ENCYCLOPEDIA OF

Physical Science
AND Technology

THIRD EDITION

Cl–Cp

Volume 3

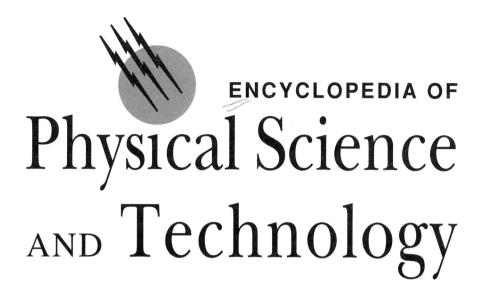

ENCYCLOPEDIA OF
Physical Science
AND Technology

THIRD EDITION

Cl–Cp

Volume 3

Editor-in-Chief

Robert A. Meyers, Ramtech, Inc.

ACADEMIC PRESS
A Harcourt Science and Technology Company

San Diego   San Francisco   Boston   New York   London   Sydney   Tokyo

This book is printed on acid-free paper.

Copyright © 2002 by ACADEMIC PRESS

Academic Press
A Harcourt Science and Technology Company
525 B Street, Suite 1900, San Diego, California 92101-4495, USA
http://www.academicpress.com

Academic Press
Harcourt Place, 32 Jamestown Road, London NW1 7BY, UK
http://www.academicpress.com

Library of Congress Catalog Card Number: 2001090661

International Standard Book Number:
          0-12-227410-5 (Set)   0-12-227420-2 (Volume 10)
          0-12-227411-3 (Volume 1)   0-12-227421-0 (Volume 11)
          0-12-227412-1 (Volume 2)   0-12-227422-9 (Volume 12)
          0-12-227413-X (Volume 3)   0-12-227423-7 (Volume 13)
          0-12-227414-8 (Volume 4)   0-12-227424-5 (Volume 14)
          0-12-227415-6 (Volume 5)   0-12-227425-3 (Volume 15)
          0-12-227416-4 (Volume 6)   0-12-227426-1 (Volume 16)
          0-12-227417-2 (Volume 7)   0-12-227427-X (Volume 17)
          0-12-227418-0 (Volume 8)   0-12-227429-6 (Index)
          0-12-227419-9 (Volume 9)

PRINTED IN THE UNITED STATES OF AMERICA
01   02   03   04   05   06   MM   9   8   7   6   5   4   3   2   1

SUBJECT AREA EDITORS

Contents

Foreword

The editors of the *Encyclopedia of Physical Science and Technology* had a daunting task: to make an accurate statement of the status of knowledge across the entire field of physical science and related technologies.

No such effort can do more than describe a rapidly changing subject at a particular moment in time, but that does not make the effort any less worthwhile. Change is inherent in science; science, in fact, seeks change. Because of its association with change, science is overwhelmingly the driving force behind the development of the modern world.

The common point of view is that the findings of basic science move in a linear way through applied research and technology development to production. In this model, all the movement is from science to product. Technology depends on science and not the other way around. Science itself is autonomous, undisturbed by technology or any other social forces, and only through technology does science affect society.

This superficial view is seriously in error. A more accurate view is that many complex connections exist among science, engineering, technology, economics, the form of our government and the nature of our politics, and literature, and ethics.

Although advances in science clearly make possible advances in technology, very often the movement is in the other direction: Advances in technology make possible advances in science. The dependence of radio astronomy and high-energy physics on progress in detector technology is a good example. More subtly, technology may stimulate science by posing new questions and problems for study.

The influence of the steam engine on the development of thermodynamics is the classic example. A more recent one would be the stimulus that the problem of noise in communications channels gave to the study of information theory.

As technology has developed, it has increasingly become the object of study itself, so that now much of science is focused on what we have made ourselves, rather than only on the natural world. Thus, the very existence of the computer and computer programming made possible the development of computer science and artificial intelligence as scientific disciplines.

The whole process of innovation involves science, technology, invention, economics, and social structures in complex ways. It is not simply a matter of moving ideas out of basic research laboratories, through development, and onto factory floors. Innovation not only requires a large amount of technical invention, provided by scientists and engineers, but also a range of nontechnical or "social" invention provided by, among others, economists, psychologists, marketing people, and financial experts. Each adds value to the process, and each depends on the others for ideas.

Beyond the processes of innovation and economic growth, science has a range of direct effects on our society.

Science affects government and politics. The U.S. Constitution was a product of eighteenth century rationalism and owes much to concepts that derived from the science of that time. To a remarkable extent, the Founding Fathers were familiar with science: Franklin, Jefferson, Madison, and Adams understood science, believed passionately in empirical inquiry as the source of truth, and felt that government should draw on scientific concepts for inspiration. The concept of "checks and balances" was borrowed from Newtonian physics, and it was widely believed that, like the orderly physical universe that science was discovering, social relations were subject to a series of natural laws as well.

Science also pervades modern government and politics. A large part of the Federal government is concerned either with stimulating research or development, as are the National Aeronautics and Space Administration (NASA) and the National Science Foundation, or with seeking to regulate technology in some way. The reason that science and technology have spawned so much government activity is that they create new problems as they solve old ones. This is true in the simple sense of the "side effects" of new technologies that must be managed and thus give rise to such agencies as the Environmental Protection Agency (EPA). More importantly, however, the availability of new technologies makes possible choices that did not exist before, and many of these choices can only be made through the political system.

Biotechnology is a good example. The Federal government has supported the basic science underlying biotechnology for many years. That science is now making possible choices that were once unimagined, and in the process a large number of brand new political problems are being created. For example, what safeguards are necessary before genetically engineered organisms are tested in the field? Should the Food and Drug Administration restrict the development of a hormone that will stimulate cows to produce more milk if the effect will be to put a large number of dairy farmers out of business? How much risk should be taken to develop medicines that may cure diseases that are now untreatable?

These questions all have major technical content, but at bottom they involve values that can only be resolved through the political process.

Science affects ideas. Science is an important source of our most basic ideas about reality, about the way the world is put together and our place in it. Such "world views" are critically important, for we structure all our institutions to conform with them.

In the medieval world view, the heavens were unchanging, existing forever as they were on the day of creation. Then Tycho Brahe observed the "new star"—the nova of 1572—and the inescapable fact of its existence forced a reconstruction of reality. Kepler, Galileo, and Newton followed and destroyed the earth- and human-centered universe of medieval Christianity.

Darwin established the continuity of human and animal, thus undermining both our view of our innate superiority and a good bit of religious authority. The germ theory of disease—made possible by the technology of the microscope destroyed the notion that disease was sent by God as a just retribution for unrepentant sinners.

Science affects ethics. Because science has a large part in creating our reality, it also has a significant effect on ethics. Once the germ theory of disease was accepted, it could no longer be ethical, because it no longer made sense, to berate the sick for their sins. Gulliver's voyage to the land of the Houynhyms made the point:

By inventing a society in which individuals' illnesses were acts of free will while their crimes were a result of outside forces, he made it ethical to punish the sick but not the criminal.

Knowledge—most of it created by science—creates obligations to act that did not exist before. An engineer, for instance, who designs a piece of equipment in a way that is dangerous, when knowledge to safely design it exists, has violated both an ethical and a legal precept. It is no defense that the engineer did not personally possess the knowledge; the simple existence of the knowledge creates the ethical requirement.

In another sense, science has a positive effect on ethics by setting an example that may be followed outside science. Science must set truth as the cardinal value, for otherwise it cannot progress. Thus, while individual scientists may lapse, science as an institution must continually reaffirm the value of truth. To that extent science serves as a moral example for other areas of human endeavor.

Science affects art and literature. Art, poetry, literature, and religion stand on one side and science on the other side of C. P. Snow's famous gulf between the "two cultures." The gulf is largely artificial, however; the two sides have more in common than we often realize. Both science and the humanities depend on imagination and the use of metaphor. Despite widespread belief to the contrary, science does not proceed by a rational process of building theories from undisputed facts. Scientific and technological advances depend on imagination, on some intuitive, creative vision of how reality might be constructed. As Peter Medawar puts it: [Medawar, P. (1969). Encounter 32(l), 15-23]: All advance of scientific understanding, at every level, begins with a speculative adventure, an imaginative preconception of what might be true—a preconception that always, and necessarily, goes a little way (sometimes a long way) beyond anything that we have logical or factual authority to believe in.

The difference between literature and science is that in science imagination is controlled, restricted, and tested by reason. Within the strictures of the discipline, the artist or poet may give free rein to imagination. Although we may critically compare a novel to life, in general, literature or art may be judged without reference to empirical truth. Scientists, however, must subject their imaginative construction to empirical test. It is not established as truth until they have persuaded their peers that this testing process has been adequate, and the truth they create is always tentative and subject to renewed challenge.

The genius of science is that it takes imagination and reason, which the Romantics and the modern counterculture both hold to be antithetical, and combines them in a synergistic way. Both science and art, the opposing sides of the "two cultures," depend fundamentally on the creative use of imagination. Thus, it is not surprising that many mathematicians and physicists are also accomplished musicians, or that music majors have often been creative computer programmers.

Science, technology, and culture in the future. We can only speculate about how science and technology will affect society in the future. The technologies made possible by an understanding of mechanics, thermodynamics, energy, and electricity have given us the transportation revolutions of this century and made large amounts of energy available for accomplishing almost any sort of physical labor. These technologies are now mature and will continue to evolve only slowly. In their place, however, we have the information revolution and soon will have the biotechnology revolution. It is beyond us to say where these may lead, but the implications will probably be as dramatic as the changes of the past century.

Computers affected first the things we already do, by making easy what was once difficult. In science and engineering, computers are now well beyond that. We can now solve problems that were only recently impossible. Modeling, simulation, and computation are rapidly be-

coming a way to create knowledge that is as revolutionary as experimentation was 300 years ago.

Artificial intelligence is only just beginning; its goal is to duplicate the process of thinking well enough so that the distinction between humans and machines is diminished. If this can be accomplished, the consequences may be as profound as those of Darwin's theory of evolution, and the working out of the social implications could be as difficult.

With astonishing speed, modern biology is giving us the ability to genuinely understand, and then to change, biological organisms. The implications for medicine and agriculture will be great and the results should be overwhelmingly beneficial.

The implications for our view of ourselves will also be great, but no one can foresee them. Knowledge of how to change existing forms of life almost at will confers a fundamentally new power on human beings. We will have to stretch our wisdom to be able to deal intelligently with that power.

One thing we can say with confidence: Alone among all sectors of society and culture, science and technology progress in a systematic way. Other sectors change, but only science and technology progress in such a way that today's science and technology can be said to be unambiguously superior to that of an earlier age. Because science progresses in such a dramatic and clear way, it is the dominant force in modern society.

Erich Bloch
National Science Foundation
Washington, D.C.

Preface

We are most gratified to find that the first and second editions of the *Encyclopedia of Physical Science and Technology* (1987 and 1992) are now being used in some 3,000 libraries located in centers of learning and research and development organizations world-wide. These include universities, institutes, technology based industries, public libraries, and government agencies. Thus, we feel that our original goal of providing in-depth university and professional level coverage of every facet of physical sciences and technology was, indeed, worthwhile.

The editor-in-chief (EiC) and the Executive Board determined in 1998 that there was now a need for a Third Edition. It was apparent that there had been a blossoming of scientific and engineering progress in almost every field and although the World Wide Web is a mighty river of information and data, there was still a great need for our articles, which comprehensively explain, integrate, and provide scientific and mathematical background and perspective. It was also determined that it would be desirable to add a level of perspective to our Encyclopedia team, by bringing in a group of eminent Section Editors to evaluate the existing articles and select new ones reflecting fields that have recently come into prominence.

The Third Edition Executive Board members, Stephen Hawking (astronomy, astrophysics, and mathematics), Daniel Goldin (space sciences), Elias Corey (chemistry), Paul Crutzen (atmospheric science), Yuan Lee (chemistry), George Olah (chemistry), Melvin Schwartz (physics), Edward Teller (nuclear technology), Frederick Seitz (environment), Benoit Mandelbrot (mathematics), Allen Bard (chemistry) and Klaus von Klitzing (physics)

concurred with the idea of expanding our coverage into molecular biology, biochemistry, and biotechnology in recognition of the fact that these fields are based on physical sciences. Military technology such as weapons and defense systems was eliminated in concert with present trends moving toward emphasis on peaceful uses of science and technology. Aaron Klug (molecular biology and biotechnology) and Phillip Sharp (molecular and cell biology) then joined the board to oversee their fields as well as the overall Encyclopedia. The Advisory Board was completed with the addition of John Bollinger (engineering), Michael Buckland (library sciences), Jean Carpentier (aerospace sciences), Ludwig Faddeev (physics), Herbert Friedman (space sciences), R. A. Mashelkar (chemical engineering), Karl Pister (engineering) and Gordon Slemon (engineering).

A 40 page topical outline of physical sciences and technology was prepared by the EiC and then reviewed by the board and modified according to their comments. This formed the basis for assuring complete coverage of the physical sciences and for dividing the science and engineering disciplines into 50 sections for selection of section editors. Six of the advisory board members decided to serve also as section editors (Allen Bard for analytical chemistry, Elias Corey for organic chemistry, Paul Crutzen for atmospheric sciences, Yuan Lee for physical chemistry, Phillip Sharp for molecular biology, and Melvin Schwartz for physics). Thirty-two additional section editors were then nominated by the EiC and the board for the remaining sections. A listing of the section editors together with their section descriptions is presented on p. v.

The section editors then provided lists of nominated articles and authors, as well as peer reviewers, to the EiC based on the section scopes given in the topical outline. These lists were edited to eliminate overlap. The Board was asked to help adjudicate the lists as necessary. Then, a complete listing of topics and nominated authors was assembled. This effort resulted in the deletion of about 200 of the Second Edition articles, the addition of nearly 300 completely new articles, and updating or rewrite of approximately 480 retained article topics, for a total of over 780 articles, which comprise the Third Edition. Examples of the new articles, which cover science or technology areas arising to prominence after the second edition, are: molecular electronics; nanostructured materials; image-guided surgery; fiber–optic chemical sensors; metabolic engineering; self-organizing systems; tissue engineering; humanoid robots; gravitational wave physics; pharmacokinetics; thermoeconomics, and superstring theory.

Over 1000 authors prepared the manuscripts at an average length of 17-18 pages. The manuscripts were peer reviewed, indexed, and published. The result is the eighteen volume work, of over 14,000 pages, comprising the Third Edition.

The subject distribution is: 17% chemistry; 5% molecular biology and biotechnology; 11% physics; 10% earth sciences; 3% environment and atmospheric sciences; 12% computers and telecommunications; 8% electronics, optics, and lasers; 7% mathematics; 8% astronomy, astrophysics, and space technology; 6% energy and power; 6% materials; 7% engineering, aerospace, and transportation. The relative distribution between basic and applied subjects is: 60% basic sciences, 7% mathematics, and 33% engineering and technology. It should be pointed out that a subject such as energy and power with just a 5% share of the topic distribution is about 850 pages in total, which corresponds to a book-length treatment.

We are saddened by the passing of six of the Board members who participated in previous editions of this Encyclopedia. This edition is therefore dedicated to the memory of S. Chandrasekhar, Linus Pauling, Vladimir Prelog, Abdus Salam, Glenn Seaborg, and Gian-Carlo Rota with gratitude for their contributions to the scientific community and to this endeavor.

Finally, I wish to thank the following Academic Press personnel for their outstanding support of this project: Robert Matsumura, managing editor, Carolan Gladden and Amy Covington, author relations; Frank Cynar, sponsoring editor; Nick Panissidi, manuscript processing; Paul Gottehrer and Michael Early, production; and Chris Morris, Major Reference Works director.

Robert A. Meyers, Editor-in-Chief
Ramtech, Inc.
Tarzana, California, USA

FROM THE PREFACE TO THE FIRST EDITION

In the summer of 1983, a group of world-renowned scientists were queried regarding the need for an encyclopedia of the physical sciences, engineering, and mathematics written for use by the scientific and engineering community. The projected readership would be endowed with a basic scientific education but would require access to authoritative information not in the reader's specific discipline. The initial advisory group, consisting of Subrahmanyan Chandrasekhar, Linus Pauling, Vladimir Prelog, Abdus Salam, Glenn Seaborg, Kai Siegbahn, and Edward Teller, encouraged this notion and offered to serve as our senior executive advisory board.

A survey of the available literature showed that there were general encyclopedias, which covered either all facets of knowledge or all of science including the biological sciences, but there were no encyclopedias specifically in the physical sciences, written to the level of the scientific community and thus able to provide the detailed information and mathematical treatment needed by the intended readership. Existing compendia generally limited their mathematical treatment to algebraic relationships rather than the in-depth treatment that can often be provided only by calculus. In addition, they tended either to fragment a given scientific discipline into narrow specifics or to present such broadly drawn articles as to be of little use to practicing scientists.

In consultation with the senior executive advisory board, Academic Press decided to publish an encyclopedia that contained articles of sufficient length to adequately cover a scientific or engineering discipline and that provided accuracy and a special degree of accessibility for its intended audience.

This audience consists of undergraduates, graduate students, research personnel, and academic staff in colleges and universities, practicing scientists and engineers in industry and research institutes, and media, legal, and management personnel concerned with science and engineering employed by government and private institutions. Certain advanced high school students with at least a year of chemistry or physics and calculus may also benefit from the encyclopedia.

Robert A. Meyers
TRW, Inc.

Guide to the Encyclopedia

R eaders of the *Encyclopedia of Physical Science and Technology (EPST)* will find within these pages a comprehensive study of the physical sciences, presented as a single unified work. The encyclopedia consists of eighteen volumes, including a separate Index volume, and includes 790 separate full-length articles by leading international authors. This is the third edition of the encyclopedia published over a span of 14 years, all under the editorship of Robert Meyers.

Each article in the encyclopedia provides a comprehensive overview of the selected topic to inform a broad spectrum of readers, from research professionals to students to the interested general public. In order that you, the reader, will derive the greatest possible benefit from the *EPST*, we have provided this Guide. It explains how the encyclopedia was developed, how it is organized, and how the information within it can be located.

LOCATING A TOPIC

The *Encyclopedia of Physical Science and Technology* is organized in a single alphabetical sequence by title. Articles whose titles begin with the letter A are in Volume 1, articles with titles from B through Ci are in Volume 2, and so on through the end of the alphabet in Volume 17.

A reader seeking information from the encyclopedia has three possible methods of locating a topic. For each of these, the proper point of entry to the encyclopedia is the Index volume. The first method is to consult the alphabetical Table of Contents to locate the topic as an article title; the Index volume has a complete A-Z listing of all article titles with the appropriate volume and page number.

Article titles generally begin with the key term describing the topic, and have inverted word order if necessary to begin the title with this term. For example, "Earth Sciences, History of" is the article title rather than "History of Earth Sciences." This is done so that the reader can more easily locate a desired topic by its key term, and also so that related articles can be grouped together. For example, 12 different articles dealing with lasers appear together in the La- section of the encyclopedia.

The second method of locating a topic is to consult the Contents by Subject Area section, which follows the Table of Contents. This list also presents all the articles in the encyclopedia, in this case according to subject area rather than A-Z by title. A reader seeking information on nuclear technology, for example, will find here a list of more than 20 articles in this subject area.

The third method is to consult the detailed Subject Index that is the essence of the Index volume. This is the best starting point for a reader who wishes to refer to a relatively specific topic, as opposed to a more general topic that will be the focus of an entire article. For example, the Subject Index indicates that the topic of "biogas" is discussed in the article Biomass Utilization.

CONSULTING AN ARTICLE

The First Edition of the *Encyclopedia of Physical Science and Technology* broke new ground in scholarly reference publishing through its use of a special format for articles.

The purpose of this innovative format was to make each article useful to various readers with different levels of knowledge about the subject. This approach has been widely accepted by readers, reviewers, and librarians, so much so that it has not only been retained for subsequent editions of *EPST* but has also been adopted in many other Academic Press encyclopedias, such as the *Encyclopedia of Human Biology*. This format is as follows:

- Title and Author
- Outline
- Glossary
- Defining Statement
- Main Body of the Article
- Cross References
- Bibliography

Although it is certainly possible for a reader to refer only to the main body of the article for information, each of the other specialized sections provides useful material, especially for a reader who is not entirely familiar with the topic at hand.

USING THE OUTLINE

Entries in the encyclopedia begin with a topical outline that indicates the general content of the article. This outline serves two functions. First, it provides a preview of the article, so that the reader can get a sense of what is contained there without having to leaf through all the pages. Second, it serves to highlight important subtopics that are discussed within the article. For example, the article "Asteroid Impacts and Extinctions" includes subtopics such as "Cratering," "Environmental Catastophes," and "Extinctions and Speciation."

The outline is intended as an overview and thus it lists only the major headings of the article. In addition, extensive second-level and third-level headings will be found within the article.

USING THE GLOSSARY

The Glossary section contains terms that are important to an understanding of the article and that may be unfamiliar to the reader. Each term is defined in the context of the article in which it is used. The encyclopedia includes approximately 5,000 glossary entries. For example, the article "Image-Guided Surgery" has the following glossary entry:

Focused ultrasound surgery (FUS) Surgery that involves the use of extremely high frequency sound targeted to highly specific sites of a few millimeters or less.

USING THE DEFINING STATEMENT

The text of most articles in the encyclopedia begins with a single introductory paragraph that defines the topic under discussion and summarizes the content of the article. For example, the article "Evaporites" begins with the following statement:

EVAPORITES are rocks composed of chemically precipitated minerals derived from naturally occurring brines concentrated to saturation either by evaporation or by freeze-drying. They form in areas where evaporation exceeds precipitation, especially in a semiarid subtropical belt and in a subpolar belt. Evaporite minerals can form crusts in soils and occur as bedded deposits in lakes or in marine embayments with restricted water circulation. Each of these environments contains a specific suite of minerals.

USING THE CROSS REFERENCES

Though each article in the *Encyclopedia of Physical Science and Technology* is complete and self-contained, the topic list has been constructed so that each entry is supported by one or more other entries that provide additional information. These related entries are identified by cross references appearing at the conclusion of the article text. They indicate articles that can be consulted for further information on the same issue, or for pertinent information on a related issue. The encyclopedia includes a total of about 4,500 cross references to other articles. For example, the article "Aircraft Aerodynamic Boundary Layers" contains the following list of references:

Aircraft Performance and Design ● Aircraft Speed and Altitude ● Airplanes, Light ● Computational Aerodynamics ● Flight (Aerodynamics) ● Flow Visualization ● Fluid Dynamics

USING THE BIBLIOGRAPHY

The Bibliography section appears as the last element in an article. Entries in this section include not only relevant print sources but also Websites as well.

The bibliography entries in this encyclopedia are for the benefit of the reader and are not intended to represent a complete list of all the materials consulted by the author in preparing the article. Rather, the sources listed are the author's recommendations of the most appropriate materials for further research on the given topic. For example, the article "Chaos" lists as references (among others) the works *Chaos in Atomic Physics, Chaos in Dynamical Systems,* and *Universality in Chaos.*

Climatology

J. E. Oliver
G. D. Bierly
Indiana State University

Hans A. Panofsky
University of California, San Diego

GLOSSARY

Albedo Fraction of solar radiation reflected; earth–atmosphere system albedo is approximately 30%.

Doldrums Calms of the intertropical convergence located between the northeast and southeast trade winds.

Eccentricity Ratio of focus–center distance to semimajor axis of ellipse.

Ecliptic Plane of the earth's orbit about the sun.

El Niño Almost periodic phenomenon involving warming of the equatorial Pacific.

Front Surface separating air masses of differing properties.

GCM Computer-derived general-circulation model of the earth's atmosphere.

Hadley cell Circulation cell with updrafts near the Equator and downdrafts near 30° latitude.

Horse latitudes Regions of sinking air and light winds located around 30° to 35° latitude, the subtropical high-pressure belts.

Isopleth Lines joining equal values on a map or chart, e.g, isotherm—equal temperatures; isobar—equal pressures.

Jet stream Rapidly flowing, narrow airstream in the upper troposphere or lower stratosphere.

Latent heat Energy released or required by change of phase (usually for water in the climate context).

Obliquity of the ecliptic Angle between the earth's orbit and the Equator.

Precession Movement of the earth's axis with a period of 26,000 years.

Solar constant Solar energy received at right angles to solar beam at the earth's mean distance from the sun. Current value: 1370 W/m^2.

Encyclopedia of Physical Science and Technology, Third Edition, Volume 3

Trades Tropical easterly winds, northeast in the Northern Hemisphere and southeast in the Southern, between the subtropical highs and the intertropical convergence.

Tropopause Top of the troposphere.

Troposphere Region of the atmosphere from the surface to 9 to 16 km, which is heated from below and is fairly well mixed.

Westerlies Mid-latitude westerly winds which flow as a train of waves perpendicular to the hemispheric pressure gradient force in the upper atmosphere.

WE SHALL DEFINE CLIMATE as weather statistics over periods of the order of 30 years as measured at particular points on the globe. Thus, climate includes not only averages of weather elements, but also measures of their variation. These might include variations throughout the day, throughout the year, from year to year, and extremes that may occur only once in several years or decades. Other definitions of climate involve statistics over different periods.

Temperatures and precipitation are the most common elements used to define climate, but wind, amount of snowfall, and other variables are also included. Climate information of all types is available from the National Climate of the National Oceanographic and Atmospheric Administration in Asheville, North Carolina, and also from the World Meteorological Office in Geneva, Switzerland.

I. INTRODUCTION

Climate data are required for planning many types of human activity, such as agriculture, architecture, and transportation. For example, in the construction of a house information is needed about the averages and extremes of temperature, precipitation, wind, and snowfall. Also, obviously, different types of agriculture are possible in different climates.

A. Brief History of Climate Study

An appreciation of climate existed long before any formal history of the discipline was initiated. People have always been aware of their environment, and if a person resides in a place for a decade or more, an image of the climate of that location is formed. Given the large seasonal changes in many world areas and their significance to human survival, such an image must have been formed in early times. However, it remained for the ancient Greeks to formalize its study; in fact, the word *climate* is derived from the Greek *klima,* meaning "slope." In this context, the word applies to the slope or inclination of the earth's axis and is

applied to an earth region at a particular elevation on that slope, that is, the location of that place in relation to the parallels of latitude and the resulting angle of the sun in the sky.

Apart from establishing the geometric relationships of climatology, Greek scholars wrote treatises on climate. The first climatography (descriptive study of climate) is attributed to Hippocrates, who wrote "Airs, Waters, and Places" in 400 BC. In 350 BC, Aristotle wrote the first meteorological treatise, "Meteorologica." The Greeks gave names to winds, described marine climates, and examined the role of mountains in determining the climate of a location.

With the decline of ancient Greece, it remained for Chinese scholars to make major efforts to understand their atmospheric environment. For example, they used seasonal rainfall to estimate harvests and tax revenues and noted the relative severity of floods and cold winters. While the Chinese continued their efforts, the Western world entered a period in which scientific inquiry was not encouraged, and climatic observations were not accorded importance. It was not until the middle of the 15th century, the Age of Discovery, that long-term observation of the atmosphere was again of interest. With the extended sea voyages and development of new trading routes, descriptive reports of world climates became available, especially those concerning prevailing wind systems.

During the 17th century, scientific analysis of the atmosphere got underway when instruments were developed. This set the scene for the advent of the modern era. The availability of instruments and the formulation of basic laws of gases led to a new era of climatic observation and analysis. Table I shows a partial listing of some of the significant events.

B. Subdisciplines of Climatology

Descriptive climatology describes the climates of the world. It is subdivided into regional climatology, which deals on a broad scale with the climates of large portions of the world, and microclimatology, the modifications of local climates by local factors, such as topography and land–water contrasts. Physical climatology attempts to explain the properties of climates by the earth–sun geometry, atmospheric composition, ground characteristics, and the laws of physics. Synoptic climatology draws linkages between physical and dynamic aspects of the large-scale atmospheric circulation and surface weather tendencies at regional and local scales. Historical climatology deals with climate change and the many hypotheses suggested to account for them. It also attempts to predict future climate change, produced both naturally and by human action.

TABLE I Significant Events in the Development of Climatology

ca. 400 BC	Influence of climate on health is discussed by Hippocrates in "Airs, Waters, and Places."
ca. 350 BC	Weather science is discussed in Aristotle's "Meteorologica."
ca. 300 BC	"De Ventis" by Theophrastus describes winds and offers a critique of Aristotle's ideas.
ca. 1593	Thermoscope is described by Galileo; the first thermometer is attributed most likely to Santorre, 1612.
1622	Significant treatise on the wind is written by Francis Bacon.
1643	Barometer is invented by Torricelli.
1661	Boyle's law on gases is propounded.
1664	Weather observations begin in Paris; although often described as the longest continuous sequence of weather data available, the records are not homogeneous or complete.
1668	Edmund Halley constructs a map of the trade winds.
1714	Fahrenheit scale is introduced.
1735	George Hadley's treatise on trade winds and effects of earth rotation is written.
1736	Centigrade scale is introduced. (It was formally proposed by du Crest in 1641.)
1779	Weather observations begin at New Haven, CT; they represent the longest continuous sequence of records in the United States.
1783	Hair hygrometer is invented.
1802	Lamark and Howard propose the first cloud classification system.
1817	Alexander von Humboldt constructs the first map showing mean annual temperature over the globe.
1825	Psychrometer is devised by August.
1827	The period begins during which H. W. Dove developed the laws of storms.
1831	William Redfield produces the first weather map of the United States.
1837	Pyrheliometer for measuring insolation is constructed.
1841	Movement and development of storms are described by Espy.
1844	Gaspard de Coriolis formulates the "Coriolis force."
1845	First world map of precipitation is constructed by Berghaus.
1848	First of M. F. Maury's publications on winds and currents at sea is written.
1848	Dove publishes the first maps of mean monthly temperatures.
1862	First map (showing western Europe) of mean pressure is drafted by Renou.
1869	Supan publishes a map showing world temperature regions.
1892	Systematic use of balloons to monitor free air begins.
1900	Term *classification of climate* is first used by Köppen.
1902	Existence of the stratosphere is discovered.
1913	Ozone layer is discovered.
1918	V. Bjerknes begins to develop his polar front theory.
1925	Systematic data collection using aircraft begins.
1928	Radiosondes are first used.
1940	Nature of jet streams is first investigated.
1956	First computer model explains general-circulation techniques.
1960	First meteorological satellite, *Tiros I,* is launched by the United States.
1966	First geostationary meteorological satellite is launched by the United States.
1968	Global atmospheric research program begins.
1978	National Climate Program is adopted by the U.S. government.
1974	Rowland and Molina formulate the theoretical linkage between CFCs and stratospheric ozone depletion.
1985	British Antarctic Survey at Halley Bay discovers Southern Hemisphere ozone hole.
1987	Montreal Protocol established.
1992	Rio Framework Convention on Climatic Change.
1997	Kyoto Climate Conference.

Because of the profound effect of climate on many aspects of human existence, there are myriad applications of climatology: bioclimatology, the effects of climate on human, animal, and plant life; architectural climatology, the impact of climate on architecture; and so on. Since climatologists cannot foresee all the applications of their knowledge, applied climatology is best left to the users of climate information.

II. FUNDAMENTALS OF CLIMATE THEORY

A. Reasons for Climate Differences

Weather and climate are produced by solar heating. The reasons for variations in climate around the earth are primarily of four types:

1. Incoming solar radiation (insolation) varies with latitude, being largest at the equator and smallest at the poles; hence, climate varies with latitude.
2. Energy transformations at the ground are completely different over water, ice, and land; hence, climates depend critically on the proximity of such surfaces.
3. Hills and mountains modify atmospheric variables.
4. Differences in land use and ground cover affect climate.

B. Solar Source

The *solar constant* is defined as the radiation intensity received at right angles to the sun's radiation at the earth's mean distance, 1.49×10^8 km. This is the average between largest distance (aphelion), in July, and smallest distance (perihelion), on January 4. Aphelion distance exceeds perihelion, at present, by ~3%. The solar constant is ~1370 W/m². It is a measure of the sun's heat output

and has been remarkably constant during the period of its measurement, the 20th century. Only a decrease of 0.1% from 1981 to 1986, observed by satellites, may be significant, and there is no reason why this small trend should continue.

Given that the solar constant has not been noted to vary much, some of the basic properties of climate can be inferred from the simple geometric laws for insolation I (radiation intensity on a horizontal surface) without an atmosphere:

$$I = (S/r^2)\cos z \qquad (1)$$

Here, S is the solar constant, r is the earth–sun distance divided by the mean distance, and z is the zenith angle of the sun—that is, the angle between the sun and the zenith.

According to a basic formula of practical astronomy, $\cos z$ depends only on time of day, time of year, and latitude. The zenith angle, and hence I, depends on time of year (seasonal variation) because the angle between the earth's Equator and its orbit around the sun, the obliquity of the ecliptic ε, is not zero. Its present value is 23.5° (Fig. 1). The larger ε is, the larger are the seasonal contrasts.

As mentioned earlier, the distance r currently varies by only 3% throughout the year. This variation is not of great importance in explaining the properties of the current climate.

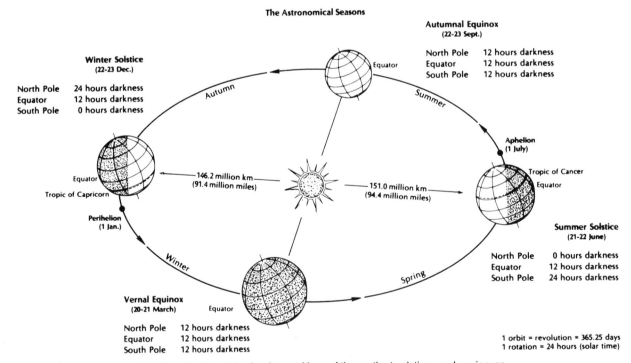

FIGURE 1 Earth's orbit, showing positions of the earth at solstices and equinoxes.

The presence of the atmosphere means that the actual insolation is less than that given by Eq. (1), even without clouds. The difference is largest at high latitudes, where the light path through the air is longest. The difference also depends on the transparency of the atmosphere, which is quite variable. Clouds are the most variable factors that affect the transparency of the atmosphere, since they both reflect and absorb sunlight. The net result of the presence of the earth's atmosphere is that the actual energy available at the surface is only about half that indicated by Eq. (1).

Even when we are given accurate radiation intensity estimates everywhere, we still need to calculate the meteorological variables such as temperature, wind, and precipitation. This is done by climate models and general-circulation models.

C. Energy Budget

Given the distribution of insolation at the surface, we shall briefly follow the fate of the energy after it has been absorbed at the ground—but averaged over the entire earth's surface. Energy received and released must almost balance on a globally and yearly averaged basis, since climate changes are slow and small in magnitude.

In this brief survey, we shall consider the average insolation at the atmosphere to be one-fourth of the solar constant S; the reason is that the earth intercepts $\pi R^2 S$ of solar radiation (R is the earth's radius), but the earth loses radiation to space over its entire surface of area $4\pi R^2$.

For convenience, we shall consider $S/4$ to be 100%. As mentioned before, about 50% is absorbed at the surface, 30% is reflected by the atmosphere (mostly by clouds), and 20% is absorbed. The ground loses energy through evaporation ($\sim$22%) and direct conduction to the atmosphere (6%). The evaporated water vapor condenses in the lower atmosphere and is an important heat source for it.

In addition, the ground loses more than 100% in infrared radiation. If this were emitted into space, the ground would lose more energy than it receives and would cool rapidly. Actually, much of this radiation is intercepted by clouds, water vapor, and carbon dioxide in the atmosphere and returned to the ground, leading to a near balance of energy at the ground. This "trapping" of infrared radiation is often called the "greenhouse effect" even though greenhouse glass also inhibits convection, which the atmosphere does not.

To complete the balance at the top of the atmosphere, the atmosphere radiates $\sim$50% to space, while $\sim$20% is emitted directly from the earth's surface; 30% is reflected solar radiation.

Energy is approximately balanced only for the globe averaged for the year, not separately for each latitude. Figure 2 shows the latitudinal distribution of incoming and outgoing radiation at the top of the atmosphere. The incoming radiation varies rapidly with latitude, being by far the largest at the Equator. The outgoing radiation is much more uniform. This is because part of this radiation originates near the tropopause ($\sim$9 km height at the

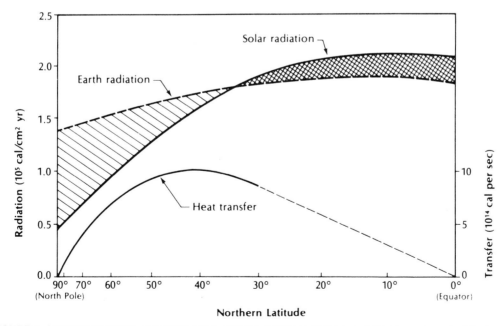

FIGURE 2 Latitudinal variation of incoming and outgoing radiations at the top of the atmosphere. Hatched area, surplus; striped area, deficit.

poles to (~16 km at the equator), which is actually cooler near the Equator than either to the north or south. As a result, low latitudes receive an excess of energy, and high latitudes a deficit. These inequalities must be made up by north–south heat exchange. This occurs in three ways:

1. Heat transport in the atmosphere
2. Heat transport in the oceans
3. Latent heat transport (evaporation at one latitude and condensation at another)

According to current estimates, factors 1 and 2 are about equally important, both transporting heat poleward. Latent heat is transported poleward and equatorward from latitudes ~30% north and south, latitudes of little rainfall and large evaporation.

The strong latitudinal variation of insolation is responsible for the "general circulation of the atmosphere," which in turn interacts with the temperature field. The atmosphere can be viewed as a giant heat engine: Heat is added preferentially in the tropics and removed at high latitudes, thereby driving the winds (providing work). Climate models and general-circulation models treat these subjects quantitatively. Since the models have become progressively more complex and are the subject of a huge worldwide research effort, they will be treated in a separate section (see Section V).

D. Hydrological Cycle

World-average precipitation and evaporation must approximately balance on an annually averaged basis. Each averages ~1 m, but whereas ~90% of the evaporation takes place over the ocean, only ~67% of the precipitation falls there. The difference is made up by river flow.

Evaporation and precipitation show large latitudinal, seasonal, and diurnal variations, which are closely related to the characteristics of the general circulation. Over the oceans, air generally sinks near latitudes 30° and rises near the Equator and near latitudes 60° north and south. Hence, most evaporation occurs into the dry air near latitudes 30°, and most precipitation, occure near the Equator and at high latitudes. All those zones move northward and southward with the sun throughout the year. Superimposed on these latitudinal migrations, evaporation over the oceans also shows seasonal cycles, with strongest evaporation in the winter, particularly in the western oceans.

Over continents, the general circulation shows strong seasonal reversals. This leads to strong seasonal variations in continental interiors, with maximum precipitation in the summer. Coastal areas are dominated by winter storms, except in regions of strong hurricane activity, which are basically late-summer phenomena.

III. REGIONAL CLIMATOLOGY

Mean fields of atmospheric variables are usually represented by isopleths, that is, lines along which a given variable is constant. For example, Figs. 3 and 4 show global isotherms, lines of constant temperature in July and January, respectively. The lines run basically east–west, showing the prime importance of latitude. The region of warmest air has a tendency to move north of the Equator in the northern summer. Otherwise, seasonal contrasts are small over water and very large over land. Also, seasonal contrasts over land grow from the Equator toward the poles. Owing to the prevalence of water in the Southern Hemisphere, seasonal differences are smaller in this hemisphere.

Over the oceans, in middle latitudes, the isotherms tend to bend poleward on the western side of the oceans and equatorward on the eastern side. These bends are caused by ocean currents, which are warm in the west and cold in the east of the ocean basins.

Variability of temperature from time to time cannot be shown in any generality. Of course, as we have seen, seasonal variation is indicated by maps drawn for different months or seasons. The largest diurnal variations in temperature are found at low latitudes in midcontinental and desert areas.

Year-to-year changes in ocean climate have been of special interest since the 1970s. One aspect of these changes in the climate system is that large portions of the oceans may develop temperature anomalies in the order of 1° to 2° (sometimes more), which are quite persistent and associated with anomalous circulation patterns. A special case of this is the El Niño–southern oscillation (ENSO) phenomenon, which occurs irregularly, roughly every 3 to 7 years. It was especially strong in the period 1997–1998 when the event received widespread coverage in the media. All ENSO events result from a weakening of the easterly trade winds in the equatorial Pacific Ocean. When these weaken, there is a build up of warm surface water and a sinking of the thermocline in the eastern Pacific. Modified circulation patterns result in drought in Indonesia and Australia and storms and floods in Peru, Columbia, and Bolivia. Worldwide repercussions also occur. For example, in the 1997–1998 event, the United States experienced stormy weather on the west coast and drier weather in the east; over most areas, warmer than normal temperatures occurred, with resulting diminished snowfall totals.

Pressure patterns are related to patterns of wind and precipitation, but the latter is more affected by local topography and other local peculiarities. Thus, pressure maps are often shown to describe the circulation. Wind is nearly parallel to the isobars (lines of constant pressure) in midlatitudes, but blows across the isobars near the Equator

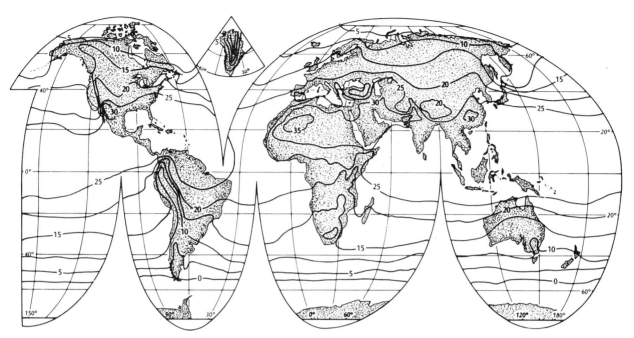

FIGURE 3 Distribution of average July temperature over the world. [From Oliver, J. E., and Hidore, J. J. (1983). "Climatology: An Introduction," by permission of Merrill, Columbus, OH.]

toward low pressure. In the Northern Hemisphere, if one puts one's back to the wind, low pressure is to the left. The reverse is true in the Southern Hemisphere. Close spacing of isobars indicates strong winds and vice versa.

Winds tend to converge into low-pressure centers (lows). Therefore, the air rises there, yielding precipitation. High-pressure ridges indicate sinking air and generally little precipitation. over the oceans, the sea level

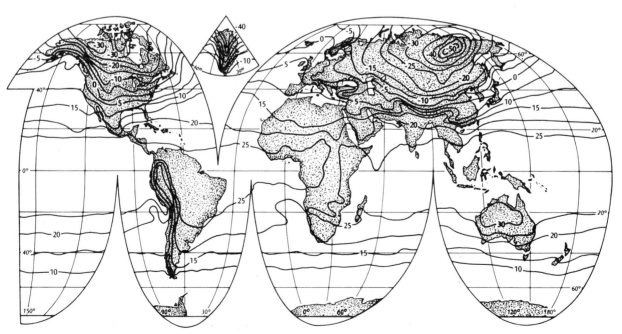

FIGURE 4 Distribution of average January temperature over the world. [From Oliver, J. E., and Hidore, J. J. (1983). "Climatology: An Introduction," by permission of Merrill, Columbus, OH.]

pressure pattern changes little between seasons, except for a northward movement in summer.

North and south of the Equator, the winds are generally from the east and are called the trades. In the Northern Hemisphere, the trades blow from the northeast; to the south, from the southeast. Thus, a convergence zone is formed near the Equator, the Intertropical Convergence Zone (ITCZ). Here air tends to rise, and precipitation is plentiful. It is a region of tropical rain forests and is also called the doldrums. Near latitude 30°, there are strong pressure ridges over the oceans with little wind and negligible precipitation. This is the belt of the horse latitudes, where most of the world's deserts are found. Between latitudes 30° north and south and 60° north and south lies the region of prevailing westerlies. Here, winds have more often westerly then easterly directions, and pressure systems generally move from west to east. This is a region of active weather, where cold air masses from polar regions meet warm air masses from the tropics at fronts. New low-pressure centers are formed at fronts, and thus precipitation is quite plentiful in this region. Near latitude 60° is the location of permanent, intense low-pressure systems with especially stormy and wet weather.

The permanent pressure and wind systems are most clearly visible over the ocean. Over the continents, there is seasonal reversal of pressure patterns. In the winter, with cold, heavy air masses over the continents, high pressure predominates in these areas, with diverging winds (clockwise in the Northern Hemisphere, counterclockwise in the Southern Hemisphere). Winter precipitation is light. In summer, pressure over the continents is low, with pre-

cipitation in midcontinent at a maximum. However, precipitation then is more spotty, in the form of showers and thunderstorms. Local topography affects some of the precipitation.

At the edges of the continents, then, there often is a reversal of winds from summer to winter; for example, in East China, winds are from the northwest in winter and southeast in summer. Such a reversal is called a monsoon and is often associated with an equally dramatic change in precipitation patterns. For example, the summer monsoon in India drives hot, moist air against the Himalayas, causing heavy precipitation. In contrast, the offshore, cold and dry winds of winter lead to little precipitation.

Although pressure patterns give an indication of general precipitation patterns, precipitation regions are often indicated separately, since not only pressure distribution, but also topography and ground characteristics contribute to the variability of precipitation. Figure 5 gives a more detailed picture of global precipitation climatology.

Sometimes, climatologists are interested in combinations of variables—for example, the simultaneous occurrences of certain temperature and precipitation ranges, which make certain types of life possible. This leads to the definitions of climate classes, the distribution of which around the globe is then shown. The most famous of these is the Köppen classification system, the details of which are given in many climatology texts.

Although regional climatology has traditionally concentrated on conditions close to the ground, upper-air climatology has increased in importance in the second half of the 20th century. There are several reasons

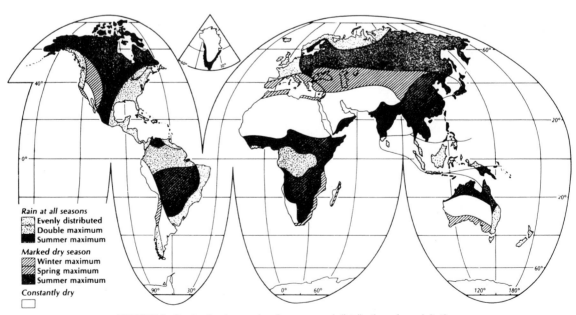

FIGURE 5 Generalized map showing seasonal distribution of precipitation.

for this: increased availability of observations; the importance of upper-air flow for steering surface storms and for weather forecasting generally; and increased air traffic.

Pressure falls more rapidly with height in cold than in warm air. Hence, up to ~20 km height, pressure tends to be lower at the poles than at the Equator. The pressure differences are largest at about tropopause height and decrease above. This is due to the fact that the horizontal temperature differences reverse near the tropopause.

In the upper air, friction is unimportant. Newton's second law shows that, in unaccelerated air, the wind is parallel to the isobars, with low pressure to its left in the Northern Hemisphere and to its right in the Southern Hemisphere.

Figure 6 shows the equivalent of a mean January pressure map at an average of 5500 m height. Actually, it is a map for conditions at 500 mbar; about half the atmosphere is above that surface and half is below. This surface is not horizontal. Here, the lines are "contour lines," of equal geopotential height, which, for practical purposes, equals ordinary geometric height. Heights are given in tens of meters. It turns out that these contour lines have the same relation to the direction of frictionless, unaccelerated flow as isobars on horizontal charts. Also, the closer the contours, the stronger the wind. Hence, Fig. 6 shows that the flow is generally from west to east, except near the Equator and the poles. This is why airplanes generally travel faster eastward than westward.

The contours are not exactly east–west, and thus average winds are not everywhere exactly from the west but have small components from the north or south. Such deviations from a strict westerly flow are associated with east–west variations of temperature in the lower atmosphere. Above 3 km, isotherms are generally approximately parallel to the contours.

Summer circulations tend to be weaker than winter circulations, since north–south temperature variations are weaker in summer. Hence, surface pressure systems move eastward more slowly in summer than in winter, and so do aircraft. Otherwise, wind patterns aloft are similar in winter and summer, except that the westerlies do not extend quite so far to the south.

As we get to the top of the troposphere (the tropopause), winds get stronger, without change of pattern. Of course, this description covers average flow only, but even winds on individual days show the same dominant west–east flow; however, the strongest winds may occur at different latitudes and may be strongly concentrated, especially above surface fronts (jet streams). Wavelike perturbations may have much larger amplitudes, with occasional closed circulation patterns even in middle latitudes.

IV. MICROCLIMATOLOGY

Regional climatology can delineate only the average climate over large areas, while local climates can be extremely varied due to local topography, local water–land and ice–snow distribution, or land use. The area of microclimatology is large, and huge books, most notably the classical work by Geiger, have been written about the subject. We shall give only a few examples here.

Even small hills influence the temperature patterns dramatically, particularly on nights with light winds. The temperatures over the slopes of shallow valleys may be 10°C or more lower than at the same heights over the center of valleys, so that cold air drains into the valley. Thus, even on calm nights there is downslope flow and general drainage along the valleys. In general, winds are channeled by valleys, so that winds in valleys often deviate from winds on nearby plateaus.

Even small water bodies moderate temperatures, particularly on light-wind winter nights. Before the water bodies freeze, the temperatures at their shores remain near freezing even though temperatures only a few kilometers away may be 10°C, even 20°C colder.

Of special interest in the 1970s was the impact of cities in local climate. First, there is the heat-island effect. Surface temperatures in cities are always higher than those in the surrounding countryside. The difference is largest in big cities, depending approximately on the fourth root of the population; it is largest at night, in winter, and with light wind speeds, depending inversely on the square root of wind speed. Cities produce a warm plume of air, slightly elevated, downstream. This is responsible for a local maximum of showers and thunderstorms on the lee side of cities.

Also, as part of microclimatology, we should mention the extremely rapid vertical variation of most meteorological variables close to the ground (except for pressure). The greatest variation is in the lowest meter, and then the variables always change more slowly. The reason is that the air very close to the surface is poorly mixed because the turbulent "eddies" there are small. In fact, in the lowest millimeter or so, eddies have no room, and vertical exchange is through molecules, which produce little transport because of their small mean free paths. Only in the area of vertical transport of momentum (wind) over rough terrain is turbulent transfer possible right into the ground. As an example of extreme vertical variation, it is possible for the surface temperature of a black asphalt road to be 80°C, whereas it is only 45°C above the surface. Similarly, cold layers can form on the ground on cold, light-wind nights. The wind speed is zero just at the surface, but it may rise to

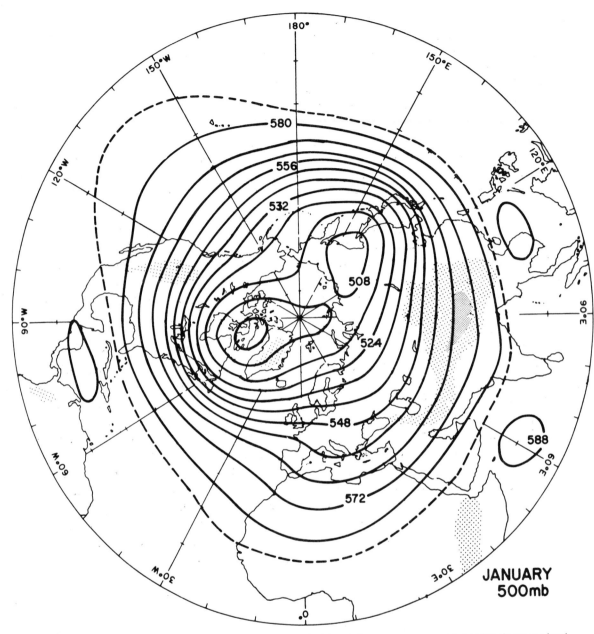

FIGURE 6 Distribution of mean January geopotential height (almost the same as geometric height) above sea level at 500 mbar. Contours labeled in tens of meters. [Based on Palmen, E., and Newton, C. W. (1969). "Atmospheric Circulation Systems," copyright Academic Press, New York.]

many meters per second in the first meter and then vary gradually.

V. CLIMATE AND GENERAL-CIRCULATION MODELS

The basic purpose of climate models originally was to explain features of current climate and general circulation in terms of the geometry of the earth–sun systems and basic physical principles. Once this was accomplished, it was possible to experiment with climate models by changing various input parameters to account for the features of past climates and to speculate about future climates.

Relatively simple "climate models" are concerned entirely with explaining the temperature distribution; the effect of circulation, if included at all, is prescribed in terms of the temperature distribution. Such climate models range from simple zero dimensional (averaged over the whole atmosphere) to models allowing for vertical and horizontal variations of input parameters. The models differ in the way boundary conditions and physical processes are handled. Sometimes clouds are prescribed, or they may be generated by the models. Ocean temperatures may be given, or the atmospheric model may be combined with an oceanic model. One of the important effects in many climate models is the ice–albedo feedback: If the model predicts a cooling, the ice sheets expand, causing increased albedo and more cooling— a "positive" feedback. In early models, this feedback was so severe that only a small cooling led to an ice-covered earth. In spite of such excesses, climate models have explained most features of the vertical temperature distributions and effects of land–water differences and topography.

General-circulation models (GCMs) attempt to duplicate the distribution of wind as well as of temperature and moisture. They are usually three dimensional; however, the earliest models had little resolution in the vertical; for example, the pioneering model by Norman Phillips in 1956 consisted of only two layers. In the meantime, much more complete models have been developed, many requiring the fastest computers available.

In the models, generally seven basic equations are solved for the seven basic variables of meteorology: pressure, density, temperature, moisture, and three velocity components. The seven equations are the gas law, the first law of thermodynamics, equations of continuity for air and water substance, and three components of Newton's second law. If the stratosphere is modeled explicitly, ozone concentration must be added as a variable and at least one equation must be included to describe the ozone budget. Since the ozone budget depends on concentrations of many other trace gases, many more variables and equations are sometimes added.

Typically, the initial state of the atmosphere is simple, with no motion. Then radiation is introduced, and the models are integrated numerically for several model years. Eventually, the models acquire many of the properties of the actual atmosphere.

In the United States alone, there are at least five centers studying the properties of GCMs. The models differ in horizontal and vertical resolution, in vertical extent, in whether clouds are assumed or generated by the model, and in the characteristics of the lower boundary. Some of the atmospheric models are coupled with GCMs of the ocean. Such coupled models produce problems because of the differences in time and space scales of important atmospheric and oceanic circulation characteristics. Other differences involve the prescription (parameterization) of motion on scales too small to be resolved but that nevertheless affect the resolved motions by their capacity to mix air with different properties.

The first real GCM in 1956 proved successful in explaining some of the basic features of the general circulation and of climate: the basic three-cell structure including the strong Hadley cell with its rising motion at the ITCZ, the sinking near latitudes of $30°$, and the equatorial flow at the surface in between. In addition, the model duplicated a persistent characteristic of atmospheric flow: the nonuniform temperature gradients, with relatively homogeneous air masses separated by sharp fronts and, associated with the fronts, narrow strems of strong west winds aloft, the jet streams.

Since Phillips's original model, more recent GCMs have become more realistic in many details: Vertical and horizontal resolution has increased; oceanic models have been combined with the GCMs; realistic hydrological cycles have been added, as have realistic prescriptions for changing clouds and snow cover; and the models have been extended to include the stratosphere with its relatively high ozone concentration.

Although the GCMs do not represent climate perfectly, they are sufficiently realistic to make experimentation with past and future climates very promising indeed.

VI. PAST CLIMATES

Climate has varied in the past on many time scales. There have been long periods ($\sim$50 million years) of relatively undisturbed climates, generally warmer than the current climate, interrupted by shorter periods (a few million years) of quite variable climates. For about the past 2 million years, we have been in such a disturbed period, with ice ages alternating with somewhat milder interglacials. The last ice age ended $\sim$10,000 years ago; since then there have been minor climate fluctuations—for example, a warm period $\sim$1000 AD and a "little ice age" in the second half of the 17th century. In the 20th century, the atmosphere warmed $\sim$0.5°C from 1880 to 1940 and cooled from 1940 to at least 1970.

Many reasons for historical climate changes have been suggested: changes in the earth's crust (e.g., migration of the continents, mountain buildings, and volcanic eruptions); changes in atmospheric composition, particularly the amount of carbon dioxide; changes in earth–sun geometry; and changes in solar heat output.

Of course, it is quite possible that some of these factors worked together, but most important climatic variations on different time scales do not have the same causes. At least, past climates have not been affected by human activity; the same cannot be said about future climates.

Climate changes on the longest time scales considered here (50 million years) almost certainly were affected by changes in the earth's crust; for example, it is possible that the active periods coincided with periods of mountain building; and perhaps, ice ages occurred when there were landmasses near the poles.

The cause of climate changes *within* the disturbed periods, which have time scales of 10,000 to 100,000 years, is now fairly well understood. We have good records from undersea and underice cores, and we have a good quantitative theory that agrees with these records. This theory was proposed in 1926 by Milankovich but has been taken seriously only since about 1970. It is known from quantitative celestial mechanics (and from observations) that several characteristics of the earth–sun geometry are variable. For example, the earth's axis does not point in the same direction in space at all times, but precesses with a period of 26,000 years. At the same time, the major axis of the earth's orbit spins in the opposite direction. The net result is that, about every 20,000 years, the sun and earth are closest in northern winter (as they are now). So right now, northern winters are relatively mild and southern winters are relatively cold. Ten thousand years from now, this situation will be reversed. A second factor is the eccentricity of the earth's orbit. Right now, the difference between aphelion and perihelion distance is ~3%, but the distance difference has been as large as 10% and at other times the orbit has been circular. The period of this variation is ~100,000 years. Finally, the angle between the earth's Equator and the earth's orbit has varied from about 22° to 24.5°, with a period of ~40,000 years. When this angle is largest, seasonal contrasts are largest.

None of these factors affect significantly the total radiation received by the earth, but they all affect seasonal contrast. Milankovich suggested that ice ages result when there are cool summers and mild winters, for snowfall can be heavy in mild winters, and less snow melts in cool summers.

The chronology of observed ice ages in the past million years agrees well with predictions of the Milankovich theory; in particular, the various periodicities have been found in the geological records. Climate models have some difficulties with the phases of the largest cycles, but there are satisfactory theories to explain these difficulties. As a result, the basic theory is generally accepted as the explanation for climate changes on the scale of 10,000 to several hundred thousand years.

There is no generally accepted hypothesis to explain more recent climate changes. Solar activity has periods of 11, 21, and 80 years, but statistical analyses of relationships between solar activity and atmospheric varia-

tions have not been convincing, even though a period of low sunspot activity coincided with the little ice age, and U.S. western droughts are correlated somewhat with the sunspot cycle. However, there are no known mechanisms connecting solar activity with changes of climate.

There is some indication of cooling at the surface following major volcanic eruptions, but the effects are not large and it is not clear whether volcanoes produce long-lasting changes. In short, there is no agreement concerning the causes of recent climate changes.

VII. FUTURE CLIMATES

Since the reliability of the Milankovich theory has been established from past records, it should also provide some indication of future climate change, in the absence of complications produced by human activity. According to this theory, a cooling trend has already set in, will intensify several thousand years from now, and will produce peak glaciation in ~20,000 years.

Superimposed on this scenario are natural short-period fluctuations, which we cannot predict because we do not understand their causes, and man-made changes. The most important of these are produced by increased CO_2 in the atmosphere, which is due to burning of fossil fuels and clearing of forests. Increases in atmospheric CO_2 have been measured and amount to about half of the CO_2 known to be emitted into the atmosphere; the rest is presumably absorbed by the ocean. Some time after the year 2050, the amount of CO_2 in the atmosphere is expected to be double that before the onset of the industrial age.

Increasing atmospheric CO_2 concentration enhances the natural "greenhouse effect," thus causing surface warming. As soon as warming begins, increasing evaporation increases the amount of water vapor in the atmosphere, producing even more warming. Many climate models have been run with increased CO_2 concentrations, typically double the normal. A consensus is starting to emerge among modelers that doubling CO_2 concentration results in a global average warming of 2° to 3°C and polar warming of more than twice that amount. Warming due to increased CO_2 is not yet large enough to be observed in view of irregular temperature variations. It is expected that CO_2 warming will be detected in the future, if the models are correct.

However, most of the models accounting for increased CO_2 are unrealistic in the specification of clouds and possibly in the lower boundary conditions. For example, it is not known whether warming will increase cloudiness or change the physical characteristics of clouds. In particular, clouds may thicken and reflect more sunlight, limiting the warming effect.

Other long-term atmospheric impacts resulting from human activity are also being monitored. Of note is the role of chlorofluorocarbons (CFCs), multiuse chemicals best known as refrigerants. Chemical breakdown of CFCs in the ozone layer of the stratosphere results in a chemical reaction leading to the diminution of ozone. The decrease modifies the role of ozone in screening solar radiation, resulting in an increase in shortwave ultraviolet radiation reaching the earth's surface.

Few environmental problems have received the media attention of the problem of global warming. The continued rise of atmospheric CO_2 content, together with the warmth of the 1980s and particularly the 1990s, which had the three warmest years on record (1998, 1997, 1995), caused some climatologists to suggest that the global warming suggested by computer models is already underway. Forecasts of increasing warmth, melting ice caps, rising sea levels, and modified environments have been widely reported. Given such forecasts, attempts have been made to introduce legislature to limit the burning of fossils fuels and, hence, CO_2 production. Notably, the Earth Summit in Rio de Janerio, Brazil, marked an important global moment of recognition of the significance of potential human impacts on climate. Further, the Climate Conference in Kyoto, Japan, in 1998 provided a substantive basis for the reduction of greenhouse emissions and a procedure for monitoring the success of mitigation efforts. The latter document lists the Intergovernmental Panel on Climate Change (IPCC) as the authority for all scientific decisions. No definitive action has yet been taken.

The relative lack of political action, particularly by the developed nations, reflects the economic costs involved and the varying interpretations of the global warming signal. Models do not provide a uniform interpretation of the future. While most climatologists agree that the earth will experience a warming trend, there is disagreement on how much the temperature will rise and when the impacts will be felt by the human population. While most models predict modest warming, an alternative scenario suggest that high latitude warming could initiate a cooling episode. The mechanism for this scenario is fairly complex, involving adjustments in the North Atlantic Deep Water oceanic circulation, and overall implications are unclear. As global models become more refined, the answers to the problem will become clearer.

In contrast to the paucity of action concerning the CO_2 problem, appreciable headway has been made in combatting the ozone depletion problem. Most industrial nations that produce CFCs have agreed to curb production and examine substitute chemicals. Such action will take time, however, and the depletion of ozone continues to be of concern. Atmospheric scientists researching the problem are now examining the dynamics that cause "holes" that periodically occur in the polar ozone layers.

Given the current interpretation of climate, speculations concerning future climates are possible. If the GCMs are correct and no preventative measures are taken, substantial global warming will take place over the next few hundred years. Eventually, perhaps after 1000 years, most fossil fuels will have been used and excess CO_2 will be taken up by the oceans. The Milankovitch mechanisms will then take over to lead, potentially, to a global cooling. Climatic change has occurred in the past and will occur in the future.

SEE ALSO THE FOLLOWING ARTICLES

GREENHOUSE EFFECT AND CLIMATE DATA • GREENHOUSE WARMING RESEARCH • HYDROLOGIC FORECASTING • METEOROLOGY, DYNAMIC (TROPOSPHERE) • METEOROLOGY, DYNAMIC (STRATOSPHERE) • POLLUTION, AIR • WEATHER PREDICTION, NUMERICAL

BIBLIOGRAPHY

Barry, R. G., and Chorley, R. J. (1998). "Atmosphere, Weather and Climate," Routledge, London.

Hidore, J. J., and Oliver, J. E. (1993). "Climatology: An Atmospheric Science," Macmillan Co., New York.

IPCC. (1991). "Climate Change: The IPCC Scientific Assessment" (Houghton, J. T., Jenkins, G. J., and Ephraums, J. J., eds.), Cambridge Univ. Press, New York.

Linacre, E. (1992). "Climate Data and Resources: A Reference and Guide," Routledge, London.

Oke, T. R. (1987). "Boundary Layer Climates," Methuen, New York.

Oliver, J. E., and Fairbridge, R., eds. (1987). "The Encyclopedia of Climatology," Van Nostrand–Reinhold, New York.

Schneider, S. H., ed. (1996). "Encyclopedia of Climate and Weather," Oxford Univ. Press, New York.

Thompson, R. D., and Perry, R., eds. (1997). "Applied Climatology: Principles and Practice," Routledge, London.

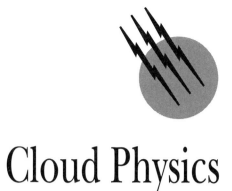

Cloud Physics

Andrew J. Heymsfield
National Center for Atmospheric Research*

GLOSSARY

Adiabatic process Thermodynamic change of state of a system in which there is no transfer of heat or mass across the boundaries of the system.

Cirriform cloud Cloud with a relatively transparent and white or silky appearance that forms at high altitudes and is composed of small particles, typically ice crystals.

Cloud condensation nucleus (CCN) Small, solid particle, typically containing a salt, on which condensation of water vapor begins in the atmosphere.

Cloud droplet Particle of liquid water from a few micrometers to ~200 μm in diameter.

Cloud seeding Any technique carried out with the intent of adding to a cloud certain particles that will alter the natural development of the cloud.

Cumuliform cloud Cloud whose principal characteristic is vertical development in the form of rising mounds, domes, or towers.

Graupel Spherical or conical ice particle about 2–5 mm in diameter, consisting of a white or opaque snowlike structure, formed by the collection of cloud droplets.

Hailstone Spherical, conical, or irregularly shaped ice particle, from ~5 mm to more than 5 cm in diameter, formed by the collection of cloud droplets and drops.

*The National Center for Atmospheric Research is supported by the National Science Foundation.

Ice crystal Any one of a number of macroscopic crystal forms in which ice appears in the atmosphere.

Ice nucleus Any particle that serves as a nucleus for the formation of ice crystals in the atmosphere.

Stratiform cloud Cloud of extensive horizontal development.

Supercooled cloud Cloud composed of liquid water droplets at temperatures below 0°C.

Supersaturation Condition existing in a given portion of the atmosphere when the relative humidity is greater than 100%.

CLOUD PHYSICS is a discipline within meteorology concerned with the properties of atmospheric clouds and the processes that operate within them, the diversity of phenomena intrinsic to natural clouds, the interactions of clouds with the atmosphere, and the effects of clouds on climate. The discipline covers the range from single clouds to large-scale weather systems and even weather on a global scale.

Cloud physicists draw on the well-developed sciences of chemistry, physics, and fluid dynamics to study these phenomena. Such topics as the thermodynamics of moist air, the physics of the growth of water droplets and ice particles, radiation, effects of clouds on climate, electrification, and chemical conversion processes are all part of this discipline. Major research tools include computers for numerical simulation and aircraft and radars for observation, along with wind tunnels and cold rooms for the study of the properties of cloud and precipitation particles.

I. CLOUD FORMATION MECHANISMS AND CLOUD CLASSIFICATION

A complete description of the many genera, species, and varieties of clouds is given in the "International Cloud Atlas" (1956) of the World Meteorological Organization. This international classification evolved from the experience of ground observers over many years, depending primarily on appearance. From a physical point of view, the distinctions among types of clouds arise from the vertical motion characteristics that produce them and from their microphysical properties—the presence or absence of ice particles and the sizes and concentrations of each type of cloud particle.

There are three fundamental classes of clouds: cirrus, cumulus, and stratus. *Cirrus* are high clouds with a silken appearance, because they are composed of ice crystals. *Cumulus* are detached, dense clouds that rise in mounds or towers from a level base. *Stratus* is the name given to an extensive layer or flat patches of low clouds showing hardly any well-defined detail. These names are some-

times used in conjunction. For example, when cirrus is in layered form, it is termed cirrostratus; a low-level layer cloud that is broken up into a wavy pattern is called stratocumulus; and so on. Similar clouds at intermediate levels are called altocumulus, and a thick-layer cloud at these levels is called an altostratus. A class of lesser fundamental importance comprises the lenticular cloud, having a lens or almond shape.

Most types of clouds are formed as a result of vertical motions produced in the following ways:

1. *Widespread gradual lifting.* Upglide motion of air at a frontal surface occurs in the cyclones of temperate latitudes and gives rise to expansive layers of deep and often layered altostratus and nimbostratus clouds. The vertical component of air velocity is of the order of a few to a few tens of centimeters per second, and the lifting results in steady precipitation of long duration.

2. *Widespread irregular stirring.* When air is cooled at the ground, fogs may form. Over land at night, the cooling may be due to the radiation of heat from the ground, but fogs also occur over land and sea when air flows slowly into regions of lower surface temperature.

3. *Convection.* Heating at the ground either by sunshine or when cool air undercuts warmer air can cause masses of air from the surface layers to ascend through a relatively undisturbed environment, often producing clouds of cumuliform type. Above the level of the cloud base the liberation of latent heat due to the conversion of water vapor to droplets usually increases the buoyancy and vertical velocity of the rising masses. In settled weather, "fair weather" cumulus clouds are well scattered and small, with horizontal and vertical dimensions of only 1 or 2 km and vertical motions from about 1–5 m sec^{-1}. In disturbed weather, cumulus congestus and cumulonimbus (thunderhead) clouds form. The tops of the latter can reach up to 20 km above the ground, spreading out into a flat-topped, anvil shape. Thunderclouds can have updrafts of more than 40 m sec^{-1} in the most extreme cases and can produce heavy rain, hail, and tornadoes.

4. *Orographic lifting.* When an air mass moves against a mountain barrier, some of the air is forced to rise. This can result in an extensive sheet of deep stratiform and cumuliform clouds. Lenticular clouds can also form high above mountains and in their lee (downwind locations) within mountain-generated waves.

II. COOLING OF MOIST AIR

A. Water Vapor in the Atmosphere

The amount of water vapor present in the atmosphere is dependent in a complex way on (1) the amount that enters the atmosphere by evaporation and sublimation, (2) its transport by air motions throughout the troposphere (lower portion of the atmosphere) and lower stratosphere, and (3) the amount precipitated in the form of rain, snow, and hail.

Two factors account for the observed decrease in the concentration of water vapor (termed the *water vapor mixing ratio*, grams of water vapor per gram of air) with height. First, the earth's surface is the primary source of water vapor. Second, the air temperature decreases with height in the troposphere. Since the maximum possible mixing ratio, termed *saturation*, decreases with decreasing temperature, water is squeezed out as parcels ascend in the atmosphere.

B. Cooling of Air in the Absence of Water Vapor Condensation

When a parcel of air is lifted in the atmosphere, it expands and cools. Heat may be added to the parcel through such effects as radiation and friction, but, in most cases, the resulting changes in the temperature of the parcel are secondary to the expansion process. It is a reasonable and useful idealization to assume, then, that the expansion is *adiabatic*, that is, that there is no transfer of heat or mass into or out of the parcel, and it is a reversible process. Furthermore, in the absence of water vapor condensation, no heat is added within a rising parcel. From the first law of thermodynamics for an ideal gas, the decrease in temperature with height in the absence of condensation, termed the *dry adiabatic lapse rate*, works out to be $\sim 10°C\,km^{-1}$, or about a cooling of $\sim 1°C$ for every 100 m of lift. The cooling rate in a rising parcel cannot be higher than dry adiabatic.

C. Cooling of Air with Water Vapor Condensation

If a parcel of air being lifted dry adiabatically achieves a relative humidity of 100% (i.e., becomes saturated), water vapor condenses to form a cloud, and latent heat of condensation is released. (This discussion does not consider the latent heat that can be released when ice is present.) Thus, the parcel now cools at a rate less than dry adiabatic, the rate depending on whether all or part of the condensate stays within the parcel. If all of it remains, the first law of thermodynamics can again be used to derive the moist adiabatic lapse rate. The resulting lapse rate is not constant as is the dry adiabatic lapse rate, but is dependent on pressure and temperature. For 1000 kPa and 20°C, this lapse rate is $\sim 4°C\,km^{-1}$, while at the same pressure and a temperature of 0°C, it increases to $\sim 6°C\,km^{-1}$. At temperatures below about $-30°C$, the moist adiabatic lapse rate approximates the dry adiabatic rate.

D. Cooling with Entrainment

Numerous in-cloud measurements, along with theoretical and laboratory studies, have indicated that *entrainment* takes place; that is, air in the environment surrounding a parcel of air is mixed into the parcel and becomes part of the rising current. A rising parcel of cloudy air into which dry air is entrained cools at a faster rate than moist adiabatic, because heat is required (1) to evaporate sufficient condensate, increasing the mixing ratio of the air to saturation, and (2) to warm the air from its original temperature to the parcel temperature. The lapse rate in an entrained parcel of air falls somewhere between the moist and dry adiabatic rates.

The adiabatic model of cooling in a rising parcel has been modified by several cloud physicists to take entrainment into account. These researchers have proposed that entrained air originates primarily either from the sides of clouds (lateral entrainment) or from their tops (cloud-top entrainment). The latter model is gaining fairly widespread acceptance.

E. Summary of Cooling Processes During Cumulus Cloud Formation and Resulting Thermal Instability

The vertical temperature distribution in a rising column of air associated with a cumulus cloud is given by the solid line in Fig. 1. The segment *AB* represents dry adiabatic ascent, and the in-cloud segment *BD* represents moist ascent with entrainment. In the absence of entrainment, temperatures would be moist adiabatic, given by dotted line *BC*. The environmental temperature distribution is given by the dashed line. Along segment *AB*, vertical velocities may be of the order of a few meters per second. Along segment *BD*, the parcel becomes increasingly warmer relative to the environment and thus more buoyant, possibly increasing the vertical velocity more than 10 m sec^{-1}. Beyond point *D*, the parcel is negatively buoyant, and the velocities decrease to 0 m sec^{-1}, producing the cloud top.

III. FORMATION OF CLOUD DROPLETS

As a parcel of air is cooled toward saturation, the relative humidity approaches 100%, and water vapor begins to condense, or *nucleate*, on small particles of airborne dust, or *cloud condensation nuclei* (CCNs). These small particles usually contain a soluble component, often a salt such as sodium chloride or ammonium sulfate. CCNs of different sizes and compositions are present at each position in the atmosphere. Some of the nuclei become wet at relative humidities below 100% and form haze, while the

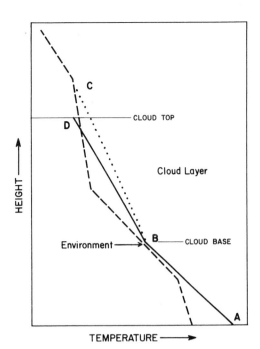

FIGURE 1 Temperature vs height distributions. Solid line, in a rising parcel of air: *AB*, below cloud base, dry adiabatic; *BD*, cloud base to cloud top, entrained ascent. Dotted line: *BC*, moist adiabatic. Dashed line: environmental temperatures.

relatively large CCNs are the most likely ones to grow and become cloud droplets.

This initial stage of droplet formation deserves a careful explanation. Over a flat, pure water surface at 100% relative humidity (saturation with respect to water), water vapor is in equilibrium, which means that the number of water molecules leaving the water surface is balanced by the number arriving at the surface. Molecules at water surfaces are subjected to intermolecular attractive forces exerted by the nearby molecules below. If the water surface area is increased by adding curvature, molecules must be moved from the interior to the surface layer, in which case energy is required to oppose the cohesive forces of the liquid. As a consequence, for a pure water droplet to be at equilibrium, the relative humidity has to exceed the relative humidity at equilibrium over a flat, pure water surface, or be *supersaturated*. The flux of molecules to and from a surface produces what is known as *vapor pressure*. The equilibrium vapor pressure is less over a salt solution than it is over pure water at the same temperature. This effect balances to some extent the increase in equilibrium vapor pressure caused by the surface curvature of small droplets. Droplets with high concentrations of solute can then be at equilibrium at subsaturation.

It follows from the surface energetics of a pure or solution droplet that once it achieves a certain "critical" radius, at a time at which the supersaturation in its environment

achieves a certain "critical" value, the droplet will grow spontaneously and rapidly.

Equations have been developed to express the equilibrium supersaturation of a droplet of a given radius in terms of the composition and size of the CCNs. A family of curves showing the fractional equilibrium relative humidity (relative humidity/100%) of water droplets containing differing masses of sodium chloride salt is shown in Fig. 2. A curve for pure water droplets is also shown. Asterisks at the peak of each curve show the critical radius and supersaturation of the droplets and indicate that the higher the mass of the salt, the lower the critical supersaturation. A family of curves can also be generated for other salts found in the atmosphere, such as ammonium sulfate.

Figure 2 can be used to understand the initial formation of droplets. One can see that a spectrum of droplet sizes will be produced in a rising parcel of air because of the diversity of sizes and composition of CCNs. When the supersaturation in the parcel has increased to such an extent that the droplets that have achieved their critical supersaturation are consuming more of the vapor than is being produced by cooling of the air, the supersaturation begins to decrease, and below a certain point no additional droplets are produced. In marine environments, where abundant sodium chloride nuclei of relatively large masses (about 10^{-15}–10^{-14} g) are produced through sea spray, the condensed water deposits on the nuclei with large masses and the peak saturation ratio achieved in a parcel is comparatively low. In continental areas, where relatively few large sodium chloride nuclei are present, the condensed water deposits on nuclei with small masses

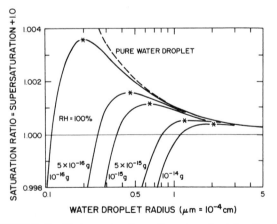

FIGURE 2 Equilibrium saturation ratio versus droplet radius for different masses of sodium chloride nuclei (solid lines). Asterisks show droplet radius where solution drop containing indicated mass of sodium chloride will continue to grow without further increase in the saturation ratio. Dashed curve is for a pure water droplet. RH, Relative humidity. [Adapted from Byers, H. R. (1965). "Elements of Cloud Physics," courtesy of the University of Chicago Press and the author.]

(about 10^{-15}–10^{-16} g) and higher supersaturation occurs. Given the same condensed liquid water content, maritime areas will have fewer and larger droplets than continental areas.

IV. PHYSICS OF THE GROWTH OF CLOUD DROPLETS

A. Diffusional Growth

After a solution droplet has been nucleated, it enters a stage of growth by diffusion of vapor to it. This growth is maintained as long as the saturation ratio, or supersaturation for it to be at equilibrium, is exceeded (see Fig. 2).

Consider a solution droplet of radius r in a supersaturated environment in which the concentration of vapor molecules at distance R from the droplet center is denoted by $n(R)$. This vapor diffuses toward the droplet and condenses on it. At any point in the vapor field the concentration of vapor molecules must satisfy the equation representing diffusion, assuming certain approximations that are not discussed here:

$$\nabla^2 n(R) = 0 \qquad (1)$$

When $R = \infty$, n must have the value n_0, the concentration of vapor at a great distance from the droplet. At the droplet's surface n must equal n_r, the equilibrium vapor concentration over the droplet surface. The solution to this equation is

$$n(R) = n_0 + (n_r - n_0)r/R \qquad (2)$$

The flux of molecules, each of mass m, on the surface of the droplet is equal to $D(\partial n/\partial R)$, where D is the diffusion coefficient. With the vapor density in the environment being ρ_v, given by mn_0, and at the droplet's surface ρ_{vr}, given by mn_r, the rate of mass increase of the droplet is

$$dm/dt = 4\pi r D(\rho_v - \rho_{vr}) \qquad (3)$$

and since the droplet is spherical, with a radius r and a density ρ_L (1 g cm^{-3}), the time rate of radius increase is

$$dr/dt = D(\rho_v - \rho_{vr})/\rho_L r \qquad (4)$$

Equations (3) and (4) can be expressed in terms of the supersaturation since the term ρ_{vr} can be found from the CCN composition and mass.

The growth histories of individual droplets at a constant supersaturation of 1% and an air temperature of 20°C have been calculated in Fig. 3 according to the principles used in deriving Eq. (4). Of considerable importance is that the drops with smaller initial radii catch up to the size of the drops with larger initial radii.

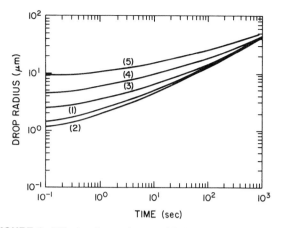

FIGURE 3 Diffusional growth rate of individual solution drops containing indicated salt mass as a function of time at 1% supersaturation and 20°C. (1) Sodium chloride, $m = 10^{-12}$ g; (2) ammonium sulfate, $m = 10^{-12}$ g; (3) sodium chloride, $m = 10^{-11}$ g; (4) sodium chloride, $m = 10^{-10}$ g; (5) sodium chloride, $m = 10^{-9}$ g. [From Pruppacher, H. R., and Klett, J. D. (1978). "Microphysics of Clouds and Precipitation," courtesy of D. Reidel Publishing Co. and the lead author.]

B. Growth through Droplet Collisions

Most of the earth's precipitation reaches the ground as "drops" of radius 100 μm or larger, many of which form in clouds whose tops do not extend to a level where ice particles are produced. Figure 3 suggests that drops cannot be produced solely through diffusion in any type of cloud. Growth through diffusion becomes very slow after a radius of $\sim$10 μm is reached, and the mechanism that then takes over consists of collisions and merging (coalescence) of cloud droplets.

A spectrum of droplet sizes is formed at each position in a cloud for reasons discussed previously, among others. Collisions between droplets, and between drops and droplets, can occur because droplets and drops of different sizes have differing fall velocities; large droplets or drops fall faster than smaller ones, overtaking and collecting some fraction of those lying in their paths. Electrical fields may promote additional collections.

As a drop falls, it collides with only a fraction of the droplets in its path because some are swept aside in the airstream moving around the drop. Also, some droplets that collide rebound and do not merge. The ratio of the actual number of collisions to the number for geometric sweep-out is called the *collision efficiency*, E, and depends primarily on the size of the collecting drop, R, and the sizes of the collected droplets, r. Collisions first occur for drops with radii as small as 10 μm, though with small efficiency. Collision efficiency E is generally an increasing function of R and r and has values greater than 0.5 when R is greater than 100 μm for most cloud situations.

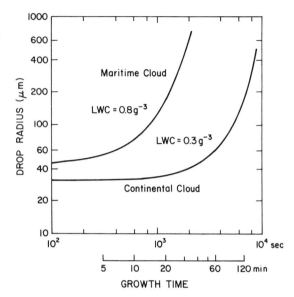

FIGURE 4 Size variation of a drop growing by collision and coalescence in a maritime- and a continental-type cloud. LWC, liquid water content. [From Braham, R. R. (1968). *Bull. Amer. Meteorol. Soc.* **49**, 343–353. Courtesy of the American Meteorological Society and the author.]

It follows from the nature of the geometric sweepout process that the change in drop radius with time is approximately equal to

$$dr/dt = \bar{E} \times \text{LWC} \times V_{\text{T}}/(4\rho_{\text{L}}r), \qquad (5)$$

where $\bar{E}$ is the effective average value of collection efficiency for the droplet population, LWC is the liquid water content, and V_{T} is the fall velocity of the collecting particle. Figure 4 shows the calculated size variations of a drop growing by collision and coalescence in a cloud in a maritime environment (containing relatively few droplets, each of fairly large size) and in a continental environment (containing a relatively large number of droplets, each of relatively small size). Note that this growth by collection is much more rapid than diffusional growth (Fig. 3).

Even if a cloud is spatially homogeneous or "well mixed" with the same average droplet concentration throughout, there will be local variations in droplet concentrations. These follow the Poisson probability law. Equation (5) does not take into account the possibility of statistical fluctuations in the droplet spectrum but applies only to average droplet growth. Some "fortunate" drops fall through regions of locally high droplet concentrations, encountering more than the average number of droplets early in their growth and subsequently can grow more rapidly. Equations have been developed to take into account the statistical, or *stochastic*, nature of the droplet and drop growth process, and it is now recognized as crucial in the early stages of coalescence. It explains why some drops form relatively rapidly and why a distri-

bution of drop sizes is produced in a relatively uniform cloud.

Two major processes limit the growth of drops in otherwise favorable conditions. First, drops larger than ~0.1 cm that collide with drops greater than 300 μm often break up and produce a multitude of smaller drops. Second, raindrops are limited to sizes of less than ~0.5 cm because at larger sizes they spontaneously break up from aerodynamic forces acting on their surface.

V. OBSERVATIONS OF CLOUD DROPLETS AND DROPS

A. In-Cloud Measurement Devices

Currently, a total of approximately 25 aircraft in the United States, Canada, and Europe are equipped to collect cloud physics data. Three primary instruments on these aircraft acquire data on the concentrations of cloud droplets as a function of droplet diameter (the cloud droplet size spectrum) and the cloud liquid water content. The size spectrum is obtained primarily from an electrooptical device known as the forward-scattering spectrometer probe (FSSP), which sizes droplets over the diameter range 1.5–46.5 μm in 3 μm increments, although other sizing bounds can be used. The size spectrum is also obtained using another electrooptical device, which produces two-dimensional images of particles; this optical array probe is described in more detail in Section IX. The liquid water content can be obtained indirectly, from the size spectrum data, by integration of the measured droplet spectrum. The liquid water content can be measured directly using "hot-wire" devices. Droplets impinge on a hot wire, and because they evaporate, they cool the wire. The temperature of the wire or the current required to maintain it at a constant temperature is used to derive the liquid water content.

B. Observations of Droplets

Most drop-size distributions, even though measured in many different types of clouds formed under differing meteorological conditions, exhibit a characteristic shape. The concentrations generally increase with size abruptly from a low value to a maximum somewhere between 10 and 20 μm, and then decrease gradually toward larger sizes, causing the distribution to be skewed with a long tail toward the larger sizes. A log normal or γ distribution function is used to approximate this characteristic shape. When coalescence growth is involved, the spectra contain a characteristic secondary peak in concentration at diameters usually between 20 and 40 μm.

Significant differences are found between spectra formed in maritime air and those formed in continental

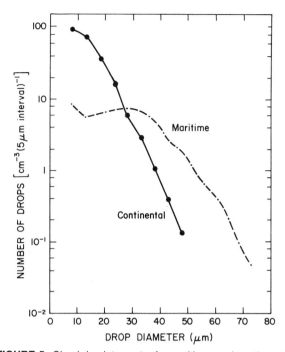

FIGURE 5 Cloud droplet spectra for maritime- and continental-type cumulus clouds. These curves are used for the drop growth rate computations in Fig. 4. [From Braham, R. R. (1968). *Bull. Amer. Meteorol. Soc.* **49,** 343–353. Courtesy of the American Meteorlogical Society and the author.]

air masses because of the previously discussed nature of the CCNs. Figure 5 illustrates these differences between typical spectra in a maritime and a continental cumulus cloud. Typical concentrations and mean diameters in maritime convective clouds are 50–100 cm^{-3} and 15–20 μm, respectively, while in continental convective clouds they are 500–1000 cm^{-3} and 10–15 μm, respectively.

Liquid water contents are dependent on the cloud type, cloud base temperature, and height above cloud base. In stratiform clouds, the values are comparatively low, usually ~0.1 g m^{-3}. In cumulus clouds, typical peak values are ~0.5 g m^{-3}, which increase with increasing cloud intensity to more than 3 g m^{-3} for severe thunderstorms. (As the cloud base temperature increases, a cloud of a given type tends to have a higher liquid water content, increasing with height from cloud base to near cloud top and then falling abruptly to zero at cloud top.)

C. Drop-Size Spectra in Rain

A multitude of measurements of drop-size spectra in rain have been made at the ground. These measurements indicate that drop-size distributions are of an approximately negative exponential form, especially in rain that is fairly steady. The so-called Marshall–Palmer distribution appears to describe the rain spectrum in most meteorological situations. This spectrum has the form

$$N(D) = N_0 e^{-\Lambda D}, \tag{6}$$

where the product $N(D)\,dD$ is the number of drops per unit volume with diameters between D and $D + dD$. The slope factor Λ of the spectrum is given by

$$\Lambda = 41 R^{-0.21} \tag{7}$$

where R is the rainfall rate (in millimeters per hour), and N_0, the intercept of the spectrum, is given by

$$N_0 = 0.08 \text{ cm}^{-4} \tag{8}$$

VI. ICE FORMATION MECHANISMS

Ice crystals form in the atmosphere in three ways: (1) on solid particles, often airborne soil particles, referred to as *ice nuclei*; (2) by secondary processes that multiply primary ice crystals formed by (1); and (3) by *homogeneous nucleation*, that is, pure water droplets freezing spontaneously when a temperature of −40°C is reached, probably typical of cirrus clouds.

A. Ice Nuclei and Primary Ice Crystals

It was realized many years ago that clouds of liquid water droplets can persist at temperatures below 0°C (supercooled) unless suitable ice nuclei are present to help ice crystals form. Ice nuclei provide a surface having a structure geometrically similar to that of ice, thereby increasing the probability of formation of the ice structure required for stability and thus causing microscopic droplets to freeze relatively rapidly at temperatures higher than −40°C.

Currently, there is much uncertainty about the mechanisms of ice nucleation in the atmosphere, but it is thought that ice nuclei operate by three basic modes. In one mode, water is absorbed from the vapor phase onto the surface of the ice nucleus, and at sufficiently low temperatures, the adsorbed vapor is converted to ice. In another mode, the ice nucleus, which is inside a supercooled droplet either by collection or as a result of its participation in the condensation process, initiates the ice phase from inside the droplet. In the third mode, the ice nucleus initiates the ice phase at the moment of contact with a supercooled droplet (such nuclei are known as contact nuclei). The relative importance of these different modes of operation is not known with certainty, but the latter two modes are thought to be much more common than the first.

Much effort has been made to measure the concentrations of ice nuclei in the air and the variation with temperature. While considerable variability in concentration is found with time and location, a fairly representative worldwide concentration is ~1 per liter active at −20°C, changing with temperature in the way shown in Fig. 6.

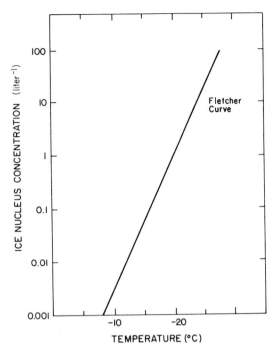

FIGURE 6 Worldwide "average" of measured ice nucleus concentrations versus temperature (the Fletcher curve).

(This is the so-called Fletcher worldwide average ice nucleus curve. It must be emphasized, however, that there is a great deal of scatter of the data about the Fletcher curve, and it may not provide a useful representation of the concentrations in the atmosphere.)

When the temperatures in-cloud are near or below $-40°$C and the ambient relative humidity is close to 100% with respect to water saturation—conducive to cirrus cloud formation—homogeneous ice nucleation is likely to be an important mechanism for ice production. The theory of homogeneous nucleation of ice has shown that when molecules of liquid water comprising a pure droplet a few microns in diameter bind to form a stable ice structure at an observable rate only when temperatures near $-40°$C is reached. This theory has been confirmed in the laboratory. What has been confirmed in studies in orographic wave clouds, and in cirrus cloud as well, is that when water condenses on a cloud condensation nucleus at temperatures near and below $-40°$C and grow to a sizes of 1 μm, the droplets freeze homogeneously. This mechanism is thought to be important in the formation of cirrus clouds.

B. Secondary Ice Crystals

Discrepancies between concentrations of ice crystals and ice nuclei have been clearly demonstrated in clouds containing air of maritime origin. These clouds, often with tops no colder than $-10°$C, contain concentrations of ice crystals that can exceed that of ice nuclei by a factor of 10^4 and typically contain droplets larger than 25 μm. Whatever process is operating seems to be most effective at a temperature of $-6°$C.

Various mechanisms have been proposed to explain this important ice nucleation process. In one, drops that are in the process of freezing build up high internal pressures, shatter, and produce secondary ice particles. Laboratory studies indicate that this mechanism applies to droplets with diameters greater than 400 μm; it is not likely to be important in most clouds, however. In another, secondary ice particles are thought to be ejected when supercooled drops freeze onto an ice particle at temperatures between $-3°$ and $-8°$C. Experiments have demonstrated that this mechanism is operative in the laboratory when the particle collecting the droplets is falling at a velocity of 1–3 m sec^{-1}, when some of the droplets being collected are larger than 25 μm in diameter and when some are about half this diameter. The nature of this process has been inferred from laboratory experiments, and it appears likely to explain the initiation of many of the secondary crystals observed in the atmosphere. Collections of ice particles at ground level and in clouds often contain fragments of crystals, which indicates that a third mechanism of producing secondary ice particles is the fracturing of ice particles on collisions with other ice particles. Ice multiplication by this process (mechanical fracture) is still not very well understood, nor has its importance been documented. A fourth mechanism involves the aerodynamic breakup of particles in subsaturated layer, where components of crystals can break off from the parent particles.

VII. OBSERVATIONS OF THE GROWTH MECHANISMS OF ICE PARTICLES

Observations of ice particle shapes and sizes from their origin in clouds through their fallout at the ground indicate that growth proceeds by the chain of processes illustrated in Fig. 7. Following nucleation, crystals grow first through vapor diffusion. Observations indicate that they must achieve a diameter of ~200 μm before they begin to collect, or *accrete*, water droplets. Water drops that freeze in the air at a diameter of 200 μm or larger collect water drops immediately. In the absence of liquid water, ice crystals grow into snow crystals and often clump together to form snowflakes. Rimed crystals or rimed snowflakes that continue to rime can grow into spherical or conical low-density ice particles (graupel) and in some instances continue to grow to diameters larger than 0.5 cm (hail) and are then nearly solid ice. Frozen drops grow immediately into graupel and can also grow into hail. Ice particles then reach the ground in the form of snow, rain, hail, or graupel.

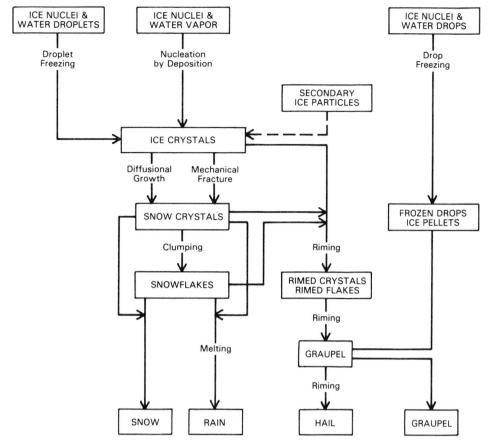

FIGURE 7 Flow chart showing processes of ice particle growth. [Adapted from Braham, R. R., and Squires, P. (1974). *Bull. Amer. Meteorol. Soc.* **55,** 543–556. Courtesy of the American Meteorological Society and the authors.]

Observations also show that ice and snow crystals grow in a large variety of shapes, or *habits*. The major habits illustrated in Fig. 8 are plates, dendrites, columns, needles, bullets, and combinations of bullets. Laboratory and theoretical studies reveal that snow crystals have one common basic shape, that of a sixfold, symmetric (hexagonal) prism composed of two basal planes (e.g., the faces of the crystals; see crystals in Figs. 8a–8c) and six prism planes (the crystal edges, see crystals in Figs. 8d–8f). Laboratory experiments reveal that the rate of propagation of the basal faces (growth along the *c* axis) relative to that of the prism faces (*a* axis) varies with temperature and supersaturation in a characteristic manner. The results of experimental studies of crystal habits are shown as a function of temperature and relative degree of water vapor supply in Fig. 9. A dashed, horizontal line in the figure shows water saturation conditions (100% relative humidity). A solid, horizontal line at the bottom of the figure shows ice saturation conditions. (In terms of relative humidity with respect to water, ice saturation equals 100% at 0°C and decreases nearly linearly with temperature, equaling ~60% at −40°C.) The important points to note from this figure

are twofold. There are changes in habit from columns to plates to columns that occur at precise temperatures, and at a particular temperature, the relative humidity controls some important features of snow crystal growth.

Measurements have illustrated that ice crystal growth rates depend strongly on temperature. Figure 10 shows the results of laboratory measurements of the growth rates along the *a* and *c* axes, designated in the figure as da/dt and dc/dt. These data represent averages over a ~3-min period, which started with crystals ~20 μm in diameter. This figure illustrates that there is a marked maximum in the growth rate along the *a* axis maximum at −15°C and that a secondary broader maximum occurs along the *c* axis at −6°C.

VIII. PHYSICS OF THE GROWTH OF ICE PARTICLES

The ice crystal growth processes presented in the previous section (Fig. 7) are discussed from a physical point of view in this section.

FIGURE 8 Major shapes of snow crystals: (a) simple plate, (b) dendrite, (c) crystal with broad branches, (d) solid column, (e) hollow column, (f) sheath, (g) bullet, and (h) combination of bullets. [From Pruppacher, H. R., and Klett, J. D. (1978). "Microphysics of Clouds," who used photographs from Nakaya, U. (1954). "Snow Crystals," Courtesy of D. Reidel Publishing Co., H. R. Pruppacher, and Harvard University Press, copyright 1954 by the President and Fellows of Harvard College.]

A. Growth of Ice Crystals by Diffusion

On formation of ice embryos, whether by sublimation directly from the vapor, by freezing of supercooled droplets, or by a secondary production mechanism, diffusional growth commences because the embryo is in an environment that is probably at or close to water saturation. The growth rate equations can be derived using theory analogous to that presented in Section IV, but since ice crystals are generally not spherical, some alterations must be made.

The approach that has been used to calculate ice crystal growth starts with an analogy between the governing equations and the boundary conditions for electrostatics and diffusion problems. Poisson's equation in electrostatics and Green's theorem lead to the following equation for the ice crystal diffusional mass growth rate,

$$dm_d/dt = 4\pi C D(\rho_v - \rho_{iR}) \tag{9}$$

where ρ_{iR} is the vapor density at the ice particle surface, C is the "capacitance" of the ice crystal, and the other terms have been previously defined. In order to apply Eq. (9) to a particular crystal form, C is specified as a function of the crystal geometry. In the simplest case of a spherical crystal of radius R, $C = R$, and Eq. (9) then has a form similar to the mass growth rate of drops [Eq. (3)]. A slightly more complex case is a simple thin hexagonal plate, for which

$C = 2R/\pi$. Laboratory experiments of the capacitances of brass models of snowflakes have confirmed the validity of this approach for calculating the ice crystal growth rate.

The linear rate of growth of a crystal can be derived from Eq. (9), given knowledge of the crystal geometry, thickness-to-diameter ratio (axial ratio, AR), and bulk density ρ_i. For example, a hexagonal plate crystal has a mass given by $m = 5.2R^3 \times AR \times \rho_i$, and so its growth rate along its major axis is

$$\frac{dR}{dt} = \frac{0.51D(\rho_v - \rho_{iR})}{AR \times \rho_i R} \tag{10}$$

For columnar and needle crystals, growth along the crystal length L is given by

$$\frac{dL}{dt} = \frac{3.22D(\rho_v \times \rho_{iR})}{\ln(2AR)\rho_i L} \times AR^2 \tag{11}$$

It is important to point out here that as crystals become increasingly large, their linear growth rate becomes increasingly small: From Eq. (10), dR/dt is proportional to $1/R$, and from Eq. (11), dL/dt is proportional to $1/L$. Data on the crystal axial ratios and bulk density are necessary to solve Eqs. (10) and (11). The data given in Fig. 10 can be used to compute the axial ratio: $AR = (da/dt)/(dc/dt)$. Additional axial ratio information is available from field data. Laboratory and field data indicate that bulk densities have typical values of between 0.5 and 0.8 g cm^{-3}.

Solving for R and L as a function of time in Eqs. (10) and (11) also requires values for D, ρ_v, and ρ_{iR}. Meteorological tables or equations can be used to derive the value of D and ρ_v if the temperature, pressure, and relative humidity of the growth environment are known. Now consider the vapor density at the crystal surface, ρ_{iR}. As the ice crystal grows, its surface is heated by the latent heat of vaporization, and because of this warming, the value of ρ_{iR} is effectively raised above the value that would apply without heating. Under stationary growth conditions, the value of ρ_{iR} is determined from the rate of latent heating and the rate of heat transfer away from the surface. If the crystal has grown to a size of more than a few hundred micrometers, at which time it has an appreciable fall velocity, it is necessary to take into account the effect of "ventilation" on the diffusion of water vapor and heat. Fairly simple expressions have been derived to solve for ρ_{iR} in Eqs. (9) to (11). Values for the term $D(\rho_v - \rho_{iR})$ as a function of temperature are shown in Fig. 11 for water saturation, for 100-μm crystals, and for pressures of 1000 (sea level) and 400 mbar ($\sim$6 km). The important points to note from this figure are that (1) this term is at a maximum at a temperature of about $-15°$C, suggesting that particles can potentially grow most rapidly at this temperature, and (2) the term is higher at 400 than at 1000 mbar, indicating

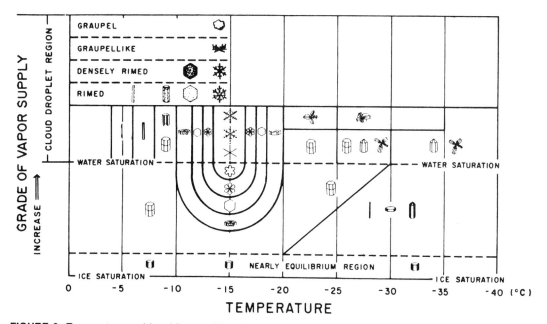

FIGURE 9 Temperature and humidity conditions for the growth of natural snow crystals of various types. [From Magono, C., and Lee, C. W. (1966). *J. Fac. Sci. Hokkaido Univ. Ser.* **7**, 2(4). Courtesy of the Faculty of Sciences, Hokkaido University, Japan.]

that the growth rate increases with altitude, given a constant temperature.

It still remains to be shown that the growth rate equations provide a reasonable representation of the experimentally measured growth rates in Fig. 10. To simulate the laboratory conditions, growth has been calculated over a ~3-min period beginning with crystals ~20 μm in diameter. The average values of da/dt and dc/dt are shown in Fig. 12. The growth equations appear to emulate the salient features of the observed growth in Fig. 10, at least during these early stages of crystal growth. Data at later stages are not available for comparison.

Ice crystals are often carried or descend into regions where the relative humidity is below saturation with respect to ice and begin to evaporate; common ice subsaturated regions are near the bases of thunderstorm anvils and at the base of cirrus clouds. The growth rate equations given by Eqs. (9) to (11) can be used to compute the rate

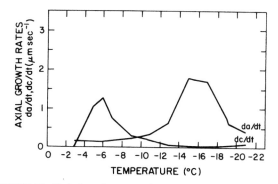

FIGURE 10 Variation of measured crystal axial growth rates with temperature over a period of ~3 min following ice crystal nucleation. [Adapted from Ryan, B. F., Wishart, E. R., and Shaw, D. E. (1976). *J. Atm. Sci.* **33**, 842–851. Courtesy of the American Meteorological Society and the lead author.]

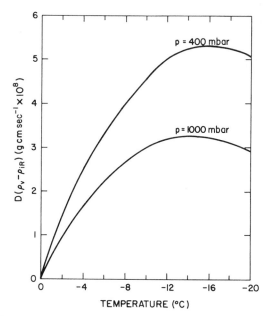

FIGURE 11 Variation of a term in the ice crystal growth rate equation with temperature at two atmospheric pressures.

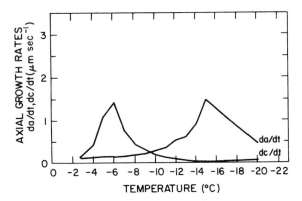

FIGURE 12 Variation of calculated crystal axial growth rates with temperature over a period of ~3 min following ice crystal nucleation.

at which crystal mass and linear dimensions decrease in these regions.

B. Growth through Collisions between Ice Particles

Snowflakes rather than individual ice crystals account for most of the precipitation reaching the ground as snow. As crystals become increasingly large, they are more likely to bump into one another, and some stick together or clump to form an aggregate. The physics of aggregation is summarized below.

Theoretical approaches to snowflake growth compute the number of collisions that can occur between the crystals in a given volume of air. Consider particles of radius R_1 and terminal velocity V_1 in concentration N_1 in close proximity to those of radius R_2, velocity V_2, and concentration N_2. The number of the faster falling particles R_1 that collide with R_2 in time dt is given by the product of the following three terms: (1) the volume that they collectively sweep out, $\pi(R_1 + R_2)^2 V_1 N_1 N_2\, dt$; (2) their relative terminal velocity, $V_1 - V_2$; and (3) the efficiency at which they collide, E. The evolution of the size spectrum of crystals and aggregates is obtained by calculating the collisions that could occur between all sizes of crystals, aggregates and crystals, and aggregates and aggregates that are present in a given volume of air, taking into account their respective collection efficiencies E.

The theoretical studies have demonstrated that aggregates can grow at a rate of up to 10 μm sec^{-1}, in contrast to single ice crystals of the same size, which grow at a rate of 1–10% of this value. Small variations in the velocity of particles of the same size can lead to an even more rapid growth rate, and aggregation leads to the development of a size distribution in which the concentration decreases exponentially with increasing size, all in agreement with observations.

C. Growth by Droplet Collection (Accretion)

An ice crystal in a cloud of supercooled water droplets grows by accretion into a rimed crystal, a graupel particle, or a hailstone. Water drops can grow by an analogous process of drop–droplet collection, as previously discussed (Section IV.B). Theoretical treatments of the accretion process are much more complex than those for drops, because they must consider a variety of complex shapes (e.g., Fig. 8) changing with time during the accretion process, the release of latent heat at the crystal surface by droplet freezing, and the density of droplets accreted on the surface of the ice particle since the density depends on the characteristics of the ice particle (these densities vary from 0.1–0.91 g cm^{-3}).

The basic equation for the particle mass growth rate during accretion can be written as the sum through diffusion and accretion,

$$dm/dt = (dm_d/dt) + A \times E \times \text{LWC} \times V_T, \qquad (12)$$

where dm_d/dt is the diffusional growth rate term, A is the cross-sectional area of the particle normal to the airflow, and the other terms are as previously defined. The diffusional growth rate term must be calculated by considering the effects of latent heating due to sublimation and droplet freezing versus conduction of heat away from the particle. During the early stages of riming, when a crystal is beginning to grow into graupel, the growth rate dm_d/dt is positive, while in later periods, for example, graupel and hail growth stages, dm_d/dt is negative. The particle terminal velocity V_T and cross-sectional area A are the two factors that control the number of droplets a particle sweeps out per unit time, and the terminal velocity is highly dependent on the particle shape (e.g., whether crystal or graupel), the diameter, and the bulk density (Fig. 13).

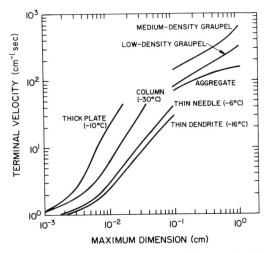

FIGURE 13 Terminal velocities of ice particles as a function of maximum crystal dimension, calculated for particles of different types.

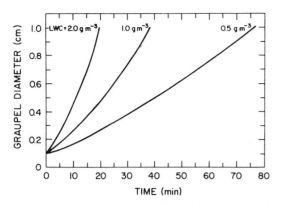

FIGURE 14 Calculated diameters of graupel particles as a function of time for several liquid water contents (LWC). Growth temperature is −15°C.

The collection efficiency is highly dependent on particle terminal velocity and cross-sectional area as well.

Assuming spherical particle growth, the linear accretional growth rate can be written analogously to Eq. (5),

$$dR/dt = \bar{E} \times \text{LWC} \times V_T/(4\rho_A R) \qquad (13)$$

such that ρ_A is the accretional density. Calculations of the growth of a spherical (graupel) ice particle from 1 mm to 1 cm in diameter (hail size) at −15°C and liquid water contents commonly observed in clouds (0.5, 1.0, and 2.0 g m⁻³) are given as a function of time in Fig. 14. Note that hail can grow from very small particles in a period as short as 20 min. The liquid water contents and temperature in Fig. 14 were chosen so that all accreted liquid water was frozen (dry growth). When growth temperatures or liquid water contents are higher, the surface temperature of the particle may rise to 0°C, and not all accreted water is frozen (wet growth). Wet growth can be very important for hailstones growing at temperatures above about −10°C.

IX. IN-CLOUD OBSERVATIONS OF ICE PARTICLES

A. Measurement Devices

Electrooptical devices are currently the primary tools used on aircraft to probe ice particles in clouds. The two probes used most frequently are one-dimensional (1-D) and two-dimensional (2-D) optical array probes, both operating on essentially the same principle. These probes project a laser beam through ∼10 cm of a cloud onto a linear array of photodiode elements. When a particle intervenes near the object (in-focus) plane of the beam, its imaged shadow momentarily occults a number of optical array elements. For the 1-D probe, the particle size is the measured shadow size divided by the optical magnification. Such probes are widely used in the size ranges 20–300 and 300–4500 μm, obtaining information on concentrations in

15 equally spaced size intervals, which is then recorded on magnetic tape. The 2-D probe uses a photodiode array and electronics similar to the 1-D probe. However, the 2-D probe contains electronics to record many pieces of shadow information for an individual particle as it passes across the photodiode array. As the particle's transit shadows the array, image slices are obtained across the shadow to develop a 2-D image. The 2-D probes are used to size particles in the same ranges as the 1-D probes, but the 2-D probes have twice the size resolution.

Several new probes on aircraft and balloons are revolutionizing the way that cloud physicists observe ice particles. The cloud particle imager (CPI) makes digital images with 2.3 μm size resolution. It uses a high-power laser pulsed at a high frequency to freeze images of particle on a CCD camera (Fig. 15). The high-volume particle sampler is a version of a 2-D probe with a very large sampling volume to provide better sampling statistics for precipitation-size particles. The cloudscope is essentially an airborne microscope that obtains images of particles as they impact a window exposed to the airstream. A balloon-borne ice crystal replicator and the hydrometeor-video-sonde (HYVIS) are instruments which capture and either preserve or obtain video images of particles as the balloon ascends through a cloud. The counterflow virtual impactor is an instrument which obtains the water content of the condensate above a size of about 10 μm.

B. Summary of In-Cloud Measurements

A fairly large number of measurements made in clouds with all droplets smaller than 25 μm show that ice crystal concentrations are scattered on either side of the Fletcher ice nucleus curve (Fig. 6), but there is a great deal of variability. Conversely, in clouds where droplets are greater than 25 μm (usually in maritime areas), measured concentrations are often 10⁴ μm times higher than those expected from ice nucleus measurements. A secondary ice crystal production mechanism that operates at about −6°C appears to account for these enhancements (see discussion in Section VI.B).

Measured ice particle size spectra typically exhibit a gamma form, where concentrations first increase with size up to tens to 100 μm, then decrease exponentially with size thereafter. In fairly quiescent clouds such as cirrus, ice particles at a given location span a size range from 1–1000 μm, while in vigorous clouds such as cumulonimbus, sizes can span a range from 1 μm to as much as 10⁵ μm (10 cm) when large hail is present. The observations of an exponential "tail" to the size distribution is not surprising, since ice particles found at a particular location could have originated from vastly different locations wherein they experienced different temperatures and humidities during their growth. Processes such as

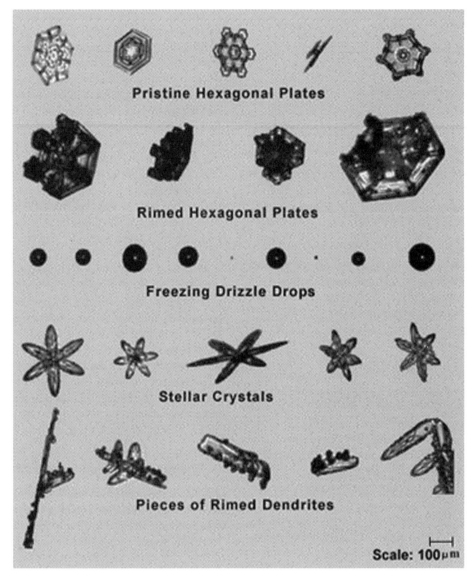

Pristine Hexagonal Plates

Rimed Hexagonal Plates

Freezing Drizzle Drops

Stellar Crystals

Pieces of Rimed Dendrites

Scale: 100µm

FIGURE 15 Images of ice particles and water droplets from the SPEC, Inc. cloud particle images. Scale is shown in the lower right-hand corner. (Courtesy of R. Paul Jawson.)

aggregation contribute further to the development of exponential spectra.

Ice crystal shapes observed in clouds vary with temperature in a way consistent with the findings from the laboratory experiments (Fig. 9). Exceptions occur when, for example, columnar ice particles fall into a planar crystal growth regime; in such a case, columnar crystals with end plates (capped columns) are observed. There is still a great deal to learn about the dependence of ice crystal shapes on ambient temperature and relative humidity below $-25°C$. While the probability of ice particle aggregation is a maximum near $0°C$, recent observations indicate that it can occur at any temperature, leading to the development of the exponential tail of the size distribution. Rimed ice crystals and graupel form in clouds that

contain both ice crystals and supercooled drops. In such clouds, ice crystals, aggregates, and frozen drops can serve as graupel and hail embryos. In-cloud and ground-based studies indicate that both columnar ice crystals and ice crystal plates have to grow by diffusion to a certain size before they can grow by riming. It turns out that the columnar crystal's minor dimension (width) has to exceed about $50 \ \mu m$ by diffusion first, while that of plates must exceed $\sim 300 \ \mu m$.

X. EFFECTS OF CLOUDS ON CLIMATE

Over the past decade, cloud physics research has been driven by uncertainties related to the role of clouds on climate. Cloud absorb the outgoing long-wave radiation emitted by the surface and the troposphere, and re-emit this energy to space at the much colder cloud-top temperatures. Clouds also tend to cool the surface by reflecting solar radiation back to space. The sum of these two effects lead to the net effect of clouds on climate, but since there is such a wide range of cloud sizes, microphysical properties, and height in the atmosphere, it is very difficult to assess the role of clouds on climate.

Uncertainties about the effects of clouds on climate has provided the impetus to conduct focused field campaigns aimed at understanding the effects of cirrus and stratocumulus clouds on incoming short-wave radiation and outgoing long-wave radiation. These experiments involve aircraft *in situ* measurements, ground-based measurements using radar and laser radar (lidar), and overflying aircraft and satellites to provide information on cloud radiative properties. Focused experiments have been conducted in North America, Europe, Japan, Brazil, and in the central and western Pacific. Long-term monitoring stations equipped with a wide range of remote sensing equipment have been placed in Oklahoma, Alaska, and the western Pacific to examine cloud properties over periods of decades.

Two cloud types, stratocumulus and cirrus, cover the greatest portion of the earth's surface and are therefore the most important cloud types from the standpoint of the effect of clouds on climate. The focused observations reveal that stratocumulus have a net cooling effect on the planet. However, the net effect of cirrus on climate is still unclear. The details of the cirrus cloud microphysics, including the mean, maximum, and effective (radiatively important) particle size, crystal shape, temperature, and height in the atmosphere, affect their radiative properties. This is an active area of research.

Satellite observations are now beginning to play a pivotal role in measuring not only the size and spatial distribution and extent of clouds, but their microphysical properties as well. The International Satellite Cloud Climatology Project (ISCCP) has been analyzing combined geostationary and sun-synchronous or polar orbiting satellite datasets since 1983. These datasets have been used to develop a near-global survey of the effective particle size in liquid water clouds and also ice clouds. Satellites are also being used to characterize the size of ice crystals in cirrus clouds and to map out the spatial distributions of aircraft-produced condensation trails (contrails). The Tropical Rainfall Measuring Mission (TRMM) is an important cloud physics platform that carries a radar for measuring precipitation rates and amounts over land and ocean areas in the tropics. The Moderate Resolution Imaging Spectroradiometer on the recently launched Terra spacecraft will enhance the capability for monitoring cirrus cloud microphysical properties. Several satellites to be launched over the next five years will further improve our ability to measure cloud microphysical properties over global scales.

The modeling of cloud physics processes in climate models has progressed significantly in recent years. Cloud physics processes are now represented in the more advanced climate models, although still rather simplistically. Processes such as rain formation, ice crystal nucleation, growth, advection, and fallout are now being included. A major process has been linking the microphysics to the radiative properties in climate models, although the parameterizations are still rather simplistic.

Over the past five years, there have been a number of studies which have sought to examine the effect of man's activities on clouds and climate. Modification of the microphysical structure of clouds as a result of biomass burning and the potential impact of the affected clouds on rainfall has been a topic of major interest. Field programs in the tropics have shown that biomass burning modifies the cloud condesation nuclei, and therefore there will be a change in the microphysical properties of clouds in the area. Increases in the CCN concentration results in more cloud droplets produced, with associated increases in cloud reflectivity. Assessing the impact of man's activities on cloud properties and reflectivity is an area of intense activity.

A. Aircraft-Induced Condensation Trails (Contrails)

Contrails form when saturation with respect to water is temporarily reached in the plume behind an aircraft, and they persist in ice supersaturated air masses. The ambient temperature necessary for contrail formation can be predicted accurately. Contrail particles consist of ice crystals. Young persistent contrails are composed of more, but smaller, ice crystals than typical cirrus clouds. Persistents contrails develop towards cirrus clouds in the course of time. The average contrail coverage exhibits a value around 0.5% over Europe. The mean global contrail cover is estimated to be of order of 0.1%. The net radiation effect of contrails is believed to enhance warming of the troposphere on average.

XI. CLOUD SEEDING

The main purposes of cloud seeding are (1) to increase precipitation, (2) to dissipate cloud or fog, and (3) to suppress hail. These purposes and their scientific basis are discussed in order below.

1. Experiments that attempt to increase precipitation are based on three main assumptions. (a) The presence of

ice crystals is necessary to produce precipitation in a super-cooled cloud, or the presence of fairly large water drops is required to initiate coalescence growth. In earlier sections it was shown that these particles are necessary for producing precipitation. (b) Some clouds precipitate inefficiently or not at all, because the particles listed in (a) are naturally deficient. The Fletcher curve (Fig. 6) shows that ice nuclei are active in concentrations of less than 1 liter^{-1} down to a temperature of $\sim 20°C$. For most clouds, a concentration of about 1 liter^{-1} is considered to be necessary to convert water vapor and cloud droplets efficiently to precipitation. Therefore, in clouds whose temperatures at cloud top do not fall below about $-20°C$, seeding can potentially increase precipitation. Clouds that would obviously not benefit from seeding are those of the maritime type in which large droplets ($>25~\mu m$) are present and high concentrations of secondary ice crystals are naturally produced. In clouds that contain small droplets, mechanisms are often not present to initiate coalescence growth. In such cases, addition of salt particles can lead to broader cloud droplet size distributions and initiate precipitation by coalescene growth. (c) Seeding the clouds artificially to produce ice crystals or water drops can alleviate the deficiency of precipitation embryos. Laboratory experiments have shown that two materials are highly effective in producing abundant ice crystals in supercooled clouds: dry ice, in the form of pellets that nucleate a large number of droplets in their path homogeneously and rapidly freeze them, and silver iodide, in the form of small (submicrometer-size) particles, which have a lattice structure very similar to that of ice and are effective ice nuclei. Coalescence growth can be initiated by introducing water drops into clouds or by seeding summertime convective clouds below cloud base with pyrotechnic flares that produce small salt particles in an attempt to broaden the cloud droplet spectrum and accelerate the coalescence process (see Section III). Dry ice is dispersed in clouds by aircraft, while silver iodide is dispersed either by aircraft or by ground-based aerosol generation units.

To produce the desired results, a cloud obviously must be seeded so as to produce an optimal concentration of ice crystals or water drops for the particular cloud conditions. For example, introducing too many particles will result in "overseeding," which will cause small, nonprecipitating particles to develop. Similarly, introducing too few particles will "underseed" the cloud and will result in too few precipitation particles.

Many problems have been encountered in the precipitation-enhancement seeding experiments conducted to date. In particular, it has been extremely difficult to deliver the optimal amount of seeding material at the correct time and place in a cloud.

2. Fogs and low clouds often pose hazards around airports. The concept for seeding these clouds is to introduce large salt particles or ice nuclei to sweep out the cloud droplets, thus clearing an area temporarily. Some success has been achieved, particularly in supercooled clouds. However, the clearing is often of short duration, and operations can be fairly expensive.

3. Two main concepts are employed in the design of hail suppression experiments. In the first, it is argued that by adding a large amount of seeding material to a thunderstorm so as to freeze most of the cloud liquid water at temperatures below about $-15°C$, accretional growth will be effectively eliminated, and large hail will not grow. The second approach is to produce artificially many embryos within the regions of a storm where hail growth is occurring, and through competition among particles for the available water, it will be unlikely that any hailstone will grow to a large size.

Many hail suppression experiments are being conducted throughout the world, notably in Russia. Some positive responses have been claimed in hail suppression experiments, but the extent of the result has been difficult to document.

XII. CONCLUDING REMARKS

Technological advances are contributing to rapid increases in the understanding of the key physical processes operative in clouds. For example, sophisticated measurements now obtained by aircraft and radar in clouds are being used in concert to increase the understanding of the symbiotic relationship between cloud dynamics and cloud microphysics. Detailed computer models that account for the dynamics, thermodynamics, and microphysics of clouds are being used to simulate the complex processes that occur in clouds from initial cloud formation through collapse. Laboratory experiments in the carefully controlled environments of wind tunnels and cold rooms are now performed to investigate key hydrometeor growth processes. The role of clouds in the earth's energy balance and in climate change is the focus of intensive research.

This article has not addressed several subjects in the field of cloud physics. For cloud chemistry and cloud electrification, see Atmospheric Chemistry and Atmospheric Electricity. The topic of cloud radiation and optics is almost as broad as that of cloud physics itself. The reader is referred to the Bibliography for information on this topic.

SEE ALSO THE FOLLOWING ARTICLES

CLIMATOLOGY • GREENHOUSE EFFECT AND CLIMATE DATA • IMAGING THROUGH THE ATMOSPHERE • METEOROLOGY, DYNAMIC • THUNDERSTORMS, SEVERE

BIBLIOGRAPHY

Cotton, W. R., and Anthes, R. A. (1989). "Storm and Cloud Dynamics," Academic Press, San Diego.

Gadsen, M., and Schroder, W. (1989). "Noctilucent Clouds," Springer-Verlag, New York.

Hobbs, P., and Deepak, A., eds. (1981). "Clouds: Their Formation, Optical Properties and Effects," Academic Press, New York.

Huschke, R. E., ed. (1970). "Glossary of Meteorology," 2nd printing, Am. Meteorol. Soc., Boston, MA.

Knight, C., and Squires, P., eds. (1982). "Hailstorms in the Central High Plains," Vols. 1 and 2, Colorado Associated Univ. Press, Boulder.

Pruppacher, H. R., and Klett, J. D. (1978). "Microphysics of Clouds and Precipitation," Reidel, Dordrecht, Netherlands.

Rogers, R. R. (1976). "A Short Course in Cloud Physics," Pergamon, Oxford.

Scorer, R. (1972). "Clouds of the World," Lothian Publ. Co., Melbourne, Australia.

Cluster Computing

Thomas Sterling

California Institute of Technology
and NASA Jet Propulsion Laboratory

GLOSSARY

Cluster A computing system comprising an ensemble of separate computers (e.g., servers, workstations) integrated by means of an interconnection network cooperating in the coordinated execution of a shared workload.

Commodity cluster A cluster consisting of computer nodes and network components that are readily available COTS (commercial off-the-shelf) systems and that contain no special-purpose components unique to the system or a given vendor product.

Beowulf-class system A commodity cluster implemented using mass-market PCs and COTS network technology for low-cost parallel computing.

Constellation A cluster for which there are fewer SMP nodes than there are processors per node.

Message passing A model and methodology of parallel processing that organizes a computation in separate concurrent and cooperating tasks coordinated by means of the exchange of data packets.

CLUSTER COMPUTING is a class of parallel computer structure that relies on cooperative ensembles of independent computers integrated by means of interconnection networks to provide a coordinated system capable of processing a single workload. Cluster computing systems achieve high performance through the simultaneous application of multiple computers within the ensemble to a given task, processing the task in a fraction of the time it would ordinarily take a single computer to perform the same work. Cluster computing represents the most rapidly growing field within the domain of parallel computing due to its property of exceptional performance/price. Unlike other parallel computer system architectures, the core computing elements, referred to as *nodes*, are not custom designed for high performance and parallel processing but are derived from systems developed for the industrial, commercial, or commodity market sectors and applications. Benefiting from the superior cost effectiveness of the mass production and distribution of their *COTS* (commercial off-the-shelf) computing nodes, cluster systems exhibit order-of-magnitude cost advantage with respect to

Encyclopedia of Physical Science and Technology, Third Edition, Volume 3

their custom-designed parallel computer counterparts delivering the same sustained performance for a wide range of (but not all) computing tasks.

I. INTRODUCTION

Cluster computing provides a number of advantages with respect to conventional custom-made parallel computers for achieving performance greater than that typical of uniprocessors. As a consequence, the emergence of clusters has greatly extended the availability of high-performance processing to a much broader community and advanced its impact through new opportunities in science, technology, industry, medical, commercial, finance, defense, and education among other sectors of computational application. Included among the most significant advantages exhibited by cluster computing are the following:

- *Performance scalability.* Clustering of computer nodes provides the means of assembling larger systems than is practical for custom parallel systems, as these themselves can become nodes of clusters. Many of the entries on the Top 500 list of the world's most powerful computers are clusters and the most powerful general-purpose computer under construction in the United States (DOE ASCI) is a cluster to be completed in 2003.
- *Performance to cost.* Clustering of mass-produced computer systems yields the cost advantage of a market much wider than that limited to the high-performance computing community. An order of magnitude price-performance advantage with respect to custom-designed parallel computers is achieved for many applications.
- *Flexibility of configuration.* The organization of cluster systems is determined by the topology of their interconnection networks, which can be determined at time of installation and easily modified. Depending on the requirements of the user applications, various system configurations can be implemented to optimize for data flow bandwidth and latency.
- *Ease of upgrade.* Old components may be replaced or new elements added to an original cluster to incrementally improve system operation while retaining much of the initial investment in hardware and software.
- *Architecture convergence.* Cluster computing offers a single general strategy to the implementation and application of parallel high-performance systems independent of specific hardware vendors and their product decisions. Users of clusters can build software application systems with confidence that such systems will be available to support them in the long term.
- *Technology tracking.* Clusters provide the most rapid path to integrating the latest technology for high-

performance computing, because advances in device technology are usually first incorporated in mass market computers suitable for clustering.
- *High availability.* Clusters provide multiple redundant identical resources that, if managed correctly, can provide continued system operation through graceful degradation even as individual components fail.

Cluster computing systems are comprised of a hierarchy of hardware and software component subsystems. Cluster hardware is the ensemble of compute nodes responsible for performing the workload processing and the communications network interconnecting the nodes. The support software includes programming tools and system resource management tools. Clusters can be employed in a number of ways. The master–slave methodology employs a number of slaved compute nodes to perform separate tasks or transactions as directed by one or more master nodes. Many workloads in the commercial sector are of this form. But each task is essentially independent, and while the cluster does achieve enhanced throughput over a single processor system, there is no coordination among slave nodes, except perhaps in their access of shared secondary storage subsystems. The more interesting aspect of cluster computing is in support of coordinated and interacting tasks, a form of parallel computing, where a single job is partitioned into a number of concurrent tasks that must cooperate among themselves. It is this form of cluster computing and the necessary hardware and software systems that support it that are discussed in the remainder of this article.

II. A TAXONOMY OF CLUSTER COMPUTING

Cluster computing is an important class of the broader domain of parallel computer architecture that employs a combination of technology capability and subsystem replication to achieve high performance. Parallel computer architectures partition the total work to be performed into many smaller coordinated and cooperating tasks and distribute these tasks among the available replicated processing resources. The order in which the tasks are performed and the degree of concurrency among them are determined in part by their interrelationships, precedence constraints, type and granularity of parallelism exploited, and number of computing resources applied to the combined tasks to be conducted in concert. A major division of parallel computer architecture classes, which includes cluster computing, includes the following primary (but not exhaustive) types listed in order of their level of internal communication coupling measured in terms of bandwidth (communication throughput) and latency (delay in transfer of data). This taxonomy is illustrated in Figure 1. Such a delineation is, by necessity, somewhat idealized because many actual parallel computers may incorporate multiple forms of parallel structure in their specific architecture. Also, the terminology below reflects current general usage but the specific terms below have varied in their definition over time (e.g., "MPP" originally was applied to fine-grain SIMD computers, but now is used to describe large MIMD computers).

1. *Vector processing.* The basis of the classical supercomputer (e.g., Cray 1), this fine-grain architecture pipelines memory accesses and numeric operations through one or more multistage arithmetic units supervised by a single controller.

2. *Systolic.* Usually employed for special-purpose computing (e.g., digital signal and image processing), systolic systems employ a structure of logic units and physical communication channels that reflect the computational organization of the application algorithm control and data flow paths.

3. *SIMD.* This *Single instruction stream, multiple data stream* or *SIMD* family employs many fine- to medium-grain arithmetic/logic units (more than tens of thousands), each associated with a given memory block (e.g., Maspar-2, TMC CM-5). Under the management of a single system-wide controller, all units perform the same operation on their independent data each cycle.

4. *MPP.* This *multiple instruction stream, multiple data stream* or *MIMD* class of parallel computer integrates many (from a few to several thousand) CPUs (central processing units) with independent instruction streams and flow control coordinating through a high-bandwidth, low-latency internal communication network. Memory blocks associated with each CPU may be independent of the oth-

ers (e.g., Intel Paragon, TMC CM-5), shared among all CPUs without cache coherency (e.g., CRI T3E), shared in SMPs (symmetric multiprocessors) with uniform access times and cache coherence (e.g., SGI Oracle), or shared in DSMs (distributed shared memory) with nonuniform memory access times (e.g., HP Exemplar, SGI Origin).

5. *Cluster computing.* Integrates stand-alone computers devised for mainstream processing tasks through local-area (LAN) or system-area (SAN) interconnection networks and employed as a singly administered computing resource (e.g., Beowulf, NOW, Compaq SC, IBM SP-2).

6. *Distributed Internet computing.* Employs wide-area networks (WANs) including the Internet to coordinate multiple separate computing systems (possibly thousands of kilometers apart) under independent administrative control in the execution of a single parallel task or workload. Previously known as *metacomputing* and including the family of GRID management methods, this emergent strategy harnesses existing installed computing resources to achieve very high performance and, when exploiting otherwise unused cycles, superior price/performance.

Cluster computing may be distinguished among a number of subclasses that are differentiated in terms of the source of their computing nodes, interconnection networks, and dominant level of parallelism. A partial classification of the domain of cluster computing includes commodity clusters (including Beowulf-class systems), proprietary clusters, open clusters or workstation farms, super clusters, and constellations. This terminology is emergent, subjective, open to debate, and in rapid transition. Nonetheless, it is representative of current usage and practice in the cluster community.

A definition of commodity clusters developed by consensus is borrowed from the recent literature and reflects their important attribute; that they comprise components

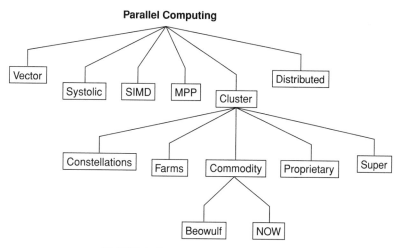

FIGURE 1 Taxonomy of cluster computing.

that are entirely off-the-shelf, i.e., already developed and available for mainstream computing:

A *commodity cluster* is a local computing system comprising a set of independent computers and a network interconnecting them. A cluster is *local* in that all of its component subsystems are supervised within a single administrative domain, usually residing in a single room and managed as a single computer system. The constituent *computer nodes* are commercial-off-the-shelf, are capable of full independent operation as is, and are of a type ordinarily employed individually for stand-alone mainstream workloads and applications. The nodes may incorporate a single microprocessor or multiple microprocessors in a symmetric multiprocessor (SMP) configuration. The *interconnection network* employs COTS LAN or SAN technology that may be a hierarchy of or multiple separate network structures. A cluster network is dedicated to the integration of the cluster compute nodes and is separate from the cluster's external (worldly) environment. A cluster may be employed in many modes including but not limited to high capability or sustained performance on a single problem, high capacity or throughput on a job or process workload, high availability through redundancy of nodes, or high bandwidth through multiplicity of disks and disk access or I/O channels.

Beowulf-class systems are commodity clusters employing personal computers (PCs) or small SMPs of PCs as their nodes and using COTS LANs or SANs to provide node interconnection. A Beowulf-class cluster is hosted by an open source Unix-like operating system such as Linux. A Windows-Beowulf system runs the mass-market widely distributed Microsoft Windows operating systems instead of Unix.

Proprietary clusters incorporate one or more components that are custom-designed to give superior system characteristics for product differentiation through employing COTS components for the rest of the cluster system. Most frequently proprietary clusters have incorporated custom-designed networks for tighter system coupling (e.g., IBM SP-2). These networks may not be procured separately (unbundled) by customers or by OEMs for inclusion in clusters comprising other than the specific manufacturer's products.

Workstation farms or open clusters are collections of previously installed personal computing stations and group shared servers, loosely coupled by means of one or more LANs for access to common resources, that, although primarily employed for separate and independent operation, are occasionally used in concert to process single coordinated distributed tasks. Workstation farms provide superior performance/price over even other cluster types in that they exploit previously paid-for but otherwise unused computing cycles. Because their interconnection network is shared for other purposes and not op-timized for parallel computation, these open clusters are best employed for weakly interacting distributed workloads. Software tools such as Condor facilitate their use while incurring minimum intrusion to normal service.

Super clusters are clusters of clusters. Principally found within academic, laboratory, or industrial organizations that employ multiple clusters for different departments or groups, super clusters are established by means of WANs integrating the disparate clusters into a single more loosely coupled computing confederation.

Constellations reflect a different balance of parallelism than conventional commodity clusters. Instead of the primary source of parallelism being derived from the number of nodes in the cluster, it is a product of the number of processors in each SMP node. To be precise, a constellation is a cluster in which there are more processors per SMP node than there are nodes in the cluster. While the nodes of a constellation must be COTS, its global interconnection network can be of a custom design.

Of these, commodity clusters have emerged as the most prevalent and rapidly growing segment of cluster computing systems and are the primary focus of this article.

III. A BRIEF HISTORY OF CLUSTER COMPUTING

Cluster computing originated within a few years of the inauguration of the modern electronic stored-program digital computer. SAGE was a cluster system built for NORAD under an Air Force contract by IBM in the 1950s based on the MIT Whirlwind computer architecture. Using vacuum tube and core memory technologies, SAGE consisted of a number of separate stand-alone systems cooperating to manage early warning detection of hostile airborne intrusion of the North American continent. Early commercial applications of clusters employed paired loosely coupled computers with one performing user jobs while the other managed various input/output devices.

Breakthroughs in enabling technologies occurred in the late 1970s, both in hardware and software, that were to have a significant long-term effect on future cluster computing. The first generations of microprocessors were designed with the initial development of VLSI technology and by the end of the decade the first workstations and personal computers were being marketed. The advent of Ethernet provided the first widely used LAN technology, creating an industry standard for a modest cost multidrop interconnection medium and data transport layer. Also at this time, the multitasking Unix operating system was created at AT&T Bell Labs and extended with virtual memory and network interfaces at UC Berkeley. Unix was adopted in its various commercial and public domain forms by the scientific and technical computing community as the principal environment for a wide range of computing system classes from scientific workstations to supercomputers.

During the decade of the 1980s, increased interest in the potential of cluster computing was marked by important experiments in research and industry. A collection of 160 interconnected Apollo workstations was employed as a cluster to perform certain computational tasks by the NSA. Digital Equipment Corporation developed a system comprising interconnected VAX 11/750s, coining the term *cluster* in the process. In the area of software, task management tools for employing workstation farms were developed, most notably the Condor software package from the University of Wisconsin. The computer science research community explored different strategies for parallel processing during this period. From this early work came the communicating sequential processes model more commonly referred to as the *message-passing model*, which has come to dominate much of cluster computing today.

An important milestone in the practical application of the message passing model was the development of PVM (parallel virtual machine), a library of linkable functions that could allow routines running on separate but networked computers to exchange data and coordinate their operation. PVM, developed by Oak Ridge National Laboratory, Emory University, and University of Tennessee, was the first major open distributed software system to be employed across different platforms. By the beginning of the 1990s, a number of sites were experimenting with clusters of workstations. At the NASA Lewis Research Center, a small cluster of IBM workstations was used to simulate the steady-state behavior of jet aircraft engines in 1992. The NOW (Network of Workstations) project at UC Berkeley began operation of the first of several clusters there in 1993 that led to the first cluster to be entered on the Top 500 list of the world's most powerful computers. Also in 1993, one of the first commercial SANs, Myrinet, was introduced for commodity clusters, delivering improvements in bandwidth and latency an order of magnitude better than the Fast Ethernet LAN most widely used for the purpose at that time.

The first Beowulf-class PC cluster was developed at NASA's Goddard Space Flight Center in 1994 using early releases of the Linux operating system and PVM running on 16 Intel 100-MHz 80486-based PCs connected by dual 10-Mbps Ethernet LANs. The Beowulf project developed the necessary Ethernet driver software for Linux and additional low-level cluster management tools and demonstrated the performance and cost effectiveness of Beowulf systems for real-world scientific applications. That year, based on experience with many other message-passing software systems, the parallel computing community set

out to provide a uniform set of message-passing semantics and syntax and adopted the first MPI standard. MPI has become the dominant parallel computing programming standard and is supported by virtually all MPP and cluster system vendors. Workstation clusters running the Sun Microsystems Solaris operating system and NCSA's PC cluster running the Microsoft NT operating system were being used for real-world applications.

In 1996, the Los Alamos National Laboratory and the California Institute of Technology with the NASA Jet Propulsion Laboratory independently demonstrated sustained performance of more than 1-Gflops for Beowulf systems costing under $50,000 and was awarded the Gordon Bell Prize for price/performance for this accomplishment. By 1997 Beowulf-class systems of more than 100 nodes had demonstrated sustained performance of greater than 10 Gflops with a Los Alamos system making the Top 500 list. By the end of the decade, 28 clusters were on the Top 500 list with a best performance of more than 500 Gflops. In 2000, both DOE and NSF announced awards to Compaq to implement their largest computing facilities, both clusters of 30 and 6 Tflops, respectively.

IV. CLUSTER HARDWARE COMPONENTS

Cluster computing in general and commodity clusters in particular are made possible by the existence of cost-effective hardware components developed for mainstream computing markets. The capability of a cluster is determined to first order by the performance and storage capacity of its processing nodes and the bandwidth and latency of its interconnection network. Both cluster node and cluster network technologies evolved during the 1990s and now exhibit gains of more than two orders of magnitude in performance, memory capacity, disk storage, and network bandwidth and a reduction of better than a factor of 10 in network latency. During the same period, the performance-to-cost ratio of node technology has improved by approximately 1000. In this section, the basic elements of the cluster node hardware and the alternatives available for interconnection networks are briefly described.

A. Cluster Node Hardware

The processing node of a cluster incorporates all of the facilities and functionality necessary to perform a complete computation. Nodes are most often structured either as uniprocessor systems or as SMPs although some clusters, especially constellations, have incorporated nodes that were distributed shared memory (DSM) systems. Nodes are distinguished by the architecture of the microprocessors employed, the number and organization of the microprocessors, the capacities of the primary and secondary storage, and the internal interconnect logic structure. The nodes of commodity clusters marketed primarily for mainstream computing environments must also incorporate standard interfaces to external devices that ensure interoperability with myriad components developed by third-party vendors. The use of the high-bandwidth interface allows clusters to be configured with little or no change to the node subsystem, minimizing any additional costs incurred on a per-node basis. The key elements of a node are briefly discussed below. It must be understood that this technology is evolving rapidly and that the specific devices that are provided as examples are likely to be upgraded in operational characteristics or to be replaced altogether in the near future.

1. *Central processing unit.* The CPU is a single VLSI integrated circuit microprocessor, possibly merged on an MCM (multichip module) with one or more cache chips. The CPU executes sequences of binary instructions operating on binary data, usually of 32- or 64-bit length. While many instructions are performed on internal data stored in registers, acquiring new data from the memory system is an important aspect of microprocessor operation, requiring one or more high-speed cache memories to minimize the average load/store access times. Both 32-bit and 64-bit architectures are used in clusters with the most popular based on the 32-bit Intel X86 family and the highest performance clusters based on the 64-bit Compaq Alpha family or IBM RS6000. The first Beowulf-class commodity clusters incorporated Intel 80486 microprocessors operating at 100 MHz. Today, descendents of this chip including the Intel Pentium III and the AMD K7 Athelon have clock rates in excess of 1 GHz. The CPU connects to an internal memory bus for high-speed data transfers between memory and CPU and to an external I/O bus that provides interfaces to secondary storage and networking control modules.

2. *Main memory.* Stores the working data and program instructions to be processed by the CPU. It is a part of a larger memory hierarchy that includes high-speed cache memories closer to the CPU and high-density persistent mass storage from which it acquires its initial data and stores its final results. For the last two decades, main memory has been dominated by DRAM technology, closely packed arrays of switched capacitive cells embedded on silicon wafers. DRAM chips containing 256 Mbits of data are available with gigabit chips to become common place in the near future. Typical cluster nodes support main memory capacities between 64 Mbytes and 1 Gbytes although large SMP or DSM nodes provide more. DRAM has undergone significant advances in recent years pro-

viding more rapid throughput as well as higher density, reducing if not closing the bottleneck between CPU and its main memory.

3. *Secondary storage.* Comprises a set of devices that provides persistent storage of a large amount of data. Secondary storage serves several purposes as a function of the usage of the data it contains. It provides all of the functions, both user applications and operating system tools, that govern the operation and computation of the CPU. It provides the data sets on which the user tasks are to operate and is the primary repository for the final results of user computations. It maintains configuration data concerning the setup and operational parameters of the computing node as well as information concerning the rest of the cluster devices and their relational roles. Because most memory systems support the virtual memory abstraction, providing a logical memory many times larger than the actual physical main memory installed, secondary storage temporarily holds those segments of the logical address space and associated data that do not fit in the existing physical main memory. Unlike main memory, data stored on secondary storage devices are retained, even when system power is disrupted. This nonvolatile property allows data to be archived indefinitely. The primary component type providing secondary storage is the venerable hard disk with its early genesis in the late 1950s based on magnetic storage (like a cassette tape) of one or more disks rotating on a single spindle at high speed and accessed by a magnetic detection head moved radically in and out across the disk surface, reminiscent of the arm of an old record turntable. Modern disk drives provide many tens of gigabytes at moderate cost and access times on the order of a few milliseconds. Other technologies are employed to provide more specialized forms of secondary storage, particularly for data portability and safe permanent archival storage. CD-ROMs developed from the original digital musical recording media provide approximately 600 Mbytes of storage at less than $1 a disk and read–write capability is now becoming commonplace, although this is of less importance to cluster systems. The long-lived and relatively diminutive floppy disk holding a mere 1.4 Mbytes is still employed, even on clusters, primarily for initial installation, configuration, and boot up.

4. *External interfaces.* Serve three important roles related to the operation and management of clusters. They provide direct user interactive access and control, they permit application data input and results to be conveyed with devices outside the system, and they connect to the cluster interconnection network and thereby to other nodes in the cluster. While there are many different types of interfaces (just look at the number of sockets on the back of a typical PC), PCI is universal from PCs to mainframes, connecting

the CPU to a plethora of interface control devices. The PCI bus has four different configurations employing 32- or 64-bit connections running at 33- or 66-MHz clock speed and a peak data throughput of 4 Gbps. The majority of network interface controllers (NIC) are compatible with one or more of these PCI forms. In the future, it is likely that a new external interface standard, Infiniband, will eventually replace PCI to deliver higher throughput and lower latency between the CPU and external devices.

B. Cluster Network Hardware

A model, but not necessarily the only possible model, of parallel processing with cluster systems involves each of the cluster nodes performing one or more tasks on local data and then exchanging the computed results with other nodes within the cluster. Networks make this possible. They provide physical channels between nodes by which data are transported and logical protocols that govern the flow and interpretation of the transferred data. Networks are employed in a broad range of integrated systems from the Internet spanning the globe requiring possibly as much as a hundred milliseconds for a message packet to reach its destination to a data bus internal to a computer integrating its various components supporting data transfers in 100 nanosec or less, a ratio of a million in network latency. Networks for commodity clusters fall in between with the initial use of Ethernet exhibiting on average approximately 100 nanosec latency falling in the middle (logarithmically speaking).

Network technology determines the potential value of cluster computing. Its principal properties are bandwidth, latency, scale, and cost. Bandwidth imposes an upper bound on the amount of data that can be transferred in unit time (e.g., Mbps, Gbps). Latency is the amount of time it takes for a message packet to transit the diameter of a system measured in microseconds. Cost is usually considered as the percentage of the total price of the hardware system. Scale is the largest number of nodes that a network can connect effectively. Together, they establish a cluster's capability, applicability, and user accessibility. Different applications exhibit varying global data access patterns that may be suitable for some networks rather than others. Higher bandwidth networks ordinarily will have greater generality of application than those networks of lower bandwidth. Similarly, for applications using short messages or involving frequent global synchronization, lower latency networks will be more general purpose than high-latency networks. But superior behavioral properties often come at additional cost that may preclude their use in many environments, where cost is a significant factor in the choice to employ clusters in the first place. Thus the

selection of a specific network is dependent on how the cluster is to be used and by whom.

A cluster network includes NICs that connect the cluster node to the network, transport layer links to carry the data, and switches that route the data through the network. NICs move data from message buffers filled by the node processor to signal packets sent out to the transport layer performing a number of translation functions on the data in the process. The data links may comprise one or more parallel channels and may be implemented with metal coaxial cable or optical fiber (advanced development of free-space optical networks is under way). Switches accept messages at their multiple input ports, determine their required routing, switch as many as possible simultaneously sending them out the appropriate output ports, and arbitrating where contention for shared resources (ports, channels) occurs. The earliest Beowulf-class systems used low-cost hubs, rather than the more expensive switches, but these permitted only one transfer to occur at a time on the entire network. Switches deliver much closer to the peak bi-section bandwidth of the network as they isolate separate disjoint paths from each other.

Together these network components can be structured to form a number of different topologies. Most simple among these and used frequently for small clusters is the star configuration using a single switch of degree n (the number of separate ports) connecting n nodes. Larger systems can be formed with a hierarchy of switches to form a tree structure. The scale of tree-based clusters is limited by the bi-section bandwidth of the root node of the tree topology. More complex network structures permit the implementation of larger systems. Among them is the CLOS network (also referred to as the fat-tree) that overcomes the deficiency of the tree topology by providing multiple channels in parallel, balanced to keep the cross-section bandwidth equal at each level in the tree. Mesh and toroidal topologies provide scalable bandwidth and locality of interconnect with fixed degree nodes but may experience relatively high latency across the system diameter. Variations on these and other network topologies are possible and depend on requirements of a given system. A few of the most widely used network technologies used in commodity clusters are described next.

1. *Ethernet* is the most widely used network for clusters, even today, although devised as a LAN and originated in the late 1970s. Its success is due in part to its repeated reinvention, which takes advantage of technology advances while meeting expanding requirements. The 10-Mbps Ethernet that was first used in Beowulf clusters in the early 1990s superceded early Ethernet at 3 Mbps. Fast Ethernet provided 100 Mbps and with low-cost switches is the mainstay of small low-cost Beowulf-class systems.

Gigabit Ethernet, as the name implies, provides a peak bandwidth of approximately 1 Gbps. But its per-node cost remains high and it suffers from the relatively long latencies of its predecessors.

2. *Myrinet* was one of the first networks to be developed expressly for the SAN and cluster market. With a cost of approximately $1600 per node, Myrinet was initially reserved for the more expensive workstation clusters. But with its superior latency properties of 20 μsec or less, it permitted some classes of more tightly coupled applications to run efficiently that would perform poorly on Ethernet-based clusters. More recently, reduced pricing has expanded its suitability to lower cost systems and has proven very popular.

3. *VIA* is a recent advance in cluster network technology involving improvements in both hardware and software to further reduce data communication latency. Typically, message packets are copied from the user application space into the operating system space or vice versa. VIA (virtual interface architecture) employs a zero-copy protocol, avoiding the O/S intermediate stage and moving the packets directly between the network transport layer and the application. Giganet's cLAN and Compaq's Server net II both implement the VIA standard, delivering best case latencies well below 10 μsec.

4. *SCI* was perhaps the first SAN to achieve IEEE standardization and has very good bandwidth and latency characteristics. Existing implementations provide between 3.2- and 8-Gbps peak bandwidth with best latencies below 4 μsec. The SCI standard includes protocol for support of distributed shared memory operation. However, most clusters employing SCI use PCI-compatible network control cards (e.g., Dolphin) that cannot support cross-node cache coherence. Nonetheless, even in distributed memory clusters, it provides an effective network infrastructure.

5. *Infiniband* is the next-generation interconnection technology to extend the capabilities of SANs. Although not yet available, an industrial consortium of major computer technology (hardware and software) manufacturers has developed and released an extensive specification that will lead first to reference implementations, and eventually to widely distributed products. Bandwidths up to 12 Gbps (employing optical channels) and latencies approaching 1 μsec will become possible with Infiniband, which replaces previous I/O buses (e.g., PCI) and migrates the network interconnect closer to the memory bus of the compute node.

V. CLUSTER SOFTWARE COMPONENTS

The earliest use of commodity clusters involved little more software than the original node operating system and basic support for a network interface protocol such as sockets in Unix. Application programmers running a single problem on a small, dedicated cluster would hand craft the parallel program and painstakingly install the code and necessary data individually on every node of the cluster system. Good results were obtained for real-world problems on Beowulf-class systems and other such clusters, motivating continued advances in cluster hardware and methodology. Today, with commodity clusters contending for dominance of the high-performance computer arena, such primitive frontier techniques can no longer be justified and, indeed, would present a serious obstacle to wider usage of commodity clusters. During the intervening period, significant advances in software support tools have been developed for cluster computing. These are in the two critical areas of programming environments and resource management tools. Together, they provide the foundation for the development of sophisticated and robust cluster system environments for industrial, commercial, and scientific application.

The environments and tools described below engage the system as a global ensemble, treating its processing nodes as a set of compute and storage resources to be managed, allocated, and programmed. But each node is itself a complete and self-sustaining logical as well as physical entity, hosting its own environment: the node operating system. While some experimental clusters incorporate custom operating systems derived expressly for use within the cluster context, the vast majority of commodity clusters employ nodes hosting conventional operating systems. Many operating systems have been used in support of clusters. The IBM AIX operating system used on their SP-2 and Compaq True64 used on their Alpha-based SC series are two examples of vendor software migrated to use with clusters. However, the dominant operating systems employed with commodity clusters are Linux and Microsoft Windows. Linux emerged as the software of choice as a result of the Beowulf Project, which implemented the first clusters using Linux and running real-world science and technical applications. Linux gained prominence because of its Unix-like structure, which was consistent with the technical computing community's environments from scientific workstations to supercomputers and because of its free open source code policy. Microsoft Windows, the world's single most widely used operating system, has been favored in business and commerce environments for clusters using ISV applications software developed for Windows such as distributed transaction processing. Windows has also been used effectively for technical computing clusters at NSCA and Cornell Theory Center. Both IBM and Compaq as an alternative cluster node operating system to their proprietary software have adopted Linux.

A. Programming Environments

Parallel programming of clusters involves a sequence of steps that transforms a set of application requirements into a set of cooperating concurrent processes and data sets. Although actual programming styles may vary significantly among practitioners, a representative methodology may be the following process:

1. Capture the application in a set of ordered routines.
2. Partition the global data into separate approximately equal regions.
3. Define tasks to be performed on each data partition.
4. Determine precedence constraints between tasks of different regions.
5. In the programming language of choice, write the sequence of statements that encodes the tasks to be performed.
6. In the global communication medium of choice, set up synchronization conditions that will govern guarded program execution.
7. Devise procedures for exchanging necessary intermediate results among concurrent tasks.
8. Create a minimalist test data set and debug compile and runtime program errors.
9. Monitor program behavior and optimize code for best performance.
10. Partition real-world data set.
11. On selected cluster, allocate nodes to data partitions.
12. Install data and tasks on designated nodes.
13. Initiate execution and acquire result values.

These steps are rarely performed in such rigid lock-step manner but all of the actions described must be accomplished prior to successful completion of executing a real parallel problem on a cluster. The effectiveness achieved in programming a cluster is difficult to measure (although some metrics have been devised to this end). Nonetheless, the ease of parallel programming is strongly influenced by the execution model assumed and the tools available to assist in the process.

Many models have been conceived in the last two decades (or more) to provide a conceptual framework for parallel program execution and programming. These have been strongly influenced by the assumptions of the characteristics of the underlying parallel computer. Pipelined supercomputers used vector models, SIMD machines used fine-grain data parallel programming, SMP systems used coarse-grain multiple threads with shared memory synchronization (e.g., open MP), and large MPPs used single-program, multiple data stream (SPMD) style (e.g., HPF) with either put/get shared memory primitives or message passing for interprocessor cooperation. Because of the relatively long global latencies and constrained network bandwidth characteristic of clusters, the programming paradigms of widest usage have been the master–slave model for embarrassingly parallel job streams of independent tasks (e.g., transaction processing, web search engines) and the message-passing model for cooperating interrelated processes. Where clusters consist of SMP nodes, hybrid models are sometimes used employing message passing between nodes and multiple threads within the nodes.

Efficient programming practices demand effective programming environments that incorporate a set of sophisticated tools to support the steps listed above. A partial list of the desired tools based on a message-passing approach might include these:

- A core language and compiler (e.g., C, Fortran)
- A language sensitive editor (e.g., Emacs)
- A linkable message-passing library [e.g., MPICH (*http://www-unix.mcs.anl.gov/mpi/mpich*), LAM (*http://www.mpi.nd.edu/lam*)]
- Numeric libraries [e.g., Scalapack (*http://www.netlib. org/scalapack*)]
- Debuggers [e.g., gdb, Totalview (*http://www.etnus. com*)]
- Performance profilers [e.g., jumpshot (*http://www-unix.mcs.anl.gov/mpi/mpich*), XPVM (*http://epm.ornl. gov/pvm*)]
- Loaders and process distribution [e.g., Scyld (*http:// www. scyld.com*), Rocks (*http://slic01.sdsc.edu*), vasystemimager (systemimager.org), OSCAR (*http:// openclustergroup.org*), etc.]
- Schedulers [e.g., LSF (*http://www.platform.com*), PBS (*http://www.openpbs.org*), Condor (*http://www.cs.wisc. edu/condor*), etc.]

Such environments and tools are in a state of flux with a combination of free open-source and commercial offerings in continuous development yielding constant improvements in functionality, performance, and reliability. But new tools that provide more complete support, especially for parallel debugging, are still required. One important trend is toward the development of PSEs or problem-solving environments that target specific application domains and provide the programmer with a high-level framework within which to cast the problem. Other programming styles for clusters such as HPF, BSP, Split-C, and UPC are also being pursued and applied by some communities, although it is unclear which if any of these will become dominant.

One challenge that complicates programming clusters is the use of nodes comprising more than one processor in an SMP configuration. Such nodes view their local memory as common, sharing the name space within the node through hardware support of cache coherence. In principle, this allows computing within each node to employ a threaded shared memory model rather than the message-passing model. One would expect that such mixed-mode programming could yield superior performance. Surprisingly, this is not the mainstream practice. The majority of users of clusters of SMP nodes program intranode operation with message-passing operations such as those provided by MPI as they do internode processing. One widely used SMP programming methodology is Open MP, which provides the added constructs necessary for parallel programming on a shared memory multiprocessor. Some programmers, striving to take advantage of both clusters and their SMP nodes, employ a hybrid programming methodology consisting of both MPI and Open MP. This has not become common practice but is expected to grow in usage, in spite of its difficulties.

B. Resource Management Software

The management of cluster computers includes many responsibilities from initial assembly and software installation to possible dynamic load balancing of user application modules. Originally, users of moderate-scale low-cost Beowulf-class performed most of these chores manually, ignoring some as unnecessary for dedicated use. But modern clusters supporting multiple users and a range of application and workload types on systems scaled to hundreds of gigaflops require sophisticated environments and tools to manage the plethora of system resources. Programming tools, while still in transition, have achieved a level of community-wide standardization. Such is not the case for resource management tools. These are still in a state of experimentation although there is general consensus on the basic requirements. The principal capabilities needed for commercial grade resource management include the following:

- *Assembly, installation, and configuration.* Setting up a cluster, whether assembled on site by staff or vendor provided, can benefit from a set of low-level tools that organize the task of installing the large suite of software and configuring the large number of system parameters. Maintaining consistency across all of the nodes can be facilitated by routines that search and validate all copies and their version numbers.
- *Scheduling and allocation.* Loading a parallel application program on to a cluster shared by other users and jobs requires software tools that determine which resources will be employed to perform what jobs and when. Far more complicated than on a conventional uniprocessor, cluster scheduling involves space sharing where a system is physically partitioned into multiple subsystems to run as many jobs.
- *System administration.* The management of user accounts, job queues, security, backups, mass storage, log journaling, operator interface, user shells, and other housekeeping activities are essential elements of a commercial-grade computing system but impose added burden due to the multiplicity of computing resources and the diverse ways in which they may be used. PBS is an example of one software system that brings much of this capability to cluster computing.
- *Monitoring and diagnosis.* The complex state of a cluster, its operational status, tasks being performed, and its performance are all constantly changing. Operator tools are required to continuously monitor the behavior of the many components comprising a cluster system and quickly diagnosing hardware or software failures when they occur. A number of such tool sets have been developed by many cluster installations although no single suite has been adopted by the community as a whole.
- *Parallel mass storage.* Almost all computations require access to secondary storage including both local and remote disk drives for support of file systems. Commercial applications frequently use their own in-house distributed software for file management optimized around the specific needs of the application. Examples of general-purpose parallel file systems used for clusters include PPFS (*http://www-pablo.cs.uiac.edu/Project/PPFS /PPFSII/PPFSIIOverview.htm*), PVFS (*http://parlweb. parl.Clemson.edu/pvfs*), and GPFS (*http://gfs.lcse.umn. edu*).
- *Reliability.* Checkpoint and restart support software allows large programs with long run times to survive transient or hard failures of system components. Individual organizations and some vendors have developed some support for this capability although a general solution is not widely available. More difficult is the detection of errors. Software fault tolerance is a field in which strides are being made but which is still largely experimental.

While substantial advances have been made during the early 2000s, continued research and development are required to produce a common cluster environment that satisfies the requirements of a broad user community and exhibits sufficient reliability to garner confidence in the robustness and therefore utility of commodity clusters for commercial and industrial grade processing. A number of efforts are under way to synthesize a number of tools into common frameworks including the Oscar, Grendel, and RWCP projects. More than one ISV offers collections of cluster middleware including PGI and Scyld. In the long term, one or more advanced programming models and their complementing runtime execution models will have to be developed before commodity clusters become significantly easier to program and perhaps more efficient as well for a broader range of application algorithms. The development of PSE for a range of widely used application classes may provide partial solution to the challenge of programming, at least for those special cases. PSE software presents a template to the user who then fills in the parameters with data relevant to the specific problem to be performed. This eliminates the need for detailed program development and reduces the time to solution as well as providing improved efficiency of operation. But for general-purpose cluster computing, significant improvements in ease of use may depend on next-generation parallel programming formalisms.

VI. SUMMARY AND CONCLUSIONS

Commodity cluster computing is growing rapidly both for high-end technical and scientific application domains and for business and commerce. The low cost, high flexibility, and rapid technology tracking are making this class of computing the platform of choice for many user domains requiring scalability and excellent price/performance. The extraordinary rate of growth in capability for commodity clusters in general and Beowulf-class systems in particular is anticipated to continue for at least the next 5 years. By 2005 to 2006, price/performance may reach $0.10 per megaflops with systems as large as 50 teraflops operational at a few sites. There is a strong likelihood that Linux and Microsoft Windows will be the mainstream operating systems, with one or the other offered by virtually every system vendor. Both are also likely to incorporate advancements that directly enhance cluster scalability and efficiency by eliminating bottlenecks and reducing overhead. Network bandwidths of 10 Gbps will have become commonplace with network latency approaching $1\ \mu\text{sec}$ through the implementation of Infiniband. MPI-2 will be the ubiquitous programming model for parallel applications and .NET or an equivalent tool set will be employed for loosely coupled workloads, primarily in the commercial sector. Packaging will become cheaper and more compact to reduce footpad and overall system cost. Finally, systems administration tools will have reached the sophistication of mainstream servers. In 10 years, the first petaflops-scale commodity clusters will have been installed.

SEE ALSO THE FOLLOWING ARTICLES

COMPUTER ARCHITECTURE • COMPUTER NETWORKS • DATABASES • DATA STRUCTURES • PARALLEL COMPUTING • PROJECT MANAGEMENT SOFTWARE

BIBLIOGRAPHY

Buyya, R. (1999). "High Performance Cluster Computing," Vol. 1, Prentice Hall, Upper Saddle River, NJ.

Gropp, W., Lusk, E., and Skjellum, A. (1999). "Using MPI: Portable Parallel Programming with the Message-Passing Interface," The MIT Press, Cambridge, MA.

Pfister, G. F. (1998). "In Search of Clusters," 2nd ed., Prentice Hall, Upper Saddle River, NJ.

Seifert, R. (1998). "Gigabit Ethernet: Technology and Applications for High-Speed LANs," Addison Wesley Longman, Reading, MA.

Sterling, T. L., Salmon, J., Becker, D. J., and Savarese, D. F. (1999). "How to Build a Beowulf: a Guide to the Implementation and Application of PC Clusters," The MIT Press, Cambridge, MA.

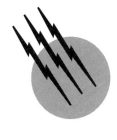

Coal Geology

Colin R. Ward

University of New South Wales

I. Chemical and Physical Characteristics of Coal
II. Petrology of Coal
III. Geology of Coal
IV. Coal Exploration and Mining Geology
V. Coal Preparation and Use

GLOSSARY

Ash Inorganic residue after incineration of coal or coke under standard conditions.

Banded coal Coal with visible stratifications, bands, or layers of bright, semibright, and dull material along its bedding planes due to the incorporation of a heterogeneous mixture of plant debris.

Carbonization Process of decomposing coal by heating in the absence of air or oxygen to produce a carbon-rich solid (coke or char), hydrogen-rich liquid (tar), and gaseous products.

Coalification Series of physical and chemical changes by which the vegetable matter in peat is transformed to coal under the influence of increased temperatures and pressures associated with burial throughout some interval of geological time.

Coke Porous, or vesicular, fused carbonaceous solid derived from the carbonization of coal.

Combustion Series of processes, accompanied by the emission of heat energy, brought about by the rapid interaction of coal with oxygen or air at high temperatures.

Grade Relative freedom of a coal from inorganic con-

taminants, such as mineral matter (which forms ash), moisture, and sulfur.

Humic coal Coal developed from plant debris that was exposed, before burial, to partial degradation by interaction with the atmosphere. In high-rank deposits, such material is typically banded coal.

Lithotype Any of the macroscopically recognizable types of bands or discrete layers seen in humic coals.

Maceral Any of the elementary, homogeneous microscopic constituents of coal, analogous to the "minerals" in rocks.

Microlithotype Microscopic band, at least 50 μm in width, made up of a particular association of macerals.

Mineral matter Inorganic material in coal, which gives rise to ash and certain volatile components when the coal is burned.

Proximate analysis Analysis of coal or coke, using standardized techniques, to determine the proportions of moisture, volatile matter, ash, and fixed carbon. (Fixed carbon is referred to as "residue" in coke analysis.)

Rank Stage of the coalification process reached by a given coal.

Sapropelic coal Coal developed by anaerobic putrefaction of algal remains, spores, and similar fine organic debris.

Encyclopedia of Physical Science and Technology, Third Edition, Volume 3

In high-rank deposits, such coals are usually massive or free of visible stratification in hand specimen.

Type Variety of coal distinguished from other varieties by features that depend on the nature of the plant debris from which it was derived, including the composition, origin, and conditions of deposition of that plant debris.

Ultimate analysis Analysis of coal or coke to determine the proportions of carbon, hydrogen, oxygen, nitrogen, and sulfur present.

COAL is a combustible sedimentary rock composed essentially of lithified plant debris. The plant debris, in various stages of preservation, initially accumulates in a bog, marsh, or swampy depositional environment (mire) to form a water-saturated, spongy sediment called peat. Burial by other strata over a long period of time compresses the peat and exposes it to higher than surface temperatures, giving rise to a progressive series of physical and chemical changes, known as coalification or rank advance, that transform it to harder, drier, and more durable material such as lignite, bituminous coal, or anthracite (Fig. 1).

Coal geology is the application of geological sciences to the study of the nature and origin as well as to the extraction and use of coal and coal deposits. It is concerned with the chemical and physical characteristics of coal, with the appearance, properties, and origin of coal constituents when studied under the microscope, and with the age, distribution, and features of coal and coal-bearing rock strata in the field. Geological techniques are used by the coal industry to evaluate coal resources and delineate features that may affect coal mining, including the impact of mining or use of coal on the environment. They are also used, in conjunction with the techniques of other sciences, to determine the suitability of coal for various forms of use.

I. CHEMICAL AND PHYSICAL CHARACTERISTICS OF COAL

The properties of a particular coal depend mainly on its rank, the degree to which it has been affected by metamorphism in the coalification process. However, there is also a significant variation in properties among coals of the same rank, brought about by differences in the assemblage of plant constituents that formed the original peat accumulation. This variation, expressed by differences in physical appearance at a megascopic and a microscopic scale, is referred to as a variation in coal type. Although parameters such as bulk chemistry are traditionally used to classify coals, rank and type are the fundamental factors that control these and most other coal characteristics. Coupled with the proportion of inorganic impurities (moisture, mineral matter and sulfur), a factor sometimes referred to as the grade of the coal, they reflect the origin of the different classes of material and ultimately determine the optimum use to which each coal can be put.

A. Proximate and Ultimate Analysis

From the chemical point of view, coal can be regarded as a mixture of solid hydrocarbon compounds, together with a certain amount of moisture and mineral matter. Although the compounds present can be investigated by sophisticated chemical methods if required, the general characteristics of a particular sample can be found by the relatively simple techniques of proximate analysis and ultimate analysis.

Proximate analysis involves heating crushed coal to specified temperatures under controlled conditions and determining proportional changes in mass produced in the process. It is used to determine the relative proportions of volatile and nonvolatile organic material produced from the coal by heating in the absence of air, referred to as volatile matter and fixed carbon, respectively, as well as the amount of moisture present and the proportion of noncombustible residue, or ash, left behind after the organic material is burned.

Ultimate analysis involves determining the proportions of the main chemical elements in the coal: carbon, hydrogen, oxygen, nitrogen, and sulfur. These elements are mostly determined using elemental analyzers. They can

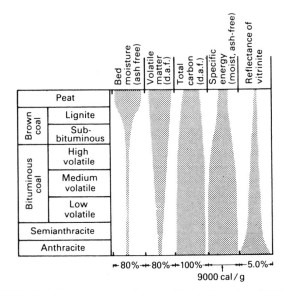

FIGURE 1 Changes in coal composition with rank advance. [From Ward, C. R. (1984). In "Coal Geology and Coal Technology." Reproduced by permission of Blackwell Scientific Publications, Melbourne.]

also be determined by burning the coal under controlled conditions or, for some tests, digesting the sample in an acid or other reagent and converting the products to simple chemical compounds for titration or gravimetric analysis. Details of the procedures normally followed are given in publications by the national standards organizations of the major coal-producing countries, such as the International Standards Organization or the American Society for Testing and Materials.

B. Mineral Matter

Material in coal, other than moisture and organic compounds, is usually described as mineral matter. Such material includes not only the discrete, crystalline particles (minerals in the geological sense) that are associated, often intimately, with the plant debris, but also any salts and other substances dissolved in the pore waters of the coal and the inorganic elements, apart from those determined by ultimate analysis (carbon, hydrogen, oxygen, nitrogen, and sulfur), that are often components of the organic compounds themselves. Much of the mineral matter in peats and lignites occurs as dissolved salts, as exchangeable ions, or as elements in the hydrocarbon compounds, but in higher rank coals, such as bituminous coal and anthracite, crystalline particles usually make up almost all of the mineral matter.

The actual minerals most commonly found in coal include quartz, clay minerals, pyrite and other sulfides, carbonates such as siderite, calcite, and dolomite, and sulfate minerals of iron and calcium. They originate in a number of ways, some being washed or blown directly into the peat swamp, some forming, along with the original organic debris, as crystalline masses (phytoliths) within the plant tissues, and some occurring as skeletal components, such as diatom frustules and sponge spicules, derived from other swamp organisms. Minerals may also form by reactions or precipitation processes in the pore waters of the peat, either as the organic matter accumulates or shortly after it is buried, or may be introduced by percolating waters into the cracks and fissures of the coal during or after the rank-advance process.

Like the organic matter, many of the minerals in coal, such as carbonates, clays, and sulfides, break down when the coal is burned. Some of the components of the mineral fraction escape with the volatile matter in proximate analysis, leaving only a residue of these minerals in the ash. A number of the elements found in the organic matter also occur in some of the minerals, such as carbon and oxygen in carbonates, sulfur in pyrite, and hydrogen and oxygen in clay minerals. Unless appropriate allowance is made—for example, by making separate determinations of the carbonate CO_2, pyritic sulfur, and sulfate sulfur content—

ultimate analysis may not always reflect the composition of the organic material in the coal.

C. Trace Elements

As well as the major elements in their hydrocarbon and mineral constituents, coals also contain a wide range of elements in trace amounts. Some of these elements, such as arsenic, selenium, and boron, are typically more abundant in coal or coal ash than they are in crustal rocks generally. Some may represent possible sources of pollution; others may represent potentially valuable by-products when the coal is used.

Many trace elements appear to be preferentially associated with the organic matter in the coal, either adsorbed onto or incorporated within the organic compounds. Others, however, are apparently associated to a greater extent with the mineral fraction, occurring as minor components of abundant minerals in the coal or as major components of minerals present themselves in only minor proportions. Boron, which appears to be associated mainly with the organic matter, is of particular interest to coal geologists, since its abundance reflects the degree of influence exerted by marine conditions on coal accumulation.

D. Combustion Properties

Combustion involves a series of exothermic chemical reactions that take place when coal at high temperature is brought into contact with air or oxygen. Although, in the past, it has provided a direct source of heat energy for processes such as domestic heating or for transport by ships and railway locomotives, coal combustion is now used mainly for steam raising at centralized plants dedicated to electric power generation or regional heat distribution (Fig. 2). Other uses include the provision of heat for cement manufacture and various industrial processes.

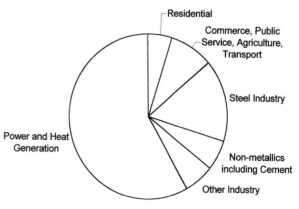

FIGURE 2 Global usage of hard (higher rank) coal in 1996. [World Coal Institute data.]

Combustion may take place in a bed of relatively large sized coal particles, a process known as solid coal combustion, or it may be brought about by the injection of finely powdered or pulverized coal, mixed with a jet of air, into the furnace. Relatively coarse coal particles may also be borned while being lifted in an up-current of air, a process known as fluidized-bed combustion.

Coal combustion by any of these methods involves two essentially simultaneous processes. The volatile matter is first driven off from the coal to burn in the hot furnace atmosphere, after which the nonvolatile fraction burns to leave behind a noncombustible ash residue.

1. Specific Energy

The amount of heat energy released on combustion per unit mass of coal is referred to as the specific energy of the coal. This property is often measured in SI units of megajoules per kilogram (MJ/kg). It is also, however, widely referred to as calorific value or heating value and measured in calories per gram (cal/g), kilocalories per kilogram (kcal/kg), or British thermal units per pound (Btu/lb). Conversion factors for these units are as follows:

$$1.00 \, MJ/kg = 429.923 \, Btu/lb$$

$$= 238.846 \, cal/g \quad or \quad kcal/kg$$

Specific energy is determined in the laboratory by sealing a sample of coal with an excess of oxygen in a gas-tight pressure vessel and igniting it by an electrical firing circuit while the vessel is immersed in a known mass of water. The heat released by the combustion process is transferred to the water surrounding the pressure vessel, and the temperature increase in this water, after a series of corrections, gives the specific energy value.

In an operating furnace, the water derived from evaporation of the coal's moisture and combustion of its organic compounds escapes as a vapor into the atmosphere. In laboratory tests for specific energy, however, this vapor is allowed to condense after combustion, increasing the amount of energy obtained by the latent heat given off in the condensation process. The specific energy value determined by laboratory testing, known as the gross specific energy, must therefore be adjusted to determine the actual amount of heat, or the net specific energy, obtained in a combustion plant where condensation of the water vapor does not take place.

If allowance is made for mineral matter and moisture, which act as diluants and reduce the amount of heat otherwise available, the specific energy increases with the rank of the coal concerned. The proportion of volatile matter, however, is also important in combustion applications, since it determines the ease with which the coal will ignite and actually release this energy. Although they have a high specific energy, anthracites, with a low proportion of volatile matter, are usually difficult to ignite, while lignites and subbituminous coals, with abundant volatile matter, often ignite readily. These lower rank coals are often liable to spontaneous combustion while in transit or being held in stockpiles.

2. Other Factors in Coal Combustion

Other materials in coal that significantly affect its use in combustion applications include sulfur, chlorine, nitrogen, and several other elements. These can combine with the combustion products to form compounds that either corrode the water tubes and other surfaces within the furnace or coat them with deposits that reduce the rate of heat transfer for processes such as stream raising. Many of these elements, along with toxic trace elements such as lead or arsenic, may also escape into the atmosphere to form potential pollutants in gases or fine ash particles from the furnace stack.

The behavior of the ash produced after combustion can also be important when a coal is used as furnace fuel. Most furnaces are designed so that the ash is collected and removed as a solid, either from the base of the furnace as bottom ash or from suspension in the stack gases as fly ash, and ash that melts in the hot part of the furnace and solidifies on reaching cooler sections is difficult to handle in these circumstances. The temperatures at which the ash goes through various stages of the melting process, the ash fusion temperatures, are determined in the laboratory by noting the behavior of molded ash specimens through an observation window in a high-temperature furnace. The chemical composition of the ash is also a guide to its behavior, while the electrical properties of fine ash particles, along with their composition, indicate the ease with which fly ash from the coal can be recovered by electrostatic precipitators from the stack gases.

When a coal is to be burned by pulverized-fuel processes, the ease with which it can be ground to fine powder may have to be determined. This is assessed by grinding the coal in a special mill with a given input of energy and evaluating a parameter called the Hardgrove grindability index from the proportion of fine particles produced. A coal with a low index value is stronger than one with a high index value, and may necessitate the installation of additional grinding equipment if a given output of pulverized coal is to be maintained.

E. Carbonization and Coking Characteristics

When coal is heated at several hundred degrees centigrade in the absence of oxygen or air—a process known as carbonization—the volatile matter is driven off and a solid, carbonenriched residue left behind. After cooling, the volatile products can be separated to form a liquid hydrocarbon fraction (called tar), aqueous solutions rich in ammonium compounds (ammoniacal liquor), and a range of hydrocarbon gases. Depending on the rank and type of the coal as well as the carbonization conditions, the solid residue may be a fused, porous solid (coke) or a noncoherent, powdery substance known generally as char.

All of these products have potential for economic use in some way, but most carbonization is currently directed toward the production of coke for iron and steel manufacture. The tars, liquors, and gases from the coal, however, are usually recovered as by-products of this coking process and used for various purposes. The coke itself is heated with iron ore and a fluxing agent such as limestone in a blast furnace, reducing the oxides in the ore to iron metal and allowing mineral contaminants to be drawn off as a silicate slag. In some cases powdered coal is also injected into the blast furnace without coking, a process known as pulverized coal injection (PCI), to perform a similar function and reduce the overall coke requirement.

An indication of the yield of coke and other products from carbonization of a particular coal can be obtained in the laboratory by a carbonization assay. The coal is heated in an inert atmosphere over a prescribed temperature range, and the products collected for further evaluation. Coals with a low proportion of volatile matter generally give a higher coke yield, and, provided that the coke is of suitable quality, are usually preferred over high-volatile coals on economic grounds. However, high-volatile coals produce a greater proportion of by-products, and the value of these may partly compensate for the lower yields of coke obtained.

In order to function properly when used, the coke must be both porous and permeable, with a large surface area to allow the various reactions to take place. However, it must also be well fused and strong, so that it will not break up into fine particles on handling or be crushed by the weight of the other materials in the blast furnace.

The nature of the coke produced from a coal can be found, in general terms, by determining the crucible swelling number (or free-swelling index) of the coal in a simple laboratory test. A small quantity of crushed coal is heated in a closed crucible until all of the volatiles have been driven off, and the shape of the coke button formed in the crucible is examined. The swelling number is graded from zero for a coal that produces a noncoherent powdery char in these circumstances to 9, where a swollen, porous coke is formed that completely fills the crucible space (Fig. 3).

The crucible swelling number depends on both the rank and the type of the coal concerned. If a single coal is to be used for coke manufacture, an intermediate value (4–6) for

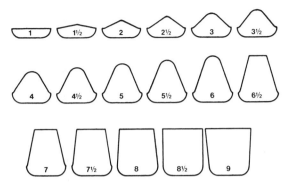

FIGURE 3 Cross-sectional profiles of coke buttons for different values of the crucible swelling number (or free swelling index). [From British Standard 1016, part 12, 1980. Reproduced by permission of the British Standards Institution, London.]

this parameter is probably desirable, since a coal with a low swelling number will not have an adequate porosity, while one with a high swelling number will not have adequate strength. In most large-scale operations, however, the coke is made from blends of several coals, each of which may have different swelling characteristics.

Other coking tests with a similar function include determination of the Grey–King coke type and determination of the Roga index. The strength of a trial batch of coke prepared from the coal in a small-scale laboratory coke oven may also be measured if required, usually by tumbling it in a specially designed drum for a prescribed period and determining the proportion of fines produced.

Some of the organic components of the coal fuse during carbonization. After the volatile matter has escaped, however, the residues of these constituents resolidify, binding the remaining particles together into a coherent mass. The changes that take place during coke formation can be studied by monitoring the length of a pressed specimen of powdered coal in a dilatometer, such as the Audibert–Arnu or the Ruhr dilatometer, or by measuring the viscosity, or fluidity, of a pressed coal powder in a Gieseler or Hoehne plastometer as the coal is heated over its carbonization range. Note is taken of the temperatures at which critical stages of the carbonization process are reached and of the degree of change, such as contraction, expansion, or fluidity, that develops in the material as the coke is formed. These data are then used in various ways to assess the suitability of a coal or a blend of coals for coke manufacture.

F. Chemical Properties and Coal Rank

Proximate and ultimate analysis, as well as most other tests on coal, are usually carried out on samples dried in the laboratory atmosphere. The results can also be recalculated, and supplemented by additional tests if necessary,

TABLE I Coal Classification by the American Society for Testing and Materials[a]

Class	Group	Fixed carbon limits, % (dry, mineral-matter-free basis)		Volatile matter limits, % (dry, (mineral-matter-free basis)		Calorific value limits, Btu/lb (moist,[b] mineral-matter-free basis		Agglomerating character
		Equal to or greater than	Less than	Greater than	Equal to or less than	Equal to or greater than	Less than	
I. Anthracitic	1. Meta-anthracite	98	—	—	2	—	—	Nonagglomerating
	2. Anthracite	92	98	2	8	—	—	
	3. Semianthracite[c]	86	92	8	14	—	—	
II. Bituminous	1. Low-volatile bituminous coal	78	86	14	22	—	—	Commonly agglomerating[e]
	2. Medium-volatile bituminous coal	69	78	22	31	—	—	
	3. High-volatile A bituminous coal	—	69	31	—	14,000[d]	—	
	4. High-volatile B bituminous coal	—	—	—	—	13,000[d]	14,000	
	5. High-volatile C bituminous coal	—	—	—	—	11,500	13,000	
						10,500	11,500	Agglomerating
III. Subbituminous	1. Subbituminous A coal	—	—	—	—	10,500	11,500	Nonagglomerating
	2. Subbituminous B coal	—	—	—	—	9,500	10,500	
	3. Subbituminous C coal	—	—	—	—	8,300	9,500	
IV. Lignitic	1. Lignite A	—	—	—	—	6,300	8,300	
	2. Lignite B	—	—	—	—	—	6,300	

[a] This classification does not include a few coals, principally nonbanded varieties, which have unusual physical and chemical properties and which come within the limits of fixed carbon or calorific value of the high-volatile bituminous and subbituminous ranks. All of these coals either contain less than 48% dry, mineral-matter-free fixed carbon or have more than 15,500 moist, mineral-matter-free British thermal units per pound. From American Society for Testing and Materials Standard D 388. Reprinted with permission from the Annual Book of ASTM Standards. Copyright, ASTM 1916 Race Street, Philadelphia, Pennsylvania 19103.

[b] *Moist* refers to coal containing its natural inherent moisture but not including visible water on the surface.

[c] If agglomerating, classify in low-volatile group of the bituminous class.

[d] Coals having 69% or more fixed carbon on the dry, mineral-matter-free basis shall be classified according to fixed carbon, regardless of calorific value.

[e] It is recognized that there may be nonagglomerating varieties in these groups of the bituminous class, and there are notable exceptions in high volatile C bituminous group.

to express them as percentages of dry, dry ash-free, or dry mineral-matter-free coal, rather than as percentages of the sample as received or in the air-dried state.

When corrected to allow for ash or mineral matter, as well as variations in coal type, the percentages of moisture and volatile matter decrease with rank advance. If allowance is made for moisture as well, then this is also true of the percentages of oxygen and hydrogen in the organic matter. Fixed carbon and total carbon, in contrast, increase. The rates of increase or decrease are not uniform, however, and different components change in proportion at different stages of the rank-advance process. Together with the specific energy or calorific value, which also increases with rank advance (Fig. 1), and in some cases an indication of the coal's behavior on coking, such as its crucible swelling number, these properties form the basis for a number of coal classifications. An example of such a classification is given in Table I.

II. PETROLOGY OF COAL

Detailed examination of coal, either in hand specimen or under the microscope, shows it to consist of discrete masses, aggregates, or layers of different organic materials. In peats and low-rank coals, these can be clearly identified as variously degraded components of the original plant debris. In higher rank coals, however, they are often considerably modified by compaction and other processes, with only remanant botanical structures visible under the microscope providing a general indication of their origin.

Coal petrology is the study of these coal constituents, including their composition, morphology, and optical properties, as well as their relation to other coal characteristics. Quantitative study of their relative abundance or quantitative evaluation of key properties of selected components, a branch of the science sometimes referred to as coal petrography, represents another tool for the assessment of coal quality and provides a geological framework to supplement coal classification by traditional chemical means.

A. Coal Lithotypes

Most coal seams are made up of heterogeneous accumulations of large and small fragments of plant debris. This debris is inherently deposited under conditions that produce at least some degradation due to interaction with the oxygen of the atmosphere before the material is buried. A range of organic compounds known as humic acids are characteristic products of this degradation, and coals originating from peat formed in this way are known as humic coals.

Some coal, however, originates from homogeneous accumulations of algae, plant spores, and related fine organic debris. Material of this type is usually deposited in an anaerobic or oxygen-poor environment and undergoes a fermentation or putrefaction process to form a different range of organic products. Coals derived from such accumulations have chemical and physical properties that are different from those of the more common humic coals and are known in coal petrology as sapropelic coals.

When humic coals of bituminous rank are examined in hand specimen, such as in field outcrops or mine exposures, up to four classes of material may be seen:

1. *Vitrain* takes the form of thin, homogeneous bands of black, glassy material, usually ranging from 3 to ~10 mm in thickness. It is generally very brittle and characterized by a close-spaced fracture pattern cutting across the layers.

2. *Fusain* occurs as thin bands of soft, powdery material, not unlike charcoal in physical appearance. Some fusain, however, is harder due to impregnation of the pores between the particles with mineral matter.

3. *Clarain* is a finely laminated coal with an overall silky luster. It generally contains very fine vitrain bands set in a matrix of more homogeneous material.

4. *Durain* is a black, homogeneous, relatively tough material with a characteristic dull, earthy luster. It resembles impure coal and carbonaceous shale in this regard but is distinguished from stony coal by having a lower density and a lower ash yield.

Together with material intermediate in character between durain and clarain, these materials form the macroscopically recognizable bands seen in most bituminous coal seams. Such bands are formally known in coal petrology as lithotypes.

Two varieties, or lithotypes, are also recognized in sapropelic coals. Material made up mainly of algal remains is known as boghead coal, while material consisting of abundant spores and similar debris is referred to as cannel coal.

The appearance of humic and sapropelic coals at lower rank (i.e., in subbituminous coals and lignites) is quite different from that of the lithotypes recognized in high-rank deposits. Although different kinds of material can be seen in hand specimen, a universal system of lithotype nomenclature for these coals has not yet been developed.

Vitrain and fusain are bands of essentially homogeneous character, but clarain, durain, and, where recognized, the intermediate forms are intimate mixtures of different, yet megascopically recognizable organic components. Clarain, durain, and related lithotypes, therefore, are often referred to as attrital coals, emphasizing their fragmental nature. It is not always practical to use the formal lithotype terms in many field situations, however, and humic coals or coal seam subsections are often described simply as interbedded mixtures of bright (vitrain or clarain) and dull constituents (durain or fusain), with an indication of the relative proportions of each (Table II).

B. Macerals

Coal can be studied microscopically by the use of either very thin sections viewed in transmitted light or highly polished sections viewed in reflected (or incident) light. The optical density of coal, particularly in high-rank

TABLE II Terms Used to Describe Banded Bituminous Coals in Hand Specimen

United States[a]		Australia[b]
Vitrain		Coal bands, bright
Attrial coal	Bright	Coal, bright with dull
	Moderately bright	Coal, dull and bright, interbanded
	Midlustrous	Coal, mainly dull with numerous bright bands
	Moderately dull	Coal, dull with minor bright bands
	Dull	
Fusain		Coal, dull

[a] After Schopf, J. M. (1960). *U.S. Geological Survey Bull.* **1111B,** 25–69.

[b] After Australian Standard 2916 (1986). "Symbols for the Graphic Representation of Coal Seams and Associated Strata," Standards Association of Australia, Sydney.

materials, requires thin section slices to be prepared to about one-third the thickness required for other rocks. Even at this thickness, however, many of the components are still opaque, and little can be seen of their constitution in transmitted-light studies. Polished sections are much easier to prepare and, particularly with oil immersion optical systems, allow a wider range of components to be examined. Although much of the pioneering work in coal petrology was conducted using thin sections and thin-section methods are still useful in some research applications, modern coal petrology is now based almost entirely on polished-section techniques.

The fundamental constituents of coal recognizable at a microscopic scale are known in coal petrology as macerals. These are discrete bands or organic particles distinguished from one another by their morphology and optical properties in much the same way as inorganic particles (minerals) are distinguished from one another under the microscope in noncoal rocks.

Macerals in coal are characterized by such properties as their reflectance in incident light, their hardness or relief on a polished surface, and their shape or inherent botanical structure (Fig. 4). Some macerals also exhibit visible fluorescence when illuminated by high-intensity blue-violet or ultraviolet light.

Coal macerals differ from one another in their chemical characteristics and aspects of their behavior during coal utilization. The chemical constitution of the different macerals may be studied in polished sections using a special electron microprobe on Fourier-transform intrared (FTIR) microanalysis techniques. The chemical and optical features of macerals also change with rank, so that distinctions among them in low-rank coals may be all but obliterated in materials of higher rank.

The most widely used classification of macerals is the Stopes–Heerlen system. This system recognizes three groups, principally on the basis of their optical properties and chemical reactivity (Table III).

1. Vitrinite Group

In bituminous and higher rank coals, vitrinite occurs as relatively homogeneous bands of various thicknesses, with a medium gray color or moderate reflectance in polished section and a red-brown to orange color in thin section. The macerals of this group are derived from wood, stems, and other plant tissue, often impregnated with organic gels, that has been preserved with little oxidation or decay in the peat swamp. Materials of similar origin in low-rank coals may have different features and are sometimes referred to as members of the huminite group. These, however, are usually regarded simply as precursors of the vitrinite macerals.

TABLE III Classification of Coal Macerals[a]

Maceral group	Maceral (or submaceral)	Principal mode of origin
Vitrinite	Telinite	Mummified cell walls of woody tissues
	Telocollinite (collotelinite)	Gel-impregnated woody tissues
	Vitrodetrinite	Small vitrinitic fragments
	Desmocollinite (collodetrinite)	Gelified smaller plant tissues
	Gelinite	Organic gels, mainly as cell infillings
Liptinite (or exinite)	Sporinite	Remains of spore coatings
	Cutinite	Remains of leaf cuticles
	Resinite	Resin bodies
	Alginite	Remains of algal masses
	Suberinite	Remains of waxy, cork-like tissues
	Liptodetrinite	Fine fragments of liptinite macerals
	Fluorinite	Lenticles of plant oils or fats
	Bituminite	Bitumen-like residues from algae, etc.
	Exsudatinite	Fluorescent cavity infilling
Inertinite	Fusinite	Oxidized cellular woody tissue
	Semifusinite	Partly oxidized tissue
	Inertodetrinite	Oxidized cell wall fragments
	Micrinite	Fine-grained, probable decay product
	Macrinite	Massive, possibly fusinized gels
	Funginite	Cellular fungal remains
Secretinite	Fusinized	Vesicular resin bodies

[a] Modified from Davis, A. (1984). In "Coal Geology and Coal Technology" (Ed. C. R. Ward), Blackwell Scientific Publications, Melbourne.

The principal members of the vitrinite group are telocollinite and desmocollinite. Telocollinite occurs as bands of homogeneous material, usually 3–12 mm in thickness, and represents the mummified remains of roots, stems, bark, and other fragments of large woody tissues. It forms the vitrain bands visible in most coal hand specimens. Desmocollinite is derived from smaller plant components, such as reed and grass fragments, and forms much thinner, but still homogeneous bands. It has slightly different optical properties than telocollinite and is characteristically found in the attrital coal lithotypes. It is particularly abundant in clarites.

2. Liptinite Group

Also referred to as the exinite group, the liptinite group is a diverse assemblage of small organic particles

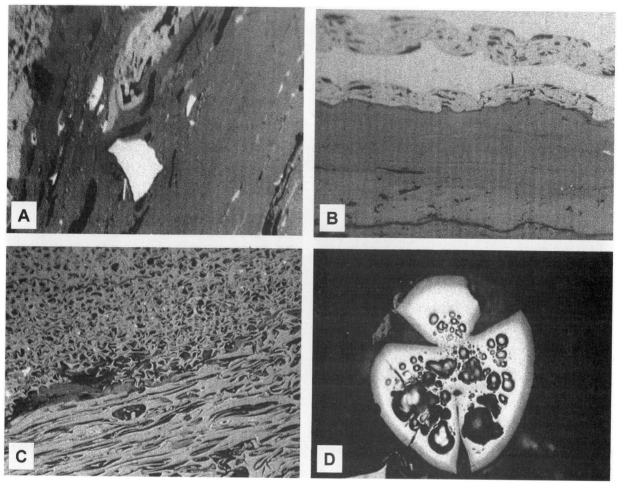

FIGURE 4 Photomicrographs of bituminous coal in polished section under oil immersion. (Photos: L. W. Gurba). Field width approximately 0.2 mm. (A) Sporinite (dark gray), inertodetrinite (white angular fragments) and semifusinite (light gray cellular masses) in vitrinite (mid-gray). The lighter colored vitrinite on the right is telocollinite and the darker vitrinite on the left is desmocollinite. (B) Semifusinite (top) and vitrinite (bottom), with a thin dark gray cutinite band near base of image. (C) Fusinite (lower half) and semifusinite. (D) Secretinite (also known as resino-sclerotinite or pseudosclerotinite).

characterized by a very dark gray color or a low reflectance in polished section and a bright yellow color, due to their greater transparence, in thin section. The macerals of this group, at least in coals of up to high-volatile bituminous rank, typically have a high hydrogen content and give a high yield of volatile matter in proximate analysis. They also exhibit a range of characteristic fluorescence effects when viewed under blue or ultraviolet light irradiation.

Liptinite macerals include small particles derived from the outer coating of spores and pollens (sporinite) and long, thin bands representing the outer skin or cuticle of leaf tissues (cutinite), as well as various types of resin bodies (resinite) and algal masses (alginite). When studied under illumination that produces fluorescence effects, a number of other liptinite macerals can be seen that are not normally visible in white light. These include fluorinite (usually found as lenticular, yellow-fluorescing bodies), bituminite (irregular shredlike aggregates), and exudatinite (fluorescent infillings of cracks and cavities in other coal macerals), each of which has optical properties different from other members of the group as well as their own individual fluorescence characteristics.

Liptinite macerals, chiefly sporinite and cutinite, are found in relatively small proportions in most humic coals. They tend to lose their identity, however, as their reflectance increases in higher rank materials. Liptinites are the dominant macerals in sapropelic coals, with alginite being abundant in bogheads and a range of related oil shales, and sporinite forming the dominant constituent of cannel coals.

3. Inertinite Group

The inertinite group takes its name from the fact that its members characteristically undergo little reaction or alteration when the coal in which they occur is carbonized (i.e., heated to high temperature in the absence of oxygen). The macerals of this group have a higher reflectance than those of the other groups, giving them a light gray color in polished section. They are opaque to transmitted light, however, and appear as black bands or particles in thin-section studies.

The inertinite macerals are mainly derived from plant tissues similar to those from which the macerals of the vitrinite group are derived. However, the material has been preserved under different conditions and suffered more extensive oxidation in the peat swamp before burial. The most common members of the group are fusinite, which occurs as highly reflecting bands with well-defined cell wall remains, and semifusinite, a moderate to highly reflecting material, cellular but generally somewhat more homogeneous than fusinite, and intermediate between fusinite and vitrinite in character. Other members of the group are inertodetrinite (detached cell wall fragments within the coal), micrinite (layers of fine, powdery, high-reflecting material), macrinite (small, rounded, homogeneous bodies), and sclerotinite (cellular masses representing fungal remains).

The inertinite macerals contain more carbon and less hydrogen than members of the other maceral groups and, all else being equal, give a lower yield of volatile matter. Fusain, of course, represents megascopic bands of fusinite, while durain and other dull lithotypes are typically rich in inertinite macerals. Although they are quite satisfactory for use in combustion processes, coals with a high inertinite content are usually of limited value in coking operations, unless they also contain, or can be blended with, vitrinite-rich materials.

C. Microlithotypes

When examined at a microscopic scale, the macerals in coal, particularly humic coal, are organized into discrete bands, lenticles, or layers, each layer being made up of either a single maceral type or an intimate mixture of different maceral components. The natural associations of macerals that make up these layers are known as microlithotypes.

Seven combinations of the three maceral groups are recognized (Table IV), and these constitute the main microlithotype groups. Three of these groups are made up of representatives of only one maceral group (monomaceral microlithotypes), three are mixtures of macerals from two different groups (bimaceral microlithotypes), and one is a

TABLE IV   Components of Coal Microlithotypes[a]

Microlithotype	Microlithotype group	Components
Monomaceral	Vitrite	>95% vitrinite
	Liptite	>95% liptinite
	Inertite	>95% inertinite
Bimaceral	Clarite	>95% (vitrinite + liptinite)
	Durite	>95% (inertinite + liptinite)
	Vitrinertite	>95% (vitrinite + inertinite)
Trimaceral (trimacerite)	Duroclarite	Vitrinite > (inertinite + liptinite)
	Clarodurite	Inertinite > (vitrinite + liptinite)
	Vitrinertoliptite	Liptinite > (vitrinite + inertinite)

[a] After Stach, E., Mackowsky, M.-Th., Teichmüller, M., Taylor, G. H., Chandra, D., and Teichmüller, R. (1982). In "Stach's Textbook of Coal Petrology," Gebruder Borntraeger, Stuttgart.

mixture of components from all three groups (trimaceral microlithotype, or trimacerite). Vitrite, for example, is a natural association or band made up entirely of vitrinite, while clarite is a mixture of vitrinite and liptinite macerals (Table IV). Duroclarite and clarodurite are made up of vitrinite, liptinite, and inertinite, with vitrinite being dominant in the former and inertinite dominant in the latter.

Components with an abundance of less than 5% in the mixture are normally disregarded in the identification of microlithotypes, as are natural bands less than 50 μm in width. Because they represent the natural assemblages of organic constituents in the peat, at a microscopic scale, microlithotypes often give information on the mode of origin of the coal. They also indicate the way in which the macerals are packed together in the coal and thus are useful in assessing, in detail, how the coal might behave when used, particularly in coke manufacture.

D. Maceral Reflectance

The proportion of the incident light reflected by all macerals increases progressively as the rank of the coal increases. This proportion, usually measured with a sensitive photometer using monochromatic light under oil immersion conditions, is referred to as the reflectance of the maceral. It can be measured for any maceral, but most reflectance studies concentrate on the relatively large, homogeneous vitrinite components.

When polarized light is used, the reflectance measured for a given vitrinite particle depends on the orientation of the polarization plane with respect to the bedding of the vitrinite. The optical anisotropy, or contrast between the maximum and minimum reflectance displayed by the vitrinite, increases markedly in high-rank coals, but each vitrinite particle still displays a maximum reflectance value when the polarization is parallel to a trace of the bedding plane. The mean of these maximum values for a number of particles in a crushed, mounted, and polished coal specimen is referred to as the mean maximum reflectance in oil ($\bar{R}_{o\,max}$). This value increases steadily from $\sim 0.3\%$ for vitrinite in lignites to more than 3.0% for vitrinite in anthracites (Fig. 1), providing a very useful petrographic index of coal rank. Together with an index of coal type based on the relative percentages of the different macerals, it is an alternative to chemical parameters in coal classification.

Where they occur together in the same coal sample, or in coals of equivalent rank, the desmocollinite member of the vitrinite group typically has a slightly lower reflectance than the telocollinite member. Use of vitrinite reflectance as a rank indicator is therefore more precise if measurements are restricted to the telocollinite component. The reflectance of vitrinite, including telocollinite, is also lower, in coals of equivalent rank, where the coal has been influenced by marine conditions during or shortly after deposition.

III. GEOLOGY OF COAL

Although carbonaceous rocks have been found in deposits from as far back as Precambrian times, true coals are not found in the rock record until plants started to appear on earth during the Silurian period. Some coal accumulations have appeared in Devonian strata, but most of the world's economically valuable deposits are from the Carboniferous or later periods.

Carboniferous coals, mainly of bituminous rank, are found in Europe and central Asia, as well as the eastern part of North America. Somewhat younger Permian coals also occur in Europe and parts of Asia, but the most significant resources of this age are found in Australia, India, Africa, South America, and Antarctica, the regions that made up, at that time, the single land mass known as Gondwanaland.

Triassic and Jurassic coals also occur in some areas (e.g., Australia), while Cretaceous coals of bituminous to subbituminous rank cover large areas of western Canada and the western United States. Low-rank brown coals (lignites) of Tertiary age occur in relatively undisturbed geological settings in Europe (particularly in Germany), the United States, and the southern part of Australia. Higher rank materials of similar age occur in more folded strata in Japan and the Southeast Asian region.

A. Peat Formation

The formation of peat, especially peat with sufficient quality, thickness, and extent to form a workable coal seam, requires plants to grow and plant debris to form at a faster rate than organic matter is destroyed by the oxidation and decay that arise from exposure to the atmosphere. It also requires, however, that the accumulating plant debris be contaminated to no more than a minimal extent by the introduction of inorganic material, such as silt and clay, to the depositional site.

Prolific plant growth takes place in swampy areas, and the debris shed from these plants is protected from oxidation beneath the stagnant waters of the swamp itself. The growing plants also provide a pattern of baffles and filters that prevent suspended sediment from penetrating very far into the swampy terrain. The water level in such swamps is very critical to peat formation, since excessive oxidation may occur if the swamp dries out or the plants may be killed off and additional noncoal sediment accumulate if the water becomes too deep.

Peat-forming swamps, bogs, and marshes may develop in a wide range of sedimentary environments, including river floodplains, deltas, lakes, and coastal lagoons. Other environments favorable for peat formation include muskeg bogs in poorly drained, high-latitude areas and localized swamps in volcanic craters or limestone sinkholes. Not all of these, however, are likely to produce peat deposits that will be preserved in a sedimentary succession to become a recognizable coal bed.

In some areas high rainfall encourages vegetation to grow and peat to build up above the local ground level, such as that of any surrounding river floodplain. Environments formed in this way are referred to as raised bogs. Peat accumulating in such deposits foms at a higher level than the silt and clay introduced by river action, and thus inherently has minimal contamination by other sedimentary material.

Depending on such factors as water depth, salinity, and possibly substrate type, different plant communities develop in different parts of the peat-forming mires. The different components in the debris released by these communities, coupled with differences in the degree of oxidation inherent in different mire environments, determine the assemblage of macerals and microlithotypes in the resultant coal (i.e., the coal type) and give rise to variations in certain properties, such as coking characteristics, in different parts of the seams ultimately produced.

Considerable debate often arises as to whether the vegetation from which the peat was formed to make a particular coal seam grew at the actual site of deposition (*in situ* or autochthonous peat) or was transported to that site (drifted or allochthonous peat). The presence of root traces in the sediments immediately beneath the seam is usually taken to indicate *in situ* growth, but the absence or lack of preservation of such structures does not necessarily imply transportation. Because of the large volume of organic matter involved and the virtual absence of other sedimentary

FIGURE 5 Coal seams and associated strata exposed in a gently dipping sedimentary sequence of the Cretaceous Age (Utah). (Photo: C. R. Ward.)

contaminants, most seams of commercial significance appear to have been derived mainly from *in situ* or essentially *in situ* accumulations of plant debris.

B. Depositional Environments of Coal Seams

Coal seams occur as beds within layered successions of other sedimentary strata, associated with such rocks as sandstone, shale, conglomerate, and possibly limestone (Fig. 5). Individual coal beds may be 100 m or more in thickness or extend over areas of several thousand square kilometers, but they are commonly much thinner and more restricted in lateral extent. Depending on their geological history after deposition, they may range from flat-lying or gently dipping beds to units in highly folded and faulted stratigraphic successions.

The depositional environments of coal formation are traditionally divided into limnic and paralic with respect to their geological settings. Limnic coals are deposited in inland areas, usually in landlocked lakes, bogs, and swampy regions of intracontinental basins, while paralic coals are formed in coastal areas with the depositional environment open, at various times, to incursion by the sea. Because such areas are more likely to be associated with continued subsidence and buildup of substantial thicknesses of sediment generally, most of the world's major coal deposits appear to have been formed under paralic depositional conditions.

Present-day examples of paralic coal-forming environments include the coastal plains of southern New Guinea, the western part of Africa, and the eastern and southern seaboards of the United States. These areas include extensive swamps that have been formed around major river deltas, behind coastal barrier systems, and at the foot of alluvial fan accumulations. Since the earth has not yet adjusted to recent changes in global sea level, however, erosion of the land and associated production of clastic sediment is taking place at a relatively high rate, compared with the conditions that might be expected to have existed at the principal times of coal formation in the geological past. Although a useful guide to the general geological framework of coal formation, these environments do not provide an exact parallel to ancient coal-forming conditions, because the extent of swampy terrain is probably more limited than it was at those times.

1. Rivers and Upper Delta Plains

Rivers and the onshore portions of river deltas are areas where streams or distributary channels traverse a fairly flat, laterally extensive plain. The channels themselves are generally too swiftly flowing and carry too much sediment to allow plant growth, but vegetation may be prolific on the levees alongside the channels and over much of the surrounding floodplain. The levees in many river areas, and at least part of the floodplain, are often too far above the water table to allow the debris from these plants to be preserved without oxidation or are subject to too much sediment influx, with flooding of the channels, to allow a clean bed of peat to build up. However, low-lying swampy areas (back-swamps) in river floodplains, abandoned and partly infilled channel segments, and the extensive swamps and marshes that surround the distributaries of river deltas, including in some cases raised bogs, are ideal sites for peat and coal accumulation.

Coal seams deposited around rivers and on the upper part of delta plains are generally interbedded with layers of sandstone and shale, built up as the stream channels change their course across the region. The sandstones tend to be lenticular in cross section, with an erosional basal contact, internal cross-stratification, and a progressive decrease in particle size from bottom to top of the bed. Coal development takes place above and to the side of these channel-fill bodies, with maximum seam development some distance from the main sandstone mass. The seam may be expected to thin out or split due to an influx of levee and overbank sediment as the sandstone body is approached (Fig. 6a).

2. Lower Delta Plain Coals

At the lower, or seaward, end of a river delta, the channels terminate in sandy shoals or distributary mouth bars built up at the points where the various river branches meet the sea. If river action is strong, relative to waves and tides, such as it is in the Mississippi River delta, the distributary channels may project, flanked by levees, out into the sea, with shallow marine embayments located in between. Swamps and marshes flank the projecting channels and interdistributary bays, forming extensive layers of peat on top of the mouth bar and bay-fill sediments (Fig. 6b).

In contrast to the coals found in the upper delta plain, lower delta plain coals tend to occur as thin, blanket-like bodies interbedded with shales that are at least partly of marine origin. The strata coarsen progressively upward from shale, often with marine fossils or evidence of burrowing by bay-dwelling organisms, through closely interlaminated sandstone and shale sequences to massive sandstones formed by the shoals at the river mouth.

Diversion of river action, combined with a continuation of basin subsidence, often results in coals formed on lower delta plains being directly overlain by marine strata. The sea water from which these strata were deposited also permeates the peat beneath, carrying with it a certain proportion of dissolved sulfate ions. Bacterial action in the peat converts some of the sulfate ions to pyrite and possibly to various organic sulfur compounds, giving the coal a high total sulfur content that ultimately causes pollution problems during mining and use. The marine influence

may also affect the processes of peat preservation, leading ultimately to the development of lower vitrinite reflectance than might be expected with rank advance. Coals overlain by freshwater deposits, on the other hand, such as those parts of the seam located close to the distributary channels, are impregnated with sulfate-bearing solutions to a lesser extent and are less liable to problems of this type. They also tend to have more normal vitrinite reflectance values for the rank of the coal concerned.

3. Back-Barrier Coal Deposits

In coastal areas without a major supply of river sediments or in river deltas where wave action is dominant over river processes, sand may be built up offshore as barrier bars and islands parallel to the general coastline. Lagoonal areas of protected, shallow water form between these barriers and the main landmass, and the shallow fringes of these areas, if the tide range is low, provide an ideal site for peat accumulation.

Back-barrier coals generally take the form of thin, discontinuous seams, often resting on burrowed, organic-rich shales deposited deeper in the lagoonal area. The seams are interbedded, at least at their seaward edge, with sandstones formed by material carried in from the barrier due to storm washover or tidal inflow processes (Fig. 6c). On the landward side, however, they tend to be interbedded with river delta deposits.

Like other seams intimately associated with marine strata, back-barrier coals generally have anomalously low vitrinite reflectance and a high total sulfur content. With some notable exceptions, such as the Beckley seam and several others in North America, they are generally too limited in extent or too poor in quality to be of commercial value.

4. Coals Associated with Alluvial Fans

Alluvial fans are lobate wedges of coarse clastic sediment built up at the foot of mountain ranges and in other areas where stream gradients undergo a sudden change. The upper parts of the fan are generally sites of rapid sedimentation and oxidation of plant debris, but the lower parts of the fan, in humid or possibly in arctic climates, may become swampy areas suitable for peat accumulation (Fig. 6d).

Coals formed in association with alluvial fans are typically interbedded with conglomerates, deposited by debris flows and braided streams running off the fan wedge. If the fan terminal occurs in a coastal area, wave action may give rise to barrier sands, with coal forming in lagoons between the fan and the coastal deposits.

Areas such as southern New Guinea or the southeastern part of Alaska are modern examples of peat formation in an

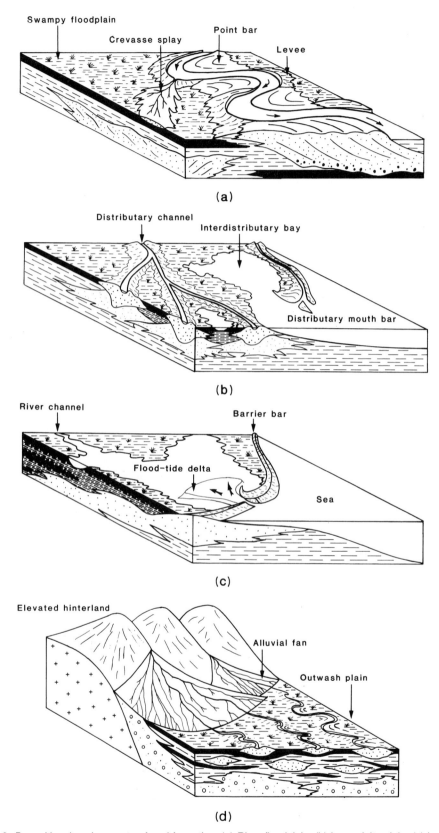

FIGURE 6 Depositional environments of coal formation. (a) River floodplain; (b) lower delta plain; (c) back-barrier lagoon; (d) alluvial fan.

alluvial fan environment. Ancient examples include parts of the Karoo Basin in South Africa and the Sydney Basin in Australia, as well as coal deposits in northern Spain and the Canadian Cordillera region.

C. Geological Features of Coal Seams

Depending, to a certain extent, on the depositional environment in which it was formed, the coal within the seam may be homogeneous in appearance from top to bottom, or it may contain subsections or layers of different coal type. Thin layers of noncoal material, sometimes referred to as "bands" or "partings," may also occur at different levels within the seam. In many cases, the noncoal bands can be traced over large areas, allowing correlation of the host seams from one region to the next.

A number of other features are found within coal seams, some of which are caused by depositional processes associated with peat formation and some by geological process during or after burial. Many of these features are of great significance in mining, limiting the extent or providing a break in continuity of workable coal, while others may be associated with roof instability, gas outbursts, and other hazards to the safety of the mining operation.

1. Splits

The lateral continuity of a coal seam may be limited by several geological features, many of which are the result of processes associated with peat accumulation. A seam may be divided, for example, by a wedge of noncoal strata into two separate coal beds, each of which is referred to as a "split" (Fig. 7a). In some cases, the process produces only a thin "rider" of coal above or below the main seam, but in others a workable coal bed can be separated into two layers, several meters apart, neither being of sufficient thickness to allow an economic mining operation.

Splits are produced by the development of a river channel, lake, or other localized sediment influx within the peat swamp. When the channel is diverted, the lake fills up or the supply of sediment is otherwise exhausted, swampy conditions are reestablished, and peat is formed once again above the mass of noncoal material.

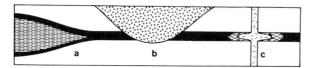

FIGURE 7 Cross section showing some geological features associated with coal seams. (a) Split in coal seam; (b) washout; (c) cindered coal around an igneous intrusion.

2. Washouts and Channel-Fill Structures

Washouts are linear belts along which the top, and in some cases the whole, of the coal seam has been replaced by sandstone or some other clastic sediment (Fig. 7b). They represent erosion of the peat and deposition of noncoal material in river channels, either at the time the organic matter was accumulating or shortly afterward, as part of the process that buried the peat deposit.

Some washouts are relatively minor structures and are responsible only for localized removal of coal from the top of the seam. Others, particularly those contemporaneous with peat deposition, can be major features up to several kilometers wide and some hundreds of kilometers in length.

Because of the contrast in physical properties between coal and channel-fill material, the rock strata around a washout, including the beds that form the normal working roof of the seam, are often cut by slickensided fractures formed by compaction processes. As well as problems with seam thinning or removal due to the washout itself, mining of the coal around washouts is affected by roof falls and other instabilities in underground operations due to these fracture patterns.

3. Igneous Intrusions and Cindered Coal

Intrusions of molten igneous rock are common in many coal-bearing successions, cooling and solidifying to form dykes, sills, plugs, and related geological features. High heat flow from large bodies of this type can give rise to local increases in coal rank, even if the coal is at some distance from the heat source. If the temperature of the coal becomes too high, however, such as might be expected if even a small intrusion was actually in contact with the seam, the igneous body may carbonize the coal and transform it into a natural cokelike material known as cinder or cindered coal.

Hot, molten magma replaces coal more readily than other rocks, occupying the space created by loss of moisture and volatile matter as the cindered material is formed. The intrusion often widens in lateral extent when it intersects the coal bed, forming a sill-like body within the seam surrounded by a cindered zone (Fig. 7c). Mineralized fluids associated with the intrusion subsequently impregnate the pores, cracks, and fissures in the cindered coal and, in many cases, interact with the magma itself to transform the igneous rock to a clay-rich alteration product.

Because of their wide extent when they intersect a coal seam, even small igneous intrusions, such as narrow dykes, can give rise to considerable loss of otherwise workable coal reserves. In most cases the surrounding cindered material is also too rich in mineral matter to be of any commercial use. In some areas, however, increases in local rank without cindering, brought about by nearby igneous intrusions, have actually increased the economic value of the associated coal deposits.

IV. COAL EXPLORATION AND MINING GEOLOGY

Energy derived from coal has long been the basis of modern industrial society, and coal still represents the world's largest single source of conventional energy. It currently provides around 27% of our primary energy needs and is expected to continue in this strategic role for many years. Coal, however, is a finite, nonrenewable resource. The amount available must be properly assessed and the material extracted and used in the most effective way in the light of this assessment if adequate quantities are to be conserved to meet future demands.

Coal exploration involves determining the nature, location, and extent of the coal resources available in a particular situation and delineating the geological features that may affect their safe and efficient extraction. Once it has been established that material of adequate quantity and quality is present to justify mine development, however, further geological studies are generally needed to assist with mine design, production scheduling, and environmental impact assessment.

A. Exploration Programs

Like other activities in mineral exploration, the discovery and geological assessment of a new coal field area involves the following components:

1. Obtaining legal title to explore the area
2. Evaluating the geological and other information already available
3. Exploring the ground surface and subsurface using geological and geophysical techniques
4. Collecting and analyzing samples and delineating quality trends
5. Evaluating coal resources and factors affecting their recovery
6. Communicating results to other members of the project team.

The ownership of coal resources and the procedures involved in gaining the right to explore and mine them vary considerably throughout the world. Most projects, however, require some form of legal agreement, usually involving government authorities, before any significant geological field work can be commenced.

Depending on the area under investigation, the preliminary stages of the exploration program may include ground and photogrammetric surveying to prepare suitable base maps for the compilation of geological data. This work is of great importance to the exploration program and can also be used as the basis for detailed mine plans if and when actual mine development finally takes place.

Systematic environmental data may also be gathered at this stage, with observations, analyses, and measurements of surface and ground-water hydrology, wind patterns, and prevailing dust and noise levels in different parts of the area, as well as data on flora and fauna patterns and sociological factors in the region. This helps to establish the conditions prevailing at the site before any mine development begins and generally permits a more objective assessment of the impact of the project on the environment to be made than if base data collection were delayed until the exploration program were well under way.

The nature of the exploration program itself depends partly on the geology of the area under investigation and partly on such factors as climate, topography, legal constraints and accessibility, as well as the time and resources available for the task. Detailed geological mapping is an essential part of many programs, measuring stratigraphic sections through coal-bearing sequences and tracing the outcrop pattern of key marker beds in the field. This may be supplemented by geophysical studies, such as gravity, magnetic, and seismic surveys, and by the drilling of a series of exploratory boreholes to determine the distribution of the coal seams and other rock units in the subsurface.

B. Geophysical Exploration

Geophysical methods have been used for many years in the search for metallic ore bodies and petroleum fields and are increasingly being applied to coal exploration programs. If the coal basin is underlain by rocks that are denser than or have different magnetic properties from those associated with the coal seams, maps showing the pattern of variation across the area in the earth's gravitational attraction or its magnetic field can be used to assess variations in the depth to this basement and hence indicate the most favorable sectors for follow-up investigation. Detailed magnetic studies can also be used, in some cases, to locate igneous intrusions within the coal-bearing sequence and hence suggest zones where cindered coal may be developed in the seams.

Seismic reflection studies involve the generation of a shock wave at or just below the ground surface, using either a falling weight or an explosive charge, and reflection of this wave from a suitable horizon in the sequence of strata beneath. On its return to the surface, the reflected wave energy is recorded by a series of sensitive receivers, called geophones, connected to a computer-based processing system. The time taken for the wave energy to travel from its source to the reflecting horizon and back again to the geophone array is then used, in conjunction with the velocity of the wave in the strata through which it has passed, to give an estimate of the depth to the reflecting layer.

A series of such tests set out along a traverse line gives a cross-sectional profile of the reflecting horizons in the sequence. Such profiles can be used to show the basic geological structure of the exploration area and also to detect any major fault displacements or other breaks in continuity of the coal seams.

Coalfield studies are concerned mainly with structures having a relief of less than 10 m located at depths of less than 1000 m. The total travel time for reflected seismic waves in these circumstances is less than 1 sec, and this means that the broad-scale techniques of seismic investigation used by the petroleum industry have to be modified, sometimes considerably, for use in coalfield programs.

Special techniques for high-resolution seismic study have been developed to provide the detail needed to delineate structures that may affect coal-mining operations. These involve the generation of a high-frequency, or short-wave-length, energy pulse using a much smaller explosive charge than normal and special high-frequency geophones. The shot point and, if possible, the geophones are placed in drill holes below the weathered layer to minimize attenuation of the signal and to eliminate minor deviations in travel time associated with passing through the loose, near-surface materials. The cost of such studies, however, is generally somewhat higher than might otherwise be the case, and much useful information can also be gained without adoption of these special methods.

C. Exploratory Drilling

Drilling of boreholes through the subsurface strata and recovery of samples in the form of broken rock cuttings or a cylindrical drill core provides the most reliable source of information on the depth, thickness, and quality of the coal and the nature of the overlying noncoal rocks at any point in the study area (Fig. 8). Rock samples, particularly those from core-drilling operations, can be examined in detail to ascertain the origin of the sequence and assess the distribution of the individual coal seams, while cores of the coal seams themselves can be subjected to a comprehensive analytical program. The mechanical properties of noncoal rock cores can also be tested to provide data on such aspects as their strength, abrasiveness, and weathering characteristics, which can facilitate mine design.

Core drilling for coal exploration is based mainly on diamond drilling techniques. A hollow cylindrical bit impregnated around one end with industrial diamonds is attached to a cylindrical core barrel and a series of hollow

FIGURE 8 Drilling an exploratory borehole in a coal-bearing sequence. (Photo: C. R. Ward.)

metal drill rods and rotated, under controlled pressure, against the base of the hole. The diamonds in the bit grind away a circle of rock, with the cuttings removed by water flushing, and a cylindrical core of the strata penetrated is left intact in the center of the core barrel.

When the barrel is full, the core is brought to the surface for examination and description by the site geologist. It may also be photographed to provide a convenient visual record of the strata present. If a coal seam has been penetrated, the core of that interval is sent, after a detailed, on-site description, to the laboratory for analysis.

The cores taken in drilling programs for bituminous coals are normally between 45 and 85 mm in diameter. Such cores are large enough to prevent undue breakage of coal and other weak materials in the drilling process and the associated loss of these beds from the core finally recovered. At the same time, they provide a sufficient quantity of coal, from most workable seams, to facilitate a comprehensive analytical program. Larger cores, 150–200 mm in diameter, may be used at a more advanced stage of the exploration program to provide bulk samples for testing, in detail, such aspects such as the preparation characteristics or coking properties of the coal. They are also used, in place of smaller-diameter or slim cores, in some lower rank deposits.

Special techniques allow a particular depth interval to be redrilled by forcing the hole to deviate off to one side,

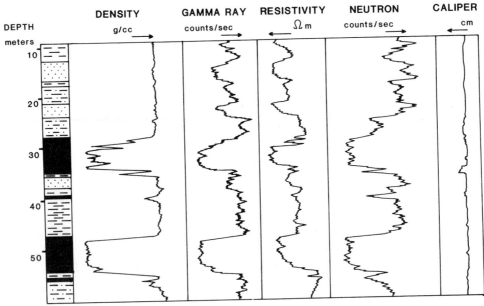

FIGURE 9 Records from down-hole geophysical logs through a coal-bearing sequence. Rock types interpreted from the data are shown in graphic form on the left-hand side.

on a second pass, from a point a short distance above the top of the zone in question. If the strata are steeply dipping, the hole may be drilled at an angle rather than vertically into the ground or even forced to follow a curved path to intersect the coal seam or other horizon in a particular direction. Other techniques make it possible to take oriented cores, such as might be needed for petrofabric or geomechanical studies, while down-hole cameras and similar imaging systems may be used to study the fracture pattern and other features exposed in the walls of the hole.

The number and location of drill holes for each program depends on the amount of information available from other sources, the complexity of the area, and the stage that has been reached in the mine development project. A regional appraisal of a large area to select targets for further study may be based on boreholes more than 10 km apart, while a spacing of 1 to 2 km is often adequate, in favorable areas at least, to provide the geological data necessary to plan a major underground mine. If open-cut mining is anticipated, however, the boreholes may have to be as few as 100 m apart to delineate adequately the seam and overburden thickness and the coal quality trends. Closer spacing, in both cases, is also necessary in problem areas, such as to map out subsurface faults and igneous intrusive bodies in the coal seam.

D. Down-Hole Geophysical Logging

Down-hole geophysical logging involves lowering a series of sensors, or sondes, down an open borehole and record-

ing signals from them that reflect the physical properties of the materials intersected at different depths (Fig. 9). The combination of properties measured by the different sondes can be used to identify the rock types present over each depth interval and, in some cases, to evaluate such properties as rock porosity and groundwater salinity. They also help to determine, to some extent at least, the thicknesses and certain quality parameters of the coal seams.

For many years, geophysical logging has been the primary tool for analyzing the strata penetrated by holes drilled for petroleum exploration. Recent developments in miniaturization, however, have also made equipment available for small-diameter boreholes that gives sufficient resolution of thin rock units to make it possible to use the data in coal exploration projects.

Geophysical logging can be used in conjunction with studies of drill cuttings to evaluate the geological sequence penetrated in noncored boreholes through coal-bearing strata. Since such holes, even with the added cost of down-hole logging, are generally less expensive than fully cored holes, this may significantly reduce the cost of the overall exploration program. It can also be used, however, in conjunction with cored boreholes, partly to build up a data bank for interpretation when no core is available and partly to resolve any problems with lithologic data or the position of stratigraphic boundaries arising from core loss. Although valuable, especially in guiding interpolation between holes with more reliable data, the logs cannot replace a core itself for coal quality determination, detailed lithologic analysis, or quantitative geotechnical investigations.

The principal down-hole logging techniques used in coal exploration are described in the following subsections.

1. γ-Ray Logging

This is a technique that measures the natural level of γ radiation emitted from the strata around the hole. High levels of activity are generally due to the presence of potassium (the ^{40}K isotope is slightly radioactive), which occurs in mica, feldspar, and some of the clay minerals. Intervals with a high γ count are mostly interpreted as shales or, in some cases, felsic volcanic sequences, while low γ radiation is commonly associated with quartzose sandstones and with coal seams.

2. Density Logging

This technique involves the input of γ rays into the rock around the hole and measurement of the amount of backscattered radiation generated from the strata. Elements with a low atomic number, such as the carbon and hydrogen dominant in coal seams, produce more backscattering than those with a higher atomic number, such as aluminum, silicon, and iron, which are most abundant in noncoal rock types. The count rate is almost inversely proportional to the density of the surrounding rocks, with coal giving a high count rate (low density value), noncoal rocks a low rate (higher density), and impure or shaley coal an intermediate count rate and density value. All else being equal, the density log value gives a rough indication of the proportion of ash and hence the overall quality of each individual coal ply.

The degree of resolution that can be obtained in density logging, and thus the precision of fixing seam boundaries or identifying thin noncoal bands within coal seams, depends on the separation between the γ-ray source and the backscattered-ray receiver on the density sonde. With a short spacing, much of the γ radiation from the source penetrates only the mud on the sides of the hole rather than the rock itself and causes the backscattered radiation to give a poor indication of the actual rock density. A longer source-to-detector spacing, however, provides a better indication of rock density but measures a larger depth interval in the process and gives a lesser resolution of boundaries between contrasting rock types. In practice, sondes with both short spacings and long spacings are used in coal exploration, the former to fix boundary positions and identify thin beds and the latter to measure rock density within intervals of uniform lithologic constitution identified by the short-spacing unit.

3. Neutron Logging

Neutron logging involves the input of neutrons into the strata and the recording of either the reduction in energy of these particles (neutron—neutron logging) or the γ rays emitted from the strata as a result of neutron capture (neutron—γ logging). Atomic nuclei of low mass, such as hydrogen, produce the greatest energy loss, and the response of the strata in neutron logging is therefore taken as an index of the total hydrogen content of the rock.

The hydrogen may be present as water in the pores of the rock and as moisture or hydrocarbon compounds in the coal. For coal seams, neutron logging may therefore indicate the amount of moisture or volatile matter and give a guide to coal rank or possibly to coking properties. Neutron logging may also help to distinguish hydrogen-deficient heat-affected or cindered coal from unaltered, hydrogen-rich material.

4. Resistivity Logging

This technique is based on a pattern of current and potential electrodes in the sonde that make contact with the walls of the hole and measure the apparent resistivity of the surrounding beds. Porous rocks with saline fluids, such as shale, generally have low resistivities, while coal and freshwater sandstone commonly have high resistivities. Resistivity logging is therefore used, as it is in the petroleum industry, to identify noncoal rock types rather than coal seams.

A close-spaced pattern of electrodes is needed for resistivity logging if thin rock units are to be identified or boundaries between rock units are to be located with adequate precision. In such cases, however, the mud cake on the walls of the hole, with a low resistivity value, tends to short-circuit the measurement process and prevent testing of the rock materials themselves. A special focused resistivity sonde is often used in coal field studies to overcome this problem, directing a greater proportion of the current through the mud cake and into the surrounding rock mass.

5. Other Down-Hole Logs

A number of other down-hole logs are available to supplement the techniques listed above. These include caliper logs, used to measure variations in hole diameter, and sonic logs, which give a measure of the seismic velocity in the various rock strata. Caliper logs show the location of zones of weak or caved strata in the hole and are also essential in correcting data from short-spaced density logs for minor irregularities in the borehole wall. Sonic log data may be useful in the interpretation of seismic records and also provide an indication of rock strength. Temperature logs also assist planning of the ventilation requirements for deep underground mines or, in some cases, indicate the location of water seepages into the bore hole.

E. Coal Resources and Coal Reserves

Although the terminology is by no means universal, the word *resources* is generally used in the coal industry to describe all of the coal that may conceivably be of value to humankind, either now or in the future. As well as economically mineable coal, this all-embracing definition may include seams that are too thin, too deep, too remote, or too poor in quality to be worked under present economic or technical conditions, as well as material about which too little is known for a detailed assessment of these factors to be made. In contrast, the term *reserves* is more restrictive, referring only to coal that has been explored in detail and that has been found, in the process, to be suitable at the time of the assessment for economic extraction and use (Fig. 10).

The mass or tonnage of coal present in an exploration area is generally calculated by the geologist from the thickness and density of the workable section of the seam combined with the area of the holding over which that sec-

tion extends. Depending on the spacing of data points for this calculation (e.g., boreholes), this gives an estimate of the amount of coal actually available in the ground, a mass referred to as the *in situ* (or in-place) resources or reserves.

Not all of this coal, however, is necessarily recovered in the course of mining operations. Minor amounts are left in the ground or removed with the overlying strata during open-cut or surface mining processes, while a large proportion of the *in situ* coal, often more than 50%, is left in underground mines due to factors such as the need to provide roof support. The mass of coal actually expected to be won from the ground when mining is established is referred to as the *recoverable reserves* (or resources) of the area in question.

The coal recovered from the ground may still contain significant amounts of mineral matter, such as pyrite concentrations or bands of noncoal material that occur within the seam. Together with any contamination that may have been introduced from admixed roof or floor strata, these materials can be removed before sale of the coal by one or more cleaning processes in a preparation plant. If such processes are used, the amount of coal from the area finally

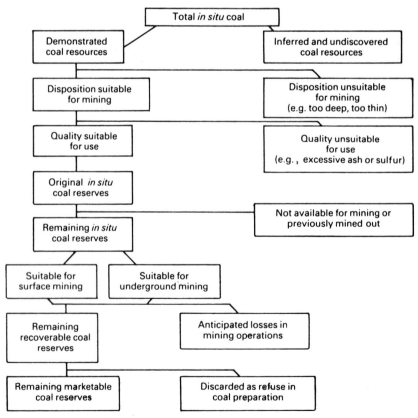

FIGURE 10 Flow sheet showing different categories of material identified in the assessment of coal resources and coal reserves. [From Ward, C. R. (1984). In "Coal Geology and Coal Technology." Reproduced by permission of Blackwell Scientific Publications, Melbourne.]

available for use, a quantity known as the marketable reserves, will represent only part of the recoverable material, the balance forming a refuse fraction to be disposed of in some way.

The evaluation of coal resources is not a simple process, since coal seams may split, thin, or vary considerably in quality over remarkably short distances or may be locally affected by such features as washouts and igneous intrusions. Displacement of seams by faulting, while not greatly affecting the total mass of coal present, may also give rise to severe limitations, or at least higher costs, in the extraction process. Statements of reserves or resources generally include a term, such as *measured*, *indicated*, or *inferred*, that describes the degree of certainty to which these factors have been delineated, the term *reserves* usually being applied only to material that has been assessed to a *measured* or *indicated* level.

Coal resources are often expressed in terms of the energy contained within the deposits rather than the actual mass of material present. A given mass of subbituminous coal, all else being equal, gives off a significantly lesser amount of heat, on combustion, than the same mass of bituminous coal or anthracite, while an equivalent mass of lignite produces an even lower output of heat energy. Such differences in specific energy can be allowed for by expressing coal resources or, indeed, resources of any other fuel in terms of "coal equivalents," where 1 metric ton coal equivalent (tce) represents the heat energy contained in 1 metric ton of bituminous coal with a specific energy of 29.30 MJ/kg (7000 cal/g). An assessment of coal resources in selected countries is given in Table V.

The total amount of known coal recoverable from the ground under present and expected economic conditions and with existing technology is estimated at 0.9×10^{12} metric tons (Table V).

The full extent of *in situ* coal in the world, including coal that might become economically recoverable in the foreseeable future and coal that is not yet fully evaluated, is difficult to determine, due to different degrees of geological knowledge and mining development, as well as to the application of different limiting criteria in the various national assessments.

By far the greatest proportion of the world's coal resources is located in three countries: the United States, the Russian Federation, and the People's Republic of China. Australia, India, South Africa, and Germany, however, also have substantial economically recoverable or potentially recoverable resources.

F. Geology in Coal Mining

Coal is mined either by surface methods (open-cut or open-cast mining) or by underground techniques. In surface mining, all of the rock from the ground surface to the top of the seam (material referred to as overburden) is removed and set aside to expose the coal, which is then taken out and transported from the mine for preparation and use (Fig. 11). With underground mining, in contrast, a series of openings is made from the surface to the coal seam, after which coal is won from excavations almost entirely confined within the coal bed.

Because of the cost of removing overburden in relation to the value of the coal produced, surface mining is restricted to relatively thick seams, or aggregations of seams, lying at shallow depth. The technique has many advantages over underground mining, including higher productivity, greater safety, and better recovery of *in situ* coal resources. It principal disadvantage, however, is the greater level of unfavorable impact, at least in the short term, on the surface environment generally.

Coal is extracted and used in large quantities, with many individual mines producing several millions or even tens of millions of metric tons per year. Compared with the value of many other ores, however, the price per unit mass of this coal is relatively low, and production costs must be kept to a minimum if mining is to remain economically viable. Modern coal mines are mostly large-scale, high-productivity operations, with layout, equipment, and production scheduling tailored to suit the geological characteristics of the particular coal deposit. The delineation of these characteristics is as much a part of an exploration program as assessment of the amount of coal actually available for use.

The geological factors affecting mine design include the thickness, depth, and structure of the seam or seams to be worked; the strength, abrasiveness, and other mechanical properties of the coal and its roof and floor strata; and the location, orientation, and extent of any major breaks in seam continuity, such as faults, splits, washouts, and igneous intrusions. As well as detailed studies undertaken to identify and map these features, quantitative assessments of the geotechnical properties of the strata are usually included in exploration programs to determine, in advance, such factors as the cost of shaft sinking or overburden removal, the stability of underground or surface mine excavations, and the ventilation or drainage requirements of the mine workings. In conjunction with detailed studies of coal quality characteristics, such data can be used to test the technical and economic feasibility of a particular mine design and determine the most appropriate schedul of production from different parts of the deposit at various stages of mine development.

The role of the geologist does not cease, however, once the mine has opened and production is under way. Ongoing monitoring of geological conditions is essential while the coal is being extracted in order to evaluate the predictions

TABLE V World Coal Resources, 1996

Country or region	Proved recoverable reserves[a]		Proved in place reserves[b]		Estimated additional resources[c]	
	Bituminous + anthracite (Mt)	Subbituminous + lignite (Mt)	Bituminous + anthracite (Mt)	Subbituminous + lignite (Mt)	Bituminous + anthracite (Mt)	Subbituminous + lignite (Mt)
Botswana	4,313	—	7,189	—	205,253	—
South Africa	55,333	—	121,218	—	5,000	—
Canada	4,509	4,114	6,435	12,905	26,045	31,990
United States	111,338	135,305	239,675	240,538	456,153	665,887
Brazil	—	11,950	—	17,072	—	15,319
Colombia	6,368	381	—	—	—	—
China	62,200	52,300	—	—	—	—
India	72,733	2,000	—	—	—	—
Indonesia	770	4,450	—	1,331	—	6,299
Japan	785	—	8,277	—	—	—
Kazakhstan	31,000	3,000	—	—	—	—
Pakistan	—	2,928	—	—	—	—
Thailand	—	2,000	—	2,315	—	3,000
Turkey	449	626	—	7,339	—	110
Uzbekistan	1,000	3,000	—	—	—	—
Czech Republic	2,613	3,564	6,401	2,547	4,928	5,284
Germany	24,000	43,000	44,000	78,000	186,000	—
Greece	—	2,874	—	—	—	—
Hungary	596	3,865	1,407	8,306	702	—
Poland	12,113	2,196	60,185	14,184	—	—
Russian Federation	49,088	107,922	75,753	124,823	1,582,479	2,358,085
Serbia, Montenegro	64	16,408	—	—	—	—
Ukraine	16,388	17,968	21,850	23,958	5,406	5,819
United Kingdom	1,000	500	—	—	—	—
Australia	47,300	43,100	65,900	48,000	125,000	215,000
Other Countries	5,531	11,269	—	—	—	—
Total world	**509,491**	**474,720**	—	—	—	—

Not all countries report categories under notes [b] and [c], and not all use the same limiting criteria.

[a] Carefully measured and assessed as exploitable under present and expected economic conditions and with existing technology. [Data from world Energy Council.]

[b] Proved amont in place that can be recovered (extracted from the earth in raw form) under present and expected conditions.

[c] Inlicated and inferred coal additional to the proved amount in place, including extensions to known deposits and as yet undiscovered resources.

made from the exploration program and modify the mining operation if necessary, taking new data or further experience into account. Such work is somewhat different from that of finding and evaluating a new coal deposit and is often the responsibility of specialist mine geologists rather than exploration-oriented personnel.

G. Mining Hazards

Coal mining, particularly by underground methods, is traditionally regarded as a dangerous occupation. Although many of the dangers have declined with the advent of safer mine equipment and improved working practices, there are still a number of hazards, primarily of a geological nature, that can impede production, prevent access to coal resources, or even lead to mine disasters with loss of life or serious injury to the work force.

Rock falling from the roof of underground workings is one of the most common problems (Fig. 12). Such falls may be initiated by excessive stresses in the rock around the mine opening, due to such factors as the pressure exerted by the overlying strata or tectonic forces within the earth's crust. They may also arise, however, from the presence of weak materials or preexisting discontinuities such as joints and bedding planes at critical locations in the rock mass. They can be prevented, to some extent, by appropriate support devices, such as timber posts and beams or steel arches around the excavation or patterns of rock

FIGURE 11 Coal mining by open-cut methods. A dragline is used to remove overburden from a gently dipping coal seam, placing it in spoil piles on the up-dip side. The coal can then be removed by a power shovel or similar equipment and placed into trucks for haulage from the mine. (Photo: C. R. Ward.)

bolts drilled and anchored into the roof strata. A detailed study of the regional stress pattern or the delineation of structures associated with falls may help to give warning of difficult ground in advance of the mine working face.

Another hazard in underground mines is the presence of gas trapped in pores, joints, and other fissures in the coal bed and its adjacent strata. This gas, often under high pressure, is usually made up of methane and/or carbon dioxide, that was liberated during rank advance or by processes associated with the emplacement of igneous intrusions. It may be discharged slowly into the mine atmosphere, creating a risk of either explosion or asphyxiation if not adequately diluted by the ventilation system. If trapped very tightly, it may be released violently as the retaining strata are removed, producing very dangerous outbursts of coal, rock, and gas, sometimes called blowouts, from around the working face.

The presence of such gas can be tested by studies of the gas released from drill holes or coal seam cores. Detailed study of structures associated with local gas pockets may also be useful, and monitoring of minor shocks produced in the strata leading up to individual outbursts (microseismic studies) can be of value in identifying dangerous sites. The gas may be allowed to bleed off through strategically placed boreholes ahead of the mine workings, and if the gas is mainly methane, it can be tapped and used to ad-

vantage as separate energy sources. In some areas, such as the San Juan Basin of the United States, methane in coal seams may be extracted as a significant energy source without any associated coal mining operation.

Although usually less dangerous than the hazards of underground mining, difficulties may be encountered with unstable slopes in the rock walls of open-cut excavations and the associated overburden emplacements or spoil piles. Slides, slumps, and other movements may occur at such sites, due to the presence of low-strength materials in the overburden sequence or degradation of particular components with atmospheric exposure. Sliding may also occur along unfavorably oriented joints, faults, and bedding or cross-bed planes.

Movements of this type may take place over many hundreds of meters or even several kilometers along the open-cut excavation. They can bury a large amount of newly exposed coal, disrupting production and adding to the cost of overburden removal. They may also reduce the amount of *in situ* coal that can be economically recovered from the mine. Unless equipment and personnel are removed from the area in time, such movements endanger mine plant and operating staff.

The incidence of such problems can be reduced in many cases by adopting less steeply sloping faces for the pit walls and spoil piles. This results, however, in the need

FIGURE 12 Roof fall in an underground mine. (Photo: J. Shepherd.)

to remove a greater volume of overburden per metric ton of coal mined and, depending on the method of removing the overburden, can also reduce the efficiency of the emplacement process.

H. Environmental Impact of Coal Mining

Like the introduction of agriculture to a previously undeveloped area, the extension of a road network, or the growth of towns and cities, the mining of coal can have a considerable and often adverse effect on the environment. Although people may be aware of the ultimate need for the energy derived from coal, the standards of modern society require that its extraction not disrupt or unduly interfere with other activities in the surrounding region. Indeed, proposals for new mines or major increases in mining activity must often be accompanied by fully researched statements of their environmental impact, so that these aspects can be assessed as part of the approval process.

One of the first factors to be considered in such proposals is the existing use of the land surface in and around the intended mine area. The value of a tract of prime agricultural land, a township, or an undisturbed wilderness, for example, and the cost of its loss or reestablishment elsewhere must be weighed against the value of the coal

beneath and the benefits to be gained from its extraction. Open-cut mines affect larger areas of the surface than underground operations, where only a small area near the pit top is generally required for mine facilities, and are therefore more likely to be limited in their development.

Detailed recording of rainfall, wind patterns, water quality, and prevailing noise and dust levels should be commenced at a very early stage in the exploration program, before the environment is significantly changed by drilling, land clearing, trial excavations, or simple increased vehicle movements. Other activities at this stage may include flora and fauna surveys, sociological investigations, and searches for significant archaeological sites. As well as establishing details of these aspects before the start of the project, such records may provide valuable data for the mine design process.

The interaction of the mining operation with the supply and movement of surface and underground water in the area is a critical part of any environmental impact study. Coal mines use considerable quantities of water for dust suppression and coal preparation, as well as for the personal amenities of the work force at the site, and, ideally, provision of this water should not interfere with other activities, such as farming, in the surrounding region. Special dams and pipelines may have to be built to ensure that an adequate supply of water is available when required.

In some mines, either open-cut or underground, large quantities of underground water must be pumped from the workings to prevent flooding and ensure safe working conditions. At least part of this water, however, may be diverted for use elsewhere on the mine site. The amount of water removed in this dewatering process and the effect of its removal on water supplies in the adjacent region must be taken into account and so, too, must the method of discharging any surplus water from the site back to the region's hydrological system.

As well as the quantities involved, the quality of the water pumped from the mine workings and of the water runoff from the pit-top area, stockpiles and overburden or refuse emplacements must also be considered in any environmental study. Such water may become contaminated with acid solutions derived from the oxidation of pyrite in some of the coal or associated strata or with suspensions of fine solids in the form of clay or coal dust. Treatment of the water may be necessary, at some mines, before it is returned to the surrounding environment.

There are many other geological factors, apart from those concerned with water, that must be considered in an assessment of the environmental impact of a coal-mining operation. These include the weathering characteristics of the overburden strata in open-cut areas or of refuse from coal preparation plants and the shaping of emplacements of these materials required to give a free-draining, aesthetically pleasing ground surface. Topsoil from the area should also be set aside, as mining proceeds, to cover these emplacements and allow early reestablishment of vegetation in the mine area.

Removal of the coal in underground mines may give rise to subsidence of the ground at the surface above the workings. The extent of the area affected by this subsidence depends on the depth of the coal, the area where full extraction has taken place, and the nature of the strata above the worked-out seam, while the amount of subsidence finally produced depends mainly on the seam thickness and the mining practice used. Underground mining may be prohibited, or at least severely restricted in some areas, due to the need to prevent subsidence of built-up areas and other key surface features or man made installations.

V. COAL PREPARATION AND USE

Coal is chiefly used as a combustion fuel (steaming coal) for power generation and other purposes or to produce coke (coking coal) for metallurgical processes such as iron and steel manufacture. Other uses that have been significant in the past, and with changes in technology could again become significant in the future, include the manufacture of gaseous and liquid hydrocarbon fuels (gasification and liquefaction) to supplement or perhaps even replace petroleum supplies.

Most equipment that uses coal is designed to handle only material with a particular combination of properties and will often not work efficiently with material of either higher or lower grade. Indeed, the design of many coal-using installations, such as power stations and coke plants, is based very heavily on the nature of the coal deposits from which they will be supplied. A process of coal preparation is often employed either at the mine or the user facility, to minimize day-to-day fluctuations in the quality of the mined material. However, such plants can also be used to upgrade the coal product, if necessary, by removing high-ash or high-sulfur components from the crushed run-of-mine coal.

A. Coal Preparation

The coal extracted from a modern mechanized mine, either open-cut or underground, typically contains material derived from intraseam bands of noncoal rock and large lenticles or other masses of mineral matter, as well as a certain amount of contamination from the roof and floor strata. The product won from the working face also contains a wide range of particle sizes, many of which are too large for the equipment in which the coal is to be used.

Coal preparation represents a series of processes aimed at improving the quality of this coal for the marketplace. Depending on the run-of-mine product and the use to which the coal will be put, it may involve any combination of the following processes:

1. *Crushing:* reducing the size of the broken particles in the run-of-mine product
2. *Sizing:* sorting the crushed coal into fractions in which the particles have specific size limits
3. *Cleaning:* separating the high-ash or mineral-rich particles from the particles of low ash, or clean coal
4. *Dewatering:* removing excess water used in the cleaning process from the coal products
5. Blending, storing, and, when necessary, loading the products of the preparation plant for transport to the coal user.

1. Size Reduction

The particles of broken coal from the working face may be up to 1 m in diameter, particularly in open-cut mines. Many of these particles, especially those in the larger size fractions, may be composite particles made up partly of clean coal and partly of shale or some other inferior material. They generally have to be reduced in size, depending on the plant, to a maximum of somewhere between 25 and 150 mm, to facilitate handling and to separate the different components in these composite particles (a process known as liberation).

Size reduction is often achieved by means of rotary breakers, which tumble the coal in large rotating drums and cause breakage of the particles on impact. Other processes, however, include pick breakers, jaw crushers, gyratory crushers, and roll crushers or a process involving some combination of these techniques. It is usually desirable to minimize the amount of fines (particles finer than ~0.5 mm) produced in size reduction, since these are very difficult to handle in the subsequent cleaning and dewatering operations.

2. Size Separation

Most of the techniques used in coal cleaning are effective only with particles falling within a certain size range; many markets also have size distribution requirements. The crushed coal is usually separated into size classes by a process in which it is passed over a series of screens in the form of perforated metal plates or woven wire meshes having openings or apertures of particular sizes. An alternative, however, is to make use of the different settling rates of coarse and fine particles in water using a settling cone or a trough or cyclone classifier.

3. Coal Cleaning

Clean coal particles, at least in the coarser size fractions, are usually separated from inferior coal or shale particles on the basis of their different relative density values. This is accomplished in a preparation plant by such equipment as the following:

1. Jigs, which separate the particles on a perforated deck in an up-and-down pulsating column of water

2. Dense medium baths, in which the light particles (clean coal) float and the dense (inferior or shaley) particles sink in a dense liquid, usually made up of a suspension of magnetite (Fe_3O_4) in water

3. Dense medium cyclones, where the particles are separated by a rapidly rotating vortex of dense liquid in a special, conical-shaped vessel

4. Shaking tables, where the particles are separated by a film of water flowing over a riffled, sideways-shaking deck

5. Spirals, where the particles are separated by a film of water flowing down and around a helical-shaped trough

Apart from separation by density, coal particles can also be separated from shale and other mineral-rich constituents by differences in their surface properties. The technique of froth flotation, widely used to clean coal fractions below ~0.5 mm in size, is based on selective attachment of the nonpolar surfaces of the organic components (coal particles) to air bubbles that rise through a suspension of fine coal and mineral solids in water. The coal particles are lifted with the bubbles and gather together in a froth at the water surface.

4. Float-Sink Testing

The behavior of coal in density-based cleaning processes can be predicted from a series of float-sink (or washability) tests carried out on a crushed sample in the laboratory. The sample is placed in a heavy liquid of known density (e.g., 1.40 g/cm^3) and the floating fraction skimmed off for weighing and analysis. The sinking fraction is then placed in a liquid of somewhat higher density and the process repeated, with further tests in still denser liquids until the desired range of separation densities has been covered (Table VI).

Data from float-sink testing are used to determine the yield and quality of both clean coal and refuse fractions produced by separation in coal cleaning equipment, such as a dense medium bath, adjusted to split the stream at a particular density value. Such data may also give the density required in the separating medium, as well as the yield of product, the yield of refuse, and the quality of

TABLE VI Float-Sink Test Data for a Crushed Coal Sample

Relative density fraction	Mass (%)	Ash (%)	Cumulative floats Mass (%)	Cumulative floats Ash (%)	Cumulative sinks Mass (%)	Cumulative sinks Ash (%)
Floats, 1.30	5.0	5.0	5.0	5.0	100.0	20.7
1.30–1.35	16.1	10.1	21.1	8.9	95.0	21.6
1.35–1.40	39.4	12.9	60.5	11.5	78.9	23.9
1.40–1.45	15.0	17.2	75.5	12.6	39.5	34.9
1.45–1.50	7.2	21.7	82.7	13.4	24.5	45.7
1.50–1.55	4.1	27.1	86.8	14.1	17.3	55.7
1.55–1.60	2.0	33.9	88.8	14.5	13.2	64.5
Sinks, 1.60	11.2	70.0	100.0	20.7	11.2	70.0

refuse that will be obtained from the material tested if clean coal of a particular quality is produced. Float–sink testing is invaluable in the design of coal preparation plants, the development of marketing strategies, and, of course, the estimation of marketable coal reserves for a given coal deposit.

B. Coal Combustion

Coal combustion processes are traditionally divided into those involving burning of solid coal (coarsely crushed or broken coal) in a permeable, granular bed and those involving combustion of finely ground or pulverized coal, blown with air through a jet into the furnace chamber. An additional process is based on a strong updraft of air through a layer of crushed coal to produce combustion in a fluidized coal bed. Stabilized slurries of fine coal in oil or water, known respectively as coal–oil mixtures (COM) and coal–water mixtures (CWM), may also be burned in combustion operations, using similar techniques to those used for combustion of fuel oil and other liquid fuels.

Most coal combustion is used to convert water to high-pressure steam, which is then used for purposes such as the driving of turbines in electric power generation (Fig. 13). The furnace, in such cases, is usually integrated with a boiler of some type to ensure efficient heat exchange for steam production. Other uses of combustion, however, without the need for a boiler, include the heating of limestone and other ingredients in a kiln for cement manufacture and the production of heat, char, and carbon monoxide for the processing of iron ores by direct reduction methods.

1. Combustion Equipment

Although furnaces fueled by hand-stoking methods (i.e., the traditional shovel) are used in some small installations, most large equipment for coal combustion incorporates

FIGURE 13 Coal-fired power station. (Photo: C. R. Ward.)

some sort of mechanical facility to introduce the coal to the combustion chamber and mix it with the supply of air necessary for an efficient combustion process. For combustion of solid coal, such mechanical stoking facilities include the following:

1. Underfeed stokers, where the coal is forced up from beneath the combustion chamber by a screw feed arrangement, and the ash left after combustion is allowed to spill over to collection points around the furnace walls.

2. Traveling grate and similar stokers, where the coal is burned in a granular bed carried across the combustion chamber on an endless chain grate or worked across the chamber on a vibrating or reciprocating metal plate. The air required for combustion passes through the grate and the coal bed, and ash is discharged from the end of the grate when passage across the furnace has been completed.

3. Sprinkler stokers, where the coal is thrown into the furnace from the blades of a rotating coal distributor in a manner similar to that introduced from a shovel by hand. The coal falls onto a grate, either stationary or traveling, and ash is discharged either through the grate or at the end of the traveling unit.

Combustion of pulverized fuel is often used in preference to solid coal combustion processes at large-scale installations such as power plants. The coal is first ground to fine powder, with most particles smaller than 75 μm in diameter, in a series of pulverizing mills, the excess moisture having been removed, if required, in a previous drying process. After grinding, the coal is mixed with preheated air and blown through jets designed to ensure mixing by turbulent flow into the combustion chamber (Fig. 14).

The ash left after combustion travels out of the furnace mainly as fine particles (fly ash) suspended in the waste gases from the combustion process. It is recovered from these gases either by electrostatic precipitators, which attract the particles to electrically charged plates in the furnace stack, or by mechanical filtration techniques. Some ash also settles directly to the lower part of the combustion chamber, from which it is removed separately as bottom ash.

2. Combustion Problems

Furnace operation is chiefly governed by such factors such as the net heat availability from the coal after moisture and ash-forming constituents are taken into account, the particle size distribution of the crushed coal, its resistance to fine grinding, and the fusion characteristics or precipitation behavior of the ash. Coarse particles of hard minerals in the coal, such as quartz, can also give rise to abrasion of the grinding equipment or of exposed surfaces in the combustion path. Other factors to be considered are the possible formation of nonconductive coatings on the furnace walls and the water tubes of the associated boiler (fouling and slagging) by condensed combustion products or corrosion of these surfaces by acids and other gases formed in the combustion process.

Sulfur dioxide and related compounds are also formed by combustion of the sulfur within the coal, and if the coal has a particularly high sulfur content, these components may give rise to atmospheric pollution when the gases are released from the furnace stack. The level of sulfur emissions can be reduced by scrubbing the stack gas with lime or magnesia slurry before it is released, while if fluidized bed combustion is used, a certain amount of crushed limestone may be added to the fuel bed to absorb the sulfur oxides from the combustion products.

Nitrogen oxides are also produced in the course of coal combustion, partly from conversion of the nitrogen compounds in the coal and partly (at least at high temperatures) by incorporation of nitrogen from the air into the combustion process. Such oxides may cause atmospheric pollution from the stack gases. Control of the air flow in the furnace may reduce the formation of nitrogen compounds, while their emission can be reduced, if necessary, by processes similar to those used to ameliorate sulfur pollution problems.

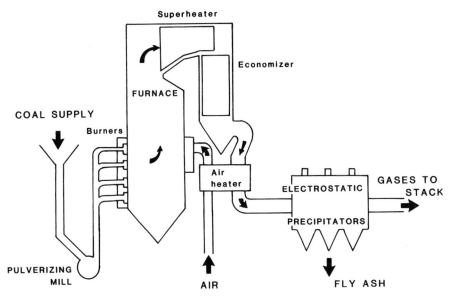

FIGURE 14 Schematic view of a pulverized-fuel boiler. Steam is produced in water tubes around the furnace walls and in superheater and economizer facilities in the combustion gas path. Fly ash is collected in this installation by electrostatic precipitators before the stack gases are released to the atmosphere.

C. Carbonization and Coke Production

Coke is prepared by heating crushed coal to high temperatures in the absence of air. This is carried out in an enclosed coke oven. The volatile matter is driven off in the process; it may be recovered and used as a heat source or refined into a range of marketable by-products from the coking operation.

Most coke is used in the blast furnace for the production of cast iron from iron oxide ores. In this process, it acts as both a heat source and a reducing agent. The coke must be strong enough to withstand the loads imposed on it by the other constituents in the blast furnace yet at the same time be sufficiently porous and permeable to allow gases to flow through and the necessary reduction reactions to take place.

1. Coke Oven Processes

Coke ovens can be classified into two basic types: those from which the volatile by-products are collected for subsequent use (by-product coke ovens) and those in which no attempt is made to recover by-products, apart from using some of the released volatiles as a heat source for the coking operation (nonrecovery coke ovens).

The simplest type of coke oven, now largely obsolete, consists of an almost-enclosed brick kiln known as a beehive oven. The coal to be coked is spread across the floor of the oven, and the top of this bed ignited just before the door to the oven is sealed. Heat from combustion of this top layer drives off the volatile matter from the rest of the coal, and combustion of the volatiles provides fur-

ther heat for the coking process. Excess volatile matter is discharged through a chimney at the top of the oven and usually burned off as waste. When the process is complete and all of the coal has been carbonized, the oven is allowed to cool, the door opened, and the coke removed for use.

The principal type of coke oven in general use is the slot oven. This equipment, which can be used either with or without a by-product recovery process, consists of a narrow refactory chamber closed by removable doors at each end. A charge of coal is introduced through a hole in the top, leveled, and then heated, after the oven is sealed, by combustion of coal gas in heating flues at the sides. The gases, tars, and liquors are collected from the oven, if required, and separated in an adjacent by-products plant. When carbonization is complete, the heating is stopped, the doors removed, and the hot coke discharged from one end by pushing at the other. After discharge the hot coke is cooled quickly, usually by drenching with a large volume of water, both to break up the mass and prevent it from burning away in the air.

Nonrecovery ovens, particularly those of the beehive design, tend to give lower coke yields, all else being equal, than those of the slot oven type. They also produce a greater degree of atmospheric pollution. The sale of by-products from slot ovens, however, can offset the cost of processing any materials that have inherently low coke yields, such as high-volatile coals or coal blends.

2. Coal Quality and Coke Characteristics

The behavior of different coals in coke manufacture depends partly on the design and operation of the coke oven and partly on the rank and type of the coal concerned. Vitrinite, liptinite, and, in some cases, part of the inertinite in bituminous coals react during carbonization to an extent depending on rank. These constituents soften and then fuse to produce a solid carbonaceous binding material punctuated with vesicles left by escaping volatile matter. However, most if not all of the inertinite and even the vitrinite and liptinite in coals that are too high or too low in rank undergo little visible change, remaining as solid particles throughout the carbonization process.

A mixture of both reactive and nonreactive macerals is required to form a strong porous coke. The nonreactive or inert components fill much the same role as the aggregate particles in concrete and form the structural framework that gives the coke its strength. Like cement in the same concrete, the fused reactive macerals then bind this aggregate of inert particles together to form a rigid mass.

A number of coal seams contain material that, when carbonized by itself, produces a good-quality coke. Such coals, in fact, are often referred to as prime coking coals. In most large-scale operations, however, coke production is based on blends of coals from several different sources, with some of these coals providing mainly inert or aggregate particles and some acting as suppliers chiefly of binding materials.

The relative proportions of reactive and inert macerals in a coal, or even in a blend of coals, and the rank of the reactive components, as indicated by vitrinite reflectance, can be used, in some cases, to predict such properties as crucible swelling number and coke strength. Petrographic data can also be used, with some coals at least, to predict such aspects as the degree of volumetric expansion of the coal in the coke oven and the reactivity to carbon dioxide in the coke produced.

Apart from its petrographic properties, the bulk chemistry of the coal is also an essential consideration in coking operations. Coals with a high proportion of sulfur or phosphorus are generally unacceptable, since these elements are transferred to the coke and ultimately incorporated in the iron and steel, where they seriously affect the metallurgical properties of the product. The composition of the coke ash, and hence that of the coal, must also be taken into account, since this affects the production of slag from the nonferrous impurities in the blast furnace charge.

In many steel-making operations pulverized coal is injected into the blast furnace as a partial replacement for some of the coke requirement. Coal used for this purpose (pulverized coal injection) does not have the same requirements for porosity and strength development as conventional coking coal, but does need to meet similar specifications for chemical parameters, such as sulfur and phosphorus content.

TABLE VII Chemical Reactions in Coal Gasification

Primary reactions	
Water–gas reaction	$C + H_2O \rightarrow CO + H_2$
Boudouard reaction	$C + CO_2 \rightarrow 2CO$
Partial combustion	$C + O_2 \rightarrow CO$
Hydrogasification	$C + 2H_2 \rightarrow CH_4$
Secondary reactions	
Shift reaction	$CO + H_2O \rightarrow H_2 + CO_2$
Methanation	$3H_2 + CO_2 \rightarrow CH_4 + H_2O$

D. Coal Gasification

Coal gasification is the conversion of coal, coke, or char to gaseous fuels by interaction with agents such as steam, air, or oxygen. Although a certain amount of gas is liberated from the coal by carbonization processes (i.e., heating in the absence of air or oxygen), complete gasification involves conversion of all of the carbonaceous material in the coal to gaseous products by reactions such as those shown in Table VII. The only residue left from complete gasification is that derived either in solid or in liquid (slag) form from the mineral matter of the coal concerned.

Gas produced from coal has been widely used for a range of commercial, residential, and industrial heating applications. It has, however, been largely if not entirely replaced in many areas by natural gas supplies. Research into gasification processes has been continuing, nevertheless, stimulated at different times by threats to future availability of adequate natural gas resources. Gasification is also part of some of the processes involved in the use of coal for liquid fuel production.

1. Gasification Processes

Depending on the type of coal used and the actual gasification technique, the products of gasification may include CO, CO_2, H_2, and CH_4 (Table VII), diluted by H_2S and atmospheric nitrogen. Many of the impurities, however, particularly pollutants such as H_2S or noncombustible materials such as CO_2 and N_2, may be removed in subsequent purification steps. Gas with only some of its diluants removed may be used directly as a fuel gas of low specific energy, but most modern gasification is directed toward the manufacture of two higher specific energy products:

1. Synthesis gas (syngas), a mixture of CO and H_2, with a specific energy of $\sim$9 kJ/m^3
2. Synthetic natural gas (SNG), made up essentially of CH_4, with a specific energy of $\sim$30 kJ/m^3.

Synthesis gas can be used to make a number of products, including methanol, ammonia, and liquid fuels, while synthetic natural gas can be used as a direct substitute for natural gas in domestic and industrial reticulation systems.

Coal gasifiers are basically enclosed pressure vessels in which crushed or pulverized coals is brought into contact with the gasifying agent. One of the most widely used is the Lurgi gasifier, with a water-jacketed vessel ∼4 m in diameter that typically operates at a pressure of up to 3 MPa. Coarsely crushed coal is fed in through a lock-hopper system at the top and passes slowly downward against an upward flow of oxygen and steam. The hot gasification products are taken off at the top, while ash is drawn off through a second lock arrangement at the base.

Although most of the reactions involved in coal gasification are exothermic, two of the most fundamental, the water–gas reaction and the Boudouard reaction, are endothermic and require an input of heat energy if they are to take place. The necessary heat may be provided by the other reactions occurring in the gasification process, such as partial combustion with oxygen to form carbon monoxide (authothermal gasifiers), or may be produced by some external heat source (allothermal gasifiers). External heat, if required, may be obtained by burning other coal or coal carbonization products or, in some units under investigation, by means of a high-temperature nuclear reactor attached to the gasification plant.

2. *In Situ* Gasification

Coal gasification can also be carried out on seams that are still in place, the coal being converted to gaseous fuels without an intermediate mining operation. This process, known as *in situ* or underground gasification, typically involves drilling boreholes or sinking shafts to intersect the seam in at least two places and providing a passage for gas through the seam from one intersection to the other using such techniques as cross-drilling, hydraulic fracturing, explosive rock breakage, or underground mining methods. The coal can also be ignited at the base of one borehole and forced, by means of compressed air, to create a narrow, fire-formed channel through the coal bed to the other.

Once the holes are linked, the coal is ignited at the base of one hole (or other entry point) and a gasification agent, such as steam or compressed air, injected into the seam. Heat from the zone of burning coal causes carbonization and gasification reactions in the seam around it, while the injection process drives the products of these reactions toward the other borehole (or boreholes) where they are taken off for use (Fig. 15).

The principal product of *in situ* gasification is a low specific energy fuel gas, made up of a mixture of CO, CO_2, N_2, and H_2. The process normally continues until the zone of burning coal reaches the second borehole and all the available coal between the holes is consumed. Although extensively used in the former Russia until the early 1950s, when natural gas became available, the process is liable to some technical problems. These include leakage of gas from Russia points other than the extraction holes, blockage of the flow of gasification agent or gas products by swelling of the coal or collapse of overlying strata, excessive heat losses, and incomplete gasification generally in some parts of the system. Except in special situations, it does not appear to represent an economically viable alternative to mining and processing by more conventional means, or to extraction of coal-seam methane resources.

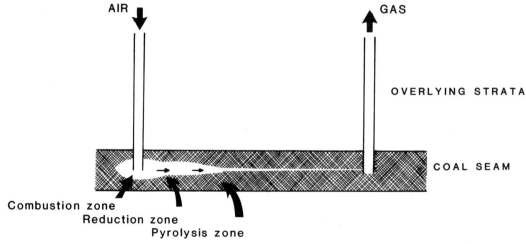

FIGURE 15 Schematic cross section of an *in situ* gasification process. [After Berkowitz, N. (1979). In "An Introduction to Coal Technology." Reproduced by permission of Academic Press, New York.]

E. Coal Liquefaction

Coal liquefaction is the process of making liquid hydrocarbon fuels and related chemical feedstocks from coal, either directly or indirectly, rather than taking them from petroleum sources. Liquefaction products derived from coal range from a semisolid, almost mineral-free hydrocarbon material known as solvent-refined coal (SRC), which is useful as a low-sulfur fuel or a coke oven blend component, to a synthetic crude oil (syncrude) that can be used as a heavy fuel oil or refined into gasoline and other light hydrocarbon fractions.

Liquefaction of coal is usually regarded as an economically viable operation only in circumstances where supplies of liquid petroleum are not readily available. It was used extensively in Germany, Britain, and several other countries before and during World War II, and a large-scale operation to produce liquid fuels from coal was subsequently established in South Africa during the early 1950s. A number of major investigations into liquefaction processes were commenced in the early 1970s in response to price increases and associated supply restrictions in the petroleum industry, and several demonstration plants were constructed as a result. With a significant amount of expansion in recent years, however, the operation in South Africa still remains the world's only major liquefaction plant.

From the chemical point of view, coal contains an excess of carbon and is deficient in hydrogen relative to crude oil and liquid petroleum products. It also contains too much undesirable mineral matter. The ratio of hydrogen to carbon has to be increased in some way, and the mineral matter removed for a successful liquefaction process. This can be accomplished in one of three basic ways (Fig. 16):

1. *Pyrolysis or partial liquefaction.* This is essentially equivalent to carbonization, whereby the hydrogen-rich fractions of the coal are driven off at high temperature in the absence of oxygen and a carbon residue (char) is left behind along with the mineral matter.

2. *Synthesis of gasification products.* Also known as indirect liquefaction, this is a process whereby the coal is gasified and the gases converted, by catalytic reactions, to liquid hydrocarbon compounds.

3. *Hydrogenation or direct liquefaction.* This covers a range of processes in which hydrogen is added to the coal from some external source, usually after the coal has been dissolved in a suitable organic solvent.

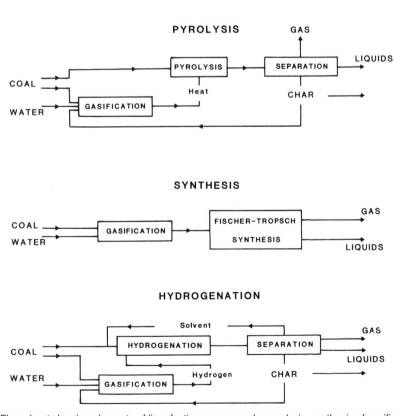

FIGURE 16 Flow sheet showing elements of liquefaction processes by pyrolysis, synthesis of gasification products, and hydrogenation techniques. [After White, N. (1979). *BHP Techn. Bull.*, **23** (2), 70–78.]

1. Pyrolysis

Pyrolysis processes resemble, in principle, those used to make coke and coke by-products. They involve introducing crushed or even pulverized coal to a closed vessel at relatively high temperatures, or a series of vessels at progresively increasing temperatures, to liberate the volatile components. The principal products of the process are tar and gas, which are extracted and separated, plus a quantity of solid residue, or char, together with the remains of the mineral matter. The tar is refined to produce a synthetic crude oil fraction, while the gas and char are used to provide heat for the process or other suitable industrial operations.

Unlike the carbonization process used in coke manufacture, the coal is generally brought to high temperature relatively quickly, since slow heating gives rise to excessive decomposition of the liquid hydrocarbon products. Flash pyrolysis, involving rapid heating of fine coal particles, gives the best tar yields for pyrolysis-based liquefaction techniques, but yields of liquid products per metric ton of coal are still lower than for other liquefaction methods.

Variations on the pyrolysis process include hydropyrolysis and flash hydropyrolysis, in which the coal is pyrolyzed under a pressurized atmosphere of hydrogen. The hydrogen may be derived in turn from gasification of the char produced by the pyrolysis, using a water–gas reaction process.

2. Synthesis of Gasification Products

A series of catalytic reactions known as Fischer–Tropsch synthesis can be used to convert the carbon monoxide and hydrogen from coal gasification (synthesis gas) to a wide range of organic products. Depending on such factors as the temperature and pressure at which the reaction takes place and the nature of the catalyst used, the process can be adapted to produce methane, alcohols, waxes, or several different types of hydrocarbon fuels. It is, in fact, the basis for the commercial liquefaction plant at Sasolburg in the South Africa.

The coal is first gasified, using oxygen and steam in a Lurgi or similar gasifier, and the gas treated to remove water, dust, hydrogen sulfide, and other contaminants. The gas is then passed (at the South African plant) to one of two different conversion processes, each based on Fischer–Tropsch reactions, converting it to either a gasoline-rich fraction or a series of products suitable for processing into diesel fuel and heavy fuel oil.

3. Hydrogenation

Hydrogenation involves increasing the ratio of hydrogen to carbon and converting the coal to liquid fuels by adding hydrogen from some external source. The first step in the process, in many operations, is to dissolve at least part of the organic matter of the coal in a suitable solvent, such as coal tar oil at several hundred degrees centigrade. In addition to making the coal more reactive, the solvent may act as the hydrogen donor for the conversion process, the external hydrogen being used to replenish that in the solvent and allow further reaction to take place.

Hydrogen is added to the mixture of coal and solvent under high temperature and pressure in the presence of a catalyst. The unreacted solid particles and mineral residues are removed and the solvent recovered for reuse, leaving a semisolid or a liquid hydrocarbon product. This product may be either used directly (as solvent-refined coal) or refined into a wide range of fuels and chemical materials.

Considered on a dry, ash-free basis, the proportion of the coal actually converted to liquid products is inherently greater than that obtained from other liquefaction techniques. However, in any commercial operation, there is also a need to provide the hydrogen as well, and this may require that additional coal be used in a subsidiary gasification plant. All else being equal, greatest yields per metric ton of coal are obtained from material with a high proportion of reactive macerals (vitrinite and liptinite) and a rank no higher than the middle of the high-volatile bituminous range. Excessive amounts of elements such as oxygen, nitrogen, chlorine, and sulfur, on the other hand, tend to combine with the hydrogen and reduce the efficiency, if not the yield, of the liquefaction process.

SEE ALSO THE FOLLOWING ARTICLES

COAL PREPARATION • COAL STRUCTURE AND REACTIVITY • COMBUSTION • ELEMENTAL ANALYSIS, ORGANIC COMPOUNDS • GEOCHEMISTRY, ORGANIC • GEOENVIRONMENTAL ENGINEERING • FOSSIL FUEL POWER STATIONS: COAL UTILIZATION • MINING ENGINEERING • PETROLEUM GEOLOGY • POLLUTION, AIR

BIBLIOGRAPHY

American Society for Testing and Materials (1996). *Annual Book of ASTM Standards,* Vol 05.05, American Society for Testing and Materials, Philadelphia.

Diessel, C. F. K. (1992). "Coal-Bearing Depositional Systems," Springer-Verlag, Berlin.

Falcon, R. M. S., and Snyman, C. P. (1986). "An introduction to coal petrography: atlas of petrographic constituents in the bituminous coals of southern Africa, *Geol. Soc. S. Afr. Rev. Paper* **2**.

Gayer, R. A., and Harris, J., eds. (1996). "Coalbed methane and coal geology," *Geol. Soc. (London) Spec. Pub.* **109**.

Gayer, R. A., and Pasek, J. (1997). "European coal geology and technology," *Geol. Soc. (London) Spec. Pub.* **125**.

Karr, C. Jr., ed. (1978). "Analytical Methods for Coal and Coal Products," Volumes 1-3, Academic Press, New York.

Lyons, P. C., and Rice, C. L., eds. (1986). "Palaeoenvironmental and tectonic controls in coal-forming basins of the United States," *Geol. Soc. Amer. Spec. Paper* **210**.

Rahmani, R. A., and Flores, R. M., eds. (1984). "Sedimentology of Coal and Coal-bearing Sequences," International Association of Sedimentologists, Special Publication **7**, Blackwell Scientific Publications, Oxford.

Rapp, A. R., Hower, J. C., and Peters, D. C. (1998). "Atlas of Coal Geology," American Association of Petroleum Geologists, *Stud. Geol.* **45** (CD-ROM).

Scott, A. C., ed. (1987). "Coal and Coal-bearing Strata: Recent Advances," Geological Society (London) Special Publication **32**, Blackwell Scientific Publications, Oxford.

Stach, E., Mackowsky, M.-Th., Teichmuller, M., Taylor, G. H., Chandra, D., and Teichmuller, R. (1982). "Stach's Textbook of Coal Petrology," 3rd edition, Gebruder Borntrager, Stuttgart.

Standards Australia (1993). "Guide to the Technical Evaluation of Hard Coal Deposits," Australian Standard, **2519**.

Taylor, G. H., Teichmuller, M., Davis, A., Diessel, C. F. K., Little, R., and Robert, P. (1998). "Organic Petrology," Gebruder Borntrager, Berlin.

Thomas, L. (1994). "Handbook of Practical Coal Geology," J. Wiley and Sons, London.

Ward, C. R., ed. (1984). "Coal Geology and Coal Technology," Blackwell Scientific Publications, Melbourne.

Ward, C. R., Harrington, H. J., Mallett, C. W., and Beeston, J. W., eds. (1995). "Geology of Australian Coal Basins," Geological Society of Australia Coal Geology Group, Special Publication **1**.

Coal Preparation

Robert A. Meyers
Ramtech Limited

Anthony D. Walters
Kilborn Engineering Limited

Janusz S. Laskowski
University of British Columbia

GLOSSARY

Dense media (or heavy liquids) Fluids used in dense-medium separation of coal and gangue particles by their relative densities. The medium can be any suitable fluid, but in commercial coal preparation operations, it is usually a suspension of fine magnetite in water.

Density (gravity) separation Separation methods based on differences in density of separated minerals, such as dense-medium separation and jigging.

Float-and-sink tests Tests carried out to determine coal washability, by which the coal is separated into various density fractions.

Floatability Description of the behavior of coal particles in froth flotation.

Hardgrove grindability index Measure of the ease by which the size of a coal can be reduced. Values decrease with increasing resistance to grinding.

Metallurgical coal Coal used in the manufacture of coke for the steel industry.

Near-density material Percentage of material in the feed within ± 0.1 density range from the separation density.

Organic efficiency (recovery efficiency) Measure of separating efficiency, calculated as

$$\frac{\text{actual clean coal yield}}{\text{theoretical clean coal yield}} \times 100$$

with both yields at the same ash content.

Probable error E_p (Ecart probable moyen) Measure of separating efficiency of a separator (jig, cyclone, etc.). Calculated from a Tromp curve.

RRB particle size distribution Particle size distribution developed by Rossin and Rammler and found by Bennett to be useful in describing the particle size distribution of run-of-mine coal.

Separation cut point δ_{50} Density of particles reporting equally to floating and sinking fractions (heavy media overflow and underflow in the dense media cyclone). The δ_{50} is also referred to as partition density.

Separation density δ_s Actual density of the dense medium.

Thermal coals Coals used as a fuel, mainly for power generation. These are lower-rank coals (high volatile bituminous, subbituminous, and lignites).

Tromp curve (partition curve, distribution curve) Most widely used method of determining graphically the value of E_p as a measure of separation efficiency. Synonyms: partition curve, distribution curve.

Washability curves Graphical presentation of the float-sink test results. Two types of plots are in use: Henry–Reinhard washability curves and M-curves.

Washing of coal (cleaning, beneficiation) Term denoting the most important coal preparation unit operation in which coal particles are separated from inorganic gangue in the processes based on differences in density (gravity methods) or surface properties (flotation). Cleaning then increases the heating value of raw coal.

COAL PREPARATION is the stage in coal production—preceding its end use as a fuel, reductant, or conversion plant feed—at which the run-of-mine (ROM) coal, consisting of particles, different in size and mineralogical composition, is made into a clean, graded, and consistent product suitable for the market; coal preparation includes physical processes that upgrade the quality of coal by regulating its size and reducing the content of mineral matter (expressed as ash, sulfur, etc.). The major unit operations are screening, cleaning (washing, beneficiation), crushing, and mechanical and thermal dewatering.

I. COAL CHARACTERISTICS RELATED TO COAL PREPARATION

Coal, an organic sedimentary rock, contains combustible organic matter in the form of macerals and inorganic matter mostly in the form of minerals.

Coal preparation upgrades raw coal by reducing its content of impurities (the inorganic matter). The most common criterion of processing quality is that of ash, which is not removed as such from coal during beneficiation processes, but particles with a lower inorganic matter content are separated from those with a higher inorganic matter content. The constituents of ash do not occur as such in coal but are formed as a result of chemical changes that take place in mineral matter during the combustion process. The ash is sometimes defined as all elements in coal except carbon, hydrogen, nitrogen, oxygen, and sulphur.

Coal is heterogeneous at a number of levels. At the simplest level it is a mixture of organic and inorganic phases, but because the mineral matter of coal originated in the inorganic constituents of the precursor plant, from other organic materials, and from the inorganic components transported to the coal bed, its textures and liberation characteristics differ. The levels of heterogeneity can then be set out as follows (Table I):

1. At the seam level, a large portion of mineral matter in coal arises from the inclusion during mining of roof or floor rock.
2. At the ply and lithotype level, the mineral matter may occur as deposits in cracks and cleats or as veins.
3. At the macerals level, the mineral matter may be present in the form of very finely disseminated discrete mineral matter particles.
4. At the submicroscopic level, the mineral matter may be present as strongly, chemically bonded elements.

Even in the ROM coal, a large portion of both coal and shale is already liberated to permit immediate concentration. This is so with heterogeneity level 1 and to some extent with level 2; at heterogeneity level 3 only crushing and very fine grinding can liberate mineral matter, while at level 4, which includes chemically bonded elements and probably syngenetic mineral matter, separation is possible only by chemical methods.

Recent findings indicate that most of the mineral matter in coal down to the micron particle-size range is indeed a distinct separable phase that can be liberated by fine crushing and grinding.

The terms *extraneous mineral matter* and *inherent mineral matter* were usually used to describe an ash-forming material, separable and nonseparable from coal

TABLE I Coal Inorganic Impurities[a]

Type	Origin	Examples	Physical separation
Strongly chemically bonded elements	From coal-forming organic tissue material	Organic sulphur, nitrogen	No
Adsorbed and weakly bonded groups	Ash-forming components in pure water, adsorbed on the coal surface	Various salts	Very limited
Mineral matter a. Epiclastic	Minerals washed or blown into the peat during its formation	Clays, quartz	Partly separable by physical methods
b. Syngenetic	Incorporated into coal from the very earliest peat-accumulation stage	Pyrite, siderite, some clay minerals	Intimately intergrown with coal macerals
c. Epigenetic	Stage subsequent to syngenetic; migration of the minerals-forming solutions through coal fractures	Carbonates, pyrite, kaolinite	Vein type mineralization; epigenetic minerals concentrated along cleats, preferentially exposed during breakage; separable by physical methods

[a] Adapted from Cook, A. C. (1981). *Sep. Sci. Technol.* **16**(10), 1545.

by physical methods. Traditionally in coal preparation processes, only the mineral matter at the first and, to some extent, the second levels of heterogeneity was liberated, the rest remained unliberated and, left with the cleaned coal, contributed to the inherent mineral matter. Recent very fine grinding, which also liberates the mineral matter at the third level of heterogeneity, has changed the old meanings of the terms *inherent* and *extraneous*. The content of the "true" inherent part of ash-forming material (i.e., the part left in coal after liberating and removing the mineral matter at the first, second, and third levels of heterogeneity) is usually less than 1%.

In recent years, the emphasis in coal preparation has been placed on reducing sulfur content of coal and on recovering the combustible material. Sulfur in coal is present in both organic and inorganic forms. The dominant form of inorganic sulfur is pyrite, but marcasite has also been reported in many coals. Pyrite occurs as discrete particles, often of microscopic size. It comprises 30–70% of total sulfur in most coals. Other forms of inorganic sulfur that may be present are gypsum and iron sulfates. The sulfate level in fresh unoxidized coals is generally less than 0.2%.

Organic sulfur in coal is believed to be contained in groups such as

THIOPHENE

R—S—R'   R—SH   R—SS—R'

ORGANIC   MERCAPTAN   ORGANIC
SULFIDE   DISULFIDE

The organic sulfur content in coals range from 0.5 to 2.5% w/w.

Physical cleaning methods can remove inorganic sulfates (gypsum) and most of the coarse pyrite; the finely disseminated microcrystalline pyrite and organic sulfur are usually not separable by such processes. This means that in the case of coal containing 70% of sulfur in pyritic form and 30% as organic sulfur, the physical cleaning can reduce the sulfur content by about 50%.

II. BREAKING, CRUSHING, AND GRINDING

The primary objectives of crushing coal are

1. Reduction of the top size of ROM coal to make it suitable for the treatment process
2. liberation of coal from middlings, and
3. size reduction of clean coal to meet market specification.

Size reduction of coal plays a major role in enabling ROM coal to be used to the fullest possible extent for power generation, production of coke, and production of synthetic fuels. ROM coal is the "as-received" coal from the mining process. Because the types of mining processes are varied, and size reduction actually begins at the face in the mining operation, it is quite understandable that the characteristics of the products from these various processes differ widely. The type of mining process directly affects the top size and particle size distribution of the mine product.

During the beneficiation of coal, the problem of treating middlings sometimes arises. This material is not of sufficient quality to be included with the high-quality clean coal product, yet it contains potentially recoverable coal. If this material is simply recirculated through the cleaning circuit, little or no quality upgrading can be achieved. Liberation can be accomplished if the nominal top size of the material is reduced, which permits recovery of the coal in the cleaning unit. Hammer mills, ring mills, impactors, and roll crushers in open or closed circuits can be used to reduce the top size of the middlings. Normally, reducing the nominal top size to the range 20–6 mm is practiced.

Breaking is the term applied to size operations on large material (say +75 mm) and *crushing* to particle size reduction below 75 mm; the term *grinding* covers the size reduction of material to below about 6 mm. However, these terms are loosely employed. A general term for all equipment is *size reduction equipment*, and because the term *comminution* means size reduction, another general term for the equipment is *comminution equipment*.

A. Breaking and Crushing

Primary breakers that treat ROM coal are usually rotary breakers, jaw crushers, or roll crushers; some of these are listed below.

Rotary Breakers (Fig. 1). The rotary breaker serves two functions—namely, reduction in top size of ROM and rejection of oversize rock. It is an autogenous size-reduction device in which the feed material acts as crushing media.

Jaw Crusher. This type of primary crusher is usually used for crushing shale to reduce it to a size suitable for handling.

Roll Crusher. For a given reduction ratio, single-roll crushers are capable of reducing ROM material to a product with a top size in the range of 200–18 mm in a single pass, depending upon the top size of the feed coal. Double-roll crushers consist of two rolls that rotate in opposite directions. Normally, one roll is fixed while the other roll is movable against spring pressure. This permits the passage of tramp material without damage to the unit. The drive units are normally equipped with shear pins for overload protection.

Hammer Mills. The swinging hammer mill (instead of having teeth as on a single-roll crusher) has hammers that are mounted on a rotating shaft so that they have a swinging movement.

B. Grinding

The most common grindability index used in conjunction with coal size reduction, the Hardgrove Grindability Index, is determined by grinding 50 g of 16×30 mesh dried coal in a standardized ball and race mill for 60 revolutions at an upper grinding ring speed of 20 rpm. Then the sample is removed and sieved at 200 mesh to determine W, the amount of material passing through the 200 mesh sieve. The index is calculated from the following formula:

$$\text{HGI} = 13.6 + 6.93W. \tag{1}$$

From the above formula, it can be deduced that as the resistance of the coal to grinding increases, the HGI decreases. The HGI can be used to predict the particle size distribution, that is, the Rossin–Rammler–Bennett curve [see Eq. (4)].

The distribution modulus (m) and the size modulus $d_{63.2}$ must be known to determine the size distribution of a particular coal.

It has been shown that for Australian coals, the distribution modulus can be calculated from the HGI by the following equation:

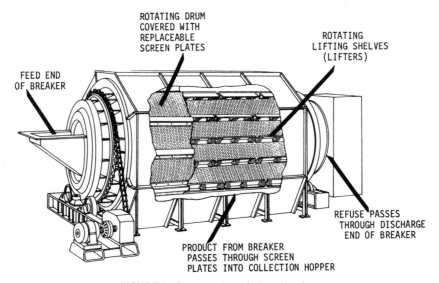

FIGURE 1 Cutaway view of rotary breaker.

$$HGI = 35.5m^{-1.54}. \qquad (2)$$

The value of $d_{63.2}$ is a function of the degree of breakage that the coal has undergone during and after mining. Having selected $d_{63.2}$ and the value of m from the HGI, one can predict the size distribution.

For the use of coal at power stations and for treatment by some of the newer beneficiation techniques (e.g., for coal/water slurries), grinding is employed to further reduce the top size and produce material with a given particle size distribution. The principal equipment used for coal grinding is the following:

1. air-swept ball mills,
2. roll or ball-and-race type mills,
3. air-swept hammer mills, and
4. wet overflow ball mills.

III. COAL SCREENING

Sizing is one of the most important unit operations in coal preparation and is defined as the separation of a heterogeneous mixture of particle sizes into fractions in which all particles range between a certain maximum and minimum size. Screening operations are performed for the following purposes:

1. Scalping off large run-of-mine coal for initial size reduction
2. Sizing of raw coal for cleaning in different processes
3. Removal of magnetic (dense medium) from clean coal and refuse
4. Dewatering
5. Separation of product coal into commercial sizes

The size of an irregular particle is defined as the smallest aperture through which that particle will pass.

In a practical screening operation the feed material forms a bed on the screen deck (Fig. 2) and is subjected to mechanical agitation so that the particles repeatedly approach the deck and are given the opportunity of passing

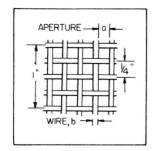

FIGURE 2 Four-mesh wire screen.

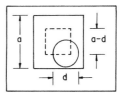

FIGURE 3 Passage of a spherical particle through a square aperature.

through. In the simplest case (Fig. 3), a spherical particle of diameter d will only pass through if it does not touch the sides of the aperture. The condition for passing is that the center of the sphere falls within the inner square, side $a - d$. The probability P of passing is thus given by the ratio between the areas of the inner and outer squares, that is,

$$P = (a - d)^2/a^2 = (1 - d/a)^2. \qquad (3)$$

The assumption that passage will be achieved only if there is no contact with the aperture sides is too restrictive. They can and do collide with the screen deck while passing through. In particular, the following factors contribute to the probability of passing:

1. Percentage open area
2. Particle shape
3. Angle of approach
4. Screen deck area
5. Bed motion
6. Size distribution of feed.

To select the correct screen for an application in a coal preparation plant, a detailed knowledge of the size distribution of the feed is necessary. Size distributions are usually presented graphically, and it is useful to use a straight-line plot, because curve fitting and subsequent interpolation and extrapolation can be carried out with greater confidence. In addition, if a function can be found that gives an acceptable straight-line graph, the function itself or the parameter derived from it can be used to describe the size distribution. This facilitates data comparison, transfer, or storage, and also offers major advantages in computer modeling or control. The particle size distribution used most commonly in coal preparation is the Rossin–Rammler–Bennett,

$$F(d) = 100(1 - \exp[(-d/d_{63.2})^m], \qquad (4)$$

where $F(d)$ is the cumulative percent passing on size d, $d_{63.2}$ is the size modulus ($d_{63.2}$ is that aperture through which 63.2% of the sample would pass), and m is the distribution modulus (the slope of the curve on the RRB graph paper). A sample of the RRB plot is shown in Fig. 4.

In Fig. 4 one can read off the slope m; the scale for $S \cdot d_{63.2}$ is also provided. If $d_{63.2}$ is the particle size (in

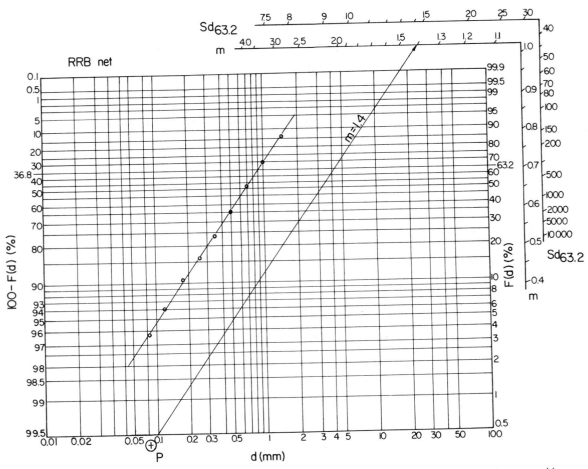

FIGURE 4 Rossin–Rammler–Bennett net with additional scales. Specific surface S is in square meters per cubic decimeter, $d_{63.2}$ in millimeters.

millimeters) belonging to $F(d) = 63.2\%$, then the scales furnish the specific surface S in m^2/dm^3.

The types of screens used in coal preparation plants generally fall into the following categories:

1. Fixed screens and grizzlies (for coarse coal).
2. Fixed sieves (for fine coal). These are used for dewatering and/or size separation of slurried coal or rejects. The sieve bend is the most commonly used in coal preparation plants for this application (Fig. 5).
3. Shaking screens. These are normally operated by camshafts with eccentric bearings. They can be mounted horizontally or inclined, and operate at low speeds with fairly long strokes (speeds up to 300 rpm with strokes of 1–3 in).
4. Vibrating screens. These are the most commonly used screens in coal preparation and are to be found in virtually all aspects of operations. A summary of their application is shown in Table II. A recent development in vibrating screens has been the introduction of very large screens (5.5 m wide, 6.4 m long). Combined with large tonnage [1000 metric tons per hour (tph) per screen], screens have been developed (banana screens) with varying slope: first section 34°, second section 24°, and third section 12°.

5. Resonance screens. These have been designed to save energy consumption. The screen deck is mounted on flexible hanger strips and attached to a balance frame, which is three or four times heavier than the screen itself.
6. Electromechanical screens. This type of screen operates with a high-frequency motion of very small throw. The motion is usually caused by a moving magnet which strikes a stop.

A. Recent Developments

A new process for the screening of raw coal of high moisture content at fine sizes has been developed by the National Coal Board. Their Rotating Probability Screen

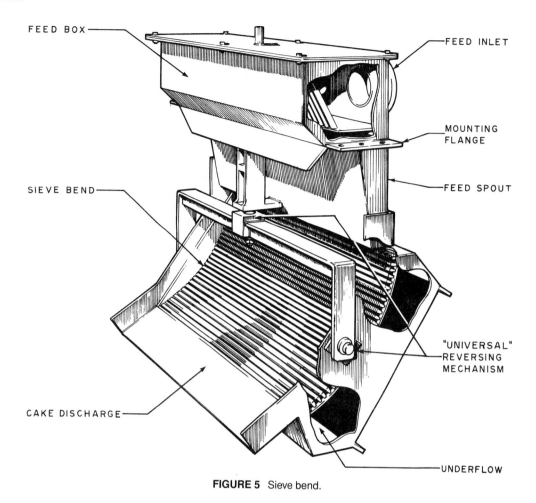

FEED BOX

FEED INLET

MOUNTING FLANGE

FEED SPOUT

SIEVE BEND

"UNIVERSAL" REVERSING MECHANISM

CAKE DISCHARGE

UNDERFLOW

FIGURE 5 Sieve bend.

(Fig. 6) has the advantage that the effective screening aperture can be changed while the screen is running. The screen "deck" is made up of small diameter stainless steel rods radiating from a central hub. The hub is rotated, and the coal falls onto the rotating spokes. The undersize coal passes through the spokes; the oversize coal is deflected over them. The speed of rotation dictates the screening aperture. The ability to screen coal with high levels of moisture has taken precedence over the accuracy of size separation. A screening operation by which the proportion of underflow product can be controlled while the machine is in operation represents an important advance in the preparation of blended coals in treatment plants where the fines are not cleaned.

IV. FLOAT-AND-SINK ANALYSIS

Most coal cleaning processes that are used to remove inorganic impurities from coal are based on the gravity separation of the coal from its associated gangue. ROM

coal consists of some proportion of both coal and shale particles, already sufficiently liberated, together with coal particles with inclusions of gangue (i.e., bands of shale). Commercially cleaned coal contains only very disseminated impurities and has a density ranging from 1.2 to 1.6. Carbonaceous shale density ranges from 2.0 to 2.6, and pure shale, clay, and sandstone have a density of about 2.6. The density of pyrite is about 5.0. The difference in density between pure coal and these impurities, in a liberated state, is sufficient to enable an almost complete separation to be achieved fairly easily. However, it has been shown that the inorganic impurity content, and hence, the ash content, ranges from pure coal containing only microscopic impurities to shale, which is entirely free from carbonaceous matter. Generally speaking, the mineral matter content of any coal particle is proportional to its density and inversely proportional to its calorific value.

For the prediction of concentration results (design of flowsheet) and for the control of preparation plant operations, a measure of concentrating operations is needed.

TABLE II Vibrating-Screen Applications in Coal Preparation Plants

Type	Number of decks	Installation angle	Aperture[a]	Screen deck type	Accessories
Run-of-mine scalper	Single	17°–25°	6 in.	Manganese skid bars, AR perforated plate with skid bars	Feed box with liners, extra high side plates, drive guard enclosures
Raw-coal sizing screen	Double	17°–25°	1 in.	AR steel perforated polyurethane, rubber	Dust enclosures, drive plate, guard enclosures
			$\frac{5}{16}$ in.	Polyurethane, wire, 304 stainless steel profile deck, rubber	Feed box with liners
Pre-wet screen	Double	Horizontal	1 in.	Wire, polyurethane, rubber	Water spray bar, side plate drip angles, drive guard enclosures, feed box liners
			1 mm	Stainless steel profile deck, polyurethane	
Dense-medium drain and rinse screen (coarse coal)	Double	Horizontal	1 in.	Wire, polyurethane, rubber	Side plate drip angles, spray bars, shower box cross flow screen or sieve bend, drip lip angles, drive guard enclosures
			1 mm	304 stainless steel profile deck, polyurethane	
Dewatering screen (coarse coal)	Single	Horizontal	1 mm	304 stainless steel profile deck, polyurethane	Sieve bend or cross flow screen, dam, discharge drip lip angles, drive guard enclosures
Desliming screen	Single	Horizontal	0.5 mm	304 stainless steel profile deck, polyurethane	Sieve bend or cross flow screen, spray bars, shower box, drive guard enclosures
Classifying screen (fine coal)	Single	28°	100 mesh	Stainless steel woven wire sandwich screens	Three-way slurry distributor and feel system
Dense-medium drain and rinse screen (fine coal)	Single	Horizontal	0.5 mm	304 stainless steel profile deck, polyurethane	Sieve bend or cross flow screen, spray bars, shower box, drip lip angles, drive guard enclosures
Dewatering screen (fine coal)	Single	Horizontal or 27°–29°	0.5 mm	304 stainless steel profile deck or woven wire, rubber, polyurethane	Sieve bend or cross flow screen, dam, drip lip angles, drive guard enclosures

[a] Typical application.

The best known means of investigating and predicting theoretical beneficiation results are the so-called washability curves, which represent graphically the experimental separation data obtained under ideal conditions in so-called float–sink tests. Float-and-sink analysis is also used to determine the Tromp curve, which measures the practical results of a density separation. The practical results of separation can then be compared with the ideal and a measure of efficiency calculated.

The principle of float–sink testing procedure is as follows: A weighed amount of a given size fraction is gradually introduced into the heavy liquid of the lowest density. The floating fraction is separated from the fraction that sinks. The procedure is repeated successively with liquids extending over the desired range of densities. The fraction that sinks in the liquid of highest density is also obtained. The weight and ash contents of each density fraction are determined.

In the example shown in Fig. 7, five heavy liquids with densities from 1.3 to 1.8 are used. The weight yields of six density fractions are calculated ($\gamma_1, \gamma_2, \ldots, \gamma_6$), and their ash contents are determined ($\lambda_1, \lambda_2, \ldots, \lambda_6$). The results are set out graphically in a series of curves referred to as washability curves (Henry–Reinhard washability curves or mean-value curve, M-curve, introduced by Mayer).

The construction of the primary washability curve (Henry–Reinhard plot) is shown in Fig. 8. It is noteworthy that the area below the primary curve (shaded) represents ash in the sample. The shaded area, changed into the rectangle of the same surface area, gives the mean-ash content in the sample ($\alpha = 16.44\%$ in our example).

The shape of the primary curve reveals the proportions of raw coal within various limits of ash content and so shows the proportions of free impurities and middlings present. Because the relative ease or difficulty of cleaning a raw coal to give the theoretical yield and ash content

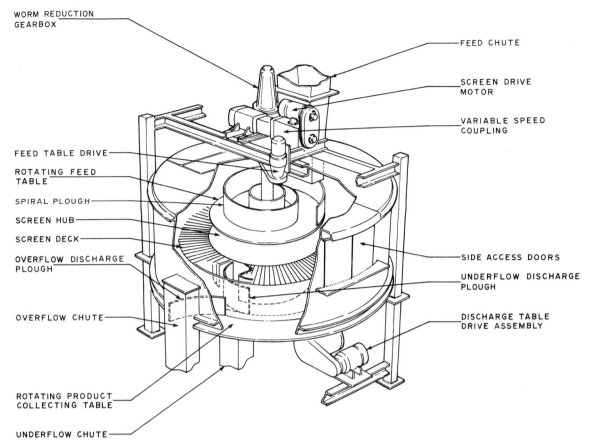

FIGURE 6 Rotating probability screen.

depends on the proportion of middlings, the shape of the primary curve also indicates whether the coal is easy or difficult to clean.

The construction of the mean-value curve (M-curve), also referred to as Mayer's curve, is shown in Fig. 9. The point where the curve intersects the abscissa gives the average ash content of the raw coal (α).

The shape of the primary washability curve and the M-curve is an indication of the ease or difficulty of cleaning the coal. The more the shape approximates the letter

L the easier the cleaning process will be. (This is further illustrated in Fig. 10.)

Figure 10 shows four different cases: (a) ideal separation, (b) easy cleanability, (c) difficult cleanability, and (d) separation impossible.

The primary washability curve for the difficult-to-clean coal (c) exhibits only a gradual change in slope revealing a large proportion of middlings.

It can be seen from the example that the further the M-curve is from the line connecting the zero yield point

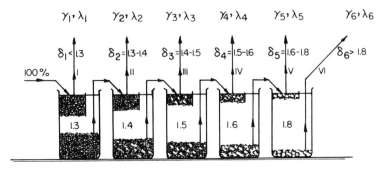

FIGURE 7 Float–sink analysis procedure.

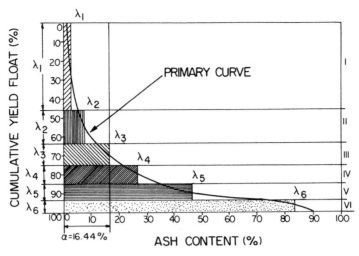

FIGURE 8 Construction of the primary washability curve (Henry–Reinhard washability diagram).

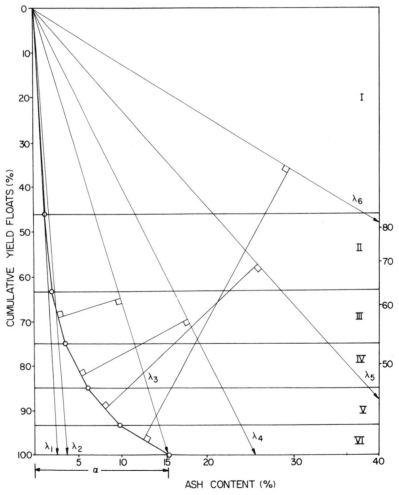

FIGURE 9 Construction of the M-value curve.

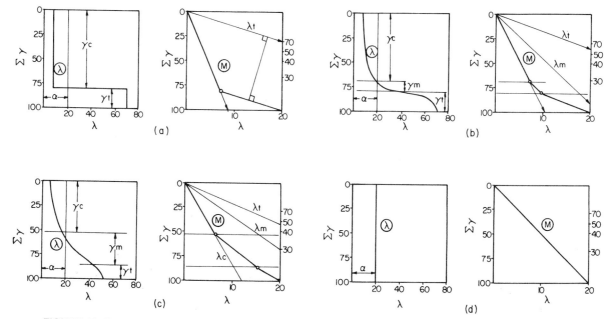

FIGURE 10 Primary washability and M-value curves for (a) ideal separation, (b) easy-to-clean coal, (c) difficult-to-clean coal, and (d) impossible separation.

with α on the λ axis, the greater the range of variation of ash content in the coal. The more gradual change in slope of an M-curve indicates more difficult washability characteristics.

The washability curve can be used in conjunction with the specific-gravity-yield curve (curve D) and the ± 0.1 near-density curve (curve E) as demonstrated in the example given in Table III and plotted in Fig. 11.

Curve A is the primary washability curve. Curve B is the clean-coal curve and shows the theoretical percent ash of the clean-coal product at any given yield. Curve C (the cumulative sink ash) shows the theoretical ash content of the refuse at any yield. Curve D, plotted directly from the cumulative percent yield of floats versus density, gives the yield of floats at any separation density. Curve E, the curve of near-density material, gives the amount of ma-

terial within ± 0.1 specific-gravity unit, and is used as an indication of the difficulty of separation. The greater the yield of the ± 0.1 fraction, the more difficult will be the separation at this separation density. For the example given, the separation process, as indicated by the ± 0.1 density curve, will be easier at higher densities than at lower ones.

V. DENSE-MEDIUM SEPARATION

The use of liquids of varying density in a simple float–sink test leads to separation of the raw coal into different density fractions. In a heavy liquid (dense medium), particles having a density lower than the liquid will float, and those having a higher density will sink. The commercial dense-media separation process is based on the same principle. Since the specific gravity of coal particles varies

TABLE III Set of Float–Sink Test Results

Specific gravity (g/cm³) (1)	Direct		Weight of ash of total (%) (4)	Cumulative weight of ash (%) (5)	Cumulative floats		Sink weight of ash (%) (8)	Cumulative sinks		±0.1 Specific gravity distribution	
	Yield γ (wt. %) (2)	Ash λ (%) (3)			Yield (%) (6)	Ash (%) (7)		Yield (%) (9)	Ash (%) (10)	Specific gravity (g/cm³) (11)	Yield (%) (12)
<1.3	46.0	2.7	1.24	1.24	46.0	2.7	15.2	54.0	16.44	—	—
1.3–1.4	17.3	7.3	1.26	2.50	63.3	3.9	13.94	36.7	28.22	1.4	63.3
1.4–1.5	11.7	16.3	1.87	4.37	75.0	5.8	12.07	25.0	38.00	1.5	29.0
1.5–1.6	10.0	26.7	2.67	7.04	85.0	8.3	9.4	15.0	48.28	1.6	14.15
1.6–1.8	8.3	46.0	3.82	10.86	93.3	11.6	5.58	· 6.7	62.66	1.7	8.3
>1.8	6.7	83.3	5.58	16.44	100.0	16.44	—	—	83.29	1.8	10.85
	100.0		16.44								

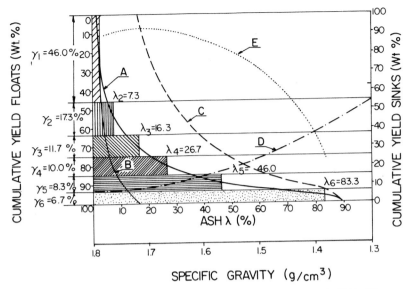

FIGURE 11 Complete set of washability curves (for data given in Table III).

from about 1.2 g/cm³ for low-ash particles to about 2.0 g/cm³ for the high-ash particles, the liquids within this range of density provide conditions sufficient for the heavy-medium separation of coals.

Of the four types of dense media that can be considered in such a process—namely, organic liquids (carbon tetrachloride, bromoform), aqueous solutions ($ZnCl_2$, $CaCl_2$), aqueous suspensions of high-density solids (ferrosilicon, magnetite, barite, quartz sand), and air fluidized bed suspension (sand)—the first two are used only in laboratory washability studies, and only the third has found wide industrial applications.

The main difference between the former two and the latter two is stability: The first two are homogeneous liquids, while the latter are composed of fine solid particles suspended in water (or air), and as such are highly unstable. Magnetite has become the standard industrial dense medium.

To achieve densities in the range 1.5–1.9, the concentration of magnetite in water must be relatively high. At this level of concentration, the suspension exhibits the characteristics of a non-Newtonian liquid, resulting in a formidable task in characterizing its rheological properties.

The principles of separation in dense media are depicted in Fig. 12, which shows a rotating drum dense-medium separator. The classification of static dense-medium separators offered 40 years ago by T. Laskowski is reproduced in Table IV.

Theoretically, particles of any size can be treated in the static dense-medium separators; but practically, the treatable sizes range from a few millimeters to about 150 mm.

TABLE IV Classification of Dense-Medium Separators[a]

	Static					
	Shallow			Deep		
Rotating-drum separates	Separation products removal with paddles	Separation products removal by belt conveyor	Separation products removal with scraper chains	Hydraulic removal of Separation products	Separation products removal by airlift	Mechanical removal of Separation products
			Examples			
Wemco drum	Link belt; SKB-Teska; Neldco (Nelson–Davis); Disa; Norwalt; Drewboy	Ridley–Sholes bath; Daniels bath	Tromp shallow; Dutch State Mines bath; McNally Tromp	McNally static bath; Potasse d'Alsace	Wemco cone; Humboldt	Chance; Barvoys; Tromp

[a] After Laskowski, T. (1958). "Dense Medium Separation of Minerals," Wyd. Gorn. Hutnicze, Katowice. Polish text.

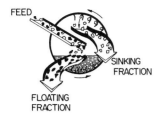

FIGURE 12 Principles of separation in dense media.

Since this process is unable to treat the full size range of raw coal, it is most important to remove the fines from the feed, because their presence in the circuit can increase the medium viscosity and increase the loss of media per ton of treated coal. The loss of magnetite in the coarse-coal cleaning is usually much below 1 kg/metric ton of washed product.

Dense-media static separators have a capability of handling high throughputs of up to 1000 metric ton/h.

Dynamic separators, which are the most efficient devices for cleaning intermediate size coal (50 down to 0.5 mm), include the dense-medium cyclone, two- and three-product Dynawhirlpool separators, the Vorsyl dense-medium separator, and the Swirl cyclone.

The most common are dense-medium cyclones (DMS) (Fig. 13a). Because the raw coal and medium are introduced tangentially into the cyclone, and the forces acting on the particles are proportional to V^2/r, where V is the tangential velocity and r the radius of the cylindrical section, the centrifugal acceleration is about 20 times greater than the gravity acceleration acting upon particles in the static dense-medium separator (this acceleration approaches 200 times greater than the gravity acceleration at the cyclone apex). These large forces account for the high throughputs of the cyclone and its ability to treat fine coal.

A DMS cyclone cut point δ_{50} is greater than the density of the medium and is closely related to the density of the cyclone underflow. The size distribution of the magnetite particles is very important for dense-medium cyclones, and magnetite that is 90% below 325 mesh (45 μm) was

found to provide the best results. Even finer magnetite (50% below 10 μm) was found to be essential for treating $-0.5 + 0.075$ mm coal. In treating such fine coal in a DMS, which technically is quite possible, the main problem that still awaits solving is medium recovery at reasonable cost.

Innovative dynamic separators are now being introduced that are able to treat the full size range (100–0.5 mm) in a single separating vessel. The Larcodems (*LA*rge, *CO*al, *DE*nse, *ME*dium, *SE*parator) is basically a cylindrical chamber that is mounted at 30° to the horizontal. Feed medium is introduced under pressure by an involute inlet at the bottom, and the raw coal enters separately through an axial inlet at the top. Clean coal is discharged through an axial inlet at the bottom end and reject expelled from the top via an involute outlet connected to a vortextractor.

The Tri-flo separator is a large-diameter three-product separator similar in operation to a Dyna Whirlpool.

A. Dense Media

Any suitable fluid can be used as a dense medium, but only fine solid particles in water suspensions have found wide industrial applications. A good medium must be chemically inert, and must resist degradation by abrasion and corrosion, have high inherent density, be easily recovered from the separation products for reuse, and be cheap.

There is a close relationship between the medium density and viscosity: the lower the medium density (i.e., the solids content in suspension), the lower its viscosity. This, in turn, is related to medium stability (usually defined as the reciprocal of the settling rate) through the size and shape of solid particles: the finer the medium particles, the lower the settling rate (hence, the greater stability), but the higher the viscosity. Medium recovery is easier for coarser medium particles. Thus the medium density cannot be changed without the viscosity and stability being affected. For constant medium density, a higher specific gravity solid used as the medium reduces solid concentration in the dense medium and then reduces viscosity and stability; both are also affected by the selection of a spherical medium type (i.e., granulated ferrosilicon).

The term *viscosity*, as used above, is not accurate. Dense media at higher solid concentrations exhibit characteristics of non-Newtonian liquids, namely, Bingham plastic or pseudo-plastic behavior, as shown in Fig. 14.[1]

[1]The use of the modified rheoviscometer, which enables handling of unstable mineral suspensions, has recently revealed that the Casson Equation fits the flow curve for the magnetite suspension better than the typically used Bingham plastic model [see the special issue of *Coal Preparation* entirely devoted to magnetite dense media: [*Coal Preparation* (1990). **8**(3–4).]

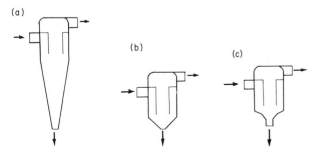

FIGURE 13 Cyclones in coal preparation (a) dense-medium cyclone; (b) water-only cyclone; (c) water-compound cyclone (Visman tricone).

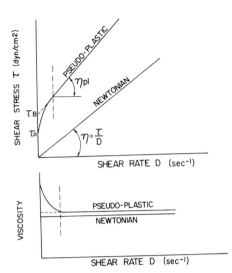

FIGURE 14 Rheological curves for Newtonian and pseudo-plastic liquids.

As the rheological curves for magnetite dense-medium show, for such systems the shear stress is not proportional to the shear rate; therefore, such systems are not characterized by one simple value of viscosity, as in the case of Newtonian liquids.

The plastic and pseudo-plastic systems are described by the Bingham equation,

$$\tau = \tau_B + \eta_{pl}D, \qquad (5)$$

and so both values τ_0 (τ_B) and η_{pl}, which can be obtained only from the full rheological curves, are needed to characterize dense-medium viscosity.

Figure 15 shows the effect of magnetite particle size and concentration on the medium rheology. As seen, the finer

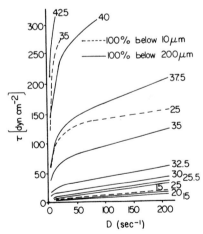

FIGURE 15 Rheological curves for magnetite dense media. Solid line, magnetite particle size 100% below 200 μm; broken line, magnetite size 100% below 10 μm. [Adapted from Berghöfer, W. (1959). *Bergbauwissenschaften*, **6**(20), 493.]

magnetite suspension exhibits much higher viscosity. It is also seen that while magnetite dense-medium behaves as a Newtonian liquid at magnetite concentration lower than 15% by volume, it is clearly pseudoplastic at higher concentrations.

In a heavy-medium separation process, coal particles whose density is higher than the medium sink while the lower-density coal particles float; the separation efficiency and the through put of the separation device depend on the velocity of coal particles in a dense medium. The viscosity of a medium has little effect on the low-density coal particles or the high-density gangue particles, but becomes critical in the separation of material of a density equal to or near that of the medium; hence, a low viscosity must be maintained to separate near-density material at a high rate of feed.

The efficiency of separation (E_p) in dense-medium baths was found to depend on the plastic viscosity of the media, and a high yield stress (τ_0) of the medium is claimed to cause elutriation of the finer particles into the float as they (and near-density particles) are unable to overcome the threshold shear stress required before the movement takes place. When a particle is held in a suspension, the yield stress is responsible for it; but when it is moving, its velocity is a function of plastic viscosity. It is understandable then that dispersing agents used to decrease medium viscosity may improve separation efficiency quite significantly. They may also decrease magnitite loses.

As the research on coal/water slurries shows, the suspension viscosity can also be reduced by close control of the particle size distribution; bimode particle distributions seem to be the most beneficial.

VI. SEPARATION IN A WATER MEDIUM

A. Jigs

Jigging is a process of particle stratification in which the particle rearrangement results from an alternate expansion and contraction of a bed of particles by pulsating fluid flow. The vertical direction of fluid flow is reversed periodically. Jigging results in layers of particles arranged by increasing density from the top to the bottom of the bed.

Pulsating water currents (i.e., both upward and downward currents) lead to stratification of the particles. The factors that affect the stratification are the following:

1. *Differential Acceleration.* When the water is at its highest position, the discard particles will initially fall faster than the coal.

2. *Hindered Settling.* Because the particles are crowded together, they mutually interfere with one another's

settling rate. Some of the lighter particles will thus be slowed by hindrance from other particles. This hindered settling effect is an essential feature, in that it helps to make separation faster; once the separation has been achieved, it helps to ensure that the material remains stratified.

3. *Consolidation Trickling.* Toward the end of the downward stroke, the particles are so crowded together that their free movement ceases, and the bed begins to compact. The further settling of the particles that leads to this compaction is referred to as *consolidation.* The larger particles lock together first, then the smaller particles consolidate in the interstices. Therefore, the larger particles cease to settle first, while the smaller particles can trickle through the interstices of the bed as consolidation proceeds.

1. Discard Extraction Mechanism

The method of discard extraction varies in different types of jig. Discard is discharged into the boot of the bucket elevator, and the perforated buckets carry it farther out. Some fine discard passes through the screen deck and is transported to the boot by screw conveyor, but the major portion is extracted over the screen plate. There are many methods of automatic control of discard.

2. Baum Jig

The most common jig used in coal preparation is the Baum jig, which consists of U-shaped cells (Fig. 16a). One limb

of the cell is open and contains the perforated deck; the other terminates in a closed chamber, to and from which air is repeatedly admitted and extracted by rotary valves coupled to an air compressor. Particle separation is effected on the perforated deck.

3. Batac Jigs

For many years, jig technology was largely unchanged (i.e., Baum type), except for advances in the design of air valves and discard extraction mechanisms. A new concept, the Batac jig, in which the air chamber is placed beneath the screen plate, was developed initially in Japan and later in Germany. The principal feature of the design (Fig. 16b) is that instead of the box being U-shaped, the air chambers are placed beneath the washbox screen plates. In this way, the width of the jig can be greatly increased and its performance improved by more equal jigging action over the width of the bed.

Jigs traditionally treat coals in the size range 100–0.6 mm; however, for optimum results, coarse and fine coals should be treated in separate jigs. Fine-coal jigs have their own "artificial" permanent bed, usually feldspar.

B. Concentrating Tables

A concentrating table (Fig. 17) consists of a riffled rubber deck carried on a supporting mechanism, connected to a head mechanism that imparts a rapid reciprocating motion

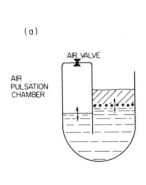
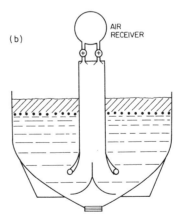

FIGURE 16 Comparison of (a) Baum and (b) Batac jigs. [After Williams, D. G. (1981). *In* "Coal Handbook" (Meyers, R. A. ed.), p. 265, Marcel Dekker, New York.]

Type	Baum jig	Batac jig
Year introduced	1892	1965
Present width	2.1 m	7.0 m
Present area	8.5 m^2	42.0 m^2
Present capacity	450 metric tons/hr	730 metric ton/hr.

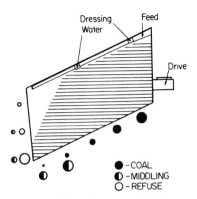

FIGURE 17 Distribution of concentrating table products by particle size and specific gravity. ●, coal; ◐, middling; ○, refuse.

in a direction parallel to the riffler. The side slope of the table can be adjusted. A cross flow of water (dressing water) is provided by means of a launder mounted along the upper side of the deck. The feed enters just ahead of the water supply and is fanned out over the table deck by differential motion and gravitational flow. The particles are stratified in layers by three forces:

1. Friction or adhesion between deck and particle
2. Pressure of the moving water
3. Differential acceleration due to the table action.

The clean coal overflows the lower side of the table, and the discard is removed at the far end. The device cleans efficiently over the size range 5–0.5 mm.

C. Water-Only Cyclones

The water-only cyclone (WOC) (first developed by the Dutch State Mines) is a cylindro-conical unit with an included apex angle up to 120° (Fig. 13b). Unlike the dense-medium cyclone, it uses no medium other than water; for this reason it is sometimes referred to as the autogenous cyclone. Geometrically, the principal differences from the dense-medium cyclone are the greater apex angle and the much longer vortex finder of the water-only cyclone.

The compound water cyclone (Visman Tricon) (Fig. 13c) is a variation of the water-only cyclone. Its distinguishing features are the three sections of the cone part with different apex angles, the first being 135°, the second 75°, and the third 20°.

Water-only cyclones do not make a sharp separation, but are commonly used as a preconcentrator or "rougher." Their efficiency can be increased by the addition of a second stage. There is also a sizing effect with water-only cyclones, and particles finer than 100 μm tend to report to the overflow regardless of ash content or relative density.

D. Spirals

Spiral separators have been used for many years, since the introduction of the Humphrey unit. The early spirals had simple profiles and were used in easy separations. Later development of Reichert spirals, suitable for a wide variety of applications, increased the use of spirals.

The effective size range for coal spiral operation (−3 mm +0.075 μm) coincides largely with the sizes that lie between those most effectively treated by heavy-media cyclones and those best treated by froth flotation. Thus, the principal areas of application might be substitution for water-only cyclones, for fine heavy-media cyclone separation, and for coarse froth flotation.

The capacity of Reichert coal spirals is about 2 tph. E_p values for Reichert Mark 9 and Mark 10 spirals have recently been reported in the range of 0.14–0.18, but they were found to be much lower for the combined product plus middlings than for the product alone. Reichert spirals offer simple construction requiring little maintenance plus low capital cost and low operating costs. Reichert-type LD Mark 10 and Vickers spirals have recently been shown to be able to achieve lower E_p values for the finer fractions (0.1 mm) but higher E_p for the coarse fractions (1 mm) than water-only cyclone. In general, the spiral is less affected by a change in particle size than WOC, but most importantly, the spirals tolerate increasing amounts of clay in the feed with little or no change in separation efficiency, in contrast to the WOC performance, which deteriorates with increasing amounts of clays. This indicates that the spiral is less influenced by increase in viscosity and is better suited for raw coals with significant proportions of clays.

VII. FLOTATION

Typically, coal fines below 28 mesh (0.6 mm) are cleaned by flotation. In some plants the coarser part of such a feed [+100 mesh (150 μm)] is treated in water-only cyclones, with only the very fine material going to flotation.

The coal flotation process is based on differences in surface properties between hydrophobic low-ash coal particles and hydrophilic high-ash gangue. Coals of different rank have various chemical composition and physical structure, and therefore their surface properties and floatability change with coalification. While metallurgical coals float easily and may require only a frother, flotation of lower-rank subbituminous coal and lignites (as well as high-rank anthracites) may be very difficult. In such a process one may need not only large amounts of oily collector but also a third reagent, the so-called promoter (Table V).

TABLE V   Coal Flotation Reagents[a]

Type	Examples	Remarks
Collectors	Insoluble in water, oily hydrocarbons, kerosene, fuel oil	Used in so-called emulsion flotation of coal, in which collector droplets must attach to coal particles
Frothers	Water-soluble surfactants; aliphatic alcohols; MIBC	To stabilize froth; adsorb at oil/water interface and onto coal; have some collecting abilities
Promoters	Emulsifiers	Facilitate emulsification of oily collector and the attachment of the collector droplets to oxidized and/or low-rank coal particles
Depressants/dispersants	Organic colloids: dextrin, carboxymethyl cellulose, etc.	Adsorb onto coal and make it hydrophilic
Inorganic salts	NaCl, $CaCl_2$, Na_2SO_4, etc.	Improve floatability; in so-called salt flotation process may be used to float metallurgical coals, even without any organic reagents; are coagulants in dewatering.

[a] Adapted from Klassen, V. I. (1963). "Coal Flotation," Gogortiekhizdat, Moscow. (In Russian.)

The surface of coal is a hydrophobic matrix that contains some polar groups as well as hydrophobic inorganic impurities. Coal surface properties are determined by

1. The coal hydrocarbon skeleton (related to the rank)
2. The active oxygen content (carboxylic and phenolic groups)
3. Inorganic matter impurities

Only the third term, the content of inorganic matter, is related to coal density. Therefore, in general there is no relationship between coal washability and coal floatability; such a relationship can be observed only in some particular cases.

Results obtained by various researchers agree that the most hydrophobic are metallurgical, bituminous coals. The contact angle versus volatile matter content curve, which shows wettability as a function of the rank (Fig. 18), is reproduced here after Klassen. As seen, lower-rank coals are more hydrophilic, which correlates quite well with oxygen content in coal, shown here after Ihnatowicz (Fig. 19). Comparison of Fig. 18 with Fig. 19 also indicates that the more hydrophilic character of anthracites cannot be explained on the same basis.

Two types of reagents are traditionally used in the flotation of coal:

Water-insoluble, oily hydrocarbons, as collectors
Water-soluble surfactants, as frothers

Insolubility of the collector in coal flotation requires prior emulsification or long conditioning time. On the other hand, the time of contact with the frother should be as short as possible to avoid unnecessary adsorption of frother by highly porous solids, such as coal.

In the flotation of coal, the *frother* (a surface-active agent soluble in water) adsorbs not only at the water/gas interface but also at the coal/water and the oil/water interfaces; it may facilitate, to some extent, the attachment of

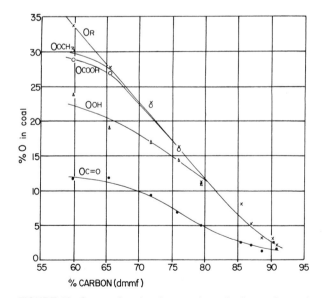

FIGURE 19   Oxygen functional groups in coals. (Large O stand for oxygen.) [From Ihnatowicz, A. (1952). *Bull. Central Research Mining Institute*, Katowice, No. 125. (In Polish).]

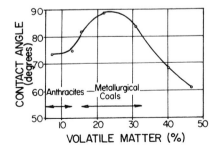

FIGURE 18   Effect of the rank on coal wettability. [From Klassen, V. I. (1963). "Coal Flotation," Gosgortiekhizdat, Moscow. (In Russian).]

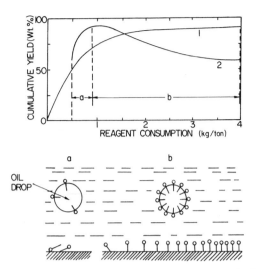

FIGURE 20 Effect of surface-active agent (frother) on the flotation of coal with water-insoluble oily collector. Curve 1 is for flotation with kerosene; curve 2 for kerosene and *n*-octyl alcohol. [From Melik-Gaykazian, V. I., Plaksin, I. N., and Voronchikhina, V. V. (1967). *Dokl. Akad. Nauk SSSR*, **173**, 883.]

oily droplets to coal particles. This can lead to a substantial improvement in flotation, as shown in Fig. 20.

In order to float oxidized and/or low-rank coals, new reagents called *promotors* have recently been tested and introduced into the industry. These are emulsifying agents that facilitate emulsification of the oily collector and the attachment of oily droplets to coal particles.

The reduction of sulfur is one of the principal benefits of the froth flotation process. Of the two types of sulfur in coal (i.e., inorganic and organic), only the inorganic sulfur (mainly represented by pyrite) can be separated from coal by physical methods. The floatability of coal-pyrite is, however, somewhat different from that of ore-pyrite, and its separation from coal presents a difficult problem. The research based on the behavior of ore-pyrite in the flotation process—namely, its poor flotability under alkaline conditions—did not result in a development of a coal-pyrite selective flotation process. The two-stage reverse flotation process proved to be much more successful. In this process, the first stage is conventional flotation with the froth product comprising both coal and pyrite. This product is reconditioned with dextrin, which depresses coal in the second stage, and is followed by flotation of pyrite with xanthates.

The fine-coal cleaning circuits in the newest plants frequently comprise water-only cyclones and flotation (Fig. 21). Such an arrangement is especially desirable in cleaning high-sulfur coals, because the difference in specific gravity between coal particles (1.25–1.6) and pyrite (5) is extremely large, and flotation does not discriminate well between coal and pyrite. Advantages of such an arrangement are clearly seen in Table VI, quoted after Miller, Podgursky, and Aikman.

VIII. SEPARATION EFFICIENCY

The yield and quality of the clean-coal product from an industrial coal preparation plant and the theoretical yield and quality determined from washability curves are known to be different. In the ideal cleaning process, all coal particles lower in density than the density of separation would be recovered in the clean product, while all material of greater density would be rejected as refuse. Under these conditions the product yield and quality from the actual concentration process and the yield and quality expected from the washability curves would be identical.

The performance of separators is, however, never ideal. As a result, some coal particles of lower than the separation density report to rejects, and some high-ash particles of higher than the separation density report to clean coal. These are referred to as *misplaced material*.

Coal particles of density well below the density of separation and mineral particles of density well above the density of separation report to their proper products: clean coal and refuse. But as the density of separation is approached, the proportion of the misplaced material reporting to an improper product increases rapidly.

Tromp, in a study of jig washing, observed that the displacement of migrating particles was a normal or

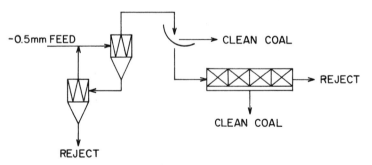

FIGURE 21 Fine-coal cleaning circuit.

TABLE VI Effectiveness of Hydrocycloning and Flotation in Ash and Sulfur Removal[a]

Particle size, mesh	Ash removal (%)		Sulfur removal (%)	
	Hydrocyclone	Flotation (single stage)	Hydrocyclone	Flotation (single stage)
28 × 100	60–65	50–60	70–80	50–60
100 × 200	40–45	40–45	60–70	30–40
200 × 325	15–18	40–45	35–40	25–30
325 × 0	0–5	50–55	8–10	20–25

[a] After Miller, F. G., Podgursky, J. M., and Aikman, R. P. (1967). *Trans. AIME* **238**, 276.

near-normal frequency (gaussian curve), and from this observation the partition curve (distribution, Tromp curve) in the form of an ogive was evolved.

The partition curve, the solid line in Fig. 22a, illustrates the ideal separation case ($E_p = 0$), and the broken-line curve represents the performance of a true separating device. The shaded areas represent the misplaced material.

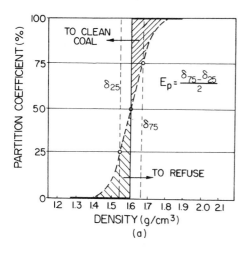

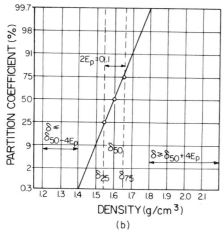

FIGURE 22 (a) Partition curve plotted on linear graph paper and (b) its anamorphosis plotted on a probability net.

The curves are plotted according to European convention and represent the percent of feed reporting to reject. In American practice, Tromp curves usually give the percent of feed reporting to washed coal.

The Tromp curve from the mathematical point of view is a cumulative distribution curve and as such can be linearized on probability graph paper. Such anamorphosis is produced by plotting the partition coefficients on a probability scale versus specific gravity on a linear scale (Fig. 22b) for dense-media separation, and versus log ($\delta - 1$) for jigs.

To determine the partition curve for a cleaning operation, one needs the yield of clean coal from this operation and the results of float–sink tests for both products—that is, for the clean coal and the refuse. Such data allow the reconstituted feed to be calculated, and from this can be found the partition coefficients, which give the percentage of each density fraction reporting to reject. As seen in Fig. 22b, the particles with densities below $\delta_{50} - 4E_p$ and the particles of density above $\delta_{50} + 4E_p$ report entirely to their proper products. The density fractions within $\delta_{50} \pm 4E_p$ are misplaced. Material of density very close to δ_{50} (near-density material) is misplaced the most. As postulated by Tromp, 37.5% of fractions within $\delta_{50}-\delta_{25}$ and $\delta_{75}-\delta_{50}$ are misplaced (this corresponds to $\delta_{50} \pm E_p$), and this percentage falls off drastically with the distance of the actual density fraction from the δ_{50} density.

Figure 23 shows partition curves for the major U.S. coal-cleaning devices. As seen, the sharpness of separation in dense-media separators is much better than in jigs or water-only hydrocyclones. Figure 24 shows E_p values plotted versus the size of treated particles for various cleaning devices. E_p values for the dense-medium bath and dense-medium cyclone are in the range of 0.02–0.04 for particles larger than 5 mm; E_p values of jigs and concentrating tables are in the range of 0.08–0.15; and for water-only cyclones, E_p values exceed 0.2. As seen, in all cases the efficiency of separation as given by E_p values decreases sharply for finer particles.

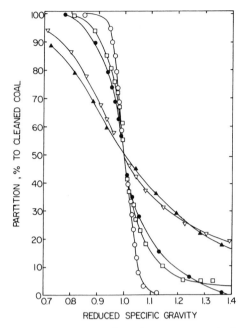

FIGURE 23 Performance of gravity separators. ▲, hydrocyclones, $\frac{1}{4}$ in. × 200 mesh; ▽, air tables, 2 in. × 200 mesh; ●, jigs: Baum, 6 in. × $\frac{1}{4}$ in.; Batac $\frac{3}{4}$ in. × 28 mesh; □, concentrating tables, $\frac{3}{4}$ in. × 200 mesh; ○, dense-medium separators: cyclone, $\frac{3}{4}$ in. × 28 mesh; vessel, 6 in. × $\frac{1}{4}$ in. [From Killmeyer, R. P. "Performance characteristics of coal-washing equipment: Baum and Batac jigs," U.S. Department of Energy, RI-PMTC-9(80).]

Another index developed to characterize the sharpness of separation is the imperfection I.

$$I = E_p/\delta_{50} \qquad (6)$$

for dense-medium separation, and

$$I = E_p/\delta_{50} - 1 \qquad (7)$$

for jigs.

According to some authors, I values vary from 0.07 for dense-medium cyclones and 0.175 for jigs to above 0.2 for flotation machines.

Conventional float and sink methods for the derivation of the partition curve are expensive and time-consuming. In the diamond and iron ore industries, the density tracer technique has been developed to evaluate the separation efficiency. This technique has also been adopted for studies of coal separation.

The tracers, usually plastic cubes prepared to known specific gravities rendered identifiable through color coding, are introduced into a separating device, and on recovery from the product and reject streams, are sorted into the appropriate specific gravity fractions and counted. This allows the points for the partition curve to be calculated. The technique, as described above, was adopted at the Julius Kruttschnitt Mineral Research Centre, Brisbane, while tracers made from plastic–metal composites that can

be detected with metal detectors mounted over conveyer belts, known as the Sentrex system, were developed in the United States.

IX. ANCILLARY OPERATIONS

There are many ancillary operations in coal preparation that are similar to operations in the mineral processing industry—namely

1. Mechanical dewatering
   a. Vibrating basket centrifuges
   b. Screen bowl centrifuges
   c. Solid bowl centrifuges
   d. Disk filters
   e. Drum filters
   f. Belt filters
   g. Plate and frame filter presses
   h. Thickeners, conventional and high capacity
2. Thermal dewatering
   a. Fluidized bed drying
   b. Rotary dryers
3. Stocking and blending systems
4. Automatic product loadout systems
5. Computer startup and process monitoring

X. COAL PREPARATION FLOWSHEETS

In ROM coal, a large portion of both coal and shale is already liberated sufficiently to permit immediate concentration. Because dewatering of coarse coal is much more efficient and cheaper than that of fines, and because many users prefer larger size fractions for their particular processes, the preparation of feed before cleaning has traditionally consisted of as little crushing as possible. Therefore, coal preparation consists mostly of gravity concentration and to some extent flotation, complemented by screening and solid/liquid separation auxiliary processes.

The precrushing needed to reduce the top size of the material is widely applied by means of rotary breakers. This process, as already pointed out, is based on the selective breakage of fragile coal and hard inorganic rock, and combines two operations: size reduction and preconcentration (because it rejects larger pieces of hard rock). A further development in coal preparation has been the application of the three-product separators to treat coarse and intermediate size fractions. The recrushed middlings produced in such a cleaning are reprocessed, together with the fine part of the feed, increasing the total coal recovery.

This complete concentration sequence is typical for processing a metallurgical coal.

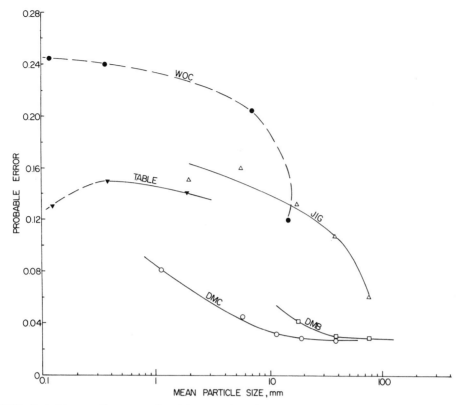

FIGURE 24 Probable error E_p vs mean size of the treated coal for dense-media baths (DMB), dense-media cyclones (DMC), jigs, concentrating tables, and water-only cyclones (WOC). [After Mikhail, M. W., Picard, J. L., and Humeniuk, O. E. (1982). "Performance evaluation of gravity separators in Canadian washeries." Paper presented at *2nd Technical Conference on Western Canadian Coals*, Edmonton, June 3–5.].

Coal preparation plants are classified according to the extent to which coal is cleaned, as follows (see Fig. 25).

Type I. Only the large size coal is reduced and screened into various size fractions.

Type II. The plant cleans coarse coal by gravity methods. The gangue content in coal is reduced in the form of coarse reject from the Bradford breaker and high-gravity rejects. The yield of saleable product is 75–85%.

Type III. The plant cleans coarse and intermediate size coal, leaving the fines untreated.

Type IV. These modern coal preparation plants clean the full size-range of raw coal.

Type V. The plant produces different saleable products that vary in quality, including deep-cleaned coal that is low in sulfur and ash.

Type VI. The plant incorporates fine grinding to achieve a high degree of liberation, using froth flotation (oil agglomeration) and water-only cyclones (or spirals).

Type VI represents a very important stage in the development of coal preparation strategy. Types I–V are all predominantly based on coarse coal cleaning, with the amount of fines reduced to a minimum; but in the type VI plant all coarse coal cleaning operations are aimed at eliminating coarse hard gangue and constitute a pretreatment stage, with the final cleaning following after fine crushing and grinding. This future development in coal preparation technology is aimed at cleaning and utilization of thermal coal, and is mainly a result of the new environmental-protection restrictions.

In the future, type VI coal preparation plants will be combined with coal/water slurry technology, and such slurries will to some extent replace liquid fuels.

XI. ON-LINE ANALYSIS

On-line measurements of parameters such as ash, moisture, and sulfur use an inferential technique in which some property can be measured that in turn relates to the parameter in the material.

Most on-line techniques use some form of electromagnetic radiation. When radiation passes through matter, there is a loss of intensity by absorption and scattering.

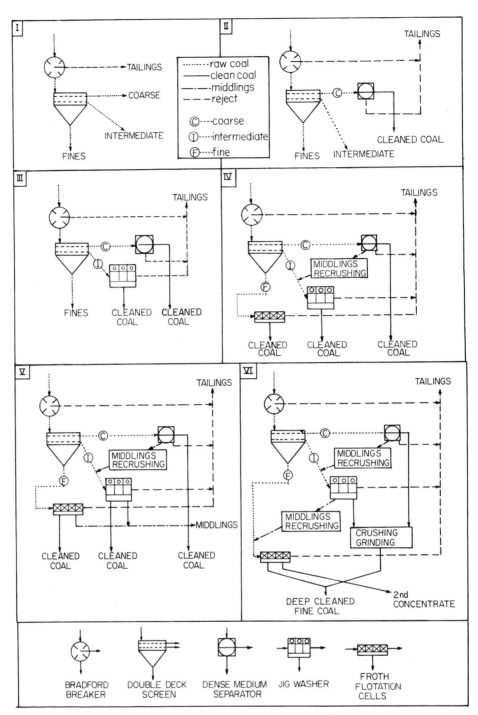

FIGURE 25 Classification of coal preparation circuits.

A. Ash Analysis

Several devices irradiate the coal with γ- or X-rays, and the backscatter is measured. At relatively low energies (<100 keV) the change in scattering coefficient with atomic number more than offsets the change in mass absorption coefficient; thus the backscatter can be correlated to mean atomic number and to the ash for a given coal. The backscattered radiation intensity is a function of both bulk density and mean atomic number. Thus, instruments using back scatter must be designed to ensure that a consistently packed sample of coal is presented for analysis. Some of the gauges using this principle are the Gunston Sortex, the Berthold–Humbolt–Wedag, and the Wultex.

When γ radiation above energy levels of 1022 keV interacts with matter, positron–electron pairs are generated. The positron is rapidly annihilated by collision with an electron and releases gamma rays, the intensity of which is strongly dependent upon atomic number. The gauge that uses the pair production principle is the COALSCAN Model 4500, which uses Radium (226) emitting γ radiation at both the 1760 and 2200 keV levels. The devices that use absorption use a low energy level source that is sensitive to the mean atomic number (2) of the material. There is an eightfold difference in the mass absorption coefficient between the organic portion of coal ($Z = 6$) and the mineral matter ($Z = 12$). By comparing the differences in absorption between low and high energy beams, the mean atomic number can be determined, and from this the ash composition can be inferred. This is the underlying principle behind such ash gauges as the COALSCAN Model 3500, the SAI 400, and the ASHSCAN slurry gauge.

Another form of radiation used for coal analysis is neutron activation. This method, known as Prompt Gamma Neutron Activation Analysis (PGNAA), has the distinct advantage of being able to conduct a complete elemental analysis rather than just total ash. This method permits the determination of the sulfur, carbon, and hydrogen composition, which also permits the calculation of specific energy. Several commercial PGNAA gauges are available—SAI Models 100 and 200, MDH Motherwell, and the Gamma Metrics.

The determination of ash in coal slurries uses the basic principle of measurement of those for solid coal. The extra problem is that the entrained air could seriously affect the accuracy of these devices. Various methods are used to compensate for this, such as pressurizing the system to collapse the air bubbles (ASHSCAN) or using a neutron attenuation measurement to correct for possible air entrainment (AMDEL).

B. Moisture Analysis

There are several methods of on-line moisture analysis:

1. Microwave attenuation/phase shift: Involves transmitting a microwave beam through a given layer of coal (conveyor belt). The maximum absorption of the energy occurs at 20 GHz and the energy absorbed by the material is calibrated against moisture content. However, simple attenuation is found to be sensitive to other properties such as particle size, bulk density, and salt content. To overcome these problems, the phase shift of the microwave energy is measured after passage through the sample or the attenuation is measured at more than one frequency.

2. Capacitance measurements: Based on the principle that electrical conductivity of coal increases with the

quantity of moisture. However, it is only applicable for the measurement of surface moisture.

3. Nuclear magnetic resonance: This method measures the hydrogen in the coal, and is able to distinguish between hydrogen in water and in hydrocarbons.

4. Neutron moisture gauges: Neutron transmission and scatter methods can be used to measure total hydrogen content, and assumes a constant hydrocarbon hydrogen.

C. Sulfur Analysis

The PGNAA method is generally used to measure elemental sulfur. Commercial gauges available are the MDH-Motherwell ELAN, Science Application International-Nucoalyzer, and the Gamma-Metrics.

XII. RESEARCH INTO NEW BENEFICIATION PROCESSES

A. Gravity

1. Otisca Process

The Otisca process is a waterless separation process employing CCl_3F (with certain additives) as a heavy liquid. In this process, raw coal with surface moisture up to 10% is subjected to separation in a static bath. The medium is recovered by evaporation from the separation products and is recycled. The Otisca process plant is completely enclosed, and the chemicals used are nonflammable, odor free, and noncorrosive.

The E_p calculated from the 8-hr plant tests for different size feeds gave the results (shown in Table VII). As seen, E_p values increase for very fine material (28×100 mesh and 100×325 mesh) but are still better than for many other methods. This indicates very favorable low-viscosity conditions in the process, which were confirmed by rheological measurements. The favorable rheology of the true organic heavy-liquid slurries has further been confirmed in experiments on the use of Freon-113 (1,1,2-trichloro-1,2,2-trifluoro ethane) in dense-medium

TABLE VII E_p **Values for Otisca Process Calculated for Different Size Feeds**[a]

Composite feed	Organic efficiency %	Probable error, E_p	Specific gravity
$\frac{3}{8} \times \frac{1}{4}$ in.	100	0.008	1.48
$\frac{1}{4}$ in. $\times 28$ mesh	99	0.015	1.49
28×100 mesh	98	0.175	1.57
100×325 mesh	96	0.260	1.80
$\frac{3}{8}$ in. $\times 325$ mesh	98	0.023	1.49

[a] After Keller, D. V. (1982). *In* "Physical Cleaning of Coal" (Liu, Y. A., ed.), p. 75, Marcel Dekker, New York.

cyclones. The separation in cyclones, with use of Freon-113 as a heavy medium, gave $E_p = 0.009$ for 28 mesh × 0 coal, $E_p = 0.045$ for 200 mesh × 0, and $E_p = 0.100$ for 400 mesh × 0 coal.

B. Magnetic Separation

This process is being investigated in the separation of coal and pyrite. The difference in magnetic susceptibility of coal and pyrite is very small; therefore, the most promising application is for high-gradient magnetic separators (HGMS) or the pretreatment of coal, which could increase the susceptibility of coal mineral matter.

Because pyrite liberation usually requires very fine grinding (e.g., below 400 mesh), high-gradient magnetic separation, the process developed to upgrade kaolin clays by removing iron and titanium oxide impurities, seems to be particularly suitable for coal desulfurization. Wet magnetic separation processes offer better selectivity for the fine dry coal that tends to form agglomerates. Experiments on high-gradient wet magnetic separation have so far indicated better sulfur rejection at fine grinding.

Perhaps the most interesting studies are on the effect of coal pretreatment, which enhances the magnetic properties of the mineral matter, on the subsequent magnetic separation. It has, for example, been shown that microwave treatment heats selectively pyrite particles in coal without loss of coal volatiles, and converts pyrite to pyrrhotite, and as a result, facilitates the subsequent magnetic separation.

In the Magnex process, crushed coal is treated with vapors of iron carbonyl [Fe(CO)$_5$], followed by the removal of pyrite and other high-ash impurities by dry magnetic separation. Iron carbonyl in this process decomposes on the surfaces of ash-forming minerals and forms strongly magnetic iron coatings; the reaction with pyrite leads to the formation of pyrrhotite-like material.

In 1970, Krukiewicz and Laskowski described a magnetizing alkali leaching process, which was applied to convert siderite particles into magnetic γ-Fe$_2$O$_3$ and Fe$_3$O$_4$. A similar process has recently been tested for high-sulfur coal, and it was reported that, after pretreatment of the coal in 0.5-M NaOH at 85°C and 930 kPa (135 psi) air pressure for 25 min, more than 50% of the coal sulfur was rejected under the same conditions in which the same high-gradient magnetic separation had removed only 5% sulfur without the pretreatment.

C. Processes Based on Surface Properties

1. Oil Agglomeration

As in coal flotation, oil agglomeration takes advantage of the difference between the surface properties of low-ash coal and high-ash gangue particles, and can cope with even finer particles than flotation. In this process, coal particles are agglomerated under conditions of intense agitation. The following separation of the agglomerates from the suspension of the hydrophilic gangue is carried out by screening.

The most important operating parameters in this process are the amount of oil, intensity of agitation (agitation time and agitation speed), and oil characteristics.

The amount of oil that is required is in the range of 5–10% by weight of solids. Published data indicate that the importance of agitation time increases as oil density and viscosity increase, and that the conditioning time required to form satisfactory coal agglomerates decreases as the agitation is intensified. Because the agitation initially serves to disperse the bridging oil to contact the oil droplets and coal particles, higher shear mixing with a lower viscosity bridging liquid is desirable in the first stage (microagglomeration), and less intense agitation with the addition of higher viscosity oil (macroagglomeration) is desirable in the second stage. Viscous oil may produce larger agglomerates that retain less moisture. With larger oil additions (20% by weight of solids), the moisture content of the agglomerated product can be well below 20% and may be reduced even further if tumbling is used in the second stage instead of agitation.

The National Research Council of Canada developed the spherical agglomeration process in the 1960s. This process takes place in two stages: First, the coal slurry is agitated with light oil in high shear blenders where microagglomerates are formed; then the microagglomerates are subjected to dewatering on screen and additional pelletizing with heavy oil.

Shell developed a novel mixing device to condition oil with suspension. Application of the Shell Pelletizing Separator to coal cleaning yielded very hard, uniform in size, and simple to dewater pellets at high coal recoveries. The German Oilfloc Process was developed to treat the high-clay, −400-mesh fraction of coal, which is the product of flotation feed desliming. In the process developed by the Central Fuel Research Institute of India, coal slurry is treated with diesel oil (2% additions) in mills and then agglomerated with 8–12% additions of heavy oil.

It is known that low rank and/or oxidized coals are not a suitable feedstock for beneficiation by the oil agglomeration method. The research carried out at the Alberta Research Council has shown, however, that bridging liquids, comprising mainly bitumen and heavy refinery residues are very efficient in agglomeration of thermal bituminous coals. Similar results had earlier been reported in the flotation of low rank coals; the process was much improved when 20% of no. 6 heavy oil was added to 2 fuel oil.

2. Otisca T-Process

This is another oil agglomeration process that can cope with extremely fine particles. In this process, fine raw coal, crushed below 10 cm, is comminuted in hammer crushers to below 250 μm and mixed with water to make a 50% by weight suspension; this is further ground below 15 μm and then diluted with water to 15% solids by weight. Such a feed is agglomerated with the use of Freon-113, and the coal agglomerates and dispersed mineral matter are separated over screen. The separated coal-agglomerated product retains 10–40% water and is subjected to thermal drying; Freon-113, with its boiling point at 47°C, evaporates, and after condensing is returned liquified to the circuit. The product coal may retain 50 ppm of Freon and 30–40% water.

Various coals cleaned in the Otisca T-Process contained in most cases below 1% ash, with the carbonaceous material recovery claimed to be almost 100%. Such a low ash content in the product indicates that very fine grinding liberates even micromineral matter (the third level of heterogeneity); it also shows Freon-113 to be an exceptionally selective agglomerant.

3. Selective Agglomeration During Pipelining of Slurries

In some countries, for example in Western Canada, the major obstacles to the development of a coal mining industry are transportation and the beneficiation/utilization of fines. Selective agglomeration during pipelining offers an interesting solution in such cases. Since, according to some assessments, pipelining is the least expensive means for coal transportation over long distances, this ingenious invention combines cheap transportation with very efficient beneficiation and dewatering. The Alberta Research Council experiments showed that selective agglomeration of coal can be accomplished in a pipeline operated under certain conditions. Compared with conventional oil agglomeration in stirred tanks, the long-distance pipeline agglomeration yields a superior product in terms of water and oil content as well as the mechanical properties of the agglomerates. The agglomerated coal can be separated over a 0.7-mm screen from the slurry. The water content in agglomerates was found to be 2–8% for metallurgical coals, 6–15% for thermal coals (high-volatile bituminous Alberta), and 7–23% for subbituminous coal. The ash content of the raw metallurgical coal was 18.9–39.8%, and the ash content of agglomerates was 8–15.4%. For thermal coals the agglomeration reduced the ash content from 19.8–48.0 to 5–12.8%, which, of course, is accompanied by a drastic increase in coal calorific value. Besides transportation and beneficiation, the agglomeration also facilitates material handling; the experiments showed that the agglomerates can be pipelined over distances of 1000–2000 km.

4. Flotation

Flotation has progressed and developed over the years; recent trends to achieve better liberation by fine grinding have intensified the search for more advanced means of improving selectivity. This involves not only more selective flotation agents but also better flotation equipment. Since the froth product in conventional flotation machines contains entrained fine gangue, which is carried into the froth with feed water, the use of froth spraying was suggested in the late 1950s to eliminate this type of froth "contamination." The flotation column patented in Canada in the early 1960s and marketed by the Column Flotation Company of Canada, Ltd., combines these ideas in the form of wash water supplied to the froth. The countercurrent wash water introduced at the top of a long column prevents the feed water and the slimes that it carries from entering an upper layer of the froth, thus enhancing selectivity.

Of many variations of the second generation columns, perhaps the best known is the Flotaire column marketed by the Deister Concentrator Company.

The microbubble flotation column (Microcel) developed at Virginia Tech is based on the basic premise that the rate (k) at which fine particles collide with bubbles increases as the inverse cube of the bubble size (D_b), i.e., $k \sim 1/D_b^3$. In the Microcel, small bubbles in the range of 100–500 μm are generated by pumping a slurry through an in-line mixer while introducing air into the slurry at the front end of the mixer. The microbubbles generated as such are injected into the bottom of the column slightly above the section from which the slurry is with drawn for bubble generation. The microbubbles rise along the height of the column, pick up the coal particles along the way, and form a layer of froth at the top section of the column. Like most other columns, it utilizes wash water added to the froth phase to remove the entrained ash-forming minerals. Advantages of the Microcel^{TM} are that the bubble generators are external to the column, allowing for easy maintenance, and that the bubble generators are nonplugging. An 8-ft diameter column uses four 4-in. in-line mixers to produce 5–6 tons of clean coal from a cyclone overflow containing 50% finer than 500 mesh.

Another interesting and quite different column was developed at Michigan Tech. It is referred to as a static tube flotation machine, and it incorporates a packed-bed column filled with a stack of corrugated plates. The packing elements arranged in blocks positioned at right angles to each other break bubbles into small sizes and obviate the need for a sparger. Wash water descends through the same flow passages as air (but countercurrently) and removes entrained particles from the froth product. It was shown in both the laboratory and the process demonstration unit that this device handles extremely well fine below 500-mesh material.

The Jameson Column developed at the University of Newcastle is considerably shorter, and its height may be only slightly greater than its diameter.

Another novel concept is the Air-Sparged Hydrocyclone developed at the University of Utah. In this device, the slurry fed tangentially through the cyclone header into the porous cylinder to develop a swirl flow pattern intersects with air sparged through the jacketed porous cylinder. The froth product is discharged through the overflow stream.

D. Chemical Desulfurization

Chemical desulfurization has been investigated as a method to selectively remove inorganic sulfur (which con-

sists of pyritic sulfur with trace amounts of sulfate) and organic sulfur without need to discard part of the coal rich in sulfur and ash as with conventional coal preparation. If this could be accomplished without loss of coal product heat content, a clean burning coal alternative to flue gas desulfurization would be at hand. Several methods have been tested for the chemical removal of pyritic or organic sulfur from coal. Treatment with aqueous sodium hydroxide or aqueous sodium hydroxide with added lime at temperatures of 200–325°C and pressures up to 12500 psi removes only 45–95% of the inorganic sulfur and none of the organic sulfur, while significantly reducing the heat content of the product coal due to hydrothermal oxidation. Sulfur dioxide, potassium nitrate with complexing agents, nitrogen dioxide, nitric acid, air oxidation with and without the bacterium *Thiobacillus ferrooxidans*, hydrogen peroxide, chlorine, and organic solvents have all been tested. In each case, there was only minimal sulfur removal (based on sulfur content per unit heat content of

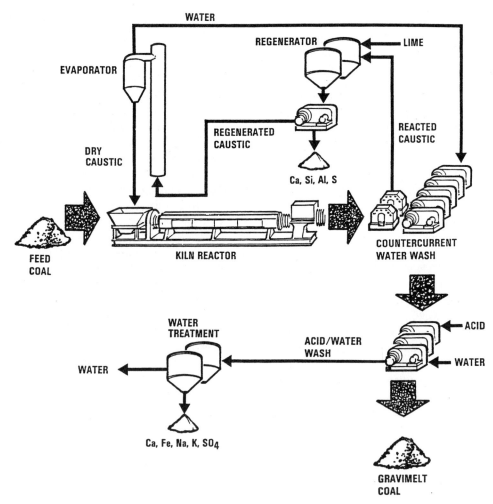

FIGURE 26 Molten–Caustic–Leaching (Gravimelt Process) flow schematic.

product coal) and considerable loss of coal heat content and/or dilution of the organic coal matrix by bonding with reactant chlorine, nitrate, etc.

However, the use of aqueous ferric sulfate has proved to be effective in selectively removing virtually all of the inorganic sulfur while the heat content increases in direct proportion to the amount of iron pyrite removed. The ferric sulfate sulfur removal technology, known as the Meyers process, was developed through pilot plant demonstration under the auspicies of the Environmental Protection Agency. Typically, 90–95% of the pyritic sulfur was removed and the ferric salt was simultaneously regenerated with air oxidation. Elemental sulfur is produced as a byproduct, and must be extracted from the coal by use of an organic solvent or steam distillation. The process has not yet been utilized commercially as removal of pyritic sulfur alone (usually only half of the total sulfur, the remainder being mainly organic

sulfur) does not meet present air pollution control standards.

Subsequently, simultaneous removal of organic sulfur and pyritic sulfur was demonstrated in the laboratory and then in a test plant utilizing molten sodium and potassium hydroxide followed by a countercurrent water wash to recover sodium, potassium, and about half of the original coal mineral matter. A dilute sulfuric acid wash removes the remainder of the coal mineral matter (ash). This process, termed Molten–Caustic–Leaching, or the Gravimelt Process, results in removal of 90–99% of the coal inorganic and organic sulfur as well as 95–99% of the coal mineral matter. The product coal meets all current air pollution control standards for sulfur and ash emissions, and has an increased heat content, usually in the range of 14000 btu/lb. This coal is also very efficient as a powdered activated carbon for water purification. This is due to the high internal pore structure

FIGURE 27 Rotary kiln reactor in operation at Gravimelt Test Plant Facility. The reactor consists of a 12-ft long furnace section, a 3-ft cooling zone, and a discharge area.

introduced by removal of sulfur and ash. The coal also functions as a cation exchange medium for water cleanup due attachment of carboxy groups during the caustic treatment.

A schematic of the Gravimelt Process is shown in Fig. 26. Feed coal is premixed with anhydrous sodium hydroxide or mixed sodium and potassium hydroxides, and then fed to a rotary kiln reactor where the mixture is heated to reaction temperatures of 325–415°C, causing the caustic to (1) melt and become sorbed in the coal matrix, (2) react with the coal sulfur and mineral matter, and (3) dissolve the reaction products containing sulfur and inorganic components. The kiln reactor can be seen in operation in Figure 27.

The reaction mixture is cooled in the cooling section of the kiln such that the mixture exits the rotary kiln in the form of pellets that are then washed with water in a seven-stage countercurrent separation system consisting of two filters and five centrifuges. As an example, four weights of wash water per weight of coal may be used, resulting in a 50% aqueous reacted caustic solution exiting the first stage of filtration. A large part of the coal-derived iron and a small amount of sodium and/or potassium remain with the coal cake exiting this countercurrent washing system.

The coal cake is next washed to remove the residual iron and caustic in a countercurrent three-stage centrifuge system. Sulfuric acid is added to the first stage to produce a pH of about 2 in order to dissolve the residual alkali and iron hydroxide. As an example, three weights of wash water per weight of coal may be used, resulting in an acid extract leaving the first stage containing 5900 ppm of mixed sodium and potassium sulfates and 12,500 ppm of iron sulfate. The product coal exiting this acidwashing system contains about one weight of sorbed water per weight of coal.

All of the iron sulfate and a major portion of the alkali sulfates can be removed from the acid extract water by a sequence of heating and lime treatment to form insoluble minerals such as gypsum. This water can then be recycled to the water wash train. Under some (very carefully controlled) conditions, the precipitate formed can be a jarosite-like double salt, which can be easily disposed of or land-filled.

A newer process configuration, with potentially large cost savings to the process, was tested in which the effluent from the acid wash train (without acid addition) was used as the makeup wash water for the water wash train. This configuration resulted in a product coal with low sulfur and moderate ash content suitable for those commercial applications that do not require low ash coal.

The aqueous caustic recovered from the first countercurrent washing circuit is limed to remove the coal-derived mineral matter, sulfur compounds, and carbonates. The mixture is provided with sufficient residence time to permit precipitation of the impurities before being centrifuged. The purified liquid is preheated and sent to a caustic evaporator where the water is recovered for recycle to the first wash train, while producing anhydrous caustic (as either a molten liquid or as flakes) for reuse in the initial leaching of the coal.

SEE ALSO THE FOLLOWING ARTICLES

COAL GEOLOGY • COAL STRUCTURE AND REACTIVITY • FOSSIL FUEL POWER STATIONS, COAL UTILIZATION • FROTH FLOTATION • FUELS • MINING ENGINEERING • PETROLEUM REFINING

BIBLIOGRAPHY

Deurbrouck, A. W., and Hucko, R. E. (1981). *In* "Chemistry of Coal Utilization" (M. A. Elliot, ed.), 2nd Suppl. vol., Chap. 10, Wiley, New York.

Gochin, R. J., and Smith, M. R. (1983). *Min. Mag.*, Dec., p. 453–467.

Horsfall, D. W., ed. (1980). "Coal Preparation for Plant Operators," South African Coal Processing Society, Cape Town.

Klassen, V. I. (1963). "Coal Flotation," Gosgorttiekhizdat, Moscow (Russian text).

Laskowski, T., Blaszczyński, S., and Slusarek, M. (1979). "Dense Medium Separation of Minerals," 2nd ed. Wyd. Slask, Katowice (Polish text).

Leonard, J. W., ed. (1979). "Coal Preparation," 4th ed., American Institute of Mechanical Engineering, New York.

Liu, Y. A., ed. (1982). "Physical Cleaning of Coal," Dekker, New York.

Meyers, R. A. (1977). "Coal Desulfurization," Marcel Dekker, New York.

Misra, S. K., and Klimpel, R. R., eds. (1987). "Fine Coal Processing," Noyes Publications, Park Ridge.

Osborne, D. G. (1988). "Coal Preparation Technology," Volumes 1 and 2, Graham and Trotman, London.

Speight, J. G. (1994). "The Chemistry and Technology of Coal," 2nd., revised and Expanded, Marcel Dekker, New York.

TRW Space and Electronics Group, Redondo Beach, CA Applied Technology Division. (1993). "Molten–Caustic–Leaching (Gravimelt) system integration project, final report," Department of Energy, Washington, DC, NTIS Order Number: DE930412591NZ, Performing Organization's Report Number: DOE/PC/91257-T18.

U.S. Department of Energy (1985). "Coal Slurry Fuels Preparation and Utilization," U.S. Department of Energy, Pittsburgh.

Wheelock, T. D., ed. (1977). "Coal Desulfurization," ACS Symp. Ser. No. 64, American Chemical Society, Washington, DC.

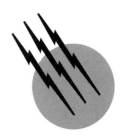

Coal Structure and Reactivity

John W. Larsen

Lehigh University

Martin L. Gorbaty

Exxon Research and Engineering Company

I. Introduction
II. Geology: Origin and Deposition
III. Coal Classifications
IV. Physical Properties
V. Physical Structure
VI. Chemical Structure
VII. Coal Chemistry

GLOSSARY

Anthracite Coals with less than ∼10% volatile matter; the highest coal rank.
Bituminous Coals that have between about 10 and 30% volatile matter and coals that agglomerate on heating.
Diagenesis Initial biological decomposition of coal-forming plant material.
Grade Suitability of a coal for a specific use.
Lignite Stage of coalification following peat; high-oxygen, low-heating-value, young coals.
Lithotypes Macroscopic rock types occurring in coals, usually in banded structures.
Macerals Homogeneous microscopic constituents of coals, analogous to minerals in inorganic rocks.
Metamorphism Chemical changes responsible for the conversion of fossil plant materials to coal.

Rank Degree of maturation of a coal.
Type Origin and composition of the components of a coal.

IN THE BROADEST TERMS, coal may be defined as a naturally occurring, heterogeneous, sedimentary organic rock formed by geological processes (temperature, pressure, time) from partially decomposed plant matter under first aerobic and then anaerobic conditions in the presence of water and consisting largely of carbonaceous material and minor amounts of inorganic substances. The main use for coal worlwide is combustion as a source of heat for generating steam to make electricity. Coals used for this purpose are called steam coals. Certain other coals are used to produce the metallurgical coke required for making iron from iron ore. Coal can also serve as the raw material for making clean, usable liquid fuels, such as gasoline, and chemicals.

TABLE I Worldwide Coal Deposits[a]

Country	Estimated resources (billions of tons)	Proven resources (billions of tons)
Australia	600	32.8
Canada	323	4.2
Federal Republic of Germany	247	34.4
India	81	12.4
People's Republic of China	1,438	98.9
Poland	139	59.6
South Africa	72	43
United Kingdom	190	45
United States of America	2,570	167
Russia	4,860	110
Other	229	55.7
	10,749	663

[a] Data from Wilson, C. L. (1980). "Coal—Bridge to the Future, Report of the World Coal Study," p. 161, copyright Ballinger Publishing Co., Cambridge, MA.

I. INTRODUCTION

Coal is perhaps the most abundant fossil fuel, and large deposits are distributed worldwide. As with other natural resources, the amounts are estimated on two bases: *estimated resources* and *proven reserves*. Estimated resources are based on geological studies (e.g., seismic surveys and exploratory drilling) and are the best estimate of the amount of coal believed to be present in a given area. Proven reserves are those amounts of coal known to be economically recoverable by current technology in a given area. Proven reserves represent a small fraction of estimated resources. Table I contains resource and reserve data for some of the major worldwide coal deposits expressed in billions of tons. It must be recognized that these estimates vary from year to year, as do the methods for obtaining the data. Therefore, the data in the table should not be viewed as being absolute. According to these data, ~663 billion tons, or 6% of worldwide resources of over 10 trillion tons, are recoverable by current technology. The worldwide reserves of coal have an energy content equivalent to ~2 trillion barrels of petroleum.

II. GEOLOGY: ORIGIN AND DEPOSITION

Coals may be viewed as fossilized plant matter. The processes by which plant matter was converted to coal are complex and involve at least three steps: accumulation, diagenesis, and metamorphism. The starting material for coal was a diversity of plant types and parts. For the plant matter to accumulate, a specific set of conditions had to prevail. The deposition rate of plant matter had to be approximately equal to the subsidence of the land on which it grew. Furthermore, there had to be water present to cover and partially preserve the plant matter. Initially, aerobic organisms began to decompose the cellulosic parts of the plant, as well as proteinaceous and porphyrin materials. This biological decomposition is called *diagenesis*. As the accumulated, partly decayed biomass was buried deeper, aerobic activity ceased, and under the increasing temperature and pressure conditions of burial it was chemically transformed to coal. This chemical transformation is called *metamorphism*. From the least serve to the most severe conditions, the coals that resulted are arranged as follows: peat → lignite → subbituminous → bituminous → anthracite. Chemically, the transformations can be viewed as combinations of deoxygenation, dehydrogenation, aromatization, and oligomerization. These transformations are inferred from the elemental analyses shown in Table II. The data in this table are typical; each class includes a range of values.

All coals contain minor amounts of inorganic material, called *mineral matter*, ranging from about 2 to 30% and averaging ~10% by weight. Inorganics were incorporated into coal as the inorganic matter present in the original vegetation, mineral detritus deposited by water or wind flowing through the decaying biomass (detrital), and by precipitation or ion exchange of soluble ions from water percolating through the coal bed during metamorphism (authigenic). The former two are called syngenetic, the latter epigenetic to differentiate the processes occurring during diagenesis from those occurring during and after metamorphism.

The major components of mineral matter include aluminosilicate clays, silica (quartz), carbonates (usually of calcium, magnesium, or iron), and sulfides (usually as pyrite and/or marcasite). Many other inorganics are present, but only in trace quantities of the order of parts per million.

TABLE II Elemental Analysis Characterizing Metamorphism as a Deoxygenation/Aromatization Process

Coal or precursor	Carbon (%)	Hydrogen (%)	Oxygen (%)
Wood	49.3	6.7	44.4
Peat	60.5	5.6	33.8
Lignite	69.8	4.7	25.5
Subbituminous	73.3	5.1	18.4
Bituminous	82.9	5.7	9.9
Anthracite	93.7	2.0	2.2

It is important to differentiate between mineral matter and ash. The former consists of inorganics present in the coal as mined; the latter is an oxide residue from combustion.

III. COAL CLASSIFICATIONS

The diverse nature of deposited plant and mineral matter and the burial conditions make it clear that coal is not one substance but a wide range of heterogeneous materials, each with considerably different chemical and physical properties from the others. The heterogeneity of coals can be seen on the macroscopic and microscopic levels, and the science of classifying coals on the basis of these difference is called *coal petrography*.

On the macroscopic level, coals, particularly bituminous coals, appear to be layered or banded. These layers, called *lithotypes*, fall into four main classes: vitrain (coherent, black, and high gloss), clarain (layered, glossy), durain (granular, dull sheen), and fusain (soft, friable, charcoal-like).

On the microscopic level, the lithotypes just described are seen to be composites made up of discrete entities called *macerals*. Optically differentiated by microscopic examination of coals under reflected light or by the study of thin sections of coal using transmitted light (Fig. 1), macerals are classified into three major groups: vitrinite,

liptinite (often termed exinite), and inertinite. The *vitrinite* class appears orange-red in transmitted light and is derived from fossilized lignin. *Liptinites* appear yellow and are the remains of spores, waxy exines of leaves, resin bodies, and waxes. *Inertinites* appear black and are believed to be the remnants of carbonized wood and unspecified detrital matter. Chemically, the liptinites are richest in hydrogen, followed by the vitrinites. Both the liptinites and vitrinites are reactive and give off a significant amount of volatiles when pyrolyzed, while the inertinites are relatively unreactive. It is not possible to generalize with accuracy the proportional maceral composition of all coals; however, it is fair to say that most bituminous coals contain about 70% vitrinite, 20% liptinite, and 10% inertinite.

It is generally believed that maceral composition ultimately determines a coal's reactivity in a particular process. For example, coals are blended together on a commercial scale to make optimum feedstocks for metallurgical coke based largely on their maceral contents. In the past, macerals were separated tediously by hand under a microscope. Today, new techniques are available to separate a coal into its component macerals much more easily, and this has renewed interest in maceral characterization and reactivity. It may even be possible one day to tailor coal feedstocks for specific processes such as pyrolysis, gasification, or liquefaction by separation and blending of macerals.

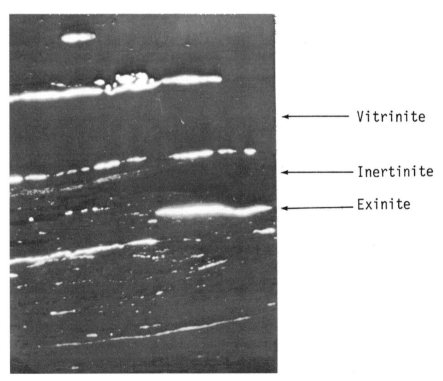

FIGURE 1 Transmission photomicrograph of bituminous coal. (Courtesy of D. Brenner, Exxon Research and Engineering Company.)

Another method of classifying coals is by the degree of metamorphism they have undergone. This classification is called *rank*—the higher the rank, the more mature the coal. Since metamorphism is a deoxygenation–aromatization process, rank correlates generally with carbon content and heat content of the coal, although there are many other, more specific methods of determining rank. Lignites and subbituminous coals are generally considered to be low rank, while bituminous coals and anthracite are considered to be high rank. The term "*rank*" must be distinguished from type and grade. Type refers to a coal's composition, the origin of the organic portion, and its organic components. Grade refers to the quality or suitability of a coal for a particular purpose. These distinctions are important because coal properties are not solely attributable to metamorphic alteration; the nature of the source material and its diagenetic alterations are also important in determining coal properties.

In general, carbon content, aromaticity (i.e., the percentage of carbon that is aromatic), number and size of condensed rings, and calorific content increase with increasing rank, while volatile matter, oxygen content, oxidizability, and solubility in aqueous caustic decrease with increasing rank. Hardness and plastic properties increase to a maximum, then decrease with increasing rank, while porosity (i.e., moisture-holding capacity) and density decrease to a minimum, then increase with increasing rank. Typical values are shown in Table III. Within each rank classification, a range of values is found.

IV. PHYSICAL PROPERTIES

Physical properties are an important consideration in all uses of coals. Most vary with rank and coal type. A brief discussion of the most important physical properties follows.

A. Mechanical Properties

1. Elasticity

Because of the heterogeneous nature and the natural occurrence of cracks and pores, determining elastic moduli is experimentally difficult. Under normal conditions, coals are brittle solids. For bituminous coals, the elastic properties are independent of rank and similar to those of phenol–formaldehyde resins. At ∼92% carbon, the moduli increase rapidly and become anisotropic.

2. Hardness (Grindability)

Grindability is normally measured by determining the number of revolutions in a pulverizer that are required to achieve a given size reduction. This is an indirect measure of the amount of work required for that size reduction. Grindability increases slowly to a maximum at ∼90% carbon and decreases rapidly beyond this point. Hardness is ascertained by measuring the size of the indentation left by a penetrator of specified shape pressed onto the coal with a specified force for a specified time. It changes only slightly between 70 and 90% carbon and increases rapidly beyond 90% carbon.

3. Plastic Deformation

Coals heated to above 400°C plastically deform, as do those that have been swollen with pyridine and other solvents that are good hydrogen bond acceptors. Very small coal particles flow and deform at room temperature under very high stresses.

B. Thermal Properties

1. Heat of Combustion

Heat of combustion is normally measured calorimetrically, but it can also be calculated with good accuracy from

TABLE III Properties of Coals of Different Rank

Property[a]	Lignite	Subbituminous	Bituminous	Anthracite
Moisture capacity (wt%)	40	25	10	<5
Carbon (wt%), DMMF	69	75	83	94
Hydrogen (wt%), DMMF	5.0	5.1	5.5	3.0
Oxygen (wt%), DMMF	24	19	10	2.5
Volatile matter (wt%), DMMF	53	48	38	6
Aromatic carbon (mol fraction)	0.7	0.78	0.84	0.98
Density (helium) (g/cm^3)	1.43	1.39	1.30	1.5
Grindability (Hardgrove)	48	51	61	40
Heat content (btu/lb), DMMF	11.600	12.700	14.700	15.200

[a]DMMF: dry, mineral-matter-free.

the elemental composition of the coal. It is the heat given off when a given amount of coal is completely burned.

2. Specific Heat

The specific heat records the heat necessary to cause a given temperature rise in a coal. It is rank dependent, decreasing from ~0.35 cal/g at 50% carbon to ~0.23 cal/g at 90% carbon. Above 90% carbon, it decreases quite rapidly.

3. Thermal Conductivity

The capacity of a coal to conduct heat is strongly dependent on the pore structure and moisture content. It has been widely studied. When coals soften thermally, between 400 and 600°C, the thermal conductivity increases as the pores collapse.

C. Electrical Properties

1. Dielectric Constant

The dielectric constant is a sensitive function of the amount of water present in a coal. Bituminous coals are insulators with static dielectric constants between 3 and 5. Above 88% carbon, the static dielectric constant increases very rapidly.

2. Electric Conductivity

The most interesting feature here is that anthracites are natural semiconductors. Conductivity increases sharply above 88% carbon, presumably due to increasing graphitic character.

3. Magnetic Susceptibility

Coals contain numerous unshared electrons, of the order of 10^{18} spins per gram. This gives rise to the bulk property called paramagnetic susceptibility, the magnitude of the force exerted on a sample in a strong magnetic field.

4. Reflectance

Coal samples are often geologically characterized by very accurate measurements of the reflectance of light from their components. There is a vast literature dealing with this empirical technique.

D. Density

The measurement of coal densities is complicated by the presence of pores in coals. A number of techniques for

dealing with this problem have been developed, and accurate data are available for whole coals and macerals. Density is rank dependent in a regular way. Vitrinite density decreases slowly from ~1.5 g/cm^3 at 50% carbon to a minimum of 1.27 at 88% carbon, then increases rapidly, reaching values as high as 1.6–1.8. Fusinites and micrinites are more dense than vitrinites, while liptinites are less dense. The density differences provide a good way of separating macerals.

V. PHYSICAL STRUCTURE

Coals are very complex mixtures of several macromolecular components and a number of inorganic materials. This fact makes the discussion of physical properties difficult. Each individual component can be isolated and its properties determined, but will the property of the whole coal be the sum or some weighted average of the individual component properties? Perhaps the easiest approach is the phenomenological one: simply to report the properties measured for the whole coal while realizing that these are the result of some complex addition of the properties of many different components. This approach is often perfectly satisfactory. If the heat of combustion of a coal is desired, it makes little difference what the individual component contributions to that heat are. The approach used here will be phenomenological, but the reader should know that this obscures the many differences among individual components.

A. Pore Structure

Coals are highly porous materials, many having void volumes as high as 20% of thier total volume. To describe the pore structure we need to know the size distribution of the pores, the size ranges of the pores, and the population of each range. The shape of the pores is important, as is the total surface area of the coal.

A reagent coming in contact with a coal first encounters its surface. Often, it reacts there. To understand reactivity patterns, we need to know the surface area. Even if the reagent must diffuse into the coal, the rate at which it does this is strongly influenced by the amount of surface it contacts. The size and shape of the pores control whether a molecule can enter and contact the surface and so are an important part of any description of the pore network.

Most information about pores comes from adsorption measurements. A sample of coal that has been evacuated to clean its surface is exposed to vapors of a gas (e.g., CO_2), which adsorb and cover the surface. By means of a very sensitive balance, the amount of adsorption is measured at several pressures. Thus, the amount of gas required to

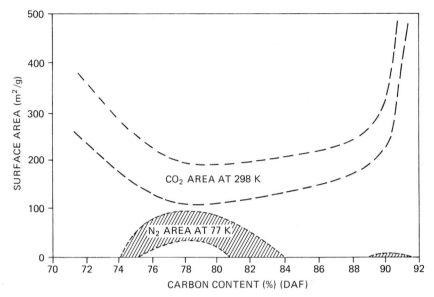

FIGURE 2 Dependence of coal surface area for dry, ash-free (DAF) coal on coal rank and technique. [Reproduced with permission from Gan, H., Nandi, S. P., and Walker, P. L. (1972). *Fuel* **51**, 272.]

cover the surface is determined. The cross-sectional area of the gas is known. This is the amount of surface that each gas molecule will cover. From these two quantities, the surface area can be calculated.

Figure 2 shows how the surface areas of coals vary with carbon content. Carbon dioxide measures the entire surface area, pore surfaces as well as external surfaces. Nitrogen cannot penetrate the pores (it does not have enough energy at these low temperatures), and so only the external surface area is measured. The surface areas are enormous, demonstrating that the coals must be very porous.

The pore volume can be estimated from some elaborate gas adsorption studies or measured directly. If the density of coal particles is measured in a vacuum and in a gas that fills the pores, different values are obtained. Density is mass over volume, and for both measurements the volume is the same. In one case (vaccum), the mass of the coal is recorded. In the other case, the mass of the coal plus the mass of the gas in the pores is recorded. If the density of the gas is known, the pore volumes are easily obtained. Some typical values are given in Table IV.

An analysis of the weight increase due to gas adsorption as the pressure of the gas increases can give the size distribution of the pores. At low pressures, pores of very small size fill by capillary condensation, while larger pores are still only surface coated. The results of such an analysis for a series of coals are given in Table IV. For bituminous coals and anthracites, most of the surface area is due to micropores, which also make the largest contribution to the pore volume. Lignites have a much more open pore structure, with the largest pores predominating. Penetration of

the pores by liquid mercury at high pressures can also provide information about the size distribution of macropores and some of the transitional pores. We shall return briefly to pore structure later to discuss micropores as packing imperfections between macromolecules.

There remains only an analysis of pore shape. This issue has been probed mainly by looking at the dependence of molecular absorption on the shape of the absorbed

TABLE IV   Illustrative Gross Pore Distributions in Coal[a]

Rank[b]	Pore volume (total), 4–30,000 Å (cm^3/g)	Pores Macro, 300–30,000 Å (cm^3/g)	Transitional, 12–300 Å (cm^3/g)	Micro, 4–12 Å (cm^3/g)
Anthracite	0.076	0.009	0.010	0.057
LV bituminous	0.052	0.014	0.000	0.038
MV bituminous	0.042	0.016	0.000	0.026
HVA bituminous	0.033	0.017	0.000	0.016
HVB bituminous	0.144	0.036	0.065	0.043
HVC bituminous	0.083	0.017	0.027	0.039
HVC bituminous	0.158	0.031	0.061	0.066
HVB bituminous	0.105	0.022	0.013	0.070
HVC bituminous	0.232	0.043	0.132	0.070
Lignite	0.114	0.088	0.004	0.022
	0.105	0.062	0.000	0.043
	0.073	0.064	0.000	0.009

[a] Reproduced with permission from Gan, H., Nandi, S. P., and Walker, P. L. (1972). *Fuel* **51**, 272.

[b] LV, Low volatile; MV, medium volatile; HVA, high volatile A; HVB, high volatile B; HVC, high volatile C.

molecules. There are fewer data here than there are on pore volume or surface area. Nevertheless, there is general agreement that the pores are often slits, oblate elipsoids, that often have constricted necks.

Coals are penetrated by an extensive network of very tiny pores and, because of this, have enormous surface areas. The smaller pores are about the same size as small molecules, so coals are molecular sieves, capable of trapping small molecules in their pores while denying access to larger molecules.

B. Macroscopic Structure

The best analogy for the macroscopic structure of coal is a fruitcake. The dough corresponds to the principal maceral, vitrinite, while the various goodies correspond to the different macerals spread throughout. The ground nuts play the role of the mineral matter. The composition of the fruitcake is not rank dependent, but rather is due to the original coal deposition and depositional environment. The macerals become more like one another as rank increases, but their proportions are not rank dependent. In fact, their relative proportions vary with both vertical and horizontal positions in a coal seam, with the vertical variation occurring over a much shorter distance scale.

To make our fruitcake a more accurate model, we must alter it somewhat. Coals tend to have banded structures, with macerals grouped in horizontal bands. A fruitcake made by a lazy cook who did not stir the batter throughly is an improved model. The nuts must also be ground to cover an enormous size range, from large chunks to micrometer-size particles. Further more, they must stick to the cake. Even extremely fine grinding in a fluid energy mill will not entirely separate the mineral matter from the organics. To separate the macerals on the basis of their different densities, the mineral matter must first be dissolved in acid. We know almost nothing about the binding forces or binding mechanism between the mineral matter and the coal.

C. Extractable Material

As much as 25% of many coals consists of small molecules that will dissolve in a favorable solvent and can thereby be removed from the insoluble portion. The amount of coal that dissolves is a function of the nature of the coal, the solvent used, and the extraction conditions. First, let us consider the solubility of coals in various solvents at their boiling points at atmospheric pressure. Typical data are contained in Table V.

Certain solvents stand out as being unusually effective, pyridine and ethylenediamine being the most noteworthy. They are not unique, however, and any solvent having a Hildebrand solubility parameter of ~11 and the capacity to serve as an effective hydrogen bond acceptor will be a

TABLE V Yields of Coal Extracts in Solvents at Their Boiling Points[a]

Solvent	Yield of coal extract (wt% of dry, ash-free coal)
n-Hexane	0.0
Water	0.0
Formamide	0.0
Acetonitrile	0.0
Nitromethane	0.0
Isopropanol	0.0
Acetic acid	0.9
Methanol	0.1
Benzene	0.1
Ethanol	0.2
Chloroform	0.35
Dioxane	1.3
Acetone	1.7
Tetrahydrofuran	8.0
Pyridine	12.5
Dimethyl sulfoxide	12.8
Dimethylformamide	15.2
Ethylenediamine	22.4
1-Methyl-2-pyrrolidone	35.0

[a] Data from Marzec, A., Juzwa, M., Betlej, K., and Sobkowiah, M. (1979). *Fuel Proc. Technol.* **2,** 35, and refer to a specific coal; they are illustrative but not representative.

good solvent. The dependence on coal rank of the amount that can be extracted by pyridine is shown in Fig. 3. The pyridine extract seems to resemble the parent coal quite closely, and the structural characteristics of the insoluble coal can be taken as present in the soluble portion, which is much easier to analyze.

Other solvents generally dissolve less of the coal, with solvents like toluene dissolving only a small percentage and hexane being ineffective. Less material can be extracted from coals by chloroform than the portion of the pyridine extract which is soluble in chloroform. If a coal and the pyridine extract of a coal are extracted with chloroform, different amounts of material go into solution. The greater percentage of the starting coal is dissolved by first extracting with pyridine. This will be explained in the next section.

The solvent power of dense, supercritical fluids is often greater than that of the corresponding liquid. Supercritical organic fluids at temperatures above 350°C are often quite good solvents for coals, and, indeed, a conversion process based on supercritical fluid extraction has received extensive development in Britain. It seems probable that some chemical decomposition is occurring that leads to the high

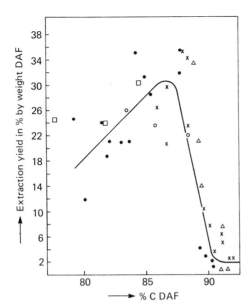

FIGURE 3 Extraction yield in pyridine at 115°C vs rank. [Reprinted with permission from Van Krevelen, D. W. (1981). "Coal," Elsevier, New York.]

extractabilities observed, often approaching 40% of the coal when toluene is the solvent.

D. Macromolecular Nature of Coals

Coals are believed to be three-dimensionally cross-linked macromolecular networks containing dissolved organic material that can be removed by extraction. This model offers the most detailed and complete explanation of the chemical and mechanical behavior of coals. It is a relatively recent model and is somewhat controversial at this writing. The insoluble portion of the coal comprises the cross-linked network, one extraordinarily large molecule linked in a three-dimensional array. This network is held together by covalent bonds and hydrogen bonds, the weak interactions that play such a large role in the association of biological molecules. The extractable portion of the coal is simply dissolved in this solid, insoluble framework. A solvent like pyridine, which forms strong hydrogen bonds, breaks the weaker hydrogen bonds within the coal structure. In the presence of pyridine, the network is less tightly held together. Molecules not tied into the network by covalent bonds can be removed. A solvent like chloroform, which does not break hydrogen bonds in coals, cannot extract all of the material that is soluble in it, because some of the soluble molecules are too large to move through the network and be removed. They do come out of the coal in pyridine, which expands the network. Once removed from the network, they readily dissolve in chloroform.

The macromolecular structure controls many important coal properties but has been relatively little studied. The native coal is a hard, brittle, viscoelastic material. This is because it is a highly cross-linked solid. The network is held together by both covalent bonds and a large number of hydrogen bonds, which tie otherwise independent parts of the network together. The network is frozen in place and is probably glassy. Diffusion rates are low because the network cannot move to allow passage of the diffusing molecule.

If the internal hydrogen bonds are destroyed, either by chemical derivatization or by complexation with a solvent that is a strong hydrogen bond acceptor, the properties of coals change enormously. The network is now quite flexible, and the coal is a rubbery solid, which is no longer brittle. Diffusion rates are enormously enhanced. All mechanical properties have been altered. All of this is due to the freeing of those portions of the network that were held to one another by the hydrogen bonds. This greatly increased freedom of motion changes the fundamental character of the solid.

VI. CHEMICAL STRUCTURE

It is impossible to represent accurately and usefully a material as complex as coal using classical chemical structures. A model large enough to show the distribution of the very many structures present in a coal would be too large and unwiedly to be useful. We shall discuss the occurrence of the various coal functional groups while ignoring their immediate environment. A phenolic hydroxyl will be listed as a phenolic hydroxyl irrespective of whether it is the only functional group on a benzene ring or whether it shares an anthracene nucleus with several other hydroxyl groups. Since the distribution of hydroxyl environments is unknown, there is no alternative.

The elements of interest are, in increasing order of structural importance, nitrogen, sulfur, oxygen, and carbon. The chemistry of hydrogen, the remaining major component, is governed by its position on one of these elements and need not be discussed separately. Trace elements will be ignored, and general mineral mater composition is covered in Section II.

A. Nitrogen

Nitrogen is a minor constituent of most coals, being less than 2% by weight. Its principal importance is that it may lead to the formation of nitrogen oxides on combustion. Its origin is believed to be amino acids and porphyrins in the plant materials from which coals were derived. Recently, X-ray methods have been developed which allowed

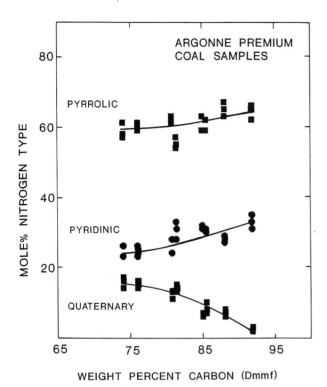

FIGURE 4 Organic nitrogen forms as a function of rank for Argonne Premium Coals. [Reprinted with permission from Kelemen, S. R., Gorbaty, M. L., and Kwiatek, P. J. (1994). *Energy Fuels* **8**, 896. Copyright American Chemical Society.]

the accurate speciation and quantification of the forms of organically bound nitrogen, including the basic pyridines and quinolines and the weakly acidic or neutral indoles and carbazoles. X-ray photoelectron spectroscopy (XPS) data showing the quantitative distributions of nitrogen types in a suite of well-preserved coals representative of many ranks (Argonne Premium Coals) are shown in Fig. 4 as a function of weight percent carbon. Pyrrolic nitrogen was found to be the most abundant form of organically bound nitrogen, followed by pyridinic, and then quaternary types. It is clear from Fig. 4 that the distributions are rank dependent, with the highest abundance of quaternary nitrogen in the lowest rank coals. The quaternary species are attributed to protonated pyridinic or basic nitrogen species associated with hydroxyl groups from carboxylic acids or phenols. The concentrations of quaternary nitrogen decrease, while pyridinic or basic nitrogen forms appear to increase correspondingly as a function of increasing rank.

B. Sulfur

The presence of sulfur in coals has great economic consequences. When coal is burned, the resulting SO_2 and SO_3 must be removed from the stack gasses at great expense or discharged to the atmosphere at great cost to the environment. The presence of sulfur in coals has been the driving force for a great deal of research directed toward its removal. Numerous processes for removing mineral matter (including inorganic sulfur compounds) from coal are in use, as are processes for removing sulfur oxides from stack gasses. Much work has been done on chemical methods for removing sulfur from coals before burning, but none are in large-scale use.

The origin of sulfur in most coals is believed to be sulfate ion, derived from seawater. During the earliest stages of coal formation, bacterial decomposition of the coal-forming plant deposits occurs. Some of these bacteria reduce sulfate to sulfide. This immediately reacts with iron to form pyrite, the principal inorganic form of sulfur in coals. It is also incorporated into the organic portion of the coal. The amount and form of sulfur in coals depend much more on the coal's depositional environment than on its age or rank. In this sense, it is largely a coal-type parameter, not a rank parameter.

Coals contain a mixture of organic and inorganic sulfur. The inorganic sulfur is chiefly pyrite (fool's gold). Exposure of coal seams to air and water results in the oxidation of the pyrite to sulfate and sulfuric acid, causing acid mine drainage from open, wet coal seams. Much of the pyrite can often be removed by grinding the coal and carrying out a physical separation, usually based on the difference in density between the organic portion of the coal and the more dense mineral matter. Coal cleaning is quite common, is often economically beneficial, and is increasing. The organic sulfur in coals, that is, the sulfur bonded to carbon, is very difficult to remove. Although many processes have been developed for this purpose, none has been commercially successful.

In the last decade, major advances have been made in speciating and quantifying forms of organically bound sulfur in coals into aliphatic (i.e., dialkylsulfides) and aromatic (i.e., thiophenes and diarylsulfides) forms. Representative results on a well-preserved suite of coals obtained from XPS and two different sulfur K-edge X-ray absorption near edge structure spectroscopy (S-XANES) analysis methods are plotted together in Fig. 5. The accuracy of the latter methods is reported to be ±10 mol%. In Fig. 5, aromatic sulfur forms are plotted against weight percent carbon. It is seen that in low-rank coals there are relatively low levels of aromatic sulfur (i.e., significant amounts of aliphatic sulfur) and that levels of aromatic sulfur increase directly as a function of increasing rank. The X-ray methods have been used to follow the chemistry of organically bound sulfur under mild oxidation in air, where it was shown that the aliphatic sulfur oxidizes in air much more rapidly than the aromatic sulfur, *ex situ*

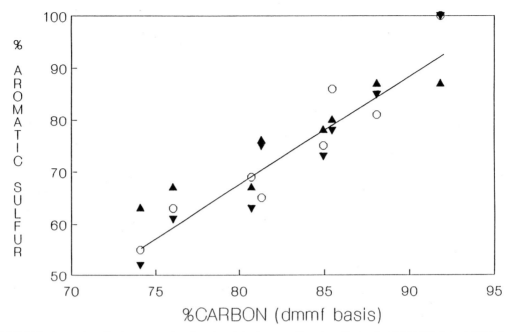

FIGURE 5 Aromatic sulfur forms as a function of rank for Argonne Premium Coals: ▲, ▼ XANES data; ○ XPS data. [Reprinted with permission from Gorbaty, M. L. (1994). *Fuel* **73,** 1819.]

and *in situ* pyrolysis, single electron transfer, and strong base conditions.

C. Oxygen

Oxygen is a major component of the plant materials that form coals. The principal precursor of coals is probably lignin, the structural material of plants. Most of the cellulose and starch is probably destroyed in the earliest stages of deposition and diagenesis, although a controversy over this continues. The oxygen content of coals decreases steadily as the rank of the coals increases. Good analytical techniques exist for the important oxygen functional groups, so their populations as a function of rank are known (Fig. 6). The detailed chemistry of the loss and transformations of these functional groups remains largely unknown.

The carboxylate (—COOH) group is the major oxygen functionality in lignites but is absent in many bituminous coals. The mechanism for its rapid loss is unknown. Its presence contributes much to the chemistry of lignites and low-rank coals. It is a weak acid and in many coals is present as the anion (—COO⁻) complexed with a metal ion. Low-rank coals are naturally occurring ion-exchange resins. Certain lignites are respectable uranium ores, having removed the uranium from ground water flowing through the seam by ion exchange.

Lignites contain water and can be considered to be hydrous gels. Removing the water causes irreversible changes in their structure. The water is an integral part of the native structure of the lignites, not just something complexed to the surface or present in pores. The nature of the water in the structure is unknown. Its presence is undoubtedly due to the presence of a large number of polar oxygen functional groups in the coal.

Hydroxyl (—OH) is present in phenols or carboxylic acids. It is present in all coals except anthracites. It is slowly lost during the coalification process. The phenols are much less acidic than the carboxylic acids and remain un-ionized in native coals.

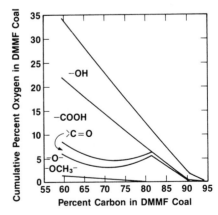

FIGURE 6 Distribution of oxygen functionality in coals: DMMF, dry, mineral-matter-free. [Reprinted with permission from Whitehurst, D. D., Mitchell, T. O., and Farcasiu, M. (1980). "Coal Liquefaction: The Chemistry and Technology of thermal Processes," Academic Press, New York.]

The methoxyl group (—OCH$_3$) is a minor component whose loss parallels that of carboxylate. The variety of groups containing the carbonyl function ($>$C$=$O) comprise a small portion of the oxygen.

In some ways, the ether functionality (—O—) is the most interesting. Its loss during coalification is not continuous. The pathway by which it is created is unknown. It is also uncertain how much of the ether oxygen is bound to aliphatic carbon and how much to aromatic carbon, but techniques capable of providing an answer to this question have just arrived on the scene.

D. Carbon

There are two principal types of carbon in coals: aliphatic, the same type of carbon that is found in kerosene and that predominates in petroleum, and aromatic, the kind of carbon found in graphite. Many techniques have been used to estimate the relative amounts of the two types of carbon in coals, but only recently has the direct measurement of the two been possible. Their variation with rank is illustrated in Fig. 7.

The carbon in coals is predominantly aromatic and is found in aromatic ring structures, examples of which are

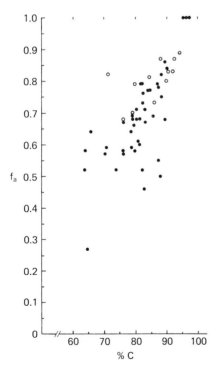

FIGURE 7 Plot of fraction of aromatic carbon (f_a) vs percent carbon for 63 coals and coal macerals. [Reprinted with permission from Wilson, M. A., Pugmire. R. J., Karas, J., Alemany, L. B., Wolfenden, W. R., Grant, D. M., and Given, P. H. (1984). *Anal. Chem.* **56,** 933.]

given below. It is very desirable to know the frequency of the occurrence of these ring systems in coals, and many techniques have been applied to this problem. Most of the techniques involve rather vigorous degradation chemistry, advanced instrumental techniques, and many assumptions.

The distribution of aromatic ring systems in coals has been estimated for the same Argonne Premium Coal samples discussed previously, based on ^{13}C NMR experiments, and these data, expressed as number of aromatic carbons per cluster (error $\approx \pm 3$), are shown in Fig. 8. For samples in this suite varying from lignite to low volatile bituminous, the results show that the average number of aromatic carbons per ring cluster ranged from between 9 (lignite) and 20 carbons (lvb), with most subbituminous and bituminous coals of this suite containing 14. This indicates an average size of 2–3 rings per cluster for most of these coals, with the lignite having perhaps an average of 1–2 and the highest rank coal having 3–4.

Our knowledge of the aliphatic structures in coals is currently inadequate. In many low-rank coals, there is a significant amount of long-chain aliphatic material. There is a debate as to whether this material is bound to the coal macromolecular structure or whether it is simply tangled up with it and thus trapped.

Much of the aliphatic carbon is present in ring systems adjacent to aromatic rings. Some is bonded to oxygen and in short chains of —CH$_2$— groups linking together aromatic ring systems. No accurate distribution of aliphatic carbon for any coal is known.

E. Maceral Chemical Structure

There are four principal maceral groups. The vitrinite macerals are most important in North American coals, comprising 50–90% of the organic material. They are derived primarily from woody plant tissue. Since they are the principal component of most coals, vitrinite maceral chemistry usually dominates the chemistry of the whole coal. Inertinite maceral family members comprise between 5 and 40% of North American coals. These materials are not at all chemically inert, as the name implies. They are usually less reactive than the other macerals, but their chemical reactivity can still be quite high. They are thought to be derived from degraded woody tissue. One member of this family, fusinite, looks like charcoal and may be the result of ancient fires at the time of the original deposition of the plant material. The liptinite maceral group makes up 5–20% of the coals in question. Its origin is plant resins, spores, and pollens, resinous and waxy materials that give rise to macerals rich in hydrogen and aliphatic structures. Terpenes and plant lipid resins give rise to the varied group of resinite macerals. From some unusual western

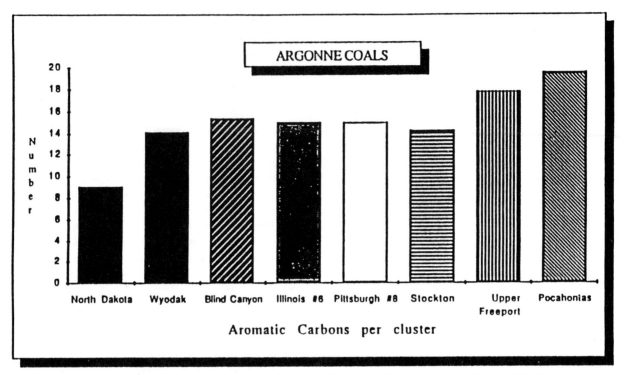

FIGURE 8 Aromatic cluster sizes in Argonne Premium Coals. [Reprinted with permission from Solum, M. S., Pugmire, R. J., Grant, D. M. (1989). *Energy Fuels* **3,** 187. Copyright American Chemical Society.]

U.S. coals, resinites have been isolated and commercially marketed.

The rank dependence of the elemental composition of the maceral groups is shown in Fig. 9. The inertinites are the most aromatic, followed by the vitrinites and then the liptinites. Their oxygen contents decrease in the order vit-

rinite > liptinite > inertinite. It is probably impossible to come up with a universally accurate reactivity scale for these materials. The conditions and structures vary too widely. *Reactivity* refers to the capacity of a substance to undergo a chemical reaction and usually refers to the speed with which the reaction occurs. The faster the reaction, the greater is the reactivity. The reactivity of different compounds changes with the chemical reaction occurring and thus must be defined with respect to a given reaction or a related set of reactions. The reactivity of macerals as substrates for direct liquefaction and their reactivity as hydrogen donors are parallel and in the order liptinite > vitrinite > inertinite > resinite. The solubility of the resinites is generally high, so their low reactivity makes little difference for most coal processing.

F. Structure Models

A number of individuals have used the structural information available to create representative coal structures. All these representations contain the groups thought to occur most frequently in coals of the rank and composition portrayed. Formulating any such structure necessarily involves making many guesses. Taken in the proper spirit, they are useful for predicting some types of chemistry and are a gauge of current knowledge. One such structure is shown in Fig. 10. It is typical of recent structure proposals

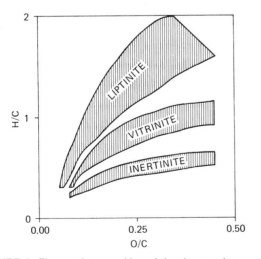

FIGURE 9 Elemental composition of the three main maceral groups. [Reprinted with permission from Winans, R. E., and Crelling, J. C. (1984). "Chemistry and characterization of coal macerals." *ACS Symp. Ser.* **252,** 1.]

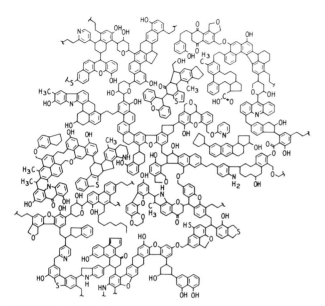

FIGURE 10 Molecular structure for a bituminous coal. [Reprinted with permission from Shinn, J. (1984). *Fuel* **63,** 1284.]

Name	Reaction	ΔH_{298}	Reaction temp. (°C)
Carbon–steam	$2C + 2H_2O \xrightarrow{\text{catalyst}} 2CO + 2H_2$	62.76	950
Water–gas shift	$CO + H_2O \xrightarrow{\text{catalyst}} CO_2 + H_2$	−9.83	325
Methanation	$CO + 3H_2 \longrightarrow CH_4 + H_2O$	−49.27	375
	$2C + 2H_2O \longrightarrow CH_4 + CO_2$	3.66	

In this scheme, hydrogen from water is added to carbon, and the oxygen of the water is rejected as carbon dioxide.

To make this chemistry a viable technology, many other steps are involved, a key one being acid gas and ammonia removal from raw syngas. Coal is not made only of carbon, hydrogen, and oxygen, but also contains sulfur and nitrogen. During gasification the sulfur is converted to hydrogen sulfide and the nitrogen to ammonia. Carbon dioxide is also produced. These gases must be removed from the raw syngas because they adversely affect the catalysts used for subsequent reactions.

Another issue in coal gasification is thermal efficiency. The thermodynamics of the reactions written above indicate that the overall $C \rightarrow CH_4$ conversion should be thermoneutral. However, in practice, the exothermic heat of the water–gas shift and methanation reactions is generated at a lower temperature than the carbon–steam reaction and therefore cannot be utilized efficiently. Recent catalytic gasification approaches are aimed at overcoming these thermal inefficiencies.

Coal gasification plants are in operation today and are used mainly for the production of hydrogen for ammonia synthesis and the production of carbon monoxide for chemical processes and for syngas feed for hydrocarbon synthesis.

VII. COAL CHEMISTRY

The basic chemical problem in coal conversion to gaseous or liquid products can be viewed as management of the hydrogen-to-carbon atomic ratio (H/C). As Fig. 11 shows, conventional liquid and gaseous fuels have atomic H/C ratios higher than that of coal. Thus, in very general terms, to convert coal to liquid or gaseous fuels one must either add hydrogen or remove carbon.

A. Gasification

Coal can be converted to a combustible gas by reaction with steam,

$$C + H_2O \longrightarrow CO + H_2,$$

reports of which can be traced back to at least 1780. The reaction is endothermic by 32 kcal/mol and requires temperatures greater than 800°C for reasonable reaction rates.

The major carbon gasification reactions are shown in Table VI, and selected relative rates are shown in Table VII. The products of the carbon–steam reaction, carbon monoxide and hydrogen, called *synthesis gas* or *syngas,* can be burned or converted to methane:

B. Liquefaction

As already discussed, coal is a solid, three-dimensional, cross-linked network. Converting it to liquids requires the breaking of covalent bonds and the removal of carbon or the addition of hydrogen. The former method of producing

TABLE VI Carbon Gasification Reactions

Name	Reaction	ΔH_{298} (kcal/mol)
Combustion	$C + O_2 \rightarrow CO_2$	−94.03
	$C + \frac{1}{2}O_2 \rightarrow CO$	−26.62
	$CO + \frac{1}{2}O_2 \rightarrow CO_2$	−67.41
Carbon–steam	$C + H_2O \rightarrow CO + H_2$	31.38
Shift	$CO + H_2O \rightarrow CO_2 + H_2$	−9.83
Boudouard	$C + CO_2 \rightarrow 2CO$	41.21
Hydrogenation	$C + 2H_2 \rightarrow CH_4$	−17.87
Methanation	$CO + 3H_2 \rightarrow CH_4 + H_2O$	−49.27

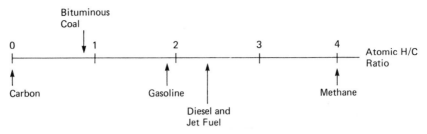

FIGURE 11 Relationship of resource to end products on the basis of the ratio of atomic hydrogen to carbon.

liquid is called *pyrolysis*, the latter *hydroliquefaction*. Coal pyrolysis, or destructive distillation, is an old technology that started on a commercial scale during the industrial revolution. Hydroliquefaction, the reaction of hydrogen with coal to make liquids, was first reported in 1869 and was developed into a commercial process in the period from 1910 to 1920.

In general, both pyrolysis and hydroliquefaction reactions begin with the same step, thermal homolytic bond cleavage to produce free radicals, effecting a molecular weight reduction of the parent macromolecule (Fig. 12). If the radicals are healed, smaller neutral molecules result, leading to liquid and gaseous products. Those radicals that are not healed recombine to form products of the same or higher molecular weight than the parent, leading eventually to a highly cross-linked carbonaceous network called *coke* (if the material passed through a plastic phase) or *char*. In pyrolysis, radicals are healed by whatever hydrogen is present in the starting coal. In hydroliquefaction, excess hydrogen is usually added as molecular hydrogen and/or as molecules (such as 1,2,3,4-tetrahydronaphthalene) that are able to donate hydrogen to the system. Thus, hydroliquefaction produces larger amounts of liquid and gaseous products than pyrolysis at the expense of additional hydrogen consumption. As shown in Table VIII, conventional pyrolysis takes place at temperatures higher than those of hydroliquefaction, but hydroliquefaction requires much higher pressures.

Many techniques have been used in hydroliquefaction. They all share the same thermal initiation step but differ in how hydrogen is provided: from molecular hydrogen, either catalytically or noncatalytically, or from organic donor molecules. Obviously, rank and type determine a particular coal's response to pyrolysis and hydroliquefaction (Table IX), and the severity of the processing conditions determines the extent of conversion and product selectivity and quality.

A large number of pyrolysis and hydroliquefaction processes have been and continue to be developed. Pyrolysis is commercial today in the sense that metallurgical coke production is a pyrolysis process. Hydroliquefaction is not commercial, because the economics today are unfavorable.

C. Hydrocarbon Synthesis

Liquids can be made from coal indirectly by first gasifying the coal to make carbon monoxide and hydrogen, followed by hydrocarbon synthesis using chemistry discovered by Fischer and Tropsch in the early 1920s. Today, commercialsize plants in South Africa use this approach for making liquids and chemicals. Variations of hydrocarbon synthesis have been developed for the production of alcohols, specifically methanol. In this context, methanation and methanol synthesis are special cases of a more general reaction. Catalyst, temperature, pressure, and reactor design (i.e., fixed or fluid bed) are all critical variables in determinig conversions and product selectivities (Table X).

$$n\,CO + (2n+1)H_2O \xrightarrow{\text{catalyst}} C_nH_{2n+2} + n\,H_2O$$

$$2n\,CO + (n+1)H_2 \xrightarrow{\text{catalyst}} C_nH_{2n+2} + n\,CO_2$$

$$n\,CO + 2n\,H_2 \xrightarrow{\text{catalyst}} C_nH_{2n} + n\,H_2O$$

$$n\,CO + 2n\,H_2 \xrightarrow{\text{catalyst}} C_nH_{2n+1}OH + (n-1)H_2O$$

TABLE VII Approximate Relative Rates of Carbon Gasification Reactions (800°C, 0.1 atm)

Reaction	Relative rate
$C + O_2 \longrightarrow CO_2$	1×10^5
$C + H_2O \longrightarrow CO + H_2$	3
$C + CO_2 \longrightarrow 2\,CO$	1
$C + 2\,H_2 \longrightarrow CH_4$	3×10^{-3}

TABLE VIII Typical Coal Liquefaction Conductions

Reaction	Temperature (°C)	Pressure (psig)
Pyrolysis	500–650	50
Hydroliquefaction	400–480	1500–2500

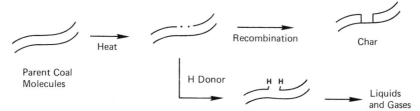

FIGURE 12 Schematic representation of coal liquefaction.

Clearly, a broad spectrum of products can be obtained, and research today is aimed at defining catalysts and conditions to improve the product selectivity. One recent advance in selectivity improvement involves gasoline synthesis over zeolite catalyst from methanol, which is made selectively from syngas. The good selectivity to gasoline is ascribed to the shape-selective zeolite ZSM-5, which controls the size of the molecules synthesized. Called MTG (methanol to gasoline), the process operates at about 400°C and 25 psig. Large-scale demonstration plants are currently under construction in New Zealand and the Federal Republic of Germany.

D. Coke Formation

The most important nonfuel use of coal is the formation of metallurgical coke. Only a few coals form commercially useful cokes, while many more agglomerate when heated. Good coking coals sell at a premium price. The supply of coals suitable for commercial coke making is sufficiently small so that blends of good and less good coking coals are used. Desirable coke properties are also obtained by blending. The coking ability of the blends can be estimated from the maceral composition and rank of the coals used in the blends. The best coking coals are bituminous coals having ~86% carbon. Other requirements for a commercial coking coal or coal blend are low sulfur and mineral matter contents and the absence of certain elements that would have a deleterious effect on the iron or steel produced.

In coke making, a long, tall, narrow (18-in.) oven is filled with the selected coal or coal blend, and it is heated very slowly from the sides. At ~350°C, the coal starts to soften and contract. When the temperature increases to between 50 and 75°C above that, the coal begins to soften and swell. Gases and small organic molecules (coal tar) are emitted as the coal decomposes to form an easily deformable plastic mass. It actually forms a viscous, structured plastic phase containing highly organized liquid crystalline regions called mesophase. With further heating, the coal polymerizes to form a solid, which on further heating forms metallurgical coke. The chemical process is extraordinarily complex.

TABLE IX Variation of Hydroliquefaction Conversion with Rank

Coal rank	Batch hydroliquefaction	
	Conversion to liquids and gas (% of coal, dry ash-free basis)	Barrels oil/ton coal
Lignite	86	3.54
High-volatile bituminous-1	87	4.68
High-volatile bituminous-2	92	4.90
Medium-volatile bituminous	44	2.10
Anthracite	9	0.44

TABLE X Hydrocarbon Synthesis Product Distributions (Reduced Iron Catalyst)

Variable	Fixed bed	Fluid bed
Temperature	220–240°C	320–340°C
H_2/CO feed ratio	1.7	3.5
Product selectivity (%)		
Gases (C_1–C_4)	22	52.7
Liquids	75.7	39
Nonacid chemicals	2.3	7.3
Acids	—	1.0
Gases (% total product)		
C_1	7.8	13.1
C_2	3.2	10.2
C_3	6.1	16.2
C_4	4.9	13.2
Liquids (as % of liquid products)		
C_3–C_4 (liquified petroleum gas)	5.6	7.7
C_5–C_{11} (gasoline)	33.4	72.3
C_{12}–C_{20} (diesel)	16.6	3.4
Waxy oils	10.3	3.0
Medium wax (mp 60–65°C)	11.8	—
Hard wax (mp 95–100°C)	18.0	—
Alcohols and ketones	4.3	12.6
Organic acids	—	1.0
Paraffin/olefin ratio	~1:1	~0.25:1

The initial softening is probably due to a mix of physical and chemical changes in the coal. At the temperature where swelling begins, decomposition is occurring and the coal structure is breaking down. Heat-induced decomposition continues and the three-dimensionally linked structure is destroyed, producing a variety of mobile smaller molecules that are oriented by the large planar aromatic systems they contain. They tend to form parallel stacks. Eventually, these repolymerize, forming coke.

SEE ALSO THE FOLLOWING ARTICLES

COAL GEOLOGY • COAL PREPARATION • GEOCHEMISTRY, ORGANIC • FOSSIL FUEL POWER STATIONS—COAL UTILIZATION • MINING ENGINEERING • POLLUTION, AIR

BIBLIOGRAPHY

Berkowitz, N. (1979). "As Introduction to Coal Technology," Academic Press, New York.
Elliott, M. A., ed. (1981). "Chemistry of Coal Utilization," 2nd suppl. vol., Wiley (Interscience), New York.
Francis, W. (1961). "Coal: Its Formation and Composition," Arnold, London.
George, G. N., Gorbaty, M. L., Kelemen, S. R., and Sansone, M. (1991). "Direct determination and quantification of sulfur forms in coals from the Argonne Premium Sample Program," *Energy Fuels* **5,** 93–97.
Schobert, H. H., Bartle, K. D., and Lynch, L. J. (1991). "Coal Science II," ACS Symposium Series 461, American Chemical Society, Washington, DC.
Gorbaty, M. L. (1994). "Prominent frontiers of coal science: Past, present and future," *Fuel* **73,** 1819.
Van Krevelen, D. W. (1961). "Coal," Elsevier, Amsterdam.

Coastal Geology

Richard A. Davis, Jr.

University of South Florida

GLOSSARY

Distributary Channels on a delta that distribute the water to the receiving basin; the opposite of tributary.

Holocene Current period of geologic time that is defined on the basis of the age of what paleontologists consider to be a fossil—10,000 years. It is arbitrary and does not fit well with the so-called Holocene rise in sea level, which actually began when the glaciers started to melt about 18,000–20,000 years ago.

Overbank Area on a delta or stream system that is out of the channel. This includes natural levees, crevasse splays, and the interdistributary areas. They are all over-bank environments that receive overbank sediments as the result of flooding.

Ridge and runnel Small, shore-parallel, and generally intertidal bar and trough immediately adjacent to the beach. The ridge (bar) is formed during waning storm conditions and then migrates landward over the runnel during calm conditions. It eventually welds onto the beach.

Tidal prism Water budget for a coastal area, typically a tidal inlet. The prism is the volume of water that is exchanged during a single tidal cycle. It is equal to the tidal range times the area of the coastal bay being served by the inlet.

Washover fan Thin, fan-shaped accumulation of sediment (usually sand) that is deposited landward of the beach and dunes as the result of storm activity. The storm surge, when combined with large waves, carries sediment from the beach and nearshore area and deposits sand on the marsh or vegetated flats. This is very common on barrier islands.

Wave-cut terrace (wave-cut bench) A rather smooth and essentially horizontal surface cut into the rocky coast as the result of prolonged wave action.

FOR PURPOSES OF this article, the coast will include the transition zone between the terrestrial environments and the open marine environments. Included are a complex set of sedimentary environments and their related processes and sediment responses. The major sedimentary environments that are considered in this article are deltas, coastal bays including estuaries and lagoons, barrier island systems, and rocky coasts. Each of these contains numerous and distinct environments within it, and each may be associated with any of the others depending upon the local coastal setting.

The conditions which tend to have major effects on the coastal environments, and in some cases control them, are tectonic and geomorphic setting, climate including waves, tidal regime, sediment flux, and hydrodynamics.

I. GENERAL COASTAL SETTING

The coast is the complicated transition zone between the terrestrial or land regime which includes fluvial (river), desert, lacustrine (lake), and glacial environments, and the marine regime which includes the continental margin, reefs, and deep ocean environments. Because the coast receives influence from both of these regimes and because it does so in differing relative amounts, there is great variety in the nature of coasts in general and especially in the specific coastal environments.

A. Major Environments and Their Definitions

In this article, the coast will be subdivided into four major categories, each of which contains numerous specific environments. These are deltas, coastal bays, barrier island complexes, and rocky coasts. It should be noted that even these major systems may occur together on a particular coast.

A **delta** is an accumulation of sediment deposited by a stream at its mouth, which protrudes into the basin of deposition. Deltas come in all sizes and shapes but have only a few major factors that control their presence and characteristics. First, it is necessary to have a place where the sediment can accumulate, so deltas form in shallow water areas of the continental shelf. Second, there must be more sediment supplied than can be carried away by marine processes such as tides, waves, and wave-generated currents. The overall size and morphology of the delta is controlled by the sediment provided by the stream and its interaction with tidal and wave processes.

Coastal bays include a wide variety of types of embayments in the general coastal trend. They may originate in a variety of ways, including tectonic activity and related crustal movement, drowning of drainage basins during rising sea level, excavation by glaciers, or by formation of a sandy barrier seaward of the mainland shore.

Barrier island complexes are dominated by the barrier island, a shore-parallel accumulation of sand that includes

the beach and adjacent surf zone. They also include coastal dunes and sandy back-island flats and wetlands that grade into the adjacent coastal bay. Inlets typically interrupt barrier islands and serve as pathways for tidal currents and migrating organisms as well as for boats.

The **rocky coast** is, as the name implies, one of irregular and typically high relief which is characterized by bedrock of almost any type. Rocky coasts are dominated by erosion, whereas the above-mentioned types are depositional coasts. The dominant process along rocky coasts is waves.

II. CLASSIFICATION OF COASTS

There are many ways to classify coasts, including morphology, dominant features, and dominant processes. One of the most important aspects of any coast, regardless of its type, is whether it is rising, sinking, or remaining stationary relative to sea level. Another important factor is the tectonic setting and history that tends to dictate the regional character of the coast. Both of these factors are related to plate tectonics that control the broad-scale coastal morphology.

A. Tectonic Classification

The morphologic classification of coasts by Inman and Nordstrom of the Scripps Institution of Oceanography offers the best and simplest organization of coastal types that is related to plate tectonics. In this classification there are three major categories: (1) leading edge or collision coasts, (2) trailing edge coasts, and (3) marginal sea coasts. A **collision coast** is formed where two lithospheric crustal plates converge. This may involve a continental margin such as on the west coasts of North and South America, or it may be along island arcs such as the Philippines or the Aleutian Islands. In either situation the coast is one of high relief, lacks a shallow shelf to accumulate sediments, and has deep water near the shore that tends to result in high wave energy. Consequently, such coasts are typically erosional in nature.

Both of the other types tend to be dominated by depositional coasts. **Trailing edge coasts** face a spreading zone. These may result from the development of new spreading centers such as on the Red Sea or the Gulf of California, they may be on continental margins where the opposite margin is also a trailing edge such as on both oceanic coasts of Africa, or they may be developed where one side is trailing and the other is a collision coast such as in North and South America.

Marginal sea coasts are similar to trailing edge coasts in their morphology but differ in their tectonic setting.

These coasts develop along marginal seas that face an island arc. The coasts of eastern Asia, such as Korea, China, and Vietnam, are modern examples. The Gulf of Mexico is also a marginal sea that is bounded on the south by a plate boundary and the volcanic province of the Caribbean Sea.

B. Process Classification

Whereas the tectonic classification described above gives a good overall approach to the broad- and large-scale nature of coasts, it does not include the specific coastal morphologies that respond to coastal processes that impact the coasts. For practical purposes, it is appropriate to consider waves and tides as the dominant processes operating on all coasts regardless of tectonic setting. It is also appropriate to consider that all coasts are dominated by deposition or by erosion over a geologically significant period of time.

Rocky coasts are erosional and wave dominated, except in some embayed settings where tides may dominate. Wave-dominated rocky coasts are typical of collision coasts. Under this scheme, depositional coasts can be either wave dominated or tide dominated. Coasts with smooth shorelines such as long barrier islands tend to be wave dominated, and those with irregular shorelines and many embayments are tide dominated.

A classification of coasts based on tidal range (Davies, 1964) produced a good attempt to categorize depositional coasts. Three arbitrary categories were designated: (1) microtidal coasts (<2 m tidal range), (2) mesotidal coasts (2–4 m), and (3) macrotidal coasts (>4 m). This was further considered by Hayes (1975, 1979), who expanded the classification and related it to wave- and tide-dominated coastal morphology (Fig. 1).

Initial application of this classification was to consider microtidal coasts to be wave dominated and macrotidal coasts to be tide dominated with mesotidal coasts a result of mixed tidal and wave energies. This has been demonstrated to be an incorrect approach to classification of coasts because the primary factor is the *relative* impact of waves and tides, not the absolute values of either one.

Tide-dominated coasts include the Florida coast on the northeast corner of the Gulf of Mexico and on Andros Island, Bahamas, where tidal ranges are less than 1 m, as well as along the German Bight of the North Sea, where ranges are up to 4 m (Fig. 2), and in the Minas Basin of the Bay of Fundy, where tidal range exceeds 10 m. In these examples, the morphology is similar with tidal currents being the dominant process and the morphology being characterized by features that are perpendicular to the overall trend of the coast.

Wave-dominated coasts follow the same pattern in that they may develop in areas with both microtidal and macrotidal ranges. For example, the barrier island coast

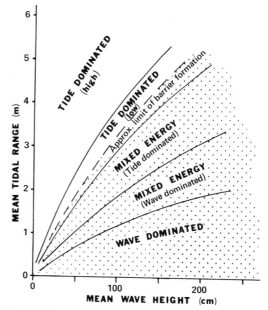

FIGURE 1 Diagram showing relationships between wave height and tidal range with various examples. [From Davis, and Hayes (1984).]

of Texas is a classic example of a microtidal, wave-dominated coast, but there is also a wave-dominated coast within the Bay of Fundy where tidal range is about 9 m (Fig. 3). The Oregon and Washington coasts are also macrotidal with spring ranges of 4 m, but they display distinctly wave-dominated morphologies. All of these examples display smooth, arcuate shorelines typical of wave-dominated coasts.

III. HOLOCENE SEA LEVEL RISE

The vast majority of the present coast of the world was shaped during the past few thousand years. The sea level

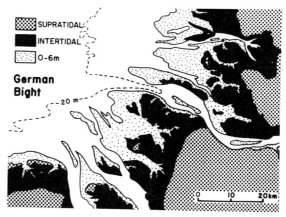

FIGURE 2 Generalized map of the tide-dominated coast along the German Bight where the tidal range is up to 4 m. [Modified from Reineck and Singh (1980).]

FIGURE 3 Photographs of the coast at Cape Split in the Bay of Fundy, Canada, where wave-dominated conditions prevail although the tidal range is 9 m.

rose quite rapidly for the first 10,000 years or so after the melting of glaciers began. This was followed by a period of slow rise. It is quite possible that, because of the geologically recent and marked changes in sea level, the present coastal configuration is atypical of most of geologic time.

Glaciation has characterized much of the higher latitude areas of the Earth during the past few million years. As the glaciers develop and increase in size, the oceans shrink because of the water volume necessary to form the ice sheets. The result is a drastic lowering of the sea level on a global scale. During periods of glacial melting, or interglacial periods as they are known, the sea level rises due to the influx of meltwater and the expansion of the volume in the oceans caused by rising temperatures of global warming.

We are currently in an interglacial period. The maximum advance of the last glacial period, the Wisconsinan, ended about 18,000 years ago. The global rise in sea level from that time to the present has been great: more than 120 m in most places.

This means that during periods of glacial advance and low sea level, the shoreline was at or near the edge of the present continental shelf. It seems very likely that under those circumstances the coast looked much different at most locations than the one we see now. Similar conditions probably existed during each of the several major advances and retreats of the ice sheets during the Pleistocene epoch.

A. Holocene Sea Level Rise

Much of the period of time since the Wisconsinan advance, during which the sea level has been rising, is referred to as the Holocene. It is defined as the last 10,000 years. The sea level has actually been rising for about 18,000 years and is continuing to do so as glaciers melt. The rates of sea level rise are interpreted to have been quite rapid for about the first 10,000–11,000 years, about 10 mm/year (Fig. 4). At about 7000 years before present, the sea level was about 10 m below the present position and the rate of rise slowed to about 2.0–2.5 mm/year in many places. What happened during the last 3000 years is still not agreed upon. Three scenarios have been proposed by various authors: (1) the sea level reached its present position at 3000 YBP and has remained essentially constant, (2) the sea level has moved above and below its present position about 1 m or so over this period, and (3) the sea level has risen gradually at a rate of about 1 mm/year over this period.

During the past century, however, there has been a recorded rate of sea level rise of about 2.5 mm/year (Hicks, 1981). It is generally believed that this increase is due, in part, to the impact of humans on the Earth, especially in the form of the greenhouse effect that is warming the global climate.

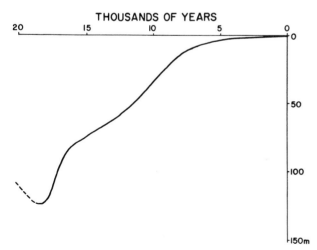

THOUSANDS OF YEARS

FIGURE 4 Generalized sea level curve since the melting of glaciers began showing the rapid rise during the early portion of this time and the marked decrease during the past few thousand years. (Modified from various sources.)

B. Influence of Sea Level Rise on the Continental Margin

The change in sea level and the related shoreline migration that has taken place during the past 18,000 years or so has had a significant impact on the continental margin as a whole and particularly on the shelf. At the time that the sea level was at its lowest, the shoreline was at or near the present shelf-slope break, meaning that there was essentially no continental shelf. As a result, rivers were emptying onto rather steep gradients and moving down the slope in the form of turbidity currents. Deltas were small or nonexistent due to the lack of shallow shelf areas on which they could develop.

During the initial rapid rise in sea level, the shoreline migrated so fast that there was little time for coastal environments to develop, especially wave-dominated ones. The shoreline tended to be quite irregular as rivers were drowned by the rapidly rising water. The broadly irregular coast caused tidal processes to dominate for nearly 10,000 years. Coastal sedimentary deposits that accumulated during the lowering of sea level associated with the advance of the Wisconsinan ice sheet were simply inundated during this rise with little or no reworking by coastal processes. As a result, there is much *relict* sediment on the present continental shelf; about two-thirds of the area is covered by such deposits.

When the sea level slowed about 7000 years ago, the reworking of coastal deposits and the influx of land-derived sediment to the coast and adjacent shelf began to have a major influence. It was at this time that older sediments were thoroughly reworked and Holocene sediments were deposited on the inner continental shelf under wave-dominated conditions.

C. Influence of Sea Level Rise on Coastal Development

Virtually the entire present coast is geologically quite young, probably less than 5000 years in most places. This is the result of the situation described above. Although coastal environments and their sediment bodies can develop in short periods of time, they do require relative stability of the shoreline in order to accomplish this development.

For example, if the shoreline is moving landward rather rapidly as the sea level rises, it is essentially impossible to develop even modest-sized river deltas. First, the rise in sea level continually reduces relief in the drainage basin and thereby tends to reduce the rate of erosion that, in turn, reduces the sediment discharge to the river mouth. Second, and most important, it takes at least a modest amount of time to accumulate a quantity of sediment large enough to develop a delta. A rapid rise in sea level is counterproductive to delta development on both counts.

Barrier islands also require some stability of the shoreline in order for wave and wave-generated currents to build a significant sand accumulation that will develop a barrier island. It takes time for waves to not only remove the unconsolidated sediment that may have accumulated, but also to erode the bedrock present along erosional coasts.

Coastal bays are the least affected by the sea level rise of all the major coastal environments. As the sea level rises and drowns valleys, the coastal bay changes its size and shape to accommodate the water level. The rate of rise does have an effect on the ability of the embayment to be modified by coastal processes in that the rapid rise permits little or no change and slow rise does allow for modification.

IV. DELTAS

A **river delta** is a large accumulation of sediment at the mouth of a stream where it discharges into a standing body of water. The term "delta" comes from the third letter of the Greek alphabet and was first applied to the Nile Delta which has a generally triangular shape although most deltas do not exhibit this outline. Regardless of their size, shape, and geographic location, river deltas have several factors in common. These include their general hydrology, major sedimentary environments, and the fact that all are the result of some combination of stream and

coastal processes of the basin into which the sediment is deposited.

A. General Morphodynamics of River Deltas

River deltas are the result of the interaction between the discharge of water and sediment from the stream, and the waves, tides, and currents associated with both of these processes. Large river deltas are generally restricted to trailing edge or marginal sea coasts because they provide large drainage basins and also an appropriate shallow shelf for accumulation of deltaic sediment. The global map of major deltas shows this relationship well with very few deltas located on leading edge coasts (Fig. 5).

Within a given delta there may be much variation in the balance between the major processes acting upon the delta. Most important is the control of discharge exerted by climate. During the wet season it is quite high, perhaps even causing major flood events, whereas during the dry season it may be considerably less. Similar effects can be imparted by coastal processes, especially waves. The wave climate of a given coastal area is also strongly influenced by seasonal variation. The high-energy time of year, typically winter, brings about large waves that cause much erosion or at least reworking of delta sediments. Conversely, the low-energy time of year may permit much progradation or building of the delta.

Short-term cycles also influence deltas, with tides being the primary one. As the tide floods, river discharge is greatly reduced or, in some cases, may even be negative

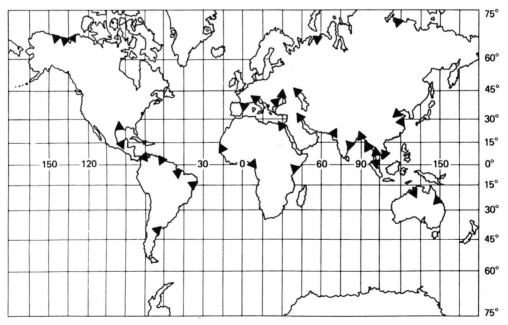

FIGURE 5 Map of the world showing the distribution of major deltas. Note that nearly all are associated with trailing edge continental margins. [From Wright (1985)].

due to the tidal influence. This is reversed during ebb tides. Regular tidal cycles of a diurnal or semi-diurnal nature tend to balance out, with the overall impact being dominated by the discharge or wave energy.

Deltas typically have both constructive and destructive phases. This may occur over a period of time such as seasonal or long-term changes, or it may take place geographically throughout the delta with a particular area experiencing progradation while another area is being eroded. The latter situation may also vary through time with the Mississippi Delta being a good example. Over the past few thousand years, there have been several shifts in the locus of deposition of the Mississippi Delta. The active lobes over this period have shifted over distances of more than 100 km (Fig. 6) during this time. The present active lobe is quite young, with most of its accumulation being since the time of the Civil War.

Recently, human activities have had a severe impact on some deltas, especially as the result of impounding and diverting the discharge. Two good examples are the Colorado River in the southwestern United States and the Nile River in Egypt. Several dams have been constructed along the Colorado River, and much of its discharge has been diverted to desert areas and to the megalopolis in southern California. As a consequence, there is little sediment discharge being carried to the mouth of the river in the Gulf of California. The result is that the delta is experiencing destruction, largely by tidal currents.

The Nile Delta is being impacted by the decrease in discharge caused by the construction of the Aswan Dam in the desert of Egypt. As a consequence, little sediment is being carried to the delta, and wave action in the Mediterranean Sea is eroding the delta.

Considerable land loss is being experienced by the Mississippi Delta due to both anthropogenic and natural factors. The combination of sea level rise, compaction of sediments, withdrawl of fluids by the petroleum industry, and damming of much of the sediment in the drainage system have resulted in a sea level rise of 1 cm/year, more than four times the global average. This causes coastal wetlands to be drowned and eroded at a rate of about 60 acres/week.

1. Major Elements of a Delta

There are three major parts of river deltas: (1) the delta plain, (2) the delta front, and (3) the prodelta. Each has certain features and environments that distinguish it from

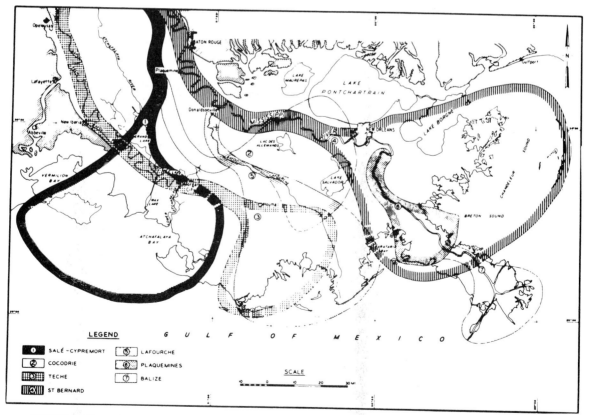

FIGURE 6 Skematic map of the Mississippi Delta showing the numerous lobes that have been the primary focus of sediment accumulation during the past few thousand years. The present active lobe is only a few hundred years old. [From Kolb and Van Lopik (1966).]

the others, and each is developed to varying proportions in individual deltas.

The **delta plain** is the landward portion and is the most complex of the three major parts of a delta. It is partially subaerial and partially subaqueous, truly the transitional part of the delta with considerable influence from the fluvial environment. Included are distributaries with their point bar deposits, natural levees, and crevasse splays that are essentially fluvial elements. The interdistributary areas consist of bays, tidal flats, and marshes which are heavily influenced by marine factors but also receive sediment from overbank deposition during high-discharge events. The delta plain is generally a rather widespread portion of the delta with accumulations that are typically thin, except for point bar sand deposits that can be as thick as the channels are deep. Sediments of the delta plain are dominated by mud and very fine sand, except for the channel deposits that may be medium to coarse sand.

The **delta front** environment is dominated by sand and tends to be the most geographically restricted of the three major delta elements. It may be comprised of distal bars, distributary-mouth bars, and tidal ridges, depending upon the dominant coastal processes that influence the delta. The sand bodies that occupy the delta front typically display well-developed cross-stratification, scour- and-fill structures, plane beds, and few bioturbation structures due to the relatively high energy that this zone experiences.

The **prodelta** generally comprises most of the volume of a delta, especially along low-energy coasts. This portion of the delta is dominated by mud. The sediments of the prodelta become finer toward the basin and grade into shelf mud of the open marine environment. Bioturbation is more prominent than in the delta front, and gravity structures such as convolute bedding, slump structures, and soft-sediment faulting are common.

B. Classification of Deltas

Deltas are generally classified by the dominant process that controls their morphology and sediment content. The major processes are river discharge, tides, and waves. Needless to say, there is a transition from one extreme to the other, and any combination of the three processes may result in a characteristic delta morphology. River-dominated deltas display a morphology that generally has several distributaries and a very irregular, digitate outline. Wave-dominated deltas tend to have a relatively small delta plain and a proportionally large delta front. Wave action reworks sediment by removing the mud and developing extensive sand bodies, typically with an orientation that is essentially parallel to the overall coastal trend. The tide-dominated delta is actually quite similar to a tide-dominated estuary (see Section V.B). The volume of sedi-

ment that accumulates is limited, and the dominat features are tidal sand ridges with orientation essentially perpendicular to the coastal trend, reflecting the orientation of tidal currents.

The most convenient approach to delta classification is that of William Galloway of The University of Texas who used a triangular diagram with each corner representing one of the three major processes (Fig. 7). By evaluating the relative roles of these major processes, it is possible to place any river delta into this classification scheme.

The Mississippi Delta with its digitate morphology and its protrusion far into the Gulf of Mexico is a classic river-dominated delta (Fig. 8). The Ebro Delta on the Spanish coast of the Mediterranean Sea exhibits a combination of riverine and wave influence with little tidal effects. The Nile and Orinoco Deltas also display a morphology that is the result of both riverine and wave influence.

Wave-dominated deltas may have a morphology that is the result of the direct impact of the waves such as the Sao Francisco Delta. This example shows a single channel with numerous wave-formed sandy beach ridges giving an overall cuspate configuration (Fig. 9). Another deltas of this type where waves generally fall into the wave-dominated category but produce a different morphology. The Senegal Delta in Africa is a good example (Fig. 10). The dominating longshore current has developed a spit across the mouth of the river, resulting in a marked displacement of the river mouth.

Tide-dominated deltas tend to have a somewhat funnel-shaped estuarine configuration. Deposition is dominated by elongated sand bodies that may be isolated, as in the Ord Delta on the northwest coast of Australia (Fig. 11), or they may be a part of a large, subparallel distributary complex such as in the Ganges-Brahmaputra Delta of Asia (Fig. 12).

C. River-Dominated Deltas

River-dominated deltas exhibit the most variety of sedimentary environments of the three major categories of deltas. The delta plain is well developed, the delta front is generally influenced primarily by wave action, and the prodelta is thick and extensive.

1. Delta Plain

The delta plain is generally extensive and has a large subaerial component, including natural levees, crevasse splays, and interdistributary marshes. All of these environments are composed of overbank deposits that accumulate as the result of flood events. The natural levees are linear ridges of sediment on the edge of the channel. They may rise as much as a few meters above the surrounding

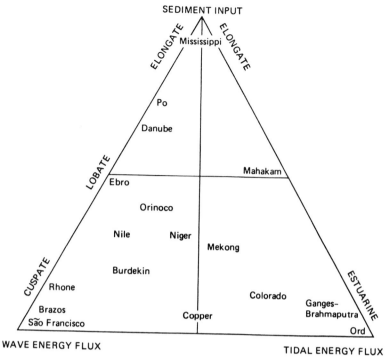

FIGURE 7 Ternary diagram of delta classification using the three major processes that control deltaic morphology. Various examples are placed in their respective places. [Modified from Galloway, W. (1975).]

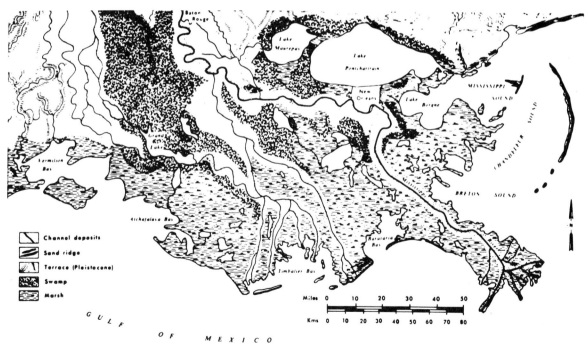

FIGURE 8 Skematic map of the Mississippi Delta, a classic example of a river-dominated delta. Observe the irregular outline of the modern lobe and also the relative smooth coast in areas that are currently experiencing destruction or erosion. [From Wright and Coleman (1973).]

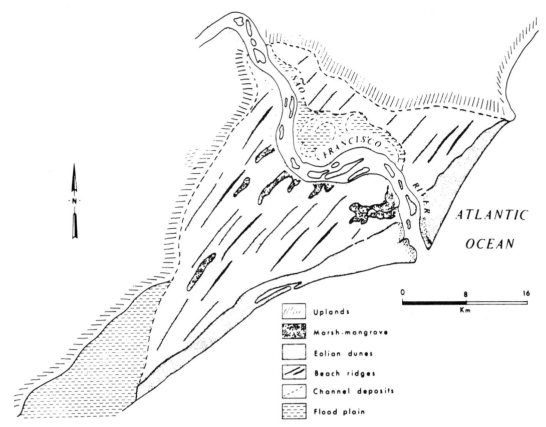

FIGURE 9 General morphology of the Sao Francisco Delta in Brazil, an excellent example of a wave-dominated delta. Note the lack of distributaries and the well-developed beach ridge complex. [From Wright and Coleman (1973).]

area in very large systems such as the Mississippi, but may be only tens of centimeters high in smaller deltas. Natural levee sequences fine upward and outward with sediments ranging from fine sand to mud. Planar bedding and scattered ripple cross-stratification are present.

Crevasse splays are fan-shaped sediment accumulations that are deposited as the result of a breaching of the natural levee. When this happens, a pulse of sediment-laden water is carried onto the interdistributary area and accumulates in a thin, fining-upward and -outward layer of fine sand and mud. These features are typically single-event accumulations, and large ones can be related to specific flood events (Fig. 13). They may cover over 100 km² and be over 1 m thick.

The marshes are formed by interdistributary sediments accumulating up to a level at or near neap high tide. It is at this elevation that *Spartina*, the common marsh grass, can become established. As flooding of distributaries, combined with tidal currents, carries fine sediment into these areas, the marshes become established. As the delta matures, the marsh becomes more widespread and eventually may occupy all of the interdistributary area of the delta plain.

The subaqueous delta plain is largely distributary channel deposits, which are dominated by point bars, and interdistributary bays. The channel deposits are thick, discontinuous sand bodies that are quite similar to their fluvial counterparts in the terrestrial regime. They contain the same type of fining-upward sequence with its related sedimentary structures.

Interdistributary bays are shallow, low-energy embayments between major distributaries. They receive fine sediment from flooding of the channels and from tidal transport. They are the subaqueous equivalent of the interdistributary marshes. As the delta matures and sediments accumulate, this environment gives way to the marshes. Some deltas also have a significant tidal flat environment which occupies the niche between the bays and marshes and which receives sediments in the same way.

2. Delta Front

The delta front on river-dominated deltas is dominated by distributary-mouth bars and distal bars. Both are influenced by waves. The distributary-mouth bars are the result of the interaction of the channel discharge and the wave

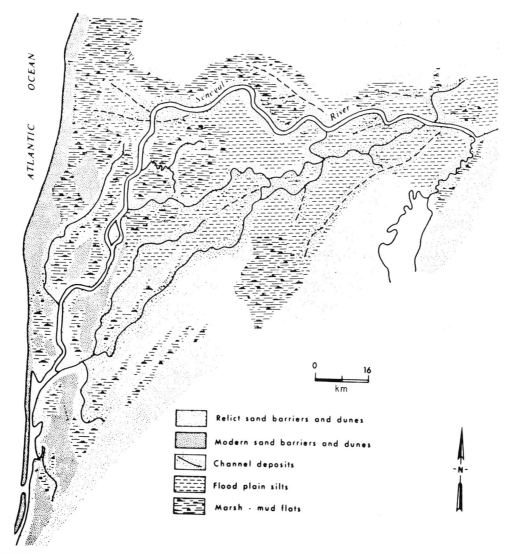

FIGURE 10 Map of the Senegal Delta, another type of wave-dominated delta, on the west coast of Africa. Here, the longshore drift diverts the main distributaries, and the outer margin is dominated by sandy beaches and beach ridges. [From Wright and Coleman (1973).]

climate of the basin. These bars may be discontinuous and somewhat elongated in the direction of the channel discharge, as on the Mississippi Delta, or they may be more shore parallel and continuous, as on the Amazon Delta.

The distal bars represent a transition between the sand-dominated delta front and the mud-dominated prodelta. They are well developed in river-dominated deltas and may be up to 25 m thick on the Mississippi Delta.

3. Prodelta

The prodelta is quite pronounced in river-dominated deltas. The fine sediment settles from suspension into the distal parts of the delta and accumulates a thick mud se-

quence. Because of the variation in sediment and water discharge, the rate of sedimentation may vary greatly for an individual delta. The rate also decreases in the seaward direction. In basins of low wave energy such as the Gulf of Mexico, the mud accumulates rapidly and causes slumping and other gravity-related phenomena.

D. Wave-Dominated Deltas

In general, wave-dominated deltas tend to be relatively small and have few distributaries and a small delta plain with an extensive delta front. The morphology is cuspate and smooth with only modest protuberance into the basin of accumulation.

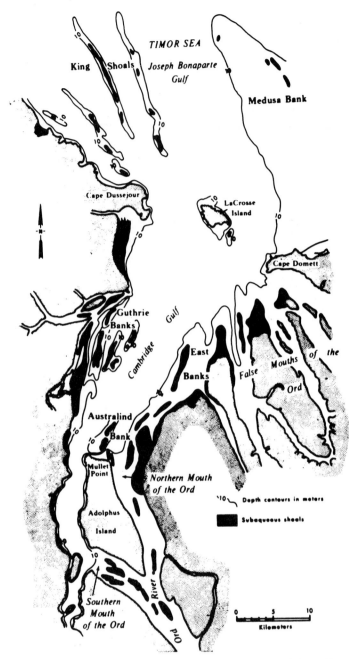

FIGURE 11 General physiography of the Ord Delta on the northwest coast of Australia, an example of a tide-dominated delta. Note that the linear sand ridges are oriented perpendicular to the coast. [From Wright, et al. (1975).]

1. Delta Plain

This environment is modest to small relative to the other primary elements. In deltas where waves are reworking sediment and developing cuspate ridges, such as on the Sao Francisco Delta, there is no distributary system and the subaerial part of the delta is essentially a complex of

beach ridges with marshy swales between them. This may be considered as a delta plain, but because it is a result of wave action it is perhaps more appropriately a delta front. Overbanking is not common, but it may carry some sediment to the swale areas between the ridges.

The straight type of wave-dominated delta where long-shore currents prevail contains a larger delta plain. It also

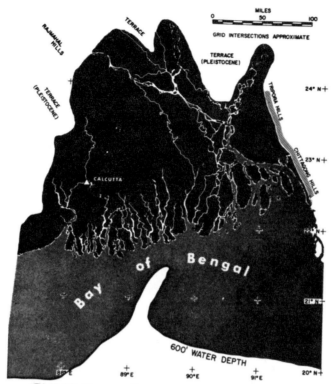

Ganges—Brahmaputra

COUNTRY : EAST PAKISTAN AND INDIA
SUBAERIAL AREA : 35,000 SQ. MILES
DRAINAGE AREA : 769,000 SQ. MILES
AVERAGE WATER DISCHARGE : 1,360,000 CU. FT/SEC.
ANNUAL SEDIMENT DISCHARGE : 700,000,000 TONS

FIGURE 12 Skematic map of the Ganges-Brahmaputra Delta in Pakistan, an example of a tide-dominated delta which has much more runoff than the Ord and therefore has a rather well-developed distributary system. Observe that the distributaries are perpendicular to the coast, a trait of tide-dominated deltas. [Modified from Shirley (1966).]

has little in the way of a distributary network, but there may be extensive flood plain or overbank accumulations in the form of marshes or swamps. Sand ridges or spits tend to be limited to the seaward part of the subaerial delta and are best considered as a delta front where wave domination occurs. The Senegal Delta of Africa is a good example of this situation.

rents. The coastal beach/dune ridge may reach over 10 m in thickness. Nearshore sand bars are also common and develop much like they do along nondeltaic coasts. In fact, the outer coast of the wave-dominated delta is much like the nearshore-beach-dune complex of most barrier island coasts.

2. Delta Front

The delta front of a wave-dominated delta is commonly the most prominent of the three major elements. Waves rework sediment into both subaerial and subaqueous sand bodies with a dominantly shore-parallel orientation. This occurs as the result of both direct wave attack and longshore cur-

3. Prodelta

The prodelta of a wave-dominated delta is generally similar to that of the river-dominated delta but much more restricted in area and thickness. It contains sand stringers and much bioturbation, especially in its distal areas. The sand layers represent intense events during which wave action extends to fairly deep water, reworking the prodelta area.

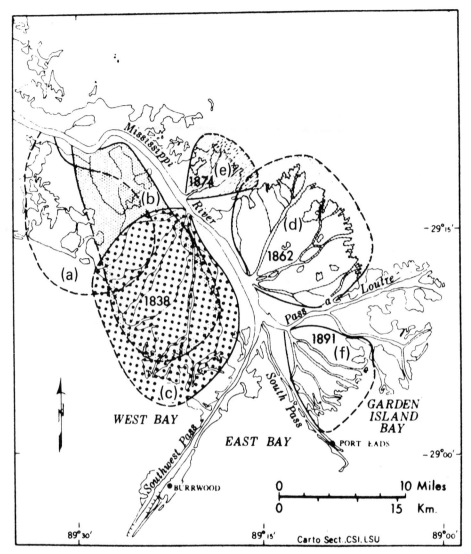

FIGURE 13 Map of large crevasse splays that have developed on the modern active lobe of the Mississippi Delta during historical time. These splays deposits result from extreme discharge such as that which would be associated with spring flooding. [From Coleman (1976).]

E. Tide-Dominated Deltas

Deltas that are dominated by tides tend to occur not only along coasts that have high tidal ranges, but also at the mouths of streams that have either a low discharge or great seasonal variability in discharge. Among the best examples are the Ord Delta in northwest Australia and the Ganges-Brahmaputra Delta in Bangladesh. Both experience high tidal ranges, and both are affected by monsoon climates that bring tremendous discharge in a short period of time with little discharge over the remainder of the year.

1. Delta Plain

In tide-dominated deltas, the delta plain is primarily composed of intertidal environments consisting mostly of tidal flats with some mangrove swamps or marshes, depending

upon latitude. Because of the generally high tidal range, these tidal flats may be quite extensive; in some places, they are a few kilometers wide. Natural levees may be well developed because of the high discharge events and the accompanying overbank conditions. The interdistributary bay area tends to be occupied by shallow, isolated basins that receive sediments during flooding, but that are typically not influenced by regular tidal activity. Channel deposits tend to display bidirectional cross-stratification because of the great influence of tides, especially during low discharge periods.

2. Delta Front

The delta front of a tide-dominated delta has a broad extent and contrasts markedly from the rather elongated and shore-parallel delta front of both the river- and wave-dominated deltas. Distributary-mouth bars may be present, but tidal ridges are the dominant sand body in this type of delta. These ridges are elongated perpendicular to the trend of the coast and are aligned with the tidal current direction. They may be several meters high and kilometers in length (Fig. 11).

3. Prodelta

The prodelta of tide-dominated deltas differs little from that of the other delta types. It is dominated by mud with local sand stringers, there is considerable influence from marine benthic organisms, and it grades into shelf mud. The prodelta sequence tends to be somewhat thinner in the tide-dominated deltas due to the tendency for strong tidal currents to carry fines away from the delta.

V. COASTAL BAYS

Coastal bays include a broad spectrum of embayments along the coastal regime with differing origins, physiography, hydrology, and dominant process. Coastal bays occur in all geologic, geomorphic, and climatic provinces of the globe. The Holocene sea level rise has resulted in what is probably an anomalously high number of coastal bays now as compared to the average situation during geologic time.

A. Types of Coastal Bays and Their Formation

Although coastal bays come in all sizes, shapes, and hydrologic styles, they can be organized into a small number of basic types: estuaries, lagoons, and polyhaline bays. Those that do not fall into any of these three categories can simply be called coastal bays.

Estuaries are coastal bays that have a significant influx of freshwater and that have good tidal circulation with the open marine environment. As a consequence, they tend to have low salinities ranging from near zero to almost normal marine concentrations (35 ppt). Estuaries may form as the result of tectonic activity, glacial erosion (fjords), drowned river systems, or formation of sediment spits across an embayment.

Lagoons are coastal bays that have no significant freshwater influx and have restricted tidal circulation with the open marine environment. These water bodies tend to be located landward of barrier islands, and most are elongated in a shore-parallel direction. In arid to semi-arid climates they may be drowned drainage systems as well. Lagoons tend to exhibit higher than normal marine salinities.

Polyhaline bays are those that display a great range in salinity and hydrologic condition. Typically, this re-flects seasonal changes in climate, especially precipitation. Shallow coastal bays that have little or no interaction with marine waters or that receive no continuous freshwater influx are controlled by the weather within the immediate area of the water body. These bays are generally shallow. During the wet season, precipitation causes the water to be nearly fresh, but the excess of evaporation during the dry season results in hypersaline conditions.

B. Estuaries

Estuaries themselves may be classified according to various parameters such as tidal range, hydrology, dominant process, or mode of origin. The classification of J. L. Davies of Australia that is based on tidal range can be applied to estuaries as discussed by Miles O. Hayes, who related tidal range to estuarine morphology.

The more classic approach used hydrology and the interaction of fresh and salt water to classify estuaries. There are three types in his classification: stratified, partially stratified, and totally mixed. A **stratified estuary** is one in which the freshwater overrides the salt water with little or no mixing, and there is an abrupt, nearly horizontal boundary (Fig. 14a). A **partially mixed estuary** is one

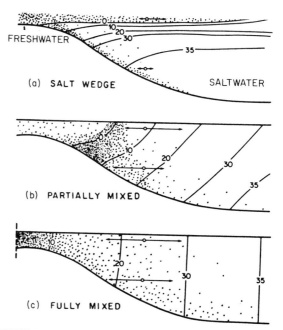

FIGURE 14 Skematic profile sections of the three primary types of estuarine circulation as described by Postma. (a) The salt wedge estuary is one that maintains a layer of freshwater overlying a layer of salt water. This is a low-energy type of estuary with isohalines essentially horizontal. (b) The partially mixed type of estuary is one where there is a modest mixing zone between the fresh and salt water masses with isohalines inclined. (c) The fully mixed estuary is a relatively high-energy system with essentially homogenized water. [From Postma (1980).]

where some mixing takes place between the fresh and salt water masses with isohaline contours displaying a diagonal orientation (Fig. 14b). **Totally mixed estuaries** show no vertical stratification, and isohalines are vertical with the water column; fresh near the headwaters of the estuary and normal marine near the mouth (Fig. 14c).

In this discussion, estuaries will be considered as consisting of two types: low-energy or wave-dominated estuaries and high-energy or tide-dominated estuaries. This is in keeping with excellent research on estuarine morphodynamics that has also provided stratigraphic models for estuarine sediment accumulations. **Wave-dominated estuaries** have barrier islands across their seaward side, are characterized by mud, and typically have low tidal ranges. **Tide-dominated estuaries** are typically open to the sea, are characterized by sand, and may develop in any tidal range.

1. Wave-Dominated Estuaries

Estuaries that are protected from open water waves by a barrier and have only modest tidal flux tend to accumulate large quantities of mud. The stream or streams that enter this embayment deposit their sediment load partly in the form of a delta and partly dispersed through the bay. The physical energy in the estuarine system is not capable of winnowing away the fines and redistributing the sediment. As a consequence, the estuary tends to slowly fill with sediment. Such an estuary is surrounded by poorly developed beaches, if any. Tidal flats are limited in extent because this type of estuary is characterized by low tidal ranges, generally less than 1 m. Marshes may be common if the sediment substrate reaches the neap high-tide level where marsh vegetation can thrive.

Good examples of low-energy estuaries are the coastal embayments along the entire U.S. Gulf of Mexico. These shallow bays are the result of the Holocene transgression flooding coastal plain drainage systems. They experience tidal ranges of less than 60 cm and are fed by a variety of streams. In some, such as San Antonio Bay and Matagorda Bay in Texas and Mobile Bay in Alabama, there are large river deltas.

These low-energy bays are characterized by mud substrates with small amounts of terrigenous sand and shell material. They contain large and numerous oyster reefs which thrive in the low-energy, brackish, and nutrient-rich waters of these bays. Only in areas where there are constrictions and tidal currents are relatively strong is terrigenous sand a significant constituent of the bottom sediments (Fig. 15). These conditions cause fine sediment to be winnowed, leaving the coarser particle behind.

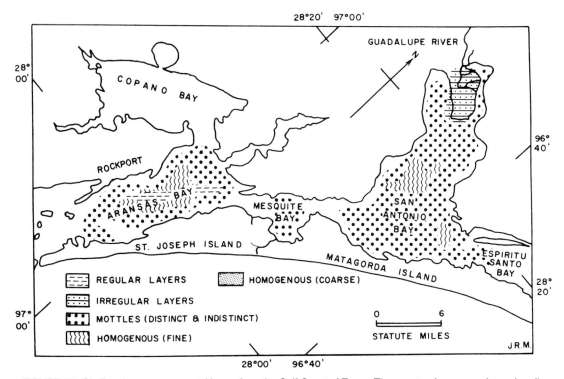

FIGURE 15 Shallow, low-energy coastal bays along the Gulf Coast of Texas. These estuaries accumulate primarily mud with some terrigenous sand and biogenic skeletal debris. Bioturbation of the bay sediments is widespread, as shown on the map. [From Shepard and Moore (1960).]

2. Tide-Dominated Estuaries

Dominance by strong tidal currents characterizes these estuaries and results in a predominantly sand substrate. It is common that a portion of the estuary, typically the margin, be dominated by wave action. There are a few situations where embayments fed by streams are also exposed to high wave energy. Actually much of the Bay of Fundy falls into this category. Although the tidal range is quite high (about 10 m), the seaward part of this estuary is wave dominated due to its orientation relative to the movement of the dominant weather systems in this area.

Most high-energy estuaries are the result of strong tidal currents that remove fine sediments and rework the sand-sized sediment into large sand bodies. Strong tidal currents do not require high tidal ranges to develop. As long as a large amount of water is moved in and out of the estuary during each tidal cycle, the work on bottom sediments can be accomplished. This amount of water, the **tidal prism**, is like a water budget and is dependent upon the tidal range and the size of the estuary. Thus, although a high tidal range makes it easier to develop a high-energy estuary, it also can be accomplished by low range but a high volume of water.

These tide-dominated estuaries tend to develop extensive tidal flats and marsh complexes. This is, in part, due to the tendency toward high tidal range but also is due to the work of tidal currents in redistributing sediment into the intertidal environment. The bottom sediments are typically fashioned into various bedforms such as ripples, megaripples, and sand waves which are commonly superimposed on large sand bodies that are oriented with the tidal currents. These estuaries tend to receive sediment from the discharge of streams and also from the open marine environment via tidal currents. Estuaries are sediment sinks and tend to have a limited lifetime. Eventually, they fill in with sediment, are colonized by marsh or mangrove communities, and then accumulate enough sediment to support a terrestrial plant community.

C. Lagoons

Lagoons tend to be low-energy coastal environments because there is no significant circulation mechanism from outside the coastal bay and they are commonly too narrow to generate enough internal circulation to significantly rework sediments. These coastal bays contrast with estuaries in several additional ways. They typically are long and narrow with a coast-parallel orientation (Fig. 16), whereas estuaries may exhibit almost any shape. Lagoons receive little terrigenous sediment influx and, therefore, have a slow rate of sediment accumulation. Salinity conditions dictate a very different type of fauna and flora from that of

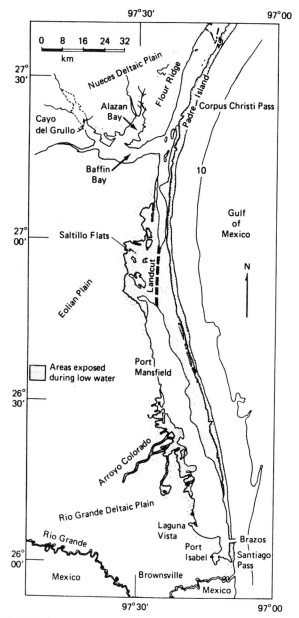

FIGURE 16 General map of the south Texas coastal region showing Laguna Madre which separates the mainland from Padre Island, the longest barrier island in the world. Laguna Madre shows a range of lagoonal conditions from near-normal marine salinities on the northern end to very hypersaline conditions in the south. Baffin Bay, adjacent to the west, is also a hypersaline lagoonal bay. [From Rusnak (1960).]

estuaries. Lagoons also tend to be in low to mid-latitudes due to climate restrictions, whereas estuaries may occur throughout the globe.

Coastal lagoons receive terrigenous sediment from washover and blowover of the barrier island that typically marks the seaward boundary of the lagoon and also from occasional runoff from intermittent streams that are activated by rainfall. Chemical precipitation is widespread in many coastal lagoons due to high salinities that are commonly achieved. Various forms of calcium carbonate or calcium-magnesium carbonate precipitate in lagoons such as Laguna Madre in Texas (Rusnak, 1960), the Coorong in South Australia, and Lake Reeve in Victoria (Australia). If salinities reach 200 ppt or more, it is possible to precipitate gypsum and halite.

D. Polyhaline Bays

There are coastal bays that have only nominal influence from freshwater runoff and tidal circulation but are markedly affected by seasonal conditions. This is due to distinct seasonality in precipitation. During the wet season the bay may have salinities that are significantly below normal marine levels, sometimes almost fresh. By contrast, during the dry season they may be hypersaline to the extent that evaporites are precipitated. These bays are characterized by a community that is quite euryhaline; that is, the organisms can tolerate marked changes in salinity.

Sediment accumulation of any type is quite slow in these bays and tends to be dominated by skeletal material and organic matter, although the latter tends to oxidize. In this respect, the polyhaline bays differ from estuaries, which receive abundant terrigenous sediment, and from lagoons, which may accumulate significant chemical sediment. Polyhaline bays are present along the Texas coast, along some areas of the Mediterranean Sea, and around the Indian Ocean where monsoon climates prevail.

VI. BARRIER ISLAND SYSTEMS

Barrier island systems are quite extensive along trailing edge coasts where coastal plains exist. They are complex depositional systems that are generally wave dominated or of mixed wave and tidal energy, and they contain a spectrum of sedimentary depositional environments. Barrier island systems occur throughout the world regardless of climate, but they must have a fairly shallow shelf on which to accumulate and they require a large amount of sediment plus significant wave influence.

A. Major Components

From the seaward to the landward location, the barrier island system consists of the nearshore zone, beach, dunes, washover fans, and various types of coastal bays with marsh and/or tidal flat margins. Barrier islands are typically interrupted by tidal inlets that permit circulation between the coastal bays landward of the barrier and the open marine environment.

The **nearshore zone** includes the bar and trough topography adjacent to the beach and extends landward to the low-tide position. It is a wave-dominated environment that is characterized by shore-parallel sand bars and intervening troughs. The number differs from location to location but depends upon the general slope of the bottom and the amount of sediment available. Rip current systems are common in this environment, as water moved landward by oncoming waves travels seaward through lows or saddles in the longshore bars.

The **beach** is the dynamic zone of unconsolidated sediment between low tide and the next landward geomorphic element. The landward boundary is typically the base of the dunes or whatever landward change in morphology or composition may exist. In nonbarrier situations it may be bedrock cliffs, glacial drift, or even man-made structures such as seawalls.

Dunes are eolian accumulations of sand that form generally linear sediment bodies immediately landward of the beach. Their size and the number of dune ridges are dependent upon the sediment available. Some barriers have such a limited sediment supply that dunes are absent or very small.

When storms cause the barrier to be temporarily inundated by waves, sediment is carried from the beach and nearshore zone to the landward side of the barrier and deposited in the form of **washover fans**. These fan-shaped sediment bodies may occur as rather isolated accumulations or they may be nearly continuous, forming washover aprons.

Breaks in barrier islands that serve as passageways for tidal currents to move between the open sea and coastal bays are called **tidal inlets**. These features may form in a variety of ways but tend to have a common morphology that is controlled primarily by the interactions between tidal currents and waves. Inlets tend to have sediment accumulations called **tidal deltas** at both the landward and seaward ends.

B. Origin of Barrier Islands

Barrier islands are widespread throughout the globe, and their morphodynamics are well understood; however, there is still much to learn about their origin. There have been three primary scenarios ascribed to the origin of barrier islands: (1) upward shoaling of subtidal sand bars due to wave action, (2) generation of long spits due to littoral drift and then breaching to form inlets, and (3) drowning of coastal ridges. Subsequent research on

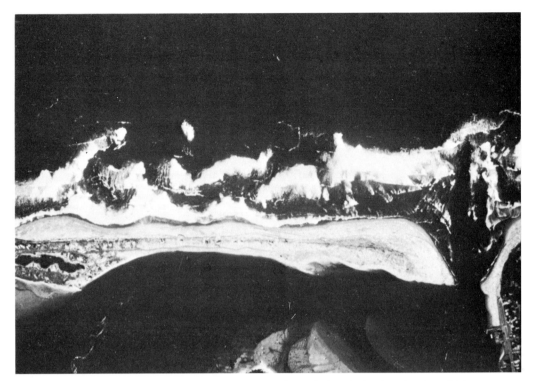

FIGURE 17 Short barrier spit (Siletz Spit) on the Oregon Coast. This spit developed under a high-energy wave climate.

the topic has supported one of these modes of origin or has called upon some combination of these.

This article will not dwell upon the merits of the various proposed mechanisms for barrier island development, but will simply refer the reader to selected important works on the subject and provide a few examples of modern barriers that seem to have readily assignable origins. The spit origin of barriers is easily recognizable and tends to be common along coasts that have moderate to high relief, especially leading edge coasts. These barriers are typically attached to some type of a headland commonly a bluff of bedrock or glacial drift. They are also characteristic of wave-dominated conditions because their formation is dependent upon a significant amount of sediment flux in a littoral drift system that is either dominated by transport in one direction alongshore and/or by significant onshore movement of sand. This type of barrier is common along the northwest coast of the United States where most of the barriers have not been breached (Fig. 17).

Most of the disagreement about the origin of barrier islands centers around the barriers that occur adjacent to coastal plains and are quite long with scattered inlets. The majority of the literature on the subject has favored the idea of drowned coastal ridges. Some research indicates that once formed, the barriers migrated long distances as

the Holocene sea level rose. Work in the Gulf of Mexico has supported the upward-shoaling idea. Recently formed barriers on the Gulf Coast of peninsular Florida have conclusively documented the origin of barriers by the upward-shoaling mode of origin. To date, no modern barriers have been documented to have been formed by the mechanism of drowned beach ridges.

There is little question that barrier islands can originate through various mechanisms. It is likely that the barriers in a given coastal area are characterized by their own dominant mode of origin. This can be determined only through extensive research that must include coring and detailed stratigraphic analysis.

C. Beach and Nearshore Zone

Probably the most dynamic of all coastal environments is that area of mobile sediment that includes the beach and the adjacent nearshore zone where waves interact with the substrate. Regardless of the specific geographic location, this environment displays a similar topographic profile (Fig. 18). The bar and trough topography may include several longshore sand bars or it may contain none at all. This represents the most obvious difference in morphology from one location to another and is dependent upon

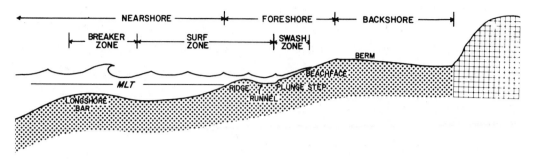

FIGURE 18 Diagrammatic profile of the beach and nearshore zone showing the major elements. [From Davis, R. A., Jr., ed. (1985). "Coastal Sedimentary Environments," Springer-Verlag, New York.]

three major factors: (1) the slope of the nearshore bottom, (2) the availability of sediment, and (3) the wave climate. Most coastal areas have two or three longshore bars (Fig. 19).

Waves begin to interact with the bottom and eventually steepen and break over the crests of the longshore bars. The wave climate at a given instant in time may be such that breaking waves develop over all bars present, such as during storms, or none may break, such as under calm conditions. Storms tend to erode the longshore bars and cause them to migrate seaward. During the intervening quiescent conditions, the bars return to their prestorm profile and migrate back to their prestorm location.

Well-sorted sand with scattered shell gravel dominates the nearshore zone. Ripples and planar beds are the most common bedding surfaces, with occasional megaripples formed in the troughs when storms generate strong longshore currents. Bioturbation structures are uncommon due to the nearly constant reworking of bottom sediments by waves and wave-generated currents.

The beach environment displays more temporal and spatial variety than the nearshore zone. The size and geometry are controlled by the sediment, the energy level,

and the morphology of the coast. Typically, a beach consists of two major parts: the **foreshore (forebeach)**, which is the intertidal portion, and the **backshore (backbeach)**, which is the supratidal part (Fig. 18). The foreshore is a fairly planar surface that slopes seaward with the gradient directly related to mean grain size; gravel beaches are quite steep (Fig. 20a) and fine sand beaches have gentle gradients (Fig. 20b). The backshore part of the beach is nearly horizontal or dips landward a few degrees (Fig. 18). This environment is typically subjected to eolian processes, except during storms when waves reach this part of the beach.

Beaches display cycles to their morphology that are related to the energy imparted on the coast. High wave energy during the storm season causes erosion and removes some or, in extreme circumstances, all of the sediment. During the quiescent periods of low wave energy, the beach is able to rebuild itself and return to a profile similar to that which preceded the storm period. These beach cycles may be seasonal or they may be related to single storm events. In either case, the morphology is similar. The prestorm profile shows a well-developed beach with a broad backshore and a relatively steep foreshore. The poststorm profile has little or no backshore, and the gradient is fairly steep.

Also associated with the poststorm profile is a **ridge and runnel** morphology. This is a small, typically intertidal, sand bar (ridge) with a low trough (runnel) separating it from the foreshore beach (Fig. 21). During the quiescent period between storms, wave-generated currents with the aid of the tidal fluctuations cause the ridge to migrate onto the beach and thereby repair damage caused by erosion during the storm (Fig. 22). This welding of the ridge or ridges enables the beach to recover to its prestorm profile. If storms interrupt this repair process, there may be a net loss of beach sediment. This would be the case during a storm season (commonly winter) with the result being little or no beach until the low-energy season (summer) when the beach can be rebuilt.

FIGURE 19 Oblique aerial photograph showing location and definition of longshore sand bars by the waves breaking over their crests. (Photo taken along Mustang Island, TX.)

FIGURE 20 Photographs of the foreshore zone of the beach showing (a) a steep slope on a gravel beach (Maine) and (b) a gentle slope on a fine sand beach (Texas).

D. Coastal Dunes

Barrier islands or other depositional coasts, where there is abundant sediment available that is not removed by storms, tend to develop dunes. Wind patterns are such that most coasts experience winds with at least some onshore component. This wind coupled with extensive beaches results in dunes.

Dunes consist of mound-like accumulations of fine to medium, well-sorted sand. These piles of sand tend

FIGURE 21 Beach area shortly after a storm showing the presence of a ridge and runnel as seen at low tide.

to form in linear ridges just landward of the beach. In some areas the ridges have nearly uniform and continuous crests (Fig. 23a), whereas in others the dunes are individually distinguishable although they may form linear ridges (Fig. 23b). The presence of dunes and their size are a function of sediment availability.

The onshore winds carry sediment from the typically dry backshore portion of the beach and transport it landward. Various items behind the beach, such as plants, rocks, debris, or other objects, may cause the sediment to become trapped or to accumulate in a shadow zone behind them (Fig. 23). This is the beginning of dune development. Continued accumulation will eventually result in a dune ridge immediately landward of the beach. These

are commonly called **foredunes** because of their position. As more and more sediment is supplied, it is common for additional ridges to develop as the beach prograles seaward (Fig. 24). Opportunistic vegetation may develop on these dunes and cause them to stabilize.

The sediment in coastal dunes displays large-scale, trough cross-stratification which appears in wedge sets when exposed (Fig. 25) by erosion, generally by storm waves and related storm tides.

E. Washover Fans

Storms and their related high waves and storm surge can overtop low barrier islands or cut through low areas on those barriers with high dunes. When this phenomenon occurs, the surging water carries abundant sediment. This sediment is deposited on the landward side of the barrier in the form of fan-shaped sediment accumulations called washover fans (Fig. 26).

The sediment that forms the washover fans is derived primarily from the surf zone and the beach. It is sand sized and generally moderately well sorted. This sediment is transported under upper flow regime conditions as sheet flow and accumulates dominantly as planar beds (Fig. 27).

Individual fans may cover from hundreds of square meters to a few square kilometers. Their thickness is typically tens of centimeters. Successive fans may accumulate in a vertical sequence and can be recognized by a subtle fining-upward trend and by vegetation debris separating the occurrences. Although a washover fan may completely bury the vegetation on the backbarrier, the marsh grasses

FIGURE 22 Several days after the development of the ridge and runnel morphology, the ridge has migrated onto the storm beach and is welding onto it, thus repairing much of the erosion that was caused by the storm.

FIGURE 23 Coastal dune ridges showing (a) an essentially continuous ridge on the Ninety-Mile Beach, Australia, and (b) discontinuous coastal foredunes along the Texas coast.

or other plants are quite hearty and can grow up through the fan sediment in a matter of months.

Intense events may result in nearly continuous washover fans, forming a washover apron. The sediment deposited by such events may reach into the coastal bay that borders the barrier on the landward side. This is quite common as the result of hurricanes (Fig. 28) and is a primary mechanism for the landward migration of the barrier island (see Section VI.G).

Wind also causes much landward transport of sediment so that blowover of sediment is an important phenomenon. This is particularly true of unvegetated dunes. The wind may cause entire dune complexes to migrate landward across a barrier island.

F. Tidal Inlets

Barriers are interrupted by channels through which tidal currents pass during each tidal cycle and through which much sediment may be transported. These tidal inlets occur in a variety of forms and are the result of multiple origins. They do, however, have a common general appearance, and they react similarly to coastal processes regardless of origin.

FIGURE 24 Oblique aerial photograph showing multiple parallel and continuous dune ridges near Seasport, Victoria, Australia.

The tidal inlet is a pathway for the tidal prism between the open marine environment and the coastal bay system between the barrier and the mainland. In this fashion, inlets act much like a bidirectional river with a predictable current pattern. Because these systems carry sediment also, they tend to deposit and rework sediments at the mouths of the inlet. These sediment accumulations are called **tidal deltas** because of their similarity to riverine deltas; they form at the mouth of a channel where it empties into a basin. In the case of inlets, the channel is quite short

and has a mouth at each end. These tidal deltas may be large or small depending upon the sediment available and the nature of the wave and current processes acting on them.

The tidal delta on the landward side of the inlet is called a **flood tidal delta** due to its accumulation as the result of flood tidal currents passing through the inlet and losing competency at the landward end of the barrier. The sediment body at the mouth of the inlet on the seaward side is called an **ebb tidal delta** for similar reasons.

FIGURE 25 Close-up view of cross-stratification in a coastal dune. The typical appearance is of wedge-shaped sets of trough cross-stratification.

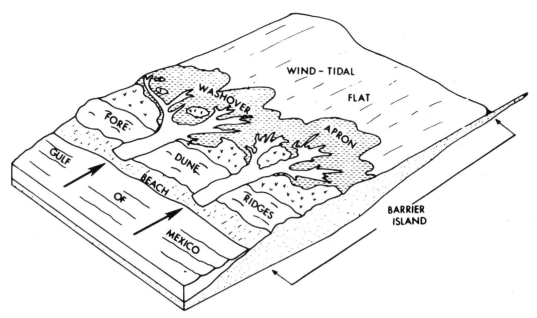

FIGURE 26 A plan view of washover fans which have formed as the result of a storm carrying sand from the beach and nearshore over the dunes and onto the back-island flats. [From Scott and McGowan, (1969)].

1. Origin of Tidal Inlets

Tidal inlets tend to originate in two primary ways; they develop as spits which form across coastal bays or as the result of storms which cut through barriers. A third mode of origin that has an unknown level of importance is where the inlet forms as the barrier develops. In all of these situations, inlets are kept open by the tidal currents that pass in and out of the coastal bay landward of the barrier.

2. Tidal Deltas

Tidal deltas may form gradually as the result of tidal currents through constricting channels as a spit migrates or they may form quickly as the consequence of a storm. In either case the tidal deltas eventually become modified by waves and currents, eventually achieving equilibrium between their morphology and the processes operating on them.

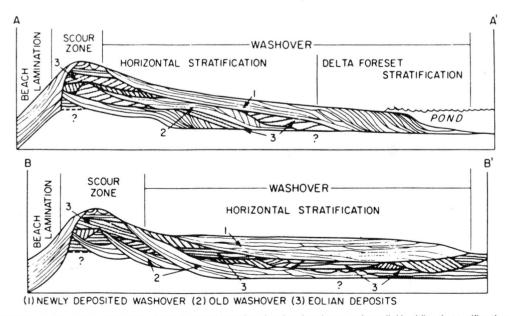

FIGURE 27 Diagrams of trenches through a washover fan showing the planar and parallel bedding that typifies these deposits. [From Schwartz, (1972)].

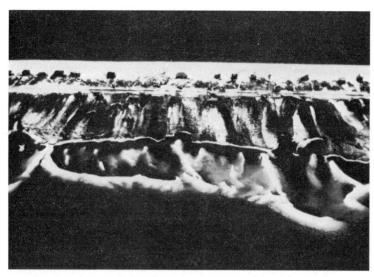

FIGURE 28 The combination of storm intensity and the narrow island caused washover deposits to be carried completely across Dauphine Island, AL, during Hurricane Frederick and to accumulate in the sound landward of the island. (Photo by Dag Nummedal.)

Flood tidal deltas are generally fan shaped and are protected from wave attack by their location in coastal bays where waves are small. These sediment bodies may be multilobate and are at least partly intertidal, allowing for marsh or mangrove vegetation (Fig. 29). Flood deltas commonly have their outer margins reworked by ebb-tidal currents, with the impact of flood currents limited to the inner and seaward portion.

Ebb deltas may have a wide variety of morphologies because of their location in the seaward side of the barrier where they are exposed to wave attack. The overall configuration of the ebb delta is controlled by the balance between the ebb-tidal currents coming out of the inlet channel and the wave climate of the open marine environment. A coast which is distinctly wave dominated may not have ebb-tidal deltas due to the removal of sediment by waves and longshore currents. Tide-dominated ebb deltas have large, elongated sediment bodies that are perpendicular to the trend of the barrier (Fig. 30). Mixed-wave and tidal energies produce ebb deltas which are distinct sediment

FIGURE 29 Vertical aerial view of a flood tidal delta showing the fan shape with a rather irregular outline. Ebb-tidal currents may cause a smoothing of the outer portion of such a flood tidal delta.

FIGURE 30 An ebb-tidal delta showing elongated sand bodies that are perpendicular to the coast. This tide-dominated ebb delta is on the coast of South Carolina. [From Oertel, (1985)].

FIGURE 31 The ebb-tidal delta at Matanzas Inlet in Florida displays a rather smoothed outer margin due to modification by waves.

bodies but which have smooth outer margins due to modification by waves (Fig. 31).

3. Inlet Dynamics

The morphology of the inlet-tidal delta system is dependent upon the interaction of tidal currents, waves, and longshore currents with the inlet. Several distinct morphologic types can be produced, giving rise to a classification of inlets (Fig. 32). A tide-dominated inlet has a straight channel that is rather stable and an ebb-tidal delta that extends a considerable distance into the open marine environment with shore-normal sediment bodies. A wave-dominated inlet tends to have little or no ebb-tidal delta.

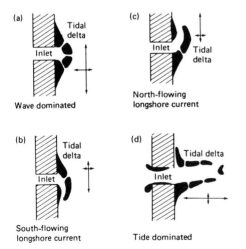

FIGURE 32 The interaction of tides, waves, and wave-generated longshore currents permits a classification of ebb-tidal deltas based on the dominant process that controls the morphology.

Wave-dominated inlets that experience considerable longshore current in a dominant direction will have an unstable channel. Longshore currents displace the inlet in the direction of the littoral drift and may even result in the inlet being closed (Fig. 33). Closure results from the tidal prism being small and the tidal currents being over powered by longshore currents and the associated sediment transport.

G. Barrier Island Morphodynamics

Barrier islands develop along wave-dominated and mixed-energy coasts regardless of tidal range. The determination of these coastal types is predicated on the relationship between tidal energy and wave energy, not the absolute size of waves or tidal range. The distinctly wave-dominated barrier island coast tends to consist of long, narrow barriers with widely spaced inlets. The islands have abundant washover deposits (Hayes, 1979). The mixed-energy barrier coast is characterized by short, stubby barriers with numerous inlets, some of which are tide dominated. The Texas coast and the Outer Banks of North Carolina are good examples of wave-dominated coasts, and the Georgia Bight area in the southeastern United States and the peninsular Gulf Coast of Florida are good examples of mixed-energy barrier coasts.

Barriers on wave-dominated coasts migrate landward as beach erosion and washover/blowover take place. The present situation of a relatively rapid sea level rise facilitates this landward migration. Inlets tend to close due to dominanting longshore drift but are reopened or new inlets are cut as the result of intense storms. These coasts are dominated by extreme events such as hurricanes.

FIGURE 33 Under circumstances of extreme littoral drift of sediment by longshore currents in combination with small tidal prism through the inlet, there may be closure such as that shown for Midnight Pass in Florida.

Mixed-energy coasts are characterized by drumstick barrier islands and fairly stable inlets. The drumstick barrier is the result of local coastal processes that cause most sediment to accumulate at one end of the barrier, thus depriving the other end of its fair share. The resulting barrier has numerous accretionary ridges at one end and the other end is narrow and low (Fig. 34). The shape resembles the drumstick of a chicken and thus the name.

This type of barrier may exhibit some migration, but it is quite different from the wave-dominated barrier. Drum-stick barriers prograde seaward at the wide end and are washed over, thus causing landward migration at the narrow end. The result is a reorientation of the island rather than landward migration.

VII. ROCKY COASTS

All of the coastal types and major elements discussed above are characterized by deposition. Rocky coasts are

FIGURE 34 Caladesi Island in Florida is a good example of a drumstick barrier island with a narrow, low-lying updrift end and a wide, beach-ridge-dominated downdrift end.

characterized by erosion. This type of coast displays a rugged morphology of high relief with bedrock being the dominant substrate; there is little unconsolidated sediment.

Rocky coasts occur throughout the world without regard to climate; however, their most common tectonic setting is along the leading edge of lithospheric plates. The west coasts of most of North and South America are characterized by rocky coasts. Climate may be a factor in the formation of rocky coasts such as glaciation along the coast of Maine and reefs in the Florida Keys.

A. Coastal Morphology

The profile of rocky coasts differs greatly from depositional coasts, both above and below the sea level. The shoreline itself is also quite different. In both cases there is considerable relief and rugged irregularity. The typical profile is one of cliffed bedrock that rises at least a few meters above sea level. The base of the cliff may have a small beach depending upon sediment availability, wave energy, and a site for accumulation. Beaches are typically restricted to small embayments in the rocky coast where there is some protection from waves (Fig. 35).

Some locations have wave-cut terraces or benches (Fig. 36) that reflect some combination of high wave energy and a rather stable sea level. Because of the tectonics associated with leading edge coasts, the relative sea level may change in rather large increments instead of gradually as it does along trailing edge coasts. As a result, the benches and terraces may be rapidly drowned or uplifted.

The beaches that do occur on this type of coast are local and narrow. They are generally only thin veneers of unconsolidated sediment resting upon a bedrock bench. It is not uncommon for the sediment and, therefore, the

FIGURE 36 Wave action has eroded a flat terrace called a wave-cut platform almost exactly at sea level along this coast in Tasmania, Australia.

beach to be seasonal in its development. Winter storms wash the sediment away into deeper water, and the relatively low wave energy during the summer permits it to return. A good example of this occurs along part of the Oregon coast. Spring tides are 3–4 m, and the beach is quite wide, up to 300 m. In November and December the beach sediment is removed, leaving an irregular bedrock surface. The sand begins to return to the beach in pulses that start in early March.

Erosion of rocky coasts takes place at irregular rates due to variation of wave energy, lithology, and resistance of bedrock and bottom profile that may attenuate waves. As a result, there are various erosional features that characterize the rocky coast. **Sea stacks** are erosional remnants of bedrock that have a vertical character and stand alone somewhat offshore from the coast itself (Fig. 37). **Arches** may also form due to the removal of nonresistant rock surrounded by resistant rock.

The irregular shoreline and nearshore topography causes great variety in the wave energy that reaches the coast. The general situation is that headlands receive a relatively high level of wave energy and the embayments receive a low amount, thus the appearance of pocket beaches in the embayments (Fig. 35). Another result of this variation in wave energy is the development of **tombolos** which are sandy strips of sediment that connect a rocky island to the mainland (Fig. 38). Wave refraction permits sediment

FIGURE 35 Small pocket beaches are fairly common along irregular rocky coasts. They form in embayments where the sand is allowed to accumulate, such as this example on the coast of Oregon. (Photo by W. T. Fox.)

FIGURE 37 Irregular resistance and subsequent erosion may leave isolated rock columns called sea stacks along the coast. These are in the famous Twelve Apostles area of southern Victoria, Australia.

to accumulate in the lee of an island, and in combination with locally generated longshore currents, it will become connected to the mainland.

B. Processes on Rocky Coasts

It is obvious that waves are the dominant process on rocky coasts. This may be the result of extensive fetch and large waves, such as along the entire eastern edge of the Pacific Ocean, but it may also be from other factors. One important contributor is the steep nearshore gradient that characterizes rocky coasts. This permits waves to move to and impact the coast without any loss of energy such as

that which would occur on gently sloping nearshore areas. The steep gradient also inhibits beach development, and beaches gradually absorb wave energy. On the rocky coast the waves hit the steep or vertical shoreline with all of their energy. Longshore currents are essentially absent along rocky coasts because of the irregular nature of the shoreline in combination with the steep nearshore gradient.

Tides have little effect on rocky coasts, except for the passive role they play in changing the focus of impact of the waves as the sea level changes. If the tidal range is small, there is little change and wave energy tends to be quite concentrated. If the tidal range is 3–4 m, then the wave energy can be spread out a bit and perhaps slow the rate of erosion relative to the coast with a low tidal range.

VIII. SUMMARY

It is readily apparent that the coastal zone is geologically complex. There are interactions of both terrestrial and marine processes with sediments and rocks. Such global factors as sea level change and plate tectonics also play an important role in coastal development. Although the coastal zone is narrow and covers a small area as compared to either the terrestrial and marine systems, it contains much variety in processes, morphology, and relief.

Most sediment is carried to the coast by rivers and empties into estuaries or is deposited on deltas. The coast also received sediment from the marine environment through the action of waves and tides. Rapid and tectonic local

FIGURE 38 Wave refraction and related sediment transport may cause rocky islands to become connected to the mainland, forming a tombolo. This example is near Cape Ann in Massachusetts.

changes in sea level as well as slow and eustatic changes overprint on these coastal processes.

Change characterizes all coasts regardless of type or geographic location. All coastal environments are dynamic. Some change rapidly in response to extreme events such as hurricanes or floods, whereas others respond to more common and slower but continual processes.

SEE ALSO THE FOLLOWING ARTICLES

CONTINENTAL CRUST • COASTAL METEOROLOGY • GLACIAL GEOLOGY AND GEOMORPHOLOGY • HYDRODYNAMICS OF SEDIMENTARY BASINS • HYDROGEOLOGY • OCEANIC CRUST • PLATE TECTONICS • SEDIMENTARY PETROLOGY

BIBLIOGRAPHY

Bird, E. C. F. (1984). "Coasts," Blackwell, Oxford.
Carter, R. W. G., and Woodroofe, C. D. (eds.) (1994). "Coastal Evolution," Cambridge Univ. Press, Cambridge.
Davies, J. L. (1980). "Geographical Variation in Coastal Development," 2nd ed., Longmans, New York.
Davis, R. A., Jr., ed. (1985). "Coastal Sedimentary Environments," Springer-Verlag, New York.
Davis, R. A., Jr., ed. (1994). "Geology of Holocene Barrier Islands," Springer-Verlag, Heidelberg.
Davis, R. A., Jr. (1994). "The Evolving Coast," Scientific American Library, New York.
Eisma, D. (1998). "Intertidal Deposits," CRC Press, Boca Raton, FL.
Komar, P. D. (1998). "Beach Processes and Sedimentation," 2nd ed., Prentice-Hall, Englewood Cliffs, NJ.
Short, A. D., ed. (1999). "A Handbook of Beach and Shoreface Morphodynamics," Wiley, Sydney.

Coastal Meteorology

S. A. Hsu
Louisiana State University

I. Coastal Weather Phenomena
II. Local Winds
III. Boundary-Layer Phenomena
IV. Air–Sea Interaction
V. Hurricanes

GLOSSARY

Aerodynamic roughness length Measure of the roughness of a surface over which wind is blowing.

Ageostrophic wind Vector difference between measured wind and assumed geostrophic wind.

Antitriptic wind Wind in which the pressure gradient force exactly balances the viscous force.

Baroclinic State of atmosphere in which surfaces of constant pressure intersect surfaces of constant density or temperature.

Brunt–Väisälä frequency Frequency of vertical oscillation of an air parcel released after displacement from its equilibrium position.

Coriolis force Apparent force on moving particles caused by the Earth's rotation.

Density current Intrusion of colder air beneath warmer air caused mainly by hydrostatic forces arising from gravity and density or temperature differences.

Frontogenesis Initial formation of a front as a result of an increase in the horizontal gradient of the air mass property.

Geostrophic wind Horizontal wind in which the Coriolis and pressure-gradient forces are balanced.

Gradient wind Horizontal wind in which the Coriolis acceleration and the centripetal acceleration together exactly balance the pressure-gradient force.

Inversion Increase in temperature with height as opposed to the normal condition, when temperature decreases with height.

Mesoscale Scale in which atmospheric motions occur within 2000 km of horizontal length and 24 hr of time.

Millibar (mb) Meteorological unit of pressure that equals 100 N m^{-2}.

Potential temperature Temperature a parcel of dry air would have if it were brought adiabatically (at approximately 1°C per 100 m) from its initial state to 1000 mb.

Rossby radius of deformation Length scale that is equal to $c/|f|$, where c is the wave speed in the absence of rotation effects and f is the Coriolis parameter caused by Earth's rotation.

Significant wave height Average of the highest one-third waves on the sea surface.

Stability parameters Dimensionless ratio of height, Z, and the Monin–Obukhov stability length, L, which relates to the wind shear and temperature (buoyancy) effect. Thus, the term Z/L represents the relative importance of heat convection and mechanical turbulence. The Richardson number is also a dimensionless ratio of wind shear and buoyancy force.

Supergeostrophic wind Type of wind with speed in excess of the local geostrophic value.

Synoptic scale Scale of atmospheric motion with horizontal length scale greater than 2000 km and time scale longer than 24 hr.

Thermal wind Vertical shear of geostrophic wind directed along the isotherms with cold air to the left in the Northern Hemisphere and to the right in the Southern Hemisphere.

Upwelling Slow, upward motion of deeper water.

Vorticity Vector measure of fluid rotation.

Wet-bulb potential temperature Wet-bulb temperature after an air parcel is cooled from its initial state adiabatically and then brought to 1000 mb.

COASTAL METEOROLOGY is an integral part of the total-system approach to understanding coastal environments. Since meteorology is the study dealing with the phenomenon of the atmosphere, coastal meteorology may be defined as that part of meteorology that deals mainly with the study of atmospheric phenomena occurring in the coastal zone. This description includes the influence of atmosphere on coastal waters and the influence of the sea surface on atmospheric phenomena, that is, the air–sea interaction.

The behavior of the atmosphere can be analyzed and understood in terms of basic laws and concepts of physics. The three fields of physics that are most applicable to the atmosphere are radiation, thermodynamics, and hydrodynamics. Owing to limitations of space, only a few topics in coastal meteorology are covered here. More information is available through the bibliography.

I. COASTAL WEATHER PHENOMENA

A. Coastal Fronts

1. Synoptic Scale Phenomena

The coastal front marks the boundary between cold continental air and warm oceanic air. A thermally direct frontal circulation exists, resulting in local enhancement of precipitation on the cold side of the frontal boundary. The coastal front is also a low-level baroclinic zone in which upward vertical motion and temperature advection in a narrow area along the coast exist. Along the U.S. Atlantic coast, both geostrophic and observed wind deformation play a role in coastal frontogenesis. The frontogenetical process involves a weak cyclone, which strengthens the preexisting temperature gradient as it moves northward. A moist baroclinic zone remains in place along the coast in the absence of strong cold advection in the wake of the

weak cyclone. The residual moisture-enhanced baroclinicity and surface vorticity are important factors contributing to a second disturbance, which intensifies as it moves northeastward parallel to the coast along the frontal zone.

2. Mesoscale Phenomena

Frictional convergence at coastlines. The coastline generally represents a marked discontinuity in surface roughness. The resulting mechanical forcing leads to a secondary circulation in the boundary layer and, consequently, to a vertical motion field that may have a strong influence on the weather in the coastal zone. In one example, heavy shower activity along the Belgium and Netherlands coasts was caused by frictional uplifting and frontogenesis occurring when a maritime polar air mass hit the coastline at a critical angle. By utilizing numerical models, scientists from the Netherlands recently found that upward motion is most pronounced when the geostrophic wind makes a small (about 20°) angle with the coastline (in a clockwise direction) and not when the geostrophic wind is perpendicular to the coastline, as is sometimes mentioned. The asymmetry relative to normal to the coastline is caused by Coriolis acceleration and not by a nonlinear effect.

Boundary-layer fronts at sea. These fronts or convergence lines develop in the cold air at sea when there are large bends or kinks in the shape of the upstream land or ice boundary from which the cold air is flowing. The air on one side of the convergence line has had a different over-water trajectory than that on the other side. The existence of such convergence in the cold air may, under the right conditions, very well be a factor in the genesis of polar vortices, which may later be intensified into polar (arctic) lows.

Orographically forced cold fronts. A coastal mountain range, for example, in southeastern Australia, can sometimes block shallow cold fronts with a northwest/southeast orientation. The violent behavior of some cold-front passages, or southerly busters, is found by Australian scientists to be at least orographically initiated. The head of the front has the character of an evolving density current, and its propagation is well predicted by density current theory over more than half of its lifetime. The horizontal roll vortex just behind the front is found to be accelerating relative to the rate of advection of cold air behind the front. This evolution is governed by warm-air entrainment.

Sea-breeze front. A sea-breeze front has the properties of the head of a gravity current. The front head has the shape of a lobe at its nose and a cleft behind the nose above the denser air from the sea. The cleft can engulf overriding lighter air from the land. The circulation at the head may form a cutoff vortex.

Island-induced cloud bands. The development of cloud bands induced by an island is a complex interaction between the airflow and the geometry of the island. First, the upwind surface flow forms a separation line with an associated stagnation point. Then, a low-level convergence zone develops along this line, resulting in an updraft line. If the updrafts are strong enough, a band cloud forms. To characterize such a flow, the Froude number, Fr, is often employed (Fr $= U/NH$, where U is the upstream windspeed, N is the Brunt–Väisälä frequency, and H is the characteristic height of the island mountain). As an example, Fr ≈ 0.2 for the island of Hawaii, where a well-defined band cloud was observed offshore of Hilo. A higher Fr tends to induce a stronger band that forms closer to shore. When Fr ≈ 1, orographic clouds may form. On the other hand, when Fr ≈ 0.1, the convergence zone moves offshore and the cloud band may be weak or may even disappear.

Cold-air damming. Cold-air damming exists when the cold air over land becomes entrenched along mountain slopes that face a warmer ocean. For example, along the eastern slopes of the Appalachians, the temperature difference can exceed 20°C between the damming region and the coast, a distance of approximately 150 km. This Appalachian cold-air damming was investigated in detail most recently by G. D. Bell and L. F. Bosart, who reveal that damming events occur throughout the year but peak in the winter season, particularly December and March, when three to five events per month might be expected. Cold-air damming is favored in late autumn and early winter when the land is coldest relative to the ocean. The event is critically dependent upon the configuration of the synoptic-scale flow. The presence of a cold dome is indicated by a U-shaped ridge in the sea-level isobar pattern, as well as in the 930-mb (about 700–800 m) height field, which is near the top of the cold dome. The potential temperature contours are also pronouncedly U shaped in the damming region, indicating relatively uniform cold air in the dome bounded by a strong baroclinic zone just to the east.

At the onset of the cold dome, warm air advection was observed over the surface-based layer of cold air advection. This differential vertical thermal advection pattern aided in generating and rapidly strengthening an inversion at the top of the cold dome, resulting in decoupling of the northeasterly flow in the cold dome from the southeasterly flow just above it. Force balance computation indicated that the acceleration of the flow to the speed of the low-level wind maximum at 930 mb was governed by the mountain-parallel component of the large-scale height (or pressure) gradient force. After the formation of the cold dome, the force balance on the accelerated flow was geostrophic in the cross-mountain direction and antitriptic in the along-mountain direction. Cold dome drainage occurs with the advection of the cold air toward the coast in response to synoptic-scale pressure falls accompanying coastal cyclogenesis.

Cyclogenesis. Cyclogenesis is defined as any development or strengthening of cyclonic circulation in the atmosphere. In certain coastal regions, cyclogenesis is a very important phenomenon, for example, along the mid-Atlantic coast of the United States and in the northwestern Gulf of Mexico. The cyclones that develop over the Yellow Sea and East China Sea often cause strong gales, and the cold air west of the cyclones spreads southward as an outbreak of the winter monsoon. Main features and processes contributing to coastal cyclogenesis along the U.S. East Coast are significant sensible heat transport over the ocean and latent heat release along the East Coast, coastal frontogenesis, and a polar jet streak propagating eastward. In a case of East Asian coastal cyclogenesis, a numerical experiment that included all physical processes simulated the development of a cyclone that developed rapidly in a way similar to that observed. In an experiment without latent heat feedback, only a shallow low appeared when the upper short-wave trough approached the inverted surface trough situated on the coast, but no further development took place. This suggests that the baroclinic forcing was enhanced by the feedback of physical processes. The latent heating had a profound impact on the amplifying jet streak circulation and the vertical coupling within the system, which appeared to prime the rapid cyclogenesis along the coast. Sensible heating contributed nearly 18% to the surface development. It helped to build a potential temperature contrast along the coast below 900 mb. Without sensible heating, the model-latent heat release was reduced. Thus, the impact of sensible heating was partly through the moist processes rather than direct heating.

Cyclogenesis in the western Gulf of Mexico is contributed in most cases by a mountain-induced standing wave developed on a climatological surface baroclinic zone over the Gulf. The effect of surface-layer baroclinicity on cyclogenesis is shown in Fig. 1. A close relationship is found between the frequency of occurrence of frontal overrunning over New Orleans, LA, and the air temperature difference between the shelf (shallow) water and deeper (warm) ocean water. Because the surface-based vorticity is directly and linearly proportional to the temperature difference across the cold shelf water and warmer Gulf water, Fig. 1 indicates that there is definitely a correlation between the surface-based, vorticity-rich air and

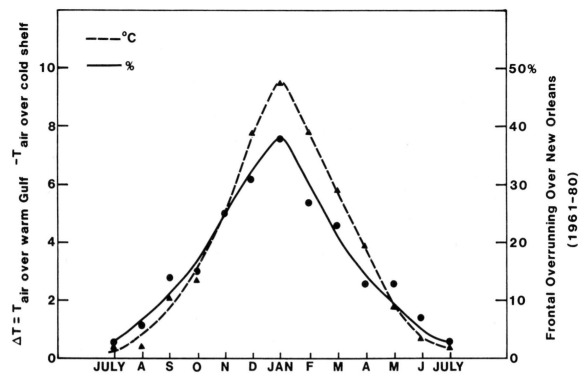

FIGURE 1 Correlation between the frequency of frontal overrunning over New Orleans, LA, and the difference in air temperature over warmer Gulf of Mexico and colder shelf waters.

the temperature difference or baroclinic zone occurring between the colder land/shelf water and the warmer deep-ocean water. A larger scale baroclinic or solenoidal field from Key West, FL, to Del Rio, TX, via Victoria, TX, and the Gulf of Mexico is shown in Fig. 2. The cold pool or cold-air damming, as discussed previously, over the cold shelf water off the coast of southern Texas and the Victoria region is clearly delineated. On the other hand, a warmer region over the Loop Current west of Key West is also illustrated.

An example of the cyclogenesis over the Western Gulf of Mexico including its effect and classification is provided in Figs. 3–6.

B. Fog and Stratiform Clouds

Fog and stratus are clouds, but the base of fog rests on the Earth's surface and stratus clouds are above the surface. Although the substance of fog and cloud is the same, their processes of formation are different. Clouds form mainly because air rises, expands, and cools. Fog results from the cooling of air that remains at the Earth's surface. Scientists at the University of Nevada recently revealed the following characteristics for convection over oceans: when warm air flows over cooler water, the air layer over the sea will usually convect even when the water surface

is 10° or more colder than the initial air temperature. An inversion at stratus cloud tops can be created by the stratus. Such inversions persist after subsidence evaporates the cloud. Radiation heat exchange does not play an essential role in stratus formation or maintenance and can either heat or cool the cloud. Dry air convection does not erode inversions at the top of the convective layer. Fogs are most likely to form at sea, where the water is coldest and needs no radiation effects to initiate cooling, or a boost from patches of warmer water, to begin convection. Both stratus cloud growth and the evaporation of clouds by cloud-top entrainment readjust the vertical structure of the air to leave a constant wet-bulb potential temperature with height.

The stratocumulus cloud deck over the east Pacific has large cloud variability, on 1–5 km scales. The cloud deck slopes upward from 700 to 1000 m in a northeast–southwest direction over a distance of 120 km. In the examples studied, vertical cloud top distributions were negatively skewed, indicating flat-topped clouds. The dominant spectral peak of the cloud-top variations was found at 4.5 km, which is 5–7 times the depth of the local boundary layer. The cloud layer was stable with respect to cloud-top entrainment instability. Structural properties of stratocumulus clouds observed off the coast of southern California, near San Francisco, and in the Gulf of Mexico are

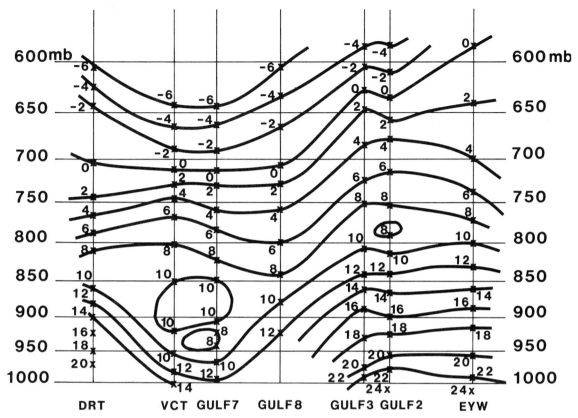

FIGURE 2 An example of a baroclinic (or solenoidal) field from Key West, FL, to Del Rio, TX, via the Gulf of Mexico and Victoria, TX, on February 22, 1986, during a special experiment. Based on radiosonde ascents from weather stations and radiosonde drops from airplanes.

similar. Marine stratocumulus cloud fields composed primarily of small cells have very steep slopes and reach their asymptotic values at short distances from the origin. As the cells composing the cloud field grow larger, the slope becomes more gradual and the asymptotic distance increases accordingly.

II. LOCAL WINDS

A. Land and Sea Breezes

The best example of local winds in the coastal zone is perhaps the land–sea breeze system. This coastal air-circulation system brings fresh air from the sea in the afternoon to cool coastal residents, whereas farther inland hot and still air is the general rule. On coasts and shores of relatively large lakes, because of the large diurnal temperature variations over land as compared to that over water, a diurnal reversal of onshore (sea breeze) and offshore (land breeze) wind occurs.

A sea breeze develops a few hours after sunrise, continues during the daylight hours, and dies down after sunset. Later, a seaward-blowing land breeze appears and contin-

ues until after sunrise. The sea breeze may extend up to 50 or 100 km inland, but the seaward range of the land breeze is much smaller. In the vertical, the sea breeze reaches altitudes of 1300–1400 m in tropical coastal areas, with a maximum speed at a few hundred meters above the ground. In contrast, the nocturnal land breeze is usually rather shallow, being only a few hundred meters deep. Typical horizontal speeds of the sea breeze are of the order of meters per second, while the vertical components are only a few centimeters per second. At specific locations, large and abrupt temperature and relative humidity changes can occur with the passage of the sea-breeze front. An example of the land- and sea-breeze system is shown in Fig. 7. The onshore and offshore wind components are shown at 3-hr time intervals during the day. The lower portion of the onshore flow is the sea breeze and that of the offshore flow is the land breeze. The maximum wind speed and its approximate height in each current are depicted by arrows. The elliptical shapes in the figure illustrate the horizontal and vertical extent of the land–sea breeze circulation. The dashed horizontal line represents the 900-mb pressure surface (approximately the convective condensation level). At 0900 LST (local standard time), the air

FIGURE 3 An example of cyclogenesis which took place over the Gulf of Mexico on February 16, 1983. This shot was taken from the GOES satellite. Notice the comma-shaped whirlpool cloud pattern and also the fact that this system was not linked to other larger scale systems. This was one of the top five cyclones generated over the Gulf of Mexico during the 1982–1983 El Niño period. Not only do surface conditions, such as sea surface temperatures, play an important part in the development and intensification of these storms, but the upper atmospheric conditions are critical as well.

FIGURE 4 An enlargement of Fig. 3 over the western Gulf of Mexico.

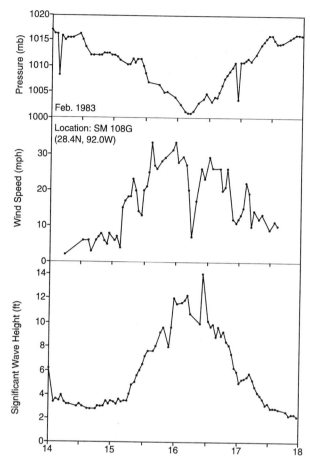

FIGURE 5 The time series analysis for this storm (Fig. 4) was made from a data buoy for atmospheric pressure, wind speed, and significant wave height during the period of cyclogenesis. Note the relationship between pressure and winds. The maximum wind speed does not usually occur at the time of the lowest pressure, but in general the lower the pressure the stronger the wind will be. This particular time series very much resembles a typical tropical cyclone plot as the wind speed would be expected to drop off dramatically in the eye or center of lowest pressure.

breeze becomes well developed by 0300 LST and reaches its maximum intensity near 0600 LST. A weak land-breeze convergence line and associated line of cumulus clouds develop offshore near sunrise. The land breeze continues until midmorning, when the sea-breeze cycle starts over again. It is interesting to note that in this model the maximum strength of the land breeze in the near-surface layer is comparable to that of the sea breeze. Because of day–night differences in stability and frictional effects, however, the observed strength of the daytime sea breeze at the surface is considerably greater than the nighttime land breeze.

The importance of the effect of latitude on the sea-breeze circulation has been investigated numerically by scientists at the U.S. National Center for Atmospheric Research. They show that at the equator the absence of the Coriolis force results in a sea breeze at all times. At the other latitudes, the Coriolis force is responsible for producing the large-scale land breeze. At 20°N, the slower rotation of the horizontal wind after sunset produces a large-scale land breeze that persists until several hours after sunrise. At 30°N, the inertial effects produce a maximum land breeze at about sunrise, and the land breeze is strongest at this latitude. At 45°, the rotational rate of the horizontal wind after sunset is faster, so that the maximum land breeze occurs several hours before sunrise. These results indicate that the Coriolis force may be more important than the reversal of the horizontal temperature gradient from day to night in producing large-scale land breezes away from the equator.

Onshore penetration of the sea breeze varies with latitude also. In midlatitude regions (generally above 40°N), even under favorable synoptic conditions, the sea breeze may extend to 100 km. At latitudes equatorward of about 35°, much greater penetrations have been reported, for example, 250 km inland of the Pakistan coastline. In tropical regions of Australia, the sea breeze can penetrate to 500 km. In such cases, the sea breeze traveled throughout the night before dissipation occurred shortly after sunrise on the second day.

The effect of the sea breeze on the long-range transport of air pollutants to cause inland nighttime high oxidants has recently been investigated by Japanese scientists. On clear nights with weak gradient winds, a surface-based inversion layer often forms between sunset and sunrise. A strong inversion forms mainly in basin bottoms in the inland mountainous region between Tokyo on the Pacific Ocean and Suzaka near the Japan Sea. An air mass that passes over the large emission sources along the coastline can be transported inland by the sea breeze in the form of a gravity current. In the case studied, a high-concentration layer of oxidants was created in the upper part of the gravity current. It descended at the rear edge of a gravity-current head because of the internal circulation within the

temperature over land is still cooler than over the sea and the land breeze is still blowing. By 1200 LST, the land has become warmer than the water, and the circulation has reversed. At this time, a line of small cumulus may mark the sea-breeze front. At 1500 LST, the sea breeze is fully developed, and rain showers may be observed at the convergence zone, 30 to 40 km inland. Because of a low-level velocity divergence, there is a pronounced subsidence and thus a clear sky near the coastal area at this time. At 1800 and 2100 LST, the sea breeze is still clearly present but is gradually weakening in intensity. By midnight or 0000 LST, the sea breeze is barely evident aloft, and the surface wind is nearly calm over land. At this time, a temperature inversion and occasionally fog appear over land. After land again becomes cooler than the water, a land

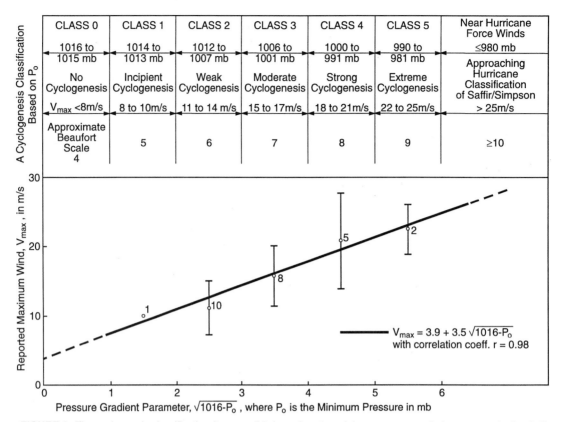

FIGURE 6 The cyclogensis classification (top panel) is based on the minimum pressure of winter storms in the Gulf of Mexico. The bottom panel shows the number of storms studied, the relationship between the pressure gradient parameter and the reported maximum winds, while the vertical bars are the standard deviations. [Reprinted with permission from Hsu, S. A. (1993). *Mariners Weather Log* **37**(2), 4.]

head, thus yielding the highest concentration of oxidants near the ground.

An example of the sea-breeze system along the coasts of Texas and Louisiana is presented in Figs. 8 and 9.

B. Low-Level Jets

Low-level jets have long been known to meteorologists. There are many manifestations of low-level jets around the world and many mechanisms for their formation. For instance, in the Northern Territory of Australia the mechanism for a jet would seem to be an inertial oscillation set up when the turbulent shearing stress falls dramatically with the formation of the nocturnal inversion. Observations from that region indicated that for geostrophic winds in the range of 10–20 m/sec ageostrophic wind magnitudes of 5–10 m/sec were common above the surface layer near sunset, with cross-isobar flow angles of above 40°. The jet that then developed by midnight was probably the result of these large ageostrophic winds, strong surface cooling, and favorable baroclinicity and sloping terrain.

Low-level jets usually have a well-marked super-geostrophic maximum in the boundary-layer wind speed

profile within a few hundred meters above the ground. They are modified by thermal stratifications, baroclinicity of the lower atmosphere, advective accelerations, and nonstationarity of the boundary layer. Some low-level jets have a diurnal life cycle with pronounced maxima during the night. An improved numerical model developed recently by German scientists is able to simulate the low-level jet. The improvements stem from the incorporation of a diurnally varying drag coefficient rather than a constant value for the entire 24 hr.

In certain regions, heavy rainfall is closely related to the low-level jet. For example, during the early summer rainy season in subtropical China and Japan, extremely heavy rainfall (at least 100 mm/day) is one of the most disastrous weather phenomena. As in other parts of East Asia, extremely heavy rainfall is also found by meteorologists in Taiwan to be closely associated with a low-level jet. For example, there was an 84% likelihood that a low-level jet of at least 12.5 m/sec would be present at 700 mb (about 3 km) 12 hr before the start of the rainfall event. They concluded that the low-level jet may form to the south of heavy rainfall as part of the secondary circulation driven by convective latent heating.

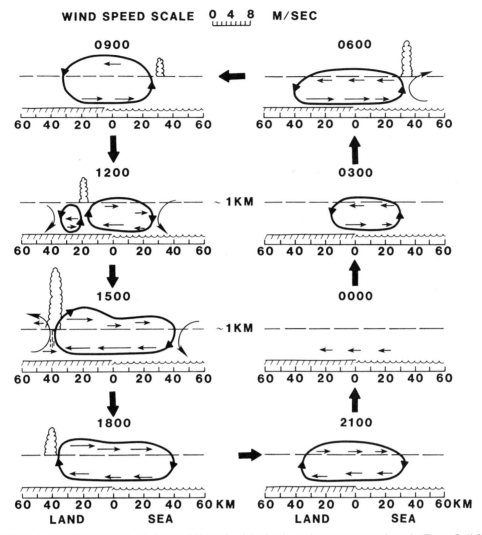

FIGURE 7 A simplified synthesized observed life cycle of the land–sea breeze system along the Texas Gulf Coast. Arrow lengths are proportional to wind speed. See text for explanation.

Along the coast of California during north-westerly up-welling favorable winds, the marine atmospheric boundary layer is characterized by a low-level jet, with peak wind speeds of as much as 30 m/sec at elevations of a few hundred meters. The vertical structure is marked by an inversion, usually at or near the elevation of the wind speed maximum. Above the inversion, the stratification is stable, and the wind shear is caused primarily by baroclinicity (thermal wind) generated by the horizontal temperature gradient between the ocean and land. Below the inversion, the flow is turbulent.

C. Other Coastal Winds

1. Wind Reversals along the California Coast

North winds along the northern California coast in summer may be interrupted by southerly winds. At the start of the particular event studied, the marine layer thickened in the southern California bight. A couple of days later the marine layer thickened from Point Conception to Monterey. Then, the marine layer thickness increase surged to the north along the coast to Point Arena, where progression stopped and an eddy formed. In this surging stage, winds switched to southerlies as the leading edge of the event passed. A day later, the leading edge surged farther to the north. Inshore winds were southerly, and the lifted marine layer extended to Cape Blanco in Oregon. This event is interpreted by C. Dorman as a gravity current surging up the coast.

2. Offshore-Directed Winds over the Gulf of Alaska

The strong thermal contrast between relatively warm off-shore waters and frigid air over the interior plateau of Alaska creates a region of hydrostatic pressure contrast

FIGURE 8 The sea-breeze system along the Gulf coast of Mexico, Texas, and Louisiana. This visible imagery from the GOES satellite shows the sinking air (or subsiding and thus clearing) on both sides of the shoreline. Notice the existence of the sea-breeze front or the convergence line displaced inland from the shore.

along the southern gulf coast of Alaska during the cold season. As a result, there is frequent regional offshore flow and nearly continuous drainage flow through mountain gaps in this region during the winter months. Scientists from the U.S. Pacific Marine Environmental Labora-

tory found that coastal mountains around Prince William Sound contribute to offshore winds in three ways: (1) by forming a physical barrier to low-level coastal mixing of cold continental and warm marine air, (2) by providing gaps through which dense continental air may be

FIGURE 9 Infrared imagery from the GOES satellite for Fig. 8. Two lines across the sea-breeze front are delineated, one in south Texas and the other in west Louisiana. They provide the horizontal temperature distribution of the cloud top at B (south Texas) and F (west Louisiana), sea surface (at D and H), and ground (at A, C, E, and G). Note that both B and F are located on the sea-breeze front. Also, there is 4°C difference between A and C as well as between E and G, indicating the advancing of cooler air onshore associated with the sea-breeze system.

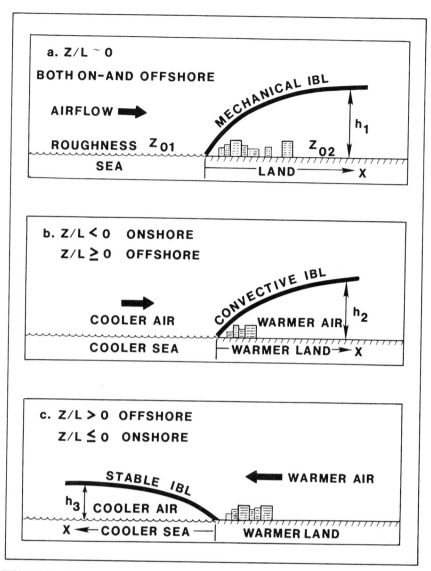

FIGURE 10 Schematics of the development of internal boundary layers across a shoreline.

channeled, and (3) by exciting mountain-lee waves. Nearshore winds are perturbed and highly localized. Cold drainage winds are eroded by heat and momentum transfer within a few tens of kilometers of the coast. Geotriptic adjustment of the regional surface-wind field is nearly achieved within a distance from the barrier corresponding to the Rossby radius of deformation. Note that in a geotriptic flow the acceleration term in the horizontal equations of motion is balanced by the pressure gradient force, Coriolis force, and frictional force.

III. BOUNDARY-LAYER PHENOMENA

A. Internal Boundary Layer

When air flows across a shoreline, its boundary layer undergoes significant modification in aerodynamic rough-ness, potential temperature, mixing ratio, and aerosol concentration. Because these changes in atmospheric layering occur within the planetary or atmospheric boundary layer, the modified layer is dubbed the internal boundary layer (IBL). There are two major IBLs, one thermal and the other mechanical. The thermal IBL can be further classified as convective or stable, depending upon the wind direction with respect to the temperature contrast between the land and sea. Figure 10 shows these IBLs. A review of the IBL is given in Garratt (1990).

B. Mechanical Internal Boundary Layer

The mechanical IBL develops across the shoreline because of roughness changes. Mechanical turbulence over-powers thermal contrast to make the stability parameter, Z/L, close to zero, or neutral. As shown in Fig. 10a, the height of the mechanical IBL, h_1, is given by the general form

$$h_1 = a \times Z_{02} \times (x/Z_{02})^b. \qquad (1)$$

Experimental results have shown that

$$a = 0.75 + 0.03\ln(Z_{01}/Z_{02}),$$

where Z_{01} and Z_{02} are the aerodynamic roughness lengths over the water (upwind) and land (downwind), respectively; X is the fetch downwind from the shoreline; and the power, b, is equal to approximately 0.8. Typical values of roughness length are about 0.01 cm for the water surface and about 10 cm for a relatively flat coast.

C. Convective Internal Boundary Layer

A convective IBL develops when cooler air flows from the sea to warmer land. It is modified by the temperature contrast, as shown in Fig. 10b. The height of the convective IBL, h_2, as derived by A. Venkatram, is

$$h_2 = \left[\frac{2C_d(\theta_{\text{land}} - \theta_{\text{sea}})X}{\gamma(1 - 2F)}\right]^{1/2}, \qquad (2)$$

where C_d is the drag coefficient inside the convective IBL; γ is the lapse rate above the boundary layer or upwind conditions; F is an entrainment coefficient, which ranges from 0 to 0.22; θ_{land} and θ_{sea} are the potential air temperatures over land and water, respectively; and X is the distance or fetch downwind from the shoreline. The dependency of h on $X^{1/2}$ has been predicted by dimensional analysis and by the modynamic approaches.

Equation (2) can be rewritten as

$$h_2 = AX^{1/2}. \qquad (3)$$

In flat coastal regions, $A \approx 60$ if h_2 is in meters and X is in kilometers.

Over urban areas in the coastal zone, however, the following formulas for the height of a daytime convective IBL have been recommended by S. R. Hanna for operational application:

$$h_2 = 0.1X \qquad \text{for } X \leq 200 \text{ m} \qquad (4a)$$

$$h_2 = 200 \text{ m} + 0.03(X - 2000 \text{ m}) \qquad \text{for } X > 2000 \text{ m}, \qquad (4b)$$

where X is the distance from the shoreline.

Under convective IBL conditions, the phenomenon of fumigation near the shoreline is a common occurrence. Recent studies have shown that for maximum ground-level concentrations

$$X_{\max} \simeq 0.3Q/(uh^2)$$

which occur at a downwind distance of

$$X_{\max} \simeq 10(u/W_*)h,$$

where W_* is the convective velocity scale,

$$W_* = (gHh/C_p\rho T)^{1/3},$$

where H is the sensible heat flux, T is the air temperature near the surface, u is the wind speed, h is the IBL height, and C_p is the specific heat at constant pressure, all inside the convective IBL.

A similar phenomenon exists when cooler air flows from a colder sea toward an oceanic warm front. For example, over the Gulf Stream when the wind direction is approximately perpendicular to the edge of the stream, a convective IBL may develop. At approximately 70 km downwind from the edge of the Gulf Stream, the height of the convective IBL, h_2, was observed at 300 m. This says that over the Gulf Stream when conditions are right

$$h_2 \approx 36X^{1/2}. \qquad (5)$$

Comparison of Eqs. (3) and (5) indicates that the height of the convective IBL over a coast is higher than that over an ocean for a given fetch and temperature difference. This is mainly due to the higher drag coefficient ($\approx 10 \times 10^{-3}$) across the shoreline on land than that ($\approx 1.5 \times 10^{-3}$) across an oceanic front at sea.

D. Stable Internal Boundary Layer

Contrary to the convective internal boundary layer, a stable IBL develops when warmer air is advected from an upstream warmer land (or sea) surface to a cooler sea downstream. This situation is shown in Fig. 10c. The main difference between convective and stable IBLs is that the heat flux is directed upward (from a warm sea to cooler air) for a convective IBL and downward (from warm air to a cooler sea) for a stable IBL. A two-dimensional numerical mesoscale model was used by J. R. Garratt to investigate the internal structure and growth of a stably stratified IBL beneath warm continental air flowing over a cooler sea. An analytical model was also used by Garratt to study a stable IBL, and excellent agreement with the numerical results was found. This analytical model states that

$$h_3 = BX^{1/2}U(g\Delta\theta/\theta)^{-1/2}, \qquad (6)$$

where h_3 is the depth of the stable IBL, which relates to X, the distance from the coast; U, the large-scale wind (both normal to the coastline); $g\Delta\theta/\theta$, in which $\Delta\theta$ is the temperature difference between continental mixed-layer air and sea surface; the mean potential temperature, θ; the acceleration caused by gravity, g; and other parameters combined as B.

From numerical results, Garratt suggests that B is a constant with a value of 0.014. Actually, B relates several other parameters, such as the flux Richardson number, the geostrophic drag coefficient, and the angle of the geostrophic wind measured counterclockwise from the positive X direction.

Comparing Eqs. (2) and (6) indicates that, for a convective IBL, $h \propto |\Delta\theta|^{1/2}$ and, for a stable IBL, $h \propto |\Delta\theta|^{-1/2}$. This difference is important, since both equations are consistent with energetic considerations.

For operational purposes, Eq. (6) may be simplified. Appropriate experiments in a flat coastal zone show that (see Fig. 10c)

$$h_3 \approx 16X^{1/2}, \qquad (7)$$

where h_3 is in meters and X is in kilometers.

IV. AIR–SEA INTERACTION

A. Meteorological Fluxes

Meteorological transport processes in coastal marine environments are important from several points of view. For example, the wind stress or momentum flux is one of the most essential driving forces in water circulation. Heat and convection are the origin of some localized coastal weather systems. Sensible heat and water vapor fluxes are necessary elements in radiation and heat budget considerations, including computation of evaporation and salt flux for a given estuarine system.

In the atmospheric surface boundary layer, the vertical turbulent transports are customarily defined as, for practical applications,

$$Momentum\ flux = \rho u_*^2 = \rho C_d U_{10}^2 = \tau$$

$$Sensible\ heat\ flux = \rho C_p C_T (T_{sea} - T_{air})U_{10} = H_s$$

$$Latent\ heat\ flux = L_T \rho C_E (q_{sea} - q_{air})U_{10} = H_l$$

$$Moisture\ flux = \rho C_E (q_{sea} - q_{air})U_{10} = E$$

$$Buoyancy\ flux = C_T U_{10}(T_{sea} - T_{air})\left(1 + \frac{0.07}{B}\right)$$

$$Bowen\ ratio = B = \frac{H_s}{H_l}$$

where

ρ = air density
u_* = friction velocity
C_d = the drag coefficient
U_{10} = wind speed at 10 m above the sea surface
C_p = the specific heat capacity at constant air pressure
C_T = the sensible heat coefficient
T_{sea} = the "bucket" seawater temperature in the wave-mixed layer (in °C)
T_{air} = the mean air temperature at the 10 m reference height (in °C)
L_T = the latent heat of vaporization
C_E = the latent heat flux coefficient

q_{sea} = the specific humidity for the sea
q_{air} = the specific humidity for the air
E = evaporation

Operationally, according to the WAMDI Group (1988), $u_* = U_{10}\sqrt{C_d}$, where

$$C_d = \begin{cases} 1.2875 * 10^{-3}, & U_{10} < 7.5\ \text{m sec}^{-1} \\ (0.8 + 0.065U_{10}) * 10^{-3}, & U_{10} \geq 7.5\ \text{m sec}^{-1} \end{cases}$$

According to Garratt (1992), $C_T = C_E \simeq 1.1 * 10^{-3}$ ($\pm 15\%$); and according to Hsu (1999),

$$B = 0.146(T_{sea} - T_{air})^{0.49}, \qquad T_{sea} > T_{air}.$$

The atmospheric stability parameter, Z/L, at height Z over the sea surface is defined as

$$\frac{Z}{L} = -\frac{\kappa g Z C_T (T_{sea} - T_{air})\left(1 + \frac{0.07}{B}\right)}{(T_{air} + 273)U_Z^2 C_d^{3/2}},$$

where L = Monin-Obukhov stability length
κ = Von Karman constant = 0.4
g = gravitational acceleration = 9.8 m sec^{-2}
U_Z = wind speed at height Z, normally set to 10 m

Thus, Z/L represents the relative importance between (or simply the ratio of) the buoyancy effect (or thermal turbulence) and the wind-shear or mechanical turbulence. Note that if $T_{sea} > T_{air}$, Z/L is negative, this stands for an unstable condition. On the other hand, if $T_{air} > T_{sea}$, Z/L is positive, the stability is said to be stable. When $T_{air} \simeq T_{sea}$, $Z/L \simeq 0$, the stability is near neutral.

B. Wind–Wave Interaction

Ocean surface waves are primarily generated by the wind. Because the water surface is composed randomly of various kinds of waves with different amplitude, frequency, and direction of propagation, their participation is usually decomposed into many different harmonic components by Fourier analysis so that the wave spectrum can be obtained from a wave record. Various statistical wave parameters can then be calculated. The most widely used parameter is the so-called "significant wave height," $H_{1/3}$, which is defined as the average height of the highest one-third of the waves observed at a specific point. Significant wave height is a particularly useful parameter because it is approximately equal to the wave height that a trained observer would visually estimate for a given sea state.

In the fetch-limited case (i.e., when winds have blown constantly long enough for wave heights at the end of the fetch to reach equilibrium), the parameters required for wave estimates are the fetch F and U_{10}. The interaction among wind, wave, and fetch are formulated for operational use as simplified from the U.S. Army Corp of Engineers (1984):

$$\frac{gH_{1/3}}{U_{10}^2} = 1.6 * 10^{-3} \left(\frac{gF}{U_{10}^2}\right)^{1/2}$$

$$\frac{gT_p}{U_{10}} = 2.857 * 10^{-1} \left(\frac{gF}{U_{10}^2}\right)^{1/3}.$$

The preceding equations are valid up to the fully developed wave conditions given by

$$\frac{gH_{1/3}}{U_{10}^2} = 2.433 * 10^{-1}$$

$$\frac{gT_p}{U_{10}} = 8.134,$$

where g is the gravitational acceleration, $H_{1/3}$ is the spectrally based significant wave height, T_p is the period of the peak of the wave spectrum, F is the fetch, and U_{10} is the adjusted wind speed.

In order to estimate hurricane-generated waves and storm surge, the following formulas are useful operationally, provided the hurricane's minimum (or central) pressure near the surface, P_0, in millibars is known (see Hsu, 1988, 1991, and 1994).

$$H_{max} = 0.20(1013 - P_0)$$

$$\frac{H_r}{H_{max}} = 1 - 0.1\frac{r}{R}$$

$$\Delta S = 0.069(1013 - P_0) = 0.35 H_{max},$$

where H_{max} (in meters) is the maximum significant wave height at the radius of maximum wind, R (in kilometers); H_r (in meters) is the significant wave height at the distance r away from R; and ΔS (in meters) is the maximum open-coast storm surge (i.e., above the astronomical tide) before shoaling. The mean R value for hurricanes is approximately 50 km.

V. HURRICANES

Hurricanes are tropical cyclones that attain and exceed a wind speed of 74 mph (64 knots or 33 m sec^{-1}). A hurricane is one of the most intense and feared storms in the world; winds exceeding 90 m sec^{-1} (175 knots or 200 mph) have been measured, and rains are torrential. The Saffir–Simpson damage-potential scale is used by the U.S. National Weather Service to give public safety officials a continuing assessment of the potential for wind and storm-surge damage from a hurricane in progress. Scale

FIGURE 11 Hurricane Florence over the north-central Gulf of Mexico at 9:05 AM (CDT) on September 9, 1988. (Courtesy of Oscar Huh and David Wilensky, Louisiana State University.)

FIGURE 12 This pre-Florence photo was taken on July 16, 1988, along the north-central section of Curlew Island, which is an undeveloped barrier island in St. Bernard Parish, LA. The photo shows a partially vegetated barrier island with a large area of washover flat deposit to the right. However, no major washover channels exist at this time period. (Courtesy of Louisiana Geological Survey, Coastal Geology Section.)

FIGURE 13 Significant hurricane impact features associated with Hurricane Florence can be seen in this photo, which was taken on October 13, 1988, at the same locality. Waves in combination with the hurricane storm surge cut several large washover channels across Curlew Island, depositing washover fans into the Chandeleur Sound. These washover channels break the barrier into smaller pieces and limit the ability of the barrier island to act as a buffer against subsequent hurricane impacts. (Courtesy of Louisiana Geological Survey, Coastal Geology Section.)

FIGURE 14 Hurricane Gilbert near Yucatan, Mexico at 4:17 PM (CDT) on September 13, 1988. (Courtesy of Oscar Huh and David Wilensky, Louisiana State University.)

FIGURE 15 This pre-Gilbert photo was taken on July 15, 1988, along the central part of Grand Isle, which is the only commercially developed barrier island in Louisiana. In 1984, the U.S. Army Corps of Engineers constructed an 11-ft-high artificial dune along the entire length of Grand Isle with wooden beach-access structures placed over the dune at regular intervals. The artificial dune provides protection against hurricane storm surge and flooding. Note the position of the shoreline in relation to the wooden beach-access structure. (Courtesy of Louisiana Geological Survey, Coastal Geology Section.)

FIGURE 16 Although Hurricane Gilbert made landfall over 400 miles away at the Mexico–Texas border. Grand Isle still experienced significant coastal erosion. This post-Gilbert photo was taken on October 12, 1988. The artificial dune was totally eroded away, and the shoreline migrated approximately 10–20 m to the landward side of the wooden beach-access structure. (Courtesy of Louisiana Geological Survey, Coastal Geology Section.)

numbers are made available to public safety officials when a hurricane is within 72 hr of landfall.

In September 1988, two hurricanes affected the coastal regions of the Gulf of Mexico. One was Florence, a minimal hurricane (scale 1), and the other was Gilbert, a catastrophic storm (scale 5). The effect of these two hurricanes on the barrier islands of Louisiana is shown in Figs. 11–16. Figure 11 is an NOAA-10 satellite image made at 9:05 AM (CDT) on September 9, 1988, as received at Louisiana State University, Baton Rouge. Wind

speeds at the eye of Florence, about 30 miles wide, were 80 mph. The eye passed 30 miles east of New Orleans around 3 AM then headed across Lake Pontchartrain, slightly west of the predicted path, over the city of Slidell, LA. The effect of Florence on a barrier island is shown in Figs. 12 (prestorm) and 13 (poststorm). Waves and surges produced by Florence physically cut several large washover channels, which broke the barrier island into smaller pieces and limited its ability to act as a buffer against subsequent hurricane impacts.

FIGURE 17 Visible imagery from the GOES satellite while Hurricane Floyd was still over the Bahama region.

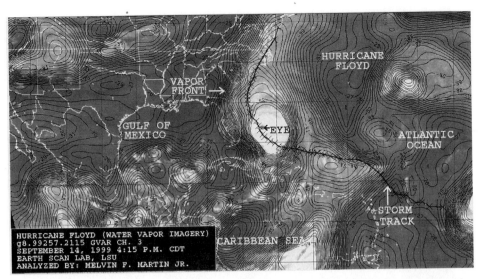

FIGURE 18 Water vapor imagery at the same time as the visible shown in Fig. 17. Note that the tongue of moisture (or the water vapor plume) extended northeastward over North Carolina. The hurricane track appears to follow this vapor plume. Note also that the vapor front which was located west of the vapor plume acts as a blocker to prohibit possible landfall of Floyd over Florida, Georgia, and South Carolina.

Hurricane Gilbert (Fig. 14), one of the most powerful (category 5) storms on record, devastated the Yucatan Peninsula, Mexico, on September 14, 1988. Two days earlier, it had destroyed an estimated 100,000 of Jamaica's 500,000 houses. Two days later, it again made landfall, striking northeastern Mexico and causing more than 200 people to perish.

The damage-potential scale categories of the Saffir–Simpson scale of hurricanes range from category 1, a minimal-size hurricane with central pressure equal to or greater than 980 mbar, up to category 5, a catastrophic strom with central pressure lower than 920 mbar. Two other category 5 hurricanes have affected the Gulf of Mexico region in this century: one in September 1935, which devastated Key West, FL, and Hurricane Camille in August 1969, which caused extensive damage along the Mississippi and Louisiana coasts.

The NOAA satellite advanced very high resolution radiometer imagery of Hurricane Gilbert was acquired with a 1.2-m program track antenna of the SeaSpace Co., Terascan System, NOAA satellite Earth station, established in the summer of 1988 at Louisiana State University, Baton Rouge. The hurricane photographs are of the 1.1-km resolution channel 2 (3.55–3.93 μm NIR) imagery. At the time of this NOAA-9 overpass, on September 13 at 4:17 PM (CDT), Gilbert was a category 5 hurricane nearly the size of the Gulf of Mexico, with the eye located at latitude 19°25.51′N, longitude 83°15.96′W. The eye, as measured with the channel 4 data, was 21 km in diameter. Channel 4 radiation temperatures ranged from 23.5°C in the center of the eye to below −83°C on the surrounding cloud tops.

Photographs taken before (Fig. 15) and after (Fig. 16) Gilbert at Grand Isle, LA, show that, even though Gilbert made landfall over 400 miles away, at the Mexico–Texas border, Grand Isle experienced significant coastal erosion.

Hurricane Floyd (Figs. 17 and 18) in 1999 caused extensive damage in North Carolina. Because its track was along the southeast coast of the United State, the exact landfall position was a challenge to forecast. With the aid of hourly water vapor imagery from the GOES satellite, improvements in earlier warnings can be made.

SEE ALSO THE FOLLOWING ARTICLES

ATMOSPHERIC TURBULENCE • CLOUD PHYSICS • COASTAL GEOLOGY • METEOROLOGY, DYNAMIC • OCEAN-ATMOSPHERIC EXCHANGE

BIBLIOGRAPHY

Bell, G. D., and Bosart, L. F. (1988). *Mon. Weather Rev.* **116**, 137.

Beyrich, F., and Klose, B. (1988). *Boundary-Laver Meteorol.* **43**, 1.

Boers, R., Spinhirne, J. D., and Hart, W. D. (1988). *J. Appl. Meteorol.* **27**, 797.

Brook, R. R. (1985). *Boundary-Layer Meteorol.* **32**, 133.

Chen, G. T. J., and Yu, C.-C. (1988). *Mon. Weather Rev.* **116**, 884.

Chen, S.-J., and Dell'Osso, I. (1987). *Mon. Weather Ree.* **115**, 447.

Coulman, C. P., Colquhoun, J. R., Smith, R. K., and Melnnes, K. (1985). *Boundary-Layer Meteorol.* **32**, 57.

Dorman, C. (1987). *J. Geophys. Res.* **92**(C2), 1497.

Garratt, J. R. (1985). *Boundary-Layer Meteorol.* **32**, 307.

Garratt, J. R. (1987). *Boundary-Layer Meteorol.* **38**, 369.

Garratt, J. R. (1990). *Boundary-Layer Meteorol.* **50**, 171.

Garratt, J. R. (1992). "The Atmospheric Boundary Layer," Cambridge Univ. Press, Cambridge, U.K.

Hanna, S. R. (1987). *Boundary-Layer Meteorol.* **40,** 205.

Hsu, S. A. (1988). "Coastal Meteorology," Academic Press, San Diego, CA.

Hsu, S. A. (1991). *Mariners Weather Log* **35**(2), 57.

Hsu, S. A. (1993). *Mariners Weather Log* **37**(2), 4.

Hsu, S. A. (1994). *Mariners Weather Log* **35**(1), 68.

Hsu, S. A. (1999). *J. Phys. Oceanogr.* **29,** 1372.

Kotsch, W. J. (1983). "Weather for the Mariner," 3rd ed., U.S. Naval Institute, Annapolis, MD.

Macklin, S. A., Lackmann, G. M., and Gray, J. (1988). *Mon. Weather Rev.* **116,** 1289.

McIlveen, J. F. R. (1986). "Basic Meteorology, A Physical Outline," Van Nostrand-Reinhold, Berkshire, England.

Roeloffzen, J. C., Van den Berg, W. D., and Oerlemans, J. (1986). *Tellus* **38A,** 397.

Smolarkiewicz, P. K., Rasmussen, R. M., and Clark, T. L. (1988). *J. Atmos. Sci.* **45,** 1872.

Stull, R. B. (1988). "An Introduction to Boundary Layer Meteorology," Kluwer, Dordrecht, The Netherlands.

Telford, J. W., and Chai, S. K. (1984). *Boundary-Layer Meteorol.* **29,** 109.

The WAMDI Group (1988). *J. Phys. Oceanogr.* **18,** 1775.

Toba, Y., Okada, K., and Jones, I. S. F. (1988). *J. Phys. Oceanogr.* **18,** 1231.

U.S. Army Corps of Engineers (1984). "Shore Protection Mannual," Visksburg, MS.

Ueda, H., and Mitsumoto, S. (1988). *J. Appl. Meteorol.* **27,** 182.

Venkatram, A. (1986). *Boundary-Layer Meteorol.* **36,** 149.

Welch, R. M., Kuo, K. S., Wielick, B. A., Sengupta, S. K., and Parker, L. (1988). *J. Appl. Meteorol.* **27,** 341.

Yan, H., and Anthes, R. A. (1987). *Mon. Weather Rev.* **115,** 936.

Zemba, J., and Friehe, C. A. (1987). *J. Geophys. Res.* **92**(C2), 1489.

Coatings, Colorants, and Paints

Peter K. T. Oldring

The Valspar Corporation

GLOSSARY

This glossary is for paints and coatings and is limited to paints and coatings.

Alkyd Polyester resin containing vegetable oil (or derived fatty acids) and phthalic anhydride. Alkyds can be air drying or non-air drying.

Binder Film-forming material which "binds" the pigment in the coating composition and to the substrate. Binders are part of a paint formulation and are in essence the non-volatile portion of the medium, colorants, additives, and fillers excluded.

Coalescence Fusing of soft, resin particles of an emulsion to form a continuous film.

Coating Pigmented or unpigmented film containing organic materials applied to a substrate such as wood, metal, plastic, paper, ceramic, or textile for decoration or protection. The term "coating" also includes paints, enamels, varnishes, lacquers, sizes, primers, basecoats, topcoats, tiecoats, and undercoats. It excludes adhesives, inks, and preformed films.

Colorant Pigment or dye which adds color to an ink or coating.

Crosslinking Chemical reactions which form a three-dimensional network in the binder, thereby giving the coating film its properties. Applicable only to thermoset coatings.

Crosslinking agent A chemically reactive species, often

Encyclopedia of Physical Science and Technology, Third Edition, Volume 3

a low molecular weight substance (often a resin) or a catalyst added to a thermoset resin in order to achieve crosslinking of the thermoset resin.

Cure The physical state of the film when it has achieved its desirable properties. Only applicable to thermoset coatings.

Dispersion Organic (e.g., polymer) or inorganic (e.g., pigments) particles dispersed in a medium, either aqueous or non-aqueous.

Drier Organometallic salt(s) added to air drying coatings which contain oils (oleoresinous or alkyds) to induce and promote the drying of the film through catalytic oxidation of fatty acid unsaturation of an alkyd or oleoresinous material. Different metal salts may have different functions, such as promoting surface or through drying.

Dye Colorant, which is soluble in the solvents and resin system, normally used in inks or textiles to produce transparent colored films.

Emulsion Suspension of fine particles of a liquid in a liquid, such as oil in water. Emulsion is frequently used to refer to polymeric dispersions in water, often prepared by emulsion polymerization.

Enamel Normally refers to a high-gloss pigmented coating on metal, glass, or ceramics. Enamels are coatings.

Film Dry or wet and not preformed. A liquid coating forms a wet film on the surface of the substrate when applied. When it becomes a coating, a continuous layer, (dry film) covering the required parts of the substrate, is formed.

Filler Insoluble particles in a coating which do not add color, but are added to improve the barrier effect of the coating film.

Gum Dried exudate of plants.

Ink Pigment dispersions in resins to provide external decoration. They are applied by printing processes directly onto substrates or onto coated substrate. Inks can be overvarnished.

Lacquer The term lacquer has different meanings in different industries and in different parts of the world. Originally, lacquers were based upon solutions of natural resins, which may or may not have been pigmented. Nowadays, lacquers are predominately transparent coatings based upon synthetic resins. Lacquers are coatings, being a continuous film formed by evaporation of solvent from a solution or dispersion.

Latex/Latices Originally referred to a natural rubber dispersion. Today, it is also used for synthetic polymer dispersions and emulsions, such as poly(vinyl acetate).

Lining Lining normally refers to can coatings and is the internal coating or lacquer (i.e., the coating in contact with foodstuffs). Lining is also used for interior tank coatings.

Medium The liquid phase of a paint.

Oil Fatty acid triglyceride, normally of vegetable origin. Oils are classified as drying, non-drying, and semi-drying.

Oil, drying Drying oils (normally vegetable oils) contain sufficient fatty acid unsaturation for free radical polymerization in the presence of air and a dryer to take place. An example is linseed oil.

Oil, non-drying Non-drying oils contain fatty acids with little or no unsaturation. An example is coconut oil.

Oil, semi-drying Semi-drying oils are intermediate to drying and non-drying. An example is soya bean oil.

Oleoresinous Containing a drying oil, usually in conjunction with a natural or synthetic resin.

Paint Mixture of pigment(s), resin(s), additives, and solvent(s) which form film when dry. Paint is a coating.

Pigment Synthetic or natural colorant, which is insoluble in the coating resins and solvents. A pigment needs dispersing and normally produces opaque-colored films. Other pigments include anticorrosive pigments, zinc dust, and aluminium flakes.

Polymer dispersion Defined in ISO 12000 as liquids or semi-liquid materials, usually milky white, containing the polymeric material in a stable condition finely dispersed in a continuous liquid phase, normally water (aqueous dispersion).

Primer First coat applied to a substrate to (1) provide corrosion resistance, (2) improve adhesion to the substrate, and (3) provide specific performance (such as barrier coat for migrants from the substrate or to block gas permeation through plastics). A primer is of a much higher film weight than a size coat. A primer is a coating.

Resin Originally, natural macromolecular products derived from residues of trees, etc. Today, resins are mainly of synthetic origin, and those containing natural products, such as alkyds, need reaction with synthetic materials to improve their performance. Resins can be either thermoplastic or thermoset. All coatings contain resins.

Resin, thermoplastic These form film primarily through evaporation of solvent. Generally, they are higher molecular weights than thermoset resins.

Resin, thermoset (chemically cured) These form film through crosslinking reactions. Although defined as thermoset, application of heat is not necessary for reaction, with some systems curing at ambient. They are generally of low molecular weight and contain functional groups, such as hydroxyl, through which crosslinking occurs. A crosslinking agent is frequently present in the coating.

Size (size coat) Initially, size coats were used to seal porous substrates and provide adhesion for subsequent layers of coatings. Nowadays, size coats refer to low film weight coatings applied over porous or non-porous substrates to improve the adhesion and mechanical and chemical resistance of subsequent coatings. A size coat is a coating.

Varnish External transparent (unpigmented) coating, used either for decoration or protection.

Vehicle Liquid component of a paint, that is, resins and solvents (i.e., a solution of the binder in the solvents).

I. HISTORY OF SURFACE COATINGS

The extent of the use of coatings, colorants, and paints can be used as an index of progress toward modern civilization. Primitive prehistoric paintings, such as those in the Altimira (Spain) and Lascaux (France) caves, were probably related to hopes of good hunting. The primitive painting of the grand bison on the walls of the Altimira cave in Spain is more than 15,000 years old. The painting of the Chinese horse on the walls of the cave at Lascaux, France, is less primitive than paintings at Altimira. The aborigines made pictures of game animals in their Obiri rock painting in Arnhem Land in northern Australia.

The first decorative coatings used on the walls of caves were based on colored mud. These water-dispersed coatings usually contained crushed berries, blood, eggs, plant sap, or milk. The Egyptians used distemper to decorate their walls in the period from 3000 to 1500 BC. The term *distemper* was coined by the British to describe waterborne coatings containing egg whites, glue, vegetable gums, or casein. The ancient Hebrews used milk or "curd paints" to decorate their homes.

The first paints were made by using binders (such as egg whites, milk curd, or pitch) to bind fillers and colorants (such as chalk, charcoal, ashes, earth colors, and vegetable dyes). The early Egyptians used pitch and balsam to caulk ships, and artists adopted these binders for paints. The first naturally based drying oil (unsaturated vegetable oil) was linseed oil obtained from flaxseeds.

The first colorants were white chalk, dark ashes, charred wood, and brighter hued earth colors. White lead [$2PbCO_3$ $Pb(OH)_2$], the oldest synthetic pigment, was produced by Pliny, who placed lead sheets in vinegar in the presence of carbon dioxide more than 2500 years ago. This primitive pigment not only provided a white paint, but also accelerated the drying (curing) of vegetable-oil-based binders.

It has been reported (Mattiello, 1941) that there was a time when the surface coatings industry dealt with a relatively small number of materials and processes for making paints and varnishes. Indeed, it was claimed (Mattiello, 1941) that between 1736 and 1900, Watin's book on varnish formulations was reprinted 14 times with only minor modifications. This was claimed to be the industry standard (Mattiello, 1941). Compare that with today's situation. A book which lasted for 200 years to the turn of the century was followed by different books which have been superceded at decreasing time intervals. It is not the writing which is inadequate, but it is a true reflection of the increasing rate of change of the surface coating industry.

The early oleoresinous paints have been largely displaced by synthetic resins, either in solution, dispersion, or solvent free. A spirit (ethanol)-soluble phenolic resin (Bakelite) was introduced as a replacement for shellac at the beginning of the 20th century. Alkyd resin, essentially an oil-modified Glyptal resin, was introduced by R. H. Kienle in the 1920s. Vinyl acetate and vinyl chloride copolymers, aminoplasts, epoxy resins, and acrylic resins were introduced as paint resins during the 1930s–1950s. Paints based on aqueous polymeric emulsions and powder coatings were introduced in the l950s. The last 10–20 years has seen a move toward more environmentally friendly coating systems, such as waterborne, high solids, radiation curable, and powder coatings.

II. NATURAL RESINS

Lacquers were introduced in the Chow dynasty (1027–256 BC) and were perfected in the Ming period (1368–1644 AD). The word *lacquer* is derived from *lac*, the name of a resin (shellac) secreted by an insect, *Coccos lacca*, that feeds on the lac tree. It has been estimated that 1,500,000 insects are required to produce sufficient lac for 0.5 kg of shellac (Mattiello, 1941). Shellac is the principal, non-vegetable, naturally occurring resin. The dried resin, called stick lac, is ground and washed to remove the red lac dye and is then melted and poured into molds. Shellac is a complex mixture of anhydrides, esters, and lactones of aliphatic and aromatic polyhydroxycarboxylic acids such as 9,10,16-trihydroxypalmitic acid. Because of its insolubility in varnish media, shellac has been used to cover knots in pine wood. Solutions of shellac are known as French polish for finishing wood. Shellac continues to be used in printing inks and as a spirit-soluble coating for candy, pills, and fruit.

However, shellac is no longer an important resin in the paint industry, due to cost and variability of supply.

Dammar resins are classified as recent fossil resins; that is, they are obtained from living plants and contain relatively large amounts of volatile solvents. These soft turpentine-soluble resins are obtained from trees in Borneo and Sumatra and are named after the export cities (i.e., Batavia, Singapore, or Penang). Dammar is used principally for its wax content.

Semi-fossil resins (i.e., those with relatively low solvent content) are obtained from dead trees. The principal East

Indies resins are batu and East India. These dark resins have been used in oleoresinous paints.

Melenket, loba, and Phillipine Manila resins are of recent origin, but pontianak is a semi-fossil resin and boea is a hard, oil-soluble fossil resin. Congo, a fossil resin from Africa, and kauri, a fossil resin from New Zealand, are also classified as copals.

Gum rosin (colophony) is obtained by tapping living pine trees or by the solvent extraction of pine tree stumps. The principal constituent of rosin is abietic acid, which can be isomerized by heat to produce levopimaric acid.

Gutta percha is a poly(trans-isoprene) obtained from the leaves of *Palaquium*, which is grown on plantations in Malaysia. Balata is a similar product obtained from the latex of *Minusops globosa*, which grows wild in Panama and northern South America. These resins have been used, to a limited extent, as paint resins but are no longer important in the coatings industry.

A naturally occurring fossil resin, amber, was used as an ingredient of varnishes as early as 250 BC. Actually, the name *varnish* is derived from *vernix*, the latin word for "amber." Amber is a fossil resin derived from an extinct variety of pine trees.

Accroides (Botany Bay gum), also called Accroides gum, is an exudate of the Australian grass tree (*Xanthorrhea*). Elemi is a soft, yellow-brown recent resin obtained from *Canarium* and *Amyris* trees in the Phillipines. Sandarac gum (juniper gum) is a yellow recent resin obtained from *Calitris quadrivavis* trees in Morocco. Mastic gum is a turpentine-soluble exudate of the *Pistacia lintiscus* tree found in the Greek archipelago.

The first lacquers were produced from the exudate of the sumac tree (*Rhus vernicifera*).

III. OILS

A. Vegetable Oils

Vegetable oils are complex mixtures of triglycerides of fatty acids, ranging from about C_{12} to C_{20}. Because climate, soil, and other natural factors affect the composition of oils, the same oil may vary in composition of its fatty acids from region to region and year to year. The most commonly used oils are C_{18} fatty acids with different levels of unsaturation, typically none, one, two, or three double bonds. Apart from a few rare examples, these fatty acids possess unbranched chains and an even number of carbon atoms. When double bonds are present, they are usually in the *cis* form and are not conjugated.

Oils are classified into drying, non-drying, and semidrying oils, depending upon the type and amount of unsaturation present in their fatty acid components. Examples of drying oils are linseed, tung (China wood), oticicia, and perilla. Examples of semi-drying oils are soya bean, tall oil fatty acids, dehydrated castor, and safflower. Examples of non-drying oils are coconut, groundnut, olive, and hydrogenated castor.

Linseed oil has been used as a drying oil in coatings for more than eight centuries. Linseed oil, obtained from the seeds of flax (*Linium usitatissimum*) by expression or solvent extraction, consists of glycerides of linolenic, oleic, linoleic, and saturated fatty acids. This oil may be heated to 125°C with air (blown) to produce a drying oil that yields a harder film than unreacted linseed oil. "Boiled" linseed oil is oil that has been heated with traces of dryers.

Tung or China wood oil is obtained by roasting and pressing the seeds of *Aleurites cordata*. Tung oil contains the triple unsaturated eleostearic acid. The tung tree is indigenous to China and Japan, but has been grown in the Gulf Coast region of the United States.

Oiticica oil is obtained by pressing the seeds from the Brazilian oiticica tree (*Licania rigida*). Its principal constituents are glycerides of α-licanic acid (4-keto-9,11, 13-octodecatrienoic acid).

Dehydrated castor oil is obtained by heating oil from the castor bean (*Ricinus communis*) with sulfuric or phosphoric acid. During dehydration a double bond is formed.

Soybean oil is obtained by crushing the seeds of the legume *Glycine max*, *Soya hispide*, *S. japanica*, or *Phaseolus hispide*. This vegetable oil contains linoleic acid (51%), linolenic acid (9%), oleic acid (24%), and saturated oils (16%).

Tall oil fatty acids (TOFA) are derived from the stumps of pine trees. Scandanavian tall oil differs in composition to North American tall oil.

Coconut, palm, and groundnut oils are non-drying, primarily consisting of saturated fatty acids. They give good color retention to any system containing them.

B. Fish Oils

Menhaden oil is obtained from North Atlantic sardines (*Clupea menhaden*), and pilchard oil is obtained from European sardines (*C. pilchardus*). Oil from North American herring (*Clopea harengus*) is similar to sardine oil, and oil from Pacific herring (*C. caerulea*) resembles pilchard oil. About 70% of these fish oils are unsaturated oils. The use of fish oils has declined.

C. Air Drying of Oils

Drying oils such as linseed, soybean, and safflower oils have *cis*-methylene-interrupted unsaturation. Other oils, such as tung oil, contain conjugated double bonds.

The principal film-forming reaction of drying oils is oxidation, which includes isomerization, polymerization, and cleavage. These reactions are catalyzed by dryers such as zirconium, cobalt, and manganese (organometallic salts).

It has been shown that oleic acid forms four different hydroperoxides at C-8, C-11, C-9, and C-10 and that unsaturated double bonds are present on C-9, C-10, and C-8. It is believed that these products undergo typical chain reaction polymerization via the standard steps of initiation, propagation, and termination.

Conjugated oils for use in oleoresinous paints have been produced by the dehydration of castor oil and the isomerization of linoleic and linolenic esters.

Linoleic and linolenic acids or esters can be converted to conjugated unsaturated acids or esters by heating with alkaline hydroxides or catalysts, which shift the methylene-interrupted polyunsaturation to conjugated unsaturations.

In addition to the dehydration of castor oil and isomerization of methylene-interrupted unsaturated oils such as linseed oil, conjugated succinyl adducts can be formed by the reaction of maleic anhydride with non-conjugated unsaturated oils. The succinyl adduct, which is attached at the 9 and 12 positions on the chain, undergoes rearrangement to produce a conjugated unsaturated acid. Additional rearrangement to a *trans-trans* form occurs at 200°C.

The reaction of 5–8% of maleic anhydride with soybean oil produces an unsaturated oil that is competitive with linseed oil and dehydrated castor oil.

D. Driers for Oils (and Alkyds)

Before 1840, white lead was used exclusively for the production of white oil paints. Initial attempts to replace the toxic lead pigment with zinc oxide were unsuccessful because the lead forms fatty acid salts, which act as dryers, whereas zinc salts are not primary dryers.

Oil-soluble salts of lead, manganese, and cobalt with fatty acids were introduced as dryers for oil-based paints in 1805. The most widely used acid for forming salts from dryers is now naphthenic acid, which is obtained in the refining of petroleum. Salts of tall oil from paper manufacture and salts of other metals such as vanadium, iron, zinc, calcium, and zirconium are also used. The effect of dryers is reinforced by the addition of 1,10-phenanthroline.

A combination of dryers is normally used. They can be classified into oxidation catalysts, polymerization catalysts, and auxiliary catalysts.

Oxidation catalysts, such as those based upon cobalt, manganese, cerium, vanadium, and iron(III) are primary dryers. They help the film absorb oxygen and participate in the formation and decomposition of peroxides. They greatly affect the surface hardness of the dried film.

Polymerization catalysts, based upon metals such as lead, zirconium, rare earths, and aluminium, assist drying to completion and are called through or secondary dryers.

Auxiliary catalysts based upon calcium, lithium, and potassium do not act alone as a dryer. They do, however, play a synergistic role with cobalt or zirconium. Calcium is the most widely used. Zinc is also in this category and is used to reduce the rate of surface drying, thereby permitting better oxygen absorption by the film as it dries.

IV. ESTER GUMS

Ester gum is produced by the esterification of rosin by glycerol. The ester obtained from pentaerythritol has a higher melting point than ester gum. Adducts of maleic anhydride and rosin are also used in varnishes and printing inks.

The most important maleic resin (ester gum) is obtained by the reaction of rosin with maleic anhydride followed by esterification with a polyol such as glycerol. Maleic acid resins are non-yellowing and can be used in oleoresinous varnishes or with alkyds.

V. OLEORESINOUS VARNISHES

Oleoresinous varnishes contain vegetable oils and are made by blending natural or synthetic resins with drying oils at high temperatures (500°F or more) and subsequently adding dryers and thinners. The original resins were fossil resins (copals) from various parts of the world. These resins were augmented by exudates of pine trees (balsams). Rosin is obtained when turpentine is distilled from the balsam.

Rapidly drying copolymers are produced by heating linseed oil with dicyclopentadiene at elevated temperatures. Syrenated oils are also obtained from linseed, soybean, dehydrated castor, and tung oils. There is some question as to whether copolymerization occurs in these reactions, but the products do give excellent films. Superior coatings are produced when the styrene or *p*-methylstyrene is grafted onto the unsaturated polymer chain.

VI. ALKYDS

Berzelius prepared glyceryl tartrate resins in 1847, and Smith prepared glyceryl phthalate resins (Glyptal) in 1901. Alkyds are a sub-class of polyesters, but due to their importance in coatings they are normally treated separately.

Patents were granted to Kienle for oil-modified polyesters in 1933. Kienle coined the euphonious name alkyd from the names of the reactants: alcohol and acid. Alkyds, the most widely used coating resins, are produced by substituting some of the phthalic anhydride in the Glyptal formulation by an oil containing saturated or unsaturated fatty acids.

Oil-modified glyceryl phthalate resins are typical alkyd resins. The term *alkyd* also describes resins in which pentaerythritol, sorbitol, or mannitol has been substituted for glycerol. Likewise, while phthalic anhydride is normally used, isophthalic, terephthalic, succinic, adipic, azelaic, and sebacic acids as well as tetrachlorophthalic anhydride have been substituted for phthalic anhydride or phthalic acid in alkyd resins.

Alkyds are prepared by reacting oils with a polyol, such as glcerol, to form monoglycerides of the oil. These are then reacted with phthalic anhydride. Alternatively, fatty acids (derived from oils) can be reacted directly with polyols and phthalic anhydride.

Alkyds can be either air drying or non-air drying (normally baking), depending upon the type (conjugated or non-conjugated) and amount of carbon—carbon unsaturation present in the oils used. Different oils have different drying characteristics, and it is the combination of type of oil and oil length (medium to long) which determines whether an alkyd will air dry. Semi-drying oils can be used in air drying alkyd paints, but it depends upon the amount of oil present and the molecular weight of the alkyd.

Alkyds are classified as short-, medium-, and long-oil alkyds based on the relative amount of oil in the formulation. While definitions may vary slightly, short-oil alkyds typically contain <45% oil, medium-oil alkyds contain 45–55% oil, and long-oil alkyds contain >55% oil.

Non-drying alkyds can be used as a plasticizing resin in some coatings (e.g., nitrocellulose wood lacquers). In others, they need to be crosslinked to form a film (industrial stoving alkyd). Air drying alkyds have been used for many years in household gloss paints and have found extensive use.

A wide range of modified alkyds are available including styrenated, vinylated, and siliconized as well as urethane alkyds.

New generation alkyds, such as waterborne or high solids, will ensure the continued use of alkyd-based paints and coatings for many years to come.

VII. OIL-FREE POLYESTERS

Polyesters include alkyds. Normally, the differentiation used is alkyds and oil-free polyesters. The latter do not contain oils or fatty acids derived from oils. They consist of polyols and polyacids. Typical polyols include neopentyl glycol, trimethylol ethane (or propane), butane (or hexane) diol, and ethylene (or propylene) glycol. Typical polyacids include iso and terephthalic, adipic, and trimelitic anhydride. Phthalic anhydride finds limited usage in polyesters.

Polyester resins are used extensively in powder coatings and find use in certain types of industrial coatings. Siliconized polyesters are used for architectural claddings, coil applied.

Polyesters can contain unsaturated components such as maleic anhydride. These polyesters can be crosslinked by a free radical initiator. When dissolved in styrene or other unsaturated monomer, they find use in glass fiber laminates.

VIII. ACRYLIC RESINS AND STYRENE

Polyacrylic acid was synthesized in 1847, but this polymer and its esters were not investigated extensively until the early 1900s, when Otto Rohm wrote his Ph.D. dissertation on acrylic resins. Rohm and his associate, Otto Haas, produced acrylic esters commercially in Germany in 1927, and Haas brought this technology to the United States in the early 1930s.

Because the α-methyl group hinders segmental chain rotation, methacrylates are harder than their corresponding acrylate ester. Poly(methyl acrylate) has a T_g of $-9°C$ compared to $105°C$ for poly(methyl methacrylate). The flexibility increases and T_g decreases as the size of the alkyl ester group increases. Poly(2-ethyl hexyl acrylate) has a T_g of $-70°C$, and poly(butyl methacrylate) has a T_g of $22°C$.

Acrylic coatings are transparent and have good solvent, corrosion, and weather resistance.

Styrene is often copolymerized with acrylic monomers. This has technical benefits in some cases, as well as a cost reduction. Styrene has poor exterior durability characteristics, due to its aromatic ring. Polystyrenes are available, but find little usage as the sole coating resin. Styrene is nearly always copolymerized with acrylic monomers for use in coatings.

Acrylic monomers generally do not contain additional functionality. There are, however, a range of acrylate and methacrylate functional monomers available. This functionality includes hydroxyl, amino, amino-amide, acid, and epoxy. These monomers are incorporated into acrylic resins for thermoset applications or thermoplastic resins which may need the improved performance they bring, such as improved substrate adhesion. Thermoset acrylic coatings can be crosslinked by crosslinking agents, such as isocyanates, melamine, benzoguanamine, or urea formaldehyde resins. They may also be self-crosslinking as a result of functionality incorporated into the ester group. Acrylamide-containing resins are one example.

In 1956, General Motors Company replaced some of its cellulose nitrate automobile finishes with acrylic coatings. The labor-intensive polishing step used for both cellulose nitrate and acrylic finishes was eliminated by a "bake, sand, bake" technique in which all sand scratches were filled in during the final bake. Thus, these coatings can replace cellulose nitrate and alkyd or melamine formaldehyde automobile finishes. Both solvent-based and waterborne acrylic coatings are available.

High-quality acrylic latices, typically made from methyl methacrylate and ethyl acrylate or related monomers, have found extensive use in water-based paints.

Acrylic resins find widespread usage in the coatings industry.

IX. PHENOLPLASTS

Phenol and substituted phenols can react with formaldehyde under basic or acidic conditions to form phenolic resins, sometimes called phenolplasts. Phenolic resins can be non-reactive (NOVOLACS) or reactive (RESOLES). Novolacs are the result of an excess of phenol under acidic conditions, while resoles are the result of an excess of formaldehyde under basic conditions. Novolac resins can participate in cross-linking reactions if a formaldehyde-generating material, such as hexamine, is added.

Phenolplasts are complex mixtures due to self-condensation, which occurs during their manufacture. Normally, the methylol groups are partially substituted by reaction with an alcohol, such as iso-butanol, to improve solubility, compatibility, and adjust reactivity.

The condensation products of phenol and formaldehyde developed by Leo Baekeland in 1909 were not suitable for oil-based coatings. However, condensates of formaldehyde with substituted phenols, such as p-octylphenol, are oil soluble and can be used with drying oils for making oleoresinous varnishes. Phenolic-modified oleoresinous spar varnishes have superior resistance properties, particularly water resistance.

Phenolic resins, per se, find limited usage in coatings, being primarily used for structural applications. Phenolic resins can be used as the sole binder in drum coatings, for example, but frequently other resins are present. Resoles are used to crosslink other resins, such as epoxy resins for can linings. Phenolic resins are used as binders for the sand in foundry moulds.

Substituted phenol-formaldehyde resins are also admixed with neoprene to produce industrial adhesives.

X. AMINOPLASTS

Urea, melamine, or benzoguanamine can react with formaldehyde to form amino resins sometimes called aminoplasts. Further reaction with monohydric alcohols produces etherified amino resins. They are complex mixtures and undergo some degree of self-condensation during their manufacture. A major usage in coatings is as cross-linking agents.

Tollens condensed urea with formaldehyde in 1884. These resins are now used for paper coatings, wash-and-wear finishes on cotton textiles, and general-purpose coatings. The last-named are produced by the condensation of urea and formaldehyde in the presence of butanol. Butanol-etherified urea resins are used as modifiers for alkyd or cellulose nitrate resins in furniture coatings.

Melamine resins are produced by the condensation of melamine and larger proportions of formaldehyde than those used for urea resins. Like urea resins, they can be used as butylated resins. Also like the urea resins, they are used as crosslinking agents for alkyds, acrylics, epoxies, and other coating resins. Some melamine-based aminoplasts are water soluble, for example, hexamethoxymethyl melamine (HMMM).

Benzoguanamine formaldehyde resins are less functional and, hence, less reactive than their corresponding melamine formaldehyde resin.

XI. EPOXY RESINS

Epoxy resins, under the name of ethoxyline resins, were patented in Europe by P. Schlack in 1939 and introduced in the United States in the late 1940s. These resins, which are produced by the condensation of bisphenol A and epichlorohydrin, contain terminal epoxy groups and may contain many hydroxyl pendant groups, depending on molecular weight.

Epoxy resins are not used alone for coatings, normally being crosslinked.

Epoxy resins, based upon bisphenol A (or F) and epichlorhydrin, cured at room temperature by aliphatic polyfunctional amines and polyamides are used in heavy duty coatings for ships, oil rigs, and storage tanks, as well as water pipes. The epoxy resin-containing component of the paint is mixed with the polyamine-containing component prior to application. This is a two pack system. They also form the basis of the two pack adhesive (Araldite) available from most hardware stores.

Epoxy resins also react at elevated temperatures with aromatic amines, cyclic anhydrides, aminoplasts, and phenolplasts. In these applications, epoxy resin is present with the cross-linking agent in the coating and no mixing prior to application is required. This is a one pack coating. Epoxy resins have been used with acrylic coating resins, as powder coatings, and as high solids coatings. Epoxy resins are used as intermediates in UV coatings. Electrodepostion automobile primers are based upon epoxy resins crosslinked with polyisocyanate. Epoxy resins are used as additive resins in some coatings to improve properties such as adhesion and resistance.

Cycloaliphatic epoxy resins are used in cationic UV curable coatings and inks.

Epoxy resins reacted with fatty acids derived from oils produce epoxy esters, which can be air drying or non-air drying. The latter find usage in machinery coatings and are normally crosslinked with an aminoplast.

XII. POLYURETHANES

Urethane resins, introduced by Otto Bayer of I. G. Farbenindustrie in 1937, were produced in Germany during World War II. These polymers, which contain the characteristic NHCOO linkage, are produced by the condensation of a diisocyanate, such as tolylene diisocyanate, and a polyol, such as a polyether or polyester with two or more hydroxyl groups. Both aromatic and aliphatic isocyanates are used commercially, but great care must be exercised when handling them, particularly the more volatile isocyantes, due to their toxicity and potential for sensitization.

Polyurethanes can be produced *in situ* by reacting a diisocyanate and a polyether or polyester-containing hydroxyl group. These resins are thermoplastic and are used in inks for polyolefin films.

A major use of polyisocyanates is as a cross-linking agent for resins containing hydroxyl, or other, functionality. The resulting cross-linked network is a polyurethane. Because isocyanates are normally very reactive, they are sometimes used in two pack coatings.

One pack baked (stoving) coatings can be produced from blocked or capped isocyanates such as phenol adducts, which release diisocyanates when heated. These then react with any resins containing hydroxyl functionality.

One-component coatings based on the reaction products of drying oils and isocyanates are available. Monoglycerides are condensed with diisocyanates in a urethane-alkyd coating system. Moisture-cured polyurethane coatings are used for wood floor coatings.

Epoxy resins can be used as the polyol in cationic electrodeposited primers for automobiles. Polyurethane clear coats are used as an automobile topcoat.

XIII. VINYL ACETATE POLYMERS

Waterborne paints based on dispersions of casein were introduced commercially in the 1930s, but they lacked washing and scrub resistance. Hence, they were displaced by latices of polyvinyl acetate, stabilized by polyvinyl alcohol, and plasticized by dibutyl phthalate, which became available in the 1950s. More recent developments include copolymers of vinyl acetate with ethylene and vinyl chloride to improve performance. Other soft comonomers, such as butyl acrylate, which act as internal plasticizers for the harder (higher glass-transition temperature T_g homopolymer) vinyl acetate, are also used.

As a general rule, vinyl acetate-based latices are restricted to internal usage, due to their poor exterior durability characteristics.

XIV. VINYL CHLORIDE AND VINYLIDENE CHLORIDE POLYMERS

A. Vinyl Chloride Containing Polymers

Polyvinyl chloride was synthesized by Baumann in the 1870s. However, because of its insolubility and intractability, it remained a laboratory curiosity until it was plasticized by Waldo Semon in the 1930s. In 1927 chemists at Carbon and Carbide, duPont, and I. G. Farbenindustrie produced copolymers of vinyl chloride and vinyl acetate that were more flexible and more readily soluble than polyvinyl chloride.

The most popular copolymer (Vinylite VYHH) contains 13% of the vinyl acetate comonomer and is soluble in aliphatic ketones and esters. A terpolymer (VYMCH) produced by the addition of maleic anhydride to the other monomers before polymerization has better adhesion than VYHH, and a partially hydrolyzed VYHH (VAGH) is more soluble than VYHH.

It is customary to add plasticizers and stabilizers to these coating resins. VAGH is compatible with alkyd, urea, and melamine resins. Copolymers with high vinyl acetate content (385E, VYCC) are compatible with cellulose nitrate.

Polyvinyl chloride (PVC) suffers from dechlorination when heated, particularly in the presence of iron. Therefore, it is necessary to incorporate hydrochloric acid scavengers in PVC-containing coatings which will undergo thermal treatment, particularly if they are on metal. Examples of scavengers include epoxy resins and epoxidized oils.

B. Vinylidene Chloride-Containing Polymers

Polyvinylidene chloride was produced by Regnault in the 1830s, but this product was not commercialized until the 1930s. Chemists at Dow and Goodyear reduced the crystallinity of this polymer by copolymerizing the vinylidene chloride with acrylonitrile (Saran) or vinyl chloride (Pliovic). These copolymers, which are soluble in methyl ethyl ketone, are characterized by good resistance to gaseous diffusion and are used as barrier coatings. These copolymers are also used as latices for coating paper and textiles.

XV. CELLULOSE DERIVATIVES

Because of strong intermolecular hydrogen bonding, cellulose is insoluble in ordinary organic solvents. However, it is soluble in N,N-dimethylacetamide in the presence of lithium chloride, in dimethyl sulfoxide in the presence of formaldehyde, in Schweitzer's reagent (cuprammonium hydroxide), and in quaternary ammonium hydroxide. However, derivatives of cellulose are soluble in commonly used solvents.

A. Cellulose Nitrate—Nitrocellulose

Cellulose nitrate was synthesized in 1845 by Schonbein, who, because he believed it to be a nitro compound instead of an ester of nitric acid, mistakenly called it nitrocellulose. Solutions of cellulose nitrate (Pyroxylin) were patented by Wilson and Green in 1884.

Modern lacquers, which were introduced in 1925, are solutions of cellulose nitrate. These coatings dry by evaporation of the solvent. Pigmented nitrocellulose lacquers were used as finishes for automobiles in 1913.

Cellulose nitrate lacquers of higher solids content were hot-sprayed in the 1940s. In order to meet competition from alkyd-amino resin coatings, cellulose nitrate coatings were improved in the 1950s by the development of multicolor lacquer enamels and superlacquers based on cellulose nitrate-isocyanate prepolymers.

Wood lacquers were based upon nitrocellulose, but that usage is declining. Alkyd resins compatible with nitrocellulose have been used to plasticize nitrocellulose lacquers.

B. Cellulose Acetate

The first cellulose acetate was a triacetate produced by Schutzenberger in 1865. A more readily soluble cellulose diacetate was produced in 1903 by Miles, who reduced the acetyl content by partial saponification of the triacetate. The diacetate is soluble in acetone. More than 20 organic acid esters of cellulose are marketed in the United States, but only the nitrate and acetate are produced in large quantities. Cellulose acetate, plasticized by acetyl triethyl citrate, is used as a coating for paper and wire screening and as a cast film. Cellulose acetate butyrate (CAB) finds widespread usage as a modifying resin in industrial coatings.

C. Cellulose Ethers

Organosoluble ethylcellulose and ethylhydroxyethylcellulose are used in coatings and inks. These tough polymeric ethers have been approved by the U.S. Food and Drug Administration for use as food additives.

Methylcellulose, which is water soluble, was produced by Suida in 1905 and by Denham, Woodhouse, and Lilienfield in 1912. Water-soluble hydroxyethylcellulose was produced by Hubert in 1920, and carboxymethylcellulose was produced by Jansen in 1921. However, these polymeric ethers were not commercialized until the late 1930s, when they were used as thickeners in waterborne coatings systems.

XVI. RUBBER AND ITS DERIVATIVES

Natural rubber exists as a colloidal dispersion (latex) in certain shrubs and trees, the principal source being *Hevea braziliensis*. In 1823, Charles Macintosh coated cloth with a solution of rubber in a hydrocarbon solvent to produce the rainwear that bears his name. Polychloroprene (Neoprene), produced commercially by duPont in 1932, was also used as a protective coating for cloth and metal.

Latices of natural rubber, styrene-butadiene rubber and acrylonitrile-butadiene rubber, have been used as coatings for textiles (e.g., rug backing), concrete, and steel. It is customary to add an aqueous dispersion of curing and stabilizing additives to the latex before application.

Chlorinated rubber containing ~67% chlorine and sold under the trade names Tornesit and Parlon has been used for several decades as a protective coating. Chlorinated rubber is insoluble in aliphatic hydrocarbons and alkanols, but is soluble in most other organic solvents. Chlorinated rubber coatings have good adhesion to Portland cement and metal surfaces and are resistant to many corrosive materials.

Cyclized rubber (Pliolite) is produced by the chlorostannic acid isomerization of natural rubber. Cyclized rubber is widely used in paper coatings. It is compatible with paraffin wax and many hydrocarbon resins. This resin is soluble in aliphatic and aromatic hydrocarbon solvents and is usually applied as a solution in toluene, xylene, or naphtha.

The trade name Pliolite S-5 is also used for a copolymer of styrene and butadiene. This resin is more economical and is often used in place of cyclized rubber.

XVII. SILICONES

Kipping synthesized polysiloxanes in the early 1900s, but failed to recognize their commercial potential. He believed these products to be ketones and, hence, called them silicones. These polymers, produced by the hydrolysis of dichlorodialkylsilanes, were commercialized in the 1930s by General Electric Co. and Dow Corning Co. Silicones have excellent resistance to water and elevated temperatures. Aluminum powder-filled silicone coatings have been used at 250°C. Silicone-modified polyesters, alkyds, epoxy, and phenolic resins are also used as coatings. A modified silicone coating with a non-stick surface is used as a finish for other coating systems.

XVIII. POLYOLEFINS

Amorphous polyolefins, such as polyisobutylene, atactic polypropylene, and copolymers of ethylene and propylene, are soluble in aliphatic solvents, but crystalline polyolefin coatings must be applied as melts. Chlorosulfonated polyethylene produced by the reaction of polyethylene with chlorine and sulfur dioxide is soluble in aromatic solvents and is used as a protective coating. Because of the presence of the SO_2Cl pendant groups, this polymer can be cured by lead oxide in the presence of a trace of water or a diamine. The bonding of polyolefin coatings to metal substrates has been improved by the introduction of mercaptoester groups and by passing the polyolefin through a fluorine-containing gaseous atmosphere.

XIX. PIGMENTS AND COLORANTS

Pigments can be divided into organic and inorganic.

A. Inorganic Pigments

Basic lead carbonate [$2PbCO_3 Pb(OH)_2$] was used by the early Greeks 2400 years ago, and other inorganic pigments such as zinc oxide (ZnO) and zinc sulfide (ZnS) were used in the late 1700s. A white pigment consisting of barium sulfate ($BaSO_4$) and zinc sulfate ($ZnSO_4$) (lithopone) was used widely in place of white lead. Ilmenite was used as a pigment by J. W. Ryland in 1865. Titanium dioxide, a white pigment, was extracted from this ore in the 1920s. It was synthesized in the 1950s and is now the most widely used white pigment. Lithopone and basic sulfate white lead pigments were introduced in the 1850s.

1. Titanium Dioxide

In 1908, Jebson and Farup produced white titanium dioxide (TiO_2) by the action of sulfuric acid on black ilmenite ($FeTiO_3$). In spite of its superior hiding power and because of its high price, this pigment was not used to any great extent until the 1930s.

In the past, the anatase form of the pigment was the most widely used, but now the more stable (exterior durability) rutile form predominates. The hiding power and tinting strength of titanium dioxide is about twice that of other white pigments.

2. Iron Oxide

Earth pigments were used by the artists who painted on the walls of the Altamira caves in prehistoric times. These inexpensive pigments are still in use today.

a. Ochres. The coloring matter in ochres is exclusively iron oxide (Fe_2O_3). The pale golden French ochre contains from 20 to 50% iron oxide (Fe_22O_3). Yellow ochre, which is greenish yellow to brownish yellow, is hydrated iron oxide [$FeO(OH)$]. The coloring material in red ochre is anhydrous ferric oxide. Burnt ochre is produced by the calcination of yellow ochre.

b. Sienna. This pigment was mined originally in Siena, Italy, but is now found throughout Italy and Sardinia. Raw sienna is yellow-brown and contains 50–70% iron oxide. Burnt sienna, which is reddish orange, is obtained by the calcination of raw sienna.

c. Umber. Raw umber, which is found in Cyprus, is greenish brown. It is contaminated with 12–20% manganese dioxide (MnO_2). Burnt umber, which is deep brown, is produced by the calcination of raw umber.

d. Other iron oxides. Haematite, which is found around the Persian Gulf, is α-ferric oxide, and this red pigment is called Persian red oxide. Limonite, a hydrated ferric oxide found in India and South Africa, is yellow. Magnetite (Fe_3O_4) is black. Many synthetic iron oxides are produced by the controlled calcination of copperas or green vitriol ($FeSO_4 \cdot 7H_2O$).

3. Iron Cyanides

The generic name *iron blue* describes insoluble salts containing an iron hexacyanide anion. Prussian blue (ferrous ferricyanide) was discovered in Berlin in 1764. Prussian brown (ferric ferricyanide) is rapidly converted to Prussian blue by acidic surfaces such as clay.

4. Chromates

a. Chromic oxides. Guignet's green [$CrO(OH)$] is obtained by heating an alkali metal bichromite with boric acid at 500°C. Chromium oxide green (Cr_2O_3) is obtained by the calcination of potassium dichromate in the presence of carbon.

b. Lead chromate.

Chrome yellow is lead chromate ($PbCrO_4$). Lead oxide chromate ($Pb_2O \cdot CrO_4$) is orange-red. Actually, lead chromate may exist as a lemon yellow rhombic form, a reddish yellow monoclinic form, and a scarlet tetragonal form, but only the monoclinic form is stable at room temperature. Chrome orange and chrome red ($PbCrO_4 \cdot PbO$) are produced by heating lead chromate with alkali. The difference in the two colors is related to particle size. Chrome green is a mixture of chrome yellow and Prussian blue. Zinc yellow is zinc chromate loosely combined with zinc hydroxide.

5. Lead Pigments

Basic sulfate white lead ($PbSO_4 \cdot PbO$) can be produced by vaporizing galena (lead sulfide, PbS) and burning the vapor with a controlled amount of air or by spraying molten lead into a furnace with sulfur dioxide and air. Leaded zinc oxide is a mixture of white lead and zinc oxide. Lead tungstate ($PbWO_4$) and lead vanadate [$Pb(VO_3)_2$] are yellow pigments. Because all lead compounds are extremely poisonous, their use is being discontinued.

6. Zinc Pigments

a. Zinc oxide.

Zinc oxide (ZnO) has been used as a white pigment since 1835. Zinc oxide occurs in the United States as franklinite, willemite, and zincite ores, which are roasted to produce zinc oxide. The oxide is reduced to zinc by heating with carbon. In both the synthetic and mineral processes, the zinc metal is volatilized and oxidized by air to zinc oxide. On a weight basis, the hiding power of zinc oxide is superior to that of white lead. It acts as a moldicide and an ultraviolet absorber.

b. Lithopone and zinc sulfide.

Zinc sulfide (ZnS) is used directly as a white pigment and also with barium sulfate ($BaSO_4$) as a component of lithopone. Lithopone, which is used as a white pigment in waterborne paints, has better hiding power than zinc oxide.

7. Antimony Pigments

The white pigment antimony oxide (Sb_2O_3), sold under the trade name Timonox, has been used to reduce chalking of anatase-pigmented paints. However, it is not needed when the rutile form of titanium dioxide is used. The principal use of antimony oxide is as a flame retardant.

8. Cadmium Sulfide

Cadmium sulfide (CdS), called cadmium yellow, lacks acid resistance but is acceptable for indoor use in neutral or alkaline environments. This pigment tends to react with carbon dioxide and form less yellow cadmium carbonate. Cadmium sulfide can be admixed with barium sulfate to produce cadmopone. Because cadmium compounds are extremely toxic, their use is being discontinued.

9. Cobalt Blue

Because cobalt blue ($CoO.Al_2O_3$) was introduced as a pigment by Thenard and Proust in the late 1800s, it is often called Thenard's blue. Because of its low refractive index, it is transparent in oily media. It is used as a blue pigment in ceramics, but is not competitive with phthalocyanine blue in coatings.

10. Ultramarine

The ancient Egyptians obtained natural ultramarine by pulverizing the precious stone lapis lazuli. The name, which means "beyond the sea," refers to the asiatic origin of this semiprecious stone. In 1828, Guimet and Gmelin independently synthesized this pigment in their attempts to win a 6000 franc prize from the Societé d'Encouragement in Paris.

In the 1870s, more than 30 plants were producing synthetic ultramarine by heating clay, sulfur, and sodium sulfate with charcoal. Through further treatment, one obtains ultramarine in shades ranging from green to red. Ultramarine is used as a pigment for paper, textiles, plastics, and rubber.

11. Metallic Pigments

Finely divided metals and alloys are used as pigments in paints. The principal pigments are obtained from aluminum, zinc, and bronze. Aluminum powder, in both leafing and non-leafing forms, is produced by atomization of molten aluminum. The atomized powder is converted to lamellar particles by ball milling with stearic acid.

Metallic powders and lamellar pigments are produced from bronze by techniques similar to those employed for aluminum. These pigments are used for decorative effects. Zinc dust is used in the manufacture of anticorrosive primers. Metallic flakes are used for the production of antistatic coatings.

B. Organic Pigments

Natural organic colorants, such as carmine, logwood, indigo, and madder, which were used by ancient people, have been replaced by synthetic pigments and dyes. The first synthetic dyestuff was prepared in 1856 by Perkin, who oxidized aniline.

The search for these synthetic colorants was catalyzed in 1876 by Witt, who identified azo (—N=N—) and other common chromophores. Organic pigment dyestuffs are insoluble, while soluble toners are precipitated by the addition of metallic salts.

1. Carbon Black

The first carbon black, called lamp black, was produced 5000 years ago by allowing a flame to impinge on the surface of an inverted bowl. More recently, bone black has been prepared by the carbonization of degreased bones. Carbon black is now produced in large quantities by the incomplete combustion of hydrocarbons. The tinting strength of carbon black pigment is inversely proportional to its particle size.

Channel black has a "blackness or jetness" that is superior to that of furnace blacks. The latter, whose production is more acceptable to environmentalists, is produced by the incomplete combustion of natural gas or oils in a special furnace. Channel black is produced by an impingement process similar to that used to produce lamp black. Furnace blacks are used as reinforcing agents in rubber, but as pigments they are inferior to other blacks.

2. Aniline Black

Aniline black (nigrosine), which is obtained by the oxidation of aniline in the presence of Cu(II) ions, is the most widely used organic pigment. Unlike carbon black, the hiding power of aniline black is due to absorption rather than the scattering of light.

3. Phthalocyanines

Phthalocyanines, which were synthesized in 1927 by de Diesbach and van der Weid, have to a large extent replaced alkali-sensitive iron blues and acid-sensitive ultramarine. Gaertner has called the discovery of phthalocyanines "the turning point in pigment chemistry."

This pigment is produced by dissolving the heterocyclic phthalocyanine in concentrated sulfuric acid and precipitating it by the addition of water. Copper phthalocyanines are stable to heat and most corrosives. Green phthalocyanine is obtained by the chlorination of blue phthalocyanine. Bromination produces a yellow product.

4. Azo Colors

Insoluble azo pigments are produced by coupling diazonium compounds with a phenolic compound such as 2-naphthol, acetoacetanilide, or pyrazolone derivatives. Soluble azo pigments, which are characterized by the presence of acidic groups, are precipitated from aqueous solution by barium, calcium, iron, copper, or manganese cations. Typical azo pigments are toluidine red, Hansa yellow, benzidine yellow, pyrazolone red, and the barium salts of lithol red and pigment scarlet.

5. Vat Pigments

Vat pigments are an important class of textile colorants. Naturally occurring indigo and madder were known to the earliest civilizations but are made synthetically today. The name vat *dye* is derived from the vat in which the reduction of color and the immersion of cloth occurs. Reoxidation takes place in the atmosphere after removal from the vat.

6. Quinacridones

Quinacridone and its derivatives are stable solvent-resistant pigments. The color of these pigments ranges from red-yellow to violet, scarlet, and maroon.

XX. SOLVENTS

Ethanol, which was used for dissolving shellac, was produced a few thousand years ago by the fermentation of starch from grains. Ethyl ether, which was produced in the 16th century by the reaction of sulfuric acid and ethanol, was called sulfuric ether. It was perhaps fortuitous that a mixture of the few available solvents, such as ethyl ether and ethanol, was used to produce a solution of cellulose nitrate (collodion) in the middle of the 19th century.

Turpentine was produced by the distillation of gum from pine trees in the 18th century and became the solvent of choice for the paint industry. The general use of ethanol as a solvent was delayed until 1906, when tax-free denatured alcohol became available.

Weizman produced acetone and butanol by the bacillus fermentation of corn during World War I. Methanol, produced by the destructive distillation of wood, has been available for more than a century, but this process has been replaced by the conversion of synthesis gas (CO and H_2) to methanol. Isopropyl alcohol, produced by the oxidation of propylene, has been available commercially since 1917. Aromatic hydrocarbon solvents, toluene and naphtha, were introduced in 1925. There is a trend away from some aromatic solvents toward aliphatic ones for toxicity reasons.

Oxygenated solvents became available about 60 years ago to meet the solvency requirements of emerging synthetic resins and some industrial coating application processes. Some of the early oxygenated solvents were shown to possess adverse health effects, some causing abnormalities in the unborn child. A modern generation of oxygenated solvents have been developed to minimize health risks.

In spite of the wide variety of solvents now available, the use of solvents in coatings and paints is decreasing because of restrictions on their use, as recommended by regulatory agencies such as the U.S. Environmental Protection Agency and European Union Directives. Hence, the trend is toward the use of coatings with a higher solids content (less solvent), waterborne coatings, and radiation curable and powder coatings.

XXI. RADIATION CURABLE COATINGS

The use of electromagnetic radiation to crosslink or cure polymeric coatings has been called radiation curing. The applied radiation may be ionizing radiation (i.e., α, β, or γ rays) or high-energy electrons and non-ionizing radiation, such as UV, visible, IR, microwave, and radiofrequency wavelengths of energy.

Most workers refer to radiation curable coatings as those which cure by exposure to UV radiation or an electron beam. There are two distinct types of radiation curable coatings. One is a free radical mechanism utilizing acrylic usaturation. The other is cationic. Both are characterized by being solvent free, so they meet solvent emissions and any potential carbon dioxide taxes.

Relatively low molecular weight oligomers normally containing acrylic unsaturation are normally dissolved in "monomers" containing acrylic unsaturation. The oligomers are often reaction products of acrylic acid with epoxy resins, hydroxy acrylates with polyisocyanates (or a low molecular weight polyester urethane containing residual isocyanate functionality), or a low molecular weight polyester reacted with an acrylic monomer, such as acrylic acid. Methacrylate unsaturation is not normally used, because it is too unreactive at the speeds at which modern equipment runs. The monomers are either multifunctional or monofunctional. Most are multifunctional and are obtained by reacting a polyol with acrylic acid, for example, TPGDA—tripropylene glycol diacrylate.

The "wet" coating contains a photoinitiator package. This may be one or more materials which decompose on exposure to UV radiation of certain wavelengths generating free radicals. Synergists may also be present to improve free radical generation of the photoinitiator, and some synergists also reduce oxygen inhibition of curing. Different photoinitiators are essential, partly to overcome pigments absorbing the UV radiation at their absorption spectrum. A mixture of photoinitiators with different absorption spectra can be used. The free radicals, once formed, initiate polymerization of the coating film through the unsaturation in the monomers and oligomers.

The light source for UV curable coatings is a medium-pressure mercury arc lamp enclosed in quartz. In the last 30 years or so, much progress has been made with the technology and chemistry. Radiation curable coatings find widespread usage for paper and board, plastics, optical fibers, digital optical recording and laser vision discs, adhesives, dental fillings, circuit board coatings, and wood. There are limitations for metal at present. UV curable inks are an important area for this technology.

Electron beam cured coatings are similar to the UV ones mentioned above, but they do not contain a photoinitiator. The electron beam has sufficiently high energy to fragment the molecules in the coating and initiate polymerization. There are only a few installations. The capital cost of the equipment is high, and, due to oxygen inhibition, an inert gas blanket is required.

Cationic UV curable coatings, which have been commercial since about 1980, contain monomers, oligomers, and photoinitiators. Typically, the monomers are vinyl or propenyl ethers and the oligomers are aliphatic epoxies. The photointiators, often aryl sulphide or phosphonium halogen salts, degrade on exposure to UV radiation. Cationic coatings do not suffer from oxygen inhibition of cure, but moisture in the atmosphere is a major problem. Cationic coatings have good performance on metals. The penetration of UV cationic coatings is less than some imagined when this technology was originally commercialized.

XXII. HIGH SOLIDS COATINGS

The solvent content of many classical coatings has been reduced in order to comply with environmental protection regulations. The first solvent emission law (Rule 66) was enacted in California in 1966. Subsequent regulations on state and national levels have resulted in a reduction of solvent contents. Europe also has controls on solvent emissions. High solids coatings refer to those coatings which contain solvent, or volatile materials, but at a much reduced level than their equivalents a few years ago. There are different amounts of solvents in different types of coatings which are all classified as high solids. A particular industry and geographic location may have different definitions to the same industry elsewhere or to other industries in the same location. As a general rule, high solids is not used by industrialists to refer to solvent-free coatings.

Application viscosity is a limiting parameter for coatings. As a general rule, the higher the molecular weight of a resin, the greater the solution viscosity. Therefore, lower molecular weight resins and reactive diluents have been developed. Too low a molecular weight of the resin will result in poor film performance, so a compromise is required. There have been many developments directed toward retaining high molecular weights, while keeping the viscosity relatively low. The other development has been toward reactive diluents with improved performance characteristics.

Thermoset coatings containing relatively low molecular weight resins are ideal candidates for high solids, particularly if a high solids aminoplast is used as the cross-linking agent. It is difficult to obtain high solid solutions of high molecular weight thermoplastic resins, with dispersion being one of the few viable routes.

High solids alkyds, polyesters, acrylics, aminoplasts, and other resins are commercially available. If solvent content cannot be reduced, then alternatives, such as waterborne, are often used.

XXIII. POWDER COATINGS

The first powder coatings were based on low-density polyethylene, which is too insoluble to be used as a solution coating. The first powder-coating process was based on flame spraying, and this was replaced by a process in which the article to be coated was heated and dipped in a fluidized bed of powdered polymer. The loosely coated article was then placed in an oven in order to sinter the powder and obtain a continuous film.

Thinner coatings of thermosetting resins, such as epoxy resins, are now applied by an electrostatic spray gun and then cured by heat. Epoxy, acrylic, polyester, and polyurethane chemistries, as well as hybrids of some, all find use in powder coatings. The majority of powder coatings are thermoset systems, thus the cross-linking agents must also be a solid at room temperatures. A sharp melting point is a prerequisite for a good flowing powder coating to give good gloss and appearance. The melting point has to be well above ambient to ensure that the powder particles do not sinter during storage before application.

The use of electrostatic spraying and the need to store above $100°C$ (often $150°C+$) make powder coatings ideally suited to coating metal articles. Therefore, it is not surprising that this technology has replaced traditional wet coating chemistries in many industrial applications, such as "white goods."

Substrates must be able to withstand the temperatures necessary for flow and cure, although newer developments are aimed at overcoming these deficiencies. An example would be the emerging generation of powder coatings for wood.

XXIV. WATERBORNE COATINGS

Waterborne coatings were used by the Egyptians 5000 years ago, and "distemper," which refers to waterborne coatings containing casein, glues, and other components, continues to be used as artists' colors. The white wash used by Tom Sawyer in Mark Twain's book was water-slaked lime. The adhesion and durability of this dispersion of lime were improved by the addition of casein.

Pigmented casein paints were introduced in the 1930s. An improved film deposit was obtained by the addition of a drying oil, such as linseed oil, and brightly colored casein paints were widely used at the Century of Progress Exposition in Chicago in 1933.

The use of casein-dispersed waterborne coatings in the 1940s was catalyzed by the introduction of roller coating techniques in "do it yourself" applications. Polyvinyl acetate, styrene-butadiene copolymer, and acrylic latices were introduced after World War II.

Waterborne coatings cover many categories of resin. There are only a few truly water-soluble resins used in coatings. Some cross-linkers, such as HMMM, are truly water soluble. One of the drawbacks of being water soluble is a sensitivity to moisture and poor water resistance properties of the dried paint film. Most waterborne coatings are derived from water-insoluble resins, by either polymerizing their monomers in water (polyvinylacetate, styrene-butadiene copolymers, and acrylics latices) or preparing a resin with either acid or basic functionality, neutralizing this functionality and dispersing in water. This process is sometimes called inverting a resin (from solvent based to water dispersed). These coatings may be (1) anionic (i.e., have a neutralizable functional carboxyl group), (2) cationic (i.e., have a neutralizable functional amine group), or (3) non-ionic (i.e., have hydroxyl soulblizing groups). During curing or the drying process, the neutralizing agent leaves the film, thereby restoring the water insolubility of the original resin. Most of the dispersed waterborne resin systems contain water-soluble solvents (butyl glycol, for example), sometimes referred to as cosolvents. Dispersed pigments and additives, such as defoamers, flow aids, emulsifiers, and coalescening agents (for emulsions), are usually present in waterborne formulations.

It is possible to emulsify other resins by dispersing in water using emulsifiers. Emulsions of epoxy resin and curing agents are used for waterborne concrete coatings and some heavy duty applications, where solvent fumes would be hazardous.

The electrodeposition process, in which charged paint particles are electrically plated out of water suspensions onto a metal article, was developed by Crossee and Blackweld in England in the 1930s and used by Brewer to coat Ford automobile frames in 1963. Electrodeposited coatings are theoretically of uniform thickness, regardless of the shape and complexity of the substrate. They can penetrate areas which would be very difficult, if not impossible, to coat by any other means, such as automobile chassis.

Originally, most water-soluble coatings used in electrodeposition were anionic, having carboxyl groups in the polymer chain, which were neutralized by the addition of sodium hydroxide or an organic amine. Some of the resins used were maleinized and styrenated drying oils, alkyds, polyesters, and acrylics.

Cationic electrodeposition has largely superceded anionic electrodeposition, mainly because better corrosion resistance is obtained. Automobile primers are based upon epoxy resin reacted with amines and neutralized by acid. Frequently, these coatings are crosslinked by a polyisocyanate.

Other applications for electrodeposited paints include central heating radiators and industrial metallic parts such as shelving.

There is a continuing growth in waterborne coatings as many industries are moving away from solvent-borne coatings.

XXV. TOXICITY

Some solvents, pigments, and binder components used in coatings are injurious to human organisms, and, hence, appropriate precautions must be taken to prevent accidental contact with these toxic ingredients. Extreme care must be used in the application of coatings containing lead, cadmium, or chrome(VI) pigments. The use of these pigments has substantially decreased. Inhalation of organic solvents, such as toluene and chlorinated solvents, must be avoided during the application of coatings. Extraction or well-ventilated areas must be used during the application of a coating containing solvent. Likewise, care must be exercised when polyurethane containing volatile isocyanates and epoxy resins containing free amines are applied. Sensitization by epoxies and isocyanates is another problem, and skin contact should be avoided with any coating. Solvent-containing paints must be applied in a well-ventilated, flame-free area.

XXVI. FUTURE TRENDS

The primary functions of paint are decoration and protection. Neither of these qualities is required for pigmented plastics and inherent decorated surfaces such as polyvinyl chloride siding and composite wallboard.

The high capital cost of cars, equipment, ships, and storage vessels necessitates their adequate protection. This can only be achieved by the application of a coating. In some other cases, such as food cans, the coating is not only protective, but is essential for the manufacture and filling of the can.

The use of traditional paints and coatings containing high levels of noxious solvents will continue to reduce.

Replacement paints and coatings will be more environmentally and user friendly. Aliphatic solvents will replace aromatic ones wherever possible. The use of toxic pigments will continue to decrease. People applying coatings and the end consumers will become more aware of any environmental or health problems associated with those coatings and changes will be enforced to overcome them.

The penetration of waterborne, powder, and high solids coatings will continue. It is probable that the move toward radiation curable inks and coatings will slow down, due to the fact that most applications suited to this technology have converted to it, with further technological developments being required before further inroads are made.

The proposals for carbon dioxide taxes in some European countries will drive the market. The increasing costs of energy will drive thermally cured coatings to lower cure schedules. Economics will also play a major role, and coatings and coated articles must remain competitive with other forms of materials, for example, laminated metal.

Perhaps the biggest change will be in those countries (many being considered as Third World) which have largely not moved from traditional coatings. They will eventually move to safer and more environmentally coatings. The only question is when?

SEE ALSO THE FOLLOWING ARTICLES

BIOPOLYMERS • COLOR SCIENCE • KINETICS (CHEMISTRY) • POLYMERS, SYNTHESIS • RUBBER, NATURAL • RUBBER, SYNTHETIC • SURFACE CHEMISTRY

BIBLIOGRAPHY

History of surface coatings

Mattiello, J. J. (1941). "Protective and Decorative Coatings," Vol. I, Wiley, New York.
Mattiello, J. J. (1941). "Protective and Decorative Coatings," Vol. II, Wiley, New York.
Mattiello, J. J. (1943). "Protective and Decorative Coatings," Vol. III, Wiley, New York.
Mattiello, J. J. (1941). "Protective and Decorative Coatings," Vol. IV, Wiley, New York.
Mattiello, J. J. (1946). "Protective and Decorative Coatings," Vol. V, Wiley, New York.

More detailed and recent treatises

Bates, D. A. (1990). "Powder Coatings," SITA Technology, ISBN 0947798005.
Holman, R., and Oldring, P. K. T. (1988). "UV & EB Curing Formulation for Printing Inks, Coatings & Paints," SITA Technology, ISBN 0947798021.
Landbourne, R., ed. (1987). "Paint and Surface Coatings: Theory and Practice," Ellis Horwood, Chichester, UK.

Lee, L. H., ed. (1988). "Adhesives, Sealants, and Coatings," Plenum, New York.

Lowe, C., and Oldring, P. K. T. (1994). "Test Method for UV & EB Curable Systems," SITA Technology, ISBN 0947798072.

Martens, C. R. (1981). "Water Borne Coatings," Van Nostrand-Reinhold, New York.

Morgans, W. M. (1984). "Outlines of Paint Technology," Vols. I and 2, Charles Griffin and Co., High Wycomb, UK.

Oldring, P. K. T., ed. (1986). "Resins for Surface Coatings," Vol. I, SITA Technology, ISBN 0947798048.

Oldring, P. K. T., ed. (1987). "Resins for Surface Coatings," Vol. II, SITA Technology, ISBN 0947798056.

Oldring, P. K. T., ed. (1987). "Resins for Surface Coatings," Vol. III, SITA Technology, ISBN 0947798064.

Oldring, P. K. T., ed. (1991). "Chemistry & Technology of UV & EB Formulation for Coatings, Inks & Paints," Vol. I, SITA Technology, ISBN 0947798110.

Oldring, P. K. T., ed. (1991). "Chemistry & Technology of UV & EB Formulation for Coatings, Inks & Paints," Vol. II, SITA Technology, ISBN 0947798102.

Oldring, P. K. T., ed. (1991). "Chemistry & Technology of UV & EB Formulation for Coatings, Inks & Paints," Vol. III, SITA Technology, ISBN 0947798161.

Oldring, P. K. T., ed. (1991). "Chemistry & Technology of UV & EB Formulation for Coatings, Inks & Paints," Vol. IV, SITA Technology, ISBN 0947798218.

Oldring, P. K. T., ed. (1994). "Chemistry & Technology of UV & EB Formulation for Coatings, Inks & Paints," Vol. V, SITA Technology, ISBN 0947798374.

Oldring, P. K. T., and Lam, P., eds. (1996). "Waterborne & Solvent Based Acrylics and Their End User Applications," SITA Technology, ISBN 0947798447.

Oldring, P. K. T., ed. (1997). "Waterborne & Solvent Based Epoxies and Their End User Applications," Wiley, New York.

Oldring, P. K. T., ed. (1999). "The Chemistry and Application of Amino Crosslinking Agents or Aminoplasts," Wiley, New York.

Oldring, P. K. T., ed. (1999). "The Chemistry and Application of Phenolic Crosslinking Agents or Phenolplasts," Wiley, New York.

Sanders, D., ed. (1999). "Waterborne & Solvent Saturated Polyesters and Their end User Applications," Wiley, New York.

Seymour, R. B. (1982). "Plastics vs Corrosives," Wiley, New York.

Seymour, R. B., and Carraher, C. E. (1988). "Polymer Chemistry: An Introduction," 2nd ed., Dekker, New York.

Seymour, R. B. (1988, 1989, 1990). "Advances in coatings science technology," *J. Coat Technol.* **60** (7S9), 57; **61** (776), 73; **62** (781), 62, 63.

Seymour, R. B., and Mark, H. F., eds. (1990). "Organic Coatings: Their Origin and Development," Elsevier, Amsterdam, The Netherlands.

Seymour, R. B., and Mark, H. F., eds. (1990). "Handbook of Organic Coatings," Elsevier, Amsterdam, The Netherlands.

Solomon, D. H., and Hawthorne, D. G. (1983). "Chemistry of Pigments and Fillers," Wiley, New York.

Tess, R. W., and Poehlein, C. W., eds. (1985). "Applied Polymer Science," 2nd ed., Am. Chem. Soc., Washington, DC.

Thomas, P., ed. (1998). "Waterborne & Solvent Based Polyurethanes and Their end User Applications," Wiley, New York.

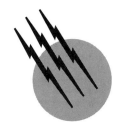

Cognitive Sciences

M. Gopnik
McGill University

GLOSSARY

Brain Physical set of neurons and neuronal connections that are responsible for cognitive processing.

Cognitive science Theories that account for intelligent behavior in both organisms and machines.

Innate knowledge Inborn assumptions that guide and constrain perception, memory, and thinking.

Memory Ability to store and retrieve information.

Mental representation Form in which information is encoded for cognitive processing.

Mind That which embodies all cognitive functions such as perception, memory, thinking, consciousness, and intentionality.

Perception Ability of an organism to apprehend various aspects of the physical world around it.

Thinking Ability to draw inferences and solve problems on the basis of information from perception and memory.

COGNITIVE SCIENTISTS want to answer some very old questions in new ways. The crucial question that they want to address is this: What is the mind and how does it work? The question itself is not new. It has been around at least since Socrates, and each age of humankind has examined the question of mind in its own ways. It is one of those odd questions that seem perfectly clear at first glance but become more and more difficult to understand as more thought is given to them. The initial question gives rise to a host of subsidiary questions: If there is such a thing as the mind, where is it? How can it be conscious (or unconscious) of its own operation? Does it operate by using just incoming data from the world or does it have some initial internal structure? Philosophical debates over the last 2000 years have raised challenging questions about the mind. These philosophical questions have set the agenda for cognitive science. What is exciting, however, is that new ways are being found to address the old question, ways that appeal to knowledge from neurology, psychology, computer sciene, linguistics, and anthropology, as well as philosophy.

Stated a little differently, it could be said that cognitive scientists want to explain intelligent behavior, both natural and artificial. In an intuitive sense, the sorts of behaviors that should be counted as intelligent are well known. In

Encyclopedia of Physical Science and Technology, Third Edition, Volume 3

general, these behaviors involve solving new problems in ways that cannot be completely accounted for simply by looking at the variables of the situation and the preprogrammed instinctual responses of the organism.

I. HISTORY

Although the problems that cognitive scientists want to deal with are old, the field of cognitive science is relatively new. At the beginning of the twentieth century it was thought that *real* scientists should talk only about things that were observable. The problem is that the mind, whatever else it is, is not directly observable. So for a long time no one talked about the mind. In psychology, the behaviorists banished the word *mind* from their lexicon. The only terms that were allowable were terms referring to things that could be directly observed. The behaviorists built their theories around observables like behavior, stimulus, and response. The mind and its allied activity, thinking, became disreputable and were ignored. In the behaviorists, view, people did not think or consider, they simply responded with behavior A when they were under stimulus condition B. In linguistics, the story was much the same. The American structuralists, explicitly acknowledging their debt to behaviorism, and wanting very much to be real scientists, declared the only acceptable evidence in linguistics to be directly observable utterances. Speakers' intuitions about their language were not to be trusted. The structuralists spent a great deal of attention on developing procedures for extracting interesting facts about language from the restricted data that they allowed themselves to look at. All this supposed rigor would have been fine if it had yielded these investigators what they wanted, an account of the behavior of organisms on the one hand or a description of language on the other. But it did not. However, their efforts were not a total failure. We did know better how pigeons could be taught to waltz and what the surface facts of many exotic languages were. These theories, in fact, told us many things, but nothing about the questions that now seem to be the crucial ones. In psychology, although behaviorists were better able to describe how pigeons learned to waltz, they could not explain how birds learned to sing or build nests. In linguistics, although the structuralists could describe strange and exotic languages, they could not answer the question of why those languages had the shape they did.

The interest in cognitive science came from many directions at once. People working in psychology, computer science, and linguistics all began to realize that it was crucial for their work to start thinking about thinking. And there was no way of finding out about intelligence without resurrecting the mind. So they took the risky step of starting to talk about the mind, intractable, immeasurable, and unob-

servable as it is. All this started in the mid-1950s. There was a Symposium on Cognition in Boulder, Colorado, in 1955. In 1956, the Second Symposium on Information Theory was held at Massachusetts Institute of Technology (MIT). It was there that many of the issues about intelligence in minds and machines were raised. The interest in these issues was fostered over the next few years by seminars, meetings, and shared research. By 1960, there was so much excitement about the new questions that Bruner and Miller asked the Carnegie Corporation to fund a Center for Cognitive Studies at Harvard University. The corporation did and the center was established. There are now many such centers, several of them funded by the Alfred P. Sloan Foundation. These centers are places where people working on the problems from different fields can come together and exchange ideas and knowledge and participate in joint research. There are programs in cognitive science at both the undergraduate and graduate levels at several universities, with more programs being established every year. What started out as a radical proposal, to look at the mind, has become received wisdom.

II. OVERVIEW

Chess playing is often cited as the paradigm example of intelligent behavior in humans, and a great deal of work has been done in trying to understand just how people do play chess successfully. Early on, this problem of playing chess became a challenge to computer scientists. They reasoned that if chess playing were the paradigm case of intelligent behavior, then if they could teach a machine to play chess they would have demonstrated that machines were capable of intelligent behavior. Computers can now play chess, better than most humans, they win because they have the ability to perform a prodigious number of calculations in a very short time. They can calculate all the consequences of a set of moves much faster than any human could even begin to do so. And that is how they win. They simply use their speed of calculation to look ahead. The interesting question is how do people, who cannot do these calculations, play so well. They must be using a totally different mechanism for representing and solving the problem. Although chess playing is one area in which computers can outdo most people, there are areas in which even the most sophisticated machine cannot begin to match the performance of even the youngest baby. For example, babies have a richness in their representation of the visual world around them that we are only just beginning to understand. No machine can "see" as well. One of the surprising discoveries that has come out of the studies in cognitive science is just how complex the seemingly simple operations like seeing, hearing, and using language really are.

When we say that something is a science, we expect that it can provide us with a unified set of laws, or at least is on the trail of such a set. That is not the case with cognitive science. There is no set of laws that can account for all cognitive behavior, nor are we on the brink of finding them. In fact, the situation is quite the contrary. As the serious study of cognitive systems continues, the specificity and uniqueness of each of the systems become more apparent. And yet there are clear connections. The way the questions are being posed and the sorts of answers that are being found are similar. The details differ, but the broad outlines are the same.

III. LEVELS OF EXPLANATION

One of the basic tenets of cognitive science is that cognitive processes in animals have evolved to fulfill the needs of the organism. These processes therefore are highly ordered and selective systems that "know" how to look for the parts of the world that are important for the functioning of the organism. Processes like seeing, hearing, and using language are not simply the result of general intelligence making sense of the world around it, but rather biologically determined systems, each with its own set of initial rules and restrictions that imposes an order on the world it encounters. For instance, the visual system of a frog is fundamentally different from the visual system of a human being. Frogs and people live in functionally different worlds and their visual systems are built to take account of this difference. Frogs visually respond primarily to small objects moving swiftly across their visual field. If the same small object is stationary in the frog's visual field, it is not perceived by the frog. This is a handy and efficient system for a creature whose main aim in life is to catch insects as they fly by.

To understand cognition from this perspective, we must assume that each cognitive system has its own rules and restrictions that impose an order on the incoming data. This means that these systems must be studied in special ways. The first is to understand what the system is trying to do, that is, we have to find out what constraints the system itself is imposing on the data. We might say metaphorically that each cognitive system has its own theory about the way the world is built and the way it is likely to change. This innate theory is usually right because it evolved in a particular world in which it was almost always successful. For example, the visual system in human beings assumes that objects have a constant size. If the visual field shows an object getting larger and larger, we interpret the information as meaning that the object is actually staying the same size but coming closer and closer. In the world in which we live and through which we move, objects change their apparent size much more often because of relative motion

than because they are suddenly growing or shrinking. Our visual system is designed to take account of these facts. When we build a theory to account for vision, this assumption of the visual system is one of the things that the theory must incorporate. The same is true for any cognitive system. We know, for example, that we do not learn language simply by listening to the language around us and finding out which parts are alike and which parts are different by some general compare-and-match procedure. Rather, we start with the ability, even the necessity, to recognize certain parts of the input as highly significant and other parts as nonsignificant even when there is no apparent reason for this distinction on the basis of the difference in the amount of information in the signal itself. For example, we know that young babies pay much greater attention to the properties of sound that are likely to be significant in the structure of language than they do to other properties of the sound signal. The first level of explanation, therefore, must describe what we have referred to as the theory of the world that the cognitive system itself imposes on the data. It must tell us what sorts of things the system expects to find in the world and what sorts of relationships they enter into.

Once we know the answer to these questions, we can ask what are the algorithms and evidence that are used to establish these entities and relationships. How does our visual system extract information about motion from changes in size of the object in the visual field. Does it do so by measuring the rate of change of the object itself relative to other surrounding objects? Or by looking at the pattern of the changing occlusion of other objects? Or changes in texture correlated with changes in size? Or a combination of all three? Or some other way altogether? The first level of explanation establishes just what it is that the system does. An explanation at the second level is concerned with finding out just how the cognitive system goes about doing what it has to do.

The third level of explanation looks at the hardware that performs these procedures. It looks at neurons and neuronal connections. It looks at brain architecture and brain chemistry. How much information does the eye itself pass on to the brain about motion? What neural pathways does the information follow? How much processing is done before the information reaches the visual cortex? Are different parts of the processing done at different locations in the visual cortex?

IV. COGNITIVE PROCESSES

Cognition is often said to be the result of three processes: perception, memory, and thinking. Any account of cognitive processes must explain how the system in question makes sense of its input, how it relates this input to its prior knowledge or experience, and how it decides to act. In rough outline the first process is related to perception, the second is related to memory, and the third is related to thinking. In order to have intelligent behavior, you have to be able to represent the information that is in the environment. The first step in this process is to take in information from the environment by using sensory receptors. The second step is to evaluate and organize this information by relating it to other information you already have stored in memory. Third, the information from the receptors and the information from memory must be related by means of some mechanisms of inference and comparison. While these three functions should be treated separately for purposes of clarity, research has shown that they are always in a state of dynamic interaction, some times in surprising ways. Every act of perception involves memory and thinking. Every memory is the result of some prior perception that has been turned into a storable representation by thinking. All thought operates on both perceptions and memories. One of the goals of cognitive scientists is to make explicit what these processes and interactions are for each specific cognitive task, such as perceiving objects, recognizing faces, understanding language, hearing melodies, and answering questions.

A. Perception

All organisms have sensory receptors that allow them to gather specific information about the world. The range of information that is available to them varies from species to species. Organisms have sensors for light intensity, color, polarization, pressure, sound over a wide range of frequencies, gravity, chemical compounds of various kinds, heat, the magnetic field of the earth, and many other aspects of the environment. In general, organisms perceive aspects of the environment which are functionally important to them. The same sensor can be used by different organisms for different ends, and the same end may be accomplished by different organism's using different sensors. For example, certain kinds of ocean-dwelling bacteria, as well as homing pigeons, have a sensor that responds to the magnetic field of the earth. The bacteria are able to use this sensor to orient themselves with respect to up-and-down position, because near the equator where these bacteria live the lines of force of the magnetic field of the earth are nearly perpendicular to the surface of the earth. Humans accomplish this same orientation task, in part, by being sensitive to gravity. The environment provides several different kinds of clues to determining up and down. Organisms have evolved that have different but appropriate means of gathering this information that is so important for their functioning. In homing pigeons, the information about the magnetic field of the earth is not used

for up-and-down orientation but rather for finding their way over long distances. Perception, therefore, cannot be understood simply by looking at the sensory systems of organisms and the kinds of information to which the systems respond. It is important also to understand these systems as they function in the ongoing life of the organism.

Very interesting work has been done on the perceptual systems of various organisms. This work has shown that even in seemingly simple organisms, perception involves more than just storing raw sensation. It involves organizing and integrating the raw sensations so that they represent objects and events which are salient to the organism. The assumption in the simplest model of perception is that it is a system in which information in the world is stored directly in a one-to-one mapping so that every piece of information that can be sensed is equally valued in the internal representation the organism constructs. This simple model does not work. For example, under this model, people looking at the world would just have to use their eyes to measure the light intensity at every point and store this information. If we wanted to build a computer that could see, this would certainly be the easiest way to do it. But that is not how people see. They do not simply see a scene composed of varying light intensities. People see separate objects in a three-dimensional space that are casting shadows and partially hiding objects that are behind them.

To understand how this works, we start with a seemingly easy question. How do we segment the visual world into objects in the first place? This question has no direct and easy answer. Dividing the world into objects is so familiar and automatic that it is difficult at first to realize that it is a complex process that relies on several assumptions and many kinds of implicit knowledge about the world. If we look at babies, we can see that they do not respond to objects in the same way as adults. As they get older their responses with respect to objects change and become more like that of the adults around them. Some investigators have interpreted this to mean that the concept of "object" evolves as babies mature; others that the concept is intact to begin with, but that the baby has to learn how to behave appropriately toward objects in the world.

However we want to explain this phenomenon, looking at the cognitive development of "objectness" can help us to understand all the seperate assumptions that we must make in order to recognize that something is an object. One assumption is that something is likely to be an object if it is contained within a single boundary and it is not an object if the boundaries are discontinuous. This rules out two blocks being considered as one object since they can each be delimited within separate boundaries, except of course if they are right beside each other. Then another test comes into play that even very young babies know. Something is likely to be an object if all the parts that are within the boundary move together. If the pieces of one block all move together in one direction while the pieces of the other block move together in another direction, then the first hypothesis that they were a single object has to be discarded. Early on, children learn that objects generally have constant properties; that is, objects most often do not change their shape, color, or size. As we refine this concept of the constancy of properties, we learn that objects may differ in which properties are constant and which are changeable. A face, for example, can change its expression, but not the relative location of the eyes, nose, and mouth. A tree can change its shape, but not its location. Children also assume that objects retain their original shape even when some piece of the object is hidden by an object in front of it. At the end of their first year, children finally behave as if objects continue to exist even when the objects disappear from their visual field when they blink, turn their head, or leave the room; or when the objects are covered or are behind or inside another object. By the time a child is 2 years old, all these separate assumptions are united into a single concept of "object" that guides the visual system into literally seeing the world as made up of objects. We do not seem to see just light intensities and then determine which parts belong to which objects. We just see the objects. Moreover, we cannot choose not to see the objects in our visual field. They are just there before us.

In order to do this, we must be processing visual information in a special way. For example, we pay selective attention to some parts of the visual field as opposed to other parts. We know, for example, that the slight changes in light intensities along the surface of a wall are not nearly as important as are the sharp changes in intensity at its edges. This information about the importance of edges has already been incorporated in computer vision systems. These systems have been designed to pay more attention to large changes in light intensities than to small ones. And some systems have been constructed so that if they encounter an unfamiliar shape, they assume that it is produced by two more familiar shapes one of which is behind the other, and, if it is possible, construct a representation of the scene in these terms. The problem of how people recognize objects they have never seen before as being instances of some general class is very difficult to make explicit in the general case. How do we sort the world we see, even when we have never before encountered this particular part of it, into houses, trees, and leaves? It is presumed that we must do it by having some prototypical representation of houses, trees, and leaves that we match against the new visual information. What goes into these prototypes and how we construct them is still far from clear.

One surprising new finding is that we seem to be built to preserve some kinds of constancy of properties even when the real world conspires to make it appear that these properties have changed. The two best examples of this are size constancy and color constancy. As previously mentioned, we judge objects to stay the same size whether they are near to us or far away. We say that when they are farther away they may appear smaller, but they really are the same size as they were originally. The same thing is true about color. We perceive objects as retaining their original color even under varying colors of illumination. It had been proposed that size constancy was learned by experiencing that as you move through the three-dimensional world objects change size. But now it seems as if children see the world as three-dimensional from the start and have size constancy from an extremely young age. The phenomenon of color constancy has made us revise our notions of how color is perceived. For a long time it had been thought that we perceived color by directly sensing the frequencies of the light reflected from colored surfaces. But if this were true then the color of an object should be perceived to change significantly as we move from incandescent light to sunlight. This, however, does not happen. Perceived colors stay constant over wide ranges of changes in illumination. It now appears that our color vision does not look at absolute frequencies but rather the relative reflectance of the light from various surfaces. The relative reflectance of different colors does stay constant under different illuminations, and therefore a system working on these relative values rather than on absolute values preserves color constancy.

These same sorts of phenomena are found not just in the visual field but also in the auditory system. Our ear receives a single complex acoustic signal. But it does not perceive the input to be a single signal. Rather it segments the signal into parts and interprets these parts as coming from different sound sources. At the very same moment I can hear people talking in the next office, my computer humming, the keys clattering, someone walking down the hall, and someone moving a chair in the adjoining room. I have extracted all this information from a single acoustic signal, and I have done it automatically. In fact, I cannot choose not to distinguish the sound sources. Cognitive scientists have been investigating what properties in the acoustic signal give us the cues we use for organizing the chaos of sound into recognizable sources and patterns. One cue we use, for example, is random variation. If two frequencies have the same sort of variation at the same point in time, then they are judged to belong to the same sound source. If a frequency in the same acoustic signal does not exhibit this variation, it is judged not to belong to the same sound source.

At the level of the neural organization that allows the tasks to be performed, we know somewhat more about cats than we do about people simply because the necessary experiments can be done on lower animals but not on people. We know that some of the organization of the perceptual world of a cat is actually done at the level of the receptor organ. The neurons in the retina are so organized that they can directly encode information about vertical and horizontal lines and some kinds of motion. The nerve fibers in the cochlea of the inner ear sort sound with reference to its frequencies and may actually group some of these frequencies. We also know that appropriate neural pathways are established early in life in response to incoming stimuli. If a normal newborn cat is kept in the dark for several months after birth, it will not develop the normal pattern of neural connections. If it is then brought into the light, it will not be able to see. On autopsy we can see that the characteristic visual neural pathways have not been established in these light-deprived cats. Although we cannot perform experiments on humans, we are able to gain insights into the neural organization of perception by studying patients with brain injuries. It is well known that damage to the brain can cause very specific deficits in cognitive functioning.

In discussing the visual system we have talked about the way in which cognitive development has given us some insights into the organization of vision. This is also true in audition. But there has been another very fruitful avenue of investigation. This has involved the use of computers to generate perceptual stimuli with known and controllable properties. By using computers to produce auditory or visual displays, we can make up stimuli having properties and relationships that never occur in the real world. In this way we can isolate particular variables and see what effect they have on perception. We can generate sound with only one frequency. We can lengthen or shorten the onset of a sound. We can vary intensities. In short, we can build a sound with exactly the properties we want to test, a sound that might never be possible in the real world. By looking at the way people perceive these stimuli, we can distinguish variables that are inextricably connected in the natural world.

B. Memory

Storing information and then accessing it at some later time is one of the fundamental processes in any cognitive system. Learning, for example, depends upon recognizing that some new experience is similar to some past experience; therefore, what worked in the last case is likely to work in the present case. For learning to take place, a representation of the first event must be constructed and stored. There also must be some means of accessing this representation at some later time and judging that the earlier and later representations are of the same sort. Explaining how these representations are constructed, organized, stored, and accessed is the job of a theory of memory.

As in perception, it is now believed that memory is different in different organisms. Parts of the world that are crucially important for one organism to encode in a representation of an event may be unimportant for another kind of creature. The first job in developing a theory of memory for an organism is to understand just what role memory plays in the organism's various behaviors. For example, we know that juvenile Atlantic salmon can learn to associate certain events with feeding. However, the same sorts of learning are beyond the capacities of older salmon. Moreover, whatever is learned as a juvenile is retained in adulthood, but no new learning can take place after a certain age. A theory of memory for the salmon would have to account for the kinds of representations that would allow this learning to take place and further would have to explain why early learning is remembered forever and why no new memories can be formed.

Human beings, however, can learn throughout their lifetime and they do not remember anything that happened approximately before the age of 2 years. Memory, not surprisingly, seems to be very different in humans than in salmon. Humans have all sorts of things stored in memory, some of which are very specific like telephone numbers, names, faces, dates, and events, and others which are more general like what dogs look like, how to play chess, and how to get to the office. Some of these things seem to be remembered extraordinarily accurately and easily. For example, we seem to have an exceptionally good memory for recognizing faces. Tests have shown that people can recognize faces they have not seen for many years. They can recognize that they have seen a face before, even if they have seen it for only a few brief moments. The representation of a face in memory seems to be accurately encoded on the basis of very brief exposure and seems to last for a very long time. We are not so good, however, at other sorts of memories. While we might remember that we have seen a face before, we often forget the context in which we saw it. Did we meet the person at a party, or in a store, or in class? And even more frequently we forget the name we should associate with the face. Almost everyone has forgotten the name that goes with even a very familiar face. The ability to recognize faces appears to be a likely candidate for being provided for in our biological makeup. From almost their first moments, newborns seem to sort out faces as warranting particular and special attention. Within 2 weeks after birth, they are able to recognize familiar faces like their mother's. There is some tentative evidence from a neurological disorder (prosopagnosia) that this ability to recognize faces can be selectively impaired. There are reported cases of people who, after cerebral injury, can no longer remember that they previously saw a particular face. They cannot recognize husbands, children, or even themselves in a mirror. In evolutionary terms it is not surprising that a creature that regulates much of its interactions on the basis of social alliances would find it important to recognize individuals.

Not everything is as easy to remember as faces. People have a difficult time remembering random lists of items. This is very different than in the case of machines. For machines, faces are extremely difficult to code and remember; random lists, such as sequences of numbers, are very easy. One of the fascinating things about human memory for lists is that what can count as a single item can very dramatically in length and in complexity. For example, people can remember sequences of about five syllables, or five words, or five sentences. Yet, clearly, in remembering five sentences you are remembering many more than five words, and in remembering five words you are remembering more than five syllables. For example, it is very difficult to remember this list of words: dog, boy, all, yellow, my, barked, the, fly, birds, not, good, is. Yet if we organize the same list of words into the following three sentences, these words are now easier to remember:

The boy is good.
Not all birds fly.
My yellow dog barked.

This phenomenon of chunking can also be seen in remembering sequences of numbers. A 12-digit sequence like 916144926839 would be very difficult to remember. However, if the sequence is interpreted as a telephone number, 9-1-614-492-6839, then it becomes much easier to remember. The ability to remember something is therefore not directly related to any simple measure of the amount of information to be remembered, but rather is a function of both the information and some higher-order theory that can impose an organization on this material. This phenomenon leads to the seeming paradox that it is sometimes easier to remember something if you actually add information to it. Traditional memory experts know that they can remember more if they organize the given information by locating it in an imagined physical location. For example, if you want to remember an arbitrary list of objects you should imagine yourself walking down a familiar street and placing each object in a specific location, the more bizarre the location the better. A dog might be put on top of a mailbox, a boy placed on a window ledge. To remember the list, you merely have to stroll down the street again and you will "see" the dog on top of his mailbox and the boy on the window ledge. It seems to be easier to remember a dog on the top of a mailbox than just to remember a dog by itself.

Theories of memory have to explain not only what gets remembered and how, but also how long these memories last. We know that some memories can last a lifetime and others disappear in a few minutes. Everyone can remember his or her wedding day or the birth of his or her children. Most people can remember some incident prior to 6 years of age. People vary in how detailed these memories are. Some can remember the details of all their birthday parties since they were 4 years old, including the presents they received and the kind of cake they had. Others just have faint glimpses of brief moments in the past. Nobody remembers all the details of all the events in his or her life. It is still not entirely clear how particular events are selected to be remembered, although it seems that the significance of the event in our lives is certainly a factor. Some memories last only a few moments. If someone tells us a telephone number, we can usually remember it long enough to pick up the telephone and dial, but after the call we may have totally forgotten the number. There is evidence that these different kinds of memories can be selectively impaired and are therefore neurologically instantiated in different ways. There are memory deficits in which a person can easily recall events from more than 50 years ago but cannot remember if he or she had lunch today. Conversely, there are cases in which the ability to remember present events is unimpaired, but the memories of a past life are gone. It is often the case that patients cannot remember any of the events which occurred just prior to sustaining a head injury, although their memory about the more distant past and about the present remains unimpaired. It used to be thought that there were two sorts of memory, short term and long term. Short-term memory was thought of as temporary storage that could hold about five items at a time. On some occasions this store was transferred to a single location in long-term memory. We now know that the situation is more complex than that. It now appears that all the information in a memory is stored in several locations. The neurological details of this process are not clear. We know that certain neurotransmitters seem to be absent or at low levels in people with memory disorders, but we do not know the exact function these substances perform. And it seems that we must posit more than two levels of storage in memory. Moreover, it may be wrong to posit a unitary memory function that is the same no matter what is being remembered. It may be that faces are remembered in a very different way than are telephone numbers, events in another way, and language in still another way. Memory is simple in juvenile Atlantic salmon. It appears as if their memories are actually hard-wired into the organism. New experiences affect the neural connections themselves. Once they are established they cannot be changed. Older fish cannot establish any new connections. That is why the young salmon can learn and old salmon can never forget. Hard wiring is certainly not the answer in human beings. Our memory seems to be very flexible, at least in some respects. We can both remember and forget. Cognitive scientists want to explain all this.

C. Thinking

Thinking about thinking started with the Greeks. They wondered whether new truths are inferred from old truths or whether all truths are known to us, in some way or another, at the outset. If we can infer new truths from old ones, how do we do it? If they are all there from the start, what and where are they? These questions are still with us in cognitive science and there are still partisans of both positions. Most people agree that some aspects of thinking about the world must be there from the start. Everyone can see that there is no way that the concept of "similarity" could be learned, since all learning is founded on judgments of similarity in the first place. Learning the details of which particular things are similar to which other particular things might come from experience, but the ability to make similarity judgments must be there at the start. Even similarity is not going to be enough. There have to be some parameters on which similarity is judged, some property or properties that must be constant in order for two things to be judged to be the same event or object. The real question, therefore, is not whether there are any initially assumed truths with which the organism starts out, but rather how rich this system is and exactly which ones they are for each organism.

Even if some or most of the fundamental categories of reasoning exist from the start, there are some things that must be learned. We still must be able to derive new truths about the world from what we already know and our new experience. The question is, how do we do this? One way of inferring new truths from old ones is by using a logico-deductive system in which the truth of the conclusion is guaranteed by the truth of the premises and the inheritability of truth under the rules of inference. If you start with true propositions and you know that your inferential rules preserve truth, then you can be confident in the truth of the conclusion. For the past 2000 years, logicians have been developing and refining such logical systems. We all are familiar with the rules of syllogistic reasoning:

All men are mortal.
Socrates is a man.

Therefore,

Socrates is mortal.

Early syllogistic reasoning has been formalized in the propositional and predicate calculi. Probability theory allows us to calculate the chances of a particular event on the basis of our knowledge of the probabilities of other events. The truth of the conclusions is guaranteed within each system. The question is, do these systems capture the way people really construct inferences? And if they do not, is it because the logics are wrong or the people are wrong? We may hold the prescriptive position that logic describes how inferences *should* and *must* be drawn, but people, imperfect creatures that they are, simply do not always behave rationally. An alternative conclusion is that logics, although they may be elegant and useful formal systems in their own right, simply are not accurate models of the way people think. Many experiments have shown that people do not draw inferences in the same way that formal systems do. Logical systems have no trouble drawing a correct inference from these two statements:

None of the artists are painters.
All of the painters are chemists.

However, ordinary people not trained in logic simply do not infer the following from these two statements: Some of the chemists are not artists.

Sometimes logics go wrong as models because they have used ordinary words like "and" and "not" and "all" in specifying their rules, but they have not retained the ordinary language meaning of these terms. For example, logicians define "if" so that propositions connected with "if" are false only if the premise is true and the conclusion is false. But that is not how "if" works in ordinary language. It would be strange to say, "If John comes then I will have a party," if I intend to have a party even if John does not come. Yet under the interpretation in logic the sentence would be true if John did not come and I had a party anyway, because implications are true when the premise is false and the conclusion is true.

Or suppose I ask you to estimate the probability that I will have a student with one blue eye. Most people would estimate that the chances are very low. Yet, the same people would estimate that the chances are very high that I will have a student with two blue eyes. Strictly speaking the set of students with one blue eye is a subset of the set of students with two blue eyes and therefore the probability is greater that I will have a student with one blue eye than that I will have one with two blue eyes. This seems to be a perfectly clear case in which the correct reading of the phrase "with one blue eye" is "with one and only one blue eye." There are many similar examples in which the probability of a subset is estimated to be larger than the probability of the larger set. For example, if we say that someone is a civil rights worker and also supports significant arms reduction and is for abortion on demand, then the probability of that person's being a bank teller is estimated to be very low. The probability of the same person's being a bank teller and an ardent feminist is rated much higher. In general, people's behavior with respect to probabilities is complex and is not in accord with theoretical expectations. But that does not mean that they are irrational. We are just beginning to understand the variables which go into real inferencing. They include prior beliefs, utility, and reasonableness. Grice's theory of conversational implicatures would account for some of these seeming irrationalities by saying that the hearer expects the speaker to be relevant, truthful, and comprehensive in his or her utterances. Therefore if a speaker refers to a student with one blue eye then the hearer has the right to assume that the fact that he says "one blue eye" is significant and salient. Since the hearer believes that the speaker knows that students normally have two blue eyes, he or she is justified in assuming that the speaker is intending to communicate "one and only one blue eye" in saying "one blue eye." The same holds in the bank teller example. The speaker in this case, under normal rules for interpreting conversations, is correct in interpreting "bank teller" to mean "a bank teller with no significant other properties." However the details of particular cases work out, the goal of cognitive scientists is to construct a model of thinking that would specify the content of our innate endowment and construct a psychologically real model of inference.

V. FIELDS OF STUDY

To understand cognitive science as a whole, we must see how each of the separate fields which compose it work. Each of the fields has its own problems and solutions. Each builds its own theoretical framework to account for the data which it is responsible for. But, at the same time, there is a general awareness in each field of the advances and setbacks in the other cognitive domains and an intention to coordinate the work and theories across fields so that the knowledge and insights in one domain can be brought to bear on the problems in another.

A. Linguistics

One of the puzzles that contributed to the genesis of cognitive science arose in linguistics. Noam Chomsky, in the 1950s, began to rethink the questions which linguists were asking about language. In any field the most radical changes come when the questions, not the answers, change. Linguists of the twentieth century had more or less agreed that language was a learned social behavior. They supposed that we learned our language in much the same way as we learn any of the other norms of our culture. Learning language was like learning how to dance or determining what to wear. Although almost all cultures have dances and clothing, their particular form might vary in any way from culture to culture. Each new generation must be taught the particular norms of the culture by the preceding generation. Moreover, any similarity between the forms in one culture and the forms in another, if not due to direct or indirect influence, was purely coincidental.

This was assumed to be true with language. Language was a product of culture, and languages could vary in any way from one another. This model of language was directly challenged by Chomsky. He proposed that language is a biologically determined, species-specific attribute of all human beings. All normal children are biologically predisposed to acquire human language. The only trigger that is needed to initiate the process is the presence of language in the child's environment. If language is present in the environment, then the maturational path of language acquisition is more or less the same in all normal children in all cultures. Any variance is the result of individual difference and is not correlated with anything in the culture or the environment of the child. From Chomsky's point of view, acquiring language is like learning to walk upright. Although language is not present at birth, it develops naturally as the normal child matures. No explicit instruction by the child's guardians is necessary. In fact it can be shown that in the case of language, explicit instruction is simply ignored by the child. Language, like walking upright, is learned, but it is not taught.

This view of language directly raises questions about the mind. First, in order to acquire language in this way, the child must come equipped with some particular initial mental structures particularly appropriate for language learning. If this is true, then it must also follow that these properties of the mind must be appropriate for learning any human language to which the child might be exposed. Therefore, there must be some fundamental universal properties all languages share. The job of linguistics is to specify these universals and to provide some account of where they come from. Recent studies of children in whom language development is impaired have provided evidence that these impairments are caused by an autosomally dominant gene. These impairments in language can be shown to affect only certain subparts of the grammar, in particular morphological features and phonology. This finding suggests that language is probably not an evolutionarily discrete entity but is composed of several constituents, each with a different evolutionary history. In order to be able to provide an account of these parts and how they interact, linguists must provide an account of the universal properties of language.

It appears that the sound system of every language is built from the same set of elements and is constructed along the same principles. Although humans can make and hear many distinctions among sounds, only some of these distinctions can serve to signal differences of meaning in languages. The other differences can convey information about whether the speaker is a man, woman, or child, or whether the person is angry or happy. From the point of view of the language system, however, they are just noise. The sound system of every possible human language is organized into discrete elements varying from one another in a limited number of ways. These elements are combined into higher-level structures such as the syllable. The rules that govern the relationships among elements at each level are universal. That is not to say that the rules and elements of each language are the same. Languages do differ from one another, but they differ in predictable ways. This is really not surprising. Because every sound signal varies from every other, the infant must have some way of determining constancies in the sound signal. There are experimental results showing that some sound differences make an infant perceive the sounds as being the same, and other differences make the infant perceive the sounds as being different. The differences that we hypothesize as being significant for language are just the ones the infant perceives as making a difference.

At the level of the organization of words (morphology) and the organization of sentences (syntax), the same seems to be true. There seem to be some universal principles constraining the kinds of grammars that languages can have. In order to see these regularities, we have to postulate an abstract level of linguistic organization of the sentence. Two sentences that appear the same on the surface may have a very different underlying structure. For instance, let us consider the following two sentences:

Mary persuaded John to leave.
Mary promised John to leave.

We can see that in one case John is leaving and in the other case Mary is leaving.

The goal of current linguistic theory is to account for the facts of each language and to construct this account in terms of some universal principles.

B. Psychology

The revolution in linguistics was certainly one of the important factors in the development of cognitive science. But the same sort of reassessment of fundamental beliefs was going on in many allied fields. In psychology, even at the time when the behaviorists were dominating the field, there were perceptual and developmental psychologists who realized that to account for visual perception or conceptual development, some internal structure had to be posited. This tied in tangentially with what was being learned about the natural development of animals. For example, ducks do not, learn to follow their mother by trial and error; they simply open their eyes and follow the first thing that they see. In the natural state this would inevitably be their mother, but in experimental situations they can be made to follow anyone or anything. If ducks have some initial states guiding their expectations and experience in the world, it would not be surprising that humans do, too.

The behaviorists' hypothesis that the newborn baby is an empty slate on which experience writes was no longer tenable. Psychologists, instead of assuming that babies know nothing and have everything to learn, actually began to look at babies. The newborns obliged by demonstrating that they came into the world knowing a great deal. It is still being demonstrated that babies know a great deal more about their world from the start than we had ever suspected. As psychologists began to frame new kinds of questions, they found that they were led to new kinds of answers, answers that had to posit minds.

C. Neurology

While psychologists were finding that explanations of the fundamental, natural behavior of organisms had to include hypotheses about the way the mind worked, neurologists were finding new ways to study the working brain. For 100 years, the primary information about neural processing had come from autopsy and from observation of patients with brain damage. In the 1950s, procedures were developed to allow the brains of awake patients undergoing neurosurgery to be electrically stimulated. The responses of the patients to this stimulation could then be charted. Not only could gross motor responses be studied, but since the patient was awake, cognitive processing could also be studied. Memory, language, perception, and thinking can all be stimulated electrically. Procedures for investigating the cognitive functioning of the brain have been developed at a dizzying pace since the 1950s. In animal studies using implanted electrodes, scientists have been able to study the behavior of individual neurons in the more or less normal functioning of the organism. Split-brain studies have given us information on the interdependence of the two hemispheres. All these studies have given us information primarily about the architecture of the brain. But the brain also has a very complex neurochemistry. Many new kinds of neurotransmitters have been discovered. The particular function of each of them is only now being understood. This new knowledge lets us probe the relationship between the presence of certain neurochemicals and associated cognitive processing. The inability to transfer information from short-term to long-term memory, for example, seems to be associated with low levels of acetylcholine. Moreover, the complexity of these chemical processes in the brain and the differential rates of firing of neural networks make it appear unlikely that the brain is operating as a simple two-state system. Machines that allow us to image the brain in greater detail than was ever possible before have been developed. The CAT (computerized axial tomography) scan gives us successive cross-sectional views of the brain that allow us to diagnose the location and extent of brain damage with greater precision than ever before. The PET (positron-emission tomography) scan allows us to monitor the pathways taken by radioactive tracers as they are metabolized in the brain. This gives us important information about some aspects of the working chemistry of the brain.

In addition to answering important questions about the architecture and the chemistry of the brain, we are now gaining some insights into the development of the brain. It is clear that the brain is not static. From birth, it is constantly forming and eliminating connections. There seems to be a dynamic interaction between the initial state of the brain at birth and experience with the world. During infancy, experience actually shapes the neural connections. These new connections in turn constrain what can be experienced. In this way the initial plasticity of the young brain is lost. The localization of function we see in the adult brain does not seem to be there initially. Even though language is normally located in the left hemisphere, a young child who has the language centers in his left hemisphere removed will be able to learn language by coding it on his right hemisphere. It is not clear how much of the brain must be intact in the infant for normal functioning to be possible. There are reports of people who function perfectly normally, yet a CAT scan reveals that they are missing 80% of their brain tissue. There are other cases in which very slight birth traumas having no detectable physical effects are correlated with very specific cognitive deficits. We simply do not know yet what accounts for these differences.

Both the new technology and the new knowledge about the brain have reawakened the old questions about the relationship of the brain to the mind. It looks as if we can, for the first time, begin to provide serious answers at the level of the actual hardware in which cognitive processing is carried out.

D. Artificial Intelligence

It is clear that the computer does not literally operate like the brain. For instance, it does not change its hard-wired connections in response to its early learning environment. It does not have prewired programs constraining the sorts of information it can consider. It clearly operates as a two-state system. In short, the computer is not a brain. But the question in artificial intelligence is not whether the computer is a brain, but can the computer, using this different kind of hardware and different kinds of processing, simulate intelligent behavior? This question has both theoretical and practical consequences. From the theoretical point of view, seeing just what the computer can do and what it cannot do can give us important insights into both theories of human intelligence and computational modeling. From the practical point of view, getting the computer to

behave intelligently would allow it to do interesting new tasks for us. In the beginning there was great optimism about what computers would be able to do. It was thought that it would be easy to get computers to translate from one natural language to another, to distinguish the spoken numbers, to answer questions about a database, to recognize pictures, and to read handwriting. All of this proved not to be easy at all. In general, the things that people do naturally and easily like speaking, understanding, seeing, hearing, and thinking are the most difficult things for the computer to do. Sometimes bits and pieces of a task can be done, but the bits that we have are not generalizable. For example, there are computer programs able to translate weather reports from English to French, but no one supposes that by adding a few more words and a few more syntactic rules the program could translate news reports. And yet humans have no more trouble understanding news reports than we do understanding weather reports. To understand weather-reports, we have to know about only a few states—snow, rain, sun, sleet, hot, and cold—and a few relationships among them. The language to encode these facts can be simple because there are no complex interrelationships that have to be qualified carefully. News reports are a totally different story. To understand the news, we must know about the physical, geographical, historical, and political world. We also have to know about people and their beliefs, intentions, goals, strategies, and possible actions, both rational and irrational. In order to represent the complex interactions held among these different factors, we use the full complexity of our syntactic and semantic power.

We might hypothesize that if we could tell the computer what we knew then it could do the same job. The problem is that we do not always know what we know. We do not yet even know how language, which we use so automatically and fluently, works. We know even less about beliefs, intentions, goals, and actions. However, even though our knowledge of what we know is not perfect, there are some areas, especially those founded on learned knowledge rather than on natural processes, that we can make more or less explicit. There have been very successful programs that have tapped the knowledge of experts in order to build systems that could simulate the behavior of these experts. These systems work best when the knowledge is about a set of constrained facts about which there is general consensus in the field. Given a problem in the field, the expert system attempts to specify a step-by-step procedure that will lead the computer to the same solution as the expert would reach. A prototypical application of expert systems is in medical diagnosis. The doctor, on the basis of reported symptoms and test results, forms an opinion about the probable cause, the expected course, and the recommended treatment of the illness. The first step in building an expert system is to ask an expert, in this case a diagnostician, to specify every step to be used in reaching a conclusion. This includes what facts to count as significant and which to ignore, what further tests to order, and how to make valid inferences on necessarily partial knowledge. Even in constrained and explicit fields, it is often very difficult for the expert to tell the programmer everything he or she needs to know. The expert often can come to the right conclusion without knowing exactly why. One of the side benefits of the development of an expert system is to make the field of knowledge in question as clear and explicit as possible.

But such a direct approach will not work for everything. We are all experts at seeing, but no one can develop an expert computer vision system by directly interrogating us, the experts. The development of computer vision systems and the development of knowledge about how vision works have cross-fertilized each other. Information about the way people actually process the visual signal has directly influenced the way some of the computer models have been constructed, and the failures of the computer models have, in turn, led to experiments about the visual system.

The standard way to build computer models of cognitive processes has been to provide a representation of the properties of the input in detail and to specify both the exact rules and the sequential ordering of these rules that act on this input. In the past few years a totally different approach to computer modeling has been introduced—connectionism. Connectionist models are composed of networks of parallel computing elements. Instead of using precise rules, these systems work by having activation values for elements in a network that are calculated from the activation values of neighboring elements. The input is also represented in terms of activation values. The system is trained on sets of input/output pairs on the basis of which the system constructs its own internal activation values. Proponents of connectionist models believe that this form of computer modeling more accurately reflects the kind of learning that goes on in humans and therefore is more likely to be successful in modeling human intelligence. Opponents of these models argue that connectionist models have not solved the fundamental inadequacies of computer models in general. These systems have provided an interesting new way of modeling cognitive functions on the computer. It is not clear, however, that this new approach will, in practice, be able to solve higher-level cognitive problems. Nor is it clear that it will be able, in principle, to provide insights into human cognitive function.

Artificial intelligence has been central in the development of cognitive science. The attempts to simulate intelligent behavior, partial as they have been, have made us realize the full complexity of what appears at first glance to be simple cognitive processes. Our understanding of cognitive processes in humans has made it possible to build more sophisticated and intelligent computer models.

E. Anthropology

One of the crucial questions in cognitive science is to determine which properties of cognition are necessary and universal. One approach to this question, which we discussed already, is to try to determine a genetic or neurological correlate of the phenomenon. Another important source of information is to look at variations among cultures. A property of the mind occurring everywhere is likely to be a manifestation of a cognitive universal. At one time anthropologists believed that cultures could vary in any way. What might seem natural in one culture would seem unnatural in another. Now we know that while some things can vary widely from culture to culture, there are also some constancies. It is these constancies that interest the cognitive scientist.

The focus of anthropology has traditionally been to document the diversity of cultures. It has searched for and found unusual and striking differences in all aspects of culture from food to dress to social organization. This is not surprising. Variation is easier to perceive than sameness. It is unremarkable that someone in a new culture walks upright. It is what we expect, and what we fully expect to see often becomes transparent. We simply do not notice things that are perfectly familiar. But this emphasis on diversity has led anthropologists away from the questions that interest cognitive scientists. If we look at the traditional explanations anthropologists have given to explain culture, the gap is even wider. Diversity or similarity between cultures has been explained either by functional necessity or by the effects of technology. Functional necessity, for example, is used to account for the organization of the family. Because women bear and nurse children it is functionally efficient for women to stay at home and care for the children and perform household chores. Because men are on the average larger and stronger and more aggressive it is functionally efficient for them to be warriors and hunters. The division of labor between men and women is seen to be the result of constraints on the organization of culture designed to maximize the efficient functioning of the system as a whole. If there is a thread of similarity among cultures, it can be accounted for by postulating that the functional demands are the same everywhere. Attributing differences in culture to the effects of technology is an old theme. For Rousseau, the advent of technology was a corrupting influence. Technological humans were essentially different from their primitive counterparts, and clearly worse. The same sort of explanation is still with us, but now technology enriches. It is claimed that technological humans can think more complexly. Without technology, complex thought is not possible. Neither of these explanations is in accord with the cognitivists' model. The view from cognitive science is that the cognitive structure of humans is universal and independent of particular functional constraints and certainly independent of short-term technological innovations. Of course functional efficiency, as one of the driving forces of evolution, probably did have a hand in shaping our cognitive structures, but that was long before the cultural variation we now see.

There are two other approaches to anthropology that do make assumptions about regularity and universality. One of these is French structuralism and the other is semiotics. French structuralism provides a model of description which organizes cultural facts into a coherent system. Each aspect of the system has value only in relation to the rest of the system. What is allowed takes its meaning from what is forbidden. Semiotics assumes that culture can best be understood as a system of signs by means of which communication among the members of the society is maintained. Both approaches tie in tangentially to cognition, but they are both closer to descriptive systems than to explanatory systems. Work in anthropology that asks specific cognitive questions is just beginning. For example, it had been believed that people in different cultures literally had different ways of seeing the world. The evidence of this difference is that people in some cultures draw very different kinds of pictures than we do to depict the world around them and when they look at our pictures they do not find them to be good representations of the world. These early results have been replaced by more subtle tests that show that the fundamental cognitive concepts of representation are the same everywhere. However, there are culturally determined levels of style and taste in pictures. Everyone agrees about which picture "looks" most like an elephant, but they differ on which picture is the "best" picture. Bringing the hypotheses that have been developed in cognitive science to bear on questions in anthropology will undoubtedly lead to new insights about humans.

F. Philosophy

Philosophical speculation about the nature of the mind has been fundamental to the development of cognitive science and, in turn, the results of cognitive science have helped rephrase and refocus some of the old questions. There are three questions of particular interest that involve the close interaction of the empirical results from the various fields of cognitive science with overall philosophical considerations. The first question concerns the organization of the cognitive faculties in the mind. Are these faculties separate and autonomous, or are they simply specific manifestations of some underlying uniform intelligence? The second question concerns the relationship between

our minds and the minds of other species. Are the minds of all creatures essentially alike, differing only in the details, or are there emergent properties of the mind that produce incommensurate differences among the species? The third question concerns the relationship between computer models of intelligence and real minds. Are the differences between them only differences in degree which can be eliminated with the advent of different sorts of machines and more sophisticated programs, or are these differences in principle unresolvable? The answers to these questions are crucial for the understanding of the mind. But they are not the sort of questions that can be left until the end, to be resolved once all the facts are in. These questions are so fundamental that they shape both the course of research itself and the kinds of explanatory theories built to account for the results of this research.

1. Modularity

The first question, about the organization of the cognitive faculties in the mind, actually questions the possibility of ever having a single theory of cognition. If each of the separate aspects of cognition is autonomous and is processed in its own module with its own internal organization and constraints, then the study of one aspect of cognition will give us little insight into any another aspect of cognition. A theory of cognition under the modularity hypothesis would simply be an amalgam of independent theories of the autonomous cognitive modules. Under the contrasting general intelligence hypothesis, there are generalized processes of perception, memory, and thinking that are the same for all the different cognitive processes. The input to the cognitive systems comes from the appropriate receptors—ears, eyes, skin, nose—which is then sent to the general processor, which organizes it, compares it with stored memories, draws inferences, and initiates action. This model would lead directly to a unified theory of cognition. Anything we found out about one cognitive system could be relevant to some other cognitive system.

The modularity of mind hypothesis has to be distinguished from the localization hypothesis. The modularity hypothesis is a hypothesis about the way the mind organizes and processes the various cognitive domains. In terms of levels of explanation we discussed before, it is a level two question. The localization hypothesis supposes that there are separate areas in the brain devoted to each cognitive system. It is a hypothesis about the hardware in which these processes are carried out. These two hypotheses are not necessarily linked. The processes could be radically different from one another, but instead of being localized they could be instantiated in complex networks in the brain. It could also be the case that each of the processes is essentially the same, but the location for each cognitive system might be different. Although there is no necessary reason for modularity to be linked with localization, there is a suspicion that it would be, empirically, an efficient way to organize things.

For a long time the general intelligence model reigned, but now there is reason to believe that there are some great differences among the various cognitive systems. For example, it looks as if language, at least in some respects, has very specific constraints. The underlying organization of language at the level of sound and syntax seems peculiar to language and unlike anything we see in the other cognitive systems. Just how finely grained these modules are is still a matter of debate, but there are hypotheses suggesting that they are very specific. It is likely that the visual system can be usefully divided into smaller modules such as color, three-dimensionality, face recognition, and shape recognition, to name just a few. These are not primarily philosophical problems, but rather empirical ones. Criteria for deciding when two systems are best thought of as autonomous have been developed and are being refined. Evidence that can decide the case in specific instances is being collected. Cognitive scientists are in the forefront of the investigation, which seems to suggest that what unites them is a single question and point of view rather than a single explanatory theory of mind.

2. Species Specificity

The second question, about species specificity, has consequences for the generalizability of data about one animal to any other. If there are emergent properties of mind that make the cognitive systems of one species radically different from the cognitive systems of another species, each species must be said to live in its own special cognitive world. Evidence about cognitive systems in one kind of organism is absolutely irrelevant to discussions of processes in another type of organism, even if the systems seem to share characteristics. Although the problem of cross-species comparison is true of all organisms, the real heat of the issue is generated when we try to go from animals, no matter what their kind, to people. Are humans really different in fundamental ways from other creatures, or are they just fancier versions of the same thing? There is no one involved in this debate who does not subscribe to the evolutionary hypothesis. Our lungs and hearts developed from some simpler lungs and hearts. But what about our minds? We know there have been moments in evolution when properties emerged that were different in kind from previous properties. We now would like to know whether certain cognitive processes are the result of such emergent properties or whether all cognitive processes can be derived from simpler processes.

The prime candidate for an emergent cognitive system is language. All normal human babies acquire language at about the same time and in about the same way. Other animals do not. It is true that animals have communication systems that can be intricate, but, it is argued, they differ from human language in several fundamental ways. First, they are very constrained in what they can talk about. For example, while the language of bees can effectively communicate about the location and the distance of the food source, it does not and cannot talk about the scenery along the way, or the likely predators to be encountered, or the similarity of this field to the one visited last week. These systems are not creative in the sense that they do not combine elements in new ways to produce entirely new utterances. Language can do this. In general, the formal properties which we find in all natural languages are absent from animal communication systems. But, second, and more important, humans can use language intentionally. Not only can language be used to inform, it can also be used to deceive. We can choose to speak or to remain silent. Animals are not free to communicate or not. In animals when the conditions for the system to be activated are present, the communication must take place. Bees returning to the hive do their dance no matter what. Speakers not only can choose to communicate, they *know* that they can choose to do so and that the hearer knows that they intend to communicate. The speaker knows that the hearer knows that the speaker intends to communicate, and so on. Humans do not just have intentions, they are conscious of their own intentions and can reflect on them. Even if we want to say that some actions in some creatures seem to be intentional, we cannot find any behavior indicating this level of reflection. But even if animals do not naturally acquire language, they might still have untapped cognitive capacities to learn such systems. Several experiments designed to teach language to apes have been mounted. The results have been mixed. The creatures have shown themselves capable of learning to use symbols to refer to objects and actions. They have been able to master simple rules for ordering these symbols. They have used these symbols not only in training situations, but also in novel situations to communicate with other creatures. In no sense, though, have they achieved the formal complexity or the levels of intentionality we see in the language of children. Given these fundamental failures, it is not clear that their accomplishment should be considered to be rudimentary language.

These theoretical considerations that argue for language being an emergent property have recently been supplemented by empirical data which indicate that certain properties of language may be controlled by an autosomally dominant gene. Several recent studies of the pattern of occurrence of specific developmental language impairments in twins and in families have indicated that these impairments are likely to be associated with an autosomally dominant gene. The language of these subjects has been studied and it can be shown that the deficit is in a specific part of their underlying grammar. Therefore, both theoretical considerations and empirical data seem to indicate that at least certain parts of language may be a species-specific attribute of humans.

When we look at the other cognitive domains, the answer to the question is less clear. Animals do have visual systems, auditory systems, memories, and some kinds of inferencing. We know that these systems may differ from one another in their basic organizing principles. Whether these differences are best thought of as creating incommensurate systems or as being continuous with one another is still a vexed question. The answer awaits the discovery of some unifying theory or the proof that none exists.

3. Computer Models

The third important question concerns the relationship between computers and minds. Are computers really "electronic brains" substituting electronic circuitry for neurons to perceive, remember, and think, or are they just a series of on–off switches that *we interpret* as performing these tasks? The answer depends on what you think is crucially important about cognitive systems and what you think is merely extra decoration. If what is important is getting the same output, then some computers can accomplish some cognitive tasks such as playing chess and proving theorems, although none as of yet can identify the objects in a picture chosen at random. It is clear that the computer does not use the same processes to accomplish the same ends as do people. Processes that computers are good at, humans are not, and vice versa. For example, computers are extraordinarily good at remembering and recalling detailed information. They can easily remember everyone's name and address and telephone number in Montreal. What they do not seem to be able to do is to reorganize this memory storage depending on the situation. Humans seem to be able to do this. Or perhaps it might be better to say that human memory seems to follow different principles than does computer memory. We are best at remembering episodes, not facts. My memory of a picnic in Paris might one time be triggered by a taste of a baguette, by the sound of an accordian, or by the need for an example of a memory. I can access it through any modality and even at the meta level of simply being an example of a memory. Moreover, this list does not begin to be exhaustive. There is seemingly no end to the ways in which this single memory can be recalled. This means that both the rules for representing memories and those for accessing memories must be very rich. How rich and how they are organized, we still do not know. But all this is just detail. Perhaps if we knew how to build larger memories, then dynamic, episodic memory would be possible. The crucial question is whether there are dif-

ferences that no amount of tinkering can fix. To find likely candidates for such differences, we have to look at some of the deeper properties of the human mind. Humans are capable of not only perceiving, remembering, and thinking, but also knowing that they are doing so. They can not only follow some course of action, but also intend to do so. Humans can not only say something; but also mean it. Machines cannot know, intend, or mean. It is only humans who design and interpret the action of computers that can. And we can even go up another level. We can be conscious that we are knowing, intending, or meaning, and conscious that we are conscious. No machine can come close.

VI. CONCLUSION

Cognitive science as a discipline is young, born in the 1950s. In terms of the questions that cognitive scientists want to answer, however, the field is very old. What it brings to these old questions is a new way of looking at the problem and a new set of tools for answering them. New knowledge about cognitive development and function has progressed at an ever-increasing rate. We have new knowledge about intelligence in animals, humans, and machines. But more important, we have found out that what we thought would be easy to know is difficult. The questions we are asking now are more intelligent. We still do not know if a creature with our intelligence is capable of answering them.

SEE ALSO THE FOLLOWING ARTICLES

ARTIFICIAL INTELLIGENCE • ARTIFICIAL NEURAL NETWORKS • COMPUTER ALGORITHMS • CYBERNETICS AND SECOND ORDER CYBERNETICS • DIGITAL SPEECH PROCESSING • GAME THEORY • HUMAN-COMPUTER INTERACTION • IMAGE PROCESSING • MATHEMATICAL LOGIC • SELF-ORGANIZING SYSTEMS

BIBLIOGRAPHY

Gardner, H. (1985). "Mind's New Science: The Cognitive Revolution in the Computer Age," Basic Books, New York.
Gopnik, I., and Gopnik, M., eds. (1986). "From Models to Modules: Studies in Cognitive Science," Ablex, Norwood, NJ.
Hirst, W., ed. (1988). "The Making of Cognitive Science: Essays in Honor of George Miller," Cambridge Univ. Press, New York.
Penrose, R. (1989). "The Emperor's New Mind: Concerning Computers, Minds and the Laws of Physics," Oxford Univ. Press, New York.
Posner, M., ed. (1989). "Foundations of Cognitive Science," MIT Press, Cambridge, Mass.
Pylyshn, Z. (1984). "Computation and Cognition: Towards a Foundation for Cognitive Science," Bradford Books, U.K./MIT Press, Cambridge, Mass.

Coherent Control of Chemical Reactions

Robert J. Gordon

University of Illinois at Chicago

Yuichi Fujimura

Tohoku University

GLOSSARY

Coherent control Control of the motion of a microscopic object by using the coherent properties of an electromagnetic field. *Coherent phase control* uses a pair of lasers with long pulse durations and a well-defined relative phase to excite the target by two independent paths. *Wave packet control* uses tailored ultrashort pulses to prepare a wave packet at a desired position at a given time.

Coherent population transfer Transfer of population from one quantum mechanical level to another using coherent radiation. The radiation may be provided by either continuous or pulsed lasers. Using the method of adiabatic passage (see *STIRAP*), 100% population transfer has been achieved.

Genetic algorithm A learning algorithm used to maximize the adaptability of a system to its environment. The method, based on the genetic processes of re-

production, crossover, and mutation, has been used to optimize the amplitudes and phases (the "genes") of the frequency components of a laser pulse in order to generate a wave packet with desired chemical properties.

Mode-selective chemistry The use of laser beams to control the outcome of a chemical reaction by exciting specific energy states of the reactants.

Optimal control theory A method for determining the optimum laser field used to maximize a desired product of a chemical reaction. The optimum field is derived by maximizing the objective function, which is the sum of the expectation value of the target operator at a given time and the cost penalty function for the laser field, under the constraint that quantum states of the reactants satisfy the Schrödinger equation.

Pendular state Superpositions of field-free rotational eigenstates in which the molecular axis librates about the field direction. Pendular states are eigenstates of the rotational Hamiltonian plus the dipole potential.

STIRAP Transfer of population by means of Stimulated Raman Adiabatic Passage, using a pump and Stokes laser. Population in a three-level system is completely transferred without populating the intermediate state if the Stokes laser precedes the pump laser in a "counter-intuitive" order.

Wave packet A localized wave function, consisting of a non-stationary superposition of eigenfunctions of the time-independent Schrödinger equation.

COHERENT CONTROL refers to a process in which the coherent properties of an electromagnetic field are used to alter the motion of a microscopic object such as an electron, atom, or molecule. The controlled process may be categorized according to the degree of freedom that is manipulated. For example, a laser beam with carefully tailored properties might be used to control the motion of electrons within an atom or molecule, thereby populating specific eigenstates, or to create electronic wave packets with interesting spatial and temporal properties. Another possibility is the use of coherent light to control the stretching and bending modes of a molecule, thereby altering its chemical reactivity. These are both examples of the control of *internal* degrees of freedom. Alternatively, a coherent light source might be used to orient a molecule in space so that a particular bond is pointing in a chosen direction. Another possibility is to use a focused laser beam to control the translational motion of an atomic or molecular beam, perhaps focusing the particles to a small volume or steering them in a new direction. In a condensed phase, a laser might be used to alter the direction of an electric or ion current. These are illustrations of control of *external* degrees of freedom. In all of these examples, the coherence of a light wave is transferred to a material target so as to alter the dynamical properties of the target in a controlled way.

I. OVERVIEW

In this article we approach the topic of coherent control from the perspective of a chemist who wishes to maximize the yield of a particular product of a chemical reaction. The traditional approach to this problem is to utilize the principles of thermodynamics and kinetics to shift the equilibrium and increase the speed of a reaction, perhaps using a catalyst to increase the yield. Powerful as these methods are, however, they have inherent limitations. They are not useful, for example, if one wishes to produce molecules in a single quantum state or aligned along some spatial axis. Even for bulk samples averaged over many quantum states, conventional methods may be ineffective in maximizing the yield of a minor side product.

A number of strategies have been developed over the past few decades to overcome the limitations of bulk kinetics. One method, known as *mode-selective chemistry*, exploits the idea that a molecule may have an eigenstate that strongly overlaps a desired reaction coordinate. Depositing energy into that degree of freedom may selectively enhance the reaction of interest. In the following section we give a number of examples using localized nuclear or electronic motion to enhance a particular process. Although this approach does not depend intrinsically on the coherent properties of light and therefore lacks one of the characteristic features of coherent control, it has played an historic role in the development of control techniques.

An inherent limitation of mode-selective methods is that Nature does not always provide a local mode that coincides with the channel of interest. One way to circumvent the natural reactive propensities of a molecule is to exploit the coherence properties of the quantum mechanical wave function that describes the motion of the particle. These properties may be imparted to a reacting molecule by building them first into a light source and then transferring them to the molecular wave function by means of a suitable excitation process.

Two qualitatively different (though fundamentally related) strategies for harnessing the coherence of light were developed in the mid-1980s. The first, proposed and developed by Paul Brumer and Moshe Shapiro, is a molecular analogue of Young's two-slit experiment, in which two coherent excitation paths promote the system to a common final state. Variation of the relative phase of the two paths produces a modulation of the excitation cross section. This method does not rely on the temporal properties of the light source and may in principle use a continuous laser. We refer to this approach as *coherent phase control* and describe it in detail in Section III.

The second approach, proposed by David Tannor and Stuart Rice, and further developed by them and others including Ronnie Kosloff and Herschel Rabitz, uses very short pulses of light to prepare a wave packet that evolves in time after the end of the pulse. After a suitable delay, an interrogating pulse projects out the product of interest. *Wave packet control* may be thought of as a generalization of mode-selective chemistry in which a short (and therefore broadband) pulse of light produces a localized non-stationary state that evolves in a predetermined fashion. Wave packet methods have been used with considerable success to control electronic, vibrational, and rotational motion of a variety of simple systems. One of the very powerful properties of this approach is that it is possible to use automated learning algorithms to tailor the laser pulses to create wave packets with desired properties. Details of wave packet control are given in Section IV.

Coherent radiation may also be used to control the *external degrees of freedom* of a molecule. For example, it is possible to create a quivering "pendular state" in which a molecule having an anisotropic polarizability is aligned along the electric field vector of a laser beam. It is also possible to use a focused laser beam to deflect a beam of molecules, perhaps focusing them to a point or steering them towards a target. Control of external degrees of freedom is discussed in Section V. We conclude this article with a brief discussion of future directions that the field is likely to take.

II. MODE-SELECTIVE CHEMISTRY

The central concept of mode-selective chemistry is illustrated in Fig. 1, which depicts the ground and excited state potential energy surfaces of a hypothetical triatomic molecule, ABC. One might wish, for example, to break selectively the bond between atoms A and B to yield products A+BC. Alternatively, one might wish to activate that bond so that in a subsequent collision with atom D the products AD+BC are formed. To achieve either goal it is necessary to cause bond AB to vibrate, thereby inducing motion along the desired reaction coordinate.

Direct excitation to the continuum usually (but not always, *vide infra*) results in rupture of the weakest bond. In order for the experimenter to have control over which bond is broken, it is helpful first to excite motion along the bond of interest. This process, known as vibrationally mediated photodissociation, preselects the desired degree of freedom before the reaction takes place. This method is illustrated in Fig. 1, where a low energy photon excites

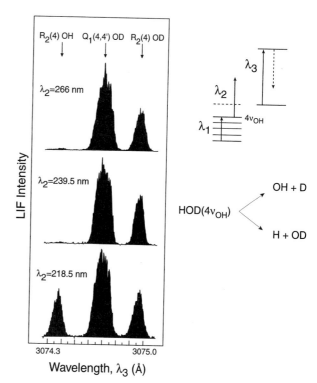

FIGURE 2 Vibrationally mediated photodissociation of water. (Provided by the courtesy of Fleming Crim.)

a bound state of the molecule causing the A–B bond to stretch. A high energy photon then promotes the molecule to the A+BC product valley of an excited potential energy surface.

The possibility of such preselection depends on the local mode character of the molecule. Typical narrow-band light sources excite a stationary eigenstate of the Hamiltonian. Such an eigenstate may be written as a linear combination of zero-order states, ϕ_i, which correspond to localized motion such as stretches of individual bonds or simple bending motion. Our goal is to excite one of those zero-order states, such as the A–B stretch. In a favorable case, that zero-order state will carry most of the oscillator strength and will also be a major component of the excited eigenstate. Designating the wave function of the zero-order "bright" state by ϕ_s and all the other zero-order states as "dark," we may write the excited state wave function as

$$\psi = c_s\phi_s + \sum_{i \neq s} c_i\phi_i. \qquad (1)$$

The bright-state character corresponding to localized vibration of the AB bond equals $|c_s|^2$. For anharmonic molecules it is not uncommon to find eigenstates with large local mode character, i.e., with $|c_s|^2 \cong 1$.

Figure 2 illustrates the vibrationally mediated bond-specific photodissociation of isotopically labeled water,

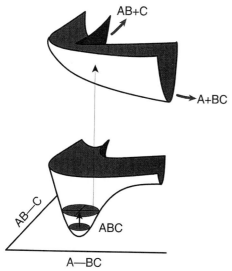

FIGURE 1 Illustration of mode-selective control of the dissociation of a triatomic molecule. (Provided by the courtesy of Fleming Crim.)

HOD. A 722.5 nm laser pulse (λ_1) excites the third overtone stretch of OH. After a short delay, a pulse of ultraviolet radiation of frequency ν_2 (wavelength λ_2) dissociates the molecule, and a third pulse with a wavelength near 308 nm (λ_3) probes the OH or OD fragments by laser-induced fluorescence. It is observed that with a dissociation wavelength of 266 or 239.5 nm, the products are almost exclusively H + OD,

$$HOD(4\nu_{OH}) + h\nu_2 \rightarrow H + OD, \qquad (2)$$

as seen in the $Q_1(4,4')$ and $R_2(4)$ fluorescence lines of OD, whereas with 218.5 nm equal amounts of OH and OD are formed. Because a stationary state of the molecule is excited by the first laser, the bond remains energized indefinitely until it collides with another particle. The excited molecule can then react, breaking preferentially the activated bond. For example, collision of HOD($4\nu_{OH}$) with a chlorine atom produces primarily HCl rather than DCl:

$$HOD(4\nu_{OH}) + Cl \rightarrow HCl + OD. \qquad (3)$$

The same principles have been applied to molecules with four atoms. For example, Fleming Crim and coworkers showed that excitation of the NH stretch of isocyanic acid enhances its reaction with Cl atoms,

$$HNCO(3\nu_1) + Cl \rightarrow HCl + NCO, \qquad (4)$$

whereas excitation of the bending mode inhibits the reaction. For the reactions of ammonia ions with neutral ammonia molecules,

$$NH_3^+ + ND_3$$

$$\rightarrow NH_3D^+ + ND_2 \qquad \text{(hydrogen extraction)} \quad (5a)$$

$$\rightarrow NH_3 + ND_3^+ \qquad \text{(charge transfer)} \quad (5b)$$

$$\rightarrow NH_2 + ND_3H^+, \qquad \text{(proton transfer)} \quad (5c)$$

Richard Zare and coworkers found that excitation of the umbrella mode of NH_3^+ selectively enhances the proton transfer reaction; however, in this case the projection of the nuclear motion onto the reaction coordinate is not as obvious.

Vibrational mode selectivity can also be used to promote electronic processes. Vibrational autoionization is a process whereby a bound electron acquires sufficient energy to escape by extracting one quantum of vibrational energy from the ionic core of the molecule. For such an energy transfer to occur, the electron must first collide with the core. Scattering of the electron with the core can be promoted if the amplitude of the nuclear motion overlaps the electronic charge density. An example of this process studied by Steven Pratt is vibrational autoionization of the 3d Rydberg electrons of ammonia, which is enhanced

by the umbrella vibration of the molecule. The d_{z^2}, d_{xz}, and d_{yz} orbitals have lobes perpendicular to the plane of the $\tilde{C}'$ Rydberg state of NH_3. Excitation of the out-of-plane umbrella mode of the molecule promotes vibrational autoionization of electrons in these orbitals but has little effect on the poorly overlapping $d_{x^2-y^2}$ and d_{xy} electrons.

It is also possible to use localized *electronic* excitation to promote reactions selectively. An example studied by Laurie Butler and coworkers is the ultraviolet photodissociation of CH_2IBr. This molecule has absorption maxima at 270, 215, and 190 nm, corresponding to localized excitation of a nonbonding iodine electron to an antibonding orbital localized on the C–I bond, ($n_I \rightarrow \sigma_{C-I}^*$), promotion of a nonbonding bromine electron ($n_{Br} \rightarrow \sigma_{C-Br}^*$), and a Rydberg transition, respectively. Photodissociation at 248.5 nm, at the edge of the $n_I \rightarrow \sigma_{C-I}^*$ transition, yields 60% I atoms and 40% Br. At 210 nm only Br atoms are formed, even though the C–I bond is the weakest bond in the molecule. In addition, some concerted IBr elimination occurs. At 193 nm all three products are formed.

Another example of electronic control studied by Butler is the photodissociation of methyl mercaptan, CH_3SH. Although the CH_3–SH bond is the weakest bond in the molecule, CH_3S + H are the primary photodissociation products at 193 nm. Bond selectivity in this case occurs even though the initially excited state is not repulsive along the reaction coordinate. Here selectivity results from non-adiabatic coupling of the initially excited metastable $2\ {}^1A''$ Rydberg state to the dissociative $n_I \rightarrow \sigma_{S-H}^*$ state. Another case where nonadiabatic coupling results in bond-selective chemistry is the photodissociation of bromoacetyl chloride, $BrCH_2COCl$. It was found for this molecule that at 248 nm the C–Cl bond is preferentially broken, even though the barrier for C–Br scission is lower than that for C–Cl scission. The reason for bond selectivity in this case is that the splitting between the adiabatic potential energy surfaces is much smaller for C–Br scission, so that nonadiabtic crossing and recrossing without reaction is much faster in this channel as compared with adiabatic motion along the C–Cl reaction coordinate. These examples illustrate that, although bond-selective photoexcitation is a general phenomenon, its mechanism depends strongly on the details of the potential energy surfaces. Studies of mode-selective reactions are, therefore, a valuable source of information about the structure of potential energy surfaces and their interactions.

III. COHERENT PHASE CONTROL

The underlying principle of coherent phase control is that the probability of an event occurring is given by the square of the sum of the quantum mechanical amplitudes

(a)   **(b)**

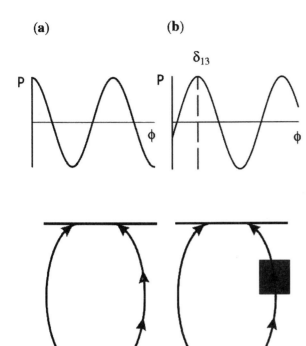

FIGURE 3 Illustration of coherent phase control by one- and three-photon excitation in the cases of (a) no intermediate resonances and (b) a quasi-bound state at the two-photon level that introduces a channel phase shift of δ_{13}. [Reproduced with permission from Fiss, J. A., Khachatrian, A., Truhins, K., Zhu, L., Gordon, R. J., and Seideman, T. (2000). *Phys. Rev. Lett.* **85**, 2096.]

associated with each independent path connecting the initial and final states. In the most commonly studied scenario, illustrated in Fig. 3a, two independent paths are the absorption of three photons of frequency ω_1 and one photon of frequency $\omega_3 = 3\omega_1$. Denoting the j-photon dipole operator by $D^{(j)}$, the probabilities of independent one- and three-photon transitions from the ground state, $|g\rangle$, to a bound excited eigenstate, $|e\rangle$, are given respectively by

$$P_3 = \left| \langle g | D^{(1)} | e \rangle \right|^2 \qquad (6)$$

and

$$P_1 = \left| \langle g | D^{(3)} | e \rangle \right|^2. \qquad (7)$$

In the chemically interesting case that the excited state is a continuum leading asymptotically to a product channel S at total energy E, the one-photon transition probability, integrated over all product scattering angles, $\hat{k}$, is given by

$$P_3^s = \left| \int d^3\hat{k} \langle g | D^{(i)} | E, S, \hat{k} \rangle \right|^2, \qquad (8)$$

with an analogous equation for P_1^s. The one- and three-photon dipole operators are given by

$$D^{(1)} = -\mu \cdot \varepsilon \qquad (9)$$

and

$$D^{(3)} = \sum_{i,j} \frac{D^{(1)} |i\rangle \langle i | D^{(1)} | j \rangle \langle j | D^{(1)}}{(E - \omega_1)(E - 2\omega_1)}, \qquad (10)$$

where μ is the electronic dipole, ε is the electric field, and the sum is over all states of the molecule.

For a single excitation path (i.e., one or three photons), the only possibility for controlling the outcome of the reaction is to select the excited eigenstate by varying E, as is normally done in mode-selective processes. A completely new form of control becomes possible, however, if both excitation paths are simultaneously available. In that case, the reaction probability is

$$P^s = \left| \int d^3\hat{k} \langle g | D^{(1)} + D^{(3)} | E, S, \hat{k} \rangle \right|^2 \qquad (11)$$

We assume that the electric field is a plane wave linearly polarized in the x direction,

$$\varepsilon_s(t) = \varepsilon_{s0} \hat{x} e^{i(k_{s,z}z - \omega_s t + \varphi_s)} \qquad (12)$$

where $k_{s,z}$ is the wave number, φ_s is an arbitrary phase, and $S = 1$ or 3. It is essential that there be a definite phase relation between the two laser fields, such that

$$\varphi = \varphi_3 - 3\varphi_1 \qquad (13)$$

is constant during an experimental run. Inserting Eqs. (9), (10), (12) and (13) into Eq. (11) and expanding the square, one obtains for the reaction probability

$$P^s = P_1^s + P_3^s + 2 \left| P_{13}^s \right| \cos\left(\delta_{13}^s + \varphi\right), \qquad (14)$$

where the cross term is given by

$$\left| P_{13}^s \right| e^{i\delta_{13}^s} = e^{-i\varphi} \int d^3\hat{k} \langle g | D^{(1)} | ES\hat{k} \rangle \langle ES\hat{k} | D^{(3)} | g \rangle. \qquad (15)$$

We refer to δ_{13}^s as a *channel phase*, which is a channel-specific property of the continuum.

The relative phase of the lasers, φ, is a new experimental tool. It is evident from Eq. (14) that the yield of each channel varies sinusoidally with φ. More importantly, one may maximize the relative yield of channel S by setting $\varphi = -\delta_{13}^s$. An experimental signature of phase control is a *phase lag* between the yields from any pair of channels,

$$\Delta\delta(S, S') = \delta_{13}^s - \delta_{13}^{s'}. \qquad (16)$$

A theoretical calculation of the branching ratio for the reaction

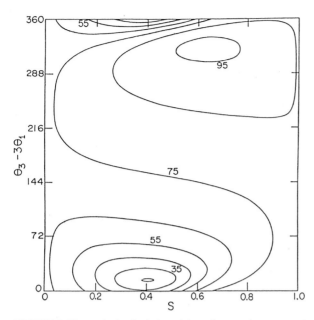

FIGURE 4 Theoretical calculation of the coherent phase control of the photodissociation of IBr by one- and three-photon excitation. [Reproduced with permission from Chan, C. K., Brumer, P., and Shapiro, M. (1991). *J. Chem. Phys.* **94,** 2688. Copyright American Institute of Physics.]

$$IBr + \omega_3, 3\omega_1 \rightarrow I + Br\left(^2P_{3/2}\right) \qquad (17a)$$

$$\rightarrow I + Br\left(^2P_{1/2}\right) \qquad (17b)$$

as a function of relative laser phase and intensity is given in Fig. 4. The contours show the fraction of $Br(^2P_{1/2})$ produced from an initial rovibrational level with quantum numbers $v = 0$, $J = 42$, averaged over initial M_J, with a photon energy of $\omega_1 = 6635.0$ cm^{-1}. The abscissa is the dimensionless ratio $x^2/(1 + x^2)$, where $x = |\varepsilon_1|^3/(\bar{\varepsilon}|\varepsilon_3|)$, and $\bar{\varepsilon}$ is defined as a single unit of electric field.

An apparatus used for coherent control experiments is depicted in Fig. 5. A molecular beam intersects the two laser beams in a plane lying between the repeller and extractor electrodes of a time-of-flight mass spectrometer. A uv laser beam of frequency ω_1 is focused into a cell containing a rare gas such as xenon. Third harmonic generation (THG) by the rare gas produces coherent vacuum ultraviolet (vuv) radiation of frequency ω_3 with a definite relative phase, φ, defined by Eq. (13). The two laser beams enter a phase-tuning cell containing a transparent gas such as hydrogen. Because the difference between the indices of refraction at ω_1 and ω_3 is proportional to the pressure of the phase-tuning gas, an increase in the gas pressure produces a linear increase of φ. Ions produced in the reaction region are repelled into a field-free flight tube and detected by a microchannel plate (MCP).

Typical experimental results are shown in Fig. 6 for the reaction

$$HI + \omega_3, 3\omega_1 \rightarrow H + I^*, \qquad (18a)$$

$$\rightarrow HI^+ + e^-. \qquad (18b)$$

The excited iodine atom produced in reaction (18a) absorbs one or two photons to yield the I^+ ion. The Xe pressure in the third harmonic cell is adjusted so that the one- and three-photon signals are approximately equal. Variation of the H_2 pressure in the phase-tuning cell produces the sinusoidal variation of the ion signals shown in Fig. 6. Evident in this figure is a phase lag of 150° between the two products, HI^+ and I. Also shown is modulation of the signal produced by photoionization of H_2S, which provides a reference phase for the HI^+ and I^+ signals.

Coherent phase control has been used to populate both bound and continuum eigenstates. Bound-to-bound state control has been demonstrated for many molecules, including HCl, CO, NH_3, CH_3I, $N(CH_3)_3$, $N(C_2H_5)_3$, $(CH_3)_2N_2H_2$, and c-C_8H_8. Bound-to-continuum control has been achieved for the photoionization of Hg, HI, DI, H_2S, and D_2S, and for the photodissociation of HI, DI, and CH_3I. In all of these studies, the use of one- vs three-photon excitation ensures that the parity changes for the two paths are the same. If the parities for the two paths are not equal, as for one- vs two-photon excitation, the average over scattering angles in Eq. (15) causes the cross term to vanish. In this case the *differential* cross section (i.e., the distribution of recoil angles) may still be controlled. One- vs two-photon control of angular distributions has been demonstrated for the photoionization of Rb and NO and for the photodissociation of HD$^+$. This method has also been used to control the direction of an electric current in a GaAs/AlGaAs quantum well and in an amorphous GaAs semiconductor.

It is possible to design multipath control schemes in which the laser phase cancels out of the interference term. One possibility is a "diamond" path configuration, $\omega_1 + \omega_2$ vs $\omega_2 + \omega_1$, with a resonance near ω_1 contributing a phase to the first path and a resonance near ω_2 contributing a phase to the second path. As before, the total probability is the square of the sum of the amplitudes for each path, but here the phases of the two laser beams appear in both paths and cancel in the cross term. In this case the control parameters are the laser frequencies, which determine the detuning from the resonances. This technique was used by Daniel Elliott and coworkers to control the differential cross sections for the ionization of Ba and NO.

Another example of phase-insensitive control utilizes the "lambda" scheme depicted in Fig. 7. In this case a strong coupling (ω_2) field mixes an excited state with

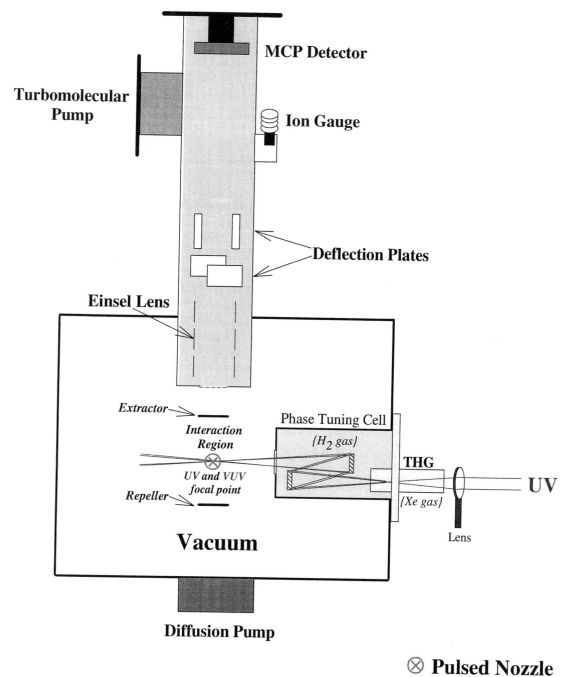

FIGURE 5 Apparatus used for coherent phase control.

the continuum, so that the two paths are $2\omega_1$ and $2\omega_1 - \omega_2 + \omega_2$. (The second path may be viewed as excitation of the continuum by $2\omega_1$ followed by emission and reabsorption of an ω_2 photon.) In the example shown in Fig. 7, a pulsed dye laser (5 ns pulse width) was used to dress the continuum of Na_2 with the $v = 35$, $J = 38$ and $v = 35$, $J = 36$ resonances of the $A^1\Sigma_u/^3\Pi_u$ manifold. A second dye laser (ω_1) induced two-photon disso-

ciation of the molecule, and spontaneous emission from Na(3p) and Na(3d) was used to monitor the branching ratio of Na(3s) + Na(3p) vs Na(3s) + Na(3d) fragments as a function of ω_2 detuning. The experimental control of the Na(3d) product shown in Fig. 8 is in excellent agreement with theory. Comparable results were obtained for the control of Na(3p), which reached a maximum near 13,317 cm^{-1}.

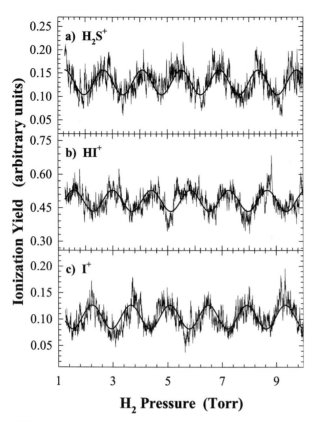

FIGURE 6 Experimental coherent phase control of the photodissociation and photoionization of HI. The three panels show the signals for the ionization of H_2S (top), which is used as phase reference, the ionization of HI (middle), and the dissociation of HI (bottom). [Reproduced with permission from Fiss, J. A., Zhu, L., Gordon, R. J., and Seideman, T. (1999). *Phys. Rev. Lett.* **82**, 65.]

The value of coherent control experiments lies not only in their ability to alter the outcome of a reaction but also in the fundamental information that they provide about molecular properties. In the example of phase-sensitive control, the channel phase reveals information about couplings between continuum states that is not readily obtained by other methods. Examination of Eq. (15) reveals two possible sources of the channel phase—namely, the phase of the three-photon dipole operator $D^{(3)}$, and that of the continuum function, $|ES\hat{k}\rangle$. The former is complex if there exists a metastable state at an energy of ω_1 or $2\omega_1$, which contributes a phase to only one of the paths, as illustrated in Fig. 3b. In this case the channel phase equals the Breit–Wigner phase of the intermediate resonance (modulo π),

$$\delta = -\cot\epsilon, \qquad (19)$$

where the ϵ is the reduced energy,

$$\epsilon = 2(E - E_{res} - \Delta)/\Gamma, \qquad (20)$$

and E_{res}, Δ, and Γ are, respectively, the unperturbed resonance center, the shift of the resonance, and its width. An example of this effect is illustrated in Figs. 9 and 10 for the photoionization of HI. The potential energy curves in Fig. 9 display the $b^3\Pi_1$ Rydberg state at approximately two-thirds of the ionization threshold. This quasibound state is predissociated by the $A^1\Pi$ continuum state. The structure evident in the phase lag spectrum in Fig. 10a is produced by the rotational levels of the $b^3\Pi_1$ state. The same rotational structure is evident in the conventional $2+1$ resonance-enhanced multiphoton ionization (REMPI) spectrum (Fig. 10b). The absence of any structure in the single-photon ionization spectra of HI (Fig. 10c) and H_2S (Figure 10d) confirms that the phase lag is produced by an intermediate resonance of HI.

The other source of a channel phase is the complex continuum wave function at the final energy E. At first it would appear from Eq. (15) that the phase of $|ES\hat{k}\rangle$ should cancel in the cross term. This conclusion is valid if the product continuum is not coupled either to some another continuum (i.e., if it is elastic) or to a resonance at energy E. If the continuum is coupled to some other continuum (i.e., if it is inelastic), the product scattering wave function can be expanded as a linear combination of continuum functions,

$$|ES\hat{k}\rangle = c_1|ES_1\hat{k}\rangle + c_2|ES_2\hat{k}\rangle, \qquad (21)$$

producing a nonzero channel phase that is only weakly energy dependent. The presence of a resonance at energy E produces an extremum in the energy dependence of $|\delta_{13}^s|$. If the underlying continuum is elastic, $|\delta_{13}^s|$ reaches a maximum on resonance, whereas if it is inelastic $|\delta_{13}^s|$ reaches a minimum on resonance. In the limiting case of an isolated resonance coupled to an elastic continuum, with both direct and resonance-mediated transitions to the medium, the channel phase has a Lorentzian energy dependence,

$$\tan\delta_{13}^s = \frac{2(q^{(1)} - q^{(3)})}{\left[\epsilon - \frac{1}{2}(q^{(1)} + q^{(3)})\right]^2 + \left[4 - \frac{1}{4}(q^{(1)} - q^{(3)})^2\right]}, \qquad (22)$$

where $q^{(j)}$ is the j-photon Fano shape parameter.

Examples of the latter two sources of the channel phase are illustrated in Fig. 11. From an independent knowledge that the channel phase for ionization of H_2S is zero (or π), it is deduced that the phase lags $\Delta\delta(I, H_2S^+)$ and $\Delta\delta(HI^+, H_2S^+)$ are equal, respectively, to the channel phases (modulo π) for the dissociation (δ_{13}^I) and ionization ($\delta_{13}^{HI^+}$) of HI. The nearly flat, nonzero values of δ_{13}^I (triangles in Fig. 11a) is indicative of coupling in the dissociation continuum, whereas the peak in δ_{13}^{HI} (diamonds)

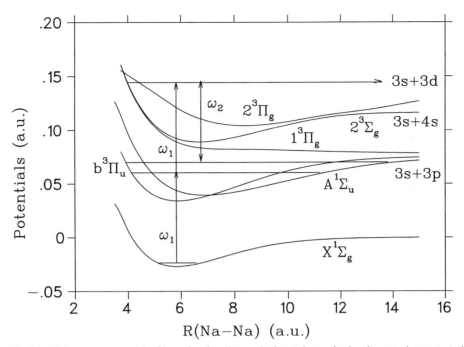

FIGURE 7 Potential energy curves for Na$_2$, showing the excitation scheme for incoherent phase control of the photodissociation of the molecule. [Reproduced with permission from Chen, Z., Shapiro, M., and Brumer, P. (1993). *J. Chem. Phys.* **98,** 6843. Copyright American Institute of Physics.]

at $\lambda_1 = 356.2$ nm (the three-photon wavelength) is caused by the 5d(π,δ) resonance of HI. This resonance is evident in the one-photon ionization spectrum (Fig. 11b), but is absent in the $2+1$ REMPI spectrum (Fig. 11c).

The secondary maximum observed in the phase lag near $\lambda_1 = 355.4$ nm has been attributed to a weak transition to a vibrationally excited Rydberg state not visible in the ionization spectrum. Examples of a minimum in $|\delta_{13}^s|$,

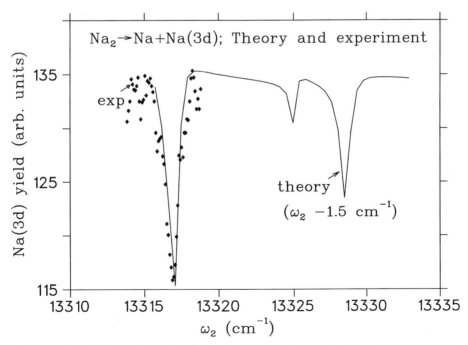

FIGURE 8 Comparison of the experimental and theoretical yields for the reaction Na$_2 \rightarrow$ Na(3s) + Na(3d), obtained by incoherent phase control. [Reproduced with permission from Chen, Z., Shapiro, M., and Brumer, P. (1993). *J. Chem. Phys.* **98,** 6843. Copyright American Institute of Physics.]

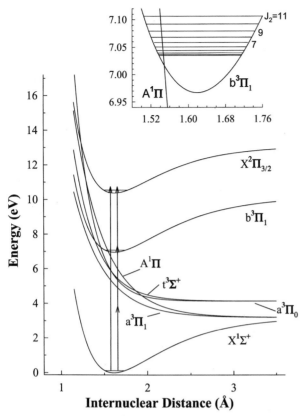

FIGURE 9 Potential energy curves of HI. The insert shows the rotational structure of the $b^3\Pi_1$ Rydberg state, which is predissociated by the $A^1\Pi$ valence state. [Reproduced with permission from Fiss, J. A., Khachatrian, A., Truhins, K., Zhu, L., Gordon, R. J., and Seideman, T. (2000). *Phys. Rev. Lett.* **85**, 2096.]

caused by inelastic coupling to a continuum, have also been observed.

IV. WAVE PACKET CONTROL

A. Introduction

Recent progress in laser technology has led to the widespread use of ultrafast lasers with pulse widths shorter than the vibrational periods of most chemical bonds. A localized state, called a nuclear wave packet, is created on a potential surface by exciting a molecule with ultrashort pulses of radiation. The time-evolution of such wave packets can be directly utilized to observe the transition states of chemical reactions. This development is one of the major accomplishments of femtosecond chemistry.

To understand how wave packets are created by ultrashort pulses, consider a molecule interacting with a pulsed laser field $\varepsilon(t)$. Figure 12 shows the time evolution of a

wave packet in a two-electronic state model. In this diagram, X_{g0} represents the lowest vibrational state of the ground electronic state, $X_e^{(1)}(0)$ is the excited-state wave packet created from X_{g0} by an optical excitation, $X_e^{(1)}(\tau)$ is the wave packet at time τ, and $X_g^{(2)}(\tau)$ represents the ground-state wave packet created by stimulated emission from $X_e^{(1)}(\tau)$. Superscripts of X denote the order of the photon-molecule interactions that are used in calculating these wave packets.

The total Hamiltonian $H(t)$ is given within the semiclassical treatment of the molecule-laser field interaction as

$$H(t) = H_0 - \boldsymbol{\mu} \cdot \boldsymbol{\varepsilon}(t). \tag{23}$$

Here H_0 is the molecular Hamiltonian, and $\boldsymbol{\mu} \cdot \boldsymbol{\varepsilon}(t)$ is the interaction between the molecule and the laser field in the dipole approximation, where $\boldsymbol{\mu}$ is the transition dipole moment of the molecule. Time evolution of the system is determined by the time-dependent Schrödinger equation,

$$i\hbar \frac{\partial}{\partial t}|\Psi(t)\rangle = H(t)|\Psi(t)\rangle. \tag{24}$$

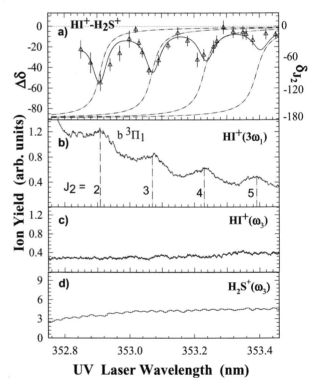

FIGURE 10 The effect of an intermediate resonance on the phase lag for the photoionization of HI. Shown are (a) the phase lag of HI$^+$ relative to H$_2$S$^+$, (b) the three-photon ionization spectrum of HI, and the one-photon ionization spectra of (c) HI and (d) H$_2$S. [Reproduced with permission from Fiss, J. A., Khachatrian, A., Truhins, K., Zhu, L., Gordon, R. J., and Seideman, T. (2000). *Phys. Rev. Lett.* **85**, 2096.]

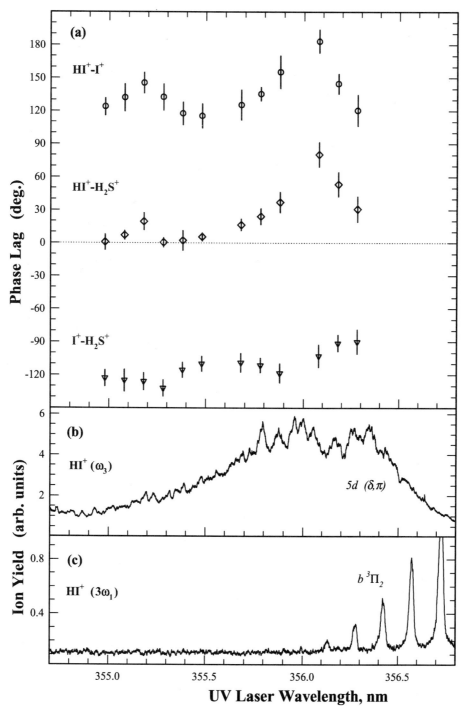

FIGURE 11 Phase lags for the photodissociation and photoionization of HI in the vicinity of the $5d(\pi,\delta)$ resonance. [From Fiss, J. A., Khachatrian, A., Zhu, L., Gordon, R. J., and Seidman, T. (1999). *Disc. Faraday. Soc.* **113,** 61. Reproduced by permission of the Royal Society of Chemistry.]

The solution of this equation $\Psi(t)$ is with the initial condition $\Psi(0)$ at $t = 0$.

Within the Born–Oppenheimer approximation, the initial wave function is expressed as $\Psi(0) = \Phi_g(r, R)X_{gv}(R)$, where $\Phi_g(r, R)$ denotes the electronic wave function and $X_{gv}(R)$ denotes the nuclear wave function. Here r and R are the coordinates of the electrons and nuclei, respectively. In the case of a weak field in which the

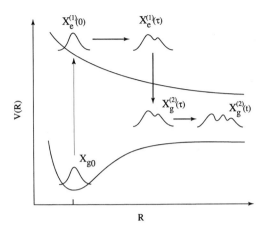

FIGURE 12 Time evolution of a wave packet, based on a perturbative treatment of a two-electronic state model.

population change is negligibly small, first-order time-dependent perturbation theory is sufficient for evaluating the time evolution of the molecular system. The nuclear wave packet $X_e^{(1)}(t)$ created on an electronic potential energy surface e from the lowest ground state, $\Phi_g(r, R)X_{g0}(R)$ in the vibrationless (low temperature) limit, is expressed, as shown in Fig. 12, as

$$
\left|X_e^{(1)}(t)\right\rangle = \frac{i}{\hbar} \int_0^t dt_1 \exp\left[-\frac{iH_e}{\hbar}(t - t_1)\right]\mu_{eg}(R)
$$
$$
\times \exp\left[-\frac{iH_g t_1}{\hbar}\right]|X_{g0}(R)\rangle \varepsilon(t_1), \quad (25)
$$

where H_e, the nuclear Hamiltonian of the electronic state e, is given by

$$
H_e = T_R + V_e(R). \quad (26)
$$

Here T_R is the kinetic energy of the nuclei and $V_e(R)$ is the potential energy. (The ground state Hamiltonian H_g, is given by an equivalent expression with V_e replaced by V_g.) In Eq. (25), $\mu_{eg}(R)$ is the electronic transition moment at the nuclear configuration R. Dephasing effects have been omitted in this equation. They can be taken into account by introducing an effective Hamiltonian, $H_{eff} = H_e - i\Gamma_{eg}$, in which $\Gamma_{eg} = \frac{1}{2}(\Gamma_{gg} + \Gamma_{ee}) + \Gamma'_{eg}$ is a dephasing constant. Here, Γ_{gg} and Γ_{ee} are the decay widths of the ground and excited states, respectively, and Γ'_{eg} is a pure dephasing constant produced by elastic collisions between the molecule and its reservoir.

Let the laser field be expressed as $\varepsilon(t) = \varepsilon_0(t)\sin(\omega_L t)$, where $\varepsilon_0(t)$ is the pulse envelope function, including polarization, and ω_L is the central frequency. For simplicity, consider a δ function excitation. In the rotating wave approximation, the field is expressed as $\varepsilon(t) = \frac{1}{2}\varepsilon_0\delta(t)\exp(-i\omega_L t)$ with field strength ε_0. Integration over t_1 in Eq. (25) gives

$$
\left|X_e^{(1)}(t)\right\rangle = \exp\left[-\frac{iH_e t}{\hbar}\right]\left|X_e^{(1)}(0)\right\rangle, \quad (27)
$$

where $X_e^{(1)}(0)$ is the nuclear wave packet created on the electronically excited state e just after the delta function excitation, given by

$$
\left|X_e^{(1)}(0)\right\rangle = \frac{i\varepsilon_0}{2\hbar}\mu_{eg}(R)|X_{g0}\rangle. \quad (28)
$$

The time evolution of the wave packet given by Eq. (27) can be evaluated by an eigenfunction expansion or by a split-operator technique. With the latter technique, the wave packet $X_e^{(1)}(t + \delta t)$ after a small increment of time δt can be expanded approximately as

$$
\left|X_e^{(1)}(t + \delta t)\right\rangle = \exp\left[-\frac{iV_e(R)\delta t}{2\hbar}\right]\exp\left[-\frac{iT_R\delta t}{\hbar}\right]
$$
$$
\times \exp\left[-\frac{iV_e(R)\delta t}{2\hbar}\right]\left|X_e^{(1)}(t)\right\rangle. \quad (29)
$$

This expansion is valid to second order with respect to δt. This is a convenient and practical method for computing the propagation of a wave packet. The computation consists of multiplying $|X_e^{(1)}(t)\rangle$ by three exponential operators. In the first step, the wave packet at time t in the coordinate representation is simply multiplied by the first exponential operator, because this operator is also expressed in coordinate space. In the second step, the wave packet is transformed into momentum space by a fast Fourier transform. The result is then multiplied by the middle exponential function containing the kinetic energy operator. In the third step, the wave packet is transformed back into coordinate space and multiplied by the remaining exponential operator, which again contains the potential.

Evolution of the wave packet on the excited state potential energy surface is described by Eq. (27). In the case of a bound potential energy surface, the wave packets are initially localized in the Franck–Condon region but eventually become delocalized because of vibrational mode-mixing processes produced by anharmonicities or because of kinetic couplings. In contrast, if the excited state is unbound, the wave packet rapidly departs from the Franck–Condon region. In both cases time evolution of the wave packet is observed by applying a second pulse, called a probe pulse, which induces stimulated emission or ionization after a selected time delay. This spectroscopic method, known as the pump–probe technique, is used to study the transition state on a femtosecond time scale. From an analysis of the pump–probe spectrum, information about the excited-state dynamics as well as structural properties of the excited potential energy surface may be obtained.

In the weak field limit, the time evolution of wave packets in pump–probe experiments can be evaluated by second-order time-dependent perturbation theory. The second-order solution of Eq. (24) is expressed as

$$\left|X_g^{(2)}(t)\right\rangle = -\frac{1}{\hbar^2}\int_0^t dt_2 \int_0^{t_2} dt_1 \exp\left[-\frac{iH_g}{\hbar}(t-t_2)\right]$$

$$\times \mu_{ge}(R)\exp\left[-\frac{iH_e}{\hbar}(t_2-t_1)\right]\mu_{eg}(R)$$

$$\times \exp\left[-\frac{iH_g t_1}{\hbar}\right]\left|X_{g0}(R)\right\rangle\varepsilon(t_2)\varepsilon(t_1). \quad (30)$$

If both the pump and probe pulses are assumed to be δ functions, Eq. (30) can be expressed as

$$\left|X_g^{(2)}(t)\right\rangle = \exp\left[-\frac{iH_g}{\hbar}(t-\tau)\right]\left|X_g^{(2)}(\tau)\right\rangle \quad (31)$$

where τ is the delay time between the pump and probe pulses, and $\left|X_g^{(2)}(\tau)\right\rangle$, the ground-state wave packet created just after irradiation by the probe pulse, has the form

$$\left|X_g^{(2)}(\tau)\right\rangle = \frac{i\varepsilon_0}{2\hbar}\mu_{ge}(R)\exp\left[-\frac{iH_e\tau}{\hbar}\right]\left|X_e^{(1)}(0)\right\rangle. \quad (32)$$

In the discussion so far, instantaneous excitation or de-excitation by a δ function pulse has been assumed to transfer wave packets from one electronic state to another state. For realistic pulses, the wave packets may be obtained by numerically integrating Eqs. (25) and (30).

B. Controlling Wave Packets with Tailored Laser Pulses

1. Perturbative Treatment

An intuitive method for controlling the motion of a wave packet is to use a pair of pump–probe laser pulses, as shown in Fig. 13. This method is called the pump–dump control scenario, in which the probe is a controlling pulse that is used to create a desired product of a chemical reaction. The controlling pulse is applied to the system just at the time when the wave packet on the excited state potential energy surface has propagated to the position of the desired reaction product on the ground state surface. In this scenario the control parameter is the delay time τ. This type of control scheme is sometimes referred to as the Tannor–Rice model.

There are many other variables in addition to τ that may be used to control the reaction products by manipulating the motion of wave packets. These include the time-dependent frequency, amplitude, and phase functions of the laser pulse. The use of tailored laser fields to alter the shape of a wave packet is a very general method for controlling the outcome of a chemical reaction.

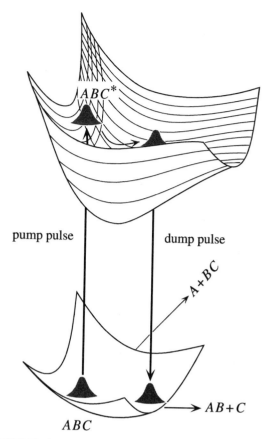

FIGURE 13 A pump–dump control scheme, used to control the branching ratio of the dissociation of a triatomic molecule, ABC.

In order to control a wave packet with tailored laser pulses, we introduce a target operator W, which is a projection operator ($W^2 = W$) that localizes the wave packet at a target position on an electronically excited potential energy surface. The target operator is one of the fundamental quantities in control problems. There are several kinds of target operators, depending on the type of object that is to be controlled. For population control, if a vibronic state X_{ef} is chosen as the target, then its target operator is expressed as $W = |X_{ef}\rangle\langle X_{ef}|$. For wave packet shaping, if a Gaussian wave packet characterized by its average position $\bar{R}$ and average momentum $\bar{P}$ is placed on an electronically excited state e, its target operator is expressed as $W_G = |X_{eG}\rangle\langle X_{eG}|$, with a coordinate representation expressed as $\langle R|W_G|R\rangle = \langle R|X_{eG}\rangle\langle X_{eG}|R\rangle$. Here $\langle R|X_{eG}\rangle$ is given as

$$\langle R|X_{eG}\rangle = (2\pi a^2)^{-\frac{1}{4}}\exp\left[i\frac{\bar{P}}{\hbar}(R-\bar{R})-\frac{(R-\bar{R})^2}{4a^2}\right], \quad (33)$$

where a, the square root of the variance, is the uncertainty in the position of the wave packet.

Applying W to Eq. (25), we obtain

$$W\big|X_e^{(1)}(t)\big\rangle = \frac{1}{\hbar}\int_0^1 dt_1 W\exp\left[-\frac{iH_e}{\hbar}(t-t_1)\right]\mu_{eg}(R)$$
$$\times\exp\left[-\frac{iH_e t_1}{\hbar}\right]|X_{g0}\rangle\varepsilon(t_1). \quad (34)$$

For a weak field, the probability $\langle W(t)\rangle$ of the wave packet existing at the target position at time t is given by

$$\langle W(t)\rangle \approx \big|W\big|X_e^{(1)}(t)\big\rangle\big|^2 = \big\langle X_e^{(1)}(t)\big|W\big|X_e^{(1)}(t)\big\rangle$$
$$= \frac{1}{\hbar^2}\int_0^{t_f}dt_1\int_0^{t_f}dt_2\langle X_{g0}|\exp\left[\frac{iH_g t_2}{\hbar}\right]\mu_{ge}(R)$$
$$\times\exp\left[\frac{iH_e}{\hbar}(t-t_2)\right]W\exp\left[-\frac{iH_e}{\hbar}(t-t_1)\right]\mu_{eg}(R)$$
$$\times\exp\left[-\frac{iH_g t_1}{\hbar}\right]|X_{g0}\rangle\varepsilon(t_2)\varepsilon(t_1). \quad (35)$$

Consider the problem of wave packet control in a weak laser field. Here "wave packet control" refers to the creation of a wave packet at a given target position on a specific electronic potential energy surface at a selected time t_f. For this purpose, a variational treatment is introduced. In the weak field limit, the wave packet can be calculated by first-order perturbation theory without the need to solve explicitly the time-dependent Schrödinger equation. In strong fields, where the perturbative treatment breaks down, the time-dependent Schrödinger equation must be explicitly taken into account, as will be discussed in later sections.

In the case of a weak field, the variational method is used to determine the properties of the laser pulses required to reach a specified target. For example, consider the shaping of a Gaussian wave packet in which the target is localized at an average position $\bar{R}$ with an average momentum $\bar{P}$. The target operator is given as W_G. To achieve the desired shape of the wave packet, we define an objective function,

$$J = \langle W_G(t_f)\rangle - \frac{1}{2}\int_0^{t_f}dt\,\lambda(t)|\varepsilon(t)|^2, \quad (36)$$

where $\langle W_G(t_f)\rangle$ is the expectation value of the wave packet localized near a given Gaussian target. The second term on the right-hand side of Eq. (36) is the constraint on the laser pulses, where $\lambda(t)$ is a time-dependent Lagrange multiplier.

Applying the variational procedure to Eq. (36), we obtain for the optimal control pulse the equation,

$$\int_t^{t_f}dt_1\langle X_{g0}|W_G^S(t_f;t,t_1)|X_{g0}\rangle\,\varepsilon(t_1) = \lambda(t)\varepsilon(t), \quad (37)$$

where $W_G^S(t_f;t_2,t_1)$ is a symmetrized operator defined as

$$W_G^S(t_f;t_2,t_1) = W_G(t_f;t_2,t_1) + W_G(t_f;t_1,t_2), \quad (38)$$

with

$$W_G(t_f;t_2,t_1) = \exp\left[\frac{iH_g t_2}{\hbar}\right]\mu_{ge}(R)\exp\left[\frac{iH_e}{\hbar}(t_f-t_2)\right]$$
$$\times W_G\exp\left[-\frac{iH_e}{\hbar}(t_f-t_1)\right]\mu_{eg}(R)$$
$$\times\exp\left[-\frac{iH_g t_1}{\hbar}\right]. \quad (39)$$

Because W_G^s in Eq. (39) is a Hermitian operator, the eigenvalues $\lambda(t)$ are real and express the yield of a given target.

This procedure is illustrated by the example of an outgoing wave packet of I_2 on the $B^3\Pi_{0^+}$ potential energy surface. The wave packet is assumed to be created from the lowest vibrational level in the ground $X^1\Sigma^+$ state. The potential energy curves for the ground and excited states are shown in Fig. 14. The target is defined as a wave packet on the B surface centered at $\bar{R} = 5.84$ Å, with the center of the outgoing momentum corresponding to a kinetic energy of 0.05 eV. The optimal field $\varepsilon(t)$ is a single pulse with a full width at half-maximum of $\sim$225 femtoseconds. The time- and frequency-resolved optimal field is shown in Fig. 15. A Wigner transform of the optimal field $F_w(t,\omega)$, given as

$$F_w(t,\omega) = 2\mathrm{Re}\int_0^\infty dt'\varepsilon^*\left(t+\frac{t'}{2}\right)\varepsilon\left(t-\frac{t'}{2}\right)g(t'), \quad (40)$$

is used. Here $g(t)$ is a window function for smoothing of a spectrum originated from a finite time width. The time- and frequency-resolved spectrum indicates the presence of positive chirp, i.e., a frequency increasing with time. This effect can be seen from the fact that the lower energy components of the continuum wave packet take relatively longer times to reach the target position, and the higher energy components take shorter times.

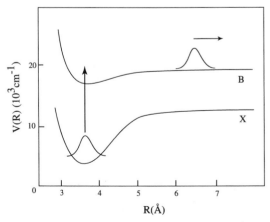

FIGURE 14 Potential energy curves for the ground ($X^1\Sigma^+$) and excited ($B^3\Pi_{0^+}$) states of I_2 vapor. Both the initial and outgoing wave packets are shown.

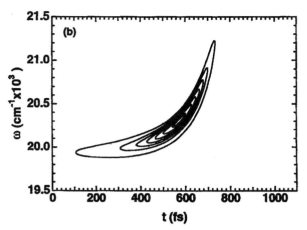

FIGURE 15 Wigner representation of the optimal electric field for I_2 wave packet control. [Reproduced with permission from Krause, Whitnell, J. L., Wilson, R. M., K. R., and Yan, Y. (1993). *J. Chem. Phys.* **99**, 6562. Copyright American Institute of Physics.]

2. Optimal Control

In Section IVB1, a perturbative treatment for wave packet control in a weak field was presented. In this section, a general theory based on an optimal control theory is presented. The resulting expression for laser pulses is applicable to strong as well as weak fields.

The expression for the optimal laser pulse is derived by maximizing the objective function J, defined as

$$J = \hbar\langle W(t_f)\rangle - \frac{1}{2}\int_0^{t_f}\frac{dt}{A(t)}|\varepsilon(t)|^2$$
$$+ 2\text{Re}\left\{i\int_0^{t_f}dt\langle\xi(t)|i\hbar\frac{\partial}{\partial t} - H(t)|\psi(t)\rangle\right\}. \quad (41)$$

The first term on the right-hand side of this equation, $\langle W(t_f)\rangle = \langle\psi(t_f)|W|\psi(t_f)\rangle$, is the expectation value of the target operator W at the final time t_f. The second term represents the cost penalty function for the laser pulses with a time-dependent weighting factor $A(t)$. The third term represents the constraint that the wave function $\psi(t)$ should satisfy the time-dependent Schrödinger equation with a given initial condition. Here $\xi(t)$ is the time-dependent Lagrange multiplier.

Carrying out the integration of the third term by parts, the objective function can be rewritten as

$$J = \hbar\langle W(t_f)\rangle - \frac{1}{2}\int_0^{t_f}dt\frac{|\varepsilon(t)|^2}{A(t)}$$
$$- 2\hbar\text{Re}\left\{\langle\xi(t)\,|\,\psi(t)\rangle\big|_0^{t_f}\right\}$$
$$+ 2\text{Re}\int_0^{t_f}dt\left\{\hbar\left\langle\frac{\partial}{\partial t}\xi(t)\,\bigg|\,\psi(t)\right\rangle\right.$$
$$\left. - i\langle\xi(t)|H(t)|\psi(t)\rangle\right\}. \quad (42)$$

By varying both $\psi(t) \to \psi(t) + \delta\psi(t)$ and $\varepsilon(t) \to \varepsilon(t) + \delta\varepsilon(t)$ in the above equation, the objective function J is expressed as $J + \delta J$, where δJ has the form

$$\delta J = \int_0^{t_f}dt\left[2\text{Re}\{i\langle\xi(t)|\mu|\psi(t)\rangle\} - \frac{\varepsilon(t)}{A(t)}\right]\delta\varepsilon(t)$$
$$+ 2\text{Re}\int_0^{t_f}dt\left[\hbar\left\langle\frac{\partial}{\partial t}\xi(t)\,\bigg|\,\delta\psi(t)\right\rangle\right.$$
$$\left. - i\langle\xi(t)|H(t)|\delta\psi(t)\rangle\right] + 2\hbar\text{Re}\{\langle\psi(t_f)|W|\delta\psi(t_f)\rangle$$
$$- \langle\xi(t_f)\,|\,\delta\psi(t_f)\rangle\}. \quad (43)$$

From the optimal condition, $\delta J = 0$, the expression for the optimal laser pulse,

$$\varepsilon(t) = -2A(t)\text{Im}\langle\xi(t)|\mu|\psi(t)\rangle, \quad (44)$$

is obtained. Here $\xi(t)$ satisfies the time-dependent Schrödinger equation,

$$i\hbar\frac{\partial}{\partial t}|\xi(t)\rangle = H(t)|\xi(t)\rangle, \quad (45)$$

with the final condition at $t = t_f$,

$$|\xi(t_f)\rangle = W|\psi(t_f)\rangle. \quad (46)$$

The optimal pulse can be obtained by solving the time-dependent Schrödinger equation iteratively with initial and final boundary conditions. First, assuming an analytical form for $\varepsilon(t)$, the time-dependent Schrödinger equation is solved to obtain $\psi(t)$ by forward propagation of the molecular system. Second, solving Eq. (45) with the same form of $\varepsilon(t)$ as before, but with the final condition, Eq. (46), the backward propagated wave function $\xi(t)$ can be obtained. A new form of the laser field $\varepsilon(t)$ can then be constructed by substituting these two wave functions, $\psi(t)$ and $\xi(t)$, into Eq. (44). These procedures are repeated until convergence is reached. This is a general procedure for obtaining optimal pulse shapes, and is called the global optimization method. By using this method, one can obtain the true optimal solution of systems having many local solutions. Convergence problems sometimes arise when global optimization is applied to real reaction systems. Several numerical methods for carrying out global optimization, such as the steepest descent method and a genetic algorithm, have been developed.

Another approach is known as the local optimization method. Here "local" means that maximization of the objective function J is carried out at each time, i.e., locally in time between 0 and t_f. There are several methods for deriving an expression for the optimal laser pulse by local optimization. One is to use the Ricatti expression for a linear time-invariant system in which a differential equation of a function connecting $\psi(t)$ and $\xi(t)$ is solved, instead of directly solving for these two functions. Another method is to solve an inverse problem for the path in a functional space of J. There are two essential points in the local optimization method. The first is to divide the time interval t_f from the initial time into infinitesimally short time intervals. The second point is to impose the final condition at the end of each time interval, i.e., $|\xi(t)\rangle = W|\psi(t)\rangle$. Following this procedure, the optimized pulse at time t is expressed as

$$\varepsilon(t) = -2A(t)\,\text{Im}\langle\psi(t)|W\mu|\psi(t)\rangle. \quad (47)$$

The simplest method for obtaining this expression is to substitute Eq. (46) into Eq. (44) after changing t_f to t in Eq. (46). This means that a (virtual) target is set just after each infinitesimally small time increment, and then the virtual target is moved toward the final true target position. The necessary condition for local optimization is therefore the assurance of an increase in the population of the target state. The merit of local optimization is that only one-sided propagation, i.e., forward or backward propagation, is needed. Its algorithm is, therefore, quite simple. Once the initial condition is specified, the time-dependent Schrödinger equation with a seed pulse as its initial pulse form can be solved to obtain a wave function after an infinitesimally increased propagation time. Next, substituting the resulting wave function into Eq. (47), an expression for the pulse at the increased time is obtained. With this pulse form, the time-dependent Schrödinger equation is solved again. This cycle of calculation described above is repeated until convergence is reached. This is a form of feedback control, because the wave function and laser pulse are related by Eq. (47). Because the local optimization method described above is nonperturbative, this method can be applied to wave packet control in intense fields. In such an intense field case, Eq. (47) can be used by letting the time increment become smaller and smaller.

As an example of optimal control, we consider the local control of a ring-puckering isomerization such as that of trimethyleneimine. The coordinate of the puckering motion q is defined as the displacement of the line joining the carbon and nitrogen atoms. The adiabatic potential energy expressed as a function of q is a double minimum potential. Figure 16 shows the adiabatic potential energy function together with several vibrational eigenfunctions. A linear dipole moment with respect to q was assumed. The time evolution of the probability density of the wave packet, $|\langle q|\psi(t)\rangle|^2$, produced by the locally optimized laser field is shown in Fig. 17. Starting from isomer A, the wave packet is almost completely transferred to the well of isomer B within 10 ps. Figure 18 shows the time variation of the locally optimized electric field. By analyzing the electric field with the help of a window Fourier transform, the optimized field may be regarded as four

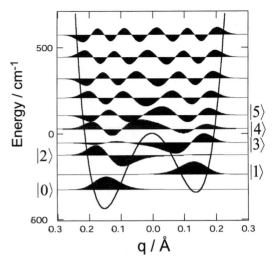

FIGURE 16 Adiabatic potential energy of trimetyleneimine as a function of the puckering coordinate q. The vibrational eigenfuctions are superimposed on the potential energy curve. [Reproduced with permission from Sugawara, M., and Fujimura, Y. (1994). *J. Chem. Phys.* **100**, 5646. Copyright American Institute of Physics.]

successive pulses with carrier frequencies that correspond to transition frequencies of the molecular eigenstates.

An advantage of the local control method described above is that it can be applied to wave packet propagation starting from an initial, nonstationary state, in contrast to ordinary wave packet control, which begins with the initial condition of a stationary state. An example where starting from such an initial condition is useful is the control of a localized state of a double-well potential. In this case, by propagating the final-state wave packet backward to the initial state, pulses that are optimized for forward

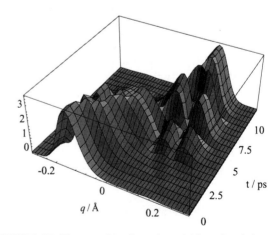

FIGURE 17 Wave packet dynamics of trimetyleneimine produced by an optimized laser field. [Reproduced with permission from Sugawara, M., and Fujimura, Y. (1994). *J. Chem. Phys.* **100**, 5646. Copyright, American Institute of Physics.]

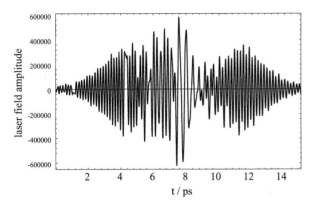

FIGURE 18 Time variation of the optimized electric field. [Reproduced with permission from Sugawara, M., and Fujimura, Y. (1994). *J. Chem. Phys.* **100**, 5646. Copyright American Institute of Physics.]

propagation can be constructed by time reversal, because the time-dependent Schrödinger equation is unitary in the case of no dissipation.

So far we have treated only the case of wave packets constructed from pure states. Consider now the control of a molecular system in a mixed state in which the initial states are distributed at a finite temperature. The time evolution of the system density operator $\rho(t)$ is determined by the Liouville equation,

$$i\hbar \frac{\partial}{\partial t} \rho(t) = L(t)\rho(t), \tag{48}$$

where the Liouville operator $L(t)$ is given by

$$L(t)\rho(t) = H(t)\rho(t) - \rho(t)H(t). \tag{49}$$

It is convenient to introduce a Liouville space, or double space, that is a direct product of cap and tilde spaces. In Liouville space, operators are considered to be vectors and Hilbert-space commutators are considered to be operators. Equation (48) is then expressed as

$$i\hbar \frac{\partial}{\partial t} |\rho(t)\rangle\rangle = L(t)|\rho(t)\rangle\rangle, \tag{50}$$

where $|\rho(t)\rangle\rangle$ is a vector, and

$$L(t) = L_0 - M\varepsilon(t) \tag{51}$$

is an operator in Liouville space. Here $L_0 = \hat{H}_0 - \tilde{H}_0$, and $\hat{H}_0$ and $\tilde{H}_0$ are molecular Hamiltonians in the cap and tilde spaces, respectively. Similarly, $M = \hat{\mu} - \tilde{\mu}$. In the Liouville representation, the objective function is rewritten as

$$J = \hbar \langle\langle W_G \,|\, \rho(t_f)\rangle\rangle - \frac{1}{2} \int_0^{t_f} dt \frac{|\varepsilon(t)|^2}{A(t)} + 2\mathrm{Re}\, i \int_0^{t_f} dt_1$$

$$\times \left\{ \langle\langle \Xi(t)| \, i\hbar \frac{\partial}{\partial t} \rho(t)\rangle\rangle - \langle\langle \Xi(t)|L(t)|\rho(t)\rangle\rangle \right\}. \tag{52}$$

Equation (52) has the same structure as that of Eq. (41). An expression for the optimal control pulse in a mixed case can therefore be obtained as

$$\varepsilon(t) = -2A(t)\mathrm{Im}\, \langle\langle \Xi(t)|\hat{\mu}|\rho(t)\rangle\rangle, \tag{53}$$

where the time-dependent multiplier $\Xi(t)$ satisfies the Liouville equation,

$$i\hbar \frac{\partial}{\partial t} |\Xi(t)\rangle\rangle = L^\dagger(t) \,|\, \Xi(t)\rangle\rangle, \tag{54}$$

with the final condition $|\Xi(t_f)\rangle\rangle = |W\rangle\rangle$.

A fundamental limitation to coherent population control is that it is impossible to transfer 100% of the population in a mixed state. That is, the maximum value of the population transferred cannot exceed the maximum of the initial population distribution of a system without any dissipative process such as spontaneous emission. This result can be simply verified using the unitary property of the density operator, $\rho(t) = U(t, t_0)\rho(t_0)U^\dagger(t, t_0)$, where $\rho(t_0)$ is the diagonalized density operator at $t = t_0$, $U(t, t_0)$ is the time-evolution operator given by

$$U(t, t_0) = \hat{T} \exp\left[-\frac{i}{\hbar} \int_{t_0}^t dt' \, V_I(t') \right], \tag{55}$$

$\hat{T}$ is a time-ordering operator, and $V_I(t')$ is the interaction between the molecules and the controlling pulses in the interaction representation. The eigenvalues of $\rho(t)$ are thus invariant with respect to unitary transformation. The population of a target state $|k\rangle$ at time t, $\langle k|\rho(t)|k\rangle$, satisfies the condition that the minimum eigenvalue of $\rho(t_0) \leq \langle k|\rho(t)|k\rangle \leq$ the maximum eigenvalue of $\rho(t_0)$. That is, the maximum population in a target state at time t_f is equal to the maximum eigenvalue of $\rho(t_0)$. Therefore, in the mixed state case, one must choose a target operator appropriate for this restriction.

3. Experimental Examples of Wave Packet Control

The key technological advance that has made optical pulse shaping widely available is the pulse modulator depicted in Fig. 19. For a Gaussian laser pulse the product (full-width at half-maximum) of duration τ and radial frequency bandwidth $\delta\omega$ is 0.44. (For a sech² pulse, the product is 0.32.) For such transformed-limited pulses the group velocity is the same for all frequencies. The properties of a laser pulse can be tailored by dispersing the pulse, filtering the frequency components, and finally reconstituting the modified pulse. This method is illustrated in Figure 19, where grating G_1 is placed at the focal point of lens L_1. A multipixel spatial light modulator (SLM) placed in the Fourier plane is programmed to alter the

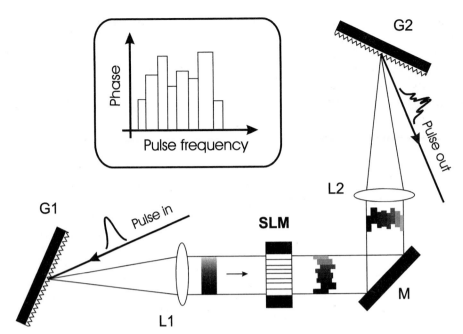

FIGURE 19 Illustration of pulse shaping using a liquid crystal spatial modulator. [Uberna, R., Amitay, Z., Loomis, R. A., and Leone, S. R. (1999). *Disc. Faraday Society* **113**, 385. Reproduced by permission of the Royal Society of Chemistry.]

phases and/or amplitudes of the light in each frequency interval. To control both amplitudes and phases, two modulators are used back-to-back. The two types of devices that have been used for this purpose are a liquid crystal spatial light modulator and an acousto-optic modulator. The modified pulse is finally recompressed by lens L_2 and grating G_2.

The phase and amplitude spectrum of the laser pulse are tailored to create a wave packet with selected properties. The various eigenstates that comprise the wave packet are populated by different frequency components of the laser pulse, each with its specified amplitude and phase. For example, rovibrational wave packets of Li_2 in the $E^1\Sigma_g^+$ state were created, consisting of vibrational levels $v = 12–16$ and rotational levels $J = 17, 19$. The phases and amplitudes of the pump pulse shown in Fig. 20 were generated with a 128-pixel liquid crystal SLM. The pulse was tailored to optimize the ionization signal at a delay time of 7 ps. The phases used to maximize or minimize the ionization signal are shown by solid and dashed lines, respectively, and the intensities at the eigenfrequencies of the wave packet are indicated by circles.

C. Genetic Algorithms

There are many cases in which the molecular Hamiltonian and the interactions with the photon fields are not completely known. For many isolated polyatomic molecules or for molecules in condensed phases, for ex-

ample, only the initial and final states of the control system are specified; the multidimensional potential energy surfaces and the reaction coordinates connecting the initial and final states cannot be determined either theoretically or experimentally. In addition, there exist experimentally unavoidable uncertainties associated with the controlling fields. The optimal control methods described in previous sections are not well suited for such systems. In such cases, a genetic algorithm (GA), which is a global optimization method, may be employed.

A GA has three fundamental operations: reproduction, crossover, and mutation. By means of these three operations, an ensemble of individuals (i.e., a population) is adapted to fit its environment. In the first operation, groups of individuals that have a higher probability of reproducing as a consequence of their better adaptation to their environment pass their genetic information onto succeeding generations. In the crossover operation, splicing of the parents' genes transfers a mixture of their genetic material to their offspring. Genetic diversity of the offspring cannot be created by crossover. By mutation, however, the genetic material is altered to avoid premature convergence to an undesirable trait.

The GA procedure has an interactive feedback (closed) loop structure without any intervention by the experimentalist. An initial guess of multiple sets of control fields is evaluated for its success in achieving the target. The most successful members of this population are transformed by the three GA operators, and their offspring are evaluated.

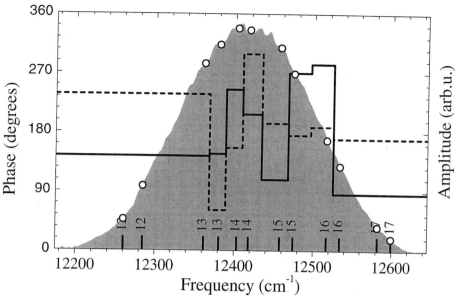

FIGURE 20 Amplitude and phase of a laser pulse optimized either to maximize or minimize the ionization of Li_2. [Uberna, R., Amitay, Z., Loomis, R. A., and Leone, S. R. (1999). *Disc. Faraday Society* **113**, 385. Reproduced by permission of the Royal Society of Chemistry.]

The entire process is repeated for many generations until convergence is achieved.

An experimental illustration of the GA is shown in Fig. 21. The molecule cyclopentadienyl–iron–dicarbonyl–chloride was irradiated with pulses of 800 nm radiation that were initially 80 ns long before entering the pulse shaper. The phases of the laser pulses were modified with a liquid crystal SLM and optimized with a GA either to maximize or minimize the ratio of $C_5H_5FeCOCl^+$ to $FeCl^+$. Convergence was achieved typically after 100 generations. The optimum yield ratio and the pulse shapes used to achieve them are shown. (A GA was similarly used in the example of Li_2 ionization illustrated in Fig. 20.)

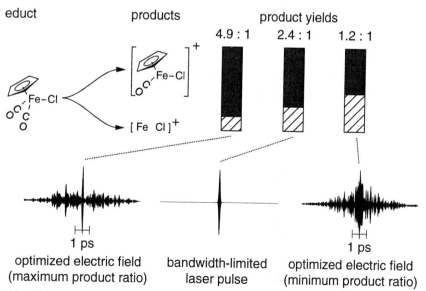

FIGURE 21 Use of the genetic algorithm to control the dissociative ionization of cyclopentadienyl–iron–dicarbonyl chloride. Shown on the bottom are the optimized electric fields generated to either maximize or minimize the ratio of $C_5H_5FeCOCl^+$ to $FeCl^+$. [Provided by the courtesy of T. Brixner, and adapted from Assion, A., Baumert, T., Bergt, M., Brixner, T., Kiefer, B., Seyfried, V., Strehler, M., and Gerber, G. (1998). *Science* **282**, 919.]

D. Coherent Population Transfer

So far, we have considered various theoretical treatments of time-dependent wave packets controlled by laser pulses to produce a desired product in a chemical reaction. Another type of problem, based on adiabatic behavior of wave functions, is the transfer of population from one state to another.

Suppose that a laser pulse interacts adiabatically with a molecular system. By the term "adiabatic" is meant that an eigenstate $\psi_\ell(t)$ at time t satisfies the time-independent Schroedinger equation:

$$H(t)\psi_\ell(t) = E_\ell(t)\psi_\ell(t). \tag{56}$$

In the adiabatic limit, t is considered to be a parameter, and $\psi_\ell(t)$ is called an adiabatic state. One of the interesting properties of this limit is that a population can be inverted by evolving the system adiabatically. This process is called adiabatic passage. Population transfer induced by a laser is generally called "coherent population transfer." For a two-level system, the complete population inversion is produced by a π-pulse or by adiabatic rapid passage.

Population transfer in a three-level system can be achieved by using one laser (known as the "pump laser," which may be either continuous wave or pulsed) to connect the ground and intermediate levels, and a second laser (the "Stokes laser") to connect the intermediate and final levels. This method, known as stimulated Raman adiabatic passage or STIRAP, is illustrated in Fig. 22. In this example, the three levels have a Λ-type configuration, where

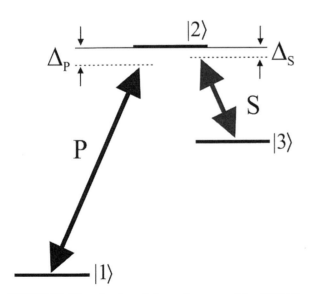

FIGURE 22 Three-level excitation scheme used for STIRAP. [Reproduced with permission from Bergmann, K., Theuer, H., and Shore, B. W. (1998). *Rev. Mod. Phys.* **70**, 1003.]

ϕ_1, ϕ_2, and ϕ_3 are the wave functions of the initial, intermediate, and final states, respectively, (denoted by $|1\rangle$, $|2\rangle$, and $|3\rangle$ in Fig. 22), subscripts P and S label the pump and Stokes processes, respectively, and $\Delta_P(\Delta_S)$ denotes the detuning of the pump (Stokes) laser. In the discussion that follows, the two laser fields are assumed to be pulsed.

The interaction matrix elements of the total Hamiltonian are expressed in the three-state model as

$$\frac{\hbar}{2} \begin{bmatrix} -2\Delta & \Omega_P(t) & 0 \\ \Omega_P^*(t) & 0 & \Omega_S(t) \\ 0 & \Omega_S^*(t) & -2\Delta \end{bmatrix}, \tag{57}$$

where $\Omega_P(t)$ is the Rabi frequency of the pump pulse, $\Omega_S(t)$ and is the Rabi frequency of the Stokes pulse. By diagonalizing the determinant of this interaction matrix, the adiabatic states ψ_0 and $\psi_\pm$, with eigenfrequencies ω_0 and $\omega_\pm$, are analytically derived as

$$\psi_0(t) = \cos[\Theta(t)]\phi_1 - \sin[\Theta(t)]\phi_3 \tag{58}$$

with eigenfrequency

$$\omega_0(t) = -\Delta(t), \tag{59}$$

$$\psi_+(t) = \sin[\eta(t)]\sin[\Theta(t)]\phi_1 + \cos[\eta(t)]\phi_2$$
$$+ \sin[\eta(t)]\cos[\Theta(t)]\phi_3 \tag{60}$$

with eigenfrequency

$$\omega_+(t) = -\frac{1}{2}\Big[\Delta(t) - \sqrt{|\Omega_P(t)|^2 + |\Omega_S(t)|^2 + \Delta^2(t)}\Big], \tag{61}$$

and

$$\psi_-(t) = \cos[\eta(t)]\sin[\Theta(t)]\phi_1 - \sin[\eta(t)]\phi_2$$
$$+ \cos[\eta(t)]\cos[\Theta(t)]\phi_3 \tag{62}$$

with eigenfrequency

$$\omega_-(t) = -\frac{1}{2}\Big[\Delta(t) + \sqrt{|\Omega_P(t)|^2 + |\Omega_S(t)|^2 + \Delta^2(t)}\Big]. \tag{63}$$

The mixing angle $\Theta(t)$ in Eqs. (58), (60), and (62) is given by

$$\sin[\Theta(t)] = \frac{|\Omega_P(t)|}{\sqrt{|\Omega_P(t)|^2 + |\Omega_S(t)|^2}} \tag{64}$$

$$\cos[\Theta(t)] = \frac{|\Omega_S(t)|}{\sqrt{|\Omega_P(t)|^2 + |\Omega_S(t)|^2}}, \tag{65}$$

and

$$\tan[\eta(t)] = \frac{\sqrt{|\Omega_P(t)|^2 + |\Omega_S(t)|^2}}{\sqrt{|\Omega_P(t)|^2 + |\Omega_S(t)|^2 + \Delta^2(t)} + \Delta(t)}. \tag{66}$$

We note that the lowest adiabatic state, Eq. (58), is expressed in terms of the initial and the final states without the intermediate state. This property implies that the system in the initial state is transferred to the final state adiabatically, with no population in the intermediate resonant state. That is, under the condition for the Rabi frequencies $|\Omega_P(t)| \ll |\Omega_S(t)|$, we can see from Eqs. (64) and (65) that $\psi_0(0) = \phi_1$ and $\psi_0(\infty) = \phi_3$ if the Stokes pulse comes before the pump pulse. The ordering of these pulses in STIRAP is counterintuitive, compared with conventional, stimulated Raman scattering processes. Figure 23 shows the time evolution of the Rabi frequencies, mixing angle, dressed-state eigenvalues, and the population of the initial and final levels. The counterintuitive pulse sequence is evident in Figure 23a. One of the properties of STIRAP

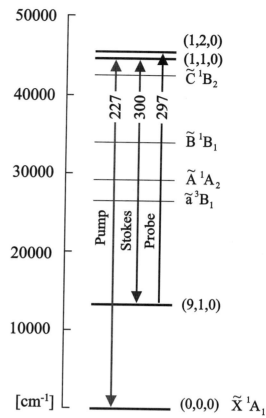

FIGURE 24 Energy levels of SO_2, showing the pumping scheme used to transfer population from the ground level to the (9,1,0) vibrationally excited level by STIRAP. [Reproduced with permission from Halfmann, T., and Bergmann, K. (1996). *J. Chem. Phys.* **104**, 7068. Copyright American Institute of Physics.]

is its robustness with respect to parameters such as Rabi frequency and time delay between the Stokes and pump pulses. The STIRAP technique can also be applied to a system with more than three levels.

An example of a STIRAP simulation and experiment is illustrated in Figs. 24 and 25 for SO_2. Figure 24 shows the energies of the laser pulses used to transfer population from the vibrationless level to the (9, 1, 0) level of the ground electronic state. Figure 25 shows the experimentally measured and numerically simulated fraction of the population transferred to the excited state as a function of the time delay between the pump and Stokes pulses. The greater efficiency of a counterintuitive pulse sequence is evident.

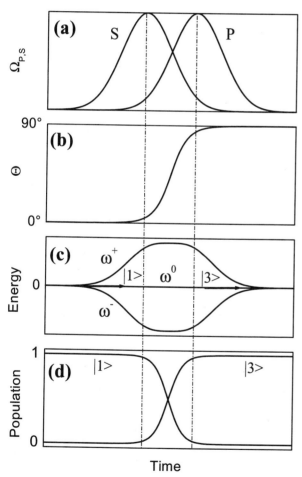

FIGURE 23 Illustration of the STIRAP technique used for coherent population transfer. Shown are the time evolution of (a) the Rabi frequencies of the pump and Stokes lasers, (b) the mixing angle, (c) the dressed-state eigenvalues, and (d) the populations of the initial and final levels. [Reproduced with permission from Bergmann, K., Theuer, H., and Shore, B. W. (1998) *Rev. Mod. Phys.* **70**, 1003.]

V. CONTROL OF EXTERNAL DEGREES OF FREEDOM

In the discussion so far, emphasis has been placed on controlling the internal degrees of freedom of an atom or

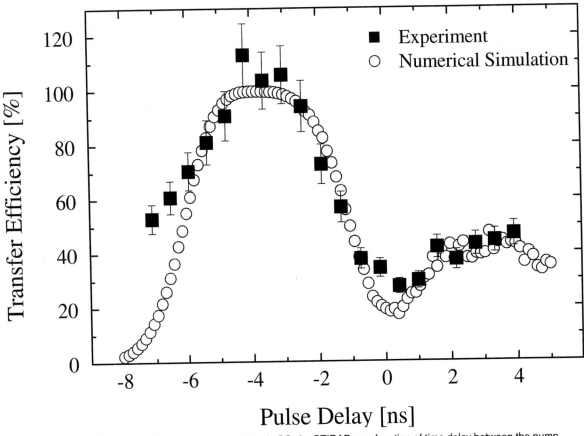

FIGURE 25 Efficiency of the transfer of population in SO$_2$ by STIRAP as a function of time delay between the pump and Stokes laser pulses. [Reproduced with permission from Halfmann, T., and Bergmann, K. (1996). *J. Chem. Phys.* **104,** 7068. Copyright American Institute of Physics.]

molecule. In this section we show how the dipole force can be used to manipulate the external motion of a molecule. A particle with an electric dipole moment μ and a polarizability α exposed to an electric field ε has a potential energy

$$V(\theta) = -\mu\varepsilon\cos\theta - \tfrac{1}{2}\varepsilon^2\big(\alpha_\parallel\cos^2\theta + \alpha_\perp\sin^2\theta\big), \quad (67)$$

where θ is the angle between μ and ε, and $\alpha_\parallel$ and $\alpha_\perp$ are the parallel and perpendicular components of the polarizability tensor, respectively. For a static electric field the first term dominates, and the potential energy has a minimum at $\theta = 0$. Static field strengths on the order of 10^4–10^5 V/cm are typically required to align or orient a molecule. (*Alignment* refers to the direction of the molecular axis without regard to which end is "up," whereas *orientation* treats a molecule as a single-headed arrow.) For example, Wei Kong used a field strength of 58 kV/cm to orient a molecular beam of BrCN. This "brute force" method of orientation has been used to study steric effects in bimolecular reactions. For example, the differential cross section for the reaction

$$K + C_6H_5I \rightarrow C_6H_5 + KI \quad (68)$$

was measured by Loesch and coworkers for different orientations of phenyl iodide. If a focusing hexapole field is used instead of a uniform dc field, it is possible to generate a beam of symmetric top molecules in a single rotational eigenstate.

Although the brute force method has produced many beautiful results, it is limited to molecules with a permanent dipole moment and is hampered by the requirements of high voltages, and, for a multipole focusing field, a complex apparatus. An alternate method of producing a strong electric field is to use a focused laser beam. In the high frequency limit, the time average of the ac electric field gives $\langle\varepsilon(t)\rangle = 0$ and $\langle\varepsilon^2(t)\rangle = \tfrac{1}{2}\langle\varepsilon_0^2\rangle$, where ε_0 is the amplitude of the field. The potential energy of the molecule in this case is

$$V(\theta) = -\tfrac{1}{4}\varepsilon_0^2\big(\Delta\alpha\cos^2\theta + \alpha_\perp\big), \quad (69)$$

where $\Delta\alpha = \alpha_\parallel - \alpha_\perp$ is the anisotropy of the polarizability. A dimensionless quantity that is useful for scaling the alignment of different molecules is given by

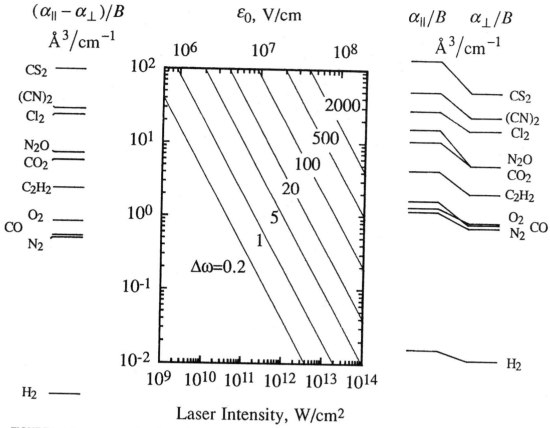

FIGURE 26 Nomogram of the dimensionless alignment parameter $\Delta\omega$ as a function of laser intensity, field strength, and polarizability anisotropy. [Reproduced with permission from Freidrich, B., and Herschbach, D. R. (1995). *J. Phys. Chem.* **99**, 15686. Copyright American Chemical Society.]

$$\omega_{\|(\perp)} = \alpha_{\|(\perp)}\varepsilon_0^2/4B, \qquad (70)$$

where B is the rotational constant. Noting that $I = 0.00265\varepsilon_0^2$, where I is the laser intensity in W/cm², and ε_0 is in V/cm, we obtain

$$\omega_{\|(\perp)} = 5.28 \times 10^{-12}\,\alpha_{\|(\perp)}[\mathring{A}^3]I[\text{W/cm}^2]/B[\text{cm}^{-1}]. \qquad (71)$$

The condition for strong alignment is $\Delta\omega = \omega_{\|} - \omega_{\perp} \cong 10$, which, for $\alpha = 10\ \mathring{A}^3$ and $B = 1\ \text{cm}^{-1}$, corresponds to $I = 2 \times 10^{11}\ \text{W/cm}^2$. A nomogram for $\Delta\omega$, plotted in Fig. 26, shows that many molecules may be aligned with fields readily achieved in the laboratory.

To align a molecule, it is necessary that the intensity of the aligning laser pulse lie below the threshold for multiphoton ionization. From Eq. (70) it is evident that the required intensity varies inversely with polarizability. Because molecules with low polarizabilities generally have higher ionization potentials, the conditions for laser alignment are fairly robust. An experiment demonstrating alignment utilized three laser pulses: a linearly polarized aligning pulse ($\tau = 3.5\ \text{ns}$, $I = 1.4 \times 10^{12}\ \text{W/cm}^2$,

$\lambda = 1.064\ \mu\text{m}$), a circularly polarized dissociating pulse (100 fs, $3 \times 10^{12}\ \text{W/cm}^2$, 688 nm), and a circularly polarized ionizing pulse (100 fs, $7 \times 10^{13}\ \text{W/cm}^2$, 800 nm). The fragment ions were accelerated toward a microchannel plate, and an image produced by electrons ejected onto a phosphor screen was captured by a charge-coupled device (CCD) camera. Typical images demonstrating alignment of C_6H_5I are shown in Fig. 27. The image on the left shows the anisotropic recoil of the iodine atom produced by a linearly polarized dissociation laser in the absence of an aligning pulse. The middle image shows isotropic recoil produced by a circularly polarized dissociation laser in the absence of an aligning pulse. Finally, the image on the right shows the highly anisotropic recoil produced by a circularly polarized dissociation laser in the presence of an aligning pulse. Other molecules that have been so aligned include I_2, ICl, CS_2, and CH_3I. If instead the aligning pulse is elliptically polarized, it is possible to align all three axes of a molecule, as was demonstrated for 3,4-dibromothiophene ($C_4H_2Br_2S$).

The average value of $\cos^2\theta$ is a measure of the extent of alignment. Values of $\langle\cos^2\theta\rangle > 0.9$ obtained with

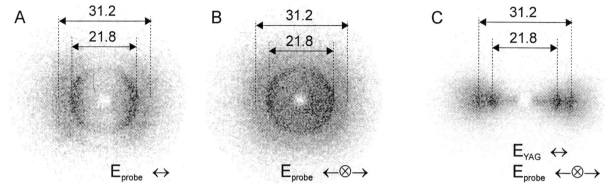

FIGURE 27 Velocity map images illustrating alignment of C_6H_5I. The images show the angular distribution of the iodine fragment for (a) a linearly polarized probe without an aligning laser, (b) a circularly polarized probe without an aligning laser, and (c) a circularly polarized probe in the presence of a linearly polarized aligning laser pulse. [Reproduced with permission from Larsen, J. J., Sakai, H., Safvan, C. P., Wendt-Larsen, I., and Stapelfeldt, H. (1999). *J. Chem. Phys.* **111**, 7774. Copyright American Institute of Physics.]

the dipole force of a focused laser beam are considerably greater than what has been achieved by the brute force method. For example, pyridazine molecules, which have a permanent dipole moment of 4 Debye, when cooled to 2 K and placed in a 60 kV/cm dc field, are oriented with approximately half of the molecules restricted within a cone of 45° half-angle. In contrast, iodine molecules at the same rotational temperature and placed in a focused laser beam with an intensity of 5×10^{11} W/cm^2 (equivalent to 1.4×10^7 V/cm) have an induced dipole moment on the order of 1 Debye, and are aligned with half the molecules restricted to a cone of 12° and 98% of the molecules within a 45° cone.

The experiments described above used nanosecond laser pulses, which are much longer than the rotational period of the molecules. At the termination of the pulse, the pendular state that is formed relaxes adiabatically to a free-rotor eigenstate. If instead picosecond laser pulses are used, a rotational wave packet is formed by successive absorption and re-emission of photons during the laser pulse. Such wave packets are expected to display periodic recurrences of the alignment after the end of the pulse.

Laser alignment of molecules can be used to control their chemical reactions. In one example, alignment of the iodine molecule was used by Stapelfeldt and coworkers to control the spin-orbit branching ratio of its photofragments. For the I_2 molecule aligned parallel to the electric field vector of the photodissociation laser, the fragments were primarily $I(^2P_{3/2}) + I(^2P_{3/2})$, whereas for perpendicular alignment the primary products were $I(^2P_{3/2}) + I(^2P_{1/2})$. In another example studied by Corkum and coworkers, the aligned molecule was "grabbed" by the rotating polarization vector of the aligning laser and forced to move with it. Using a pair of counterrotating, circularly polarized, chirped laser pulses, the rate of rotation of a chlorine molecule was accelerated from 0 to 6 THz in 50 ps, going from near rest to angular

momentum states with $J \sim 420$. At the highest spinning rate the molecule overcame the centrifugal barrier and dissociated.

The dipole force can also be used to control the motion of the center of mass of a molecule. A focused laser beam has a radial intensity dependence $I(r)$, so that the potential energy function in Eq. (69) has a minimum at the focal point. The depth of the induced-dipole well can be orders of magnitude greater than what is commonly obtained in magneto-optic traps. For example, at an intensity of 10^{12} W/cm^2, the well depth for I_2 molecules is 256 K. Atoms or molecules encountering such a potential well will be deflected towards the high intensity region. As illustrated in Fig. 28, a Nd:YAG (IR) laser focused to a spot size 7 μm in diameter at an intensity of 9×10^{11} W/cm^2 was used to deflect a molecular beam of CS_2. An intense $(8 \times 10^{13}$ W/cm$^2)$ colliding-pulse mode-locked (CPM)

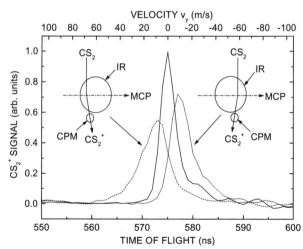

FIGURE 28 Deflection of a molecular beam of CS_2 using the dipole force of a focused laser beam. [Reproduced with permission from Stapelfeldt, H., Sakai, H., Constant, E., and Corkum, P. B. (1997). *Phys. Rev. Lett.* **79**, 2787.]

laser ionized the molecules, which were then detected by a microchannel plate (MCP). Shown in the figure are the time-of-flight spectra of the undeflected beam (solid curve) and the deflected beams (dashed and dotted curves). Molecules deflected towards the detector display an earlier arrival time than those deflected away from the detector.

VI. CONCLUDING REMARKS

Since its conception in the mid-1980s, there has been significant progress in the field of coherent control of chemical reactions. Theoretical tools for computing the optimal laser pulses for manipulating the motion of vibrational and electronic wave packets have been developed, and laboratory methods for tailoring the shapes of laser pulses are now available. These techniques have been applied to control the branching ratios of simple unimolecular ionization and dissociation reactions. Genetic algorithms have been developed and implemented for controlling more complex unimolecular reactions. The method of phase control via competing quantum mechanical paths has likewise been used to control elementary branching ratios, and a theory relating the phase lag to fundamental molecular quantities has been developed. Adiabatic passage methods have been used to achieve 100% population transfer between quantum mechanical states. Coherent methods have been developed also to control external degrees of freedom. These methods have been used to align molecules and to alter their center-of-mass translational motion.

Despite these notable successes, much remains to be done before coherent control can become a practical tool. Virtually all the successes to date have involved very simple molecules. Although learning algorithms may prove to be useful for controlling complex molecules, they have so far shed little light on the dynamics involved. Two very important problems where experiments lag far behind theory are the selective control of molecules with different chirality and the control of bimolecular reactions. A major

goal in controlling external degrees of freedom is the orientation of asymmetric molecules. In view of the major advances in laser and pulse-shaping technology over the past decade, which were barely anticipated when coherent control schemes started to emerge, it appears likely that at least some these frontier problems will be solved in the coming decade.

SEE ALSO THE FOLLOWING ARTICLES

CHEMICAL THERMODYNAMICS • DYNAMICS OF ELEMENTARY CHEMICAL REACTIONS • KINETICS (CHEMISTRY) • LASERS • NUCLEAR CHEMISTRY • PHOTOCHEMISTRY BY VUV PHOTONS • PHOTOCHEMISTRY, MOLECULAR • PROCESS CONTROL SYSTEMS • QUANTUM MECHANICS

BIBLIOGRAPHY

Bergmann, K., Theuer, H., and Shore, B. W. (1998). "Coherent population transfer among quantum states of atoms and molecules," *Rev. Modern Phys.* **70,** 1003–1025.

Crim, F. F. (1999). "Vibrational state control of bimolecular reactions: Discovering and directing the chemistry," *Acc. Chem. Res.* **32,** 877–884.

Gordon, R. J., and Fujimura, Y., eds. (2001). "Advances in Multiphoton Processes and Spectroscopy, Volume 14. Quantum Control of Molecular Reaction Dynamics: Proceedings of the US—Japan Workshop," World Scientific, Singapore.

Gordon, R. J., and Rice, S. A. (1997). "Active control of the dynamics of atoms and molecules," *Ann. Rev. Phys. Chem.* **48,** 601–641.

Gordon, R. J., Zhu, L., and Seideman, T. (1999). "Coherent control of chemical reactions," *Acc. Chem. Res.* **32,** 1007.

Kohler, B., Krause, J. L., Raksi, F., Wilson, K. R., Yakovlev, V. V., Whitnell, R. M., and Yan, Y. J. (1995). "Controlling the future of matter," *Acc. Chem. Res.* **28,** 133–140.

Rice, S. A., and Zhao, M. (2000). "Optical Control of Molecular Dynamics," Wiley Interscience, New York.

Shapiro, M., and Brumer, P., (2000). "Coherent control of atomic, molecular, and electronic processes," *Adv. Atomic Mol. Optical Phys.* **42,** 287–345.

Tannor, D. (2002). "Introduction to Quantum Mechanics: A Time-dependent Perspective," University Science Books, Sausalito.

Warren, S., Rabitz, H., and Dahleh, M. (1993). "Coherent control of quantum dynamics: The dream is alive," *Science* **259,** 1581–1589.

Cohesion Parameters

Allan F. M. Barton

Murdoch University

GLOSSARY

Chameleonic behavior Capacity of the molecules of a compound to assume a character similar to their environment by intermolecular or intramolecular association.

Cohesion parameter Quantity with dimensions of square root of pressure used to characterize the cohesive and adhesive properties of materials.

Cohesive energy Thermodynamic quantity describing the sum of molecular effects in a material that cause it to remain in a condensed state.

Cohesive pressure (cohesive energy density) Ratio of the cohesive energy to the volume for a material.

Dispersion forces Intermolecular forces present in all materials, whether polar or not, arising from the fluctuating molecular dipoles that result from the positive nucleus and negative electrons.

Hansen parameters Cohesion parameters resulting from the subdivision of the Hildebrand parameter into dispersion (nonpolar), polar, and hydrogen bonding contributions.

Hildebrand parameter, solubility parameter Square root of the cohesive pressure of a liquid.

Homomorph The homomorph of a polar molecule is a nonpolar molecule having the same size and shape.

Interaction cohesion parameters Cohesion parameters resulting from the subdivision of the Hildebrand parameter into dispersion (nonpolar), orientation, induction, acid, and base contributions.

Regular solution A solution (solvent plus solute) that has a completely random molecular distribution despite possible specific interactions between solute and solvent molecules.

THE TERM COHESION PARAMETER came into use in the early 1980s as a general term for a class of quantities with dimensions of (pressure)$^{1/2}$. In the simplest application of the concept, the cohesion parameter values of each of two pure materials are multiplied together to yield a numerical value for the cohesive pressure expected when the materials are mixed or the adhesive pressure expected when they are in contact.

Cohesion parameters provide a simple method of correlating and predicting the cohesive and adhesive properties of substances from a knowledge of the properties of the individual components only. There are, of course, numerous more sophisticated theories and techniques for this purpose, but none is as easy to use in practical applications.

Cohesion parameters provide estimates of the enthalpy changes on mixing and do not explicitly treat any deviations from ideal entropy changes. They are based on J. H. Hildebrand's definition of a special class of "regular" solutions, those involving an ideal entropy change on mixing despite the existence of enthalpy changes, but are now applied far more widely.

The simplest cohesion parameter is the Hildebrand parameter, commonly called the *solubility parameter*, values of which are now being included by manufacturers of polymers and solvents in the specification sheets of their products.

I. MAIN CLASSES OF COHESION PARAMETERS

Pure materials exist as condensed phases (e.g., liquids, crystals, glasses, rubbers) over certain ranges of temperature and pressure because in some circumstances these states are more stable than the gaseous state: There are energetic advantages in the molecules being packed closely together. In these condensed phases, strong attractive forces exist between the molecules. The *cohesive energy* $-U$ is positive (equal and opposite in sign to the internal energy), and the cohesive energy per unit volume V of the material is defined as the *cohesive pressure* or cohesive energy density, given the symbol c.

Similarly, many mixtures also exist as homogeneous, condensed phases for the same reason: There is a significant cohesive pressure maintaining that state in existence. The basis of the cohesion parameter approach to mixtures is that a material with a high cohesive pressure requires more energy for dispersal than is gained by mixing it with a material of low cohesive pressure, so immiscibility (separation of phases) results, but two materials with similar cohesive pressure values gain cohesive energy on dispersal, so mixing occurs.

The change in cohesive pressure associated with the process of mixing two components i and j with their respective cohesive pressures ^ic and ^jc is given by the *interchange cohesive pressure*, ^{ij}A,

$$^{ij}A = {}^ic + {}^jc - 2^{ij}c, \qquad (1)$$

where ^{ij}c is the cohesive pressure characteristic of the intermolecular forces acting between molecules of type i

and type j. This equation can be understood in a simple way by considering what happens when unit volumes of components i and j are mixed: Two $i-j$ interactions are formed for each pair of $i-i$ and $j-j$ interactions broken.

A. Hildebrand Parameter

At pressures below atmospheric pressure (i.e., for temperatures below the boiling point), the molar cohesive energy $-U$ of a liquid can be taken as equal to the molar enthalpy of vaporization ΔH less the pressure–volume work (which for a vapor with ideal gas behavior is RT per mole, where R is the molar gas constant and T the absolute temperature). This is the basis of the original definition by Joel H. Hildebrand and Robert L. Scott of the solubility parameter, hereafter called the *Hildebrand parameter* δ:

$$\delta = c^{1/2} = (-U/V)^{1/2} = [(\Delta H - RT)/V]^{1/2}. \qquad (2)$$

This parameter was intended for use only with nonpolar, nonassociating liquid systems that form regular solutions, but its use has been extended to all types of material. For *all* liquids, the value of the Hildebrand parameter may be determined directly from thermodynamic data by means of this equation, but it is only for regular solutions that predictions using the Hildebrand parameter are reliable.

A regular solution has an ideal entropy of formation, that is, a random molecular distribution, despite the existence of interactions that lead to a nonideal enthalpy of formation (heat of mixing). This means that regular mixtures are restricted to those systems in which only dispersion forces are acting. (Dispersion forces, or London forces, arise from the fluctuating dipoles that result from the positive nucleus and negative electron "cloud" in every atom. They occur in all systems and are distinct from forces associated with molecular polarity and "chemical" interactions between molecules.) For systems like this, without the orientation and ordering effects of polar molecules, the cohesive pressure between unlike molecules is given to a good approximation by the geometric mean of the cohesive pressures of the individual components,

$$^{ij}c = ({}^ic\,{}^jc)^{1/2}. \qquad (3)$$

From a combination of Eqs. (1), (2), and (3), it follows that the interchange cohesive pressure is given by Eq. (4):

$$^{ij}A = ({}^ic^{1/2} - {}^jc^{1/2})^2 = ({}^i\delta - {}^j\delta)^2. \qquad (4)$$

By means of this fundamental equation, it is possible to describe the thermodynamics of the mixing process in terms of Hildebrand parameters. For example, the mole fraction activity coefficient f_x at infinite dilution of component j with molar volume ^jV in component i is given by

$$RT \ln {}^{j}f_x^{\infty} = {}^{j}V^{ij}A = {}^{j}V({}^{i}\delta - {}^{j}\delta)^2. \quad (5)$$

Equations such as this hold exactly for regular solutions and are good approximations for many other useful systems.

On the basis of the assumtions made in the derivation of cohesion parameter expressions, the effective Hildebrand parameter $\bar{\delta}$ of a binary solvent mixture is volumewise proportional to the Hildebrand parameter values of its components, so

$$\bar{\delta} = {}^{i}\phi^{i}\delta + {}^{j}\phi^{j}\delta, \quad (6)$$

where ϕ is the volume fraction, defined by

$$^{i}\phi = \frac{{}^{i}V^{i}x}{{}^{j}V^{i}x + {}^{i}V^{i}x}, \quad (7)$$

where x is the mole fraction and V is the molar volume.

B. Component Cohesion Parameters

In addition to the dispersion forces described above for regular solutions, most chemical systems also exhibit polar interactions and specific interactions (e.g., hydrogen bonding). To be generally useful, theories and models aiming to systematize and predict the behavior of matter must deal with molecular interactions on the basis of their natures or origins as well as their strengths. The cohesive properties characteristic of the condensed states of matter are produced by various intermolecular forces, and the cohesive pressures ${}^{i}c$, ${}^{j}c$, and ${}^{ij}c$ represent the resultant effect of all these forces acting between molecules of types i and j.

For this reason, five interaction cohesion parameters have been introduced to describe the properties of materials in greater detail than is possible with the Hildebrand parameter. These are dispersion, orientation, induction, Lewis acid, and Lewis base cohesion parameters.

Dispersion forces, occurring in all molecules, whether polar or not, give rise to a dispersion cohesive pressure ${}^{i}c_d$ and a corresponding dispersion cohesion parameter ${}^{i}\delta_d$ in a pure material i:

$$^{i}c_d = {}^{i}\delta_d^2. \quad (8)$$

The nonpolar, dispersive interactions between unlike molecules of type i and type j provide a contribution to the cohesive pressure that is based on the geometric mean of the individual values:

$$^{ij}c_d = \left({}^{i}c_d{}^{j}c_d\right)^{1/2} = {}^{i}\delta_d{}^{j}\delta_d. \quad (9)$$

A simple interpretation of this geometric-mean behavior is that the interaction is of a "symmetrical" nature: Each member of a pair of molecules interacts by virtue of the same molecular property (the polarizability). It follows that

$$^{ij}A_d = {}^{i}\delta_d^2 + {}^{j}\delta_d^2 - 2{}^{i}\delta_d{}^{j}\delta_d = \left({}^{i}\delta_d - {}^{j}\delta_d\right)^2. \quad (10)$$

Orientation effects result from dipole–dipole (or Keesom) interactions and occur between molecules that have permanent dipole moments. The orientation cohesive pressure of a pure material i is denoted ${}^{i}c_o$, and the corresponding orientation cohesion parameter ${}^{i}\delta_o$ is defined by

$$^{i}c_o = {}^{i}\delta_o^2. \quad (11)$$

Like dispersion forces, these are symmetrical interactions, depending on the same property of each molecule, which in this case is the dipole moment. It follows that the geometric-mean rule is obeyed closely for orientation interactions between unlike molecules. For "ideal" polar molecules, which may be represented by spherical force fields with small ideal dipoles at their centers, this contribution to the cohesive pressure in mixtures of i and j molecules is

$$^{ij}c_o = \left({}^{i}c_o{}^{j}c_o\right)^{1/2} = {}^{i}\delta_o{}^{j}\delta_o \quad (12)$$

and the interchange cohesive pressure due to orientation is

$$^{ij}A_o = \left({}^{i}\delta_o - {}^{j}\delta_o\right)^2. \quad (13)$$

Dipole induction effects arise from dipole-induced forces (Debye interactions) occurring between molecules with permanent dipole moments and any other neighboring molecules, whether polar or not, and resulting in an induced nonuniform charge distribution. In contrast to dispersion and orientation interaction, dipole induction interactions are "unsymmetrical," involving the dipole moment of one molecule and the polarizability of the other. Thus, the cohesive pressure term for induction in a pure material i involves the product ${}^{i}\delta_i{}^{j}\delta_d$, where ${}^{i}\delta_i$ is the induction cohesion parameter, and in a mixture of i and j,

$$^{ij}c_i = {}^{i}\delta_i{}^{j}\delta_d + {}^{j}\delta_i{}^{i}\delta_d. \quad (14)$$

It can be shown, therefore, that the interchange cohesive pressure due to induction is

$$^{ij}A_i = 2{}^{i}\delta_i{}^{i}\delta_d + 2{}^{j}\delta_i{}^{j}\delta_d - 2{}^{j}\delta_i{}^{i}\delta_d - 2{}^{i}\delta_i{}^{j}\delta_d \quad (15)$$

$$^{ij}A_i = 2\left({}^{i}\delta_d - {}^{j}\delta_d\right)\left({}^{i}\delta_i - {}^{j}\delta_i\right). \quad (16)$$

Lewis acid–base or electron donor–acceptor interactions can be denoted

$$\text{A} \quad + \quad \text{:D} \quad \rightleftarrows \quad \overset{\delta-}{\text{A}} \cdots \overset{\delta+}{\text{D}} \quad (17)$$

Lewis acid   Lewis base
(electron pair   (electron pair
acceptor)   donor)

The Lewis acid–base complex is formed by an overlap between a filled electron orbital of the donor with a vacant orbital in the acceptor and differs from a "normal" chemical bond in that only one molecule supplies the pair of electrons. These interactions are unsymmetrical, involving donor and acceptor with different roles, so it is necessary to use two separate parameters to characterize these interactions, a Lewis acid cohesion parameter δ_a and a Lewis base cohesion parameter δ_b. The acid–base interchange cohesion pressure is

$$^{ij}A_{ab} = 2\left(^i\delta_a - {}^j\delta_a\right)\left(^i\delta_b - {}^j\delta_b\right). \quad (18)$$

Hydrogen bonding interactions are a special type of Lewis acid–base reactions with the electron acceptor being a Brönsted acid (proton acid):

$$\text{—X—H} + \quad \text{:Y—} \quad \rightleftarrows \text{—X—H} \cdots \text{Y—}$$

Electron pair   Electron pair   (19)
acceptor;   donor; proton
proton donor   acceptor

One of the assumptions central to the cohesion parameter approach to the properties of materials is that the various contributions to the cohesive pressure of a substance (either pure or mixed) are additive, so the interchange cohesive pressure for a mixing process is

$$^{ij}A = {}^{ij}A_d + {}^{ij}A_o + {}^{ij}A_i + {}^{ij}A_{ab}. \quad (20)$$

For a pure substance, the total cohesion parameter is

$$^i\delta^2 = {}^i\delta_d^2 + {}^i\delta_o^2 + 2^i\delta_i{}^i\delta_d + 2^i\delta_a{}^i\delta_b. \quad (21)$$

It is clear that this total cohesion parameter is identical to the Hildebrand parameter, which can be determined from Eq. (2).

C. Hansen Parameters

A much simpler three-component cohesion parameter method was developed by C. M. Hansen on an empirical basis, the parameters being determined either experimentally or by semiempirical equations. This assumes that the cohesion pressure is made up of a linear combination of contributions from nonpolar or dispersion interactions (δ_d^2), polar interactions (δ_p^2), and hydrogen bonding or similar specific association interactions (δ_h^2). The Hildebrand parameter (which can be determined from thermodynamic properties) for any material i is related to the Hansen parameters by

$$^i\delta^2 = {}^i\delta_d^2 + {}^i\delta_p^2 + {}^i\delta_h^2. \quad (22)$$

The interchange cohesive pressure associated with the mixing of i and j is

$$^{ij}A = \left(^i\delta_d - {}^j\delta_d\right)^2 + \left(^i\delta_p - {}^j\delta_p\right)^2 + \left(^i\delta_h - {}^j\delta_h\right)^2. \quad (23)$$

It is important to note that the Hansen parameter method ignores the unsymmetrical nature of the induction and acid–base components of the cohesive pressure. In particular, with hydrogen bonding there is no means of separating the proton donor and proton acceptor capabilities of any material. (Of course, this is true also of the Hildebrand parameter, but Hansen parameters *appear* to take hydrogen bonding into account, in a manner that is often misleading; see Section II.C.)

Despite these theoretical shortcomings, Hansen parameters have been fairly widely used in the polymer and coatings industries. Frequently, the three-component parameters have been plotted on three mutually perpendicular coordinates. A solubility "volume" in Hansen space is then drawn up for each solute and compared with the point locations in this space of each solvent. Equation (23) has been modified by doubling the scale on the dispersion axes with the aim of providing "spheres" of solubility for each solute. The distance of the solvent coordinates ($^i\delta_d$, $^i\delta_p$, $^i\delta_h$) from the center point ($^j\delta_d$, $^j\delta_p$, $^j\delta_h$) of the solute sphere of solubilities then is

$$^{ij}R = \left[4\left(^i\delta_d - {}^j\delta_d\right)^2 + \left(^i\delta_p - {}^j\delta_p\right)^2 + \left(^i\delta_h - {}^j\delta_h\right)^2\right]^{1/2} \quad (24)$$

or

$$^{ij}A = \left(^i\delta_d - {}^j\delta_d\right)^2 + 0.25\left[\left(^i\delta_p - {}^j\delta_p\right)^2 + \left(^i\delta_h - {}^j\delta_h\right)^2\right]. \quad (25)$$

This distance ^{ij}R can be compared with the radius ^iR of the solute solubility sphere, and if $^{ij}R < {}^jR$ the likelihood of the solvent i dissolving the solute j is high. The incorporation of the numerical factor 4 in Eq. (24) does not appear to be necessary to provide a spherical interaction volume, and an equation based on Eq. (23) is just as satisfactory:

$$^{ij}r = {}^{ij}A^{1/2} = \left[\left(^i\delta_d - {}^j\delta_d\right)^2 + \left(^i\delta_p - {}^j\delta_p\right)^2 + \left(^i\delta_h - {}^j\delta_h\right)^2\right]^{1/2}. \quad (26)$$

In some applications, only two of the three Hansen parameters are used, so that the location of solvents can be displayed on two-dimensional maps (e.g., Fig. 1) and compared with solute solubility regions.

D. Other Cohesion Parameters

There are several variations on Hildebrand parameters, Hansen parameters, and interaction cohesion parameters, introduced in attempts to achieve a compromise between simplicity of operation and validity of prediction.

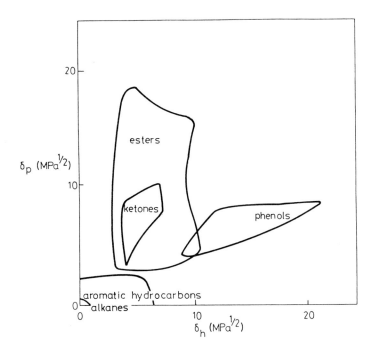

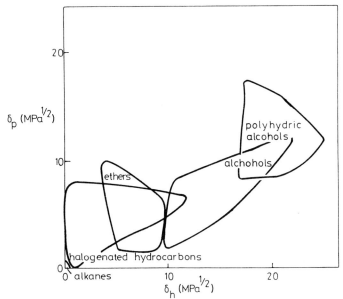

FIGURE 1 Hansen parameter δ_p–δ_h locations for various classes of organic compound. [Adapted from Klein, E., Eichelberger, J., Eyer, C., and Smith, J. (1975). *Water Res.* **9,** 807.]

Hansen parameters must be represented in three dimensions, but it is possible to use fractional cohesive pressures plotted on a triangular chart,

$$c_d = \delta_d^2/\delta^2; \quad c_p = \delta_p^2/\delta^2; \quad c_h = \delta_h^2/\delta^2 \qquad (27)$$

where δ^2 is given by Eq. (22). It is also possible to use triangular representation with fractional cohesion parameters,

$$f_d = \frac{\delta_d}{\delta_d + \delta_p + \delta_h};$$

$$f_p = \frac{\delta_p}{\delta_d + \delta_p + \delta_h}; \qquad (28)$$

$$f_h = \frac{\delta_h}{\delta_d + \delta_p + \delta_h}$$

Triangular representations make the excessively simplifying assumption that the total or Hildebrand parameter

δ is constant for all materials and that it is the relative magnitude of the three contributions (dispersion forces, polar interactions, hydrogen bonding) that determines the extent of miscibility.

The Hildebrand parameter has been subdivided in other ways. One approach was to divide it into two main components, defining a nonpolar cohesion parameter δ_d and a polar parameter δ_o. These are related by

$$\delta^2 = \delta_d^2 + \delta_o^2. \tag{29}$$

This approach neglects both induction interactions and, more important, specific interactions. The induction "correction" can be taken into account to some extent by means of the factor ^{ij}b, as shown in the following expression for the interchange cohesive pressure:

$$^{ij}A = \left(^i\delta_d - {}^j\delta_d\right)^2 + {}^{ij}b\left(^i\delta_o - {}^i\delta_o\right)^2. \tag{30}$$

This is equivalent to Eq. (25) with the δ_h term omitted.

II. LIMITATIONS OF COHESION PARAMETERS

Cohesion parameters provide one of the simplest methods of correlating and predicting the cohesive and adhesive properties of interacting materials from a knowledge of the properties of the individual components. It is therefore to be expected that there will be severe limitations on their use. What is surprising is that this simple correlation method works as well as it does in practice. It is really a semiquantitative version of the statement "like dissolves like." In fact, of course, whether any particular correlation or prediction method is seen to "work" depends on the precision that is expected in the application being considered.

A. Burrell Hydrogen Bonding Classes

In the use of Hildebrand parameters, the existence of hydrogen bonding is the most obvious cause of discrepancies from "regular" behavior. H. Burrell was one of the first to attempt to deal with the hydrogen bonding factor in the application of Hildebrand parameters to practical systems by the simple expedient of dividing solvents into three classes according to their hydrogen bonding capacities, on the assumption that complete miscibility can occur only if the degree of hydrogen bonding is comparable in the components:

1. Liquids with weak or poor hydrogen bonding capacity, including hydrocarbons, chlorinated hydrocarbons, and nitrohydrocarbons

2. Liquids with moderate hydrogen bonding capacity, including ketones, esters, ethers, and glycol monoethers
3. Liquids with strong hydrogen bonding capacity, such as alcohols, amines, acids, amides, and aldehydes.

This classification is still widely used in practical applications, and a few typical examples are presented in Tables I and II (Section VI).

B. Geometric-Mean Corrections

One of the specific assumptions in the development of cohesion parameter expressions is the geometric-mean approximation [Eq. (3)]. There are several equations that permit some correction to be made for derivations from this behavior, for example,

$$^{ij}c = ({}^ic^jc)^{1/2}(1 - {}^{ij}l), \tag{31}$$

where ^{ij}l is a dimensionless constant of the order of 0.01 to 0.1, characteristic of a given pair of materials. When this is incorporated, the empirical expression for interchange cohesive pressure corresponding to Eq. (4) becomes

$$^{ij}A = ({}^i\delta - {}^j\delta)^2 + 2^{ij}l{}^i\delta{}^j\delta. \tag{32}$$

The value of the correction term can be estimated from either liquid- or gas-phase data.

C. Chameleonic Behavior

The most important situation requiring caution in the use of cohesion parameters, particularly Hildebrand parameters and Hansen parameters, is that in which electron donor–acceptor interactions within components are very different from those between components. Common examples are systems involving hydrogen bonding: alcohols (particularly methanol), carboxylic acids, water, primary and secondary amines, and glycol ethers. Most commonly, it is hydrogen bonding within the pure component that causes a user the most problems, as one tends to overlook this, whereas one is more on the alert for new interactions when components are mixed.

Although this has long been understood in a general way, it was put most clearly by K. L. Hoy, who proposed the term *chameleonic* for those compounds that have the capacity to assume a character similar to their environment.

By dimerization or intramolecular association, what would otherwise be a polar material can behave in a nonpolar manner, thus minimizing the energy. Examples are carboxylic acids (structure I), glycol ethers (II and III), and diols (IV). It is clear that the cohesion parameters

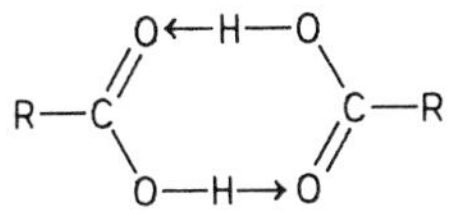

SCHEME 1

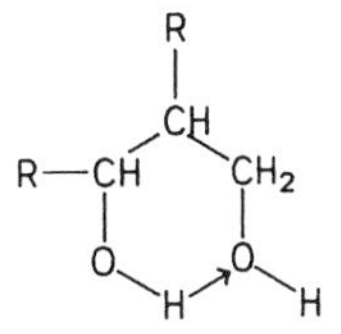

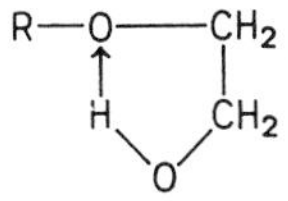

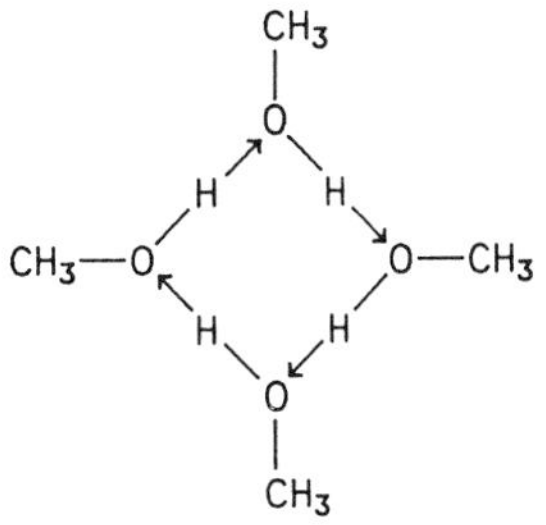

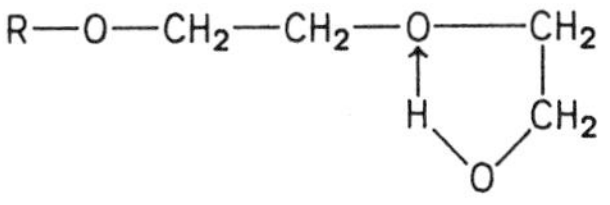

of associated and dissociated forms are very different, so the particular value exhibited will depend on the situation in which the compound occurs. Structures such as tetramers (V) and chains (VI) have been proposed for methanol and methanol–ethanol mixtures.

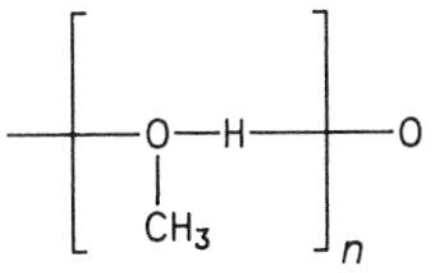

In the case of water, it is necessary to distinguish between systems of water with low-permittivity organic liquids (i.e., associated water, with limited interaction between water and organic) and systems of water with hydrogen bonding organic liquids, where there is more intermolecular association at the expense of intramolecular association.

The difficulty in all such examples is that Hildebrand parameters and Hansen parameters are trying to indicate two or more things with one number or one set of numbers:

1. The extent of cohesion within the pure compound
2. The potential for cohesion in a mixture.

When one says that the Hansen hydrogen bonding parameter δ_h of ethanol is 20 MPa$^{1/2}$, it means that within pure ethanol $\delta_h^2 = 400$ MPa is the extent of cohesion due to hydrogen bonding, but it is also inferred that δ_h reflects the hydrogen bonding capability in a mixture.

It is clear that this will work best when the two components of the mixture are *similar*, and as the differences increase, situations arise where the effective δ_h may be much less than or much greater than that in a pure liquid.

As far as hydrogen bonding is concerned, some pure compounds are both proton donors and proton acceptors. Here, the extent of cohesion due to association in a mixture generally can have any value up to that in a pure compound. Examples are alcohols, carboxylic acids, water, and primary and secondary amines.

In contrast, many compounds have a dominant capacity to accept protons: ketones, aldehydes, esters, ethers, tertiary amines, aromatic hydrocarbons, and alkenes. Here, the potential for cohesion in mixtures due to hydrogen bonding is *greater* than that in pure compounds. There are also some that are proton donors only, such as trichloromethane.

It is clear that neither Hildebrand parameters nor Hansen parameters are adequate to handle this problem in a quantitative way. There are two possible approaches:

1. To treat each associated species in both the pure liquids and the mixtures as new compounds, with formation constants that can be evaluated. This type of approach is traditional and well established, but rather cumbersome.

2. To use a full set of interaction cohesion parameters with separate Lewis acid and base contributions as well as dispersion, induction, and orientation terms. This method has the greater practical potential.

On reference to Eq. (20), it can be seen that when $^{ij}A_{ab}$ is large and negative, exothermic mixing (with evolution of heat) may be explained, in contrast to the restriction to athermic or endothermic processes, when only dispersion and polar forces exist. Thus, the answer to the criticism that Hildebrand parameters cannot cope with molecular association or exothermic interactions is that they should be expressed in the form of interaction cohesion parameters. The price to be paid is greater complexity in both evaluation and use.

III. COHESION PARAMETERS AND OTHER SOLVENT SCALES

Parameters describing and correlating the solvent capacities of liquids have been based on a great variety of chemical and physical properties. Some are measures of liquid "basicity." Others are direct determinations of the solubility of a representative solute in a range of liquids, for example, the solubility of hydrogen chloride in the solvents at 10°C.

Once a solubility scale has been established, it is necessary to determine the position on it of any required solvent. If the solubility scale has a theoretical basis, it may be possible to calculate values from information on other properties, but if it is an empirical scale, direct testing is usually required.

Because Hildebrand parameter values can be determined readily and are widely available, they have been used in conjunction with various other theoretical or empirical parameters to provide more effective predictions of solvent properties. Several examples of the parameters that have been used in this way to form "hybrid" cohesion parameter scales are now described.

A. Hydrogen Bonding Parameters

The Burrell hydrogen bonding classification (Section II.A) has been developed further by assigning quantitative values to the hydrogen bonding capabilities of liquids and plotting graphs of Hildebrand parameters against hydrogen bonding capacities.

An alternative quantitative measure of the capacity of organic liquids to donate and accept hydrogen bonds is sound velocity in solvent–paper systems. Paper fibers are held together largely by hydrogen bonds, and on wetting paper with a liquid, most disruption of the fiber bonding occurs in the presence of those solvents that preferentially form fiber–liquid hydrogen bonds. As a result, the velocity of sound through the paper, which depends on the degree of bonding, decreases as the liquid hydrogen bonding capability increases. Water, as the most common solvent for disrupting fiber–fiber bonds in paper, was chosen as a reference standard, and the hydrogen bonding parameter was defined as

$$\frac{100 \times \text{sound velocity in water-soaked paper}}{\text{sound velocity in liquid-soaked paper}}$$

B. Electrostatic Parameters

The cohesive properties of materials are closely related to the various electrostatic properties, such as dipole moment, relative permittivity, polarizability, ionization potential, and refractive index. Both dipole moment and relative permittivity are of particular importance: In the absence of specific association, the dipole moment tends to determine the orientation of solvent molecules around molecular solutes, while dissolution of ions is promoted by high relative permittivity of the solvent. The *electrostatic factor*, which is defined as the product of dipole moment and relative permittivity, takes both effects into account and provides a basis for the classification of solvents. The *fractional polarity* is also noteworthy: The natures of molecular interactions have been discussed in terms of the fraction of the total interactions due to dipole–dipole or orientation effects. This fractional polarity can be calculated from the dipole moment, polarizability, and ionization potential once certain assumptions have been made.

C. Spectroscopic Parameters

Numerous parameters have been developed on the basis of spectroscopic measurements. One direct hydrogen bonding parameter is based on the effect that a liquid has on a small amount of alcohol introduced into that liquid: The greater the extent of hydrogen bonding between the liquid and the alcohol hydroxyl groups, the weaker the O—H bond and the lower the frequency of the infrared radiation absorbed by that bond. If deuterated methanol or ethanol is used, the O–D stretch band is in a spectroscopic region with little interference, permitting detection at low alcohol concentrations. The extent of the shift to lower frequencies of the O–D stretching infrared absorption of deuterated alcohol in the liquid under study thus provides a measure of its hydrogen bond acceptor capability. The spectrum of a solution of deuterated methanol or ethanol is compared with that of a solution in benzene or other reference liquid, and the hydrogen bonding parameter is defined as 10% of the O–D absorption shift expressed in wave numbers. The choice of the reference liquid is important. Benzene is not an inert solvent, the aromatic electron system having some hydrogen acceptor properties, and there is even the possibility of some hydrogen bonding between methanol and tetrachloromethane, another reference solvent. It appears that alkanes such as cyclohexane, heptane, and isooctane may be preferable as standards.

Spectroscopic hydrogen bonding parameters form a special case of a more general type of parameter described by such names as "electron donating power" and "electron accepting power."

D. Empirical Solvent Scales

There are many empirical tests in common use for quantifying solvent behavior. One group of tests describes the "solvent power" or "strength" of liquid hydrocarbons. The *aniline point* or *aniline cloud point*, which is based on the fact that aniline is a poor solvent for aliphatic hydrocarbons and an excellent one for aromatics, is defined as the minimum equilibrium solution temperature for equal volumes of aniline and solvent. An approximately quantitative correlation exists between Hildebrand parameters and the aniline point for hydrocarbons, but it is of limited value. The *kauri–butanol number* (KB) is a measure of the tolerance of a standard solution of kauri resin in 1-butanol to added hydrocarbon diluent. There is an approximately linear relationship between the Hildebrand parameter and the KB number for hydrocarbons with KB > 35:

$$\delta/\text{MPa}^{1/2} = 0.040\,\text{KB} + 14.2. \tag{33}$$

The solvent power of "chemical" or "oxygenated" liquids (such as alcohols, ketones, esters, and glycol ethers) is much greater than that of hydrocarbons, and different scales are necessary for their description. The dilution ra-

tio is widely used since it is a direct measure of the tolerance of a solvent–resin mixture to added diluent, and qualitative correlations exist between dilution ratios and Hildebrand parameters. The heptane number of hydrocarbon solvents is a measure of the relative solvent power of high-solvency hydrocarbons in the presence of resins not soluble in heptane. The wax number and other miscibility numbers may also be correlated with the solvent Hildebrand parameter.

Another type of empirical solvent classification scheme has been developed in connection with chromatography, where it is useful to distinguish solvent strength or "polarity" from solvent "selectivity." Gas–liquid chromatographic methods are particularly convenient for quantitative characterization of the solvent properties of the stationary phase, whether this is a liquid or a polymer (see Section V.A). One approach is to define a polarity index, a measure of the capacity of a liquid to dissolve or interact with various polar test solutes. There is, in general, a good correlation between the polarity index values and Hildebrand parameter values, but liquids such as diethyl ether and triethylamine that are strong proton acceptors but have no proton donor capacity have Hildebrand parameter values similar to those of alkanes, although they show up as moderately polar on the polarity index scale. This is because Hildebrand parameters are based on pure liquid properties, while the polarity index is based on the interactions between different liquids.

E. Hybrid Scales

Often cohesion parameters are included in hybrid solvent scales which incorporate several kinds of parameters. For example, a "universal solubility" treatment includes the Hildebrand parameter as a measure of the energy necessary to create a cavity in the bulk liquid to accommodate a solute molecule. This contribution is combined with parameters allowing for solute–solvent dipole interactions, hydrogen-bond donor acidity, hydrogen-bond acceptor basicity, and an empirical coordinate covalency parameter.

IV. APPLICATIONS OF COHESION PARAMETERS

A. Liquids

For a pair of regular liquids, the infinite dilution activity coefficient expression in terms of Hildebrand parameters is given by Eq. (5). For dilute solutions with specific interactions and size effects, resulting in nonzero enthalpies of mixing and nonideal entropies of mixing, the expressions become more complex but are still useful.

For systems composed of two liquids that have substantial but incomplete mutual miscibility, cohesion parameters may be used in combination with other empirical expressions. Although satisfactory data correlation is possible, prediction of the exact extent of mutual solubility is difficult.

Cohesion parameter methods have been used for the correlation of activity coefficients of a wide range of systems. One of the most flexible approaches to the correlation and prediction of activity coefficients has been to include terms for some or all of the following:

- Regular enthalpy of mixing from Hildebrand parameters
- Polar orientation and induction effects
- Entropy effects from hydrogen bonding equilibria
- Gibbs free energy term for the breaking of hydrogen bonds in association interactions

Each component is thus characterized by several parameters. The overall average error in prediction for 845 literature data points for ~300 systems was 25% in the infinite dilution activity coefficient, falling to better than 9% for all saturated hydrocarbons in all solvents, adequate for screening and for some design purposes.

One of the early applications of regular-solution theory was the discussion of activity coefficient ratios and equilibrium constants for complex formation in terms of Hildebrand parameters. There has been considerable debate as to the correct basis of the equilibrium constant (concentration, mole fraction, volume fraction, or molality), different methods proving preferable for various systems. Extension to complexes of ionic species (as in solvent extraction systems) has also occurred.

A common application of solvents is in the separation of the components of mixtures, by countercurrent liquid–liquid extraction or by extractive distillation or by azeotropic distillation. These solvent-aided separation processes can be planned with the aid of cohesion parameters. The selectivity $^kS_{ij}$ of a solvent k toward a dilute mixture of i and j, derived from relationships such as Eq. (5), is given by

$$RT \ln {}^kS_{ij} = {}^iV^{ik}A - {}^jV^{jk}A, \qquad (34)$$

with contributions from dispersion, orientation, induction, entropy, and acid–base terms. In general, the choice of a noninteracting solvent with a cohesion parameter that differs significantly from the cohesion parameters of the components to be separated enhances the solvent selectivity, although it reduces the solvent capacity. It is therefore necessary to reach a compromise such that the liquid has an adequate capacity but retains good selectivity and immiscibility between the phases. However, if a liquid can be obtained that specifically interacts with the component to be extracted, then both the selectivity and the solvent capacity are increased. For example, in the case of an alkane i (for which both $^i\delta_a$ and $^i\delta_b$ are negligible), an aromatic hydrocarbon j ($^j\delta_a$ negligible), and an electron-accepting liquid k ($^k\delta_b$ negligible), the term for the specific interaction is given by $2^jV^j\delta_b{}^k\delta_a$.

Another common situation in which cohesion parameters have been employed involves solvent extraction of an electrically neutral ion pair or complex used to transport ionic species distributed in relatively dilute solution between two very dissimilar phases such as hydrocarbon and water. As in all other applications of cohesion parameters, it should not be expected that distribution ratios can be predicted in detail by means of cohesion parameters, and even if the infinite dilution values are approximately correct, these have limited applicability. Rather, the emphasis is on formulating broad trends and establishing correlations.

B. Gases

In many situations, such as the dissolution of oxygen in water, there are negligible specific interactions when gases dissolve in liquids, and it is reasonable to expect cohesion parameters, even simple Hildebrand parameters, to provide methods of correlation and prediction.

J. H. Hildebrand and others have long pointed out the close relationship that exists between the logarithm of the mole fraction gas solubility, $\log {}^jx_s$, and the solvent Hildebrand parameter $^i\delta$. For example, the values of $\log {}^jx_s$ for nitrogen, argon, methane, ethylene, and ethane were found to be linear functions of $^i\delta$ for the normal primary alcohols as solvents, but it was subsequently shown that some curved lines became straight when $^i\delta^2$ was used in place of $^i\delta$:

$$-\ln {}^jx_s = -\ln {}^jx_{ideal} + ({}^j\bar{V}/RT)({}^i\delta - {}^j\delta)^2, \qquad (35)$$

where $\bar{V}$ is the average partial molar volume of the gas in a range of liquids. There are several similar equations, all tacitly or explicitly assuming that the gaseous solute is condensed to a hypothetical "liquid" state (with hypothetical liquid–state Hildebrand parameter and molar volume) before mixing with the solvent.

The solubilities of gases such as tetrafluoromethane, sulfur hexafluoride, and carbon dioxide deviate from straight lines because of various properties.

Another simple application of cohesion parameters to gases is the plotting of gas solubility directly against solvent Hildebrand parameter values. A relatively sharp maximum is usually obtained at a Hildebrand parameter value corresponding to that of the hypothetical liquid form of the dissolved gas.

As well as this situation of a gas possessing a hypothetical liquid-like molar volume when it dissolves in a liquid at low pressures, there is also the case of gases that liquefy or achieve liquid-like molar volumes because of low temperatures and/or high pressures. The solvent properties of compressed gases, especially carbon dioxide, for biochemicals and polymers have been receiving more attention recently because of their application in high-pressure gas chromatography and supercritical fluid chromatography.

Cohesion parameter concepts are being utilized for studies of cryogenic liquids, refrigerants, and aerosol propellants, as well as fundamental dense gas and vapor–liquid properties. For example, it is reported that even at 25 K above their critical temperatures, compressed gaseous helium ($\delta \approx 8\,\text{MPa}^{1/2}$) and xenon ($\delta \approx 16\,\text{MPa}^{1/2}$) at 200 MPa pressure separate into two phases because of their divergent cohesion properties.

C. Solids

The regular-solution theory and the Hildebrand parameter are based on the enthalpy changes occurring when liquids are mixed. In order to extend these concepts to solutions of crystalline solids in liquids, it is necessary to estimate the thermodynamic activity of the solid referred to a hypothetical liquid subcooled below its melting point. If this is done, expressions such as Eq. (5) can be used to evaluate cohesion parameters of nonvolatile solutes or to estimate their solubilities.

The Hildebrand parameter, defined in Eq. (2), has a liquid state basis so this hypothetical liquid reference state is necessary when the method is extended to solids. Use of the vaporization enthalpy or sublimation enthalpy of a solid at 25°C in Eq. (2) uncorrected for the crystal–liquid transition enthalpy change does not yield a Hildebrand parameter. It is another type of cohesion parameter that should not be confused with any of those defined in Section I.

Despite this complication, the solubility ($^j x_s$ on the mole fraction scale) of a solid j in a liquid i has been shown to vary regularly with the Hildebrand parameter of the solvent, plots of log $^j x_s$ against $^i\delta$ being nearly linear or only slightly curved.

It is possible to derive more informative thermodynamic expressions that involve various assumptions, and several of these predict that, for solutions that approach regular behavior, plots of log $^j x_s$ against $(^i\delta - {}^j\delta)^2$ should be approximately linear, and this relatively simple method is probably the most widely used method of correlating solid solubilities in terms of Hildebrand parameters.

If $^i\delta$, $^j\delta$, and ^jV (for the subcooled liquid j corresponding to the solid of interest) as well as the melting point and

entropy of melting are known for approximately regular systems, it is possible to calculate the data necessary for phase diagrams, either temperature against $^i x_s$ plots for binary systems or triangular diagrams with temperature contours for ternary systems.

The more detailed and informative interaction cohesion parameters have not been widely used in the correlation of solid–liquid solubilities. Hansen parameters have been extended to ionic solids, but the determination of their values has not been pursued to any great extent. Any extension of interaction cohesion parameters to ionic systems in the future would be very valuable in dealing with solid–liquid systems.

D. Polymers

The parameter most commonly used in the discussion of polymer solutions is the polymer–liquid interaction parameter χ. This parameter has been associated with various theoretical treatments, but it can be considered a general, dimensionless parameter reflecting intermolecular forces between a particular polymer and a particular liquid. It was introduced as an enthalpy of dilution with no entropy component but has been considered subsequently to be a Gibbs free energy parameter. As originally formulated, the polymer–liquid interaction parameter was expected to be inversely dependent on absolute temperature and independent of polymer concentration, but as now empirically defined it depends in an unspecified way on temperature, solution composition, and polymer chain length.

It can be shown that the enthalpy part of the interaction parameter, χ_H, is related to the interchange cohesive pressure by

$$\chi_H = {}^{ij}A\,{}^iV/RT. \tag{36}$$

If Hildebrand parameters are used, from Eq. (4),

$$\chi_H = (^iV/RT)(^i\delta - {}^j\delta)^2 \tag{37}$$

and in terms of Hansen parameters

$$\chi_H = (^iV/RT)\left[\left(^i\delta_d - {}^j\delta_d\right)^2 + \left(^i\delta_p - {}^j\delta_p\right)^2 + \left(^i\delta_h - {}^j\delta_h\right)^2\right]. \tag{38}$$

The enthalpy component of the polymer–liquid interaction parameter is therefore closely related to the cohesion parameters of the components of the polymer solution.

Cohesion parameter predictions of thermodynamic property values must be corrected to allow for the substantial size differences between polymer and solvent molecules. However, if only semiquantitative "compatibility" information is required, cohesion parameters of amorphous polymers may be used in the same way as those of liquids. When crystalline polymers are considered, it is

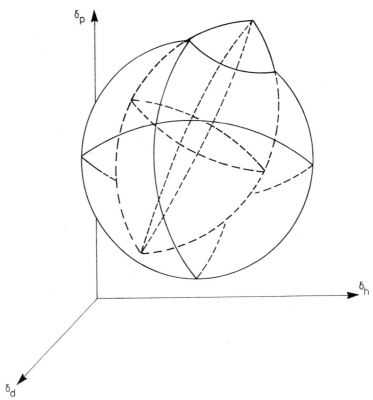

FIGURE 2 Hansen parameter three-dimensional model showing the extent of overlap between poly(phenylene oxide phosphonate ester) (spherical) and cellulose acetate. [Adapted from Cabasso, I., Jaguer-Grodzinski, J., and Vofsi, D. (1977). In "Polymer Science and Technology" (D. Klempner and K. C. Frisch, eds.), Vol. 10, p. 1, Plenum, New York.]

in principle necessary to calculate the activity of the crystalline solid relative to the real or hypothetical amorphous material at the same temperature. In practice, the distinction between amorphous and crystalline polymers is not clear-cut, and nonthermodynamic empirical methods are frequently used in the estimation of polymer cohesion parameters.

It is not correct to assume that the "best" solvent for a polymer is necessarily the corresponding monomer or low molecular weight liquid polymer ("oligomer") made up of the same repeating units. Such liquids usually have cohesion parameter values lower than those of the polymers because their molar volumes are higher. Rather, the best solvent is another liquid with a cohesive energy rather higher than that of the monomer but with a cohesive pressure the same as that of the polymer.

In one of the original applications of cohesion parameters to polymer solutions, the liquid Hildebrand parameter was combined with the Burrell hydrogen bonding classification (Section II.A). Hansen's three-component cohesion parameter system was also first developed for polymer–liquid systems.

As well as the assumption of spherical Hansen parameter "volumes" of polymer–liquid interaction, as described

in Section I.C and illustrated in Fig. 2, it is possible to present this information in the form of irregular volumes. Three-dimensional models, stereographs, sets of projections, and contour maps have all been used. In many cases the solubility behavior of a polymer can be adequately represented in two dimensions by δ and δ_h or by δ and δ_p. Another useful type of plot involves the subdivision of the Hildebrand parameter into a "volume-dependent" part $\delta_v = (\delta_d + \delta_p)^{1/2}$ and a "residual" or hydrogen bonding part δ_h. Triangular diagrams also provide a method of conveying information in three properties on two-dimensional plots by making the excessively simple assumption that the Hildebrand cohesion parameter is uniform for all materials and that it is the relative magnitude of the three contributions that determines the extent of miscibility. This is illustrated in Fig. 3.

Also widely used for polymers are hybrid maps of cohesion parameters with other quantities such as hydrogen bonding parameter, dipole moment, and fractional polarity.

Cohesion parameters are widely used to predict polymer solubility or swelling in liquids, but some liquids also affect the mechanical properties of polymers by environmental stress cracking or crazing. Whatever the cause of

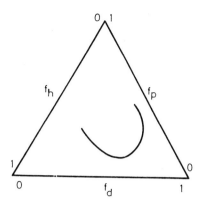

FIGURE 3 Limiting boundary of solubility for cellulose nitrate in terms of fractional cohesion parameters for dispersion (f_d), polar (f_p), and hydrogen bonding (f_h) effects. [Adapted from Gardon, J. L., and Teas, J. P. (1976). In "Treatise on Coatings" (R. R. Myers and J. S. Long, eds.) Vol. 2, Chap. 8, Dekker, New York.]

these effects, there is a surprisingly close correlation with the liquid Hildebrand parameter (Fig. 4).

E. Rate and Transport Properties

Many varied theoretical and empirical scales are available for the description of the effect of solvents on the rates of chemical processes in solution, cohesion parameters forming only one. Because the transition state theory can be interpreted in terms of pseudothermodynamic properties, including volume and enthalpy, it is also possible to attribute cohesion parameter values to the activated state. Cohesion parameters are closely related to the internal pressures in the system, and for nonpolar reactions (as well as for polar reactions in nonpolar solvents) the internal pressure of solvents influences reaction rates in the same direction as does external pressure.

The activated complex in a chemical reaction is considered to have properties that approach those of the products of the reaction, and there is a general rule that, if the reaction is one in which the products are of higher cohesion than the reactants, it is accelerated by solvents of high cohesive pressure. Conversely, if the solvent is similar to the reactants in cohesion properties, the rate tends to be lower. For polar or ionic processes the effect of the solvent is more complex, and the cohesion parameter concept is less useful.

Viscous flow can be regarded as a rate process in which molecules move into holes or voids in the liquid. According to this model, the activation energy for viscous flow is related to the energy required to form such a hole and hence to the cohesion parameter of the liquid. One successful application was a relationship between Hildebrand parameters and the initial slopes of plots of dilute so-

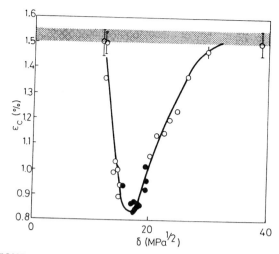

FIGURE 4 Critical strain ε_c for crazing (open circles) or cracking (filled circles) of poly(2,6-dimethyl-1,4-phenylene oxide) as a function of the Hildebrand parameter δ of the solvent. The ε_c values in air are in the band near $\varepsilon_c = 1.5\%$. [Adapted from Bernier, G. A., and Kambour, R. P. (1968). *Macromolecules* **1**, 393.]

lution viscosities against mole fraction composition. In a very different kind of system, the dropping points of greases made from lithium stearate soap and oils have been shown to be a function of the Hildebrand parameters of the oils.

The limiting viscosities of dilute polymer solutions have also received attention from this point of view. The viscosity of a dilute solution is a maximum in the best solvent, that is, in one in which the cohesion parameters of solvent and polymer are comparable. In a "good" solvent the polymer molecules are "unfolded" or "uncoiled," obtaining to the maximum extent the more favorable polymer–liquid interactions (and therefore resulting in the greatest viscosity), while in a "poor" solvent the polymer molecules remain folded because of the more favorable intramolecular interactions. Viscosity–cohesion parameter correlations have proved successful for such dilute polymer solutions.

Although the viscosity of a dilute polymer solution is a minimum in poor solvents, as the polymer concentration is increased there is a changeover in behavior: The viscosity exhibits a maximum in poor solvents at higher concentration. There is an aggregation or clustering of polymer molecules as a preliminary to phase separation when solvents are used that have cohesion parameters near the limits of the miscibility range for the polymer. This rather complex behavior can be described in terms of entropy and free volume effects, which are not present in nonpolymer systems, and it should not be expected that it can be fully correlated by means of cohesion parameters that provide direct information only on the enthalpy aspect of the interactions.

Other transport properties have also been correlated by means of cohesion parameters, including gas, liquid, and solid diffusion, permeation in polymers, reverse osmosis in membranes, and the time-dependent mechanical properties of polymers.

F. Surfaces

As a result of the close relationship between solubility or phase separation and surface activity, cohesion parameters are useful for the characterization of heterogeneous as well as homogeneous systems. Interfacial free energy ("surface tension") is the excess free energy due to the existence of an interface, arising from imbalanced molecular forces. These forces in the bulk of the material are the origin of cohesive properties, so the close link between adhesion and cohesion is easy to understand.

Various empirical or semiempirical equations have been used to link cohesion parameters with surface tension. There is considerable theoretical and experimental justification for subdividing surface free energy into additive components analogous to the cohesion parameter components on the basis of the types of molecular interaction. In fact, Hansen considered the characterization of surfaces in terms of the Hansen parameters of the liquids that spread spontaneously on them. Ideas such as these have been extended to practical problems of wetting, dewetting, adhesion, lubrication, adsorption, colloids, emulsions, and foams.

G. Biological Systems

Solubility is of major importance in biochemical processes, and correlations with cohesion properties have been explored for such purposes as transport of molecules through biological tissues, rationalization of physicochemical influences on biological responses, formulation of drugs in liquid dosage form, design of insecticides, and properties of biocompatible materials.

One particularly interesting example is the action of fluorinated ethers as inhalation convulsants, on the one hand, or as anesthetics, on the other. Those such as hexafluorodiethyl ether with Hildebrand parameters of less than 15 MPa$^{1/2}$ are powerful convulsants, whereas those with $\delta > 15$ MPa$^{1/2}$ such as methoxyflurane are anesthetics. Although all the ethers dissolve equally well in bulk lipids and have similar octanol–water partition values, they dissolve differentially into specific local microenvironments or subregions, which can be considered to have Hildebrand parameter values different from that of the bulk. This indicates that the cohesion parameter concept is valuable both in models concerned with general partition, concentration, or activity and in models that assume more specific mechanisms.

V. EVALUATION OF COHESION PARAMETERS

A. Thermodynamic Calculations and Inverse Gas Chromatography

From the definition of the Hildebrand parameter [Eq. (2)], it is apparent that it is necessary to determine both the enthalpy of vaporization and the molar volume for its evaluation. There is rarely much difficulty in finding a reliable value for the molar volume of a liquid, and solids can be treated as subcooled liquids with molar volumes extrapolated from the liquid state values. Frequently, the main problem is obtaining the enthalpy of vaporization at the temperature of interest, usually 25°C. Direct experimental information is frequently unavailable, and extrapolation methods or even empirical calculations are often necessary, based on such properties as boiling point, corresponding states, activity coefficients, and association constants. The "RT" correction in Eq. (2) assumes that the vapor is ideal, and although gas law corrections may be applied, even at the normal boiling point the correction is usually negligible.

Polymers and solids pose particular problems because the enthalpy of vaporization is unavailable. Interactions of polymers or noncrystalline solids (particularly those used as plasticizers) with liquids can be studied conventionally by using the polymer or solid as the stationary phase in gas chromatography columns. The activity coefficients at infinite dilution of volatile liquids in the polymer may be determined by this "inverse gas chromatography" using mobile phases to investigate the properties of stationary phases, rather than the reverse. From these activity coefficients, cohesion parameters may be estimated for polymers and some organic solids and liquids.

Hildebrand parameters, Hansen parameters, and the more detailed interaction cohesion parameters can be evaluated from inverse gas chromatography results, but to date this has not been widely practiced.

The thermodynamic quantity internal pressure, given by

$$\pi = (\partial U/\partial V)_T = (\partial p/\partial T)_V - p \qquad (39)$$

and directly accessible from experiment for such nonvolatile materials as polymers [to which Eq. (2) cannot be applied] as well as mixed systems, can also provide cohesion parameter values.

B. Empirical Methods

A list of liquids can be compiled with a gradation of Hildebrand parameter values to form a "solvent spectrum." In its most common form it includes subdivision into categories of hydrogen bonding capacity, as indicated in Table I. The Hildebrand parameter of a solute can then be taken as the midpoint of the range of solvent Hildebrand parameters that provides complete miscibility or the particular value that provides maximum solubility, or maximum swelling in the case of a cross-linked polymer. An ASTM test method for polymer solubility ranges uses mixtures of solvents to provide a spectrum of closely spaced Hildebrand parameters. Other physical properties that can be used as well as solubility and swelling include viscosity and related properties such as grease dropping points. A range of semiempirical equations is also available for correlation and prediction of cohesion parameters.

C. Homomorph Methods for Hansen Parameters

In any multicomponent cohesion parameter system such as Hansen parameters, there arises the problem of evaluating the components of the Hildebrand parameter separately. One obvious approach is to compare the properties of compounds that differ only in the presence or absence of a certain group. Here, the homomorph concept is useful: The homomorph of a polar molecule is a nonpolar molecule having very nearly the same size and shape.

For liquids, the Hansen dispersion parameter obtained by homomorph methods can be subtracted from the total cohesion pressure using Eq. (22), with the remainder being split into Hansen hydrogen bonding and polar parameters so as to optimize the description of the solubility and swelling behavior of a range of liquids and polymers. Both empirical methods and group methods (see next section) can be used. Once the three Hansen parameters for each liquid are evaluated, the Hansen parameters for each polymer can be obtained.

This method may distort the relative magnitudes of the intermolecular forces, but as pointed out in Section I.C, the theoretical bases of Hansen parameters are not good in any case.

Interaction cohesion parameters could, in principle, be evaluated in a similar way, but there has been little activity in this area.

D. Group Contribution Methods

Many properties of materials change in a regular way with increasing chain length in a homologous series, and some properties are conveniently linear. The miscibility behavior of materials depends to a large extent on the cohesive and volume properties, specifically the molar cohesive energy $-U$ and the molar volume V, and these quanti-

ties, together with their ratios and products, can be estimated in terms of standard contributions from groups of atoms.

The molar cohesive energy can be represented by the summation of atomic or group contributions:

$$-U = -\sum_z {}^z U. \qquad (40)$$

Hildebrand parameters can be calculated from

$$\delta = \left(\frac{-U}{V}\right)^{1/2} = \left(-\sum_z {}^z U \bigg/ \sum_z {}^z V\right)^{1/2}. \qquad (41)$$

Also useful are the group molar attraction constants ${}^z F$ defined by

$$\sum_z {}^z F = -\sum_z {}^z U \sum_z {}^z V = \delta V \qquad (42)$$

so

$$\delta = \sum_z {}^z F \bigg/ \sum_z {}^z V. \qquad (43)$$

Values of ${}^z F$ and ${}^z V$ have been tabulated for the most common organic molecular groups.

VI. SELECTED VALUES

It is neither appropriate nor practicable to provide here a comprehensive compilation of values, but exhaustive tables are included in the *Handbook of Solubility Parameters and Other Cohesion Parameters* listed in the bibliography. Rather, listed in Table I are typical values for some liquids whose Hildebrand parameter values, Hansen parameter values, and interaction cohesion parameter values are known with reasonable reliability. The Burrell hydrogen bonding classification (Section II.A) is also included. There is considerable variation in the Hansen parameters reported for water. A study of the solubilities of a range of *organics in water* suggests $\delta_d = 20$, $\delta_p = 18$, $\delta_h = 18$, and $\delta_t = 32 \, \text{MPa}^{1/2}$ rather than the results in Table I, which are more consistent with the behavior of *water in organic liquids*. This variability in Hansen parameter values is a fundamental problem associated with the use of the single parameter δ_h, rather than the pair of acid and base parameters.

For polymers, interaction cohesion parameters are yet to be determined in any detail, so the values given in Table II are restricted to ranges of polymer Hildebrand parameters (for use with solvents of specified Burrell hydrogen bonding class) and sets of Hansen parameters [together with the interaction radius ${}^i R$ of Eq. (25)]. Table III presents preferred Hildebrand parameter values for some well-studied

TABLE I Typical Hildebrand Parameter, Hansen Parameter, and Component Cohesion Parameter Values

Liquid	Hildebrand parameter (δ_t/MPa$^{1/2}$)	Component cohesion parameters (MPa$^{1/2}$)					Hansen parameters (MPa$^{1/2}$)			Burrell hydrogen bonding class	Molar volume (cm^3 mol^{-1})
		δ_d	δ_o	δ_i	δ_a	δ_b	δ_d	δ_p	δ_h		
Pentane	14.5	14.5	0	0	0	0	14.5	0.0	0.0	Poor	115
Hexane	14.9	14.9	0	0	0	0	14.9	0.0	0.0	Poor	131
Diethyl ether	15.3	13.7	5	1	0	6	14.5	2.9	5.1	Moderate	105
Cyclohexane	16.8	16.8	0	0	0	0	16.8	0.0	0.2	Poor	108
Ethyl acetate	18.2	14.3	8	2	0	6	15.8	5.3	7.2	Moderate	98
Toluene	18.2	18.2	0	0	0	1	18.0	1.4	2.0	Poor	107
Tetrahydrofuran	18.6	15.5	7	2	0	8	16.8	5.7	8.0	Moderate	82
Benzene	18.8	18.8	0	0	0	1	18.4	0.0	2.0	Poor	89
Acetone	19.6	13.9	10	3	0	6	15.5	10.4	7.0	Moderate	74
Chlorobenzene	19.8	18.8	4	0.6	0	2	19.0	4.3	2.0	Poor	102
Bromobenzene	20.2	19.6	3	0.4	0	2	20.5	5.5	4.1	Poor	105
1,4-Dioxane	20.7	16.0	11	2	0	2	19.0	1.8	7.4	Moderate	86
Pyridine	21.7	18.4	8	2	0	10	19.0	8.8	5.9	Strong	81
Acetophenone	21.7	19.6	6	1	0	7	19.6	8.6	3.7	Moderate	117
Benzonitrile	21.9	18.8	7	2	0	5	17.4	9.0	3.3	Poor	103
Propionitrile	22.1	14.1	14	4	0	4	15.3	14.3	5.5	Poor	71
Quinoline	22.1	21.1	4	0.6	0	9	19.4	7.0	7.6	Strong	118
N,N-Dimethylacetamide	22.1	16.8	10	3	0	9	16.8	11.5	10.2	Moderate	92
Nitroethane	22.5	14.9	12	5	0	2	16.0	15.5	4.5	Poor	71
Nitrobenzene	22.7	19.4	7	2	0	2	20.0	8.6	4.1	Poor	103
N,N-Dimethylformamide	24.1	16.2	13	5	0	9	17.4	13.7	11.3	Moderate	77
Dimethylsulfoxide	24.5	17.2	13	4	0	11	18.4	16.4	10.2	Moderate	71
Acetonitrile	24.7	13.3	17	6	0	8	15.3	18.0	6.1	Poor	53
Nitromethane	26.4	14.9	17	6	0	3	15.8	18.8	5.1	Poor	54
Water (see text)	48	13	31	21	34	22	16	16	42	Strong	18

polymers. For solid materials, in general only Hildebrand parameters are available (Table IV).

VII. CURRENT STATUS

Theoretical development of this topic appears to have reached something of a plateau, but there is an increasing number of practical applications of cohesion parameters and of computational methods simplifying the process.

An excellent example of the determination and application of Hildebrand parameters to a "new" solvent and its compatibility with polymers is provided by 1,8-cineole. This compound, present at high levels in the leaf oils of some eucalypts, is proposed as a replacement for the solvent 1,1,1-trichloroethane (which is now known to cause

TABLE II Typical Hildebrand Parameter and Hansen Parameter Values for Polymers, δ (MPa$^{1/2}$)

Polymer (manufacturer)	Hansen parameter				Hildebrand parameter ranges in liquids of Burrell hydrogen bonding class		
	δ_d	δ_p	δ_h	^jR	Poor	Moderate	Strong
Pentalyn® 255 alcohol-soluble resin (Hercules)	17.6	9.4	14.3	10.6	18–21	15–22	21–30
Pentalyn® 830 alcohol-soluble rosin resin (Hercules)	20.0	5.8	10.9	11.7	17–19	16–22	19–23
Cellulose nitrate, 0.5 sec	15.4	14.7	8.8	11.5	23–26	16–30	26–30
Cellolyn® 102 pentaerythritol ester of rosin, modified (Hercules)	21.7	0.9	8.5	15.8	16–21	17–22	21–24
Versamid® 930 thermoplastic polyamide (General Mills)	17.4	−1.9	14.9	9.6	—	—	19–23
Poly(methyl methacrylate) (Rohm and Haas)	18.6	10.5	7.5	8.6	18–26	17–27	—

TABLE III Preferred Hildebrand Parameter Values for Selected Polymers

Polymer	δ/MPa$^{1/2}$
Polyacrylonitrile	26
Polybutadiene	17.0
Poly(butyl acrylate)	18.5
Cellulose acetate	24
Cellulose nitrate	21
Polychloroprene	18.5
Poly(dimethylsiloxane)	15.5
Ethyl cellulose	20
Polyethylene	17.0
Poly(ethylene oxide)	24
Poly(ethyl methacrylate)	18.5
Polyisobutylene	16.5
Polyisoprene, natural rubber	17.0
Poly(methyl acrylate)	20.5
Poly(methyl methacrylate)	19.0
Polypropylene	16.5
Polystyrene	18.5
Poly(tetrafluoroethylene)	13
Poly(vinyl acetate)	20
Poly(vinyl chloride)	19.5

stratospheric ozone depletion). On the basis of calculations such as those described in Section V, a Hildebrand parameter of 18 MPa$^{1/2}$ for cineole was deduced, which is within the range of values suggested by the polymer solubilities. This is close to the value for trichloroethane (17 MPa$^{1/2}$), successfully predicting the efficacy of cineole as a replacement solvent.

Further, in developing a new solvent or solvent blend, it is also necessary to determine what polymers are likely to be affected adversely if exposed to the liquid or vapor. Table V shows that those polymers having Hildebrand parameters within 1 MPa$^{1/2}$ unit of the cineole value are soluble, and within 3 MPa$^{1/2}$ units there can be significant swelling, which is useful as an initial guide. However, the Hildebrand parameters of polymers (such as polyethylene) showing good resistance to cineole, despite having similar Hildebrand parameter values, demonstrate the limitations of such predictions.

A demonstration of the great variety and extent of applications of cohesion parameters is provided by the results of an internet search using Alta Vista, <http://www.altavista.com/cgi-bin/query?pg=aq>, conducted in June 1999 for the expression ("solubility parameter*" or "cohesion parameter*" or "Hildebrand parameter*" or "Hansen parameter*"), yielding 460 hits.

The information resulting from this search has now been organized and collected at the two sites, <http://www.mallee.com/parameters.html> and <http://wwwchem.murdoch.edu.au/staff/barton/parameters.html>, with convenient hyperlinks to most of the sites found in the search.

Motivation for providing free internet information of the kind seen in these sites is determined by commercial considerations through the opportunity for attracting potential clients:

- Computer modeling and simulation software for chemical systems often include the ability to estimate cohesion parameters, a major growth area.
- Chemical manufactures incorporate cohesion parameter values (of either the Hildebrand or Hansen variety) in material safety data sheets.
- A few educational institutions provide cohesion parameter information as a component of chemical or polymer science, for example, University of Missouri—Rolla <http://www.umr.edu/~wlf/> and discussion lists such as <http://www.hmco.com/college/chemistry/resourcesite/digests/chemedl/cedjun96/ msg00149.htm>.
- Some publishers and conference organizers providing titles or abstracts of papers to be presented or published include cohesion parameter topics, but full texts are rarely available.

TABLE IV Typical Hildebrand Parameter Values for Predominantly Covalent Crystalline Solids (Assumed to be Subcooled Liquids)

Solid	δ/MPa$^{1/12}$
Alcohol: 1-hexadecanol	20
Aliphatic acids	18–22
Amines, anilines, amides	20–30
Aromatic hydrocarbons	20–22
Barbituric acid derivatives	23–28
Benzoic acid, substituted benzoic acids	23–29
Cholesterol	19
Choleseryl esters	15–19
Cortisone and related compounds	27–30
Halogen compounds of Sn, As, Sb, Bi	23–30
Iodine	29
Lipids	18–27
Metal soaps	18–19
Methyl xanthines, including caffeine	24–29
Norethindrone and derivatives	20–22
Phenols, including antioxidants and nitrophenols	19–22
Phosphorus	27
Sulfonamides	25–30
Sulfur	26
Testosterone and derivatives	19–20

TABLE V Relative Resistance of Polymers to 1,8-Cineole ($\delta = 18$ MPa$^{1/2}$)

Effect of cineole on polymer (4-month continuous exposure)	Polymer	δ (MPa$^{1/12}$)
Soluble	Natural rubber	17.0
	Polystyrene	18.5
	Styrene-butadiene elastomer	—
Strongly swollen (>100%)	Neoprene	18.5
	Polyurethane rubber	20–21
	Silicone rubber	15.5
Little swelling (<10%)	Acrylic-styrene-acrylonitrile	~22
	Polyester urethane	~20
	Polyester urethane	~20
	Polycarbonate/acrylic-styrene-acrylonitrile	—
	Fluoro elastomer	—
Resistant	Acrylonitrile-butadiene-styrene	—
	Acetal	21–22
	Nylon-6	24–32
	Nylon-6,6	28–32
	Polyethylene	17
	Poly(butylene terephthalate)	21–22
	Poly(ethylene terephthalate)	21–22
	Poly(methyl methacrylate)	20.5
	Polypropylene	16.5
	Polycarbonate	19–22
	Styrene-acrylonitrile	—
	Poly(tetrafluoroethylene)	13
	Poly(vinyl chloride)	19.5

- Scientific and technical consultancies use cohesion parameters to promote their services.
- Institutions and individuals include reference to their publications or research activities in resumes and bibliographies, such as the University of Geneva, <http://www.unige.ch/sciences/pharm/fagal/barra~4.html#pubi>, and Princeton University, <http://www.princeton.edu/~chemical/faculty/pubs.html>.

As with most other areas of the physical sciences, the amount of internet-accessible material describing the **theoretical background** is limited. Probably the most comprehensive source is that written by John Burke (Oakland Museum of California), <http://sul-server-2.stanford.edu/byauth/burke/solpar/>.

For values of individual **liquids**, by far the greatest amount of quantitative information is provided in the pages of Charles Tennant & Co, <http://www.ctennant.co.uk/tenn04.htm>, with smaller numbers of compounds be-

ing listed by companies such as Shell, DuPont, Monsanto, Aeropres, Lambiotte, and Eastman.

A smaller but increasing number of sites deal with **compressed gases and supercritical fluids**, for example, *Polymerizations in Supercritical CO$_2$* from the University of Groningen, <http://polysg2.chem.rug.nl/>.

Solvent applications of fluids have always been an important use of cohesion parameters, such as *A Review of Supercritical Carbon Dioxide Extraction of Natural Products from Engineering World*, <a href="http://www.exicom.org/cew/oct97/awasthi.htm">.

Chemical properties, as well as physical properties, are now being correlated by means of cohesion parameters, for example, *Solution Effects on Cesium Complexation with Calixarene—Crown Ethers from Liquid to Supercritical Fluids* (University of Idaho), <http://www.doe.gov/em52/65351.html>.

Liquid crystal applications are still poorly represented, and studies of **solids** are mostly limited to **pharmaceuticals**, such as *Partial-Solubility Parameters of Naproxen and Sodium Diclofenac* from the *Journal of Pharmacy and Pharmacology*, <http://dialspace.dial.pipex.com/town/avenue/ax60/j98020.htm>.

As in the printed literature, applications to **polymers** dominate internet cohesion parameter sites, for example, American Polywater Corporation on polycarbonate stress cracking, <http://www.polywater.com/cracking.html>; as well as Millipore, <http://millispider.millipore.com/micro/mieliq/MAL104.htm>; and *Evaluating Environmental Stress Cracking of Medical Plastics* from Medical Device Link, <http://www.devicelink.com/mpb/archive/98/05/001.html>.

Polymer design is represented by IF/Prolog, <http://www.ifcomputer.com/Products/IFProlog/Applications/PolymerDesign/home_en.html>.

Surfaces receive less attention than homogeneous fluid and solid phases, although coatings and adhesives are important, with typical sites being University of Missouri—Rolla, <http://www.umr.edu/~wlf/Adhesion/young.html>, and Shell Chemicals, <http://www2.shell-chemical.com/CMM/WEB/GlobChem.NSF/Searchv/SC:2023-94?OpenDocument>.

Chromatographic properties, such as *Relationship Between Retention Behavior of Substituted Benzene Derivatives and Properties of the Mobile Phase in RPLC* in the *Journal of Chromatographic Science*, <http://www.j-chrom-sci.com/353sun.htm>, provide additional possibilities for applications of cohesion parameters to complex systems.

Nanoparticles are beginning to be discussed in terms of cohesion parameters, for example, on the formation of gliadin nanoparticles: the influence of the solubility parameter of the protein solvent, <http://link.springerny

.com/link/service/journals/00396/bibs/8276004/ 82760321.htm>, but **dyes** are surprisingly under-represented.

Among **environmental** applications, the area of materials substitution is the most potentially productive use of cohesion parameters, for example, *Toluene Replacement in Solvent Borne Pressure Sensitive Adhesive Formulations* (Shell Chemicals), <http://www2.shellchemical.com/CMM/WEB/GlobChem.NSF/Searchv/ SC:2023-94? OpenDocument>; *Solvents—the Alternatives* from the Waste Reduction Resource Center, Raleigh, NC, <http://www.p2pays.org/ref/01/00023.htm/ index.htm>; and *Solvent Substitution Data Systems* (U.S. EPA), <http://es.epa.gov/ssds/ssds.html>.

Database and compilations include ThermoDex, <http://thermodex.lib.utexas.edu/search.jsp>, and Infochem, <http://www.infochem.demon.co.uk/data.htm>.

Examples of **modeling and simulation sites** are *SciPolymer* from ScienceServe, <http://www.scienceserve. com/wwwscivision/scipolymer/prediciti.htm>; *Molecular Analysis Pro*™, <http://www.sge.com/software/soft/ 13051.htm>; *The Askadskii Approach* from Million-Zillion Software, <http://www.millionzillion.com/cheops/ metho1/askadskii.htm>; *Cerius2* from Molecular Simulations Inc., <http://www.man.poznan.pl/software/msi-doc/cerius38/TutMats/tut_synthia.doc.htm1#740585>; <http://www.msi.com/solutions/polymers/misc.html>; and *ProCAMD* from Capec, <http://www.capec.kt.dtu. dk/capec/docs/software/procamd/procamd.html>.

QSAR (quantitative structure activity relationships) and QSPR (quantitative structure property relationships), which study relationships between useful chemical and physical attributes and the molecular properties, are ideal candidates for this kind of approach, for exam-ple, Molecular Analysis Pro™, <http://www.chemsw. com/13051.htm>.

While internet sources do not provide the comprehensive and integrated information available in the *Handbook of Solubility Parameters and Other Cohension Parameters* (Barton, 1992) or even the convenient summary in the *Encyclopedia of Physical Science and Technology*, they should not be overlooked for information on recent developments.

SEE ALSO THE FOLLOWING ARTICLES

BONDING AND STRUCTURE IN SOLIDS • GAS CHROMATOGRAPHY • HYDROGEN BONDS • LIQUIDS, STRUCTURE AND DYNAMICS • MOLECULAR HYDRODYNAMICS • SURFACE CHEMISTRY

BIBLIOGRAPHY

Barton, A. F. M. (1983). In "Polymer Yearbook" (H. G. Elias, and R. A. Pethrick, eds.), p. 149, Harwood, Chur, Switzerland.

Barton, A. F. M. (1992). "Handbook of Solubility Parameters and Other Cohesion Parameters," 2nd ed., CRC Press, Boca Raton, FL.

Barton, A. F. M. (1990). "Handbook of Polymer-Liquid Interaction Parameters and Solubility Parameters," CRC Press, Boca Raton, FL.

Barton, A. F. M., and Knight, A. R. (1996). *J. Chem. Soc. Faraday Trans.* **92,** 753.

Burrell, H. (1975). In "Polymer Yearbook" (J. Brandrup, and G. H. Immergut, eds.), 2nd ed., IV-337, Wiley (Interscience), New York.

Hansen, C. M. (1969). *Ind. Eng. Chem. Prod. Res. Dev.* **8,** 2.

Hansen, C. M., and Beerbower, A. (1971). In "Kirk-Othmer Encyclopedia of Chemical Technology," 2nd ed. (A. Standen, ed.), Suppl. Vol., p. 889, Wiley (Interscience), New York.

Hoy, K. L. (1970). *J. Paint Technol.* **42,** 76.

Karger, B. L., Snyder, L. R., and Eon, C. (1978). *Anal. Chem.* **50,** 2126.

Rowes, R. (1985). "Chemists propose universal solubility equation," *Chem. Eng. News* (March 18), 20.

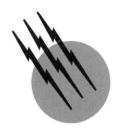

Collider Detectors for Multi-TeV Particles

C. W. Fabjan
CERN

GLOSSARY

Accelerator, circular A machine which increases the kinetic energy of particles (e.g., electrons, protons, and their antiparticles); magnetic fields are used to guide them on a circular path many times through the same accelerating system.

Accelerator, linear A machine which accelerates particles in a straight line.

Calorimeter A particle detector which measures the energy of an elementary particle by absorbing the particle and converting its energy into a measurable signal.

Collider, circular A machine in which two circular accelerators are combined to accelerate and store beams moving in opposite directions. It is used to produce very high energy phenomena in the collisions between particles in the beams.

Collider, linear A machine consisting of two linear accelerators which accelerate beams in opposite directions.

Collider detector (CD) A complex instrument used to measure the momenta and energies of the particles produced in a collider collision. Layers of detectors dedicated to specific measurement tasks surround the collision point.

Electron volt (eV) Unit of energy used in high-energy and accelerator physics. It is defined as the kinetic energy imparted to an electron passing through a potential difference of 1 V.

Hadron electron ring accelerator (HERA) An electron–proton collider, 6.3 km in circumference, at Deutsches Elektronen-Synchrotron (DESY) in Germany.

Higgs particle Putative particle introduced to describe one mechanism by which elementary particles acquire masses in their interaction with the all-pervasive Higgs field.

Large electron–positron (LEP) collider The world's largest particle accelerator, 27 km in circumference, at the European Organization for Nuclear Research, or CERN.

Large hadron collider (LHC) A particle accelerator under construction in the tunnel of the LEP machine at

Encyclopedia of Physical Science and Technology, Third Edition, Volume 3

CERN. The LHC will collide proton beams with a combined energy of 14 TeV.

Relativistic heavy-ion collider (RHIC) A circular particle accelerator, 3.8 km in circumference, at the U.S. Department of Energy's Brookhaven National Laboratory. It provides collisions between protons and between ions as heavy as gold.

Standard model (SM) A model that describes our present understanding of nature's electroweak and strong forces. It is beautiful, in agreement with almost all observations, but not the ultimate truth. Arguably, new physics horizons will be opened when the TeV frontier is reached.

Tevatron A particle accelerator at the U.S. Department of Energy's Fermi National Accelerator Laboratory (Fermilab), 6.3 km in circumference, which can accelerate protons to energies of almost 1 TeV. It is also used as a proton–antiproton collider.

Tracking detectors Instruments which measure the trajectories of charged particles with a spatial resolution ranging from 0.01 to 1 mm. Typically, these detectors consist of thin layers of noble-gas mixtures or silicon, in which thin wires or metal electrodes collect the ionization produced by the passage of the particle signaling the particle position.

COLLIDER DETECTORS for multi-TeV particles are complex instruments that detect and measure simultaneously the parameters of thousands of particles in a detector volume of 10,000–20,000 m^3. This splash of particles is produced in ultra-energetic collisions between two protons in counterrotating beams in a particle collider. The next generation of these instruments, at present under construction, aims at the study of physics phenomena at energies of 10^{11} to 10^{12} eV or more, such as the mechanism generating the masses of fundamental particles, the concept unifying the fundamental forces, or the process through which the matter–antimatter symmetry was violated during the first microsecond after the big bang. These instruments are composed of several layers of detectors, each with a specific measurement task, and are subdivided into $\sim 10^8$ detection cells. Collisions exhibiting novel physics signals are expected to occur only rarely, typically at the rate of 1 in 10^{10} collisions or less; therefore, these instruments also push the data rates to limits exceeding 10^{15} bits/sec. Worldwide collaborations of thousands of physicists and engineers are building research facilities, in concert with industry, to develop these instruments.

I. OVERVIEW

Exploring the structure of matter at the microscopic scale, exposing the laws of physics which shape the evolution of matter in our universe, grasping the forces which hold particles together: these are some of the aims of elementary particle physics. Several tools are used in this research, but foremost among them are particle accelerators called colliders and their associated experimental apparatus (collider detectors) which open the way into this invisible and microscopic world.

This overview will set the stage: The collider machines are being developed to create this new (physics) world; collider detectors (CDs) live with them in a symbiotic relationship. They "see" the new physics, capturing the fleeting signals and transforming them into bytes—terabytes—of information. Thousands of scientists around the globe are decoding this data in search of deeper or new insights into our physics understanding.

The "actors" in a CD will be a number of different detectors, each playing a specialized measurement role. In Section II we will highlight features of modern instrumentation, which will set the foundation for designing a collider detector (Section III). Finally, in Section IV we will turn the blueprint into physics and engineering reality.

A. What Is a Collider?

A collider accelerates and stores beams of elementary particles (e.g., protons or electrons). Two beams, traveling at relativistic speed close to the speed of light in opposite directions, are guided to intersect inside the center of the CD, and particles will, with a certain probability, collide. In such an interaction the energy carried by the particles will be concentrated in a tiny volume, which will create a state of very high energy density. This state will materialize into a variety of particles (pions, kaons, protons, etc.) to be captured and identified in the CD. Occasionally, these signals may reflect a hitherto unknown state of matter or a novel fundamental particle.

In such a collision the particles also "feel" each other intimately and sense each other's internal structure. Much as an electron microscope reveals the structure of matter at the 10^{-8}-cm scale, the most energetic particle colliders probe matter at the 10^{-17}-cm scale.

In short, the higher the energy of the colliding beams, the higher the microscopic resolution, the energy density, and the sensitivity for signals of new physics. The energy of these particle beams is measured in electron volts (eV), with present colliders operating between giga electron volts (1 GeV = 10^9 eV) and tera electron volts (1 TeV = 10^{12} eV). In today's colliders different stable particles—electron–positron (e^-e^+), proton–proton (pp), proton–antiproton (p$\bar{\text{p}}$), electron–proton (ep), ion–ion—are being used, the choice depending foremost on the research emphasis. Equally decisive, however, is the technology available to accelerate and store the particles.

B. Why Colliders?

In the collision of two particles of energies (momenta, masses) $E_1(\bar{p}_1, m_1)$ and $E_2(\bar{p}_2, m_2)$ the total center-of-mass energy (E_{CMS}) can be expressed in the form

$$E_{CMS} = \left[(E_1 + E_2)^2 - (\bar{p}_1 + \bar{p}_2)^2\right]^{1/2}.$$

For two relativistic, counterrotating beams colliding head-on,

$$E_{CMS} \approx [2E_1 \cdot E_2 \cdot (1 + \beta_1\beta_2)]^{1/2}$$

or

$$E_{CMS} \approx 2E \quad \text{for} \quad E_1 = E_2, \beta_1 = \beta_2 \approx 1.$$

Contrast this result with the energy E_{CMS} made available in the collision of an accelerated beam of particles, energy E_1, colliding with nucleons, mass m_N, on a stationary target ($E_2 = m_N, \beta_2 = 0$):

$$E_{CMS} \approx \left[2E_1 \cdot m_N c^2\right]^{1/2} \approx [2E_1(\text{GeV})]^{1/2},$$

with $m_N c^2 \approx 1$ GeV. It would need an energy $E_1 = 2E^2/m_N c^2$ to match the E_{CMS} achievable in a collider with two beams of energy E. As an example, the 27-km-long large hadron collider (LHC) ring would turn into an impossibly large 200,000-km accelerator ring.

Obviously, if energy is at a premium, one should build a collider. These machines are honed to extend the energy frontier for physics research; see Table I. A few explanations will help an individual to understand the shorthand of this table. Perhaps the most striking feature is the difference in E_{CMS} between the e^+e^- collider, or large electron–positron (LEP) collider, and $p\bar{p}$ collider (Tevatron), and the future LHC. Although the LEP collider ring has a circumference of 27 km and a bending radius $r = 3100$ m, the synchrotron radiation loss dW/dt,

$$dW/dt \sim (E/mc^2)^4/r^2,$$

is so high that for electrons, with their small mass m, it is unrealistic to go beyond the LEP collider energy with circular machines. Today, higher energies can be reached only with a $pp(p\bar{p})$ machine for which synchrotron losses are not significant. The LHC, under construction in the LEP collider tunnel, will be able to handle beams 70 times the LEP collider energy.

Dramatically new ways are being pursued to prepare colliders with energies beyond those of the LEP collider and the LHC. Any future energy increase in e^+e^- machines will require novel linear collider geometries. The pp route implies very high field superconducting magnet technologies and very large (100-km) circumference rings. A remarkable compromise between these two extremes could be a $\mu^+\mu^-$ collider, which promises the

TABLE I Today's Energy Frontier of Collider Detectors[a]

	BNL RHIC	CERN LEP collider	CERN LHC	DESY HERA	Fermiab Tevatron 1	Fermiab Tevatron 2
Colliding particles	pp Au–Au	e^+e^-	pp Pb–Pb	ep	$p\bar{p}$	$p\bar{p}$
Beam energy (GeV)	250 (proton) 100/nucleon (Au)	≤104	7,000 (proton) 2,760/nucleon (Pb)	28 + 920	900	1000
E_{CMS} (GeV)	500 (pp) 200 (NN)	≤208	14,000 (pp) 5,520 (NN)	320	1800	2000
Collision rate (Hz)	$<10^6$ (p) $\sim10^3$ (Au)	<10	10^9 (p) 10^4 (Pb)	Few 10^2	$\sim10^6$	$\sim10^7$
Discovery reach, (GeV)	10 GeV/fm^3 in $\sim10^3$ fm^3	208	5,000	100	<300	~500
Major physics emphasis (examples)	Quark–gluon plasma	Standard model (SM)	Origin of particle masses	Proton structure	SM top quark	Top quark
		Higgs search	Physics beyond the SM	Particle searches		Search for symmetry breaking

[a] BNL, the U.S. Department of Energy's Brookhaven National Laboratory; RHIC, relativistic heavy-ion collider; CERN, the European Organization for Nuclear Research; LEP, large electron–positron; LHC, large hadron collider; DESY, the German laboratory Deutsches Elektronen-Synchrotron; HERA, hadron electron ring accelerator; Fermilab, the U.S. Department of Energy's Fermi National Accelerator Laboratory; NN, nucleon–nucleon.

attractiveness of a circular e^+e^- machine without its bane of synchrotron radiation.

Table I also shows that pp machines can be used to collide fully ionized nuclei, up to Pb^{92+}. A dedicated nuclear collider, the relativistic heavy-ion collider (RHIC), has been constructed to study nuclear matter under extreme conditions (e.g., in the quark–gluon plasma phase). Starting in 2006 this program will also be pursued at the LHC.

C. Collider Detectors

Table I also hints at an intimate relation between the type of collider and its collision rate, which determines the key features of the CD.

In every e^+e^- annihilation all the energy is converted into the physics state under study. For this reason the discovery reach for new particles of mass M approaches approximately $M \approx E_{CMS}$, with rather clean signals and little disturbing background.

The contrast to pp(pp̄) machines is striking. Protons are made of three quarks, held together with gluons, which share in a fluctuating way the energy of the proton. These pp(pp̄) colliders are actually quark–quark (quark–antiquark) and gluon–gluon colliders at reduced energy. In most cases the quarks collide at glancing angles, transferring little energy. Only rarely, typically in 1 collision in 10^6 or less, is there a head-on encounter with most of the energy of both particles made available. In pp collisions, the effective energy reach for discovery is therefore much smaller than E_{CMS}. Furthermore, most of the time much energy is squandered to produce the familiar collision products, which is a big nuisance. The new, exciting physics happens excruciatingly rarely. Some of the "most wanted" new particles might be discovered only at the rate of 1 in 10^{14}.

In these colliders the particles in the beams are grouped into bunches: at the LHC, every 25 nsec, pairs of intense proton bunches will sweep through each other; this produces typically 20 collisions and some 3000 particles.

For these reasons CDs at pp colliders, such as the LHC, must function at very high collision rates, must select the rare interesting physics phenomena with lightning speed, and need to measure them in the presence of thousands of background particles. These are the challenges for the new generation of TeV collider detectors under construction for the LHC, as will be described in the following sections.

II. COLLIDER DETECTOR COMPONENTS: A PRIMER

The particles created in the wake of a collision are kinematically described by their momentum $\bar{p}$ and energy E.

While in principle a measurement of $\bar{p}$ and E of all the particles produced characterizes the original state, sometimes more specific quantities, such as the identity or mass of a particle, are measured with specialized instrumentation.

We present the principal detector components used for these measurements and argue that the laws of physics shape the basic configuration and dimensions of a CD, conceptually shown in Fig. 1.

How are momenta and energies measured?

A. Momentum Measurement of Charged Particles

For the momentum measurement there is only one recipe: immerse the volume around the collision in a magnetic field, and instrument this volume with tracking detectors, such that the curved particle trajectory may be reconstructed and hence its momentum inferred. The minimal requirements for such a measurement are as follows:

- At least three position measurements must be made along the trajectory to deduce the curvature.
- Tracking detectors which disturb the trajectory ever so gently must be used to minimize instrumental "blurring" of the trajectory.

Given the radius of curvature r [m] of a particle with momentum p [GeV/c] and unit charge in a magnetic field B [T], with

$$r \approx 10p/3B,$$

one derives an approximate expression for the momentum accuracy $\delta p/p$ of such a detector:

$$\frac{\delta p}{p} = \frac{\sigma}{\sqrt{N_m}} \cdot \frac{80p}{3L^2B}, \qquad (1)$$

where σ [m] is the accuracy in the position measurement of the tracking detector and N_m counts the number of measurements along the track. Including the spectrometer length L [m] and the magnetic field strength B [T], there are four parameters available for the spectrometer optimization.

There are two main classes of tracking detectors used in CDs. They are discussed next.

1. Solid-State Detectors

Solid-state detectors (see Fig. 2) provide a signal by collecting the charge liberated in the passage of the particle through a semiconductor. Suitably implanted electrodes, typically strips, apply an electric drift field, in which the ionization charges are collected and induce a detectable signal. The workhorse for this type of detector is silicon

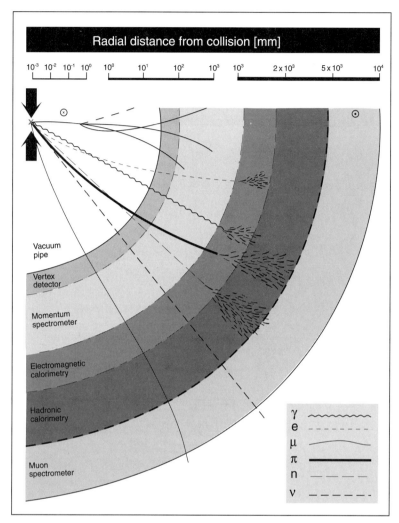

FIGURE 1 Cross section through a conceptual configuration of a collider detector. Like nested Russian dolls, a sequence of detectors enclose the collision point. In each detector layer, the particle is subjected to a specific measurement, shedding, layer by layer, its physics information. Note the three logarithmic scales of the radial dimension.

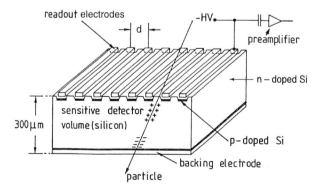

FIGURE 2 Schematic view of a silicon microstrip detector (not to scale). The charges liberated in the passage of the charged particle are collected at the readout electrodes, from which the position is derived. The distance (*d*) between strips is typically in the range of 20–50 μm. HV.

(Si), which has profited from the technological advances of the computer-chip industry.

It is easy to estimate the strength of the signal. In Si, an energy of $\epsilon \sim 3.6$ eV is required to move an electron into the conduction band. Industry can process, relatively easily, Si disks 300 μm thick. With an average energy loss of ~ 4 MeV/cm, about 30,000 electron–hole pairs are created, which is large compared with the effective thermal noise $N_e \sim 1000$ e in the preamplifier, attached to electrodes. If each electrode is connected to an electronic channel to register the signal, the spatial resolution σ is comparable to the strip spacing d: $\sigma = d/\sqrt{12}$.

Typical sizes of such detectors range from 5×5 to $\sim 10 \times 10$ cm^2. A modern development is the pixel detector, made possible through the prowess of the electronics industry. A checkerboard of electrodes is used, with

a)

b)

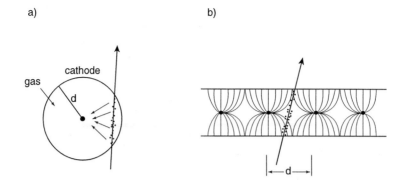

c)

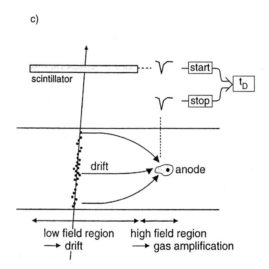

FIGURE 3 Three generations of wire-chamber tracking detectors: (a) proportional tube (also called, when operated at very high gas amplification, the Geiger-Müller counter); (b) multiwire proportional chamber. Registration of the ionization electrons measures the track with a resolution $\sigma = d/\sqrt{12}$. Measuring the drift time interval t_D improves the spatial accuracy; typical values in a drift chamber (c) are $\sigma \sim 100 \ \mu$m.

dimensions of $\sim$50 × (few) hundred square microns, allowing a two-dimensional measurement capability. Modern versions have the readout electronics piggybacked to and matching the size of the tiny electrodes. Such detectors allow tracks to be followed with a precision of $\leq$20 μm even in collisions with hundreds or thousands of tracks. They are the key component to identifying particles which decay after a few tens or hundreds of microns, such as the enigmatic beauty particles.

2. Gaseous Detectors

The second category of tracking detector works on ionization produced in layers of suitable gas mixtures. Again,

electrodes collect the charge for further signal processing. The primary signal is very feeble: in 1 cm of argon at atmospheric pressure, a favorite detector gas, only about 100 electron–ion pairs are created, which is not enough for convenient processing. However, a clever trick is used: in sufficiently strong electric fields, the ionization electrons can gain enough energy on their migration to ionize in turn the gas, which initiates an avalanche of electron–ion pairs. Physicists have found gas mixtures in which this avalanche amplification can, in a controlled way, reach values up to 10^6, which makes the signal processing straightforward. The ancestor of such devices is the Geiger-Müller counter (Fig. 3a). A thin wire, d 50 μm, is the collecting anode. Thanks to the characteristic cylindrical electric field, $E \sim V_o/r$, conditions for avalanche amplification can be adjusted to occur within a few microns of the wire surface.

Experimentation in particle physics was revolutionized in the late 1960s with the advent of the multiwire proportional chamber (MWPC; see Fig. 3b), a deceptively simple extension of the proportional tube. Simple, robust, and cheap, it allows large detector areas to be instrumented with high-quality tracking detectors and tracking information to be processed at megahertz event rates.

The drift chamber (Fig. 3c) is an ingenious derivative. A measurement of the drift time interval t_D of the ionization electrons between creation and arrival at the anode wire gives the distance $d = v_D t_D$, if the drift velocity v_D is known. In suitable gas mixtures typical velocities are $v_D \sim 1$–5 cm/μsec. The achievable spatial resolution is ~ 100 μm.

The invention of the MWPC and the drift chamber was the start of an explosive development of gaseous wire chambers: it is relatively easy to shape electric fields and cajole the electrons to drift in line with the demands of physicists. The pinnacle in purity and elegance is the time projection chamber (Fig. 4). A gas-filled cylinder, with a central electrode, drifts the electrons to and projects them onto the detector disks at the end, which are wire chambers. They provide a two-dimensional view of the projected tracks, with the drift time giving the third dimension. It is an ideal imaging tracker for high-multiplicity topologies occurring at low event rates. This technique was optimal for and reached its apogee in CDs at the LEP collider. It is also the technique of choice for tracking in experiments at ion colliders (RHIC, LHC heavy ions).

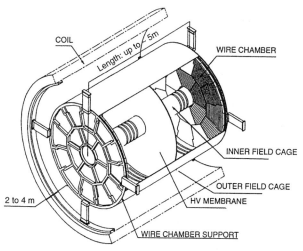

FIGURE 4 Schematic view of a time projection chamber. The large cylinder acts as a drift chamber, projecting the track ionization onto the wire chambers at the two end faces.

B. Energy Measurement

Energy measurement of particles is a second, complementary technique. In contrast to the gentle interactions in tracking detectors, the energy measurement relies on complete absorption of the particles. Massive detectors are instrumented, in which the particles interact either electromagnetically or by means of the strong (nuclear) interaction. Sufficiently energetic particles (E 100 MeV) produce a cascade of secondary particles with increasingly lower energy. These instruments are called calorimeters in analogy to the instruments which measure the total absorption of mechanical or chemical energy through a temperature rise. For calorimeters employed in CDs, an intermediate step in the absorption process—ionization or excitation of the detector molecules—is used to derive a conveniently measurable signal. On average this signal S is proportional to the number of cascade particles N produced, which in turn is proportional to the incident particle's energy, $S = aN = bE$. In individual measurements, the number N fluctuates because the cascade is a statistical sequence of approximately independent collisions. The fluctuations ΔN of cascade particles, $\Delta N \sim \sqrt{N}$, will generally determine the ultimate limit to the accuracy (resolution) of the energy measurement. The relative resolution,

$$\Delta E/E \sim \frac{\Delta N}{N} \sim \frac{\sqrt{N}}{N} \sim 1/\sqrt{N} \sim 1/\sqrt{E},$$

improves with increasing energy, which makes calorimetry more precise than momentum spectroscopy for energies around and beyond ~ 100 GeV. A second attractive and distinctive feature is the capability of measuring neutral particles (high-energy gammas, neutrons, etc.). This is a further reason why calorimetry is essential in modern particle physics, where increasingly we need to obtain rather complete information about a collision event. Two categories of calorimeters are being used. They are described next.

1. Electromagnetic Calorimeters

For the energy measurement of electrons, positrons, and gammas the absorption proceeds essentially by means of the electromagnetic interaction. Bremsstrahlung of photons from electrons and positrons, and e^+e^- production in the interaction of gammas in the Coulomb field of the nuclei of the detector material are the principal processes driving and propagating the electromagnetic (e.m.) cascade. These instruments are therefore called electromagnetic calorimeters. A pictorial view of such a particle shower is shown in Fig. 5.

The correlation between the measured energy and the number of cascade electrons and positrons is very tight,

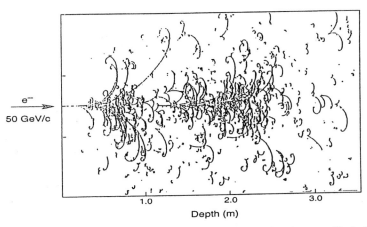

e⁻
50 GeV/c

1.0   2.0   3.0

Depth (m)

FIGURE 5 A bubble-chamber picture of an electromagnetic shower. The chamber was filled with a liquid Ne–H$_2$ mixture, which has a long characteristic absorption length for electrons. It was immersed in a 3-T magnetic field. The curved lines are the images of electron and positron tracks in the electromagnetic cascade. In collider detectors, materials for which this absorption length is at least 10 times shorter are chosen.

which results in a potentially very high quality energy measurement; the practical limit to the energy resolution in e.m. calorimeters is at the level of $\sigma(E)/E \approx 0.001/\sqrt{E(\text{GeV})}$. That is, for a 1-GeV electron (gamma) the intrinsic resolution can be as good as 1% (i.e., 10 MeV). At 100 GeV it could reach 0.1% (i.e., 100 MeV), although in present-day detectors instrumental effects mask this stupendous precision.

Practical examples of such high-quality instruments are certain types of crystals, in which the e^+e^- cascade excites the crystal molecules with subsequent emission of light flashes. A famous example of such a crystal is bismuth germanate, which was developed for an LEP collider experiment. It has become the detector of choice for positron-emission tomography (PET) used in medical diagnostics.

2. Hadron Calorimeters

Hadron calorimetry is more complex. In the nuclear and strong interactions a very wide spectrum of cascade particles is produced: part of the energy is channeled into energetic hadron production, and part is deviated into nuclear collisions accompanied by soft neutron and photon emission, nuclear breakup, spallation, and so forth. This multitude and complexity of reaction channels severely disturbs the energy–signal correlation and hence the intrinsic resolution is much worse. In the best devices, in which certain tricks are used to improve this correlation, the intrinsic resolution is measured to be at the level of $\sigma(E)/E \sim 0.2/\sqrt{E(\text{GeV})}$, although in many practical detectors the resolution is closer to $\sigma(E)/E \sim 0.5/\sqrt{E(\text{GeV})}$. Nevertheless, at the high-energy frontier where typical energies are well beyond 100 GeV, these

devices have a very respectable accuracy at the percent level.

3. Calorimeter Dimensions

In both types of calorimeters the instrumental dimensions are determined by a characteristic mean free path between the collisions in the cascade. For the e.m. devices this length X_0 (radiation length) is determined by the density of atomic electrons; one finds approximately

$$X_0 \approx 180 A/Z^2 [\text{g cm}^{-2}].$$

For lead, with density $\delta = 11.4$ g cm^{-3}, one obtains $X_0 \approx 5.5$ g cm^{-2}; hence X_0 (Pb) ≈ 0.5 cm. Approximately 50 X_0 (i.e., 25 cm Pb) are enough to absorb $\approx 99\%$ of a 100-GeV particle.

For hadrons, the density of nuclear scatterers determines the nuclear interaction length λ. For Pb again, we find $\lambda(\text{Pb}) = 194$ g cm^{-2} or ~ 17 cm. Because of the higher multiplicity in secondary-particle production, fewer mean free paths are needed for complete absorption. Approximately 10 λ are needed to absorb a 100-GeV hadron: hadron calorimeters therefore typically have a depth of ~ 2 m:

4. Calorimeter Optimization

We have broadly sketched the physics underlying the energy measurement. Practical devices need to be finely instrumented to extract a signal: it may be ionization charge left in a suitable detector medium (e.g., liquid argon), or it may be a light flash in a plastic scintillator or a scintillating crystal. Such instruments may be sensitive in the full detector volume (e.g., a crystal) or divided into separate layers of absorber and instrumentation (sampling calorimeters). It is the choice of this instrumentation which will tailor the calorimeter to a specific measurement requirement. This quest for optimization has led many research groups over the past 20 yr to explore an incredible range of instrumentation possibilities. In particular, the next generation of CDs is characterized by excellent e.m. calorimetry, combined with hadron calorimetry optimized for very good high-energy measurements.

C. Particle Identification and Event Topologies

Inspection of Fig. 1 also indicates that the combined p and E measurements contain a wealth of—for the physics program—essential additional information on the types of particles produced:

1. *Short-lived particles.* Particles containing a charm or a beauty quark are short-lived, as is the heavy tau-lepton: they decay within the beam pipe. Reconstruction of the decay vertices with high-resolution tracking detectors allows this very interesting group of particles to be identified.

2. *Electrons and photons.* These particles deposit their energy in the e.m. calorimeter. A matching track would identify an electron; the neutral photon remains invisible in the tracking detectors.

3. *Hadrons.* Most of the energy is deposited in the hadronic calorimeter. Information on the charge is obtained from the tracker.

4. *Jets.* Free quarks do not exist but manifest themselves as "jets" of hadrons, measured both in the tracker and in the calorimeter.

5. *Muons.* These are as essential as electrons. Because they weigh approximately 200 times more than electrons

they cannot be absorbed in a calorimeter and are therefore identified as charged particles, traversing the detector and leaving only a relatively faint signal of ionization.

6. *Neutrinos.* Although they traverse the detectors, almost never leaving any direct signal, their production can still be inferred. They do carry with them E and $\bar{p}$ and reveal themselves, provided all the other particles are adequately well measured. In practice, the energy component projected on a plane transverse to the collision axis, E_T, reveals most tellingly the missing energy carried away by the neutrino.

D. Performance Limits of Detectors

In Table II, the performance limits of the various detectors are summarized. These limits of individual detector performance set the scale and determine the performance of the CD.

We have gained an understanding about the basic detector components. The physics of the detection processes imposes a natural arrangement and sets the scale for the size of the detector components, as illustrated with a few benchmark numbers:

1. *Momentum measurement:*
   a. The typical magnetic field reached with superconducting magnets is $B \sim 2$ T.
   b. The typical spatial resolution with Si detectors is $\sigma = 20 \ \mu$m.
   c. A 1% precision measurement at 100 GeV/c would require the following:
      (1) Four layers of position measurement with $\sigma = 20 \ \mu$m (alternatively, 100 drift-chamber measurements with $\sigma \approx 100 \ \mu$m).
      (2) Length of spectrometer $L \approx 100$ cm.

TABLE II Performance Limits of Collider Detector Components

Measurement	Detector	Limit to performance	Process limiting performance	Practical performance
Position of charged particles	Silicon-strip detectors	σ 5 μm	Spread of ionization electrons along track	$\sigma = 10$–20 μm
	Drift chamber	σ 50 μm	Diffusion of ionization electrons	$\sigma = 50$–150 μm
			Thermal noise in electronics	
Energy measurement of electrons or photons	Crystals	$\sigma \sim 10$ MeV at 1 GeV	Signal fluctuations Signal sampling	$\sigma = 10$–100 MeV at 1 GeV
	Fine sampling calorimetry	$\sigma \sim 1$ GeV at 100 GeV		$\sigma \approx 1$ GeV at 100 GeV
of hadrons	Sampling calorimeters	$\sigma \sim 3$–5 GeV at 100 GeV	Fluctuations in absorption process	$\sigma \sim 5$ GeV at 100 GeV

2. *Energy measurement* (surrounding the momentum spectrometer, starting with the thin e.m. calorimeter):
   a. For 100-GeV photon or electron absorption, approximately 50 X_0 are needed, which require typically 50 cm.
   b. For hadron containment, $E \sim$ few hundred GeV; 12–14 λ are required. In relatively compact instruments values of $\lambda \sim 20$ cm can be reached: a total of 250–300 cm are required.

3. *Muon measurement* (performed in the last instrumentation layer surrounding the hadron calorimeter; very large magnetic-field volumes are required):
   a. Technologically practical field levels in air are limited to 0.5–1 T; they reach 1.8 T in saturated iron.
   b. Typically 2–5 m of momentum spectroscopy are needed.

In conclusion, it is clear that the laws of physics shape the basic configuration of a collider detector: A "Russian doll" sequence of application-specific detectors enclose the collision point, peeling off the physics information carried by the particle, step-by-step; see once more Fig. 1.

In the next section, we will show how the aims of a specific physics research program and the ingenuity and tastes of the researchers lead to a design of such a research facility, the collider detector.

III. DESIGNING A COLLIDER DETECTOR

A panoply of measurement techniques have been developed to track and capture particles. The laws of physics shape the generic CD. We are now equipped to partake in this exciting intellectual adventure: the creation of a new detector and research facility. Intellect, as well as experience seasoned with emotion, shapes the design, as will be illustrated with the largest effort undertaken to date: the LHC general-purpose CD facilities.

A. Intellect: Design Shaped by Physics Potential

Research since the 1970s, and in particular during the 1990s at the LEP collider, has culminated in and supported the Standard Model of Electroweak and Strong Interactions (SM) as a remarkably good approximation of the world experienced so far. However, the clarity and precision of the results is casting a shadow of new physics not yet observed directly or contained in the SM:

- According to the SM a new, fundamental field (the Higgs field) permeates all space. Particles acquire masses through interaction with this field. If this is true, one particle, the Higgs particle, will be observable at the LHC. Research at the LEP collider has placed a lower bound on the Higgs mass of $M_H > 121 \; m_p$ (m_p is the proton mass) and an upper bound of, very likely, $M_H < 200 \; m_p$.

- Nature has operated with, until now, mysterious violations of symmetries. One such mechanism created an imbalance between matter and antimatter, such that a minute amount of matter, 1 quark in 10^9, survived. Stars, galaxies, and life owe their existence to this tiny violation.

- Neutrinos appear to have small masses, which the SM has not predicted. Experiments must show the way out of this conundrum.

- Our universe is dominated by invisible dark matter. Circumstantial evidence favors the existence of unknown elementary particles, as postulated, for example, in one popular extension of the SM: supersymmetry. If this is true, supersymmetric particles should be discovered at the LHC.

This catalog is incomplete, but it gives a flavor of the breadth of physics to be studied and provides the yardstick to gauge the performance of the CD design. As an example, the various signatures through which a Higgs particle might reveal its existence are shown in Fig. 6. Books have been written, journals filled, and conferences organized to develop, debate, and anticipate this physics. The consequences and challenges for the CD are synthesized in Table III.

Measurement accuracies are typically a factor of 2–5 better than for CDs at the LEP collider. However, the dramatically novel feature is operation at a collision rate of 10^9 sec^{-1}, which is needed to observe the new phenomena with adequate significance. As each collision typically produces $\sim$100 particles, the CD must see approximately 10^{15} particles for each detected Higgs particle: if the proverbial haystack is made up of some 10^7 straws, then searching for the Higgs particle is like looking for a needle in not 1 but 10^8 of these haystacks.

B. Experience and Emotion: Design Driven by Physicists

More than 2000 physicists and engineers are mobilized to design and construct an LHC CD. Early in the program the important decision was made to construct two such general facilities in view of the enormous physics stakes. This would ensure competition, cross-fertilization of ideas

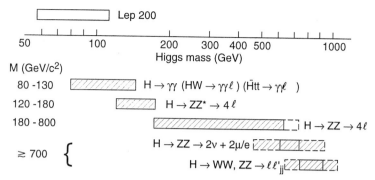

FIGURE 6 Diagrammatic representation of the experimentally most suitable decay modes in the search for the Higgs particle at the large hadron collider (LHC), as a function of the Higgs mass. The large electron–positron (LEP) collider experiments have provided a lower bound of 114 GeV/c^2.

and methods, and cross-checking on the quality of the detectors and the results.

However, with the physics program, the experimental requirements, and the measurement techniques all in the public domain, might the two CDs not look like two clones? The answer is an emphatic no: scientists, not computers with a universal operating system, are designing the research facilities. Two examples may illustrate the process.

High-quality muon spectroscopy is recognized as a cornerstone of a CD. These measurements must be done in the last shell outside the hadron calorimeter, therefore dominating the overall size and cost. The two collaborations, ATLAS (a toroidal LHC apparatus) and CMS (compact muon solenoid), have chosen very different solutions for the muon spectrometer magnet (Fig. 7). ATLAS opted for three dedicated toroidal magnetic fields. Muons are measured twice: in the inner tracker and in the toroidal muon field. An audacious design was developed, with eight 25-m-long superconducting coils forming the central toroid. CMS pursued a one-magnet concept: a high-field (4-T) and large ($R = 3$ m) solenoid encloses the central tracker and calorimetry. Muons are measured with high precision in the central tracker and are reconfirmed in the return yoke of the magnet, albeit with considerably less accuracy. Although compact and elegant, it is somewhat more risky than the ATLAS approach, as the high-quality muon measurement must succeed in the presence of thousands of charged particles flooding the central tracker.

The equally important measurement of electrons and photons is another example. Members of ATLAS were experienced with calorimetry based on the liquid-argon ionization technique. The prime motivation was stability and uniformity of measurement response. They decided to "breed" it into a form able to survive the ferocious LHC environment. A totally new accordion geometry (Fig. 8) had to be invented. CMS boasted experts who had participated in developing the revolutionary crystal calorimeter built for the L3 experiment at the LEP collider. The elegant, high-performance method used for L3 had to be ruggedized for LHC survival. A remarkable worldwide collaboration between particle physicists, material scientists, crystallographers, and industry succeeded in growing a novel crystal, PbWO$_4$, fit for LHC research.

TABLE III Discovery Physics and Corresponding Measurements at the LHC

Discovery signal	Measurement	Measurement accuracy	Signal rate per collision
Higgs $\rightarrow \gamma\gamma$	$m_H < 2m_Z$: precise γ energy measurement	1% at 100 GeV	$\sim 10^{-13}$
$\rightarrow ZZ$	$m_H > 2m_Z$: momentum of e and μ; ν (= missing energy)	~ 2% at 100 GeV/c	$\sim 10^{-11}$
Other mechanisms of symmetry breaking	Precision μ, e spectroscopy at large momenta	~ 10% at 1 TeV	$\sim 10^{-14}$
Supersymmetric particles	Jets of particles; missing energy; e, μ	Energy with ~ 1% at 1 TeV	$\sim 10^{-13}$
New gauge bosons	e, μ; ν (= missing energy)	Missing energy with few % at 1 TeV	$\sim 10^{-14}$
Quark constituents	Precision jet measurements	~ 1% linearity up to few TeV	$\sim 10^{-14}$

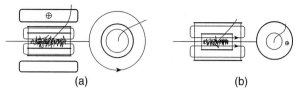

(a)   (b)

FIGURE 7 Conceptual muon spectrometer magnet configurations adopted by the two LHC collaborations: (a) ATLAS (a toroidal LHC apparatus) configuration and (b) CMS (compact muon solenoid) configuration. For both experiments, a longitudinal view (left) and a transverse view (right) are shown. Arrows indicate the direction of the magnetic-field lines. Note the different curvature of the muon tracks (long heavy line) in the two projections. For ATLAS only the central toroidal magnet is drawn.

These examples show two groups of physicists with two different stagies to meet the same challenge. Ultimately it is scientists—teams with different experiences, know-how, and tastes—who drive the final decisions. Without these strong, personal involvements the CDs would never be built.

C. Complexity of the Design

Much as the physical size of the CD is a direct consequence of the physics and technology of the detection process, its complexity is driven by the physics research.

Complexity reflects the number of independent detection cells (channels), the associated signal-processing rate, and the staggering level of event selectivity.

1. Number of Detection Cells

The central tracker is constructed of enough independent detector channels such that the probability of simultaneous occupancy of one element by more than one particle is very small (e.g., of the order of 0.1%). The consequence of this approach is that a few 10^6 channels are needed.

Different considerations prevail in the vertex detector, whose role is the reconstruction of decay vertices produced by telltale short-lived particles. The decay topology must be reconstructed in three dimensions with a resolution of $10 \times 10 \times 50~\mu m^3$; silicon detectors are divided into pixels of typically $50 \times 300~\mu m^2$. Three layers surrounding the 30-cm-long collision zone are divided into a total of $\sim 10^8$ pixel elements.

In the calorimetry, the concept of very fine subdivision is limited by the natural scale of the transverse dimensions of the particle cascade, and low occupancy cannot be achieved. Remember, however, that most collisions are glancing ones, transferring little energy to the collision products: a redeeming feature. These thousands of low-energy particles just produce a background

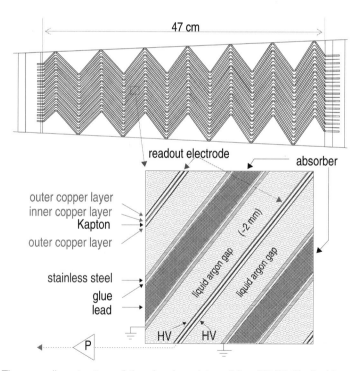

FIGURE 8 (Top) The accordion structure of the absorber plates of the ATLAS Pb–liquid-argon electromagnetic calorimeter. The particle enters from the left. (Bottom) Details of the electrode structures needed to collect the ionization charge.

noise and do not significantly disturb the measurement of the valuable energetic particles producing a strong signal.

Dividing the e.m. calorimeter into cells of natural dimensions results in typically ~2000 cells/m^2, subdivided longitudinally a few times. A few 10^5 channels will suffice for such a detector. Similar arguments lead to a subdivision of the hadronic calorimeter into approximately 10^4 channels.

For the muon system, considerations on occupancy similar to those for the inner tracker also apply. Therefore, approximately 0.5×10^6 tracking cells are needed.

2. Signal-Processing Rate and Event Selectivity

In an LHC detector, every 25 nsec more than 2000 particles will be produced. What is more, the collision products induce an intense level of background radiation comparable to, sometimes much greater than, the particle rates. The signals left in the various detectors will linger for up to a few hundred nanoseconds. At any given moment there are several waves of event information racing through the CD and its signal-processing system. Several strategies are needed to digest this simultaneity of event information.

A new form of signal processing has to be used to deal with this staggering amount of information (Fig. 9). The primary signals of hundreds of such events are stored sequentially in "pipelines." Detectors combined with ultrafast signal processors decide on the potential physics interest of each event, typically in less than 2 μsec. After this initial preselection the compacted data rate is reduced to ~10^{13} bits/sec, which corresponds to the data volume of approximately 10^8 simultaneous telephone calls. Subsequently more refined algorithms are applied to this data to retain events of potential physics interest. We expect that only 1 event in 10^7 collisions (i.e., 100 events/sec) will meet such event-selection criteria and will be archived. These CDs require communication links and a processing power similar to those of a large telecommunications company.

3. Engineering Complexity

The CDs are the size of a six-story office building. They are composed of some 20,000 detector elements, constructed with 0.01- to 1-mm tolerances. Many of these units are supported by lightweight frames, familiar in the space industry. The detector components, together with the required services, are being fitted together with millimeter clearances; they are surveyed with tens of thousands of optical rays, which locate them in the 20,000-m^3 detector volume with an accuracy of 10–30 μm.

The 10^8 signal sensors, some of which work at cryogenic temperatures, are exposed to a ferocious radiation

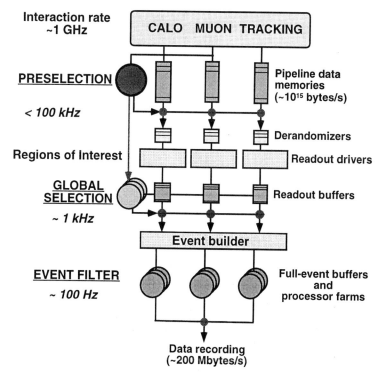

FIGURE 9 Conceptual diagram of the data-processing and event-selection system. Signals from each of the 10^8 detector channels are processed through one of these pipelines needed to store the data awaiting a selection decision.

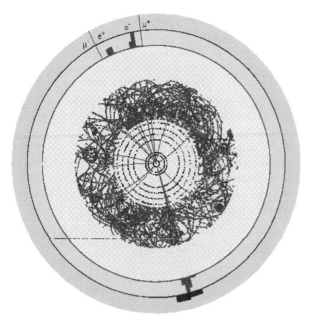

FIGURE 10 Computer simulation of an H → eeμμ event in the ATLAS detector. The electrons are identified as charged particles in the inner tracker with a matching energy deposit in the electromagnetic calorimeter. The muon tracks are clearly visible in the outer spectrometer. Although thousands of particles crowd the tracking detectors, the electron and muon tracks can be identified and measured in the computer reconstruction.

environment of up to 10^6 particles/cm^2 sec reminiscent of certain military applications. Many of the detectors are practically inaccessible and are therefore designed to reliability standards of the space industry.

These detectors use military- and space-type technologies, and engineering methods usually associated with big space projects, but are constructed with university personnel and budgets. Remarkably, the cost of the LHC experimental program has not shown the same level of inflation as performance and specifications: it is comparable to the past LEP collider program.

D. The Design in Virtual Reality

Long before any metal is cut or any amplifier produced, the CD is given a rigorous "health check" in computer-simulated reality. Fortunately, the developments in hardware detectors have been paralleled by equally intensive collaborations to provide the necessary simulation tools. Suffice it to say that we have learned to simulate the detector performance, the influence of engineering choices, and the optimization of materials, services, and so forth at least to the level at which these features influence the final detector performance. These complex detectors sim-

ply could not be built without these faithful tools. The importance of such tools can perhaps best be gauged by realizing that the most modern version, dubbed GEANT, an object-oriented code in C^{++}, is being developed by more than 100 physicists and informatics engineers over several years in a worldwide collaboration. Besides the world's particle physics laboratories, the European Space Agency (ESA), NASA, and medical establishments are using this code. Applying the more reliable GEANT estimates on radiation doses in tumor therapy, doctors estimate that approximately 10,000 more lives per year will be saved in the United States alone. As a preview, the GEANT simulation of an event pp → particles + Higgs → particles + 2e + 2μ is shown in Fig. 10.

IV. CONSTRUCTING A COLLIDER DETECTOR

In the previous sections the basic design features of the new generation of CDs, and the motivation for them, were explained. But can such a physics and detector concept be transformed into engineering reality? Today we know that the answer is yes, for the following reasons:

- During the 1990s many groups developed these novel detectors and demonstrated their performance in LHC-like beam tests.
- The groups have learned to address the new management aspects of a project of such unprecedented size.

The evolution of scale and complexity of the new CDs is indicated in Table IV.

In addition to the technological and engineering complexity (see Section III.C.3), several new issues have to be successfully addressed:

TABLE IV Scale and Complexity of Collider Detectors

	Tevatron	LEP collider	LHC
Measurement volume of CD [m^3]	2000	2000	10,000/20,000
Measurement channels	Few ×10^6	Few ×10^6	10^8
Interaction rate [sec^{-1}]	10^6–10^7	10	10^9
Size of collaboration			
Physicists/engineers	500	500	2000
Institutions	50	50	150
Countries	15	20	50
Time scale (concept to first data) [yr]	~10	~10	~20

TABLE V Tools for Managing the Construction of Collider Detectors

Tool	Purpose
Computer-aided design (CAD) systems[a]	• Construct virtually the complete CD • Check fit of all components • Design layout of cables, cryogenic lines, pipes, other services (total length ~100,000 km)
Virtual-reality software	Simulate dynamic procedures: • Installation of detector components • Access to components
Engineering data management system (EDMS)	Central depository of all information:[b] • Text files • Drawings • Videos Total data volume of documentation for one CD: Tbyte
Project management tools	Variety of programs used for • Approval process for drawings and documents • Procurement • Scheduling • Project progress monitoring • Production database of quality control and detector operational data

[a] The centrally used CAD must interface with the different CADs used around the world.

[b] To provide a worldwide collaboration with the unique detector baseline design.

• *Technology*. The detectors are pushing the limits of realizable technologies. Never before was performance demanded at the physics limits for such large systems, nor was industry asked in many-year-long collaborations to develop technologies to LHC demands. One major stimulus of this development is the healthy fight for funds, in which particle physics is joined by other sciences (biology, astronomy, space-base experiments). Never before have funding agencies obtained such an outstanding performance/price ratio.

• *Time scale*. The increase in the time scale is significant. On a purely technical level, new concepts of documentation need to be implemented to retain the memory of the CD design, construction, and modification over several decades.

• *Globalization*. The scale of these instruments is such that the responsibility for a given subdetector (e.g., the tracker) is typically shared by 10 countries, with their own national priorities and funding-agency conditions.

Fortunately, new worldwide networked tools have become available to support this globalization; see Table V. The toolbox is the World-Wide Web: a CERN invention, originally developed to help in the analysis of physics data.

For the LHC, construction of the detector components, the experimental infrastructure, and underground halls has started. A computer-generated view of the two LHC detectors, as they will be in place in 2005, is shown in Figs. 11 and 12.

V. CLOSING REMARKS

A collaboration of physicists, engineers, and technicians, in concert with industrial partners, conceives, designs, and builds these physics cathedrals. With these instruments, new worlds are explored, the laws of physics are extended, and a deeper understanding of our universe, our own origin, is reached. Like all human expeditions to new frontiers it is a fascinating amalgam of human curiosity, intellect, and passion. Like past explorers we reach these new frontiers with the most appropriate, state-of-the-art

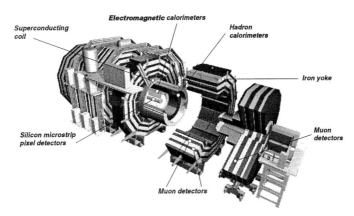

FIGURE 11 A three-dimensional exploded view of the CMS facility, under construction by a worldwide collaboration. In operation it will have a length of almost 30 m and a diameter of 15 m.

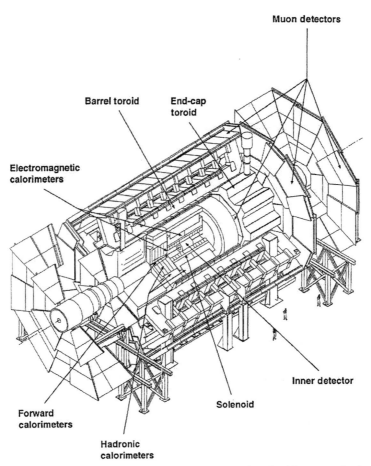

FIGURE 12 A three-dimensional view of the ATLAS facility. The total length of the apparatus is close to 50 m, the diameter is approximately 22 m, and the total weight exceeds 7000 tons.

technologies. As befits the twenty-first century, this expedition to the beginning of our universe is a global voyage of spaceship Earth.

ACKNOWLEDGMENTS

I am grateful for advice and critical comments from W. Blum, N. Ellis, F. Gianotti, and T. S. Virdee. I am indebted to the CERN desktop publishing service for the thoughtful editing of the manuscript and for very professionally transcribing my notes and sketches into printable form.

SEE ALSO THE FOLLOWING ARTICLES

ACCELERATOR PHYSICS AND ENGINEERING • ATOMIC AND MOLECULAR COLLISIONS • COLLISION-INDUCED SPECTROSCOPY • ENERGY TRANSFER, INTRAMOLECULAR • NUCLEAR PHYSICS • PARTICLE PHYSICS, ELEMENTARY

BIBLIOGRAPHY

ATLAS (1994). "Technical Proposal for a General-Purpose pp Experiment at the Large Hadron Collider at CERN," CERN/LHCC/94-43, CERN, Geneva. (Electronic link to CERN Library existing.)

Brandt, D. *et al.* (2000). "Accelerator physics at LEP," *Rep. Prog. Phys.* **63,** 939–1000.

CMS (1994). "The Compact Muon Solenoid, Technical Proposal," CERN/LHCC/94-38, CERN, Geneva. (Electronic link to CERN Library existing.)

Ellis, N., and Virdee, T. S. (1994). "Experimental challenges in high-luminosity collider physics," *Annu. Rev. Nucl. Part. Sci.* **44,** 609–653.

Fabjan, C. W. (1994). LHC: Physics, machine, experiments. *In* "AIP Conference Proceedings 342" (A. Zepeda, ed.), Am. Inst. of Phys., New York. (Electronic link to CERN Library existing.)

Lefèvre, P., and Petterson, T., eds. (1995). "The Large Hadron Collider," CERN/AC/95-05 (LHC), CERN, Geneva. (Electronic link to CERN Library existing.)

Virdee, T. S. (1990). "Experimental techniques," *Rep. CERN 99-04.* (Electronic link to CERN Library existing.)

Williams, H. H. (1986). "Design principles of detectors at colliding beams," *Annu. Rev. Nucl. Part. Sci.* **36,** 361–417.

Collision-Induced Spectroscopy

Lothar Frommhold

University of Texas

I. Brief Overview
II. Collision-Induced Absorption
III. Collision-Induced Light Scattering
IV. Virial Expansions
V. Significance for Science and Technology
VI. Conclusions

GLOSSARY

Absorption coefficient The natural logarithm of the ratio of incident and transmitted intensity divided by the optical path length, $\alpha(\omega) = \log_e(I_0/I)/L$. It is a function of frequency ω, temperature, and gas density.

Allowed and forbidden transitions Atomic and molecular systems exist in a variety of states. Spectroscopic lines arise from transitions between such states, but not all possible transitions are "allowed" for emission or absorption of a photon. Selection rules exist which determine which of the transitions are optically allowed. Forbidden transitions may, however, take place without the emission or absorption of a photon, e.g., in collisional interactions.

Bound vs free van der Waals systems A certain, usually small fraction of the atoms or molecules ("monomers") of virtually any gas exist as van der Waals molecules, i.e., systems of two (or more) monomers, bound together by the weak van der Waals intermolecular forces ("dimers," "trimers," . . .). Below, we will be concerned mainly with complexes of two (or more) unbound monomers which exist only for the very short duration of a fly-by encounter. Free and bound van der Waals systems have many properties in common, but only the latter possess the relative stability of a molecule. Properties of bound and free van der Waals systems will be referred to as supramolecular properties.

Dipole Molecules have a permanent dipole moment if the centers of positive and negative charge do not coincide. Dipole moments can also be induced by external fields (polarization) or momentarily by collisional interactions.

Frequency units Frequency f is measured in cycles per second, or hertz (Hz). Spectroscopists have good reasons to express frequencies ν in cycles per centimeter, or wavenumbers (cm^{-1}), so that $\nu = f/c$, with c being the speed of light in vacuum. Theorists usually prefer angular frequencies $\omega = 2\pi f$; units are radians per second.

Infrared spectroscopy Molecular spectra are observed at frequencies ranging roughly from the microwave region ($\approx 10^{10}$ Hz) to the soft X-ray region ($> 10^{16}$ Hz) or more. While the high-frequency spectra arise from electronic transitions that are of lesser interest here, the low-frequency spectra occur in the microwave,

Encyclopedia of Physical Science and Technology, Third Edition, Volume 3

infrared, and visible region of the electromagnetic spectrum and arise typically from internal rotation and vibration and from molecular encounters.

Raman spectroscopy If monochromatic light falls on a molecular target, polarization and light scattering results. The scattered light is basically of the same frequency as the incident light, but modulated with the rotovibrational frequencies characteristic of the molecule.

Spectroscopic notation Examples are $S_0(0)$ and $Q_3(1)$. The meaning of $X_n(j)$ is as follows: $j = 0, 1, \ldots$ is the rotational quantum number of the initial state; the subscript $n = v' - v$ is the difference of the vibrational quantum numbers of initial (v) and final (v') vibrational state; and X stands for one of these letters $O, P, Q, R,$ S, etc., each specifying a different rotational transition: $j' - j = -2, -1, 0, 1, 2,$ etc., respectively. We note that the subscript n is often omitted when it is clear what vibrational band it refers to.

Stimulated emission For every photon absorption process an inverse process exists, called stimulated emission. Stimulated emission is important when the population of states of sufficiently high energy are significant. For example, in the far infrared, where photon energies are comparable to the mean thermal energy of collisional pairs of molecules, stimulated emission forces the absorption to zero as frequencies approach zero.

Virial expansion At gas densities that are substantially smaller than those of liquids and solids, certain properties of a gas may be described by a power series of density. The leading term is typically linear in density and reflects the contributions of the (non-interacting) monomers. The next term is quadratic in density and reflects the induced contributions of exactly two interacting monomers, etc. The most familiar virial expansion is that of the equation of state, which relates pressure, volume, and temperature of a *real* (as opposed to an *ideal*) gas. There are, however, several other and, for our present focus, more relevant examples of virial expansions that are related to the dielectric properties of gases, to be mentioned below in some detail.

I. BRIEF OVERVIEW

Rarefied molecular gases absorb and emit electromagnetic radiation if the individual molecules are *infrared-active,* i.e., if the structure of individual molecules is consistent with the existence of an electric dipole moment. Homonuclear diatomic molecules (H_2, N_2, ...) are infrared-inactive, but characteristic rotovibrational absorption bands and certain continuous spectra exist in the rarefied gases composed of polar molecules (HCl, NO, ...). Com-

pressed molecular gases, on the other hand, show quite generally a variety of *additional* absorption bands—even if the individual molecules are infrared-inactive. These are the *collision-induced* absorption spectra that arise from fluctuating dipole moments induced momentarily when molecules collide. Collision-induced dipole moments are of a *supramolecular* nature; they are properties of *complexes* of two or more interacting molecules and are foreign to the individual molecules of the complex, as long as these are separated most of the time from all the other molecules by distances amounting to several molecular diameters or more—as may be thought of being the case in rarefied gases.

Similarly, if monochromatic laser light is incident on a molecule, the molecule is polarized by the electric field. This field-induced dipole will emit (or "scatter") radiation of the frequency of the incident light. It will also emit at other frequencies that are shifted relative to the laser frequency by certain rotovibrational transition frequencies of the individual molecules, if the molecule is Raman-active, i.e., if the invariants of the polarizability tensor are nonzero for certain rotational and/or vibrational transitions of the molecule. Compressed gases show quite generally a variety of *additional* Raman bands, the collision-induced Raman bands, even if the individual molecules are Raman-inactive. For example, in rarefied monatomic gases, the scattered laser light will be strictly at the laser frequency; no shifted Raman lines or bands exist. However, in the compressed rare gases, Raman continua exist which are due to collisional and, to some extent, to bound van der Waals pairs (and, at higher gas densities, triples, ...) of interacting atoms. Collision-induced Raman spectra of the common gases are well known and will be discussed in greater detail below.

II. COLLISION-INDUCED ABSORPTION

A. A Discovery

In his famous dissertation, J. D. van der Waals argued compellingly in 1873 that the forces between two molecules of a gas must be repulsive at near range and attractive at larger separations. Ever since, theorists conjectured the existence of what today would be called *van der Waals molecules* or *dimers*, that is, weakly bound systems of two argon atoms (Ar_2) in a gas consisting almost purely of argon atoms (Ar) or of two oxygen molecules (O_2)$_2$ bound together by the weak van der Waals forces in a gas that otherwise consists purely of O_2 molecules; etc. In spectroscopic laboratories around the world efforts ensued to discover characteristic dimer bands and thus demonstrate directly the existence of dimers. However, it took almost 100 years of dedicated research efforts before some such dimer bands of the elusive van der Waals molecules were actually discovered. Today, we know that such dimers exist in small concentrations in virtually all gases, almost without exception.

In 1949 H. L. Welsh and his associates had hoped to demonstrate the existence of dimers in compressed oxygen gas but failed—just like all other efforts elsewhere had failed at the time. However, in the course of that work a new and arguably more significant type of absorption spectra was discovered instead: the much stronger and truly universal spectra of *unbound* pairs of molecules. In other words, fluctuating dipoles induced in collisionally interacting, free molecules, e.g., O_2–O_2 [to be distinguished from the bound pairs $(O_2)_2$], etc., actually absorb more radiation and absorb over a greater frequency band than the bound dimers do. One simple reason for the stronger absorption of light by collisional pairs is that, typically in compressed gases, at any instant one counts many more collisional pairs than bound dimers. In short, collision-induced absorption of compressed oxygen gas was discovered. In quick succession similar absorption bands were seen in virtually all common molecular compressed gases—a truly universal, new, *supramolecular* spectroscopy was thus discovered: collision-induced absorption.

This discovery of collision-induced absorption was accomplished at infrared frequencies, where the rotovibrational bands of the common molecules typically are found. The new spectroscopy is, however, not limited to such frequencies; it is now known to extend from the microwave region throughout the infrared and well into the visible—and in a few known cases actually beyond.

A quantitative knowledge of the absorption of light by the Earth's atmosphere is essential to scientists, especially to astronomers who need to correct their observational data for such absorption as much as possible. Since the atmosphere absorbs very little visible light, in 1885 Janssen attempted to measure absorption by the atmospheric gases in a high-pressure cell. He found a number of absorption bands of oxygen, unknown from previous studies conducted at much lower gas densities. Absorption in these bands could be enhanced by the addition of nitrogen, but pure nitrogen did not show any absorption bands in the visible and near ultraviolet regions of the electromagnetic spectrum. A telling feature of these new absorption bands is that the absorption coefficient of pure, pressurized oxygen increases with increasing gas density as the *square* of density when the expectation at the time would have been a linear dependence; the enhancement of the absorption coefficient by the addition of nitrogen was found to be proportional to the product of O_2 and N_2 densities. These density dependences suggest a kind of absorption that requires two interacting O_2 molecules, or an O_2–N_2 pair, as

opposed to just one O_2 or one N_2 molecule for every absorption process. The new absorption bands had early on been called *interaction-induced* bands by some prominent spectroscopists. However, the process seemed somewhat mysterious because this type of absorption appeared to be limited to situations involving oxygen; others of the most common gases do not have such striking pressure-induced bands in the visible region of the spectrum. Today, we understand that these interaction-induced absorption bands of oxygen are collision-induced bands involving *electronic* (as opposed to purely rotovibrational) transitions of the O_2 molecules; further details may be found below.

B. Monatomic Gases

1. Pure Monatomic Gases

Collision-induced absorption by pure monatomic gases has not been observed. Of course, it is clear that collisional *pairs* of like atoms cannot develop a collision-induced dipole, owing to their inversion symmetry which is inconsistent with the existence of a dipole moment. However, triatomic and higher complexes of like atoms theoretically could absorb infrared radiation, but apparently these absorption coefficients are so small that thus far a measurement has been impossible. Even at the highest densities, e.g., in liquefied rare gases, only a very small upper limit of the infrared absorption coefficient could be established for a few rare gases. Pure monatomic gases are probably the only gases that do not show significant collision-induced absorption at any frequency well below X-ray frequencies.

2. Mixtures

However, *mixtures* of monatomic gases absorb in the far infrared. Whereas like collisional pairs such as Ar–Ar do not support a dipole moment, dissimilar pairs such as He–Ar generally do. The collisional complex of two dissimilar atoms lacks the inversion symmetry that precludes the existence of an electric dipole moment. Absorption by dissimilar pairs is now well known, even if gas densities are well below liquid state densities.

At a given frequency, the intensity of a beam of light falls off exponentially with increasing path length x,

$$I(x) = I_0\, e^{-\alpha x},$$

if absorption occurs (Lambert's law). The measurement determines the absorption coefficient α as function of frequency, temperature, and the densities ρ_1, ρ_2, of atoms of both species (assuming a binary mixture). The so-called absorption spectrum α is preferably presented by the "normalized" absorption coefficient, $\alpha/(\rho_1\rho_2)$, which is invariant under variation of the densities ρ_1, ρ_2, as long as

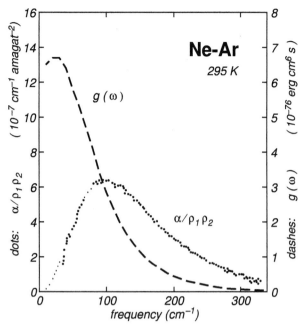

FIGURE 1 The dots represent the measurement of the binary collision-induced translational absorption coefficient of the mixture of neon and argon gas at 295 K by Bosomworth and Gush [(1965). *Can. J. Phys.* **43**, 735]. The dashed curve is the so-called spectral density function derived from the above by correcting the measurement for stimulated emission; it is a diffuse "line" that may be thought to be centered at zero frequency.

no three-body or higher order interactions interfere with the measurement (i.e., at intermediate gas densities that are well below liquid state densities). An example of the absorption spectrum of mixtures of neon and argon is shown in Fig. 1. Densities employed in that measurement were in the order of roughly 10 amagats (1 amagat $\approx 2.7 \times 10^{19}$ atoms/cm^3). The frequencies are shown in wavenumber units (cm^{-1}); they are in the far infrared region of the electromagnetic spectrum and are commensurate with the reciprocal time scales of the atomic collisions. The mean relative speed of the Ne–Ar pair at the temperature of 295 K is about 750 m/s, and the size of the collision diameter for Ne–Ar collisions is in the $\approx 3 \times 10^{-11}$ m range, so that the duration of an average Ne–Ar collision amounts to $\Delta t \approx 4 \times 10^{-14}$ s. According to Heisenberg's uncertainty relation,

$$\Delta t \, \Delta \omega > 0.5, \tag{1}$$

the spectral frequency band $\Delta \omega$ of the average collision amounts to at least $\Delta \omega \approx 1 \times 10^{13}$ rad/s, or >70 cm^{-1}. Actually, absorption over a much greater range of frequencies was observed (Fig. 1). The estimated spectral width in wavenumber units, $\Delta \omega / (2\pi c) \approx 70$ cm^{-1}, is actually very close to the half-width of the so-called spectral density function $g(\omega)$ which is also sketched in Fig. 1 (dashed

curve). The spectral density function is obtained by dividing the measured, normalized absorption coefficient $\alpha/(\rho_1 \rho_2)$ by the photon energy $\hbar \omega$ and by $[1 - \exp(-\hbar \omega / kT)]$ to correct for stimulated emission. Both factors force the absorption to zero as the frequencies approach zero. The spectral density function may be considered (half of) a spectral line centered at zero frequency, with a half-width of roughly 70 cm^{-1} as was estimated from Heisenberg's uncertainty relation.

The translational absorption band is the only collision-induced absorption spectrum in mixtures of monatomic gases at frequencies well below the ultraviolet.

According to quantum mechanics, the absorption of a photon corresponds to a transition of the colliding pair from a state of relatively low energy of relative motion to a higher such state. The spectrum shown in Fig. 1 is therefore called a "translational" spectrum.

3. Ternary and Many-Body Interactions

In the measurement (Fig. 1), the density variation of the normalized absorption coefficient $\alpha/(\rho_1 \rho_2)$ was carefully checked and found to be independent of either density. This density invariance indicates the binary nature of the collision-induced spectrum in the range of gas densities employed in that measurement. We mention that if the density of either gas is further increased, a point is reached where the three-body interactions—and eventually many-body interactions—manifest themselves by a breakdown of the invariance of the normalized absorption coefficient. In a few cases ternary absorption spectra have actually been separated from the binary contributions for detailed analyses. At even higher densities ($\approx$liquid state densities), true many-body effects control the spectroscopy. All these many-body spectra differ from the binary spectra in characteristic ways; the spectra observed at the highest densities must be considered superpositions of supramolecular spectra of binary, ternary, . . . , many-body systems.

C. Molecular Gases

In molecular (as opposed to monatomic) gases much richer collision-induced absorption spectra are observed in a number of spectral bands, because of the increased degrees of freedom of the collisional complexes. Any collisional pair possesses the degrees of the translational motion and an associated kinetic energy of relative motion. Additional degrees of freedom and thus of energies are associated with rotational and vibrational motion of one or more molecules of the complex, if molecular collisions are considered. Accordingly, photons are absorbed over a much greater range of frequencies, in the vicinity of the various rotovibrational bands of the molecules, and at sums and differences of such rotovibrational frequencies, if two or more molecules interact. In other words, the energy of a collisional pair involving at least one molecule is given by the sum of translational and rotovibrational energies of the interacting molecules. Absorption of a photon corresponds to a transition of the transient "supermolecule" from a state of lower to one of higher energy, and because there are typically many different rotovibrational states accessible, photons of varying energy (frequency) can be absorbed by the pair. One speaks of collision-induced rotovibrational spectral bands. It is important to remember that these induced bands are observable even if the corresponding bands of the individual molecules are "forbidden," i.e., if the individual molecules are infrared-inactive at such frequencies.

1. Rotovibrational Spectra

As an example, Fig. 2 shows the collision-induced fundamental band of the H_2 molecule in compressed hydrogen gas, which is "forbidden" in isolated (i.e., non-interacting)

H_2 molecules. A photon is absorbed in the near infrared, in the broader vicinity of the transition frequency from the ground state to the lowest vibrationally excited state ($v = 0 \rightarrow 1$), which occurs at about $4155\ cm^{-1}$. Especially at the lower temperatures, three broad lines labeled $Q(1)$, $S(0)$, and $S(1)$ are noticed which correspond in essence to transitions of the H_2 molecule involving rotational quantum numbers $j = 1 \rightarrow 1$, $j = 0 \rightarrow 2$, and $j = 1 \rightarrow 3$, subject to the vibrational transition $v = 0 \rightarrow 1$, combined with a change of the translational energy of relative motion of the collisional pair of H_2 molecules. These lines are very diffuse (i.e., not "sharp," like a bright, thin line of light against a dark background, which elsewhere would be called a spectroscopic line), reflecting the short duration of the collisional encounter [Heisenberg's uncertainty relation, Eq. (1)].

Especially at the higher temperatures in Fig. 2, other lines may be discovered in careful analyses of the measurements, such as lines associated with higher rotational states ($j = 2, 3, \ldots$). Perhaps more surprisingly, so-called double transitions can also be discovered (but are not immediately discernible in Fig. 2) which correspond to

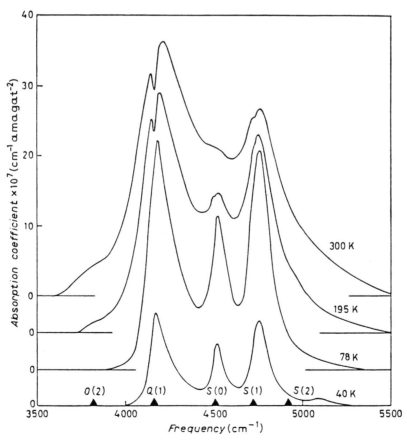

FIGURE 2 The collision-induced absorption spectrum of gaseous hydrogen of moderate density at infrared frequencies near the fundamental band of H_2. [After Hunt, J. L., and Welsh, H. L. (1964). *Can. J. Phys.* **42**, 873.]

simultaneous rotovibrational transitions in both H_2 molecules, e.g., a purely rotational transition in one molecule and a purely vibrational transition in the other. The latter cause absorption near sums and differences of the rotovibrational transition frequencies of the H_2 molecule, duly broadened by the short duration of the collision.

At the four temperatures shown in Fig. 2 the spectra are quite similar, except that at the higher temperatures the lines are broader, reflecting the decreasing duration of the average collision with increasing temperature, i.e., with increasing relative speed of the collisional encounters, Eq. (1).

The three diffuse lines seen so clearly in the Fig. 2 are called the Q and S lines of the H_2 molecule. However, one must keep in mind that none of these lines can be due to single, non-interacting H_2 molecules, owing to the inversion symmetry of H_2 which is inconsistent with the existence of an electric dipole moment. Hydrogen molecules are infrared-inactive, and rarefied hydrogen gas shows virtually no absorption at such frequencies. The spectrum shown (along with others shown below in Figs. 3, 4, and 9) are observable in compressed gases only and are collision induced. Significantly, there must be at least one interacting atom or molecule nearby for these lines to appear. An important point to be made here is that this type of rotovibrational absorption spectra occurs universally in virtually all molecular gases and in mixtures of atomic and molecular gases. Moreover, the absorption of pairs of molecules is

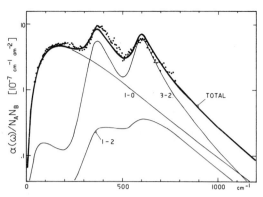

FIGURE 4 The enhancement of the binary collision-induced rototranslational absorption spectrum of hydrogen obtained by the addition of helium to hydrogen at the temperature of 77.4 K. The dots represent the measurement. The heavy solid curve represents a calculation based on first principles. The spectra of several contributing dipole components (marked 1-2, 1-0, 3-2) are sketched lightly. The $S_0(0)$ and $S_0(1)$ lines of H_2 are seen near 354 and 585 cm^{-1}. A translational spectrum much stronger than in Fig. 3 is also noticeable, which arises from the dissimilar nature of the collision partners. [After Meyer, W., and Frommhold, L. (1986). *Phys. Rev. A* **34**, 2237.]

not limited to the fundamental band of H_2 (or of any other molecule in collisional interaction with atoms or another molecule). Instead, it occurs also at the other vibrational bands, e.g., the overtone bands where vibrational quantum numbers v change by 2 or some larger integer, at frequencies in the near infrared and even the visible, and in the rototranslational band in the far infrared (see Figs. 3, 4, and 9).

2. Mixtures of Gases

The collision-induced spectral features may quite generally be enhanced by admixtures of other atomic or molecular gases. In binary gas mixtures, say of hydrogen and helium, at gas densities where binary interactions prevail, one may quite naturally distinguish the supramolecular spectra of H_2–H_2 and H_2–He pairs; absorption by the former varies as the hydrogen density squared and absorption by the latter varies as the product of hydrogen and helium densities. If binary mixtures of molecular gases, say of hydrogen and nitrogen, three types of collision-induced spectra can be distinguished: those of H_2–H_2, H_2–N_2, and N_2–N_2 pairs, again on the basis of their density dependences. Mixtures of more than two gases will quite generally show contribution of all possible pairs, and at higher densities triples, . . . that are consistent with the existence of electric dipole moments during interaction. An example of such enhancement of collision-induced absorption in the mixture of hydrogen and helium is given below (Fig. 4).

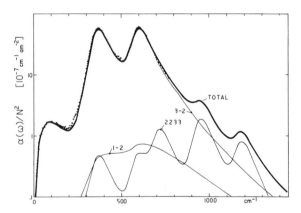

FIGURE 3 The binary collision-induced rototranslational absorption spectrum of hydrogen at the temperature of 77.4 K. The dots represent the measurement. The heavy solid curve represents a calculation based on first principles. That result is actually composed of contributions from mainly three different induced dipole components, which are sketched lightly and marked 1-2, 2233, 3-2. The main features are the rotational $S_0(0)$ and $S_0(1)$ lines centered at 354 and 585 cm^{-1}. A translational peak near 100 cm^{-1}, arising from orientational transitions, is also discernible. The theoretical structures at 940 and 1170 cm^{-1} arise from double transitions $S_0(0) + S_0(1)$ and $2S_0(1)$. [After Meyer, W., Frommhold, L., and Birnbaum, G. (1989). *Phys. Rev. A* **39**, 2434.]

3. Rototranslational Spectra

Figure 3 shows the collision-induced absorption of compressed hydrogen gas in the far infrared (dots), i.e., at much lower frequencies than the absorption spectra shown in Fig. 2. As we noted above, in the discussion related to Fig. 1, as the frequencies approach zero, absorption falls off to zero for several reasons, one of them being stimulated emission. The rapid increase of absorption with frequency increasing from zero and the first broad peak correspond to translational absorption. The next two peaks near 354 and 585 cm^{-1} are the collision-induced $S_0(0)$ and $S_0(1)$ lines of H_2. The remaining two broad peaks are double rotational transitions of the type $S(0) + S(1)$ and $S(1) + S(1)$, combined with a change of the translational state of the pair *by absorption of a single photon*—truly a supramolecular feature. We note that besides the measurement (dots), a calculation based on first principles is shown in Fig. 3; theory reproduces all aspects of the measurement closely and on an absolute frequency scale and with precision.

If helium is added to the hydrogen, collision-induced absorption is enhanced in proportion to the helium partial density. The enhanced absorption is also proportional to the hydrogen density, indicating as its origin the H_2–He pair. Figure 4 shows the enhancement spectrum, which is recorded in the same frequency range as Fig. 3. The absorption by H_2–He pairs is again zero at zero frequency, rises to a first translational peak and two further peaks, the $S_0(0)$ and $S_0(1)$ lines of H_2, and falls off at higher frequencies; note the absence of double transitions in this case. Another noteworthy fact is the very strong translational peak which is characteristic of enhancement spectra of virtually all mixtures of gases.

4. Many-Body Effects

The collision-induced absorption spectra mentioned thus far were all obtained at densities substantially less than liquid state densities; the observed absorption is primarily due to exactly two interacting atoms or molecules. Of course, when we go to densities approaching liquid or solid state densities (e.g., several 100 amagats), molecular complexes of more than two molecules will also shape the observable spectra. While initially collision-induced absorption becomes rapidly more important with increasing density, many-body contributions will quickly modify the shapes and (normalized) intensities, relative to the binary spectra shown above.

One such many-body effect that is very striking actually appears at moderate or even small gas densities: the so-called intercollisional interference. However, at low density that effect is limited to a small number of relatively narrow frequency regions so that the binary character of

the bulk of low-density spectra is undeniable. Specifically, dipoles induced in subsequent collisions tend to have a significant anticorrelation, resulting in partial cancellations. At the higher photon frequencies, this phase relationship is scrambled and amounts to virtually nothing, but at the lowest frequencies absorption dips results, e.g., at zero frequency and at the Q and (less strikingly) S line centers. Such dips are discernible in Fig. 2 at the higher temperatures. The dips may be viewed as inverted Lorentzian line profiles which broaden roughly in proportion to the density variation (see Fig. 5). The intercollisional interference dips are often narrow in comparison to the collision-induced Q or S lines on which they reside, because their width is given by the reciprocal *mean time between collisions*, when the widths of the intracollisional lines are given by the reciprocal *mean duration* of a collision. Under conditions where binary spectra can be recorded, the former is generally substantially longer than the latter. (However, at liquid densities the intercollisional dips may be very broad.) The intercollisional dip is a true many-body feature.

Figure 6 attempts to demonstrate the similarities and differences of the hydrogen spectra of highly compressed gas, of the liquid gas, and of the solid gas. Whereas the curve labeled *50 atm, 78 K* is still similar to the 78 K curve shown in Fig. 2, the uppermost curve in Fig. 6 was recorded at a much higher pressure (4043 atm) than the 300 K curve of Fig. 2. The intercollisional dip is now much broader (from the points marked Q_p to Q_r); the normalized absorption coefficient $\alpha/(\rho_1\rho_2)$ is no longer invariant under density variation. When we compare the spectra of the gas at 78 K with those of the liquid (at 17.5 K) and solid (at 11.5 K), we notice not only a sharpening of the lines with decreasing temperature, but a few new features may also be noticed. The spectrum of the solid shows broad bands that arise from combination tones of molecular frequencies and lattice frequencies (phonon spectra). The long extension of the phonon spectra toward higher frequencies is probably due to multiple phonon generation. Three weak double transitions of the type $S_1 + S_0$ are also discernible. At higher spectral resolution than was employed in Fig. 6, the $S(0)$ and $S(1)$ groups show weak single transitions, $S_1(0)$ and $S_1(1)$, and much stronger double transitions of the type $Q_1(j) + S_0(j)$, with $j = 0$ and 1, and the Q branch shows fine structures related to orientational transitions of two ortho-H_2 molecules.

5. Infrared-Active Gases

The emphasis thus far was on infrared-inactive gases. Spectral lines that are allowed in the individual molecules will at high gas densities be accompanied by a broad collision-induced background. In such a case, it will often be difficult to separate allowed and induced components

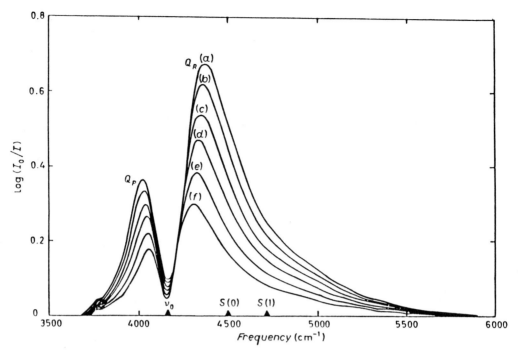

FIGURE 5 The intercollisional dip in the Q branch of H_2, here due to H_2–He collisions, at various densities. The absorption pathlength was 4 cm, the temperature was 298 K, and the mixing ratio of hydrogen and helium was fixed to 1:18. Helium densities from (a) to (f) are 1465, 1389, 1304, 1204, 1088, and 950 amagats. [After Gush, H. P., et al. (1960). *Can. J. Phys.* **38**, 180.]

if their intensities differ widely. Nevertheless, a number of very careful measurements exist in deuterium hydride (HD) and in mixtures of HD with rare gases at densities where binary interactions prevail.

The HD molecule is infrared-active, because the zero-point vibrational motion of the proton is slightly greater than that of the deuteron—a non-adiabatic effect. As a consequence, one side of the HD molecule is more

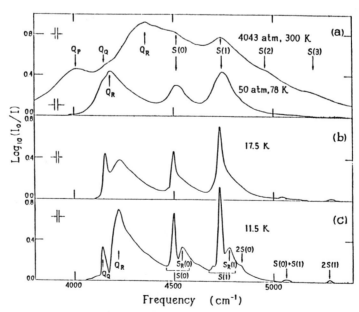

FIGURE 6 The collision-induced absorption spectrum of hydrogen in the fundamental band of the H_2 molecule in (a) the compressed gas, (b) the liquid gas, and (c) the solid gas. [After Hare, W. F. J., and Welsh, H. L. (1957). *Can. J. Phys.* **36**, 88.]

positively charged than the other; thus, a weak dipole moment exists—rotovibrational molecular bands thus occur. Collision-induced bands also occur, basically at the same frequencies as the allowed lines but more diffuse. The two dipole components are of comparable magnitude so that quite striking interference effects result. We note that the permanent dipole moments of the more common polar molecules are generally much stronger than the induced dipoles so that such interferences are harder to detect. Nevertheless, clear indications of such an interference have been discovered in the far wings of allowed spectral lines in compressed gases.

6. Electronic Collision-Induced Spectra

All spectra mentioned thus far arise from supramolecular transitions that leave the electronic states of the collisional partners unchanged. The possibility of the electronic state changing is, of course, finite if the photon energies are sufficiently large. Collision-induced electronic spectra typically occur at higher photon energies (e.g., in the ultraviolet region of the electromagnetic spectrum) than we considered above. However, a few of the common molecules possess electronic states with excitation energies low enough so that collision-induced spectra in the infrared, visible, and near ultraviolet region occur.

Absorption of a photon $\hbar\omega$ by an interacting molecular pair $A + B$ can symbolically be written as

$$A + B + \hbar\omega \rightarrow A' + B' + \Delta E,$$

where ΔE represents a change of the (e.g., translational) energy of the pair and primes indicate a possible change of the internal states of A and/or B. We may distinguish two different supramolecular collision-induced absorption processes: one affects the rotovibrational energies of A, B only and the other involves electronic transitions. Electronic collision-induced absorption typically requires visible and ultraviolet photons. Rotovibrational collision-induced absorption occurs normally in the microwave and infrared regions of the spectrum. The boundaries are, of course, not rigid and depend very much on the specific systems under consideration.

Collision-induced rotovibrational transitions are often studied at atmospheric and higher densities. Collision-induced electronic transitions, on the other hand, are usually studied at much lower densities, where absorption would be difficult to measure. In that case, other detection schemes, such as the ensuing product fluorescence, are employed.

As mentioned above, it has long been established that compressed oxygen has several diffuse absorption bands in the visible and infrared that are unknown from studies of rarefied oxygen and air (Janssen 1885). The intensities of these bands vary as density squared, indicating absorption by *pairs* of O_2 molecules. These bands are now understood to correspond to the "forbidden" electronic transitions from the ground state, $X^3 \sum_g^-$, to the electronically excited states $a^1\Delta_g$ and $b^1 \sum_g^+$ of O_2; simultaneous vibrational transitions of one or both of the interacting O_2 molecules may also occur in the process. Individuals who have handled liquid air are familiar with the blue tint of liquid oxygen which is caused by these electronic collision-induced absorption bands in the red portion of the visible spectrum. Simultaneous electronic transitions in *both* interacting O_2 molecules are also observed at shorter wavelengths. Similar bands of O_2–N_2 pairs have long been known, too. A few other molecules are known to have quite analogous, collision-induced electronic absorption bands in the visible and near ultraviolet regions of the spectrum in a compressed gas environment.

7. Line Profiles

The individual line shapes of the collision-induced spectra have previously been described by a Lorentzian profile, as for pressure-broadened lines. The width of the former is approximately given by the mean reciprocal duration of a collision, the ratio of root mean square speed v and range ρ of the induction mechanism (which is roughly of the order of the collision diameter, $\rho \approx \sigma$). In the low-density limit, v/ρ is density independent. However, the wings of collision-induced lines are roughly exponential and fall off much faster than the Lorentzian profile. Collision-induced spectral features are strikingly asymmetric (due to the principle of detailed balance), whereas Lorentzian profiles are not. For all these reasons, generalized spectral model profiles have been found that describe collision-induced profiles quite well. They are analytical and nearly as simple as the Lorentzian function, but without its limitations. These usually include the Lorentzian profile as a limiting case.

8. Charged Systems

So far, the collisional systems considered were all neutral. It is noteworthy that, in collisions involving one electrically charged particle, induction occurs, which in fact is much stronger than the quadrupole-induced dipoles discussed above. For example, induced spectra involving ion-neutral collisions and electron-neutral collisions (polarization bremsstrahlung) have been observed. It turns out that much stronger allowed electronic spectra usually prevail under conditions in which significant concentrations of free electrons or charged particles are observed. These often mask the weaker, induced spectra.

9. Nonlinear Interactions

With the advent of the laser it became possible to investigate non-linear interactions of light with matter. We just mention hyper Raman, stimulated Raman, coherent anti-Stokes Raman gain (SRG) spectroscopies, along with various other multi-photon processes, which have widened the scope of molecular spectroscopy dramatically. Most of these also have a collision-induced counterpart, much as the cases discussed above.

10. van der Waals Molecules

van der Waals forces are weak when compared to chemical forces that bind the common molecules—so weak that most dimer–monomer collisions will destroy (dissociate) the dimer. There is a certain destruction rate of dimers which balances the formation rates and keeps bound dimer concentrations typically at a low level. The average lifetime of bound dimers amounts, in many cases, to just a few mean free times between monomer collisions, or roughly $\approx 10^{-9}$ s in air at standard temperature and pressure; with increasing density and temperature the lifetimes decrease correspondingly. For comparison, we note that a collisional complex may be thought to "exist" for the duration of a fly-by collision, or roughly 10^{-12} s or so, nearly independent of density. According to Heisenberg's uncertainty relation [Eq. (1)], typical dimer bands may be much "sharper" than the collision-induced lines, if they are not severely pressure broadened.

The infrared spectra of van der Waals molecules are due to the same collision-induced dipole moments that generate the Q, S, etc. lines of the collision-induced spectral profiles shown above. Dimer features thus appear near the centers of these ("forbidden") lines in the collision-induced spectra, i.e., near zero frequency (where they are difficult to record), and at the Q, S, etc. lines of the (forbidden) rotovibrational bands where the dimer features were actually discovered. We note that collision-induced Raman spectra show similar dimer signatures in the vicinity of the Rayleigh line and at the other line centers discernible in the collision-induced Raman spectra.

van der Waals dimer bands are known to be highly susceptible to pressure broadening; this is one reason why it took so long to actually record dimer bands. Moreover, if one wants to resolve dimer bands spectroscopically, relatively high spectral resolution must be employed. We note that the spectra shown above were recorded with low resolution and relatively high pressures so that the dimer features seem absent. However, in recent years in several collision-induced vibrational spectral bands, dimer bands could be recorded and analyzed, and invaluable new knowledge concerning intermolecular interactions

has been obtained in this way. The structures attributed to the $(H_2)_2$ dimer, which are indicated by the small rectangle near the $S_0(0)$ line center in Fig. 9 below, were seen in the Voyager spectra of Jupiter and Saturn, in which the conditions are more favorable than in the older laboratory measurements (low density and long absorption path lengths). For most binary systems, similar dimer structures must be expected. Spectral transitions involving dimers are from bound-to-bound and bound-to-free states of the pair, while typical collision-induced spectra correspond to free-free transitions of the pair.

D. Collision-Induced Emission

Any gas that absorbs electromagnetic radiation will also emit; supramolecular absorption and supramolecular emission are inseparable. However, cold gases will emit in the far infrared, which will often go unnoticed. However, the emission spectra of the outer planets are well known (see Fig. 9 below). Striking supramolecular emission occurs in hot and dense environments, e.g., shockwaves, "cold" stars, etc.

E. Collision-Induced Dipoles

In most cases when absorption or emission of electromagnetic radiation occurs, one can identify an electric dipole moment that is responsible for such spectroscopic processes. Rotating or vibrating electric dipoles emit (and absorb) at the frequencies of rotation and vibration; translationally accelerated charges (dipoles) emit a continuum. Emission and absorption take place in *transitions* between certain quantum states. For example, molecules with a permanent electric dipole moment, such as HCl or H_2O, emit in transitions between rotovibrational states if certain selection rules (i.e., conservation of energy and angular momentum) are satisfied. Even if no permanent dipole moment exists, molecules may emit in transitions between *electronic* states, but these transitions typically require more energy than the rotovibrational ones: a higher photon energy for absorption and higher excitation energy (e.g., higher temperatures) for emission. At room temperature, at frequencies in the infrared and in rarefied gases, the common homonuclear diatomic gases (hydrogen, nitrogen, etc.) do not undergo electronic transitions; no absorption is observed because their inversion symmetry is inconsistent with the existence of a permanent dipole moment.

Supramolecular systems, on the other hand, usually do possess a "permanent" dipole moment during their short lifetime. Four mechanisms are known that induce an electric dipole moment in two or more interacting molecules: (1) multipole induction, (2) exchange force interactions, (3) dispersion interaction, and (4) molecular frame distortion—the same mechanisms that are familiar from the studies of the intermolecular forces.

Multipole induction arises from the fact that all molecules are surrounded by an electric field. While molecules are electrically neutral, the electric field surrounding each molecule is set up by the internal electronic and nuclear structure of the molecule. It may be described by a multipole expansion, i.e., by a superposition of dipole, quadrupole, octopole,... fields. For example, the monopole and dipole terms are zero for all neutral homonuclear diatomic molecules; in this case the lowest order multipole is a quadrupole. When two such molecules interact, the collisional partners are polarized and thus possess momentarily—for the duration of the collision—dipole moments that interact with electromagnetic radiation. In the case of pure compressed hydrogen gas, quadrupolar induction provides nearly 90% of the total induced absorption. Since the quadrupole field rotates with the molecule, collision-induced rotational S lines are quite prominent in the spectra of compressed hydrogen and, of course, of all similar molecules.

Exchange forces control the repulsive part of the intermolecular interactions. In a collision at near range, when the electronic charge clouds of the collisional partners overlap, a momentary redistribution of electric charge occurs that is caused by electron exchange (Pauli exclusion principle). Especially when dissimilar atoms or molecules are involved, a dipole moment results from this redistribution. In most cases, the partner with fewer electrons temporarily assumes a positive charge, and the other assumes a negative one. This mechanism is usually the dominant one when dissimilar particles collide (as is the case for the spectra shown in Figs. 1 and 4). Exchange force-induced dipoles in molecules can also have a certain anisotropy of quadrupolar or higher symmetry. Examples of spectral features induced by anisotropic overlap are shown in Figs. 3 and 4 (components marked 1-2) and also implicitly in most of the other figures.

Dispersion forces control the attractive part of the intermolecular interactions. Over moderately wide separations, atoms or molecules interact through dispersion forces that are of an electric nature and arise from electronic intercorrelation. For dissimilar pairs, these are associated with a dipole moment whose asymptotic strength is proportional to the inverse seventh power of the intermolecular separation, and the polarity is typically the opposite of the overlap-induced dipole. The dispersion dipole is usually weaker than multipole-induced and overlap-induced dipoles, but is usually discernible in discriminating analyses.

Molecular frame distortion by collisions may break temporarily the high symmetry many molecules possess.

For example, the unperturbed CH_4 molecule, owing to its tetrahedral symmetry, has a zero dipole moment, even though the C–H bond is strongly polar: the vector sum of these four dipoles is zero as long as the exact tetrahedral symmetry persists. However, the momentary displacement of one of the hydrogen atoms by a collision will immediately produce a non-zero dipole moment which then may interact with radiation.

Typically, three or all four of the interaction-induced dipole moments are present at the same time. They cause collision-induced absorption, as well as the absorption of bound van der Waals systems. The point to be made here is that supramolecular complexes may have properties very different from those of the (non-interacting) monomers, such as a dipole moment even if the non-interacting constituents may be without.

We note that collision-induced dipoles are roughly one to three orders of magnitude weaker than the permanent dipoles of ordinary molecules. Correspondingly, the collision-induced spectra of individual pairs of molecules are typically much less intense than analogous spectra of ordinary molecules. Nevertheless, at high density the spectra can be intense. Observable intensities, when integrated over a line or a spectral band, are proportional to the number of molecules, that is, to the density, if the spectra of polar molecules are considered. For binary-induced spectra, on the other hand, integrated intensities are proportional to the number of *pairs* in a sample. If N molecules exist in a box, we have $N(N-1)/2$ pairs. Since, in all practical cases, N is a very large number, N and $(N-1)$ are nearly indistinguishable, and we can approximate the number of pairs by $N^2/2$. This quadratic density dependence may, at high densities, generate substantial intensities from pairs of molecules even if the individual collision-induced dipole moments are weak.

1. Ternary Dipoles

Recently, semi-empirical models of the most significant dipole components of three interacting homonuclear diatomic molecules have been obtained. Ternary-induced dipoles are the vector sum of the pairwise-additive dipoles (which are often well known) and the (previously essentially unknown) irreducible ternary dipoles. The model is consistent with three different experimental manifestations of the irreducible dipole component, in this case of compressed hydrogen gas: the third virial coefficient of the collision-induced, integrated absorption spectrum of the fundamental band of H_2; the triple transition $3Q_1$ by absorption of a single photon at $12,466\,cm^{-1}$; and certain features of the intercollisional dip of the Q_1 line, all of which are significantly shaped by the irreducible induced dipole.

When two molecules collide at near range, exchange forces redistribute electronic charge. As a result, a collision-induced dipole may thus be created, along with higher electric multipole moments. The strongest of these is the exchange-force-induced quadrupole moment. In the electric field of that quadrupole a third molecule will be polarized and an irreducible ternary dipole is thus created. There are several other mechanisms that contribute some to the irreducible dipole moment, but the exchange quadrupole-induced dipole (EQID) appears to be by far the most significant contribution to the irreducible ternary-induced dipole of three H_2 or similar molecules.

2. Dipoles and Dense Matter

In dense systems (e.g., in liquids and solids) the three-body and probably higher order cancellations due to destructive interference are most important. We distinguish between two components of translational spectra: one due to the diffusive and the other to the oscillatory ("rattling") motions of the molecules in a liquid. The latter is the analog of the intercollisional spectrum and consists of a dip to very low intensities near zero frequencies. While translational spectra of monatomic liquids are rather well understood, this cannot be said of those of the molecular liquids.

3. Electronic-Induced Dipoles

The induction proceeds, for example, via the polarization of molecule B in the multipole field of the electronically excited atom A, by long-range transition dipoles which vary with intermolecular separation R as R^{-3} or R^{-4}, depending on the symmetry of the electronic excited state involved. At near range, a modification of the inverse power dependence due to electron exchange is often quite noticeable, much as this is known for the rotovibrational collision-induced spectra discussed above.

III. COLLISION-INDUCED LIGHT SCATTERING

If a molecular substance is irradiated by the intense monochromatic light of a laser, light of the same frequency ν_0 is scattered (the Rayleigh line). Besides, at lower frequencies ($\nu < \nu_0$, Stokes wing), as well as at higher ones ($\nu > \nu_0$, anti-Stokes wing), other lines may appear that are shifted relative the incident frequency by certain rototranslational molecular transition frequencies. Many such states normally exist, certainly for the Stokes wing. This basically describes the Raman spectrum of ordinary molecules.

If laser light is directed at a sample of a rarefied monatomic (rather than a molecular) gas, no Raman spectra besides the Rayleigh line are observed, owing to the absence of rotovibrational states. However, at high enough densities, continuous Stokes and anti-Stokes wings appear that are collision induced. These arise because the polarizability of two interacting atoms differs slightly from the sum of polarizabilities of the (non-interacting) atoms: the collisional complex possesses an *excess* polarizability which generates these wings. The wings are continuous because the translational energies of relative motion of the two atoms are continuous, i.e., unlike the rotovibrational energies of bound diatomic molecules which are quantized like all periodic motion. Collision-induced Raman spectra of the rare gases were predicted and soon after demonstrated in actual measurements by G. Birnbaum and associates in 1967.

It has long been known that two types of light scattering by gases must be distinguished; so-called polarized scattering maintains the polarization of the incident laser beam, but depolarized scattering does not. Depolarized scattering arises from the anisotropic part of the polarization tensor of a molecule; it rotates the polarization plane of the incident laser beam randomly so that nearly no memory of the polarization of the incident beam remains in the scattered light at a given frequency.

Figures 7 and 8 show the Raman spectra of two interacting He atoms as obtained at gas densities corresponding to 10 or 20 times their density at standard temperature and pressure (1 atm and 273 K). Similar spectra are obtained in a gas of the rare isotope, ^{3}He. In an actual measurement the depolarized (Fig. 7) and the polarized (Fig. 8) spectra are superimposed and must be separated artificially, on the basis of their different polarizations, which introduces considerable uncertainty in the weaker of the two (Fig. 8). Similar collision-induced Raman spectra are known for the other rare gases and for mercury vapor, another monatomic gas.

The profiles of the Raman spectra, Figs. 7 and 8, resemble the spectral density function $g(\omega)$ seen in collision-induced absorption (Fig. 1). In fact, Heisenberg's uncertainty principle, Eq. (1), will directly relate the observed half-widths of these profiles and the mean duration of the collision, just as this was pointed out above in Fig. 1. Note that the range of spectroscopic interaction of the trace is shorter than that of the anisotropy of the polarizability tensor, so that the polarized spectra are actually more diffuse than the depolarized ones, under otherwise comparable conditions. Furthermore, we note that the reader may find the profiles shown in Figs. 7 and 8 to look different from the dashed profile in Fig. 1. The apparent differences of shape are, however, largely due to the logarithmic intensity

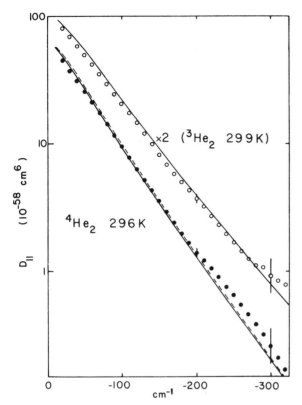

FIGURE 7 The Stokes side of the binary collision-induced depolarized Raman spectrum of the helium diatoms of the abundant and the rare isotope of helium at room temperature. The intense Rayleigh line (at zero frequency shift) was suppressed. Dots and circles represent the measurement; the solid lines represent a calculation based on the fundamental theory. For clarity of the display, the spectrum of the rare isotope was multiplied by a factor of 2. [After Dacre, P. D., and Frommhold, L. (1982). *J. Chem. Phys.* **76**, 3447.]

scales used in Figs. 7 and 8, when a linear scale was used in Fig. 1.

Similar work with compressed *molecular* gases demonstrates that the collision-induced Raman spectra are as universal as their collision-induced absorption counterpart discussed above. In molecular gases, additional rotovibrational lines and bands appear, much like those seen in the infrared, regardless of whether the gases are Raman-active or not.

Just as in the case of collision-induced absorption, collision-induced Raman spectra of binary and many-body complexes differ significantly. In the low-density limit in Raman-inactive gases, the binary collision-induced spectra are dominant, and the intensities vary as the square of density. At increasing densities a point is reached where many-body cancellations of the binary intensities are observed, owing to ternary (and higher order) interactions. At

still higher densities, and in the case of liquids and solids, many-body collisional complexes contribute to the observable spectra which further modifies the induced spectra seen in the low-density limit.

A. Collision-Induced Polarizabilities

All molecules are polarized in an external electric field, i.e., the field pulls the electrons slightly off to one side and the nuclei to the other so that a "field-induced dipole moment" **d** results,

$$\mathbf{d} = \mathbf{A} \cdot \mathbf{F}, \qquad (2)$$

where **F** is the electric field strength. If an alternating field (laser) is used, the field-induced dipole moment oscillates and thus radiates at the same frequency as the incident light (Rayleigh scattering). Since the polarizability **A**, a 3×3 tensor with two invariants, trace, and anisotropy, depends on the molecular orientation and the nuclear structure of the molecule, the scattered light is also modulated by certain molecular rotation and vibration frequencies. These modulations make up the familiar Raman spectra of the ordinary (unperturbed) molecules.

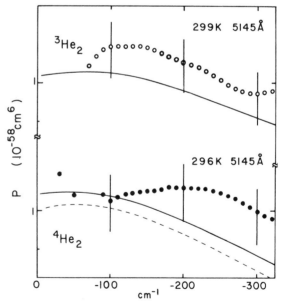

FIGURE 8 The Stokes side of the collision-induced polarized Raman spectrum of the helium diatoms at room temperature. The intense Rayleigh line has been suppressed. Dots and circles represent the measurements. The size of the error bars give an impression of the uncertainty of the measurement, which was substantial, owing to the low scattered intensities in the presence of the depolarized background. The solid lines are calculations based on the fundamental theory. [After Dacre, P. D., and Frommhold, L. (1982). *J. Chem. Phys.* **76**, 3447.]

Supramolecular Raman spectra are generated by the *excess* of the polarizability invariants of the interacting complex over the sum of the corresponding invariants of the non-interacting (i.e., widely separated) members of the supramolecular complex. A simple example will illustrate this: The atoms of the monatomic gases are completely isotropic; the field-induced dipole will always be exactly parallel to the applied field, $\mathbf{d} \parallel \mathbf{F}$, so that the anisotropy (which would rotate the dipole relative to the field) is zero—as long as the atoms are sufficiently spaced apart (rarefied gas). At higher densities, one cannot ignore the likelihood of another atom B sailing by close enough so that the local field at the position of atom A is perturbed: it is now the vector sum of the external field and the electric field set up by the dipole induced in atom B. For example, the resulting pair dipole moment and thus the polarizability is typically greater than the sum of atomic polarizabilities if the internuclear axis is parallel to the field; it is smaller if the internuclear axis is perpendicular to the field. At the other orientations, the plane of polarization of the scattered light is actually rotated relative to that of the incident beam. In other words, the anisotropy of the atom pair is no longer zero and a collision-induced depolarized Raman spectrum exists. To a large extent, this is simply due to the fact that the effective exciting field in the former case is enhanced by the vicinity of the dipole in the collisional partner, while in the latter case it is weakened (this is the dipole-induced-dipole interaction, often abbreviated DID). To some lesser extent, electronic overlap at near range and dispersion forces at more distant range have an effect on the polarizability of the pair, and the DID model must be considered an approximation that describes roughly 90% of the pair polarizability of most systems of interest; it is thus quite useful.

The field-induced, excess dipole moment of the pair goes from very small values at large initial separations to a maximum at the point of nearest approach, whereupon it falls off again. During this time, the dipole is oscillating with the very high frequency of the exciting field. The Raman spectra (e.g., the Fourier transform of the autocorrelation function of the time-varying excess dipole moment) consist of the Rayleigh line and the Stokes and anti-Stokes wings. While the Rayleigh line arises mostly from the scattering of monomers so that its width is not affected by the short lifetime of the collisional complexes, the Stokes and anti-Stokes wings are of a width consistent with the reciprocal lifetime of the collisional complex.

We note that the pressure-induced depolarization of scattered light by dense, isotropic gases is closely related to the mechanism that produces the depolarized collision-induced spectra.

IV. VIRIAL EXPANSIONS

A. Equation of State

A gas in thermal equilibrium obeys the equation of state,

$$p = kT \left\{ \frac{N_A}{V} + \frac{B(T)}{V^2} + \frac{C(T)}{V^3} + \cdots \right\},$$

where p, T, V, k designate pressure, temperature, molar volume, and Boltzmann's constant, respectively; N_A is Avogadro's number, and $B(T)$, $C(T)$, ... are the second, third, ... virial coefficients of the equation of state. The infinite series to the right is called a virial expansion.

In the limit of a highly diluted gas ($V \to \infty$), the right-hand side of this expression is equal to ρkT, where $\rho = N_A/V$ is the density of the gas. This is the *ideal gas approximation* of the equation of state which approximates the gas as a collection of non-interacting point particles—which is quite a reasonable model for rarefied gases. The second and higher terms in the virial expansion represent the effects of the intermolecular interactions. In particular, the second virial coefficient $B(T)$ expresses the effect of the strictly binary interactions upon the pressure of the gas. The third coefficient $C(T)$ describes the effect of the ternary interactions, and so forth. With increasing gas densities (e.g., in compressed gases) these virial coefficients become more and more important. At densities as high as those of liquids, the virial expansion becomes meaningless. Under such conditions every particle of the fluid is in simultaneous interaction with quite a number of other particles nearby.

The second virial coefficient is expressed in terms of the pair-interaction potential $V(R)$,

$$B(T) = -2\pi N_A^2 \int_0^\infty \left(e^{-V(R)/kT} - 1 \right) R^2 \, dR.$$

Similarly, $C(T)$ can be expressed in terms of ternary interactions, of both the pairwise and the irreducible kind. Valuable information on intermolecular interactions has been obtained from measurements of the virial coefficients. The second, third, ... virial coefficients are functions of temperature only and can be calculated in terms of the interactions of two, three, ... molecules in the volume V. In other words, the N_A-body problem of the imperfect gas has been reduced to a series of one-, two-, three-body, ... problems which are much more tractable than the very difficult many-body problem of an amorphous fluid.

B. Other Virial Expansions

It is plausible that besides the equation of state there should be other thermodynamic functions, i.e., other properties of matter that may be described by superposition of the effects of unitary, binary, ternary, . . . , molecular interactions, that is, by a virial expansion. In fact, the discussion of collision-induced spectral intensities above suggests such a virial expansion of the observed line shapes and the integrated absorption coefficients.

C. Collision-Induced Spectra

Rarefied gases interact with electromagnetic radiation in proportion to density variations: if at a given frequency absorption exists, the absorption coefficient will, in general, double if gas densities (pressures) are doubled. However, with increasing number density of the gas, as we approach roughly $\approx 1\%$ of the liquid state densities, supramolecular absorption, emission, and light scattering becomes increasingly important: the contributions from *pairs* of molecules will increase as density squared; molecular *triples* contribute proportional to density cubed; etc. In other words, at not too high densities and at most (but not all!) frequencies of a given collision-induced band, a virial expansion of spectral intensities is possible and permits a separation of the contributions of monomers, dimers, trimers, ... It is clear that with increasing gas densities the collision-induced contributions must be of increasing importance—regardless of whether the dimers, trimers are van der Waals molecules or collisionally interacting complexes with fluctuating dipole moments.

The leading term of any virial expansion is due to the non-interacting monomer contributions. Contrary to the leading coefficient of the equation of state, the corresponding spectroscopic first virial coefficient vanishes if the molecules are infrared- or Raman-inactive; in that case the virial series starts with the second spectroscopic coefficient, expressible in terms of the pair-interaction potential and the induction operator (i.e., induced dipole or induced polarizability invariants), depending on whether we consider absorption and emission or the Raman process.

We note that certain sum formulas, e.g., the integrated intensity of a collision-induced band, may at intermediate densities be represented by a virial expansion.

D. Dielectric and Refractive Virial Expansions

Other equilibrium properties of gases and liquids are known that possess a virial expansion and are intimately related to the collision-induced spectroscopies. The density dependence of the relative dielectric constant ϵ of a gas is given by the Clausius-Mossotti equation,

$$\frac{\epsilon - 1}{\epsilon + 2} = \frac{A_\epsilon(T)}{V} + \frac{B_\epsilon(T)}{V^2} + \frac{C_\epsilon(T)}{V^3} + \cdots,$$

where $A_\epsilon, B_\epsilon, \ldots$ are the first, second, . . . dielectric virial coefficients. The dielectric coefficient provides a measure of the polarization of matter, and in principle there are two mechanisms that can be distinguished: orientation of existing (permanent) dipoles in the external electric field and generation of dipoles by field-induced polarization, $\mathbf{d} = A\mathbf{F}$. In other words, the first virial coefficient is given by the Debye expression,

$$A_\epsilon(T) = \frac{4\pi N_A}{3(4\pi \epsilon_0)} \left(A + \frac{d_p}{3kT} \right),$$

which is the sum of the field-induced and the permanent dipole contribution. Accordingly, the second dielectric virial coefficient, which represents the leading term describing the lowest order deviations from the dielectric ideal gas behavior, is written as

$$B_\epsilon(T) = \frac{2\pi N_A^2}{3\Omega(4\pi \epsilon_0)}$$
$$\times \iint \left[A_{12}(R) + \frac{d_{12}^2}{3kT} \right] \mathrm{e}^{-V(R)/kT} \, d^3R \, d^2\omega_{12}.$$

The integration is over all positions and orientations of molecule 2 relative to molecule 1 (here assumed to be identical). The quantity Ω is defined by $\int d^3R \, d^2\omega_{12} = \Omega V$. The excess isotropic polarizability of the pair, $A_{12}(R)$, also called the collision-induced trace of the pair polarizability, is a function of intermolecular separation R. Similarly, the squared collision-induced dipole moment, d_{12}, more precisely the *excess* of the squared dipole moment of the pair above those of the non-interacting molecules, is also a function of separation and orientation.

Static dielectric properties are measured with static electric fields. If the frequencies of alternating fields approach those of visible light, the refractive index n, which is related to the dynamic (frequency-dependent) dielectric constant by $\epsilon = n^2$, becomes important. In this case an equation that is completely analogous to the Clausius-Mossotti expression, and which is commonly called the Lorentz-Lorenz equation, is of interest. It has formally the same virial coefficients as the former, only the polarizabilities are now measured at the frequencies of visible light. Furthermore, when at high frequencies the rotational inertia of polar molecules does not permit the molecules to orient fast enough in response to the applied alternating field, the orientational terms simply disappear from the expression for $B_\epsilon(T)$.

The isotropic pair polarizability A_{12} mentioned here (the trace of the pair polarizability tensor) is the very same quantity that controls the purely polarized collision-induced Raman process. Furthermore, the collision-induced dipole moments, d_{12}, whose squares occur in the expression for the second virial dielectric coefficient, are clearly the same that cause collision-induced absorption (and emission), and intimate relationships exist between the second virial spectroscopic coefficient of absorption and the part of the dielectric coefficient $B_\epsilon(T)$ that arises from the orientational dependence.

E. Kerr Constant

If a uniform, strong electric field is applied to a fluid, it becomes birefringent. In that case the refractive index is no longer isotropic (independent of the direction of propagation of light). Instead, we distinguish refractive indices $n_\parallel$ and $n_\perp$ for propagation of light in a direction parallel and perpendicular to the electric field vector, which increasingly differ as the field strength increases (Kerr effect). Every substance has a Kerr constant K which determines how much $n_\parallel$ and $n_\perp$ will differ for a given field strength, F,

$$K = \lim_{F \to 0} \left\{ \frac{6n(n_\parallel - n_\perp)V}{(n^2 + 2)^2(\epsilon + 2)^2 F^2} \right\}.$$

This effect is related to the optical anisotropy of molecules. In optically anisotropic gases, the molar Kerr constant varies linearly with gas density. However, as the density is increased, collision-induced optical anisotropies arise, which can be accounted for by a virial expansion such as

$$K = A_K(T) + B_K(T) \frac{1}{V} + C_K(T) \frac{1}{V^2} + \cdots.$$

The $A_K(T), B_K(T), \ldots$ are the first, second, $\ldots$ virial Kerr coefficients. A_K is the ideal gas value of the molar Kerr constant; for monatomic gases it is related to the hyperpolarizability. The second Kerr coefficient is given by

$$B_K(T) = \frac{8\pi^2 N_A^2}{405kT(4\pi\epsilon_0)} \int_0^\infty \beta_v(R)\beta_0(R)e^{-V(R)/kT}R^2\,dR,$$

where $\beta_0(R)$ and $\beta_v(R)$ are the (nearly equal) collision-induced anisotropies at zero frequency and the frequency v of the incident light, respectively. It is exactly this same collision-induced anisotropy that generates the depolarized collision-induced Raman spectra.

V. SIGNIFICANCE FOR SCIENCE AND TECHNOLOGY

A. Molecular Physics

Throughout this article numerous remarks have been made that illustrate the significance of the collision-induced spectroscopies for the study of molecular interactions. We summarize these by stating that complete binary spectra can be reproduced in all detail by a rigorous, quantum mechanical procedure if two functions of the molecular interaction are known: the intermolecular interaction potential and the pair induction operator (i.e., the collision-induced dipole surface if infrared absorption and emission are considered, or the collision-induced polarizability invariants for polarized or depolarized, collision-induced Raman spectra, respectively). Inversely, we may say that measurements of such collision-induced spectra *define* these functions of the interaction, certainly if accurate spectra over a wide frequency and temperature range are obtained. At present, no entirely satisfactory procedure is known to obtain these functions from spectroscopic measurements. Nevertheless, reasonably successful inversions of measurements exist which have generated valuable information concerning intermolecular interactions.

While ternary spectra are known and valuable pioneering work with various third virial coefficients exists, the precision of the data is usually somewhat limited. Consequently, the analyses have not always been as discriminating as one would like. The problem is, of course, the presence of the dominating binary process, combined with the contributions from four-body (and higher) interactions. A subject of considerable interest, namely, the separation of the irreducible parts of the three-body interactions from the pairwise component, has recently shown great promise.

B. Atmospheric Sciences

The interest of the planetary scientist in collisional absorption comes as no surprise. The most abundant molecules and atoms in space are non-polar (H_2, H, and He). In the dense and cool regions of space (i.e., in planetary atmospheres and "cool" stars), the most significant spectroscopic signatures that can be observed in the infrared are of the collision-induced nature. The atmospheres of the outer planets are opaque in the far infrared because of collision-induced absorption of H_2–H_2 and H_2–He pairs.

The exploration of the solar system by infrared spectroscopy has been an exciting and productive area of astronomical research, especially in recent times. Major goals are the detection of atmospheric constituents and their elemental and isotopic abundance ratios, which are so important to the understanding of the evolutionary process of the solar system, the establishment of thermal properties (i.e., brightness temperature and effective temperature), and the vertical thermal structure of the atmospheres ($p - T$ profiles). In all these endeavors collision-induced spectroscopy is an essential ingredient. While we have emphasized the hydrogen (H_2–H_2) collision-induced spectral contributions, other systems such as H_2–He, H_2–CH_4, and H_2–N_2 are also important under many conditions. The two systems mentioned last are, for example, of interest in

Titan's atmosphere, which like the Earth's atmosphere is composed primarily of nitrogen. It has, therefore, received much attention in recent years. The related spectra of the van der Waals dimers of such systems [e.g., $(H_2)_2$ and H_2N_2] appear to be of considerable interest for astrophysics for the same reasons.

The atmospheres of the outer planets are thought to be of a composition that resembles that of the primordial solar nebula. Therefore, the study of the composition might provide important answers to the ancient scientific problem of the origin of the solar system. Collision-induced spectroscopy can help to determine one of the most important parameters of primordial matter, the helium-to-hydrogen abundance ratio, which may be determined from the distinct features of the collision-induced spectra of H_2–H_2 and H_2–He and of the associated $(H_2)_2$ dimer spectra. (It is noteworthy that the H_2–He system is one of the few that do not form a bound dimer.)

Figure 9 shows the far infrared part of the emission spectra of the north equatorial region of Jupiter, recorded by the Voyager I IRIS spectrometer during the 1979 fly-by encounter. The frequency axis (abscissa) ranges from 200 to 600 cm^{-1}. Intensities are measured in units of *brightness temperature*, i.e., the temperature a black body has that emits the same intensity at the given wavelength. It is clear that high brightness temperature corresponds to high emission intensity. However, one should keep in mind the fact that brightness temperature is a highly non-linear measure of intensity.

At low frequencies, we notice structures that have been identified as molecular bands of the NH_3 molecule. Simi-

lar bands of other molecules, such as CH_4, occur at higher frequencies (>600 cm^{-1}) and are not shown in the figure. These strong (allowed) bands come from the deep interior of the atmospheres where temperatures are high. Of special interest here are the broad, relatively unstructured regions of the smallest intensities, extending from about 250 to beyond 600 cm^{-1}. These are the collision-induced rotational absorption lines of H_2 molecules that are collisionally interacting with other H_2 molecules or with He atoms. (These lines are the same shown in Figs. 3 and 4.) We note that helium atoms are the second most abundant species after H_2 in the atmosphere. Contrary to the NH_3 and CH_4 bands mentioned, these lines are forbidden in the non-interacting H_2 molecules. The broad dips of the emission spectra are centered near 354 and 585 cm^{-1}. The H_2 $S_0(0)$ and $S_0(1)$ lines are actually "dark fringes" in the thermal emission spectrum of Jupiter; their origin is completely analogous to the well-known dark Fraunhofer lines in the solar spectrum. The relatively cool outer regions of Jupiter's atmosphere are opaque at these frequencies, owing to collision-induced absorption; emitted radiation reflects the temperatures of these opaque regions. In contrast to the Fraunhofer lines, the collision-induced features are very broad because of the short durations of typical H_2–H_2 and H_2–He collisions [Eq. (1)]. We note that the hydrogen densities in the outer regions where collision-induced absorption takes place amount to about 0.4 amagat; the mean absorption pathlength amounts to roughly 20 km.

C. Astrophysics

In 1952, a few years after the discovery of collision-induced absorption, G. Herzberg pointed out the collision-induced $S_3(0)$ overtone structure of hydrogen in the spectra of Uranus and Neptune. This was the first direct evidence for the existence of molecular hydrogen (H_2) in the atmospheres of the outer planets, which consist of roughly 90% H_2 molecules! This direct detection of H_2 had to await the discovery of collision-induced absorption.

D. Applied Sciences

The liberation of observational data for astronomy, satellite-supported meteorology, and remote atmospheric sensing from the aggravating influence of the Earth's atmosphere has been a classical problem in the applied sciences. Precise quantitative knowledge of the coefficients of continuous absorption, especially in the far wings of spectral lines, and of their temperature dependence is indispensable for the solution of the inverse problem in satellite meteorology and weather prediction. The inverse problem attempts to reproduce accurately the distribution curves

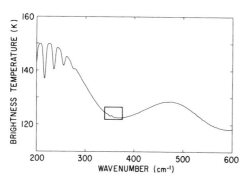

FIGURE 9 Emission spectrum of Jupiter's north equatorial belt obtained with the Voyager 1 IRIS far infrared spectrometer in the fly-by mission. The relatively sharp, striking structures at the lowest frequencies are NH_3 bands. The collision-induced $S(0)$ line of the H_2 molecule ranges from roughly 280 to 420 cm^{-1} as a broad, inverted feature. A similar dark and broad feature at higher frequencies is partially discernible, which is due to the $S(1)$ line of H_2, with a center near 585 cm^{-1}. The small rectangle near the center of the $S(0)$ line points out an interesting structure arising from bound-to-free transitions involving the van der Waals molecule $(H_2)_2$. [After Frommhold, L., Samuelson, R., and Birnbaum, G. (1984). *Astrophys. J.* **283**, L82.]

of physical parameters of the atmosphere from measurements of spectral composition and emission. For these tasks the collision-induced spectra of the atmospheric constituents (e.g., N_2, O_2, and H_2O) are essential.

The propagation of laser beams through the atmosphere is affected by atmospheric extinction from the scattering and absorption of light, both of which have a significant collision-induced component. Long-range monitoring of various physical and chemical parameters of the atmosphere (LIDAR) is a promising new direction in science and engineering; it is affected by collisional spectroscopies. To some extent all laser communication and information transmission systems, locating and telemetering systems, and mapping and navigational systems require access to quantitative data describing the effect of a dense atmosphere on the parameters of laser beams, which serve as the carriers of information.

Photoattenuation at wavelengths in the extreme red wings of resonant lines of electronic transitions have a strong collision-induced component. The degree of attenuation increases rapidly with increasing temperature, which has a detrimental effect on the performance of gas lasers. Since every scattering process has a stimulated counterpart, the coefficient of collision-induced scattering is likely to increase with increasing laser power (stimulated collision-induced scattering) to severely limit the highest attainable internal power density of high-power lasers. Collision-induced dipoles are known to be the prime cause of far-wing absorption of radiation in excimer lasers and multiphoton processes. Volumetric heating of non-polar gases, liquids, and even solids is possible by utilizing collision-induced absorption lines of the systems involved. Other applications in laser physics and chemistry have been proposed that attempt to control collisional processes and involve collision-induced spectroscopic transitions and lasers.

Frozen deuterium-tritium mixtures may be used as nuclear fuel for inertial confinement fusion reactors. Collision-induced, vibrational-rotational spectra of liquid and solid mixtures of deuterium are known, which are the isotopic analogs of the hydrogen spectra shown in Fig. 5. However, new infrared lines in the tritiated solid hydrogens below about 11 K were observed, which are due to tritium molecules perturbed by the electrostatic field of nearby ions that were formed by the beta rays of a decaying tritium nucleus. This is a form of collisional induction by a charged particle. The new lines are apparently those of the fundamental band but are shifted by the strong field of the electric monopoles (Stark shift).

In recent years a considerable technological interest in the non-linear optical properties of liquids has evolved. It centers around the third-order susceptibility, which controls many aspects of optical signal processing, image processing, stimulated scattering, and so on. Molecular susceptibility is related to the polarizabilities that determine the Rayleigh and induced Raman spectra of the fluids.

VI. CONCLUSIONS

The examples of collision-induced spectra shown in this article are chosen for their relative simplicity. The spectra were those of complexes of atoms and simple molecules, recorded under well-defined laboratory conditions and reproduced from the fundamental theory with precision for a demonstration of the basic principles involved. These choices, however, do not indicate the scope of collisional induction, which actually encompasses (1) quite large molecules as well as the smallest ones; (2) virtually any gas or mixtures of gases, liquids, and solids; (3) spectra in virtually any frequency band of the electromagnetic spectrum, up to X-ray frequencies; and (4) optical phenomena observable at any temperature, from near absolute zero to tens of thousands of kelvin.

Collision-induced spectroscopy is the extension of the spectroscopy of ideal gases to one of real gases and to important aspects of the condensed state. It is thus a very practical science that continues to provide new understanding of molecular interactions. Almost from the moment of their discovery, the collision-induced spectroscopies have had an enormous impact in astrophysics and other disciplines. Their significance for science and technology seems to be ever increasing. The field is diverse and has prospered through the furtherance of many disciplines and technologies. Not only has the full extent of microwave, infrared, and Raman spectroscopy with low and high resolutions been mobilized, but these techniques had to be paired with other advanced technologies (e.g., ultrahigh pressure capabilities and laser and cryogenic technologies) before the now familiar, very general statements concerning the collision-induced spectroscopies could be made. New theoretical thinking, combining the elements of statistical mechanics, liquid state theory, thermodynamics, quantum chemistry, and molecular dynamics studies, had to be developed and supported by modern supercomputers for the simulation of measurements and quantitative tests of the assumptions made. Perhaps because of the great diversity of interests and resources that have been essential for all work in the collision-induced spectroscopies, only recently have a few major attempts been known to review the field and to collect the existing knowledge in a few conference proceedings and monographs. These are quoted below.

Under conditions that are not favorable for the occurrence of electronic or molecular spectra, that is, at low temperatures and if non-polar molecules are considered, collision-induced spectra can be quite prominent, especially at high gas densities, in liquids and solids. Best known are the rototranslational absorption spectra in the far infrared and microwave regions of the non-polar gases and liquids; vibrational absorption bands in the near infrared, analogous Raman spectra, especially of the Raman-inactive gases; and various simultaneous transitions in pairs and triples of interacting molecules. Collisionally induced spectra are ubiquitous in dense environments in almost any gas; the only exception of such absorption spectra seem to be the pure monatomic gases.

Collision-induced spectra and the spectra of van der Waals molecules (of the same monomeric species) are due to the same basic dipole induction mechanism. An intimate relationship of the induction mechanisms responsible for the collision-induced spectra with the dielectric virial properties of matter exists.

SEE ALSO THE FOLLOWING ARTICLES

ATOMIC AND MOLECULAR COLLISIONS • ENERGY TRANS-FER • INFRARED SPECTROSCOPY • MICROWAVE SPEC-TROSCOPY, MOLECULAR • PLANETARY ATMOSPHERES • RAMAN SPECTROSCOPY

BIBLIOGRAPHY

Frommhold, L. (1994). "Collision-induced Absorption of Gases," Cambridge Univ. Press, Cambridge.

Birnbaum, G., Borysow, A., and Orton, G. S. (1996). "Collision-induced absorption of H_2–H_2 and H_2–He in the rotational and fundamental bands for planetary applications," *Icarus* **123**, 4.

Borysow, A., Jørgensen, U. G., and Zheng, C. (1997). *Astron. Astrophys.* **324**, 185.

Herman, R. M., ed. (1999). "Spectral Line Shapes," Vol. 10, Am. Inst. Physics, Woodbury, NY. See the articles in that volume by Borysow, A., Le Duff, Y., Moraldi, M., Tipping, R. H., and Zoppi, M., and others.

Tabisz, G. C., and Neuman, M. N., ed. (1995). "Collision- and Interaction-Induced Spectroscopy," NATO ASI Series C, Vol. 452, Kluwer Academic, Dordrecht, Boston, London.

Bibliographies of collision-induced absorption and light scattering exist which are being updated every few years; these are quoted in the literature mentioned above.

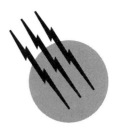

Color Science

Robert M. Boynton

University of California, San Diego

GLOSSARY

Chromaticity Ratios x, y, z of each of the tristimulus values of a light to the sum of the three tristimulus values X, Y, Z, these being the amounts of three primaries required to match the color of the light.

Chromaticity diagram Plane diagram formed by plotting one of the three chromaticity coordinates against another (usually y versus x).

Color Characteristics of sensations elicited by light by which a human observer can distinguish between two structure-free patches of light of the same size and shape.

Colorant A substance, such as a dye or pigment, that modifies the color of objects or imparts color to otherwise achromatic objects.

Colorimetry Measurement and specification of color.

Color matching Action of making a test color appear the same as a reference color.

Color order System of reference whereby the relation of one color to another can be perceived and the position of that color can be established with respect to the universe of all colors.

Color rendering General expression for the effect of a light source on the color appearance of objects in comparison with their color appearance under a reference light source.

Color temperature Absolute temperature of a blackbody radiator having a chromaticity closest to that of a light source being specified.

Metamerism (1) Phenomenon whereby lights of different spectral power distributions appear to have the same color. (2) Degree to which a material appears to change color when viewed under different illuminants.

Optimal colors Stimuli that for a given chromaticity have the greatest luminous reflectance.

Primaries (1) Additive: Any one of three lights in terms of which a color is specified by giving the amount of each required to match it by combining the lights. (2) Subtractive: Set of dyes or pigments that, when mixed in various proportions, provides a gamut of colors.

Radiance Radiant flux per unit solid angle (intensity) per unit area of an element of an extended source or reflecting surface in a specified direction.

Reflectance Ratio of reflected to incident light.

Reflection Process by which incident flux leaves a surface or medium from the incident side, without change in wavelength.

COLOR SCIENCE examines a fundamental aspect of human perception. It is based on experimental study under controlled conditions susceptible to physical measurement. For a difference in color to be perceived between two surfaces, three conditions must be satisfied: (1) There must be an appropriate source of illumination, (2) the two surfaces must not have identical spectral reflectances, and (3) an observer must be present to view them. This article is concerned with the relevant characteristics of lights, surfaces, and human vision that conjoin to allow the perception of object color.

I. PHYSICAL BASIS OF PERCEIVED COLOR

The physical basis of color exists in the interaction of light with matter, both outside and inside the eye. The sensation of color depends on physiological activity in the visual system that begins with the absorption of light in photoreceptors located in the retina of the eye and ends with patterns of biochemical activity in the brain. Perceived color can be described by the color names white, gray, black, yellow, orange, brown, red, green, blue, purple, and pink. These 11 basic color terms have unambiguous referents in all fully developed languages. All of these names (as well as combinations of these and many other less precisely used nonbasic color terms) describe colors, but white, gray, and black are excluded from the list of those called hues. Colors with hue are called chromatic colors; those without are called achromatic colors.

Although color terms are frequently used in reference to all three aspects of color (e.g., one may speak of a sensation of red, a red surface, or a red light), such usage is scientifically appropriate only when applied to the sensation; descriptions of lights and surfaces should be provided in physical and geometrical language.

II. CIE SYSTEM OF COLOR SPECIFICATION

A. Basic Color-Matching Experiment

The most fundamental experiment in color science entails the determination of whether two fields of light such as those that might be produced on a screen with two slide projectors, appear the same or different. If such fields are abutted and the division between them disappears to form a single, homogeneous field, the fields are said to match. A match will, of course, occur if there is no physical difference between the fields, and in special cases color matches are also possible when substantial physical differences exist between the fields. An understanding of how this can happen provides an opening to a scientific understanding of this subject.

Given an initial physical match, a difference in color can be introduced by either of two procedures, which are often carried out in combination. In the first instance, the radiance of one part of a homogeneous field is altered without any change in its relative spectral distribution. This produces an achromatic color difference. In the second case, the relative spectral distribution of one field is changed such that, for all possible relative radiances of the two fields, no match is possible. This is called a chromatic color difference.

When fields of different spectral distributions can be adjusted in relative radiance to eliminate all color difference, the result is termed a metameric color match. In a color-matching experiment, a test field is presented next to a comparison field and the observer causes the two fields to match exactly by manipulating the radiances of so-called primaries provided to the comparison field. Such primaries are said to be added; this can be accomplished by superposition with a half-silvered mirror, by superimposed images projected onto a screen, by very rapid temporal alternation of fields at a rate above the fusion frequency for vision, or by the use of pixels too small and closely packed to be discriminated (as in color television). If the primaries are suitably chosen (no one of them should be matched by any possible mixture of the other two), a human observer with normal color vision can uniquely match any test color by adjusting the radiances of three monochromatic primaries. To accomplish this, it sometimes proves necessary to shift one of the primaries so that it is added to the color being matched; it is useful to treat this as a negative radiance of that primary in the test field. The choice of exactly three primaries is by no means arbitrary: If only one or two primaries are used, matches are generally impossible, whereas if four or more primaries are allowed, matches are not uniquely determined.

The result of the color-matching experiment can be represented mathematically as $t(T) = r(R) + g(G) + b(B)$, meaning that t units of test field T produce a color that is matched by an additive combination of r units of primary R, g units of primary G, and b units of primary B, where one or two of the quantities r, g, or b may be negative. Thus any color can be represented as a vector in R, G, B space. For small, centrally fixated fields, experiment shows that the transitive, reflexive, linear, and associative properties of algebra apply also to their empirical counterparts, so that color-matching equations can be manipulated to predict matches that would be made with a change in the choice of primaries. These simple relations break down for very low levels of illumination and also with higher levels if the fields are large enough to permit significant contributions by rod photoreceptors or if the fields are so bright as to bleach a significant fraction of cone photopigments, thus altering their action spectra.

Matches are usually made by a method of adjustment, an iterative, trial-and-error procedure whereby the observer manipulates three controls, each of which monotonically varies the radiance of one primary. Although such settings at the match point may be somewhat more variable than most purely physical measurements, reliable data result from the means of several settings for each condition tested. A more serious problem, which will not be treated in this article, results from differences among observers. Although not great among those with normal color vision, such differences are by no means negligible. (For those with abnormal color vision, they can be very large.) To achieve a useful standardization—one that is unlikely to apply exactly to any particular individual—averages of normal observers are used, leading to the concept of a standard observer.

In the color-matching experiment, an observer is in effect acting as an analog computer, solving three simultaneous equations by iteration, using his or her sensations as a guide. Although activity in the brain underlies the experience of color, the initial encoding of information related to wavelength is in terms of the ratios of excitations of three different classes of cone photoreceptors in the retina of the eye, whose spectral sensitivities overlap. Any two physical fields, whether of the same or different spectral composition, whose images on the retina excite each of the three classes of cones in the same way will be indiscriminable. The action spectra of the three classes of cones in the normal eye are such that no two wavelengths in the spectrum produce exactly the same ratios of excitations among them.

B. Imaginary Primaries

Depending on the choice of primaries, many different sets of color-matching functions are possible, all of which describe the same color-matching behavior. Figure 1 shows experimental data for the primaries 435.8, 546.1, and

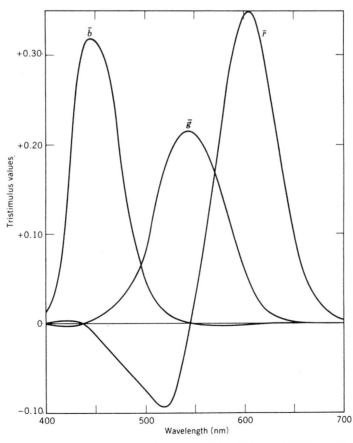

FIGURE 1 Experimental color-matching data for primaries at 435.8, 546.1, and 700.0 nm. [From Billmeyer, F. W., Jr., and Saltzmann, M. (1981). "Principles of Color Technology," 2nd ed. Copyright ©1981 John Wiley & Sons, Inc. Reprinted by permission of John Wiley & Sons, Inc.]

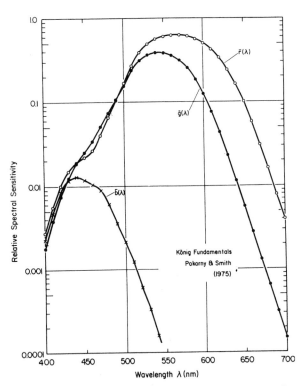

FIGURE 2 Estimates of human cone action spectra (König fundamentals) derived by V. Smith and J. Pokorny. [From Wyszecki, G., and Stiles, W. S. (1982). "Color Science: Concepts and Methods, Quantitative Data and Formulate," 2nd ed. Copyright ©1982 John Wiley & Sons, Inc. Reprinted by permission of John Wiley & Sons, Inc.]

700.0 nm. Depicted in Fig. 2 are current estimates of the spectral sensitivities of the three types of cone photoreceptors. These functions, which have been inferred from the data of psychophysical experiments of various kinds, agree reasonably well with direct microspectrophotometric measurements of the absorption spectra of outer segments of human cone photoreceptors containing the photopigments that are the principal determinants of the spectral sensitivity of the cones.

The cone spectral sensitivities may be regarded as color-matching functions based on primaries that are said to be imaginary in the sense that, although calculations of color matches based on them are possible, they are not physically realizable. To exist physically, each such primary would uniquely excite only one type of cone, whereas real primaries always excite at least two types.

Another set of all-positive color-matching functions, based on a different set of imaginary primaries, is given in Fig. 3. This set, which makes very similar predictions about color matches as the cone sensitivity curves, was adopted as a standard by the International Commission on Illumination (CIE) in 1931.

By simulating any of these sets of sensitivity functions in three optically filtered photocells, it is possible to remove the human observer from the system of color measurement (colorimetry) and develop a purely physical (though necessarily very limited) description of color, one that can be implemented in automated colorimeters.

C. Chromaticity Diagram

A useful separation between the achromatic and chromatic aspects of color was achieved in a system of colorimetry adopted by the CIE in 1931. This was the first specification of color to achieve international agreement; it remains today the principal system used internationally for specifying colors quantitatively, without reference to a set of actual samples.

The color-matching functions $\bar{x}(\lambda)$, $\bar{y}(\lambda)$, and $\bar{z}(\lambda)$ are based on primaries selected and smoothed to force the $\bar{y}(\lambda)$ function to be proportional to the spectral luminous efficiency function $V(\lambda)$, which had been standardized a decade earlier to define the quantity of "luminous flux" in lumens per watt of radiant power. The $\bar{x}(\lambda)$, $\bar{y}(\lambda)$, and $\bar{z}(\lambda)$ functions were then scaled to equate the areas under the curves, an operation that does not alter the predictions they make about color matches.

To specify the color of a patch of light, one begins by integrating its spectral radiance distribution $S(\lambda)$ in turn with the three color-matching functions:

$$X = k \int S(\lambda)\bar{x}(\lambda)\,d\lambda,$$

$$Y = k \int S(\lambda)\bar{y}(\lambda)\,d\lambda,$$

$$Z = k \int S(\lambda)\bar{z}(\lambda)\,d\lambda.$$

The values X, Y, and Z are called relative tristimulus values; these are equal for any light having an equal-radiance spectrum. Tristimulus values permit the specification of color in terms of three variables that are related to cone sensitivities rather than by continuous spectral radiance distributions, which do not. Like R, G, and B, the tristimulus values represent the coordinates of a three-dimensional vector whose angle specifies chromatic color and whose length characterizes the amount of that color.

Chromaticity coordinates, which do not depend on the amount of a color, specify each of the tristimulus values relative to their sum:

$$x = X/(X + Y + Z);$$

$$y = Y(X + Y + Z);$$

$$z = Z/(X + Y + Z)$$

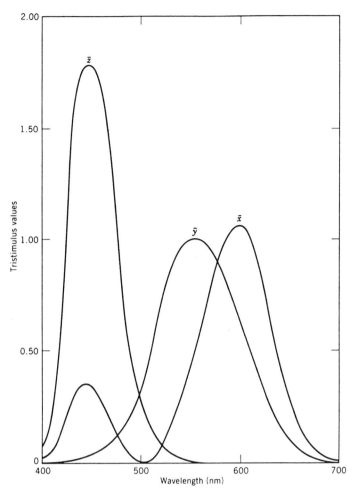

FIGURE 3 Tristimulus values of the equal-energy spectrum of the 1931 CIE system of colorimetry. [From Billmeyer, F. W., Jr., and Saltzmann, M. (1981). "Principles of Color Technology," 2nd ed. Copyright ©1981 John Wiley & Sons, Inc. Reprinted by permission of John Wiley & Sons, Inc.]

Given any two of these, the third is determined (e.g., $z = 1 - x - y$). Therefore, full information about chromaticity can be conveniently represented in a two-dimensional diagram, with y versus x having been chosen by the CIE for this purpose. The resulting chromaticity diagram is shown in Fig. 4. If one wishes to specify the quantity of light as well, the Y tristimulus value can be given, allowing a color to be fully specified as x, y, and Y, instead of X, Y, and Z. The manner in which the quantity of light Y is specified is determined by the normalization constant k.

Depending on the choice of primaries for determining color-matching functions, many other chromaticity diagrams are possible. For example, the set of color-matching functions of Fig. 1 leads to the chromaticity diagram of Fig. 5. This so-called *RGB* system is seldom used.

The affine geometry of chromaticity diagrams endows all of them with a number of useful properties. Most fundamental is that an additive mixture of any two lights will fall along a straight line connecting the chromaticities of the mixture components. Another is that straight lines on one such diagram translate into straight lines on any other related to it by a change of assumed primaries. The locations of the imaginary primaries X, Y, and Z are shown in Fig. 5, where one sees that the triangle formed by them completely encloses the domain of realizable colors. The lines X–Y and X–Z of Fig. 5 form the coordinate axes of the CIE chromaticity diagram of Fig. 4. Conversely, the lines B–G and B–R in Fig. 4 form the coordinate axes of the chromaticity diagram of Fig. 5. The uneven grid of nonorthogonal lines in Fig. 4, forming various angles at their intersections, translates into the regular grid of evenly spaced, orthogonal lines in Fig. 5. This illustrates that angles and areas have no intrinsic meaning in chromaticity diagrams.

The CIE in 1964 adopted an alternative set of color-matching functions based on experiments with large ($10°$) fields. Their use is recommended for making predictions about color matches for fields subtending more than $4°$ at the eye.

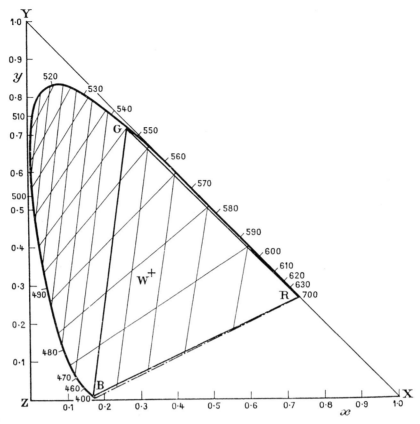

FIGURE 4 The *XYZ* chromaticity diagram showing the locations of the *RGB* primaries of Fig. 5 and the projection of the rectilinear grid of that figure onto this one. [From LeGrand, Y. (1957). "Light, Colour, and Vision," 2nd ed., Wiley Interscience, New York.]

Table I lists values of the CIE color-matching functions for 2° and 10° fields at 10-nm wavelength values. Tables for 1-nm wavelength values for 2° and 10° fields are available in *Color Measurement*, the second volume in the series *Optical Radiation Measurements*, edited by F. Grum and C. J. Bartleson.

III. COLOR RENDERING

From an evolutionary viewpoint, it is not surprising that sunlight is an excellent source for color rendering. Its strong, gap-free spectral irradiance distribution (Fig. 6) allows the discrimination of a very large number of surface-color differences. Color appearance in sunlight provides the standard against which the adequacy of other sources for color rendering is often judged.

A. Best and Worst Artificial Sources for Color Rendering

Of the light sources in common use today, low-pressure sodium is one of the poorest for color rendering, coming very close to being one of the worst possible. This illuminant consists mainly of the paired sodium lines that lie very close together (at 589.0 and 589.6 nm) in the "yellow" region of the spectrum; although some other spectral lines are also represented, these are present at such low relative radiances that low-pressure sodium lighting is for practical purposes monochromatic.

For a surface that does not fluoresce, its spectral reflectance characteristics can modify the quantity and geometry of incident monochromatic light, but not its wavelength. Viewed under separate monochromatic light sources of the same wavelength, any two surfaces with arbitrarily chosen spectral distributions can be made to match, both physically and visually, by adjusting the relative radiances of incident lights. Therefore, no chromatic color differences can exist under monochromatic illumination.

The best sources for color rendering emit continuous spectra throughout the visible region. Blackbody radiation, which meets this criterion, is shown for three temperatures in Fig. 7. These curves approximate those for tungsten sources at these temperatures.

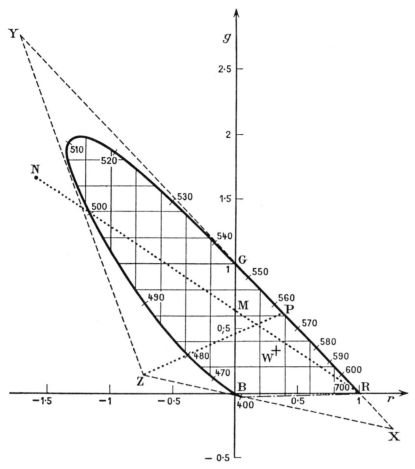

FIGURE 5 The *RGB* chromaticity diagram showing the locations of the *XYZ* primaries of Fig. 4. [From LeGrand, Y. (1957). "Light, Colour, and Vision," 2nd ed., Wiley Interscience, New York.]

B. Intermediate Quality of Fluorescent Lighting

Much of the radiant flux produced by incandescence emerges as infrared radiation at wavelengths longer than those visible; this is not true of fluorescent light, which is more efficiently produced, accounting for its widespread use. This light results from the electrical energizing of mercury vapor, which emits ultraviolet radiation. Although itself invisible, this radiation elicits visible light by causing the fluorescence of phosphors suspended in a layer coating the inside of a transparent tube.

Fluorescent lamps emit energy at all visible wavelengths, which is a good feature for color rendering, but their spectra are punctuated by regions of much higher radiance whose spectral locations depend on the phosphors chosen and the visible radiations of mercury vapor. Radiant power distributions of six types of fluorescent lamps are shown in Fig. 8.

C. Efficacy

The amount of visible light emitted by a source is measured in lumens, determined by integrating its radiant power output $S(\lambda)$ with the spectral luminous efficiency function $V(\lambda)$. The latter, which is proportional to $\bar{y}(\lambda)$, peaks at ~555 nm. Therefore, the theoretically most efficient light source would be monochromatic at this wavelength, with the associated inability to render chromatic color differences. Efficacy does not include power lost in the conversion from electrical input to radiant output, which may vary independently of the efficacy of the light finally produced.

D. Correlated Color Temperature

A blackbody, or Planckian radiator, is a cavity within a heated material from which heat cannot escape. No matter what the material, the walls of the cavity exhibit a

TABLE I Spectral Tristimulus Values for Equal Spectral Power Source

| a. CIE 1931 Standard Observer | | | | b. CIE 1964 Supplementary Observer | | | |
Wavelength (nanometer)	$\bar{x}(\lambda)$	$\bar{y}(\lambda)$	$\bar{z}(\lambda)$	Wavelength (nanometer)	$\bar{x}_{10}(\lambda)$	$\bar{y}_{10}(\lambda)$	$\bar{z}_{10}(\lambda)$
380	0.0014	0.0000	0.0065	380	0.0002	0.0000	0.0007
385	0.0022	0.0001	0.0105	385	0.0007	0.0001	0.0029
390	0.0042	0.0001	0.0201	390	0.0024	0.0003	0.0105
395	0.0076	0.0002	0.0362	395	0.0072	0.0008	0.0323
400	0.0143	0.0004	0.0679	400	0.0191	0.0020	0.0860
405	0.0232	0.0006	0.1102	405	0.0434	0.0045	0.1971
410	0.0435	0.0012	0.2074	410	0.0847	0.0088	0.3894
415	0.0776	0.0022	0.3713	415	0.1406	0.0145	0.6568
420	0.1344	0.0040	0.6456	420	0.2045	0.0214	0.9725
425	0.2148	0.0073	1.0391	425	0.2647	0.0295	1.2825
430	0.2839	0.0116	1.3856	430	0.3147	0.0387	1.5535
435	0.3285	0.0618	1.6230	435	0.3577	0.0496	1.7985
440	0.3483	0.0230	1.7471	440	0.3837	0.0621	1.9673
445	0.3481	0.0298	1.7826	445	0.3687	0.0747	2.0273
450	0.3362	0.0380	1.7721	450	0.3707	0.0895	1.9948
455	0.3187	0.0480	1.7441	455	0.3430	0.1063	1.9007
460	0.2908	0.0600	1.6692	460	0.3023	0.1282	1.7454
465	0.2511	0.0739	1.5281	465	0.2541	0.1528	1.5549
470	0.1954	0.0910	1.2876	470	0.1956	0.1852	1.3176
475	0.1421	0.1126	1.0419	475	0.1323	0.2199	1.0302
480	0.0956	0.1390	0.8130	480	0.0805	0.2536	0.7721
485	0.0580	0.1693	0.6162	485	0.0411	0.2977	0.5701
490	0.0320	0.2080	0.4652	490	0.0162	0.3391	0.4153
495	0.0147	0.2586	0.3533	495	0.0051	0.3954	0.3024
500	0.0049	0.3230	0.2720	500	0.0038	0.4608	0.2185
505	0.0024	0.4073	0.2123	505	0.0154	0.5314	0.1592
510	0.0093	0.5030	0.1582	510	0.0375	0.6067	0.1120
515	0.0291	0.6082	0.1117	515	0.0714	0.6857	0.0822
520	0.0633	0.7100	0.0782	520	0.1177	0.7618	0.0607
525	0.1096	0.7932	0.0573	525	0.1730	0.8233	0.0431
530	0.1655	0.8620	0.0422	530	0.2365	0.8752	0.0305
535	0.2257	0.9149	0.0298	535	0.3042	0.9238	0.0206
540	0.2904	0.9540	0.0203	540	0.3768	0.9620	0.0137
545	0.3597	0.9803	0.0134	545	0.4516	0.9822	0.0079
550	0.4334	0.9950	0.0087	550	0.5298	0.9918	0.0040
555	0.5121	1.0000	0.0057	555	0.6161	0.9991	0.0011
560	0.5945	0.9950	0.0039	560	0.7052	0.9973	0.0000
565	0.6784	0.9786	0.0027	565	0.7938	0.9824	0.0000
570	0.7621	0.9520	0.0021	570	0.8787	0.9556	0.0000
575	0.8425	0.9154	0.0018	575	0.9512	0.9152	0.0000
580	0.9163	0.8700	0.0017	580	1.0142	0.8689	0.0000
585	0.9786	0.8163	0.0014	585	1.0743	0.8526	0.0000
590	1.0263	0.7570	0.0011	590	1.1185	0.7774	0.0000
595	1.0567	0.6949	0.0010	595	1.1343	0.7204	0.0000
600	1.0622	0.6310	0.0008	600	1.1240	0.6583	0.0000
605	1.0456	0.5668	0.0006	605	1.0891	0.5939	0.0000
610	1.0026	0.5030	0.0003	610	1.0305	0.5280	0.0000

continues

TABLE I (*continued*)

a. CIE 1931 Standard Observer				b. CIE 1964 Supplementary Observer			
Wavelength (nanometer)	$\bar{x}(\lambda)$	$\bar{y}(\lambda)$	$\bar{z}(\lambda)$	Wavelength (nanometer)	$\bar{x}_{10}(\lambda)$	$\bar{y}_{10}(\lambda)$	$\bar{z}_{10}(\lambda)$
615	0.9384	0.4412	0.0002	615	0.9507	0.4618	0.0000
620	0.8544	0.3810	0.0002	620	0.8563	0.3981	0.0000
625	0.7514	0.3210	0.0001	625	0.7549	0.3396	0.0000
630	0.6424	0.2650	0.0000	630	0.6475	0.2835	0.0000
635	0.5419	0.2170	0.0000	635	0.5351	0.2283	0.0000
640	0.4479	0.1750	0.0000	640	0.4316	0.1798	0.0000
645	0.3608	0.1382	0.0000	645	0.3437	0.1402	0.0000
650	0.2835	0.1070	0.0000	650	0.2683	0.1076	0.0000
655	0.2187	0.0816	0.0000	655	0.2043	0.0812	0.0000
660	0.1649	0.0610	0.0000	660	0.1526	0.0603	0.0000
665	0.1212	0.0446	0.0000	665	0.1122	0.0441	0.0000
670	0.0874	0.0320	0.0000	670	0.0813	0.0318	0.0000
675	0.0636	0.0232	0.0000	675	0.0579	0.0226	0.0000
680	0.0468	0.0170	0.0000	680	0.0409	0.0159	0.0000
685	0.0329	0.0119	0.0000	685	0.0286	0.0111	0.0000
690	0.0227	0.0082	0.0000	690	0.0199	0.0077	0.0000
695	0.0158	0.0057	0.0000	695	0.0318	0.0054	0.0000
700	0.0114	0.0041	0.0000	700	0.0096	0.0037	0.0000
705	0.0081	0.0029	0.0000	705	0.0066	0.0026	0.0000
710	0.0058	0.0021	0.0000	710	0.0046	0.0018	0.0000
715	0.0041	0.0015	0.0000	715	0.0031	0.0012	0.0000
720	0.0029	0.0010	0.0000	720	0.0022	0.0008	0.0000
725	0.0020	0.0007	0.0000	725	0.0015	0.0006	0.0000
730	0.0014	0.0005	0.0000	730	0.0010	0.0004	0.0000
735	0.0010	0.0004	0.0000	735	0.0007	0.0003	0.0000
740	0.0007	0.0002	0.0000	740	0.0005	0.0002	0.0000
745	0.0005	0.0002	0.0000	745	0.0004	0.0001	0.0000
750	0.0003	0.0001	0.0000	750	0.0003	0.0001	0.0000
755	0.0002	0.0001	0.0000	755	0.0001	0.0001	0.0000
760	0.0002	0.0001	0.0000	760	0.0001	0.0000	0.0000
765	0.0002	0.0001	0.0000	765	0.0001	0.0000	0.0000
770	0.0001	0.0000	0.0000	770	0.0001	0.0000	0.0000
775	0.0001	0.0000	0.0000	775	0.0000	0.0000	0.0000
780	0.0000	0.0000	0.0000	780	0.0000	0.0000	0.0000
Totals	21.3714	21.3711	21.3715	Totals	23.3294	23.3324	23.3343

characteristic spectral emission, which is a function of its temperature. The locus of the chromaticity coordinates corresponding to blackbody radiation, as a function of temperature, plots in the chromaticity diagram as a curved line known as the Planckian locus (see Fig. 4). The spectral distribution of light from sources with complex spectra does not approximate that of a Planckian radiator. Nevertheless, it is convenient to have a single index by which to characterize these other sources of artificial light. For this purpose the CIE has defined a correlated color temperature, determined by calculating the chromaticity coordinates of the source and then locating the point on the blackbody locus perceptually closest to these coordinates.

E. Color-Rendering Index

The CIE has developed a system for attempting to specify the quality of color rendering supplied by any light source. The calculations are based on a set of reflecting samples specified in terms of their reflectance functions. The

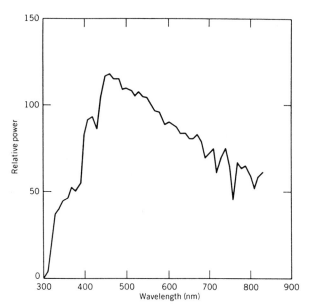

FIGURE 6 Spectral power distribution of typical daylight. [From Billmeyer, F. W., Jr., and Saltzmann, M. (1981). "Principles of Color Technology," 2nd ed., Copyright ©1981 John Wiley & Sons, Inc. Reprinted by permission of John Wiley & Sons, Inc.]

calculations begin with the choice of a reference illuminant specified as a blackbody (or daylight) radiator having a color temperature (or correlated color temperature) as close as possible to the correlated color temperature of the test illuminant; the choice of reference illuminant depends on the correlated color temperature of the test illuminant (daylight is used as a reference above 5000 K). For each of the samples, defined by their spectral reflectance functions, the amount of color shift ΔE introduced in going from reference to test illuminant is determined using the CIELUV formula described in Section VIB. There are 14 reference samples in all. A special color-rendering index R_i, peculiar to each sample, is calculated as $100-4.6\Delta E$. Most commonly a single-number index is calculated from the mean of a subset of eight special color-rendering indices to provide a final value known as the general color-rendering index R_a. The factor 4.6 was chosen so that a standard warm white fluorescent lamp would have an R_a of ~50; tungsten-incadescent sources score very close to 100. Table II gives R_a values for several commonly used artificial sources. Despite its official status, R_a is of limited value because of its many arbitrary features, especially its dependence on so limited a set of color samples. It is most useful for distinguishing large differences in color rendering, but not so useful for discriminating among sources of very high color-rendering properties. Individual values of R_i can be useful for determining the manner in which light sources differ in their color-rendering properties.

The intermediate color-rendering properties of most fluorescent light sources are closer to the best than to the worst. Mercury vapor and high-pressure sodium sources, widely used for street lighting, have poor color-rendering properties that fall between those of fluorescent and low-pressure sodium illumination.

IV. GLOBAL SURFACE PROPERTIES

The term *reflection* characterizes any of a variety of physical processes by which less than 100% of the radiant energy incident on a body at each wavelength is returned without change of wavelength. Reflection is too complicated for detailed specification at a molecular level for most surfaces and wavelengths of light. For this reason and because the molecular details are unimportant for many practical purposes, methods have been devised for measuring the spectral reflectance of a surface—the spectral distribution of returned light relative to that which is incident. Reflectance depends on the wavelength and angle of incidence of the light, as well as the angle(s) at which reflected light is measured.

A. Specular and Diffuse Reflectance

A familiar example of specular reflectance is provided by a plane mirror, in which the angles of light incidence and reflectance are equal. An ideal mirror reflects all incident light nonselectively with wavelength. If free of dust and suitably framed, the surface of an even less than ideal real mirror is not perceived at all; instead, the virtual image of an object located physically in front of the mirror is seen as if positioned behind.

Although specular reflectance seldom provides information about the color of a surface, there are exceptions. In particular, highly polished surfaces of metals such as gold, steel, silver, and copper reflect specularly. They also reflect diffusely from within but do so selectively with wavelength so that the specular reflection is seen to be tinged with the color of the diffuse component. More often, because highlights from most surfaces do not alter the spectral distribution of incident light, specular reflection provides information about the color of the source of light rather than that of the surface.

Diffuse reflectance, on the other hand, is typically selective with wavelength, and for the normal observer under typical conditions of illumination it is the principal determinant of the perceived color of a surface. A surface exhibiting perfectly diffuse reflectance returns all of the incident light with the distribution shown in Fig. 9, where the luminance (intensity per unit area) of the reflected light decreases a cosine function of the angle of reflection

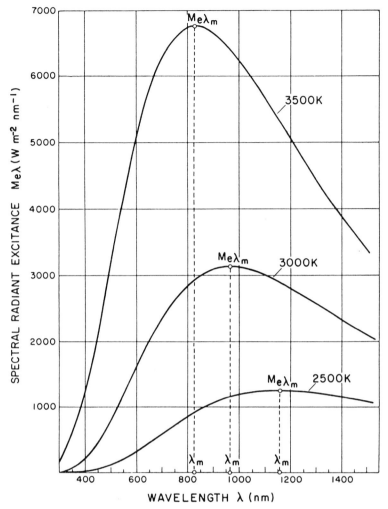

FIGURE 7 Spectral radiance distributions of a blackbody radiator at three temperatures. [From Wyszecki, G., and Stiles, W. S. (1982). "Color Science: Concepts and Methods, Quantitative Data and Formulae," 2nd ed. Copyright © 1982 John Wiley & Sons, Inc. Reprinted by permission of John Wiley & Sons, Inc.]

relative to normal. As such a surface is viewed more and more obliquely through an aperture, a progressively larger area of the surface fills the aperture—also a cosine function. The two effects cancel, causing the luminance of the surface and its subjective counterpart, lightness, to be independent of the angle of view.

No real surface behaves in exactly this way, although some surfaces approach it. Some simultaneously exhibit specular and diffuse reflectance; that of a new automobile provides a familiar example. The hard, highly polished outer surface exhibits specular reflectance of some of the incident light. The remainder is refracted into the layers below, which contain diffusely reflecting, spectrally selective absorptive pigments suspended in a binding matrix. Light not absorbed is scattered within this layer with an intensity pattern that may approximate that of a perfectly diffuse reflector. Because of the absorptive properties of

the pigments, some wavelengths reflect more copiously than others, providing the physical basis for the perceived color of the object.

Many intermediate geometries are possible, which give rise to sensations of sheen and gloss; these usually enable one to predict the felt hardness or smoothness of surface without actually touching them.

B. Measuring Diffuse Surface Reflectance

The diffuse spectral reflectance of a surface depends on the exact conditions of measurement. To some extent these are arbitrary, so that in order for valid comparisons of measurements to be made among different laboratories and manufacturers, standard procedures are necessary. To agree on and specify such procedures has been one of the functions of the CIE, which has recommended four

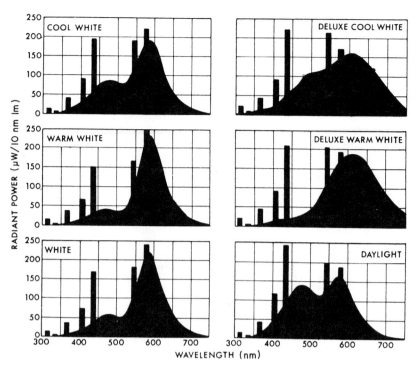

FIGURE 8 Spectral radiance distributions of six typical fluorescent lamps. [From Kaufman, J. E., ed. (1981). "IES Lighting Handbook; Reference Volume," © 1981 Illuminating Engineering Society of North America.]

procedures for measuring diffuse spectral reflectance, the most sophisticated of which is illustrated at the bottom left in Fig. 10. It makes use of an integrating sphere painted inside with a highly reflecting and spectrally nonselective paint made from barium sulfate. When light is admitted into an ideal integrating sphere, the sphere "lights up" uniformly as a result of multiple diffuse internal reflections. The size of the sphere does not matter so long as the ports cut into it, which permit entry and exit of the incident and reflected light, do not exceed 10% of the total surface area.

The surface to be measured fills an opening at the bottom, oriented horizontally. The incident light, which should be of limited cross section in order to be confined to the sample being measured, enters at an angle of 5° to normal. To eliminate the specular component of reflection from the measurement, a light trap is introduced, centered at an angle of 5° on the other side of normal; the remaining diffusely reflected component illuminates the sphere. Ideally, the exit port could be located almost anywhere. In practice, it is located as shown, so that it "sees" only a small opposite section of the sphere. As an added precaution, a small baffle is introduced to block the initially reflected component of light, which otherwise would strike the sphere in the area being measured. When the cap, shown in the lower left of Fig. 10 is black, it eliminates the specular (or direct) component, and only the diffuse reflectance is measured. When the cap is white (or when the sphere's surface is continuous, as at the bottom right

of the figure), both specular and diffuse components contribute, and the measurement is called total reflectance. Measurements are made, wavelength by wavelength, relative to the reflectance of a calibrated standard of known spectral reflectance. The spectral sensitivity of the detector does not matter so long as it is sufficiently sensitive to allow reliable measurements.

The arrangement of Fig. 10 ensures that all components of diffusely reflected light are equally weighted and that the specular component can be included in or eliminated from the measurement. Often, however, there is no true specular component, but rather a high-intensity lobe with a definite spread. This renders somewhat arbitrary the distinction between the specular and diffuse components. Operationally, the distinction depends on the size of exit port chosen for the specular light trap. Reflectance measurements are usually scaled relative to what a perfectly diffuse, totally reflecting surface would produce if located in the position of the sample. Figure 11 shows the diffuse spectral reflectance curves of a set of enamel paints that are sometimes used as calibration standards.

C. Chromaticity of an Object

The chromaticity of an object depends on the spectral properties of the illuminant as well as those of the object. A quantity $\phi(\lambda)$ is defined as $\rho(\lambda)S(\lambda)$ or $\tau(\lambda)S(\lambda)$, where $\rho(\lambda)$ symbolizes reflectance and $\tau(\lambda)$ symbolizes

TABLE II Color and Color-Rendering Characteristics of Common Light Sources[a]

Test lamp designation	CIE chromaticity coordinates		Correlated color temperature (Kelvins)	CIE general color rendering Index, R_a	CIE special color rendering indices R_i													
	x	y			R_1	R_2	R_3	R_4	R_5	R_6	R_7	R_8	R_9	R_{10}	R_{11}	R_{12}	R_{13}	R_{14}
Fluorescent lamps																		
Warm White	0.436	0.406	3020	52	43	70	90	40	42	55	66	13	−111	31	21	27	48	94
Warm White Deluxe	0.440	0.403	2940	73	72	80	81	71	69	67	83	64	14	49	60	43	73	88
White	0.410	0.398	3450	57	48	72	90	47	49	61	68	20	−104	36	32	38	52	94
Cool White	0.373	0.385	4250	62	52	74	90	54	56	64	74	31	−94	39	42	48	57	93
Cool White Deluxe	0.376	0.368	4050	89	91	91	85	89	90	86	90	88	70	74	88	78	91	90
Daylight	0.316	0.345	6250	74	67	82	92	70	72	78	82	51	−56	59	64	72	71	95
Three-Component A	0.376	0.374	4100	83	98	94	48	89	89	78	88	82	32	46	73	53	95	65
Three-Component B	0.370	0.381	4310	82	84	93	66	65	28	94	83	85	44	69	62	68	90	76
Simulated D_{50}	0.342	0.359	5150	95	93	96	98	95	94	95	98	92	76	91	94	93	94	99
Simulated D_{55}	0.333	0.352	5480	98	99	98	96	99	99	98	98	96	91	95	98	97	98	98
Simulated D_{65}	0.313	0.325	6520	91	93	91	85	91	93	88	90	92	89	76	91	86	92	91
Simulated D_{70}	0.307	0.314	6980	93	97	93	87	92	97	91	91	94	95	82	95	93	94	93
Simulated D_{75}	0.299	0.315	7500	93	93	94	91	93	93	91	94	91	73	83	92	90	93	95
Mercury, clear	0.326	0.390	5710	15	−15	32	59	2	3	7	45	−15	−327	−55	−22	−25	−3	75
Mercury improved color	0.373	0.415	4430	32	10	43	60	20	18	14	60	31	−108	−32	−7	−23	17	77
Metal halide, clear	0.396	0.390	3720	60	52	84	81	54	60	83	59	5	−142	68	55	78	62	88
Xenon, high pressure arc	0.324	0.324	5920	94	94	91	90	96	95	92	95	96	81	81	97	93	92	95
High pressure sodium	0.519	0.418	2100	21	11	65	52	−9	10	55	32	−52	−212	45	−34	32	18	69
Low pressure sodium	0.569	0.421	1740	−44	−68	44	−2	−101	−67	29	−23	−165	−492	20	−128	−21	−39	31
DXW tungsten halogen	0.424	0.399	3190	100	All 100 except for $R_6 = R_{10} = 99$														

[a] Lamps representative of the industry are listed. Variations from manufacturer to manufacturer are likely, especially for the D series of fluorescent lamps and the high-intensity discharge lamps. A high positive value of R_i indicates a small color difference for sample i. A low value of R_i indicates a large color difference.

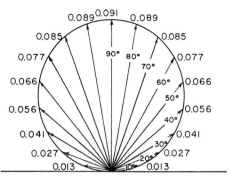

FIGURE 9 Intensity distribution of light reflected from a perfectly diffuse (Lambertian) surface, showing proportion of reflected light within 5° of each indicated direction. [From Boynton, R. M. (1974). *In* "Handbook of Perception" (E. C. Carterette and M. P. Friedman, eds.), Vol. 1. Copyright 1974 Academic Press.]

transmittance. Whereas reflectance ρ is the fraction of incident light returned from a surface, transmittance τ is the fraction of incident light transmitted by an object. Tristimulus values are then calculated as follows:

$$X = k \int \phi_\lambda \bar{x}(\lambda)\, d\lambda,$$

$$Y = k \int \phi_\lambda \bar{y}(\lambda)\, d\lambda,$$

$$Z = k \int \phi_\lambda \bar{z}(\lambda)\, d\lambda.$$

Calculation of chromaticity coordinates then proceeds as described above for sources. If an equal-energy spectrum is assumed, the source term $S(\lambda)$ can be dropped from the definition of $\phi(\lambda)$. When the chromaticity of a surface is specified without specification of the source, an equal-energy spectrum is usually implied.

D. Fluorescence

The practice of colorimetry so far described becomes considerably more complicated if the measured surface exhibits fluorescence. Materials with fluorescing surfaces, when excited by incident light, generally both emit light at a longer wavelength and reflect a portion of the incident light. When making reflectance measurements of nonfluorescent materials, there is no reason to use incident radiation in the ultraviolet, to which the visual mechanism is nearly insensitive. However, for fluorescent materials these incident wavelengths can stimulate substantial radiation at *visible* wavelengths. The full specification of the relative radiance (reflection plus radiation) properties of such surfaces requires the determination, for *each* incident wavelength (including those wavelengths in the ultraviolet known to produce fluorescence), of relative radiance at *all* visible wavelengths, leading to a huge matrix of measurement conditions. As a meaningful and practical shortcut, daylight or a suitable daylight substitute can be used to

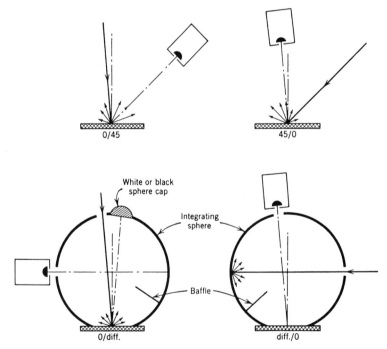

FIGURE 10 Schematic diagram showing the four CIE standard illuminating and viewing geometries for reflectance measurements. [From Wyszecki, G., and Stiles, W. S. (1982). "Color Science: Concepts and Methods. Quantitative Data and Formulae," 2nd ed. Copyright © 1982 John Wiley & Sons, Inc. Reprinted by permission of John Wiley & Sons, Inc.]

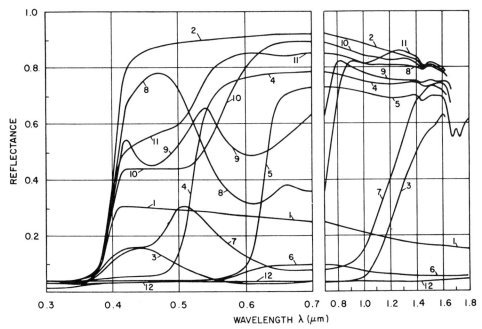

FIGURE 11 Diffuse spectral reflectance curves of a set of enamel paints having the following color appearances: (1) medium gray, (2) white, (3) deep blue, (4) yellow, (5) red, (6) brown, (7) medium green, (8) light blue, (9) light green, (10) peach, (11) ivory, and (12) black. [From Wyszecki, G., and Stiles, W. S. (1982). "Color Science: Concepts and Methods, Quantitative Data and Formulae," 2nd ed. Copyright © 1982 John Wiley & Sons, Inc. Reprinted by permission of John Wiley & Sons, Inc.]

irradiate the sample, and a spectrophotometer can be located at the exit port to register the spectral distribution of the reflected light, which will include the component introduced by fluorescence. Daylight substitutes are so difficult to obtain that the most recent standard illuminant sanctioned by the CIE, called D-65, has been specified only mathematically but has never been perfectly realized. (A suitably filtered, high-pressure xenon arc source comes close.)

E. Optimal Colors

Because of the broadband characteristics of the cone spectral sensitivity functions, most of the spectrum locus in the chromaticity diagram is very well approximated by wave bands as broad as 5 nm. A reflecting surface that completely absorbed all wavelengths of incident broadband (white) light and reflected only a 5-nm wave band would have a chromaticity approximating the midpoint of that wave band along the spectrum locus. Such a surface would also have a very low reflectance because almost all of the incident light would be absorbed. Extending the wave band would increase reflectance, but at the cost of moving the chromaticity inward toward the center of the diagram, with the limit for a nonselective surface being the chromaticity of the illuminant. For any particular reflectance, the domain of possible chro-

maticities can be calculated; the outer limit of this domain represents the locus of optimal surface colors for that chromaticity.

The all-or-nothing and stepwise reflectance properties required for optimal surface colors do not exist either in nature or in artificially created pigments (see Fig. 11), which tend to exhibit instead gently sloped spectral reflectance functions. For any given reflectance, therefore, the domain of real colors is always much more restricted than the ideal one. Figure 12 shows the CIE chromaticity diagram and the relations between the spectrum locus, the optimal colors of several reflectances, and the real surface colors of the Optical Society of America Uniform Color Scales set.

F. Metamerism Index

As already noted, *metamerism* refers to the phenomenon whereby a color match can occur between stimuli that differ in their spectral distributions. In the domain of reflecting samples the term carries a related, but very different connotation; here the degree of metamerism specifies the tendency of surfaces to change in perceived color, or to resist doing so, as the spectral characteristics of the illuminant are altered. Surfaces greatly exhibiting such changes are said to exhibit a high degree of metamerism, which from a commercial standpoint is undesirable.

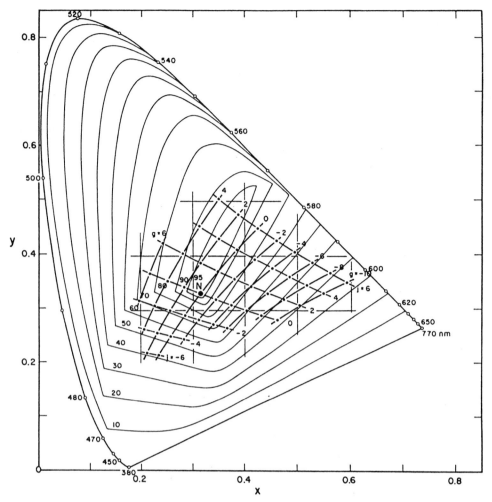

FIGURE 12 Locus of optimal colors of reflectances indicated, also showing the locations of 47 surface colors of ~30% reflectance developed by the Optical Society of America to be equally spaced perceptually. [From Wyszecki, G., and Stiles, W. S. (1982). "Color Science: Concepts and Methods, Quantitative Data and Formulae," 2nd ed. Copyright © 1982 John Wiley & Sons, Inc. Reprinted by permission of John Wiley & Sons, Inc.]

Most indices of metamerism that have been proposed depend either on the assessed change in color of specific surfaces with change in illuminant, calculated by procedures similar to those used for the color-rendering index of illuminants, or on the number of intersections of the spectral reflectance functions of the samples being assessed. For two samples to be metameric, these functions must intersect at least three times; in general, the more intersections, the lower is the degree of metamerism that results, implying more resistance to color change with a change in illuminant. In the limiting case, where the curves are identical, there is no metamerism and the match holds for all illuminants.

Except for monochromatic lights on the curved portion of the spectrum locus in the chromaticity diagram, the number of possible metamers is mathematically infinite. Taking into account the limits of sensitivity of the visual

system for the perception of color differences, the number of possible metamers increases as one approaches the white point on the chromaticity diagram, moving inward from the outer limits of realizable reflecting colors.

V. PHYSICAL BASIS OF SURFACE COLOR

The physical basis of the color of a surface is related to processes that alter the spectral distribution of the returned light in the direction of an observer, relative to that of the incident illumination.

A. Color from Organic Molecules

The action of organic molecules, which provide the basis for much of the color seen in nature, has been interpreted with accelerating precision since about 1950 in the context of molecular orbital theory. The interaction of light with small dye molecules can be completely specified, and although such detailed interpretation remains impractical for large dye molecules, even with currently available supercomputers, the origin of color in organic molecules is considered to be understood in principle at a molecular level.

Most organic dyes contain an extended conjugated chromophore system to which are attached electron donor and electron acceptor groups. Although the wavelength of a "reflected" photon is usually the same as that of the incident one, the reflected photon is not the same particle of light as the incident one. Instead, the surface is more properly regarded as a potential emitter of light, where (in the absence of incandescence, fluorescence, or phosphorescence) incident light is required to trigger the molecular reaction that produces the emitted radiation. Except for fluorescent materials, the number of emitted photons cannot exceed *at any wavelength* the number that are incident, and the frequency of each photon is unchanged, as is its wavelength if the medium is homogeneous. In considering nonfluorescent materials, the subtle exchange of reflected for incident photons is of no practical importance, and the term *reflection* is often used to describe the process as if some percentage of photons were merely bouncing off the surface.

B. Colorants

A colorant is any substance employed to produce reflection that is selective with wavelength. Colorants exist in two broad categories: dyes and pigments. In general, dyes are soluble, whereas pigments require a substrate called a binder. Not all colorants fall into either of these categories. (Exceptions include colorants used in enamels, glasses, and glazes.)

C. Scatter

Because pigments do no exist as dissociated, individual molecules, but instead are bound within particles whose size and distribution may vary, the spectral distribution of the reflected light depends only partly on the reaction of the dye or pigment molecules to absorbed light. In addition to the possibility of being absorbed, reflected, or transmitted, light may also be scattered. Particles that are very small relative to the wavelength of light produce Rayleigh scattering, which varies inversely as the fourth power of wavelength. (Rayleigh scatter causes the sky to appear blue on a clear day; without scatter from atmospheric particles, the sky would be black, as seen from the moon.) As scattering particles become larger, Mie scattering results. Wave-

length dependence, which is minimal for large-particle scatter, becomes a factor for particles of intermediate size. The directionality of scattered light is complex and can be compounded by multiple scattering. There are two components of Rayleigh scattering, which are differentially polarized. Mie scattering is even more complicated than the Rayleigh variety, and calculations pertaining to it are possible to carry out only with very large computers.

Scatter also occurs at object surfaces. For example, in "blue-eyed" people and animals, the eye color results mainly from scatter within a lightly pigmented iris. As a powder containing a colorant is ground more finely or is compressed into a solid block, its scattering characteristics change and so does the spectral distribution of the light reflected from it, despite an unchanging molecular configuration of the colorant. Often such scatter is nonselective with wavelength and tends to dilute the selective effects of the colorant. A compressed block of calcium carbonate is interesting in this respect because it comes very close to being a perfectly diffuse, totally reflecting, spectrally nonselective reflector.

D. Other Causes of Spectrally Selective Reflection

The spectral distribution of returned light can also be altered by interference and diffraction. Interference colors are commonly seen in thin films of oil resting on water; digital recording disks now provide a common example of spectral dispersion by diffraction.

Light is often transmitted partially through a material before being scattered or reflected. Various phenomena related to transmitted light per se also give rise to spectrally selective effects. In a transmitting substance, such as glass, light is repeatedly absorbed and reradiated, and in the process its speed is differentially reduced as a function of wavelength. This leads to wavelength-selective refraction and the prismatic dispersion of white light into its spectral components.

The most common colorants in glass are oxides of transition metals. Glass may be regarded as a solid fluid, in the sense of being a disordered, noncrystalline system. The metal oxides enter the molton glass in true solution and maintain that essential character after the glass has cooled and hardened. Whereas much colored glass is used for decorative purposes, color filters for scientific use are deliberately produced with specific densities and spectral distributions caused by selective absorption (which usually also produces some scatter), by reflection from coated surfaces, or by interference.

The visual characteristics of metals result from specular reflection (usually somewhat diffused) which, unlike that from other polished surfaces, is spectrally selective.

If the regular periodicity of their atoms is taken into account, the reflectance characteristics of metals can also be understood in terms of the same molecular orbital theory that applies to organic colorants. In this case it serves as a more fundamental basis for band theory, in terms of which the optical and electrical conductance properties of metals and semiconductors have classically been characterized.

E. Subtractive Color Mixture

The addition of primaries, as described in Section IA, is an example of what is often termed additive color mixture. Four methods of addition were described, all of which have in common the fact that photons of different wavelengths enter the eye from the same, or nearly the same, part of the visual field. There are no significant interactions between photons external to the eye; their integration occurs entirely within the photoreceptors, where photons of different wavelengths are absorbed in separate molecules of photopigments, which for a given photoreceptor are all of the same kind, housed within the cone outer segments.

Subtractive color mixing, on the other hand, is concerned with the modification of spectral light distributions external to the eye by the action of absorptive colorants, which, in the simplest case, can be considered to act in successive layers. Here it is dyes or pigments, not lights, that are mixed. The simplest case, approximated in some color photography processes, consists of layers of nonscattering, selectively absorptive filters. Consider the spectral transmittance functions of the subtractive primaries called cyan and yellow in Fig. 13 and the result of their combination: green. The transmittance function for the resulting green is simply the product, wavelength by wavelength, of the transmittance functions of cyan and yellow. When a third subtractive primary (magenta) is included in the system, blue and red can also be produced by the combinations shown. If all three subtractive primaries are used, very little light can pass through the combination, and the result is black.

If the filters are replaced by dyes in an ideal nonscattering solution, transmittance functions of the cyan, yellow, and magenta primaries can be varied quantitatively, depending on their concentration, with little change of "shape"—that is, each can be multiplied by a constant at each wavelength. By varying the relative concentrations of three dyes, a wide range of colors can be produced, as shown by the line segments on the CIE chromaticity diagram of Fig. 14. Subtractive color mixtures do not fall along straight lines in the chromaticity diagram.

Dichroic filters, such as those used in color television cameras, ideally do not absorb, but instead reflect the component of light not transmitted, so that the two components are complementary in color. By contrast, examination of an ordinary red gelatin filter reveals that the appearance of light reflected from it, as well as that transmitted through it, is red. The explanation for the reflected component is similar to that for colors produced by the application of pigments to a surface.

Consider elemental layers within the filter and a painted surface, each oriented horizontally, with light incident downward. In both cases, the light incident at each elemental layer consists of that not already absorbed or backscattered in the layers above. Within the elemental layer, some fraction of the incident light will be scattered upward, to suffer further absorption and scatter before some of it emerges from the surface at the top. Another fraction will be absorbed within the elemental layer, and the remainder will be transmitted downward, some specularly and some by scatter. In the case of the painted surface, a fraction of the initially incident light will reach the backing. In the case of the red gelatin filter, light emerging at the bottom constitutes the component transmitted through the filter. For the painted surface, the spectral reflectance of the backing will, unless perfectly neutral, alter the spectral distribution of the light reflected upward, with further attenuation and scattering at each elemental layer, until some of the light emerges at the top.

The prediction of color matches involving mixtures of pigments in scattering media is, as the example above suggests, not a simple matter. For this purpose, a theory developed by Kubelka and Munk in 1931, and named after them, is (with variations) most often used. Complex as it is, the theory nevertheless requires so many simplifying assumptions that predictions based on it are only approximate. Sometimes Mie scattering theory is applied to the problem of predicting pigment formulations to match a color specification, but more often empirical methods are used for this purpose.

VI. COLOR DIFFERENCE AND COLOR ORDER

Paradoxically, exact color matches are at the same time very common and very rare. They are common in the sense that any two sections of the same uniformly colored surface or material will usually match physically and therefore also visually. Otherwise, exact color matches are rare. For example, samples of paints of the same name and specification, intended to match, seldom do so exactly if drawn from separate batches. Pigments of identical chemical specification generally do not cause surfaces to match if they are ground to different particle sizes or suspended within different binders. Physical matches of different materials, such as plastics and fabrics, are usually impossible because differing binders or colorants must be used.

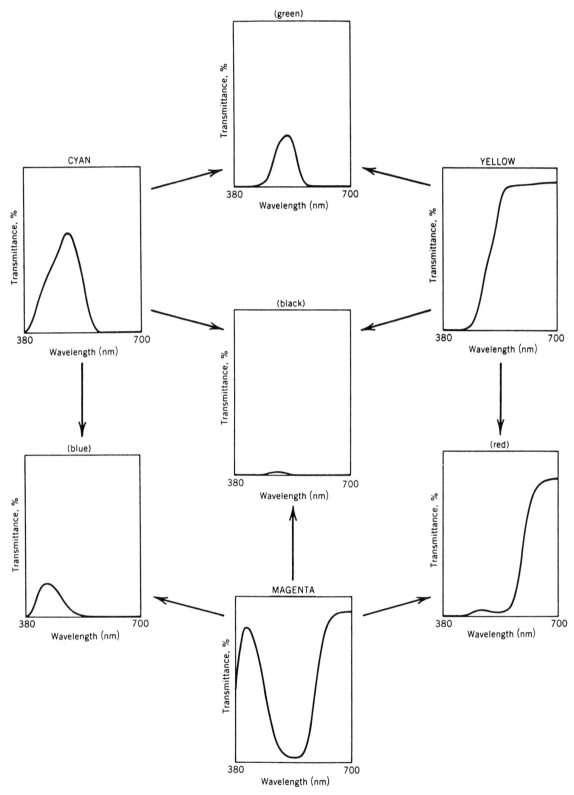

FIGURE 13 Spectrophotometric curves of a set of subtractive primary filters and their mixtures, superimposed in various combinations. [From Billmeyer, F. W., Jr., and Saltzmann, M. (1981). "Principles of Color Technology," 2nd ed. Copyright © 1981 John Wiley & Sons, Inc. Reprinted by permission of John Wiley & Sons, Inc.]

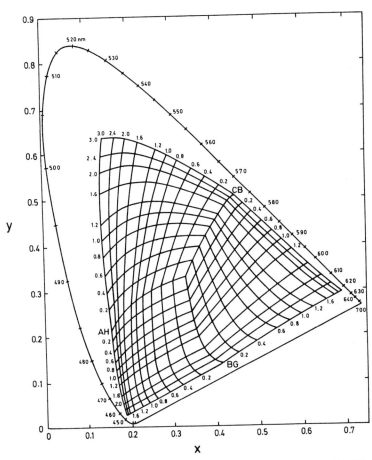

FIGURE 14 Chromaticities in a white light (~6500 K) of indicated combinations of dyes AH, BG, and CB in various concentrations. (Courtesy D. L. MacAdam and Springer-Verlag.)

In such cases—for example, matching a plastic dashboard with the fabric of an automobile seat—metameric matches must suffice; these cannot be perfect for all viewing conditions and observers. Given the difficulty or impossibility of producing perfect matches, it is important to be able to specify tolerances within which imperfect matches will be acceptable.

The issue of color differences on a more global scale will also be considered. Here concern is with the arrangement of colors in a conceptual space that will be helpful for visualizing the relations among colors of all possible kinds—the issue of color order.

A. Color-Difference Data

Figure 15 shows the so-called MacAdam discrimination ellipses plotted in the CIE chromaticity diagram. These were produced more than 40 years ago by an experimental subject who repeatedly attempted to make perfect color matches to samples located at 25 points in chromaticity space. The apparatus provided projected rather than sur-

face colors, but with a specified achromatic surround. For a set of settings at a given reference chromaticity, the apparatus was arranged so that the manipulation of a single control caused chromaticity to change linearly through the physical match point in a specified direction while automatically keeping luminance constant. Many attempted matches were made for each of several directions, as an index of a criterion sensory difference. The standard deviations of which were plotted on both sides of each reference chromaticity. Each of the ellipses of Fig. 15 was fitted to collections of such experimental points. MacAdam developed a system for interpolating between the measured chromaticities, and further research extended the effort to include luminance differences as well, leading to discrimination ellipsoids represented in $x–y–Y$ space. By this criterion of discrimination, there are several million discriminable colors.

Early calculational methods required the use of graphical aids, some of which are still in widespread use for commercial purposes. Very soon it was recognized that, if a formula could be developed for the prediction of

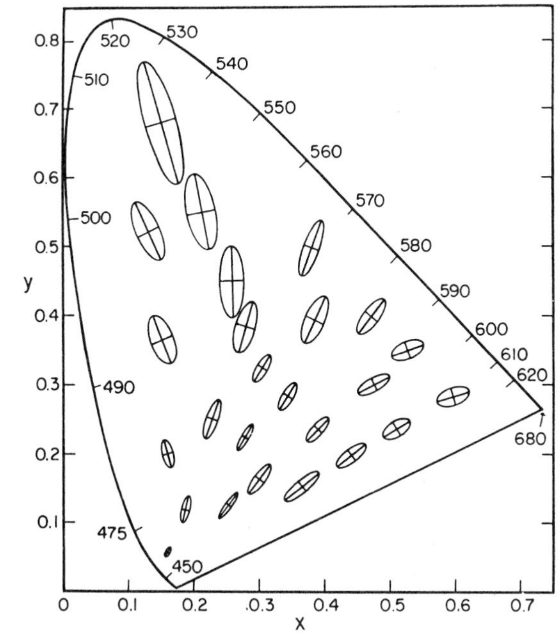

FIGURE 15 MacAdam discrimination ellipses, 10 times actual size. (Courtesy D. L. MacAdam and Springer-Verlag.)

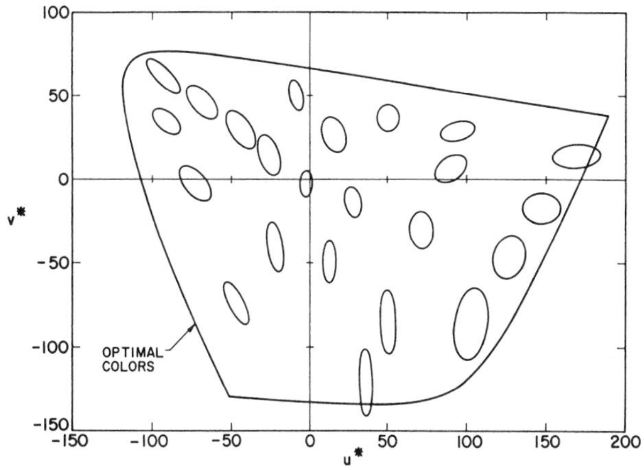

FIGURE 16 MacAdam ellipses ($L^* = 50$) shown in the CIE u^*, v^* diagram. [From Wyszecki, G., and Stiles, W. S. (1982). "Color Science: Concepts and Methods, Quantitative Data and Formulae," 2nd ed. Copyright ©1982 John Wiley & Sons, Inc. Reprinted by permission of John Wiley & Sons, Inc.]

just-discriminable color differences, measurements of color differences could be made with photoelectric colorimeters. Many such formulas have been proposed. To promote uniformity of use, the CIE in 1976 sanctioned two systems called CIELAB and CIELUV, the second of which will be described here.

B. CIELUV Color Difference Formulas

It has long been recognized that the 1931 CIE chromaticity diagram is perceptually nonuniform, as revealed by the different sizes and orientations of the MacAdam ellipses plotted thereon. For the evaluation of chromatic differences, the ideal chromaticity space would be isotropic, and discrimination ellipsoids would everywhere be spheres whose cross sections in a constant-luminance plane would plot as circles of equal size.

Many different projections of the chromaticity diagram are possible; these correspond to changes in the assumed primaries, all of which convey the same basic information about color matches. The projection of Fig. 16 is based on the following equations:

$$u' = 4X/(X + 15Y + 3Z);$$

$$v' = 9Y/(X + 15Y + 3Z).$$

The CIELUV formula is based on the chromatic scaling defined by this transformation combined with a scale of lightness. The system is related to a white reference object having tristimulus values X_n, Y_n, Z_n, with Y_n, usually taken as 100; these are taken to be the tristimulus values of a perfectly reflecting diffuser under a specified white illuminant.

There quantities are then defined as

$$L^* = 116Y/Y_n - 16,$$

$$u^* = 13L^*(u' - u'_n),$$

$$v^* = 13L^*(v' - v'_n).$$

These attempt to define an isotropic three-dimensional space having axes L^*, u^*, and v^*, such that a color difference ΔE^*_{uv} is defined as

$$\Delta E^*_{uv} = \Delta L^* + \Delta u^* + \Delta v^*.$$

The MacAdam ellipses, plotted on the u^*, v^* diagram, are more uniform in size and orientation than in the CIE diagram. Without recourse to nonlinear transformations, this is about the greatest degree of uniformity possible. These data and the CIE's color difference equations are recommended for use under conditions in which the observer is assumed to be adapted to average daylight; they are not recommended by the CIE for other conditions of adaptation.

It is not difficult to write a computer program that will calculate the ΔE^*_{uv} values appropriate to each member of a pair of physical samples. Starting with knowledge of the spectral reflectance distributions of the samples and the spectral irradiance distribution of the illuminant, one calculates the tristimulus values X, Y, and Z. From these, the L^*, u^*, and v^* values for each sample are calculated and inserted into the final formula. Given that voltages proportional to tristimulus values can be approximated using suitably filtered photocells and electronics, it is a short step to the development of fully automated devices that, when aimed in turn at each of two surfaces, will register a color difference value.

The CIELUV formula is only one of more than a dozen schemes that have been suggested for calculating color differences, some simpler but most more elaborate. None of these performs as well as would be desired. Correlations of direct visual tests with predictions made by the best of these systems, including CIELUV, account for only about half the experimental variance. Different formulas make predictions that correlate no better than this with one another. In using CIELUV to predict color differences in self-luminous displays, agreement is lacking concerning the appropriate choice of reference white. For industrial applications in which reflecting materials are being evaluated, differential weighting of the three components entering into the CIELUV equation may be helpful, and to meet the demands of specific situations, doing so can significantly improve the predictive power of the system. For example, when samples of fabrics are being compared, tolerance for luminance mismatches tends to be greater than for mismatches along the chromatic dimensions.

Despite these problems and limitations, calculations of chromatic differences by formula has proved useful, automated colorimeters for doing so exist, and the practice may be regarded as established, perhaps more firmly than it should be. Room for improvement at a practical level and for a better theoretical understanding of the problem certainly exists.

C. Arrangement of Colors

In the years before the development of the CIE system of color specification, which is based on radiometric measurement, colors could be described only by appeal to labeled physical samples. Any two people possessing a common collection of samples could then specify a color by reference to its label. Whereas in principle such sets of colors could be randomly arranged, an orderly arrangement is clearly preferable in which adjacent colors differ by only a small amount.

For more than 100 years, it has been thought that colors can be continuously arranged in a domain composed of two cones sharing a common base, as shown in Fig. 17. A line connecting the cone apices defines the axis of achromatic colors, ranging from white at the top to black at the bottom. A horizontal plane intersecting one of the cones, or their common base, defines a set of colors of equal lightness. Within such a plane colors can be represented in an orderly way, with gray at the center and colors of maximum saturation on the circumference. The hues on the circumference are arranged as they are in the spectrum, in the order red, orange, yellow, green, blue, and violet, with

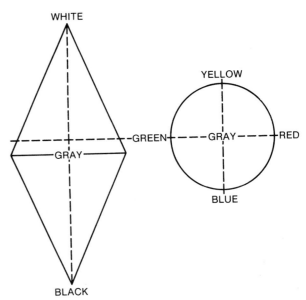

FIGURE 17 Representation of three-dimensional domain of surface colors.

the addition of a range of purples shading back to red. In moving from gray at the center toward a saturated hue, the hue remains constant while the white content of the color gradually diminishes as its chromatic content increases, all at constant lightness. As the intersecting horizontal plane is moved upward, the represented colors are more highly reflecting and correspondingly lighter in appearance, but their gamut is more restricted as must be so if only white is seen at the top. As the intersecting horizontal plane is moved down ward, colors become less reflecting and darker in appearance, with progressively more restricted gamuts, until at the bottom only a pure black is seen.

Considering a typical cross section, a chromatic gray is at its center. Significant features, first observed by Newton, pertain to the circumferential colors. First, they represent the most highly saturated colors conceivable at that level of lightness. Second, adjacent hues, if additively mixed, form legitimate blends; for example, there is a continuous range of blue-greens between blue and green. Third, opposite hues cannot blend. For example, yellow and blue, when additively mixed, produce a white that contains no trace of either component, and the sensations of yellow and blue are never simultaneously experienced in the same spatial location. Two colors that when additively mixed yield a white are called complementary colors, and in this kind of color-order system they plot on opposite sides of the hue circle.

There are several such systems in common use, of which only one, the Munsell system, will be described here. In this system, the vertical lightness axis is said to vary in *value* (equivalent to lightness) from 0 (black) to 10 (white). At any given value level, colors are arranged as described

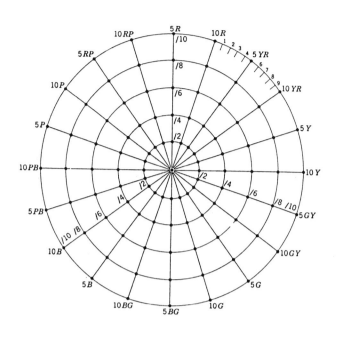

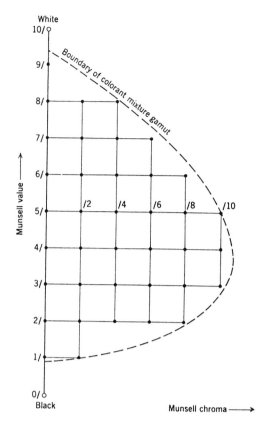

FIGURE 18 Organization of colors in the Munsell system. [From Wyszecki, G., and Stiles, W. S. (1982). "Color Science: Concepts and Methods, Quantitative Data and Formulae," 2nd ed. Copyright ©1982 John Wiley & Sons, Inc. Reprinted by permission of John Wiley & Sons, Inc.]

above but labeled circumferentially according to the hues blue, green, yellow, red, purple (and adjacent blends), and radially according to their saturation (called chroma). Figure 18 illustrates the system.

CIE chromaticity coordinates and Y values have been determined for each of the Munsell samples—the so-called Munsell renotations. A rubber-sheet type of transformation exists between the locations of Munsell colors at a given lightness level and their locations the CIE diagram, as shown in Fig. 19, for Munsell value 5. This figure illustrates that color order can also be visualized on the CIE diagram. A limitation of the CIE diagram for this purpose, in addition to its perceptually nonuniform property, is that it refers to no particular lightness level. The dark surface colors, including brown and black, do not exist in isolated patches of light. These colors are seen only in relation to a lighter surround; in general, surface colors are seen in a complex context of surrounding colors and are sometimes called related colors for this reason.

Although surrounding colors can profoundly influence the appearance of a test color in the laboratory situation, these effects are seldom obvious in natural environments and probably represent an influence of the same processes responsible for color constancy. This concept refers to the fact that colors appear to change remarkably little despite changes in the illuminant that materially alter the spectral distribution of the light reaching the retina. In other words, the perceived color of an object tends, very adaptively, to be correlated with its relative spectral reflectance, so that within limits color seems to be an unchanging characteristic of the object rather than triply dependent, as it actually is, on the characteristics of the illuminant, object, and observer.

VII. PHYSIOLOGICAL BASIS OF COLOR VISION

Progress in modern neuroscience, including a progressively better understanding of sensory systems in physical and chemical terms, has been especially rapid since about 1950 and continues to accelerate. The following is a very brief summary of some highlights and areas of ignorance related to color vision.

The optical system of the eye receives light reflected from external objects and images it on the retina, with a

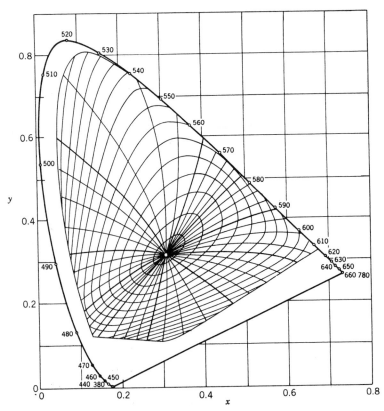

FIGURE 19 Location of circles of constant chroma and lines of constant hue from the Munsell system plotted on the CIE chromaticity diagram. [From Billmeyer, F. W., Jr., and Saltzmann, M. (1981). "Principles of Color Technology," 2nd ed. Copyright © 1981 John Wiley & Sons, Inc. Reprinted by permission of John Wiley & Sons, Inc.]

spectral distribution that is altered by absorption in the eye media. By movements of the eyes that are ordinarily unconsciously programmed, the images of objects of interest are brought to the very center of a specialized region of the retina, known as the fovea centralis, where we enjoy our most detailed spatial vision. The color of objects in the peripheral visual field plays an important role in this process.

The spectral sensitivities of the three classes of cone photoreceptors shown in Fig. 2 depend on the action spectra of three classes of photopigments, each uniquely housed in one type of cone. The cones are very nonuniformly distributed in the retina, being present in highest density in the fovea and falling off rapidly to lower density levels across the remainder of the retina.

The colors of small stimuli seen in the periphery are not registered so clearly as in foveal vision, but if fields are enlarged sufficiently, the periphery is capable of conveying a great deal of information about color. In the fovea there are few if any short-wavelength-sensitive (S) cones, and the long-wavelength-sensitive (L) and middle-wavelength-sensitive (M) cones are present in roughly equal numbers and very high density.

The L and M cones subserve spatial vision and also provide chromatic information concerned with the balance between red and green. Outside the fovea, the proportion of S cones increases but is always very small. The coarseness of the S-cone mosaic makes it impossible for them to contribute very much to detailed spatial vision; instead, they provide information concerned almost exclusively with the second dimension of chromatic vision.

As noted earlier, color is coded initially in terms of the ratios of excitation of the three kinds of cones. A single cone class in isolation is color-blind, because any two spectral distributions can excite such cones equally if appropriate relative intensities are used. The same is true of the much more numerous rod photoreceptors, which can lead to total color-blindness in night vision, where the amount of light available is often insufficient to be effective for cones. At intermediate radiance levels, rods influence both color appearance and, to a degree, color matches. Interestingly, vision at these levels remains trichromatic in the sense that color matches can be made using three primaries and three controls. This suggests that rods feed their signals into pathways shared by cones, a fact documented by direct electrophysiological experiment.

Light absorption in the cones generates an electrical signal in each receptor and modulates the rate of release of a neurotransmitter at the cone pedicles, where they synapse with horizontal and bipolar cells in the retina. The latter deliver their signals to the ganglion cells, whose long, slender axons leave the eye at the optic disk as a sheathed bundle, the optic nerve. Within this nerve are the patterns of impulses, distributed in about a million fibers from each eye, by means of which the brain is exclusively informed about the interaction of light with objects in the external world, on the basis of which form and color are perceived.

Lateral interactions are especially important for color vision. The color of a particular area of the visual scene depends not only on the spectral distribution of the light coming from that area, but also on the spectral distribution (and quantity) of light coming from other regions of the visual field. In both retina and brain, elaborate lateral interconnections are sufficient to provide the basis for these interactions.

Figure 20 summarizes, in simplified fashion, a current model of retinal function. The initial trichromacy represented by L, M, and S cones is transformed within the retina to a different trichromatic code. The outputs of the L and M cones are summed to provide a luminance signal, which is equivalent to the quantity of light as defined by flicker photometry. The L and M cone outputs are also differenced to form a red—green signal, which carries information about the relative excitations of the L and M cones. An external object that reflects long-wavelength light selectively excites the L cones more than the M, causing the red–green difference signal to swing in the red direction. A yellow–blue signal is derived from the difference between the luminance signal and that from the S cones.

The color appearance of isolated fields of color, whether monochromatic or spectrally mixed, can be reasonably well understood in terms of the relative strengths of the red–green, yellow–blue, and luminance signals as these are affected in turn by the strength of the initial signals generated in the three types of cones. In addition, increasing S-cone excitation appears to influence the red–green opponent color signal in the red direction.

There has been a great deal of investigation of the anatomy and physiology of the brain as it relates to color perception, based on an array of techniques that continues to expand. The primary visual input arrives in the striate cortex at the back of the head; in addition, there are several other brain centers that receive visual input, some of which seem specifically concerned with chromatic vision. The meaning of this activity for visual perception is not yet clear. In particular, it is not yet known exactly which kinds or patterns of activity immediately underlie the sensation of color or exactly where they are located in the brain.

SEE ALSO THE FOLLOWING ARTICLES

BONDING AND STRUCTURE IN SOLIDS • COATINGS, COLORANTS, AND PAINTS • GLASS • HOLOGRAPHY • LIGHT SOURCES • OPTICAL DIFFRACTION • RADIOMETRY AND PHOTOMETRY • SCATTERING AND RECOILING SPECTROSCOPY

BIBLIOGRAPHY

Billmeyer, F. W., Jr., and Saltzman, M. (1981). "Principles of Color Technology," 2nd ed., Wiley, New York.

Boynton, R. M. (1979). "Human Color Vision," Holt, New York.

Grum, F., and Bartleson, C. J. (1980). "Optical Radiation Measurements," Vol. 2, Academic Press, New York.

Kaufman, J. E., ed. (1981). "IES Lighting Handbook: Reference Volume," Illuminating Engineering Society of North America, New York.

MacAdam, D. L. (1981). "Color Measurement: Theme and Variations," Springer-Verlag, Berlin and New York.

Marmion, D. M. (1991). "Handbook of U.S. Colorants," 3rd ed., Wiley, New York.

Mollon, J. D., and Sharpe, L. T., eds. (1983). "Colour Vision: Physiology and Psychophysics," Academic Press, New York.

Nassau, K. (1983). "The Physics and Chemistry of Color," Wiley, New York.

Wyszecki, G., and Stiles, W. S. (1982). "Color Science: Concepts and Methods, Quantitative Data and Formulae," 2nd ed, Wiley, New York.

Zrenner, E. (1983). "Neurophysiological Aspects of Color Vision in Primates," Springer-Verlag, Berlin and New York.

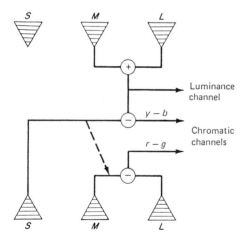

FIGURE 20 Opponent-color model of human color vision at the retinal level.

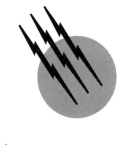

Combustion

F. A. Williams
University of California, San Diego

GLOSSARY

Adiabatic flame temperature Temperature achieved if all the heat of combustion is used to increase the temperature of the combustion products at constant pressure.

Burning velocity Velocity of propagation of a deflagration.

Deflagration Reaction front propagating into a combustible at a subsonic velocity, in which forward diffusion raises the temperature to a value at which reaction proceeds at an appreciable rate.

Detonability limits Limiting values of chemical composition or pressure beyond which a combustible system cannot be made to experience detonation.

Detonation Reaction front propagating into a combustible at a supersonic velocity, in which a leading shock raises the pressure and temperature to high values at which reaction proceeds at an appreciable rate.

Diffusion flame Combustion process in which fuel and oxidizer diffuse to a reaction layer from opposite sides.

Extinction Process in which a combustible system reacting at an appreciable rate is brought to a condition in which it is reacting at a negligible rate.

Flammability limits Limiting values of chemical composition or pressure beyond which ignition cannot be achieved.

Heat of combustion The energy released when a given amount of fuel reacts with oxidizer to form specified combustion products at constant pressure and temperature.

Ignition Process in which a combustible system reacting at a negligible rate is brought to a condition in which it is reacting at an appreciable rate.

Ignition delay Time interval required for ignition to occur after beginning the application of an ignition stimulus.

Ignition energy Energy that must be supplied to a combustible system to achieve ignition.

Ignition temperature Temperature to which a combustible system must be raised to achieve ignition.

Quenching distance For a combustible confined between

Encyclopedia of Physical Science and Technology, Third Edition, Volume 3

two parallel plates, the minimum distance between the plates below which ignition cannot be achieved.

COMBUSTION is a chemical process that liberates heat. Typically, it involves finite-rate chemistry in fluid flow with heat and mass transfer. The technology of combustion finds application in equipment for heating, for power production, and for propulsion, as well as in techniques for creating explosions, for destruction of toxic wastes, and for mitigating fire hazards. The science of combustion is focused on obtaining fundamental descriptions of combustion phenomena by experimental and mathematical methods. Basic principles of combustion are sufficiently well developed for the subject to be qualified as an applied science.

I. APPLICATIONS OF COMBUSTION

A. The Role of Combustion in the Development of Science and Technology

Use of fires by mankind for warmth and protection and in connection with food predates recorded history. Although the origins of furnaces are obscure, we know that they preceded scientific knowledge of combustion, as did the development of pyrotechnic rockets and also of guns (forged in Germany during the 14th century), which require combustion for their construction and operation. Applications of combustion played an essential role in the industrial revolution, beginning with the development of the steam engine in the early 17th century. In general, use of the art of combustion for technological advancement has preceded establishment of the corresponding science of combustion. For example, when Otto built his successful internal combustion engine in 1876, the principles of flame propagation within the device were unknown. Combustion processes are complicated enough that even today trial-and-error plays a prominent role in their technological applications.

Experience of combustion exerted a dominant influence on early but misdirected concepts of matter. During the Middle Ages, fire was thought to be one of the four basic elements of matter. The phlogiston theory of Stahl, which impeded progress in chemistry during the 18th century, relied heavily on observations of effects of combustion. However, combustion also played an important and beneficial role in the establishment of the modern science of chemistry. Weight and volume changes produced by combustion were investigated by Jean Ray, Boyle, Hooke, and Mayow during the 17th century. Later, the combustion-related studies of Lavoisier, Priestley, and Scheele (which

began in the 1770s) led by 1800 to knowledge of the overall chemical changes that occur in common combustion processes and thereby aided in laying the foundations of modern chemistry.

The sciences of chemical kinetics and of molecular transport processes did not begin to flourish until the middle of the 19th century. The science of combustion benefited from these developments, as well as from earlier advances in thermodynamics and in fluid mechanics. The experiments on combustion waves by Bunsen (1866), Berthelot and Vieille (1881), and Mallard and le Chatelier (1881, 1883) and the associated theoretical explanations offered by Mikhel'son (1890), Chapman (1899), and Jouguet (1905, 1917) led to the establishment of combustion as a scientific discipline early in the 20th century. Research in combustion has been growing ever since and continues to contribute to the understanding of fluid dynamics, heat and mass transfer, and various aspects of chemistry, especially chemical kinetics. In particular, flames offer laboratory tools for investigating rates of chemical processes.

B. Combustion in Furnaces and Boilers

Furnaces may be classified by type of fuel; gas-fired (e.g., natural gas, largely methane, CH_4), liquid-fired (e.g., fuel oil), or solid-fired (e.g., coal or wood). In typical furnace designs there is a combustion chamber, either enclosed or open to the air, in which fuel combines with the oxygen, O_2, in air to form combustion products (largely carbon dioxide, CO_2, and water, H_2O) and to release heat. The heat can be transferred to a working fluid (e.g., air, water, or steam), which distributes it as required or expands (e.g., against a piston) to produce work. In steam engines, the heat from the furnace is transferred to a boiler in which water is converted into steam, the working fluid. There are many applications of combustion (e.g., in the steel industry and in welding) in which a working fluid is not employed: Heat released by combustion is applied directly to the object that is to be heated.

Design of the combustion chamber is a central aspect of furnace design. Objectives are to obtain efficient and clean combustion and efficient and trouble-free heat transfer. Typically, fuel is metered into the combustion chamber where it meets air entrained by forced or natural convection. In gas-fired furnaces, fuel jets at the periphery of the chamber admit the fuel, possibly premixed with air, in flow patterns (often involving swirl) that are designed to achieve efficient combustion. In oil-fired furnaces, atomization of the liquid by the fuel injector (atomizer) and liquid-jet penetration are critical aspects in producing fine droplets properly distributed in air. Types of solid-fired furnaces may be listed as fixed-bed (in which the solid fuel is supported on a grate as it burns), fluidized-bed (in which gas flow through the bed of fuel agitates the solid elements of the bed to give them a fluid-like behavior), and pulverized-fuel (in which finely ground solid fuel is transported by a gas stream and injected in much the same way as the fuels of gas-fired furnaces). Coal–oil or coal–water slurries (solid suspensions in liquids) transport coal as a liquid to be burned in furnaces of liquid-fired design. Thus coal, the world's most plentiful fuel, can be used in furnaces of all three basic types.

C. Combustion in Piston Engines

The purpose of a piston (or reciprocating) engine, the production of power, is achieved through exertion of pressure on a moving piston by a working fluid. In an external combustion engine (e.g., a steam engine) the working fluid and the fluid in which the combustion occurs are not the same, whereas in an internal combustion engine they are the same. The two principal types of internal-combustion piston engines are spark-ignition engines and compression-ignition (diesel) engines. Both types operate in a cyclic process in which every cycle involves intake of fuel and air, compression, ignition, combustion, expansion of the gas against the piston, and exhaust of combustion products. In the spark-ignition engine, an electrical spark ignites the combustible mixture. In the diesel engine, autoignition occurs (without a spark) as a consequence of the temperature and pressure rise associated with compression. In spark-ignition engines, the fuel and air are often mixed prior to intake, although increasing use of direct injection of fuel (after air intake) is being made; in diesel engines fuel injection usually occurs during compression of the air. These engines usually employ liquid fuels, although gaseous and solid fuels also can be used; for example, pulverized coal was the fuel in Diesel's (1892) original design. There is increasing interest in dual-fuel engines, which employ mainly natural gas, for example, but utilize liquid fuel, usually in a diesel mode, for controlled ignition.

Combustion in internal-combustion reciprocating engines involves propagation of flame from the points of spark or compression ignition through the rest of the combustible mixture in the combustion chamber. Since the gas in the chamber is in turbulent motion, this process is one of turbulent flame propagation. The velocity of propagation and the chamber size affect the time required for combustion to occur, although for diesel combustion the time needed for burnout of remaining fuel pockets, after completion of propagation, also influences the total combustion time. The combustion time is significant in engine design, because power production is greatest if combustion begins after compression is completed and is completed before expansion begins. Design objectives include achieving high engine efficiency (power produced, divided by the product of the heat of combustion per unit mass of fuel and the rate of fuel consumption) and low rates of emission of pollutants (oxides of nitrogen, mainly nitric oxide, NO; carbon monoxide, CO; unburnt hydrocarbons; odoriferous compounds; and particulate matter). Among the approaches to achieving these objectives are exhaust-gas recirculation (in which some of the products of combustion are mixed into the intake air) and stratified-charge combustion (in which attempts are made to achieve tailored, nonuniform distributions of fuel and air throughout the chamber to produce improved combustion characteristics). One of many classes of designs employing such approaches is the direct-injection, stratified-charge engine.

The combustion principles that apply to piston engines also apply to rotary engines, in which a suitably designed rotating element is substituted for the reciprocating piston.

D. Combustion in Gas Turbines and Jet Engines

In contrast to the cyclic character of reciprocating engines, gas turbines and most jet engines involve continuous flow of fuel and air (an exception is the pulsejet which operates on an unsteady gas-dynamic wave-propagation principle). These are internal combustion engines in which compression, combustion, and expansion occur in separate sections of the device. The purpose of the gas turbine is to produce power, which is extracted by the turbine as the heated working fluid expands through it. The purpose of the jet engine is to produce thrust, force (associated with the high-velocity exhaust) that can propel a vehicle. The basic types of jet engines are turbojets and ramjets, the latter being restricted mainly to propulsion at high speeds. Turbojets employ the same types of combustion chambers as gas turbines, while ramjets require different designs because of the higher gas velocities in the chamber. To augment thrust, turbojets sometimes are equipped with afterburners, whose combustion chambers resemble those of ramjets.

In gas turbines and turbojets, fuel (usually liquid) is injected into the combustion chamber through atomizers, and air from a compressor is directed toward the fuel through a perforated liner (or can) designed to facilitate stable mixing and combustion in the flow. There may be a number of tubular combustion chambers around the shaft that connects the compressor and the turbine, or an annular region around the shaft may be employed as a combustion chamber, with an annular liner inside an annular casing. Annular designs facilitate combustion by providing larger areas and therefore lower gas velocities in the chamber, but they present difficulties in flow control (e.g., in achieving uniformity of temperature at the chamber exit). Can-type liners around injectors can be employed in annular chambers to improve control while retaining other advantages of annular designs.

In ramjets and afterburners, flames typically are stabilized by bluff bodies (e.g., rods or baffles), ahead of which fuel is injected and behind which there is flow separation with recirculation zones that anchor flames. The flames spread from the recirculation zones at angles determined by the turbulent burning velocity. Recirculation for anchoring flames also can be produced by wall recesses or by reverse jets (fuel injected in the upstream direction, opposite to the direction of the gas flow). The dump combustor, in which a sudden expansion of the cross-sectional area of the combustion chamber produces flow separation and recirculation, is an example of a ramjet in which flames are stabilized by wall recesses.

Most current jet aircraft engines have air streams that bypass the turbojet combustor and turbine and to which fuel often may be added and burnt in ramjet fashion, so that these bypass engines possess attributes of both turbojets and ramjets. For propulsion at high supersonic speeds, the temperature increase associated with deceleration of air to subsonic velocities can impose unacceptably large heat loads on the vehicle, and consideration of combustion in supersonic flows therefore becomes attractive. This has led to designs of supersonic-combustion ramjets in which combustion occurs in a supersonic mixing layer as the fuel and air mix. Such "scramjets" are being considered for propulsion of the aerospace plane, designed to fly from Earth to orbit. Detonative-combustion engines, which operate on principles of standing or traveling detonations, also have been considered for these types of applications. One such concept under consideration and experimentation is the pulse-detonation engine, consisting of a number of tubes sequentially filled with a detonable fuel-air mixture, having detonations periodically initiated in each tube at controlled frequencies to achieve tailored thrust. For flight propulsion at hypersonic speeds, it may be impractical to bring air into an internal combustion chamber, and therefore concepts of external-burning ramjets have been explored, in which fuel is injected into the air outside the vehicle, where it ignites and burns, producing thrust on external surfaces.

E. Combustion in Rocket Engines

The applications discussed thus far have been concerned with air-breathing engines, in which the oxidizer for the fuel is the O_2 in the air. In rocket engines, the oxidizer instead is carried within the vehicle; this is essential for propulsion in space. The principle of the rocket is that the hot products of combustion from the combustion chamber are expanded through a nozzle to a high velocity for producing thrust. By employing oxidizers other than air, rocket engines can achieve higher temperatures of combustion products and greater thrust per unit mass of propellant consumed (called specific impulse) than if air were the oxidizer.

There are two main types of rockets—solid-propellant and liquid-propellant. In solid-propellant rockets, a combustible solid material—the propellant "grain" (e.g., a mixture of nitrocellulose and nitroglycerine, called a *double-base propellant*, or crystals of an oxidizer such as ammonium perchlorate, NH_4ClO_4, dispersed in a matrix of a hydrocarbon fuel polymer to form a composite material called a *composite propellant*)—is contained in the combustion chamber. After ignition (e.g., by electrical heating or by initiation of a small explosive or pyrotechnic charge), the grain burns at its surface, causing hot gas to issue into the portion of the chamber not occupied by solid. The burning rate or regression velocity of the propellant, that is, the velocity at which the grain surface recedes, is important in engine design and is a property of the combustion process.

In liquid-propellant rockets, the combustibles are liquids, which are carried in tanks and injected into the combustion chamber through intricate injectors that atomize the fuel and promote mixing. Two types of liquid-propellant rockets are (a) monopropellant rockets, which employ a single liquid fuel (e.g., hydrazine, N_2H_4) that is capable by itself of combustion with heat release, and (b) the more common bipropellant rockets, which employ two liquids, a fuel and an oxidizer (e.g., cryogenically cooled liquid hydrogen, H_2, and fluorine, F_2), that burn with heat release when brought into contact with each other. The time required for completion of combustion of the injected liquids is relevant to the design of liquid-propellant rockets; this time often is related to droplet burning times.

The hybrid rocket engine employs a solid-fuel grain in the combustion chamber and an injected liquid or gaseous oxidizer. The regression rate of the grain surface in hybrid rockets depends on the rates of heat and mass transfer in the gas layer adjacent to the surface where the fuel reacts with the oxidizer in the combustion process.

F. Relationship of Combustion to Fire and Explosion Hazards

Deflagration is the intended mode of combustion for nearly all rockets and air-breathing engines. In explosives (e.g., trinitrotoluene, TNT) the intended mode is detonation because the much higher associated pressures enable greater forces to be applied. Most combustible mixtures are capable of detonation, and in most applications, detonation (which produces the most damaging type of explosion) must be avoided. Studies of safety in the handling of combustible mixtures address problems of preventing ignition, pressure buildup, and detonation. Safety standards for working with fuels and combustible mixtures were developed through investigations of the combustion of these materials. Design of equipment (safety valves, flame arresters, etc.) for safety in combustible handling relies on knowledge of combustion properties.

Unwanted fires involve the combustion of fuels in air. Combustion studies contribute to the development of methods for fire prevention, fire detection, fire hazard evaluation, fire damage assessment, and fire suppression. In particular, investigations of mechanisms for extinction of combustion suggest elements of operation for fire extinguishers, which are employed in fire suppression. Identification of fire retardant materials is aided by combustion knowledge. Strategies for controlling large fires employ estimates of combustion behavior. In general, much of the field of fire-control research concerns combustion research.

G. Waste Incineration and Underground Combustion

Combustion provides a major method for disposal of waste materials. It is especially attractive for destruction of hazardous or toxic wastes, because thorough conversion of toxic materials to nonharmful molecules can be achieved with appropriate incinerator designs. Combustion at high pressures, above the critical point at which liquid–gas distinctions disappear, exhibits promising attributes for some of these purposes and therefore is being studied. In other situations, beneficial power production may accompany waste destruction through suitable incinerator design.

Controlled combustion in underground deposits of coal and oil shale has been investigated as a promising means for extracting useful fuels and for production of power. On the other hand, uncontrolled underground combustion in coal and peat deposits can be difficult to extinguish and can threaten human life and pose severe environmental problems.

II. SCIENTIFIC SUBDISCIPLINES OF COMBUSTION

A. Thermodynamics of Combustion

The first and second laws of thermodynamics are basic to combustion processes. Thermodynamic cycles for production of work from heat play roles in evaluating efficiencies of piston engines and gas turbines. The thermodynamics of

open chemical systems is the most relevant to combustion. In particular, heats of reaction and chemical equilibria are of prominent importance. For many combustion purposes, use of the thermodynamics of ideal gas mixtures is sufficient, but sometimes nonideality and condensed phases must be considered—the former in detonation of solid explosives and the latter in droplet burning, for example.

Heats of combustion of some common fuels, per unit mass of fuel, to form gaseous CO_2 and liquid H_2O as products are given in Table I, along with representative adiabatic flame temperatures for mixtures initially at room temperature (298.15 K). Adiabatic flame temperatures can be estimated from heats of combustion by equating the product of an average specific heat at constant pressure for combustion products and the difference between the adiabatic flame temperature and the initial temperature to the product of the heat of combustion and the fraction of the mass of the initial mixture that is fuel. Improved calculations of adiabatic flame temperatures account for chemical equilibria in products (equilibrium dissociation) and usually employ available computer routines. The entries in the last three rows of Table I are average estimates because of variations in the fuels.

B. Chemical Kinetics of Combustion

In combustion processes, the chemistry proceeds at finite rates and generally involves a number of elementary reaction steps. Two principal aspects of the chemical kinetics of combustion are the reaction mechanism (the sequence of elementary steps involved) and the rate of each elementary step. The relative rates of the steps vary with conditions, because they depend on the temperature, pressure, and composition of the system. Thus, steps that are important for some conditions become unimportant for others, which leads to changes in mechanisms as conditions change. The reaction mechanisms are known for many but not all combustion processes.

Reaction mechanisms in combustion often involve chain reactions, in which a reactive intermediate species (such as hydroxyl, OH), called a chain carrier, which is created in some steps and destroyed in others, accelerates the overall rate of conversion of fuel to products. During much of the combustion process these intermediates can achieve steady-state concentrations (in which their total rate of creation equals their total rate of destruction), or they can be involved in reaction steps that attain partial equilibrium (in which the forward and backward rates of the step are equal). If steady-state or partial-equilibrium approximations can be verified for a particular combustion process, then improved understanding of its chemical kinetics is obtained, and simplified descriptions of the overall rate of conversion of fuel to products may be

TABLE I Selected Heats of Combustion and Flame Temperatures for Various Fuels

Fuel	State	Formula	Standard heat of combustion at 298.15 K (kJ/g)	Oxidizer	Pressure (atm)	Adiabatic flame temperature, maximum (K)
Carbon	Solid (graphite)	C	32.8	—	—	—
Hydrogen	Gas	H_2	141.8	Air	1	2400
				O_2	1	3080
Carbon monoxide	Gas	CO	10.1	Air	1	2400
Methane	Gas	CH_4	55.0	Air	1	2220
				Air	20	2270
				O_2	1	3030
				O_2	20	3460
Ethane	Gas	C_2H_6	51.9	Air	1	2240
Ethene	Gas	C_2H_4	50.3	Air	1	2370
Acetylene	Gas	C_2H_2	48.2	Air	1	2600
				O_2	1	3410
Propane	Gas	C_3H_8	50.4	Air	1	2260
Heptane	Liquid	C_7H_{16}	48.4	Air	1	2290
Dodecane	Liquid	$C_{12}H_{26}$	47.7	Air	1	2300
Benzene	Liquid	C_6H_6	41.8	Air	1	2370
Methanol	Liquid	CH_3OH	22.7	—	—	—
Fuel oil	Liquid	—	42–47	Air	1	2300
Coal	Solid	—	20–36	Air	1	2200
Wood	Solid	—	19–23	Air	1	2100

developed. Therefore, methods for identifying and verifying these approximations in combustion are continually evolving. The resulting descriptions are now called reduced chemical-kinetic mechanisms.

When reaction mechanisms are not fully known, but measurements of overall rates of fuel destruction or of heat release can be made, then empirical, one-step, overall, reaction-rate expressions can be obtained. In terms of the fuel concentration f (g/cm^3), the oxidizer concentration g (g/cm^3), pressure p (atm), and temperature T (K), the empirical rate w (g/cm^3 sec) of the reactant mixture conversion often can be written as

$$w = A f^l g^m p^n e^{-E/RT}, \qquad (1)$$

where R (8.31 J/mol K) is the universal gas constant, and the overall activation energy E, pressure exponent n, reaction orders l and m, and prefactor A are constants. Typically, 40 kJ/mol $\leq E \leq$ 200 kJ/mol, $0 \leq n \leq 2$, $0 \leq l \leq 2$, and $0 \leq m \leq 2$, although values outside these ranges can occur. The exponential factor in Eq. (1) is called the Arrhenius factor, and the rate expression is said to be an Arrhenius expression, because Arrhenius employed the exponential form in early work (1899).

An example of a reaction for which Eq. (1) sometimes can be used is the oxidation of methane, whose overall reaction is

$$CH_4 + 2O_2 \rightarrow CO_2 + 2H_2O. \qquad (2)$$

Some elements in the mechanism of reaction (2) are

$$CH_4 + O_2 \rightarrow CH_3 + H + O_2 \qquad (3)$$

$$H + O_2 \rightleftarrows OH + O \qquad (4)$$

$$CH_4 + OH \rightarrow CH_3 + H_2O \qquad (5)$$

$$CH_3 + O \rightarrow CH_2O + H \qquad (6)$$

$$CH_2O + OH \rightarrow HCO + H_2O \qquad (7)$$

$$HCO + O_2 \rightarrow H + CO + O_2 \qquad (8)$$

$$CO + OH \rightarrow CO_2 + H \qquad (9)$$

$$H + OH \rightleftarrows O + H_2 \qquad (10)$$

$$H_2 + OH \rightleftarrows H_2O + H \qquad (11)$$

$$2OH \rightleftarrows H_2O + O \qquad (12)$$

$$H + 2O_2 \rightarrow HO_2 + O_2 \qquad (13)$$

Here H, OH, and O are some of the chain carriers, and reaction (4) is an example of chain branching in which the number of carriers is increased. Reactions (4) and (10)–(13) also occur in H_2 oxidation. During portions of the combustion history, steady-state approximations (e.g., for HCO) and partial-equilibrium approximations [e.g., for reaction (11)] may apply. The former would enable reactions (7) and (8) to be replaced by

$$CH_2O + OH \rightarrow H_2O + H + CO, \qquad (14)$$

the rate being that of (7). The establishment of Eq. (14) is an example of a step toward derivation of reduced chemical-kinetic mechanisms. Derivation of Eq. (1) from the detailed mechanism and from the rates of each step by simplification of this type has not been attained for reaction (2), although a four-step reduced mechanism has become quite successful for methane flames. There has been significant recent advancement in reduced chemistry for combustion, including the chemistry of production of pollutants such as NO in flames. A further objective in the chemical kinetics of combustion is to obtain unknown rates of elementary steps from measurement and theory.

C. Fluid Mechanics of Combustion

Fluids can be treated as continua for nearly all purposes in combustion; an exception occurs in structures of gaseous detonations that begin with shocks so strong that the kinetic theory of gases is needed to describe them. The conservation equations of fluid mechanics describe combustion processes. In addition to the usual equations for conservation of mass, momentum, and energy, equations for conservation of chemical species are needed in combustion. Rates of chemical reactions [e.g., Eq. (1)] appear in these last equations. For most purposes, mixtures of ideal gases can be considered and treated as Newtonian fluids, so that the Navier–Stokes equations apply (augmented by chemistry). For some purposes (e.g., in spray combustion) conservation equations for multiphase flows must be considered. The fluid mechanics of turbulence is relevant to turbulent combustion. However, there are many laminar combustion processes in which the complexities of turbulence do not arise.

A distinguishing characteristic of the fluid mechanics of combustion is the decrease in the density of the fluid that is associated mainly with the temperature increase produced by the heat release. Density changes of a factor of five are typical and introduce flow effects (e.g., strong gas expansion) that are not commonly encountered in other areas of fluid mechanics.

D. Transport Phenomena in Combustion

Molecular diffusion, heat conduction, and viscosity play important roles in many combustion processes. For example, interdiffusion of fuel and oxidizer often must precede chemical heat release. The transport coefficients (diffusion coefficients, thermal conductivity, and coefficient of viscosity) that are multiplied by the gradients to obtain the fluxes of chemical species, heat, and momentum, as well as the transport equations that relate the fluxes to the gradients of concentration, temperature, and velocity, can be obtained from the kinetic theory of gases and experiment for ideal gas mixtures. Although the coefficients differ by as much as a factor of 10 for different species under different conditions, and although small differences sometimes are important in combustion processes, it may be assumed as a rough approximation that all molecular and thermal diffusivities, as well as the kinematic viscosity, vary inversely with pressure and at atmospheric pressure are about 0.1 cm^2/sec at room temperature and 1 cm^2/sec near flame temperatures. The ratio of the kinematic viscosity to the thermal diffusivity (the Prandtl number) and the ratios of the thermal diffusivity to diffusion coefficients of various species (Lewis numbers) are relevant numbers near unity in reacting mixtures of ideal gases.

Turbulent transport processes are important in turbulent combustion, and sometimes turbulent transport coefficients can be used. The values of these coefficients depend on the turbulence levels and generally are much larger than the molecular values just given.

Many combustion processes in both laminar and turbulent flows occur at high Reynolds numbers (large ratios of intertial to viscous forces) as a consequence of high velocities. Under these conditions, the boundary-layer theory of fluid mechanics often applies. That is, viscous and transport effects are important only in thin layers of the fluid, called boundary layers. The theory of heat and mass transfer across boundary layers therefore is relevant to many combustion problems. Bases for describing chemically reacting boundary layers in ideal gas mixtures are available. Transfer theory provides a general framework in which transport processes at all Reynolds numbers can be described, without restriction to ideal gas mixtures, and both with and without combustion. Transfer theory thus finds applications in combustion.

III. PRINCIPLES OF COMBUSTION

A. Self-Accelerating Character of Combustion

A distinguishing attribute of many combustion processes that gives them a qualitative difference from other chemical rate processes is the tendency for the combustion rate to increase as the combustion proceeds. In chemistry, this property often is called autocatalysis. There are two causes for the progressive increase in the rate of combustion. One is chain branching, exemplified in reaction (4). Since the combustion rate increases as the concentrations of chain carriers increase, branching can accelerate the process. The other is the increase in rate with temperature, seen in the exponential factor in Eq. (1). Since combustion is exothermic (releases heat), the temperature tends to increase as the process proceeds, and therefore the rate increases. The self-acceleratory character of the reaction introduces features such as explosions, categorized as branched-chain explosions, if chain branching dominates, or thermal explosions, if the increase in rate with temperature dominates.

B. Premixed Combustion

In many respects, the most fundamental principle in the study of combustion is the need to distinguish between two basically different arrangements of fuel; premixed and nonpremixed, also called homogeneous and heterogeneous. Premixed systems are those in which the fuel and oxidizer are intimately mixed before combustion begins. An example of premixed combustion is the flame cone of a Bunsen burner, in which gaseous fuel is thoroughly mixed with air at the base of the burner tube that holds the flame at its upper exit. Fuels that burn without additional oxidizers, such as monopropellants and many explosives, are intrinsically premixed. Premixed combustion can proceed in transient processes nearly homogeneously throughout the entire system or in waves (deflagrations and detonations); thin fronts that propagate into the unburnt combustibles. Thus, self-acceleratory phenomena and combustion waves are characteristics of premixed combustion.

C. Combustion in Heterogeneous Mixtures

The opposite extreme from premixed combustion is combustion in systems in which the fuel and oxidizer are separated initially and mix as they burn—for example, a wood fire. Most liquid and solid fuels that require an oxidizer burn in this nonpremixed (initially unmixed) fashion, as do gaseous fuels in many applications (e.g., in industrial flares). In these combustion processes, the chemical heat release occurs in a flame into which fuel and oxidizer are transported from opposite sides. Diffusion is essential to this type of combustion; fuel and oxidizer diffuse into the reaction zone while heat and reaction products diffuse out. Therefore, these flames are called diffusion flames. The processes tend to be nonexplosive and nonpropagating because the heat-release rates are limited by diffusion rates.

There are combustion processes in heterogeneous mixtures that possess diffusion-flame substructures while exhibiting premixed-combustion behavior at larger scales; examples are combustion waves in fuel sprays or dust suspensions (e.g., pulverized coal) in air. There also are diffusion flames with premixed-flame substructures. For example, under appropriate conditions the fuel and oxidizer

constituents in composite solid propellants burn in diffusion flames, but the oxidizer NH_4ClO_4 behaves as a monopropellant and decomposes first in a premixed flame. In this example, the collection of diffusion flames at the propellant surface can be viewed as a substructure of the overall propellant deflagration (a premixed-combustion concept); thus there are three successive levels—premixed, diffusion, and premixed—each at a larger scale than its predecessor. Finally, in nonhomogeneous gas mixtures of fuel and oxidizer there may be variations in the fuel–oxidizer ratio, which give rise to separate regions of premixed flames and diffusion flames or to partially premixed diffusion flames (if the mixing is insufficient to support purely premixed flames). Despite the complexities of fuel arrangement that can occur in heterogeneous mixtures, the two extreme concepts of diffusion flames and premixed combustion remain helpful principles in bringing order to descriptions of the combustion processes.

IV. INITIATION OF COMBUSTION

A. Critical Conditions for Ignition

Premixed combustibles that are not reacting rapidly always are metastable in that they react at small but nonzero rates. Often (e.g., for CH_4–air mixtures at room temperature) the rates are so small that in hundreds of years the extent of reactant depletion would not be detectable. Ignition is the process by which these mixtures, reacting at negligible rates, are caused to begin to react rapidly. In combustion there are many situations in which necessary conditions for achieving ignition can be defined accurately. When they exist, these critical conditions can readily be identified in the graph of rates of heat release and of heat loss shown in Fig. 1. If the dependence of the rate of heat loss on temperature is sufficiently strong, because the system has a large thermal conductivity or a large surface-to-volume ratio (small dimensions), then there is only one intersection giving a balance of rates of generation and loss at a low (negligible) rate of heat release. At lower conductivities (larger systems), there are three intersections, the middle one being an unstable balance. The loss line on which a critical ignition condition is identified corresponds to a system size above which the only balance is the high-temperature intersection with a rapid rate of heat release. Ignition can be achieved by external stimuli when the high-rate intersection exists. It occurs spontaneously without external stimuli (spontaneous combustion) when the only balance is the high-rate balance. The spontaneous process occurs when the system is big enough for the loss rate to depend sufficiently weakly on the temperature of the system.

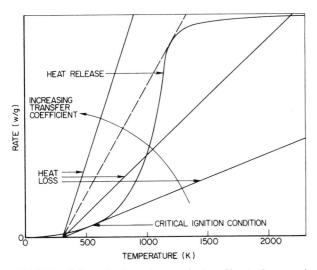

FIGURE 1 Schematic dependences of rates of heat release and of heat loss on temperature, illustrating criticality.

B. Concept of Thermal Runaway

In the limit of no heat loss (adiabatic systems), there are no intersections in Fig. 1, and the slow combustion causes the temperature to increase with time. The consequent increase in the rate of heat release often leads eventually to a rapid rate of temperature increase, called thermal runaway, many examples of which have been seen in experiment. Mathematical theories for ignition and for thermal explosion may employ asymptotic methods in which the rate of temperature increase, at the time of ignition, approaches (but does not reach) infinity. Also, thermal runaway is often characteristic of ignitions (when they occur) in systems with heat loss.

C. Ignition Energies and Ignition Temperatures

Conditions for achieving ignition by application of thermal stimuli can be expressed as ignition energies if the rates of application are rapid or as ignition temperatures (variously called spontaneous ignition temperatures or autoignition temperatures) if the rates of application are slow. In the former case, all the energy applied may contribute to ignition, while in the latter, some of it is lost, and steady conditions are approached in which the applied rate equals the loss rate. There are several methods of ignition; exposure to a sufficiently hot surface, to hot inert gas, to a small flame (a pilot), to an electrically heated wire, to a radiant energy source, to an explosive charge, or to an electrical spark discharge. Ignition criteria for the last two can be expressed best in terms of ignition energy and for the first two in terms of ignition temperature; the others fall in between. The energy that must be supplied to a system to achieve ignition usually exceeds the ignition energy because of losses; for example, in spark ignition the electrical energy that must be supplied to the spark (the spark-ignition energy) exceeds the ignition energy because of heat conduction to the electrodes and energy carried away in the gas by shocks. Ignition energies and ignition temperatures for a few fuels in air at atmospheric pressure are given in Table II. Values of these quantities can vary appreciably with chemical composition and experimental conditions; minimum values appear in Table II. Among practical fuels, ignition temperatures tend to be around 700 K for coal and 600 K for newspaper, dry wood, and gasoline.

Representative dependences of the spontaneous ignition temperature on pressure are provided by the explosion-limit curves shown in Fig. 2. If in a given experiment, ignition is identified with a constant value of w

TABLE II Ignition and Flammability Properties of Selected Fuels in Air

Fuel	Minimum ignition energy (mJ)	Spontaneous ignition temperature (K)	Lower flammability limit (% by volume of gaseous fuel in mixture)	Upper flammability limit (% by volume of gaseous fuel in mixture)	Minimum quenching distance (cm)	Maximum burning velocity (cm/sec)
H_2	0.02	850	4	75	0.06	300
CO	—	900	12	74	—	45
CH_4	0.29	810	5	15	0.21	44
C_2H_6	0.24	790	3	12	0.18	47
C_2H_4	0.09	760	2.7	36	0.12	78
C_2H_2	0.03	580	2.5	100	0.07	160
C_3H_8	0.24	730	2.1	9.5	0.18	45
C_7H_{16}	0.24	500	1.1	6.7	0.18	42
$C_{12}H_{26}$	—	480	0.6	—	0.18	40
C_6H_6	0.21	840	1.3	8	0.18	47
CH_3OH	0.14	660	6.7	36	0.15	54

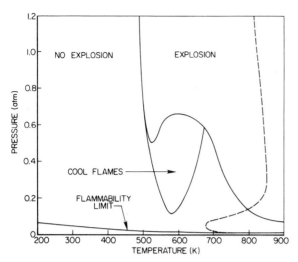

FIGURE 2 Schematic representation of explosion limits in a pressure–temperature diagram for C_3H_8–O_2 mixtures having 50% fuel (solid curves) and for H_2–O_2 mixtures having 67% fuel (dashed curve).

in Eq. (1), then the explosion limit in Fig. 2 should show a gradual increase in temperature with decreasing pressure. This behavior, characteristic of thermal explosions, is seen to apply at the higher pressures for both examples shown. At very low pressures there is a similar behavior, but anomalies occur at pressures between about 0.1 and 0.5 atm in these examples.

For H_2 in O_2 there is an explosion peninsula, its point having an ignition temperature below 700 K at a pressure somewhat below 0.1 atm. Peninsulas of this kind are found for other fuels as well (e.g., for wet CO in O_2) and are explained by the branched-chain character of the explosion. Chain carriers can be removed more rapidly, thereby reducing the rate, either by recombination steps involving collision of three molecules [e.g., reaction (13)] if the pressure is high or by diffusion to the wall of the vessel containing the mixture (where they can be lost by recombination) if the pressure is low. At intermediate pressures, corresponding to the tip of the peninsula, both of these carrier-loss mechanisms are relatively inefficient, and the mixture is prone to spontaneous combustion. According to this explanation, the limit curve should depend on the size of the vessel, on the wall material, and so on; experimentally, it does, and variations of ignition temperature (at a given pressure) of more than 100 K are not uncommon.

A different type of anomaly is observed for hydrocarbon oxidation and is exemplified by the propane results shown in Fig. 2. In the region identified as cool flames, a few blue flames are observed to propagate sequentially through the reactant mixture in the vessel without causing ignition or explosion. For the reactant mixture in this example, as

many as six cool flames were observed, but different fuels can exhibit more or less; cool-flame phenomena have also been seen in CO oxidation. Since the temperature rise in a cool flame is small, chemical kinetics are believed to be responsible for the phenomena. Chain carriers can be produced and then consumed by products at a later stage in the mechanism before ignition develops. Cool flames have been suggested to be relevant to knock (explosion ahead of the flame) in spark-ignition engines; antiknock compounds may suppress cool flames. It has also been suggested that they are responsible for the will-o'-the-wisp—fleeting blue lights occasionally seen in swamps at night—although ordinary blue methane flames in air (from the CH_4 released by decay in the bog) may provide a better explanation.

D. Ignition Delay Time

Once conditions for ignition are established, there is a time delay prior to thermal runaway. The delay is controlled by the chemical kinetics, although in various applications, such as in spark-ignition engines, fluid mechanics also may play a role in the delay. The ignition delay is relevant in the design of many types of combustion engines. The delay depends on the temperature–time history of the reactant mixture. At constant temperature the delay often can be expressed as ρ/w, where ρ (g/cm^3) is the density of the initial reactant mixture and w possesses a form like Eq. (1). The delay is seen to decrease as the temperature increases.

Results of this kind can be used to estimate ignition temperatures by assigning the delay a reasonable value for the experiment to be performed. Because of the Arrhenius form, the ignition temperature is not a precisely defined quantity but depends on the time scale of the experiment. Ignition would occur even at room temperature if the time were long enough. Typical overall activation energies are large enough that the ignition temperature depends only weakly on the time scale; restricting attention to the wide range of time scales encountered in practice may still lead to relatively small variations in the ignition temperature. The ignition delay in adiabatic homogeneous systems is called the adiabatic thermal explosion time.

In ignition by deposition of radiant energy, for example, the ignition delay depends on the incident radiant energy flux. At low fluxes, the delay is long, and ignition can be characterized in terms of the critical flux (or power) needed for ignition; the energy required approaches infinity. At high fluxes the delay is short, and the critical energy for ignition is useful instead; the power approaches infinity. At intermediate fluxes, a critical relationship between power and energy must be satisfied to achieve ignition, with both power and energy exceeding their minimum requirements.

E. Flammability Limits and Quenching Distances

In gas mixtures of fuel and oxidizer, it has been found experimentally that ignition cannot be achieved if the fuel or oxidizer concentration is too low. The critical concentration of either fuel or oxidizer, below which ignition is impossible, is the flammability limit of the mixture. These limits can be expressed as minimum and maximum fuel percentages, between which the fuel percentage must lie for the mixture to be combustible. The minimum percentage is called the lower flammability limit (LFL) and the maximum is the upper flammability limit (UFL). Flammability limits for a few fuels in air at atmospheric pressure and room temperature are listed in Table II. If the intent is to burn the mixture, then it is necessary to keep the fuel percentage between the LFL and the UFL. To handle mixtures safely, without the possibility of combustion occurring, the fuel percentage should be kept below the LFL or above the UFL. For example, in partially filled gasoline tanks of automobiles, the fuel percentage in the air above the liquid is above the UFL (which is about 6%).

The LFL and the UFL vary with pressure and temperature and normally tend to approach each other as either of these quantities is decreased. Thus, as the pressure is reduced, the LFL and UFL converge and meet each other at a critical pressure, the minimum pressure limit of flammability. At pressures below this critical value, the mixture does not burn at any fuel percentage. The pressure limit of flammability for any given mixture depends on temperature; an indication of this dependence is shown in Fig. 2 for the C_3H_8–O_2 mixture, where it is seen to lie well below the explosion-limit curve. Between the flammability limit and the explosion limit the mixture can be ignited by an external stimulus but does not ignite spontaneously.

A standard apparatus in which the LFL and UFL are measured is a vertical glass tube 5 cm in diameter and 100 cm in length, with its ends either closed or open to the atmosphere. The tube is filled with the gas mixture, and a strong spark is discharged either at the top or at the bottom of the tube. According to one criterion, the mixture is judged to be flammable if a flame propagates to the other end of the tube. The results of the experiment depend on whether the tube is open or closed and on whether the spark is at the top (downward propagation) or at the bottom (upward propagation). The limits usually are the widest for upward propagation in a closed tube. The limits usually reported are the widest because the results are used most often in connection with safety.

Since the limits depend on the experiment, they are properties not only of the gas mixture but also of the configuartion of the system. Limits generally widen as the size of the gas container increases. Tabulations of limits are useful only if the dependence on the configuration is not too great. This condition usually is satisfied because differences in the UFL for upward or downward propagation in closed or open tubes seldom exceed a few percent; changes in the LFL seldom reach a factor of two. In addition, when tubes greater than 5 cm in diameter are employed, the widening of the limits is found to be comparably small. One reason for selecting 5 cm is that when smaller diameters are employed, the flammable range begins to narrow appreciably. If the tube diameter is too small, then combustion cannot be achieved at any fuel percentage. The critical diameter below which combustion is impossible is the quenching diameter of the mixture.

If a deflagration propagating through a gas meets a tubular restriction with a diameter less than the quenching diameter, then combustion fails to penetrate into the tube. Similarly, if the deflagration encounters parallel plates whose separation is less than a critical value (the quenching distance), then combustion fails to proceed between the plates. Experimentally, the quenching diameter is about 20–50% greater than the quenching distance. Quenching distances for a few fuels in air at atmospheric pressure and room temperature are given in Table II. The values listed are those for the fuel percentage that gives the minimum value; the quenching distance increases significantly as the fuel percentage departs from the optimum. There is a corresponding dependence of the ignition energy on the fuel percentage. Values of quenching diameters are employed in designs of flame arresters in which fine grids are placed in gas lines to prevent flames from propagating through them.

Flammability limits and quenching distances are manifestations of the same general phenomena and are associated with heat loss and with the strong dependence of heat-release rates on temperature. Because of this strong dependence, rates of heat loss that are only roughly 10% of the rate of heat release can cause extinction of a deflagration. Qualitatively, it is as if a system at an upper intersection (see Fig. 1) were subjected to heat-loss lines with slopes that gradually increase until a tangency condition corresponding to the dashed line is reached, beyond which only a slow-reaction intersection is possible. (However, the figure is not quite applicable because it excludes the energy used in raising the temperature of the mixture as it burns.)

Since it is the ratio of the rate of heat loss to the rate of heat release that is of significance in extinction, reduction of the rate of heat release is an alternative way to extinguish deflagrations. This can be achieved by adding flame extinguishants to the combustible to reduce the rate by reducing the flame temperature (through dilution, for instance, by addition of water or of an inert gas) or by slowing the chemical kinetics at a constant flame temperature (through chemical inhibition, for instance, by addition of materials such as bromine-containing species that combine with chain carriers to reduce their concentrations in the reaction zone). An extreme version of slowing the chemical kinetics can arise if a change in conditions produces a somewhat abrupt change in the kinetic mechanism to one that releases heat at a much slower rate. An example of this may be found in H_2–O_2 combustion, in which the influence of nonreactive HO_2 as a sink for chain carriers through reactions such as Eq. (13) is known to become dominant at temperatures below about 900 K.

Approximate correlations between quenching distances and minimum ignition energies have been developed. It can be stated that to ignite a combustible, an amount of energy must be deposited locally that is roughly sufficient to raise the temperature to the adiabatic flame temperature in a disk-shaped volume of diameter equal to the quenching diameter and of thickness equal to the deflagration thickness. This ignition criterion can be expressed solely in terms of either quenching diameter or deflagration thickness by using the result that the quenching diameter generally is between 20 and 60 times the deflagration thickness (essentially because heat-loss rates that are roughly 10% of heat-release rates cause extinction, as indicated earlier). The basis of the ignition-energy criterion is that if it is not satisfied, then heat loss to the cool combustible surrounding the ignition point causes extinction after the energy deposition.

V. DEFLAGRATION

A. Establishment of Laminar Flames

When mass, momentum, and energy conservation equations are written for planar, steady, adiabatic combustion waves through which reactants are converted to products, it is found that a one-parameter family of waves may exist. The possiblilities are conveniently illustrated in the pressure–volume diagram shown in Fig. 3. In the initial combustible, the pressure is p_0 and the volume per unit mass is v_0; corresponding quantities in the burnt gas are p_∞ and v_∞. The locus of burnt-gas states is called the Hugoniot curve, and the negative of the slope of the straight line connecting the burnt state to the initial state is proportional to the square of the propagation velocity of the wave. Therefore, the slope must be negative; this divides the Hugoniot curve into two distinct branches, the upper and the lower. Detonations have final states on the upper branch and deflagrations on the lower branch. There is a minimum propagation velocity for detonations corresponding to the tangency at the upper Chapman–Jouguet point, and a maximum propagation velocity for

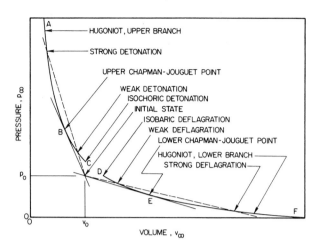

FIGURE 3 Schematic locus of burnt-gas states for combustion waves.

deflagrations corresponding to the tangency at the lower Chapman–Jouguet point. The two dashed lines in Fig. 3 illustrate representative intermediate conditions, and each has two intersections, corresponding to weak and strong waves, as indicated in the figure. The waves encountered in combustion are weak (in fact, nearly isobaric) deflagrations and strong or often Chapman–Jouguet detonations. As the heat of combustion is increased, the separation between the initial state and the Hugoniot increases. Some additional properties of these waves are summarized in Table III, where the subscripts + and − refer to upper and lower Chapman–Jouguet conditions, respectively, and the subscripts 1, max, and min identify additional constants determined by the initial state of the mixture. Initiation of combustion typically produces a deflagration that later may undergo a transition to a detonation.

A common way to initiate a deflagration (also called a laminar or turbulent flame depending on whether the flow is laminar or turbulent) is to discharge a spark inside a tube containing a combustible mixture. A planar deflagration may develop and travel at a nearly constant speed (burning velocity) for a distance roughly between 2 and 50 tube diameters from the spark. A spherical chamber can be used with the spark at the center to initiate a spherical flame. For experimental investigations of deflagrations, it is helpful to have a stationary rather than a propagating flame in the laboratory. This can be achieved by use of

1. A Bunsen burner
2. A slot burner (Bunsen-type burner with an elongated rectangular outlet to produce nearly planar, two-dimensional flow)
3. A flat-flame burner (in which the combustible is passed through a cooled, porous flat plate, and a

TABLE III Summary of Types of Combustion Waves and Their Properties

	Section in Fig. 3	Pressure ratio $p = (p_\infty/p_0)$	Volume ratio $v = (v_\infty/v_0)$	Propagation Mach number M_0	Burnt-gas Mach number M_∞	Remarks
Strong detonations	Line A–B	$p_+ < p < \infty$	$v_{min} < v < v_+$ $(v_{min} > 0)$	$M_{0+} < M_0 < \infty$	$M_\infty < 1$	Seldom observed; requires special experimental arrangement
Upper Chapman–Jouguet point	Point B	$p = p_+$ $(p_+ > 1)$	$v = v_+$ $(v_+ < 1)$	$M_0 = M_{0+}$ $(M_{0+} > 1)$	$M_\infty = 1$	Usually observed for waves propagating in tubes
Weak detonations	Line B–C	$p_1 < p < p_+$ $(p_1 > 1)$	$v_+ < v < 1$	$M_{0+} < M_0 < \infty$	$M_\infty > 1$	Seldom observed; requires very special gas mixtures
Weak deflagrations	Line D–E	$p_- < p < 1$	$v_1 < v < v_-$ $(v_1 > 1)$	$0 < M_0 < M_{0-}$	$M_\infty < 1$	Often observed; $p \approx 1$ in most experiments
Lower Chapman–Jouguet point	Point E	$p = p_-$ $(p_- < 1)$	$v = v_-$ $(v_- > 1)$	$M_0 = M_{0-}$ $(M_{0-} < 1)$	$M_\infty = 1$	Not observed
Strong deflagrations	Line E–F	$0 < p < p_-$	$v_- < v < v_{max}$ $(v_{max} < \infty)$	$M_{0\,min} < M_0 < M_{o-}$ $(M_{0\,min} > 0)$	$M_\infty > 1$	Not observed; forbidden by cosiderations of wave structure

flame is ignited and stabilized in a planar, one-dimensional configuration adjacent to the plate)

4. A stagnation-flow burner (in which counterflowing streams of reactants and products or inerts from opposed ducts serve to stabilize a planar flame in a flow with strain, which is said to stretch the flame)

When the flame is stationary, the velocity of the combustible entering the flame (or the normal component of velocity in non-one-dimensional configurations) is the burning velocity. Many measurements of burning velocities and of temperature and composition profiles through laminar deflagrations have been made with stationary flames.

B. Structures of Laminar Flames

A representative flame structure is illustrated in the upper part of Fig. 4. Pressure decreases with distance through the flame (see also Fig. 3), but the fractional change is very small (of the order of the square of the Mach number of propagation), typically 10^{-6} to 10^{-3}, and it has been exaggerated in Fig. 4 for illustration. The rate of heat release and the rate of fuel consumption [w in Eq. (1)] follow approximately the same curve, which peaks sharply near the burnt-gas end of the flame. Most of the reaction occurs near this peak, where the fuel concentration is significantly

below its initial value. In the figure the fuel concentration is represented in terms of its mass fraction, the mass of fuel per unit mass of mixture. Although the fuel is flowing toward this reaction zone, it enters it mainly by diffusion rather than by convection, because the concentration gradient is appreciable but the level is small (roughly 10% of the initial concentration). Temperature and velocity both increase, typically by a factor of 5 to 8, and they are near their maximum values in the reaction zone. The portion

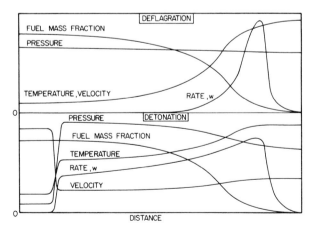

FIGURE 4 Schematic illustration of representative structures of planar deflagrations and detonations.

of the flame ahead of the reaction time typically is thicker than the reaction zone and often involves convection and diffusion, with little reaction. Thus, there is approximately a two-zone structure, a preheat zone in which heat conduction increases the temperature until the w of Eq. (1) becomes appreciable, followed by a reaction zone. It is because of this two-zone character that the deflagration propagates at a well-defined burning velocity.

Variations in this structure may occur in different flames. Chain carriers may be produced mainly on the hot side and diffuse forward to react and produce appreciable heat release in the preheat zone. Calculations of flame structures with detailed chemical kinetics show that various chemical species and reactions appear and then disappear in different parts of the flame. Study of deflagration structure is a continuing research topic which has experienced appreciable advancement in recent years.

The overall thickness δ of the deflagration can be estimated (e.g., by dimensional analysis) as

$$\delta = \sqrt{\alpha\rho/w}, \tag{15}$$

where α (cm^2/sec) is the thermal diffusivity. The definition of a thickness is somewhat arbitrary because of the continuous variation in properties. Since most of the thickness often is occupied by the preheat zone, definitions based on the half-width of the peak of w tend to underestimate thicknesses defined in terms of temperature changes. Thicknesses based on temperature changes tend to be proportional to $p^{-n/2}$ [see Eq. (1)]. That is, they decrease as pressure increases, and at atmospheric pressure they have values in the range of 10^{-2} to 10^{-1} cm.

C. Laminar Burning Velocities

Unlike ignition temperatures, laminar burning velocities have more well-defined values. They can be estimated (again by dimensional analysis) as

$$V = \sqrt{\alpha w/\rho}, \tag{16}$$

where V denotes the burning velocity. From Eq. (1) it is seen that their dependence on the flame temperature is strong (through the Arrhenius factor). Since values of n typically are near 2, their dependence on pressure (proportional to $p^{n/2-1}$) is weak; they may increase or decrease slowly as the pressure is increased. The dependence on flame temperature causes them to increase as the initial temperature of the combustible increases and to depend strongly on the composition of the combustible mixture. The composition dependence for a few hydrocarbon fuels (in air at atmospheric pressure and initially at room temperature) is shown in Fig. 5, where the equivalence ratio is defined as the ratio of the initial fuel concentration to that required for a stoichiometric mixture. [In such a mixture,

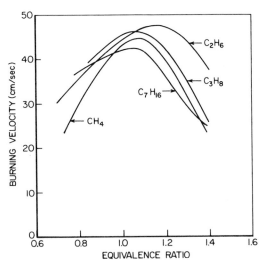

FIGURE 5 Dependence of the burning velocity on the equivalence ratio for four hydrocarbon fuels in air at standard conditions.

if all the fuel were to burn to CO_2 and H_2O, as in reaction (2), for example, there would be no fuel or O_2 remaining in the products.] Maximum burning velocities for various fuels (in air at atmospheric pressure and initially at room temperature) are listed in Table II; minimum values occur at the flammability limits and tend to be around 5 or 10 cm/sec. The maximum value occurs near stoichiometric conditions (an equivalence ratio ϕ of unity). The flame temperature is a maximum very near stoichiometry [exactly at stoichiometry if reaction (2), for example, is an exact representation of the overall process], but V often achieves its maximum somewhat farther from stoichiometry (usually at fuel-rich conditions, $\phi > 1$, rather than fuel-lean conditions), because of differing diffusivities for different chemical species and variations in chemical kinetics with ϕ.

Values of burning velocities measured by different techniques typically differ from each other by a few percent. Techniques include observations of both propagating and stationary flames. Rough values of burning velocities can be obtained from pressure–time histories in closed chambers if the flame shape can be estimated. In the soap-bubble method, the combustible is contained in an essentially constant-pressure spherical bubble and ignited by a spark at the center; the burning velocity is calculated from the observed rate of expansion of the spherical flame by subtracting the expansion associated with the density decrease across the flame. Use of Bunsen and slot burners involves small uncertainties in calculating the normal component of gas velocity from the observed flame shape and the measured combustible flow rate to the burner. Flat-flame burners can stabilize flames over a range of burning velocities, because as the fuel flow rate is decreased, the flame moves closer to the burner, causing the rate of heat loss to the burner to increase and therefore the flame temperature to decrease, and thus maintaining a balance between the flow velocity and the burning velocity; the flow velocity above which the flame no longer can be stabilized by the burner and travels away with the flow (i.e., the velocity at which flame blowoff occurs) corresponds to nearly adiabatic conditions and therefore is comparable to the burning velocities found with the other experimental techniques. In stagnation-flow burners, flame stretch generally modifies the burning velocity, and values to be compared with results of other methods can be obtained by extrapolation to low rates of strain. Recent most reliable results for CH_4, obtained by a variant of this last method, are approximately 10% below those of Table II and Fig. 5.

D. Laminar Flames in Nonuniform Flows

In nonuniform flows, flame stretch affects the internal structure of the flame and can change the flame temperature from its normal adiabatic value by modifying relative diffusion rates of heat and various chemical species. Small differences in Lewis numbers can cause large differences in the influence of stretch on burning velocities, because of the strong dependence of the burning velocity on the flame temperature. For deflagrations stabilized at the exit of a duct, the boundary layer at the duct wall is a region of strain and therefore stretches the flame that is anchored there. The consequent flame-structure modifications usually are not of great significance in determining the conditions for blowoff (the average exit flow velocity above which the flame is blown away) or flashback (the average exit flow velocity below which the flame propagates into the duct); these conditions can be estimated by evaluating the flow velocity and burning velocity at a distance from the wall that is equal to the quenching distance (since flame propagation at scales less than the quenching distance generally does not occur). Flame-structure modifications by stretch can, however, be significant for deflagrations away from the walls. For combustible mixtures in which stretch tends to reduce the flame temperature (e.g., those having thermal diffusivities exceeding the average reactant diffusivities, such as in fuel-rich H_2–air flames or fuel-lean C_3H_8–air flames), abrupt flame extinction can be produced by stretch alone. In the opposite case (e.g., for fuel-lean H_2–air flames or fuel-rich C_3H_8–air flames), heat loss is needed for abrupt extinction of single planar stretched flames.

E. Turbulent Flame Propagation

In most practical combustion chambers, flames are turbulent. The burning velocity generally is observed to increase as the intensity of the turbulence increases. Although the mechanisms of turbulent flame propagation are not well understood, there exist many ideas on the subject and various methods for estimating turbulent burning velocities. A fact that has been established is that there is more than one regime of turbulent flame propagation. Regimes can be distinguished on the basis of two dimensionless numbers, a Reynolds number that is characteristic of the turbulence and a Damköhler number that is characteristic of the chemistry. The latter, the ratio of a flow time to a chemical time, first introduced by Damköhler in 1936, is of paramount importance in combustion and may be defined as

$$D = \tau V/\delta, \qquad (17)$$

where τ is a flow time, and δ and V are given by Eqs. (15) and (16). For use in turbulent combustion, the flow time τ may be taken as the ratio of the size of the largest turbulent eddies to an intensity defined as the average value of the magnitude of the turbulent velocity fluctuation (the difference between the local instantaneous velocity and the mean velocity). With this definition, two limiting regimes of turbulent flame propagation are those of large and small values of D.

At sufficiently large Damköhler numbers, laminar flames are thin compared with all scales of turbulence, and turbulent flames consist of wrinkled laminar flames—the reaction-sheet regime. At sufficiently small Damköhler numbers, laminar flames are thick compared with all scales of turbulence and therefore cannot exist in the turbulent flow; turbulent combustion occurs in the manner in which reactions occur in well-stirred chemical reactors in the chemical industry—the distributed-reaction regime. At sufficiently large turbulence Reynolds numbers, there may well exist additional (currently unknown) regimes of turbulent flame propagation at intermediate Damköhler numbers, between those corresponding to the reaction-sheet and the distributed-reaction regimes. Methods for estimating turbulent burning velocities are most well developed for the reaction-sheet regime; here the ratio of the turbulent burning velocity to the laminar burning velocity is seen geometrically to be the ratio of the wrinkled flame area to the cross-sectional area of the turbulent flame and depends on the intensity U of the turbulence, but not on its scale. A rough formula for the turbulent burning velocity, often used, is $U + V$, for the reaction-sheet regime.

Many practical turbulent-combustion processes—for example, those in spark-ignition engines—are widely believed to occur in the reaction-sheet regime. At high turbulence intensities (large values of U/V), these flames involve many reaction sheets and pockets of unburnt gas that have been cut off by the wrinkled laminar flame propagation. At low intensities (small U/V), only single sheets occur, which are slightly wrinkled. If D becomes too small, then the previously indicated influences of flow nonuniformities on laminar-flame structure arise; and under appropriate conditions, there are local extinctions of the wrinkled flames, causing holes to develop in the reaction sheets.

F. Heterogeneous Deflagrations

Combustion waves in fuel sprays, earlier identified as premixed-combustion processes with diffusion-flame substructures, provide examples of heterogeneous deflagrations. Since the droplets in heterogeneous spray deflagrations burn as diffusion flames, Eq. (1) cannot be used in Eqs. (15) and (16) to obtain the thickness of the heterogeneous deflagration and its burning velocity. Instead, w may be estimated as the product of the number of droplets per unit volume and the burning rate (mass per unit time) of a droplet. The resulting burning velocity decreases (and deflagration thickness increases) as the droplet size increases, if the equivalence ratio is held fixed (i.e., if the droplets are made large by combining the same amount of fuel into a smaller number of droplets). In sprays of sufficiently small droplets, the evaporation time of a droplet is short compared with the passage time (δ/V) through a gaseous deflagration. The combustion process thus becomes essentially the same as that which has been described for gaseous fuels [so Eq. (1) again can be used, for example]. For typical hydrocarbon fuels in air at atmospheric pressure and room temperature, this homogeneous mechanism occurs at initial droplet diameters below about $10 \, \mu m$ and therefore requires finely atomized sprays. With coarser atomization, deflagration thicknesses greater than 10^{-1} cm are not uncommon.

VI. DETONATION

A. Transition from Deflagration to Detonation

Deflagrations propagating along tubes for distances greater than roughly 50 tube diameters often experience a transition to detonation. The transition tends to occur sooner in tubes closed at one or both ends than in open tubes. The process of transition is not instantaneous and usually not one dimensional; often it is preceded by a period of turbulent flame propagation. The expansion of the gas on its passage through the deflagration leads to gas motion in the direction of propagation, as the combustion products are brought to rest at the walls, and to associated increased pressures in the burnt gas. These higher pressures give rise to pressure waves that propagate at sound velocities or greater in the direction of motion of the deflagration, quickly overtaking the deflagration and

passing into the combustible ahead of it, compressing the mixture and consequently raising its temperature. The temperature increase can be great enough at positions of shock convergence to generate local explosions that develop into new propagating deflagrations moving outward from points that are often located at walls. From these ignition sites, deflagrations travel in both directions along the tube, increasing the overall rate of pressure-wave generation. Eventually, pressures in excess of final detonation pressures develop, and a detonation propagates through the rest of the combustible in the tube.

This sequence is favored by confinement, because the walls reflect the pressure waves back into the combustible. Openings allow relief by transmitting pressure waves to the surroundings of the system. In open combustible clouds, transition to detonation is rare, although it can occur in sufficiently reactive combustibles. The transition time in the open appreciably exceeds that in an enclosure. In general, the greater the degree of partial confinement the sooner detonation develops. In this respect, confined combustibles are more hazardous than combustibles open to the atmosphere.

B. Detonation Velocities

Planar detonations differ from planar deflagrations in that, in a reference frame in which the combustible gas ahead of the wave is at rest, the gaseous products just behind the wave are moving toward the combustible. Since this reference frame is the laboratory frame for detonations traveling in closed tubes, for example, a region of gas expansion must develop in the product gas between the detonation and the end wall. This expansion generates expansion waves that tend to reduce the pressure just behind the detonation. Therefore, strong detonations are weakened by the expansion, and their burnt-gas state tends to approach the upper Chapman–Jouguet point (see Fig. 3). Since $M_\infty > 1$ for weak detonations (see Table III), the expansion waves cannot overtake them or affect them; in fact weak detonations do not occur in combustion for reasons associated with their structures. Thus, detonations in tubes evolve toward Chapman–Jouguet waves and are observed to propagate at the Chapman–Jouguet velocity C. At the large values of heat release, representative of most detonations, this velocity can be approximated as

$$C = \sqrt{2(\gamma^2 - 1)Q}, \qquad (18)$$

where γ is the specific-heat ratio (a number typically between 1.2 and 1.3) and Q is the heat released per unit mass of mixture (expressed in units of velocity squared).

An approximation to the C of Eq. (18) can be shown to be $[2\gamma_\infty(\gamma_\infty + 1)T_\infty R/W_\infty]^{1/2}$, where the subscript again identifies the fully burnt gas, W (g/mol) is the average molecular weight of the gas, and R (here 8310 g m^2/ sec^2 mol K) has been defined earlier. In this form, the dependence of C on the final temperature and molecular weight is clearer. Values of C for detonable gas mixtures initially at room temperature and atmospheric pressure typically lie between 1000 and 4000 m/sec; for stoichiometric H$_2$–O$_2$ mixtures the value is about 2830 m/sec and for stoichiometric CH$_4$–O$_2$ mixtures about 2500 m/sec. Thus, representative detonation velocities are on the order of 1000 times representative deflagration velocities.

The boundary layer at the wall of a tube decreases the velocity of propagation of a detonation along the tube through non-one-dimensional interaction with the wave; the decrease varies inversely as the tube diameter. Decreases in detonation velocities also may be associated with incomplete combustion, since decreasing Q decreases C [see Eq. (18)]; there have been observations of unusually low-velocity detonations that must involve incomplete combustion. There also are overdriven detonations in tubes that propagate at velocities greater than C; these are strong detonations (initiated by a strong shock wave) that have not had time to decay to Chapman–Jouguet waves. In addition, standing detonations (stationary in the laboratory frame) have been produced in highly supersonic flows of combustible gas mixtures; these too are strong detonations, in that the component of gas velocity normal to the wave exceeds C.

C. Structures of Planar Detonations

The lower part of Fig. 4 illustrates detonation structure. A strong shock wave at the front of the detonation increases the pressure and temperature and initiates chemical heat release, which thereafter proceeds to deplete the fuel, in reactions that are essentially uninfluenced by molecular transport (because of the high velocities prevailing everywhere). This structure was found independently by Zel'dovich (1940), von Neumann (1942), and Döring (1943). The thickness of the shock (through which all properties except the mass fractions of the chemical species change abruptly) is only a few molecular mean free paths and has been expanded greatly in the figure for purposes of illustration. The thickness of the detonation therefore is controlled by the rates of the chemical reactions behind the shock and is approximately equal to the length l of the reaction zone, which is roughly

$$l = (\rho/w)C(\gamma - 1)/(\gamma + 1) \qquad (19)$$

since the velocity behind the strong leading shock is $C(\gamma - 1)/(\gamma + 1)$. Here, the reaction time ρ/w, with w given by Eq. (1), is to be evaluated at the pressures and temperatures prevailing behind the leading shock. Detonation thicknesses vary strongly with equivalence ratios, for

example, and decrease with increasing pressure, in proportion to p^{1-n}. For gas mixtures initially at room temperature and atmospheric pressure, thicknesses ranging roughly from 10^{-3} to 1 cm are representative.

Since transport properties are unimportant behind the leading shock, accurate calculations of detonation structures with detailed chemical kinetics can be made comparatively easily. It is often found that there is a relatively long induction period of nearly constant temperature, followed by a period of rapid heat release, as suggested by the w curve in Fig. 4 which, for clarity of presentation, is shown with a much longer jump across the initial shock, and a correspondingly smaller increase thereafter, than is typical in real gaseous detonations. The properties just behind the leading shock and in the burnt gas far downstream can be calculated even more simply, because they do not depend on the reaction rates. The pressures are of the greatest practical interest from the viewpoint of producing damage. There is a pressure peak behind the leading shock; thereafter, the pressure decreases toward its final value. In the notation of Table III, the peak pressure is approximately $p_0 M_0^2 2\gamma/(\gamma + 1)$ at sufficiently large heat release and approximately twice the final pressure, which itself may typically be 20 times the initial pressure. Ordinary walls cannot withstand these pressures.

D. Transverse Structures of Detonations

Most planar detonations are unstable to both planar and nonplanar disturbances by a mechanism involving the propagation of pressure waves through the reaction zone. The instability is strongest for nonplanar disturbances and leads to a three-dimensional structure of propagating detonations that involves incident, transmitted, and reflected shock waves, with the reactions occurring behind the transmitted and (possibly) incident shocks. The triple point, where the three shocks meet, is the point of highest pressure. Detonations in tubes consist of cells with boundaries defined by moving triple points. The one-dimensional structure and propagation velocity represent an average over a large number of cells.

The cells can be seen by placing a smoked foil on the inner wall of a tube and then passing a detonation through a detonable mixture in the tube; afterward, the trajectories of the triple points are visible in the smoke layer on the foil as writings on the wall. The cell spacing can be defined as the distance between the intersections of triple points. The spacing is about 10 to 30 times the thickness l of Eq. (19). In round tubes near detonability limits, spinning modes are seen (spinning detonations), in which a triple point follows a helical path along the wall. Strong detonations that are far from the Chapman–Jouguet conditions do not exhibit cellular structures.

E. Direct Initiation, Transmission, Detonability Limits, and Failure of Detonations

In addition to developing from deflagrations, detonations can be initiated directly by rapidly depositing a sufficient amount of energy in the combustible—for instance, from an explosive charge that is large enough. There are highly sensitive liquid and solid explosives in which direct initiation can be produced relatively easily by a shock; their susceptibility to initiation is measured by impact tests in standard configurations to provide sensitivity ratings of the materials. Although there is an initial period of adjustment with direct initiation, deflagrations are not involved. The criterion for achieving direct initiation by instantaneous deposition of energy is that the energy deposited be sufficient to maintain a shock as strong as the leading shock of the Chapman–Jouguet detonation for a time as long as the chemical reaction time in the detonation. There are simplified descriptions of histories of blast waves (strong shocks) that have been used successfully with this criterion to calculate the minimum initiator energies required.

A detonation traveling in a tube can be transmitted to an unconfined combustible at an open end of the tube if certain conditions are satisfied. It is found that transmission will occur if the tube diameter exceeds 13 cell sizes (i.e., about $200l$).

There are combustible mixtures that can experience deflagration but not detonation. Thus, for example, a detonability limit curve could be drawn in Fig. 2 slightly above the flammability limit curve. The two limit curves often are found to be close together. There is no evident fundamental reason for this, because the processes controlling the two limits are very different; even the chemical kinetics usually differ in deflagrations and detonations, the former, for example, not requiring initiation chemistry in which chain carriers are produced from stable molecules because the carriers can diffuse in the hot reaction zone to start the combustion. Detonability limits involve gas-dynamics processes that cool the combustible behind the leading shock before the chemical reactions are completed. They therefore are related to processes of failure of detonations.

Detonations will fail to propagate through a combustible mixture in a tube if the tube diameter is less than a critical value related to l. Similarly, a cylinder of solid explosive cannot be detonated if its diameter is less than a critical value, the failure diameter. Failure occurs more readily in the absence of confinement, because lateral expansion (with consequent cooling) of the gas behind the leading shock is then facilitated. In the open, a combustible layer of thickness roughly less than l cannot be detonated. This critical thickness depends on the properties of the gas outside the layer. Because of the short passage time of the reaction zone in a detonation, it is the inertia rather than the strength of a wall that provides the confinement relevant to failure. If the mass per unit area of the wall is greater than roughly l times the initial density of the combustible, then the detonation will behave as if it were confined. Thin plastic sheets, totally incapable of withstanding detonation pressures or of contributing to the development of a detonation, nevertheless can cause a gaseous detonation, once established, to propagate as if it were under confinement.

F. Heterogeneous Detonations

Detonations can occur not only in homogeneous gaseous, liquid, or solid explosives, but also in heterogeneous mixtures such as fuel sprays in air, suspensions of aluminum powder or coal dust in air, and composite solid propellants. Chemical reaction mechanisms in these detonations involve vaporization or heterogeneous reactions at fuel surfaces. The length l of the reaction zone therefore tends to be longer than for homogeneous detonations and, although Eq. (19) would still apply, Eq. (1) cannot be used in it. Explosions in grain elevators have involved heterogeneous detonations in grain–air mixtures. Since detonations are the most damaging kinds of explosions, evaluation of detonation hazards is essential in safety considerations whenever combustibles are employed.

VII. DIFFUSION FLAMES

A. Experimental Configurations of Diffusion Flames

In diffusion flames, two types of feed streams can be identified; fuel streams and oxidizer streams. A common configuration is a fuel jet issuing into a quiescent oxidizing gas or into a co-flowing oxidizer stream in a duct, as studied by Burke and Schumann (1928). Flat diffusion flames can be established in counterflowing streams of fuel and oxidizer, with the flame oriented normal to the axis of the streams. The counterflow configuration is useful for scientific investigation not only because the flame is flat, but also because there is no region of quenching (where fuel and oxidizer meet each other without reacting), which occurs at the tube exit in co-flow configurations, for example.

B. Structures of Laminar Diffusion Flames

In diffusion flames, there are regions on each side of the reaction zone that involve diffusion of reactants without reaction. The reaction zone may be approximated as a thin sheet with no fuel on one side and no oxidizer on the other. Concentrations of both of these reactants are approximated as zero at the sheet, and the sheet location is determined by the requirement that the fuel and oxidizer

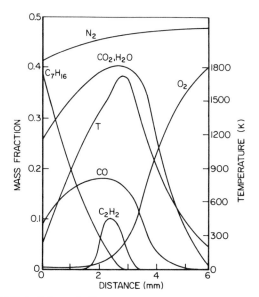

FIGURE 6 Schematic illustration of the structure of a laminar diffusion flame of C_7H_{16} in an O_2–N_2 mixture.

diffuse into it in stoichiometric proportions. Figure 6 is a sketch in which the sheet can be imagined to be in the middle (at 3 mm); this figure also illustrates some additional attributes of hydrocarbon–air diffusion-flame structure. Pressure is not shown in Fig. 6 because pressures remain nearly constant throughout diffusion flames.

Figure 6 pertains to a counterflow experiment with a liquid heptane pool below a nitrogen–oxygen duct. The distance is the height above the vaporizing surface of the fuel. The inert (nitrogen) diffuses to the fuel surface against convection without reacting. Product concentrations and especially temperature peak sharply at the blue reaction sheet where the heat is liberated and products are produced. The blue color is chemiluminescent radiation emitted from electronically excited molecules in the reaction. In these hydrocarbon flames, some CO also is produced at the reaction sheet and is oxidized more slowly in a broader O_2-rich region. On the fuel side at elevated temperatures, the hydrocarbon fuels pyrolyze (react under increased temperature to form many other hydrocarbons, represented in the figure by just one of the important ones, C_2H_2). Thus, a mixture of hydrocarbon fuels, in fact, is oxidized in the reaction zone even if the original fuel is pure. If residence times are long enough at elevated temperatures on the fuel side, then the pyrolysis results in the formation and growth of soot particles (carbon-rich particles about 1–100 nm in size that give the flame a yellow color), which also may be oxidized in the reaction zone.

The figure shows a small amount of O_2 penetrating the flame and reaching cool fuel regions where temperatures are too low for it to react. Small amounts of reactant pene-

tration occur when residence times are made short enough (e.g., by increasing the exit velocity in the oxidizer duct). Extinction occurs at a critical residence time τ, which can be defined in terms of a minimum Damköhler number,

$$D = \tau w / \rho, \qquad (20)$$

where τ can be estimated as the separation between the fuel and oxidizer ducts divided by the exit velocity.

C. Droplet Burning

A spherical fuel droplet in a quiescent, ambient oxidizing gas burns by vaporization and diffusion of the fuel to a spherical reaction sheet surrounding the droplet. The diameter of the reaction sheet, typically 10 times the droplet diameter, is determined by the requirement that the reactants diffuse into it in stoichiometric proportions; it increases as the ratio of inert to oxidizer in the ambient gas is increased. The radial profiles of temperature and species concentrations as functions of the distance from the surface of the droplet are like those shown in Fig. 6. The reaction sheet is sensitive to nonspherical motion of the gas and therefore is distorted appreciably by forced or natural convection (i.e., by droplet motion or buoyancy). In many diesel and liquid-propellant rocket engines, pressures exceed the critical point of the fuel (the pressure above which the distinction between liquid and gas disappears); the diffusion flame then involves fuel diffusion without vaporization, as if the fuel were gaseous. In multicomponent fuels, the less volatile constituents tend to accumulate at the droplet surface, and the internal temperature can rise above the vaporization or nucleation temperature, leading to a sudden disruption of the droplet, with rapid dispersion and burning of the fine mist that is thereby produced.

The burning rate of a droplet (its rate of mass loss) is controlled by the rate of heat transfer to the liquid, usually by heat conduction from the flame. Since the ratio of the flame diameter to the droplet diameter remains approximately constant with time, this rate is proportional to the droplet diameter. Therefore, the square of the droplet diameter decreases linearly with time; the constant of proportionality in this relationship, the evaporation constant K, is found to be

$$K = 8\alpha(\rho_g / \rho_l) \ln(1 + B), \qquad (21)$$

where α again is the thermal diffusivity of the gas, ρ_g / ρ_l is the ratio of densities of the gas and the liquid, and B is the transfer number. The transfer number is the ratio of the impetus for mass transfer to the resistance to mass transfer, and for droplet burning it is

$$B = [qY + c_p(T_\infty - T_l)]/L, \qquad (22)$$

where q is the heat released in combustion per unit mass of oxidizer consumed, Y is the mass fraction of oxidizer in the ambient gas, c_p is the specific heat at constant pressure for the gas, T_∞ is the ambient temperature, T_l is the liquid surface temperature, and L is the energy required for vaporization per unit mass of fuel. Here, T_l can be approximated as the boiling temperature of the liquid (generally it is slightly less), and if internal liquid heating and radiant energy loss are neglected, L can be approximated as the thermodynamic heat of vaporization of the fuel at its boiling point. Typical values of B are around 10 and of K around 10^{-2} cm^2/sec.

D. Burning of Solid Fuel Particles

Spherical fuel particles can burn with diffusion flames by mechanisms related to those of droplet burning. Often (e.g., for carbon under many conditions) the combustion occurs on the surface of the solid fuel particle. An alternative to Eq. (22), often useful in solid-particle combustion, is

$$B = (vY + X)/(1 - X), \qquad (23)$$

where v is the stoichiometric ratio of mass of fuel to mass of oxidizer and X is the mass fraction of fuel in the gas at the particle surface. For a fuel such as carbon, that normally does not vaporize, X is zero. Aluminum and magnesium are examples of fuels that vaporize during combustion. The combustion products of solid fuels, especially of metals, often tend to be solids or liquids. Condensation of reaction products then occurs in the combustion process, and a number of different combustion mechanisms are observed, depending on whether condensation occurs at the fuel surface, in a gas-phase flame, or in the oxidizing gas relatively far from the particle. The relative volatility of the fuel and its oxide has an important influence in affecting which ones of the possible combustion mechanisms will develop.

E. Combustion in Laminar and Turbulent Fuel Jets

The flame height is of practical interest in the combustion of gaseous jets of fuel in oxidizing atmospheres. Also of concern in safety, for example, is the rate of radiant energy emission from the flame. There are correlations (based on experiment and on theory) between energy emission rates and flame heights. Flame heights and shapes usually are defined from visual observations but in more scientific studies are related to the positions at which the mixture is stoichiometric. Flame heights defined in different ways typically differ by amounts on the order of 10%, although differences between average and maximum heights in turbulent situations can exceed a factor of 2. Figure 7 is a schematic illustration of the dependence of the average flame height on the exit velocity of the fuel jet; the numbers would apply roughly to a CH$_4$ jet issuing from a tube 0.5 cm in diameter into air at atmospheric pressure and room temperature.

At low exit velocities, the diffusion flame is laminar, and its height increases in proportion to the exit velocity u. Theory and experiment agree in showing the laminar height to be proportional to ud^2/α, where d is the exit diameter. Buoyancy can play an important role, and it causes departures from this type of dependence if the exit is not round. Transition to turbulence begins at exit velocities above a critical value, and the portion of the flame that is turbulent then rapidly increases with u. The dashed line in Fig. 7 marks the lower boundary of the turbulent portion. The height of the turbulent flame is independent of u and proportional to d, as predicted if the molecular diffusivity α is replaced by a turbulend diffusivity proportional to ud.

The base of the flame is observed to be lifted abruptly to an appreciable distance above the jet exit when u exceeds another critical value. This value depends somewhat on the exit geometry of the jet, and depending on the fuel and

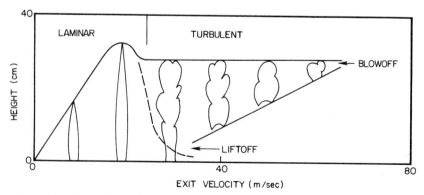

FIGURE 7 Schematic illustration of flame height for open-jet diffusion flames, showing laminar and turbulent regimes, liftoff velocity, liftoff height, and blowoff velocity.

oxidizer it may occur in the laminar or turbulent regime. The liftoff is caused by extinction of the diffusion flame in the region of high rates of strain at the base of the flame. At greater heights, the rate of strain is less because the jet has spread, and the flame again can begin. The height at which the lifted flame begins—the liftoff height—increases approximately linearly with u. At a large enough exit velocity, the liftoff height is nearly equal to the flame height, and blowoff of a roughly spherical flame occurs at a third critical exit velocity; the blowoff velocity. If u exceeds the blowoff velocity, then a diffusion flame cannot be stabiiized by the jet. These phenomena are partially understood.

F. Flame Spread Through Heterogeneous Fuels

In many accidental fires, for example, diffusion flames spread through arrays of solid or liquid fuels. The process of flame spread usually involves heating the nonburning fuel to a temperature at which it begins to give off combustible gas that can participate in the combustion process. A spread velocity can be defined as the velocity at which this heating front moves through the fuel array and can be estimated from the heat flux imparted by the combustion and the energy required to bring the virgin fuel into participation. Many different physical phenomena can be involved in flame spread (radiation, conduction, convection, buoyancy, surface tension, etc.). Spread phenomena are important not only in unwanted fires but also in processes designed to produce desirable results, such as the *in situ* combustion of coal or oil shale underground in fuel recovery.

VIII. COMBUSTION INSTABILITIES AND EXTINCTION

A. Stability of Laminar Deflagrations

Planar laminar deflagrations are subject to a number of types of instabilities. These instabilities sometimes lead to turbulence, but more often produce nonplanar propagating laminar flames, which can be regular or chaotic. The instabilities may be ordered from those of greatest importance at the largest scales to those having major influences at the smallest scales.

At large scales, the body force associated with buoyancy is a dominant instability for upward-propagating flames; the low-density burnt gas below provides an unstable configuration. Buoyancy contributes a stabilizing influence to planar flames propagating downward. Accelerating flames experience an effective body force that is destabilizing; it is stabilizing for decelerating flames. Buoyant instabilities

have been observed for flames propagating in large vented enclosures. By increasing the flame area, they increase the overall burning rate and contribute to the transition to turbulent combustion within the enclosure. Diffusion flames on ceilings have been observed to develop buoyant convection cells; similar phenomena can be expected for premixed flames on large burners pointed downward. Flames with low burning velocities propagating upward in narrow tubes develop an elongated nose, like that of a large air bubble moving upward in a liquid in a narrow tube, and travel at the same buoyant velocity, about $0.3\sqrt{dG}$, where d is the tube diameter and G is the acceleration of gravity (about 1000 cm/sec^2 at the Earth's surface).

At somewhat smaller scales, deflagrations can be dominated by the hydrodynamic instability, which was discovered independently in theoretical analyses by Darrieus (1938) and Landau (1944). Because of the density differences across the flame, approximated as a discontinuity propagating at a constant burning velocity, there is an inviscid, nonacoustic, hydrodynamic instability in the absence of body forces that causes transverse disturbances to grow at a rate inversely proportional to the wavelength. The mechanism of this instability, which is present for every deflagration, is well understood, but its consequences when the amplitude of the disturbance becomes large are still under investigation. The instability appears to lead not to turbulence but to a steady, nonplanar structure having a corrugated flame shape. The somewhat perplexing fact that these nonplanar waves are not often seen must be attributable to the stabilizing influences of buoyancy (or of container walls) at large scales and of the third potential source of instability—diffusive–thermal effects—at small scales.

Differences in diffusivities of reactants and heat (diffusive–thermal effects) can be responsible for the instability of planar deflagrations at scales comparable to the flame thickness δ. Since a protuberance of the reaction zone toward the reactants constitutes a source for heat and a sink for reactants, reactant diffusivities in excess of heat diffusivities (that is, Lewis numbers less than unity) increase the flame temperature and therefore the burning velocity at the protuberance, thus providing instability. Conversely, Lewis numbers greater than unity tend to stabilize the flame against nonplanar disturbances. For Lewis numbers less than unity, the combination of diffusive–thermal and hydrodynamic instabilities leads to cellular flames, observed, for example, for fuel-lean H_2–air flames. These are flames rounded toward the fresh combustible that have cusps pointing toward the burnt products. Cell sizes depend on the Lewis numbers. Related diffusive–thermal instabilities rarely occur for diffusion flames but sometimes are observed, with their physical mechanisms having been clarified recently.

B. Extinction of Laminar Combustion

Extinction has been seen to be associated with heat loss. In cellular flames, extinction occurs sooner at the cusps because of the lower flame temperatures there, induced by the diffusive–thermal effects. After cusp extinction, the nonburning gas there provides a heat sink for the remaining nose of the flame. As flammability limits are approached through dilution of lean H_2–air flames, an increasing fraction of the gas in the cusp regions remains unburnt. Very near the limit (just prior to nose extinction), fingers of flames wander through the combustible, leaving most of it unburnt. This phenomenon is not observed when diffusive–thermal effects are stabilizing.

C. Inherent Oscillations of Burning Solids

Combustibles with Lewis numbers sufficiently greater than unity are stable to transverse disturbances but unstable to time-dependent planar disturbances. These planar instabilities are seen most readily for solid combustibles, especially those with solid combustion products, because diffusion coefficients in solids are so small that their Lewis numbers are large. The instabilities develop into finite-amplitude, pulsating propagation of the deflagration, in which there are relatively long quiet periods somewhat resembling autoignition broken by short periods of rapid propagation during which the combustible heated in the quiet period is consumed. If mild, the pulsations are periodic; reactions with relatively large overall activation energies in systems with large Lewis numbers can produce large-amplitude, chaotic pulsating combustion. Some periodic "chuffing" instabilities in solid-propellant combustion may be caused by the pulsating mechanism. Some solid combustible cylinders with large enough diameters are more likely to experience spinning modes of deflagration (in which the leading and hottest reaction front propagates along a helical path) than planar pulsations.

D. Acoustic Instabilities in Combustion Chambers

Oscillatory combustion inside chambers containing combustibles can be produced by the response of the combustion to acoustic waves. If, on the average, an increased rate of heat addition occurs in phase with the pressure increases of acoustic waves, then amplification of the sound by the combustion results. This criteria was known to Rayleigh (1878) and explains many observations of singing and vibrating flames reported by Higgins (1777), Tyndall (1867), and others. Amplification of acoustic waves by combustion is responsible for the combustion instabilities observed in many types of combustion chambers. Solid-propellant and liquid-propellant rocket motors have been found to be especially prone to acoustic combustion instabilities; testing for the instability has become an integral part of motor development programs. The instabilities are detrimental in that they can lead to excessively high rates of heat transfer and unacceptably high average chamber pressures through the growth of the amplitudes of the oscillations into nonlinear ranges. Methods of providing enhanced damping for attenuating the oscillations have been developed.

E. System Instabilities in Combustion Equipment

There is a wide variety of instabilities that involve interactions between different parts of combustion devices. Processes occurring in intakes and exhausts as well as in the combustion chamber can be important. Acoustic wave propagation in one or more of these components may or may not be relevant. In addition to examples of occurrences in industrial furnaces, there are examples of "chugging" in liquid-propellant rocket motors (a pressure-coupled interaction between the flow rate in the propellant feed lines and the combustion in the chamber) and of low-frequency oscillations in solid-propellant rocket motors that involve a pressure-coupled resonance between the gas flowing into the chamber by combustion and that flowing out through the nozzle. Consideration of potential system instabilities is an essential aspect of design.

IX. FUELS FOR COMBUSTION AND FLAMMABILITY RATINGS

A. Combustible Materials

The three primary sources for the principal fuels currently used in combustion are coal, petroleum, and natural gas. Classified by age from oldest to youngest, coals fall under four main ranks—anthracite, bituminous, subbituminous, and lignite—with the heat of combustion generally decreasing in the order cited. Petroleum is mainly a mixture of hydrocarbons with average molecular weights ranging from very high values (giving high viscosities) for heavy crude oil to lower values for light crude oil. The principal combustible constituent of natural gas is CH_4, which is present in volume percentages ranging from roughly 70 to nearly 100 in different supplies. Refinement of these natural fuels is most important for petroleum and provides liquid fuels ranging from fuel oils to gasoline.

Numerous other materials serve as combustibles in various applications. Although some of them occur in nature, many are formulated especially for specific uses. Most materials not intended for use in combustion nevertheless can burn, either as fuels or as oxidizers, if conditions are favorable; essentially the only materials that cannot burn are combustion products in their most stable states and the noble gases. Steel and aluminum are examples of materials that are not commonly considered to be combustible but that burn with great heat release after ignition in hot, high-pressure O_2.

B. Hazard Assessments

There are many facets to the assessment of combustion hazards. In fire ratings, for example, attention is given to ignition, fire spread, rate of heat release, smoke production, oxygen depletion, and production of toxic products. Numerous standard tests of combustible materials are employed in evaluating these various aspects of flammability and combustion. Increased use of synthetic materials that are not based on cellulose complicates the task of evaluation, because wider ranges of combustion behavior are encountered. Combustion histories depend not only on the combustible materials, but also on the environments and configurations in which these materials are arranged. Increasing attention is being paid to the roles of the configurations in combustion hazards. Safe handling of fuels necessitates consideration of the entire range of fire and explosion hazards, from deflagration and diffusion flames to detonation. Transportation of liquefied natural gas in large volumes is an example of an operation that has received considerable study from the viewpoint of hazard assessment. That combustion hazards can arise in relatively unexpected situations is exemplified by the incident (Three-Mile Island, 1979) in which the reaction of hot metal with water liberated H_2, which mixed with air, ignited, and burned inside the containment vessel of a nuclear reactor, producing a significant pressure rise that threatened loss of containment. Combustion of carbon in the nuclear reactor, possibly accompanied by H_2 detonation, was a major contributor to the Chernobyl (1986) disaster. These examples illustrate how challenging the assessment of combustion hazards can be.

C. Environmental Constraints

Concern about the detrimental effects of combustion on the environment has exerted significant influences on both the technology and the science of combustion. Making realistic estimates of environmental effects can be difficult. For example, the extent of cooling of the Earth's atmosphere through high-altitude interaction of sunlight with the soot particles that might be transported there from fires ignited in a nuclear war ("nuclear winter") would be challenging to assess, even if the percentages of fu-

els converted to soot (perhaps 3%) were well known. Although global warming often is accepted as a fact and is attributed to human activities largely involving the use of combustion, the only incontrovertible conclusion that can be drawn from environmental studies is that mankind is responsible for the gradual increase of CO_2 concentration in the atomosphere. Since CO_2 buildup is likely to be detrimental, investigations have begun on technological means for removing CO_2 from combustion exhausts and suitably sequestering it, an apparently great challenge in view of the large fraction of the products that is CO_2, but one which, perhaps surprisingly, appears to possess a variety of different solutions that potentially are economically viable.

Best estimates of environmental effects are employed to establish regulations on permissible combustion emissions from stationary and mobile power plants and on open burning. Studies of emissions involving sulfur and its oxides, oxides of nitrogen, carbon monoxide, hydrocarbons, particulates (such as soot), and complex gaseous species that produce noticeable odors and may be toxic in low concentrations have led to improved knowledge of the chemistry of these compounds in combustion processes and of the regions in combustion chambers where they are produced and survive. The constraints that have been placed on some of these emissions have significantly affected fuel economy and fuel selection. For example, it has become expensive to design furnances for use of high-sulfur coals and fuel oils, and in piston-engine design there is a trade-off between efficiency of combustion and NO production. Consideration of these facts along with the previously discussed use of combustion in waste disposal and incineration of toxic materials demonstrates that there are many aspects of the interactions between combustion and environmental problems.

D. Availability of Fuels

The energy available in coal reserves, largely in the United States, the Russia, and China, appreciably exceeds the energy available in reserves of petroleum and natural gas. Oil shale is a relatively untapped source of liquid fuel (shale oil), which is comparable in abundance to petroleum. With dwindling petroleum supplies, increasing attention is being given to the development of alternative liquid fuels derived from coal and from oil shale. A number of processes for obtaining liquid fuels from these sources have long been available, but for economic reasons have enjoyed relatively little improvement or application. Renewable resources in vegetation (mainly wood) account for roughly only 5% of the energy being used and provide marginal prospects for becoming major contributors, although under special circumstances they can be important sources of liquid fuels (alcohols). Liquid fuels are so advantageous in many applications (notably in mobile power plants) that there are incentives to maximize their availability. Gaseous fuels also possess advantages, such as the ease of distribution and lack of ash. Coal gasification is a viable process for production of a fuel alternative to natural gas. The future of exotic schemes for widespread usage of other fuels (e.g., H_2) in combustion, for example, in conjunction with fuel cells, depends on economic and technological developments, as well as, in the long run, on future scientific advance.

SEE ALSO THE FOLLOWING ARTICLES

AEROSOLS • CHEMICAL KINETICS, EXPERIMENTATION • CHEMICAL THERMODYNAMICS • COAL STRUCTURE AND REACTIVITY • FIRE DYNAMICS • FLUID DYNAMICS • GAS-TURBINE POWER PLANTS • INTERNAL COMBUSTION ENGINES • JET AND GAS TURBINE ENGINES • LIQUID ROCKET PROPELLANTS • MARINE ENGINES • SOLID PROPELLANTS

BIBLIOGRAPHY

Glassman, I. (1996). "Combustion," 3rd ed., Academic Press, San Diego.

Lewis, B., and von Elbe, G. (1987). "Combustion, Flames and Explosion of Gases," 3rd ed., Academic Press, Orlando, FL.

Liñán, A., and Williams, F. A. (1993). "Fundamental Aspects of Combustion," Oxford Univ. Press, New York.

Peters, N. (1999). "Turbulent Combustion," Cambridge Univ. Press, Cambridge.

Strehlow, R. A. (1984). "Combustion Fundamentals," McGraw-Hill, New York.

Williams, F. A. (1985). "Combustion Theory," 2nd ed., Addison-Wesley, Redwood City.

Cometary Physics

W.-H. Ip

Institutes of Astronomy and Space Science, National Central University, Taiwan

GLOSSARY

a Semimajor axis of an orbit.

AU Astronomical unit $= 1.496 \times 10^{13}$ cm, which is the semimajor axis of the earth's orbit around the sun.

Blackbody An idealized perfectly absorbing body that absorbs radiation of all wavelengths incident on it.

Bow shock A discontinuity defining the interface of the transition of a supersonic flow to a subsonic flow as it encounters an obstacle.

Carbonaceous chondrites A special class of meteorites containing up of a few percent carbon.

Contact surface A surface separating two flows of different chemical and/or flow properties.

Comet showers The infrequent injection of a large number of new comets into the inner solar system by passing stars at close encounters with the solar system.

CONTOUR An American space mission to perform flyby observations of comets Encke, Schwassmann-Wachmann 3, and d'Arrest with the spacecraft to be launched in 2002.

DE (disconnection event) The major disruption of a cometary ion tail by the apparent ejection of large plasma condensation or separation of the main ion tail from the comet head.

Deep Impact An American space mission to comet 9P/Tempel 1 for active Impact experiments. The scheduled launch date is in 2004.

Deep Space 1 An American Spacecraft with ion propulsion to flyby comet Borrelly in 2002.

eV Electron volt $= 1.6 \times 10^{-12}$ ergs.

Giotto A spacecraft from the European Space Agency for flyby observations of comet Halley in March 1986.

HST Hubble Space Telescope, an optical telescope launched into geocentric orbit by NASA in 1990.

ICE International Cometary Explorer, an interplanetary spacecraft (ISEE 3), which was redirected to interception of comet Giacobini/Zinner in September 1985.

IHW International Halley Watch, an international program to coordinate the ground-based observations of comet Halley.

IMS Ion mass spectrometer that can make mass-separating measurements of ions.

ISO Infrared Space Observatory, an infrared telescope launched into geocentric orbit by ESA in 1993.

Jet force The repulsive force on a cometary nucleus due to anisotropic gas emission. The jet force was invoked to explain the nongravitational effects observed in cometary orbital evolutions.

keV Kilo electron volt = 10^3 eV.

KT boundary The Cretaceous–Tertiary boundary separating two distinct geological structures about 65 million years ago. This boundary is characterized by a thin layer of enhanced iridium deposition that might have come from an impact event of a comet or an asteroid of kilometer size.

Mass extinction Episodes of large-scale disappearances of life forms on earth. Statistical correlations suggest a periodicity of about 26 million years.

Meteoroid A small solid particle orbiting around the sun in the vicinity of the Earth.

NMS Neutral mass spectrometer, which can make mass-separating measurements of neutral gas.

nT Nanotesla (10^{-9} T) units of magnetic field strength.

Oort cloud The reservoir of new comets at large distances (10^4–3×10^4 AU) from the sun.

PAH Polycyclic aromatic hydrocarbons, which might be representative of the very small grains in interstellar space.

POM Formaldehyde polymers of the form of $(H_2CO)_n$, with $n = 10$–100. The chainlike polymers could be terminated and stabilized by the addition of monovalent atoms or ions (i.e., $HO–CH_2–O–..CH_2–CN$).

pc Parsec: 1 pc = 3.086×10^{18} cm.

Rosetta A European mission designed to collect surface material from the short-period comet P/Wirtanen with a lander to be launched from a mother spaceship orbiting around the cometary nucleus. The spacecraft will be launched in 2003.

Sakigake A Japanese spacecraft for remote-sensing measurements at comet Halley in 1986.

Solar nebula The primordial disk of gas and condensed matter befor the formation of the planetary system.

Solar wind A radial outflow of ionized gas (mostly protons) from the solar corona. The mean number density of solar wind at 1 AU solar distance is 5 cm^{-3}, average speed at earth is 400 km s^{-1}, and the mean electron temperature is 20,000 K.

Stardust An American sample return mission to flyby comet Wild 2 for dust collection. The comet encounter will be in 2004 and the sample will be parachuted back to the earth in 2006.

Suisei A Japanese spacecraft for *in-situ* plasma measurements at comet Halley in 1986.

VEGA Two Soviet spacecraft dedicated to close flyby observations of comet Halley in 1986 after performing a series of measurements at Venus swingbys.

COMETS are known to be the Rosetta Stone to decipher the origin of the solar system. The chemical composition of the volatile ice, the mineralogical properties and isotopic abundances of the dust grains, and finally the physical structure of the cometary nuclei all carry fundamental information on the condensation and agglomeration of small icy planetesimals in the solar nebula. The process of comet–solar-wind interaction is characterized by very complex and interesting plasma effects of special importance to solar system plasma physics; also, the different aspects of momentum exchange, energy transfer, and mass addition in a plasma flow are closely linked to large-scale outflows in interstellar space. For example, the generation of massive outflows from the T Tauri stars and other young stellar objects and their interaction with the surrounding medium are reminiscent of the comet–solar-wind interaction. The formation of small structures called cometary knots in the Helix planetary nebula is reminiscent of the comet–solar wind interaction (Fig. 1). The cometary plasma environment therefore could be considered a laboratory simulating many basic astronomical processes.

I. INTRODUCTION

The historical context of cometary research is an interesting document of progress in science. In Oriental records, comets were usually called guest stars or visiting stars to describe their transient appearances in the celestial sphere. One of the earliest systematic records of comets can be found in the Chinese astronomical history of 467 B.C. The historian Kou King-ting noted in 635 B.C. that cometary tails point away from the sun. In Europe, study can be traced back to Aristotle's postulation that comets should be associated with transient phenomena in the earth's atmosphere. As astronomy started to blossom in the Middle Ages, Tycho Brahe, Kepler, and Newton all paid special attention to the dynamical nature and origin of comets. The study of comets has been closely coupled with scientific and technical progress in the past few centuries. The most important example is comet Halley, which was deduced to be in elliptic orbit around the sun with a period of 75.5 years by Edmond Halley in 1682. His famous

FIGURE 1 Structures of the so-called cometary knots observed at the Helix Nebula (NGC 7293). They are formed by interstellar cloudlets being blown off by intense ultraviolet radiation from the central massive stars. Photo origin: NASA.

prediction that this comet should return in 1758 has become a milestone in astronomy. Three returns later, in March 1986, comet Halley was visited by a flotilla of spacecraft taking invaluable data revealing the true nature of the most primitive bodies in the solar system. In total, six spacecraft from the former Soviet Union, Japan, Europe, and the United States participated in this international effort (see Fig. 2). In addition, an extensive network for ground-based comet observations was organized by NASA to coordinate the study of comet Halley. This program, called International Halley Watch (IHW), proved to be successful in maintaining a high level of cometary research in all areas. The long-time coverages provided by ground-based observations are complementary to the snapshots produced by spacecraft measurements. In fact,

investigations of the nucleus rotation, coma activities, and dynamics of the plasma tail all require comparisons of the spacecraft measurements with the observational results gathered by the IHW and other similar programs.

Before the Halley encounters, a first look at the cometary gas environment was obtained by the NASA spacecraft International Cometary Explorer (ICE), at comet Giacobini–Zinner on September 11, 1985. Even though it did not have a scientific payload as comprehensive as the payloads carried by the Giotto and Vega probes, many exciting new observations pertinent to comet–solar-wind interaction were obtained. Because ICE went through the ion tail of comet Giacobini–Zinner, the experimental data are particularly useful in addressing questions about the structures of cometary plasma tails.

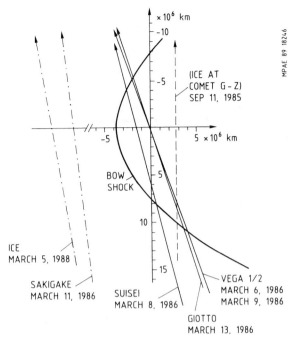

FIGURE 2 A schematic view of the encounter geometries of several spacecraft at comet Halley in March 1986 and the ICE spacecraft at comet Giacobini–Zinner in September 1985.

Because of the retrograde orbital motion of comet Halley, the encounter speeds during flyby observations were very high (68 km s^{-1} for Giotto and 79 km s^{-1} for Vega 1). High-speed dust impact at close approaches to the comet nucleus became a hazard for the spacecraft. This problem caused a temporary loss of radio link of Giotto to the earth when the probe was at a distance of about 1000 km from the cometary nucleus. As a result, no images were obtained beyond that point even though the spacecraft was targeted to approach the comet as close as 600 km on the sunward side. Several other instruments were also damaged. In spite of these calculated risks, the comet missions as a whole achieved major scientific successes beyond expectations.

In other arenas of cometary research, there has been very interesting progress as well. The new concept of an inner Oort cloud between 10^3 AU and 10^4 AU is one of them. In the present work, we shall make use of these many new results to attempt to build a concise picture of modern cometary physics. Because of the page limit, it is impossible to go into all details. For further information, the reader should consult the appropriate references as listed.

II. ORBITAL DYNAMICS

The Keplerian orbit of a comet can be characterized by several orbital elements, namely, the semimajor axis (a),

the inclination (i), the eccentricity (e), the argument of perihelion (ω) and the longitude of the ascending node (Ω) with respect to the vernal equinox (γ). Their mutual relations are shown in Fig. 3. The orbital period is given by

$$p = a^{3/2} \text{ years} \tag{1}$$

where a is in units of AU. Thus, for comet Halley, $a = 17.9$ AU and $P = 76$ years. The closest distance to the sun, the perihelion, is given by

$$q = a(1 - e) \tag{2}$$

and the largest distance from the sun, the aphelion, is given by

$$Q = a(1 + e) \tag{3}$$

The total specific angular momentum is expressed as

$$L = \sqrt{\mu a(1 - e^2)} \tag{4}$$

whereas the component perpendicular to the ecliptic plane is

$$L_z = \sqrt{\mu a(1 - e^2)} \cdot \cos i \tag{5}$$

In the foregoing equation, $\mu = \text{GM}_\odot$, where G is the gravitational constant and $\text{M}_\odot$ the solar mass. The orbit of a comet is prograde (or direct) if its inclination (i) $< 90°$, and retrograde (or indirect) if $i > 90°$.

Because of perturbations from the passing stars, planetary gravitational scattering, and/or jet forces from surface outgassing, the orbital elements can change as a function of time. The orbital elements determined at a certain time (t_0) are called osculating elements for this particular epoch. To infer the future or past orbital elements of a comet, orbital integrations taking into account all relevant perturbation effects must be carried out from $t = t_0$

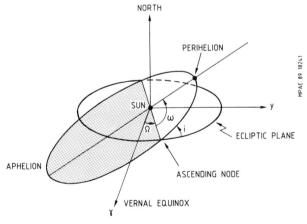

FIGURE 3 Description of a cometary orbit with i denoting the inclination; ω is the argument of perihelion; Ω is the longitude of the ascending node; and γ is the equinox.

to the time interval of interest. Because of limitations of computing machine time, computational accuracies, and observational uncertainties, such calculations usually do not cover a time period much more than a few thousand years if very accurate orbital positions are required. To investigate long-term evolutions, statistical methods are often used such that the orbital behavior of a sample of comets over a time span of millions to billions of years can be described.

Certain invariants in celestial mechanics are useful in identifying the orbital characteristics of small bodies (comets and asteroids) in the solar system. In the restricted three-body problem with the perturbing planet (Jupiter in this case) moving in a circular orbit, the Tisserand invariant is defined as

$$T_J = \frac{1}{a} + 2\left[2q\left(1 - \frac{q}{2a}\right)\right]^{1/2} \cos i \qquad (6)$$

where a and q are in units of Jupiter's semimajor axis, a_J. Most of the short period comets have $T_J < 3$.

Comets are generally classified into three orbital types according to their periods: namely, the long-period comets with $p > 200$ years; the intermediate-period comets with P between 20 and 200 years; and finally, the short-period comets with $P < 20$ years. A compilation of the orbital data of the observed comets shows that there are in total 644 long-period comets 25 intermediate-period comets, and 88 short-period comets. Although such classification is somewhat arbitrary in the divisions of the orbital periods, it makes very clear distinctions in the inclination distribution. The inclinations of the short-period comets are mostly less than $30°$, and those of the long-period comets have a relatively isotropic distribution. As discussed below, one active research topic at the present moment is whether these different inclination distributions are also indicative of different dynamic origins of these comet populations.

Note that among the long-period comets, there is a group of sun-grazers, called the Kreutz family, that might have originated from the breakup of a single large comet. A number of them have perihelion distances inside the sun. The SOLWIND and Solar Maximum Mission satellites detected 13 such comets, while the SOHO space solar observatory detected more than 200 of the Kreutz family comets (see Fig. 4).

Figure 5 shows the $1/a$ distribution for the new, long-period comets in units of 10^{-4} AU^{-1}. As first discussed by Oort (1950), the peak at $1/a < 10^{-4}$ AU^{-1} suggests that the new comets mostly come from an interstellar reservoir in the form of a spherical shell surrounding the sun. The inner radius of this so-called Oort cloud is at about 10^4 AU and that of the outer radius at about $3–5 \times 10^4$ AU. Because of the perturbing effects of passing stars, inter-

stellar molecular clouds, and the galactic tidal forces (see Section IX), a continuous influx of new comets from the Oort cloud to the inner solar system can be maintained.

In addition to the classical Oort cloud, made detectable via the injection of a constant flux of new comets, the presence of a massive inner Oort cloud between 10^3 AU and 5×10^3 AU has recently been postulated. Only the very infrequent passages of stars would scatter comets in this region into the inner solar system. Such events could give rise to the so-called comet showers lasting about 2–3 million years at irregular intervals of 20–30 million years.

For a comet in the Oort reservoir, with radial distance $r > 10^4$ AU from the sun, its angular momentum can be expressed in terms of the orbital velocity (V_c) and the angle (θ) between the comet–sun radius vector and the velocity vector (see Fig. 7), that is,

$$r V_c \sin \theta = [\mu a (1 - e^2)]^{1/2} \qquad (7)$$

which yields [cf. Eq. (4)]

$$V_c^2 \theta^2 \approx \frac{2\mu q}{r^2} \qquad (8)$$

for small θ. Assuming that the velocity vectors of comets in the Oort region are sufficiently randomized by stellar perturbations, we can express the distribution function

$$f_\theta(\theta) d\theta = \frac{\sin \theta}{2} d\theta \approx \frac{\theta d\theta}{2} \qquad (9)$$

In combination with Eq. (8) we find that the perihelion distance of new comets should be

$$f_q(q) dq = \frac{\mu dq}{2 V_c^2 r^2} \qquad (10)$$

In other words, f_q is independent of q. The population of long-period comets with $1/a \approx 10^{-4}$ AU, however, is a mixture of new comets and the evolved ones with one or more perihelion passages through the solar system. Because of planetary perturbations, the perihelion distribution of the evolved comets would be significantly modified. Numerical simulations show that the frequency distribution of perihelion distance of long-period comets should be highly depleted inside the orbit of Jupiter. It becomes constant only for $q \gtrsim 30$ AU where the planetary perturbation effects are no longer important.

Note that the velocity of a long-period comet near its aphelion in the Oort region is of the order of 200 m s^{-1}. On the other hand, the average speed (V_*) of a passing star relative to the solar system is about 30 km s^{-1}. The effect of stellar perturbation can therefore be treated using the impulse approximation as follows. With closest approach distances given as D_c and $D_\odot$, respectively, the velocity increments received by the comet and the sun can be given as

$$\Delta \underline{V}_c = \frac{2GM_*}{V_*} \frac{D_c}{D^2} \qquad (11)$$

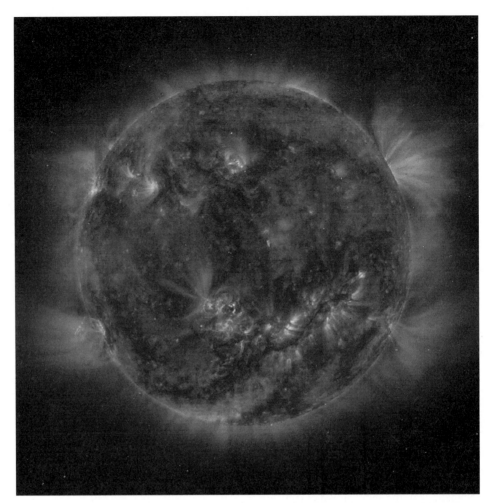

FIGURE 4 The appearance of a sun-grazing comet observed by the SOHO coronograph instrument. Photo origin: ESA.

and

$$\Delta \underline{V}_\odot = \frac{2GM_*}{V_*} \frac{D_\odot}{D_*^2}. \tag{12}$$

The overall effect in the velocity change of the comet relative to the sun is then

$$\delta \underline{V}_c = \Delta \underline{V}_c - \Delta \underline{V}_\odot \tag{13}$$

Over the age of the solar system (4.5×10^9 years), the root-mean-squared velocity increment of a comet as a result of stellar perturbations can be estimated to be $V_{RMS} \approx 100$ m s^{-1}. As this value is comparable to the orbital velocity of a comet in circular motion at 9×10^4 AU, the outer boundary of the Oort cloud may be set at this distance as a result of stellar perturbations.

As mentioned before, the binding energy of a new comet is of the order of 10^{-4} AU^{-1}. During its passages through the inner solar system, a comet will be subject to gravitational perturbations by the planets. The most important perturbing planet is Jupiter, followed by Saturn. Accord-

ing to numerical calculations, the typical energy change (ΔE) per perihelion passage is a function of the perihelion distance and orbital inclination. For a comet with i between $0°$ and $30°$, $E > 10^{-4}$ AU^{-1} if $q < 15$ AU, and for i between $150°$ and $180°$, $E > 10^{-4}$ AU^{-1} only if $q < 5$ AU. This difference is due to the larger relative velocities of retrograde comets during planetary encounters and resulting smaller orbital perturbations. The excess of retrograde orbits in the frequency distribution of the inclinations of long-period comets may be a consequence of such a selective effect in planetary encounters.

The orbital transformation of long-period comets into short-period comets is one of the possible outcomes of a sequence of random walk processes. The capture efficiency depends on the inclinations and other orbital elements of the comets in question. For example, a "capture" zone exists for long-period comets with 4 AU $< q < 6$ AU and $i < 9°$. Another possible scenario is that the majority of short-period comets are not supplied by long-period comets with isotropic inclination distribution

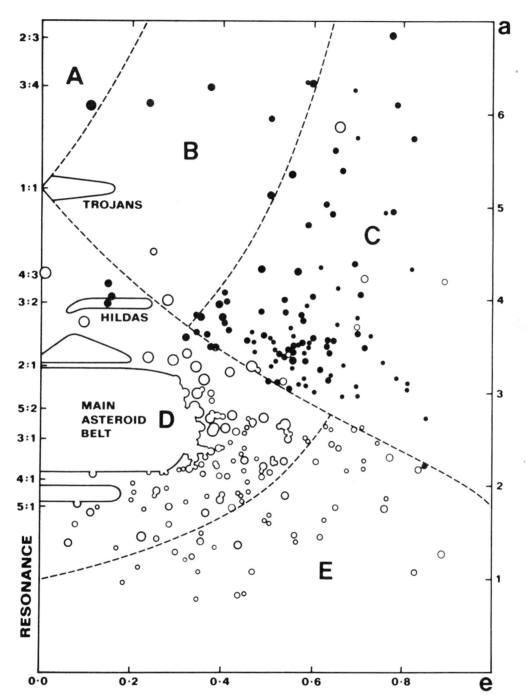

FIGURE 5 Distributions of short-period comets (solid circles) and asteroids (open circles) plotted in a diagram of semimajor axis (*a*) vs eccentricity (*e*). A indicates the transjovian region, B is the Jupiter family of weak cometary activity, C is the Jupiter family of strong cometary activity, D is the minor planet region, and E is the apollo region. The thick dashed curve denotes the critical value of $T_J = 3$ (with $\cos i = 1$) separating the cometary region (B + C) with $T_J < 3$ from the asteroidal region (D and E) with $T_J > 3$. [From Kresak, L. (1985). *In* "Dynamics of Comets: Their Origin and Evolution" (Carusi, A., and Valsecchi, G. B., eds.), IAU coll. 83, Reidel, Dordrecht, pp. 279–302.]

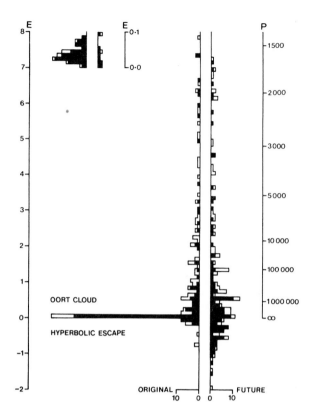

FIGURE 6 Frequency distribution of the original reciprocal semi-major axes of long-period comets with $(1/a)_{orig} < 10^{-3}$ AU^{-1}. [From Kresak, L. (1987), in "The Evolution of the Small Bodies of the Solar System," (Fulchignoni, M., and Kresak, L., eds.) *Proc. of the Intern. School of Physics* "Enrico Fermi," Course 98, Societa Italiana di Fisica, Bologna-North-Holland publ. co., pp. 10–32.]

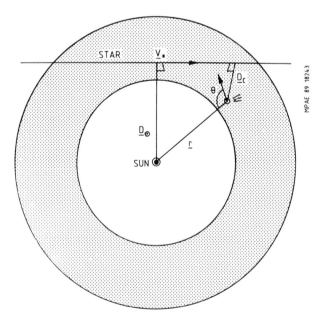

FIGURE 7 The encounter geometry of a passing star with the solar system.

[i.e., $f_i(i) \alpha \sin i$], but rather by a comet belt of low inclinations, located just outside the orbit of Neptune.

The dynamical influence of Jupiter can be recognized in the aphelion distribution of short-period comets that tend to cluster near the semimajor axis of Jupiter. Furthermore, the distribution of the longitudes of perihelion (ω) for such a Jupiter family has a mimimum near the perihelion of longitude (ω_J) of Jupiter.

The final fate of the orbital evolution of short-period comets would be determined by perturbation into escape orbit via close encounter with Jupiter or other planets, direct collision with a planet, or catastrophic fragmentation by hypervelocity impact with interplanetary stray bodies or crashing into the sun (see Fig. 4). The fragmentation process would lead to the formation of a meteor stream composed of small dust particles. Meteor streams could also be produced by partial fragmentation, surface cratering, and, of course, outgassing activities. For example, the short-period comet P/Encke is associated with the meteor stream S. Taurids, and P/Giacobini–Zinner is associated with the October Draconids.

The Geminids stream is connected with the Apollo object 3200 Phaethon. Since Apollo objects are basically defined as Small bodies in Earth-crossing orbits (and Amor objects are bodies in Mars-crossing orbits), suggestions have been made that 3200 Phaethon may in fact be a defunct cometary nucleus. There are several possible clues supporting the hypothesis that short-period comets could be a source for the Apollo–Amor objects. First, several of these Apollo–Amor objects are in relatively high inclinations ($>30°$), which are very unusual for the asteroidal population. Second, three Apollo–Amor objects have aphelia beyond Jupiter's orbit. Finally, the Apollo object 2201 Oljato has a surface UV reflectance very different from that of the asteroids. Short-period comets like P/Arend–Regaux and P/Neujmin II are almost inactive; therefore, they might have reached the turning point of becoming Apollo–Amor objects. Recent dynamical calculations show that, in addition to the injection from the main-belt asteroidal population via chaotic motion near the 3:1 Jovian commensurability and the ν_6 secular resonance in the inner boundary of the asteroid belt, a significant fraction of the Apollo–Amor objects (≈ 2400 in total) could indeed come from short-period comets.

III. GENERAL MORPHOLOGY

Despite its brilliance in the night sky, a comet has a solid nucleus of only a few kilometers in diameter. The brightness comes from the dust, gas, and ions emitted from the nucleus. For example, during its 1986 passage near the earth's orbit, comet Halley lost on the order of 3.1×10^4 kg of volatile ice per second and about an equal amount in small nonvolatile dust particles. The expansion of the neutral gas (mostly water and carbon monoxide) from the central nucleus would permit the formation of a large coma visible in optical, ultraviolet, and infrared wave-lengths. The optical emission in a cometary coma is largely from the excitation of the minor constituents, such as CN and C_2, by the solar radiation. These radicals are the daughter products from photodissociation of parent molecules such as HCN, C_2H_4, C_2H_6, and other more complicated molecules. The dirty snowball model of the cometary nucleus first proposed by Whipple (1950) describes the nucleus as a mixture of frozen ice and nonvolatile grains. The main components of the volatile ice are H_2O, CO_2, CO, H_2CO, CH_4, and NH_3, followed by minor species such as HCN, C_2H_4, C_2H_6, CS_2, and others.

A. Surface Sublimation

An icy nucleus will begin to evaporate significantly at perihelion approach to the sun once its surface temperature exceeds a certain value. Since the gas pressure in the cometary coma is much smaller than the critical pressure at the triple point, direct sublimation from the solid phase into vapor will occur. Under steady-state conditions, the energy equation in its simplest form can be written as

$$F_\odot e^{-\tau}(1 - A_v)r^{-2}\cos\theta\cos\phi$$
$$= \varepsilon_{IR}\sigma T^4 + \frac{L(T)}{N_A}\dot{Z} - K(T)\nabla T|_s \quad (14)$$

where $F_\odot$ is the solar flux at 1 AU, τ is the optical depth of the dust coma, A_v is the surface albedo, $_r$ is the heliocentric distance in AU, ϕ and θ are the local hour angle and latitude, ε_{IR} is the infrared emissivity, σ is the Stefan–Boltzmann constant, T is the surface temperature, $L(T)$ is the latent heat of sublimation, N_A is the Avogadro number, $\dot{Z}$ (in units of molecules $cm^{-2}s^{-1}$ str^{-1}) is the gas production rate, and $K(T)$ is the thermal conductivity at the surface.

Another important equation is the Clapeyron–Clausius equation. For water we have

$$\log p(\text{mm Hg}) = \frac{-2445.5646}{T} + 8.2312\log T$$
$$- 0.01677006\,T + 1.20514$$
$$\times 10^{-5}T^2 - 6.757169 \quad (15)$$

with the equilibrium vapor pressure p in units of millimeters of mercury. And for CO_2 we have

$$\log p(\text{mm Hg}) = \frac{-1367.3}{T} + 9.9082 \qquad \text{for } T > 138\text{ K} \quad (16)$$

and

$$\log p(\text{mm Hg}) = -\frac{1275.6}{T} + 0.00683T + 8.307$$
$$\text{for } T < 138\text{ K} \quad (17)$$

The latent heat of vaporization of water ice is

$$L(T) = 12420 - 4.8T \quad (18)$$

with $L(T)$ in cal mole^{-1}. For CO_2 ice we have

$$L(T) = 12160 + 0.5\,T - 0.033\,T^2 \quad (19)$$

Finally, the sublimation rate is related to the equilibrium vapor pressure by the following equation:

$$m\dot{Z}(T) = p(T)\left(\frac{m}{2\pi KT}\right)^{1/2} \quad (20)$$

Figure 8 shows the variations of $\dot{Z}(T)$ as a function of the solar distance r with $\langle\cos\theta\cos\phi\rangle = \frac{1}{4}$, $A_v = 0.02$, and $\varepsilon_{IR} = 0.4$. For CO_2 ice, a significant level of surface sublimation starts at $r \approx 10$ AU; strong outgassing would occur at $r \approx 2$–3 AU for H_2O ice.

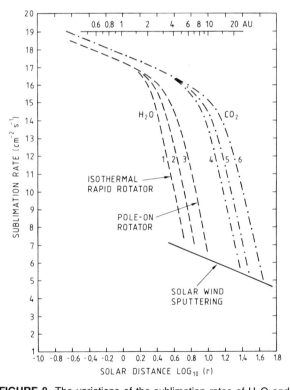

FIGURE 8 The variations of the sublimation rates of H_2O and CO_2 ices as a function of the solar distance. The surface albedo A_v is taken to be 0.02 and the infrared emissivity ε_{IR} is assumed to be 0.4. Two extreme cases of the orientation of the spin axis (isothermal rapid rotator and pole-on rotator) are shown. The surface sputtering rate via solar-wind protons is also given.

B. Gas Comas and Ion Tails

As the parent molecules (H_2O, CO, CO_2, CH_4, H_2CO, etc.) move away from the nucleus, they will be either dissociated or ionized by solar ultraviolet radiation. The photodissociation life-time for water molecules is about 10^5 s at 1 AU solar distance. With an expansion speed (V_n) of around 1 km s^{-1}, most of the water molecules will be dissociated into hydrogen atoms and hydroxide (OH) at a cometocentric distance of a few 10^5 km. The OH will also be dissociated subsequently. A similar process occurs for other parent molecules. As a result, at a cometocentric distance of a few 10^6 km, the extended neutral coma will be made up only of atomic species (H, C, O, N, S, ...). In Lyman alpha emission at 1216 Å, the atomic hydrogen cloud of a comet can become the most prominent structure in the solar system (Fig. 9). The ionization of the cometary neutrals by solar radiation and charge exchange with solar wind protons will lead to the formation of an ion tail flowing in the antisolar direction. In optical wavelengths, CO^+ and H_2O^+ are the most visible among the ions. The general features of the interaction of the cometary coma with the solar radiation and interplanetary plasma are sketched in Fig. 10.

C. Dust Coma and Mantle

The gaseous drag from the expanding outflow is able to carry dust particles away from the nuclear surface. There is, however, an upper limit on the particle size to be ejected in this manner. The critical particle size is reached when the gravitational attraction of the nucleus outweighs the gaseous drag. The stronger the surface gas sublimation rate, the larger will be the critical size (see Section VI). For comet Halley, its out-gassing rate at 1 AU solar distance would enable the emission of solid particles up to a

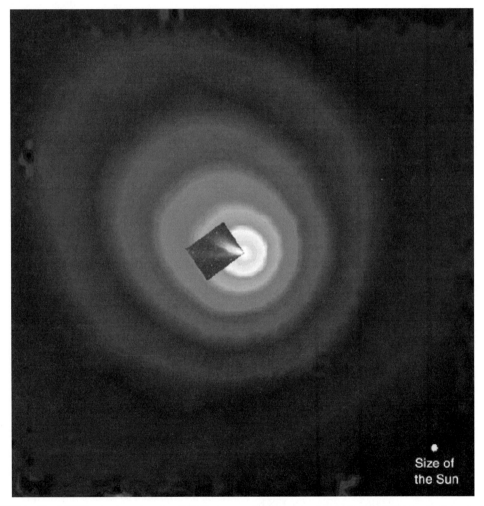

FIGURE 9 The SWAN experiment on the SOHO spacecraft observed a huge atomic hydrogen cloud surrounding Comet Hale–Bopp. The hydrogen cloud is 70 times the size of the Sun. Photo origin: ESA/NASA.

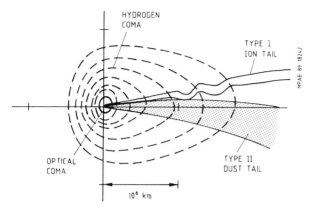

FIGURE 10 A schematic view of the interaction of the cometary coma with the solar wind. The Type I ion tail points along the radial direction; the Type II dust tail is usually more diffuse and curved.

diameter of a few tens of centimeters. A halo of centimeter-size particles in its vicinity was indeed detected by radar observations.

In the above scenario, solid particles with diameters larger than the critical value will remain on the nuclear surface, forming a sort of dust mantle. When a comet moves along a Keplerian orbit, its outgassing rate will change as a function of solar distance (see Fig. 8). The formation of a dust mantle is therefore a time-dependent process. Also, irregular structures of the nuclear surface might lead to rather patchy coverage by such a dust layer. Since a dust mantle a few centimeters thick would be sufficient to insulate the heat flow such that the effective gas sublimation rate is reduced by a factor of 100, the surface area covered by a thin dust mantle would become much less active.

As a comet goes through many perihelion returns near the sun, its gas sublimation rate will begin to subside. One consequence is that the critical size of dust grain ejection will become smaller and smaller. As a result, more and more dust particles will be accumulated on the nucleus surface; that is, the dust mantle will become thicker and more widespread as the comet ages. It is perhaps for this reason that "old" comets (such as comet Encke and comet Halley) are generally observed to have anisotropic gas emission from a few localized active regions. The rest of the surface, presumably, is covered by a thickening dust layer. On the other hand, comet Wilson (1986), which is a new comet in the Oort sense, did not display any anisotropic outgassing effect. This may be simply because its "fresh" surface still lacks an insulating dust mantle of significant size.

D. Dust Tails

The elongated dust tails are formed of micron-size dust particles being accelerated away by the solar radiation

pressure force. The ratio of the radial component of the solar radiation force (F_r) to the solar gravitation (F_g) can be expressed as

$$1 - \mu = \frac{F_r}{|F_g|} = \frac{1.2 \times 10^4}{\rho d} Q_{pr} \qquad (21)$$

where Q_{pr} is the Mie efficiency factor for radiation pressure, ρ is the density of the dust grain, and d is its size. The value of Q_{pr} depends on both the size of the grain and on its optical properties. For a dielectric particle with $d \approx 1\ \mu$m, $Q_{pr} < 0.5$, while $Q_{pr} > 1$ for conductors. In general, the $1\ \mu$ value is largest in the size range between 0.1 and 1 μm. Figure 11 shows the efficiency factor for radiation pressure, Q_{pr}, integrated over the solar spectrum for different grain properties as a function of grain radius. It can be seen that Q_{pr}, for dielectric grains (e.g., olivine), decreases sharply for $a < 0.3\ \mu$m. Submicron grains are therefore relatively invisible.

In the cometocentric frame, the instantaneous locus in the dust tail formed of solid particles with the same value of $1 - \mu$ is called syndyname. Another type of locus of interest is the so-called synchrone formed of particles of different values of $1 - \mu$ but emitted from the nucleus

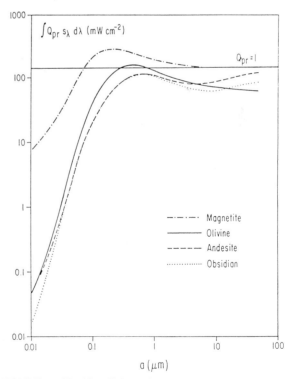

FIGURE 11 The Mie efficiency factor for radiation pressure Q_{pr} integrated over the solar spectrum for different grain materials as a function of grain radius. [From Hellmich, R. and Schwehm, G. (1982), in "Proc. Intern Conf. on Cometary Exploration" Gombosi, T., ed., Vol. I, p. 175.]

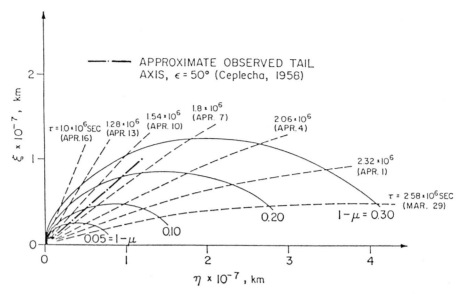

FIGURE 12 A description of the formation of the syndynames (——) and synchrones (– –) in the dust tail of comet Arend–Roland (1957 III) on April 27.8. 1957. [According to Finson, M. L., and Probstein, R. F. (1968). *Astrophys. J.* **154,** 353.]

at the same time. A classical description by Finson and Probstein (1968) of these two types of loci in the dust tail structure of comet Arend–Roland 1957 is given in Fig. 12. The brightness distribution and orientation of a cometary dust tail can be analyzed by taking into consideration (a) the size distribution of the dust particles, (b) the total production rate of dust particles as a function of time, and (c) the initial velocity of the emitted dust particles as a function of time and size. In this way, a synthetic dust tail model can be constructed and compared with observations. From parameter studies, a best fit could be obtained with a certain dust particle size distribution. Such a treatment has been used to derive useful information on the size distributions of dust particles in the dust tails of many comets (see Section VI).

IV. COMETARY SPECTRA

Figure 13 is a composite spectrum of comet Tuttle 1990 XIII between the ultraviolet and infrared wavelength range. The ultraviolet data up to 0.28 μm came from the International Ultraviolet Explorer (IUE) and the rest from ground-based observations. The figure shows emission features from various atomic and molecular species (HI, OI, CI, SI, CS, OH, NH, NH_2, CH, CN, C_2, C_3). Before spacecraft missions to comet Giacobini–Zinner in 1985 and to comet Halley in 1986, such spectra were the only means to infer the composition of the gas comas. The information on the spatial distributions of the brightness of individual species can be used to infer the corresponding abundances and life-times against photodissociation.

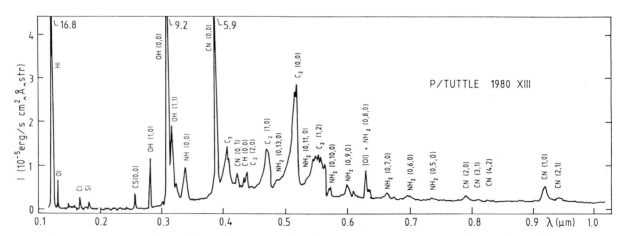

FIGURE 13 A composite spectrum of the coma emissions from comet Tuttle 1980 XIII. [From Larson, S., and Johnson, J. R. (1985).]

For example, the Lyman alpha emission of HI and the very strong emission of OH in many comets have shown that water is the main constituent of cometary ice. The CO emission at 1510 Å in several comets suggests that carbon monoxide can be a very important component as well, but its abundance relative to H_2O varies from comet to comet. Rocketborne UV observations and neutral mass spectrometer measurements at comet Halley have further shown that nearly half of the CO molecules were released in a distributed region (of radius $\approx 25,000$ km) surrounding the nucleus. Dust grains and formaldehyde polymers have been suggested as the possible parents.

A fraction of the CN radicals could be released from small dust grains. Ground-based optical observations of comet Halley detected jet-like features of CN emission in the coma. Sublimation from organic dust grains and emission from local inhomogeneities on the nucleus surface are two possible explanations. Despite their brightness, the CN, C_2, and C_3 radicals are all minor constituents with relative abundances on the order of 0.1%.

Table I lists most of the species identified in cometary spectra from observations at different wavelengths. Spacecraft measurements at comet Halley showed that the chemical composition of the gas coma is much more complex than indicated here. Remote-sensing observations, however, will continue to be a powerful tool in gathering basic chemical information about a large sample of comets. The radio observations of comet Hyakutake and comet Hale–Bopp have shown the presence of hydrocarbon molecules such as C_2H_2, C_2H_6, and HNC.

Continuum emission from dust particles could be very strong for some comets. There is a trend indicating that the dust production rate is proportional to the production rate of the CN radicals. This effect, if confirmed by statistical correlations of more comets, would indicate that a part of the CN radicals could indeed originate from the dust particles in the coma. The dust comas of several comets were observed to be highly anisotropic. As illustrated in Fig. 14, a system of dust jets could be seen in the coma of comet Halley near the time of the Giotto encounter. The CN– (and C_2–) jets detected at the same time, however, did

not align well with these dust jets. It is possible that "invisible" submicron dust grains of organic composition could release these gaseous molecules or their parent molecules via sublimation.

Infrared observations of comet Halley from the ground discovered an emission feature near 3.4 μm. The same feature was seen in the spectra taken by the infrared spectrometer experiment (IKS) on Vega 1 (see Fig. 15). Several mechanisms might be responsible for such emission, such as heating of very small organic grains of large polycyclic aromatic hydrocarbon molecules (PAH) by UV photons, thermal emission by small cometary grains with organic mantles, and resonant fluorescence of small gaseous molecules. Emissions from the OCS, CO, CO_2, H_2CO, and H_2O molecules were also detected in the IKS spectra (see Section VII).

Infrared emissions from water molecules were observed from the Kuiper Airborne Observatory (KAO) with very high spectral resolution. In addition to deducing water production rates at different times, the KAO observations are important in finding strong anisotropic outflow of the water vapor from the central nucleus and in determining the ortho–para ratio (OPR) of comet Halley. The latter can be used as a probe to the temperature at which the water was last equilibrated, that is, when it condensed into water ice. Infrared observations of water molecules of this type therefore hold the promise of mapping the temperature variations of the solar nebula during the condensation of the icy planetesimals.

V. COMETARY NUCLEI

A. Albedo

Before the spacecraft flyby observations of comet Halley, ground-based broadband photometric observations of several comets had shown that the V-J and J-K colors of their nuclei are similar to those of the Trojan D-type asteroids. This in turn suggests that the cometary nuclei must be very dark, with the albedo $p_v < 0.1$. At the same time, the mean albedo of cometary dust grains measured at 1.25 μm for a number of comets was found to be on the order of 0.02–0.04. The close-up views from the Vega and Giotto spacecraft have shown that the surface albedo ($p_v \approx 0.02$–0.05) of comet Halley is indeed among the darkest of solar system objects. One possible reason for such a dark color might be related to early irradiation effects on the hydrocarbon material mixed in the cometary ice.

B. Size

When coma activity is negligible, the sunlight reflected by the bare nucleus accounts for all of the optical emission

TABLE I Species in Cometary Optical, UV, IR and Radio Spectra

Location	Species
Coma (gas)	H, C, C_2, $^{12}C^{13}C$, C_3, O, OH, H_2O, S, S_2 CH, CH_4, CN, CO, CO_2, CS, C_2H_2, C_2H_6, HCN, HNC, HCO, H_2CO, CH_3CN, NH, NH_2, OCS Na, K, Ca, Cr, Mn, Fe, CO, Ni, Cu, V, Ti(?)
Tails (ions)	C^+, CH^+, CN^+, CO^+, CO_2^+, N_2^+, OH^+, H_2O^+, H_3O^+, Ca^+, H_2S^+

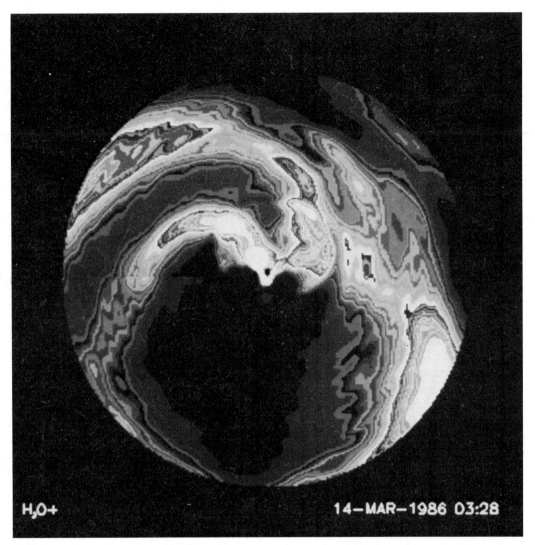

H₂O+   14-MAR-1986 03:28

FIGURE 14 A false color image-enhanced picture of the dust jet system in the coma of comet Halley taken on March 14, 1986, at the South Africa Astronomical Observatory by C. Cosmovici and P. Mack. Image processing by G. Schwarz, DLR. Wessling, FRG.

observed. Under this circumstance, the radius (a) of the cometary nucleus can be given as

$$a^2 = [p_v \phi(\theta)]^{-1} 10^{0.4(M_\odot - H + 5\log(r_\Delta))} \qquad (22)$$

where p_v is the geometrical albedo, $\phi(\theta)$ is the phase function, $M_\odot$ is the absolute magnitude of the sun, H is the apparent nuclear magnitude, r is the solar distance of the comet, and Δ is the geocentric distance. Figure 16 shows the distributions of the nucleus radii for a number of periodic comets and near-parabolic comets. In the present compilation of the data from Roemer (1966), p_v is assumed to be 0.02. As a rule, most comets have a radius of a few kilometers, except for some giant comets such as comet Humason 1962 VIII and comet Wirtanen 1957 VI, which showed signs of being new comets en-

tering the inner solar system for the first time. Near parabolic comets appear to be larger in general than the short-period ones if they all have the same albedo of 0.02.

Radar observations using large antennas have been successful in making independent determination of the radii of a few comets. For example, the periodic comet P/Encke was found to have a radius of $1.5^{+2.3}_{-1.0}$ km from the Arecibo radar observations. The observations of comet IRAS–Araki–Alcock 1983d showed that the radius ranges from 2.5 km for a solid-ice surface to 8 km for a surface of loosely packed snow. The nucleus surface seems to be very rough or porous on scales of a few meters or more.

According to the imaging experiments on the Giotto and Vega spacecraft, the physical size of comet Halley, which

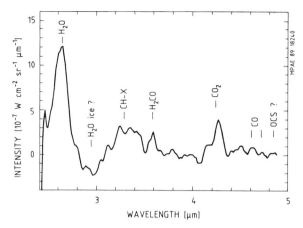

FIGURE 15 A spectrum of comet Halley between 2.5 and 5 μm, obtained from the average of five individual spectra taken at a cometocentric distance of 42,000 km. [From Combes, M, *et al.* (1989). *Icarus*, **76**, 404.]

has a 2:1 ellipsoidal shape in the first approximation, can be characterized as follows: (a) maximum length = 16 ± 1 km, (b) intermediate length = 8.2 ± 0.8 km, and (c) minimum length = 7.5 ± 0.8 km. The total volume is thus of the order of 500 km^3.

C. Mass and Density

On the basis of detailed calculations of the nongravitational effects on comet Halley's orbital motion, the mass of the nucleus has been estimated to be between 5×10^{16} g

and 1.3×10^{17} g. This would mean an average density of 0.08–0.24 g cm^{-3} for the nucleus. The uncertainties involved in the computations do allow a higher density, however. If comet Halley has already experienced 3000 orbital revolutions in similar orbits in the inner solar system, the original mass before its commencement of active mass loss could be estimated to be about 5 to 6 times its present mass.

D. Surface Activity

As indicated by the dust jet features on the sunward side of the nucleus, the surface outgassing process of comet Halley was highly an isotropic (see Fig. 17). The dust jets were confined within cones of about 120° for Vega and 70° for Giotto. The active region covered an area of no more than 10% of the total surface area. Interestingly, in spite of the strong anisotropic emission, a significant dust background was seen on the nightside with a ratio of 3:1 for the sunward–antisunward brightness variation. It is not yet clear what mechanism was responsible for this nightside dust coma. Nonradial transport of the dust particles from the jet source to the nightside would require very special lateral flow condition in the inner coma.

E. Surface Features

Radar observations have indicated that the nuclear surfaces of comets could be rough on scales of a few meters or more. The Giotto HMC showed that comet Halley's

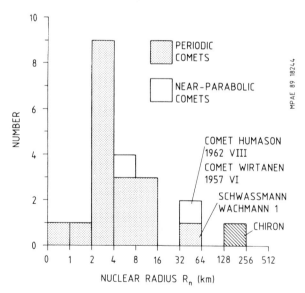

FIGURE 16 Frequency distributions of the radii of periodic comets and new comets. All comets are assumed to have a surface albedo of 0.02 [Data from Romer, E. (1966). *In* "Nature et Origine des cometes." *Pub. Institut d'Astrophysique*, Liege, Belgium, p. 15.]

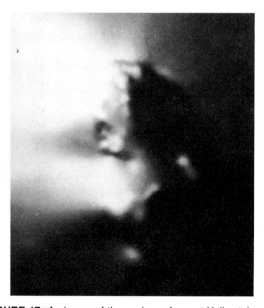

FIGURE 17 An image of the nucleus of comet Halley taken by the Halley multicolor camera onboard Giotto. [Photo courtesy of the Max-Planck-Institut für Aeronomie.]

surface morphology has a roughness of a few hundred meters. Furthermore, a number of "craters" can be identified. One of them has a depth of 200 m and a diameter of about 2 km. Some other topological structures such as "mountains" and "hills" could be delineated as well.

F. Internal Structure

Even though the subsurface structure of comet Halley's nucleus is hidden from view, there are several clues to its possible physical nature. First, the very low bulk density (<0.3 g cm^{-3}) suggests that the internal structure must be very fluffy at different scales. One concept is that, as a result of random accretion of icy planetesimals in the solar nebula, a fractal model might be a good approximation. In this way, the internal configuration of a cometary nucleus of a few kilometers in size could be very similar to the irregular share of porous interplanetary dust grains (see Fig. 18). Besides the nonvolatile building blocks, the nucleus would also consist of icy aggregates sintering or glueing the matrix together. In other words, a comet nucleus could be visualized as an assemblage of many cometesimals of different sizes (1–100 m, say).

During the long-term storage in the Oort cloud for a period of 4.5×10^9 years, comets would be constantly bombarded by galactic cosmic rays. Subsurface material down to a depth of 1–10 m would be processed by energy deposition of such energetic charged particles. Such surface irradiation might permit the buildup of a layer of very volatile cover on the surface of comets in the Oort cloud. This effect could be related to the fact that new comets such as comet Kohoutek 1973 XII, during their first inbound entry into the solar system, tend to brighten up anomalously at large heliocentric distances (>4 AU) but fizzle away at perihelion and outbound passage.

For an evolved comet, its subsurface structure may be approximated by a number of layers with different chemical compositions. The innermost core, containing pristine ice without suffering from any substantial sublimation loss, is covered by a mantle of core material with the most volatile component (i.e., CO or CO_2) already purged. In this mantle, water ice and the refractory dust grains are still intact. At the nuclear surface, a crust of dust particles with much-reduced water-ice content could exist. The exchange of mass and hence the chemical differentiation effect as a result of time variation of the temperature profile

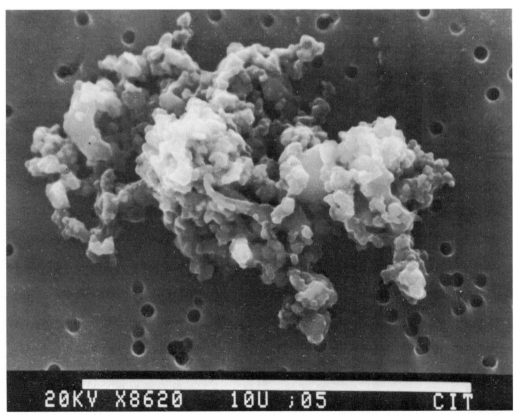

FIGURE 18 An electron microscopic picture of a highly porous chondritic interplanetary dust particle (U2-18A3B). [From Fraundorf, P., Brownlee, D. E., and Walker, R. M. (1982). *In* "Comets" (Wilkening, L. L., ed.), Univ. of Arizona Press, p. 383.]

could be determined by not just the orbital motion of the comet but also its rotational movement (i.e., day-and-night variations).

The sudden flareups of the coma activities of some comets have been suggested to be the result of phase transition of the cometary ice. If the interior of a cometary nucleus is made up of amorphous ice from low-temperature condensation in the solar nebula, a transformation into cubic ice will take place at about 153 K. The latent heat release would be an additional heat source to the cometary nucleus. Furthermore, the heat conduction coefficient of crystalline ice is about a factor of 10 higher than that for amorphous ice at 153 K; the heat balance of the nucleus consequently would be strongly influenced by the physical nature of the ice.

G. Rotation

From a number of observations such as the time variations of the light curves and the production of CN-shells in the coma of comet Halley, two periodicities of the temporal changes were found. A period of 2.2 days is superimposed on a longer period of 7.3 days. This means the nucleus of comet Halley could be spinning as well as precessing. If the ellipsoidal shape of the nucleus is approximated as a symmetric top with a major axis c and minor axes $(a = b) < c$, the rotational motion of the nucleus could be in one of two possible states, namely, one with the ratio of the spin period (P1) to the precession period (P2) being characterized by P1/P2 = 3.37 and the other with P1/P2 = 0.30. For example, arguments have been presented to support the scenario that the nucleus is rotating with a spin period of 7.4 days and a nutation period of 2.2 days. The cone angle between these two rotational vectors is about 76°. Other possible configurations have been suggested as well. Because of existing uncertainties in the ground-based and spacecraft data, no definite answer can be derived yet.

VI. DUST

A. Composition

The compositional variation of cometary dust grains had been inferred from collected samples of the interplanetary dust particles (see Fig. 18). A porous aggregate of CI carbonaceous chondrite composition, containing elements from C to Fe in solar abundance, is generally considered to be typical of cometary grains. While the dust-particle experiments on Vega and Giotto have not provided complete information on the mineralogical makeup of different kinds of grains in the coma of comet Halley, the preliminary results show that a large portion of the solid particles

TABLE II Average Atomic Abundances of the Elements in Halley's Dust Grains and in the Whole Comet, Dust and Ice

Element	Halley Dust	Halley Dust and ice	Solar system	CI-chondrites
H	2025	3430	2,600,000	492
C	814	≡940	940	70.5
N	42	76	291	5.6
O	890	1627	2,216	712
Na	10	10	5.34	5.34
Mg	≡100	100	≡100	≡100
Al	6.8	6.8	7.91	7.91
Si	185	185	93.1	93.1
S	72	72	46.9	47.9
K	0.2	0.2	0.35	0.35
Ca	6.3	6.3	5.69	5.69
Ti	0.4	0.4	0.223	0.223
Cr	0.9	0.9	1.26	1.26
Mn	0.5	0.5	0.89	0.89
Fe	52	52	83.8	83.8
Co	0.3	0.3	0.21	0.21
Ni	4.1	4.1	4.59	4.59

[From Jessberger, E. K., and Kissel, J., (1990). *In* "Comets in the Post-Halley Era" Newburn, R., Neugebauer, M., and Rahe, J., (eds.).]

indeed have compositions similar to that of the CI carbonaceous chondrites (see Table II). Furthermore, several types of dust grains can be identified. The first one is characterized by the CI chondritic composition with enrichment in the elements carbon, magnesium, and nitrogen. A small population of particles of pure silicate composition (Si, O, Mg, and Fe) constitute the second group. Organic (CHON) grains composed solely of light elements, H, C, N, and O constitute the third group. The CHON particles might be partly of interstellar grain origin with their outer mantles covered by organic polymerized material from cosmic ray and/or UV photon irradiation.

According to ground-based infrared observations, the composition of dust particles in the coma of comet Halley should be about 56% olivine, 36% pyroxene, and 8% layer lattice silicates. These estimates are not in disagreement with the *in-situ* measurements at comet Halley. Thus the PIA and PUMA dust experiments indicated that the dust of comet Halley is composed of two main components: a refractory organic phase (CHON) and a siliceous, Mg-rich phase. The CHON organic material further serves as coating of the silicate cores (see Fig. 19). The CHON-dominated grains have an average density of about 1 g cm^{-3} and the silicate-dominated grains of 2.5 g cm^{-3}. Also noteworthy is the finding that the isotopic ratio of $^{12}C/^{13}C$ in different grains varies from about 1 to 5000. Ground-based spectroscopic observations of

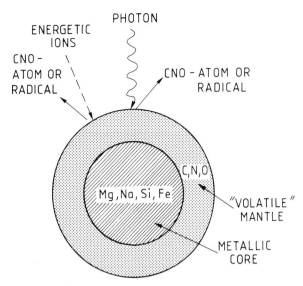

FIGURE 19 A schematic view of a possible inhomogeneous structure of cometary dust grains.

comet Halley showed that $^{12}C/^{13}C = 65$ in CN. There is therefore evidence that the isotopic ratio of carbon in comet Halley (and perhaps other comets also) is below the nominal terrestrial value of $^{12}C/^{13}C = 90$. Such isotopic anomaly might be the result of nucleosynthetic environment of the solar nebula or ion–molecule reactions in interstellar clouds.

B. Size Distribution

Before the *in-situ* measurements at comet Halley, a standard dust-size distribution had been constructed by Devine *et al.* (1986). This was given as

$$n(a)\,da \begin{cases} C_0[1 - a_0/a]^M \left[\dfrac{a_0}{a}\right]^N da & a \geq a_0 \\ O & a < a_0 \end{cases} \quad (23)$$

where $M = 12$, $N = 4$, C_0 is a normalization constant, a_0 ($= 0.1\ \mu$m) is the radius of the smallest dust grain, and a is the grain radius. In Fig. 20 this model distribution is compared to the size distribution obtained by the Vega and Giotto measurements. It is clear that while the Devine *et al.* size distribution predicted the absence of submicron grains, the spacecraft data indicated a large number of such small particles. The possible existence of a very high flux of 10^{-17} g particles (i.e., $a = 0.01\ \mu$m) was also reported. This component of very small grains might be related to the polycyclic aromatic hydrogen carbons (PAHs) in interstellar space.

The total mass production rate of the dust particles detected by the Giotto DIDSY dust particle experiment, if extrapolated to $m_{max} = 1$ g, amounts to a gas-to-dust ratio of about 7:1. But if the observed mass distribution is

extrapolated to a larger mass, say $m_{max} = 1$ kg, the corresponding gas-to-dust ratio would be about 1:1. It is interesting to note that radar observations of comet Halley in 1985 detected a weak echo that might have come from a halo of large (>2-cm radius) grains ejected from the cometary nucleus. Figure 21 shows the Doppler spectra of the full radar echo (nucleus and coma components) for comets (a) IRAS–Araki–Alcock, and (b) Hyakutake. The data show the presence of extended structures of large dust grains with sizes of a few centimeters surrounding the nuclei of comet IAA and Hyakutake.

C. Dynamics

After emission from the nuclear surface, dust particles will be accelerated by the drag force of the expanding gas. At the same time, the gravitational attraction of the nucleus itself tends to decelerate the motion of the dust particles is relatively small, such that they do not affect the expansion of the gas, we can write the equation of momentum transfer as

$$m \frac{du}{dt} = \pi a^2 \rho_{gas} C_D (u_{gas} - u)^2 - \frac{G m M_N}{r_c^2} \quad (24)$$

where $m = \frac{4}{3}\pi a^3 \delta$ is the mass of the grain with radius a and density δ, u_{gas} is the flow velocity of the gas, ρ_{gas} is the gas density, M_N is the mass of the cometary nucleus, G is the universal constant of gravitation, C_D is the drag coefficient, and r_c is the cometocentric distance. From Eq. (24) it can be seen immediately that there exists an

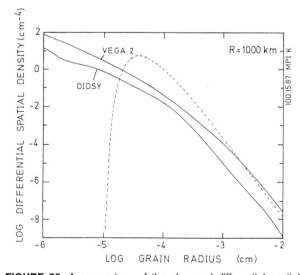

FIGURE 20 A comparison of the observed differential spatial density of cometary dust particles as a function of radius of the grains with theoretical curve determined by Devine *et al.* (1986). The experimental curve was deduced from the cumulative densities measured by the DIDSY impact detectors on Giotto. [From Lamy, Ph., *et al.* (1987). *Astron. Astrophys.* **187,** 767.]

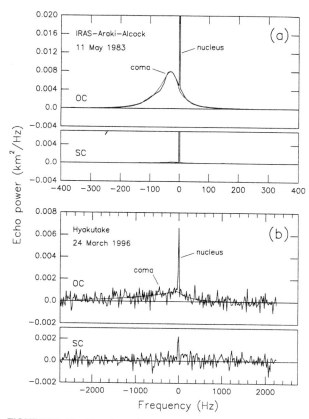

FIGURE 21 The Doppler radar spectra of comets (a) IRAS-Araki-Alcock, and (b) Hyakutake. [From J. K. Harmon et al. (1999). *Planetary and Space Science* **47**, 1409.]

upper limit of the grain radius (a_c) above which $du/dt < 0$, and hence no cometary grains of the corresponding sizes could be lifted off the nuclear surface. This critical radius can be approximated as

$$a_c \approx 3 \frac{C_D Q_m u_0}{4\pi \delta G M_N} \quad (25)$$

where u_0 is the initial velocity of the gas at the nuclear surface and Q_m is the total gas mass production rate. Thus, in the case of comet Halley at 1 AU solar distance, with $Q_m \approx 3.1 \times 10^7$ g s^{-1} and $u_0 \approx 0.3$ km s^{-1}, $C_D \approx 0.5$, $M_n \approx 10^{17}$ g, and $\delta \approx 1$ g cm^{-3}, we have $a_c \approx 50$ cm.

Dust grains with $a < a_c$ will be lifted over the surface and will move radially outward. After a distance of a few cometary radii, the gas drag effect will become insignificant and the dust particles will begin to move radially with a certain terminal velocity. For comet Halley at 1 AU solar distance, the velocity of μm-size particles is on the order of 0.3–0.5 km s^{-1}. The solar radiation pressure will continue to act on the dust particles such that, for particles emitted initially in the sunward direction, they will be eventually stopped at a parabolic envelope (see Fig. 22).

D. Thermal Emission

The solar radiation energy absorbed by cometary dust grains is mostly reemitted in the infrared wavelengths. For a spherical grain of radius a, the surface temperature of a dust particle will be determined by the following energy balance equation:

$$\int_0^\infty F_\odot(\lambda) Q_{abs}(\lambda, a)\pi a^2 \, d\lambda$$
$$= \int_0^\infty \pi B(\lambda, T) Q_{abs}(\lambda, a) 4\pi \alpha^2 \, \delta\lambda \quad (26)$$

where $F_\odot(\lambda)$ is the solar radiation at wavelength λ, $Q_{abs}(\lambda, a)$ is the absorption efficiency, and $B(\lambda, T)$ is the Planck function. With $Q_{abs} = 1$ (i.e., for a blackbody) or $Q_{abs} =$ constant (i.e., for a gray body), we have

$$T = \frac{280}{\sqrt{r}} \text{ K} \quad (27)$$

Now, for a certain size distribution $n(a)$, the thermal flux is given as

$$F_\lambda = \int_{a_{min}}^{a_{max}} \pi B(\lambda, T) Q_{abs}(a, \lambda) a^2 n(a) \, da \quad (28)$$

where a_{min} and a_{max} are the minimum and maximum radii of the dust grains. A set of infrared spectra for the central coma of comet Halley taken in March 1986 are shown in Fig. 23. Superimposed on the thermal emission of heated dust particles with a temperature of the order

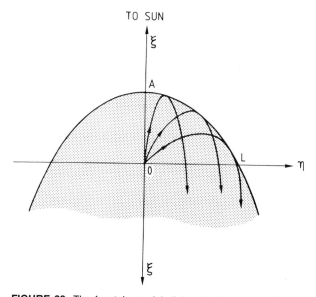

FIGURE 22 The fountain model of the dust trajectories in the sunward hemisphere near the nucleus together with their enveloping parabola. [Adapted from Mendis, D. A., Houpis, H. L. F., and Marconi, M. (1985). *Fundamental Cosmic Physics* **10**, 1.]

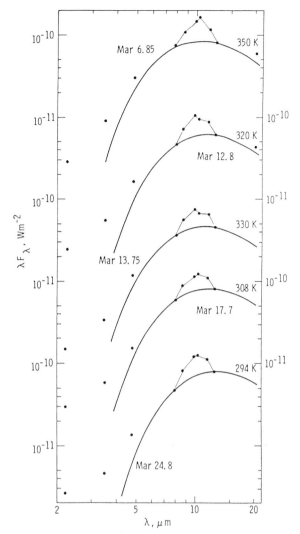

FIGURE 23 Thermal emissions from the central coma of comet Halley in March 1986, arranged chronologically. Each data set is offset by factor 10; the scales are shown on alternate sides of the figure. Blackbody temperature curves have been fit through the 7.8 μm and 12.5 μm data points, and the corresponding color temperatures are given. [From Hanner, M., *et al.* (1987). *Astron. Astrophys.* **187**, 653.]

of 300–350 K, a broad emission at 10 μm can be seen. This emission, together with another emission signature at 18 μm, was generally attributed to silicate grains; suggestions have also been made that these features could have come from polymerized formaldehyde grains in the coma.

E. Dust Jet Features

A combination of the presence of active regions on the cometary nuclear surface and rotational motion could lead to the formation of spiral dust jets in a cometary coma.

Figures 14 and 17 show the closeup views of a system of dust jets in the coma of Comet Halley. Prominent jet structures had also been observed in the coma of Comet C/1995 O1 Hale–Bopp (Fig. 24). From the dynamical behavior of the jet morphology, the rotation period of this comet has been deduced to be about 11 h.

VII. ATMOSPHERE

A. Composition

1. Comet Halley

A compilation of the molecular abundances obtained from the observations of comet Halley is given in Table III. After H_2O, CO is the second most dominant species, with an abundance of about 10–15% relative to water. According to the neutral mass spectrometer (NMS) experiment on Giotto and rocketborne ultraviolet observations, a significant fraction ($\approx$50%) of the CO was produced in a distributed region within a radial distance of about 25,000 km. The organic CHON dust grains or formaldehyde polymers could be the sources of these molecules. On the other hand, the third most abundant molecules, CO_2, appeared to have been emitted solely from the central nucleus. The heavy ion detector (PICCA) experiment on Giotto detected cometary ions with mass up to 213 AMU in the inner coma (see Fig. 25). The quasi-periodic spacing of the mass peaks has led to the suggestion that fragmentation of complex formaldehyde polymers $(H_2CO)_n$ or POMs could be responsible for the appearance of these heavy ions. But the presence of other kinds of hydrocarbons and organic molecules is also possible.

As the most primitive small bodies in the solar system, comets were generally thought to have formed at large

TABLE III Main Molecular Abundances of the Ice of Comet Halley

Molecules	Relative abundance	Remark
H_2O	1.0	
CO	0.10–0.15	Up to 50% from distributed source
CO_2	0.04	
CH_4	0.01–0.05	
H_2CO	0.04	
NH_3	0.003–0.01	Lower limit more likely
N_2	<0.001–0.1	Lower limit more likely
HCN	0.001	
OCS	<0.07	
CS_2	0.001	
S_2	<0.001	Inferred from observations of comet IRAS–Araki–Alcock (1983d)

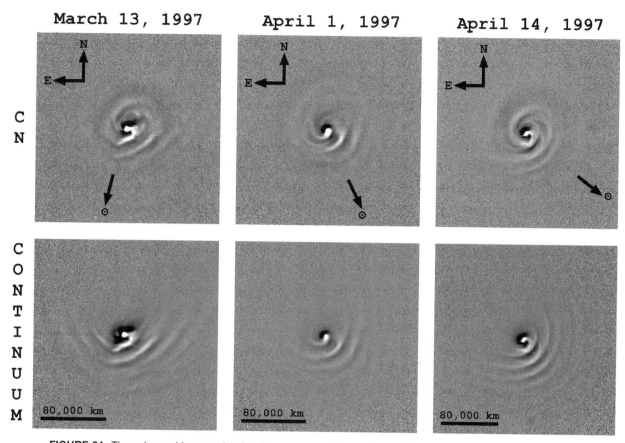

March 13, 1997 **April 1, 1997** **April 14, 1997**

C N

C O N T I N U U M

80,000 km 80,000 km 80,000 km

FIGURE 24 The enhanced images showing the coma morphology of comet Hale-Bopp in March/April, 1997. The direction to the Sun is shown with black arrows. [From B. E. A. Mueller, N. H. Samarasinha, and M. J. S. Belton, "Earth, Moon, Planets, . . .]

distances from the sun, where the condensation temperature is relatively low ($T_c \lesssim 50$ K). The bulk composition is therefore expected to be very volatile in comparison with other solar system objects, such as the asteroids. The compositional measurements of the gas and dust components at comet Halley indeed support this assessment. Figure 26 is a summary of the abundances of several major elements (C, N, O) relative to silicon. It is obvious that comet Halley's bulk composition is most similar to that of the sun.

The large amount of CO and CO_2 in comparison with the small abundance of CH_4 means that the chemistry in the solar nebula might be dominated by the oxidation of carbons, as may be found in interstellar clouds. The very small amount of N_2 may be understood in terms of the inefficiency of trapping N_2 during the low-temperature condensation of water ice. The details are still to be clarified by laboratory simulation of condensation processes.

According to the Giotto NMS experiment, the deuterium to hydrogen (D/H) ratio was determined to be $0.6 \times 10^{-4} < \text{D/H} < 4.8 \times 10^{-4}$, which brackets the cor-

responding value in the ocean water of the earth (D/H $= 1.6 \times 10^{-4}$). Both values represent a deuterium enrichment compared with interstellar space, where the D/H ratio $\approx 10^{-5}$. On the other hand, the cometary and planetary values are considerably smaller than the D/H ratios observed in several molecules (HCN, H_2O, HCO^+, etc.) found in interstellar dark clouds and in the organic inclusions of a few meteorite samples (the Chainpur meteorite shows a deuterium enrichment with D/H ratio up to 9×10^{-4}). Although ion chemistry may play a role in the deuterium enrichment of the molecules mentioned above, the D/H ratio of comet Halley could be more closely related to the ill-understood condensation process of icy planetesimals.

Two other isotope ratios deduced from the Giotto NMS experiment are $^{18}O/^{16}O = 0.0023 \pm 0.0006$ and $^{34}S/^{32}S = 0.045 \pm 0.010$, which, within the experimental errors, are both equal to the corresponding terrestrial values. Thus, the D/H, $^{18}O/^{16}O$, and $^{34}S/^{32}S$ ratios all seem to indicate that comet Halley's isotopic compositions are the same as those of the solar system material. One exception is the

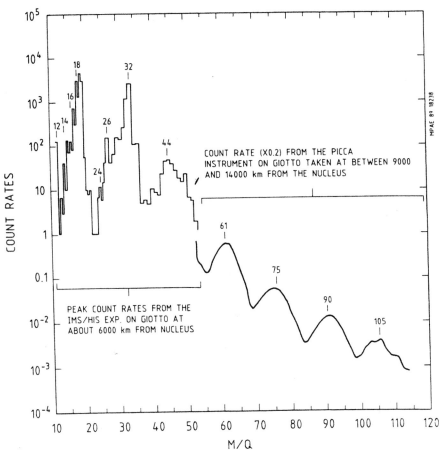

FIGURE 25 A composite ion mass spectrum taken by the IMS and PICCA instruments on Giotto in the coma of comet Halley. [From Balsiger, H., *et al.* (1986). *Nature*, **321**, 326, and A. Korth *et al.* (1986). *Nature*, **321**, 335.]

$^{12}C/^{13}C$ ratio which was determined to be ≈ 65 by ground-based high spectral resolution measurements of CN.

2. Comets Hyakutake and Hale–Bopp

The apparitions of two bright comets, C/Hyakutake and C/Hale–Bopp in 1995 and 1996, had provided new information on the chemical compositions of long-period comets. This was particularly opportune because of the availability of several advanced radio telescopes and the operation of the Infrared Space Observatory at that time, permitting the tracking of the coma developments of C/Hale–Bopp from large heliocentric distances to perihelion. Table IV shows the molecular abundance of C/Hyakutake and of C/Hale–Bopp. It can be seen that at a heliocentric distance of 1 AU, the molecular composition of C/Hyakutake is not too different from that of Comet Halley. On the other hand, the relative abundances of CO and of CO_2 are much larger for C/Hale–Bopp at 4 AU. This effect is consistent with the expectation that volatile ices such as CO and CO_2 should be the first to sublimate at large heliocentric distance when the surface

temperature is still too low for water ice to sublimate at high level of gas production rate. Figure 27 summarizes the gas production rates of molecular species observed in radio wavelengths. The crossover of the CO and H_2O values at about 4 AU during the inbound orbit of C/Hale–Bopp is most conspicuous.

B. Coma Expansion

In the inner coma, the cometary outflow is collision-dominated within a radial distance of r_c given as

$$r_c = \frac{Q\sigma}{4\pi u_{gas}} \qquad (29)$$

where Q is the total gas production rate, σ is the collisional cross section among the neutral molecules, and u_{gas} is the expansion speed of the neutral coma. For comet Halley at a solar distance of 1 AU, $Q \approx 10^{30}$ molecules s^{-1}, we obtain $r_c = 25,000$ km if $\sigma = 3 \times 10^{-15}$ cm^2 and $u_{gas} \approx 0.8$ km/s.

At sublimation, the gas molecules have an average radial velocity below the sonic value. But the expansion velocity quickly reaches supersonic value within a short

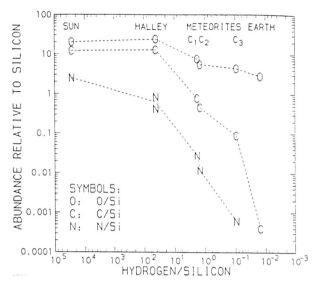

FIGURE 26 Relative abundances of elements in the material released by comet Halley. A gas/dust ration of 2 at the source was assumed. For comparison, the abundances in the solar nebula and in carbonaceous are given. Estimates for the abundance on the earth (crust plus mantle) are also included. [From Geiss, J. (1987). *Astron. Astrophys.* **187,** 859.]

distance ($\lesssim$30 m) from the nuclear surface. Several effects are important in the thermal budget and in further acceleration of the gas. First, adiabatic cooling is essential in reducing the temperature of the coma gas to a value as low as 10–30 K at a cometocentric distance of 300 km. Second, as a result of photolytic processes, the photodissociation fragments could obtain a significant amount of excess kinetic energy ($\approx$ a few eV) at dissociation. For example, hydrogen atoms from H_2O dissociation move with an initial speed of around 20 km s^{-1}. The fast H-atoms created inside the collision-dominated region will be thermalized and slowed down via collision with the background gas. Such an energy-transfer process is of importance in the photolytic heating of the coma. Since infrared emission due to rotational transitions in the highly polar molecules H_2O could be a very effective cooling mechanism, the total energy budget is then determined by a balance between the infrared cooling and photolytic heating. The exact magnitude of the infrared cooling rate, however, is still subject to debate. In Fig. 28 we show the effects of including and neglecting the IR cooling effect on the expansion velocity of the gas. A detailed fit with the gas velocity from the Giotto NMS experiment can be achieved by adjusting different parameters in the momentum-transfer process and energy budget.

The images of the near-nucleus environment of comet Halley indicate that gas and dust production were concentrated mostly on the sunlit side. Further evidence was provided by the TKS spectroscopic experiment on Vega,

which could be used to map the spatial distribution of dust and water vapor in the vicinity of the comet nucleus. According to the experimenters, most of the dust emission is confined to a narrow cone with a half-angle of about 25°. They further suggested that the angular distribution of H_2O vapor up to several thousands of kilometers from the nucleus is even narrower. Given such narrow gas jet structure, theoretical model calculations generally show that an anisotropic gas outflow should be maintained to a cometocentric distance of a few 10^4 km. An unexpected result is that the neutral gas density distribution measured by the Giotto NMS experiment can be well fitted by a spherically symmetric model.

Besides the jet structures, expanding shells of CN radicals were observed in the coma of comet Halley. Their formation could be characterized by two quasi-periods of 2.2 days and 7.4 days, respectively (cf. Section V.G). This means that the outgassing process of comet Halley was strongly modulated by the rotational movement of its nucleus. Such halo formation had already been noticed in the 1910 return of comet Halley.

The expansion velocity of the CN shells was found to vary with the heliocentric distance at production. At $R \approx 1.3$ AU, the shell velocity V_s is about 0.8 km s^{-1} and at $R \approx 0.8$ AU, V_s increases to 1.2 km s^{-1}. Such systematic variation can be explained in terms of the dependence of the photolytic heating rate on the solar distance.

TABLE IV Molecular Abundances in Comets

Molecule	T(K)	C/Hyakutake at 1 AU	Others at 1 AU	C/Hale-Bopp at 1 AU
H_2O	152	100.	100.	100.
CO	24	5.–30.	2.–20.	80.
CO_2	72	≤7.	3.–6.	30. at 4.6 AU
CH_4	31	0.7	≤0.5–2.	
C_2H_4	54	0.3–0.9		
C_2H_6	44	0.4		
CH_3OH	99	2.	1.–7.	6.
H_2CO	64	0.2–1.	0.05–4.	0.1–0.2
NH_3	78	0.5	0.4–0.9	
N_2	22		0.02	
HCN	95	0.15	0.1–0.2	0.6
HNC		0.01		
CH_3CN	93	0.01		
HC_3N	74		≤0.02	
H_2S	57	0.6	0.3	6.
OCS	57	0.3	≤0.5	
S_2		0.005	0.02–0.2	
SO_2	83		≤0.001	

[From Bockelee-Morvan, D., *in* IAU Symp. No. 178, "Molecules in Astrophysics: Probes and Processes" (van Dishoeck, E. F., ed), Leiden July 1–5, 1996.]

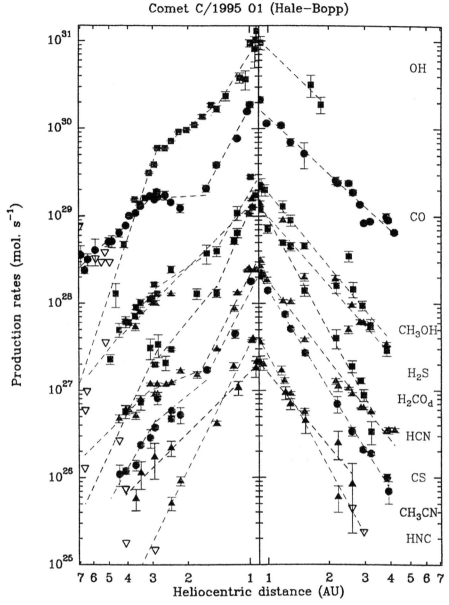

FIGURE 27 Gas production curves from observations at IRAM, JCMT, CSO, SESST and Nancay radio telescopes. [From Bockelee-Morvan, D., and Rickman, H., *Earth, Moon, and Planets*: *An International Journal of Solar System Science* (M. E. Bailey, ed.), Dordrecht, Holland.]

Another topic of interest concerns the possible recondensation of water droplets. During coma expansion, small icy grains may be dragged away from the nuclear surface together with the sublimating gas flow. The detection of an icy grain halo is quite difficult, however, as the lifetimes of icy grains—if they contain dark-colored absorbing material—should be very short, thus limiting the dimension of the icy grain halo to just a few hundred kilometers at 1 AU solar distance. Only for solar distances >3 AU would the low evaporation rate of the icy grains allow the icy grain halo to be stable against sublimation. It is perhaps for this

reason that the 3-μm absorption signature of water ice was detected for comets Bowell (1980b) and Cernis (1983f) only at large solar distances. The concept of a distributed source is still a valid one if the recondensation of water molecules into particulate droplets during the adiabatic cooling phase is possible. The icy droplets (or clusters of a few hundred H_2O molecules) so formed would probably have radii of the order of 15 Å. Depending on the fraction of water vapor that might go through the recondensation process, the latent heat released within a few nuclear radii could lead to substantial increase of the gas temperature.

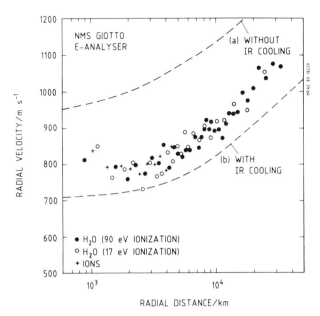

FIGURE 28 A comparison of the neutral gas radial velocity measured by the Giotto NMS experiment with theoretical model calculations assuming the presence and absence of infrared cooling effect of the water molecules in the coma. [Giotto NMS data from Lämmerzahl, P., et al. (1987). *Astron. Astrophys.* **187**, 169.]

No clear indication of the existence of icy droplets in the inner coma of comet Halley (or other comets) has been found. A possible indirect method of investigating this potentially important effect is to simulate the surface sublimation of water ice in a vacuum under controlled conditions. Laboratory simulation experiments should be designed to examine *ang* the possible production of water dimers [$(H_2O)_2$] and other water clusters of larger structures.

VIII. PLASMA

Before the space missions to comets Giacobini–Zinner and Halley, a general picture of solar-wind–comet interaction included the following steps (see Fig. 29):

(a) Ionization and pickup of the cometary ions beginning at large cometocentric distances ($r \gtrsim 10^6$ km)

(b) Heating and slowdown of the solar-wind flow due to assimilation of the heavy cometary ions

(c) Formation of a cometary shock

(d) Amplification of the magnetic field strength on the front side of the coma and draping of the field lines into a magnetic tail

(e) Stagnation of the solar-wind plasma flow by ion-neutral friction in the inner coma

(f) Formation of a contact surface shielding the outward-expanding ionospheric flow from the external plasma flow

It is satisfying that *in-situ* measurements at these two comets have confirmed this global picture. There are, however, many new and interesting effects that had not been previously considered. In this section we shall focus our attention on the major observational results from the spacecraft measurements. The survey will start from the inner region.

A. The Ionosphere

In the inner coma within a few thousand kilometers, where $r < r_c$ [see Eq. (24)], the ion temperature should be similar to the neutral temperature, since these two gases are closely coupled in this region. The ion temperature inside the contact surface of comet Halley at about 4800 km was determined to be ≈200 K by the Giotto IMS and NMS experiments (see Fig. 30). This result is in reasonable agreement with theoretical calculations. Because of the very effective cooling by the H_2O molecules via inelastic collision, the temperature of the electron gas (T_e) is expected to be nearly the same as the temperature of the neutral gas over a radial distance of about 1000 kilometers, beyond which point there is a sharp increase of kT_e to ≈2–3 eV. The exact location and gradient of T_e depends on a combination of the H_2O electron-neutral cooling function, electron transport, and thermal conductivity in the actual

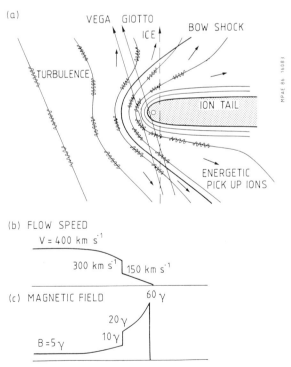

FIGURE 29 A schematic view of the comet–solar-wind interaction process.

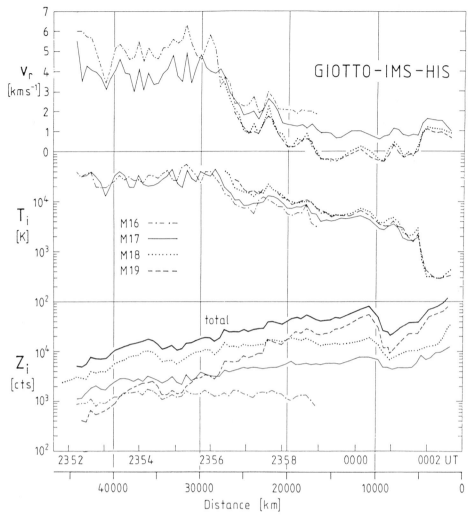

FIGURE 30 Profiles of the ion velocity (V_r) opposite to the spacecraft ram direction in the rest frame of the nucleus, temperatures (T_i), and the relative densities (Z_i) for ion masses M16, M17, M18, and M19. The measurements are from the Giotto IMS experiment. [From Schwenn, R., *et al.* (1987). *Astron. Astrophys.* **187**, 502.]

situation. Because of a lack of plasma instruments capable of measuring low-energy electrons ($kT_e < 1$ eV) the Vega and Giotto missions have not been able to shed light on the ionospheric electron temperature distributions in the coma of comet Halley. Note that the plasma wave experiment on the ICE spacecraft to comet Giacobini–Zinner measured a very dense, cold plasma in the central part of the ion tail. The electron temperature there is as low as 1.3×10^4 K (i.e., $kT_e \approx 1$ eV). Since the point of closest approach to the nucleus was at a distance of 7800 km, the electron temperature near the center of the ionosphere ($r < 1000$ km) is expected to be even lower.

The observational data from the Giotto mission for the ion temperature and the radial velocity of the ions are summarized in Fig. 30. The discontinuities at the contact surface are most notable. The ions created inside the contact surface are very cold, with a temperature of $T_i \approx 200$ K;

those outside have $T_i \approx 3000$ K. While the external plasma is nearly stagnant, the cold ionosphere plasma has a radial outward velocity of about 1 km s^{-1}. The magnetic field was found to drop from a value of 50 nT to effectively zero at this interface (see Fig. 31). The general behavior of the magnetic field variation in the vicinity of the contact surface can be understood in terms of a balance between the ion-neutral friction and Lorentz force, that is,

$$J \times B = K_{in} n_i m_i n_n u_{gas} \tag{30}$$

or

$$\frac{1}{4\pi}\left(B\frac{\partial B}{\partial R} + \frac{B^2}{R_{cur}}\right) = k_{in} n_i m_i n_n u_{gas} \tag{31}$$

where k_{in} is the ion–molecule collision rate, n_n is the neutral number density, n_i is the ion number density, u_{gas} is

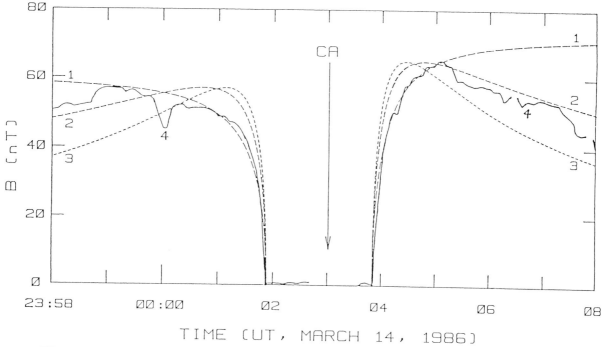

FIGURE 31 Magnetic field measurements made by the Giotto magnetometer experiment showing the inner pile-up region inbound and outbound and the magnetic cavity region. Theoretical curves following the ion-neutral friction theory are also compared. [Experimental curves from Neubauer *et al.* (1986). *Nature* **321**, 352; theoretical curves from Wu, z.-J. (1987). *Ann. Geophys.* **6**, 355.]

the neutral gas speed, and R_{cur} is the radius of curvature of the magnetic field lines.

The plasma can be taken to be stationary as a first approximation, and the cavity boundary can be assumed to be fixed. Furthermore, the ionospheric plasma inside the contact surface can be described in terms of photochemical equilibrium such that the ion number density can be expressed as

$$n_i(R) = \left(\frac{\beta Q}{4\pi \alpha u_{gas}}\right)^{1/2} \frac{1}{R} \quad (32)$$

where α is the electron dissociative recombination coefficient and β the photoionization rate. In this case Eq. (32) has a very simple solution for the magnetic field profile:

$$B(R) = B_{max} \frac{[1 - 2\ln(R/R_{max})]^{1/2}}{(R/R_{max})} \quad (33)$$

where R_{max}, proportional to $Q^{3/4}$, is the radial distance at which the piled-up magnetic field reaches its maximum value B_{max}. The essential feature of a rapid decrease of the magnetic field by about 40 to 60 nT over a distance of about 3000 km can be reproduced rather well. According to the Giotto magnetometer experiment, a final decrease of the magnetic field strength by $\Delta B \approx 25$ nT takes place over a radial distance as small as 25 km. This abrupt change

cannot be explained in terms of ion-neutral friction; and additional force such as particle pressure might be invoked to account for this effect.

At the boundary of the contact surface, a spike of ion density enhancement should be produced, accompanying the sharp rise of the magnetic field. The structure of this density jump can be described in terms of accumulation of the ionospheric plasma into a stationary thin shell. The total content of ionization is determined by the injection of ionospheric particles and their subsequent loss via electron dissociative recombination in this layer.

Outside the magnetic field-free cavity, in the region $r = 10^4$ to 2.4×10^4 km, the ion temperature $T_i \approx 10^3$–10^4 K and the ion velocity $v_i \approx 3$–6 km s^{-1}. The ion temperature was determined by a thermal equilibrium condition in which

$$T_i \approx T_n + \frac{m_i}{3k}(v_i - v_n)^2 \quad (34)$$

In this region, another interesting feature is that there was a sharp discontinuity in the ion density at $r \approx 10^4$ km (see Fig. 32). Outside the density maximum at about 3×10^4 km, the cometary ion density profile follows an r^{-2} dependence. Inside the density maximum, the density profile has an r^{-1} dependence, as predicted by photochemical equilibrium models in which the ion production

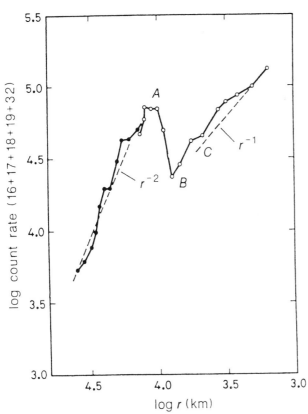

FIGURE 32 The radial profile of the sum of the ion counting rates for masses 16 to 19 and 32, showing that inside the magnetic cavity (C at $r = 4600$ km) the total counting rate tends to follow a $1/r$ dependence and that in the outer part ($r > 16,000$ km) a $1/r^2$ dependence fits the data well. The peak in the ion count rate is at A and the minimum is at B. [From Balsiger, H., *et al.* (1986). *Nature* **321**, 326.]

via photoionization is balanced by electron dissociative recombination.

A possible explanation for such a change in the ion density distribution is that there exists a gradient in the electron temperature at 10^4 km with the electron temperature T_e being much higher outside than inside. In this scenario, the plasma loss effect via electron dissociative recombination will mainly occur inside this radius. It should be mentioned that such a feature appeared to be quasi-stationary during the time interval of the spacecraft encounters with comet Halley. In addition to the Giotto measurements, both the plasma measurements on Vega and ground-based spectroscopic observations of the H_2O^+ emission in the coma revealed such a plasma structure.

One possible mechanism responsible for this electron temperature gradient is that the electron temperature could depend on the configuration of the magnetic field. In a magnetically closed region, the electron temperature is largely determined by a balance between photo-electron heating and collisional cooling by the neutral coma. In

contrast, in the case of an open-field region, there could be an extra heat input via thermal conduction along field lines from the external interplanetary medium.

B. Cometopause

Another interesting result from *in-situ* measurements by the Vega and Giotto probes at comet Halley concerns the detection of the so-called cometopause separating the plasma regimes dominated by the fast-moving solar-wind protons on one side and by the slow, cold, heavy ions of cometary origin on the other side. The Vega plasma experiment first discovered this sharp boundary of about 10^4 km thickness at a cometocentric distance of about 1.4×10^5 km (see Fig. 33). While the solar-wind proton flux quickly decreased across this boundary, the intensity of the cometary heavy ions rapidly increased. Such a plasma boundary effect was also observed during Giotto inbound at a cometocentric distance of 1.35×10^5 km, although the transition was not as sharply defined as that seen by Vega. The issue was further complicated by the fact that near this location a sudden jump of the magnetic field strength by 20 nT was detected by Giotto's magnetometer experiment, whereas the magnetic field variations measured by the Vega 1 and 2 spacecraft across the cometopause region were rather smooth. Within the framework of standard magnetohydrodynamic process in comet–solar-wind interaction, a model calculation separating the chemical processes from the two-dimensional plasma flow dynamics shows that charge exchange would be effective in eliminating solar-wind protons and hot cometary ions only at cometocentric distance ≤ 6–8×10^4 km. This theoretical result is consistent with the IMS observations on Giotto. Thus it may be concluded that the charge-exchange effect is not the immediate cause of the formation of the cometopause structure. On the other hand, plasma instabilities, such as the nonresonant firehose instability, might be important in leading to such a cometary plasma boundary under appropriate conditions.

C. Cometary Bow Shock and the Upstream Region

The transition region between the supersonic and the subsonic solar-wind flow was found to be very diffuse in the case of comets Giacobini–Zinner and Halley (see Fig. 34). The wavy and oscillatory patterns observed are somewhat similar to the quasi-parallel bow shock of the earth's magnetosphere, in which case the solar-wind flow direction is nearly parallel to the normal of the shock surface. The generation of large-amperiod of 75 to 135 s. These "100-s" waves have been suggested to correspond to ion cyclotron waves for water-group ions (i.e., H_2O^+, O^+, OH^+, etc.) ionized and picked up in the solar wind.

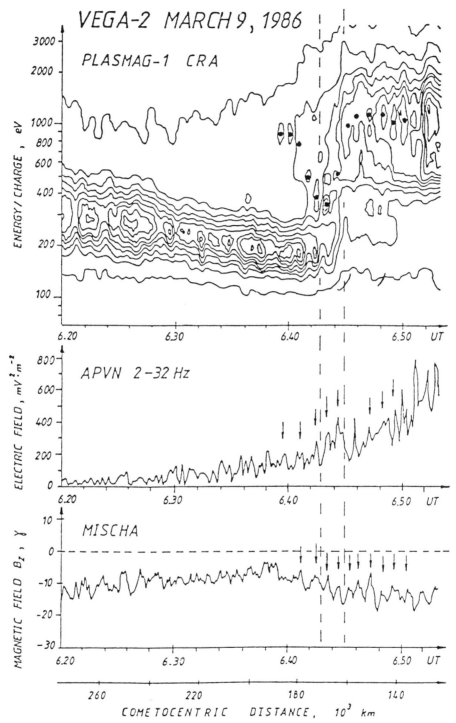

FIGURE 33 The formation of a cometopause structure (indicated by vertical dashed lines) as indicated by the ion measurements on Vega 2. The solar wind protons disappeared abruptly at about 1.5×10^5 km away from the comet nucleus. A cold population of heavy cometary ions showed up afterward with increasing fluxes. Oscillations in the ion flux, electric field, and magnetic field components (B_z pointing toward the north pole of the ecliptic plane) were also observed in the vicinity of the cometopause. Maxima are shown by dots and arrows. [From Galeev, A. A., *et al.* (1988). *J. Geophys. Res.* **93**, 7527.]

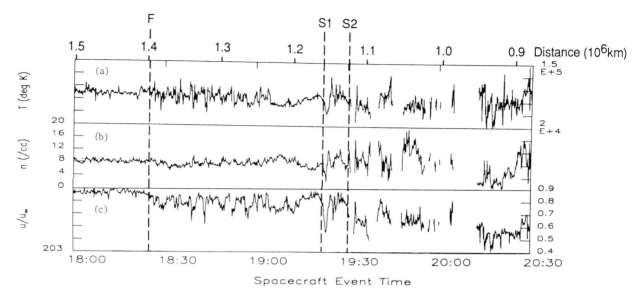

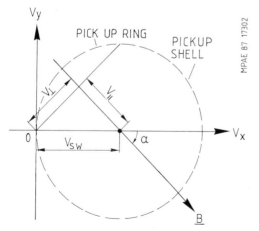

FIGURE 34 Solar wind proton bulk flow parameters in the bow shock region: (a) temperature [T], (b) number density [n], and (c) flow speed (u). The foreshock is marked F, the beginning of the low-speed dip i is marked S1, and the "permanent" speed change is marked S2. [Adapted from Coates, A. J., *et al.* (1987). *Astron. Astrophys.* **187**, 55.]

The basic cause of the plasma turbulence is associated with free energy produced by pickup of cometary ions. Initially, the velocity of these ions relative to the solar wind plasma flow has two components, $V_\parallel$ and $V_\perp$, as shown in Fig. 35. In the quasi-perpendicular case with $\alpha \approx \pi/2$, the cometary ions are accelerated by the $V_{sw} \times B$ electric field into cycloidal trajectories. In the solar-wind frame, the new ions gyrate around the magnetic-field lines with large pitch angles. Such a velocity distribution is unstable to the growths of left- and right-hand polarized Alfven (L and R) waves. The propagation direction of these waves should be sunward along the average, spiral magnetic-field direction.

FIGURE 35 The two components, $V_\parallel$ and $V_\perp$, of the initial velocity of the cometary pickup ions. Pitch angle scattering effect leads to a transformation of the ring distribution into a thin spherical shell.

If the average magnetic field is nearly parallel to the solar-wind flow direction, the new cometary ions are at first stationary with respect to the comet. However, the resultant cold beam of heavy ions having a speed of about V_{sw} relative to the solar wind leads to the nonresonant firehose instability under the following condition:

$$P_\parallel > P_\perp + \frac{B^2}{8\pi} \qquad (35)$$

where $P_\parallel$ is the thermal plasma pressure component parallel to the magnetic field and $P_\perp$ is the component in the perpendicular direction. This implies that the firehose instability, which excites long-wavelength waves, becomes possible when the cometary ion pressure exceeds the sum of the magnetic-field and solar-wind thermal pressure.

The main consequence for resonating ions of the various plasma instabilities is to rapidly isotropize the ring-beam distribution into a spherical shell. The growth rate is usually on the order of $0.1\ \Omega_i$ to Ω_i (ion gyrofrequency). The large level of magnetic-field turbulence will in turn permit stochastic acceleration of the cometary ions at large distances from the cometary bow shock. The detection of energetic heavy ions with energies up to a few hundred keV in large upstream regions where other acceleration processes are weak has been considered generally to be evidence of the second-order Fermi acceleration process. Other effects, such as first-order Fermi acceleration and lower hybrid turbulence, could be operational at localized regions at the same time.

As for high-frequency waves, one somewhat surprising discovery of cometary kilometric radiation (CKR) was made by the Sakigake spacecraft as it flew by comet Halley. The plasma wave experiment picked up discrete radio emissions in the frequency range of 30 to 195 kHz. These emissions occurring at the local plasma frequency may be the result of conversion of the electrostatic plasma waves to electromagnetic waves in the turbulent plasma environment of comet Halley. The generation of the CKR thus may be similar to the Type II solar radio bursts from coronal shock waves. In fact, the cometary bow shocks were identified as the source region of the CKR. It was further proposed that motion of the cometary bow shock could excite Alfven and Langmuir turbulences that eventually lead to the CKR emissions.

D. The Ion Tail

Combining the plasma measurements at comet Giacobini–Zinner on the tailward side of the comet at a cometocentric distance of 7800 km, we might produce a schematic model for the cometary magnetic field configuration as follows (see Fig. 36):

(a) The field draping model is basically confirmed except for the detection of the formation of a magnetosheath at the ion tail boundary. The magnetic field strength in the lobes of the ion tail is on the order of 60 nT. This relatively high field may be explained in terms of pressure balance at the tail boundary, where the total external pressure of the cometary ions was as large as the solar-wind ram pressure.

(b) A thin plasma sheet with a total thickness of about 2000 km and a width of about 1.6×10^4 km was found at the center of the ion tail. The peak electron density and an electron temperature were determined to be $n_e = 6.5 \times 10^2$ cm^{-2} and $T_e = 1.3 \times 10^4$ K by the plasma wave instrument.

(c) The plasma flow velocity gradually decreased to zero at the ion tail center. A significant amount of electron heating was seen between the ion tail and the bow shock.

We expect similar morphologies to be found in the plasma environment of comet Halley after appropriate spatial scalings. Snapshots by spacecraft flyby observations, however, do not reflect the many time-variable features seen in the ion tails of bright comets. For instance, the structure of the ion tail is often characterized by the appearance of a system of symmetric pairs of ion rays, with diameters ranging between 10^3 and a few 10^4 km, folding toward the central axis. When the ion rays are first formed, with an inclination of about 60° relative to the central axis, the angular speed is high with a linear speed $\approx$50 km s^{-1}. But near the end of the closure, the perpendicular speed is extremely low, no more than a few kilometers per second. Time-dependent MHD simulations have been applied to these phenomena with emphasis on the effect of temporal changes of the interplanetary magnetic field. No satisfactory answers to the formation of the ion rays have yet been found. A similar situation exists for the large-scale ion-tail disconnection events that had been suggested to be result of magnetic field reconnection.

E. Electrostatic Charging

Cometary dust grains, once released from the nucleus surface, will be subject to solar radiation and plasma interaction. The emission of photoelectrons and the impact of positive ions would be balanced by the surface collection

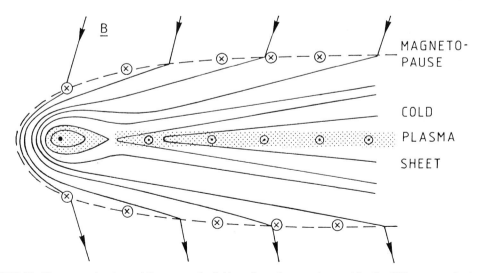

FIGURE 36 The general nature of the magnetic field configuration as observed by the ICE spacecraft at comet Giacobini–Zinner. The kinky structure of the magnetic field lines at the "magnetopause" is generated by a current layer required to couple the slow-moving cometary plasma with the external solar wind.

of electrons. The electrostatic potential of the grain surface is determined by the equilibrium condition that the net current should be zero. The electron charge on the dust grain with surface potential ϕ is

$$q = \frac{a\phi}{300} \text{ e.s.u.} \tag{36}$$

with ϕ in volts and particle radius in centimeters. In the solar wind, the Lorentz force experienced by the charged dust particle can be written as

$$f_L = q(E + V_d \times B) \tag{37}$$

where V_d is the dust velocity relative to the comet, B is the interplanetary magnetic field, e is the electronic charge, and $E = -V_{sw} \times B$ is the interplanetary motional electric field acting on the charged grain. For submicron particles, the Lorentz force for a ϕ value of a few volts is sufficient to produce strong perturbations on the orbital motion. Furthermore, during unusual episodes of enhanced energetic electron impact, the electrostatic charging effect could become strong enough to permit fragmentation of the dust particles. Some inhomogeneous structures of cometary dust tails not explained by the syndyname or synchrone formulation may be a consequence of such an effect.

At large solar distances where surface sublimation activities completely subside, the direct interaction of the solar wind plasma with the nucleus surface could lead to differential electrostatic charging of the surface. Against the weak gravitational force of the cometary nucleus, it is possible that some sort of dust levitation and surface transport process might occur. The details are very poorly understood, however.

F. X-Ray Emissions

Very intense extreme ultraviolet radiation and X-ray emissions at the coma of Comet C/Hyakutake 1996 B2 were discovered by chance by the German Roentgen X-ray satellite (ROSAT). Many more cometary X-ray extended sources were subsequently detected. Such X-ray emissions were now understood to be produced by charge transfer process between the neutral gas in the cometary comas and the solar wind minor heavy ions such as O^{5+} and C^{6+}. As a result of the charge transfer effect, a heavy ion such as O^{6+} will lose a positive charge to neutral molecules such as H_2O. The product ion, namely C^{5+}, will usually be excited to a higher electronic state. The radiative transition back to the ground state will lead to the emission of a photon of a few hundred eV. It is for this reason that the gas comas of comets have been found to be very powerful emitters of X-rays. The spatial distribution of the cometary X-ray emission is usually very symmetrical with respect to the sun–comet axis. It generally has a crescent-shaped morphology displaced toward the sunward side. This is simply because of the depletion of the solar wind ions as they stream toward the cometary center.

IX. ORIGIN

There are good dynamical reasons why the Oort cloud of new comets should have peak concentration between around 10^4 AU and 3×10^4 AU. First, with a stellar number density (n^*) of about 0.1 pc^{-3} in the solar neighborhood and an average encounter velocity (V^*) of about 30 km s^{-3}, the cumulative velocity perturbation over the age of the solar system would be large enough to release comets located at 5–10 $\times 10^4$ AU away from the sun. At the same time, stellar perturbations could maintain a steady influx of new comets only if their orbital distances are beyond 2×10^4 AU. Objects closer to the sun would be subject to stellar perturbations only rather infrequently. For example, over the age of the solar system (t_{sy}), the closest penetration distance (r) by a passing star can be approximated to be

$$\pi r_*^2 n_* v_* t_{sy} \approx 1 \tag{38}$$

Since $r_* \approx 10^3$ AU, comets with smaller orbital distances would be very stable against stellar perturbation. On the other hand, because of a lack of injection mechanism, these comets in the inner region would not be observable. It is thus possible that a rather massive inner Oort cloud may exist at solar distances between a few hundred and 1000 AU. Because particles in this system should be related to the condensation and accretion processes in the solar system without suffering from significant scattering effects by the passing stars, they should most likely be confined in a relatively flat configuration. Figure 37 illustrates the most up-to-date scenario according to current thinking. Between 10^3 AU and 2–3 $\times 10^4$ AU, there will be sporadic scatterings from time to time due to stellar passages. Once this happens, a burst of comets will be injected into the solar system that would last a few million years. During this interval, the flux of new comets could be strongly enhanced in comparison to the steady-state injection rate from the outer Oort cloud. Such a transient phenomenon is called a comet shower.

Records of biological mass extinctions show that a certain periodicity (≈ 26 Myr) may be identified. Several lines of evidence (for example, the iridium enrichment at the Cretaceous–Tertiary boundary) have seemed to indicate that mass extinctions of this sort could be related to catastrophic impact events by either comets or asteroids of kilometer size. This possible connection is strengthened by the reported evidence of a periodicity in the ages of impact craters of similar value and in phase with biological extinction events.

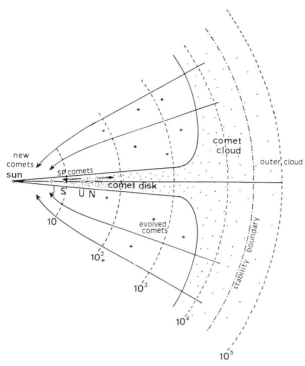

FIGURE 37 A schematic view of the structure of the cometary reservoir including the flat disk of inner Oort belt and the spherical shell of outer Oort cloud. [From Fernández, J. A., and Ip, W.-H. (1990) *in* "Comets in the Post-Halley Era" Newburn, R., Neugebauer, M., and Rahe, J., (eds.), Reidel, Dordrecht.]

Several interesting ideas have been proposed to explain the interconnection between the mass extinction/cratering events and cometary showers. These include (1) a solar companion star Nemesis with a 26-Myr orbital period, (2) periodic excursions of the solar system through the galactic plane, and (3) a trans-Neptunian planet X. The exact cause is under active investigation. In fact, whether the periodic crater events are statistically significant is still an open question. Furthermore, asteroid impacts might have contributed to the majority of these craters.

In addition to stellar perturbations, interstellar molecular clouds have been found to contribute to the destruction of the Oort cloud. The perturbation effect during the penetration passage by a giant molecular cloud with a total mass of 5×10^5 M and a radius of about 20 pc could be most dangerous to the survival of the Oort cloud. However, its occurrence is rather infrequent (approximately 1–10 encounters during the solar system lifetime). Molecular clouds of intermediate sizes with masses of a few 10^3–10^4 $M_\odot$ would be far more numerous and could produce rather strong perturbations comparable to the passing stars. The time interval between encounters of the solar system with intermediate molecular clouds has been estimated to be about 30 Myr.

One other important perturbation effect concerns the galactic tidal force. Because of the disk mass in the galaxy, the tidal force per unit mass, perpendicular to the galactic plane, acting on a comet as it moves to a galactic latitude ϕ is

$$f_{\text{tide}} = 4\pi G \rho_{\text{disk}} r \sin \phi \qquad (39)$$

where ρ_{disk} is the mass density of the galactic disk. The change in the comet's angular momentum, which is proportional to $f_{\text{tide}} \cos \phi$, thus will have a latitudinal dependence of the form

$$\Delta L \propto \sin 2\phi. \qquad (40)$$

Figure 38 shows the expected boundary of the Oort cloud as a result of the galactic tidal effect. The $\sin 2\phi$ dependence is clearly indicated. Statistical study of the frequency distribution of the galactic latitudes of new comets shows that there is indeed a minimum near the equatorial region with maxima at approximately ±45°. There are also reports of clustering of the aphelion points of new comets in the galactic coordinate system. Such an effect may be the result of close encounters of passing stars in these angular positions.

The total mass and number of comets stored in the outer and inner Oort clouds are still very uncertain. On dynamical grounds, it is estimated that there should be approximately 10^{11}–10^{12} comets in the outer Oort cloud between 1×10^4 and 5×10^4 AU. As for the inner Oort cloud between 10^3 and 10^4 AU, a similar (or larger) number of comets could exist there. With an average mass of

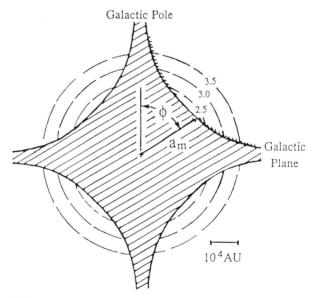

FIGURE 38 The latitudinal variation of the outer boundary of the Oort cloud as a result of galactic tidal effect. [From Morris, D. E., and Muller, R. A., (1986). *Icarus* **65**, 1.]

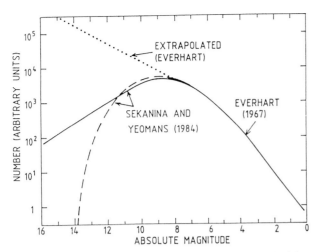

FIGURE 39 Distribution of the absolute brightness (H_{10}) of the long-period comets according to several statistical calculations. [From Everhart, E. (1967). *Astron. J.* **72**, 1002, and Sekanina, Z., and Yeomans, D. K. (1984). *Astron. J.* **89**, 154.]

10^{17} g for the comet nucleus, the total cometary mass (M_c) is thus on the order of 3–30 $M_{\oplus}$.

The actual values of M_c and number of comets (N_c) depend sensitively on the size distribution of comet nuclei. Figure 39 gives some estimates of the distribution of the absolute brightness distribution (H_{10}) of new comets. There is an apparent deficiency of comets with $H_{10} > 8$. This effect might be simply a matter of incomplete sampling, because comets in this range are relatively small. On the other hand, the cosmogonic process or physicochemical effects during the evolutionary history of the cometary nuclei could also contribute to the variations of the absolute brightness distribution. The determination of the H_{10} value of a comet depends on fitting its brightness variation as a function of power law in solar distance,

$$H = H_{10} + 5\log_{10}(\Delta) + 10\log_{10}(r) \qquad (41)$$

where r is the apparent magnitude of a comet when it is at heliocentric distance r and geocentric distance Δ. In Eq. (41) an inverse fourth-power dependence of the comet's brightness on r is assumed. Since the coma activities and surface properties of a comet could have strong effects on its brightness variation, it is by no means certain that Eq. (41) can be applied to all cases. Comprehensive surveys of distant comets at large solar distances (>3 AU), where the outgassing process generally ceases, are urgently needed to improve our knowledge in this aspect. The application of the following brightness–mass relation

$$\log_{10} m = -16.9 + 0.5 H_{10} \text{ kg} \qquad (42)$$

would mean that the cumulative number of the long-period comets has a power law dependence of $N(m) \propto m^{-1.16}$ for $m > 10^{14}$ kg. In comparison, the main-belt asteroids have

a mass distribution given by $N(m) \propto m^{-1}$. Since the assumed brightness–mass relation has large uncertainties, it is difficult to attach too much significance to the similarity of the spectral indexes for the mass distributions of comets with $H_{10} < 6$ and of the main-belt asteroids. In any event, if the cometary population has gone through extensive collisional fragmentation in the solar nebula before ejection into the distant Oort region, a cumulative mass distribution of the form of $N(m) \propto m^{-1}$ should result.

Several possible routes exist for the formation of comets in the primordial solar nebula. For example, it has been proposed that comets could have condensed and agglomerated in fragments on the outskirts of a massive solar nebula. Or, interaction of strong outflows from the young sun with the neighboring dusty environment could have facilitated the formation of icy planetesimals in the outer solar system. Finally, for a solar nebula model with small mass (<0.1 $M_{\oplus}$), gravitational instability in the central dust layer could lead to the formation of icy planetesimals outside Neptune's orbit.

One other mechanism that has been quantitatively studied by numerical methods is the gravitational scattering of small planetesimals from the Uranus–Neptune accretion zone. According to model calculations, planetesimals with a total mass of a few $M_{\oplus}$ could be ejected into near-parabolic orbits at the end of the accretion process of Uranus and Neptune. Those with aphelion distances reaching beyond 10^4 AU would have their orbital inclinations isotropized; however, the ones with smaller aphelia would tend to avoid frequent stellar perturbations and thus maintain a flat disklike distribution (see Fig. 37).

The presence of a flat cometary belt received some support from computer simulation experiments in which the orbital evolution of comets under the influence of planetary perturbations were traced. It was found that the short-period comets are most likely supplied by a trans-Neptunian source populations of low orbital inclinations. In the case of an isotropic inclination distribution for the parent population (i.e., the long-period comets) a significant fraction of the short-period comets would be captured into retrograde orbits. Since this is not observed, the obvious conclusion is that the short-period comets must be supplied by a belt (or disk) of comets outside the orbit of Neptune. The required number of comets between 40 and 200 AU is on the order of 10^8. Under steady-state condition of orbital diffusion, there should be about 10^6 such comets orbiting between Saturn and Uranus before they are captured into short-period comets.

A very interesting development from the point of view of cometary origin has to do with the discovery of Kuiper belt objects (KBOs) in trans-Neptunian orbits by David Jewitt and Jane Luu in 1992. These objects, presumably of icy composition, have sizes of a few hundred kilometers.

The total number of the KBOs with diameters >100 km has been estimated to be on the order of 10^5. Their orbits are located mostly beyond Neptune's orbit and, just like Pluto, a significant fraction of the discovered population is trapped in the 2:3 orbital resonance with Neptune. A number of the KBOs have been found to be in orbits of large semimajor axis ($\sim$100 AU). The observational statistics suggested that the total number ($\sim 3 \times 10^4$) of such scattered KBOs with diameters >100 km at the present time could be comparable to the classical KBOs. The original population of the KBOs could be as much as a factor of 100 bigger than the present one. A combination of planteary gravitational perturbations and collision destruction served to reduce the number of KBOs to the present value. It is most likely that the small collisional fragments of the KBOs could serve as an important source population of short-period comets in the past as well as the present time.

X. PROSPECTS

After the preliminary studies of comets Giacobini–Zinner and Halley via flyby reconnaissance, the future impetus of cometary research will be set by long-term investigations afforded only by rendezvous missions and, as a final goal, the return of samples of comet material in their pristine condition to Earth for laboratory analyses. Because of the potential threats of the catastrophic collisional effects of near-Earth objects (NEOs), there is also considerable interest in studying the physical properties of cometary nuclei. It is for these reasons that a number of comet missions have been planned and will be realized in the near future. These include NASA's Stardust mission for coma dust sample return, the CONTOUR mission for multiple comet flyby observations, and the Deep Impact project for active cratering experiment on a target comet. The Deep Space 1 spacecraft employing ion drive propulsion system will encounter Comet Borrelly in 2002. The title of the most daring and comprehensive comet mission belongs to the Rosetta mission of the European Space Agency. The Rosetta spacecraft will be launched in 2003 with Comet Wirtanen as its final target. After rendezvous, the spacecraft orbiting around the cometary nucleus will deploy a small spaceship for landing on the surface of the nucleus. The lander will provide *in-situ* measurements of the chemical and material properties of the cometary ice and dust mantle in great detail. Last but not least, the interior of the cometary nculeus will be probed by the radar tomography experiment on the mother spaceship.

The foregoing new initiatives, which will be among the major thrusts in planetary research, are expected to produce scientific data in the next decade. The outlook

for cometary study is therefore extremely promising. Certainly even more ambitious projects might lie ahead at the beginning of the 21st century. At this point, we should mention the potential industrial and commercial interests in cometary materials as important resources in space.

The important contributions from the International Halley Watch network during the comet Halley epoch and more recently the worldwide observations of comet Hyakutake and comet Hale–Bopp in 1996 and 1997 have shown that coordinated ground-based observations are an indispensable part of cometary research. Besides optical observations, infrared and UV measurements using satellite-borne telescopes would be very informative in monitoring the coma activities, dust emission, and gas and dust compositions of a wide variety of comets at different stages of outgassing. For this, a dedicated planetary telescope capable of performing simultaneous imaging and spectroscopic observations in the wavelength range from ultraviolet to infrared would be most desirable. For example, there are many issues related to the thermodynamics and hydrodynamics of coma expansion that require IR observations at high spatial and spectral resolution. The extremely variable structures of cometary ion tails are closely related to momentum and energy-transfer from the solar wind to cometary ions; well-planned observations capable of measuring the ion densitis, plasma velocities, and spatial configurations such as ion rays and plasmoids are urgently needed. The results from comet Halley have shown that the effects involved are extremely complicated. The correlations of remote-sensing observations and *in-situ* measurements carried out in the missions just mentioned will be most useful from this point of view.

The number density and size distribution of comets in the outer solar system are critical to our understanding of the origin of the short-period comets and the distant Oort cloud. Surveys by a new generation of space telescopes in optical and IR wavelengths will possibly provide the definitive answer. The observations of extrasolar systems exemplified by Beta Pictoris will yield complementary knowledge on the formation of the solar nebula and its disk of planetesimal population.

Laboratory simulations of ice condensation and gas trapping at low temperatures have provided very interesting insights into the physical nature of the cometary ice and the surface sublimation processes. Further progress in this area is expected.

Theoretical studies, as always, are the basic tools in laying the foundation for understanding different physicochemical phenomena and their interrelations. The five areas of ongoing theoretical efforts (i.e., dust mantle modeling, gas thermodynamics and hydrodynamics, comet–solar-wind interaction, coma chemistry, and cometary orbital dynamics) are following a trend of increasing sophistication. In combination with the opportunities of new observations and space missions in the next decade, modern cometary physics has just reached a new era.

ACKNOWLEDGMENT

This work was supported in part by NSC 90-2111-M-008-002 and NSC 89-2112-M-008-048.

SEE ALSO THE FOLLOWING ARTICLES

PRIMITIVE SOLAR SYSTEM OBJECTS: ASTEROIDS AND COMETS • SOLAR SYSTEM, GENERAL • SOLAR PHYSICS • STELLAR STRUCTURE AND EVOLUTION

BIBLIOGRAPHY

Brandt, J. C., and Chapman, R. D. (1981). "Introduction to Comets," Cambridge Univ. Press, Cambridge U.K.

"Comet Halley 1986, World-Wide Investigations, Results, and Interpretation," (1990). Ellis Horwood, Chichester, U.K.

Comet Halley Special Issue (1987). *Astron. Astrophys.* **187**(1, 2).

Crovisier, J., and Encrenaz, Th. (1999) "Comet Science," Cambridge Univ. Press, Cambridge, U.K.

Devine N., Fechtig, H., Gombosi, T. I., Hanner, M. S., Keller, H. U., Larson, M. S., Mendis, D. A., Newburn, R. L., Reinhard, R., Sekanina, Z., and Yeomans, D. K. (1986). *Space Sci. Rev.* **43**, 1.

Huebner, W. F., ed. (1990). "Physics and Chemistry of Comets in the Space Age," Springer-Verlag, Berlin.

Krishna Swamy, K. S. (1997). "Physics of Comets," 2nd ed., World Scientific, Singapore.

Mendis, D. A., Houpis, H. L. F., and Marconi, M. L. (1985). "The physics of comets," *Fundam. Cosmic Phys.* **10**, 1–380.

Newburn, R. L., Neugebauer, M., and Rahe, J. (eds.). (1991). "Comets in the Post-Halley Era," Kluwer Academic Publishers, Dordrecht.

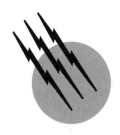

Communications Satellite Systems

Takashi Iida
Hiromitsu Wakana

Communications Research Laboratory

GLOSSARY

Asynchronous transfer mode (ATM) A network technology that efficiently provides multimedia services such as voice, video, and data by using a fixed–sized packet called a cell.

Code-division multiple access (CDMA) The resources of both frequency bandwidth and time are shared by all users using orthogonal codes. Undesired signals can be extracted by using small cross correlation.

Cross-polarization discrimination The power ratio of the copolarized wave to the cross-polarized wave converted by obstacles such as rain or ice particles.

Downlink The communication path from a satellite to an earth station.

EIRP Effective isotropically radiated power, which is the product of the transmitting power and the antenna gain.

Frequency-division multiple access (FDMA) A multiple access method to share satellite transponders simultaneously by different users using nonoverlapping frequency bands.

G/T The ratio of antenna gain to system noise temperature. It specifies the quality of receivers in satellites or earth stations.

Intermodulation Intermodulation products are produced by nonlinear amplifiers such as satellite transponders, earth stations' high-power amplifiers. Multiplication products among the input signals to the nonlinear amplifiers appear at their output and may cause interference to the output signals.

Regenerative transponder A transponder that can perform decoding, demodulation, modulation, and demodulation.

Satellite switching Redirection of the signal destination inside satellite routing devices.

Time-division multiple access (TDMA) A multiple access method to share satellite transponders by different users using nonoverlapping time slots.

Transponder Satellite onboard device that is used to receive the uplink signals and transmit the downlink signals. It consists of low-noise amplifiers, frequency converters, switches, and high-power amplifiers. A regenerative transponder includes modulator, demodulator, error-correcting encoder, and decoder.

Encyclopedia of Physical Science and Technology, Third Edition, Volume 3

Trellis-coded modulation (TCM) A modulation method that is combined with modulation and error-correction codes (convolutional coding) without degrading the power or bandwidth efficiencies.

Uplink The communication path from an earth station to a satellite.

I. COMMUNICATIONS SATELLITES

A. History and Current Status

It has been more than 40 years since the former Soviet Union launched Sputnik, the world's first artificial satellite, in 1957. The field of satellite communications then began with experiments using orbiting satellites, and the first Trans-Pacific television satellite-relay experiment occurred on November 23, 1963. Unfortunately, the images received on that day created quite a shock as they carried the news of the tragic death of John F. Kennedy, the American president, by assassination.

Satellite communications by geostationary satellites was first proposed by Arthur C. Clarke in 1945. He explained how man-made satellites in a circular stationary orbit 35,870 km above the equator could be used for global communications. Nineteen years later, in 1964, the National Aeronautics and Space Administration (NASA) in the United States launched the SYNCOM-III satellite (Fig. 1) into a geostationary orbit above the Pacific Ocean as the world's first geostationary satellite, which

was subsequently used to relay images from the Tokyo Olympics. For this purpose, the earth station on the Japan side was set up at the Kashima Space Research Center of the Radio Research Laboratory (the present CRL: Communications Research Laboratory) of the Ministry of Posts and Telecommunications (MPT). The success of SYNCOM-III led to the establishment of the International Telecommunications Satellite Organization (INTELSAT) in 1964, an international body overseeing satellite communications, and opened the way to the commercialization of space. Satellite communications is still the most successful field in space commerce. It is currently a $30 billion (about 3 trillion yen) per year enterprise and is growing rapidly in many areas. These include direct-to-home broadcast enterprises, mobile satellite services, fixed satellite services, and broadband Internet services.

1. Experimental Communications Satellites

The development of communications satellites got under way in the United States after the launch of the Vanguard artificial satellite. In particular, after the development of the ECHO-1 and ECHO-2 (passive satellites) starting in 1960 and that of the RELAY-1 and RELAY-2 satellites, NASA and the Department of Defense embarked on the SYNCOM program as a joint project to develop geostationary communications satellites.

After this, NASA began its Applications Technology Satellite (ATS) project with the aim of researching satellite and testing technologies, geostationary orbits, gravitational and attitude stabilization technologies, and satellite applications technologies. This project developed a total of six satellites. Furthermore, in the field of communications, the project undertook the development of earth-directed beam antennas, and mounted an electronic despun antenna and a mechanical despun antenna on the ATS-1 (Fig. 2) and ATS-3, respectively. For the final satellite of the

FIGURE 1 SYNCOM-III satellite. [Weight: 35 kg; diameter: 0.71 m; height: 0.39 m (cylinder section only).]

FIGURE 2 External view of ATS-1 satellite. [Weight: 352 kg (dry); height: 1.37 m; generated power: 175 W.]

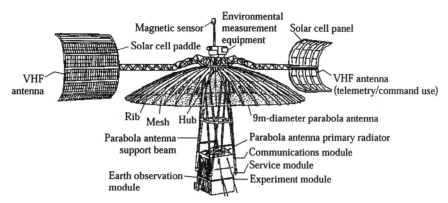

FIGURE 3 External view of ATS-6 satellite. [Weight: 1360 kg; expanded length: 16 m; generated power: 485 W.]

ATS series, the 3-axis-attitude-stabilization-type ATS-6 (Fig. 3) was developed to further improve antenna beam directionality. A joint CTS experiment was also performed with Canada in 1975 to test direct satellite broadcasting in the 12 GHz band.

The ATS would reappear in the form of the Advanced Communications Technology Satellite (ACTS) (Fig. 4) project with the aim of developing 30/20 GHz band satellite communications technology. With the launching of the ACTS satellite in 1993, a variety of experiments have come to be performed in relation to personal satellite communications, high-speed satellite communications, etc.

In Europe, space development began in 1967 with the joint development of the Symphonie communications satellite by France and the former West Germany. Satellite development has since centered about the ESA with the goal of satisfying a variety of needs associated with telephone traffic, television program switching and distri-

bution, and special services for various European countries, as well as to develop the European space industry in the world market. This work is connected with the design of future satellites such as the orbital test satellite (OTS), European region communications satellite (ECS), and large-scale multipurpose satellite (LSAT, later renamed "Olympus"). At the ESA, considerable effort is being spent on satellite development programs related to future technologies. The ARTEMIS satellite, for example, which will carry missions related to L-band mobile communications and optical intersatellite communications, is being launched. Independent development work is also being performed in France, Germany, and Italy.

Although the development of space communications technology in Japan began in the latter half of the 1960s with the acquisition of technology from the United States, Japan had earlier achieved remarkable success with its role in the satellite relay of the 1964 Tokyo Olympics. Japan's

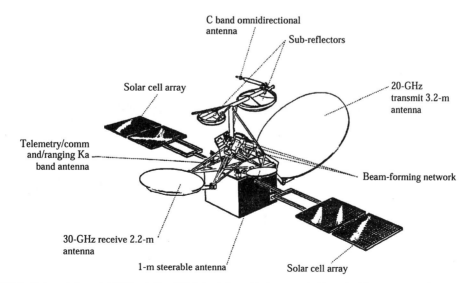

FIGURE 4 External view of ACTS satellite. [Weight: 1.4 ton (dry); expanded length: 14 m; generated power: approx. 1.8 kW.]

TABLE I   Transition of INTELSAT Satellites

Satellite	I	II	III	IV	V	VI	K	VII	VIII
Year of initial launch	1965	1967	1968	1971	1980	1986	1992	1993	1997
Size (m)	Diameter 0.72, height 0.59	Diameter 1.42 height 0.67	Diameter 1.42, height 1.04	Diameter 2.38, height 5.28	Total height 6.44, total length 15.59	Diameter 3.64, height 11.84	Total length 21.00	Total length 4.56 total length 21.84	Total length 3.44 total length 10.80
Weight at launch (kg)	68	162	293	1385	1944	4240	2500	3590	3600
Attitude stabilization system	Spin	Spin	Spin	Spin	3-axis	Spin	3-axis	3-axis	3-axis
Frequency band	C	C	C	C	C & Ku	C & Ku	Ku	C & Ku	C & Ku
Equivalent band width (MHz)	50	130	500	500	2144	3300	864	2432	2891
No. of transponders	2	1	2	12	27	48	16	46	44
No. of bidirectional + TV channels	240 or TV	240 or TV	1200 + TV	4000 + 2TV	12,000 + 2TV	24,000 + 3TV	32TV	18,000 + 3TV	22,500 + 3TV
Design life (years)	1.5	3	5	7	7	13	10	15	14–17

medium-capacity geostationary communications satellite (CS) for experimental purposes was launched in 1977. Using the CS, a variety of experiments were performed. The goal of this work was to develop the world's first satellite communications technology using 30/20 GHz bands (Ka band). In the field of mobile satellite communications, the Engineering Test Satellite ETS-V was launched in 1987. The purpose of these experiments was to establish basic technologies for mobile satellite communications and promote the use of mobile satellite services.

Experiments were also performed using a medium-scale broadcasting satellite for experimental purposes (BSE) launched in 1978. Their objective was to develop satellite-broadcasting technologies, and they greatly contributed to the growth of satellite broadcasting.

The Pan-Pacific Information Network Experiment was conducted by using the ETS-V satellite in cooperation with the University of Hawaii, a basic experiment for establishing a nonprofit communications network for tele-education in the Pacific Ocean region; following this experiment, the Pan-Pacific Regional Telecommunication Network Experiment and Research by Satellite (PARTNERS) project has been conducted by Japan, Thailand, Indonesia, Papua New Guinea, Fiji, Malaysia, and the Philippines.

2. Operational Satellite Activities

a. Fixed satellite services. As of December 1999, INTELSAT included 143 member nations and operated a total of 18 satellites over the Atlantic, Pacific, and In-

dian Oceans consisting of the INTELSAT V–VIII and K satellites. The change in INTELSAT satellite specifications over time is summarized in Table I. The table shows that the satellite becomes larger and larger and the cost per channel has been dramatically reduced. An external view of the INTELSAT VI satellite, which features a typical satellite configuration, is shown in Fig. 5, and an external view of the INTELSAT VII satellite is shown in Fig. 6.

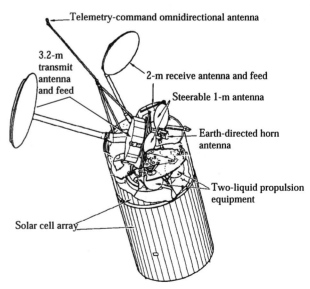

FIGURE 5 External view of INTELSAT-VI satellite. [Weight: 1910 kg (dry); total length: approx. 11.6 m; generated power: 2100 W.]

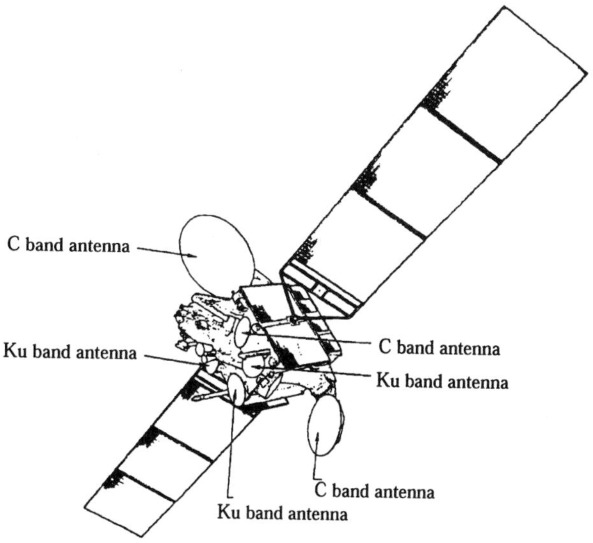
C band antenna
Ku band antenna
C band antenna
Ku band antenna
C band antenna
Ku band antenna

FIGURE 6 External view of INTELSAT-VII satellite. [Weight: 1454 kg (dry); expanded length: approx. 21.84 m; generated power: 3930 W.]

In response to INTELSAT, Intersputnik was founded in 1972 by East European countries led by the former Soviet Union. Intersputnik provides services to 24 member nations and various nonmember Western European countries.

In March 1983, the Orion Satellite Corporation submitted a business plan for an international satellite communications system separate from INTELSAT to the Federal Communications Commission (FCC) in the United States. Then, in November 1984, President Ronald Reagan officially approved a basic policy of allowing the deployment of non-INTELSAT satellite communications systems. This trend toward deregulation has accelerated the planning of non-INTELSAT systems. In addition, domestic satellite communications systems operated by various countries have begun to introduce international communications and are now competing for users by offering rates lower than those of INTELSAT. Systems of this kind include ORIONSAT, PanAmSat, COLUMBIASAT, RIMSAT, and COLUMBIASAT.

As for the domestic satellite communications, in 1972 the United States adopted an Open Sky Policy that allows communications operators to freely participate in the domestic satellite communications market. Western Union's WESTAR satellite consequently began operations in 1974, and at present, a number of satellite communications systems are in use. These systems are owned by three satellite communications operators: GE Americom, PanAmSat, and Loral Skynet. The total number of satellites provided by these systems is 41 (as of August 2000).

Turning to Canada, its domestic geostationary satellite communications system introduced in 1970 was the first of its kind in the world. Canada's domestic satellite communications are operated using the ANIK-E and F of C- and Ku band hybrid satellite. Mexico, meanwhile, operates the SATMEX satellite and the Solidaridad satellite that mounts mission devices for mobile communications; Brazil operates the BRASILSAT satellite; and Argentina the NAHUEL satellites.

The European Telecommunications Satellite Organization (EUTELSAT) became a permanent organization in 1985 and currently operates seven EUTELSAT, HotBird satellites targeting the entire European region as service area. As of August 2000, member nations totaled 50, including Russia and former East European countries. Domestic communications satellites in Europe are France's TELECOM, Germany's DFS-Copernikus, Luxembourg's ASTRA, Italy's ITALSAT, and Spain's HISPASAT.

Japan developed the CS-2 and CS-3 for operational use. Other Japanese satellites in operation include the JCSAT satellites and SUPERBIRD satellites. The N-STAR satellite, the successor to the CS-3, has been in operation since 1995.

More than 20 regional or domestic communications satellites are being operated in the Asia–Oceania region by a variety of countries, institutions, and enterprises. These satellites belong to various systems including Indonesia's PALAPA; India's INSAT; Australia's OPTUS; ARABSAT (Arab Satellite Communications Organization), a satellite communications system established in 1976 for the 22 member nations of the League of Arab States; China's CHINASAT, SINOSAT, and CHINASTAR; Korea's KOREASAT; Thailand's THAICOM; Malaysia's MEASAT; APSTAR, a joint venture among China, Hong Kong, and Thailand; and ASIASAT, based in Hong Kong. In addition, there are Singapore's ST, Philippines' AGILA, Turkey's TURKSAT, and Israel's AMOS.

b. Mobile satellite services. Typical of mobile satellite communications systems is the INMARSAT system, which was founded in 1979 with the aim of providing international maritime satellite communications services. Operations began in 1982. The INMARSAT system consisted of 87 member nations as of December 1999; it was privatized in April 1999.

The development of land mobile communications satellites for domestic use began in the 1980s in the United States and Canada. The MSAT satellite was launched in 1995. It provides mobile communications services for land vehicles, ships, and airplanes using the 1.6/1.5 GHz band (L band) for the North America region. In addition to these satellites, Australia's OPTUS-B satellite and Mexico's Solidaridad satellite have been launched. In Europe, a European Mobile System (EMS) was developed by ESA. The EMS package is to be mounted on the Italian

ITALSAT-2 communications satellite scheduled for launch in 2000. In Asia, GARUDA is opearated in the Southeast Asia area.

From about 1990 on, there have been many proposals in both the United States and Europe for low earth orbit (LEO)-based satellite communications systems to support mobile satellite communications systems using handsets. LEO systems have been classified by the FCC into "Big LEO" (large-scale LEO) and "Little LEO" (small-scale LEO). Big LEO systems can handle voice transmissions and employ 1 GHz and higher frequency bands. Little LEO systems, on the other hand can handle only data transmissions and employ frequency bands under 1 GHz.

The 66-satellite IRIDIUM system was proposed in 1990 as an international personal satellite-communications system to provide worldwide cellular telephone services. Unfortunately, IRIDIUM ceased its operation in April 2000, while another LEO satellite GLOBALSTAR began its operation in January 2000. These systems are to use the 1.6 GHz and 2.5 GHz frequency bands for user connection. The INMARSAT established ICO system is to use 10 to 12 satellites at an orbital altitude of 10,355 km using the 2.0/2.2 GHz band. Taking a look at Little LEO systems, the ORBCOMM satellite (Fig. 7) system began services in April 1995. In Europe, the Italian TEMISAT satellite performing store-and-forward communications has already been launched, and Germany has launched a similar satellite called SAFIR.

In addition to LEO systems, the former Soviet Union's MOLNIYA, having a highly elliptical orbit (HEO), is a domestic communications satellite that has been operating since 1965.

c. Broadcasting satellite services. The first genuine broadcasting satellite in the Americas was the DBS-1 launched in 1993, and DirecTV and ECHOSTAR have since been providing multichannel (200 channels) digital broadcasting services. In Canada, provisional DBS services began to be provided using the ANIK-C. The broadcasting services is now offered by NIMIQ satellite.

Direct broadcasting satellites in Europe is operated by EUTELSAT's HotBird, Sweden's SIRIUS, Norway's THOR, and Spain's HISPASAT-1 communications and broadcasting satellite. Around 42 million households in morethan 20 countries are receiving services provided by Luxembourg's ASTRA satellite. Russia, meanwhile, has been operating the EKRAN and GALS satellites.

In Japan, the BS-2 was launched in 1984 with the goal of eliminating receive impediment areas in broadcasting. This satellite became the world's first practical broadcasting satellite for individual reception by delivering real-time broadcasts to isolated islands over one channel. The second-generation BS-3 broadcasting satellite was launched in 1990, which can provide three channels of broadcast services. BSAT satellites are now operating.

Korea has launched its KOREASAT satellite for broadcasting, and India's INSAT, Australia's OPTUS, and the League of Arab States' ARABSTAT have been providing satellite broadcasting. There are also Egypt's NILESAT and Indonesia's CAKRAWARTA. ASIASAT's DTH services are spreading rapidly.

For digital audio broadcasting, World Space Corp's AFRISTAR and ASIASTAR are operated.

d. Intersatellite communications. One example of intersatellite communications is that performed by data-relay satellites that exchange data with observation satellites or other kinds of satellites orbiting the earth. The TDRS satellite (Fig. 8), developed in the United States, relays data from up to 20 satellites, which can be processed simultaneously. Europe's ARTEMIS satellite and Japan's DRTS project can be included in this category. Russia is also known to have data relay satellites, but details are scarce.

B. Regulatory Issues

1. Definition of Satellite Communications

Basic concepts required for an understanding of satellite communications systems are illustrated in Fig. 9. As shown, a satellite communications system is broadly divided into a space segment consisting of a space station (satellite) and a ground segment consisting of earth stations. Here, the technology for connecting the space segment with the ground segment includes communications-circuit design, communications schemes, and point-to-multipoint connection technology in the up- and downlinks.

Wireless communications targeting spacecraft are generally referred to as "space radio communications." On the

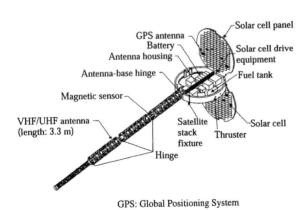

GPS: Global Positioning System

FIGURE 7 External view of ORBCOMM MicroStar satellite. [Weight: 38 kg; total length: 3.7 m; generated power: 230 W.]

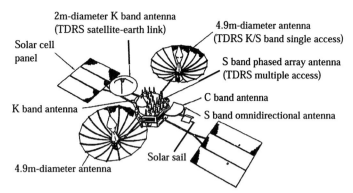

FIGURE 8 External view of TDRS satellite. [Weight: 2120 kg; expanded length: 17.42 m; generated power: 1740 W.]

other hand, communications specifically between radio stations on earth (earth stations) via a radio station set up on a spacecraft (space station) are called "satellite communications." Most satellite communications systems, moreover, currently use geostationary satellites. Satellite communications are classified according to Radio Regulations (RR) of the International Telecommunication Convention as "fixed satellite service," which provides communications between fixed stations, and "mobile satellite service," which provides communications between moving stations such as aircraft, ships, and automobiles. In addition, the communication format whereby parties on the ground directly receive radio transmissions from a space station is generally called "satellite broadcasting" and defined by RR as a "broadcasting satellite service."

We point out here that the meaning of the term "satellite communications" often includes both satellite communications and satellite broadcasting. We use "satellite communications" in this sense.

2. Frequency Usage in Satellite Communications

Frequencies used in satellite communications are allocated to various types of services by RR. Table II lists

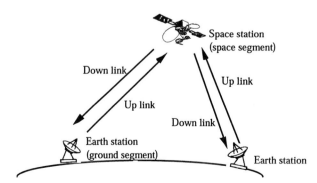

FIGURE 9 Conceptual diagram of satellite communications system.

typical frequency bands used by various services and the names given to these bands, and indicates which services employ which bands by a check mark (✔).

Most communications satellites providing fixed satellite services (such as INTELSAT satellites) use the 6/4 GHz band (where the frequency for the uplink from an earth station to the space station is given as the numerator and that for the downlink from the space station to an earth station is given as the denominator) called the "C band" and the 14/12 GHz band (Ku band). Japan's N-STAR satellite launched in 1995 uses the 30/20 GHz quasi-millimeter-wave band (Ka band). Also, the 50/40 GHz millimeter wave (Q/V band) has been highlighted recently.

With regard to mobile satellite services, satellites of the INMARSAT uses the 1.6/1.5 GHz band (L band). In Japan as well, the N-STAR satellite provides domestic mobile communications services using the S band (2.6/2.5 GHz). Finally, the 8/7 GHz band (X band) is mainly used by military satellites.

TABLE II Typical Frequency Bands Used by Various Services

Frequency Band	Name	Fixed satellite service	Broadcasting satellite service	Mobile satellite service
137 MHz	VHF			✔
400 MHz	VHF		✔	✔
1.6/1.5 GHz	L			✔
2.6/2.5 GHz	S		✔	✔
6/4 GHz	C	✔	✔	
8/7 GHz	X	✔		✔
14/12 GHz	Ku	✔	✔	
30/20 GHz	Ka or quasi-millimeter-wave	✔	✔	✔
50/40 GHz	Q, V or millimeter-wave	✔	✔	✔

Broadcasting satellite services are allocated frequencies in the 700 MHz, 2.6 GHz, 12 GHz, 22 GHz, and 42 GHz bands. Among these, the 700 MHz band and the 2.6 GHz band are used for community reception by the India's INSAT satellite, respectively. The 12 GHz band, moreover, is used for individual reception by Japan's BS-3 broadcasting satellite.

Frequency usage also concerns other elements of satellite communications, namely, tracking, telemetry, and commanding (TT&C) for satellite control and intersatellite links (ISL). The S band is normally used for TT&C. As for ISL, the S and Ku bands are used for communications between the space shuttle and the TDRS and for data relay among orbiting satellites.

Frequency is a limited resource. Therefore, to meet the increasing demand for communications and broadcasting throughout the world, devising ways to open up new frequency bands and to efficiently use previously allocated frequency bands are key technical issues in satellite communications.

3. Orbital Positions of Geostationary Satellites

The positions shown reflect the necessity of setting geostationary satellite intervals in the lattitude at or greater than a fixed value to prevent radio-wave interference that would normally occur if the interval between satellites that use the same frequency band was too close. This interference is caused in particular when radiation from one satellite falls within the antenna-beam width of an earth station belonging to another satellite. The plan for broadcasting satellites in the 12 GHz band has been to establish an interval of 6°. In the case of communications satellites, a prior announcement must be made to member nations of the International Telecommunication Convention through the Radio Regulations Board (RRB) of the ITU. In this regard, necessary adjustments will have to be made with the countries concerned with respect to frequency usage, geostationary orbital position, earth station specifications, etc. At present, C band satellites are basically allocated to an orbital position interval of 4° and Ku band satellites to one of 3°, although both of these are allocated to an interval of 2° in the United States.

4. Characteristics of Satellite Communications

Here, we will discuss general characteristics of satellite communications. We first present advantages of these forms of communication.

1. Wide-area coverage can be easily obtained.
2. Communication circuits robust against natural disasters can be established.

TABLE III Comparison between Satellite Communications and Optical Fiber

Service	Satellite	Optical fiber
Trunk line	✔	✔✔
Broadcasting	✔✔	✔
Mobile communication	✔✔	—
Network	✔	✔

3. Communication services such as multidestination delivery and broadcasting can be easily provided with wide bandwidth.
4. A communication circuit can be easily set up with any point on the ground.

On the other hand, satellite communications using geostationary orbits have the following disadvantages: (1) The round-trip time for a radio wave to travel between a satellite and the earth is about one-fourth of a second, and (2) the propagation distance of a radio wave is long, resulting in large propagation loss and generally low receiving levels.

These problems, however, will be overcome to some extent by advances in technology. We can also make a comparison between communications by satellite and that by optical fiber, its main rival. As shown in Table III, satellites can accommodate mobile communications, whereas optical fiber cannot, and satellites can also be used to construct a network over a wide area provided that earth stations can be set up. All in all, this means that it would be difficult to replace satellite communications with optical fiber.

C. Satellite Orbits

A satellite normally flies at altitudes of 300 km and higher at which atmospheric effects can be ignored. An orbit, moreover, generally takes on an elliptical shape, as shown in Fig. 10. There are six parameters that determine such an orbit, the main ones being the semimajor axis, inclination that indicates how far the orbit is tilted from the equatorial plane, and eccentricity that indicates just how elliptical the ellipse is. In addition, the points at which the altitude of the satellite with respect to the earth's surface are highest and lowest are called apogee and perigee, respectively.

At an altitude of 1000 km, the speed at which a satellite must travel so as not to fall to earth is about 8 km/s. Considering, however, that the gravitation of the earth is inversely proportional to the second power of distance, a satellite will still not fall at slower speeds provided that orbit altitude is higher. For example, satellite speed need be only about 3 km/s to balance gravitation if an altitude of about 36,000 km can be reached. At this speed, the time it

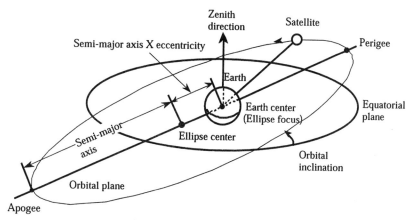

FIGURE 10 Basic parameters of a satellite orbit.

takes for the satellite to complete one orbit (period) about the earth is exactly 24 hours. This means that if the orbit of the satellite lies on the equatorial plane, the satellite s angular velocity will be the same as that of the earth, and the satellite will appear to be situated at the same position when seen from earth. This is called a geostationary orbit, and a satellite placed in such an orbit is called a geostationary satellite. The geostationary orbit is presently often referred to as simply GEO.

The distance from the ground to a satellite in a GEO is about 36,000 km, and at this altitude, the round-trip propagation time of a radio wave is about one-fourth of a second. The attenuation of signal strength due to wave propagation, moreover, is proportional to the square of the distance. It is therefore desirable that distance to the satellite be as short as possible in mobile satellite communications, especially in the case of portable terminals. An orbit at a lower altitude would therefore be more effective in this regard, and an orbit of altitude from 500 to several thousand kilometers (with most being under 1500 km) (Fig. 11) is referred to as a low earth orbit, or LEO for

short. This latitude limit of about 1500 km is designed to avoid the radiations of the lowest Van Allen belt.

The time during which a satellite placed into a LEO is visible from a point on the earth is limited. For example, at an altitude of 1000 km, the period of a LEO is about 1 hour and 45 minutes, making the satellite visible for about 12 minutes (given that the satellite's angle of elevation is at least $10°$). Accordingly, to achieve continuous communications service in this case, about 40 to 70 satellites would be needed and circuit switching would have to be performed from one satellite to another. There are also systems, moreover, that use altitudes from several 1000 to 20,000 km, slightly higher than a LEO altitude. This kind of an orbit is called a medium earth orbit (MEO). The period of a MEO at an altitude of about 10,000 km, for example, is about 5 or 6 hours, which means that only about 10 satellites would be needed to achieve continuous communications service. The preceding kinds of orbits for which satellites do not appear stationary are generally referred to as nongeostationary orbits. Again, the orbits of the MEO satellites are designed to avoid the

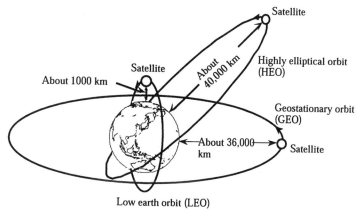

FIGURE 11 GEO, HEO, and LEO orbits.

most harmful radiation effects of the higher Van Allen belts.

GEO, MEO, and LEO are normally circular orbits with an eccentricity of nearly zero. There are also times, however, when highly elliptical orbits (HEO) are purposely used. The angle of elevation at which a geostationary satellite is seen from earth becomes lower as altitude becomes higher. Considering that communications can be disrupted when the paths of radio waves are blocked by buildings and other structures, especially in mobile communications, it is desirable that the angle of elevation be as high as possible. With this in mind, satellites of the former Soviet Union adopted orbits of this type, called "Molniya orbits," that have proved to be of practical use. A Molniya orbit has an eccentricity near 0.5, an altitude of about 40,000 km at apogee, orbital inclination near 63°, and a period of 12 hours. For the HEO, at least 2 or 3 satellites would have to be launched to provide continuous communications service, and circuit switching would have to be performed.

D. Configuration and Function of Satellites

A communications satellite is an independent system floating in space. It provides its own electric power supply, maintains its attitude, withstands the harsh environment of space, and sees that mission devices operate normally within the required mission life. The design of a satellite consists of conceptual design, preliminary design, and critical design based on a satellite communications systems plan, plus performance-requirements design and the construction of various manufacturing models (BBM, breadboard model; EM, engineering model; PFM, proto flight model; and FM, flight model). The satellite must then pass a thermal-vacuum test that simulates a space environment in a space chamber as well as vibration and other tests before being loaded onto a rocket and launched.

A communications satellite consists of basic equipment supporting a mission (called satellite bus as a whole): the attitude control subsystem, a power supply subsystem, TT&C subsystem, propulsion subsystem, the thermal control subsystem, and structure subsystem. The mission equipment is called a communication subsystem, including an antenna and transponder. Table IV shows the functions of subsystems.

Among subsystems, it is the attitude control subsystem that has the biggest influence on both configuration and performance of the communications satellite. It is necessary to maintain satellite attitude in order to always point an antenna of the satellite at the earth to communicate with the earth stations, and to point a solar array in the direction of the sun. A satellite in space overcomes radiation pressure by the sun and the influence of disturbances generated by the interaction torque of geomagnetism with the residual magnetism of the satellite in order to make the attitude stable.

Attitude control systems for maintaining appropriate satellite orientation in space are broadly divided into the spin-stabilization type and the three-axis-stabilization type. An example of the former is the INTELSAT-VI

TABLE IV Function and Typical Equipment of Subsystems of Communication Satellite

Subsystem	Function	Typical equipment
Communication	Receiving, amplification, processing and retransmitting of the signal	Transponders and antennas
Attitude control	Obtaining the information for attitude and orbit and maintaining the satellite attitude to the desired direction with certain precision as well as pointing the antenna beam at the earth	Attitude sensors (earth sensor, sun sensor, etc.), attitude control electronics, despin control electronics (spin stabilized) and wheel (3-axis stabilized)
Power supply	Supplying necessary electric power to the equipment during the mission period; generating power by battery during eclipse; controlling surplus power; battery charge, discharge, and reconditioningcontrol during eclipse	Solar battery panel, power supply control, power control, and battery
TT&C (tracking, telemetry, and commanding)	Receiving, demodulating, and distributing commands for house- and station-keeping and collecting, modulating, and transmitting telemetry and repeating ranging signals	Equipment and antenna for TT&C
Propulsion	Firing apogee motor to place on the drift orbit from transfer orbit and generating the necessary thrust for station-keeping and attitude control	Apogee motor, thrusters, propellant tanks and fuel plumbing
Thermal control	Maintaining the temperature of each subsystem within the requirement during all mission periods	Thermal insulation blanket, thermal coating, heater, and heat pipes
Structure	Supporting all subsystems against severely mechanical and thermal condition in the launching and in space and maintaining, deploying, and rotating the solar panel	Basic structure, solar battery, deployment mechanism, and rotating mechanism

shown in Fig. 5, and an example of the latter is the INTELSAT-VII shown in Fig. 6.

E. Future Prospects

President Clinton first endorsed the National Information Infrastructure (NII) for the United States in September 1993. This was followed by Vice President Gore's speech on the Global Information Infrastructure (GII) in March 1994. Also in the United States, the success of NASA's ACTS satellite has raised awareness that an information highway can be easily achieved through the use of satellites, and many proposals are thus being made for Ka band geostationary satellite system projects. A proposal based on the demand for Internet-type data communications using LEO satellites has been made for a system called TELEDESIC using the huge number of 840 LEO satellites in its initial proposal.

In Japan, the active role played by satellite communications in the aftermath of the Great Hanshin Earthquake on January 17, 1995, demonstrated once again the usefulness of this form of communications. The present consensus is that satellite communications systems should be established in harmony with terrestrial systems such as optical fiber as a foundation for the information and communications society of the 21st century. To this end, efforts are being made to initiate development of experimental satellites having transmission speeds at the gigabit level.

It is also being said that the Asia–Pacific region will play a leading role in the 21st century. Growth in this region has been remarkable, and the demand for communications has been growing faster there than in any other region in the world. The countries making up the Asia–Pacific region, however, are dispersed over an exceptionally large area, and it is no exaggeration to say that only terrestrial systems and facilities like optical fiber would not be sufficient to satisfy communications needs. This region must therefore be given particular attention as an area holding much promise for satellite communications.

The world market for satellite communications over the next 10 years is estimated to be about US $10–50 billion per year. It is therefore predicted that about 200 geostationary satellites including present projects will need to be launched during this time throughout the world. In addition, the many proposals for satellite communications systems that use nongeostationary orbits suggest that more than 1000 satellite launchings in total could take place during this period. Needless to say, satellite communications is still a growing field.

It is not certain how satellite systems and terrestrial systems will be used in the future. However, "the Pelton Merge" will probably provide some useful insight. This concept predicts that all media will be "seamlessly" merged through the use of universal digital standards to transmit information.

II. SATELLITE COMMUNICATIONS SYSTEMS

A. Communication System Design

1. Satellite Link Calculation

This section presents a calculation method for a link budget to design a satellite communication system. Figure 12 shows a schematic diagram of a typical satellite link. The link from the earth station to the satellite is called the uplink, and the link from the satellite to the earth station is called the downlink. The performance of the digital satellite communication link is specified in terms of a bit error rate (BER), which is a function of the ratio of carrier power to noise density (C/N_0) or the ratio of energy-per-bit to noise density (E_b/N_0). Main contributors in the noise density N_0 are both thermal noise in a satellite's receiver in uplink, and thermal noise in an earth station's receiver in downlink. Interference from other telecommunication systems may be assumed as equivalent to thermal noise.

This satellite link, therefore, can be modeled as a cascade circuit, as shown in the bottom of Fig. 12. The C_1 represents the signal power at the input of the satellite receiver, and the receiver noise N_1 is added to C_1. All noisy amplifiers that are included in the satellite transponder can be replaced by one noiseless amplifier and a single noise source. Then, the C/N_0 in uplink, which is referred to as $(C/N_0)_{Uplink}$, is represented by C_1/N_1. Here, G_1 is the gain of the noiseless amplifier, G_T^s is the gain of the satellite transmit antenna, L_D is the propagation loss, and G_R^e is the gain of the earth station antenna. In the earth station, the signal is received at a signal power of C_2 at the input of the receiver, and the receiver noise N_2 is added to C_2. Again, all noisy amplifiers in the earth station are replaced by a noiseless amplifier with an equivalent gain of G_2 and a noise source. The C_2/N_2 is C/N_0 in downlink, which is referred to as $(C/N_0)_{Downlink}$, and the C_3/N_3 is C/N_0 in the total satellite link, $(C/N_0)_{Total}$. Here, C_3 and N_3 are the signal power and noise density at the output of the receiver, respectively. They can be written as

$$C_3 = C_1 G_D G_2 \tag{1}$$

$$N_3 = N_1 G_D G_2 + N_2 G_2 \tag{2}$$

where $G_D = G_1 G_T^s L_D^{-1} G_R^e$.

Therefore,

$$\frac{N_3}{C_3} = \frac{N_1}{C_1} + \frac{N_2}{G_1 C_1} = \frac{N_1}{C_1} + \frac{N_2}{C_2} \tag{3}$$

This can be rewritten as

$$\left(\frac{C}{N_0}\right)_{Total}^{-1} = \left(\frac{C}{N_0}\right)_{Uplink}^{-1} + \left(\frac{C}{N_0}\right)_{Downlink}^{-1} \tag{4}$$

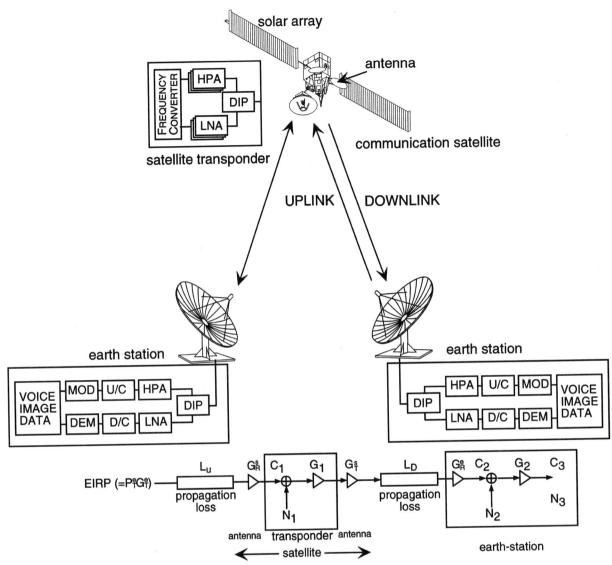

FIGURE 12 A satellite communication link.

These (C/N_0) values are represented in real values. This equation can be rewritten in unit of decibel hertz (dBHz) as

$$\left(\frac{C}{N_0}\right)_{\text{Total}} = -10\log_{10}\left[10^{-(C/N_0)_{\text{Uplink}}} + 10^{-(C/N_0)_{\text{Downlink}}}\right]$$

$$\left(\frac{C}{N_0}\right)_{\text{Uplink}} = -10\log_{10}\left[10^{-(C/N_0)_{\text{Total}}} - 10^{-(C/N_0)_{\text{Downlink}}}\right]$$

$$\left(\frac{C}{N_0}\right)_{\text{Downlink}} = -10\log_{10}\left[10^{-(C/N_0)_{\text{Total}}} - 10^{-(C/N_0)_{\text{Uplink}}}\right]$$
$$(5)$$

This is a basic equation representing the relationship between (C/N_0) values in uplink and downlink. This equation indicates that even when one of the two is much larger than

the other, the total link quality of $(C/N_0)_{\text{Total}}$ is less than that of the smaller one.

2. Power Calculation

When an antenna with a gain of G_T^e transmits a radio wave at an output power of P_T^e, the flux density in the main-beam direction of the antenna at a distance of R is given by

$$P = \frac{P_T^e G_T^e}{4\pi R^2} \tag{6}$$

The product of the transmitting power and the antenna gain is called the effective isotropically radiated power (EIRP). The signal power that is received by an antenna with an aperture area of A is

$$P_R^s = \frac{P_T^e G_T^e}{4\pi R^2} A\eta \tag{7}$$

where η is the aperture efficiency of the receiving antenna. For reflector antennas, η is typically in the range of 0.5 to 0.75, lower for small antennas and higher for large Cassegrain antennas.

Antenna gain is given by

$$G_R^s = \frac{4\pi}{\lambda^2} A\eta \tag{8}$$

where λ is the wavelength. Then, Eq. (7) can be rewritten by using Eq. (8) as follows:

$$P_R^s = \left(\frac{\lambda}{4\pi R^2}\right)^2 P_T^e G_T^e G_R^s \tag{9}$$

When the free-space path loss is defined in decibels by

$$L_u = 20\log_{10}\left(\frac{4\pi R}{\lambda}\right) \tag{10}$$

the receiving power is written in decibels by

$$P_R^s(\text{dBW}) = P_T^e(\text{dBW}) + G_T^e(\text{dB}_i) - L_u(\text{dB}) + G_R^s(\text{dB}_i) \tag{11}$$

Actual link calculation includes other effects such as attenuation in atmosphere, rain attenuation, antenna tracking errors, losses associated with transmitting and receiving antennas, etc. Figure 13 shows the free-space path loss at different satellite orbits and frequencies. At typical altitudes of 1000 km, 10,000 km, and 35,787 km for LEO, MEO, and GEO satellites, respectively, the path losses of MEO and GEO are 20.0 dB and 31.1 dB larger than that of LEO. In terms of power efficiency, LEO has definite advantages, resulting in smaller aperture and weaker transmitting powers of user terminals.

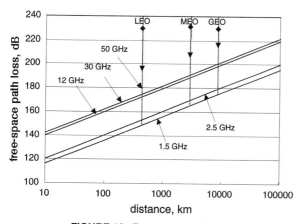

FIGURE 13 Free-space path loss.

3. Noise Temperature

The term "*noise temperature*" is generally used to present how much thermal noise is generated from active or passive devices in the receiving system. The noise power P_n is written as noise temperature as follows:

$$P_n = kT_n B \tag{12}$$

where

$k = $ Bolzmann's constant $= 1.38 \times 10^{-23}$ J/K
$\quad\;\; = -228.6$ dBW/K/Hz

$T_n = $ noise temperature (K)

$B = $ bandwidth (Hz)

The noise power density N_0 can be given, in real values and decibel values, by

$$N_0 = kT_n \;(\text{watts/Hz})$$
$$N_0 = -228.6 + 10\log T_n \;(\text{dBW/Hz}) \tag{13}$$

Figure 14 shows a typical receiving system of a satellite or an earth station, which has a cascade connection of lossy circuits and amplifiers. Let the loss in lossy circuit #n be denoted as L_n(or Ln), and the antenna noise temperature T_{ANT}, the LNA noise temperature as T_{LNA}, the noise temperature of the downconverter as $T_{D/C}$, and the noise temperature of the IF amplifier as T_{IF}. A lossy circuit not only attenuates the signal power, but also generates thermal noise. The equivalent noise temperature at the output of a lossy circuit is given by

$$T_{Loss} = T_0\left(1 - \frac{1}{L}\right) \tag{14}$$

where T_0 is the physical temperature of the lossy circuit, and L is an attenuation in real value larger than unity. The equivalent noise temperature of the whole receiving system, which is called the system noise temperature T_{System} and is defined at the input of LNA, is calculated as follows:

$$T_{System} = \frac{T_{ANT}}{L_1} + T_{L1} + T_{LNA} + \frac{(T_{D/C} + T_{L2})L_2}{G_{LNA}} + \frac{(T_{IF} + T_{L3})L_2 L_3}{G_{LNA}G_{D/C}} \tag{15}$$

where G_{LNA} and $G_{D/C}$ are gains of LNA and D/C amplifiers, respectively, and T_{Ln} is a noise temperature generated by a lossy circuit L_n. If the LNA gain is relatively large, all the terms beyond the third of Eq. (15) can be neglected. The noise temperature of the first-stage (front-end) amplifier and lossy circuit is the main contributor to the overall performance of the receiving system. To improve the performance of the receiving system, therefore, we have to keep the noise temperature of the front-end amplifier of

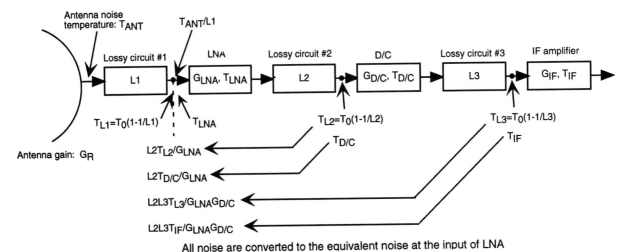

All noise are converted to the equivalent noise at the input of LNA

FIGURE 14 Noise temperature of a receiving system.

the receiver and the loss between this amplifier and the antenna as low as possible. When GaAs FET amplifiers are employed, noise temperatures from 70 K to 200 K can be achieved without physical cooling.

Noise figure (NF) is generally used to specify the noise generated from an electric device. The relation between NF and noise temperature T is given by

$$T = T_0(NF - 1) \qquad (16)$$

where T_0 is the reference temperature, usually 290 K.

4. G/T

If Eq. (9) is used to represent the received signal power, the ratio of the signal power to the noise density (C/N_0) is written as

$$\frac{C}{N_0} = \frac{P_T G_T G_S}{k T_{\text{System}}} \left[\frac{\lambda}{4\pi R} \right]^2 = \frac{P_T G_T}{k} \left[\frac{\lambda}{4\pi R} \right]^2 \frac{G_S}{T_{\text{System}}} \qquad (17)$$

where G_S is the antenna gain of the receiving terminal, and P_T and G_T are the transmitting power and antenna gain of the transmitting terminal, respectively. Here, G_S/T_{System} is a ratio that can be used to specify the quality of the receiving terminal, and it is called the figure of merit. Strictly speaking, since T_{System} is the equivalent noise temperature at the input port of LNA, G_S should be the antenna gain defined at the input port of LNA. Assuming that G_R is an actual antenna gain, $G_S = G_R/L_1$ and Eq. (15) gives

$$\frac{G_S}{T_{\text{System}}} = \frac{\dfrac{G_R}{L_1}}{\dfrac{T_{\text{ANT}}}{L_1} + T_0\left(1 - \dfrac{1}{L_1}\right) + T_{\text{LNA}}}$$

$$= \frac{G_R}{T_{\text{ANT}} + T_0(L_1 - 1) + L_1 T_{\text{LNA}}} \qquad (18)$$

5. Intermodulation Products and Interference

When a satellite transponder is operated in the nonlinear region to amplify simultaneously several input signals at different frequencies, multiplication products among input signals will cause interference with the output signals. The undesirable signals are called intermodulation products. Cochannel interference can also arise from frequency reuse operation by using orthogonal polarization or spatial isolations. When a neighboring satellite uses the same system parameters such as frequency, polarization, or service area, adjacent-satellite interference will also occur. Although these interference signals are not thermal noise, they are usually evaluated by using the carrier-to-interference densities in the link calculation as follows:

$$\left(\frac{C}{N_0}\right)^{-1}_{\text{Total}} = \left(\frac{C}{N_0}\right)^{-1}_{\text{Uplink}} + \left(\frac{C}{N_0}\right)^{-1}_{\text{Downlink}}$$

$$+ \left(\frac{C}{N_0}\right)^{-1}_{\text{IM}} + \left(\frac{C}{N_0}\right)^{-1}_{\text{I}} \qquad (19)$$

where $(C/N_0)_{\text{IM}}$ is the ratio of the signal power to the power density of intermodulation products that fall into the transmission bandwidth of this signal. The effects of the intermodulation products are regarded as equivalent thermal noise. Similarly, $(C/N_0)_{\text{I}}$ is the ratio of the signal power to the power density of interference.

6. Example of Link Budget

Table V shows an example of a link budget. It is assumed that a Ka-band geostationary satellite is located at 132°E, and the earth station is in Tokyo (35.7°N and 139.7°E). The distance between the satellite and the earth station is 37,225.5 km. The path loss is calculated from Eq. (10),

TABLE V   Link Budget

Uplink		Downlink	
Earth station		Satellite	
Frequency	29.0 GHz	Frequency	19.0 GHz
Transmit power	10.0 dBW	Transponer gain	120.0 dB
Feeder loss	0.3 dBW	Transmit power	4.1 dBW
Antenna diameter	5.0 m	Feeder loss	3.2 dB
Antenna gain	61.4 dBi	Antenna gain	27.0 dBi
Pointing loss	0.3 dB		
EIRP	70.8 dBW	EIRP	27.9 dBW
Path loss	213.1 dB	Path loss	209.4 dB
Absorption loss	1.5 dB	Absorption loss	0.8 dB
Satellite		Earth station	
		Antenna diamer	5.0 m
Antenna gain	30.0 dBi	Antenna gain	57.7 dBi
		Pointing loss	0.3 dB
Feeder loss	2.1 dB	Feeder loss	0.2 dB
Receive power	−115.9 dBW	Receive power	−125.1 dBW
T_{system}	917.1 K	T_{System}	248.9 K
T_{ANT}	290 K	T_{ANT}	69.3 K
T_{LNA}	627.1 K	T_{LNA}	169.6 K
NF	5 dB	NF	2.0 dB
G/T	−1.7 dB/K	G/T	33.6 dB/K
N_0	−199.0 dB/HZ	N_0	−204.6 dB/Hz
$(C/N_0)_{Uplink}$	83.1 dBHz	$(C/N_0)_{Down}$	79.5 dBHz
$(C/N_0)_{Total}$	78.0 dBHz		
$(C/N_0)_{Req}$	71.0 dBHz		
Bit rate	2.0 Mbps		
$(E_b/N_0)_{Req}$	11.3 dB	BER $= 10^{-/}$	
Coding gain	5.8 dB	Convolutional FEC with $R = 1/2$ and $K = 7$	
Implementation loss	2.5 dB		
Link margin	6.9 dB		

and the antenna gain is calculated from Eq. (8) with an aperture efficiency of 0.6. G/T can be also obtained from Eq. (18). The required (C/N_0) for obtaining a bit error probability of 10^{-7} is given by

$$\left(\frac{C}{N_0}\right)_{Req} = \left(\frac{E_b}{N_0}\right) - G_{code} + R + M_i \text{ (dBHz)} \quad (20)$$

where G_{code} is the coding gain of the forward error collection (FEC) in dB, R is the information rate in bit/s, and M_i is the implementation loss in dB. Here $(E_b/N_0) = 11.3$ dB at a BER of 10^{-7} for QPSK, and $G_{code} = 5.8$ dB for convolutional coding with a coding rate of $1/2$, a constraint length of 7, and Viterbi decoding (see Table VI). An information rate $R = 2$ Mbps and an implementation loss $M_i = 2.5$ dB are assumed.

B. Communications Payload Hardware

The main functions of the communication payload of a communication satellite are as follows:

(a) Receive signals that are transmitted from earth stations located within service areas.

(b) Amplify these weak-power signals with lowest noise and signal distortion.

(c) Translate the frequencies of the uplink signals to their downlink frequencies.

(d) Achieve high-power amplification and transmit the signals to destination earth stations in a given downlink frequency, polarization, and transmitting power.

(e) Using a regenerative transponder, demodulate received signals and transmit them to earth stations after remodulation as well as decoding and encoding of FEC. In multibeam satellites, an onboard switching performs interconnection between the beams.

This section describes the communications payload hardware, which consists of an antenna, transmit equipment, receive equipment, and a satellite transponder (or repeater).

TABLE VI Beamwidth, Coverage, and Antenna Diameter

Beamwidth	Earth coverage diameter (km)	Antenna diameter (cm)			
		1.5 GHz	12 GHz	30 GHz	50 GHz
17°	10,697	76	10	4	2
10°	6262	130	16	6	4
5°	3125	260	32	13	8
2.8°	1749	464	58	23	14
1°	625	1299	162	65	39
0.5°	312	2598	325	130	78

1. Antennas

Satellite antennas concentrate the satellite's transmitting power into a designated geographical region on earth and avoid interference from undesired signals transmitted from outside of the service area. The antenna radiation characteristic is usually specified by its half-power beamwidth (HPBW). For an aperture antenna, HPBW is approximately given by

$$\theta_{\mathrm{HPBW}} = 65° \frac{\lambda}{D} \qquad (21)$$

where D is the antenna diameter and λ is the wavelength. The relationship among the HPBW, the earth coverage diameter, and the antenna diameter at different frequencies is shown in Table VI. Global beam coverage from geostationary satellites needs a field of view of 17.4°. Since reflector antennas are not efficient when the diameter is less than 8λ, horn antennas or wire antennas are used for a wide-beam antenna whose beam angle is larger than about 8°. On the other hand, the larger reflector antennas are needed to produce spot-beam coverage in the lower frequencies.

Four main types of antennas are used on present communication satellites: wire antennas, horn antennas, reflector antennas, and array antennas. Wire antennas were used on the early operational satellites such as INTELSAT I and II with an antenna gain of about 4 dBi for receive and about 9 dBi for transmit. Now they are mostly used as the tracking, telemetry, and command (TT&C) antennas because of their wide beamwidth.

Horn antennas are one of the simplest directional antennas. Depending on their shape, there are several types of horn antennas: pyramidal, sectoral, conical, dual-mode, corrugated, multiflare, etc. (Fig. 15). The pyramidal horn and conical horn are simple and easy to fabricate, but they have the drawback that the radiation pattern is not circularly symmetrical and so antenna efficiency is low. They are also used as array elements of array antennas because of their small sizes. The dual-mode horn can produce a circularly symmetric pattern and lower sidelobes by adding TE_{11} and TM_{11} modes, but the disadvantage is narrow bandwidth. The corrugated horn generates a cir-

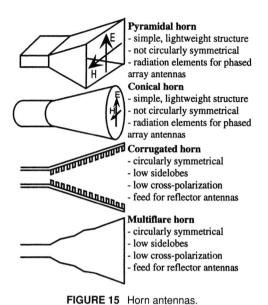

Pyramidal horn
- simple, lightweight structure
- not circularly symmetrical
- radiation elements for phased array antennas

Conical horn
- simple, lightweight structure
- not circularly symmetrical
- radiation elements for phased array antennas

Corrugated horn
- circularly symmetrical
- low sidelobes
- low cross-polarization
- feed for reflector antennas

Multiflare horn
- circularly symmetrical
- low sidelobes
- low cross-polarization
- feed for reflector antennas

FIGURE 15 Horn antennas.

cularly symmetric pattern, low sidelobes, and low cross-polarization by using the hybrid electric HE_{11} mode. These dual-mode, corrugated, and multiflare horns are also used as feeds for reflector antennas.

The reflector antenna is most frequently used in communications satellites because of its simple structure, light weight, and high gain (Fig. 16). It has one or more reflective surfaces, which are paraboloid, hyperboloid, spheroid, etc. A parabolic antenna consists of a single reflector shaped as an axially symmetric paraboloid of revolution and the feed situated at the reflector's focus. The feed may block some of arriving waves and cause antenna gain to

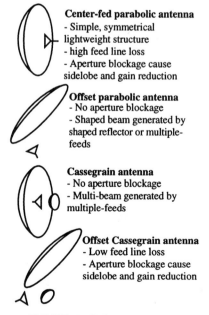

Center-fed parabolic antenna
- Simple, symmetrical lightweight structure
- high feed line loss
- Aperture blockage cause sidelobe and gain reduction

Offset parabolic antenna
- No aperture blockage
- Shaped beam generated by shaped reflector or multiple-feeds

Cassegrain antenna
- No aperture blockage
- Multi-beam generated by multiple-feeds

Offset Cassegrain antenna
- Low feed line loss
- Aperture blockage cause sidelobe and gain reduction

FIGURE 16 Reflector antennas.

drop and sidelobes to increase. The reflector of an offset-parabolic antenna, therefore, has offset to avoid the aperture blockage. A Cassegrain antenna uses a paraboloid of revolution as a main reflector and a hyperboloid of revolution as a subreflector. One focal point of the subreflector coincides with the focal point of the main reflector, while the other focal point of the subreflector coincides with the feed. Multibeam using reflector antennas can be achieved by multiple feeds situated near the focal position. An increase in the number of feeds, however, causes degradation of the radiation performance because offset of multiple feeds from the focal position increases. This type of multibeam antennas is used in offset parabolic and offset Cassegrain antennas.

An array antenna can generate steerable beams and particular radiating patterns with high directivity and low sidelobes by using a large number of radiating elements. The design of an array involves the selection of radiating elements and array geometry and the determination of the element excitations required for achieving a particular performance, which is not possible with a single radiating element. As the radiation elements in a satellite phased array antenna, horn, dipole, helix, and microstrip patch antennas are mostly used. Figure 17 shows an S-band phased array antenna, which was installed in the ETS-VI satellite to establish intersatellite links between the geostationary satellite and low-earth-orbiting satellites. The array consists of 19 radiating elements, each of which is equipped with one phase shifter for transmit beams and two phase shifters for receive beams so that each of these beams can be electrically scanned independently. This antenna is 1.8×1.8 m in size. Figure 18 shows a Ka-band active phased array antenna (APAA) of the Japanese Gigabit Satellite. Two antennas with an aperture diameter of 2.2 m for transmit and 1.5 m for receive are installed on the same surface of the satellite's earth panel without deployable structure. The APAA consists of 38 subarray units, each of which consists of 64 horn-antenna elements with a mutual spacing of 2.2λ. Figure 18 shows this subarray unit. Four beams with a scanning angle of $\pm8°$ are available.

To allow larger aperture and larger gain of satellite antennas, a deployable antenna, which uses meshed or solid reflectors folded for launch and deployed on orbit, is needed. There are two types of deployable antennas: the wrap-rib type and the umbrella type. The wrap-rib antenna consists of flat ribs that can spread out into radial directions, a center hub, and a reflective film (mesh) that extends across the ribs. The umbrella-type deployable antenna also consists of multiple ribs and a center hub, but it can achieve expansion by having the ribs open up from their base like an umbrella. Figure 19 shows multiple umbrella-type reflectors installed in the ETS-VIII satellite. The unfolded reflector size is 19.2 m × 16.7 m.

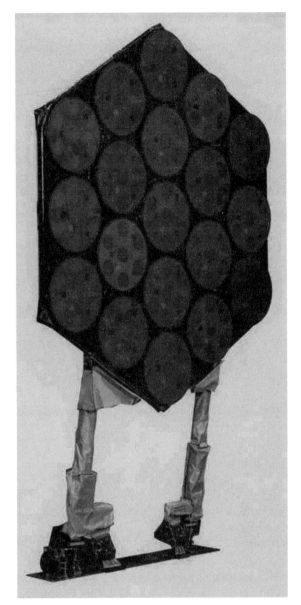

FIGURE 17 S-band phased array antenna of the ETS-VI satellite.

The reflector is assembled with 14 umbrella-type modules, each of which is deployed by using stepping motors. To achieve a multibeam phased-array antenna, the feed consists of 31 elements. The feed array is located about 1 m away from the focal point. Two separate reflectors for transmit and receive in S band are installed in the satellite to avoid signal coupling and interference by passive intermodulation.

2. Traveling-Wave Tube (TWT) Amplifier and Solid-State Amplifier

Figure 20 shows a typical block diagram of a satellite transponder. The left side shows the receiving equipment,

FIGURE 18 Ka-band phased array antenna of the Gigabit satellite.

which consists of a receiving antenna, low-noise amplifiers (LNA), and frequency converters. The frequency converter translates the radio frequency (RF) of the received signals into the intermediate frequency (IF), so this converter is called a downconverter. The right side shows the transmitting equipment, which consists of a transmitting antenna, high-power amplifiers (HPAs), and frequency converters (upconverters). When a single frequency converter is used, instead of a downconverter and upconverter, the converter translates the uplink frequency to the downlink frequency directly. Such a conversion is called single conversion; the former is called double conversion.

The HPA mostly used in satellite transponders is a TWTA or solid-state power amplifier (SSPA) such as a gallium-arsenide field-effect transistor (GaAs FET) amplifier. Since the lifetime of TWTAs is limited, and they are the least reliable component in most transponders, a TWTA is replaced by an SSPA for the lower frequencies. However, at high frequencies such as Ku- or Ka-bands, high-power and broadband TWTAs are still operating.

3. Low-Noise Amplifier (LNA) and Frequency Converter

The receive equipment consists of LNAs, downconverters, and amplifiers after frequency conversion. The LNA is the main contributor to determine the figure of merit G/T of the satellite receiver, and it is required to have a low noise temperature and a high gain. As LNAs, parametric amplifiers, FETs, and high-electron-mobility transistors (HEMTs) are used. In the C and Ku bands, FET amplifiers are dominant because of the use of GaAs and high-mobility electron techniques. The frequency converter consists of a mixer, a local oscillator, and filters.

FIGURE 19 S-band multiple umbrella-type deployable antenna of the ETS-VIII satellite. [Photograph courtesy of NASDA, Japan.]

4. On-Board Switching and Processing

Figure 21a shows a conventional satellite communication network using single-beam coverage and bent-pipe transponders. Earth stations communicate with each another within the wide beam coverage and can connect to terrestrial networks through a gateway station. Figure 21b shows a multibeam satellite that can perform interbeam connections through gateway stations via terrestrial networks. Figure 21c shows a multibeam satellite that has an on-board switching function for interbeam connections. The combination of on-board processing with beam-switching becomes very effective in increasing communi-

cations capacity. Table VII shows the onboard switching types and their features. In IF switching, routes of IF signals can be changed by connections between inputs and outputs of the microwave switch matrix (MSM). In a filter bank matrix, the signal destination can be selected by choosing the uplink carrier frequency. In switching techniques accompanied with regenerative transponders, demodulated baseband signals are switched or remodulated IF signals are switched by MSM. Since uplink and downlink are decoupled in this system, separate optimizations such as modulation and FEC schemes and multiple access are available.

For example, the advanced communications technology satellite (ACTS), launched by NASA in 1993, has two switching types: IF MSM and a baseband processor. The IF MSM achieves interconnections among three fixed spot beams for the high burst rate (HBR) system, capable of serving three time-division multiple access (TDMA) channels with a bandwidth of 900 Mbps each. The MSM is a four-by-four matrix providing crosspoint switching with a switching speed of 100 ns. It provides a 1-ms or 32-ms TDMA frame with switching at the 1-μs or 32-μs boundaries. The baseband processor, on the other hand, is used with two families of hopping spot beams for the low-burst-rate (LBR) system, intended to provide TDMA thin rout circuits of 64 kbps. It provides demodulation, convolutional decoding, storage, switching, convolutional coding, and remodulation of thin route circuits with 64 kbps

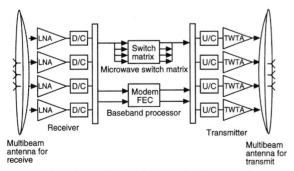

LNA: low noise amplifier, D/C: downconverter, U/C: upconverter
TWTA: travelling wave tube amplifer

FIGURE 20 Block diagram of a satellite transponder.

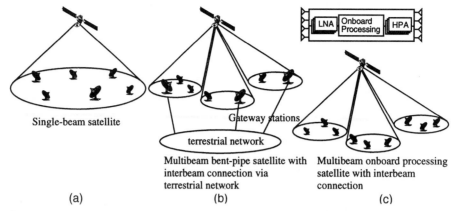

FIGURE 21 Satellite communication network using (a) a single beam and (b, c) multibeam antennas.

bandwidth. Convolutional coding and decoding is coupled with data rate reduction to achieve on-demand margin improvement of 10 dB to selected uplink and downlink earth stations experiencing rain attenuation.

The Japanese Communications and Broadcasting Engineering Test Satellite (COMETS), which was launched in 1998, has a regenerative transponder and an IF filterbank matrix, and Ka-band and millimeter-wave multibeam antennas. The input to the regenerative transponder for each beam is eight channels of single-channel-per-carrier (SCPC) signals, and its output is a remodulated binary phase-shift keying (BPSK) signal with a time-division multiplexed (TDM) format. Since user terminals use lower-data-rate SCPC signals for the uplink, compared to the TDMA scheme, it can reduce the transmitting power of terminals, and their size as well. Since only one TDM signal is used for downlink, it can make the satellite

transponder operate at its saturation point. Such effective use of the transponder power can moderate the demand for the specification of user terminals and can reduce the size of terminals.

C. Radio Propagation

To design satellite communication systems, we need quantitative knowledge of how radio waves propagate between the earth and space. Satellite communication generally uses VHF and higher frequencies to avoid reflections by the ionosphere. There is a boundary at about 10 GHz between increased ionospheric effects at lower frequencies and increased tropospheric effects at higher frequencies. Ionospheric effects, which are often inversely proportional to the square of the frequency, are listed in Table VIII. The major propagation effects at the frequencies above 10 GHz

TABLE VII Onboard Switching and Processing

Switching types	Features
IF switching without regeneration	After RF signals are downconverted to IF, the demultiplexed signals are routed through the microwave switch matrix (MSM) to their destination without demodulation or error correction. The connection pattern is controlled by groundstations.
IF switching with regeneration	After frequency conversion from RF to IF, the IF signals are demodulated, the FEC decoded, and the FEC encoded and modulated again before inputting them to the MSM. The onboard demodulation and modulation with FEC can separate (C/N_0) in uplink and (C/N_0) in downlink, and improve the total link quality.
Baseband switching	The uplink signals are downconverted, demodulated, FEC decoded, switched, and then modulated, FEC encoded, upconverted, and transmitted to the downlink. Since switching is performed at baseband, flexible connections such as different multiple access, modulation, and FEC schemes are available.
Asynchronous transfer mode (ATM)-based switching	Signals of different rates are translated to a same size packet called a cell, 53 bytes in size, and consists of virtual path identifier (VPI), virtual circuit identifier (VCI), error detection bits, and 48-byte data. Cells are routed through the satellite fast packet switch by VPI.
Fast packet switching	A packet finds its path by self-routing through a multistage switch of the banyan network. Each stage consists of smaller switching elements. Several modified switching techniques have been proposed to avoid the blocking.
Optical baseband switching	Optical switching can provide ultrahigh speed and efficient switching: presently throughput exceeding 80 Gbps will be available. Basic switching methods are TDM-based optical rings, wave division multiplexingoptical rings, and optical token rings.

TABLE VIII Ionospheric Effects

Effects	Frequency dependence	500 MHz	1 GHz	3 GHz	10 GHz
Faraday rotation	$1/f^2$	1.2 rotations	$108°$	$12°$	$1.1°$
Scintillation	$1/f^{1.5}$		$>20\,\text{dB}_\text{p-p}$	$10\,\text{dB}_\text{p-p}$	$4\,\text{dB}_\text{p-p}$
Absorption (middle latitude)	$1/f^2$	$<0.04\,\text{dB}$	$<0.01\,\text{dB}$	$<0.001\,\text{dB}$	$<10^{-4}\,\text{dB}$
Absorption (polar cap)	$1/f^2$	$0.2\,\text{dB}$	$0.05\,\text{dB}$	$6 \times 10^{-3}\,\text{dB}$	$5 \times 10^{-4}\,\text{dB}$
Direct-of-arrival variation	$1/f^2$	$48''$	$12''$	$1.32''$	$0.12''$

are rain attenuation, depolarization, gaseous attenuation, radio noise, and scintillation.

1. Rain Attenuation and Depolarization

Rainfall attenuation, which is caused by absorption and scattering from water and ice drops, can impair a satellite communication link at frequencies above 10 GHz, as well as increasing noise temperature and impairing cross-polarization discrimination. At frequencies below 10 GHz, rainfall attenuation may normally be neglected. A number of methods have been proposed to predict the rainfall attenuation statistics. In this section, we introduce one recommended by the International Telecommunications Union (ITU).

a. Long-term rainfall attenuation statistics.
The following parameters and the geometry shown in Fig. 22 are required:

$R_{0.01}$: Average annual rainfall rate corresponding to a cumulative time probability of 0.01% at a given location (integration time 1 minute) (mm/h)

h_s: Height above mean sea level of the earth station (km)

θ: Satellite elevation angle (degrees)

ϕ: Latitude of the earth station (degrees)

f: Frequency (up to 55 GHz) (GHz)

R_e: Effective radius of the earth (8500 km)

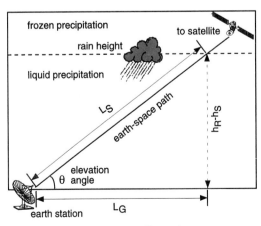

FIGURE 22 Earth–satellite path geometry.

The equivalent rain region height h_R is give by, from $\phi.$,

$$h_R\,(\text{km}) = \begin{cases} 3.0 + 0.028\phi & \text{for } 0 \le \phi \le 36° \\ 4.0 - 0.075(\phi - 36) & \text{for } 36° < \phi \end{cases}$$

(22)

The slant-path length L_S below the rain height is

$$L_S\,(\text{km}) = \frac{(h_R - h_S)}{\sin\theta} \quad \text{for } \theta \ge 5°$$

(23)

For $\theta < 5°$, accounting for the earth's curvature, it is given by

$$L_S\,(\text{km}) = \frac{2(h_R - h_S)}{\sqrt{\sin^2\theta + \dfrac{2(h_R - h_S)}{R_e}} + \sin\theta}$$

(24)

Then, the horizontal projection L_G is given by

$$L_G\,(\text{km}) = L_S \cos\theta$$

(25)

Next, we need to obtain the rainfall rate, $R_{0.01}$, using statistical data obtained at the location in question. If long-term statistic is not available, we can estimate it from the maps of rainfall rate shown in the Recommendation ITU-R. The specific rain attenuation, γ_R, can be estimated by using frequency-dependent coefficients k and α and $R_{0.01}$.

$$\gamma_R\,(\text{dB/km}) = k(R_{0.01})^\alpha$$

(26)

For linear and circular polarization, k and α can be obtained from Table IX and the following equation:

$$k = \left[k_H + k_V + (k_H - k_V)\cos^2\theta \cos 2\tau\right]/2$$

(27)

$$\alpha = \left[\alpha_H k_H + \alpha_V k_V + (\alpha_H k_H - \alpha_V k_V)\cos^2\theta \cos 2\tau\right]/2k$$

TABLE IX Regression Coefficients for the Specific Rain Attenuation

Frequency (GHz)	k_H	k_V	α_H	α_V
10	0.0101	0.00887	1.276	1.264
12	0.0188	0.0168	1.217	1.200
20	0.0751	0.0691	1.099	1.065
30	0.187	0.167	1.021	1.000
40	0.350	0.310	0.939	0.929
50	0.536	0.479	0.873	0.868

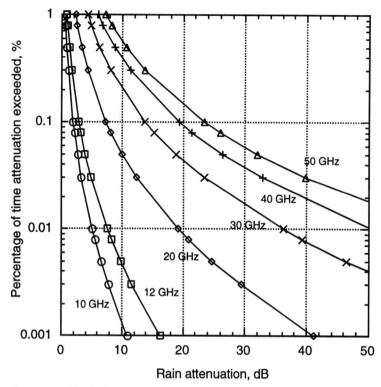

FIGURE 23 Rain attenuation distribution derived from ITU-R model, for Tokyo at an elevation angles of 45°.

where τ is the polarization tile angle relative to the horizon ($\tau = 45°$ for circular polarization).

The horizontal reduction factor, $r_{00.1}$, for 0.01% of the time is given by

$$r_{0.01} = \frac{1}{1 + (L_G/L_0)} \qquad (28)$$

where L_0 (km) $= 35 \exp(-0.015 R_{0.01})$ for $R_{0.01} < 100$ mm/h, otherwise $L_0 = 35 \exp(-1.5) = 7.810$.

The predicted rain attenuation exceeded for 0.01% of an average year is obtained from:

$$A_{0.01} \text{ (dB)} = \gamma_R L_S r_{0.01} \qquad (29)$$

The rain attenuation corresponding to cumulative time p (from 0.001% to 1%) can be determined by

$$A_p = 0.12 p^{-0.546 - 0.043 \log p} A_{0.01} \qquad (30)$$

For example, Fig. 23 shows the estimated rain attenuation distributions determined from the ITU-R model for an earth station in Tokyo of the climate zone K at an elevation angle of 45° for frequencies from 10 to 50 GHz.

b. Rain depolarization. Reuse of the same frequency by means of orthogonal polarization can increase channel capacity in satellite communications systems. A depolarized radiowave caused by rain or ice particles

has its polarization state change to an orthogonally polarized state, resulting in interference between the two orthogonally polarized channels. Degradation of cross-polarization discrimination (XPD) arises from different attenuation and phase shift of horizontally and vertically polarized waves caused by nonspherical raindrops or ice particles. The XPD is defined as

$$\text{XPD} = 20 \log \left| \frac{E_C}{E_X} \right| \text{ (dB)} \qquad (31)$$

where E_C is the received electric field in the copolarized wave and E_X is the electric field converted to the cross-polarized wave. The relationship between the rain-induced XPD and the rain attenuation A is approximately given by

$$\text{XPD} = U + (V - 20) \log L_S - V \log A \qquad (32)$$

where U and V are empirically determined coefficiencies, and L_S is the slant path length.

$$\begin{aligned} U &= 26.0 \log f + 4.1 - 20 \log|\sin 2\tau| \\ &\quad - 40 \log(\cos \theta) + 0.0053 \sigma^2 \\ V &= 12.8 f^{0.19} \quad \text{for } 10 \leq f \leq 20 \text{ GHz} \\ &= 22.6 \quad \text{for } 20 < f \leq 40 \text{ GHz} \end{aligned} \qquad (33)$$

where f is the frequency in GHz, θ is a elevation angel in degrees, τ is polarization tilt angle in degrees with respect to the horizon, and σ is the standard deviation of the raindrop canting angle.

2. Gaseous Attenuation

In the frequencies used for the satellite communications, oxygen has a series of very close absorption lines near 60 GHz and an isolated absorption line at 118.74 GHz, and water vapor has absorption lines at 22.3, 183.3, and 323.8 GHz. The ITU-R proposed a prediction method to calculate the gaseous attenuation for a given set of parameters such as frequency f, elevation angle θ, height (km) above mean sea level of the earth station h_s, surface temperature, and water vapor density at the surface. Assuming the specific attenuation for dry air γ_0 and waver vapor γ_w, and the equivalent height for dry air h_0 and water vapor h_w, the total slant path gaseous attenuation A_g is given by

$$A_g = \frac{\gamma_0 h_0 \exp(-h_s/h_0) + \gamma_w h_w}{\sin \theta} \quad \text{(dB)} \quad (34)$$

where the elevation angle θ is higher than $10°$. Then h_w is given by

$$h_w = h_{w0}\left[1 + \frac{3.0}{(f - 22.2)^2} + \frac{5.0}{(f - 183.3)^2 + 6.0} + \frac{2.5}{(f - 325.4)^2 + 4.0}\right] \quad \text{(km)} \quad (35)$$

where $h_{w0} = 1.6$ km for clear weather, and $h_{w0} = 2.1$ km for rainy weather.

3. Radio Noise

When radio waves are attenuated by absorption, thermal noise is also radiated simultaneously from the absorbing medium. The attenuation of the absorbing medium A (dB) and the noise temperature T_s (K) are related by the following formula.

$$T_s = T_a(1 - 10^{-A/10}) \quad (36)$$

where T_a is the equivalent temperature (K) of the absorbing medium, for which the following approximation formula has been proposed based on the temperature T_0 (K) at ground level.

$$T_a = 1.12T_0 - 50 \quad \text{(K)}$$

Using this result, T_a ranges from about 255 to 290 K for ground temperature ranges of 0–30°C.

4. Faraday Rotation and Scintillation

a. Faraday rotation. Faraday rotation is a phenomenon whereby the polarization plane of radio waves rotates due to magnetic flux lines and electrons in the ionosphere, resulting in different phase velocities for left- and right-hand circularly polarized waves. The rotation angle is roughly in inverse proportion to the square of the frequency, as shown in Table VIII. This rotation has no effect on received powers of circularly polarized waves, but reduces received powers for linearly polarized waves. For example, when linearly polarized waves are used, the attenuation is 9.3 dB at 850 MHz and 0.6 dB at 1.6 GHz. But at higher frequencies of several GHz and above, the amount of rotation of linearly polarized waves is so small as to make the signal attenuation practically negligible.

b. Ionospheric scintillation. Ionospheric scintillation arises from spatial and temporal fluctuations in the ionosphere's electrical characteristics and is normally observed as irregular transient fluctuations in the received signal strength. This amplitude scintillation is thought to be accompanied by fluctuations in phase and apparent elevation angle. Ionospheric scintillation is maximum near the geomagnetic equator and smallest in the middle latitude regions. The level of ionosphere scintillation appears to be inversely proportional to the frequency raised to a power of approximately 1.5, so in satellite communication systems operating at 10 GHz and above, it has a negligible effect.

c. Tropospheric scintillation. Tropospheric scintillation is typically produced by refractive index fluctuations in the first few kilometers of altitude and is caused by high-humidity gradients and temperature inversion layers. The magnitude of the scintillation depends on the magnitude and structure of the refractive index fluctuations, increasing with frequency, and with the path length through the medium, and decreasing as the antenna beamwidth decreases because of aperture averaging.

D. Communications Systems

A satellite communications system consists of the space segment and the ground segment. In the early stage (from the middle of the 1960s to the middle of the 1970s), satellites were very small and could transmit a few signals with only weak power. The first commercial satellite, INTELSAT-I (launched in 1965), had a mass of only 34 kg in orbit, and had two transponders with a total bandwidth of 50 MHz. Antennas of the earth stations, therefore, needed a large aperture of about 30 m in diameter. In the 1980s, satellites became larger and more powerful, resulting in a decrease of the antenna diameter of the earth stations to about 2 m. The INTELSAT-VII had a mass of about 1500 kg in orbit, and had 46 transponders with a total bandwidth of 2432 MHz. The larger the satellites and the more transmitting power, the smaller the earth stations could be. Such a small earth station is called a very small aperture terminal (VSAT).

A satellite network architecture is classified into three categories: point-to-point (mesh), point-to-multipoint (broadcast), and multipoint interactive. In mesh-type networks, earth stations communicate directly with each other on a one-to-one basis, either by means of user reservation or on demand. In point-to-multipoint networks, a broadcast of information by the satellite is more efficient than that by terrestrial networks with fiber-optic cables or copper wires. In multipoint interactive networks using VSATs, the satellite can work as a common connection point for the network, excluding the necessity for a separate physical link between the hub and each VSATs. In this section, we will describe multiple access techniques, and modulation and error correction techniques.

1. Multiple Access Techniques

Multiple access (MA) is a technique to use a satellite communication system efficiently by sharing satellite resources such as frequency bandwidth, power, time, and space by a large number of earth stations. Three main MA techniques have been used: frequency-division multiple access (FDMA), time-division multiple access (TDMA), and code-division multiple access (CDMA) (Fig. 24). FDMA and TDMA share the frequency bandwidth and the time of the satellite transponders, respectively. In CDMA, the earth stations share both frequency and time using a set of mutually orthogonal codes. When each earth station is dynamically assigned the satellite resources depending on its call needs, this MA is called demand-assignment multiple access (DAMA).

a. FDMA. FDMA is the most common MA technique for satellite communication systems. Especially, a single channel per carrier (SCPC), in which each carrier bears one voice or data channel, has been used in most operating satellite communication systems because of its simple implementation. The FDMA signals occupy nonoverlapping frequency bands with guard bands between signals to avoid interference to neighboring channels. When a satellite amplifier is operated close to its saturation, nonlinear amplification produces intermodulation (IM) products, which may cause interference in the signals of other users. In order to reduce IM, it is necessary to operate the transponder by reducing the total input power. This causes inefficient use of available transponder power.

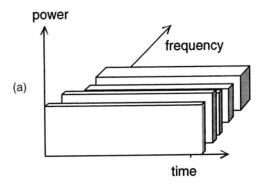

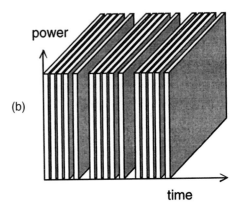

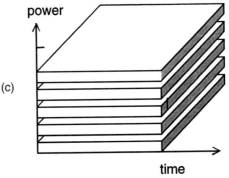

FIGURE 24 Resource sharing of (a) FDMA, (b) TDMA, and (c) CDMA.

b. TDMA. In TDMA, each user is assigned a nonoverlapping time slot with a guard time, which avoids intersymbol interference in neighboring channels. TDMA transmissions are organized in a frame structure, which contains one or two reference bursts that synchronize the network and identify the frame. Synchronization is one of major technical problems for implementing TDMA. Since TDMA bursts occupy the full bandwidth of the satellite transponder at a time, the satellite transponder can be operated in full saturation, resulting in increased channel capacity. Once synchronization is acquired, time-slot assignments for temporally accessed users are easier to adjust than frequency channel assignments in FDMA.

c. CDMA. In CDMA, all signals occupy the same frequency bandwidth and are transmitted simultaneously in time, but the different signals are distinguished from one another at the receiver by the specific spreading codes or frequency hopping pattern. Spread spectrum multiple access (SSMA), which is the most popular CDMA, is achieved by two techniques: direct sequence (DS) modulation and frequency hopping (FH) modulation. In DS, the modulated signal is multiplied by pseudorandom noise (PN) codes with a chip rate of R_c, which is much larger than an information bit rate of R_b. The resulting signal has wider frequency bandwidth than the original modulated signal. At a receiving terminal, the received signal is despread by multiplying the same PN sequence. In FH, bandwidth spreading is achieved by pseudorandom frequency hopping. The hopping pattern and hopping rate are determined by the PN code and code rate, respectively.

d. Random access. An MA method in which an earth station transmits its signal to a satellite at a random time or in bursts without coordinating its access with other earth stations is called random access. To avoid collisions between signals and increase throughput, several methods such as unslotted or slotted ALOHA, reservation ALOHA, and spread ALOHA have been proposed. When an earth station determines that its transmitted packet has collided, retransmission of this packet is performed after a random delay.

2. Modulation Method

Digital satellite communication systems send information by modulating the amplitude, phase, or frequency of a carrier. The performance of digital modulation methods can be evaluated under the following criteria: good power and bandwidth efficiencies, immunity to nonlinearity, robust synchronization, and implementation simplicity. The most popular modulation technique used in satellite communications is the family of phase-shift keying (PSK). Table X shows bandwidths and E_b/N_0 for a bit error probability of 10^{-5} for coherent detection, nonlinear immunity, and implementation simplicity for binary PSK (BPSK), quadra-

ture PSK (QPSK), offset QPSK (OQPSK), $\pi/4$-PSK, and minimum shift keying (MSK) modulations. The MSK is a special case of the OQPSK using a sinusoidal pulse instead of a rectangular pulse in the OQPSK, and it is also a special case of continuous-phase frequency shift keying (CPFSK) with a modulation index of 0.5. The MSK has a larger null-to-null bandwidth than the QPSK, but a much smaller 99%-power bandwidth than the QPSK. The $\pi/4$-QPSK has become very popular for mobile satellite communications as well as for terrestrial mobile communications because of its compact spectrum and, capability of differential detection.

The BER of the PSK family in an additive white Gaussian noise channel with coherent detection and no bandwidth limitation is given by

$$P = \frac{1}{2}\,\mathrm{erfc}\left(\sqrt{\frac{E_b}{N_0}}\right) \qquad (37)$$

where E_b/N_0 is the ratio of energy per bit to noise density, and erfc( ) is the complementary error function given by

$$\mathrm{erfc}(x) = \frac{2}{\sqrt{\pi}}\int_x^\infty \exp(-t^2)\,dt \qquad (38)$$

3. Error-Control Techniques

Error correction is achieved by automatic repeat request (ARQ) and forward error correction (FEC).

a. Automatic repeat request. In ARQ systems, if a receiving terminal does not detect errors in received data, it sends an acknowledgment (ACK) signal to the transmitting terminal. If the receiving terminal does detect errors, it send a negative acknowledgment (NACK) signal to notify that the data block was not correctly received and to request retransmission of the same block. Stop-and-wait, go-back-N, and selective-repeat request are three basic methods of ARQ. For satellite communication systems where error-correcting codes may not work reliably because of low and variable link quality or high interference, ARQ can provide adequate robustness and required error performance. Because of the long delay in satellite

TABLE X Comparison of Modulation Performance

Modulation	99%-power BW	Null-to-null BW	E_b/N_0 for 10^{-5}	Nonlinearity	Implementation
BPSK	20.56	2.0	9.6	D	A
QPSK	10.28	1.0	9.6	C	B
OQPSK	10.28	1.0	9.6	B	C
$\pi/4$-PSK	10.28	1.0	9.6	B	C
MSK	1.18	1.5	9.6	A	D

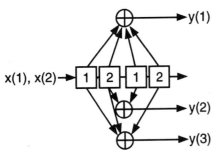

FIGURE 25 Convolutional coding for rate 2/3.

Uncoded E_b/N_0 (dB)	BER	$R = 1/3$, $K = 7$	$R = 1/2$, $K = 7$	$R = 2/3$, $K = 8$	$R = 3/4$, $K = 9$
9.6	10^{-5}	5.7	5.1	4.6	4.2
11.3	10^{-7}	6.2	5.8	5.2	4.8

TABLE XI Coding Gain for Viterbi Decoding (Soft Decision)

links, however, a long enough interval must be specified for timeout to prevent premature timeouts. If the sender performs timeout too early, while the ACK signal is still on the way, it will send a duplicate.

b. Forward error correction. The most popular FEC techniques in satellite communication systems are convolutional coding and block coding. In convolutional coding, k successive information bits are encoded continuously without their sequence being broken to a sequence of n bits ($n > k$). As shown in Fig. 25, the encoding can be achieved by linear operations of a finite-state shift register, and it is also represented by a trellis diagram (Fig. 26). In this case, the code rate $R = k/n$ is 2/3, and the constraint length K, which is defined as the number of k bits that affect formation of each n bits, is 2. Since the minimum (Hamming) distance between codewords is 3, this convolutional coding can correct up to one bit error. The most popular decoding algorithm is the maximum-likelihood decoding developed by Viterbi (known as Viterbi algorithm) to use the trellis structure for reducing the complexity of the evaluation. Table XI shows the coding gain for convolutional coding.

The most popular block code is Reed–Solomon (RS) code. A generator polynomial for a t-tuple error-correcting nonbinary RS code is given by

$$g(x) = (x - \alpha)(x - \alpha^2)(x - \alpha^3) \cdots (x - \alpha^{2t}) \quad (39)$$

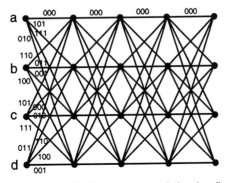

FIGURE 26 Trellis diagram for convolutional coding.

where α is a primitive element of a Galois field GF(q). Let a block of k information symbols be ($m_0, m_1, \ldots, m_{k-1}$). A code polynomial $c(x)$ is generated by multiplying the information polynomial $m(x) = m_0 + m_1 x + \cdots + m_{k-1}x^{k-1}$ by $g(x)$.

$$c(x) = m(x)g(x)$$
$$c(x) = c_0 + c_1 x + c_2 x^2 + \cdots + c_{n-1}x^{n-1}$$
$$\Rightarrow (c_0, c_1, c_2, \ldots, c_{n-1}) \quad (40)$$

A codeword ($c_0, c_1, \ldots, c_{n-1}$) can be obtained from coefficiencies of $c(x)$. The code rate is k/n. Since the minimum Hamming distance is $2t + 1$ ($= n - k + 1$), a t-tuple error can be corrected and a $2t$-tuple error can be detected.

Concatenated codes can improve the FEC performance by using two levels of FEC coding: The inner code is usually a convolutional code with the Viterbi algorithm, and the outer code is the RS code with an interleaver. After the inner code corrects most errors, the outer code corrects the burst errors that appear at the output of the inner decoder. By using the interleaving, which shuffles the bit sequence to make burst errors more random, and deinterleaving, the outer-code bits can further improve the error-correction performance.

4. Coded Modulation

In the FEC, the bit error performance can be improved by expanding frequency bandwidth. For example, the E_b/N_0 required for a BER of 10^{-5} in the case of the convolutional coding and the soft-decision Viterbi decoding with $R = 1/2$ and $K = 7$ is 5.1 dB smaller than that for uncoding PSK. On the other hand, the coded PSK requires twice the bandwidth of the uncoded PSK signal because of the increase in the symbol rate of the PSK.

This can also be done by increasing the number of phases of PSK instead of expanding the bandwidth. When we use a convolutional code of $R = 2/3$ and 8-PSK, the 8-PSK signal has the same bandwidth as uncoded 4-PSK. However, the BER performance degrades by about 4 dB because of the increase in phase. If the coding gain becomes more than 4 dB, this will be acceptable. This is the basic idea of coded modulation, which is combined with modulation and error-correcting codes without degrading power and bandwidth efficiencies. Trellis-coded

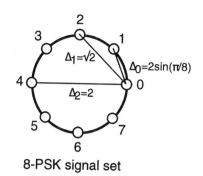

8-PSK signal set

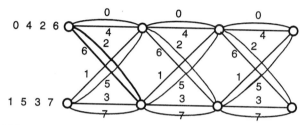

FIGURE 27 (a) Signal set of 8-PSK. (b) Trellis diagram for 8-PSK.

modulation uses a state-transition diagram, which is similar to the trellis diagram of the convolutional codes. In TCM, symbols of the modulation signal are assigned to trellis branches, although binary (or nonbinary) code symbols are assigned to branches in the convolutional codes. Figure 27a shows the signal set of 8-PSK, and Fig. 27b shows the trellis diagram, in which phase states of 8-PSK are assigned to the trellis branches. TCM can improve the bit error performance without making sacrifices for bandwidth efficiency.

III. TOPICS IN SATELLITE COMMUNICATIONS SERVICES

A. Broadband Technologies

Terrestrial networks, which are now providing various advanced multimedia services, will soon cover urban areas, but most of the population living in rural and remote areas lacking adequate telecommunications infrastructure will not have the opportunity to participate in such advanced communications because of the high cost of extending the networks. However, using broadband and global satellite systems, which will provide seamless compatibility with terrestrial fiber networks, multimedia services can be easily extended to remote areas. Table XII shows examples of Ka-band broadband GEO systems, which were accepted for filing by the Federal Communications Commission (FCC) of the United States by September 1996. The Ka band is not only expected to serve as an alternative to the crowded C- and Ku-band, but also will provide

many low-cost, global broadband interactive services such as desktop-to-desktop videoconferencing, Internet access, electronic messaging and facsimile, electronic commerce, news and information gathering, distance learning and corporate training, telemedicine, direct-to-home video, and shopping. Moreover, in December 1996, Motorola filed for the LEO system "M-Star," which uses higher frequencies of the V-band and Q-band. It was followed by Hughes with "Expressway," prompting the FCC to allow others to file for Q/V-band systems. Finally, 16 Q/V-band satellite systems were filed by U.S. companies, as shown in Table XIII. These satellite systems are proposed to use the following new technologies for broadband satellite services.

1. Multibeam Antenna

In order to achieve higher-data-rate transmission, higher EIRP, G/T, and frequencies are needed. Since antenna beamwidth becomes narrower when large aperture antennas operating at higher frequencies are used to increase EIRP and G/T, a large number of beams are needed to cover global areas. In the proposed Q/V-band systems, several tens of spot beams with a beam size of less than 1• will be installed. The NASA's ACTS satellite has hopping spot beams with a 0.3• beamwidth and a steerable beam with a 1• beamwidth in the Ka band. The CRL's Gigabit Satellite has scanning spot beams using Ka-band active phased array antennas.

2. Satellite Switching

As mentioned in Section B.4, onboard routing is effective to perform interbeam connections among multibeam coverage areas. The Q/V-band systems, as shown in Table XIII, employ IF switching using MSM, baseband switching, ATM switching (see Table VII), and satellite-switched TDMA (SS-TDMA) for routing. In SS-TDMA, interconnection between beams is performed cyclically, and each earth station transmits a TDMA burst when the required interconnection is established.

3. Networking

Because of the interoperability of satellite and terrestrial networks, in which ATM is widely accepted as the future networking technology, satellite ATM networks will play an important role in broadband satellite systems. Figure 28 shows an example of satellite ATM network architecture. The architecture can be classified into two groups: satellite ATM networks with bend-pipe (transparent) transponders and satellite ATM networks with on-board processing. In the former, all protocol processing is performed

TABLE XII Broadband GEO Systems: Ka-Band Satellite Applications Accepted for Filing by the FCC of the United States by September 1996

System	Owner	No. of satellites (orbits)	Bit rate (Mbps)	Capacity (Gbps)
Astrolink	Lockheed Martin Corp.	9 GEO (15)	8.448	7.7
CyberStar	Loral Aerospace Holding, Inc.	3 GEO	3.088	4.9
EchoStar	EchoStar Satellite Corp.	2 GEO	155.52	N/A
Galaxy	Hughes Communications Galaxy, Inc.	20 GEO (15)	6	4.4
GE*Star	GE American Communications, Inc.	9 GEO (5)	40	1.8
KaStar	KaStar Satellite Communications Corp.	2 GEO	155.52	7.4
Millennium	Comm, Inc.	4 GEO	51.84	7.5
Morning Star	Moring Star Satellite Co., L.L.C.	4 GEO	30	0.625
NetSat 28	NetSat 28	1 GEO	1.544	7.2
Orion F-2	Orion Atlantic, L. P.	1 GEO	3.088	N/A
Orion F-7/F-8/F-9	Orion Network Systems, Inc.	3 GEO	3.088	N/A
PAS-10/PAS-11	PanAmSat Corp.	2 GEO	45	N/A
VisonStar	VisionStar, Inc.	1 GEO	N/A	N/A

TABLE XIII Broadband Q/V-Band Satellite Systems Proposed in the United States

System	Owner	No. of satellites	Onboard switching MSM	Onboard switching ATM	Onboard switching Base-band	Onboard switching SS-TDMA	Satellite capacity (Gbps)	EIRP (dBW)	Transponder bandwidth, MHz (number)
Aster	Spectrum Astro, Inc.	25 GEO			✔	✔	10	66.7	470 and 980 (18)
Cyberpath	Loral Space & Comm. Ltd.	10 GEO		✔		✔	17.9	77.5	142 (100)
Expressway	Hughes Comm. Inc.	14 GEO				✔	65	55 (V) 48 (Ku)	300 (20) 250 (8)
GESN	TRW	14 GEO & 15 MEO				✔	50 & 70	83 (GEO) 78 (MEO)	300 (32) 3000 (48)
GE*Star Plus	GE American	11 GEO	✔				70	59 (V) 52 (Ku)	300 (400) 250 (16)
GS-40	Globalstar L.P.	80 LEO	✔				1	52	18/90 (30)
M-Star	Motorola	72 LEO	✔			✔	3.6	21–29	90 (104)
Orblink	Orbital Science Corp.	7 MEO	✔				75	62 60	50 (800) 1000 (20)
Pentriad	Denali Telecom. LLC	9 Molniya	✔				36	73.0 (V) 77.7 (W)	78 (80) 78 (12)
Q/V Band	Lockheed Martin	9 GEO		✔	✔		45	64.2 62.7	125 (48) 125 (8)
SpaceCast	Hughes Comm. Inc.	6 GEO				✔	64	72(V) 5 (Ku)	300 (40) 250 (16)
StarLynx	Hughes Comm. Inc.	4 GEO & 20 MEO			✔		5.9 & 6.3	70.5 (GEO) 56 (MEO)	270 (40) 270 (32)
VBS	Teledesic	72 LEO			✔		4	59	N/A
VStream	PanAmSat	12 GEO	✔				3.2	60	375 (80)

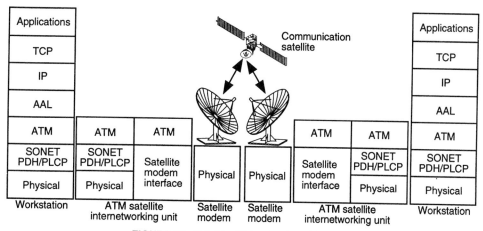

FIGURE 28 Satellite ATM network architecture.

on the ground. In the latter, the onboard switch performs ATM switching functions. The ATM satellite interworking unit interconnects ATM and satellite network and performs bandwidth allocation, network access control, traffic control, etc. ATM cells are transported by using existing digital transmission formats such as plesiochronous digital hierarchy (PDH), physical layer convergence protocol (PLCP), and synchronous digital hierarchy (SDH). As multiple access schemes, FDMA, TDMA, and CDMA are candidates, while multifrequency TDMA (MF-TDMA) was proposed to improve power efficiency for earth stations. Although full-band TDMA requires a high burst rate and high transmitting power for user earth terminals, MF-TDMA can decrease transmission rate, transmitting power, and the terminal's cost. Since ATM header error check is able to correct only single-bit errors, the burst errors cannot be corrected. An ATM adaptation layer 5 (AAL5) protocol has a 32-bit cyclic redundancy check (CRC), but FEC or ARQ techniques for improving error performance are needed.

TCP is a widely used transport protocol in terrestrial networks, but large delays in GEO, delay variations in LEO, and large delay–bandwidth product of the connection impair the throughput efficiency. The congestion control methods for broadband satellite networks may be different from those for low-latency terrestrial networks. In an acknowledgement- and timeout-based congestion control protocol such as TCP, the throughput is restricted by waiting time for receiving an ACK. The window size is used to adjust the amount of data to be sent starting at the data acknowledged. For high-bandwidth and high-delay transmission, small window size causes a problem. In Request-for-Comments (RFC) 1323, a window scale option was proposed, allowing the sender and receiver to negotiate a window scale factor for extending windows of up to 2^{30} bytes. Modifications of TCP can increase the throughput up to 15 Gbps even for the large delay of GEO

systems. Several studies are still being carried out to improve the TCP transmission performance.

New technologies such as utilization of higher frequency bands, multibeam coverage with frequency reuse, on-board fast packet switching, on-board processing, and monolithic microwave integrated circuit (MMIC) technologies increase both capacity and flexibility of the broadband satellite networks and can provide higher bandwidth-on-demand availability and quality of service (QoS) at lower cost to more users.

B. International Experiments

Many international experiments have been conducted for the GII (Global Information Infrastructure) since the G-7 Ministerial Conference on the Information Society was held in Brussels, Belgium, in 1995. The 11 pilot projects were assigned as the GIBN (Global Interoperability for Broadband Network) projects. On the other hand, the Trans-Pacific High Data Rate (HDR) Satellite Communication Network Experiment between Japan and the United States has been incubated in the Japan–U.S. Cooperation in Space project (now JUSTSAP: Japan–U.S. Science, Technology and Space Application Program) since 1993, and this project is combined with the GIBN projects. Furthermore, the international experiment was also conducted under the framework of the cooperation between developing country and developed country. The GIBN itself concluded in May 1999; however, many international experiments are still ongoing or planned as phase 2 of experiment.

1. Description of Experiments

To establish the GII, we will need many activities including international cooperation. Table XIV shows the international experiments conducted by the CRL of the MPT or MPT as Japan's representative organization.

TABLE XIV   List of International Experiments

Experiment item	Period	Destinations	Satellite	Data rate	Experiments
Japan–U.S. Trans-Pacific HDR Experiment (Phase 1)	Dec. 1996– Mar. 1997	Japan–U.S.	INTELSAT and ACTS 45	Mbps	Remote high-definition video postproduction
Japan–U.S. Trans-Pacific HDR Experiment (Phase 2)	1997–July 2000	Japan–U.S.	INTELSAT and N-STAR	45–156 Mbps	Transmitting visible human data and remote astronomical observation
Japan–Canada HDR Experiment	Oct.1997– Feb. 1998	Japan–Canada	INTELSAT and ANIK-E	45 Mbps	TV-conference, telemedicine
Japan–Europe–GA MMA Project	Sep. 1997– Nov. 1999	Japan–ESA, France, Switzerland and Italy	INTELSAT (Also Used EUTELSAT for Multipoint Distributed Conference)	2 Mbps	Multimedia and Internet applications, TV-conference, telemedicine, high-energy physics, exchange of earth observation data
Post–PARTNERS Experiment	1996–2001	Japan, Thailand, Malaysia, Indonesia, Philippines, Fiji	JCSAT-3 (1 from 2000) and SuperBird-C	2 Mbps (maximum for networking), 1.5 Mbps propagation tele-education,etc.)	Telemedicine, tele-education, networking via satellite, satellite–earth propagation
Japan–Korea HDR Satellite Communication Experiment (Phases 1 and 2)	1999–2002 (Phase 1: 1999, Phase 2: 2000–2002)	Japan–Korea	Phase 1: PanAm Sat/Ku Band; Phase 2: Koreasat 3/Ku Band	Phase 1: 45 Mbps, Phase 2: 155 Mbps	ATM LAN-to-LAN interconnection, ATM-based interactive multimedia service, HDV transmission, 3D HDV transmission, real-time VLBI, and cancer center collaboration
Gigabit test bed using planned gigabit experimental satellite	2005–2010	Japan–Asia Pacific countries	Gigabit Experimental Satellite (planning phase)	1.5–1244 Mbps	Currently inviting: telemedicine, tele-education, distributed database, high-speed Internet, large-scale trunking, etc.

a. Japan–U.S. Trans-Pacific HDR Satellite Communication Experiment.   The Trans-Pacific HDR Satellite Communication Experiment has been studied in the JUSTSAP since 1993 as mentioned earlier. Figure 29 shows the final structure of the experimental system configuration. A hybrid construction was employed for the experimental network, comprising the satellite circuits of NASA's experimental ACTS satellite and the INTELSAT satellite (with no rental fees applied), and three interconnected optical fiber networks in Japan, Hawaii, and California. The first phase of the Trans-Pacific HDR Satellite Communication Experiment was successfully conducted for postproduction high-definition video (HDV) transmission via 45 Mbps ATM (asynchronous transfer mode) circuit between Tokyo and California via Hawaii by using ACTS and INTELSAT in March 1997. The objectives of this experiment were to promote cooperation between Japan and the United States in the space communications field, to demonstrate the technological possibilities for constructing an international broadband network for satellite communications systems, and at the same time to demonstrate that satellite communications is a very practical broadband network application. It was decided to implement the High-Precision Video/Remote Post-Production Experiment proposed by the Jet Propulsion Laboratory (JPL) as the broadband network application for the experiment.

The second phase of this experiment was conducted by using a satellite network of INTELSAT and N-STAR in July 2000. The Internet was configured to use a high-speed satellite link of 45 Mbps in this experiment, and there is originality in having realized an application to make data available to users at multiple points. In a remote astronomy distance-education demonstration, a telescope at Wilson Mountain astronomical observatory was controlled remotely from CRL, NASA/JPL, KMSI (Keck Math/Science Institute), and the University of Maryland. The telescope data was displayed in a computer at hand through a dispersion filing system via a server at NASA/ARC (Ames Research Center) and participants were enabled to utilize these data simultaneously. Pupils both of Sohka High School in Tokyo and of Crossroads School in Santa Monica experienced a virtual astronomical observation classroom. In a visible human

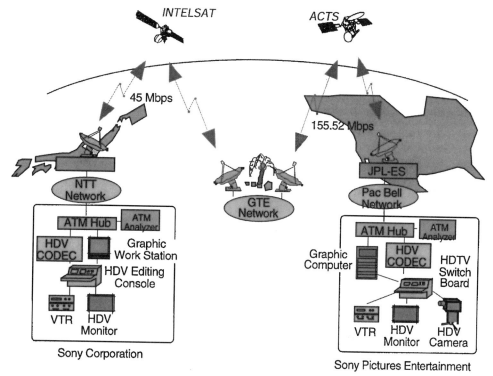

FIGURE 29 Configuration of Trans-Pacific HDR experiment.

telemedicine demonstration, data from the visible human of NLM (National Library of Medicine) was transmitted at high speed to Sapporo Medical College successfully. The information that could be transmitted in about 3 minutes conventionally was transmitted via satellite in about 10 seconds.

b. Japan–Canada HDR experiment. At the February 1995 meeting of the G7 Ministerial Conference, Japan's Posts and Telecommunications Minister and Canada's Trade and Industry Minister agreed that their countries should cooperate in an experimental linking of CRL and the Communications Research Centre (CRC) of Industry Canada. The link of 45 Mbps was used for an HDV telemedicine experiment by configuring through INTELSAT and ANIK-E satellite systems in the Japan–Canada HDR experiment for a remote workshop conducted on cardiac surgery and neuroscience, which clearly demonstrated the utility of broadband networks. To set up this network, many possible network configurations have been examined. In the initial stage, a 155-Mbps HDV conference was planned between Japan and Canada. To do that, the only method was the optical marine cable from Japan to North America, where the 155-Mbps capacity is available from Japan to the West Coast of the United States. There was no link available from the United States to Canada, because of the difficulty of coordinating both

Canada and Japan to the United States. Finally, the Intelsat satellite link is available for 45 Mbps to Vancouver.

c. Japan–Europe GAMMA (JEG) project. The Japan–Europe link experiment has been examined to secure the network to promote international satellite communications services in the Europe. In this experiment, the transmission rate was restricted to 2 Mbps. The EC (European Community) is interested in the satellite access technology rather than HDR, although the final agreement involves the HDR capability. The MPT coordinated the Japanese side and ESA (European Space Agency) started the GAMMA (Global Architecture for Multi-Media Access). The experiment is called the JEG (Japan–Europe Gamma) project. The Japan–ESA administrative meeting agreed on the JEG project in July 1997. The first experiments conducted included a satellite Internet experiment, DAVIC-VoD (Digital Audio-Visual Council—Video on Demand), and TV conferencing.

d. Post-PARTNERS project. Japan has also conducted a satellite communication network experiment with people in the Asia-Pacific countries since ISY (International Space Year) 1992. The PARTNERS project was conducted for experiments via 64 kbps link, including tele-education and telemedicine during 1992–1995. The Post-PARTNERS project with a 2 Mbps link has been conducted since FY 1996. The network configured was by using JCSAT-3 and SuperBird-C. The experiment includes telemedicine, tele-education, networking via satellite, and satellite–earth propagation.

e. Japan-Korea HDR satellite communication experiment.

Korea and Japan shared the view that the Korea–Japan HDR satellite communications experiment project should be promoted in order to strengthen collaboration between both countries for the establishment of APII (Asia–Pacific Information Infrastructure). Under the Japan–Korea HDR satellite communication experiments agreement concluded at the ministerial meeting in 1996 (the specific procedure was chosen at the director-general-level meeting in 1997), various HDR satellite communication experiments, such as three-dimensional high-definition video (3D HDV) transmission experiments, will be implemented until 2002. The experiments are planned in two phases. The first phase, which lasts until 1999, will use the 45 Mbps capacity of the Japan–Korea satellite link. The second phase, which lasts until 2002, will use a 155-Mbps capacity satellite link. PanAmSat and Koreasat-3 will be used for the first and second phases, respectively.

f. Gigabit test bed using planned gigabit satellite.

The CRL is conducting research on both system concepts and key technologies of the experimental gigabit communications satellite, including a Ka-band scanning spot beam antenna for global and high-data-rate user access, ATM-based high-speed and flexible onboard satellite switches for bandwidth-on-demand multimedia networking, HDR intersatellite communications, and interoperability protocols between the satellite and the ground networks. These technologies are also applied for the multiple utilization of the space data network system. Figure 30 shows a concept of the experimental gigabit communications satellite system. Using the experimental gigabit communications satellite, the application experiments were proposed at the APEC (Asia Pacific Economic Conference) and APT (Asia Pacific Telecommunity) in June 1998.

2. Consideration of Satellite Link for International Experiments

The international experiments mentioned earlier are reviewed in terms of why the satellite communication was/is used, what result was obtained, and what the role of satellite communication is.

a. Why satellite communications is used.

Most of the international experiments have been conducted by using satellite communication link. The reason why the satellite communications used is divided into three categories as follows:

• *Demonstration experiments on the merits of satellite communication*: Satellite communication has many merits: (i) wide coverage, (ii) broadcasting ability, (iii) immediate installation of communication link, (iv) wide bandwidth, and (v) disaster communication. Table XV shows how the experiments use the feature of satellite communications capability. Most international experiments mainly demonstrated the feature of wide bandwidth capability, while the Post-PAERTNERS experiment places more emphasis on the wide coverage capability. Although there is no experiment aiming only at using the disaster communication capability, this feature is shown in Table XV.

• *Experiments comparing satellite communications link with terrestrial link or hybrid link environment*: Experiments in this category were conducted to show feasibility and as preoperational studies for multimedia and Internet applications via satellite communications and terrestrial hybrid network environments. The experiment of the JEG project is belongs to this category. In the JEG project, the transmission characteristics through a hybrid configuration of both satellites and terrestrial lines was of interest, that is, the experiment demonstrated how well the TV conference software works on the network through Japanese domestic terrestrial links, satellite communication link via INTELSAT in the Indian Ocean, and European terrestrial links. The link throughput was 2 Mbps, not such a high speed, but the target of this experiment has been attained sufficiently. In addition to the TV conference between two points, the demonstration was successfully conducted among multiple points in the Europe as described in the previous chapter.

• *Experiments because links other than satellite are not available*: Regarding the Japan–Canada HDR experiment, the plan was to conduct the experiment from mid-1996 through the spring of 1998, using as a network link of the 155 Mbps international GIBN experimental line and a trans-Pacific undersea optical fiber cable. However, it proved to be more difficult than expected to establish lines within the United States and between the United States and Canada. Thus the INTELSAT and ANIK-E satellites were used.

b. The role of satellite communication: further considerations.

It was as easy to use the satellite communication in the international experiment as expected. This is why the satellite communications has many merits, including wide bandwidth and ease of link establishment. Especially in international experiments in which many organizations participate on a voluntary basis, satellite

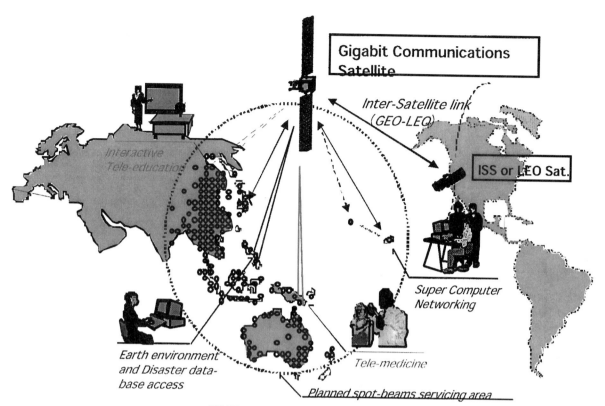

FIGURE 30 Concept of gigabit satellite.

communications could skip the tedious interface or coordination process, although it has not been easy to overcome it in actuality.

Attention should be paid to the propagation delay. The degradation of transmission efficiency in a link having a large product of delay (latency) and bandwidth (transmission rate) could increase not only in a satellite link but also in a long-distance optical fiber link, as they become operational. As far as ATM concerned, there is not so much degradation because a large product of delay and bandwidth, since the hand-shaking is not necessary in the data transmission mode. On the other hand, the TCP is a protocol of the hand-shaking type. It is well known that the transmission efficiency decreases as the delay–bandwidth product increases.

On the other hand, channel quality will be degraded because of propagation delay of the geostationary satellite in the case of interactive real-time communications applications. Furthermore, it is clear that the addition of a process delay for the digitalization of the image increases the degradation, as mentioned previously. However, for multimedia communications such as the Internet that require

TABLE XV How Experiments Use Satellite Communications Capabilities

Experiment item	Wide coverage	Broadcasting ability	Immediate installation	Wide bandwidth	Disaster communication
Japan–U.S. Trans-Pacific HDR Experiment	✔	•	✔	✔✔	✔
Post-PARTNERS Experiment	✔✔	✔	✔		✔
Japan–Korea HDR Satellite Communication Experiment	✔	✔	✔	✔✔	✔
Gigabit Test Bed using Planned Gigabit Experimental Satellite	✔	✔	✔	✔✔	✔

high speed, it must be considered that not all applications necessarily require a strictly simultaneous response. Thus we must recognize that there are some cases in which the delay is a major obstacle, and others in which it is not.

In conclusion, it has certainly been clarified that satellite communications is very useful, especially in phase 1 of the international experiment. However, it seems that satellite communications will be replaced by terrestrial fiber cable in the phase-2 experiment if such a cable system is available. Nevertheless, since satellite communications will be used more and more, especially in the initial phase of the experiment, to prepare skills in satellite communication for quick installation is preferable for future international experiments.

SEE ALSO THE FOLLOWING ARTICLES

COMMUNICATION SYSTEMS, CIVILIAN • MICROWAVE COMMUNICATIONS • NETWORKS FOR DATA COMMUNICATION • RADIO PROPAGATION • RADIO SPECTRUM UTILIZATION • SATELLITE RF COMMUNICATIONS AND ONBOARD PROCESSING • SYNTHETIC APERTURE RADAR • TELECOMMUNICATIONS • VAST—VERY SMALL APERTURE TERMINAL SATELLITE EARTH STATION

BIBLIOGRAPHY

Akylidiz, I. F., and Jeong, S.-H. (1997). "Satellite ATM Networks: A Survey," *IEEE Commun. Mag.* **35,** 30–43.

Aruga, H., Sakura, T., Nakaguro, H., Akaishi, A., Kadowaki, N., and Araki, T. (2000). "Development results of Ka-band multibeam active phased array antenna for gigabit satellite," Proc. 18th AIAA International Communications Satellite Systems Conference, 25–32.

Clarke, A. C. (1945). "Extra-terrestrial relays: can rocket stations give world-wide radio coverage?," *Wireless World* **51,** 305–308.

Evans, J. V., and Dissanayake, A. (1998). "The prospects for commercial satellite services at Q- and V-band," *Space Commun.* **15,** 1–19.

Evans, J. V. (2000). "The U.S. filings for multimedia satellites: A review," *Int. J. Satellite Commun.* **18,** 121–160.

Gargione, F., Iida, T., Valdoni, F., and Vatalaro, F. (1999). "Services, technologies, and systems at Ka band and beyond—A survey," *IEEE J. Selected Areas in Commun.* **17,** 133–144.

Gedney, R. T., Schertler, R., and Gargione, F. (2000). "The Advanced Communications Technology Satellite," Scitech Publishing, Inc.

Gibson, J. D. (ed.) (1996). "The Communications Handbook," CRC Press, Boca Raton, FL.

Helm, N. R. (2000). "Satellite communications for the Internet," 18th AIAA International Communications Satellite Systems Conference, AIAA-2000-1165.

Iida, T. (ed.) (2000). "Satellite Communications—System and Its Design Technology," Ohmsha Publishing and IOS Press, Tokyo, Japan.

Iida, T., and Suzuki, Y. (1999). "International experimentation aspect of satellite communications toward GII," 50th International Astronautical Congress (IAF'99), IAF-99-M.1.05. (in press in *Space Technology*).

Iida, T., and Pelton, J. N. (1993). "Conceptual study of advanced Asia–Pacific telecommunications satellite for future gigabit transmission," *Space Commun.* **11,** 193–203.

Maral, G., and Bousquet, M. (1993). "Satellite Communications Systems," John Wiley & Son, New York.

Ohmori, S., Wakana, H., and Kawase, S. (1997). "Mobile Satellite Communications," Artech House, Boston.

Pelton, J. N. (1996). "Satellite and Wireless Telecommunications," Prentice Hall, Englewood Cliffs, NJ.

Pelton, J. N. (1998). "Telecommunications for the 21st Century," *Sci. Am.*

Pratt, T., and Bostian, C. W. (1986). "Satellite Communications," John Wiley & Sons, New York.

Special issue (1996). *The Role of Satellites in the GII, Space Commun.* **14,** 7–14.

Special issue (May 1986). *Special Issue on Japan's CS (Sakura) Communications Satellite Experiments, IEEE Trans. Aerosp. Electron. Syst.* **AES-22,** 222–323.

Wakana, H., Saito, H., Yamamoto, S., Ohkawa, M., Obara N., Li, H.-B., and Tanaka, M. (2000). "COMETS experiments for advanced mobile satellite communications and advanced satellite broadcasting," *Int. J. Satellite Commun.* **18,** 63–85.

Communication Systems, Civilian

Simon Haykin
McMaster University

GLOSSARY

Amplitude modulation (AM) Process in which the amplitude of a carrier is varied about a mean value, linearly with a baseband signal.

Average probability of symbol error Probability that the symbol recreated at the receiver output of a digital communication system differs from that transmitted, on the average.

Baseband Bank of frequencies representing an original signal as delivered by a source of information.

Bit Acronym for binary digit.

Digital subscriber line (DSL) A system that connects a user terminal (e.g., computer) to a telephone company's central office.

Discrete multitone (DMT) A form of digital modulation using the discrete Fourier transform.

Double-sideband suppressed-carrier (DSBSC) modulation Modulated wave resulting from multiplying a baseband signal by carrier wave.

Frequency modulation (FM) Form of angle modulation in which the instantaneous frequency is varied linearly with the baseband signal.

Frequency-shift keying (FSK) Digital modulation in which the frequency of a carrier takes on a discrete value chosen from a pre-selected set in accordance with the input data.

Orthogonal frequency-division multiplexing (OFDM) Another form of multichannel modulation that differs from DMT in areas of application and some aspects of design.

Phase modulation (PM) Form of angle modulation in which the angle of a carrier is varied linearly with the baseband signal.

Phase-shift keying (PSK) Digital modulation in which the phase of a carrier takes on a discrete value chosen from a preselected set in accordance with the input data.

Prediction Filtering operation used to make an inference about the future value of a baseband signal, given knowledge of past behavior of the signal up to a certain point in time.

Pulse-analog modulation Modulation process in which some characteristic feature (e.g., amplitude, duration, or position) of each pulse in a periodic pulse train (used as carrier) is varied in accordance with the baseband signal.

Encyclopedia of Physical Science and Technology, Third Edition, Volume 3

Pulse-code modulation (PCM) Process that involves the uniform sampling of a baseband signal, quantization of the amplitude of each sample to the nearest one of a finite set of allowable values, and representation of each quantized sample by a code.

Signal-to-noise ratio Ratio of the average power of a message signal to the average power of additive noise, both being measured at a point of interest in the system (e.g., receiver output).

Single-sideband (SSB) modulation Form of amplitude modulation in which only the upper sideband or lower sideband is transmitted.

Vestigial sideband (VSB) modulation Form of amplitude modulation in which one sideband is passed almost completely, whereas just a vestige, or trace, of the other sideband is retained.

I. MODEL OF A COMMUNICATION SYSTEM

Figure 1 shows the block diagram of a communication system. The system consists of three major parts: (1) transmitter, (2) communication channel, and (3) receiver. The main purpose of the transmitter is to change the message signal into a form suitable for transmission over the channel. This modification is achieved by means of a process known as modulation.

The communication channel may be a transmission line (as in telephony and telegraphy), an optical fiber (as in optical communications), or merely free space in which the signal is radiated as an electromagnetic wave (as in wireless communications and radio and television broadcasting). In propagating through the channel, the transmitted signal is distorted because of nonlinearities or imperfections in the frequency response of the channel. Other sources of degradation are noise and interference picked up by the signal during the course of transmission through the channel. Noise and distortion constitute two basic problems in the design of communication systems. Usually, the transmitter and receiver are carefully designed so as to minimize the effects of noise and distortion on the quality of reception.

The main purpose of the receiver is to recreate the original message signal from the degraded version of the transmitted signal after propagation through the channel. This recreation is accomplished by a process known as demodulation, which is the reverse of the modulation process in the transmitter. However, owing to the unavoidable presence of noise and distortion in the received signal, the receiver cannot recreate the original message signal exactly. The resulting degradation in overall system performance is influenced by channel characteristics and the type of modulation scheme used. Some modulation schemes are less sensitive to the effects of noise and distortion than others.

In any communication system, there are two primary communication resources, namely, transmitted power and channel bandwidth. A general system design objective is to use these two resources as efficiently as possible. In most communication channels, one resource may be considered more important than the other. We therefore classify communication channels as power-limited or band-limited. For example, a telephone circuit is a typical band-limited channel, whereas a space communication link or a satellite channel is typically power-limited.

When the spectrum of a message signal extends down to zero or low frequencies, we define the bandwidth of the signal as that upper frequency above which the spectral content of the signal is negligible and, therefore, unnecessary for transmitting the pertinent information. For example, the average voice spectrum extends well beyond 10 kHz, though most of the energy is concentrated in the range of 100 to 600 Hz, and a band from 300 to 3400 Hz gives good articulation. Accordingly, telephone circuits that respond well to the latter range of frequencies give quite satisfactory commercial telephone service.

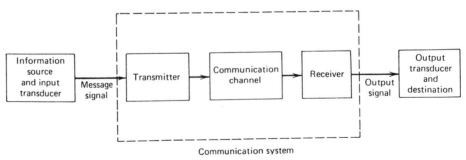

FIGURE 1 Model of an electrical communication system. [From Haykin, S. (1983). "Communication Systems," 2nd ed. Wiley, New York. © John Wiley & Sons, Inc.]

II. ANALOG MODULATION

Modulation is formally defined as the process by which some characteristic of a carrier is varied in accordance with a modulating wave. The baseband signal is referred to as the modulating wave, and the result of the modulation process is referred to as the modulated wave.

In analog or continuous-wave (cw) modulation, the modulating wave consists of an analog signal (e.g., voice signal, video signal), and the carrier consists of a sine wave. Basically, there are two types of analog modulation: amplitude modulation and angle modulation. In amplitude modulation the amplitude of the sinusoidal carrier wave is varied in accordance with the baseband signal. In angle modulation, on the other hand, the angle of the sinusoidal carrier wave is varied in accordance with the baseband signal.

A. Amplitude Modulation

Consider a sinusoidal carrier wave $c(t)$ defined by

$$c(t) = A_c \cos(2\pi f_c t) \qquad (1)$$

where A_c is the carrier amplitude and f_c the carrier frequency. For convenience, we have assumed that the phase of the carrier wave is zero in Eq. (1). Let $m(t)$ denote the baseband signal that carries the specification of the message. The carrier wave $c(t)$ is independent of $m(t)$. An amplitude-modulated (AM) wave is described by

$$s(t) = A_c[1 + k_a m(t)] \cos(2\pi f_c t) \qquad (2)$$

where k_a is a constant called the amplitude sensitivity of the modulator.

The envelope of the AM wave $s(t)$ equals

$$a(t) = A_c|1 + k_a m(t)| \qquad (3)$$

where $|\cdot|$ denotes the absolute value of the enclosed quantity. The envelope of $s(t)$ has the same shape as the baseband signal $M(t)$, provided that two requirements are satisfied:

1. The amplitude of $k_a m(t)$ is always less than unity, that is,

$$|k_a m(t)| < 1 \quad \text{for all } t$$

The absolute maximum value of $k_a m(t)$ multiplied by 100 is referred to as the percentage modulation.

2. The carrier frequency f_c is much greater than the highest frequency component of the baseband signal $m(t)$.

The spectrum of the AM wave $s(t)$ consists of a carrier, upper sideband, and lower sideband. Let W denote the highest frequency component of the baseband signal $m(t)$. For positive frequencies, the carrier is located at f_c, the upper sideband extends from f_c to $f_c + W$, and the lower sideband extends from $f_c - W$ to f_c. Hence, the transmission bandwidth of the AM wave equals $2W$, that is, exactly twice the message bandwidth. For negative frequencies, the spectrum of the AM wave is the mirror image of that for positive frequencies.

Figure 2 illustrates the amplitude modulation process. Figure 2a shows a modulating wave $m(t)$ that consists of a single tone or frequency component, Fig. 2b the sinusoidal carrier wave, and Fig. 2c the corresponding AM wave. The amplitude spectra of the modulating wave, the carrier, and the AM wave are shown on the right-hand side of the figure.

1. Double-Sideband Suppressed-Carrier Modulation

The carrier wave $c(t)$ is completely independent of the information-carrying signal or baseband signal $m(t)$, which means that the transmission of the carrier wave represents a waste of power. This points to a shortcoming of amplitude modulation, namely, that only a fraction of the total transmitted power is affected by $m(t)$. To overcome this shortcoming, we may suppress the carrier component from the modulated wave, resulting in double-sideband suppressed-carrier (DSBSC) modulation. Thus, by suppressing the carrier, we obtain a modulated wave that is proportional to the product of the carrier wave and the baseband signal.

To describe a DSBSC-modulated wave as a function of time, we write

$$s(t) = c(t)m(t) = A_c \cos(2\pi f_c t)m(t) \qquad (4)$$

This modulated wave undergoes a phase reversal whenever the baseband signal $m(t)$ crosses zero. Accordingly, unlike amplitude modulation, the envelope of a DSBSC-modulated wave is different from the baseband signal. For obvious reasons, a device that performs the operation described in Eq. (4) is called a product modulator.

The transmission bandwidth required by DSBSC modulation is the same as that for amplitude modulation, namely, $2W$. Figure 2d illustrates the DSBSC-modulated wave for single-tone modulation and the corresponding amplitude spectrum.

2. Quadrature-Carrier Multiplexing

The quadrature-carrier multiplexing or quadrature-amplitude modulation (QAM) scheme enables two DSBSC-modulated waves (resulting from the application of two independent message signals) to occupy the same

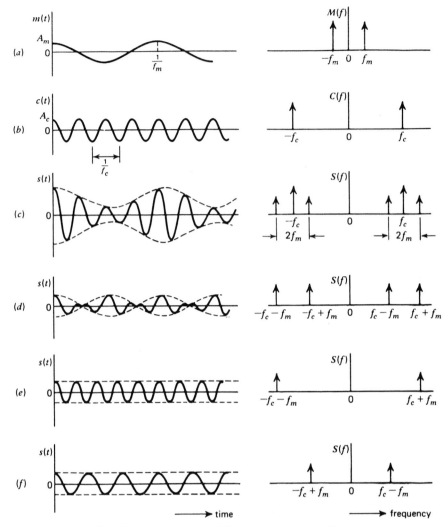

FIGURE 2 Time-domain (left) and frequency-domain (right) characteristics of different modulated waves produced by a single tone. (a) Modulating wave. (b) Carrier wave. (c) AM wave. (d) DSBSC wave. (e) SSB wave with the upper-side frequency transmitted. (f) SSB wave with the lower-side frequency transmitted. [From Haykin, S. (1983). "Communication Systems," 2nd ed. Wiley, New York. © John Wiley & Sons, Inc.]

transmission bandwidth, and yet it allows for the separation of the two message signals at the receiver output. It is therefore a bandwidth-conservation scheme. The transmitter part of the system involves the use of two separate product modulators that are supplied with two carrier waves of the same frequency but differing in phase by 90°. The transmitted signal $s(t)$ consists of the sum of these two product modulator outputs, as shown by

$$s(t) = A_c m_1(t) \cos(2\pi f_c t) + A_c m_2(t) \sin(2\pi f_c t) \quad (5)$$

where $m_1(t)$ and $m_2(t)$ denote the two different message signals applied to the product modulators. Thus, $s(t)$ occupies a transmission bandwidth of $2W$, centered at the carrier frequency f_c, where W is the message bandwidth of $m(t)$ or $m_2(t)$.

Quadrature-carrier multiplexing is used in color television. In this system, synchronizing pulses are transmitted in order to maintain the local oscillator in the receiver at the correct frequency and phase with respect to the carrier used in the transmitter.

3. Single-Sideband Modulation

Amplitude modulation and DSBSC modulation are wasteful of bandwidth because they both require a transmission bandwidth equal to twice the message bandwidth. In either case, one-half of the transmission bandwidth is occupied by the upper sideband of the modulated wave, whereas the other half is occupied by the lower sideband. However, the upper and lower sidebands are uniquely related to each other by virtue of their symmetry about the carrier frequency; that is, given the amplitude and phase spectra of either sideband, we can uniquely determine the other. This means that insofar as the transmission of information is concerned, only one sideband is necessary, and if the carrier and the other sidebands are suppressed at the transmitter, no information is lost. Thus, the communication channel has to provide only the same bandwidth as the baseband signal, a conclusion that is intuitively satisfying. When only one sideband is transmitted, the modulation system is referred to as a single-sideband (SSB) system.

The precise frequency–domain description of an SSB wave depends on which sideband is transmitted. In any event, the essential function of an SSB modulation system is to translate the spectrum of the modulating wave, either with or without inversion, to a new location in the frequency domain; the transmission bandwidth requirement of the system is one-half that of an amplitude or DSBSC modulation system. The benefit of using SSB modulation is therefore derived principally from the reduced bandwidth requirement and the elimination of the high-power carrier wave. The principal disadvantage of the SSB modulation system is its cost and complexity.

The time-domain description of the SSB wave is much more complicated than that of the DSBSC wave. Specifically, we write

$$s(t) = \tfrac{1}{2}A_c m(t)\cos(2\pi f_c t) \pm \tfrac{1}{2}A_c \hat{m}(t)\sin(2\pi f_c t) \quad (6)$$

where $\hat{m}(t)$ is the Hilbert transform of the baseband signal $m(t)$. A Hilbert transformer consists of a two-port device that leaves the amplitudes of all frequency components of the input signal unchanged, but it produces a phase shift of $-90°$ for all positive frequency components of the input signal and a phase shift of $+90°$ for all negative frequency components. The minus sign on the right-hand side of Eq. (6) refers to an SSB-modulated wave that contains only the upper sideband, whereas the plus sign refers to an SSB-modulated wave that contains only the lower sideband.

Figure 2e shows the SSB wave (with upper-side frequency) resulting from the use of single-tone modulation. Figure 2f shows the corresponding SSB wave with lower-side frequency.

4. Vestigial Sideband Modulation

Single-sideband modulation is rather well suited for the transmission of voice because of the energy gap in the spectrum of voice signals between zero and a few hundred hertz. When the baseband signal contains significant components at extremely low frequencies (as in the case of television and telegraph signals and computer data), however, the upper and lower side-bands meet at the carrier frequency. This means that the use of SSB modulation is inappropriate for the transmission of such baseband signals because of the difficulty of isolating one sideband. This difficulty suggests another scheme known as vestigial sideband (VSB) modulation, which is a compromise between SSB and DSBSC modulation. In this modulation scheme, one side-band is passed almost completely, whereas just a trace, or vestige, of the other sideband is retained. The transmission bandwidth required by a VSB system is therefore

$$B_T = W + f_v \quad (7)$$

where W is the message bandwidth and f_v the width of the vestigial sideband.

Vestigial sideband modulation has the virtue of conserving bandwidth almost as efficiently as SSB modulation, while retaining the excellent low-frequency baseband characteristics of double-sideband modulation. Thus, VSB modulation has become standard for the transmission of television and similar signals where good phase characteristics and transmission of low-frequency components are important but the bandwidth required for double-sideband transmission is unavailable or uneconomical.

It is of interest, however, that in commercial television broadcasting the transmitted signal is not quite VSB-modulated, because the shape of the transition region is not rigidly controlled at the transmitter. Instead, a VSB filter is inserted in each receiver. The overall performance is the same as VSB modulation except for some wasted power and bandwidth. Figure 3 shows the idealized frequency response for a VSB filter designed for television receivers.

B. Angle Modulation

In angle modulation the angle of the carrier wave is varied according to the baseband signal. In this method of modulation the amplitude of the carrier wave is maintained constant. An important feature of angle modulation is that it can provide better discrimination against noise and interference than amplitude modulation. However, this improvement in performance is achieved at the expense of increased transmission bandwidth; that is, angle modulation provides a practical means of exchanging transmission bandwidth for improved noise performance. Such a trade-off is not possible with amplitude modulation.

Two forms of angle modulation are distinguished: phase modulation and frequency modulation. These two methods of modulation are closely related in that the properties of one can be derived from those of the other. It is customary to define these two forms of angle modulation as follows:

1. Phase modulation (PM) is that form of angle modulation in which the angle $\theta_i(t)$ is varied linearly with the baseband signal $m(t)$, as shown by

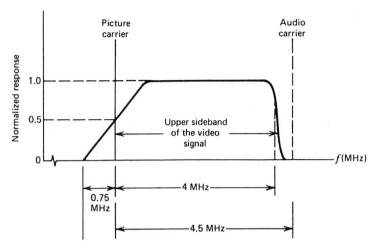

FIGURE 3 Frequency response of a VSB filter used in television receivers. [From Haykin, S. (1983). "Communication Systems," 2nd ed. Wiley, New York. © John Wiley & Sons, Inc.]

$$\theta_i(t) = 2\pi f_c t + k_p m(t) \qquad (8)$$

The term $2\pi f_c t$ represents the angle of the unmodulated carrier, and the constant k_p represents the phase sensitivity of the modulator. For convenience, we have assumed in Eq. (8) that the angle of the unmodulated carrier is zero at $t = 0$. The PM wave $s(t)$ is thus described in the time domain by

$$s(t) = A_c \cos[2\pi f_c t + k_p m(t)] \qquad (9)$$

2. Frequency modulation (FM) is that form of angle modulation in which the instantaneous frequency $f_i(t)$ is varied linearly with the baseband signal $m(t)$, as shown by

$$f_i(t) = \frac{1}{2\pi}\frac{d_i(t)}{dt} = f_c + k_f m(t) \qquad (10)$$

The term f_c represents the frequency of the unmodulated carrier, and the constant k_f represents the frequency sensitivity of the modulator. Integrating Eq. (10) with respect to time and multiplying the result by 2π, we get

$$\theta_i(t) = 2\pi f_c t + 2\pi k_f \int_0^t m(t)\, dt \qquad (11)$$

where, for convenience, we have assumed that the angle of the unmodulated carrier wave is zero at $t = 0$. The FM is therefore described in the time domain by

$$s(t) = A_c \cos\left[2\pi f_c t + 2\pi k_f \int_0^t m(t)\, dt\right] \qquad (12)$$

A consequence of allowing the angle $\theta_i(t)$ to become dependent on the message signal $m(t)$ as in Eq. (8) or on its integral as in Eq. (11) is that the zero crossings of PM wave or FM wave no longer have a perfect regularity in their spacing; zero crossings refer to the instants of time

at which a waveform changes from negative to positive value or vice versa.

The differences between amplitude-modulated and angle-modulated waves are illustrated in Fig. 4 for the case of single-tone modulation. Figures 4a and b refer to the sinusoidal carrier and modulating waves, respectively; Figs. 4c, d, and e show the corresponding AM, PM, and FM waves, respectively.

Figure 5a shows that an FM wave can be generated by first integrating $m(t)$ and then using the result as the input to a phase modulator. Conversely, a PM wave can be generated by first differentiating $m(t)$ and then using the result as the input to a frequency modulator, as in Fig. 5b.

Consider an FM wave produced by using a single-tone modulating wave of frequency f_m. Let Δf denote the frequency deviation defined as the maximum departure of the instantaneous frequency of the FM wave from the carrier frequency f_c. The ratio $\Delta f/f_m$ is called the modulation index, commonly denoted by β. When β is less than 0.5, the modulated wave is referred to as narrowband FM. When β is large, it is referred to as wide-band FM.

1. Transmission Bandwidth of FM Waves

In theory, an FM wave contains an infinite number of side frequencies so that the bandwidth required to transmit such a signal is similarly infinite in extent. In practice, however, the FM wave is effectively limited to a finite number of significant side frequencies compatible with a specified amount of distortion. We can therefore specify an effective bandwidth required for the transmission of an FM wave.

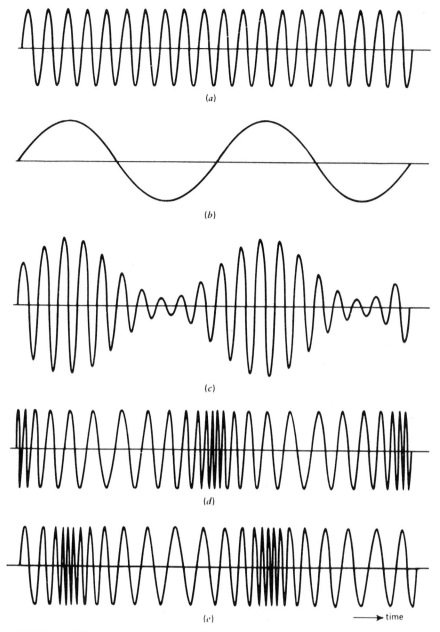

FIGURE 4 AM, PM, and FM waves produced by a single tone. (a) Carrier wave. (b) Sinusoidal modulating wave. (c) AM wave. (d) PM wave. (e) FM wave. [From Haykin, S. (1983). "Communication Systems," 2nd ed. Wiley, New York. © John Wiley & Sons, Inc.]

For the case of single-tone modulation, we calculate the transmission bandwidth of the FM wave by using Carson's rule:

$$B_T \simeq 2\Delta f + 2f_m = 2\Delta f(1 + 1/\beta) \qquad (13)$$

When β is small compared with 1 rad, as in the case of narrow-band FM, the transmission bandwidth B_T is approximately $2f_m$. On the other hand, in the case of wide-band FM for which β is large compared with 1 rad, the transmission bandwidth B_T is approximately $2\Delta f$.

C. Effects of Noise

One of the basic issues in the study of modulation systems is the analysis of the effects of noise on the performance of the receiver and the use of the results of this analysis in system design. Another matter of interest is the comparison of the noise performances of different modulation–demodulation schemes. In order to carry out this analysis, we obviously need a criterion that describes in a meaningful manner the noise performance of the modulation

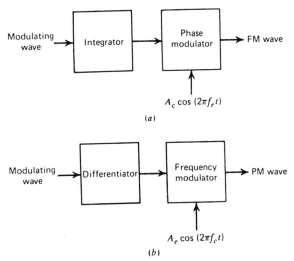

FIGURE 5 Relationship between frequency modulation and phase modulation. (a) Scheme for generating an FM wave by using a phase modulator. (b) Scheme for generating a PM wave by using a frequency modulator. [From Haykin, S. (1983). "Communication Systems," 2nd ed. Wiley, New York. © John Wiley & Sons, Inc.]

system under study. In the case of cw modulation systems, the customary practice is to use the output signal-to-noise ratio as an intuitive measure for describing the fidelity with which the demodulation process in the receiver recovers the original message from the received modulated signal in the presence of noise. The output signal-to-noise ratio is defined as the ratio of the average power of the message signal to the average power of the noise, both measured at the receiver output. Such a criterion is perfectly well defined as long as the message signal and noise appear additively at the receiver output.

In performing noise analysis, it is customary to model the noise as white, Gaussian, and stationary. Under this condition, we find that for the same average transmitted (or modulated message) signal power and the same average noise power in the message bandwidth, an SSB receiver will have exactly the same output signal-to-noise ratio as a DSBSC receiver, when both receivers use coherent detection for the recovery of the message signal. Furthermore, in both cases, the noise performance of the receiver is the same as that obtained by simply transmitting the message signal itself in the presence of the same noise. The only effect of the modulation process is to translate the message signal to a different frequency band.

Coherent detection assumes the availability of a local carrier in the receiver with the same frequency and phase as that of the carrier in the transmitter. The noise performance of an AM receiver is always inferior to that of a DSBSC or SSB receiver. This is due to the wastage of transmitter power that results from transmitting the carrier as a

component of the AM wave. Also, an AM receiver suffers from a threshold effect that arises when the carrier-to-noise ratio is small compared with unity. By threshold we mean a value of the carrier-to-noise ratio below which the noise performance of a detector deteriorates much more rapidly than proportionately to the carrier-to-noise ratio. A detailed analysis of the threshold effect in envelope detectors is complicated, however.

Since frequency modulation is a nonlinear modulation process, the noise analysis of an FM receiver is also quite complicated. Nevertheless, we can state that when the carrier-to-noise ratio is high, an increase in the transmission bandwidth B_T of the FM system provides a corresponding quadratic increase in the output signal-to-noise ratio of the system. However, this result is valid only if the carrier-to-noise ratio is high compared with unity. It is found experimentally that as the input noise is increased so that the carrier-to-noise ratio is decreased, the FM receiver breaks. At first, individual clicks are heard in the receiver output, and as the carrier-to-noise ratio decreases still further, the clicks rapidly merge into a crackling or sputtering sound. Under this condition, the FM receiver is said to suffer from the threshold effect.

1. Preemphasis and Deemphasis

The noise performance of an FM receiver can be improved significantly by the use of preemphasis in the transmitter and deemphasis in the receiver. With this method, we artificially emphasize the high-frequency components of the message signal before modulation in the transmitter and therefore before the noise is introduced in the receiver. In effect, the low-frequency and high-frequency portions of the power spectral density of the message are equalized in such a way that the message fully occupies the frequency band allotted to it. Then, at the discriminator output in the receiver, we perform the inverse operation by deemphasizing the high-frequency components, so as to restore the original signal-power distribution of the message. In such a process the high-frequency components of the noise at the discriminator output are also reduced, thereby effectively increasing the output signal-to-noise ratio of the system.

The use of the simple linear preemphasis and deemphasis filters is an example of how the performance of an FM system can be improved by utilizing the differences between characteristics of signals and noise in the system. These simple filters also find application in audio tape recording. In recent years nonlinear preemphasis and deemphasis techniques have been applied successfully to tape recording. These techniques (known as the Dolby-A, Dolby-B, and DBX systems) use a combination of filtering and dynamic range compression to reduce the effects of noise, particularly when the signal level is low.

2. Noise Comparison of Analog Modulation Schemes

Comparing the noise performance of different CW modulation schemes, we may make the following observations:

1. Among the family of AM schemes, SSB modulation is optimum.

2. The use of FM improves noise performance, but at the expense of an increased transmission bandwidth, assuming that the FM system operates above threshold for the noise improvement to be realized. The exchange of increased transmission bandwidth for improved noise performance follows a square law, which is the best that can be done with cw modulation.

III. PULSE AND DIGITAL MODULATION

In cw modulation, some parameter of a sinusoidal carrier wave is varied continuously in accordance with the message. This is in direct contrast to pulse modulation, in which some parameter of a pulse train is varied in accordance with the message. There are two basic types of pulse modulation: pulse-analog modulation and pulse-code modulation. In pulse-analog modulation systems, a periodic pulse train is used as the carrier wave, and some characteristic feature of each pulse (e.g., amplitude, duration, or position) is varied in a continuous manner in accordance with the pertinent sample value of the message. On the other hand, in pulse-code modulation, a discrete-time, discrete-amplitude representation is used for the signal, and as such it has no cw counterpart.

A. Sampling Theorem

An operation that is basic to the design of all pulse-modulation systems is the sampling process whereby an analog signal is converted to a corresponding sequence of numbers that are usually uniformly spaced in time. For such a procedure to have practical utility, it is necessary that we choose the sampling rate properly, so that this sequence of numbers uniquely defines the original analog signal. This is the essence of the sampling theorem, which can be stated in two equivalent ways:

1. A band-limited signal of finite energy, which has no frequency components higher than W Hz, is completely described by specifying the values of the signal at instants of time separated by $1/2W$ s.

2. A band-limited signal of finite energy, which has no frequency components higher than W Hz, can be completely recovered from a knowledge of its samples taken at the rate of $2W$ per second.

The sampling rate of $2W$ samples per second, for a signal bandwidth of W Hz, is often called the Nyquist rate; its reciprocal $1/2W$ is called the Nyquist interval. The sampling theorem serves as the basis for the interchangeability of analog signals and digital sequences, which is so valuable in digital communication systems.

The derivation of the sampling theorem, as described above, is based on the assumption that the signal $g(t)$ is strictly band-limited. Such a requirement, however, can be satisfied only if $g(t)$ has infinite duration. In other words, a strictly band-limited signal cannot be simultaneously time-limited, and vice versa. Nevertheless, we can consider a time-limited signal to be essentially bandlimited in the sense that its frequency components outside some band of interest have negligible effects. We can then justify the practical application of the sampling theorem.

1. Aliasing

When the sampling rate $1/T_s$, exceeds the Nyquist rate $2W$, there is no problem in recovering the original signal $g(t)$ from its sampled version $g_\delta(t)$. When the sampling rate $1/T_s$ is less than $2W$, however, the aliasing effect arises. That is, a high-frequency component in the spectrum of the original signal seemingly takes on the identity of a lower frequency component in the spectrum of its sampled version. It is evident that if the sampling rate $1/T_s$ is less than the Nyquist rate $2W$, the original signal cannot be recovered exactly from its sampled version, and information is thereby lost in the sampling process.

Another factor that contributes to aliasing is the fact that a signal cannot be finite in both time and frequency, as mentioned previously. Since this violates the strict bandlimited requirement of the sampling theorem, we find that whenever a time-limited signal is sampled, there will always be some aliasing produced by the sampling process. Accordingly, in order to combat the effects of aliasing in practice, we use two corrective measures:

1. Before sampling, a low-pass anti-alias filter is used to attenuate those high-frequency components of the signal that lie outside the band of interest.

2. The filtered signal is sampled at a rate slightly higher than the Nyquist rate.

For example, a bandwidth of 3.4 kHz is considered adequate for the transmission of voice signals over a telephone network. In this basis, a sampling rate of 8 kHz (slightly greater than twice 3.4 kHz) is considered the standard for the digital transmission of voice signals.

B. Pulse-Analog Modulation

In pulse-amplitude modulation (PAM), the amplitudes of regularly spaced rectangular pulses vary with the instantaneous sample values of a continuous message signal in a one-to-one fashion. This method of modulation is illustrated in Figs. 6a, b, and c, which represent a message signal, the pulse carrier, and the corresponding PAM wave, respectively.

The PAM wave $s(t)$ is easily demodulated by a low-pass filter with a cutoff frequency just large enough to accommodate the highest frequency component of the message signal $m(t)$. However, the reconstructed signal exhibits a slight amplitude distortion caused by the aperture effect due to lengthening of the samples.

In pulse-duration modulation (PDM), the samples of the message signal are used to vary the duration of the individual pulses. This form of modulation is also referred to as pulse-width modulation or pulse-length modulation. The modulating wave may vary with the time of occurrence of the leading edge, the trailing edge, or both edges of the pulse. In Fig. 6d the trailing edge of each pulse is varied in accordance with the message signal.

In PDM, long pulses expend considerable power during the pulse while bearing no additional information. If this unused power is subtracted from PDM, so that only time transitions are preserved, we obtain a more efficient type of pulse modulation known as pulse-position modulation (PPM). In PPM the position of a pulse relative to its unmodulated time of occurrence is varied in accordance with the message signal (Fig. 6e).

C. Pulse-Code Modulation

In PAM, PDM, and PPM, only time is expressed in discrete form, whereas the respective modulation parameters (namely, pulse amplitude, duration, and position) are varied in a continuous manner in accordance with the message. Thus, in these modulation systems, information transmission is accomplished in analog form at discrete times. On the other hand, in pulse-code modulation (PCM), the message signal is sampled and the amplitude of each sample is rounded off to the nearest one of a finite set of allowable values, so that both time and amplitude are in discrete form. This allows the message to be transmitted by means of coded electrical signals, thereby distinguishing PCM from all other methods of modulation.

The use of digital representation of analog signals (e.g., voice, video) offers the following advantages: (1) rugged-

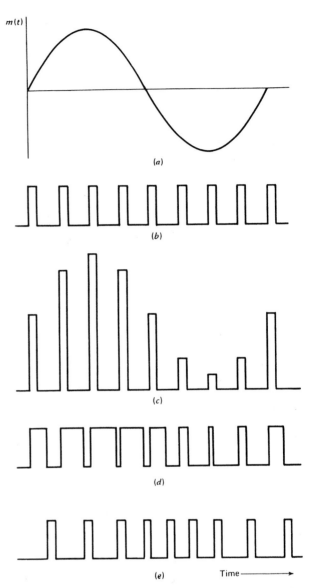

FIGURE 6 Different forms of pulse-analog modulation for the case of a sinusoidal modulating wave. (a) Modulating wave. (b) Pulse carrier. (c) PAM wave. (d) PDM wave. (e) PPM wave. [From Haykin, S. (1983). "Communication Systems," 2nd ed. Wiley, New York. © John Wiley & Sons, Inc.]

ness to transmission noise and interference, (2) efficient regeneration of the coded signal along the transmission path, and (3) the possibility of a uniform format for different kinds of baseband signals. These advantages, however, are attained at the cost of increased transmission bandwidth requirement and increased system complexity. With the increasing availability of wide-band communication channels, coupled with the emergence of the requisite device technology, the use of PCM has become a practical reality.

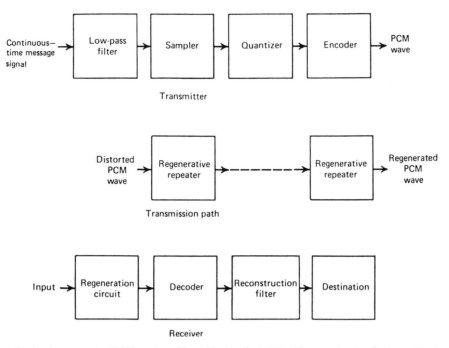

FIGURE 7 Basic elements of a PCM system. [From Haykin, S. (1983). "Communication Systems," 2nd ed. Wiley, New York. © John Wiley & Sons, Inc.]

1. Elements of PCM

Pulse-code modulation systems are considerably more complex than PAM, PDM, and PPM systems, in that the message signal is subjected to a greater number of operations. The essential operations in the transmitter of a PCM system are sampling, quantizing, and encoding (Fig. 7). The quantizing and encoding operations are usually performed in the same circuit, which is called an analog-to-digital converter. The essential operations in the receiver are regeneration of impaired signals, decoding, and demodulation of the train of quantized samples. Regeneration usually occurs at intermediate points along the transmission route as necessary.

2. Sampling

The incoming message wave is sampled with a train of narrow rectangular pulses so as to approximate closely the instantaneous sampling process. In order to ensure perfect reconstruction of the message at the receiver, the sampling rate must be greater than twice the highest frequency component W of the message wave in accordance with the sampling theorem. In practice, a low-pass filter is used at the front end of the sampler in order to exclude frequencies greater than W before sampling. Thus, the application of sampling permits the reduction of the continuously varying message wave to a limited number of discrete values per second.

3. Quantizing

A continuous signal, such as voice, has a continuous range of amplitudes, and therefore its samples have a continuous amplitude range. In other words, within the finite amplitude range of the signal we find an infinite number of amplitude levels. It is not necessary in fact to transmit the exact amplitudes of the samples. Any human sense (the ear or the eye), as ultimate receiver, can detect only finite intensity differences. This means that the original continuous signal can be approximated by a signal constructed of discrete amplitudes selected on a minimum-error basis from an available set. The existence of a finite number of discrete amplitude levels is a basic condition of PCM. Clearly, if we assign the discrete amplitude levels with sufficiently close spacing, we can make the approximated signal practically indistinguishable from the original continuous signal.

The conversion of an analog (continuous) sample of the signal to a digital (discrete) form is called the quantizing process.

4. Encoding

In combining the processes of sampling and quantizing, the specification of a continuous baseband signal becomes limited to a discrete set of values, but not in the form best suited to transmission over a line or radio path. To exploit the advantages of sampling and quantizing, we require the use of an encoding process to translate the discrete set of sample values to a more appropriate form of signal. Any plan for representing each of this discrete set of values as a particular arrangement of discrete events is called a code. One of the discrete events in a code is called a code element or symbol. For example, the presence or absence of a pulse is a symbol. A particular arrangement of symbols used in a code to represent a single value of the discrete set is called a code word or character.

In a binary code, each symbol can be either of two distinct values or kinds, such as the presence or absence of a pulse. The two symbols of a binary code are customarily denoted 0 and 1. In a ternary code, each symbol can be one of three distinct values or kinds, and so on for other codes. However, the maximum advantage over the effects of noise in a transmission medium is obtained by using a binary code, because a binary symbol withstands a relatively high level of noise and is easy to regenerate.

Suppose that in a binary code, each code word consists of n bits (bit is an acronym for binary digit). Then, using such a code, we can represent a total of 2^n distinct numbers. For example, a sample quantized into one of 128 levels can be represented by a seven-bit code word.

5. Regeneration

The most important feature of PCM systems is the ability to control the effects of distortion and noise produced by transmitting a PCM wave through a channel. This capability is accomplished by reconstructing the PCM wave by means of a chain of regenerative repeaters located at sufficiently close spacing along the transmission route.

6. Decoding

The first operation in the receiver is to regenerate (i.e., reshape and clean up) the received pulses one last time. These clean pulses are then regrouped into code words and decoded (i.e., mapped back) into a quantized PAM signal. The decoding process involves generating a pulse the amplitude of which is the linear sum of all the pulses in the code word, with each pulse weighted by its place value (2^0, 2^1, 2^2, 2^3, ...) in the code.

7. Filtering

The final operation in the receiver is to recover the signal wave by passing the decoder output through a low-pass reconstruction filter whose cutoff frequency is equal to the message bandwidth W. Assuming that the transmission path is error free, the recovered signal includes no noise, with the exception of the initial distortion introduced by the quantization process.

8. Noise in PCM Systems

The performance of a PCM system is influenced by two major sources of noise:

1. Transmission noise, which may be introduced anywhere between the transmitter output and the receiver input. The effect of transmission noise is to introduce bit errors into the received PCM wave, with the result that in the case of a binary system, a symbol 1 occasionally is mistaken for a symbol 0, or vice versa. Clearly, the more frequently such errors occur, the more dissimilar to the original message signal is the receiver output. The fidelity of information transmission by PCM in the presence of transmission noise is conveniently measured in terms of the error rate, or probability of error, defined as the probability that the symbol at the receiver output differs from that transmitted.

2. Quantizing noise, which is introduced in the transmitter and is carried along to the receiver output. Typically, each bit in the code word of a PCM system contributes 6 dB to the signal-to-noise ratio at the receiver output.

Unlike frequency modulation, PCM offers an exponential law for the trade-off between transmission bandwidth and noise performance. It is for this reason that PCM is the preferred method for the transmission of voice and video signals.

D. Differential Pulse-Code Modulation

When a voice or video signal is sampled at a rate higher than the Nyquist rate, the resulting sampled signal exhibits high correlation between adjacent samples. The meaning of this high correlation is that in an average sense, the signal does not change rapidly from one sample to the next, with the result that the difference between adjacent samples has a variance that is smaller than the variance of the signal itself. When these highly correlated samples are encoded, as in a standard PCM system, the resulting encoded signal contains redundant information. This means that symbols that are not absolutely essential to the transmission of information are generated as a result of the encoding process. By removing this redundancy before encoding, we obtain a more efficient coded signal.

Now, if we know a sufficient part of a redundant signal, we can infer the rest, or at least make the most probable estimate. In particular, if we know the past behavior of a signal up to a certain point in time, it is possible to make some inference about its future values; such a process is commonly called prediction. The fact that it is possible to predict future values of the signal $m(t)$ provides motivation for the differential quantization scheme shown in Fig. 8. The input to the quantizer is a signal that consists of the

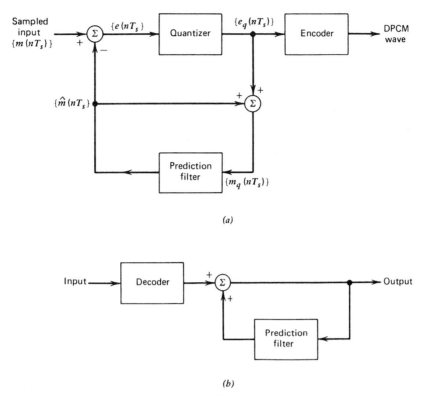

FIGURE 8 DPCM system. (a) Transmitter. (b) Receiver. [From Haykin, S. (1983). "Communication Systems," 2nd ed. Wiley, New York. © John wiley & Sons, Inc.]

difference between the unquantized input symbol $m(nT_s)$. This predicted value is produced by using a prediction filter. The difference signal $e(nT_s)$ is called a prediction error. By encoding the quantizer output, as in Fig. 8a, we obtain a variation of PCM known as differential pulse-code modulation (DPCM). This encoded signal is used for transmission.

The receiver for reconstructing the quantized version of the input is shown in Fig. 8b. It consists of a decoder to reconstruct the quantized error signal. The quantized version of the original input is reconstructed from the decoder output using the same design of prediction filter in the transmitter.

E. Delta Modulation

The exploitation of signal correlations in DPCM suggests the further possibility of oversampling a baseband signal (i.e., at a rate much higher than the Nyquist rate) purposely to increase the correlation between adjacent samples of the signal, so as to permit the use of a simple quantizing strategy for constructing the encoded signal. Delta modulation (DM), which is the one-bit (or two-level) version of DPCM, is precisely such a scheme.

In its simple form, DM provides a staircase approximation to the oversampled version of an input baseband

signal (Fig. 9). The difference between the input and the approximation is quantized into only two levels, namely, $\pm\delta$, corresponding to positive and negative differences, respectively. Thus, if the approximation falls below the signal at any sampling epoch, it is increased by δ. If, on the other hand, the approximation lies above the signal, it is diminished by δ. Provided that the signal does not change too rapidly from sample to sample, the staircase approximation remains within $\pm\delta$ of the input signal.

The principal virtue of DM is its simplicity. It can be generated by applying the sampled version of the incoming baseband signal to a modulator that involves a summer, quantizer, and accumulator interconnected (Fig. 10).

In comparing the DPCM and DM networks of Figs. 8 and 10, respectively, we note that they are basically similar, except for two important differences, namely, the use of a one-bit (two-level) quantizer in the delta modulator and the replacement of the prediction filter by a single delay element.

IV. DATA TRANSMISSION

In this section we consider some of the issues involved in transmitting digital data (of whatever origin) over a communication channel. The transmission may be in baseband

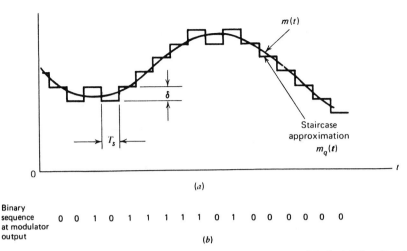

FIGURE 9 Delta modulation. [From Haykin, S. (1983). "Communication Systems," 2nd ed. Wiley, New York. © John Wiley & Sons, Inc.]

or passband form. For baseband transmission, we use discrete pulse modulation, in which the amplitude, duration, or position of the transmitted pulses is varied in a discrete manner in accordance with the digital data. However, it is customary to use discrete PAM because it is the most efficient in terms of power and bandwidth utilization.

To transmit digital data over bandpass channels (e.g., satellite channels) we require the use of digital modulation techniques whereby the amplitude, phase, or frequency

of a sinusoidal carrier is varied in a discrete manner in accordance with the digital data. In practice, the variation of phase or frequency of the carrier is the preferred method.

A. Baseband Data Transmission

The basic elements of a baseband binary PAM system are shown in Fig. 11. The signal applied to the input of the

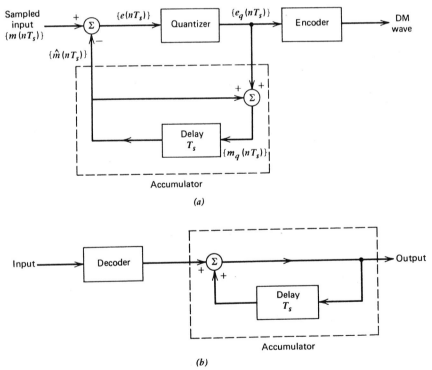

FIGURE 10 DM system. (a) Transmitter. (b) Receiver. [From Haykin, S. (1983). "Communication Systems," 2nd ed. Wiley, New York. © John Wiley & Sons, Inc.]

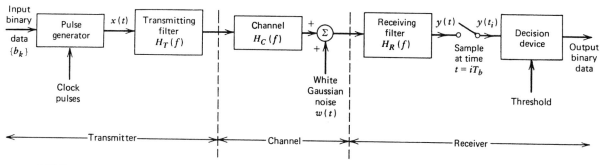

FIGURE 11 Baseband binary data transmission system. [From Haykin, S. (1983). "Communication Systems," 2nd ed. Wiley, New York. © John Wiley & Sons, Inc.]

system consists of a binary data sequence $[b_k]$ with a bit duration of T_b seconds; b_k is in the form of 1 or 0. This signal is applied to a pulse generator, producing the pulse waveform

$$x(t) = \sum_{k=-\infty}^{\infty} a_k g(t - kT_b) \qquad (14)$$

where $g(t)$ denotes the shaping pulse that is normalized, so that we write

$$g(0) = 1 \qquad (15)$$

The amplitude a depends on the identity of the input bit b; specifically, we assume that

$$a_k = \begin{cases} +a & \text{if the input bit } b_k \text{ is represented} \\ & \text{by symbol 1} \\ -a & \text{if the input bit } b_k \text{ is represented} \\ & \text{by symbol 0} \end{cases} \qquad (16)$$

The PAM signal $x(t)$ passes through a transmitting filter of transfer function $H_T(f)$. The resulting filter output defines the transmitted signal, which is modified in a deterministic fashion as a result of transmission through the channel of transfer function $H_C(f)$. In addition, the channel adds random noise to the signal at the receiver input. Then the noisy signal is passed through a receiving filter of transfer function $H_R(f)$. This filter output is sampled synchronously with the transmitter, the sampling instants being determined by a clock or timing signal that is usually extracted from the receiving filter output. Finally, the sequence of samples thus obtained is used to reconstruct the original data sequence by means of a decision device. The amplitude of each sample is compared with a threshold. If the threshold is exceeded, a decision is made in favor of symbol 1 (say). If the threshold is not exceeded, a decision is made in favor of symbol 0. If the sample amplitude equals the threshold exactly, the flip of a fair coin will determine which symbol was transmitted.

1. Baseband Shaping

Typically, the transfer function of the channel and the pulse shape are specified, and the problem is to determine the transfer functions of the transmitting and receiving filters so as to enable the receiver to recognize the sequence of values in the received signal wave.

In solving this problem, we have to overcome the intersymbol interference (ISI) caused by the overlapping tails of other pulses adding to the particular pulse $A_i p(t - iT_b)$, which is examined at the sampling time iT_b. If this form of interference is too strong, it may result in erroneous decisions in the receiver. Clearly, control of ISI in the system is achieved in the time domain by controlling the function $p(t)$.

The receiving filter output $y(t)$ is sampled at time $t = iT_b$ (with i taking on integer values), yielding

$$y(t_i) = \sum_{k=-\infty}^{\infty} A_k p[(i - k)T_b)] + n(t_i)$$

$$= A_i + \sum_{\substack{k=-\infty \\ k \neq i}}^{\infty} A_k p[(i - k)T_b)] + n(t_i) \qquad (17)$$

where $A_k p(t)$ is the response of the cascade connection of the transmitting filter, the channel, and the receiving filter, which is produced by the pulse $a_k g(t)$ applied to the input of the cascade. In Eq. (17), the first term A_i represents the ith transmitted bit. The second term represents the residual effect of all other transmitted bits on the decoding of the ith bit; this residual effect is called the intersymbol interference. The last term $n(t_i)$ represents the noise sample at time t_i.

In the absence of ISI and noise, we observe from Eq. (17) that

$$y(t_i) = A_i$$

which shows that under these conditions, the ith transmitted bit can be decoded correctly. The unavoidable presence

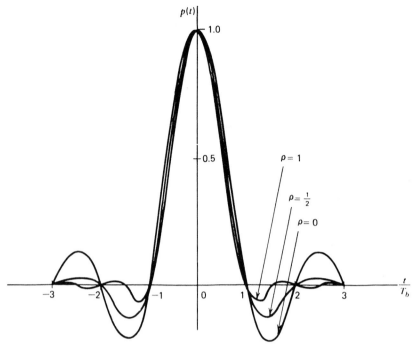

FIGURE 12 Time response $p(t)$ for varying rolloff factor ρ. [From Haykin, S. (1983). "Communication Systems," 2nd ed. Wiley, New York. © John Wiley & Sons, Inc.]

of ISI and noise in the system, however, introduces errors in the decision device at the receiver output. Therefore, in the design of the transmitting and receiving filters, the objective is to minimize the effects of noise and ISI and thereby deliver the digital data to their destination in an error-free manner as far as possible.

One signal waveform that produces zero ISI in a realizable fashion is defined by

$$p(t) = \mathrm{sinc}(2B_T t)\frac{\cos(2\pi\rho B_T t)}{1 - 16\rho^2 B_T^2 t^2} \qquad (18)$$

where $\mathrm{sinc}(2B_T t)$ is the sinc function

$$\mathrm{sinc}(2B_T t) = \frac{\sin(2\pi B_T t)}{2\pi B_T t} \qquad (19)$$

and the transmission bandwidth B_T is related to the bit duration T_b as

$$B_T = 1/2T_b$$

The parameter ρ is called the rolloff factor; its value lies within the range $0 \le p \le 1$.

The function $p(t)$ of Eq. (18) consists of the product of two factors: the factor $\mathrm{sinc}(2B_T t)$ associated with an ideal low-pass filter and a second factor that decreases as $1/|t|^2$ for large $|t|$. The first factor ensures zero crossings of $p(t)$ at the desired sampling instants of time $t = iT$, with i an integer (positive and negative). The second factor reduces the tails of the pulse considerably below that obtained from the ideal low-pass filter, so that the transmission of

binary waves using such pulses is relatively insensitive to sampling time errors. In fact, the amount of ISI resulting from this timing error decreases as the rolloff factor ρ is increased from zero to unity.

The time response $p(t)$ is plotted in Fig. 12 for $\rho = 0$, 0.5, and 1. For the special case of $\rho = 1$, the function $p(t)$ simplifies as

$$p(t) = \frac{\mathrm{sinc}(4B_T t)}{1 - 16B_T^2 t^2} \qquad (20)$$

This time response exhibits two interesting properties:

1. At $t = \pm T_b/2 = \pm 1/4B_T$, we have $p(t) = 0.5$; that is, the pulse width measured at half-amplitude is exactly equal to the bit duration T_b.

2. There are zero crossings at $t = \pm 3T_b/2$, $\pm 5T_b/2$, ... in addition to the usual zero crossings at the sampling times $t = \pm T_b$, $\pm 2T_b$, ...

These two properties are particularly useful in generating a timing signal from the received signal for the purpose of synchronization. However, this requires the use of a transmission bandwidth double that required for the ideal case corresponding to $\rho = 0$.

2. Correlative Coding

Thus far we have treated ISI as an undesirable phenomenon that produces a degradation in system performance.

Indeed, its very name connotes a nuisance effect. Nevertheless, by adding ISI to the transmitted signal in a controlled manner, it is possible to achieve a signaling rate of $2B_T$ symbols per second in a channel of bandwidth of B_T hertz. Such schemes are called correlative coding or partial-response signaling schemes. The design of these schemes is based on the premise that since ISI introduced into the transmitted signal is known, its effect can be interpreted at the receiver. Thus, correlative coding can be regarded as a practical means of achieving the theoretical maximum signaling rate of $2B_T$ symbols per second in a bandwidth of B hertz, as postulated by Nyquist, using realizable and perturbation-tolerant filters.

3. Adaptive Equalization

An efficient approach to the high-speed transmission of digital data (e.g., computer data) over a voice-grade telephone channel (which is characterized by a limited bandwidth and high signal-to-noise ratio) involves the use of two basic signal processing operations:

1. Discrete PAM by encoding the amplitudes of successive pulses in a periodic pulse train with a discrete set of possible amplitude levels.
2. A linear modulation scheme that offers bandwidth conservation (e.g., VSB) to transmit the encoded pulse train over the telephone channel.

At the receiving end of the system, the received wave is demodulated and then synchronously sampled and quantized. As a result of dispersion of the pulse shape by the channel, the number of detectable amplitude levels is often limited by ISI rather than by additive noise. In principle, if the channel is known precisely, it is virtually always possible to make the ISI (at the sampling instants) arbitrarily small by using a suitable pair of transmitting and receiving filters, so as to control the overall pulse shape in the manner described above. The transmitting filter is placed directly before the modulator, whereas the receiving filter is placed directly after the demodulator. Thus, insofar as ISI is concerned, we can consider the data transmission to be essentially baseband.

However, in a switched telephone network, two factors contribute to the distribution of pulse distortion on different link connections: (1) differences in the transmission characteristics of the individual links that can be switched together and (2) differences in the number of links in a connection. The result is that the telephone channel is random in the sense of being one of an ensemble of possible channels. Consequently, the use of a fixed pair of transmitting and receiving filters designed on the basis of average channel characteristics may not adequately reduce ISI. To realize the full transmission capability of a telephone channel, there is need for adaptive equalization. By *equalization* we mean the process of correcting channel-induced distortion. This process is said to be adaptive when it adjusts itself continuously during data transmission by operating on the input signal.

Among the philosophies for adaptive equalization of data transmission systems are prechannel equalization at the transmitter and postchannel equalization at the receiver. Because the first approach requires a feedback channel, it is customary to use only adaptive equalization at the receiving end of the system. This equalization can be achieved, before data transmission, by training the filter with the guidance of a suitable training sequence transmitted through the channel so as to adjust the filter parameters to optimum values. The typical telephone channel changes little during an average data call, so that precall equalization with a training sequence is sufficient in most cases encountered in practice. The equalizer is positioned after the receiving filter in the receiver.

4. Eye Pattern

One way to study ISI in a PCM or data transmission system experimentally is to apply the received wave to the vertical deflection plates of an oscilloscope and to apply a sawtooth wave at the transmitted symbol rate $1/T$ to the horizontal deflection plates. The waveforms in successive symbol intervals are thereby translated into one interval on the oscilloscope display. The resulting display is called an eye pattern because, for binary waves, its appearance resembles the human eye. The interior region of the eye pattern is called the eye opening.

An eye pattern provides a great deal of information about the performance of the pertinent system (Fig. 13):

1. The width of the eye opening defines the time interval over which the received wave can be sampled without

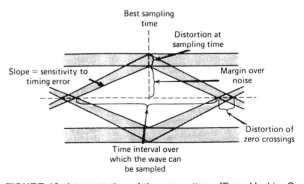

FIGURE 13 Interpretation of the eye pattern. [From Haykin, S. (1983). "Communication Systems," 2nd ed. Wiley, New York. © John Wiley & Sons, Inc.]

error from ISI. It is apparent that the preferred time for sampling is the instant of time at which the eye is open widest.

2. The sensitivity of the system to timing error is determined by the rate of closure of the eye as the sampling time is varied.

3. The height of the eye opening, at a specified sampling time, defines the margin over noise.

When the effect of ISI is severe, traces from the upper portion of the eye pattern cross traces from the lower portion, with the result that the eye is completely closed. In such a situation, it is impossible to avoid errors due to the combined presence of ISI and noise in the system.

5. Digital Subscriber Lines

A digital subscriber line (DSL) operates over a local loop (less than 1.5 km) that provides a direct connection between a user terminal (e.g., computer) and a telephone company's central office. In this way, a user is connected to a broadband backbone data network, which is based on technologies such as the asynchronous transfer mode (ATM) and Internet Protocol (IP). The information-bearing signal is thereby kept in the digital domain all the way from the user terminal to an Internet service provider, with the signal being switched or routed at regular intervals in the course of its transmission through the data network.

A major source of channel impairment in DSLs is the near-end crosstalk (NEXT), which is generated by transmitters at the same end of the cable as the receiver. The other major source of impairment is ISI that arises due to the imperfect frequency response of the channel.

A widely used line code for DSLs is the 2BIQ codes, which stands for a pair of binary digits encoded into one quaternary symbol. This code is a block code representing a four-level PAM signal.

Asymmetric digital subscriber lines (ADSL) are designed to simultaneously support three services on a single twisted-wire pair:

1. Data transmission downstream (toward the subscriber)
2. Data transmission upstream (away from the subscriber)
3. Plain old telephone service

The DSL is said to be "asymmetric" because the downstream bit rate is much higher than the upstream bit rate.

B. Passband Data Transmission

When digital data are to be transmitted over a bandpass channel, it is necessary to modulate the incoming data onto a carrier wave (usually sinusoidal) with fixed frequency limits imposed by the channel. The data may represent digital computer outputs or PCM waves generated by digitizing voice or video signals, and so on. The channel may be a microwave radio link or satellite channel, and so on. In any event, the modulation process involves switching or keying the amplitude, frequency, or phase of the carrier in accordance with the incoming data. Thus, there are three basic signaling techniques: amplitude-shift keying (ASK), frequency-shift keying (FSK), and phase-shift keying (PSK), which can be viewed as special cases of amplitude modulation, frequency modulation, and phase modulation, respectively. Figure 14 illustrates these three basic signaling techniques for binary data input.

Ideally, FSK and PSK signals have a constant envelope. This feature makes them impervious to amplitude non-linearities, as encountered in microwave radio links and satellite channels. Accordingly, in practice, FSK and PSK signals are much more widely used than ASK signals.

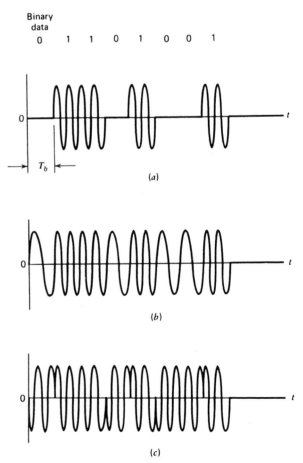

FIGURE 14 Three basic forms of signaling binary information. (a) Amplitude-shift keying. (b) Frequency-shift keying. (c) Phase-shift keying. [From Haykin, S. (1983). "Communication Systems," 2nd ed. Wiley, New York. © John Wiley & Sons, Inc.]

The system is said to be coherent when the locally generated carrier in the receiver is synchronized to that in the transmitter in both frequency and phase. It is said to be noncoherent when the phase of the incoming signal is destroyed in the receiver.

1. Coherent Binary PSK

In a coherent binary PSK system, a pair of signals, $s_1(t)$ and $s_2(t)$, are used to represent binary symbols 1 and 0, respectively; they are defined by

$$s_1(t) = \sqrt{2E_b/T_b} \cos(2\pi f_c t) \qquad (21a)$$

$$s_2(t) = \sqrt{2E_b/T_b} \cos(2\pi f_c t + \pi)$$
$$= -\sqrt{2E_b/T_b} \cos(2\pi f_c t) \qquad (21b)$$

where $0 \le t \le T_b$ and E_b is the transmitted signal energy per bit. In order to ensure that each transmitted bit contains an integral number of cycles of the carrier wave, the carrier frequency f_c is chosen equal to n_c/T_b for some fixed integer n_c. A pair of sinusoidal waves differing only in phase by 180°, as defined above, are referred to as antipodal signals.

2. Coherent Binary FSK

In a binary FSK system, the symbols 1 and 0 are distinguished from one another by the transmission of one of two sinusoidal waves that differ in frequency by a fixed amount. A typical pair of sinusoidal waves is described by

$$s_i(t) = \begin{cases} \sqrt{2E_b/T_b} \cos(2\pi f_i t), & 0 \le t \le T_b \\ 0, & \text{elsewhere} \end{cases} \qquad (22)$$

where the carrier frequency f_i equals one of two possible values: f_1 and f_2. The transmission of frequency f_1 represents symbol 1, and the transmission of frequency f_2 represents symbol 0.

3. Coherent Quadrature Signaling Techniques

Channel bandwidth and transmitted power constitute two primary "communication resources," the efficient utilization of which provides the motivation for the search for spectrally efficient modulation schemes. The primary objective of spectrally efficient modulation is to maximize the bandwidth efficiency, defined as the ratio of data rate to channel bandwidth (in units of bits per second per hertz) for a specified probability of symbol error. A secondary objective is to achieve this bandwidth efficiency at a minimum practical expenditure of average signal power or, equivalently, in a channel perturbed by additive white Gaussian noise, a minimum practical expenditure of average signal-to-noise ratio.

Examples of the quadrature-carrier multiplexing system include the following signaling techniques:

1. Quadriphase-shift keying (QPSK), which is an extension of the binary PSK. In this technique, the phase of the carrier takes on one of four possible values. Specifically, the binary pairs $10, 00, 01, 11$ are assigned the phase values $\pi/4$, $3\pi/4$, $5\pi/4$, and $7\pi/4$, respectively. This is illustrated in Fig. 15.

2. Minimum shift keying (MSK), which is a special form of continuous-phase frequency-shift keying, with the detection in the receiver being performed in two successive bit intervals. The nominal carrier frequency f_c is equal to the arithmetic mean of the two frequencies f_1 and f_2 used to represent symbols 1 and 0, respectively. Moreover, the carrier frequency f_c is chosen equal to an integral multiple of one-fourth of the bit rate in order to make the phase of the transmitted signal continuous at the bit transition instants. One other requirement is to make the deviation ratio

$$h = T_b(f_1 - f_2)$$

equal to one-half. Figure 16 illustrates the wave-form of the MSK signal.

4. M-Ary Signaling Techniques

In an M-ary signaling scheme, we send any one of M possible signals, $s_1(t), s_2(t), \ldots, s_M(t)$, during each signaling interval of duration T. For almost all applications, the number of possible signals M is 2^n, where n is an integer. The symbol duration T is equal to nT_b, where T_b is the bit duration. These signals are generated by changing

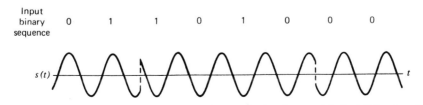

FIGURE 15 QPSK wave $s(t)$. [From Haykin, S. (1983). "Communication Systems," 2nd ed. Wiley, New York. © John Wiley & Sons, Inc.]

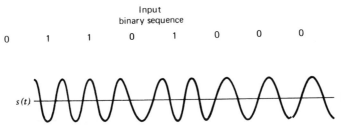

FIGURE 16 MSK wave $s(t)$. [From Haykin, S. (1983). "Communication Systems," 2nd ed. Wiley, New York. © John Wiley & Sons, Inc.]

the aplitude, phase, or frequency of a carrier in M discrete steps. Thus, we have M-ary ASK, M-ary PSK, and M-ary FSK digital modulation schemes. The QPSK system considered above is an example of M-ary PSK with $M = 4$.

M-Ary signaling schemes are preferred over binary signaling schemes for transmitting digital information over bandpass channels when the requirement is to conserve bandwidth at the expense of increasing power. In practice, a communication channel rarely has the exact bandwidth required for transmitting the output of an information source by means of binary signaling schemes. Thus, when the bandwidth of the channel is less than the required value, we use M-ary signaling schemes so as to utilize the channel efficiently.

5. Detection of Signals with Random Phase in the Presence of Noise

Up to this point, we have assumed that the information-bearing signal is completely known at the receiver. In practice, however, in addition to the uncertainty due to the additive noise of a receiver, there is often an additional uncertainty due to the randomness of certain signal parameters. The usual cause of this uncertainty is distortion in the transmission medium. Perhaps the most common random signal parameter is the phase, which is especially true for narrow-band signals.

For example, transmission over a multiplicity of paths of different, and variable lengths or rapidly varying delays in the propagating medium from transmitter to receiver may cause the phase of the received signal to change in a way that the receiver cannot follow. Synchronization with the phase of the transmitted carrier may then be too costly, and the designer may simply choose to disregard the phase information in the received signal at the expense of some degradation in the noise performance of the system.

For example, in the noncoherent detection of binary FSK signals, one can simplify receiver complexity by using a pair of frequency discriminators tuned to the frequencies f_1 and f_2, which represent binary symbols 0 and 1, respectively.

Another example is that of differential phase-shift keying (DPSK), which can be viewed as the noncoherent version of binary PSK. It eliminates the need for a coherent reference signal at the receiver by combining two basic operations at the transmitter: (1) differential encoding of the input binary wave and (2) phase-shift keying—hence the name differential phase-shift keying. In effect, to send symbol 0 we phase-advance the current signal waveform by 180°, and to send symbol 1 we leave the phase of the current signal waveform unchanged. The receiver is equipped with a storage capability, so that it can measure the relative phase difference between the waveforms received during two successive bit intervals. Provided that the unknown phase contained in the received wave varies slowly (i.e., slowly enough to be considered essentially constant over two bit intervals), the phase difference between waveforms received in two successive bit intervals will be independent of θ.

6. Noise Comparison of Digital Modulation Schemes Using a Single Carrier

It is customary to use the average probability of symbol error to assess the noise performance of digital modulation schemes. It should be realized, however, that even if two systems yield the same average probability of symbol error, their performances, from the user's viewpoint, may be quite different. In particular, the greater the number of bits per symbol, the more the bit errors will cluster together. For example, if the average probability of symbol error is 10^{-3}, the expected number of symbols occurring between any two erroneous symbols is 1000. If each symbol represents one bit of information (as in a binary PSK or binary FSK system), the expected number of bits separating two erroneous bits is 1000. If, on the other hand, there are two bits per symbol (as in a QPSK system), the expected separation is 2000 bits. Of course, a symbol error generally creates more bit errors in the second case, so that the percentage of bit errors tends to be the same. Nevertheless, this clustering effect may make one system more attractive than another, even at the same symbol error rate. In the final analysis, which system is preferable will depend on the situation.

Two systems having an unequal number of symbols can be compared in a meaningful way only if they use the same amount of energy to transmit each bit of information. It is the total amount of energy needed to transmit the complete message, not the amount of energy needed to transmit a particular symbol satisfactorily, that represents the cost of the transmission. Accordingly, in comparing the different data transmission systems considered, we shall use as the basis of our comparison the average probability of symbol error expressed as a function of the bit energy-to-noise density ratio E_b/N_0.

In Table I, we have summarized the expressions for the average probability of symbol error P_e for the coherent PSK, conventional coherent FSK with one-bit decoding, DPSK, noncoherent FSK, QPSK, and MSK, when operating over an AWGN channel. In Fig. 17 we have used these expressions to plot P_e as a function of E_b/N_0.

In summary, we can state the following:

1. The error rates for all the systems decrease monotonically with increasing values of E_b/N_0.

2. For any value of E_b/N_0, the coherent PSK produces a smaller error rate than any of the other systems. Indeed, in the case of systems restricted to one-bit decoding perturbed by additive white Gaussian noise, the coherent PSK system is the optimum system for transmitting binary data in the sense that it achieves the minimum probability of symbol error for a given value of E_b/N_0.

3. The coherent PSK and the DPSK require an E_b/N_0 that is 3 dB less than the corresponding values for the conventional coherent FSK and the noncoherent FSK, respectively, to realize the same error rate.

4. At high values of E_b/N_0, the DPSK and the noncoherent FSK perform almost as well (to within ~ 1 dB)

as the coherent PSK and the conventional coherent FSK, respectively, for the same bit rate and signal energy per bit.

5. In QPSK two orthogonal carriers $\sqrt{2/T} \cos(2\pi f_c t)$ and $\sqrt{2/T} \sin(2\pi f_c t)$ are used, where the carrier frequency f_c is an integral multiple of the symbol rate $1/T$, with the result that two independent bit streams can be transmitted and subsequently detected in the receiver. At high values of E_b/N_0, coherently detected binary PSK and QPSK have the same error rate performances for the same value of E_b/N_0.

6. In MSK the two orthogonal carriers $2\sqrt{T_b} \cos (2\pi f_c t)$ and $2\sqrt{T_b} \sin(2\pi f_c t)$ are modulated by antipodal symbol shaping pulses $\cos(\pi t/2T_b)$ and $\sin(\pi t/2T_b)$, respectively, over $2T_b$ intervals, where T_b is the bit duration. Correspondingly, the receiver uses a coherent phase decoding process over two successive bit intervals to recover the original bit stream. Thus, MSK has exactly the same error rate performance as QPSK.

7. Discrete Multitone

The ADSL, described in Section IV.A.5, is a data transmission system capable of realizing megabit per second rates over existing twisted-pair telephone lines. An elegant modulation technique well suited for this application is discrete multitone (DMT), which allows the modulator characteristics to be a function of measured channel characteristics. The basic idea of DMT is rooted in a commonly used engineering principle: divide and conquer, whereby a difficult problem is solved by dividing it into a number of simpler problems and then combining the solutions.

In DMT, a wideband channel is transformed into a set of N subchannels that operate in parallel. What makes DMT distinctive is the fact that the transformation is performed in discrete time as well as discrete frequency. Consequently, the input–output behavior of the entire communication system admits a linear matrix representation, which lends itself to implementation using the discrete Fourier transform.

Figure 18 shows a block diagram of the DMT data-transmission system. The transmitter consists of the following functional blocks:

- Demultiplexer, which converts the incoming serial data stream into parallel form.
- Constellation encoder, which maps the parallel data into $N/2$ multibit subchannel with each subchannel being represented by a specific signal constellation (e.g., M-ary quadrature amplitude modulation).
- Inverse discrete Fourier transform (IDFT), which transforms the frequency-domain parallel data at the constellation encoder output into parallel time-domain

TABLE I Summary of Formulas for the Symbol Error Probability for Different Data Transmission Systems

System	Error probability P_e
Coherent binary signaling	
Coherent PSK	$\frac{1}{2} \operatorname{erfc}(\sqrt{E_b/N_0})$
Coherent FSK	$\frac{1}{2} \operatorname{erfc}(\sqrt{E_b/2N_0})$
Noncoherent binary signaling	
DPSK	$\frac{1}{2} \exp(-E_b/N_0)$
Noncoherent FSK	$\frac{1}{2} \exp(-E_b/2N_0)$
Coherent quadrature signaling	
QPSK	$\operatorname{erfc}(\sqrt{E_b/N_0})$
MSK	$-\frac{1}{4} \operatorname{erfc}^2(\sqrt{E_b/N_0})$

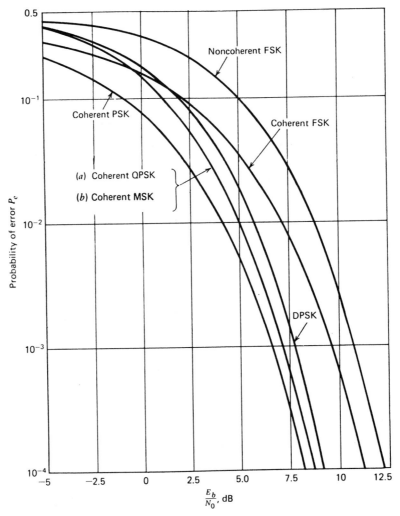

FIGURE 17 Comparison of the noise performances of different PSK and FSK systems. [From Haykin, S. (1983). "Communication Systems," 2nd ed. Wiley, New York. © John Wiley & Sons, Inc.]

data. For efficient implementation of the IDFT, it is customary to use the fast Fourier transform algorithm by choosing $N = 2^k$ where k is a positive integer.

- Digital-to-analog converter (DAC), which converts the digital data into an analog form for transmission over the channel.

The receiver consists of an analog-to-digital converter (DAC), serial-to-parallel converter, discrete Fourier transformer (DFT), decoder, and multiplexer, which perform the inverse operations of the transmitter. A reconstruction of the original data stream is achieved.

An important feature of DMT is that it permits the realization of a capacity of data transmission in accordance with Shannon's information capacity theorem:

$$C = \frac{1}{2} \log_2(1 + \text{SNR})$$

where the capacity C is measured in bits per transmission and SNR is the (dimensionless) received signal-to-noise ratio at the channel output. This remarkable performance is realized through the use of loading, which refers to the process of allocating transmit power to the N individual subchannels so as to maximize the bit rate of the entire multichannel transmission system.

8. Orthogonal Frequency-Division Multiplexing

DMT is one particular form of multichannel modulation. Another closely related form of multichannel modulation is orthogonal frequency-division multiplexing (OFDM), which differs from DMT in areas of application and some aspects of its design.

OFDM is used for data transmission over radio broadcast channels and wireless communication channels. This domain of application requires some changes to the design

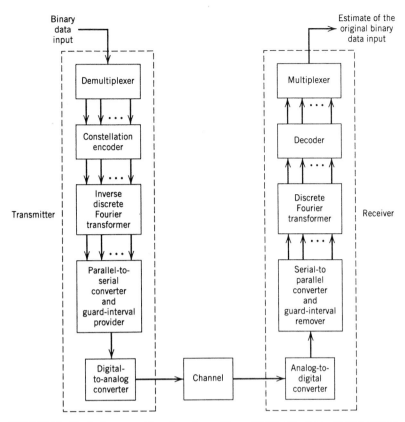

FIGURE 18 Block diagram of the discrete-multitone (DMT) data-transmission system.

of the OFDM system. Unlike DMT, which uses loading for bit allocation, OFDM uses a fixed number of bits per subchannel. This restriction is made necessary by the fact that a broadcast channel involves one-way transmission, and in a wireless communication environment the channel is varying too rapidly.

The other modification in OFDM involves the use of (1) an upconverter after the ADC in Fig. 18 to translate the transmitted frequency to facilitate signal propagation over a radio wireless channel, and (2) a downconverter before the ADC to undo the frequency translation performed by the upconverter in the transmitter.

V. SPREAD-SPECTRUM MODULATION

A major issue of concern in the study of digital communications is that of providing for the efficient utilization of bandwidth and power. Notwithstanding the importance of these two primary communication resources, there are situations where it is necessary to sacrifice this efficiency in order to meet certain other design objectives. For example, the system may be required to reject interference as in wireless communications. This requirement is catered to by a class of signaling techniques known as spread-spectrum modulation.

In the so-called direct-sequence spread-spectrum technique, two stages of modulation are used. First, the incoming data sequence is used to modulate a wide-band binary code, which transforms this narrow-band data sequence into a noiselike wide-band signal. The resulting wide-band signal undergoes a second modulation using a phase-shift keying technique (e.g., QPSK). For its operation, such a spread-spectrum modulation system relies on the availability of a noiselike spreading code called a pseudonoise sequence.

At the receiver, the channel output passes through two stages of demodulation: QPSK detection, followed by despreading. The net result is a processing gain (PG) due to spreading, which is defined by

$$PG = \frac{T_b}{T_c}$$

where T_b is the bit duration of the data sequence at the transmitter input, and T_c is the chip duration, that is, the duration of binary symbol 1 or 0 in the spreading code.

VI. MULTIPLEXING

Multiplexing is the process of combining several message signals (originating from independent users) for their transmission over a common channel. Three commonly used methods of multiplexing are as follows:

1. Frequency-division multiplexing (FDM), in which CW modulation (e.g., frequency modulation) is used to translate each message signal to reside inside a specific frequency slot inside the passband of the channel by assigning it a distinct carrier frequency. At the receiver, a bank of filters is used to separate the different modulated signals and prepare them individually for demodulation.

2. Time-division multiplexing (TDM), in which some form of pulse modulation (e.g., pulse code modulation) is used to position samples of the different message signals in nonoverlapping time slots.

3. Code-division multiplexing (CDM), in which each message signal is identified by a distinctive spreading code.

Thus, in FDM the message signals appear across the channel all the time. In contrast, in TDM each message signal occupies the full passband of the channel. CDM differs from both FDM and TDM in that each message signal occupies the full passband of the channel all the time; for demultiplexing to function satisfactorily, the spreading codes assigned to the individual message signals should ideally be orthogonal to each other.

ACKNOWLEDGMENT

The material presented in this article is summarized from the author's book, "Communication Systems" (see Bibliography). The author is indebted to John Wiley and Sons, publisher of this book, for permission to write this article.

SEE ALSO THE FOLLOWING ARTICLES

COMMUNICATION SATELLITE SYSTEMS • DATA TRANSMISSION MEDIA • NETWORKS FOR DATA COMMUNICATION • RADIO SPECTRUM UTILIZATION • SIGNAL PROCESSING, ANALOG • SIGNAL PROCESSING, DIGITAL • SIGNAL PROCESSING, GENERAL

BIBLIOGRAPHY

Bellamy, J. C. (1991). "Digital Telephony," 2nd ed., Wiley, New York.

Gitlin, R. D., Hayes, J. F., and Weinstein, S. B. (1992). "Data Communication Principles," Plenum, New York.

Haykin, S. (2001). "Communication Systems," 4th ed. Wiley, New York.

Jayant, N. S., and Noll, P. (1984). "Digital Coding of Waveforms." Prentice-Hall, Englewood Cliffs, NJ.

Jeruchim, M. C., Balaban, B., and Shanmugan, J. S. (2000). "Simulation of Communication Systems," 2nd ed. Kluwer, Dordrecht, The Netherlands.

Lee, E. A., and Messerschmitt, D. G., (1994). "Digital Communication," 2nd ed. Kluwer Academic, Dordrechat, The Netherlands.

Proakis, J. G. (1995). "Digital Communication," 3rd ed., McGraw-Hill, New York.

Schwartz, M. (1980). "Information Transmission, Modulation and Noise," 3rd ed., McGraw-Hill, New York.

Starr, T., Cioffi, J. M., and Silverman, P. J. (1999). "Understanding Digital Subscriber Line Technology," Prentice-Hall, Englewood Cliffs; NJ.

Viterbi, A. J., and Omura, J. K. (1979). "Principles of Digital Communication and Coding," McGraw-Hill, New York.

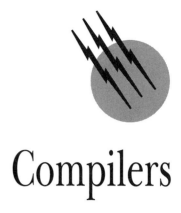

Compilers

Keith D. Cooper
Ken Kennedy
Linda Torczon
Rice University

GLOSSARY

Back end The final phase of compilation, where the program is translated from the compiler's intermediate representation into operations for the target machine.

Compiler A program that translates an executable program from one form to another.

Constant propagation An optimization that discovers, at compile time, expressions that must have known constant values, evaluates them, and replaces their run-time evaluation with the appropriate value.

Data-flow analysis A collection of techniques for reasoning, at compile time, about the flow of values at run-time.

Front end The initial stage of compilation, where the program is translated from the original programming language into the compiler's intermediate representation.

High-level transformations Transformations performed on an intermediate representation that is close to the source language in its level of abstraction.

Instruction selection The process of mapping the compiler's intermediate representation of the program into the target language produced by the compiler.

Lexical analysis That part of the compiler's front end that has the task of converting the input program from a stream of individual characters into a stream of words, or tokens, that are recognizable components of the source language. Lexical analysis recognizes words and assigns them to syntactic categories, much like parts of speech. The pass that implements lexical analysis is called a *scanner*.

List scheduling An algorithm for reordering the operations in a program to improve their execution speed. A list scheduler constructs a new version of the program by filling in its schedule, one cycle at a time. The scheduler must respect the flow of values in the original program and the operation latencies of the target machine.

Memory hierarchy management A collection of transformations that rewrite the program to change the order in which it accesses memory locations. On machines with cache memories, reordering the references can increase the extent to which values already in the cache are reused, and thus decrease the aggregate amount of tie spent waiting on values to be fetched from memory.

Optimizer The middle part of a compiler, it rewrites the program in an attempt to improve its *runtime* behavior. Optimizers usually consist of several distinct passes of analysis and transformation. The usual goal of an optimizer is to decrease the program's execution time; some optimizers try to create smaller programs as well.

Parallelization The task of determining which parts of the program are actually independent, and can therefore execute concurrently. Compilers that use these techniques usually treat them as high-level transformations, performing both analysis and transformations on a near-source representation of the program.

Programming languages Artificial (or formal) languages designed to let people specify algorithms and describe data structures, usually in a notation that is independent of the underlying target machine.

Regular expressions A mathematical notation for describing strings of characters. Efficient techniques can convert a collection of regular expressions into a scanner.

Semantic elaboration That part of a compiler's front end that checks semantic rules and produces an intermediate representation for the program. Often, semantic elaboration is coupled directly to the parser, where individual actions can be triggered as specific grammatical constructs are recognized.

Source-to-source translator A compiler that produces, as its target language, a programming language (rather than the native language of some target computer).

Syntax analysis That part of the compiler's front end that has the task of determining whether or not the input is actually a program in the source language. The parser consumes a stream of categorized words produced by the scanner and validates it against an internal model of the source language's grammatical structure (or syntax). The pass that implements syntax analysis is called a *parser*.

Vectorization A specialized form of parallelization that tries to expose computations suitable for SIMD, or vector, execution.

COMPUTER PROGRAMS are usually written in languages designed for specifying algorithms and data structures. We call such languages *programming languages*. On the other hand, the computer hardware executes operations, or *instructions*, that are much less abstract than the operations in the programming language, with set of available instructions varying from computer to computer. Before a program can execute, it must be translated from the programming language in which it is written into the machine language instructions for the computer on which it will run. The program that performs this translation is called a *compiler*.

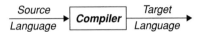

Formally, a compiler is simply a program that takes as its input an executable program and produces as its output an equivalent program. For the purpose of discussing translation, we refer to the language of the input as the *source language* and the language of the output as the *target language*. The input program is typically written in some well-known programming language, such as Fortran, C, C++, Ada, Java, Scheme, or ML. The output program is rewritten into the set of native operations of a particular computer. (Some compilers use high-level languages as their target language. Such compilers are often called *source-to-source translators*.)

I. STRUCTURE OF A COMPILER

While compilers can be built as monolithic programs, more often, they are implemented as a series of distinct phases, generally organized into three major sections: a front end, an optional optimizer (or middle end), and a back end.

The *front end* analyzes the source program to determine whether or not it is well formed—that is, if it is a valid program in the source language. To accomplish this, it must first analyze each word in the program, classifying it into a syntactic category—a task called *lexical analysis*. It must then determine if the string of classified words fits into the grammatical structure of the language, a task called *syntax analysis*. In addition, it must check a variety of extrasyntactic rules to ensure that the source program has meaning, a process called *context-sensitive analysis*. Finally, it must construct an *internal representation* (IR) of the program for use by the later phases.

The *optimizer* takes the program, expressed in the compiler's IR, and produces an "improved" version of the program, usually expressed in the same IR. A program can be improved in different ways: it might be made to run faster; it might be made more compact in memory or secondary storage; or it might be transformed to consume less power. To produce the improved program, the optimizer must reason about how the program will behave when it executes. It can use this knowledge to simplify, specialize, and rearrange the elements of the internal form of the program. At each step in the process, it must preserve the meaning of the program, as expressed in its externally visible behavior.

The *back end* maps the program, expressed in the compiler's IR, into the target language. If the target is an instruction set for a particular computer, then the back end must account for that machine's finite resources and idiosyncratic behavior. This adds significant complexity to the task of code generation. The back end must select a set of operations to implement each construct in the program. It must decide where in the computer's memory each value will reside—a task complicated by the hierarchical memory systems of modern computers, and by the fact that the fastest locations, called registers, are extremely limited in number.[1] It must choose an execution order for the operations—one that preserves the proper flow of values between operations and avoids requiring the processor to wait for results. This may necessitate insertion of null operations to ensure that no operation begins to execute before its operands are ready. Many of the most challenging problems in compiler construction occur in the back end.

II. RECOGNIZING VALID PROGRAMS

Before the compiler can translate a source language program into a target machine program, it must determine whether or not the program is well formed—that is, whether the program is both grammatically correct and meaningful, according to the rules of the source language. This task—determining whether or not the program is valid—is the largest of the front end's tasks. If the front end accepts the program, it constructs a version of the program expressed in the compiler's internal representation. If the front end rejects the program, it should report the reasons back to the user. Providing useful diagnostic messages for erroneous programs is an essential part of the front end's work.

To address these issues, the typical front end is partitioned into three separate activities: lexical analysis, syntactic analysis, and context-sensitive analysis.

A. Lexical Analysis

A major difference between a programming language and a natural language lies in the mechanism that maps words into parts of speech. In most programming languages, each word has a unique part of speech, which the compiler can determine by examining its spelling. In a natural language,

a single word can be mapped to several different parts of speech, depending on the context surrounding it. For example, the English words "fly" and "gloss" can be used as either noun or verb. The simpler rules used by programming languages permit the compiler to recognize and classify words without considering the grammatical context in which they appear.

To specify the spelling of words and their mapping into parts of speech, compiler writers use a formal notation called *regular expressions*. Regular expressions describe strings of symbols drawn from a finite alphabet, as well as ways to combine such strings into longer strings. First, any finite string of characters drawn from the alphabet is a regular expression. From these we can build longer strings by applying any of three rules. If r and s are regular expressions, then

1. $(r \mid s)$ is a regular expression, denoting a string that is either r or s
2. rs is a regular expression, denoting an occurrence of r followed immediately by an occurrence of s
3. r^* is a regular expression, denoting zero or more consecutive occurrences of r

Using regular expressions, we can compactly specify the spelling of fairly complex words. For example, since any specific word is a regular expression, we can specify the reserved words of a programming language, e.g., **define, do, if,** and **car,** simply by listing them. We can define a more complex construct, the counting numbers over the alphabet of digits, as $(1\mid2\mid3\mid4\mid5\mid6\mid7\mid8\mid9)\ (0\mid1\mid2\mid3\mid4\mid5\mid6\mid7\mid8\mid9)^*$. (We read this as "a digit from 1 through 9 followed by zero or more digits from 0 through 9.") Notice that this regular expression forbids leading zeros.

We call the pass of the compiler that recognizes and classifies words the *scanner* or *lexical analyzer*. The scanner consumes a stream of characters and produces a stream of words; each annotated with its part of speech. In a modern compiler system, the scanner is automatically generated from a set of rules, specified by regular expressions. The scanners generated by this process incur small, constant cost per character, largely independent of the number of rules. For this reason, recognizers derived from regular expressions have found application in tools ranging from text editors to search engines to web-filtering software.

B. Syntactic Analysis

The job of the *syntax analyzer* is to read the stream of words produced by the scanner and decide whether or not that stream of words forms a *sentence* in the source language. To do this, the compiler needs a formal description of the source language, usually called a *grammar*. Informally, a grammar is just a collection of rules for deriving

---

[1](Possible cross reference to cache memory, registers, . . .)

sentences in some language. For example, the following grammar describes a class of simple English sentences:

1. Sentence → Subject Predicate period
2. Subject → noun
3. Subject → adjective noun
4. Predicate → verb noun
5. Predicate → verb adjective
6. Predicate → verb adjective noun

In these rules, underlined, lowercase symbols are *terminal symbols* for the grammar—that is, they are actual parts of speech for words that can appear in a valid sentence. The capitalized symbols are syntactic variables called *nonterminal symbols*. Nonterminals are distinguished by the fact that they appear on the left-hand side of one or more rules. Each nonterminal is defined by a set of rules. The nonterminals thus provide structure to the grammar. Each rule describes some aspect of an English sentence and an interpretation of the nonterminal that appears on its left-hand side. We read the first rule as "a Sentence is a Subject, followed by a Predicate, followed by a period." The next two rules establish two ways to construct a Subject. The final three rules give three options for building a Predicate.

Consider the sentence "Compilers are programs." It fits the simple grammar. To see this, first convert each word to its part of speech: "noun verb noun period." To derive a sentence with this structure, we can start with Sentence and use the rules to rewrite it into the desired sentence:

Grammatical Form	Rule
Sentence	—
→ Subject Predicate period	1
→ noun Predicate period	2
→ noun verb noun period	4

This derivation proves that any sentence with the structure "noun verb noun period" fits within the grammar's model of English. This includes "Compilers are programs" as well as "Tomatoes are horses."

We can depict this derivation graphically, as a *derivation tree*, sometimes called a *parse tree*.

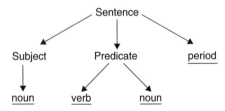

The front end of a compiler does not need to derive sentences. Instead, it must solve the inverse problem—given a

stream of words, construct its derivation tree. This process is called *parsing*. The parser uses a model of the source language's grammar to automatically construct a parse tree for the input program, if such a tree exists. If this process fails, the input program, as classified by the scanner, is not a sentence in the language described by the grammar.

Tools that construct an efficient parser from a grammar are widely available. These tools, called *parser generators*, automate most of the process of building the parser, and have simple interfaces to automatically derived scanners.

C. Context-Sensitive Analysis

The parser, alone, cannot ensure that the source program makes sense. Consider our English sentences: "Compilers are programs" and "Tomatoes are horses." Both fit the model embodied in our grammar for simple sentences; in fact, both have the same derivation. (Remember, the derivation operates on parts of speech, not the actual words.) However, the first sentence is a cogent comment on the nature of compilers while the second sentence is nonsense. The difference between them is not grammatical—it lies in properties of the words that the grammar cannot express. The verb "are" implies a relationship of equality or similarity. Compilers are, in fact, a specific kind of program. Tomatoes and horses are dissimilar enough that it is hard to accept them as equal.

Similar correctness issues arise in computer programs. Names and values have extra grammatical properties that must be respected. A correct program must use values in ways that are consistent with their definitions. Names can have distinct meanings in different regions of the program. The definition of a programming language must specify many properties that go beyond grammar; the compiler must enforce those rules.

Dealing with these issues is the third role of the front end. To succeed, it must perform *context-sensitive analysis*—sometimes called *semantic elaboration*. This analysis serves two purposes: it checks correctness beyond the level of syntax, as already discussed, and it discovers properties of the program that play an important role in constructing a proper intermediate representation of the code. This includes discovering the lifetime of each value and where those values can be stored in memory.

Specification-based techniques have not succeeded as well in context-sensitive analysis as they have in scanning and parsing. While formal methods for these problems have been developed, their adoption has been slowed by a number of practical problems. Thus, many modern compilers use simple ad hoc methods to perform context-sensitive analysis. In fact, most parser generator systems include substantial support for performing such ad hoc tasks.

III. INTERNAL REPRESENTATIONS

Once the compiler is broken into distinct phases, it needs an internal representation to transmit the program between them. This internal form becomes the definitive representation of the program—the compiler does not retain the original source program. Compilers use a variety of internal forms. The selection of a particular internal form is one of the critical design decisions that a compiler writer must make. Internal forms capture different properties of the program; thus, different forms are appropriate for different tasks. The two most common internal representations—the abstract syntax tree and three-address code—mimic the form of the program at different points in translation.

- The *abstract syntax tree* (AST) resembles the parse tree for the input program. It includes the important syntactic structure of the program while omitting any nonterminals that are not needed to understand that structure. Because of its ties to the source-language syntax, an AST retains concise representations for most of the abstractions in the source language. This makes it the IR of choice for analyses and transformations that are tied to source program structure, such as the high-level transformations discussed in Section VC.
- *Three-address code* resembles the assembly code of a typical microprocessor. It consists of a sequence of operations with an implicit order. Each operation has an operator, one or two input arguments, and a destination argument. A typical three-address code represents some of the relevant features of the target machine, including a realistic memory model, branches and labels for changing the flow of control, and a specified evaluation order for all the expressions. Because programs expressed in three-address code must provide an explicit implementation for all of the source language's abstractions, this kind of IR is well suited to analyses and transformations that attack the overhead of implementing those abstractions.

To see the difference between an AST and a three-address code, consider representing an assignment statement $a[i] \leftarrow b * c$ in each. Assume that a is a vector of 100 elements (numbered 0–99) and that b and c are scalars.

The AST for the assignment, shown on the left, captures the essence of the source-language statement. It is easy to see how a simple in-order treewalk could reprint the original assignment statement. However, it shows none of the details about how the assignment can be implemented. The three-address code for the assignment, shown on the right, loses any obvious connection to the source-language statement. It imposes an evaluation order on the statement: first b, then c, then $b * c$, then i, then $a[i]$, and, finally, the assignment. It uses the notation @b to refer to b's address in memory—a concept missing completely from the AST.

Many compilers use more than one IR. These compilers shift between representations so that they can use the most appropriate form in each stage of translation.

IV. OPTIMIZATION

A source language program can be mapped into assembly language in many different ways. For example, the expression $2 * x$ can be implemented using multiplication, using addition (as $x + x$), or if x is an unsigned integer, using a logical shift operation. Different ways of implementing an operation can have different costs. Over an entire program, these cost differences can mount up.

Compiler optimization is the process by which the compiler rewrites the internal representation of a program into a form that yields a more efficient target-language program. The word "optimization" is a misnomer, since the compiler cannot guarantee optimality for the resulting code. In practice, optimizers apply a fixed sequence of analyses and transformations that the compiler writer believes will produce better code.

Optimizers attempt to improve the program by analyzing it and using the resulting knowledge to rewrite it. Historically, optimizers have primarily focused on making the program run faster. In some contexts, however, other properties of the program, such as the size of the compiled code, are equally important.

A. Static Analysis of Programs

To improve a program, the optimizer must rewrite the code (or its IR form) in a way that produces better a target language program. Before it can rewrite the code, however, the compiler must prove that the proposed transformation

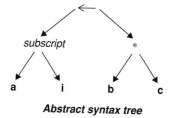

Abstract syntax tree

LOAD VAR	@b	$\Rightarrow r_1$
LOAD VAR	@c	$\Rightarrow r_2$
MULT	r_1, r_2	$\Rightarrow r_3$
LOAD VAR	@i	$\Rightarrow r_4$
MULTIMMED	$r_4, 4$	$\Rightarrow r_5$
ADDIMMED	$r_5, @a$	$\Rightarrow r_6$
STORE	r_3	$\Rightarrow r_6$

Three-address code

is *safe*—that is, the transformation does not change the results that the program computes—and likely to be *profitable*—that is, the resulting code will be an improvement over the existing code. To accomplish this, the compiler analyzes the program in an attempt to determine how it will behave when it runs. Because the compiler performs the analysis before the program runs, the analysis is considered a static analysis. In contrast, an analysis built into the running program would be a dynamic analysis.

Static analysis takes many forms, each one tailored to address a particular problem in the safety or profitability of optimization. One such problem is *constant propagation*—discovering variables whose run-time values can be determined at compile time. If the compiler can discover that **x** always has the value 2 at a particular point in the program, it can specialize the code that uses **x** to reflect that knowledge. In particular, if **x** has the value 2 in an expression **x ∗ y**, it can replace the expression with **2 ∗ y** or with **y + y**, either of which may be faster than a generic multiply. (The former avoids referring to **x**, with a possible memory reference, while the latter also replaces the multiply operation with an addition, which may be cheaper.)

Analyses such as constant propagation are formulated as problems in *data-flow analysis*. Many data-flow problems have been developed for use in optimization. These include the following:

- *Finding redundancies*: An expression is *redundant* if, along any path that reaches it, that value has already been computed. If the compiler can prove that an expression is redundant, it can replace the expression with its previously computed value.
- *Live variable analysis*: A variable is *live* at some point in the code if there exists a path from that point to a use of its value. With the results of live variable analysis, the compiler can stop preserving a variable at the point where it stops being live.
- *Very-busy expressions*: An expression is *very busy* at some point if it will be used, along every path leaving that point before one of its inputs is redefined. Moving the expression to the point where it is very busy can eliminate unneeded copies of the expression, producing a smaller target program.

These problems are formulated as systems of equations. The structure of the equations is dictated by the control-flow relationships in the program. The solutions to the equations are found by using general solvers, analogous to Gaussian Elimination, or by using specialized algorithms that capitalize on properties of the program being analyzed.

B. Classic Scalar Transformations

To apply the results of analysis, the compiler must rewrite the code in a way that improves it. Hundreds of optimizing transformations have been proposed in the literature. These techniques address a wide range of problems that arise in the translation of programming languages. Examples include reducing the overhead introduced by source-language abstractions, managing hardware features such as memory, and general strategies that use properties such as commutativity and associativity to speed expression evaluation.

Among the transformations that you will find in a modern optimizer are as follows:

- *Useless code elimination*: The compiler may discover that some part of the source code is useless—either it cannot execute or its results are never used. In this case, the compiler can eliminate the useless operations. This may make the resulting code faster (because it no longer executes the operations). The resulting code should also be smaller. The presence of other transformations, such as constant propagation, inline substitution, and redundancy elimination, can create useless code.
- *Inline substitution*: In many cases, the compiler can replace an invocation of a procedure or function with a copy of the called procedure, after renaming variables to avoid conflicts and to enforce the parameter binding and mapping of the original call. This eliminates the overhead of the call itself. It may create opportunities to specialize the inlined code to the context that calls it. Because inline substitution can significantly increase the size of the program by duplicating code, it is most attractive when the called procedure is small, as often happens in object-oriented languages.
- *Constant folding*: If the compiler can determine, at compile time, the value of an expression in the program, it can eliminate the operations that evaluate the expression and replace any references to that expression directly with its value. This speeds up the program by avoiding computation. If the expression is used in a control flow decision, folding may lead to the elimination of larger code fragments.

To design an optimizer, the compiler writer must select a set of transformations to apply. For most transformations, several implementation techniques exist. Those distinct methods may require different kinds of analysis, may operate on different IRs, and may address different cases. Thus, the choice of specific algorithms for the transformations has an impact on how well the optimizer works.

Finally, individual transformations can create opportunities for further improvement; they can also foreclose such opportunities. Thus, selecting an order in which to apply the transformations has a strong impact on the optimizer's behavior.

C. High-Level Transformations

Some compilers apply aggressive transformations to reshape the code to perform well on a specific computer architecture. These transformations depend heavily on properties of the underlying architecture to achieve their improvement. Some of the problems that compilers attack in this way are as follows:

- *Vectorization*: A vector computer applies the same operation to different elements of the same array in parallel. To support vector hardware, compilers apply a series of transformations to expose loops that can be expressed as vector computations. Vectorization typically requires the compiler to determine that the execution of a particular statement in any iteration does not depend on the output of that statement in previous iterations.
- *Parallelization*: A parallel computer uses multiple processors to execute blocks of code concurrently. To make effective use of parallel computers, compilers transform the program to expose data parallelism—in which different processors can execute the same function on different portions of the data. Parallelization typically requires the compiler to determine that entire iterations of a loop are independent of one another. With independent iterations, the distinct processors can each execute an iteration without the need for costly interprocessor synchronization
- *Memory hierarchy management*: Every modern computer has a cache hierarchy, designed to reduce the average amount of time required to retrieve a value from memory. To improve the performance of programs, many compilers transform loop nests to improve data locality—i.e., they reorder memory references in a way that increases the likelihood that data elements will found in the cache when they are needed. To accomplish this, the compiler must transform the loops so that they repeatedly iterate over blocks of data that are small enough to fit in the cache.

All of these transformations require sophisticated analysis of array indices to understand the reference patterns in the program. This kind of analysis is typically performed on an AST-like representation, where the original structure of the array references is explicit and obvious. Most compilers that perform these optimizations carry out the analysis

and transformation early in the compilation process. Because these transformations are tied, by implementation concerns, to a source-like representation of the code, they are often called *high-level transformations.*

V. CODE GENERATION

Once the front end and optimizer have produced the final IR program, the back end must translate it into the target language. The back end must find an efficient expression for each of the program's constructs in the target language. If the target language is assembly code for some processor, the generated program must meet all of the constraints imposed by that processor. For example, the computation must fit in the processor's register set. In addition, the computation must obey both the ordering constraints imposed by the flow of data in the program and those imposed by the low-level functioning of the processor. Finally, the entire computation must be expressed in operations found in the assembly language. The compiler's back end produces efficient, working, executable code. To accomplish this, it performs three critical functions: instruction selection, instruction scheduling, and register allocation.

A. Instruction Selection

The process of mapping IR operations into target machine operations is called *instruction selection.* Conceptually, selection is an exercise in pattern matching—relating one or more IR constructs into one or more machine operations.

If the details of the IR differ greatly from those of the target machine, instruction selection can be complex. For example, consider mapping a program represented by an AST onto a Pentium. To bridge the gap in abstraction between the source-like AST and the low-level computer, the instruction selector must fill in details such as how to implement a switch or case statement, which registers are saved on a procedure call, and how to address the ith element of an array. In situations where the abstraction levels of the machine and the IR differ greatly, or in situations where the processor has many different addressing modes, instruction selection plays a large role in determining the quality of the final code.

If, on the other hand, the details of the IR and the target machine are similar, the problem is much simpler. With a low-level, linear IR and a typical RISC microprocessor, much of the code might be handled with a one-to-one mapping of IR operations into machine operations. The low-level IR already introduces registers (albeit an unlimited number), expands abstractions like array indexing into their component operations, and includes the temporary names required to stitch binary operations into more complex expressions. In such situations, instruction selection is less important than high-quality optimization, instruction scheduling, and register allocation.

Instruction selectors are produced using three distinct approaches: *tree pattern matching, peephole optimization,* and *ad hoc, hand-coded solutions.*

- *Tree pattern matching* systems automatically generate a tree-matcher from a grammatical description of the tree. The productions in the grammar describe portions of the tree. Each production has an associated snippet of code and a cost for that snippet. The tree-matcher finds a derivation from the grammar that produces the tree and has the lowest cost. A postmatching pass uses the derivation to emit code, pasting together the snippets from the various productions.
- *Peephole-optimization systems* analyze and translate the IR, one small segment at a time. These matchers use a sliding window that they move over the code. At each step, the engine expands the operations into a detailed, low-level representation, simplifies the low-level code within the window, and compares the simplified, low-level IR against a library of patterns that represent actual machine operations. These matchers can work with either tree-like or linear IRs.
- *Ad hoc methods* are often used when the IR is similar in form and level of abstraction to the target machine code. Of course, this approach offers neither the local optimality of tree pattern matching nor the systematic simplification of peephole optimization. It also lacks the support provided by the tools that automate generation of both tree-matchers and peephole matchers. Still, compiler writers use this approach for simple situations.

B. Instruction Scheduling

On most computer systems, the speed of execution depends on the order in which operations are presented for execution. In general, an operation cannot execute until all of its operands are ready—they have been computed and stored in a location that the operation can access. If the code tries to execute an operation before its operands are available, some processors stall the later operation until the operand is ready. This delay slows program execution. Other processors let the operation execute. This strategy inevitably produces invalid or incorrect results.

Similarly, the compiled code might not make good use of all the functional units available on the target machine. Assume a machine that can execute one operation of type A and one operation of type B in parallel at each cycle, and that the compiled code contains the sequence AAABB-BAAABBB; then the processor can use just half of its potential. If the compiler can reorder the code into the sequence ABABABABABAB, then the computer might execute it in half the time—because at each step, it can execute an operation of type A and another of type B.

The goal of *instruction scheduling* is to reorder the operations in the target machine program to produce a faster running program. This task depends heavily on low-level details of the target machine. These details include the number of operations it can execute concurrently and their types, the amount of time it takes to execute each operation (which may be nonuniform), the structure of the register set, and the speed with which it can move data between registers and memory.

The instruction-scheduling problem is, in general, NP-complete; thus, it is likely that no efficient method can solve it optimally. To address this problem, compilers use greedy, heuristic scheduling methods. The most common of these methods is called *list scheduling.* It relies on a list of operations that are ready to execute in a given cycle. It repeatedly picks an operation from the list and places it into the developing schedule. It then updates the list of ready operations, and records when operations that depend on the just-scheduled operation will be ready. This method works on straight-line code (i.e., sequences of code that contain no branches). Many variations have been proposed; they differ primarily in the heuristics used to select operations from the ready list.

To improve the quality of scheduling, compilers use several variations on replication to create longer sequences of straight-line code. Profile-based techniques, such as *trace scheduling,* use information about the relative execution frequency of paths through the program to prioritize regions for scheduling. Loop-oriented techniques, such as *software pipelining,* focus on loop bodies under the implicit assumption that loop bodies execute more often than the code that surrounds them. Still other techniques use graphs that describe the structure of the program to select and prioritize regions for scheduling.

C. Register Allocation

Before the code can execute on the target computer, it must be rewritten so that it fits within the register set of the target machine. Most modern computers are designed around the idea that an operation will draw its operands from registers and will store its result in a register. Thus, if the computer can execute two operations on each cycle, and those operations each consume two values and produce another value, then the compiler must arrange to have four operands available in registers and another two registers available for results, in each cycle. To complicate matters further, processors typically have a small set of registers—32 or 64—and the ratio of registers per functional unit has been shrinking over the last decade.

Compilers address these issues by including a register allocator in the back end. Deferring the issue of register management until late in compilation permits the earlier phases largely ignore both the problem and its impact on code quality. Earlier phases in the compiler typically assume an unlimited set of registers. This lets them expose more opportunities for optimization, more opportunities to execute operations concurrently, and more opportunities for reordering. Once the compiler has explored those opportunities and decided which ones to take, the register allocator tries to fit the resulting program into the finite storage resources of the actual processor.

At each point in the code, the allocator must select the set of value that will reside in the processor's registers. It must rewrite the code to enforce those decisions—moving some values held in the unlimited, or virtual, register set into memory. When it moves some value into memory to make room in the register set—called *spilling* the value—it must insert code to store the value, along with code to retrieve the value before its next use. The most visible result of register allocation is the insertion of code to handle spilling. Thus, allocation usually produces a larger and slower program. However, the postallocation program can execute on the target machine, where the preallocation program may have used nonexistent resources.

The register allocation problem is NP-complete. Thus, compilers solve it using approximate, heuristic techniques. The most popular of these techniques operates via an analogy to graph coloring. The compiler constructs a graph that represents conflicts between values—two values conflict when they cannot occupy the same space—and tries to find a k-coloring for that conflict graph. If the compiler can find a k-coloring, for k equal to the number of registers on the processor, then it can translate the coloring into an assignment of those values into the k registers. If the compiler cannot find such a coloring, it spills one or more values. This modifies the code, simplifies the conflict graph, and creates a new coloring problem. Iterating on this process—building the graph, trying to color it, and spilling—produces a version of the program that can be colored and allocated. In practice, this process usually halts in two or three tries.

D. Final Assembly

The compiler may need to perform one final task. Before the code can execute, it must be expressed in the native language of the target machine. The result of instruction selection, instruction scheduling, and register allocation is a program that represents the target machine code. How-ever, it may not have the requisite form of a target machine program. If this is the case, the compiler must either convert the program into the format from which it can be linked, loaded, and executed, or it must convert it into a form where some existing tool can do the job.

A common way of accomplishing this is to have the compiler generate assembly code for the target machine, and to rely on the system's assembler to convert the textual representation into an executable binary form. This removes much of the knowledge about the actual machine code from the compiler and lets many compilers for the machine share a single assembler.

VI. SUMMARY

A long-term goal of the compiler-building community has been to reduce the amount of effort required to produce a quality compiler. Automatic generation of scanners, of parsers, and of instruction selectors has succeeded; the systems derived from specifications are as good (or better) than handcrafted versions. In other areas, however, automation has not been effective. Adoption of specification-based tools for context-sensitive analysis has been slow. Few systems have tried to automatically generate optimizers, instruction schedulers, or register allocators—perhaps because of the intricate relationship between these tools, the IR, and the target machine's architecture.

A modern compiler brings together ideas from many parts of computer science and bends them toward the translation of a source language program into an efficient target language program. Inside a compiler, you will find practical applications of formal language theory and logic, greedy heuristic methods for solving NP-complete problems, and careful use of algorithms and data structures ranging from hash tables through union-find trees, alongside time-tested, ad hoc techniques. Compilers solve problems that include pattern recognition, unification, resource allocation, name–space managemnt, storage layout, set manipulation, and solving sets of simultaneous equations. Techniques developed for compiling programming languages have found application in areas that include text editors, operating systems, digital circuit design and layout, theorem provers, and web-filtering software.

SEE ALSO THE FOLLOWING ARTICLES

BASIC PROGRAMMING LANGUAGE • CLUSTER COMPUTING • COMPUTER ALGORITHMS • COMPUTER ARCHITECTURE • DATA STRUCTURES • PARALLEL COMPUTING • SOFTWARE ENGINEERING

BIBLIOGRAPHY

Aho, A. V., Sethi, R., and Ullman, J. D. (1986)."Compilers: Principles, Tools, and Techniques," Addison-Wesley, Reading, MA.

Cooper, K. D., and Torczon, L. (2002). "Engineering a Compiler," Morgan-Kaufmann, San Francisco.

Fischer, C. N., and LeBlanc, R. J., Jr. (1991). "Crafting a Compiler with C," Benjamin Cummings.

Morgan, C. R. (1998). "Building an Optimizing Compiler," Digital Press.

Muchnick, S. S. (1997). "Advanced Compiler Design and Implementation," Morgan-Kaufmann, San Francisco.

Wilhelm, R., and Maurer, D. (1995). "Compiler Design," Addison-Wesley, Reading, MA.

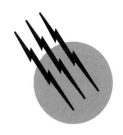

Complex Analysis

Joseph P. S. Kung

Department of Mathematics, University of North Texas

Chung-Chun Yang

Department of Mathematics, The Hong Kong University of Science and Technology

I. The Complex Plane
II. Analytic and Holomorphic Functions
III. Cauchy Integral Formula
IV. Meromorphic Functions
V. Some Advanced Topics

GLOSSARY

Analytic function A complex function with a power series expansion. An analytic function is holomorphic and conversely.

Argument In the polar form $z = re^{i\theta}$ of a complex number, the argument is the angle θ, which is determined up to an integer multiple of 2π.

Cauchy integral formula A basic formula expressing the value of an analytic function at a point as a line integral along a closed curve going around that point.

Cauchy–Riemann equations The system of first-order partial differential equations $u_x = v_y$, $u_y = -v_x$, which is equivalent to the complex function $f(z) = u(x, y) + iv(x, y)$ being holomorphic (under the assumption that all four first-order partial derivatives are continuous).

Conformal maps A map that preserves angles infinitesimally.

Disk A set of the form $\{z : |z - a| < r\}$ is an open disk.

Domain A nonempty connected open set in the complex plane.

Half-plane The upper half-plane is the set of complex numbers with positive imaginary part.

Harmonic function A real function $u(x, y)$ of two real variables satisfying Laplace's equation $u_{xx} + u_{yy} = 0$.

Holomorphic functions A differentiable complex function. See also Analytic functions.

Laurent series An expansion of a complex function as a doubly infinite sum: $f(z) = \sum_{m=-\infty}^{\infty} a_n(z - c)^m$.

Meromorphic function A complex function that is differentiable (or holomorphic) except at a discrete set of points, where it may have poles.

Neighborhood of a point An open set containing that point.

Pole A point a is a pole of a function $f(z)$ if $f(z)$ is analytic on a neighborhood of a but not at a, $\lim_{z \to a} f(z) = \infty$, and $\lim_{z \to a}(z - a)^k f(z)$ is finite for some positive integer k.

Power series or Taylor series An expansion of a complex function as an infinite sum: $f(z) = \sum_{m=0}^{\infty} a_n(z - c)^m$.

Region A nonempty open set in the complex plane.

Riemann surface A surface (in two real dimensions) obtained by cutting and gluing together a finite or infinite number of complex planes.

TO OVERSIMPLIFY, COMPLEX ANALYSIS is calculus using complex numbers rather than real numbers. Because a complex number $a + ib$ is essentially a pair of real numbers, there is more "freedom of movement" over the complex numbers and many conditions become stronger. As a result, complex analysis has many features that distingush it from real analysis. The most remarkable feature, perhaps, is that a differentiable complex function always has a power series expansion. This fact follows from the Cauchy integral formula, which expresses the value of a function in terms of values of the function on a closed curve going around that point.

Many powerful mathematical techniques can be found in complex analysis. Many of these techniques were developed in the 19th century in conjunction with solving practical problems in astronomy, engineering, and physics. Complex analysis is now recognized as an indispensable component of any applied mathematician's toolkit. Complex analysis is also extensively used in number theory, particularly in the study of the distribution of prime numbers.

In addition, complex analysis was the source of many new subjects of mathematics. An example of this is Riemann's attempt to make multivalued complex functions such as the square-root function single-valued by enlarging the range. This led him to the idea of a Riemann surface. This, in turn, led him to the theory of differentiable manifolds, a mathematical subject that is the foundation of the theory of general relativity.

I. THE COMPLEX PLANE

A. Complex Numbers

A complex number is a number of the form $a + ib$, where a and b are real numbers and i is a square root of -1, that is, i satisfies the quadratic equation $i^2 + 1 = 0$. Historically, complex numbers arose out of attempts to solve polynomial equations. In particular, in the 16th century, Cardano, Tartaglia, and others were forced to use complex numbers in the process of solving cubic equations, even when all three solutions are real. Because of this, complex numbers acquired a mystical aura that was not dispelled until the early 19th century, when Gauss and Argand proposed a geometric representation for them as pairs of real numbers.

Gauss thought of a complex number $z = a + ib$ geometrically as a point (a, b) in the real two-dimensional space. This represents the set $\mathbf{C}$ of complex numbers as a real two-dimensional plane, called the *complex plane*. The x-axis is called the *real axis* and the real number a is called

the *real part* of z. The y-axis is called the *imaginary axis* and the real number b is called the *imaginary part* of z.

The *complex conjugate* $\bar{z}$ of the complex number $z = a + ib$ is the complex number $a - ib$. The *absolute value* $|z|$ is the real number $\sqrt{a^2 + b^2}$. The (*multiplicative*) *inverse* or *reciprocal* of z is given by the formula

$$\frac{1}{z} = \frac{\bar{z}}{|z|} = \frac{a - ib}{\sqrt{a^2 + b^2}}.$$

The complex number $z = a + ib$ can also be written in the *polar form*

$$z = re^{i\theta} = r(\cos\theta + i\sin\theta),$$

where

$$r = \sqrt{a^2 + b^2}, \qquad \tan\theta = y/x.$$

The angle θ is called the *argument* of z. The argument is determined up to an integer multiple of 2π. Usually, one takes the value θ so that $-\pi < \theta \le \pi$; this is called the *principal value* of the argument.

B. Topology of the Complex plane

The absolute value defines a *metric* or distance function d on the complex plane $\mathbf{C}$ by

$$d(z_1, z_2) = |z_1 - z_2|.$$

This metric satisfies the triangle inequality

$$d(z_1, z_2) + d(z_2, z_3) \ge d(z_1, z_3)$$

and determines a topology on the complex plane $\mathbf{C}$. We shall need the notions of open sets, closed sets, closure, compactness, continuity, and homeomorphism from topology.

C. Curves

A *curve* γ is a continuous map from a real interval $[a, b]$ to $\mathbf{C}$. The curve γ is said to be *closed* if $\gamma(a) = \gamma(b)$; it is said to be *open* otherwise. A *simple closed curve* is a closed curve with $\gamma(t_1) = \gamma(t_2)$ if and only if $t_1 = a$ and $t_2 = b$.

An intuitively obvious theorem about curves that turned out to be very difficult to prove is the Jordan curve theorem. This theorem is usually not necessary in complex analysis, but is useful as background.

The Jordan Curve Theorem. The image of a simple closed curve (not assumed to be differentiable) separates the extended complex plane into two regions. One region is bounded (and "inside" the curve) and the other is unbounded.

D. The Stereographic Projection

It is often useful to extend the complex plane by adding a point ∞ at infinity. The extended plane is called the *extended complex plane* and is denoted $\bar{\mathbf{C}}$. Using a *stereographic projection*, we can represent the extended plane $\bar{\mathbf{C}}$ using a sphere.

Let $\mathcal{S}$ denote the sphere with radius 1 in real three-dimensional space defined by the equation

$$x_1^2 + x_2^2 + x_3^2 = 1$$

Let the x_1-axis coincide with the real axis of $\mathbf{C}$ and let the x_2-axis coincide with imaginary axes of $\mathbf{C}$. The point $(0, 0, 1)$ on $\mathcal{S}$ is called the *north pole*. Let (x_1, x_2, x_3) be any point on $\mathcal{S}$ not equal to the north pole. The point $z = x + iy$ of intersection of the straight line segment emanating from the north pole N going through the point (x_1, x_2, x_3) with the complex plane $\mathbf{C}$ is the *stereographical projection* of (x_1, x_2, x_3). Going backwards, the point (x_1, x_2, x_3) is the *spherical image* of z. The north pole is the spherical image of the point ∞ at infinity. The stereographic projection is a one-to-one mapping of the extended plane $\bar{\mathbf{C}}$ onto the sphere $\mathcal{S}$. The sphere $\mathcal{S}$ is called the *Riemann sphere*. The stereographical projection has the property that the angles between two (differentiable) curves in $\mathbf{C}$ and the angle between their images on $\mathcal{S}$ are equal.

The mathematical formulas relating points and their spherical images are as follows:

$$x_1 = \frac{z + \bar{z}}{1 + |z|^2}, \quad x_2 = \frac{z - \bar{z}}{i(1 + |z|^2)}, \quad x_3 = \frac{|z|^2 - 1}{1 + |z|^2}$$

and

$$z = \frac{x_1 + ix_2}{1 - x_3}$$

Let z_1 and z_2 be two points in the plane $\mathbf{C}$. The *spherical* or *chordal distance* $\sigma(z_1, z_2)$ between their spherical images on $\mathcal{S}$ is

$$\sigma(z_1, z_2) = \frac{2|z_1 - z_2|}{\sqrt{1 + |z_1|^2}\sqrt{1 + |z_2|^2}}.$$

Let $d\sigma$ and ds be the length of the infinitesimal arc on $\mathcal{S}$ and $\mathbf{C}$, respectively. Then

$$d\sigma = 2(1 + |z|^2)^{-1} ds.$$

E. Connectivity

We need two notions of connectedness of sets in the complex plane. Roughly speaking, a set is *connected* if it consists of a single piece. Formally, a set S is connected if S is not the union $A \cup B$ of two disjoint nonempty open (and hence, closed) subsets A and B. A set is *simply connected* if every closed curve in it can be contracted to a point, with the contraction occurring in the set. It can be proved that a set is simply connected in $\mathbf{C}$ if its complement in the extended complex plane is connected. If A is not simply connected, then the connected components (that is, open and closed subsets) of its complement $\bar{\mathbf{C}}$ not containing the point ∞ are the "holes" of A.

A *region* is a nonempty open set of $\mathbf{C}$. A *domain* is a non-empty connected open set of $\mathbf{C}$. When $r > 0$, the set $\{z : |z - c| < r\}$ of all complex numbers at distance strictly less than r from the *center c* is called the *open disk with center c and radius r*. Its closure is the *closed disk* $\{z : |z - c| \leq r\}$. Open disks are the most commonly used domains in complex analysis.

II. ANALYTIC AND HOLOMORPHIC FUNCTIONS

A. Holomorphic Functions

Let $f(z)$ denote a complex function defined on a set Ω in the complex plane $\mathbf{C}$. The function $f(z)$ is said to be *differentiable* at the point a in Ω if the limit

$$\lim_{h \to 0} \frac{f(a + h) - f(a)}{h}$$

is a finite complex number. This limit is the *derivative* $f'(a)$ of $f(z)$ at a. Note that the limit is taken over all complex numbers h such that the absolute value $|h|$ goes to zero, so that h ranges over a set that has two real dimensions. Thus, differentiability of complex functions is a much stronger condition than differentiability of real functions. For example, the length of every (infinitesimally) small line segment starting from a is changed under the function $f(z)$ by the same real scaling factor $|f'(a)|$, independently of the angle. The formal rules of differentiation in calculus hold also for complex differentiation.

It is possible to construct complex functions that are differentiable at only one point. To exclude these degenerate cases, we only consider complex functions that are differentiable at every point in a region Ω. Such functions are said to be *holomorphic* on Ω.

B. The Cauchy–Riemann Equations and Harmonic Functions

A complex function $f(z)$ can be expressed in the following way:

$$f(z) = f(x + iy) = u(x, y) + iv(x, y)$$

where $u(x, y)$ is the real part of $f(z)$ and $v(x, y)$ is the imaginary part of $f(z)$. The functions $u(x, y)$ and $v(x, y)$ are real functions of two real variables.

Differentiability of the complex function $f(z)$ can be rewritten as a condition on the real functions $u(x, y)$ and $v(x, y)$. Let $f(z) = u(x, y) + iv(x, y)$ be a complex function such that all four first-order partial derivatives of u and v are continuous in the open set Ω. Then a necessary and sufficient condition for $f(z)$ to be holomorphic on Ω is

$$\frac{\partial u}{\partial x} = \frac{\partial v}{\partial y}, \qquad \frac{\partial u}{\partial y} = -\frac{\partial v}{\partial x}$$

These equations are the *Cauchy–Riemann equations*. In polar coordinates $z = re^{i\theta}$, the Cauchy–Riemann equations are

$$r\frac{\partial u}{\partial r} = \frac{\partial v}{\partial \theta}, \qquad \frac{\partial u}{\partial \theta} = -r\frac{\partial v}{\partial r}$$

The square of the absolute value of the derivative is the Jacobian of $u(x, y)$ and $v(x, y)$, that is

$$|f'(z)|^2 = \frac{\partial u}{\partial x}\frac{\partial v}{\partial y} - \frac{\partial u}{\partial y}\frac{\partial v}{\partial x}$$

A *harmonic or potential function* $h(x, y)$ is a real function having continuous second-order partial derivatives in a nonempty open set Ω in R^2 satisfying the two-dimensional *Laplace equation*

$$\frac{\partial^2 h}{\partial x^2} + \frac{\partial^2 h}{\partial y^2} = 0$$

for all $z = x + iy$ in Ω. We shall see that $f(z)$ being holomorphic in Ω implies that every derivative $f^{(n)}(z)$ is holomorphic on Ω, and, in particular, the real functions $u(x, y)$ and $v(x, y)$ have partial derivatives of any order. Hence, by the Cauchy–Riemann equations, $u(x, y)$ and $v(x, y)$ satisfy the Laplace equation and are harmonic functions. The harmonic functions $u(x, y)$ and $v(x, y)$ are called the *conjugate pair* of the holomorphic function $f(z)$.

The two-dimensional Laplace equation governs (incompressible and irrotational) fluid flow and electrostatics in the plane. These give intuitive physical models for holomorphic functions.

C. Power Series and Analytic Functions

Just as in calculus, we can define complex functions using power series. A *power series* is an infinite sum of the form

$$\sum_{m=0}^{\infty} a_m(z - c)^m$$

where the *coefficients* a_m and the *center* c are complex numbers. This series determines a complex number (depending on z) whenever the the series converges. Complex power series work in the same way as power series over the reals.

In particular, a power series has a *radius of convergence*, that is, an extended real number ρ, $0 \leq \rho \leq \infty$, such that the series converges absolutely whenever $|z - c| < \rho$. The radius of convergence is given explicitly by *Hadamard's formula*:

$$\rho = \frac{1}{\displaystyle\lim_{n\to\infty} \sup \sqrt[n]{|a_n|}}$$

A function $f(z)$ is *analytic* in the region Ω if for every point c in Ω, there exists an open disk $\{z : |z - c| < r\}$ contained in Ω such that $f(z)$ has a (convergent) power series or *Taylor expansion*

$$f(z) = \sum_{m=0}^{\infty} a_m(z - c)^m$$

When this holds, $f^{(n)}(c)/m! = a_m$.

Polynomials are analytic functions on the complex plane. Other examples of analytic functions on the complex plane are the *exponential function,*

$$e^z = \sum_{m=0}^{\infty} z^n/n!$$

and the two *trigenometric functions,*

$$\cos z = \frac{e^{iz} + e^{-iz}}{2}, \qquad \sin z = \frac{e^{iz} - e^{-iz}}{2i}$$

The inverse under functional composition of e^z is the (*natural*) *logarithm* $\log z$. The easiest way to define it is to put z into polar form. Then

$$\log z = \log re^{i\theta} = \log r + i\theta$$

Since θ is determined up to an integer multiple of 2π, the logarithmic function is multivalued and one needs to extend the range to a Riemann surface (see Section V.D) to make it a function. For most purposes, one takes the value of θ so that $-\pi < \theta \leq \pi$. This yields the *principal value* of the logarithm.

III. CAUCHY INTEGRAL FORMULA

A. Line Integrals and Winding Numbers

Line integrals are integrals taken over a curve rather than an interval on a real line. Let $\gamma : [a, b] \to C$ be a piecewise continuously differentiable curve and let $f(z)$ be a continuous complex function defined on the image of γ. Then the *line integral* of $f(z)$ along the path γ is the Riemann integral

$$\int_a^b f(\gamma(t))\gamma'(t)\, dt$$

This integral is denoted by

$$\int_\gamma f(z)\, dz$$

Line integrals are also called *path integrals* or *contour integrals*. Line integrals behave in similar ways to integrals over the real line, except that instead of moving along the real line, we move on a curve.

The *winding number* or *index* $n(\gamma, a)$ of a curve γ relative to a point a is the number of time the curve winds or goes around the point a. Formally, we define $n(\gamma; z)$ using a line integral:

$$n(\gamma, a) = \frac{1}{2\pi i} \int_\gamma \frac{d\zeta}{\zeta - a}$$

This definition is consistent with the intuitive definition. One can show, for example, that (a) $n(\gamma, a)$ is always an integer, (b) the winding number of a circle around its center is 1 if the circle goes around in a counterclockwise direction, and -1 if the circle goes around in a clockwise direction, (c) the winding number of a circle relative to a point in its exterior is 0, and (d) if a is a point in the interior of two curves and those two curves can be continuously deformed into one another without going through a, then they have the same winding number relative to a.

B. Cauchy's Integral Formula

In general, line integrals depend on the curve. But if the integrand $f(z)$ is holomorphic, Cauchy's integral theorem implies that the line integral on a simply connected region only depends on the endpoints.

Cauchy's integral theorem. Let $f(z)$ be holomorphic on a simply connected region Ω in $\mathbf{C}$. Then for any closed piecewise continuously differential curve γ in Ω,

$$\int_\gamma f(z)\, dz = 0$$

One way to prove Cauchy's theorem (due to Goursat) is to observe that if the curve is "very small," then the line integral should also be "very small" because holomorphic functions cannot change drastically in a small neighborhood. Hence, we can prove the theorem by carefully decomposing the curve into a union of smaller curves. Another way to think of Cauchy's theorem is that a line integral over a curve γ of a holomorphic function on a region Ω is zero whenever γ can be continuously shrunk to a point in Ω.

Cauchy's integral formula. Let $f(z)$ be holomorphic on a simply connected region Ω in $\mathbf{C}$ and let γ be a simple piecewise continuously differentiable closed path going counterclockwise in Ω. Then for every point z in Ω inside γ,

$$f^{(m)}(z) = \frac{m!}{2\pi i} \int_\gamma \frac{f(\zeta)\, d\zeta}{(\zeta - z)^{m+1}}$$

In particular,

$$f(z) = \frac{1}{2\pi i} \int_\gamma \frac{f(\zeta)\, d\zeta}{\zeta - z}$$

Cauchy's formula for $f(z)$ follows from Cauchy's theorem applied to the function $(f(\zeta) - f(z))/(\zeta - z)$, and the general case follows similarly.

A somewhat more general formulation of Cauchy's formula is in terms of the winding number. If $f(z)$ is analytic on a simply connected nonempty open set Ω and γ is a closed piecewise continuously differentiable curve, then, for every point z in Ω,

$$f^{(m)}(z) = n(\gamma, z)\frac{m!}{2\pi i} \int_\gamma \frac{f(\zeta)\, d\zeta}{(\zeta - z)^{m+1}}$$

Cauchy's integral formula expresses the function value $f(z)$ in terms of the function values around z. Take the curve γ to be a circle $|z - a| = r$ of radius r with center a, where r is sufficiently small so that the circle is in Ω. Using the geometric series expansion

$$\frac{1}{\zeta - z} = \frac{1}{\zeta - a} \sum_{m=0}^{\infty} \left(\frac{z - a}{\zeta - a}\right)^m$$

and interchanging summation and integration (which is valid as all the series converge uniformly), we obtain

$$f(z) = \sum_{m=0}^{\infty} \frac{(z - a)^{m+1}}{2\pi i} \int_{|z-a|=r} \frac{f(\zeta)\, d\zeta}{(\zeta - a)^{m+1}}$$

This gives an explicit formula for the power series expansion of $f(z)$ and shows that a holomorphic function is analytic. Since an analytic function is clearly differentiable, being analytic and being holomorphic are the same property.

C. Geometric Properties of Analytic Functions

Analytic functions satisfy many nice geometric and topological properties. A basic property is the following.

The open mapping theorem. The image of an open set under a non-constant analytic function is open.

Analytic functions also have the nice geometric property that they preserve angles. Let γ_1 and γ_2 be two differentiable curves intersecting at a point z in Ω. The *angle* from γ_1 to γ_2 is the signed angle of their tangent lines at z. A function $f(z)$ is said to be *conformal* at the point z if it preserves the angles of pairs of curves intersecting at z. An analytic function $f(z)$ preserves angles at points z where the derivative $f'(z) \neq 0$. If an analytic function $f(z)$ is conformal at every point in Ω (equivalently, if $f'(z) \neq 0$

for every point z in Ω), then it is said to be a *conformal mapping on* Ω.

Examples of conformal mappings (on suitable regions) are Möbius transformations. Let a, b, c, and d be complex numbers such that $ad - bc \neq 0$. Then the bilinear transformation

$$T(z) = \frac{az + b}{cz + d}$$

is a *Möbius transformation*. Any Möbius transformation can be decomposed into a product of four elementary conformal mappings: translations, rotations, homotheties or dilations, and inversions. In addition to preserving angles, Möbius transformations map circles into circles, provided that a straight line is viewed as a "circle" passing through the point ∞ at infinity.

D. Some Theorems about Analytic Functions

If $f(z)$ is a function on the disk with center 0 and radius ρ and $r < \rho$, let $M_f(r)$ be the maximum value of $|f(z)|$ on the circle $\{z : |z| = r\}$.

Cauchy's inequality. Let $f(z)$ be analytic in the disk with center 0 and radius ρ and let $f(z) = \sum_{m=0}^{\infty} a_n z^n$. If $r < \rho$, then

$$|a_n| r^n \leq M_f(r)$$

A function is *entire* if it is analytic on the entire complex plane. The following are two theorems about entire function. The first follows easily from the case $n = 1$ of Cauchy's inequality.

Liouville's theorem. If $f(z)$ is an entire function and $f(z)$ is bounded on **C**, then f must be a constant function.

The second is much harder. It implies the fact that if two or more complex numbers are absent from the image of an entire function, then that entire function must be a constant.

Picard's little theorem. An entire function that is not a polynomial takes every value, with one possible exception, infinitely many times.

Applying Liouville's theorem to the reciprocal of a nonconstant polynomial $p(z)$ and using the fact that $p(z) \to \infty$ as $z \to \infty$, one obtains the following important theorem.

The fundamental theorem of algebra. Every polynomial with complex coefficients of degree at least one has a root in **C**.

It follows that a polynomial of degree n must have n roots in **C**, counting multiplicities.

The next theorem is a fundamental property of analytic functions.

Maximum principle. Let Ω be a bounded domain in **C**. Suppose that $f(z)$ is analytic in Ω and continuous in the closure of Ω. Then, $|f(z)|$ attains its maximum value $|f(a)|$ at a boundary point a of Ω. If $f(z)$ is not constant, then

$$|f(z)| < |f(a)|$$

for every point z in the interior of Ω.

Using the maximum principle, one obtains Schwarz's lemma.

Schwarz's lemma. Let $f(z)$ be analytic on the disk $D = \{z : |z| < 1\}$ with center 0 and radius 1. Suppose that $f(0) = 0$ and $|f(z)| \leq 1$ for all points z in D. Then,

$$|f'(0)| \leq 1$$

and for all points z in D,

$$|f(z)| \leq |z|$$

If $|f'(0)| = 1$ or $|f(a)| = |a|$ for some nonzero point a, then $f(z) = \alpha z$ for some complex number α with $|\alpha| = 1$.

Another theorem, which has inspired many generalizations in the theory of partial differential equations, is the following.

Hadamard's three-circles theorem. Let $f(z)$ be an analytic function on the annulus $\{z : \rho_1 \leq |z| \leq \rho_3\}$ and let $\rho_1 < r_1 \leq r_2 \leq r_3 < \rho_3$. Then

$$\log M_f(r_2) \leq \frac{\log r_3 - \log r_2}{\log r_3 - \log r_1} \log M_f(r_1)$$
$$+ \frac{\log r_2 - \log r_1}{\log r_3 - \log r_1} \log M_f(r_3)$$

It follows from the three-circles theorem that $\log M_f(r)$ is a convex function of $\log r$.

Cauchy's integral theorem has the following converse.

Morera's theorem. Let $f(z)$ be a continuous function on a simply connected region Ω in **C**. Suppose that the line integral

$$\int_{\partial \triangle} g(z)\, dz$$

over the boundary $\partial \triangle$ of every triangle in Ω is zero. Then $f(z)$ is analytic in Ω.

E. Analytic Continuation

An analytic function $f(z)$ is usually defined initially with a certain formula in some region D_1 of the complex plane. Sometimes, one can extend the function $f(z)$ to a function $\hat{f}(z)$ that is analytic on a bigger region D_2 containing D_1 such that $\hat{f}(z) = f(z)$ for all points z on D_1. Such an extension is called *analytic continuation*. Expanding the function as a Taylor (or Laurent series) is one possible way to extend a function locally from a neighborhood of a point. Contour integration is another way.

Analytic continuation results in a unique extended function when it is possible. It also preserves identities between functions.

The uniqueness theorem for analytic continuation. Let $f(z)$ and $g(z)$ be analytic in a region Ω. If the set of points z in Ω where $f(z) = g(z)$ has a limit point in Ω, then $f(z) = g(z)$ for all z in Ω. In particular, if the set of zeros of $f(z)$ in Ω has a limit point in Ω, then $f(z)$ is identically zero in Ω.

Permanence of functional relationships. If a finite number of analytic functions in a region Ω satisfy a certain functional equation in a part of Ω that has a limit point, then that functional equation holds everywhere in Ω.

For example, the Pythagorean identity
$$\sin^2 x + \cos^2 x = 1$$
holds for all real numbers x. Thus, it holds for all complex numbers.

Deriving a functional equation is often the key step in analytic continuation. A famous example is the Riemann functional equation relating the gamma function and the Riemann zeta function.

A quick and direct way (when it works) is the following method.

Schwarz's reflection principle. Let Ω be a domain in the upper half-plane that contains a line segment L on the real axis. Let $f(z)$ be a function analytic on Ω and continuous on L. Then the function extended by defining $f(z) = \overline{f(\bar{z})}$ for z in the "reflection" $\tilde{\Omega} = \{z | \bar{z} \in \Omega\}$ is an analytic continuation of $f(z)$ from Ω to to the bigger domain $\Omega \cup \tilde{\Omega}$.

We note that not all analytic functions have proper analytic continuations. For example, when $0 < |a| < 1$, the *Fredholm series*
$$\sum_{m=1}^{\infty} a^n z^{n^2}$$
converges absolutely on the closed unit disk $\{z : |z| \le 1\}$ and defines an analytic function $f(z)$ on the open disk $\{z : |z| < 1\}$. However, it can be shown that $f(z)$ has no analytic continuation outside the unit disk. Roughly speaking, the Fredholm functions are not extendable because of "gaps" in the powers of z. The sharpest "gap" theorem is the following.

The Fabry gap theorem. Let
$$f(z) = \sum_{m=0}^{\infty} a_m z^{b_m}$$
where b_m is a sequence of increasing nonnegative integers such that

$$\lim_{m \to \infty} b_m / m = \infty$$

Suppose that the power series has radius of convergence ρ. Then $f(z)$ has no analytic continuation outside the disk $\{z : |z| < \rho\}$.

IV. MEROMORPHIC FUNCTIONS

A. Poles and Meromorphic Functions

A point a is an *isolated singularity* of the analytic function $f(z)$ if $f(z)$ is analytic in a neighborhood of a, except possibly at the point itself. For example, the function $f(z) = 1/z$ is analytic on the entire complex plane, except at the isolated singularity $z = 0$. If the limit $\lim_{z \to a} f(z)$ is a finite complex number c, then we can simply define $f(a) = c$ and $f(z)$ will be analytic on the entire neighborhood. Such an isolated singularity is said to be *removable*.

If the limit $\lim_{z \to a} f(z)$ is infinite, but for some positive real number α, $\lim_{z \to a} |z - a|^{\alpha} |f(z)|$ is finite, then a is a *pole* of $f(z)$. It can be proved that if the last condition holds, then the smallest such real number α must be a positive integer k. In this case, the pole a is said to have order k. Equivalently, k is the smallest positive integer such that $(z - a)^k f(z)$ is analytic on an entire neighborhood of a (including a itself).

If an isolated singularity is neither removable nor a pole, it is said to be *essential*. Weierstrass and Casorati proved that an analytic function comes arbitrarily close to every complex number in every neighborhood of an isolated essential singularity. A refinement of this is a deep theorem of Picard.

Picard's great theorem. An analytic function takes on every complex value with one possible exception in every neighborhood of an essential singularity.

These results say that near an isolated singularity, an analytic function behaves very wildly. Thus, the study of isolated singularities has concentrated on analytic functions with poles. A complex function $f(z)$ is *meromorphic* in a region Ω if it is analytic except at a discrete set of points, where it may have poles.

The *residue* $\text{Res}(f, a)$ of a meromorphic function $f(z)$ at the isolated singularity a is defined by

$$\text{Res}(f, a) = \frac{1}{2\pi i} \int_{\gamma} f(\zeta) \, d\zeta$$

If one allows negative powers, then analytic functions can be expanded as power series at isolated singularities. The idea is to write a meromorphic function $f(z)$ in a neighborhood of a pole a as a sum of an analytic part and a singular part. Suppose the function $f(z)$ is analytic in a region containing the annulus $\{z : \rho_1 < |z - a| < \rho_2\}$.

Then we can define two functions $f_1(z)$ and $f_2(z)$ by:

$$f_1(z) = \frac{1}{2\pi i} \int_{\{\zeta : |\zeta - a| = r\}} \frac{f(\zeta)\, d\zeta}{\zeta - z}$$

where r satisfies $|z - a| < r < \rho_2$, and

$$f_2(z) = -\frac{1}{2\pi i} \int_{\{\zeta : |\zeta - a| = r\}} \frac{f(\zeta)\, d\zeta}{\zeta - z}$$

where r satisfies $\rho_1 < r < |z - a|$. The function $f_1(z)$ is analytic in the disk $\{z : |z - a| < \rho_2\}$ and the function $f_2(z)$ is analytic in the complement $\{z : |z - a| > \rho_1\}$. By the Cauchy integral formula, $f(z) = f_1(z) + f_2(z)$ and this representation is valid in the annulus $\{z : \rho_1 < |z - a| < \rho_2\}$.

The functions $f_1(z)$ and $f_2(z)$ can each be expanded as Taylor series. Using the transformation $z - a \mapsto 1/z$ and some simple calculation, we obtain the *Laurent series* expansion

$$f(z) = \sum_{m=-\infty}^{\infty} a_m (z - a)^m$$

where

$$a_m = \frac{1}{2\pi i} \int_{\{\zeta : |\zeta - a| = r\}} \frac{f(\zeta)\, d\zeta}{(\zeta - z)^{m+1}}$$

valid in the annulus $\{z : \rho_1 < |z - a| < \rho_2\}$. Note that

$$\mathrm{Res}(f, a) = a_{-1},$$

and the point a is a pole of order k if and only if $a_{-k} \neq 0$ and every coefficient a_{-m} with $m > k$ is zero. The polynomial

$$\sum_{m=1}^{k} \frac{a_m}{(z - a)^m}$$

in the variable $1/(z - a)$ is called the *singular* or *principal part* of $f(z)$ at the pole a.

B. Elliptic Functions

Important examples of meromorphic functions are elliptic functions. Elliptic functions arose from attempts to evaluate certain integrals. For example, to evaluate the integral

$$\int_0^1 \sqrt{1 - x^2}\, dx$$

which gives the area $\pi/2$ of a semicircle with radius 1, one can use the substitution $x = \sin\theta$. However, to evaluate integrals of the form

$$\int_0^1 \sqrt{(1 - x^2)(1 - k^2 x^2)}\, dx$$

we need elliptic functions. Elliptic functions are doubly periodic generalizations of trigonometric functions.

Let ω_1 and ω_2 be two complex numbers whose ratio ω_1/ω_2 is not real and and let L be the 'integer lattice'

$\{c\omega_1 + d\omega_2\}$, where c and d range over all integers. An *elliptic function* is a meromorphic function on the complex plane with two (independent) periods, ω_1 and ω_2, that is, $f(z + \omega_1) = f(z)$, $f(z + \omega_2) = f(z)$, and every complex number ω such that $f(z + \omega) = f(z)$ for all points z in the complex plane is a number in the lattice L.

A specific example of an elliptic function is the *Weierstrass $\wp$-function* defined by the formula

$$\wp(z) = \frac{1}{z^2} + \sum_{\omega \in L \setminus \{0\}} \left(\frac{1}{(z - \omega)^2} - \frac{1}{\omega^2} \right)$$

This defines a meromorphic function on the complex plane that is doubly periodic with periods ω_1 and ω_2. The $\wp$-function has poles exactly at points in L. Weierstrass proved that every elliptic function with periods ω_1 and ω_2 can be written as a rational function of $\wp$ and its derivative $\wp'$.

C. The Cauchy Residue Theorem

A simple but useful generalization of Cauchy's integral formula is the Cauchy residue theorem.

The Cauchy residue theorem. Let Ω be a simply connected region in the complex plane, let $f(z)$ be a function analytic on Ω except at the isolated singularities a_m, and let let γ be a closed piecewise continuously differentiable curve in Ω that does not pass through any of the points a_m. Then

$$\frac{1}{2\pi i} \int_{\gamma} f(z)\, dz = \sum n(\gamma, a_m) \mathrm{Res}(f, a_m)$$

where the sum ranges over all the isolated singularities inside the curve γ.

Cauchy's residue theorem has the following useful corollary.

The argument principle. Let $f(z)$ be a meromorphic function in a simply connected region Ω, let $a_1, a_2, \ldots$, be the zeros of $f(z)$ in Ω, and let $b_1, b_2, \ldots$, be the poles of $f(z)$ in Ω. Suppose the zero a_m has multiplicity s_m and the pole b_n has order t_n. Let γ be a closed piecewise continuously differentiable curve in Ω that does not pass through any poles or zeros of $f(z)$. Then

$$\frac{1}{2\pi i} \int_{\gamma} \frac{f'(\zeta)\, d\zeta}{f(\zeta)} = \sum s_m n(\gamma, a_m) - \sum t_j n(\gamma, b_j)$$

where the sum ranges over all the zeros and poles of $f(z)$ contained in the curve γ.

The name "argument principle" came from the following special case. When γ is a circle, the argument principle says that the change in the argument of $f(z)$ as z traces the circle in a counterclockwise direction, equals

$$Z(f) - P(f)$$

the difference between the number $Z(f)$ of zeros and the number $P(f)$ of poles of $f(z)$ inside γ, counting multiplicities and orders.

D. Evaluation of Real Integrals

Cauchy's residue theorem can be used to evaluate real definite integrals that are otherwise difficult to evaluate.

For example, to evaluate an integral of the form

$$\int_0^{2\pi} R(\cos\theta, \sin\theta)\, d\theta$$

where R is a rational function of $\cos\theta$ and $\sin\theta$, let $z = e^{i\theta}$. If we make the substitutions

$$\cos\theta = \frac{z + z^{-1}}{2}, \qquad \sin\theta = \frac{z - z^{-1}}{2i}$$

the integral becomes a line integral over the unit circle of the form

$$\int_{|z|=1} S(z)\, dz$$

where $S(z)$ is a rational function of z. By Cauchy's residue theorem, this integral equals $2\pi i$ times the sum of the residues of the poles of $S(z)$ inside the unit circle. Using this method, one can prove, for example, that if $a > b > 0$,

$$\int_0^{2\pi} \frac{d\theta}{(a + b\cos^2\theta)^2} = \frac{\pi(2a + b)}{a^{3/2}(a + b)^{3/2}}$$

One can also evaluate improper integrals, obtaining formulas such as the following formula due to Euler: For $-1 < p < 1$ and $-\pi < \alpha < \pi$,

$$\int_0^\infty \frac{x^{-p}\, dx}{1 + 2x\cos\alpha + x^2} = \frac{\pi \sin p\alpha}{\sin p\pi \sin\alpha}$$

E. Location of Zeros

It is often useful to locate zeros of polynomials in the complex plane. An elegant theorem, which can be proved by elementary arguments, is the following result.

Lucas' theorem. Let $p(z)$ be a polynomial of degree at least 1. All the zeros of the derivative $p'(z)$ lie in the convex closure of the set of zeros of $p(z)$.

Deeper results usually involve using some form of Rouché's theorem, which is proved using the argument principle.

Rouché's theorem. Let Ω be a region bounded by a simple closed piecewise continuously differentiable curve. Let $f(z)$ and $g(z)$ be two functions meromorphic in an open set containing the closure of Ω. If $f(z)$ and $g(z)$ satisfy

$$|f(z) - g(z)| < |f(z)| + |g(z)|$$

for every point z on the curve bounding Ω, then

$$Z(f) - P(f) = Z(g) - P(g)$$

Hurwitz's theorem. Let $(f_n(z): n = 1, 2, \ldots)$ be a sequence of functions analytic in a region Ω bounded by a simple closed piecewise continuously differentiable curve such that $f_n(z)$ converges uniformly to a nonzero (analytic) function $f(z)$ on every closed subset of Ω. Let a be an interior point of Ω. If a is a limit point of the set of zeros of the functions $f_n(z)$, then a is a zero of $f(z)$. If a is a zero of $f(z)$ with multiplicity m, then every sufficiently small neighborhood K of a contains exactly m zeros of the functions $f_n(z)$, for all n greater than a number N depending on K.

F. Infinite Products, Partial Fractions, and Approximations

A natural way to write a meromorphic function is in terms of its zeros and poles. For example, because $\sin\pi z$ has zeros at the integers, we expect to be able to "factor" it into product. Indeed, Euler wrote down the following product expansion:

$$\sin\pi z = \pi z \prod_{j=1}^\infty \left(1 - \frac{z}{n}\right)\left(1 + \frac{z}{n}\right)$$

With complex analysis, one can justify such expansions rigorously.

The question of convergence of an infinite product is easily resolved. By taking logarithms, one can reduce it to a question of convergence of a sum. For example, the product

$$\prod_{m=1}^\infty (1 + a_m)$$

converges absolutely if and only if the sum $\sum_{m=1}^\infty |\log(1 + a_m)|$ converges absolutely. Since $|\log(1 + a_m)|$ is approximately $|a_m|$, the product converges absolutely if and only if the series $\sum_{m=1}^\infty |a_m|$ converges absolutely.

The following theorem allows us to construct an entire function with a prescribed set of zeros.

The Weierstrass product theorem. Let $(a_j : j = 1, 2, \ldots)$ be a sequence of nonzero complex numbers in which no complex number occurs infinitely many times. Suppose that the set $\{a_j\}$ has no (finite) limit point in the complex plane. Then there exists an entire function $f(z)$ with a zero of multiplicity m at 0, zeros in the set $\{a_j\}$ with the correct multiplicity, and no other zeros. This function can be written in the form

$$f(z) = z^m e^{g(z)} \prod_{j=1}^\infty \left(1 - \frac{z}{a_j}\right)$$

$$\times e^{a_j/z + (1/2)(a_j/z)^2 + \cdots + (1/m_j)(a_j/z)^{m_j}}$$

where m_j are positive integers depending on the set $\{a_j\}$, and $g(z)$ is an entire function.

From this theorem, we can derive the following representation of a meromorphic function.

Theorem. A meromorphic function on the complex plane is the quotient of two entire functions. The two entire functions can be chosen so that they have no common zeros.

In particular, one can think of meromorphic functions as generalizations of rational functions.

The *gamma function* $\Gamma(z)$ is a useful function which can be defined by a product formula. Indeed,

$$\Gamma(z) = \frac{e^{-\gamma z}}{z} \prod_{m=1}^{\infty} \left(1 + \frac{z}{m}\right)^{-1} e^{z/m}$$

where γ is the *Euler–Mascheroni constant* defined by

$$\gamma = \lim_{n \to \infty} \left[\left(\sum_{m=1}^{n} \frac{1}{m}\right) - \log n\right]$$

It equals $0.57722\ldots$. The gamma function interpolates the integer factorials. Specifically, $\Gamma(n) = (n-1)!$ for a positive integer n and $\Gamma(z)$ satisfies the functional equation

$$\Gamma(z + 1) = z\Gamma(z).$$

Another useful functional equation is the *Legendre formula*

$$\sqrt{\pi}\,\Gamma(2z) = 2^{2z-1}\Gamma(z)\Gamma(z + 1/2)$$

This gives the following useful value: $\Gamma(1/2) = \sqrt{\pi}$.

Rational functions can be represented as partial fractions; so can meromorphic functions.

Mittag-Leffler's theorem. Let $\{b_j : j = 1, 2, \ldots\}$ be a set of complex numbers with no finite limit point in the complex plane, and let $p_j(z)$ be given polynomials with zero constant terms, one for each point b_j. Then there exist meromorphic functions in the complex plane with poles at b_m with singular parts $p_j(1/z - b_j)$. These functions have the form

$$g(z) + \sum_{j=1}^{\infty} \left(p_j\left(\frac{1}{z - b_j}\right) - q_j(z)\right)$$

where $q_j(z)$ are suitably chosen polynomials depending on $p_j(z)$, and $g(z)$ is an entire function.

Taking logarithmic derivatives and integrating, one can derive Weierstrass's product theorem from Mittag-Leffler's theorem.

Two examples of partial fraction exapnsions of meromorphic functions are

$$\frac{\pi^2}{\sin^2 \pi z} = \sum_{m=-\infty}^{\infty} \frac{1}{(z - n)^2}$$

and

$$\pi \cot \pi z = \frac{1}{z} + \sum_{m=1}^{\infty} \frac{2z}{z^2 - n^2}$$

Runge's approximation theorem says that a function analytic on a bounded region Ω with holes can be uniformly approximated by a rational function all of whose poles lie in the holes. Runge's theorem can be proved using a Cauchy's integral formula for compact sets.

Runge's approximation theorem. Let $f(z)$ be an analytic function on a region Ω in the complex plane; let K be a compact subset of Ω. Let $\epsilon > 0$ be a given (small) positive real number. Then there exists a rational function $r(z)$ with all its poles outside K such that

$$|f(z) - r(z)| < \epsilon$$

for all z in K.

V. SOME ADVANCED TOPICS

A. Riemann Mapping Theorem

On the complex plane, most simply connected regions can be mapped conformally in a one-to-one way onto the open unit disk. This is a useful result because it reduces many problems to problems about the unit disk.

Riemann mapping theorem. Let Ω be a simply connected region that is not the entire complex plane, and let a be a point in Ω. Then there exists a unique analytic function $f(z)$ on Ω such that $f(a) = 0$, $f'(a) > 0$, and $f(z)$ is a one-to-one mapping of Ω onto the open disk $\{z : |z| < 1\}$ with radius 1.

For example, the upper half-plane $\{\zeta : \text{Im}(\zeta) > 0\}$ and the unit disk $\{z : |z| < 1\}$ are mapped conformally onto each other by the Möbius transformations

$$\zeta = \frac{i(1 + z)}{1 - z}, \qquad z = \frac{\zeta - i}{\zeta + i}$$

The Schwarz–Christoffel formula gives an explicit formula for one-to-one onto conformal maps from the open unit disk or the upper half-plane to polygonal domains, which are sets in the complex plane bounded by a a closed simple curve made up of a finite number of straight line segments. It is rather complicated to state, and we refer the reader to the books by Ahlfors and Nehari in the bibliography. As an example, one can map the upper-half complex plane into a triangle with interior angles $\alpha_1 \pi$, $\alpha_2 \pi$, and $\alpha_3 \pi$ (where $\alpha_1 + \alpha_2 + \alpha_3 = 1$) by the inverse of the function

$$F(z) = \int_0^z t^{\alpha_1 - 1}(t - 1)^{\alpha_2 - 1}\, dt$$

The function $F(z)$ is called the *triangle function of Schwarz*.

B. Univalent Functions and Bieberbach's Conjecture

Univalent or *Schlicht* functions are one-to-one analytic functions. They have been extensively studied in complex analysis. A famous result in this area was the Bieberbach conjecture, which was proved by de Branges in 1984.

Theorem. Let $f(z)$ be a univalent analytic function on the unit disk $\{z : |z| < 1\}$ with power series expansion

$$f(z) = z + a_2 z^2 + a_3 z^3 + \cdots$$

(that is, $f(0) = 0$ and $f'(0) = 1$). Then

$$|a_n| \le n.$$

When equality holds, $f(z) = e^{-i\theta} K(e^{i\theta} z)$, where

$$K(z) = \frac{z}{(1-z)^2} = z + 2z^2 + 3z^3 + 4z^4 + \cdots$$

The function $K(z)$ is called *Koebe's function*.

Another famous result in this area is due to Koebe.

Koebe's 1/4-theorem. Let $f(z)$ be a univalent function. Then the image of the unit disk $\{z : |z| < 1\}$ contains the disk $\{z : |z| < 1/4\}$ with radius $1/4$.

The upper bound $1/4$ (the "Koebe constant") is the best possible.

C. Harmonic Functions

Harmonic functions, defined in Section II, are real functions $u(x, y)$ satisfying Laplace's equation. We shall use the notation: $u(x, y) = u(x + iy) = u(z)$. Harmonic functions have several important properties.

The mean-value property. Let $u(z)$ be a harmonic function on a region Ω. Then for any disk D with center a and radius r whose closure is contained in Ω,

$$u(a) = \frac{1}{2\pi} \int_0^{2\pi} u(a + re^{i\theta}) \, d\theta$$

The maximum principle for harmonic functions. Let $u(z)$ be a harmonic function on the domain Ω. If there is a point a in Ω such that $u(a)$ equals the maximum $\max\{u(z) : z \in \Omega\}$, then $u(z)$ is a constant function.

An important problem in the theory of harmonic function is *Dirichlet's problem*. Given a simply connected domain Ω and a piecewise continuous function $g(z)$ on the boundary $\partial\Omega$, find a function $u(z)$ in the closure $\bar{\Omega}$ such that $u(z)$ is harmonic in Ω, and the restriction of $u(z)$ to the boundary $\bar{\Omega}\backslash\Omega$ equals $g(z)$. (The *boundary* $\partial\Omega$ is the set $\bar{\Omega}\backslash\Omega$.)

For general simply connected domains, Dirichlet's problem is difficult. It is equivalent to finding a Green's function. For a disk, the following formula is known.

The Poisson formula. Let $g(z) = g(e^{i\phi})$, $-\pi < \phi \le \pi$, be a piecewise continuous function on the boundary $\{z : |z| = 1\}$ of the unit disk. Then the function

$$u(z) = u(re^{i\theta})$$

$$= \frac{1}{2\pi} \int_{-\pi}^{\pi} \left(\frac{1 - r^2}{1 + r^2 - 2r \cos(\phi - \theta)} \right) g(e^{i\phi}) \, d\phi$$

is a solution to Dirichlet's problem for the unit disk.

D. Riemann Surfaces

A *Riemann surface* $\mathcal{S}$ is a one-dimensional complex connected paracompact Hausdorff space equipped with a *conformal atlas*, that is, a set of maps or *charts* $\{h : D \to N\}$, where D is the open disk $\{z : |z| < 1\}$ and N is an open set of $\mathcal{S}$, such that

1. The union of all the open sets N is $\mathcal{S}$.
2. The chart $h : D \to N$ is a homeomorphism of the disk D to N.
3. Let N_1 and N_2 be neighborhoods with charts h_1 and h_2. If the intersection $N_1 \cap N_2$ is nonempty and connected, then the composite mapping $h_2^{-1} \circ h_1$, defined on the inverse image $h_1^{-1}(N_1 \cap N_2)$, is conformal.

Riemann surfaces originated in an attempt to make a "multivalued" analytic function single-valued by making its range a Riemann surface. Examples of multivalued functions are *algebraic functions*. These are functions $f(z)$ satisfying a polynomial equation $P(f(z)) = 0$. A specific example of this is the square-root function $f(z) = \sqrt{z}$, which takes on two values $\pm\sqrt{z}$ except when $z = 0$. This can be made into a single-valued function using the Riemann surface obtained by gluing together two sheets or copies of the complex plane cut from 0 to ∞ along the positive real axis. Another example is the logarithmic function $\log z$, which requires a Riemann surface made from countably infinitely many sheets.

Intuitively, the *genus* of a Riemann surface $\mathcal{S}$ is the number of "holes" it has. The genus can be defined as the maximum number of disjoint simple closed curves that do not disconnect $\mathcal{S}$. For example, the extended complex plane has genus 0 and an annulus has genus 1. There are many results about Riemann surfaces. The following are two results that can be simply stated.

Picard's theorem. Let $P(u, v)$ be an irreducible polynomial with complex coefficients in two variables u and v. If there exist nonconstant entire functions $f(z)$ and $g(z)$ satisfying $P(f(z), g(z)) = 0$ for all complex numbers z, then the Riemann surface associated with the algebraic equation $P(u, v) = 0$ has genus 0.

Koebe's uniformization theorem. If $\mathcal{S}$ is a simply connected Riemann surface, then $\mathcal{S}$ is conformally equivalent to

1. (Elliptic type) The Riemann sphere. In this case, $\mathcal{S}$ is the sphere.
2. (Parabolic type) The complex plane **C**. In this case, $\mathcal{S}$ is biholomorphic to **C**, **C**\{0}, or a torus.
3. (Hyperbolic type) The unit disk $\{z : |z| < 1\}$.

Complex manifolds are higher-dimensional generalizations of Riemann surfaces. They have been extensively studied.

E. Other Topics

Complex analysis is a vast and ever-expanding area. "Nine lifetimes" do not suffice to cover every topic. Some interesting areas that we have not covered are complex differential equations, complex dynamics, Montel's theorem and normal families, value distribution theory, and the theory of complex functions in several variables. Several books on these topics are listed in the Bibliography.

SEE ALSO THE FOLLOWING ARTICLES

CALCULUS ● DIFFERENTIAL EQUATIONS ● NUMBER THEORY ● RELATIVITY, GENERAL ● SET THEORY ● TOPOLOGY, GENERAL

BIBLIOGRAPHY

Ahlfors, L. V. (1979). "Complex Analysis," McGraw-Hill, New York.

Beardon, A. F. (1984). "A Primer on Riemann Surfaces," Cambridge University Press, Cambridge, U.K.

Blair, D. E. (2000). "Inversion Theory and Conformal Mappings," American Mathematical Society, Providence, RI.

Carleson, L., and Gamelin, T. W. (1993). "Complex Dynamics," Springer-Verlag, Berlin.

Cartan, H. (1960). "Elementary Theory of Analytic Functions of One or Several Complex Variables," Hermann and Addison-Wesley, Paris and Reading, MA.

Cherry, W., and Ye, Z. (2001). "Nevanlinna's Theory of Value Distribution," Springer-Verlag, Berlin.

Chuang, C.-T., and Yang, C.-C. (1990). "Fix-Points and Factorization of Meromorphic Functions," World Scientific, Singapore.

Duren, P. L. (1983). "Univalent Functions," Springer-Verlag, Berlin.

Farkas, H. M., and Kra, I. (1992). "Riemann Surfaces," Springer-Verlag, Berlin.

Gong, S. (1999). "The Bieberbach Conjecture," American Mathematical Society, Providence, RI.

Gunning, R. (1990). "Introduction to Holomorphic Functions of Several Variables," Vols. I, II, and III. Wadsworth and Brooks/Cole, Pacific Grove, CA.

Hayman, W. K. (1964). "Meromorphic Functions," Oxford University Press, Oxford.

Hille, E. (1962, 1966). "Analytic Function Theory," Vols. I and II. Ginn-Blaisdell, Boston.

Hille, E. (1969). "Lectures on Ordinary Differential Equations," Addison-Wesley, Reading, MA.

Hu, P.-C., and Yang, C.-C. (1999). "Differential and Complex Dynamics of One and Several Variables," Kluwer, Boston.

Hua, X.-H., and Yang, C.-C. (1998). "Dynamics of Transcendental Functions," Gordon and Breach, New York.

Kodiara, K. (1984). "Introduction to Complex Analysis," Cambridge University Press, Cambridge, U.K.

Krantz, S. G. (1999). "Handbook of Complex Variables," Birkhäuser, Boston.

Laine, I. (1992). "Nevanlinna Theory and Complex Differential Equations," De Gruyter, Berlin.

Lang, S. (1987). "Elliptic Functions," 2nd ed., Springer-Verlag, Berlin.

Lehto, O. (1987). "Univalent Functions and Teichmüller Spaces," Springer-Verlag, Berlin.

Marden, M. (1949). "Geometry of Polynomials," American Mathematical Society, Providence, RI.

McKean, H., and Moll, V. (1997). "Elliptic Curves, Function Theory, Geometry, Arithmetic," Cambridge University Press, Cambridge, U.K.

Morrow, J., and Kodiara, K. (1971). "Complex Manifolds," Holt, Rinehart and Winston, New York.

Nehari, Z. (1952). "Conformal Mapping," McGraw-Hill, New York; reprinted, Dover, New York.

Needham, T. (1997). "Visual Complex Analysis," Oxford University Press, Oxford.

Palka, B. P. (1991). "An Introduction to Complex Function Theory," Springer-Verlag, Berlin.

Pólya, G., and Szegö, G. (1976). "Problems and Theorems in Analysis," Vol. II. Springer-Verlag, Berlin.

Protter, M. H., and Weinberger, H. F. (1984). "Maximum Principles in Differential Equations," Springer-Verlag, Berlin.

Remmert, R. (1993). "Classical Topics in Complex Function Theory," Springer-Verlag, Berlin.

Rudin, W. (1980). "Function Theory in the Unit Ball of $\mathbf{C}^n$," Springer-Verlag, Berlin.

Schiff, J. L. (1993). "Normal Families," Springer-Verlag, Berlin.

Schwerdtfeger, H. (1962). "Geometry of Complex Numbers," University of Toronto Press, Toronto. Reprinted, Dover, New York.

Siegel, C. L. (1969, 1971, 1973). "Topics in Complex Function Theory," Vols. I, II, and III, Wiley, New York.

Smithies, F. (1997). "Cauchy and the Creation of Complex Function Theory," Cambridge University Press, Cambridge, U.K.

Steinmetz, N. (1993). "Rational Iteration, Complex Analytic Dynamical Systems," De Gruyter, Berlin.

Titchmarsh, E. C. (1939). "The Theory of Functions," 2nd ed., Oxford University Press, Oxford.

Vitushkin, A. G. (ed.). (1990). "Several Complex Variables I," Springer-Verlag, Berlin.

Weyl, H. (1955). "The Concept of a Riemann Surface," 3rd ed., Addison-Wesley, Reading, MA.

Whitney, H. (1972). "Complex Analytic Varieties," Addison-Wesley, Reading, MA.

Whittaker, E. T., and Watson, G. N. (1969). "A Course of Modern Analysis," Cambridge University Press, Cambridge, U.K.

Yang, L. (1993). "Value Distribution Theory," Springer-Verlag, Berlin.

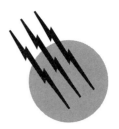

Composite Materials

M. Knight
D. Curliss
Air Force Research Laboratory

I. Characteristics
II. Constituent Materials
III. Properties of Composites
IV. Analysis of Composites
V. Fabrication of Composites
VI. Uses of Composites

GLOSSARY

Advanced composites Composite materials applicable to aerospace construction and consisting of a high-strength, high-modulus fiber system embedded in an essentially homogeneous matrix.

Anisotropic Not isotropic; having mechanical and/or physical properties that vary with direction relative to a natural reference axis inherent in the materials.

Balanced laminate Composite laminate in which all laminae at angles other than 0° and 90° occur only in ±pairs.

Constituent In general, an element of a larger grouping. In advanced composites, the principal constituents are the fibers and the matrix.

Cure To change the properties of a thermosetting resin irreversibly by chemical reaction.

Fiber Single homogeneous strand of material, essentially one-dimensional in the macrobehavior sense.

Interface Boundary between the individual, physically distinguishable constituents of a composite.

Isotropic Having uniform properties in all directions. The measured properties are independent of the axis of testing.

Lamina Single ply or layer in a laminate made of a series of layers.

Laminate Unit made by bonding together two or more layers or laminae of materials.

Matrix Essentially homogeneous material in which the reinforcement system of a composite is embedded.

Orthotropic Having three mutually perpendicular planes of elastic symmetry.

Transversely isotropic Material having identical properties along any direction in a transverse plane.

Woven fabric composite Form of composite in which the reinforcement consists of woven fabric.

1, or x, axis Axis in the plane of the laminate that is used as the 0° reference for designating the angle of a lamina.

2, or y, axis Axis in the plane of the laminate that is perpendicular to the x axis.

3, or z, axis Reference axis normal to the plane of the laminate x, y axes.

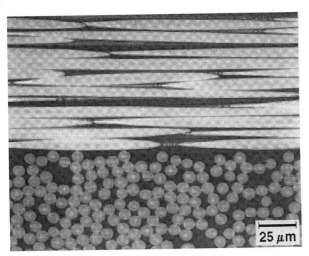

FIGURE 1 Cross section of a graphite fiber–reinforced epoxy polymer.

A COMPOSITE MATERIAL is described in this chapter as a material composed of two or more distinct phases and the interfaces between them. At a macroscopic scale, the phases are indistinguishable, but at some microscopic scales, the phases are clearly separate, and each phase exhibits the characteristics of the pure material. In this chapter, we are only describing the characteristics, analysis, and processing of high-performance structural composite materials. This special class of composites always consists of a reinforcing phase and a matrix phase. The reinforcing phase is typically a graphite, glass, ceramic, or polymer fiber, and the matrix is typically a polymer, but may also be ceramic or metal. The fibers provide strength and stiffness to the composite component, while the matrix serves to bind the reinforcements together, distribute mechanical loads through the part, provide a means to process the material into a net shape part, and provide the primary environmental resistance of the composite component. In Fig. 1, we can see the distinct cross section of graphite fibers in an epoxy matrix.

I. CHARACTERISTICS

Many materials can be classified as composites. They are composed of several distinctly different and microscopically identifiable substances. Composites are widely used in many industries and applications today, driven by the need for strong, lightweight materials. The composites reduce weight and allow for designs that tailor the mechanical properties of the material to meet the loading requirements of the structure. In addition, composites are replacing traditional engineering materials in many industrial, recreational, architectural, transportation, and infrastructure applications.

Composites occur very commonly in nature. Some of the best examples are wood, bone, various minerals, mollusk shells, and insect exoskeletons. In wood, the cellulose fibers of the cell wall are "glued" together by the lignin matrix. Bone is composed of calcium hydroxyapatite crystals in a protein matrix. Mollusk shells are composites of calcium carbonate layers in various geometries bound together by a multilayer matrix. Insect exoskeletons bear a striking resemblance to man-made fiber-reinforced composites. Some insects even exhibit apparent "layers" of fibrous chitin embedded in a protein matrix, where the orientation of the fibers varies from layer to layer, much as we might design a man-made fiber-reinforced composite. This example of a natural composite can be clearly seen in Fig. 2. Modern materials engineers have used the composite concept—reinforcement in a matrix—to create a class of modern materials that offers opportunities significantly greater than those of more common engineering materials.

Composites can be made of a such a wide variety of materials that it is impractical to discuss each one individually. One of the principal characteristics of all composites is that they have a reinforcement phase distinct from the matrix phase. The individual characteristics of the two phases combine to give the composite its unique properties.

Classes of materials commonly used for reinforcements are glasses, metals, polymers, ceramics, and graphite. The reinforcement can be in many forms, such as continuous fibers or filaments, chopped fibers, woven fibers or yarns, particles, or ribbons. The criteria for selecting the type and form of reinforcement vary in accordance with the design requirement for the composite. However, certain general qualities are desirable, including high strength, high modulus, light weight, environmental resistance, good elongation, low cost, good handleability, and ease of manufacture. By far, the most widely used reinforcement is E-glass.

FIGURE 2 Scanning electron microscope (SEM) image of a bessbeetle (*Odontotaenius disjunctus*) elytra fracture surface.

E-glass offers excellent strength, compatibility with common matrix polymers, and is very low in cost. Various types of graphite fibers are commonly used in aerospace and the recreational products industry, where light weight and maximum material performance are very important to the designer.

The matrix binds the reinforcement together and enhances the distribution of the applied load within the composite. Polymeric materials are widely used as matrix materials. Two general classes of polymers are used: thermosets and thermoplastics. Thermosets are initially low molecular weight molecules that are often viscous liquids at room temperature—what we commonly think of as "resins." Their low viscosity and fluid behavior make them very suitable to low-cost processing. The thermoset resins undergo chemical reactions when heated (or initiated by some other energy source such as UV light, electron beam, or microwave) and form a high molecular weight cross-linked polymer. In contrast, thermoplastics are high molecular weight linear polymers that are fully formed prior to processing as a composite matrix. When heated to temperatures well above their glass transition temperature, T_g, they soften and exhibit a viscosity low enough to flow and consolidate the composite. In general, they must be heated to much higher temperatures than thermosets, exhibit much higher melt viscosity, and require higher pressures to form well-consolidated composite laminates. Thermoplastics offer some advantages such as reprocessability, recyclability, and, in general, higher toughness. However, thermoplastics also have several limitations that have restricted their wider acceptance as matrix materials for fiber-reinforced composites. Thermoplastics have lower solvent resistance than thermosets and require more expensive processing equipment, there are fewer commercially available thermoplastic matrix preforms available than for thermosets, and modern toughened thermosets offer similar performance to thermoplastic matrix composites. For such economic and performance reasons, thermoplastics are not widely used as thermosets for advanced composite matrix polymers. Other matrix materials are metals, ceramics, glasses, and carbon. They perform the same function in composites as the polymer matrix. These materials (with the exception of carbon) are still experimental, and their combined fraction of the composite matrix materials market is insignificant. Carbon has been used since the 1970s for exotic high-temperature ablative applications such as rocket motor nozzles. The Properties of Composites and Analysis of Composites sections of this article are general and apply to these developmental composite materials. The Processing and Applications sections, however, are concerned only with polymer matrix composites.

The matrix influences the service temperature, service environment, and fabrication process for composites. Compatibility with the reinforcement is a consideration in selecting the matrix. The matrix must adhere to the reinforcement and be capable of distributing the loads applied to the composite.

The properties of a composite can be tailored by the engineer to provide a wide range of responses, which increases their usefulness. Composites can be made to exhibit some interesting responses when loaded: They can be designed to twist and bend when loaded in plane and to extend or contract when loaded in bending. Analysis approaches are available for predicting these responses.

There are many processes for the fabrication of composites. These often result in reduction in number of parts, reduction in production time, and savings in overall manufacturing cost. The number of industries using composites and the various uses of composites continues to grow. It is difficult to foresee what the future of this class of materials will be.

II. CONSTITUENT MATERIALS

A composite can contain several chemical substances. There are additives, for example, to improve processability and serviceability. However, the two principal constituents that are always present in advanced composites are the matrix and the reinforcement. Generally, they are combined without chemical reaction and form separate and distinct phases. Ideally, the reinforcement is uniformly distributed throughout the matrix phase. The combination of the properties of the reinforcement, the form of the reinforcement, the amount of reinforcement, and matrix properties gives the composite its characteristic properties.

The matrix phase contributes to several characteristics of the composite. The matrix provides some protection for the reinforcement from deleterious environmental conditions such as harmful chemicals. The matrix plays an important role in determining the physical and thermophysical properties of the composite. In continuous filament, unidirectionally reinforced composites, the properties transverse to the filaments are strongly influenced by the properties of the matrix. The distribution of the applied load throughout the composite is influenced by the properties of the matrix.

Table I shows typical values of selected properties of common matrix materials. The properties are tensile strength, F^{tu}, Young's modulus, E^t, total strain (or strain-to-failure), ε^t, coefficient of thermal expansion, α, and specific gravity. It can be seen that there is a wide variation in these values between types of matrix materials.

TABLE I Matrix Materials

Property	Epoxy	Polyimide	Polyester	Polysulfone	Polyether ether ketone	Al 2024	Ti 6-4
E^{tu} (MPa)	6.2–103	90	21–69	69	69	414	924
E^t (GPa)	2.8–3.4	2.8	3.4–5.6	2.8	3.6	72	110
ε^t (%)	4.5	7–9	0.5–5.0	50–100	2.0	10	8
α (10^{-6} m m^{-1} K^{-1})	0.56	0.51	0.4–0.7	0.56	0.5	24	9.6
Specific gravity	1.20	1.43	1.1–1.4	1.24	1.2	2.77	4.43

There is great variety in polymers typically used for composite matrix materials. As discussed earlier, thermosets and thermoplastics make up the two general families of engineering polymers; but there are many different polymers within each family that exhibit very diverse properties, depending on their chemical composition. Thermosets are generally named for the characteristic reactive group of the resin (e.g., epoxy, maleimide), whereas thermoplastics are generally named for either their building block ("mer" unit; e.g., polystyrene, polyethylene, polypropylene, polyvinyl chloride) or for a characteristic repeating chemical group within the thermoplastic polymer (e.g., polysulfone, polyimide). It is more appropriate to refer to the matrix polymer as a resin system, the system being a mixture of the base polymer (or thermoset resin and curing agents). Diluents, fillers, tougheners, and other modifiers are sometimes added to the resin system to alter viscosity, increase toughness, modify reactivity of the thermosets, or change other properties of the base polymer system. Because there are so many starting combinations, it is easy to see how there can be a wide variation in the properties of materials in the same general class (e.g., based on the same basic polymer, but with different additives). The other principal constituent of a composite is the reinforcement. There are several types of materials, and their various forms are used as reinforcements. The continuous fiber has been used most extensively for the development of advanced composites. This form of reinforcement provides the highest strength and modulus. It can be used to make other forms such as woven fabric, chopped fibers, and random fiber mats. These reinforcement forms typically reduce the mechanical performance compared to unidirectional fibers, but offer benefits in fabrication. Glass, graphite, and polymeric fibers are generally produced as bundles of many filaments of very small diameter. Metal, boron, and ceramic reinforcements are usually single fibers. After fabrication, fibers are processed with surface treatments for protection during handling and weaving and also for chemical compatibility with the matrix systems. After forming and treating, the filaments are typically wound on spools for use by manufacturers in fabricating composites, producing unidirectional preforms, or weaving into various geometries of textile preforms.

Table II lists the properties of some of the fibers, measured in the longitudinal direction (along the axis of the fiber), used in composite materials: tensile strength F^{tu}, Young's modulus E^t, coefficient of expansion α, strain-to-failure ε^t, diameter, and density ρ. Mechanical properties transverse to the longitudinal axis are not shown. Because of the small diameter of the fibers, transverse properties are not measured directly. Variations in the fiber properties can be caused by several factors. There can be variations in the composition of the starting material such as in E-, S-, and C-glass fibers. There can be variations in processing such as in the way the processing temperature is changed to vary the strength and modulus of graphite fibers. Also, the difficulty of performing mechanical testing on fibers contributes to uncertainty and scatter in the measured properties of fibers.

TABLE II Fiber Materials

Property	Boron	Carbon	Graphite	Aramid	Alumina	Silicon carbide	E-glass	S-glass
E^{tu} (MPa)	2.8–3.4	0.4–2.1	0.81–3.6	2.8	1.4	3.3	3.4	4.6
E^t (GPa)	379–414	241–517	34–552	124	345–379	427	69	83
α (10^{-6} m m^{-1} K^{-1})	4.9	−0.09	−0.09	−4.0	3.4	.40	5.1	3.4
ρ (g cm^{-3})	2.5–3.3	1.55	1.55	1.60	3.90	3.07	2.55	2.5
Diameter (10^{-3} m)	0.05–0.2	0.008	0.008	0.013	0.38–0.64	0.14	0.005–0.013	0.009–0.010
ε^t (%)	0.67	1.0–2.0	0.4–2.0	2.5	0.4	0.6	4.8	5.4

The reinforcement is the main load-bearing phase of the composite. It provides strength and stiffness. There is a direct relationship between an increase in volume fraction of reinforcement and an increase in strength and stiffness of the composite material. This relationship depends on the assumption of compatibility with the matrix and on the existence of good bonding to the fibers.

The reinforcement and matrix are combined either before or at the time of fabrication of the composite. This depends on the fabrication process. A common practice in making continuous-fiber-reinforced laminates is to combine the constituents before fabrication into a continuous "tapelike" preform that is used much like broadgoods in that shapes are cut out of the preform and fabricated into parts. To produce this preform product, fibers are combined with resin, typically by drawing the fiber bundle through a resin or resin solution bath. Several bundles of resin-impregnated fibers are then aligned and spread into very thin layers (0.127 mm thick) on a release ply backing. The resin is usually partially cured during production of the preform to reduce its "tackiness" and improve the handleability of the preform. This tapelike preform is known as prepreg, or unidirectional tape. It is an expensive method for producing a preform, but the preform is a continuous, well-characterized, well-controlled method to combine the matrix resin and the reinforcing fiber. After prepregging, the material is usually stored in a freezer to retard the chemical reaction until the material is used. If the matrix system is a thermoplastic polymer, then no reaction can occur, and the material may be stored indefinitely at room temperature. These layers of unidirectional fibers and resin are used to make laminates by stacking many layers in directions specified by the engineer. The number of "plies" in a laminate and the direction of fibers in each layer is dependent on the mechanical properties required for the part.

The next two sections, Properties of Composites and Analysis of Composites, describe how an engineer would design a composite laminate to have the properties needed for an application. It is exactly this tailorability that makes composites attractive for engineering applications.

III. PROPERTIES OF COMPOSITES

In many of the applications in which composite materials are used, they can be considered to be constructed of several layers stacked on top of one another. These layers, or laminae, typically exhibit properties similar to those of orthotropic materials. Orthotropic materials have three mutually perpendicular planes of material property symmetry. Figure 3 shows a lamina with its coordinate system and two of the planes of symmetry. We will first discuss the

properties of the lamina and some factors that influence them. Next, the properties of laminates will be discussed.

The lamina is made of one thickness of reinforcement embedded in the matrix. The elastic and strength properties of the reinforcement and the elastic and strength properties of the matrix combine to give the lamina its properties. In addition to the properties of the constituents, the amount of reinforcement, the form of the reinforcement, and the orientation and distribution of the reinforcement all influence the properties of the lamina.

The reinforcement provides the strength and stiffness of the composite. Increasing the amount of reinforcement increases the strength and stiffness of the composite in the direction parallel to the reinforcement. The effect of the form of the reinforcement is not as simple. However, some general observations can be made. Laminae reinforced by long, continuous, parallel fibers have greater strength and stiffness than laminae reinforced by short, randomly oriented fibers. Woven fiber–reinforced laminae usually have greater strength perpendicular to the principal fiber direction than do unwoven fiber–reinforced laminae. The strength and stiffness of laminae reinforced by unwoven continuous fibers decrease as the angle of loading changes from parallel to the fibers to perpendicular to the fibers.

Table III shows typical values for some properties of composite materials made of unwoven continuous fiber reinforcements. The table shows the strength and elastic properties of a laminate made of several laminae stacked on top of one another with all the fibers aligned in the same direction. The properties in the direction parallel to the fibers are much greater than the properties in the direction perpendicular to the fibers. This variation of properties with the orientation of the lamina axis is called anisotropy.

The single lamina serves as a building block. The engineer can select the orientation and number of each of the laminae in a laminate and design the laminate such that it has the required response. This designing of a laminate has some interesting implications that the engineer should understand. Two important factors are balance and symmetry.

Balance and symmetry simplify the analysis of the laminate and give it more conventional response characteristics. Balance in a laminate means that for each lamina with a positive angle of orientation there must be a lamina with an equal negative angle of orientation. Both laminae must have the same mechanical and physical characteristics. This is important in controlling the laminate's overall response to loading both in service and in fabrication. Symmetry means that for every lamina above the midplane of the laminate there is a lamina an equal distance below the midplane that is of the same type with the same orientation. Symmetry also influences the laminate response to loads.

TABLE III Typical Properties of Composite Materials: Laminates Reinforced With Unidirectional Continuous Fibers

Property	Unit	E-glass epoxy	Aramid epoxy	Graphite epoxy	Boron epoxy
Parallel to the fibers					
Tensile strength σ_x^T	MPa	1100	1380	1240	1296
Tensile modulus E_x^T	GPa	39.3	75.8	131	207
Poisson's ratio ν_{xy}	—	0.25	0.34	0.25	0.21
Total strain ε^T	%	2.2	1.8	1.21	0.66
Compressive strength σ_x^c	MPa	586	276	1100	2426
Compressive modulus E_x^c	GPa	39.3	75.8	131	221
Shear strength τ_{xy}	MPa	62.0	44.1	62.0	132
Shear modulus G_{xy}	GPa	3.45	2.07	4.83	6.2
Transverse to the fibers					
Tensile strength σ_y^T	MPa	34.5	27.6	41.4	62.7
Tensile modulus E_y^T	GPa	8.96	5.5	6.2	18.6
Compressive strength σ_y^c	MPa	138	138	138	310
Compressive modulus E_y^c	GPa	8.96	5.5	6.2	24.1
Specific gravity	—	2.08	1.38	1.52	2.01
Fiber volume V_f	%	~50	~60	~62	~50

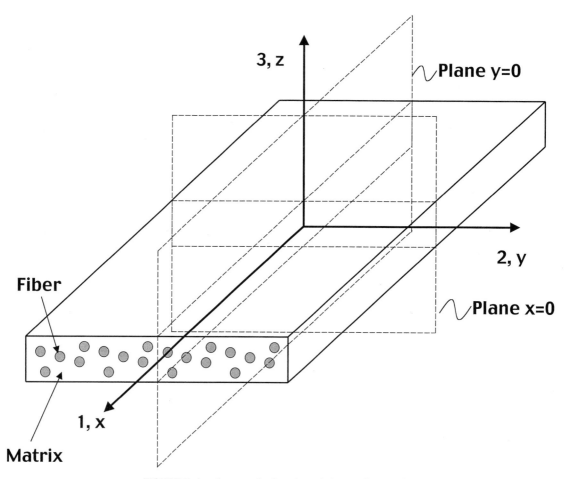

FIGURE 3 Lamina coordinate axis and planes of symmetry.

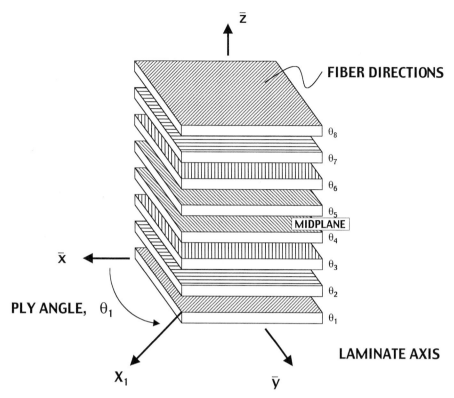

FIGURE 4 Orientation and location of laminae in a laminate.

If a laminate is not balanced and symmetrical, it will twist or bend when in-plane loads are applied. Laminates may also extend or contract when bending loads are applied. Whether the results are good or bad depends on whether they were planned or unplanned during the design of the laminate. Figure 4 shows how the laminae are oriented and stacked in a laminate.

IV. ANALYSIS OF COMPOSITES

Composite materials are complex. The properties of the constituents are different, and the fiber properties are anisotropic. The composite may also be constructed by layers, with the fiber directions varying layer to layer. Analysis of the mechanical properties of such laminates is a sophisticated process; research into better methods to predict composite performance is being pursued. However, acceptable engineering analysis methods have been developed that allow structural parts to be designed with composite materials. Further research is required to develop sound engineering methods to predict failure in composite materials, especially when subjected to severe environments that may degrade the matrix, the reinforcement, or the interfaces of the composite material. In this section, a brief summary of the currently accepted approach to performing stress analysis of composites is presented.

The emphasis has been focused on unidirectional fiber-reinforced composites. The lamina or ply form of advanced composites has been developed into the basic unit for analysis. Most of the structural applications of advanced composites involve material in a laminated form. The laminates are constructed of plies or laminae laid up to a designed configuration (see Fig. 4).

The approach to the analysis of composites starts with the lamina and its elastic properties. Then these are related to the geometry of the lay-up for the laminate. The elastic properties and orientation of the laminae are used to calculate the modulus and stiffness of the laminate. The constitutive relationship and a selected failure criterion are used to estimate failure.

In developing the analysis of the lamina, several assumptions were made. It was assumed that (1) the fibers and matrix were bonded together, (2) the lamina was void free, (3) the lamina's thickness was small in comparison with its width and length, (4) the lamina was a homogeneous orthotropic material, and (5) the fibers were uniformly distributed within the matrix.

The lamina is analyzed as a macroscopic, homogeneous, orthotropic material in a plane stress condition. If the coordinate axes for the laminate are oriented parallel and transverse to the fiber axis (see Fig. 3), the constitutive equation relating stress α and strain ε is

$$\begin{bmatrix} \sigma_1 \\ \sigma_2 \\ \tau_{12} \end{bmatrix} = \begin{bmatrix} Q_{11} & Q_{12} & 0 \\ Q_{12} & Q_{22} & 0 \\ 0 & 0 & Q_{66} \end{bmatrix} \begin{bmatrix} \varepsilon_1 \\ \varepsilon_2 \\ \gamma_{12} \end{bmatrix} \quad (1)$$

where Q is called the reduced stiffness and is defined as

$$Q_{11} = \frac{E_1}{1 - v_{12}v_{21}}; \quad Q_{22} = \frac{E_2}{1 - v_{12}v_{21}}$$
$$Q_{12} = \frac{v_{12}E_2}{1 - v_{12}v_{21}}; \quad Q_{66} = G_{12} \quad (2)$$

where E_1 is Young's modulus in the direction parallel to the fibers; E_2 is Young's modulus in the direction perpendicular to the fibers; v_{12} and v_{21} are the major Poisson's ratio and minor Poisson's ratio, respectively; and G_{12} is the in-plane shear modulus.

Equation (1) can be inverted to give the form

$$\begin{bmatrix} \varepsilon_1 \\ \varepsilon_2 \\ \gamma_{12} \end{bmatrix} = \begin{bmatrix} S_{11} & S_{11} & 0 \\ S_{12} & S_{22} & 0 \\ 0 & 0 & S_{66} \end{bmatrix} \begin{bmatrix} \sigma_1 \\ \sigma_2 \\ \tau_{12} \end{bmatrix} \quad (3)$$

where the S terms are the compliance coefficients and are defined as

$$S_{11} = 1/E_t; \quad S_{22} = 1/E_2$$
$$S_{12} = -v_{12}/E_1; \quad S_{66} = 1/G_{12} \quad (4)$$

Equation (4) relates the compliance coefficients to the engineering constants. These can be determined by mechanical testing. Also, estimates of the engineering constants can be made by using equations developed by micromechanics. In this approach, the properties of the constituents are used in equations for the engineering constants. These are

$$E_1 = E_f V_f + E_m V_m$$
$$v_{12} = v_f V_f + v_m V_m$$
$$P/P_m = (1 + \xi \eta V_f)/(1 - \eta V_f) \quad (5)$$
$$\eta = \frac{(P_f/P_m) - 1}{(P_f/P_m) + \xi}$$

where V_f, V_m are the volume fraction of the fiber and matrix, respectively; v_f, v_m are Poisson's ratio of the fiber and matrix, respectively; P is the composite modulus E_2, G_{12}, or G_{23}; P_f is the corresponding fiber modulus E_f, G_f, or v_f, respectively; P_m is the corresponding matrix modulus E_m, G_m, or v_m, respectively; and ξ is a factor related to the arrangement and geometry of the reinforcement; for square packing $\xi = 2$, and for hexagonal packing $\xi = 1$.

Because not all laminae in a laminate are oriented with the fibers parallel or transverse to the laminate coordinate axis x–y, there must be a way to find the properties of the lamina in the laminate coordinate systems. This is done through a transformation. By a combination of mathematical transformation and substitution, the following relationship between stress and strain for an arbitrary lamina k is developed:

$$\begin{vmatrix} \sigma_x \\ \sigma_y \\ \tau_{xy} \end{vmatrix}_k = \begin{vmatrix} \bar{Q}_{11} & \bar{Q}_{12} & \bar{Q}_{16} \\ \bar{Q}_{12} & \bar{Q}_{22} & \bar{Q}_{26} \\ \bar{Q}_{16} & \bar{Q}_{26} & \bar{Q}_{66} \end{vmatrix}_k \begin{vmatrix} \varepsilon_x \\ \varepsilon_y \\ \gamma_{xy} \end{vmatrix}_k \quad (6)$$

The $\bar{Q}$ terms are the components of the stiffness matrix for the lamina referred to an arbitrary axis. They are defined as

$$\bar{Q}_{11} = Q_{11}\cos^4\theta + 2(Q_{12} + 2Q_{66})\sin^2\theta\cos^2\theta + Q_{22}\sin^4\theta$$
$$\bar{Q}_{22} = Q_{11}\sin^4\theta + 2(Q_{12} + 2Q_{66})\sin^2\theta\cos^2\theta + Q_{22}\cos^4\theta$$
$$\bar{Q}_{12} = (Q_{11} + Q_{22} - 4Q_{66})\sin^2\theta\cos^2\theta + Q_{22}(\sin^4\theta + \cos^4\theta) \quad (7)$$
$$\bar{Q}_{66} = (Q_{11} + Q_{22} - 2Q_{12} - 2Q_{66})\sin^2\theta\cos^2\theta + Q_{66}(\sin^4\theta + \cos^4\theta)$$
$$\bar{Q}_{16} = (Q_{11} - Q_{12} - 2Q_{66})\sin^2\theta\cos^3\theta + (Q_{12} - Q_{22} + 2Q_{66})\sin^3\theta\cos\theta$$
$$\bar{Q}_{26} = (Q_{11} - Q_{12} - 2Q_{66})\sin^2\theta\cos\theta + (Q_{12} - Q_{22} + 2Q_{66})\sin\theta\cos^3\theta$$

where θ is the ply angle according to the Tsai convention (see Fig. 4). Counterclockwise rotations are positive and clockwise rotations are negative.

The constitutive relationships for the lamina and linear small deformation theory were used to develop the analysis for composite structures. Some assumptions that were made are as follows: (1) The laminae are bonded together, and they do not slip relative to one another when load is applied; (2) the normals to the undeformed midplane of the laminate are straight, and they remain so with no change in length after deformation; (3) the thickness of the plate is small compared with the length and width; and (4) the strain in the thickness direction is negligible. The in-plane strain is assumed constant for all the laminae. The stress varies from lamina to lamina. As a simplification, the force and moment resultants were defined. The force resultants N_x, N_y, and N_{xy} were defined as the sum of the laminae stresses per unit width. The moment resultants M_x, M_y, and M_{xy} were defined as the sum of the respective stresses, times the area over which they act, multiplied by the appropriate moment arm. The in-plane strains at the midplane, ε_x^0, ε_y^0, and γ_{xy}^0, and the curvatures, κ_x, κ_y, and κ_{xy}, are related to the resultants as shown in Eq. (8).

$$\begin{bmatrix} N_x \\ N_y \\ N_{xy} \\ --- \\ M_x \\ M_y \\ M_{xy} \end{bmatrix} \begin{bmatrix} A_{11} & A_{12} & A_{16} & \vdots & B_{11} & B_{12} & B_{16} \\ A_{12} & A_{22} & A_{26} & \vdots & B_{12} & B_{22} & B_{26} \\ A_{16} & A_{26} & A_{66} & \vdots & B_{16} & B_{26} & B_{66} \\ ---&---&---&---&---&---&--- \\ B_{11} & B_{12} & B_{16} & \vdots & D_{11} & D_{12} & D_{16} \\ B_{12} & B_{22} & B_{26} & \vdots & D_{12} & D_{22} & D_{26} \\ B_{16} & B_{26} & B_{66} & \vdots & D_{16} & D_{26} & D_{66} \end{bmatrix} \begin{bmatrix} \varepsilon_x^0 \\ \varepsilon_y^0 \\ \gamma_{xy}^0 \\ --- \\ \kappa_x \\ \kappa_y \\ \kappa_{xy} \end{bmatrix}$$

(8)

where N_x, N_y, and N_{xy} are force resultants; M_x, M_y, and M_{xy} are moment resultants; $[A]$ is the in-plane stiffness matrix for a laminate; $[B]$ is the coupling stiffness matrix for a laminate; $[D]$ is the bending stiffness matrix for a laminate; ε_x^0, ε_y^0, and γ_{xy}^0 are the strains at the laminate geometric mid-plane; and κ_x, κ_y, and κ_{xy} are the curvatures of the laminate.

Examination of Eq. (8) shows that the $[A]$ matrix is the coefficients for the in-plane strains. The $[B]$ matrix relates the curvatures to the force resultants and the in-plane strains to the moment resultants. The $[D]$ matrix relates the curvatures to the moment resultants. Equation (8) can be partially or fully inverted, depending on whether the strains, curvatures, forces, or moments are known in a given situation.

The definitions for the elements of the $[A]$, $[B]$, and $[D]$ matrices are

$$A_{ij} = \sum_{k=1}^{n} (\bar{Q}_{ij})_k (h_k - h_{k-1}) \qquad (9)$$

$$B_{ij} = \frac{1}{2} \sum_{k=1}^{n} (\bar{Q}_{ij})_k (h_k^2 - h_{k-1}^2) \qquad (10)$$

$$D_{ij} = \frac{1}{3} \sum_{k=1}^{n} (\bar{Q}_{ij})_k (h_k^3 - h_{k-1}^3) \qquad (11)$$

Figure 5 shows how k and h are defined for the laminae.

The force resultants and moment resultants are defined as

$$\begin{bmatrix} N_x \\ N_y \\ N_{xy} \end{bmatrix} = \int_{-h/2}^{h/2} \begin{bmatrix} \sigma_x \\ \sigma_y \\ \tau_{xy} \end{bmatrix} dz \qquad (12)$$

and

$$\begin{bmatrix} M_x \\ M_y \\ M_{xy} \end{bmatrix} = \int_{-h/2}^{h/2} \begin{bmatrix} \sigma_x \\ \sigma_y \\ \tau_{xy} \end{bmatrix} z\,dz \qquad (13)$$

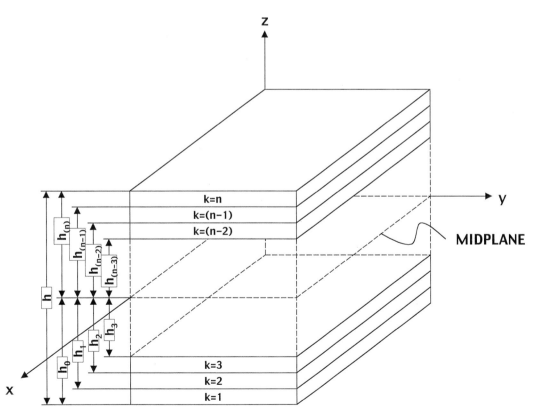

FIGURE 5 Relationship of laminae to the laminate coordinates.

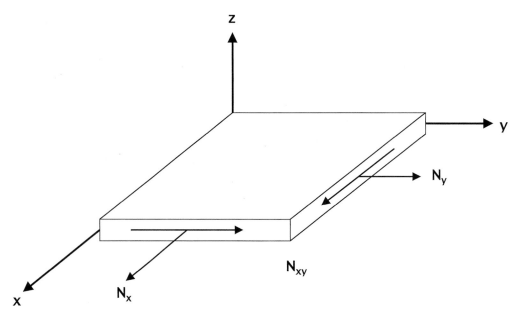

FIGURE 6 Force resultants on an element.

where σ_x, σ_y, and τ_{xy} are the stresses in the laminate co-ordinate system and z is the distance from the midplane in the direction normal to the midplane. Figures 6 and 7 show how the force and monment resultants act on an element in the laminate.

Equation (8) is the constitutive equation for a general laminated plate. Significant simplifications of Eq. (8) are possible. If the $[B]$ is made zero, the set of equations for the stress and moment resultants is uncoupled. "Uncou-pled" means that in-plane loads generate only in-plane responses, and bending loads generate only bending re-sponses. The $[B]$ can be made zero if for each lamina above the midplane there is a lamina with the same proper-ties, orientation, and thickness located at the same distance below the midplane. This is significant not only in sim-plifying the calculations but also in the physical response to load and in fabrication. If the $[B]$ is zero, the laminate will not warp when cured, and no bending will be induced

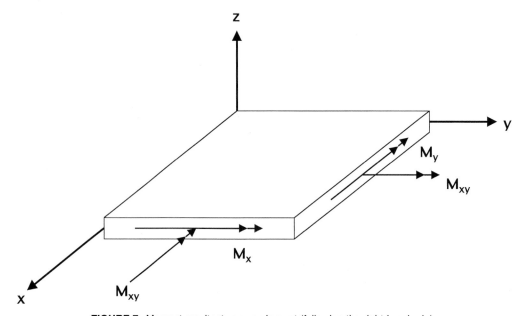

FIGURE 7 Moment resultants on an element (following the right-hand rule).

when the laminate is under in-plane loads. Equation (8) becomes

$$\begin{bmatrix} N_x \\ N_y \\ N_{xy} \end{bmatrix} = \begin{bmatrix} A_{11} & A_{12} & A_{16} \\ A_{12} & A_{22} & A_{26} \\ A_{16} & A_{26} & A_{66} \end{bmatrix} \begin{bmatrix} \varepsilon_x^0 \\ \varepsilon_y^0 \\ \gamma_{xy}^0 \end{bmatrix} \quad (14)$$

and

$$\begin{bmatrix} M_x \\ M_y \\ M_{xy} \end{bmatrix} = \begin{bmatrix} D_{11} & D_{12} & D_{16} \\ D_{12} & D_{22} & D_{26} \\ D_{16} & D_{26} & D_{66} \end{bmatrix} \begin{bmatrix} k_x^0 \\ k_y^0 \\ k_{xy}^0 \end{bmatrix} \quad (15)$$

In the preceding discussion, only the elastic properties of the laminate were considered. The elastic behavior of a laminate can be used to analyze the strength behavior of a laminate. To determine the strength of a laminate, we need a failure criterion for the lamina. It is assumed that the response of the lamina will be the same when it is in the laminate under the same stresses or strains. The strength of the laminate will be related to the strength of the individual lamina. The general approach is to determine the force and moment resultants or the mid-plane strains and curvatures for the laminate by using the laminate plate equation or an inverted form. The stress or strain is calculated for each lamina in the laminate axis system, and then it is transformed into the lamina axis system for each lamina and the failure criteria applied to determine if failure occurred in the lamina. If the first-ply failure concept for the laminates is applied, the laminate is considered to have failed when the first lamina fails. No single approach has been universally accepted for strength analysis of laminates after first-ply failure.

V. FABRICATION OF COMPOSITES

Fabrication of components from composite materials is somewhat different from that using traditional engineering materials in that the properties of a composite are highly dependent on the geometry of the reinforcement. The structural designer must consider the issues associated with processing the composite part to ensure that reinforcement volume fraction, reinforcement geometry, and other material properties can be produced economically. The diversity of composite applications has stimulated the development of a wide range of techniques for fabricating structural composites. In fact, one of the principal reasons for the success of composites is the ease of fabrication and the many different processes with widely varying levels of sophistication and cost that are available for their production. Structural and decorative composites can be fabricated with techniques ranging from very crude hand lay-up processes without molds to very

sophisticated techniques with complex molds, woven 3D reinforcement preforms, and artificial intelligence–guided computer-controlled resin infusion and curing. The configuration of the part, along with the basic manufacturing considerations such as volume, production speed, and market conditions, determine whether a part will be built in open or closed molds, by compression molding, or by an automated system. Composite fabrication technologies can be classified as either open or closed molding, the choice of appropriate technique being governed by factors mentioned earlier.

We can group most of the processes into two classes: open molding and closed molding. The main distinction is that open molds are one piece and use low pressure or no pressure, and closed molds are two pieces and can be used with higher pressure.

A. Open-Mold Processes

Open-mold processes such as spray-up, wet hand lay-up, autoclave, filament winding, vacuum infusion, pultrusion, or combinations of these techniques are the most common open-mold methods to produce composite products. Many products are suited to these manufacturing methods, including aerospace structures, tanks, piping, boat hulls and structures, recreational vehicle components, commercial truck cabs and components, structural members, and plumbing applications (e.g., tubs, showers, pools, and spas).

In spray-up and wet hand lay-up open molding, the mold surface typically has a high-quality finish and is the visual surface of the finished part. Common to all open molding techniques is mold preparation. To prepare the mold surface prior to spray-up, hand lay-up, or vacuum infusion, the mold is treated with a release agent to aid in composite part removal and then may be coated with a "gel coat" (a color-tinted layer of resin that becomes the visual surface of the finished part).

In spray-up fabrication, the thermoset resin is sprayed into the prepared mold simultaneously with chopped reinforcing fiber. The random sprayed-up mat of fiber and resin may then be compacted with hand rollers prior to cure to produce a more dense part. A hand lay-up component, the resin, and reinforcement (usually a fabric or random fiber mat) are laid into the mold, compacted with rollers, and allowed to cure. Often hand lay-up is combined with spray-up techniques depending on the structural requirements of the part. Fiber volumes of 15 to 25% are typically achieved with these techniques. There are several variations of the basic process. A vacuum bag made of a nonporous, nonadhering material can be placed over the lay-up. Then a vacuum is drawn inside the bag. The atmospheric pressure outside the bag eliminates the voids and forces out entrapped air and excess resin. Another approach is to use a pressure bag. The bag is placed against the lay-up and the mold covered with a pressure plate. Air or steam pressure is applied between the bag and the plate.

Vacuum infusion is an open molding process that is very suitable for large components for many important reasons. Vacuum infusion uses an airtight membrane over the entire part to provide vacuum pressure on the reinforcement and to prevent any volatile resin products from escaping into the atmosphere. The resin is introduced after the entire reinforcement is laid into the mold and the vacuum membrane is in place; this reduces some issues associated with the working time of the resin prior to cure. Finally, higher volume fractions of reinforcement are achievable since the reinforcement is compacted by vacuum pressure and only the minimum amount of resin necessary is added. Reinforcement volume fractions up to 70% have been reported.

An open-mold technique that is widely used in the aerospace industry and is slightly different from the preceding processes is autoclaving. One difference in this process is that the entire assembly (the lay-up and supporting unit) is placed inside an autoclave. An autoclave is a large pressure vessel that is used to provide heat and pressure to the lay-up during cure. Autoclaves are usually cylindrical, with an end that opens for full access to the interior. They have provision to pull vacuum on the lay-up assembly, and they often have multiple temperature sensors that are used to monitor the temperature of the part during cure. The curing takes place under pressure, 1–10 bar (15–150 psi), and at elevated temperature. The lay-up assembly is slightly different (Fig. 8). The top surface of the lay-up is covered with a perforated or porous release film, and if necessary bleeder plies of dry cloth are added to absorb excess resin. Then the assembly is sealed within a nonporous sheet material and placed into the autoclave. The application of pressure and control of temperature is critical. This process offers better quality control than other low- or no-pressure molding processes.

Another process that is used extensively is filament winding. The concept of wrapping filaments around articles to improve their performance is very old. The modern practice of filament winding was developed in response to the requirements for lightweight pressure vessels. Filament winding uses continuous reinforcement to maximize the use of fiber strength. Preimpregnated tape, or a single strand that has passed through a resin bath, is wound onto a mandrel in a prescribed pattern. Design and winding technique allow the maximum fiber strength to be developed in the direction desired. When the winding is completed, the assembly is cured either at room temperature or in an oven. After cure, the mandrel is removed. This process provides for a high level of fiber content.

The process of pultrusion is the opposite of extrusion. The reinforcement is passed through a resin bath and then pulled through a die that controls the resin content and final shape. The die can be heated to cure the resin, or the material can be passed through an oven for curing.

B. Closed-Mold Processes

The closed-mold processes use a two-part mold or die. When the two parts are put together, they form a cavity in the shape of the article to be molded. The molds are usually made of metal with smooth cavity surfaces. Higher pressures and temperatures than those in open molding are usually used. The processes produce very accurate moldings. Most of the processes are attractive for mass production.

Matched die molding is a closed-mold process. There are variations to this process. The main variations concern the form of the starting material and the manner in which it is introduced into the mold. In some cases, the reinforcement is first made into a preform and placed into the mold and then a metered amount of resin is added—this is known as resin transfer molding, or RTM. RTM is a widely used technique for production of components that require accurate dimensional tolerances, since the outer

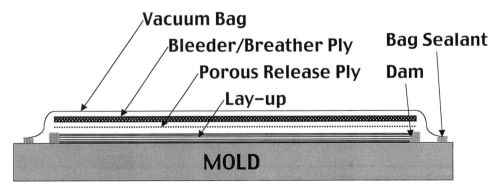

FIGURE 8 Cross section of the composite laminate lay-up and vacuum bagging processing method.

surface of the part is determined by the tool surface. In other cases, a resin–reinforcement mixture is made and a premeasured amount placed into the mold. The molding compound can be introduced automatically or manually. The molding temperatures range from 100°C (212°F) to 140°C (284°F). Pressures range from 7 to 20 bar. Cure cycles can be as short as minutes.

The selection of a fabrication process depends on several factors, including the materials to be processed, the size and design of the article, the number of articles, and the rate of production. Processes differ in their capacity to use different forms of reinforcement and to achieve the proper distribution and amount of reinforcement. The chemistry and rheology of the resin are important factors in process selection. Closed molds require higher temperatures and pressures.

The size and shape of the article to be produced affect the selection. Very large articles such as boat hulls and vehicle bodies and components are more easily and economically produced in open-mold processes. Small gears and precision electrical parts are more suitably produced in closed molds. Shapes that are surfaces of revolution are ideal for filament winding. Very large cylindrical containers have been fabricated by this process. In most open-mold processes, the molds are made of low-cost materials and are easily fabricated but have shorter lives. Autoclave processing of composites, while considered an open-mold technique, requires accurate, robust tools because of the relatively high temperatures and pressures used in the autoclave. Autoclave techniques are well suited to large structural components for aerospace applications; hence, dimensional accuracy of the tools is critical. Open-mold, hand lay-up processes have higher labor cost. If one is making a large number of parts and requires high production rates, mold life and labor cost are important factors. Open-mold processes are usually more costly in these two areas than closed-mold processes. Also, some closed-mold processes can be automated.

Automating the fabrication of advanced composites and improving processing science for composites are two current goals. The advantages of advanced composites are lighter weight, higher strength- and modulus-to-weight ratios, flexibility in design and fabrication, and usually fewer parts per component. Automating the fabrication process could result in a reduction in labor cost and an improvement in quality. The computer-aided manufacturing technology could be utilized to reduce the total labor hours. The application of higher precision control technology could improve quality and lower rejection rates. Work in processing science should result in increased understanding of the cure process, which will aid the development of resin systems and automating production cycles.

Fabrication processes for other matrix materials are important for the use and continued development of these composites. However, not as much work has been done in these areas. The use of these materials represents a small part of the overall uses of composite materials.

VI. USES OF COMPOSITES

Composite materials have been introduced into almost every industry in some form or fashion. We shall look at some of the advantages of using composites and discuss some of the industries that have made used of these materials.

The wide range of property values attained with composites and the ability to tailor the properties is an advantage. Composite materials also generally have higher strength- and modulus-to-weight ratios than traditional engineering materials. These features can reduce the weight of a system by as much as 20 to 30%. The weight savings translates into energy savings or increased performance. Advanced composites exhibit desirable dynamic properties and have high creep resistance and good dampening characteristics. In fact, the superior fatigue performance of composite materials enables them to be used to repair metallic airframes with fatigue damage.

Since composite materials can be manufactured into almost any shape, they allow great design flexibility and offer reduced parts count for articles. The opportunity to select the constituents, tailor them to obtain the required properties, and then through design make the optimum use of the properties is a situation that makes composites very attractive to many industries.

The matrix polymer can impart great chemical and corrosion resistance to composites. The transportation industry has made extensive use of composite materials. The light weight and high strength and the ability to easily manufacture aerodynamic shapes have resulted in lower fuel costs. The lack of corrosion of the materials and the low maintenance cost have reduced the cost of ownership and extended the service life of many parts and products. Examples of products in this industry include auto and truck bodies and parts, trailers, tanks, special-purpose vehicles, and manufacturing equipment.

Composites have added new dimensions to the design and construction of buildings. Their ease of manufacture, light weight, high strength, low maintenance, decorativeness, and functionality have had a significant impact on the industry. New-construction time has been reduced and more flexibility has been added to the design of structures.

Composite materials affected the marine industry very early in their development, and their influence continues to grow. Lack of corrosion, low maintenance, and design flexibility have contributed to the acceptance of composites. The ease of fabricating very large and strong articles in one piece has been another. In addition to pleasure boats, large military and commercial boats and ship hulls have been fabricated. Large tanks for fuel, water, and cargo have been used aboard ships. Composites have made the greatest impact in the sporting goods industry, replacing traditional materials at a revolutionary pace. Applications such as golf club shafts, fishing poles, tennis rackets, skiing equipment, boating applications, and many other sports equipment products are now produced almost exclusively using advanced composites. In most cases, the change in material has translated into an improvement in performance or safety for participants.

The aerospace and military markets are the two areas that have accounted for the largest effort in the development and advancement in composite technology. The need for stronger, stiffer, and lighter structures was an opportunity for composite materials to demonstrate their superiority over more commonly used materials. Durability and low maintenance are additional assets. These increase the service life and reduce the cost of maintaining systems. The development of new and the improvement of existing fabrication processes have brought about a reduction in manufacturing cost. There have been reductions in the number of parts required to construct some components by using molding and composite materials. The unique features of composites have enabled designers to formulate advanced systems that could be made only of composite materials. New military aircraft almost exclusively utilize advanced composites for structure. Rocket motor cases, nozzles, and nose cones are missile applications. Radar domes, rotor blades, propellers, and many secondary structure components such as fairings, doors, and access panels are also fabricated from advanced composites. Numerous pressure vessels, armaments, and items of space hardware are made of selected composite materials.

The use of composite materials will continue to grow. As more engineers come to understand composites, more opportunities will be recognized for their use. As the use of composites increases, more developments will take place in the areas of constituent materials, analysis, design, and fabrication. Composite materials offer tremendous for tailorability, design flexibility, and low-cost processing with low environment impact. These attributes create a very bright future composite materials.

SEE ALSO THE FOLLOWING ARTICLES

ADHESION AND ADHESIVES • BIOPOLYMERS • CARBON FIBERS • FRACTURE AND FATIGUE • METAL MATRIX COMPOSITES • POLYMERS, MECHANICAL BEHAVIOR • POLYMERS, THERMALLY STABLE • SANDWICH COMPOSITES

BIBLIOGRAPHY

Ashton, J. E., Halpin, J. C., and Petit, P. H. (1969). "Primer on Composite Materials: Analysis," Technomic Publishing Company, Stamford, CT.

Hull, D. (1981). "An Introduction to Compositive Materials," Cambridge University Press, London.

Jones, R. M. (1975). "Mechanics of Composite Materials," Scripta Book Company, Washington, D.C.

Sih, G. C., and Hsu, S. E. (1987). "Advanced Composite Materials and Structures," VNU Science Press, Utrecht, The Netherlands.

Tsai, S. W. (1985). "Composites Design—1985," Think Composites, Dayton, OH.

Tsai, S. W., and Hahn, H. T. (1980). "Introduction to Composite Materials," Technomic Publishing Company, Westport, CT.

Whitney, J. M., Daniel, I. M., and Pipes, R. B. (1982). "Experimental Mechanics of Fiber Reinforced Composite Materials," Society for Experimental Stress Analysis, Brookfield Center, CT.

Industry Overview: FRP Materials, Manufacturing Methods and Markets, (1999). *Composites Technol.* **5**, 6–20.

Computational Aerodynamics

David A. Caughey
Cornell University

GLOSSARY

Artificial viscosity Term added to a numerical approximation to provide an artificial (i.e., nonphysical) dissipative mechanism to prevent round-off errors from accumulating and destroying the accuracy of the solution.

CFL condition Constraint, first described by Courant, Friedrich, and Lewy, that limits the size of the time step by which an explicit method may advance the numerical solution of an initial-value problem to be consistent with the physics of local wave propagation.

Conservation form Divergence form of a system of partial differential equations describing conservation laws, important when solutions containing discontinuities (shocks) are to be computed.

Direct numerical simulation (DNS) Solution procedure for the Navier-Stokes equations in which all scales of turbulent eddies are fully resolved, resulting in an exact solution for turbulent flow.

Euler equations Equations describing the inviscid flow of a compressible fluid. These equations constitute a hyperbolic system of partial differential equations; the Euler equations are nondissipative, and weak solutions containing surfaces of discontinuity (which approxi-

mate the behavior of shock waves) must be allowed for many practical problems.

Explicit method Numerical method for solving initial-values problems in which the solution at a given point and time depends explicitly only upon the solution at preceding time levels.

Implicit method Numerical method for solving initial-values problems in which the solution at a given point and time depends upon the solution at neighboring points at the same time level, as well as upon the solution at preceding time levels.

Inviscid flow Approximation to the fluid-flow equations, useful at large Reynolds numbers, in which the viscous, or internal friction, forces acting within the fluid are neglected.

Large-eddy simulation (LES) Solution procedure for the Navier-Stokes equations in which the largest, energy-containing scales of a turbulent flow are resolved, while the smaller scales are modeled.

Mach number Ratio of flow velocity to speed of sound; a measure of the importance of compressibility in determining the behavior of fluid flows.

Mesh generation The process of generating a volume-filling mesh of hexahedral or tetrahedral cells as a basis for the description of the solution to a nonlinear field problem. Often a pacing item for the application of computational techniques to fluid-flow problems.

Reynolds-averaged Navier-Stokes (RANS) equations Equations for the mean properties in a turbulent flow obtained by decomposing the fields into mean and fluctuating components and averaging the Navier-Stokes equations. Solutions of these equations for the mean properties of the flow require knowledge of various correlations (the Reynolds stresses) of the fluctuating components.

Shock capturing Numerical method in which shock waves (and other discontinuities) are incorporated into the solution by smearing them out with artificial dissipative terms so that the solution remains continuous, although with very steep gradients near the shocks.

Shock fitting Numerical method in which shock waves (and other discontinuities) are treated as discontinuities and are explicitly fitted into the solution as internal boundaries, across which specified jump relations are satisfied.

Transonic flow Fluid flow that contains regions of both subsonic and supersonic flow velocities; usually occurs at flight Mach numbers near unity.

Turbulence model Phenomenological model used in Reynolds-averaged Navier-Stokes equations to relate the turbulent stresses to the mean flow properties.

Turbulent flow Flow in which unsteady fluctuations play a major role in determining the effective mean stresses in the field; regions in which turbulent fluctuations are important inevitably appear in fluid flow at large Reynolds number.

Upwind method Numerical method for CFD in which spatial asymmetry is introduced into the difference stencil to introduce dissipation into the approximation, thus stabilizing the scheme. This is a popular mechanism for the Euler equations, which have no natural dissipation, but also is effective for the Navier-Stokes equations, especially at high Reynolds number.

COMPUTATIONAL AERODYNAMICS is concerned with the development and application of numerical techniques to compute solutions to practical aerodynamic problems, such as the design of aircraft, and to pursue fundamental research into the behavior of fluid flow, especially in the turbulent regime. The general subject area includes the development of efficient numerical algorithms and their implementation and use on high-speed, digital computers. Applications include the prediction of aerodynamic forces and heat transfer rates to vehicles in flight. These computational methods have revolutionized the aerodynamic design of flight vehicles.

I. INTRODUCTION

A. Scope of Computational Aerodynamics

In 1946 John Von Neumann, a remarkable pioneer in a variety of fields spawned by the development of high-speed, digital computers, wrote the words:

"Indeed, to a great extent, experimentation in fluid dynamics is carried out under conditions where the underlying physical principles are not in doubt, where the quantities to be observed are completely determined by known equations. The purpose of the experiment is not to verify a proposed theory but to replace a computation from an unquestioned theory by direct measurements. Thus, wind tunnels are, for example, used at present, at least in part, as computing devices of the so-called analogy type ... to integrate the nonlinear partial differential equations of fluid dynamics."

Since the time these words were written, the widespread use of computers in science and engineering has had a major impact upon the design practices of modern engineers as well as upon the range of problems that is open to attack by scientists. As suggested by the above quotation, the engineering science of aerodynamics has been a forerunner in the development of computational methods. In this article, the major subject areas of computational aerodynamics will be presented, with a focus on the algorithmic approaches that have resulted in important practical advances in the state of the art. In order to describe these methods, and the algorithmic ideas upon which they are based, it is necessary to introduce enough of the relevant physics of fluid mechanics that the magnitude of the task, and the scope of the achievements to date, can be appreciated.

B. Relation to Wind Tunnel Testing

The time-honored method of aerodynamic evaluation is, of course, the wind tunnel. When used for aerodynamic design purposes (as opposed to its role in fundamental fluid mechanical research), the wind tunnel provides measurements of the aerodynamic forces and heat transfer rates acting on a model of the vehicle of interest. It is this design role of the wind tunnel, not its role in fundamental research, that Von Neumann had in mind when he referred to wind tunnels as analog computers.

Both wind tunnels and computational techniques have their advantages and limitations. The wind tunnel has the advantage that with proper scaling it is, in principle, a perfect simulation of the complete physics (although this proper scaling is rarely achieved in practice). Computational analyses, on the other hand, can be performed quite quickly and are much less energy intensive than wind tunnel testing.

The use of forces measured on a model mounted in a wind tunnel to predict forces on a full-scale vehicle is based upon modeling principles which guarantee that the forces will scale in a particular way if certain dimensionless parameters are the same for both the model and full-scale flows. For aircraft problems the relevant dimensionless parameters are the Reynolds number, which characterizes the relative importance of viscous effects, and the Mach number, which characterizes the relative importance of the compressibility of the fluid (usually air). (Both of these parameters will be defined in later sections.) It generally is not possible to match one or both of these parameters with the corresponding flight parameters in a given test, so great care and considerable insight are required to extrapolate the effects of the mismatched parameter (or parameters). In addition, the wind tunnel usually is used to approximate the flow in an unbounded medium using a flow of limited extent, and the degree to which the walls of the wind tunnel affect the flow in the vicinity of the model is another potential source of error that must be carefully assessed. Of course, these wind tunnel wall effects can be reduced by making the model very small relative to the size of the tunnel, but this makes it harder to match the Reynolds number, which, as will be seen, is proportional to the linear dimension of the model.

Computational methods, on the other hand, are limited by their inability to incorporate enough of the relevant physics into the mathematical model, while still being able to perform the computation with a reasonable expenditure of computer resources, and uncertainties about the degree to which numerical error affects the accuracy with which the equations actually are solved. The constraint on the physical models introduces an uncertainty that again calls for sound judgment to determine whether the missing physics might play an important role. In addition, it often is necessary to approximate the usually complex geometries even in a numerical calculation, and the degree to which this geometrical approximation may compromise the accuracy of the results is another source of uncertainty. The goal of computational aerodynamics is to bring increasingly accurate physical models of the fluid mechanics to bear on problems of ever greater geometrical complexity during engineering design studies, and much progress has been made toward this goal in the past three decades.

The present-day roles of wind tunnel testing and computational aerodynamics are complementary. As increasingly accurate and economical computational techniques are developed, they are used in studies to identify promising candidate designs. These are then tested in the wind tunnel to make certain that any approximations made in the computation are warranted. The best designs are frequently then used as spring boards for new parametric studies, which are again performed on the computer and validated in the wind tunnel. The net result is that, with approximately the same amount of wind tunnel testing, a much better final design is achieved when computational methods are used to screen candidate configurations. As computational methods continue to evolve, the testing of preliminary designs increasingly will be replaced by computation, but it is likely that final designs will continue to be tested in the wind tunnel for the foreseeable future.

II. FLUID MECHANICAL BACKGROUND

In this section, the fluid mechanical background necessary to understand and appreciate the algorithmic aspects of computational aerodynamics will be described.

A. The Navier-Stokes Equations

As may be inferred from the quotation of John Von Neumann, which was introduced at the beginning of this article, aerodynamics is virtually unique within the engineering sciences in having a generally accepted mathematical framework for describing most problems of practical interest. Such diverse problems as the high-speed flow of air past an airplane wing, the motions of the atmosphere responsible for our weather, and the unsteady air currents associated with the flapping of the wings of a housefly all are described by a set of partial differential equations known as the Navier-Stokes equations. These consist of a system of partial differential equations that express the fundamental conservation laws upon which the science of fluid mechanics is based. These include the conservation of mass, stating that fluid is neither created nor destroyed in any arbitrary region of space; momentum equations that express Newton's second law for the motion of the fluid; and an energy equation based on the first law of thermodynamics. In both the momentum and energy equations, the further approximation that the state of stress within the fluid is linearly related to the rate of strain also is made, but this is widely accepted as an excellent approximation in most instances. These equations can be written in a system of Cartesian coordinates (x, y) as follows. Let ρ and p denote the fluid density and pressure, respectively; u and v the components of the fluid velocity in the x and y directions; and e the total energy per unit volume. For a calorically perfect gas having a ratio of specific heats $\gamma = c_p/c_v$ (a good approximation for air from normal temperatures up to about 1700°C), the pressure is related to the total energy by the equation of state

$$p = (\gamma - 1)\left(e - \rho \frac{u^2 + v^2}{2}\right). \qquad (1)$$

The Navier-Stokes equations can then be written in the form

$$\frac{\partial \mathbf{w}}{\partial t} + \frac{\partial \mathbf{f}}{\partial x} + \frac{\partial \mathbf{g}}{\partial y} = \frac{\partial \mathbf{R}}{\partial x} + \frac{\partial \mathbf{S}}{\partial y}, \qquad (2)$$

where

$$\mathbf{w} = \{\rho,\ \rho u,\ \rho v,\ e\}^T \qquad (3)$$

is the vector of dependent variables and

$$\mathbf{f} = \{\rho u,\ \rho u^2 + p,\ \rho uv,\ (e + p)u\}^T \qquad (4a)$$

and

$$\mathbf{g} = \{\rho v,\ \rho uv,\ \rho v^2 + p,\ (e + p)v\}^T \qquad (4b)$$

are the flux vectors. The effects of viscosity are governed by the viscous flux vectors

$$\mathbf{R} = \{0,\ \tau_{xx},\ \tau_{xy},\ u\tau_{xx} + v\tau_{xy}\}^T \qquad (5a)$$

and

$$\mathbf{S} = \{0,\ \tau_{xy},\ \tau_{yy},\ u\tau_{xy} + v\tau_{yy}\}^T. \qquad (5b)$$

The viscous stresses appearing here are related to the derivatives of the components of the velocity vector by

$$\tau_{xx} = 2\mu\frac{\partial u}{\partial x} - \frac{2}{3}\mu\left(\frac{\partial u}{\partial x} + \frac{\partial v}{\partial y}\right) \qquad (6a)$$

$$\tau_{xy} = \mu\left(\frac{\partial u}{\partial y} + \frac{\partial v}{\partial x}\right) \qquad (6b)$$

$$\tau_{yy} = 2\mu\frac{\partial v}{\partial y} - \frac{2}{3}\mu\left(\frac{\partial u}{\partial x} + \frac{\partial v}{\partial y}\right), \qquad (6c)$$

where μ is the coefficient of viscosity. In the above, terms representing heat conduction have been omitted from the energy equation for simplicity. Also, here and elsewhere in this article, equations will be written in two space dimensions for the sake of economy; the extension to problems in three space dimensions is straightforward unless otherwise noted.

If the Navier-Stokes equations are nondimensionalized by normalizing lengths with respect to a length L representative of the size of the vehicle past which the flow is to be determined and by normalizing flow properties (such as density, viscosity, and flow velocities) by their values in the undisturbed free stream far from the vehicle, an important *nondimensional parameter*, the Reynolds number

$$\mathbf{Re} = \frac{\rho_\infty V_\infty L}{\mu_\infty}, \qquad (7)$$

appears. The viscous stress terms on the right-hand side of Eq. (2) are multiplied by the inverse of $\mathbf{Re}$. Physically, the Reynolds number can be interpreted as a measure of the relative importance of the viscous, or internal friction, forces in determining the behavior of the flow; when the

TABLE I The Hierarchy of Aerodynamic Approximations, after Chapman (1979)

Stage	Model	Equations	Time frame
I	Linear potential	Laplace	1960s
IIa	Nonlinear potential	Full potential	1970s
IIb	Nonlinear inviscid	Euler	1980s
III	Modeled turbulence	Reynolds-averaged Navier-Stokes	1990s
IV	LES	Navier-Stokes (+ subgrid turb. model)	1980s–
V	DNS	Fully resolved Navier-Stokes	1980s–

Reynolds number is large, viscous effects are small except in regions of very large velocity gradients.

The computational resources required to solve the complete Navier-Stokes equations numerically are enormous, especially for problems in which turbulence occurs. A computation that resolves all scales of turbulent motions is called a direct numerical simulation (DNS). In 1971, Howard Emmons of Harvard University estimated the computer time required to solve a simple, turbulent pipe-flow problem, including direct computation of all eddies containing a significant fraction of the turbulent kinetic energy. For a computational domain consisting of approximately 12 diameters of the pipe, the computation of the solution at a Reynolds number of $\mathbf{Re}_D = 10^7$ would require about 10^{17} s on a 1970 main frame computer. Of course, much faster computers are now available, but even for a machine capable of 100 Gigaflops (100×10^9 floating-point operations per second)—representative of the fastest machines currently available—such a calculation would require more than 10,000 years to complete. In fact, it probably is not necessary to resolve all the energy-containing scales in the computation. A great deal of current work is focused on developing suitable dissipative models to remove the required energy of the subgrid scales in these large-eddy simulations (LES) of turbulent flows.

More recently, Dean Chapman of the NASA Ames Research Center has divided problems in computational aerodynamics into four classes, or stages, depending on the degree to which the complete Navier-Stokes equations have been approximated. A computation in which the complete Navier-Stokes equations are solved, including direct simulation of the large-scale turbulent fluctuations, but with the subgrid scale turbulence modeled, is called a Stage IV approximation by Chapman. The computational resources required to solve a Stage IV approximation to the Navier-Stokes equations numerically for the flow past a realistic, but geometrically simplified three-dimensional wing-body configuration were estimated by Chapman to exceed the capabilities of 1980s generation supercomputers (so-called Class VI supercomputers such as the CRAY 1-S or CYBER 205) by a factor of approximately 4000. Thus, in order to make predictions for practical engineering problems, judicious approximations must be made. The first such simplification that seems reasonable is to model *all* scales of turbulence; this is termed a Stage III computation by Chapman. This can be done, for example, by solving the Reynolds-averaged Navier-Stokes (RANS) equations and modeling phenomenologically the higher order moments required for closure of the system. Such a calculation for a realistic, three-dimensional geometry has been estimated by Chapman to require about 40 times the power of Class VI supercomputers. Such computations became feasible for research purposes around 1990 and are now feasible for limited cases in the design cycle.

B. Boundary Layer Theory

Fortunately, in many aerodynamic problems the effects of viscosity are confined to very thin regions (called boundary layers) in the immediate vicinity of the body surface. Thin boundary layers are a naturally occurring feature of solutions of the Navier-Stokes equations at large Reynolds numbers when the no-slip condition is imposed at solid surfaces (resulting from thermodynamic equilibrium between the surface and the fluid) and the body is sufficiently thin that the viscous layers do not separate from the surface. Outside these viscous layers, the fluid behaves very nearly as a perfect, or inviscid, fluid, which is described by much simpler equations (that will be presented in Section II.C). In addition, since the boundary layers appearing in these flows are thin, this inviscid approximation gives a good approximation to the pressure forces acting on the body surface, and the net forces acting upon the body can be predicted quite accurately. The advantage of this inviscid model is twofold: first, the thin boundary layers no longer need to be resolved in the calculation, and the length scales that need to be resolved in the solution are those determined by the geometry of the body itself; second, since the equations themselves are considerably simpler than the Navier-Stokes equations, they also are significantly less costly (in terms of computer resources) to solve. In addition, numerical methods developed to solve the Navier-Stokes equations must also be capable of solving them accurately in regions where the viscous stresses are not important, i.e., in regions where they reduce to their inviscid counterparts. Thus, algorithms for solving the Navier-Stokes equations often are first tested on the inviscid equations of motion, and it is with attempts to solve these equations describing inviscid flow past aircraft geometries that much of computational aerodynamics is concerned. Therefore, the remainder of this article will concentrate on these aspects, returning to the question of viscous effects and of turbulence only in the closing sections.

C. Inviscid Flow Models

For many design purposes, it is sufficient to represent the flow past an aerospace vehicle as that of an ideal (or inviscid), compressible fluid. This is appropriate for flows at large Reynolds numbers that contain only negligibly small regions of separated flow. The equations describing inviscid flows can be obtained as a simplification of the Navier-Stokes equations in which the viscous terms are neglected altogether. This results in the Euler equations of inviscid flow

$$\frac{\partial \mathbf{w}}{\partial t} + \frac{\partial \mathbf{f}}{\partial x} + \frac{\partial \mathbf{g}}{\partial y} = 0. \tag{8}$$

The Euler equations comprise a hyperbolic system of partial differential equations, and their solutions contain features that do not appear in solutions of the Navier-Stokes equations. In particular, while the diffusion (viscous) terms in the Navier-Stokes equations guarantee that the solutions will be smooth, the absence of these diffusion terms from the Euler equations allows them to have solutions that are discontinuous across surfaces in the flow. Such solutions to the Euler equations must be interpreted within the context of generalized (or weak) solutions, and this theory provides the framework for developing the properties of any discontinuities that may appear. In particular, the jumps in dependent variables (such as density, pressure, and velocity) across such discontinuities cannot be arbitrary but must be consistent with the integral form of the original conservation laws upon which the differential equations are based. For the Euler equations, these jump conditions are called the Rankine-Hugoniot relations. The solution of the Euler equations for steady flows past a realistic, three-dimensional configuration is termed a Stage IIb computation by Chapman, and such computations are routinely performed in the design phase for aerospace vehicles.

If the flow can further be approximated as steady and irrotational, it is possible to introduce a velocity potential Φ such that the velocity $\mathbf{q}$ is given by

$$\mathbf{q} = \nabla \Phi \tag{9}$$

and the steady form of the Euler equations reduces to

$$\frac{\partial(\rho u)}{\partial x} + \frac{\partial(\rho v)}{\partial y} = 0, \tag{10}$$

where, from the definition of the potential in Eq. (9),

$$u = \frac{\partial \Phi}{\partial x} \qquad \text{and} \qquad v = \frac{\partial \Phi}{\partial y}$$

are the velocity components, and the density is given by the isentropic relation

$$\rho = \left\{ 1 + \frac{1}{2}(\gamma - 1)\mathbf{M}_\infty^2 [1 - (u^2 + v^2)] \right\}^{1/(\gamma-1)}, \quad (11)$$

where $\mathbf{M}_\infty$ is the Mach number of the free stream (i.e., the ratio of the velocity to the speed of sound in the free stream). Equation (11) can be used to eliminate the density from Eq. (10), which can then be expanded as

$$(a^2 - u^2)\frac{\partial^2 \Phi}{\partial x^2} - 2uv\frac{\partial^2 \Phi}{\partial x \partial y} + (a^2 - v^2)\frac{\partial^2 \Phi}{\partial y^2} = 0, \quad (12)$$

where a is the local speed of sound.

If a new (s, n) Cartesian coordinate system is introduced with the s axis aligned with the velocity vector, then at any point Eq. (12) can be written as

$$(1 - \mathbf{M}^2)\frac{\partial^2 \Phi}{\partial s^2} + \frac{\partial^2 \Phi}{\partial n^2} = 0, \quad (13)$$

where $\mathbf{M} = \sqrt{u^2 + v^2}/a$ is now the Mach number based on the local flow velocity and speed of sound. This equation shows clearly how the Mach number enters the problem for compressible flows.

Equation (13) is a second-order, quasilinear, partial differential equation whose type depends upon the sign of the coefficient $(1 - \mathbf{M}^2)$. The type of a partial differential equation is intimately related to how information about its solution is transmitted from one point to another, and, in this particular case, it is clear that this flow of information is changed dramatically when the local Mach number changes from subsonic ($\mathbf{M} < 1$) to supersonic ($\mathbf{M} > 1$). When the local Mach number is supersonic, Eq. (13) is said to be of hyperbolic type, and information is transmitted along characteristic surfaces having inclination

$$\left(\frac{dn}{ds} \right)_{\text{char}} = \pm \frac{1}{\sqrt{\mathbf{M}^2 - 1}}$$

relative to the s axis. When the local Mach number is subsonic, the equation is of an elliptic type, and the effect of a change at any point is felt at all other points in the field.

Thus, Eq. (10) contains a mathematical description of the physics necessary to predict the important features of transonic flows. It is capable of changing type, depending upon whether the local Mach number is less than or greater than unity, and the conservation form allows surfaces of discontinuity, or shock waves. Solutions at this level of approximation, while somewhat simpler and more economical than those of the Euler equations, are also considered by Chapman to belong to Stage II and are capable

of accurately describing transonic flows containing weak shock waves.

Finally, if the flow can be approximated by small perturbations to a uniform stream, the potential equation can further be simplified to

$$(1 - \mathbf{M}_\infty^2)\frac{\partial^2 \phi}{\partial x^2} + \frac{\partial^2 \phi}{\partial y^2} = 0 \qquad \text{for} \quad |u^2 + v^2 - 1| \ll 1, \quad (14)$$

where ϕ is the perturbation velocity potential defined as

$$\Phi = x + \phi \quad (15)$$

and the velocity of the free stream is assumed to be unity. Further, if the flow can be approximated as incompressible (i.e., the Mach number is everywhere negligibly small), then Eq. (10) reduces to

$$\frac{\partial^2 \phi}{\partial x^2} + \frac{\partial^2 \phi}{\partial y^2} = 0 \qquad \text{for} \quad \mathbf{M}_\infty \ll 1. \quad (16)$$

Since the speed of sound in the atmosphere (under standard sea level conditions) is approximately 340 m/s (760 mph), the effects of compressibility will be relatively unimportant for flight speeds of less than about 100 m/s (220 mph). This corresponds to a flight Mach number of about 0.3, and since it is the square of the Mach number that appears in Eq. (14), compressibility corrections are on the order of 10% or less.

Since Eqs. (14) and (16) are linear, superposition of elementary solutions can be used to construct more general solutions. Such surface-singularity methods—called panel methods in aerodynamic applications—have been highly developed for the calculation of flows past quite complex, three-dimensional configurations on relatively modest computers. These methods are included in Chapman's Stage I.

The results of any of the methods corresponding to Stages I and II can be improved by adding corrections to account for the small, but finite, thickness of the boundary layer. For flows having limited regions of separation, such patched solutions are capable of accurately predicting surface pressure distributions (including drag forces), even for transonic flows containing shock waves. The behavior of the flow in the boundary layer can be predicted once the external inviscid flow is known. Computational techniques also are required for predicting the flow within the boundary layer, but space constraints do not allow their description in this article.

III. ALGORITHMIC ASPECTS

The earliest calculations for compressible flows were based, necessarily, upon linearizations of the inviscid models presented in Section II.C. The hodograph method of analysis uses the specialized fact that, in two dimensions, if the velocity coordinates and physical coordinates are interchanged in Eq. (12), the resulting equation is linear. This hodograph transformation results in a formulation in which one finds the coordinates in the physical plane as a function of the velocity components. Although the equation is linear, the formulation of boundary conditions is sufficiently difficult that only problems with very simple geometries (such as the flow in a planar jet or past a wall having a sharp corner) can be solved without the use of numerical methods. In addition, the formulation does not handle discontinuities easily, since it is now the independent variables that must change discontinuously.

A second form of linearization is achieved by approximating the nonlinear terms in Eq. (10) by some specified function. [There is an interesting analogy with turbulence models that frequently take assumed forms for the dependence of time averages of higher order (or nonlinear) moments of the fluctuating quantities.] In the case of compressible flows, the approach must be of severely limited applicability, since the assumed form of the nonlinear terms implies *a priori* knowledge of where the sonic lines and shocks must appear in the solution.

In order to treat problems of sufficient geometrical complexity to be of practical value, numerical methods are required.

A. The Panel Method

The earliest numerical methods used for compressible flow problems were based on making compressibility corrections to solutions of the linear potential problem, defined by Eq. (14). In fact, programs for solving these problems were among the first widely used, large-scale numerical methods for engineering analysis. In addition, solutions of Eq. (16) are also of interest for low-speed, or nearly incompressible, flows. The linearity of these equations allows the construction of arbitrarily complex solutions by the superposition of relatively simple (analytically describable) elementary solutions. The elementary solutions used in these analyses correspond to distributions of singular solutions whose forms are specified over each of a number of nonoverlapping subdomains on the surface of the vehicle. The strengths of the singularities on each of these facets (or panels) is determined by satisfying the boundary condition that there be no fluid flux through the surface of the body.

A major advantage of panel methods, relative to the more general analyses that will be described in Section III.B, is that it is necessary to describe (and to discretize into panels) only the surface of the body; thus, a three-dimensional solution—such as that for the flow past a complete vehicle—requires the solution of a basically two-dimensional problem. The panel method has been developed extensively for practical problems since its first application to three-dimensional problems more than 30 years ago, but since the techniques used to solve the potential equation in this way are quite different from those used to solve nonlinear problems, and these methods are not readily extendable to the nonlinear case, they will not be discussed further here.

B. Nonlinear Methods

1. General Considerations

As is seen in Section II.C, linear methods can never predict truly transonic phenomena; calculations of transonic flow past aircraft configurations must be based on at least the full potential equation or the Euler equations. For such nonlinear problems, it is necessary to describe the solution throughout the entire domain (not just on the body surface, as in the panel method) and to use some sort of iteration (which may, or may not, actually approximate the physics of an unsteady flow process) to drive the flow variables toward the solution of the nonlinear equations. Solutions for both the full potential and Euler equations generally are based on finite-difference or finite-volume techniques. In both these techniques, a grid, or network, of points is distributed throughout the flow field, and the solution is represented by its values at these points. In a finite-difference method the derivatives appearing in the original differential equation are approximated by finite differences of the values of the solution at neighboring points (and times if the flow is unsteady), and substitution of these into the differential equation yields a system of algebraic equations relating the values of the solution at neighboring grid points. In a finite-volume method, the unknowns are taken to be representative values of the solution in the control volumes formed by the intersecting grid surfaces, and the equations are constructed by balancing fluxes across the surfaces of each control volume with the rates of change of the variables within the control volume. The algebraic equations relating the values of the solution in the neighboring cells are usually very similar in appearance to those resulting from a finite-difference formulation. A third class of numerical methods, which is more highly developed for elliptic problems (and is widely used in structural analysis), is called the finite-element method. In that method, the solution is represented using simple interpolating functions within each mesh cell, or element, and the equations for the nodal values are obtained by integrating a variational or residual formulation of the equations of motion over the elements. Such methods are less widely used for compressible flows, although significant progress has been made in the past decade.

Since the potential equation and the Euler equations are nonlinear, the algebraic equations resulting from either a finite-difference or finite-volume formulation also are nonlinear, and a scheme of successive approximation usually is required to solve them. These equations are highly local in nature, however, and efficient iterative methods usually are available to solve them.

Treatment of complex geometries. Although the equations of fluid mechanics summarized in Section II have been written in a simple Cartesian coordinate system, their solution for practical engineering problems usually requires that they be expressed on a grid that matches the shape past which the flow is to be computed. Solving the problem on such a *boundary-conforming* (or body-fitted) grid system has the advantages of (1) eliminating the need for cumbersome, and possibly inaccurate or destabilizing, interpolation formulas to enforce the boundary conditions and to obtain the solution on the body surface and (2) allowing the relatively easy and efficient clustering of mesh surfaces near the body surface where gradients in the solution usually are the largest and where high accuracy is required.

The problem may be formulated using either a *structured* or an *unstructured* grid system. Structured grids are those in which the grid points can be ordered in a regular Cartesian structure, i.e., the points can be given indices (i, j) such that the nearest neighbors of the (i, j) point are identified by the indices $(i \pm 1,\ j \pm 1)$. The grid cells for these meshes are thus quadrilateral in two dimensions and hexahedral (but having nonplanar faces) in three dimensions. Unstructured grids have no regular ordering of the points or the cells, and a connectivity table must be maintained to identify which points and edges belong to which cells. Unstructured grids most often consist of triangular cells in two dimensions and tetrahedral cells in three dimensions, or combinations of these and quadrilateral and hexahedral cells, respectively. In addition, grids having purely quadrilateral (or hexahedral) cells may also be unstructured, e.g., when multilevel grids are used for adaptive refinement.

Implementations on structured grids are generally more efficient than those on unstructured grids, since indirect addressing is required for the latter, and efficient implicit methods often can be constructed that take advantage of the regular ordering of cells in structured grids.

For structured grids, the grid surfaces can be viewed surfaces of a new, transformed, coordinate system, and the expression of the system of conservation laws in this new coordinate system thus reduces the problem to one of (apparently) Cartesian geometry. Fortunately, it is quite easy to transform systems of conservation laws to a new coordinate system. The transformation will be described here for the Euler equations [Eqs. (8) of Section II.C]. The local properties at any point of a transformation to a new coordinate system $(\xi,\ \eta)$ are contained in the Jacobian matrix $\mathbf{J}$, which is defined as

$$\mathbf{J} = \begin{pmatrix} x_\xi & x_\eta \\ y_\xi & y_\eta \end{pmatrix} \qquad (17)$$

and for which the inverse is

$$\mathbf{J}^{-1} = \begin{pmatrix} \xi_x & \xi_y \\ \eta_x & \eta_y \end{pmatrix} = \frac{1}{h} \begin{pmatrix} y_\eta & -x_\eta \\ -y_\xi & x_\xi \end{pmatrix}, \qquad (18)$$

where

$$h = x_\xi y_\eta - x_\eta y_\xi \qquad (19)$$

is the determinant of the Jacobian matrix. In these equations, subscripts denote partial differentiation with respect to the corresponding independent variable.

Introduction of the contravariant components of the flux vectors

$$\{\mathbf{F}, \mathbf{G}\}^T = \mathbf{J}^{-1}\{\mathbf{f}, \mathbf{g}\}^T \qquad (20)$$

allows the Euler equations to be written in the compact form

$$\frac{\partial h\mathbf{w}}{\partial t} + \frac{\partial h\mathbf{F}}{\partial \xi} + \frac{\partial h\mathbf{G}}{\partial \eta} = 0 \qquad (21)$$

when the transformation is independent of time [i.e., the $(\xi,\ \eta)$ coordinates are fixed in Cartesian space]. If the coordinate transformation is time dependent (as might be required if the vehicle is deflecting control surfaces or the structure is deforming due to aerodynamic loads), a similar conservation form can also be derived but includes additional fluxes when the surfaces of the control volume are moving through space.

The Navier-Stokes equations can be transformed in a similar manner, although the transformation of the viscous terms is somewhat more complicated and will not be given here. Since the potential equation is simply the continuity equation (the first of the conservation laws comprising the Euler equations), the transformed potential equation is immediately seen to be given by

$$\frac{\partial(\rho h U)}{\partial \xi} + \frac{\partial(\rho h V)}{\partial \eta} = 0, \qquad (22)$$

where

$$\{U, V\}^T = \mathbf{J}^{-1}\{u, v\}^T \qquad (23)$$

are the contravariant components of the velocity.

The generation of a suitable boundary-conforming coordinate system often is a major obstacle to solving problems of practical interest, particularly when dealing with very complex geometries. A number of techniques, including conformal mapping, algebraic transformation sequences, and the solution of systems of elliptic or hyperbolic partial differential equations for the mesh coordinates, are used. For very complex geometries, the domain may be subdivided into a number of subdomains, with separate grids constructed in each of the blocks. Such a grid is called a multiblock grid. A boundary-conforming grid used to compute the flow past a two-dimensional airfoil is shown in Fig. 1. This grid was generated using a combined algebraic-conformal mapping procedure. It contains 16,384 cells, but only a fraction of the cells—those nearest the airfoil surface—are shown for the sake of clarity. The complete grid extends to a circle having a radius equal to 30 body lengths, where a uniform freestream boundary condition is imposed on the solution.

As described above, an alternative approach is to abandon the lexicographical ordering of the grid and to use an unstructured grid. Most commonly this is done using cells of triangular (in two dimensions) or tetrahedral (in three dimensions) shape, but other shapes can be used as well. For unstructured grids, the mesh generation step is no longer usefully considered a coordinate transformation, and the equations are developed by computing the fluxes directly in the Cartesian system. There is considerably more overhead (both in storage and in CPU time) for such methods, but the greater ease of mesh generation usually makes this the method of choice for complex geometries, especially in three dimensions. Unstructured grids also can be adapted locally to features of the solution, such as large gradients and shock waves, much more efficiently than structured grids. General methods for the

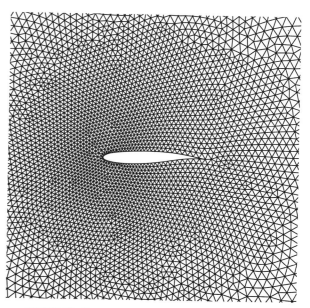

FIGURE 2 Unstructured grid for an airfoil computation.

automatic generation of unstructured grids for very complex, three-dimensional geometries have been developed, based on Delaunay triangulation techniques and so-called *advancing front* methods. An unstructured grid consisting of triangular cells in the vicinity of a two-dimensional airfoil geometry is shown in Fig. 2.

Shock capturing. Most numerical methods for solving compressible flow problems rely on the introduction of artificial (nonphysical) terms in the equations to smear out any shock waves that develop in the solution and allow shocks to be captured naturally by the numerical scheme without any special treatment at points near the shocks. This *shock capturing* approach is in contrast with that of *fitting* the shocks as surfaces of discontinuity, which then must be treated as internal boundaries in the flow calculation, across which the appropriate jump relations must be enforced as an internal boundary (or compatibility) condition. The artificial (or numerical) viscosity added in a shock-capturing scheme acts to smear out the discontinuities that the inviscid theory would predict in much the same manner as molecular viscosity smears out shock waves in the real world. The length scales of these phenomena are, or course, quite different. Under conditions typically found in the atmosphere, shock wave thicknesses are of the order of a few microns (10^{-6} m); under the action of the artificial viscosity of a numerical scheme, the shock thicknesses scale with the grid spacing, which might correspond to a physical distance of 10 cm for a reasonable mesh spacing on a full-scale wing. It is important to realize that even when the Navier-Stokes equations are solved, artificial viscosity usually is necessary when

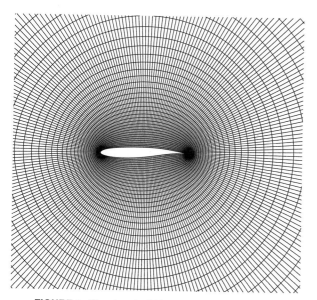

FIGURE 1 Structured grid for an airfoil computation.

the solution contains shock waves, since it is impractical to use mesh spacings fine enough to resolve the physical shock structure defined by the molecular viscosity.

2. Transonic Potential Methods

The primary advantage of solving the potential equation, rather than the Euler equations, results from the existence of a velocity potential in the former case, allowing the solution to be described in terms of a single scalar function Φ, rather than the vector of unknowns, and resulting system of equations, in the more general case. The formulation of numerical schemes to solve this single scalar equation is complicated by the fact that, as noted earlier, the equation changes type when the local Mach number changes from subsonic to supersonic and vice versa. (In contrast, the unsteady Euler equations are always of hyperbolic type.) Differencing schemes for the potential equation must, therefore, be *type dependent* (i.e., must change their form depending upon whether the local Mach number is subsonic or supersonic). These methods generally are based on central, or symmetric, differencing formulas, which are appropriate for subsonic flows in which disturbances are free to spread in all directions, with modifications to reflect the directionality of signal propagation in supersonic regions, in which disturbances introduces at a point can be felt only in a limited, conical region downstream of the point. This directional bias can be introduced into the difference equations either by adding an artificial viscosity proportional to the third derivative of the potential Φ or by replacing the density at each point in the supersonic zone by its value slightly upstream of the actual location. Algebraically, the two approaches are equivalent, since this upwinding of the density evaluation also effectively introduces a correction proportional to the third derivative of the potential.

It is important that such artificial viscosity (or compressibility) be added in such a way that the effect vanishes as the mesh spacing is refined. In this way, the numerical approximation will approach the differential equation in the limit of zero mesh spacing, and the method is said to be *consistent* with the original differential formulation. In addition, for flows with shock waves it is important to base the method on the *conservation form* of the potential equation, Eq. (10), rather than the quasilinear form of Eq. (12). This will ensure that the shock jumps are uniquely determined. The shock jump relation corresponding to Eq. (10) is different, however, from the Rankine-Hugoniot condition for the Euler equations. Since the entropy is everywhere conserved in the potential formulation, and since there is a finite entropy jump across a Rankine-Hugoniot discontinuity, it is clear the solutions must be different. For weak shocks, however, the entropy jump given by the Rankine-Hugoniot condition is of third order in the strength of the shock, so the difference is quite small, and the economies afforded by the potential theory make computations based upon this approximation attractive for many transonic problems.

The nonlinear difference equations resulting from a discrete approximation to the potential equation generally are solved using an iterative, or relaxation, technique. The equations are linearized by computing approximations to all but the highest derivatives from the preceding solution in an iterative sequence, and a correction is computed at each mesh point in such a way that the approximate equations are satisfied identically. It frequently is useful in developing these iterative techniques to think of the iterative process as a discrete approximation to a continuous, time-dependent process. Thus, the iterative process approximates an equation of the form

$$\beta_0 \frac{\partial \Phi}{\partial t} + \beta_1 \frac{\partial^2 \Phi}{\partial \xi \partial t} + \beta_2 \frac{\partial^2 \Phi}{\partial \eta \partial t} = \frac{a^2}{\rho}\left[\frac{\partial(\rho h U)}{\partial \xi} + \frac{\partial(\rho h V)}{\partial \eta}\right].$$
(24)

The parameters β_0, β_1, and β_2 can then be chosen to ensure that the time-dependent process converges to a steady state in both subsonic and supersonic regions.

Even with the values of the parameters chosen to provide stable convergence to the steady state solution, many hundreds of iterations may be necessary for the iteration to converge to within an acceptable tolerance of the exact solution to the difference equations, especially when the grid is very fine. This slow convergence on fine grids is a characteristic of all iterative schemes and is a result of the fact that the difference equations are highly local and the elimination of the global low wave number component of the error naturally requires many sweeps through the field. A powerful technique for circumventing this difficulty with the iterative solution of partial differential equations has been developed and applied with great success to aerodynamic problems. The technique is known as the multigrid method and uses the fact that after a few sweeps of any good iterative technique the error remaining in the solution is quite smooth and can be represented accurately on a coarser grid. Application of the same iterative technique on this coarser grid soon makes the error smooth on this grid as well, and the process of coarsening the grid can be repeated again and again until the grid contains only a few cells in each coordinate direction. The corrections that have been computed at all grid levels are then added back into the solution on the fine grid, and the entire multigrid cycle is repeated. The accuracy of the converged solution on the fine grid is determined by the accuracy of the approximation on that grid, since the coarser levels are used only to effect a more rapid convergence of the iterative process. In theory, for a broad class of problems the work (per mesh cell) to solve the equations in this way is independent of the number of mesh cells, which is the best order estimate one can hope for. In many practical calculations, as few as 10 multigrid cycles may be required.

One of the earliest successes of computational aerodynamics has been the design of so-called *shock-free* configurations for flight in the transonic regime. The transonic drag rise associated with the appearance of shock waves near the wings of aircraft cruising at high subsonic Mach numbers limits the speeds at which subsonic transport aircraft can fly economically. Numerical methods capable of analyzing transonic flows have led to the ability to design airfoils and wings capable of supporting large regions of locally supersonic flow, but maintaining smooth recompression of the flow back to subsonic velocities. Figure 3 shows the surface pressure distribution of one such airfoil (designed by David Korn at the Courant Institute of Mathematical Sciences of New York University) at its design condition of zero angle of attack and 0.75 Mach number. Since the airfoil is cambered (i.e., its upper and lower surfaces are not symmetrical), it produces lift even at zero angle of attack, and since there is an appreciable acceleration of the flow over the upper surface to produce this lift, there is a sizeable pocket of supersonic flow in the vicinity of the upper surface. Figure 3 shows the pressure coefficient, which is a nondimensional measure of the amount by which the pressure differs from its value if the free stream, on the upper and lower surfaces of the airfoil as a function of distance along the airfoil surface. Aerodynamicists conventionally plot negative values of pressure coefficient up, so higher values correspond to lower pressures (and, correspondingly, higher Mach numbers). For comparison, results of computations using both the potential and Euler equations are shown; for this essentially shock-free flow, the two solutions are virtually identical.

Finally, although shock-free flows are achievable, it is important to realize that they usually are quite sensitive to small changes in body geometry, as well as to changes in flight Mach number and/or angle of attack. To illustrate this (and also to show some pressure distributions containing shock waves), the surface pressure distributions are plotted for the same airfoil at Mach numbers slightly below and above the design value in Figs. 4a and 4b. At the lower Mach number, the supersonic pocket breaks into two smaller zones, each terminated by a weak shock wave; at the higher Mach number, the supersonic zone grows significantly larger and is terminated by a single (much stronger) shock. Note the larger differences between the potential and Euler solutions as the shock wave becomes stronger.

While results have been shown here only for two-dimensional computations, transonic potential calculations have become common for three-dimensional wings and wing-fuselage combinations in both industry and government research laboratories, including those of NASA. These methods were heavily used in the design of the Boeing 757 and 767 aircraft and in preliminary design work for the Boeing 777.

3. Euler Equation Methods

As noted earlier, the Euler equations of inviscid, compressible flow constitute a hyperbolic system of partial

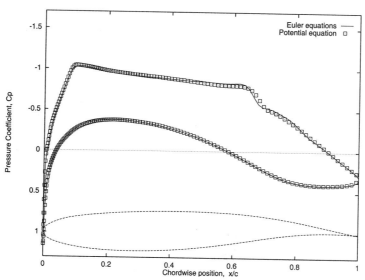

FIGURE 3 Surface pressure distribution on a shock-free airfoil at its design Mach number $M_\infty = 0.75$. Results of both the nonlinear potential and the Euler equations are shown.

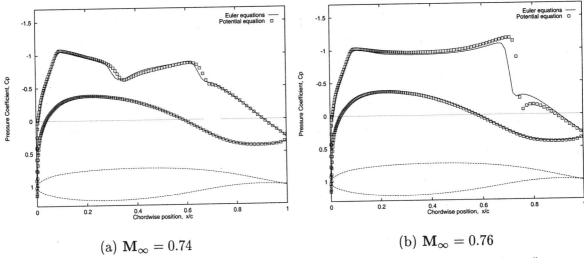

(a) $\mathbf{M}_\infty = 0.74$

(b) $\mathbf{M}_\infty = 0.76$

FIGURE 4 Surface pressure distributions for a shock-free airfoil at off-design conditions. Results of both the nonlinear potential and the Euler equations are shown.

differential equations, and numerical methods to solve them for aerodynamic applications rely heavily on the rather well-developed mathematical theory of such systems. As in the case of the nonlinear potential theory, discontinuous solutions (corresponding to shock waves) play an important role. Most methods are also of the shock-capturing type, so it is important to maintain the conservation form of the equations to ensure that the shock strengths are properly determined. In fact, the ideas of shock capturing and the importance of conservation form were developed first within the framework of the mathematical theory of hyperbolic systems of conservation laws. Hyperbolic systems describe the evolution in time of physical systems undergoing some unsteady process. Thus, if the state of the system is known at some initial instant, the hyperbolic system will determine its state at any subsequent time. This feature is frequently used even in calculations for which the solution is steady (i.e., independent of time). In this case, it is the asymptotic solution for large time that is of interest, so the equations are solved for large enough values of the time that the steady state is closely approximated. To maintain the hyperbolic character of the solution, and to keep the numerical method consistent with the physics it is trying to predict, it is necessary to determine the solution sequentially at a number of intermediate time levels between the initial and final (asymptotic) states; such a sequential process is said to be a time marching of the solution.

The simplest hyperbolic partial differential equation is the one-dimensional equation for a single scalar variable

$$\frac{\partial u}{\partial t} + \frac{\partial f(u)}{\partial x} = 0, \tag{25}$$

where $a = \partial f/\partial u$ is real. Although it is important that the numerical implementation be based on this conservation form of the equation, many questions about the behavior of the equation are more easily studied using the quasilinear form

$$\frac{\partial u}{\partial t} + a\frac{\partial u}{\partial x} = 0, \tag{26}$$

where a is assumed constant for purposes of analysis. Most methods for analyzing the behavior of numerical algorithms apply only to equations that are linear, but they provide useful information because stability of the linear equation is a necessary condition for stability of the nonlinear system.

Explicit methods. The simplest practical methods for solving hyperbolic systems are *explicit* in time. That is, the value of the solution is computed at each point at a particular level of time using only the values of the solution at the preceding time level. This is in contrast to *implicit* methods, which may also use neighboring values of the solution at the new time level and thus require the solution of an algebraic system of equations at each time level. This need to solve algebraic systems (which often are similar to the equations resulting from a discretization of an elliptic problem, such as steady, subsonic potential flow) makes implicit methods more costly in terms of computer resources per time step than explicit methods. The size of the time step that can be used in an explicit method, however, is limited by a constraint known at the Courant-Friedrichs-Lewy (or CFL) condition. If Δx is the mesh spacing, then for Eq. (26) the maximum allowable time step Δt for most common explicit schemes is determined by the constraint that the Courant number

$$C = \frac{a \Delta t}{\Delta x} \leq 1. \tag{27}$$

Broadly interpreted, the CFL condition states that the time step must be no greater than the time required for a signal to propagate across a mesh cell. Thus, if the mesh is very fine, the allowable time step also must be very small, and many time steps must be taken to reach the desired asymptotic steady state.

The most widely used explicit methods for hyperbolic systems are of the Lax-Wendroff type. The form of the difference corresponding to one of these methods will be given later in Eqs. (29a) and (29b). At this point, a physical interpretation of their characteristics will be provided by describing the modified equation corresponding to the method when applied to Eq. (26). The modified equation is the differential equation corresponding to the numerical approximation when the leading error terms, proportional to various powers of the mesh spacing, introduced by the approximation are included. For Lax-Wendroff methods applied to Eq. (26), the modified equation is

$$\frac{\partial u}{\partial t} + a \frac{\partial u}{\partial x} = a \frac{\Delta x^2}{6} (C^2 - 1) \left[\frac{\partial^3 u}{\partial x^3} + \frac{3 \Delta x}{4} \frac{\partial^4 u}{\partial x^4} \right]. \tag{28}$$

It can easily be shown that derivatives of even order appearing on the right-hand side of Eq. (28) cause changes in the amplitude of the solution, while derivatives of odd order cause changes in the speed at which the solution propagates through space. Since there are no higher order derivatives in the original equation, these terms represent errors. The former are called dissipative (or amplitude) errors, while the latter are called dispersive (or phase) errors. Note that, although there are no dissipative or dispersive terms in the Euler equations, the viscous terms in the Navier-Stokes equations are dissipative in nature. By analogy, the introduction of artificial (or numerical) dissipative terms into a numerical approximation is frequently described as the addition of an *artificial viscosity*.

It is clear from the form of the coefficients in Eq. (28) that both types of errors for the Lax-Wendroff scheme can be made arbitrarily small by reducing the size of the mesh spacing Δx, but that for any finite mesh spacing the errors will remain. In fact, some dissipative error is helpful in preventing unwanted oscillations in the solution (caused by the round-off error introduced by the finite precision arithmetic of any practical computer) from growing larger. In particular, a negative coefficient of the fourth-derivative term will result in exponential damping of these perturbations. This is ensured by requiring that the Courant number C remain less than unity. This is an alternative interpretation of the CFL condition.

The simplest and most efficient form of the Lax-Wendroff scheme, and one that has been widely applied to aerodynamic problems, is MacCormack's method. If the values at uniformly spaced mesh points i at time level $n \Delta t$ are denoted as u_i^n, MacCormack's method for Eq. (25) can be written as a two-step sequence in which provisional values $\bar{u}_i^{n+1}$ are computed according to

$$\bar{u}_i^{n+1} = u_i^n - \frac{\Delta t}{\Delta x} \left(f_i^n - f_{i-1}^n \right) \tag{29a}$$

and then corrected according to

$$u_i^{n+1} = \frac{1}{2} \left[u_i^n + \bar{u}_i^{n+1} - \frac{\Delta t}{\Delta x} \left(\bar{f}_{i+1}^{n+1} - \bar{f}_i^{n+1} \right) \right]. \tag{29b}$$

It is easily verified that this is a consistent approximation to the modified Eq. (28). For flows with shock waves, the very small dissipation provided by the term that is third order in the mesh spacing must be augmented by additional artificial viscosity terms in practical calculations.

A popular alternative approach is to treat the spatial and temporal discretizations separately. If the spatial derivative appearing in Eq. (26) is approximated using finite differences, but the time variable is left continuous, there results (for a simple centered spatial difference)

$$\frac{du_i}{dt} + a \frac{u_{i+1} - u_{i+1}}{2 \Delta x} = 0. \tag{30}$$

This is a system of ordinary differential equations relating the change in the solution at each point to the values of its neighbors. The well-developed science of numerical methods for solving systems of ordinary differential equations can now be brought to bear on this problem. In particular, conventional multistage Runge-Kutta schemes are well suited to solve the equations arising from this approach for hyperbolic problems (i.e., problems having little or no dissipation). In addition, this approach is attractive for use in conjunction with multigrid methods, as the dissipative characteristics of the time-stepping algorithm can be tailored so that it is an effective smoother of high wave number errors.

Of course, the Euler equations are a system of equations, not a single scalar equation. But, the ideas of this section can easily be extended to systems, as can the implicit methods to be described in Section III.B.3.

Implicit methods. The time-step limitation imposed by the CFL condition often is overly restrictive, especially when only the asymptotic steady-state solution is of interest. In this case, it is desirable to take very large time steps so that the asymptotic steady state is reached with a minimum of computational effort. Time steps larger than those allowed by the condition $CFL \leq 1$ are possible only with the use of implicit methods, but at the cost of having to solve systems of algebraic equations at each time step. For a particular problem, then, the implicit method will be more economical only if the advantage of the larger time step outweighs the increased cost per time step of the method.

Using the same nomenclature as for MacCormack's scheme, a general implicit method using data at two time levels for a system of equations of the form of Eq. (25) can be written as

$$\mathbf{u}_i^{n+1} = \mathbf{u}_i^n - \Delta t \frac{\theta \delta_x \mathbf{f}_i^{n+1} + (1 - \theta) \delta_x \mathbf{f}_i^n}{2 \Delta x}, \qquad (31)$$

where

$$\delta_x \mathbf{f}_i = \mathbf{f}_{i+1} - \mathbf{f}_{i-1}$$

is a central difference operator, and θ is a parameter that varies from 0 for an explicit scheme to 1 for a fully implicit one. For the Euler equations, $\mathbf{f}$ is a nonlinear function of $\mathbf{u}$, and iteration would normally be necessary to solve Eq. (31) since it is necessary to know $\mathbf{u}^{n+1}$ in order to determine $\mathbf{f}^{n+1}$. This could be very costly in terms of computer time and usually is avoided by linearizing the flux vector about the previous time level to give

$$\mathbf{f}_i^{n+1} = \mathbf{f}_i^n + \{\mathbf{a}\} \Delta \mathbf{u}_i^n + O(\Delta t^2), \qquad (32)$$

where $\{\mathbf{a}\}$ is now a square matrix, called the Jacobian of the flux vector $\mathbf{f}$ with respect to the solution $\mathbf{u}$, and

$$\Delta \mathbf{u}_i^n = \mathbf{u}_i^{n+1} - \mathbf{u}_i^n$$

is the change in the solution from the nth to the $n + 1$st time level. This linearization allows the scheme of Eq. (31) to be approximated as

$$[\mathbf{I} + \theta \{\mathbf{a}\}] \Delta \mathbf{u}_i^n + \delta_x \mathbf{f}_i^n = 0, \qquad (33)$$

which is a system of linear equations that can be solved for the changes in the solution at each point. Solution of Eq. (33) requires inversion of a system of equations that is block tridiagonal (i.e., the basic structure of the matrix corresponding to the implicit operator is tridiagonal, but the elements are square blocks of dimension equal to the number of equations in the system of differential equations). The blocks are 3×3, 4×4, or 5×5 for the one-, two-, and three-dimensional Euler equations, respectively.

A similar procedure can be followed to construct implicit schemes for multidimensional problems. In this case, however, the expense of solving the equations is greatly increased by the fact that the bandwidth of the equations is much larger since the change in the solution in each cell is related to those at its neighbors in each of the coordinate directions. The problem still can be made computationally tractable, however, by further approximating the implicit operator as the product of two or more factors, each of which has a much smaller bandwidth. The factors can be chosen to be upper and lower triangular matrices (in which case the scheme is called an approximate LU

factorization) or so that each of the factors contains differences in only one coordinate direction [in which case the method is called an alternating direction implicit (ADI) method]. For the LU methods, solution of the equations is achieved by marching through the field for each factor, sequentially inverting the blocks. For the ADI schemes, a block tridiagonal system of equations must be solved for each factor along each line of the mesh.

Flux computations. The above sections have represented the fluxes (i.e., the approximations to the spatial derivatives) in terms of symmetric, or central, difference approximations, and this approach, with carefully tailored artificial dissipative terms, has been very successful. Great strides have also been made using asymmetric, or upwind, differencing techniques as well. It can be argued that these upwind methods are more consistent with the physics and mathematics of the equations. Each of the eigenvalues of the Jacobian matrix $\{\mathbf{a}\}$ in Eq. (32) can be identified with the propagation of a wave, and the sign of the eigenvalue represents the direction of propagation of the wave, so the "upwind" direction for the various characteristic variables can be different, at least when the flow is subsonic. In this case, the flux vector must be "split" into two parts, each representing waves that propagate in one of the two directions, and each component is differenced accordingly. Higher order versions of a method, originally introduced by Godunov, also have been developed, in which a discontinuity is allowed between the two states separated by the mesh surface; such methods are particularly effective at capturing shock waves with no overshoot and a minimum of smearing.

Practical computations. Methods based on the solution of the Euler equations have been applied to very complex three-dimensional problems, including complete aircraft. These methods are widely used in the preliminary design stages for transport aircraft in the cruise condition, where very little flow separation is likely to occur.

4. Navier-Stokes Equation Methods

Laminar flows. As described in Section II, the relative importance of viscous effects is characterized by the value of the Reynolds number. For the large values of aerodynamic interest, the Reynolds number characterizes the thickness of the boundary layers when the flow remains attached and also determines whether the flow in these layers remains laminar or becomes turbulent. If the Reynolds number is not too large (typically, less than 10^6, or so), the flow in the boundary layer remains smooth, and adjacent layers (or laminae) of fluid slide smoothly past one another. When this is the case, solution of the

Navier-Stokes equations is not too much more difficult, in principle, than solution of the Euler equations of inviscid flow. Greater resolution is required for the large gradients in the boundary layers, so many more mesh cells may be necessary to achieve adequate accuracy. In most of the flow field, however, the flow behaves as if it were nearly inviscid, and the methods developed for the Euler equations are appropriate. The equations must, of course, be properly modified to include the additional terms resulting from the viscous stresses, and care must be taken to ensure that any artificial dissipation added is sufficiently small relative to the physical dissipation. The solution of the Navier-Stokes equations for laminar flows, then, is somewhat more costly in terms of computer resources, but not significantly more difficult from an algorithmic point of view than solution of the Euler equations. Unfortunately, most aerodynamic flows of practical interest occur at such large Reynolds numbers that the flow in the boundary layers becomes turbulent.

Turbulence models. When the Reynolds number exceeds a critical (transition) value, then the flow in the boundary layer develops unsteady fluctuations and the flow in the boundary layer is said to become turbulent. The flow properties in turbulent boundary layers, such as the time-averaged values of velocity and shear stresses, are dominated by the effects of the fluctuating velocities and pressures. While these fluctuations still are described by the Navier-Stokes equations, the range of length and time scales over which significant fluctuations occur is so large that to resolve them all would require many orders of magnitude more storage and computer time than will be available for the foreseeable future. (Several estimates for these requirements were presented in Section II.)

The only alternative for practical calculations of turbulent flows at the present time is to make additional approximations and to introduce models for at least some features of the turbulence. The greatest reduction in computational effort is achieved by modeling all aspects of the turbulence. This usually is done by introducing the RANS equations. This approach is based upon the assumption that the effects of the turbulence can be described statistically. The solution is decomposed into time-averaged and fluctuating components; for example, the velocity components might be given by

$$u = U + u' \quad \text{and} \quad v = V + v', \quad (34)$$

where U and V are the average values of u and v, taken over a time interval that is long compared to the turbulence time scales but short compared to the time scales of any nonturbulent unsteadiness in the flow field. If $\langle u \rangle$ denotes such a time average of the u component of velocity, then, e.g.,

$$\langle u \rangle = U.$$

If a decomposition of the form of Eq. (34) for each of the flow variables is substituted into the Navier-Stokes equations, and they are then averaged as described above, the resulting equations describe the evolution of the mean flow variables (such as U, V, etc.). These equations are nearly identical to the original Navier-Stokes equations written for the mean flow variables because terms proportional to $\langle u' \rangle$, $\langle v' \rangle$, etc. are identically zero. Because of the nonlinearity of the equations, however, several terms that involve the fluctuating variables remain. In particular, terms proportional to $\langle \rho u'v' \rangle$ appear. Dimensionally, this term is equivalent to a stress, and it is, in fact, called a Reynolds stress term. Physically, the Reynolds stresses are the turbulent counterpart of the (viscous) molecular stresses, and they appear as a result of the transport of fluid momentum by the turbulent fluctuations. The implication of the appearance of such terms in the equations describing the mean flow is that the mean flow cannot be determined without some knowledge of the effects of these fluctuating components.

The Reynolds stresses must be related to the mean flow at some level of approximation using a phenomenological model. In other words, it is necessary to *model* the effects of the turbulence in order to compute the mean flow. The simplest procedure is to relate the Reynolds stresses to the mean flow properties directly, but since the turbulence that is responsible for these stresses is a function not only of the local mean flow properties but of the flow history as well, such an approximation cannot have broad generality. A more general procedure is to develop differential equations for the Reynolds stresses themselves. This can be done by taking higher order moments of the Navier-Stokes equations. For example, if the x-momentum equation is multiplied by v and then averaged, the result will be an equation describing the evolution of the Reynolds stress $\langle \rho u'v' \rangle$. Again, because of the nonlinearity of the equations, however, yet higher moments of the fluctuating variables (e.g., terms proportional to $\langle \rho u'^2 v' \rangle$ and $\langle \rho u'v'^2 \rangle$) will appear in this equation. This is an example of the problem of *closure* of the method, represented by the fact that these third-order correlations of the fluctuating variables must be known in order to solve the equations for the Reynolds stresses. It is hoped that these higher order correlations can be modeled in a more universal fashion, but there has been no clear demonstration to date that this is the case.

A variety of turbulence models has been developed to deal with this inability to close the equation set. The simplest models, based on the original mixing length hypothesis of Prandtl, relate the Reynolds stresses to the local properties of the mean flow. Such *algebraic* models are unable to account for the history of the developing turbulence, however, and more complete models include additional partial differential equations for various properties of the turbulence. The most commonly used such models use additional equations for the turbulence kinetic energy k and the dissipation rate ε of the turbulence (or some other quantity, such as a turbulence frequency ω that determines a time scale).

Finally, even using turbulence models within the context of RANS, it is beyond the capabilities of current computers to resolve all the viscous terms in the equations for most practical problems. Since it is the gradients normal to the surface that are most important (at least for nonseparated flows), the grid spacing usually is adjusted so that only the gradients in that direction are resolved. This leads naturally to a further simplification of the equations in which the viscous terms corresponding to derivatives along the body surface are neglected altogether. Since the neglect of these terms changes the nature of the equations, in a manner similar to the boundary layer approximation, the resulting equations are called the thin-layer (or parabolized) Navier-Stokes equations.

Large-eddy simulations. The difficulty of developing generally applicable phenomenological models on the one hand, and the enormous computational resources required to resolve all the scales of the turbulent motions on the other, have led to the development of large-eddy simulation (LES) techniques. In this approach, the largest length and time scales—those that are most likely to be problem dependent—are fully resolved, but the smaller (subgrid) scales are modeled. A filtering technique is applied to the Navier-Stokes equations which results in equations having a form that is similar to the original equations, but with additional terms representing the effects of a subgrid-scale tensor on the larger scales. LES techniques date back to the pioneering work of Smagorinsky, who developed an eddy-viscosity subgrid model for use in meteorological problems. The model turned out to be too dissipative for large-scale meteorological problems in which large-scale, predominantly two-dimensional motions are affected by three-dimensional turbulence, but generalizations of Smagorinsky's model are beginning to find application in engineering problems.

Practical computations. Design computations involving the Navier-Stokes equations commonly use algebraic models for the turbulence properties, but it is an active area of research to couple the partial differential equation models to the RANS equations efficiently. The task is made difficult by the additional stiffness of the turbulence equations. The development of LES and DNS techniques remain an active research topic.

IV. CONCLUDING REMARKS

This article has described some of the ideas underlying the development of computational aerodynamics up to the present time. The emphasis has been on methods useful for determining the forces acting upon flight vehicles under relatively conventional conditions. In this situation, with the exception of the prediction of turbulent fluctuations and their effects, the physics of the processes to be predicted are well understood, and it is primarily a problem of solving known equations for practical geometries with sufficient accuracy. Great progress has been made in recent years, especially for problems for which inviscid flow models are adequate. At the present time, numerical techniques for solving better approximations to the Navier-Stokes equations are also under intense development and are beginning to see design use. These include inviscid models that allow for rotational effects (such as strong shock waves and free vortices) and the RANS equations, using a variety of turbulence models. The practical use of these methods will expand with the further development of more efficient algorithms and the continued availability of faster and more powerful computers.

Although the emphasis of this article has been on the development of engineering tools, numerical solutions of fluid mechanical problems also hold the promise of supplying fundamental knowledge in such areas as the nature and behavior of turbulence itself. Although current computers are only marginally up to the task, it should be possible in the not too distant future to compute the turbulent fluctuations themselves, at least for relatively simple flows. In fact, this is an area of intense activity, with the results of these fluid mechanics "experiments" providing detailed information under more carefully controlled situations than any physical laboratory experiment can hope to achieve, although at more limited values of the Reynolds number. The insight provided by the results of such calculations should be of great value in the development of more realistic models of turbulence for use in computations of practical interest.

SEE ALSO THE FOLLOWING ARTICLES

AIRCRAFT AERODYNAMIC BOUNDARY LAYERS • AIRCRAFT PERFORMANCE AND DESIGN • FLIGHT (AERODYNAMICS) • FLUID DYNAMICS • LINEAR SYSTEMS OF EQUATIONS • STRUCTURAL ANALYSIS, AEROSPACE

BIBLIOGRAPHY

Agarwal, R. K. (1999). "Computational fluid dynamics of whole-body aircraft," *Ann. Rev. Fluid Mech.* **31**, 125–169.

Anderson, D. A., Tannehill, J. C., and Pletcher, R. H. (1997). "Computational Fluid Mechanics and Heat Transfer," 2nd ed., Taylor & Francis, Washington, DC.

Caughey, D. A., and Hafez, M. M., eds. (1994). "Frontiers of Computational Fluid Dynamics—1994," Wiley, Chichester.

Caughey, D. A., and Hafez, M. M., eds. (1998). "Frontiers of Computational Fluid Dynamics—1998," World Scientific, Singapore.

Chapman, D. R. (1979). "Computational aerodynamics: Review and outlook," *AIAA J.* **17,** 1293–1313.

Hisrch, C. (1988). "Numerical Computation of Internal and External Flows," Vol. 1 and 2, Wiley, New York.

Jameson, A. (1989). "Computational aerodynamics for aircraft design," *Science* **245,** 361–371.

Laney, C. B. (1998). "Computational Gasdynamics," Cambridge Univ. Press, Cambridge.

Moin, P., and Kim, J. (1997). "Tackling turbulence with supercomputers," *Sci. Am.* **January**.

Moin, P., and Mahesh, K. (1998). "Direct numerical simulation: A tool in turbulence research," *Annu. Rev. Fluid Mech.* **30,** 539–578.

Roe, P. (1986). "Characteristic-based schemes for the Euler equations," *Ann. Rev. Fluid Mech.* **18,** 337–365.

Computational Chemistry

Matthias Hofmann

Ruprecht-Karls-Universität

Henry F. Schaefer III

Center for Computational Quantum Chemistry

GLOSSARY

ab initio Latin term meaning *from the beginning*. In the context of computational chemistry, computations from first principle without any empirical input except fundamental physical constants.

Density functional theory (DFT) Theory based on the electron density as the crucial property rather than the wave function in traditional ab initio methods.

Force field A set of equations describing the potential energy surface of a chemical system.

Hamiltonian operator An operator that describes the kinetic and potential energy of a system treated by wave mechanics.

Molecular mechanics Theoretical treatment of molecules by a force field based on classical mechanics and electrostatics.

Molecular modeling Branch of computational chemistry concerned with computer-aided molecular design.

Orbital Function to describe a single electron. Molecular orbitals (MOs) build the total wave function of a system and are expanded in terms of atomic orbitals, AOs (basis functions). Orbitals can be occupied or virtual.

Quantum mechanics Mathematical treatment based on the wavelike nature of small particles.

Schrödinger equation A differential equation for the quantum-mechanical treatment of a system.

Self-consistent field (SCF) method Method used to solve mathematical equations which depend on their own solution.

Semiempirical Making use of experimental results to derive parameters for approximations made in quantum-mechanical methods.

COMPUTATIONAL CHEMISTRY is the scientific discipline of applying computers to gain chemical information. It is the link between theoretical and experimental chemistry. Theoretical chemistry is mainly concerned with the development of mathematical models which allow one to derive chemical properties from calculations and to interpret experimental observations. The mathematical models developed in theoretical chemistry are usually validated by comparison with experiment. Theoretical chemistry existed before the arrival of electronic computers. Computational chemistry, however, relies heavily on powerful microelectronics to cope with huge computational tasks. It focuses on the application of theoretical methods which require calculational treatments which are by far too large to be done without fast computers.

Of course, the strict separation of theoretical, computational, and experimental chemistry is of an academic nature. In practice, theoreticians often not only develop a new method but also need to design more efficient algorithms to make the method applicable. Before computational results can be interpreted, computational chemists need to undertake benchmark studies to determine the limitations of a method. Without the knowledge about the accuracy of the applied mathematical model, any computational study is without scientific significance. Likewise, many experimentalists use computer programs to support or complement their experimental studies.

I. HISTORY OF COMPUTATIONAL CHEMISTRY

As early as 1929, only 3 years after Schrödinger's formulation of the fundamental equation that bears his name, Dirac stated correctly,

The underlying physical laws necessary for the mathematical theory of a large part of physics and the whole of chemistry are thus completely known, and the difficulty is only that the exact application of these laws leads to equations much too complicated to be soluble.

Hence, the further development of quantum chemistry was aiming at approximate solutions of the Schrödinger equation by simplifying the required mathematical treatment.

In the 1930s the basics for a wide range of computational methods based on quantum mechanics were laid by the development of the Hartree-Fock method. In 1951 Roothaan, for the first time, considered molecular orbitals as a linear combination of analytic atomic one-electron functions, shifting the mathematical task from the numerical solution of coupled differential equations to the evaluation of integrals over basis functions. Introduction of approximations for the most difficult integrals through the use of suitable parameters led to the development of semiempirical methods beginning in the 1950s. The more rigorous ab initio methods benefited from the use of Gaussian- instead of Slater-type basis functions, as pointed out by Boys in 1950 but generally accepted only two decades later. Configuration interaction (CI) was the first theoretical level used to include electron correlation and was widely applied during the 1970s. In the late 1970s many-body perturbation theory (Møller-Plesset methods), and during the 1980s coupled cluster methods, became more popular because they are more economical and more rapidly convergent, respectively, than CI. The 1990s can be considered the decade of density functional theory, which by that time had become so-phisticated enough to be useful for applications in chemistry. Molecular mechanics emerged in the mid-1960s and has become more sophisticated and more useful with time.

Due to the number of various approximations, early computations performed to try to reproduce experimental findings yielded varying degrees of success. Computational chemistry could become a recognized scientific discipline only after a real predictive power was established. Perhaps the first case in which theory proved to be accurate enough to challenge experiment was the structure determination of methylene (CH_2). From a spectroscopic investigation the ground state of this molecule was first concluded to be linear. However, this interpretation had to be revised after reliable computations predicted a significantly bent structure in 1970.

The development of different methods and their efficient implementation is only one reason for the success of computational chemistry. Another factor is the dramatic development of computer technology (i.e., computational speed as well as the amount of core memory and of disk storage). Today's personal desktop computers provide many times the computer power of early "supercomputers" at a fraction of the price. The combined development of both software and hardware allowed computational chemistry to become for chemical research an indispensable tool which allows one to plan experiments more carefully and hence to optimize the use of laboratory resources. The importance of computational chemistry was honored when the 1998 Nobel Prize in Chemistry was awarded to two pioneers of the field, J. A. Pople and W. Kohn.

II. METHODS USED IN COMPUTATIONAL CHEMISTRY

The methods used in computational chemistry can be classified according to the sophistication of the underlying model (Fig. 1). Molecular mechanics methods are based on classical mechanics and are computationally the fastest. Semiempirical methods are based on a wave function description, in which some integrals are approximated by means of parameters and many others are neglected to reduce the computational cost. Ab initio methods use only fundamental physical constants but no further experimental results; Hartree-Fock (H F) theory is the starting level, which can be improved upon by accounting for electron correlation in various ways. Density functional methods are also quantum mechanical but are based on the electron density to describe chemical systems. They are often considered "ab initio" although some empirical parameters enter the energy functionals.

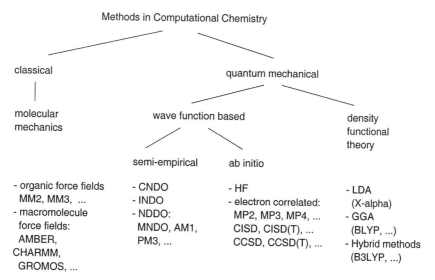

Methods in Computational Chemistry

classical — molecular mechanics

quantum mechanical — wave function based — semi-empirical / ab initio; density functional theory

- organic force fields
  MM2, MM3, ...
- macromolecule
  force fields:
  AMBER,
  CHARMM,
  GROMOS, ...

- CNDO
- INDO
- NDDO:
  MNDO, AM1,
  PM3, ...

- HF
- electron correlated:
  MP2, MP3, MP4, ...
  CISD, CISD(T), ...
  CCSD, CCSD(T), ...

- LDA
  (X-alpha)
- GGA
  (BLYP, ...)
- Hybrid methods
  (B3LYP, ...)

FIGURE 1 Classification of computational chemistry methods. AMBER, assisted model building with energy refinement; CHARMM, chemistry at Harvard molecular mechanics; GROMOS, Groningen molecular simulation; CNDO, complete neglect of differential overlap; INDO, intermediate neglect of differential overlap; NDDO, neglect of diatomic differential overlap; MNDO, modified neglect of differential overlap; AM1, Austin model 1; PM3, parametric method number 3; HF, Hartree-Fock; MP2, Møller-Plesset, second order; MP3, Møller-Plesset, third order; MP4, Møller-Plesset, fourth order; CISD, configuration interaction singly and doubly excited; CISD(T), configuration interaction singly, doubly, and triply (estimated) excited; CCSD, coupled cluster singly and doubly excited; CCSD(T), coupled cluster singly, doubly, and triply (estimated) excited; LDA, local density approximation; GGA, generalized gradient approximation; BLYP, Becke/Lee, Yang, and Parr; B3LYP, Becke three-parameter/Lee, Yang, and Parr.

A. Force Field Methods

A force field (FF) is a set of equations describing the potential energy surface of a chemical system. A molecular mechanics (MM) method uses a force field based on a classical mechanical representation of molecular forces to calculate static properties of a molecule (e.g., structure and energy of an energy minimum structure). Molecular dynamics (MD) also implements a force field but generates dynamic properties (e.g., evolution of an structure in time) by calculating forces and velocities of atoms. In MM methods atoms are treated as "balls" of different masses and sizes, and bonds are "springs" connecting the balls without an explicit treatment of electrons. The main advantage of this simple classical approach is the small computational cost, which allows one to treat very large molecules. FFs are typically constructed to yield experimentally accurate structures and relative energies. Some FFs are generated to accurately compute other properties such as vibrational spectra.

The observation that properties of chemical functional groups are normally transferable from one compound to another validates the MM approach. The most basic component in a FF is the atom type and one element usually contributes several atom types. Each bond is characterized by the atom types involved and has a "natural" bond length since the variation with the chemical environment is relatively small. Similarly, bond angles between atom types have typical values. The energy absorptions in in-

frared (IR) spectroscopy associated with a certain bond stretch or angle deformation also fall in narrow ranges, which demonstrates that the variation of force constants is also relatively small. The existence of an increment system for heats of formation, for example, shows that the energy behaves additively as well. Hence, in MM the energy is expressed classically as a function of geometric parameters.

1. Energy Terms

Advanced force fields distinguish several atom types for each element (depending on hybridization and neighboring atoms) and introduce various energy contributions to the total force field energy, E_{FF}:

$$E_{FF} = E_{str} + E_{bend} + E_{tors} + E_{vdW} + E_{elst} + \cdots,$$

where E_{str} and E_{bend} are energy terms due to bond stretching and angle bending, respectively; E_{tors} depends on torsional angles describing rotation about bonds; and E_{vdW} and E_{elst} describe (nonbonded) van der Waals and electrostatic interactions, respectively (Fig. 2). In addition to these basic terms common to all empirical force fields there may be extra terms to improve the performance for specific tasks. Each term is a function of the nuclear coordinates and a number of parameters. Once the parameters have been defined, the total energy, E_{FF}, can be computed and subsequently minimized with respect to the coordinates.

Bonded

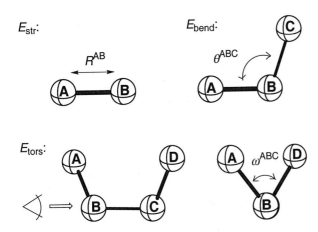

Non-Bonded

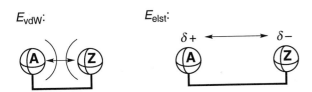

FIGURE 2 Most basic energy terms included in empirical force field (FF) methods.

$a.$ *Stretch energy.* The harmonic approximation gives the stretch energy of a bond between atom types A and B, E_{str}^{AB}, as

$$E_{str}^{AB}(\Delta R) = k^{AB}\Delta R^2,$$

where k^{AB} is the force constant and $\Delta R = R_0^{AB} - R^{AB}$ is the bond length deviation from the natural value, R_0^{AB}, for which E_{str}^{AB} is defined to be zero. Further improvement can be achieved by including higher anharmonic terms to the equation. While these expressions describe the potential well for R close to R_0, the energy goes to infinity for large distances (Fig. 3). In contrast, a morse potential allows the energy to approach the dissociation energy, D, as R increases:

$$E_{Morse}(\Delta R) = D(1 - e^{\sqrt{k/2D}\Delta R})^2,$$

but it is much more expensive in terms of computational cost.

$b.$ *Bending energy.* The harmonic approximation for the bending energy, E_{bend}^{ABC}, due to the deformation of the angle between the A–B and B–C bonds, is sufficient for most purposes:

$$E_{bend}^{ABC}(\Delta\theta) = k^{ABC}\Delta\theta^2$$

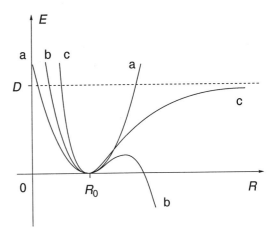

FIGURE 3 (Curve a) Harmonic potential, $E(R) = k(R_0 - R)^2$; (curve b) third-order polynomial anharmonic potential, $E(R) = k(R_0 - R)^2 + k'(R_0 - R)^3$; (curve c) Morse potential, $E(R) = D(1 - e^{\sqrt{k/2D}(R_0-R)})^2$.

with

$$\Delta\theta = \theta_0^{ABC} - \theta^{ABC}.$$

If higher accuracy is desired (e.g., for computing IR frequencies), a third-order term can be included with an anharmonicity constant set to be a fraction of k^{ABC}.

$c.$ *Torsion energy.* The torsional potential, due to the rotation of bonds A–B and C–D about bond B–C, is periodic in the torsional angle ω, which is defined as the angle between the projections of A–B and C–D onto a plane perpendicular to B–C. The torsional energy therefore is expressed as a Fourier series:

$$E_{tors}^{ABCD}(\omega) = \sum_n V_n \cos(n\omega),$$

which allows the representation of potentials with various minima and maxima (Fig. 4). Three terms are enough to model the most common torsional potentials.

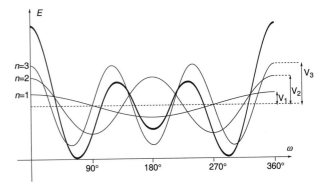

FIGURE 4 A three-minimum potential (bold line) represented as a three-term Fourier series: $E(\omega) = \sum_{n=1}^{3} V_n \cos(n\omega)$.

d. van der Waals energy. The van der Waals term, E_{vdW}, covers nonelectrostatic interactions between nonbonded atoms. The van der Waals energy is positive (repulsive) and very large at short distances, zero at large distances, but slightly negative (attractive) at moderate distances due to temporarily induced multipole attraction (dispersion force), the most important attractive contribution of which (dipole–dipole interaction) has an R^{-6} dependence. The Lennard-Jones potential for E_{vdW} includes a repulsive term, which is set proportional to $(R^{-6})^2$ to grow faster than R^{-6}:

$$E_{LJ}(R) = \varepsilon^{AB}\left[\left(\frac{R_0^{AB}}{R}\right)^{12} - 2\left(\frac{R_0^{AB}}{R}\right)^6\right].$$

ε^{AB} determines the energy depth of the minimum. Alternatively, a Buckingham or Hill potential can be used that employs an exponential function for the repulsive term.

For each atom type a van der Waals radius, R_0, and the atom softness, ε, have to be determined, from which the diatomic parameters are calculated according to

$$R_0^{AB} = R_0^A + R_0^B \quad \text{and} \quad \varepsilon^{AB} = \sqrt{\varepsilon^A \varepsilon^B}.$$

e. Electrostatic energy. The electrostatic energy, E_{elst}, is due to the electrostatic interactions arising from polarized electron distributions based on electronegativity differences. It can be modeled by Coulomb interactions of point charges associated with individual atoms:

$$E_{elst}(R^{AB}) = \frac{Q^A Q^B}{\varepsilon R^{AB}},$$

ε being a dielectric constant, which can be used to model the effect of the same or other molecules present (e.g., solvent). The atomic charges, Q, are commonly obtained by fitting to the electrostatic potential as calculated by an electronic structure method. An E_{elst} description based on dipole–dipole interactions between polarized bonds can alternatively be employed.

Hydrogen bonds are nonbonded interactions between a positively charged hydrogen atom and an electronegative atom with lone electron pairs (mostly oxygen or nitrogen) and can be adequately modeled by appropriately chosen atomic charges. Although a single hydrogen bond is a very weak interaction, the large number occurring in biomolecules (e.g., proteins) makes hydrogen bonding a very important factor.

In the large size limit, the bonded interactions increase linearly with the system size, but the nonbonded interactions show a quadratic dependence and determine the computational cost. The van der Waals interactions quickly fall off with the distance (R^{-6} dependence) and may be neglected for large separations. The electrostatic interaction (proportional to R^{-1}) is much more far reaching and

needs to be considered out to very long distances. Fast multipole methods (FMMs) can be applied to reduce the computational cost of evaluating E_{elst}.

f. Other energy contributions. So that the performance can be improved, force fields include further parameters to take care of special cases. For example, cross terms account for the interplay between different contributions (e.g., longer bonds for small angles). Correction terms may be introduced to describe substituent effects (e.g., anomeric effect). Additional terms may be introduced to adequately treat special cases like pyramidalization of sp^2 hybridized atoms. Hydrogen bonding may be treated explicitly (in addition to the electrostatic interaction) with a special set of van der Waals interaction parameters. Pseudo atoms maybe introduced to model lone pairs. In addition, atoms in unusual bonding situations (three-membered rings, molecules with linearly conjugated π-systems, aromatic compounds, etc.), which are not described adequately by the normal parameters, can be defined as new atom types.

The force field energy, E_{FF}, corresponds to the energy relative to a molecule with noninteracting fragments. Therefore, only energies for molecular structures built from the same fragments (conformers) can be compared directly. So that energy between different molecules (isomers) can be compared, the energy scale is converted to heats of formation by adding bond increments (estimated from bond dissociation energies minus the heat of formations of the atoms involved) and possibly group increments (e.g., methyl group):

$$\Delta H_f = E_{FF} + \overset{\text{bonds}}{\sum} \Delta H^{AB} + \overset{\text{groups}}{\sum} \Delta H^G.$$

2. Parametrization

Determining the parameters for a force field is a substantial task. In general, not all necessary data are available from (accurate) experiments. Modern electronic structure computations can provide unknown data relatively easily and with sufficient accuracy. Another problem is the large number of parameters: for a force field with N atom types, the number of different types of bonds, bond angles, and dihedral angles scales as N^2, N^3, and N^4, respectively, each requiring several parameters. So that the number of parameters can be reduced, the atom dependency can be reduced (e.g., the torsional parameters may be treated as dependent on the B–C central bond only and not on the atom types A and D). The parametrization effort can be reduced further by defining "generic" parameters to be used for less common bond types or when no reference data are available. This, of course, reduces the quality of a calculation. By deriving the di-, tri-, and tetra-atomic parameters from atomic data (atom radii, electronegativities, etc.), universal force fields (UFFs) allow one to include basically all elements. The performance, however, is relatively poor.

The kind of energy terms, their functional form, and how carefully (number, quality, and kind of reference data) the parameters were derived determine the quality of a force field. Accurate force fields exist for organic molecules (e.g., MM2, MM3), but more approximate force fields (e.g., with fixed bond distances) optimized for computational speed rather than accuracy [e.g., AMBER (assisted model building with energy refinement), CHARMM (chemistry at Harvard molecular mechanics), GROMOS (Groningen molecular simulation)] are the only practical choice for the treatment of large biomolecules. The type of molecular system to be studied determines the choice of the force field.

One limitation of force field methods is that they can describe only well-known effects that have been observed for a large number of molecules (this is necessary for the parametrization). The predictive power of these methods is limited to extrapolation or interpolation of known effects.

3. Quantum-Mechanical and Molecular-Mechanical (QM/MM) Method

Another limitation of MM is the inability to investigate reactions. While force field methods are capable of describing conformational changes, for which all bonds remain intact along the reaction coordinate, they are by construction not capable of treating reactions in which bonds are broken and/or formed. The classical model is not designed to describe the electronic rearrangement associated with bond breaking and bond formation. Such problems are better treated by electronic structure methods discussed below. For large systems, a combined quantum-mechanical and molecular-mechanical (QM/MM) method can be applied. In this approach the reactive part of the molecule to be studied is described by a quantum-mechanical (semiempirical, ab initio, or DFT) method while the rest of the system is treated by a force field. The problem with this approach is the "communication" between classical and quantum-mechanical potential (i.e., how the atoms close to the QM/MM border are treated).

The total energy, E_{tot}, may be computed as follows:

$$\mathbf{E}_{tot} = \mathbf{E}_{QM} + \mathbf{E}_{MM} + \mathbf{E}_{QM/MM},$$

where the quantum-mechanical contribution, $\mathbf{E}_{QM}$, and the molecular-mechanical contribution, $\mathbf{E}_{MM}$, are defined by a QM method and a MM method, respectively. The coupling term, $\mathbf{E}_{QM/MM}$, includes parameters that can be fitted to reproduce experimental results and are specific to the chosen combination of QM and MM methods.

Alternatively, the total energy, E_{tot}, may be extrapolated from QM and MM calculations on a small part and on the whole of a suitably partitioned system (IMOMM-integrated molecular orbital, molecular mechanics method)

$$\mathbf{E}_{tot} = \mathbf{E}_{QM} \text{ (small)} + \mathbf{E}_{MM} \text{ (whole)} - \mathbf{E}_{MM} \text{ (small)}$$

B. Wave Function Quantum-Mechanical Methods

The explicit treatment of electrons in atoms and molecules requires quantum mechanics, which invokes a wave function, Ψ, to describe the system of electrons and nuclei. The square of the wave function represents the probability of a particle's being at a given position. The central goal becomes the solution of the (time-independent) Schrödinger equation,

$$\mathbf{H}\Psi = E\Psi,$$

which relates the wave function, Ψ, to the energy, E, of the system. The Hamiltonian operator, $\mathbf{H}$, consists of the kinetic ($\mathbf{T}$) and the potential energy ($\mathbf{V}$) operators:

$$\mathbf{H} = \mathbf{T} + \mathbf{V}.$$

The fact that electrons instantly adjust to changes in nuclear positions due to the much greater masses of the nuclei allows the motions of electrons and nuclei to be separated (Born-Oppenheimer approximation). The electronic wave function depends on only the nuclear position, not on the nuclear momenta. The electronic Hamiltonian, $\mathbf{H}_e$, in atomic units is given by

$$\mathbf{H}_e = \mathbf{T}_e + \mathbf{V}_{ne} + \mathbf{V}_{ee} + \mathbf{V}_{nn} = -\frac{1}{2} \sum_i^{Elec.} \nabla_i^2$$

$$+ \sum_a^{Nucl.} \sum_i^{Elec.} \frac{Z_a}{|\mathbf{R}_a - \mathbf{r}_i|} + \sum_i^{Elec.} \sum_{j>i}^{Elec.} \frac{1}{|\mathbf{r}_i - \mathbf{r}_j|}$$

$$+ \sum_a^{Nucl.} \sum_{b>a}^{Nucl.} \frac{Z_a Z_b}{|\mathbf{R}_a - \mathbf{R}_b|},$$

where $\mathbf{r}$ and $\mathbf{R}$ represent the electronic and nuclear coordinates, respectively, and the Laplacian is defined as

$$\nabla_i^2 = \left(\frac{\partial^2}{\partial x_i^2} + \frac{\partial^2}{\partial y_i^2} + \frac{\partial^2}{\partial z_i^2} \right).$$

The nucleus–nucleus repulsion, $\mathbf{V}_{nn}$, is constant for a given geometry, and the kinetic energy, $\mathbf{T}_e$, and the electron–nucleus attraction, $\mathbf{V}_{ne}$, are easy to evaluate. The electron–electron repulsion, $\mathbf{V}_{ee}$, however, depends on the distances between electrons and is the reason why the Schrödinger equation cannot be solved exactly for systems with more than one electron.

The energy can be computed as the expectation value of the Hamiltonian operator:

$$E = \frac{\int \Psi^* \mathbf{H} \Psi \, d\tau}{\int \Psi^* \Psi \, d\tau} = \frac{\langle \Psi | \mathbf{H} | \Psi \rangle}{\langle \Psi | \Psi \rangle},$$

where the common *bra-ket* notation is used. The variational principle states that any trial wave function will give an energy equal to or higher than the exact value because the real system will adopt the best possible wave function (which corresponds to the exact energy). Thus, a trial wave function constructed in terms of a number of parameters can be improved by minimizing the energy with respect to the parameters (MO coefficients).

A meaningful trial wave function should approach zero as $\mathbf{r}$ goes to infinity; it should be normalized, that is,

$$\langle \Psi | \Psi \rangle = 1$$

(meaning the probability that the system is located somewhere in space is one); and it should comply with the Pauli principle. The latter states that two electrons must differ in at least one quantum number. Furthermore, the wave function should be antisymmetric (i.e., it should change sign when two electrons are interchanged). This is a characteristic property of electrons. Antisymmetry can be ensured by using Slater determinants with one-electron functions (orbitals) ϕ_i in columns and electrons $(1, 2, \ldots)$ in rows.

1. Hartree-Fock Method

Hartree-Fock theory employs a single Slater determinant. In the restricted Hartree-Fock (RHF) method, one spatial function ϕ_i is multiplied by an α (representing spin up, spin quantum number $m_s = +\frac{1}{2}$) or β (representing spin down, $m_s = -\frac{1}{2}$) spin function with the properties

$$\langle \alpha | \alpha \rangle = \langle \beta | \beta \rangle = 1; \qquad \langle \alpha | \beta \rangle = \langle \beta | \alpha \rangle = 0$$

to give spin orbitals $\phi\alpha$ and $\phi\beta$ (or ϕ and $\bar{\phi}$ for short). This is appropriate for closed-shell species with paired electrons, but open-shell species with unpaired electrons cannot be expected to have identical α and β orbitals. The unrestricted Hartree-Fock (UHF) method allows a different spatial function for each electron. However, UHF wave functions can suffer from spin contamination (i.e., the spurious mixing of higher spin states into the desired one [more formally, the expectation value of the $\mathbf{S}^2$ operator is larger than the correct value of $S(S+1)$, S being the total spin]). Restricted open-shell Hartree-Fock (ROHF)–based methods avoid the problem of spin contamination but do not allow spin polarization.

For example, a Slater determinant for the ground state of the hydrogen molecule can be written as follow:

$$\Phi(1, 2) = \frac{1}{\sqrt{2}} \begin{vmatrix} \phi_g(1) & \overline{\phi_g}(1) \\ \phi_g(2) & \overline{\phi_g}(2) \end{vmatrix}$$

$$= \frac{1}{\sqrt{2}} [\phi_g(1)\overline{\phi_g}(2) - \phi_g(2)\overline{\phi_g}(1)] = -\Phi(2, 1),$$

where ϕ_g represents the bonding molecular orbital (MO), the $1\sigma_g$ orbital.

The electronic Hamiltonian can be written as sums of one-electron ($\mathbf{h}_i$) and two-electron ($\mathbf{g}_{ij}$) operator plus the constant nuclear–nuclear repulsion:

$$\mathbf{H} = \sum_i \mathbf{h}_i + \sum_i \sum_{j>1} \mathbf{g}_{ij} + \mathbf{V}_{nn}$$

with

$$\mathbf{h}_i = -\frac{1}{2} \nabla_i^2 - \sum_a^{\text{Nucl.}} \frac{Z_a}{|\mathbf{R}_a - \mathbf{r}_i|}$$

and

$$\mathbf{g}_{ij} = \frac{1}{|\mathbf{r}_i - \mathbf{r}_j|}.$$

The Hartree-Fock energy expression becomes

$$E = \sum_i^N \langle \phi_i | \mathbf{h}_i | \phi_i \rangle + \frac{1}{2} \sum_{i=1}^N \sum_{j=1}^N (\langle \phi_i \phi_j | \mathbf{g}_{ij} | \phi_i \phi_j \rangle$$

$$- \langle \phi_i \phi_j | \mathbf{g}_{ij} | \phi_j \phi_i \rangle) + V_{nn} = \sum_i^N h_i$$

$$+ \frac{1}{2} \sum_{i=1}^N \sum_{j=1}^N (J_{ij} - K_{ij}) + V_{nn},$$

or

$$E = \sum_i^N \langle \phi_i | \mathbf{h}_i | \phi_i \rangle + \frac{1}{2} \sum_{i=1}^N \sum_{j=1}^N (\langle \phi_j | \mathbf{J}_i | \phi_j \rangle$$

$$- \langle \phi_j | \mathbf{K}_i | \phi_j \rangle) + V_{nn}.$$

J_{ij} is called a Coulomb integral because it corresponds to the electronstatic repulsion of the charge distributions due to ϕ_i^2 and ϕ_j^2; the exchange integral, K_{ij}, has no classical equivalent. $\mathbf{J}_i$ and $\mathbf{K}_i$ are Coulomb and exchange operators, respectively.

To find a minimum energy, one can vary the orbitals under the condition that they remain orthogonal by using the method of Lagrange multipliers. This leads to the definition of the Fock operator, $\mathbf{F}_i$, which describes the kinetic energy, the nuclear attraction energy, and the electron repulsion energy of one electron in the field of the other electrons:

$$\mathbf{F}_i = \mathbf{h}_i + \sum_j^N (\mathbf{J}_j - \mathbf{K}_j).$$

A set of Hartree-Fock equations is obtained,

$$\mathbf{F}_i \phi_i = \sum_j^N \lambda_{ij} \phi_j,$$

which can be simplified by a unitary transformation, which does not change the total wave function, to make the Lagrange multipliers diagonal:

$$\mathbf{F}_i \phi_i' = \varepsilon_i \phi_i'.$$

These special MOs ϕ_i' are called canonical MOs and the ε_i are the corresponding MO energies, the expectation values of the Fock operator in the MO basis:

$$\varepsilon_i = \langle \phi_i' | \mathbf{F}_i | \phi_i' \rangle.$$

According to Koopman's theorem the orbital energy corresponds to the ionization energy for a particular electron (neglecting orbital relaxation). The Hartree-Fock equations can be solved only iteratively because the Fock operator depends on all the occupied MOs by means of the Coulomb and exchange operators.

The molecular orbitals, ϕ_i, are constructed as a linear combination of atomic orbitals (LCAO), χ_α, which form the basis set (see below),

$$\phi_i = \sum_\alpha^M c_{\alpha i} \chi_\alpha.$$

This leads to the Roothaan-Hall equations, which correspond to the Fock equations in the AO basis:

$$\mathbf{FC} = \mathbf{SC}\varepsilon.$$

The elements of the overlap matrix, $\mathbf{S}$, are defined as $S_{\alpha\beta} = \langle \chi_\alpha | \chi_\beta \rangle$, and the Fock matrix elements are given by

$$F_{\alpha\beta} = \langle \chi_\alpha | \mathbf{h} | \chi_\beta \rangle + \sum_j^{\text{occ.MO}} \langle \chi_\alpha | \mathbf{J}_j - \mathbf{K}_j | \chi_\beta \rangle.$$

The energy in terms of integrals over basis functions is given by

$$E = \sum_\alpha^M \sum_\beta^M D_{\alpha\beta} \langle \chi_\alpha | \mathbf{h} | \chi_\beta \rangle$$

$$+ \frac{1}{2} \sum_\alpha^M \sum_\beta^M \sum_\gamma^M \sum_\delta^M D_{\alpha\beta} D_{\gamma\delta} (\langle \chi_\alpha \chi_\gamma | \mathbf{g} | \chi_\beta \chi_\delta \rangle$$

$$- \langle \chi_\alpha \chi_\gamma | \mathbf{g} | \chi_\delta \chi_\beta \rangle) + V_{\text{nn}},$$

which introduces the density matrix elements, $D_{\gamma\delta}$, as

$$D_{\gamma\delta} = \sum_j^{\text{occ.MO}} c_{\gamma j} c_{\delta j}.$$

The two-electron integrals are often written without the $\mathbf{g}$ operator, and for further simplification only the indices are given:

$$\langle \chi_\alpha \chi_\gamma | \mathbf{g} | \chi_\beta \chi_\delta \rangle \equiv \langle \chi_\alpha \chi_\gamma | \chi_\beta \chi_\delta \rangle \equiv \langle \alpha \gamma | \beta \delta \rangle$$

$$\equiv \iint \chi_\alpha(1) \chi_\gamma(2) \frac{1}{|\mathbf{r}_1 - \mathbf{r}_2|} \chi_\beta(1) \chi_\delta(2) \, d\mathbf{r}_1 \, d\mathbf{r}_2.$$

The Roothaan-Hall equations give the orbital coefficients as eigenvectors of the Fock matrix. So that the Fock matrix can be constructed, however, the density matrix, $\mathbf{D}$ (i.e., the orbital coefficients), has to be known. To start the iterative procedure, one must make an initial guess (e.g., from another calculation, or $\mathbf{D}$ is just set to zero), from which a Fock matrix can be derived. Diagonalization of the Fock matrix gives new (improved) orbital coefficients which allow one to build a new density matrix and a new Fock matrix. The procedure must be continued until the change is less than a given threshold and a self-consistent field (SCF) is generated (Fig. 5).

In the large basis set limit the Hartree-Fock method formally scales with the fourth power of the number of basis functions due to the two-electron integrals. In practice, computations have a more favorable scaling. Modern algorithms are close to linear scaling because the Coulomb part of the electron–electron interaction which contributes the most to the computational effort (the distances where exchange becomes negligible is relatively short) can be replaced by a multipole interaction for large distances (fast multipole method, FMM). In a conventional HF implementation the two-electron integrals are computed and stored on disk. In contrast, in the direct SCF method, the integrals are recomputed whenever they are needed

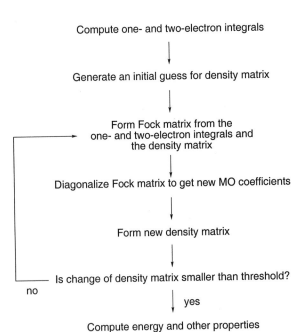

FIGURE 5 Schematic representation of the self-consistent field (SCF) procedure. MO, molecular orbital.

to avoid the storage bottleneck and the slow input–output operations. It is also possible to effectively screen for integrals which contribute only negligibly and thus can be discarded. The use of symmetry, if present, also reduces the computational cost considerably.

2. Semiempirical Methods

The HF method represents a point of departure in electronic structure theory. One direction involves improvement of the accuracy by including electron correlation (see Section II.B.3.). Semiempirical methods, however, try to provide moderate accuracy, but at much lower cost than that of ab initio methods. Therefore, only valence electrons are treated explicitly and core electrons are replaced by an effective core (covering nucleus plus core electrons) and a minimal basis of orthogonal Slater-type orbitals (usually only s and p types) is chosen to describe the valence electrons.

The two-electron integrals require the main computational effort in a HF calculation and their number is significantly reduced in semiempirical methods by the zero differential overlap (ZDO) approximation. This basic semiempirical assumption sets products of functions for one electron but located at different atoms equal to zero (i.e. $\mu_A(1)\nu_B(1) = 0$, where μ_A and ν_B are two different orbitals located on centers A and B, respectively). The overlap matrix, $\mathbf{S}$, is set equal to the unit matrix, $S_{\mu\nu} = \delta_{\mu\nu}$, and the two-electron integrals $\langle \mu\nu \mid \lambda\sigma \rangle$ are zero, unless $\mu = \nu$ and $\lambda = \sigma$, that is,

$$\langle \mu\nu \mid \lambda\sigma \rangle = \delta_{\mu\nu}\delta_{\lambda\sigma}\langle \mu\mu \mid \lambda\lambda \rangle,$$

where $\delta_{ij} = 0$ for $i \neq j$ and $\delta_{ij} = 1$ for $i = j$.

All three-and four-center two-electron integrals vanish automatically. One-electron integrals involving three centers are also set to zero. The remaining integrals are handled as parameters which partly compensate the errors introduced by the ZDO approximation. The parameters are derived from experimental data on atoms or are fitted to reproduce experimental results for molecules. The various semiempirical methods introduce different approximations for the one- and two-electron parts of the Fock matrix elements,

$$F_{\mu\nu} = \langle \mu|\mathbf{h}|\nu \rangle + \sum_{\lambda}^{AO} \sum_{\sigma}^{AO} D_{\lambda\sigma}(\langle \mu\nu \mid \lambda\sigma \rangle - \langle \mu\lambda \mid \nu\sigma \rangle),$$

with the one-electron operator

$$\mathbf{h} = -\frac{1}{2}\nabla^2 - \sum_{a} \frac{Z'_a}{|\mathbf{R}_a - \mathbf{r}|} = -\frac{1}{2}\nabla^2 - \sum_{a} \mathbf{V}_a,$$

where Z'_a denotes the charge resulting from the nucleus plus the core electrons.

a. Complete neglect of differential overlap.
The complete neglect of differential overlap (CNDO) approximation is the most rigorous: only the one- and two-center Coulomb terms among the two-electron integrals survive:

$$\langle \mu_A\nu_B \mid \lambda_C\sigma_D \rangle = \delta_{AC}\delta_{BD}\delta_{\mu\lambda}\delta_{\nu\sigma}\langle \mu_A\nu_B \mid \mu_A\nu_B \rangle.$$

$\langle \mu_A\nu_B \mid \mu_A\nu_B \rangle$ are independent of the orbital type (to guarantee rotational invariance) and there are only two parameters, $\langle \mu_A\nu_A \mid \mu_A\nu_A \rangle = \gamma_{AA}$ and $\langle \mu_A\nu_B \mid \mu_A\nu_B \rangle = \gamma_{AB}$, for the two-electron integrals. The γ_{AB} depends only on the nature of the atoms A and B and the distance between them and can be interpreted as the average electrostatic repulsion of one electron at center A and one electron at center B. The integral γ_{AA} is the average repulsion of two electrons at one atom.

The one-electron integrals are

$$\langle \mu_A|\mathbf{h}|\nu_A \rangle = -\delta_{\mu\nu} \sum_{a}^{Nucl.} \langle \mu_A|\mathbf{V}_a|\mu_A \rangle.$$

The Pariser-Pople-Parr (PPP) method is a special case of CNDO, restricted to the treatment of π electrons.

b. Intermediate neglect of differential overlap.
In the intermediate neglect of differential overlap (INDO) approximation the two-electron integrals are limited to the Coulomb integrals. One-electron integrals involving different orbitals of one center and $\mathbf{V}_a$ operator from another have to disappear to guarantee rotational invariance. The one-electron integrals are the same as in the CNDO approximation and the two-electron integrals are given by

$$\langle \mu_A\nu_B \mid \lambda_C\sigma_D \rangle = \delta_{\mu_A\lambda_C}\delta_{\nu_B\sigma_D}\langle \mu_A\nu_B \mid \mu_A\nu_B \rangle$$

and parametrized as γ_{AB} and γ_{AA}.

INDO is comparable to CNDO in computational cost but has the advantage that electronic states of different multiplicities can be distinguished.

MINDO/3 (modified intermediate neglect of differential overlap) was the first successful semiempirical method to give reasonable predictions of molecular properties. The main improvement over earlier methods was the use of molecular data rather than atomic data for the parametrization. However, the number of parameters to be determined in MINDO/3 increases with the square of the number of atoms included because one parameter depends on the type of bonded atoms.

c. Neglect of diatomic differential overlap.
Many of the shortcomings of MINDO/3 are corrected in the neglect of diatomic differential overlap (NDDO) approximation, which includes no further approximations beyond ZDO. Thus, all integrals involving any two orbitals on one

center with any two orbitals on another center are kept, which increases the number of integrals dramatically.

The one-electron integrals are

$$\langle \mu_A | \mathbf{h} | \nu_B \rangle = \langle \mu_A | -\tfrac{1}{2}\nabla^2 - \mathbf{V}_A - \mathbf{V}_B | \nu_B \rangle$$

and

$$\langle \mu_A | \mathbf{h} | \nu_A \rangle = \delta_{\mu\nu} \langle \mu_A | -\tfrac{1}{2}\nabla^2 - \mathbf{V}_A | \mu_A \rangle - \sum_{a \neq A} \langle \mu_A | \mathbf{V}_a | \nu_A \rangle.$$

The two-electron integrals are given by

$$\langle \mu_A \nu_B | \lambda_C \sigma_D \rangle = \delta_{AC} \delta_{BD} \langle \mu_A \nu_B | \lambda_A \sigma_B \rangle.$$

d. Modified NDDO. The more successful semiempirical methods, MNDO (modified neglect of differential overlap), AM1 (Austin model 1), and PM3 (parametric method number 3), are all based on NDDO but differ in the treatment of core–core repulsion and how the parameters are assigned. There are only atomic parameters, no diatomic parameters as in MINDO/3. The "modified" NDDO methods calculate the overlap matrix, **S**, explicitly rather than using the unit matrix.

The MNDO method tends to overestimate the repulsion between atoms separated by approximately the sum of their van der Waals radii. To correct for this deficiency, AM1 modifies the core–core term by Gaussian functions. PM3 is essentially equivalent to AM1 but uses (automated) full optimization of the parameter set against a much larger collection of experimental data while the AM1 parameters are tuned by hand. PM3 therefore on average gives results in somewhat better agreement with experiment.

Extending the basis set to include d functions as in MNDO/d and PM3(tm) raises the number of integrals (i.e., the number of parameters) tremendously but allows a larger variety of applications, for example, those including transition metal compounds (albeit with variable accuracy) or hypervalent molecules.

Semiempirical programs usually report heats of formation calculated from the electronic energies less the calculated energies for the atoms plus the experimental heat of formations for the atoms:

$$\Delta H_f = E_{\text{calc.}}(\text{molecule}) - \sum^{\text{atoms}} E_{\text{calc.}}(\text{atom})$$
$$+ \sum^{\text{atoms}} H_f(\text{atom}).$$

The semi ab initio model 1 (SAM1) is another modified NDDO method, but it does not replace integrals by parameters. The one- and two-center electron repulsion integrals are explicitly calculated from the basis functions [employing a standard STO-3G (Slater-type orbital from three Gaussian functions) Gaussian basis set] and scaled by a function which has to be parametrized. SAM1 calculations take about twice as long as AM1 or PM3 calculations do.

3. Electron-Correlated Methods

In the Hartree-Fock approach the real electron–electron interaction is replaced by an interaction with an averaged field. This means HF suffers from an exaggeration of electron–electron repulsion. The difference between the energy obtained at the HF level and the exact (nonrelativistic) energy (for a given basis set) is defined as the correlation energy. The name reflects that this energy difference is connected to the correlated movement of electrons which is not considered in the HF method and which reduces the electron–electron repulsion. The HF description typically allows electrons to be unrealistically resident in the internuclear region. This leads to an underestimated nuclear–nuclear repulsion and to bond lengths that are too short. As a consequence, stretching force constants and harmonic stretching frequencies computed at the HF level are too large. Likewise, the polarity of bonds is overestimated (the more electronegative atom tolerates a higher electron density in the HF picture) and computed dipole moments are often too large.

Dynamic electron correlation, which is connected to the correlated movement of electrons, can be distinguished from static (near-degeneracy) electron correlation, which deals with the insufficiency of the one-determinant approach. HF usually provides a suitable description of closed-shell molecules in their electronic ground state. However, the homolytic dissociation of such a molecule generates two electronic states which are very close in energy. This situation requires a description by more than one Slater determinant (i.e., at least a two-configuration method). The energy difference between the HF method and a multiconfigurational method is the static correlation energy. Accounting for electron correlation is essential for quantitative answers from electronic structure calculations. Different post-HF methods which attempt to recover all or part of the correlation energy are discussed in the following text.

Within the closed-shell HF picture, molecular orbitals are occupied by either exactly two or exactly zero electrons represented by the variationally best one-determinant wave function. Correlated levels give a different electron density which cannot be represented by a single Slater determinant. A logical starting point to account for electron correlation is to expand a multideterminantal wave function with the HF wave function as a starting point:

$$\Psi = a_0 \Phi_{\text{HF}} + \sum_{i=1} a_i \Phi_i,$$

where a_0 usually is close to 1. Because this is analogous to expanding one MO in terms of AOs, one speaks of the basis set as the one-electron basis (responsible for the one-electron functions, the MOs) while the number

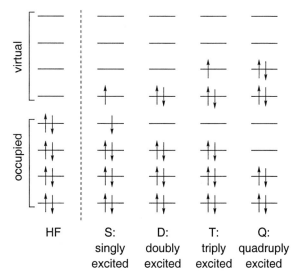

FIGURE 6 Examples of singly, doubly, triply, and quadruply excited determinants derived from a Hartree-Fock (HF) reference.

of determinants included in a correlated wave function builds the many-electron basis.

In closed-shell HF theory there is only one determinant which has the lowest MOs occupied (occupation number = 2). The remaining orbitals are empty or "virtual" (occupation number = 0). Additional determinants are generated by exciting one or more electrons from an occupied MO into an unoccupied (virtual) MO. According to the number of excited electrons, one speaks of singles, doubles, triples, quadruples, and so forth (S, D, T, Q, respectively; Fig. 6).

The larger the basis set the more virtual MOs and the more excited Slater determinants can be generated. The quality of a calculation is determined by both the size of the basis set and the number of excited determinants that are considered. If all possible determinants together with an infinite basis set could be used, one would get the exact solution of the nonrelativistic Schrödinger equation within the Born-Oppenheimer approximation. Because a different chemical environment mostly affects the valence electrons, but does not influence the core electrons, the frozen core approximation includes only determinants with excited valence electrons. Also the highest virtual orbitals may be left unoccupied in all determinants (frozen virtuals).

a. Configuration interaction. In the configuration interaction (CI) procedure the trial function is constructed as a linear combination of the ground (reference configuration) and excited Slater determinants. The MO coefficients remain fixed throughout the calculation and are usually taken from the HF orbitals. Alternatively, natural orbitals (which are defined as diagonalizing the one-electron den-

sity matrix) may be used, which promise faster convergence of the CI expansion. In general, several Slater determinants are contracted linearly to form eigenfunctions of the spin operators S_Z and S^2 and which are called spin adapted configurations, or configuration state functions (CSFs):

$$\Psi_{CI} = \sum_{i=0} a_i \Phi_i.$$

The expansion coefficients, a_i, are then determined variationally to give the minimum energy.

For a full CI (FCI) the number of determinants grows factorially with the size of the system. A full CI recovers all of the electron correlation energy (for a given basis set) but can be applied only to obtain benchmark results for very small molecules to assess the performance of more economical methods. For applications to larger molecules, the CI expansion has to be truncated to make the computation feasible. CIS, CISD, CISDT, and CISDTQ correspond to expansions through singly, doubly, triply, and quadruply excited CSFs, respectively. According to Brilluoin's theorem, the CI matrix elements of a closed-shell restricted HF wave function with singly excited CSFs vanish. Hence, CIS does not improve the description of the ground state. Doubles are found to contribute most to the correlation energy and consequently CISD (including only singly and doubly excited determinants) is the most widely applied CI method because inclusion of triples and quadruples is typically computationally too demanding.

FCI is size consistent, but truncated CI methods are not. This means the energy computed for two noninteracting molecules is not identical to the sum of the energies computed for the individual molecules. This unphysical behavior is a major drawback of any truncated CI. So that CISD can be made approximately size consistent, the Davidson correction can be applied in which the contribution of quadruples, ΔE_Q, is estimated from the correlation energy given at the CISD level, ΔE_{CISD}, and the coefficient of the reference configuration, a_0:

$$\Delta E_Q = (1 - a_0)\Delta E_{CISD}.$$

CISD was also extended to quadratic CISD (QCISD) by the inclusion of some higher-order terms to yield a size-extensive method.

b. Multiconfiguration self-consistent field. The HF method does not give a good first-order description when more than one nonequivalent resonance structure is important for the electronic structure of a molecule. A multiconfiguration self-consistent field (MCSCF) calculation may be used instead. Not only the coefficients for the determinants are optimized in MCSCF, but also the MO coefficients simultaneously. The selection of configurations is not trivial. One easy way to construct an MCSCF is the complete active space self-consistent field, or CASSCF (also called full optimized reaction space, or FORS). Instead of choosing configurations, one must select a set of "active" (occupied and unoccupied) orbitals and all possible (symmetry-adapted) configurations within this "active space" are automatically included in the MCSCF. The method is called restricted active space self-consistent field, or RASSCF, when subsets of the active orbitals are restricted to have a certain (minimum or maximum) number of electrons to reduce the computational cost.

The MCSCF provides a good first-order description covering the static electron correlation due to degeneracy problems. Dynamic electron correlation should be addressed with the MCSCF wave function as a reference. The multireference configuration interaction, or MRCI, generates excited determinants from all (or selected) determinants included in the MCSCF. The complete active space perturbation theory, second order (CASPT2) is a more economical approach. Both methods can be applied to compute excited states.

c. Many-body perturbation theory. Perturbation theory assumes that somehow an approximate solution to a problem can be found. The missing correction, which should be small, is then considered as a perturbation of the system. When the perturbation is to correct for the approximation of independent particles the method is called many-body perturbation theory, or MBPT. In electronic structure theory the Hamiltonian operator, $\mathbf{H}$, is written as a combination of a reference Hamiltonian, $\mathbf{H}_0$, which can be solved for, and a perturbation $\mathbf{H}'$:

$$\mathbf{H} = \mathbf{H}_0 + \lambda \mathbf{H}',$$

λ being the perturbation parameter ($0 \leq \lambda \leq 1$) which determines the strength of the perturbation. The energy, E, and wave function, Ψ, are expanded as Taylor series in λ:

$$E = E_0 + \lambda E_1 + \lambda^2 E_2 + \lambda^3 E_3 + \cdots$$

and

$$\Psi = \Psi_0 + \lambda \Psi_1 + \lambda^2 \Psi_2 + \lambda^3 \Psi_3 + \cdots.$$

The Schrödinger equation,

$$\mathbf{H}\Psi = E\Psi,$$

gives Ψ_0 and E_0 as the solution in the absence of any perturbation ($\lambda = 0$). E_1, E_2, etc., and Ψ_1, Ψ_2, etc., are the first-, second-, etc., order corrections to the energy and wave function, respectively. For $\lambda > 0$ the Schrödinger equation becomes

$$(\mathbf{H} + \lambda \mathbf{H}')(\Psi_0 + \lambda \Psi_1 + \lambda^2 \Psi_2 + \cdots)$$
$$= (E_0 + \lambda E_1 + \lambda^2 E_2 + \cdots)(\Psi_0 + \lambda \Psi_1 + \lambda^2 \Psi_2 + \cdots).$$

Because this equation has to be true for all values of λ, the terms connected to the same power of λ can be separated:

$$\lambda^0: \mathbf{H}_0\Psi_0 = E_0\Psi_0,$$

$$\lambda^1: \mathbf{H}_0\Psi_1 + \mathbf{H}_1\Psi_0 = E_0\Psi_1 + E_1\Psi_0,$$

$$\lambda^2: \mathbf{H}_0\Psi_2 + \mathbf{H}_1\Psi_1 + \mathbf{H}_2\Psi_0 = E_0\Psi_2 + E_1\Psi_1 + E_2\Psi_0,$$

and so forth, which gives the zeroth-, first-, second-, etc., order perturbation equations. If one chooses the intermediate normalization condition,

$$\langle \Psi_0 | \Psi_0 \rangle = 1; \qquad \langle \Psi_0 | \Psi_i \rangle = 0, \qquad i > 0,$$

simple energy expressions are obtained:

$$E_0 = \langle \Psi_0 | \mathbf{H}_0 | \Psi_0 \rangle,$$

$$E_1 = \langle \Psi_0 | \mathbf{H}' | \Psi_0 \rangle,$$

$$E_2 = \langle \Psi_0 | \mathbf{H}' | \Psi_1 \rangle,$$

and so forth. Knowledge of wave function corrections up to order i allows calculation of the energy up to order $(2i + 1)$. This relationship is known as the Wigner theorem.

In Møller-Plesset (MP) perturbation theory the unperturbed Hamiltonian, $\mathbf{H}_0$, is taken as the sum over n Fock operators (n = number of electrons) giving a total of twice the average electron–electron repulsion energy and the perturbation operator becomes the difference between the exact electron–electron repulsion and twice the average electron–electron repulsion. With this choice of $\mathbf{H}_0$ the zeroth-order energy is just the sum of MO energies and the first-order energy equals the Hartree-Fock energy. The second-order correction, $E(\text{MP2})$, is the first to contribute to the electron correlation energy and can be calculated from the two-electron integrals over MOs:

$$E(\text{MP2}) = \sum_{i<j}^{\text{occ}} \sum_{a<b}^{\text{virt}} \frac{[\langle \phi_i \phi_j | \phi_a \phi_b \rangle - \langle \phi_i \phi_j | \phi_b \phi_a \rangle]^2}{\varepsilon_i + \varepsilon_j - \varepsilon_a - \varepsilon_b}.$$

MP2 is a very economical method but often overcorrects for electron correlation effects. The MP2 energy calculation scales only with N^4, but the transformation of AO to MO integrals is an N^5 step (Table I). The next step in the series, MP3, also includes only contributions from doubly excited determinants but scales with N^6. Full MP4 involves singly, doubly, triply, and quadruply excited determinants and is an N^7 method. In applications, triples are sometimes left out (MP4SDQ is N^6) to make the calculation affordable. MPn methods with $n > 4$ are not used routinely because they are both very complex and expensive in terms of resources.

MP theory is size extensive but not variational (i.e., there is no guarantee that the correct energy is lower than

TABLE I Formal Scaling of Various Methods with the Size of the Molecular System, *N*, and Size-Extensive and Variational Properties

Method[a]	Scaling	Size extensive?	Variational?
HF	N^4	Yes	Yes
MP2	N^5	Yes	No
MP3	N^6	Yes	No
MP4SDQ	N^6	Yes	No
MP4SDTQ	N^7	Yes	No
MP5	N^8	Yes	No
MP6	N^9	Yes	No
CCSD	N^6	Yes	No
CCSD(T)	N^7	Yes	No
CCSDT	N^8	Yes	No
CISD	N^6	No	Yes
CISDT	N^8	No	Yes
CISDTQ	N^{10}	No	Yes

[a] HF, Hartree-Fock; MP, Møller-Plesset (numbers 2–6 refer to second through sixth order; S, singly excited; D, doubly excited; T, triply excited; and Q, quadruply excited; CC, coupled cluster (S, D, and T as for MP except T in parentheses is estimated); CI, configuration interaction (S, D, T, and Q as for MP).

an MP*n* energy). This is no problem because usually only relative energies are of interest. However, the MP*n* series does not necessarily converge. When the HF reference provides a poor description of the electronic structure, the MP*n* series may become divergent and produce even worse results than those of HF.

d. Coupled cluster methods. Coupled cluster (CC) theory was originally formulated for nuclear physics and only later was applied to the electron correlation problem in quantum chemistry. Today it is the method of choice for highly accurate computations. CC theory uses an exponential expansion of a reference function Φ_0, usually the Hartree-Fock determinant (in contrast with the linear expansion of CI):

$$\Psi_{CC} = e^{T}\Phi_0,$$

where the cluster operator **T** is defined as

$$\mathbf{T} = \mathbf{T}_1 + \mathbf{T}_2 + \mathbf{T}_3 + \cdots \mathbf{T}_N.$$

The excitation operators, $\mathbf{T}_i$, generate all *i*th excited Slater determinants from the reference:

$$\mathbf{T}_1\Phi_0 = \sum_i^{occ} \sum_a^{virt} t_i^a \Phi_i^a;$$

$$\mathbf{T}_2\Phi_0 = \sum_{i<j}^{occ} \sum_{a<b}^{virt} t_{ij}^{ab} \Phi_{ij}^{ab}; \qquad \cdots$$

The expansion coefficients, *t*, are called amplitudes. Substituting the exponential function by a series,

$$e^{T} = 1 + \mathbf{T} + \tfrac{1}{2}\mathbf{T}^2 + \tfrac{1}{6}\mathbf{T}^3 + \tfrac{1}{24}\mathbf{T}^4 + \cdots = \sum_{k=0}^{\infty} \frac{1}{k!}\mathbf{T}^k$$

$$= 1 + \mathbf{T}_1 + \left(\mathbf{T}_2 + \tfrac{1}{2}\mathbf{T}_1^2\right) + \left(\mathbf{T}_3 + \mathbf{T}_2\mathbf{T}_1 + \tfrac{1}{6}\mathbf{T}_1^3\right)$$

$$+ \left(\mathbf{T}_4 + \mathbf{T}_3\mathbf{T}_1 + \tfrac{1}{2}\mathbf{T}_2^2 + \tfrac{1}{2}\mathbf{T}_1^2\mathbf{T}_2 + \tfrac{1}{24}\mathbf{T}_1^4\right) + \cdots$$

shows that due to the exponential ansatz in CC,

i.) a given excitation level in general is not due to just one excitation operator (there are "disconnected" terms in addition to the "connected" term; for example, for doubles, $\mathbf{T}_1^2$ and $\mathbf{T}_2$, respectively).

ii.) restricting **T** generates not only excitations up to a given level, but also higher ones (quartets, etc.) up to infinity. For example, for $\mathbf{T} = \mathbf{T}_2$,

$$e^{\mathbf{T}_2} = 1 + \mathbf{T}_2 + \tfrac{1}{2}\mathbf{T}_2^2 + \tfrac{1}{6}\mathbf{T}_2^3 + \cdots.$$

This is in contrast with both perturbation and CI methods and therefore CC theory should provide a better description of electron correlation effects at a given truncation level.

The Schrödinger equation becomes

$$\mathbf{H}\,e^{T}\Phi_0 = E_{CC}e^{T}\Phi_0,$$

and the CC energy is given by

$$E_{CC} = \left\langle \Phi_0 \middle| \mathbf{H}\,e^{T} \middle| \Phi_0 \right\rangle.$$

Because the Hamiltonian operator contains only one- and two-electron operators, only the first few terms of the exponential series give nonzero values:

$$E_{CC} = \left\langle \Phi_0 \middle| \mathbf{H} \middle| \left(1 + \mathbf{T}_1 + \mathbf{T}_2 + \tfrac{1}{2}\mathbf{T}_1^2\right)\Phi_0 \right\rangle.$$

Further simplification leads to a CC energy expression from the two-electron integrals over MOs:

$$E_{CC} = E_0 + \sum_{i<j}^{occ} \sum_{a<b}^{virt} \left(t_{ij}^{ab} + t_i^a t_j^b - t_i^b t_j^a\right)$$

$$\times \left(\langle \phi_i\phi_j \mid \phi_a\phi_b \rangle - \langle \phi_i\phi_j \mid \phi_b\phi_a \rangle\right).$$

CCSD is the only pure CC method that can be used in routine applications. Explicit treatment of triples (CCSDT) is usually too expensive. However, the contribution of triples can be estimated perturbatively in the CCSD(T) method. Brueckner (B) theory is a variation of CC theory which uses orbitals that make the singles contribution vanish. The accuracy and computational cost of BD is comparable to that of CCSD. Excited electronic states may be treated within the CC formalism by the equation-of-motion (EOM-CC) approach.

e. R12 methods. HF is not exact because of the approximate treatment of electron–electron repulsion. Post-HF methods try to recover the electron correlation energy by expanding the N-electron wave function in terms of Slater determinants built from one-electron functions (orbitals). But the convergence toward the exact solution of the Schrödinger equation is typically slow due to the poor description of the cusp region ($r_1 - r_2 = 0$) of the wave function. Faster convergence may be achieved by adding terms that describe electron correlation effects more directly than the product of one-electron orbitals (e.g., by including a function that explicitly depends on the coordinates of two electrons rather than just one electron: the interelectronic coordinate r_{12}).

The r_{12}-dependent wave function may then be used with CI, MBPT, or CC, giving R12 methods. The formula for computing the energy includes integrals over three- and four-electron coordinates, which are difficult to evaluate and which increase with N^6 and N^8. However, insertion of a resolution of the identity allows three- and four-electron integrals to be written as sums over products of integrals involving only two-electron coordinates. The substitution is exact only for a complete basis set, but it is a very good approximation for fairly big basis sets, which have to be used for very accurate results anyway. R12 methods converge to the same basis set limit as conventional electron correlation methods do but faster.

4. Basis Sets

A basis set is used to express the unknown MOs in terms of a set of known functions. The more basis functions used, the more accurate the description of the MOs. Any type of function can be used, but most efficient are basis functions with a physical behavior (e.g., approach zero for large distances between electrons and nuclei) and which make the integral evaluation easy. Slater-type orbitals (STOs) are related to the exact solutions for the hydrogen atom, but Gaussian-type orbitals (GTOs) are preferable for computational ease:

$$\chi^{STO}_{\varsigma,n,l,m}(r, \theta, \varphi) = N Y_{l,m}(\theta, \varphi) r^{n-1} e^{-\varsigma r}$$

and

$$\chi^{GTO}_{\varsigma,n,l,m}(r, \theta, \varphi) = N Y_{l,m}(\theta, \varphi) r^{2n-2-l} e^{-\varsigma r^2};$$

$$\chi^{GTO}_{\varsigma,l_x,l_y,l_z}(x, y, z) = N x^{l_x} y^{l_y} z^{l_z} e^{-\varsigma r^2}.$$

N is a normalization constant and $Y_{l,m}$ are the spherical harmonic functions. STOs lack radial nodes, which are introduced by making linear combinations of STOs. They are primarily used for high-accuracy atomic and diatomic calculations and with semiempirical methods, which neglect all three- and four-center integrals (which cannot

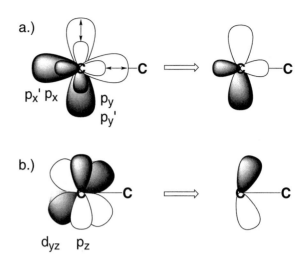

FIGURE 7 Adding flexibility to basis sets: (a) Split valence basis sets provide more and less diffuse orbitals to adjust to different bonding situations (e.g., σ- and π-bonding). (b) Higher angular momentum basis functions allow for polarization to gain better overlap.

be evaluated analytically for STOs). GTOs do not have the proper behavior near the nucleus (in contrast with the "cusp" of the STOs) and fall off too rapidly far away from the nucleus. A better description is provided by contractions of several Gaussian functions. GTOs are generally used because of the computational efficiency.

The number of basis functions employed for each atom determines the quality of the basis set. A minimum basis provides only as many shells as necessary to accommodate all electrons (i.e., 1s for H and He; 1s, 2s, and 2p for first-row elements, etc.) A double zeta basis set adds more flexibility for the description of different bonding situations (Fig. 7) by using two functions varying in the exponent ς (a measure of the diffuseness) for each orbital of an occupied shell (i.e., 1s, 1s' for H, He; 1s, 1s', 2s, 2s', 2p, 2p' for first-row elements). Because bonding involves valence electrons only, core electrons are described by a single basis function and only the valence region is split in split valence or valence double zeta basis sets (1s, 1s' for H, He; 1s, 2s, 2s', 2p, 2p' for first-row elements). Basis sets can be improved by adding more functions: triple zeta (TZ), quadruple zeta (QZ), quintuple zeta (5Z), etc., have three, four, five, etc., times the number of basis functions of a minimum basis set, respectively. The exponents are usually determined variationally for the atoms. Each basis function may consist of several "primitive" GTOs. Contraction of several GTOs to one basis function (contracted GTO) is especially useful for the inner orbital to mimic the cusp at the nucleus. Additional higher angular momentum functions are usually added as polarization functions to allow for a polarized charge distribution (e.g., one additional p set for H and He; one d set for first-row elements). For correlated wave functions even more than one set is essential if high accuracy is desired (e.g., one d and one p set for H, He; one f set and two d sets for first-row elements). Anionic species or molecules with many lone pairs have very diffuse electron distributions and require an additional set of diffuse functions (often denoted by a "+"). The basis sets by Pople and coworkers are very popular: STO-3G is a minimum basis with three GTOs contracted to represent one atomic orbital. 3-21G is a split valence basis with core orbitals contracted from three GTOs and the valence region described by one basis function contracted by two primitives and another uncontracted function. 6–31G* is constructed analogously but includes one set of polarization d functions on heavy atoms (other than hydrogen). Dunning's correlation consistent basis sets (cc-p VXZ; $X = $ D, T, Q, 5, 6, ...) represent a series of basis sets converging to the basis set limit. This allows extrapolation to the infinite basis set limit. The cc-pVDZ (correlation consistent polarized valence double zeta) basis set has [3s, 2p, 1d/2s, 1p] contracted from (9s, 4p, 1d/4s, 1p) for first-row and hydrogen atoms, respectively. The next-better basis set, cc-p VTZ, has one basis function more of each type and adds one next-higher angular momentum function: [4s, 3p, 2d, 1f/3s, 2p, 1d]. Augmentation of one extra diffuse function for each type of angular momentum (1s, 1p, 1d for cc-pVDZ) is denoted by the prefix *aug-*.

5. Optimizing the Performance/Cost Ratio

In any application a compromise between computational cost and accuracy has to be made. Accurate geometries are relatively easy to compute; MP2 with a polarized double zeta basis set usually gives satisfactory results. Accurate energetics require a better theoretical treatment and therefore are generally determined from a "single energy point calculation" at a higher level [e.g., CCSD(T) with a TZP basis set]. Whereas a single slash (/) is used to separate method and basis set specification, a double slash (//) means "at the geometry optimized at" (e.g., CCSD(T)/TZP//MP2/DZP). Zero-point vibrational energy (ZPE) corrections to relative energy are usually applied but need not be derived from a frequency calculation at the highest level of optimization. ZPEs are usually scaled by an empirical correction factor (depending on the theory level) to account for the overestimation of vibrational frequencies.

To achieve chemical accuracy (i.e., ± 1 kcal mol^{-1}) for relative energies, for molecules of chemical interest, investigators devised interpolation schemes based on an additivity assumption. Most popular is the G2 method, a general procedure based on ab initio theory for the accurate prediction of energies of molecular systems, like enthalpies of formation, bond energies, ionization potentials, electron affinities, and proton affinities. Starting from the MP4/6-311G(d,p)//MP2(fu)/6-31G* + 0.8929 ZPE(HF/6-31G*) level, corrections for diffuse functions, higher polarization functions, and a more complete electron correlation treatment, as well as a "higher level correction" depending on the number of α and β valence electrons and an empirical factor, are included to extrapolate to the QCISD(T)/6-311 + G(3df,2p) level. This procedure gives a mean absolute deviation from experimental data of 1.21 kcal mol^{-1} for the "G2 test set," a large number of various types of relative energies accurately known from experiment. Variations to the G2 method including the use of DFT methods have been proposed either to further increase the accuracy (e.g., G3) or to reduce the computational expense.

C. Density Functional Theory

The Hohenberg-Kohn theorem provides the inspiration for density functional theory (DFT): all ground-state properties of a system are functionals of the charge density. In particular the correct energy can be derived from the correct charge density. Conversely, an incorrect density will give an energy above the correct energy. In a DFT calculation the energy is optimized with respect to the density. It should be much simpler to handle the total electron density (which depends on three coordinates) than to treat all electrons explicitly (involving one spin plus three spatial coordinates per electron). Although Hohenberg and Kohn proved that the electron density determines the electronic ground-state energy, the functional to convert the electron density function into an energy value is unknown.

Each contribution to the total energy—the kinetic energy, $E_T[\rho]$; the nucleus–electron attraction, $E_{ne}[\rho]$; and the electron–electron repulsion, $E_{ee}[\rho]$—can be expressed as a functional of the total electron density:

$$E[\rho] = E_T[\rho] + E_{ne}[\rho] + E_{ee}[\rho].$$

In analogy to HF theory, $E_{ee}[\rho]$ can be divided into a Coulomb ($E_J[\rho]$) part and an exchange part. The formula for E_{ne} is exact:

$$E_{ne}[\rho] = -\sum_a^{Nucl.} Z_a \int \frac{\rho(\mathbf{r})}{|\mathbf{r} - \mathbf{R}_a|} \, d\mathbf{r},$$

and

$$E_J[\rho] = \frac{1}{2} \iint \frac{\rho(\mathbf{r}_1)\rho(\mathbf{r}_2)}{|\mathbf{r}_1 - \mathbf{r}_2|} \, d\mathbf{r}_1 \, d\mathbf{r}_2$$

holds true for electrons moving independently in the field caused by all electrons—approximations which are hoped to be corrected by a combined exchange and correlation term, E_{XC}. The major task of DFT is to develop approximate but accurate functionals for E_T and E_{XC}.

For the noninteracting uniform electron gas, the kinetic and exchange energies can be derived:

$$E_T^{TF}[\rho] = \tfrac{3}{10}(3\pi^2)^{2/3} \int \rho(\mathbf{r})^{5/3}\, d\mathbf{r}$$

and

$$E_K^{D}[\rho] = \tfrac{3}{4}\left(\tfrac{3}{\pi}\right)^{1/3} \int \rho(\mathbf{r})^{4/3}\, d\mathbf{r}.$$

In Thomas-Fermi theory the total energy is expressed as

$$E^{TF}[\rho] = E_{ne}[\rho] + E_J[\rho] + E_T^{TF}[\rho],$$

while the Thomas-Fermi-Dirac expression adds the exchange expression $E_k^{D}[\rho]$. The uniform electron gas is a model too crude to describe molecules: neither TF nor TFD gives bonding between atoms.

As a way to improve on the expression for the kinetic energy, Kohn-Sham theory calculates the kinetic energy for noninteracting electrons and corrects for the error relative to the real kinetic energy by means of the exchange correlation term $E_{XC}[\rho]$:

$$E^{DFT}[\rho] = E_T^{S}[\rho] + E_{ne}[\rho] + E_J[\rho] + E_{XC}[\rho],$$

where $E_T^{S}[\rho]$ can be computed from a Slater determinant:

$$E_T^{S}[\rho] = \sum_{i}^{N} \langle \phi_i | -\tfrac{1}{2}\nabla^2 | \phi_i \rangle,$$

and $E_{XC}[\rho]$ is usually separated into an exchange contribution ($E_X[\rho]$) and a correlation contribution ($E_C[\rho]$).

In the Kohn-Sham implementation of DFT, the density, ρ, is derived from a single Slater determinant with orthonormal orbitals, ϕ_i:

$$\rho(\mathbf{r}) = \sum_{i}^{occ} |\phi_i(\mathbf{r})|^2.$$

The energy is then optimized by solving a set of one-electron equations, the Kohn-Sham equations, but with electron correlation included:

$$\mathbf{h}^{KS}\phi_i = \varepsilon_i \phi_i,$$

where

$$\mathbf{h}^{KS} = -\frac{1}{2}\nabla^2 + \sum_{a}^{Nucl.} \frac{Z_a}{|\mathbf{R}_a - \mathbf{r}|} + \int \frac{\rho(\mathbf{r}')}{|\mathbf{r} - \mathbf{r}'|}\, d\mathbf{r}' + \mathbf{V}_{XC}(\mathbf{r}).$$

The (Kohn-Sham) orbitals, ϕ_i, which are used to represent the electron density, ρ, can be determined numerically or variationally as an expansion of basis functions.

The Kohn-Sham equations have to be solved iteratively because the Coulomb term depends on the density (i.e., the orbitals to be determined). The main advantage of DFT methods is that they include some treatment of electron correlation at a computational cost equivalent to that of the HF method. The main disadvantage of DFT, however, is that there is no hierarchy of increasingly better functionals. The performance of a given functional must be assessed by comparison with experimental data and there is no consistent way to improve the quality of a given functional. This is in contrast with wave function methods, in which a more complete treatment of electron correlation (and a more flexible basis set) means a closer approach to the exact solution.

1. Local Density Approximation

The local density approximation (LDA) assumes variations of the density to be slow and treats the local density as a uniform electron gas:

$$E_X^{LDA}[\rho] = -\tfrac{3}{4}\left(\tfrac{3}{\pi}\right)^{1/3} \int \rho^{4/3}(\mathbf{r})\, d\mathbf{r}.$$

The X_α method is an example of LDA in which the correlation energy is neglected and the exchange energy expression is multiplied by a parameter α. A fairly accurate expression for the correlation energy of the uniform electron gas, the VWN (Vosko, Wilk, Nusair) functional, was derived by fitting it to Monte Carlo results.

2. Generalized Gradient Approximation

As a way to better treat the nonuniform electron distribution of molecules, the exchange and correlation functionals were modified to include derivatives of the density in the gradient-corrected approximation, or generalized gradient approximation (GGA). Gradient-corrected exchange functionals were developed [e.g., by Perdew and Wang (PW86) and Becke (B)]. Popular correlation functionals are those of Lee, Yang, and Parr (LYP), Perdew (P86), Perdew and Wang (PW91), and Becke (B91). Hybrid methods (such as the popular Becke three-parameter functional, B3) use part of the exchange as computed by the HF method. The three parameters in B3 determine the mixing of LDA and exact exchange, as well as the gradient-corrected contributions to the exchange and correlation terms. The parameters are fitted to experimental thermodynamical data. Exchange and correlation functionals can freely be combined to give an arsenal of DFT methods (e.g., BLYP, BP86, B3LYP, etc.), but B3LYP is the most popular because of its consistently good performance.

Recently, the DFT formalism has been extended to treat excited electronic states through the implementation of time-dependent DFT (TD-DFT).

III. APPLICATIONS OF COMPUTATIONAL CHEMISTRY

When a chemical problem is to be studied computationally, an appropriate level of theory must first be chosen. Simple qualitative concepts such as the frontier molecular orbital (FMO) theory or orbital symmetry (Woodward-Hoffman) rules, which were developed on the basis of primitive computations, can often successfully predict relative reactivities and selectivities. Qualitative concepts are very useful as they can provide chemical insight and understanding. For quantitative answers more sophisticated computational methods have to be employed and a compromise between desired accuracy and computational cost has to be made. In any case, it is essential to know about the possibilities and limitations of the method to be applied.

A. The Potential Energy Surface

The methods described in Section II provide expressions for the energy as a function of the atomic coordinates [i.e., they describe the potential energy surface (PES) for a given molecular formula] (Fig. 8). This allows one to search for an atomic arrangement which makes the energy a minimum (geometry optimization). More efficient optimization algorithms can be employed when the forces and the Hessian [first and second derivatives, respectively, of the energy with respect to (w.r.t.) the nuclear coordinates] are also known. One global and usually many additional local minima exist for a given formula and relative energies for isomeric structures can be obtained. The transition structure, which is the highest point on a minimum energy path connecting two minima, can also be localized. This reveals information about chemical reactions, such as mechanistic details, activation barriers, and so forth, which is difficult or impossible to deduce from experimental investigations. Minima and transition states are stationary points (i.e., the forces are zero). All eigenvalues of the Hessian are positive for minima while transition states have one negative eigenvalue. The eigenvector corresponding to the negative eigenvalue describes the transition mode. Following the minimum energy path in mass-weighted coordinates (intrinsic reaction coordinate, IRC) allows one to confirm which minima are connected by a transition state. Harmonic vibrational frequencies can be derived from the force constants and allow one to compute entropy and enthalpy values.

The molecular geometry input can be provided in the form of a Z-matrix [i.e., be defined through internal coordinates (bond length, bond angles, and dihedral angles)]. Due to improved optimization algorithms (handling redundant internal coordinates) and the size of molecules that can now be computed routinely, the input is mostly provided as Cartesian coordinates, often generated with the help of a graphical user interface to the computational chemistry program package.

The computational speed of the more approximate methods allows one not only to treat larger molecular systems, but also to address new problems. The function of a protein is determined by its three-dimensional structure. Hence, there is considerable interest in solving the protein folding problem by predicting the three-dimensional (secondary and tertiary) structure of a protein based on the amino acid sequence (primary structure). The binding of a substrate (or inhibitor) is crucial for the catalytic activity of an enzyme. Possible binding modes are investigated by molecular docking, in which various structures of intermolecular complexes are generated and evaluated.

B. Analyzing the Wave Function

The wave function, Ψ, which describes the electron distribution around a given nuclear arrangement, not only can be used to compute the energy of the system, but also offers useful interpretation opportunities.

1. Wave Function Analysis in Terms of Basis Functions

Each MO ϕ_i (used to build Ψ) is constructed from basis functions χ_i located on the nuclei and is occupied by n_i electrons (0, 1, or 2 for HF, but any number between 0 and 2 is possible for correlated wave functions). It is therefore possible to distribute the electrons to individual atoms. However, there is no unique prescription. Mulliken population analysis, for example, employs

$$\rho_A = \sum_{\alpha \in A}^{AO} \sum_{\beta}^{AO} D_{\alpha\beta} S_{\alpha\beta}$$

$$= \sum_{\alpha \in A}^{AO} \sum_{\beta}^{AO} \left(\sum_{i}^{MO} n_i c_{\alpha i} c_{\beta i} \right) \left(\int \chi_\alpha \chi_\beta \right)$$

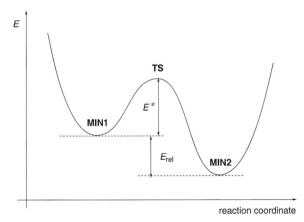

FIGURE 8 A model potential energy surface showing a transition structure, **TS**, connecting two minima, **MIN1** and **MIN2**; the activation barrier, $E^{\neq}$, for the transformation of **MIN1** to **MIN2**; and the relative energy, E_{rel}, of **MIN1** versus **MIN2**.

to determine the number of electrons ρ_A associated with atom A. The atomic charge, Q_A, is given as the difference between the nuclear charge Z_A and ρ_A:

$$Q_A = Z_A - \rho_A.$$

Bond orders b_{AB} can be defined on the basis of the sum of electrons shared between atoms A and B:

$$b_{AB} = \sum_{\alpha \in A} \sum_{\beta \in B} (\mathbf{DS})_{\alpha\beta} (\mathbf{DS})_{\beta\alpha}.$$

It is arbitrary how the electron density arising from basis functions located on different atoms is divided. The Mulliken procedure can give unphysical charges and shows a strong basis set dependence. Because of these shortcomings an atom definition which does not depend on the basis set is desirable.

2. Wave Function Analysis Based on the Electron Density

Bader's theory of atoms in molecules (AIM) is based on the electron density $\rho(\mathbf{r})$ (which can be computed by integrating the square of the wave function over the coordinates of all but one electron) and the gradient of the density, $\nabla \rho(\mathbf{r})$. Regions in space, so-called atomic basins, are defined as all points from which following the gradient of the density leads to a common attractor = nucleus. Integration of the electron density for each basin gives the number of electrons associated with the nucleus in that basin. However, AIM charges are counterintuitive in some cases. Neither AIM charges nor Mulliken charges reproduce the dipole or higher multipoles of a molecule. For points on the surface separating atomic basins, the derivative of the electron density along the normal vector equals zero. Points where the derivative perpendicular to the normal vector is zero as well are called bond critical points. They are the points of minimum electron density along the bond path, the path of maximum electron density connecting two nuclei. The electron density at the bond critical point correlates with the bond strength.

3. Localized MOs

The canonical MOs obtained as eigenfunctions of the Fock operator have contributions from all basis functions and thus are delocalized over all centers. They do not reflect the common picture of localized bonds between two atoms. However, the orbitals may be freely transformed by making linear combinations without changing the total wave function. Hence, an orbital rotation matrix can be applied to transform the canonical into localized orbitals which reflect bonds between two atoms. Several localization schemes were proposed, but the natural bond

orbital (NBO) analysis has some advantages (e.g., has no strong basis set dependence, is computationally inexpensive, and can also be applied to electron correlated methods). In a first step, natural atomic orbitals (NAOs) are generated by a diagonalization of the atomic blocks of the density matrix giving pre-NAOs which are orthogonalized in several steps applying occupancy-weighted and normal orthogonalization procedures. Diagonal values of the density matrix in the resulting NAO basis correspond to orbital populations which can be summed up to give atomic charges. Off-diagonal blocks define bonds between atoms. The resulting localized MOs give a description in agreement with chemical intuition (core orbitals, lone pairs, and bonds) and can be analyzed in terms of bond polarity, bond bending, hybridization, and so forth. Effects such as hyperconjugation can be investigated by analyzing the interactions between formally occupied and formally empty orbitals within a localized orbital picture.

C. Computing Properties

One-electron properties such as the electric dipole moment, the quadrupole moment, and the magnetic susceptibility can be evaluated from the wavefunction as the expectation value of an operator $\mathbf{O}$ which is a sum of n one-electron operators $\mathbf{o}$:

$$\langle \Psi | \mathbf{O} | \Psi \rangle = \sum_i \langle \phi_i | \mathbf{o} | \phi_i \rangle = \sum_\mu \sum_\nu D_{\mu\nu} \langle \chi_\mu | \mathbf{o} | \chi_\nu \rangle.$$

Most properties represent the response of the system to a perturbation. The effect can be calculated by derivative methods, perturbation theory, or propagator methods. For example the wave function, Ψ, is changed by an electric field which enters the Hamiltonian as part of the potential energy term. The energy in the presence of an electric field $\mathbf{F}$, $E(\mathbf{F})$, can be written as a Taylor expansion:

$$E(\mathbf{F}) = E_0 - \mu_0 \mathbf{F} - \frac{1}{2}\alpha \mathbf{F}^2 - \frac{1}{6}\beta \mathbf{F}^3 - \cdots,$$

where the permanent dipole moment, μ_0, the polarizability, α, and the (first) hyperpolarizability, β, are the first, second, and third derivatives, respectively, of the energy with respect to the field $\mathbf{F}$. The analogous expression for the presence of a magnetic field, $\mathbf{B}$, is

$$E(\mathbf{B}) = E_0 - \mathbf{m}_0 \mathbf{B} - \frac{1}{2\mu_0}\zeta \mathbf{B}^2 - \cdots,$$

where the magnetic moment, $\mathbf{m}_0$, and the magnetizability, ζ, are the first and second derivatives of the energy with respect to the magnetic field (μ_0 is the vacuum permeability).

D. Dynamics and Modeling Solvation

Standard computations treat isolated molecules (i.e., model the low-pressure gas phase situation). Chemistry in the condensed phase, however, can be significantly different because ionic and polar species are specifically stabilized. Computations can consider solvation explicitly by including the solvent molecules or as a continuous medium effect (reaction field methods).

When the solute is embedded in a bulk of solvent molecules, an "ensemble" (i.e., a large number of configurations of the molecular aggregation) has to be generated and properly averaged to derive macroscopic properties. Molecular dynamics (MD) creates a "trajectory": the configurations are obtained by following the evolution of the system in time by applying the classical equations of motion. The alternative, Monte Carlo (MC) method produces the configurations randomly: a starting configuration is perturbed and the new configuration is accepted if the new energy is lower ($\Delta E < 0$) than the old one. If $\Delta E > 0$ the new configuration is accepted with a probability proportional to the Boltzman factor, $\exp(-\Delta E/kT)$ (Metropolis method). The forces are needed for MD, but energies are sufficient for MC simulations. Meaningful averaging requires a huge number of conformations to be calculated, which leaves parametrized force fields as basically the only practible methods to be used in MD and MC simulations.

Reaction field methods model solutions by placing the solute in a cavity of a polarizable medium. The electrostatic potential due to the solute molecule polarizes the surrounding medium which in turn changes the charge distribution of the solute. Hence, the electrostatic interaction has to be evaluated self-consistently (self-consistent reaction field, SCRF). A term for creating the cavity (calculated from the surface of the cavity) has a be added to the solvation energy. Explicit treatment of solvent molecules can be combined with a reaction field method.

Dynamics can also be used to model the mechanics and rates of reactions at a fundamental level. While a complete potential energy surface is the ideal starting point of any type of accurate dynamical computation, it can be obtained for only very simple systems. Ab initio MD can be performed by repeated calculation of forces, generation of a new geometry, and convergence of the MO coefficients for the new geometry. This iterative scheme, however, is very time consuming because accurate MO coefficients are required at each point of the simulation. The Car-Parrinello method does not optimize the electronic (i.e., MO coefficients) and nuclear coordinates separately, but simultaneously, because it could be shown that the errors in the nuclear forces and in the electronic forces cancel out.

E. Spectroscopic Data

Vertical and adiabatic ionization energies are given as the energy difference between the neutral molecule and the corresponding cation in the neutral geometry and in the relaxed geometry, respectively. Analogous comparison with the anion gives electron affinities. The various ionization energies measured in PE (photoelectron) spectroscopy can be computed as differences between the neutral ground state and different electronic states of the cation. The energy difference between the electronic ground state and electronically excited states without the loss of an electron corresponds to the transitions observed in UV-VIS (ultraviolet-visible) spectroscopy.

Many other properties measured by spectroscopic methods can be computed as derivatives of the energy. The force constants (second-derivative w.r.t. the nuclear coordinates, $\mathbf{r}$) allow one to calculate harmonic vibrational frequencies and the corresponding normal modes. The derivatives of the dipole moment and of the polarizability w.r.t. the normal modes are proportional to the intensity of infrared absorptions and of Raman bands, respectively.

First and second derivatives of the energy w.r.t. nuclear magnetic spin, $\mathbf{I}$, give the hyperfine coupling constant $\mathbf{g}$ (measured by electron spin resonance, ESR, spectroscopy) and the nuclear coupling constants $\mathbf{J}$ of nuclear magnetic resonance (NMR) spectroscopy, respectively. The nuclear magnetic shielding constants, σ, are given as the mixed derivatives of the energy w.r.t. an external magnetic field $\mathbf{B}$ and the magnetic moments $\mathbf{I}$ of the nuclei. The NMR chemical shifts correspond to differences in σ for a nucleus in a given molecule and in a reference compound. Magnetic properties suffer from the "gauge origin" problem because the magnetic field is a vector potential, which is not uniquely defined. For finite basis sets, results depend on the choice for the origin. This problem can be largely overcome by using either the gauge-invariant atomic orbital (GIAO) method or the individual gauge for localized orbital (IGLO) method. Comparison of calculated and measured spectroscopic data can help to identify new molecules and to determine their structure.

The wealth of properties (Fig. 9) that can be computed as derivatives of the energy stresses the importance of the derivative methods. The ability to derive expressions for at least first and second derivatives is always desirable for a new theoretical level.

IV. OUTLOOK FOR COMPUTATIONAL CHEMISTRY

During the last decades computational chemistry has evolved into an indispensable tool for understanding and

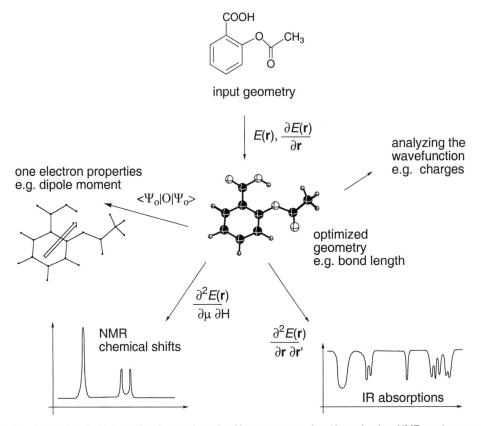

FIGURE 9 Some chemical information that can be gained from a computational investigation. NMR, nuclear magnetic resonance; IR, infrared.

predicting molecular properties and chemical behavior. The era foreseen by R. S. Mulliken in his acceptance speech for the 1966 Nobel Prize has arrived:

I would like to emphasize strongly my belief that the era of computing chemists, when hundreds if not thousands of chemists will go to the computing machine instead of the laboratory for increasingly many facets of chemical information, is already at hand.

Future goals of computational chemistry will be to determine structural, spectroscopic, and dynamic properties of even larger molecular system, also in the condensed phase, with even better accuracy and speed. This will allow computational chemistry to also play an important role in fields related to chemistry such as biology and material science.

ACKNOWLEDGMENTS

The authors are grateful to the following persons for comments and valuable suggestions: H. F. Bettinger, M. Bühl, B. Goldfuss, I. Hyla-Kryspin, K. N. Kirschner, P. v. R. Schleyer, and A. Y. Timoshkin. M.H. is grateful to Professor R. Krämer for support.

SEE ALSO THE FOLLOWING ARTICLES

MECHANICS, CLASSICAL • MOLECULAR ELECTRONICS • PHOTOELECTRON SPECTROSCOPY • PROTEIN STRUCTURE • QUANTUM CHEMISTRY • QUANTUM MECHANICS

BIBLIOGRAPHY

Dirac, P. A. M., (1929). *Proc. R. Soc. Lond. Ser. A*, **123,** 719.

Jensen, F. (1999). "Introduction to Computational Chemistry," Wiley, Chichester, UK.

Leach, A. R. (1996). "Molecular Modelling Principles and Applications," Longman, Essex, England.

Lipkowitz, K. B., and Boyd, D. B., eds. (1990–). "Reviews in Computational Chemistry," Vol. 1–, VCH, New York.

Mulliken, R. S. (1967). *Science*, **137,** 13–24.

Schleyer, P. V. R. et al., eds. (1998). "Encyclopedia of Computational Chemistry," Wiley, Chichester, UK.

Young, D. (2001). "Computational Chemistry," Wiley, Interscience, New York.

Computer Algorithms

Conor Ryan

University of Limerick

GLOSSARY

Algorithm Sequence of well-defined instructions the execution of which results in the solution of a specific problem. The instructions are unambiguous and each can be performed in a finite amount of time. Furthermore, the execution of all the instructions together takes only a finite amount of time.

Approximation algorithm Algorithm that is guaranteed to produce solutions whose value is within some prespecified amount of the value of an optimal solution.

Asymptotic analysis Analysis of the performance of an algorithm for large problem instances. Typically the time and space requirements are analyzed and provided as a function of parameters that reflect properties of the problem instance to be solved. Asymptotic notation (e.g., big "oh," theta, omega) is used.

Deterministic algorithm Algorithm in which the outcome of each step is well defined and determined by the values of the variables (if any) involved in the step.

For example, the value of $x + y$ is determined by the values of x and y.

Heuristic Rule of thumb employed in an algorithm to improve its performance (time and space requirements or quality of solution produced). This rule may be very effective in certain instances and ineffective in others.

Lower bound Defined with respect to a problem. A lower bound on the resources (time or space) needed to solve a specified problem has the property that the problem cannot be solved by any algorithm that uses less resources than the lower bound.

Nondeterministic algorithm Algorithm that may contain some steps whose outcome is determined by selecting from a set of permissible outcomes. There are no rules determining how the selection is to be made. Rather, such an algorithm terminates in one of two modes: success and failure. It is required that, whenever possible, the selection of the outcomes of individual steps be done in such a way that the algorithm terminates successfully.

Encyclopedia of Physical Science and Technology, Third Edition, Volume 3

NP-Complete problem Decision problem (one for which the solution is "yes" or "no") that has the following property: The decision problem can be solved in polynomial deterministic time if all decision problems that can be solved in nondeterministic polynomial time are also solvable in deterministic polynomial time.

Performance Amount of resources (i.e., amount of computer time and memory) required by an algorithm. If the algorithm does not guarantee optimal solutions, the term "performance" is also used to include some measure of the quality of the solutions produced.

Probabilistically good algorithm Algorithm that does not guarantee optimal solutions but generally does provide them.

Simulated annealing Combinatorial optimization technique adapted from statistical mechanics. The technique attempts to find solutions that have value close to optimal. It does so by simulating the physical process of annealing a metal.

Stepwise refinement Program development methods in which the final computer program is arrived at in a sequence of steps. The first step begins close to the problem specification. Each step is a refinement of the preceding one and gets one closer to the final program. This technique simplifies both the programming task and the task of proving the final program correct.

Usually good algorithm Algorithm that generally provides optimal solutions using a small amount of computing resources. At other time, the resources required may be prohibitively large.

IN ORDER to get a computer to solve a problem, it is necessary to provide it with a sequence of instructions that if followed faithfully will result in the desired solution. This sequence of instructions is called a computer algorithm. When a computer algorithm is specified in a language the computer understands (i.e., a programming language), it is called a program. The topic of computer algorithms deals with methods of developing algorithms as well as methods of analyzing algorithms to determine the amount of computer resources (time and memory) required by them to solve a problem and methods of deriving lower bounds on the resources required by any algorithm to solve a specific problem. Finally, for certain problems that are difficult to solve (e.g., when the computer resources required are impractically large), heuristic methods are used.

I. ALGORITHMS AND PROGRAMS

An *algorithm* can take many forms of detail. Often the level of detail required depends on the target of the algorithm. For example, if one were to describe an algorithm

on how to make a cup of tea to a human, one could use a relatively coarse (high) level of detail. This is because it is reasonable to assume that the human in question can fill in any gaps in the instructions, and also will be able to carry out certain tasks without further instructions, e.g., if the human is required to get a cup from a cupboard, it would be fair to assume that he/she knows how to do this without elaboration on the task.

On the other hand, a *program* is generally a *computer program*, and consists of a set of instructions at a very fine level of detail. A fine level of detail is required because computer programs are always written in a particular *language*, e.g., Basic, C++, Pascal, etc. Furthermore, every step in a task must be specified, because no background knowledge can be assumed. An often used distinction is that an algorithm specifies *what* a process is doing, while a program specifies *how* the process should be done. The truth is probably somewhere between these two extremes—while an algorithm should be a clear statement of what a process is doing, it is often useful to have some level of specification of functionality in an algorithm.

It is not very natural for humans to describe tasks with the kind of level of detail usually demanded by a programming language. It is often more natural to think in a *top-down* manner, that is, describe the problem in a high level manner, and then rewrite it in more detail, or even in a specific computer language. This can often help the person concerned to get a problem clear in his/her own mind, before committing it to computer. Much of this chapter is concerned with the process of *refinement*. Refinement of algorithms is (usually) an iterative process, where one begins with a very high level—that is, the *what*—and by repeatedly modifying the algorithm by adding more detail (the *how*) brings the algorithm closer and closer to being code, until the final coding of the algorithm becomes a very clear task. Ideally, when one is writing a program, one should not have to figure out any logic problems; all of these should be taken care of in the algorithm.

Algorithms are not just used as an aid for programmers. They are also a very convenient way to describe what a task does, to help people conceptualize it at a high level, without having to go through masses of computer code line by line.

Consider the following problem, which we will state first in English:

Mary intends to open a bank account with an initial deposit of $100. She intends to deposit an additional $100 into this account on the first day of each of the next 19 months for a total of 20 deposits (including the initial deposit). The account pays interest at a rate of 5% per annum compounded monthly. Her initial deposit is also on the first day of the month. Mary would like to know what the balance in her account will be at the end of the 20 months in which she will be making a deposit.

In order to solve this problem, we need to know how much interest is earned each month. Since the annual interest rate 5%, the monthly interest rate is 5/12%. Consequently, the balance of the end of a month is

$$(\text{initial balance} + \text{interest})$$
$$= (\text{initial balance}) * (1 + 5/1200)$$
$$= 241/240 \, (\text{initial balance})$$

Having performed this analysis, we can proceed to compute the balance at the end of each month using the following steps:

1. Let balance denote the current balance. The starting balance is $100, so set balance = 100.
2. The balance at the end of the month is $241/240 *$ balance. Update balance.
3. If 20 months have not elapsed, then add 100 to balance to reflect the deposit for the next month. Go to step 2. Otherwise, we are done.

This, then, is an algorithm for calculating the monthly balances. To refine the algorithm further, we must consider what kind of machine we wish to implement our algorithm on. Suppose that we have to compute the monthly balances using a computing device that cannot store the computational steps and associated data. A nonprogrammable calculator is one such device. The above steps a will translate into the following process:

1. Turn the calculator on.
2. Enter the initial balance as the number 100.
3. Multiply by 241 and then divide by 240.
4. Note the result down as a monthly balance.
5. If the number of monthly balances noted down is 20, then stop.
6. Otherwise, add 100 to the previous result.
7. Go to step 3.

If we tried this process on an electronic calculator, we would notice that the total time spent is not determined by the speed of the calculator. Rather, it is determined by how fast we can enter the required numbers and operators (add, multiply, etc.) and how fast we can copy the monthly balances. Even if the calculator could perform a billion computations per second, we would not be able to solve the above problem any faster. When a stored-program computing device is used, the above instructions need be entered into the computer only once. The computer can then sequence through these instructions at its own speed. Since the instructions are entered only once (rather than 20 times), we get almost a 20 fold speed up in the computation. If the balance for 1000 months is re-

quired, the speedup is by a factor of almost 1000. We have achieved this speedup without making our computing device any faster. We have simply cut down on the input work required by the slow human!

A different approach would have been to write program for the algorithm. The seven-step computational process stated above translates into the Basic program shown in Program 1.

PROGRAM 1

```
10 balance = 100
20 month = 1
30 balance = 241*balance/240
40 print month, "$";balance
50 if month = 20 then stop 60 month
     = month + 1
70 balance = balance + 100
80 go to 30
```

PROGRAM 2: Pascal Program for Mary's Problem

```
line program account(input,output)
1   {computer the account balance at
       the end of each month}
2   const  InitialBalance = 100;
3          MonthlyDeposit = 100;
4          TotalMonths = 20;
5          AnnualInterestRate = 5;
6   var balance, interest, MonthlyRate:
       real
7          month:integer;
8   begin
9          MonthlyRate := AnnualInterest-
           Rate/1200;
10         balance := InitialBalance;
11         writeln('Month Balance');
12         for month := 1 to TotalMonths
              do
13            begin
14              interest := balance *
                  MonthlyRage;
15              balance := balance
                  + interest;
16              writeln(month:10, ' ',
                  balance:10:2);
17              balance := balance +
                  MonthlyDeposit;
18            end; of for
19         writeln;
18  end; of account
```

In Pascal, this takes the form shown in Program 2. Apart from the fact that these two programs have been written in different languages, they represent different programming styles. The Pascal program has been written in such a way as to permit one to make changes with ease. The number of months, interest rate, initial balance, and monthly additions are more easily changed into Pascal program.

Each of the three approaches is valid, and the one that should eventually be used will depend on the user. If the task only needs to be carried out occasionally, then a calculator would probably suffice, but if it is to be executed hundreds or thousands of times a day, then clearly one of the computer programs would be more suitable.

II. ALGORITHM DESIGN

There are several design techniques available to the designer of a computer algorithm. Some of the most successful techniques are the following:

- Divide and conquer
- Greedy method
- Dynamic programming
- Branch and bound
- Backtracking

While we do not have the space here to elaborate each of these, we shall develop two algorithms using the divide and conquer technique. The essential idea in divide and conquer is to decompose a large problem instance into several smaller instances, solve the smaller instances, and combine the results (if necessary) to obtain the solution of the original problem instance. The problem we shall investigate is that of sorting a sequence $x[1], x[2], \ldots, s[n]$ of $n, n > 0$ numbers; where n is the size of the instance. We wish to rearrange these numbers so that they are in nondecreasing order (i.e., $x[1] < x[2], \ldots, x[n]$).

For example, if $n = 5$, and $(x[1], \ldots, x[5]) = (10,18, 8,12,19)$, then after the sort, the numbers are in order $(8,9,10,12,18)$. Even before we attempt an algorithm to solve this problem, we can write down and English version of the solution, as in Program 3. The correctness of this version of the algorithm is immediate.

PROGRAM 3: *First Version of Sort Algorithm*

```
Procedure sort;
Sort x[I], 1 < I < n into nondecreasing
    order;
End;{of sort}
```

Using the divide and conquer methodology, we first decompose the sort instance into several smaller instances. At this point, we must determine the size and number

of these smaller instances. Some possibilities are the following:

(a) One of size $n - 1$ and another of size 1
(b) Two of approximately equal size
(c) K of size approximately n/k each, for some integer $k, k > 2$

We shall pursue the first two possibilities. In each of these, we have two smaller instances created. Using the first possibility, we can decompose the instance (10,18,8,12,9) into any of the following pairs of instances:

(a) (10,18,8,12) (9)
(b) (10) (18,8,12,9)
(c) (10,18,8,9) (12)
(d) (10,18,12,9) (8)

and so on. Suppose we choose the first option. Having decomposed the initial instance into two, we must sort the two instances and then combine the two-sorted sequences into one. When (10,18,8,12) is sorted, the result is (8,10,12,18). Since the second sequence is of size 1, it is already in sorted order. To combine the two sequences, the number 9 must be inserted into the first sequence to get the desired five-number sorted sequence. The preceding discussion raises two questions. How is the four-number sequence sorted? How is the one-number sequence inserted into the sorted four-number sequence? The answer to the first is that this, too, can be sorted using the divide and conquer approach. That is, we decompose it into two sequences: one of size 3 and the other of size 1. We then sort the sequence of size 3 and then insert the 1 element sequence. To sort the three-element sequence, we decompose it into two sequences of size 2 and size 1, respectively. To sort the sequence of size 2, we decompose it into two of size 1 each. At this point, we need merely insert one into the other. Before attempting to answer the second question, we refine Program 3, incorporating the above discussion. The result is Program 4.

PROGRAM 4: *Refinement of Program 3*

```
line procedure sort(n)
1  {sort n numbers into nondecreasing
      order}
2  if n > 1 then begin
3      sort(n - 1); sort the first
          sequence
4      insert(n - 1, x[n]);
5      end; {of if}
6      writeln;
7  end;{of sort}
```

Program 4 is a recursive statement of the sort algorithm being developed. In a recursive statement of an algorithm, the solution for an instance of size n is defined in terms of solutions for instances of smaller size. In Program 4, the sorting of n items, for $n > 1$ items, is defined in terms of the sorting of $n - 1$ items. This just means that to sort n items using procedure sort, we must first use this procedure to sort $n - 1$ items. This in turn means that the procedure must first be used to sort $n - 2$ items, and so on. This use of recursive generally poses no problems, as most contemporary programming languages support recursive programs. To refine Program 4, we must determine how an insert is performed. Let us consider an example. Consider the insertion of 9 into (8,10,12,18). We begin by moving 9 from position 5 of the sequence and then comparing 9 and 18, since 18 is larger, it must be brought to the right of 9. So 18 is moved to position 5. The resulting sequence is as follows ("—" denoted an empty position in the sequence):

$$8\ 10\ 12\ —\ 18$$

Next, 9 is compared with 12, and 12 is moved to position 4. This results in the following sequence:

$$8\ 10\ —\ 12\ 18$$

Then 9 is compared with 10 and 10 is moved to position 3. At this time we have the sequence:

$$8\ —\ 10\ 12\ 18$$

Finally, 9 is compared with 8. Since 9 is not smaller than 8, it is inserted into position 2. This results in the sequence (8,9,10,12,18). With this discussion, we can refine Program 4 to get Program 5. Program 5 is then easily refined to get the Pascal-like code of Program 6. This code uses position 0 of the sequence to handle insertions into position 1. The recursion in this procedure can be eliminated to get Program 7. The algorithm we have just developed for sorting is called insertion sort. This algorithm was obtained using the stepwise refinement process beginning with Program 3. As a result of using this process, we have confidence in the correctness of the resulting algorithm. Formal correctness proofs can be obtained using mathematical induction or other program verification methods.

PROGRAM 5: Refinement of Program 4

line procedure *sort* (*n*)

```
1 {sort n numbers into nondecreasing
     order}
2 if n > 1 then begin
3       sort(n - 1); sort the first
             sequence
4       assign t the value x[n];
5       compare t with the xs
             beginning at x[n - 1];
```

```
6       move the xs up until the
             correct place for t is
             found
7       insert t into this place
8       end; {of if}
9 end; {of sort}
```

PROGRAM 6: Refinement of Program 5

line procedure *sort* (*n*)

```
1 {sort n numbers into nondecreasing
     order}
2 if n > 1 then begin
3       sort(n - 1); sort the first
             sequence
4       assign t and x[0]the value
             x[n];
5       assign i the value n - 1;
6       while t < x[i] do {find correct
             place for t};
7       begin
8           move x[i] to x[i + 1];
9           reduce i by 1;
10      end; {of while}
11      put t into x[i + 1]
12      end; {of if}
13 end; {of sort}
```

PROGRAM 7: Refinement of Program 6

line procedure *sort* (*n*)

```
1 {sort n numbers into nondecreasing
     order}
2 for j := 2 to n do
3 begin {insert x[j] into x[1:j - 1]}
4    assign t and x[0] the value x[j];
5       assign i the value j - 1;
6       while t < x[i] do {find correct place
             for t}
7       begin
8           move x[i] to x[i + 1];
9           reduce i by 1;
10      end; {of while}
11      put t into x[i + 1]
12      end; {of if}
13 end; {of sort}
```

Program 7 is quite close to being a Pascal program. One last refinement gives us a correct Pascal procedure to sort. This is given in Program 8. This procedure assumes that the numbers to be sorted are of type integer. In case numbers of a different type are to be sorted, the type of declaration of *t* should be changed. Another improvement of Program 8 can be made. This involves taking the statement **x[0]** := **t** out of the for loop and initializing **x[0]** to a very small number before the loop.

PROGRAM 8: *Refinement of Program 7*

line procedure *sort* (*n*)

```
 1 {sort n numbers into nondecreasing
     order}
 2 for j := 2 to n do
 3 begin {insert x[j] into x[1:j-1]}
 4     assign t and x[0] the value
         x[j];
 5     assign i the value j - 1;
 6     while t < x[i] do {find correct
         place for t}
 7     begin
 8         move x[i] to x[i+1];
 9         reduce i by 1;
10     end;{of while}
11     put t into x[i+1]
12 end;{of sort}
```

Let us consider the route our development process would have taken if we had decided to decompose sort instances into two smaller instances of roughly equal size. Let us further suppose that the left half of the sequence is one of the instances created and the right half is the other. For our example we get the instances (10,18) and (8,12,9). These are sorted independently to get the sequence (10,18) and (8.9,12). Next, the two sorted sequence are combined to get the sequence (8,9,10,12,18). This combination process is called merging. The resulting sort algorithm is called merge sort.

PROGRAM 9: *Final version of Program 8*

line procedure *sort*(*n*)

```
 1 {sort n numbers into nondecreasing
     order}
 2 var t, i, j : integer;
 2 begin
 2 for j := 2 to n do
 3 begin {insert x[j] into x[1 : j - 1]}
 4     t := x[j];x[0]:=t;i:=j-1;
 6     while t < x[i] do {find correct
         place for t}
 7     begin
 8         x[i + 1] := x[i];
 9         i := i - 1;
10     end;{of while}
11     x[i+1] := t
12 end;{of for}
13 end;{of sort}
```

PROGRAM 9: *Merge Sort*

line procedure *MergeSort*(*X*, *n*)

```
1 {sort n numbers in X}
2 if n > 1 then
3 begin
```

```
4 Divide X into two sequences A and B
    such that A contains ⌊n/2⌋
    numbers, and B the rest
5     MergeSort (A, ⌊ n/2⌋)
6     MergeSort (A, n - ⌊ n/2⌋)
7     merge (A, B);
8 end;{of if}
9 end;{of MergeSort }
```

Program 9 is the refinement of Program 3 that results for merge sort. We shall not refine this further here. The reader will find complete programs for this algorithm in several of the references sited later. In Program 9, the notation [*x*] is used. This is called the floor of *x* and denotes the largest integer less than or equal to *x*. For example, $\lfloor 2.5 \rfloor = 2$, $\lfloor -6.3 \rfloor = -7$, $\lfloor 5/3 \rfloor = 1$, and $\lfloor n/2 \rfloor$ denotes the largest integer less than or equal to $n/2$.

III. PERFORMANCE ANALYSIS AND MEASUREMENT

In the preceding section, we developed two algorithms for sorting. Which of these should we use? The answer to this depends on the relative performance of the two algorithms. The performance of an algorithm is measured in terms of the space and time needed by the algorithm to complete its task. Let us concentrate on time here. In order to answer the question "How much time does insertion sort take?" we must ask ourselves the following:

1. What is the instance size? The sort time clearly depends on how many numbers are being sorted.
2. What is the initial order? An examination of Program 7 means that it takes less time to sort *n* numbers that are already in nondecreasing order than when they are not.
3. What computer is the program going to be run on? The time is less on a fast computer and more on a slow one.
4. What programming language and compiler will be used? These influence the quality of the computer code generated for the algorithm.

To resolve the first two question, we ask for the worst case or average time as a function on instance size. The worst case size for any instance size *n* is defined as

$$T_W(n) = \max\{t(I) \mid I \text{ is an instance of size } n\}.$$

Here $t(I)$ denotes the time required for instance I. The average time is defined as:

$$(T_A)m = \frac{1}{N} \sum t(I).$$

where the sum is taken over all instances of size n and N is the number of such instances. In the sorting problem, we can restrict ourselves to the $n!$ different permutations of any n distinct numbers. So $N = n!$.

A. Analysis

We can avoid answering the last two questions by acquiring a rough count of the number of steps executed in the worst or average case rather than an exact time. When this is done, a paper and pencil analysis of the algorithm is performed. This is called performance analysis. Let us carry out a performance analysis on our two sorting algorithms. Assume that we wish to determine the worst-case step count for each. Before we can start we must decide the parameters with respect to which we shall perform the analysis. In our case, we shall obtain times as a function of the number n of numbers to be sorted. First consider insertion sort. Let $t(n)$ denote the worst-case step count of Program 6. If $n < 1$, then only one step is executed (verify that $n < 1$). When $n > 1$, the recursive call to sort $(n-1)$ requires $t(n-1)$ steps in the worst case and the remaining steps count for some linear function of n step executions in the worst case. The worst case is seen to arise when $x[N]$ is to be inserted into position 1. As a result of this analysis, we obtain the following recurrence for insertion sort:

$$t(n) = \begin{cases} a, & n <= 1, \\ t(n-1) + bn + c, & n > 1, \end{cases}$$

where a, b, and c are constants. This recurrence can be solved by standard methods for the solution of recurrences. For merge sort, we see from Program 9 that when $n < 1$, only a constant number of steps are executed. When $N > 1$, two calls to merge sort and one to merge are made. While we have not said much about the division into A and B is to be performed, this can be done in a constant amount of time. The recurrence for Merge Sort is now seen to be

$$t(n) = \begin{cases} a, & n <= 1, \\ t(\lfloor n/2 \rfloor) + t(n - \lfloor n/2 \rfloor) + m(n) + b, & n > 1, \end{cases}$$

where a and b are constants and $m(n)$ denotes the worst-case number of steps needed to merge n numbers. Solving this recurrence is complicated by the presence of the floor function. A solution for the case n is a power of 2 is easily obtained using standard methods. In this case, the floor function can be dropped to get the recurrence:

$$t(n) = \begin{cases} a, & n <= 1, \\ 2t(n/2) + m(n) + b, & n > 1. \end{cases}$$

The notion of a step is still quite imprecise. It denotes any amount of computing that is not a function of the parameters (in our case n). Consequently, a good approximate solution to the recurrences is as meaningful as an

exact solution. Since approximate solutions are often easier to obtain than exact ones, we develop a notation for approximate solutions.

Definition [Big "oh"]. $f(n) = O(g(n))$ (read as "f of n is big oh of g of n") iff there exist positive constants c and n_0 such that $f(n) \leq cg(n)$ for all n, $n \geq n_0$. Intuitively, $O(g(n))$ represents all functions $f(n)$ whose rate of growth is no more than that of $g(n)$.

Thus, the statement $f(n) = O(g(n))$ states only that $g(n)$ is an upper bound on the value of $f(n)$ for all n, $n > n$. It does not say anything about how good this bound is. Notice that $n = O(n^2)$, $n = O(n^{2.5})$, $n = 0(n^3) n = O(2^n)$, and so on. In order for the statement $f(n) = 0(g(n))$ to be informative, $g(n)$ should be a small function of n as one can come up with for which $f(n) = O(g(n))$. So while we often say $3n + 3 = O(n2)$, even though the latter statement is correct. From the definition of O, it should be clear that $f(n) = O(g(n))$ is not the same as $0(g(n)) = f(n)$. In fact, it is meaningless to say that $O(g(n)) = f(n)$. The use of the symbol $=$ is unfortunate because it commonly denoted the "equals" relation. Some of the confusion that results from the use of this symbol (which is standard terminology) can be avoided by reading the symbol $=$ as "is" and not as "equals." The recurrence for insertion sort can be solved to obtain

$$t(n) = O(n^2).$$

To solve the recurrence for merge sort, we must use the fact $m(n) = O(n)$. Using this, we obtain

$$t(n) = O(n \log n).$$

It can be shown that the average number of steps executed by insertion sort and merge sort are, respectively, $0(n^2)$ and $0(n \log n)$. Analyses such as those performed above for the worst-case and the average times are called asymptotic analyses. $0(n2)$ and $0(n \log n)$ are, respectively, the worst-case asymptotic time complexities of insertion and merge sort. Both represent the behavior of the algorithms when n is suitably large. From this analysis we learn that the growth rate of the computing time for merge sort is less than that for insertion sort. So even if insertion sort is faster for small n, when n becomes suitably large, merge sort will be faster. While most asymptotic analysis is carried out using the big "oh" notation, analysts have available to them three other notations. These are defined below.

Definition [Omega, Theta, and Little "oh"]. $f(n) = \Omega(g(n))$ read as "f of n is omega of g of n") iff there exist positive constants c and n_0 such that $f(n) \geq cg(n)$ for all n, $n \geq n_0$. $f(n)$ is $\Theta(g(n))$ (read as "f of n is theta of g of n") iff there exist positive constants c_1, c_2, and n_0 such that $c_1 g(n) \leq f(n) \leq c_2 g(n)$ for all n, $n \geq n_0$.

$f(n) = o(g(n))$ (read as "f of n is little oh of g of n") iff $\lim_{n \to \infty} f(n)/g(n) = 1$.

Example. $3n + 2 = \Omega(n)$; $3n + 2 = \Theta(n)$; $3n + 2 - o(3n)$; $3n^3 = \Omega(n^2)$; $2n^2 + 4n = \Theta(n^2)$; and $4n^3 + 3n^2 = o(4n^3)$.

The omega notation is used to provide a lower bound, while the theta notation is used when the obtained bound is both a lower and an upper bound. The little "oh" notation is a very precise notation that does not find much use in the asymptotic analysis of algorithms. With these additional notations available, the solution to the recurrence for insertion and merge sort are, respectively, $\Theta(n^2)$ and $\Theta(n \log n)$. The definitions of O, Ω, Θ, and o are easily extended to include functions of more than one variable. For example, $f(n, m) = O(g(n, m))$ if there exist positive constants c, n_0 and m_0 such that $f(n, m) < cg(n, m)$ for all $n > n_0$ and all $m > m_0$. As in the case of the big "oh" notation, there are several functions $g(n)$ for which $f(n) = \Omega(g(n))$. The $g(n)$ is only a lower bound on $f(n)$. The θ notation is more precise that both the big "oh" and omega notations. The following theorem obtains a very useful result about the order of $f(n)$ when $f(n)$ is a polynomial in n.

Theorem 1. Let $f(n) = a_m n^m + a_{m-1} n^{m-1} + \cdots + a_0$, $a_m \neq 0$.

(a) $f(n) = O(n^m)$
(b) $f(n) = \Omega(n^m)$
(c) $f(n) = \Theta(n^m)$
(d) $f(n) = o(a_m n^m)$

Asymptotic analysis can also be used for space complexity. While asymptotic analysis does not tell us how many seconds an algorithm will run for or how many words of memory it will require, it does characterize the growth rate of the complexity. If an $\Theta(n^2)$ procedure takes 2 sec when $n = 10$, then we expect it to take 8 sec when $n = 20$ (i.e., each doubling of n will increase the time by a factor of 4). We have seen that the time complexity of an algorithm is generally some function of the instance characteristics. As noted above, this function is very useful in determining how the time requirements vary as the instance characteristics change. The complexity function can also be used to compare two algorithms A and B that perform the same task. Assume that algorithm A has complexity $\Theta(n)$ and algorithm B is t of complexity $\Theta(n^2)$. We can assert that algorithm A is faster than algorithm B for "sufficiently large" n. To see the validity of this assertion, observe that the actual computing time of A is bounded from above by $c*n$ for some constant c and for all n, $n \geq n_2$, while that of B is bounded from below by $d*n^2$ for some constant d and all n, $n \geq n_2$. Since

TABLE I Values of Selected Asymptotic Functions

$\log n$	n	$n \log n$	n^2	n^3	2^n
0	1	0	1	1	2
1	2	2	4	8	4
2	4	8	16	64	16
3	8	24	64	512	256
4	16	64	256	4,096	65,536
5	32	160	1,024	32,768	4,294,967,296

$cn \leq dn^2$ for $n \geq c/d$, algorithm A is faster than algorithm B whenever $n \geq \max\{n_1, n_2, c/d\}$.

We should always be cautiously aware of the presence of the phrase "sufficiently large" in the assertion of the preceding discussion. When deciding which of the two algorithms to use, we must know whether the n we are dealing with is in fact "sufficiently large." If algorithm A actually runs in $10^6 n$ msec while algorithm B runs in n^2 msec and if we always have $n \leq 10^6$, then algorithm B is the one to use.

Table I and Fig. 1 indicate how various asymptotic functions grow with n. As is evident, the function 2^n grows very rapidly with n. In fact, if a program needs 2^n steps for execution, then when $n = 40$ the number of steps needed is $\sim 1.1 \times 10^{12}$. On a computer performing 1 billion steps per second, this would require ~ 18.3 min (Table II). If $n = 50$, the same program would run for ~ 13 days on this computer. When $n = 60$, ~ 310.56 years will be required to execute the program, and when $n = 100 \sim 4x1013$ years will be needed. So we may conclude that the utility of programs with exponential complexity is limited to small n (typically $n \sim 40$). Programs that have a complexity that is a polynomial of high degree are also of limited utility.

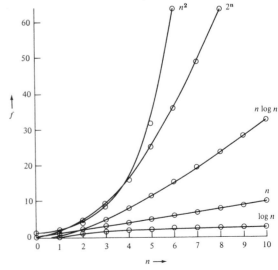

FIGURE 1 Plot of selected asymptotic functions.

TABLE II   Times on a 1 Billion Instruction per Second Computer[a]

n	Time for $f(n)$ instructions on a 10^9 instruction/sec computer						
	$f(n)=n$	$f(n)=n\log_2 n$	$f(n)=n^2$	$f(n)=n^3$	$f(n)=n^4$	$f(n)=n^{10}$	$f(n)=n^n$
10	0.01 μsec	0.03 μsec	0.1 μsec	1 μsec	10 μsec	10 sec	1μsec
20	0.02 μsec	0.09 μsec	0.4 μsec	8 μsec	160 μsec	2.84 hr	1 msec
30	0.03 μsec	0.15 μsec	0.9 μsec	27 μsec	810 μsec	6.83 day	1 sec
40	0.04 μsec	0.21 μsec	1.6 μsec	64 μsec	2.56 msec	121.36 day	18.3 min
50	0.05 μsec	0.28 μsec	2.5 μsec	125 μsec	6.25 msec	3.1 yr	13 day
100	0.10 μsec	0.66 μsec	10 μsec	1 msec	100 msec	3171 yr	4×10^3 yr
1,000	1.00 μsec	9.96 μsec	1 msec	1 sec	16.67 min	3.17×10^3 yr	32×10^{283} yr
10,000	10.00 μsec	130.3 μsec	100 msec	16.67 min	115.7 day	3.17×10^{23} yr	—
100,000	100.00 μsec	1.66 msec	10 sec	11.57 day	3171 yr	3.17×10^{33} yr	—
1,000,000	1.00 msec	19.92 msec	16.67 min	31.71 yr	3.17×10^7 yr	3.17×10^{43} yr	—

[a] 1 μsec = 10^{-6} sec; 1 msec = 10^{-3} sec.

For example, if a program needs n^{10} steps, then using our 1 billion steps per second computer (Table II) we will need 10 sec when $n = 10$; 3171 years when $n = 100$; and $3.17 * 10^{13}$ years when $n = 1000$. If the program's complexity were n^3 steps instead, we would need 1 sec when $n = 1000$; 110.67 min when $n = 10,000$; and 11.57 days when $n = 100,000$. From a practical standpoint, it is evident that for reasonably large n (say $n > 100$) only programs of small complexity (such as n, $n \log n$, n^2, n^3, etc.) are feasible. Furthermore, this would be the case even if one could build a computer capable of executing 10^12 instructions per second. In this case, the computing times of Table II would decrease by a factor of 1000. Now, when $n = 100$, it would take 3.17 years to execute n^{10} instructions, and 4×10^{10} years to execute 2^n instructions.

B. Measurement

In a performance measurement, actual times are obtained. To do this, we must refine our algorithms into computer programs written in a specific programming language and compile these on a specific computer using a specific compiler. When this is done, the two programs can be given worst-case data (if worst-case times are desired) or average-case data (if average times are desired) and the actual time taken to sort measured for different instance sizes. The generation of worst-case and average test data is itself quite a challenge. From the analysis of Program 7, we know that the worst case for insertion sort arises when the number inserted on each iteration of the **for** loop gets into position 1. The initial sequence $(n, n-1, \ldots, 2, 1)$ causes this to happen. This is the worst-case data for Program 7. How about average-case data? This is somewhat harder to arrive at. For the case of merge sort, even the worst-case data are difficult to devise. When it becomes difficult to generate the worst-case or average data, one resorts to

simulations. Suppose we wish to measure the average performance of our two sort algorithms using the programming language Pascal and the TURBO Pascal (TURBO is a trademark of Borland International) compiler on an IBM-PC. We must first design the experiment. This design process involves determining the different values of n for which the times are to be measured. In addition, we must generate representative data for each n. Since there are $n!$ different permutations of n distinct numbers, it is impractical to determine the average run time for any n (other than small n's, say $n < 9$) by measuring the time for all $n!$ permutations and then computing the average. Hence, we must use a reasonable number of permutations and average over these. The measured average sort times obtained from such experiments are shown in Table III. As predicted by our earlier analysis, merge sort is faster than insertion sort. In fact, on the average, merge sort will sort 1000 numbers in less time than insertion sort will take for 300! Once we have these measured times, we can fit a curve (a quadratic in the case of insertion sort and an $n \log n$ in the case of merge sort) through them and then use the equation of the curve to predict the average times for values of n for which the times have not been measured. The quadratic growth rate of the insertion sort time and the $n \log n$ growth rate of the merge sort times can be seen clearly by plotting these times as in Fig. 2. By performing additional experiments, we can determine the effects of the compiler and computer used on the relative performance of the two sort algorithms. We shall provide some comparative times using the VAX 11780 as the second computer. This popular computer is considerably faster than the IBM-PC and costs ~100 times as much. Our first experiment obtains the average run time of Program 8 (the Pascal program for insertion sort). The times for the V AX llnso were obtained using the combined translator and interpretive executer, pix. These are shown in

TABLE III   Average Times for Merge and Insertion Sort[a]

n	Merge	Insert
0	0.027	0.032
10	1.524	0.775
20	3.700	2.253
30	5.587	4.430
40	7.800	7.275
50	9.892	10.892
60	11.947	15.013
70	15.893	20.000
80	18.217	25.450
90	20.417	31.767
100	22.950	38.325
200	48.475	148.300
300	81.600	319.657
400	109.829	567.629
500	138.033	874.600
600	171.167	—
700	199.240	—
800	230.480	—
900	260.100	—
1000	289.450	—

[a] Times are in hundredths of a second.

TABLE IV   Average Times for Insertion Sort[a]

n	IBM-PC turbo	VAX pix
50	10.9	22.1
100	38.3	90.47
200	148.3	353.9
300	319.7	805.6
400	567.6	1404.5

[a] Times are in hundredths of a second.

V AX. This time our insertion sort program ran faster on the V AX. However, as expected, when n becomes suitably large, insertion sort on the V AX is slower than merge sort on an IBM-PC. Sample times are given in Table V. This experiment points out the importance of designing good algorithms. No increase in computer speed can make up for a poor algorithm. An asymptotically faster algorithm will outperform a slower one (when the problem size is suitably large); no matter how fast a computer the slower algorithm is run on and no matter how slow a computer the faster algorithm is run on.

IV. LOWER BOUNDS

The search for asymptotically fast algorithms is a challenging aspect of algorithm design. Once we have designed an algorithm for a particular problem, we would like to know if this is the asymptotically best algorithm. If not, we would like to know how close we are to the asymptotically best algorithm. To answer these questions, we must determine a function $f(n)$ with the following property:

PI: Let A be any algorithm that solves the given problem. Let its asymptotic complexity be $O(g(n))$. $f(n)$ is such that $g(n) = \Omega(f(n))$.

That is, $f(n)$ is a lower bound on the complexity of every algorithm for the given problem. If we develop an

Table IV. As can be seen, the IBM-PC outperformed the V AX even though the V AX is many times faster. This is because of the interpreter pix. This comparison is perhaps unfair in that in one case a compiler was used and in the other an interpreter. However, the experiment does point out the potentially devastating effects of using a compiler that generates poor code or of using an interpreter. In our second experiment, we used the Pascal compiler, pc, on the

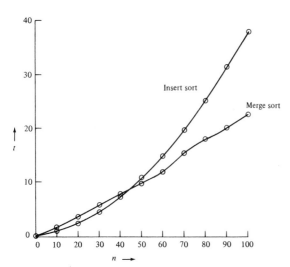

FIGURE 2   Plot of times of Table III.

TABLE V   Comparison between IBM-PC and VAX[a]

n	IBM-PC merge sort turbo	VAX insertion sort pc
400	109.8	64.1
500	138.0	106.1
600	171.2	161.8
700	199.2	217.9
800	230.5	263.5
900	260.1	341.9
1000	289.5	418.8

[a] Times are in hundredths of a second.

algorithm whose complexity is equal to the lower bound for the problem being solved, then the developed algorithm is optimal. The number of input and output data often provides a trivial lower bound on the complexity of many problems. For example, to sort n numbers it is necessary to examine each number at least once. So every sort algorithm must have complexity $\Omega(n)$. This lower bound is not a very good lower bound and can be improved with stronger arguments than the one just used. Some of the methods for obtaining nontrivial lower bounds are the following:

1. Information-theoretic arguments
2. State space arguments
3. Adversary constructions
4. Reducibility constructions

A. Information-Theoretic Arguments

In an information-theoretic argument, one determines the number of different behaviors the algorithm must exhibit in order to work correctly for the given problem. For example, if an algorithm is to sort n numbers, it must be capable of generating $n!$ different permutations of the n input numbers. This is because depending on the particular values of the n numbers to be sorted, any of these $n!$ permutations could represent the right sorted order. The next step is to determine how much time every algorithm that has this many behaviors must spend in the solution of the problem. To determine this quantity, one normally places restrictions on the kinds of computations the algorithm is allowed to perform. For instance, for the sorting problem, we may restrict our attention to algorithms that are permitted to compare the numbers to be sorted but not permitted to perform arithmetic on these numbers. Under these restrictions, it can be shown that $n \log n$ is a lower bound on the average and worst-case complexity of sorting. Since the average and worst-case complexities of merge sort is $\Theta(n \log n)$, we conclude that merge sort is an asymptotically optimal sorting algorithm under both the average and worst-case measures. Note that it is possible for a problem to have several different algorithms that are asymptotically optimal. Some of these may actually run faster than others. For example, under the above restrictions, there may be two optimal sorting algorithms. Both will have asymptotic complexity $\Theta(n \log n)$. However, one may run in $10n \log n$ time and the other in $2On \log n$ time. A lower bound $f(n)$ is a tight lower bound for a certain problem if this problem is, in fact, solvable by an algorithm of complexity $O(f(n))$. The lower bound obtained above for the sorting problem is a tight lower bound for algorithms that are restricted to perform only comparisons among the numbers to be sorted.

B. State Space Arguments

In the case of a state space argument, we define a set of states that any algorithm for a particular problem can be in. For example, suppose we wish to determine the largest of n numbers. Once again, assume we are restricted to algorithms that can perform comparisons among these numbers but cannot perform any arithmetic on them. An algorithm state can be described by a tuple (i, j). An algorithm in this state "knows" that j of the numbers are not candidates for the largest number and that $i = n - j$ of them are. When the algorithm begins, it is in the state $(n, 0)$, and when it terminates, it is in the state $(1, n - 1)$. Let A denote the set of numbers that are candidates for the largest and let B denote the set of numbers that are not. When an algorithm is in state (i, j), there are i numbers in A and j numbers in B. The types of comparisons one can perform are $A : A$ ("compare one number in A with another in A"), $A : B$, and $B : B$. The possible state changes are as follows:

$A : A$ This results in a transformation from the state (i, j) to the state $(i - 1, j + 1)$

$B : B$ The state (i, j) is unchanged as a result of this type of comparison.

$A : B$ Depending on the outcome of the comparison, the state either will be unchanged or will become $(i - 1, j + 1)$.

Having identified the possible state transitions, we must now find the minimum number of transitions needed to go from the initial state to the final state. This is readily seen to be $n - 1$. So every algorithm (that is restricted as above) to find the largest of n numbers must make at least $n - 1$ comparisons.

C. Adversary and Reducibility Constructions

In an adversary construction, one obtains a problem instance on which the purported algorithm must do at least a certain amount of work if it is to obtain the right answer. This amount of work becomes the lower bound. A reducibility construction is used to show that, employing an algorithm for one problem (A), one can solve another problem (B). If we have a lower bound for problem B, then a lower bound for problem A can be obtained as a result of the above construction.

V. NP-HARD AND NP-COMPLETE PROBLEMS

Obtaining good lower bounds on the complexity of a problem is a very difficult task. Such bounds are known for a handful of problems only. It is somewhat easier to relate the complexity of one problem to that of another using the notion of reducibility that we briefly mentioned in the last section. Two very important classes of reducible problems are NP-hard and NP-complete. Informally, all problems in the class NP-complete have the property that, if one can be solved by an algorithm of polynomial complexity, then all of them can. If an NP-hard problem can be solved by an algorithm of polynomial complexity, then all NP-complete problems can be so solved. The importance of these two classes comes from the following facts:

1. No NP-hard or NP-complete problem is known to be polynomially solvable.
2. The two classes contain more than a thousand problems that have significant application.
3. Algorithms that are not of low-order polynomial complexity are of limited value.
4. It is unlikely that any NP-complete or NP-hard problem is polynomially solvable because of the relationship between these classes and the class of decision problems that can be solved in polynomial nondeterministic time.

We shall elaborate the last item in the following subsections.

VI. NONDETERMINISM

According to the common notion of an algorithm, the result of every step is uniquely defined. Algorithms with this property are called deterministic algorithms. From a theoretical framework, we can remove this restriction on the outcome of every operation. We can allow algorithms to contain an operation whose outcome is not uniquely defined but is limited to a specific set of possibilities. A computer that executes these operations are allowed to choose anyone of these outcomes. This leads to the concept of a *nondeterministic algorithm*. To specify such algorithms we introduce three new functions:

- **choice**(S): Arbitrarily choose one of the elements of set S.
- **failure**: Signals an unsuccessful completion.
- **success**: Signals a successful completion.

Thus the assignment statement $x = $ **choice**$(1 : n)$ could result in x being assigned anyone of the integers in the range $[1, n]$. There is no rule specifying how this choice is to be made. The failure and success signals are used to define a computation of the algorithm. The computation of a nondeterministic algorithm proceeds in such a way that, whenever there is a set of choices that leads to a successful completion, one such set of choices is made and

the algorithm terminates successfully. A nondeterministic algorithm terminates unsuccessfully if there exists no set of choices leading to a success signal. A computer capable of executing a nondeterministic algorithm in this way is called a nondeterministic computer. (The notion of such a nondeterministic computer is purely theoretical, because no one knows how to build a computer that will execute nondeterministic algorithms in the way just described.)

Consider the problem of searching for an element x in a given set of elements $a[1 \ldots n]$, $n \geq 1$. We are required to determine an index j such that $a[j] = x$. If no such j exists (i.e., x is not one of the a's), then j is to be set to 0. A nondeterministic algorithm for this is the following:

```
j = choice(1:n)
if a[j] = x then print U); success endif
print ("0"); failure.
```

From the way a nondeterministic computation is defined, it follows that the number 0 can be output if there is no j such that $a[j] = x$. The computing times for **choice**, **success**, and **failure** are taken to be $O(1)$. Thus the above algorithm is of nondeterministic complexity $O(1)$. Note that since a is not ordered, every deterministic search algorithm is of complexity $\Omega(n)$. Since many choice sequences may lead to a successful termination of a nondeterministic algorithm, the output of such an algorithm working on a given data set may not be uniquely defined. To overcome this difficulty, one normally considers only decision problems, that is, problems with answer 0 or 1 (or true or false). A successful termination yields the output 1, while an unsuccessful termination yields the output 0. The time required by a nondeterministic algorithm performing on any given input depends on whether there exists a sequence of choices that leads to a successful completion. If such a sequence exists, the time required is the minimum number of steps leading to such a completion. If no choice sequence leads to a successful completion, the algorithm takes $O(1)$ time to make a failure termination. Nondeterminism appears to be a powerful tool. Program 10 is a nondeterministic algorithm for the sum of subsets problem. In this problem, we are given a multiset $w(1 \ldots n)$ of n natural numbers and another natural number M. We are required to determine whether there is a sub multiset of these n natural numbers that sums to M. The complexity of this nondeterministic algorithm is $O(n)$. The fastest deterministic algorithm known for this problem has complexity $O(2^{n/2})$.

PROGRAM 10: Nondeterministic Sum of Subsets

line procedure *NonDeterministicSumOfSubsets(W,n,M)*
2 **for** $i := 1$ to n **do**

```
3              x(i) := choice({0,1});
4              endfor
5     if ∑ⁿᵢ₌₁ w(i)x(i) = M then success
6              else failure;
7     endif;
7     end;
```

A. NP-Hard and NP-Complete Problems

The size of a problem instance is the number of digits needed to represent that instance. An instance of the sum of subsets problem is given by $(w(I), w(2), \ldots, w(n), M)$. If each of these numbers is a positive integer, the instance size is

$$\left\lceil \sum_{i=1}^{n} \log_2(w(i) + 1) \right\rceil + \lceil \log_2(M + 1) \rceil$$

if binary digits are used. An algorithm is of polynomial time complexity if its computing time is $O(p(m))$ for every input of size m and some fixed polynomial $p(-)$.

Let P be the set of all decision problems that can be solved in deterministic polynomial time. Let NP be the set of decision problems solvable in polynomial time by nondeterministic algorithms. Clearly, $P \subset NP$. It is not known whether $P = NP$ or $P \neq NP$. The $P = NP$ problem is important because it is related to the complexity of many interesting problems. There exist many problems that cannot be solved in polynomial time unless $P = NP$. Since, intuitively, one expects that $P \subset NP$, these problems are in "all probability" not solvable in polynomial time. The first problem that was shown to be related to the $P = NP$ problem, in this way, was the problem of determining whether a propositional formula is satisfiable. This problem is referred to as the satisfiability problem.

Theorem 2. Satisfiability is in P iff $P = NP$.

Let A and B be two problems. Problem A is polynomially reducible to problem B (abbreviated A reduces to B, and written as A α B) if the existence of a deterministic polynomial time algorithm for B implies the existence of a deterministic polynomial time algorithm for A. Thus if A α B and B is polynomially solvable, then so also is A. A problem A is NP-hard iff satisfiability α A. An NP-hard problem A is NP-complete if $A \in NP$. Observe that the relation α is transitive (i.e., if A α B and B α C, then A α C). Consequently, if A α B and satisfiability α A then B is NP-hard. So, to show that any problem B is NP-hard, we need merely show that A α B, where A is any known NP-hard problem. Some of the known NP-hard problems are as follows:

NP1: Sum of Subsets

Input: Multiset $W = \{w_i \mid 1 \leq i \leq n\}$ of natural numbers and another natural number M.

Output: "Yes" if there is a submultiset of What sums to M; "No" otherwise.

NP2: 0/1-Knapsack

Input: Multisets $P = \{P_i \mid 1 \leq i \leq n\}$ and $W = \{W_i \mid 1 \leq i \leq n\}$ of natural numbers and another natural number M.

Output: $x_i \in \{0, 1\}$ such that $\sum_i$ is maximized and $iWiXi\ M$.

NP3: Traveling Salesman

Input: A set of n points and distances $d(i, j)$. The $d(i, j)$ is the distance between the points i and j.

Output: A minimum-length tour that goes through each of the n points exactly once and returns to the start of the tour. The length of a tour is the sum of the distances between consecutive points on the tour. For example, the tour $1 \rightarrow 3 \rightarrow 2 \rightarrow 4 \rightarrow 1$ has the length $d(1, 3) + d(3, 2) + d(2, 4) + d(4, 1)$.

NP4: Chromatic Number

Input: An undirected graph $G = (V, E)$.

Output: A natural number k such that k is the smallest number of colors needed to color the graph. A coloring of a graph is an assignment of colors to the vertices in such a way that no two vertices that are connected by an edge are assigned the same color.

NP5: Clique

Input: An undirected graph $G = (V, E)$ and a natural number k.

Output: "Yes" if G contains a clique of size k (i.e., a subset $U \subset V$ of size k such that every two vertices in U are connected by an edge in E) or more. "No" otherwise.

NP6: Independent Set

Input: An undirected graph $G = (V, E)$ and a natural number k.

Output: "Yes" if G contains an independent set of size k (i.e., a subset $U \subset V$ of size k such that no two vertices in U are connected by an edge in E) or more. "No" otherwise.

NP7: Hamiltonian Cycle

Input: An undirected graph $G = (V, E)$.
Output: "Yes" if G contains a Hamiltonian cycle (i.e., a path that goes through each vertex of G exactly once and then returns to the start vertex of the path). "No" otherwise.

NP8: Bin Packing

Input: A set of n objects, each of size $s(i)$, $1 \le i \le n$ [$s(i)$ is a positive number], and two natural numbers k and C.
Output: "Yes" if the n objects can be packed into at most k bins of size c. "No" otherwise. When packing objects into bins, it is not permissible to split an object over two or more bins.

NP9: Set Packing

Input: A collection S of finite sets and a natural number k.
Output: "Yes" if S contains at least k mutually disjoint sets. "No" otherwise.

NP10: Hitting Set.

Input: A collection S of subsets of a finite set U and a natural number k.
Output: "Yes" if there is a subset V of U such that V has at most k elements and V contains at least one element from each of the subsets in S. "No" otherwise.

The importance of showing that a problem A is NP-hard lies in the $P = NP$ problem. Since we do not expect that $P = NP$, we do not expect. NP-hard problems to be solvable by algorithms with a worst-case complexity that is polynomial in the size of the problem instance. From Table II, it is apparent that, if a problem cannot be solved in polynomial time (in particular, low-order polynomial time), it is intractable, for all practical purposes. If A is NP-complete and if it does turn out that $P = NP$, then A will be polynomially solvable. However, if A is only NP-hard, it is possible for P to equal NP and for A not to be in P.

VII. COPING WITH COMPLEXITY

An optimization problem is a problem in which one wishes to optimize (i.e., maximize or minimize) an optimization function $f(x)$ subject to certain constraints $C(x)$. For example, the NP-hard problem NP2 (0/1-knapsack) is an optimization problem. Here, we wish to optimize (in this case maximize) the function $f(x) = \sum_{i=1}^{n} p_i x_i$ subject to the following constraints:

(a) $x_i \in \{0, 1\}$, $1 \le i \le n$
(b) $\sum_{i=1}^{n} w_i x_i \le M$

A feasible solution is any solution that satisfies the constraints $C(x)$. For the 0/1-knapsack problem, any assignment of values to the x_i'S that satisfies constraints (a) and (b) above is a feasible solution. An optimal solution is a feasible solution that results in an optimal (maximum in the case of the 0/1-knapsack problem) value for the optimization function. There are many interesting and important optimization problems for which the fastest algorithms known are impractical. Many of these problems are, in fact, known to be NP-hard. The following are some of the common strategies adopted when one is unable to develop a practically useful algorithm for a given optimization:

1. Arrive at an algorithm that always finds optimal solutions. The complexity of this algorithm is such that it is computationally feasible for "most" of the instances people want to solve. Such an algorithm is called a *usually good algorithm*. The simplex algorithm for linear programming is a good example of a usually good algorithm. Its worst-case complexity is exponential. However, it can solve most of the instances given it in a "reasonable" amount of time (much less than the worst-case time).
2. Obtain a computationally feasible algorithm that "almost" always finds optimal solutions. At other times, the solution found may have a value very distant from the optimal value. An algorithm with this property is called a *probabilistically good algorithm*.
3. Obtain a computationally feasible algorithm that obtains "reasonably" good feasible solutions. Algorithms with this property are called *heuristic algorithms*. If the heuristic algorithm is guaranteed to find feasible solutions that have value within a prespecified amount of the optimal value, the algorithm is called an approximation algorithm. In the remainder of this section, we elaborate on approximation algorithms and other heuristics.

A. Approximation Algorithms

When evaluating an approximation algorithm, one considers two measures: algorithm complexity and the quality of the answer (i.e., how close it is to being optimal). As in the case of complexity, the second measure may refer to the worst case or the average case. There are several categories of approximation algorithms. Let A be an algorithm that generates a feasible solution to every instance I of a problem P. Let $F^*(I)$ be the value of an optimal solution, and let $F'(I)$ be the value of the solution generated by A.

Definition. A is a k-absolute approximation algorithm for P iff $|F^*(I) - F'(I)| \leq k$ for all instances I. k is a constant. A is an $f(n)$-approximate algorithm for p iff $|F^*(I) - F'(I)|/F^*(I) \leq f(n)$ for all I. The n is the size of I and we assume that $|F^*(I)| > 0$. An $f(n)$-approximate algorithm with $f(n) \leq \varepsilon$ for all n and some constant ε is an ε-approximate algorithm.

Definition. Let $A(\varepsilon)$ be a family of algorithms that obtain a feasible solution for every instance I of P. Let n be the size of I. $A(\varepsilon)$ is an approximation scheme for P if for every $A(\varepsilon) > 0$ and every instance I, $|F^*(I) - F'(I)|F^*(I) \leq \varepsilon$. An approximation scheme whose time complexity is polynomial in n is a polynomial time approximation scheme. A fully polynomial time approximation scheme is an approximation scheme whose time complexity is polynomial in n and $1/\varepsilon$.

For most NP-hard problems, the problem of finding k-absolute approximations is also NP-hard. As an example, consider problem NP2 (01 1-knapsack). From any instance $(P_i, W_i, 1 \leq i \leq n, M)$, we can construct, in linear time, the instance $(k+1)p_i, w_i, \leq i \leq n, M)$. This new instance has the same feasible solutions as the old. However, the values of the feasible solutions to the new instance are multiples of $k+1$. Consequently, every k-absolute approximate solution to the new instance is an optimal solution for both the new and the old instance. Hence, k-absolute approximate solutions to the 0/1-knapsack problem cannot be found any faster than optimal solutions.

For several NP-hard problems, the ε-approximation problem is also known to be NP-hard. For others fast ε-approximation algorithms are known. As an example, we consider the optimization version of the bin-packing problem (NP8). This differs from NP8 in that the number of bins k is not part of the input. Instead, we are to find a packing of the n objects into bins of size C using the fewest number of bins. Some fast heuristics that are also ε-approximation algorithms are the following:

First Fit (FF). Objects are considered for packing in the order $1, 2, \ldots, n$. We assume a large number of bins arranged left to right. Object i is packed into the leftmost bin into which it fits.

Best Fit (BF). Let $cAvail[j]$ denote the capacity available in bin j. Initially, this is C for all bins. Object i is packed into the bin with the least $cAvail$ that is at least $s(i)$.

First Fit Decreasing (FFD). This is the same as FF except that the objects are first reordered so that $s(i) \geq s(i+1)1 \leq i leq n$.

Best Fit Decreasing (BFD). This is the same as BF except that the objects are reordered as for FFD. It should be possible to show that none of these methods guarantees optimal packings. All four are intuitively appealing and can be expected to perform well in practice. Let I be any instance of the bin packing problem. Let $b(I)$ be the number of bins used by an optimal packing. It can be shown that the number of bins used by FF and BF never exceeds $(17/10)b(I) + 2$, while that used by FFD and BFD does not exceed $(11/9)b(I) + 4$.

Example. Four objects with $s(1:4) = (3, 5, 2, 4)$ are to be packed in bins of size 7. When FF is used, object 1 goes into bin 1 and object 2 into bin 2. Object 3 fits into the first bin and is placed there. Object 4 does not fit into either of the two bins used so far and a new bin is used. The solution produced utilizes 3 bins and has objects 1 and 3 in bin I, object 2 in bin 2, and object 4 in bin 3.

When BF is used, objects 1 and 2 get into bins 1 and 2, respectively. Object 3 gets into bin 2, since this provides a better fit than bin I. Object 4 now fits into bin I. The packing obtained uses only two bins and has objects 1 and 4 in bin 1 and objects 2 and 3 in bin 2. For FFD and BFD, the objects are packed in the order 2,4, 1,3. In both cases, two-bin packing is obtained. Objects 2 and 3 are in bin 1 and objects 1 and 4 in bin 2. Approximation schemes (in particular fully polynomial time approximation schemes) are also known for several NP-hard problems. We will not provide any examples here.

B. Other Heuristics

PROGRAM 12: *General Form of an Exchange Heuristic*

1. Let j be a random feasible solution [i.e., $C(j)$ is satisfied] to the given problem.
2. Perform perturbations (i.e., exchanges) on i until it is not possible to improve j by such a perturbation.
3. Output i.

Often, the heuristics one is able to devise for a problem are not guaranteed to produce solutions with value close to optimal. The virtue of these heuristics lies in their capacity to produce good solutions most of the time. A general category of heuristics that enjoys this property is the class of exchange heuristics. In an exchange heuristic for an optimization problem, we generally begin with a feasible solution and change parts of it in an attempt to improve its value. This change in the feasible solution is called a perturbation. The initial feasible solution can be obtained using some other heuristic method or may be a randomly generated solution. Suppose that we wish to minimize the objective function $f(i)$ subject to the constraints C. Here, i denotes a feasible solution (i.e., one that satisfies C). Classical exchange heuristics follow the steps given in Program 12. This assumes that we start with a random feasible solution. We may, at times, start with a solution constructed by some other heuristic. The quality of the solution obtained using Program 12 can be improved by running this program several times. Each time, a different starting solution is used. The best of the solutions produced by the program is used as the final solution.

1. A Monte Carlo Improvement Method

In practice, the quality of the solution produced by an exchange heuristic is enhanced if the heuristic occasionally accepts exchanges that produce a feasible solution with increased $f(\ )$. (Recall that f is the function we wish to minimize.) This is justified on the grounds that a bad exchange now may lead to a better solution later. In order to implement this strategy of occasionally accepting bad exchanges, we need a probability function $\mathrm{prob}(i, j)$ that provides the probability with which an exchange that transforms solution i into the inferior solution j is to be accepted. Once we have this probability function, the Monte Carlo improvement method results in exchange heuristics taking the form given in Program 13. This form was proposed by N. Metropolis in 1953. The variables *counter* and n are used to stop the procedure. If n successive attempts to perform an exchange on i are rejected, then an optimum with respect to the exchange heuristic is assumed to have been reached and the algorithm terminates. Several modifications of the basic Metropolis scheme have been proposed. One of these is to use a sequence of different probability functions. The first in this sequence is used initially, then we move to the next function, and so on. The transition from one function to the next can be made whenever sufficient computer time has been spent at one function or when a sufficient number of perturbations have failed to improve the current solution.

PROGRAM 13: Metropolis Monte Carlo Method

1. Let i be a random feasible solution to the given problem. Set *counter* $= 0$.
2. Let j be a feasible solution that is obtained from i as a result of a random perturbation.
3. If $f(j) < f(i)$, then [$i = j$, update best solution found so far in case i is best, *counter* $= 0$, go to Step 2].
4. If $f(j) \geq f(i)$ If *counter* $= n$ then output best solution found and stop. Otherwise, $r =$ random number in the range $(0, 1)$.
   If $r < \mathrm{prob}(i, j)$, then [$i = j$, *counter* $= 0$] else [*counter* $=$ *counter* $+ 1$].
   go to Step 2.

The Metropolis Monte Carlo Method could also be referred to as a *metaheuristic*, that is, a heuristic that is general enough to apply to a broad range of problems. Similar to heuristics, these are not guaranteed to produce an optimal solution, so are often used in situations either where this is not crucial, or a suboptimal solution can be modified.

VIII. THE FUTURE OF ALGORITHMS

Computer algorithm design will always remain a crucial part of computer science. Current research has a number of focuses, from the optimization of existing, classic algorithms, such as the sorting algorithms described here, to the development of more efficient approximation algorithms. The latter is becoming an increasingly important area as computers are applied to more and more difficult problems.

This research itself can be divided into two main areas, the development of approximation algorithms for particular problems, e.g. Traveling salesman problem, and into the area of metaheuristics, which is more concerned with the development of general problem solvers.

IX. SUMMARY

In order to solve difficult problems in a reasonable amount of time, it is necessary to use a good algorithm, a good compiler, and a fast computer. A typical user, generally, does not have much choice regarding the last two of these. The choice is limited to the compilers and computers the user has access to. However, one has considerable flexibility in the design of the algorithm. Several techniques are available for designing good algorithms and determining how good these are. For the latter, one can carry out an asymptotic analysis. One can also obtain actual run times on the target computer. When one is unable to obtain a low-order polynomial time algorithm for a given problem, one can attempt to show that the problem is NP-hard or is related to some other problem that is known to be computationally difficult. Regardless of whether one succeeds in this endeavor, it is necessary to develop a practical algorithm to solve the problem. One of the suggested strategies for coping with complexity can be adopted.

SEE ALSO THE FOLLOWING ARTICLES

BASIC PROGRAMMING LANGUAGE ● C AND C++ PROGRAMMING LANGUAGE ● COMPUTER ARCHITECTURE ●

DISCRETE SYSTEMS MODELING • EVOLUTIONARY ALGO-
RITHMS AND METAHEURISTICS • INFORMATION THEORY
• SOFTWARE ENGINEERING • SOFTWARE RELIABILITY
• SOFTWARE TESTING

BIBLIOGRAPHY

Aho, A., Hopcroft, J., and Ullman, J. (1974). "The Design and Analysis of Computer Algorithms," Addison-Wesley, Reading, MA.

Canny, J. (1990). *J. Symbolic Comput.* **9** (3), 241–250.

Garey, M., and Johnson, D. (1979). "Computers and Intractability," Freeman, San Francisco, CA.

Horowitz, E., and Sahni, S. (1978). "Fundamentals of Computer Algorithms," Computer Science Press, Rockville, MD.

Kirkpatrick, S., Gelatt, C., and Vecchi, M. (1983). *Science* **220,** 671–680.

Knuth, D. (1972). *Commun. ACM* **15** (7), 671–677.

Nahar, S., Sahni, S., and Shragowitz, E. (1985). *ACM/IEEE Des. Autom. Conf.*, 1985, pp. 748–752.

Sahni, S. (1985). "Concepts in Discrete Mathematics," 2nd ed., Camelot, Fridley, MN.

Sahni, S. (1985). "Software Development in Pascal," Camelot, Fridley, MN.

Sedgewick, R. (1983). "Algorithms," Addison-Westley, Reading, MA.

Syslo, M., Deo, N., and Kowalik, J. (1983). "Discrete Optimization Algorithms," Prentice-Hall, Engle-wood Cliffs, NJ.

Computer Architecture

Joe Grimes
California Polytechnic State University

GLOSSARY

Accumulator Usually a part of the arithmetic-logic unit (ALU) of a computer that is used for intermediate storage.

Address Name identifying a location where data or instructions may be stored.

Assembler Language translator program that converts assembly language instructions into conventional machine language.

Assembly language A computer language that is more easily understood than the language that is translated into in order for the computer to understand it.

Bus A set of lines shared by one or more components of a computer system. In many computers, a common bus interconnects the memory, processor, input, and output units. Because the bus is shared, only one unit may send information at a given time.

Cache Memory that is accessed more rapidly than main memory but less rapidly than a register. Today often found in processor chip with several levels of cache present in the computer.

Combinational circuit A digital circuit whose output value(s), at a given point in time, is dependent only on the input values at that time.

Compiler Language translator program that converts application program language instruction, such as Pascal or FORTRAN, into a lower level language.

Complex instruction set computer (CISC) A class of computers whose instruction set (conventional machine language instructions) was large, instructions were of multiple lengths, instructions had variable formats, instructions took variable lengths of time to execute, etc.

Computer architecture The study of the computer as viewed by the programmer.

Control That part of the processor that controls itself and the datapath of the processor.

Datapath The components of the datapath may include registers, arithmetic and logic unit (ALU), shifter, bus, and multiplexors. It also contains the buses necessary in order to transfer information from one of the components to another.

Debugger Tool used by a programmer to locate and correct an error in a computer program.

Exception An action in the computer that will ultimately result in the suspension of the execution of a program. The exception is caused by an activity outside of the program and is handled in such a way that the program can be resumed after the exception is handled.

Exception cycle That part of the instruction cycle that checks for exceptions and handles them if they are pending. The exception is handled by saving the suspended program's state and resumes execution of instructions in the exception handler routine (program).

Input Movement of data from an input device, such as a keyboard, to the processor or another unit of the computer system.

Interrupt Special control signal(s) that diverts the flow of instruction execution to the instruction in the location associated with the particular interrupt.

Memory A unit where information (data and instructions) may be stored.

Memory consistency The same data stored in different modules is correctly stored.

Memory cycle Operations required to store or retrieve information from memory.

Microprogram Program that interprets conventional machine language instructions that are interpreted directly by the hardware.

Operating system Set of programs that monitor and operate computer hardware, and simplify such tasks as input/output (I/O), editing, and program translation. Also serves as an interface with the user.

Output Movement of data from a unit of the computer system to an output device such as a printer.

Program counter Register in which the location of the current instruction is stored.

Reduced instructions set computer (RISC) This class of computer had features that were the opposite of those of the CISC computers.

Registers Memory locations that are internal to processor and that can be accessed quickly. Some registers are available for programmer while the processor only uses others internally.

Sequential circuit A digital circuit whose output value(s), at a given point in time, is dependent on the input values as well as the state of the circuit at that time. Memory units are sequential circuits.

Virtual memory Approach used to extend main memory by using a combination of main memory and secondary memory to store a program.

COMPUTER TECHNOLOGY has come a long way from the days when a 1940s computer, without stored program capability, required the space of a family room and living room in a large home, and when turned on, would dim the lights of all the homes in the community. Computer advances have made it possible for computers to be as much a part of our lives as our brains. Like our brains we often use computers without knowing that we are doing so. Today, a computer controls the digital alarm clock, controls the microwave, assists the checkout person in a store that stocks computer coded items, partially controls vehicles with sometimes in excess of fifty computer controllers, is used for computer-assisted instruction, and in general, substantially automates the modern work environment. A computer purchased today for a thousand dollars has more performance, functionality, and storage space than a computer purchased for several million dollars in the 1960s. These computer advances result from technological as well as design breakthroughs. Over the past 30–35 years, performance advances have averaged around 25% each year. Technological advances have been more consistent than those of computer architecture. In a sense, computer architecture had been in a state of hibernation until the mid-1980s because microprocessors of the 1970s were merely miniature clones of mainframes.

I. INTRODUCTION

Trends such as the use of vendor independent operating systems (UNIX), making company proprietary environments public, and a large reduction in assembly language programming have opened a window of opportunity for new architectures to be successful. Starting in the mid-1980s, these opportunities, along with quantitative analyses of how computers are used, a new type of architecture has reared its head. Because of this and as a result of improvements in compilers, enhancements of integrated circuit technology, and new architectural ideas, performance has more than doubled each year between mid-1980s and the present.

Most textbooks related to this subject have titles that encompass both the architecture and the organization of a computer system. There is a difference between computer architecture and computer organization with computer architecture referring to the programmer model (those aspects that have a direct impact on the logical execution of a program) and computer organization referencing the operational units and their interconnections that result in the architectural specification. Architectural issues relate to such things as the instruction set, data structures, memory addressing modes, interrupt capabilities, and I/O interfaces. Organizational features include those hardware capabilities unavailable to the programmer that pertain to the implementation of the architecture and the interconnection of the computer components.

The study of computer architecture amounts to analyzing two basic principles, performance and cost for a set of computer architecture alternatives that meet functionality requirements. This appears to be straightforward when considered at a general level, but when these topics are looked at in detail the water can become muddy. If a systematic approach is used the concepts can be straightforward. Fortunately, central processing units, memory hierarchy, I/O systems, etc., are layered, and these layers may be considered as levels of abstractions. Each level of abstraction has its own abstractions and objects. By studying aspects of computer science in this fashion, it is possible to censor out details that are irrelevant to the level of abstraction being considered and to concentrate on the task at hand.

An analogy to the following effect has been made: if cost and performance improvements in automobiles had kept up with computers, a car would cost ten cents and would allow travel between Los Angeles and New York in five minutes. If this pace is to continue, it is believed that computer design must be studied in a layered fashion using a systematic approach afforded by quantitative techniques. Also, it may be appropriate to adopt new models for studying computers that are different from that of the traditional Von Neumann model (Fig. 1) developed by John Von Neumann in 1940. This Von Neumann con-

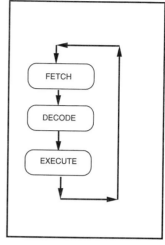

FIGURE 2 Block diagram of the instruction cycle for a computer.

cept is still the basis of many computers today with many improvements such as miniaturization and other contributions to performance improvement. Therefore, this article has been written to present this approach and create the enthusiasm in the readers so that they will be biting at the bit to chip in with their own new ideas. This section strives to introduce necessary background material and many of the concepts that will be addressed in more depth throughout the rest of the article.

It would be wonderful be able to come up with a good definition of a computer. It is impossible to give an all-encompassing definition, because a person working with computers continually refines his definition of it based on the increased understanding of its capabilities and enhancements that have occurred. A start at the definition might be Fig. 1, which gives a high-level view of the components of a Von Neumann computer with the five traditional components: datapath, control, memory, input, and output, with the former two sometimes combined and called the processor. Processors come in two flavors: imbedded processors that are internal to products such as radios, microwaves, automobiles, and standard processors that are part of a computer system. Every part of every computer can be classified under one of the five components. Computer architecture addresses how the programmer views these five components at the level of abstraction where they are programming and the computer organization addresses the implementation of the architecture in the inter- and intrarelationship of the components.

The fundamental responsibility of a computer is to execute a program, which consists of a set of instructions stored in the computer's memory. A computer operates by repeatedly performing an instruction cycle as depicted in a simple form in Fig. 2. The computer understands a simple set of instructions called an *instruction set* and cannot act

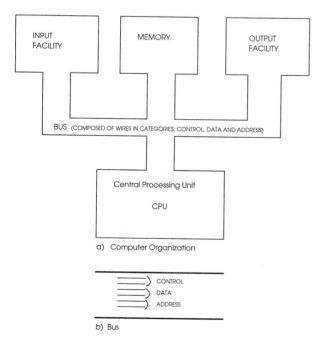

FIGURE 1 High-level Von Neumann computer organization.

on special cases that are not covered by the set of instructions. The computer would repeat this cycle over and over again for each instruction. This type of work is boring and best suited for a machine.

As mentioned previously, a computer has levels of abstractions. A person writing the instructions for the computer is a computer programmer. This view of the computer is at a high level without much detail about the way the computer operates. The person who would design a new path from the memory to the processor, or how the memory should handle the instruction cycle, would be viewing the computer at a much lower level of detail at the hardware level. An automobile would be a good analogy regarding levels of abstraction. The driver of the automobile views it at a much higher level of abstraction than the person designing any automobile system such as engine, steering, and brakes. A computer has the potential for performance improvements with performance inversely proportional to the time it takes to do a job. Performance is dependent on the architecture and organization of a computer, and should be traded off with cost.

The task of the computer architect is to participate in the phases of the life cycle, described later, of architecture that satisfies computer-user functional requirements and yet are economically implementable using available technology. All of this is quite relative, as something that is too costly in one case may be perfectly reasonable in another case.

This article will deal with these and other aspects of the computer definition in more detail.

II. HISTORICAL PERSPECTIVE

Each writer on the topic of computer history has a different perspective to present. Their presentations vary in terms of dates of events and who was responsible for certain breakthroughs. Absolute dates are not as important in the study of computer architecture as a relative perspective of what has happened and how quickly it has occurred. One good way to gain a perspective on the milestones or developments in the computer industry is through the classification of computers by generation, with generations distinguished by the technology used in the implementation. Figure 3 presents the classical generation classification of computers by the technologies used. There is no solid agreement between historians on defining generations, especially the dates. The entry of operations per second is not meant to be absolute and is meant to give only a picture of the relative performance improvements. Although the generations concept gives an overview of the historical development of computers, it is worthwhile to look at the details further. There are two interesting ways to do this: development paths of contributors could be reviewed, or the evolution of the abstraction-layered computer could be traced. The abstraction-layered computer will be reviewed here because it is important to the evolution of computer architecture.

It is widely accepted that the first operational general purpose computer was built in the 1940s. Because of recent research and legal activities, there is some controversy over whether the credit for this should be given to John Atanasoff and his graduate student Berry for the

Generation	Dates	Technology	Computers	Operations/Sec
0	300BC-1948	Mechanical or Electrical Mechanical	Abacus, Pascal Machine, Difference Engine	Few
1	1948-1957	Vacuum Tube	IBM701, UNIVAC I	40,000
2	1957-1967	Transistor	IBM 1401, CDC 6600, PDP 1	250,000
3	1967-1973	Integrated Circuits	IBM 360, ILLIAC IV, PDP 11, Data General Nova	1,000,000
4	1973-1980	Large Scale Integrated Circuits	Macintosh, Sun 2, DEC VAX	10,000,000
5	1980-Present	Very-Large-Scale Integrated Circuit		?

FIGURE 3 Computer generations.

Atanasoff/Berry Computer built in the early 1940s at Iowa State University, or J. Presper Eckert and John Mauchly for their ENIAC (Electronic Numerical Integrator and Calculator) built in the mid-1940s at the Moore School of the University of Pennsylvania, because Mauchly had briefly visited Atanasoff before he built the ENIAC. Its basic principles were enunciated in a memorandum written by von Neumann in 1945 and, largely because of this widely circulated report, von Neumann's name alone has come to be associated with concept of the stored-program computer.

One of the remarkable features of the von Neumann report was its focus on the logical principles and organization of the computer rather than on the electrical and electronic technology required for its implementation. Thus the idea of architecture appears early in the modern history of the computer. The word "architecture" itself, however, appears to have been first used in the early 1960s by the designers of the IBM System/360 series of computers. They used this word to denote the logical structure and functional characteristics of the computer as seen by the programmer. Over the years since, the theory and practice of computer architecture has extended to include not only external characteristics but also the computer's internal organization and behavior. Thus, a computer is now commonly viewed as possessing both an "outer" and an "inner" architecture.

The original digital computers had two abstractions, digital logic and something comparable to the conventional machine level (programming in binary) of today, two levels of detail. Because the digital logic level was unstructured in design and unreliable, M. V. Wilkes, in the early 1950s, proposed an intermediate abstraction, microprogramming that would perform some of the hardware tasks. A few of these computers were constructed in the 1950s. When IBM introduced the 360 family (all 360 family computer had the same architecture, but the organization was different for each family member) of computers shortly after the publication *Datamation* made the clever prediction that the concept of microprogramming would never catch on, microprogramming caught on like wild fire and by early the 1970s most computers had the abstraction 2 present. Recently things have gone full circle with true RISC computers not having the microprogramming abstraction present.

As mentioned previously, early computers had only two levels of abstraction. Each computer site had to develop their own program environment and it was a difficult as the computer manufacturer did not provide any software support. In the 1950s assemblers and compilers, supplied by computer vendors, were developed without operating systems. The 1950s programmer had to operate the computer and load the compiler or assembler along with the program

into the computer. At this point there were three levels of abstraction: the digital logic, conventional machine language, and application (in the case where compilers were present).

Starting around 1960, operating systems were supplied by the computer manufacturer to automate some of the operator's tasks. The operating system was stored in the computer at all times and automatically loaded the required compiler or assembler from magnetic tape as needed. The programmer still had to supply the program and necessary control information (request the specific compiler) on punched cards. During this era, the programmer would place the program (deck of punched cards) in a tray, and later an operator would carry this tray with many other programs into the sacred computer room (room maintained in fixed temperature and humidity range with reduced dust level), run the programs, and return the results and programs to the individual programmers. Often the time between placing the cards in the tray and receiving the results may have been several hours (even days at the end of the university term). In the meantime, if the due date were close, the programmer would sweat blood hoping that the program would run properly. This type of operating system was called a batch system.

In future years, operating systems became more sophisticated. Time-sharing operating systems were developed that allowed multiple programmers to communicate directly with the computer from terminals at local or remote sites. Other capabilities added to the operating system included such things as new instructions called "operating system calls." In the case of the UNIX operating system, it interprets the "cat" instruction but it is capable of much more than that. As compilers became much more sophisticated and assembly language programming became less common, there was an evolution from the CISC to RISC computer, and as a result a reduction in need for the microprogramming abstraction.

III. LEVELS OF ABSTRACTION

A computer architect may try to be a Don Quixote of computer science and try to view all aspects of a computer system at one time, or can proceed in a manageable fashion and view it as a set of abstractions or layers. Contemporary computers may be viewed as having at least two abstractions, with the bottom abstraction considered by computer architects being the digital logic abstraction, abstraction 0. Below abstraction 0 it would be possible to consider the device abstraction or even below that, the physics abstraction with the theory of devices. The abstraction above digital logic on some computers is microprogramming, abstraction 1. Often digital logic and microprogramming

Language/Machine	Abstraction	Abstraction	Hardware/Software Implemented
L6 (M6)	Application	A6	
L5 (M5)	Application Program	A5	
L4 (M4)	Assembly Language	A4	Software
L3 (M3)	Operating System	A3	
L2 (M2)	Conventional Machine	A2	
L1 (M1)	Microprogram	A1	
	Digital Logic	A0	Hardware

FIGURE 4

abstractions are grouped together and called the control. On almost all computers there is a conventional machine language abstraction. Figure 4 displays the levels that are present on many modern computers. A true RISC computer would not have the microprogramming abstraction present. A programmer dealing with a particular abstraction has a distinct set of objects and operations present, and need not be concerned with the abstractions below or above. Because abstractions are interdependent, a computer designer must ultimately be concerned with all levels present but need only deal with one or two at any time.

The digital logic abstraction is present on all computers and it defines the language understood by the computer. This language (L1) is the microprogramming abstraction in the case of Fig. 4, where all the abstractions are present. These instructions are usually very simple, such as move data from one storage location to another, shift a number or compare two numbers, but they are laborious, boring (some people would disagree; computer science is an art and as with any area of art not all critics agree), and very subject to programming errors. These microinstructions are directly interpreted by the hardware (digital logic abstraction). At each abstraction, except A0, there is a language that defines a machine and vice versa. The language is often called the instruction set.

The next level of abstraction is the conventional machine language and it is more people friendly, more efficient for programming, and less subject to programming errors. For a machine with all the abstractions of Fig. 4 present, the conventional machine language is not understood by the digital logic abstraction and interpreted by the microprogram.

IV. DATA STRUCTURES IN A COMPUTER SYSTEM

In the computer, a computer byte is composed of 8 binary bits, and the computer word is composed of a number of bytes with the word length dependent on the computer.

A user should read the manufacturer's documentation to determine the word size for a particular computer system. Therefore, a character, integer, or decimal number must be represented as a bit combination. There are some generally accepted practices for doing this:

- *Character*: Almost all computers today use the ASCII standard for representing character in a byte. For example "A" is represented by 65_{10}, "a" is represented by 97_{10}, and "1" is represented by 49_{10}. Most architecture textbooks will provide a table providing the ASCII representation of all character. In the past there were vendor-specific representations such as EBCDIC by IBM.
- *Integer*: The computer and user must be able to store signed (temperature readings) and unsigned (memory addresses) integers, and be able to manipulate them and determine if an error has occurred in the manipulation process. Most computers use a twos complement representation for signed numbers and the magnitude of the number to represent unsigned numbers.
- *Decimal number*: Decimal numbers are represented using a floating point representation with the most important one being the IEEE Standard 754, which provides both a 32-bit single and a 64-bit double precision representation with 8-bit and 11-bit exponents and 23-bit and 52-bit fractions, respectively. The IEEE standard has become widely accepted, and is used in most contemporary processors and arithmetic coprocessors.

The computer is a finite state machine, meaning that it is possible to represent a range of integers and a subset of the fractions. As a result, a user may attempt to perform operations that will result in numeric values outside of those that can be represented. This must be recognized and dealt with by the computer with adequate information provided to the user. Signed integer errors are called overflow errors, floating point operations can result in overflow or underflow errors, and unsigned integer errors are called carry errors.

The computers store numbers in twos complement or floating point representation because it requires less memory space. The operations are performed using these representations because the performance will always be better.

The computer architect must determine the algorithm to be used in performing an arithmetic operation and mechanism to be used to convert from one representation to another. Besides the movement of data from one location to another, the arithmetic operations are the most commonly performed operations; as a result, these arithmetic algorithms will significantly influence the performance of the computer. The ALU and Shifter perform most of the arithmetic operations on the datapath.

V. ARCHITECTURE LIFE CYCLE

A computer architecture has a life cycle that is analogous in nature to that of software. The phases might be considered:

1. *Planning*: The planning should include a needs assessment, cost assessment, feasibility analysis, and schedule.
2. *Specification*: The specification will provide a definition of the architectural requirements. Development of a test (validation) plan should be included.
3. *Architecture design*: The design phase will include such things as instruction set design and functional organization.
4. *Architecture implementation*: This portion of the life cycle is crucial to a correct instruction set. One key issue to be resolved in implementation is whether a feature should be implemented in software or hardware.
5. *Testing*: Testing will usually involve testing of units and the entire architecture.
6. *Maintenance*: Maintenance will include correcting errors and providing enhancements.

Each phase is important and should be given its due portion of the effort as a part of a process that may be cyclic rather than sequential through the phases. The life cycle may deal with one of the five major architecture components or even one of their subcomponents. As recommended by Amdahl's Law, most complete computer system developers have strived to provide a balance in the performance of the components, but memory components have lagged behind processor in performance. The computer architects strive for balance in the throughput and processing demands of the components. There are two design components that are ever changing:

A. The performance is changing in the various technology areas of the components, and rate of technology change differs significantly from one component type to another.
B. The new applications and new input/output devices constantly change the nature of the demand on the system in regard to instruction set requirements and data needs.

VI. INTERRELATION OF ARCHITECTURE COMPONENTS

The basic function of a computer is the execution of programs. The execution of a typical program requires the following:

- The movement of the program into memory from a disk or other secondary storage device
- Instructions and data are fetched by the processor from memory as needed
- Data is stored by the processor in memory as required
- Instructions are processed as directed by the program
- Communication occurs between the processor and the I/O devices as needed

All of this activity occurs as a result of instructions and is carried out by the control unit, and the five categories of activities occur (not as five sequential actions) at various points during the process of executing the program. The instructions that cause the program execution to happen may be any combination of operating system instructions and instructions in the program being processed.

A computer system consists of processor(s), memories, and I/O modules that are interrelated in an organized fashion. The interrelationship is successful if the communication between the processor(s), memories, and I/O modules is coordinated. That is, they must speak the same language and operate at a speed that is mutually acceptable. For example, two people communicating with each other can only do so if they speak at different times and in a language both understand. This section will discuss the processor, memory and I/O device and their interrelationship, and future sections will build on the topics initiated here.

A. Processor, Memory, I/O System, and the Interrelationship

A computer system is composed of components that may be classified as processor, memory, and I/O that interact or communicate with each other. There must be a path, logically depicted in Fig. 5, between these components in order for this communication to take place. Figure 5a

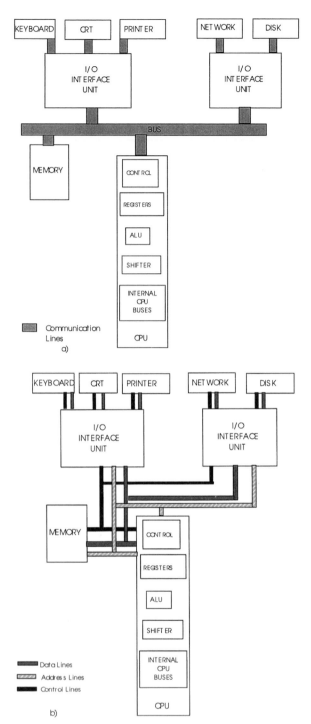

FIGURE 5 Interconnection between major computer components.

illustrates a simple single bus structure that may be used for the interaction between components, and Fig. 5b classifies the bus signals into the three possible categories of data, address, and control. The location of components relative to each other is important only in the sense that

signals travel more rapidly between components that are closer to each other. Components that communicate with each other frequently should be organized so as to minimize the distance between them.

The physical connection between the components is often termed the interconnect structure. Timing considerations must be taken into account and often the interconnection will be more complex than just a set of wires and will include digital circuits used to coordinate the timing of data, address, and control signals between the components. The reason for this is because different components may operate at different speeds. The data lines are used to transfer information, instructions, and data, and are the sole reason for the existence of the communication address and control lines. It must be possible to move information in either direction between processor and memory or between I/O devices and processor or memory. The interrelationship between the components is defined as the interconnect structure as well as the signals that are transmitted between the components over the interconnection.

Figure 5 appears to allow the exchange of all types of signals in either direction. However, in practice, memory is only capable of sending data and the I/O devices do not send address signals; otherwise all other possible exchanges may occur. The following represents a summary of the activities of each of the components:

- *Processor*: The processor is the coordinator of the computer system. It processes abstraction 2 programs by fetching their instructions from memory, decoding them, and executing them. The control portion of the processor is like a director of a play and coordinates the activities of all other portions of the processor, and the other computer system components working from a script called the instructions of a program. The control unit may be strictly hardware, or a combination of hardware and software (microprogram). The processor controls all other activities if it is only processor in the computer system. If multiple processors exist, then one must be the master or arbitration occurs between the processors for control of the other components.
- *Memory*: The memory, often called the store or storage, is the location where information is stored. *Registers*, cache, RAM, ROM, disks, and tapes are different types of memory. The early computers had only one register and it was called an accumulator. However, in the context used so far in this chapter the memory component will be either ROM or RAM. The memory component will contain N words, with each word containing a fixed number of bits and located at one of the addresses $0, 1, 2, \ldots, N - 1$. Memory locations could be compared to tract houses, because all of them look the same and can only be

distinguished by their addresses. To address N locations, it is necessary to have $\ln_2 N$ address lines between the processor and memory.

• *I/O modules*: The I/O modules are composed of the I/O interface unit and the peripheral(s). From the user's view, the I/O modules receive all the instructions and data from the computer user. From the processor point of view, the I/O module is analogous to the memory component in that it will be accessed at a specific address and information can be read/written to it. The I/O modules are an intermediary between the processor and I/O devices to off load some of the processor overhead. However, it is different from memory in that it may send control signals to the processor that will cause a processor *exception* to occur. Many computer manufacturers use the term exception to include interrupts, and the two terms will be synonymously used throughout this article. Examples of I/O devices are keyboards, printers, disk drives, CD-ROM drives, DVD drives, tape drives, and ZIP drives.

The scenario of the program cycle of a computer program illustrates the communication that must occur between the computer components. The sole purpose of a computer is to execute programs. In the following the size of the unit of transfer depends on the number of data lines in the communication lines. A possible set of computer-related steps for the program cycle include:

1. Create the program using an editor. This involves the following steps.
   a. Cause the editor to start executing.
      1. Issue editor execute command. The user types the command at the keyboard (from a window environment, the keyboard may be replaced with a mouse action) and it is transferred from the keyboard through the I/O interface unit and the communication lines to processor. This command will ultimately be recognized by the operating system and the operating system will cause the editor to start executing.
      2. Editor execute command is displayed on the monitor. When the command of (1), above, is received by the processor, it is conveyed by the processor through the communication lines and the I/O interface unit to the monitor where it is displayed.
      3. Editor program is loaded into memory. The operating system interprets the editor execute command and causes the processor to command the disk to load the editor program into memory. The processor carries this out by sending the command over the communication lines and

through the I/O interface unit to the disk. The disk then sends the editor program to memory, one unit at a time, over the communication lines and through the I/O interface unit.

   4. Processor executes the editor initialization instructions in the editor program. This includes the tasks of moving the instructions one at a time from memory to the processor using the communication lines for execution by the processor. The instructions are executed by the processor, resulting in such things as messages transferred from the processor to the CRT via the communication lines and I/O interface unit for display on the CRT.

   b. Interact with the editor and develop the program. Among other things the editor interaction includes:
      1. Issue commands to the editor. The user issues these commands to the editor through the keyboard to the processor. The editor instructions interpret the command and if it is valid carries it out, but if it is invalid the editor instructions will cause the processor to send an appropriate message over the communication lines and through the I/O interface unit to the CRT.
      2. Retrieve existing program file. This is a user-issued command to the editor, as given in (1), but is frequently used and of special interest. The editor will prompt the user for the name of the file to be retrieved by sending a message to the CRT.
      3. Enter program instructions. The request to insert program instructions is a user-issued command to the editor, as given in (1), but is frequently used and of special interest.
      4. Save the program. The request to save a program is a user-issued command to the editor, as given in (1), but is frequently used and likewise of special interest.
2. Compile the program.
3. Execute program.

Details of an instruction cycle will be addressed in the next section.

B. Central Processing Unit

The processor has the responsibility of causing instructions to be processed. The processor processes an instruction in an instruction cycle consisting of a number of small actions. Understanding the instruction cycle is important to the understanding of the processor.

Example VIB: Considering the abstraction 2 instruction, such as the assembled version of the instruction MOVE.W FOO.L, D3 (this instruction moves a word from the memory location "L" into a register named D3.), A potential instruction cycle is as follows:

1. Fetch the instruction from memory into the instruction register.
2. Increment the *program counter* by two.
3. Decode the fetched instruction.
4. Fetch the address of FOO from memory into the processor.
5. Increment the program counter by four to point to the next instruction.
6. Fetch the content of the location FOO into the processor.
7. Place the content of FOO in D3.
8. Update the content of the CCR appropriately.

It should be noted that a program could be written to perform each of the eight steps of the instruction cycle. In actuality, the microprogram is a program that carries out the instruction cycle of a conventional machine language instruction. For each of the eight steps above, there are sets of microinstructions in the microprogram that perform the specified task.

The processor repeats the instruction cycle, with variations for different instructions, for each instruction executed. The general instruction cycle for a single instruction is illustrated in Fig. 6. Steps 1, 2, and 4 correspond to instruction fetch subcycle, step 3 corresponds to the instruction decode subcycle, and steps 5–8 correspond the execute subcycle. In the MOVE.W FOO.L, D3 instruction, after decoding the first word of the instruction it is necessary to return to the fetch portion of the instruction cycle to fetch the two words corresponding to the address of FOO, but a further decode is not necessary. The in-

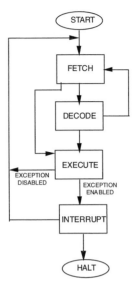

FIGURE 7 Basic instruction processing cycle with exception handling capability.

struction, MOVE.L (6, A3, D2.W), D2, is an example of an instruction in which a return to the fetch portion of the instruction cycle must occur followed by a further decode of the instruction. The information regarding the addressing mode, (6, A3, D2.W), is stored in an extension word that must be fetched and decoded after the opcode word is fetched and decoded. All instruction cycles require an interaction between the processor and memory and possibly between the processor and other computer components. For the MOVE.W FOO.L, D3 instruction, the processor must interact with memory, memory read cycle, in steps 1, 4, and 6.

An exception subcycle occurs if an exception is pending, and this addition to the instruction cycle is included in Fig. 7. The exception is caused by an event separate from the program that is being processed and is serviced in such a way that the program processing may continue after the exception has been serviced. In the case of the instruction cycle illustrated above, none of the steps are a part of the exception subcycle, and it was assumed that no exception was pending. Exception processing will not be covered extensively here, but an exception might be considered analogous to the exception of the following example. A maskable exception is one that can be ignored for some time, but an unmaskable exception is one that must be handle immediately. An I/O device that needs service by the processor causes one type of exception. For example, when a user types information at the keyboard, the keyboard may raise an exception in the processor to alert the processor that it has received information from the user and request permission to send it. The exception is transmitted from the keyboard over the control lines to the processor. The keyboard exception is an example of

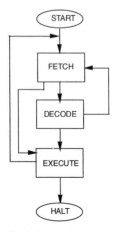

FIGURE 6 Basic instruction processing cycle.

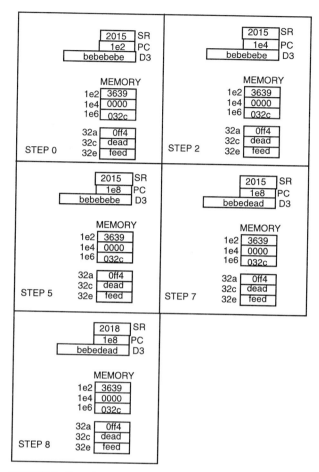

FIGURE 8 Steps in MOVE FOO.L, D3 instruction cycle in which conventional machine states are changed.

a maskable exception. A memory failure error is an example of an unmaskable exception and will be raised by the memory unit by transmitting a signal over the control lines to the processor. The signal that raises the exception is sent by the device to the processor over the communication control lines. Significant effort must go into planning the process for dealing with exceptions in order to do so in a way that will not cause performance deterioration of the processor.

Figure 8 gives the conventional machine language state changes that occur as the instruction MOVE.W FOO.L, D3 is processed. It should be noted that the conventional machine language instruction corresponding to MOVE.W FOO.L, D3 is 0x36390000032c and it is assumed that it is stored at memory location 0x1e2. At execution time the assembly language form, MOVE.W FOO.L, D3, does not exits. Only the steps given above that cause one or more conventional machine states to change are shown. The microprogram machine states will change as a result of each of the machine cycle steps above, and those steps will be carried out as a result of executing several microinstructions. All machine states are given in hexadeci-

mal, and except for the 32-bit register D3, all of the other pictured machine states are 16-bit locations. It should be noted that only the low order word of D3 is modified (Big Endian memory configuration is assumed.) as a result of this instruction and the SR N bit is set and the Z and C bits are cleared in step 8. Likewise, it is arbitrarily assumed that FOO is at address 0x32c. The assembly language or conventional machine language programmer is only aware of machine state changes occurring at steps 0, 2, 5, 7, and 8, but microprogram abstraction machine states will change at all of the steps. For example, the first word of the conventional machine language instruction, 0x36390000032c, will be moved from memory to the processor at step 2.

The processor is a part of the hardware of a computer system. It is composed of sequential and combinational circuits. The processor may be logically subdivided into components such as the control, ALU, registers, shifter, and internal processor buses. Most every computer system that was ever developed would have all of these components present. Modern processor chips also contain additional components such as cache memory. The interconnection of processor components via the internal buses may also be viewed as a network. The control component is the part of the processor that is the director of a processor activity. As discussed above, there is an interrelationship between the major computer components such as the processor, memory, and I/O devices. Also, within the processor there is an interrelationship between its registers, ALU, etc. Each interrelationship much be synchronized.

1. ALU

The ALU is a combinational circuit, and is that part of the processor that performs arithmetic, logic, and other necessary related operations. Sometimes there is a separate component, shifter, which is used to perform the shift operations on data items. The abstraction 2 programmer usually considers the shifter activities to be a part of the ALU. The Abacus, Blaise Pascal machine, and other ancestors of the computer were really just ALUs, and for the most part the other components of modern computer systems are merely there to hold data for or transfer it to the ALU. Thus, the ALU could be considered the center of the computer system. However, it does not determine its own activities. Abstraction 1 or 2 instructions determine its activity via the control. The ALU will be actively involved in steps 2, 3, and 5 of the instruction cycle of Example VIB and may be involved in other steps.

2. Control

Control coordinates all activities. As mentioned previously, unless multiple processors exist within the computer system, the uniprocessor control directs all activities of the other elements of the processor and in general all components throughout the computer system as well as its own activity. For each instruction, control performs a series of actions. From a very high level it is possible to consider the set of actions to be the three of Figure 2. Those three steps may be further refined to the eight steps for the MOVE.W FOO.l, D3. These eight steps may be further filtered so that a set of microinstructions corresponds to each of the eight steps. Of course, these microinstructions could be broken down even further. Thus the computer activities may be studied at any level of detail. The control component will contain combinational and sequential circuits. The microprogram will be placed in the control store, which is a set of sequential circuits. However, the RISC computers do not contain a microprogram and thus control store is not present, and the control unit is strictly hardware.

3. Registers

This sequential circuit is the fastest and most expensive part of the memory hierarchy. The registers are the part of the memory hierarchy that are directly a part of the processor datapath. Because these are the fastest memory it is desirable to have all of the active data present in them.

At a given time only a few are available to the programmer, because of the principle that if more were present, it would take longer to access the one desired, because the search process would take a longer period of time.

4. Exception (Interrupt) Processing

Processors vary significantly in the sophistication of their exception (interrupt) handling facilities. The exception capabilities of the Motorola 68000 family is well above the average and can be used to place all exception processing out of reach of garden variety user activities. All processors must have some degree of exception processing support, because at a minimum it must be possible to perform processor resets, which amounts to an exception.

5. Processor Organization Overview

A simplified logical view of a possible processor organization is given in Fig. 9. The registers, ALU, and shifter along with the three buses are often called the internal datapath of the processor. The datapath design should be designed to optimize the execution speed of the most frequently used abstraction 2 instructions. The datapath activity is controlled by the CONTROL component via the dashed control lines. The dashed lines represent one or more single control lines. The control component also controls its

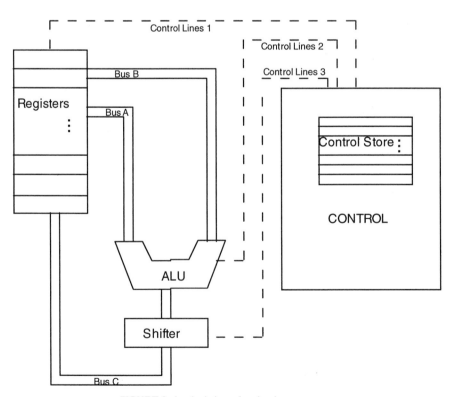

FIGURE 9 Logical view of a simple processor.

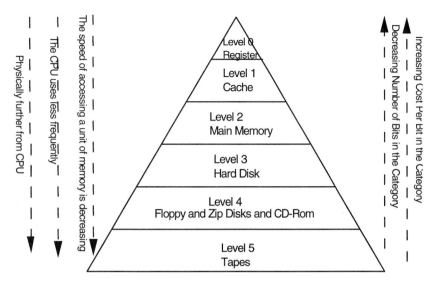

The speed of accessing a unit of memory is decreasing

The CPU uses less frequently

Physically further from CPU

Increasing Cost Per Bit in the Category

Decreasing Number of Bits in the Category

Level 0
Register

Level 1
Cache

Level 2
Main Memory

Level 3
Hard Disk

Level 4
Floppy and Zip Disks and CD-Rom

Level 5
Tapes

FIGURE 10 Hierarchal presentation of some types of memory.

own internal activity. This processor would contain more additional control lines than those shown, for example, ones that control the registers. As sure as there is water in the oceans, there is something missing, and that is the connection of the internal datapath and CONTROL signal to the main memory and I/O modules. This was done deliberately to reduce the complexity at this point. A complete dissertation of the processor can usually be found in most computer architecture textbooks. The control lines will carry out the following general tasks:

- Control lines 1: Used to select the register to be placed on Bus A and or Bus B, and/or receive data from Bus B.
- Control lines 2: Used to determine the ALU operation to be performed.
- Control lines 3: Used to determine the Shifter operation to be performed.

Control must be able to communicate commands to all of the computer components, as illustrated for the Registers, ALU, and Shifter, and the commands must be carried out in a proper sequence:

- First, the proper registers must be placed on Bus A and Bus B.
- Second, the ALU must generate the proper result.
- Third, the Shifter must produce the proper output.
- Finally, the content of Bus C must be stored back in the proper register.

For high performance, the computer architecture must be planned using the principles discussed previously and Amdahl's law.

C. Memory System Hierarchy

The memory in the memory hierarchy of a computer system is used to store information, instructions, and data that will be used by the computer system. Memory is often classified as registers, cache memory, main memory, hard disk, floppy disk, and tapes. These are pictured in a hierarchal form in Fig. 10 with locations within each type of memory randomly accessible except for tapes. Tapes are sequentially accessible, and in the long run each disk data unit is accessible in equal time, but at a given time the access time for a particular unit is dependent on the location of the disk components. The term "access" designates the memory activities that are associated with either a read or a write. Randomly accessible means that a memory location may be read or written in the same amount of time irregardless of the order of accesses of memory locations, and sequentially accessible means that the time required to access a memory location is dependent on location of the immediate prior memory access.

Example VIC: If a company stored all of their folders in a file cabinet alphabetically, then an employee would be able to find any folder in the file with roughly equal speed (randomly accessible). However, if the folders were stored flat in nonalphabetized order in a box, then the employee would have to sequentially search for the desired folder in the box. These filing systems are analogous to the randomly accessible and sequentially accessible memory storage units.

There are other forms of information storage such as CD-ROM and cassettes. The control store is a memory unit but is not considered in the memory hierarchy because it is only used to store microprogram instructions. Memory hierarchy locations are used to store abstraction

2 instructions and data. In the future it may be possible that control store will become a part of the memory hierarchy if programs are translated down to that level and the control store becomes a RAM-type memory for storage of the translated program along with the corresponding data.

In the memory hierarchy, the memory level in the higher location of the pyramid of Fig. 10 is usually physically closer to the processor, faster, more frequently accessed, smaller in size, and each bit is more expensive. Some of the characteristics of memory types are access time, bandwidth, size limit, management responsibility, and location within the computer system (within processor or on the external bus).

1. Principles of Memory Design

Size, cost, and speed are the major design parameters in the memory hierarchy. The principle behind memory hierarchy design is to keep the cost per unit of memory as close as possible to that of the least expensive memory and keep the average access time as close as possible to that of the fastest memory. To accomplish this, the design must use a minimal amount of the memory type at the top of the pyramid of Fig. 10 and attempt to keep information that will be accessed in the near future in memory at the top of the pyramid. Removable disks and tapes are of unlimited size, and that means that the user may continue to buy more of the media until his needs are met.

It would be most desirable to have every item to be used in a register. However, this is impossible, because current computers must move information into memory using level 2 instructions. In order to follow these design principles, it is necessary to study the relationship of memory references by a program and utilize those relationships that are found when possible. Studies have shown that the memory references in a small period of time tend to be clustered for both data and instructions within a few localized regions of memory, as illustrated in Fig. 11. This is known as the locality principle and predicts that there is high probability that those items referenced recently or those nearby the recently referenced memory locations will be referenced in the near future. A copy of those items that, according to the locality principle, are most likely to be used in the near future should be kept in the memory that is accessed most quickly.

Example VIC 1: Average cost and average access time.

A memory design issue is the ordering of bytes in memory that may follow either the little Endian or Big Endian definitions. Within a particular manufacturer's computers it will be either one or the other for all of their computers

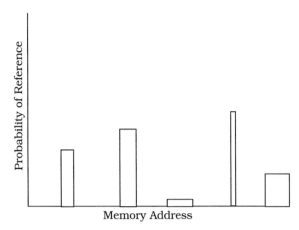

FIGURE 11 Probability of referencing memory at a snapshot of time.

within a computer family. Little Endian and big Endian are terms introduced by Cohen (1981) in an amusing article and the term "Endian" is borrowed from Jonathan Swift's *Gulliver's Travels* that poked fun at politicians who waged war over the issue of whether eggs should be cracked on the big end or little end.

D. I/O System

The I/O system is a crucial part of any computer system, but it will not be discussed in greater detail in this article.

E. Interconnection

The interconnection of computer components is critical to computer performance. The interconnection of components within a computer has been discussed, and is a key concern in the implementation of an architecture. The interconnection between computers is covered under the data and computer communications category in books and articles dealing with that topic.

VII. ENHANCING ARCHITECTURE IN THE TRADE-OFF OF PERFORMANCE AND COST

Every person who buys a computer desires one that will have maximum performance with minimum cost and meet his minimum requirements. The language of the computer is determined by the instructions that it understands. Languages of computers are quite similar in nature because computers are all built using approximately the same underlying principles. The performance of a computer system is influenced by three critical components: instruction count, clock cycle time, and clock cycles per instruction (CPI). The number of instructions required is determined by the instruction set of the computer and the efficiency of the compiler in generating the program that will be executed. The processor implementation determines the clock cycle time and the number of clock cycles per instruction. The performance of a computer can be enhanced to almost any desired level, but the performance level must be traded off with the cost. If the cost is too great, it will not be affordable and as a result will not sell. In this section, we will look at some current techniques that are used to enhance the performance of a computer.

A. Instruction Set Design

The instruction set influences the processor implementation and must be determined based on the processor efficiency. Some of the key instructions of a computer include the

1. memory reference instructions
2. arithmetic-logical instructions
3. branch instructions

Other factors in the design of an instruction set include number of instructions to be included, whether or not the instructions should have a fixed length and/or fixed format, the addressing modes available, the robustness or complexity of the instructions, and whether or not the instructions, and support data structures such as "stacks" and data types.

The guidelines—make the common case fast (make the memory access instructions fast because they occur about 50% of the time in a general purpose program) and simplicity promotes regularity (make the instructions fixed length and fixed format)—play a significant role in the definition of modern instruction sets and in particular the RISC computers. In the instruction set design, it is not always possible to have fixed length and format instructions and as a result the principle, a good design will require compromises, must be followed. The instruction set has a significant influence on the implementation, and the choice of implementation strategies will affect the clock rate and the CPI for the computer. The final principle, smaller is faster, comes into play when the number of instructions is minimized or the number of memory locations is small (in the case of registers). The number of methods and complexity of instruction access of memory (addressing modes) will play a significant role in the performance level that the computer can achieve with best performance achieved if the design principles "simplicity" and "small" are observed.

The implementation of the instruction set through the design of the datapath and control within the processor will play a major role in the performance of the computer with the technology available being an important factor. Technology will significantly influence such things as the density of transistors on a chip. The density of transistors will influence the distance between two points, and the higher density chips will provide the capacity for adding features on the processor chip as opposed to externally, thus reducing the time required to transfer information between components.

A significant influence on the instruction set design has been the past desire by computer manufacturers to provide backward compatibility (the capability of executing existing programs without modification). If this requirement were met by retaining the old instruction set in the new instruction set of the processor, it would result in greater overhead and a reduction in the potential performance of the processor. If this requirement were met by using a software interpreter, then the old programs would not achieve top performance because of the overhead of the interpreter.

B. Instruction Set Implementation Enhancements

It was mentioned above that technology advancements have played a significant role in the advancement of the performance of a processor. Implementation techniques play a major role as well with the following being some of those advancements.

1. *Pipelining*: This is an architecture implementation technique that allows multiple instructions to overlap in execution. The processor is organized as a number of stages that allow multiple instructions to be in various stages of their instruction cycle. The pipeline for instructions is analogous to an assembly line for automobiles. To implement the pipeline, additional processor resources are required. In order to achieve maximum efficiency, the instruction cycle must be divided so that approximately the same amount of work is done in each of the stages. There are a number of potential hazards that exist:

   a. *Data hazard*: This occurs if an instruction in an earlier stage of the pipeline attempts to use the content of a location that has not been written by an instruction that precedes it in the pipeline.

   b. *Branch hazard*: This occurs if instructions following the branch are brought into the pipeline when the branch will actually occur or vice versa. A method used to eliminate this problem is to provide the resources to bring both sets of possible future instructions into the pipeline. Another solution is to rearrange the order of instructions to cause instructions preceding the branch in the program to follow the branch in the pipeline. This assumes these instructions do not influence the branch decision.

c. *Exception hazard*: This hazard could occur if an exception occurs and instructions in the pipeline are not handled properly.

All hazards can be remedied but the method used may require significant processor resources. The data and branch hazards can be remedied by using the compiler to reorder the instructions.

2. *Superpipelining*: This is a processor design that divides the pipeline up into a large number of small stages. As a result, many instructions will be at various stages in the pipeline at any one point in time. There is no standard distinguishing pipelined processors from superpipelined processor, but seven has been a commonly accepted division point. The ideal maximum speedup of the processor is proportional to the number of pipeline stages. However, the required resources for fixing the hazard problem increases more rapidly as the number of stages increases.

3. *Superscalar*: This implementation of the processor provides more than one pipeline. As a result, multiple instructions may be at the same stage in their instruction cycle. A superpipeline and a superscalar implementation may exist within the same processor.

4. *Very long instruction word* (*VLIW*): This processor implementation provides for the processor to use instructions that contain multiple operations. This means that the multiple instructions are contained in a single instruction word. It is usually the responsibility of the compiler to place compatible instructions in the same word.

All of these techniques introduce parallelism into the execution of instructions within the processor.

C. Input and Output Enhancements

As computers have evolved there has been increasing complexity placed in the function of the I/O devices. The following are some of the evolutionary steps that have occurred.

1. Initially the processor handled the peripheral devices directly.
2. The I/O device became a controller of the peripheral device, but the processor performed the programmed I/O (the processor stopped processing other instructions until the I/O activity had completed and the device had signaled its completion of the required action).

3. The I/O device is the same as (2) but the processor uses interrupts to allow itself to process instructions of another program until it is interrupted by the I/O device when the interrupt activity is completed.
4. The I/O device becomes a controller that has direct memory access capability. As a result, it can place data in memory directly.
5. In the most sophisticated case to date, the I/O device becomes a complete processor with its own instruction set designed for I/O. As a result, the processor may specify a set of I/O activities and be interrupted only after all of the requests are fulfilled.

D. Memory Enhancements

Memory speed enhancements have not kept up with processor speed improvements. As a result, other techniques have been used in order to enhance the performance of the memory system. The characteristics of a memory system that are adjusted to improve performance are included in the following table:

Location of memory in computer system
Processor
Main memory
Secondary memory
Physical characteristics
Semiconductor
Magnetic
Optical
Nonvolatile vs volatile
Nonerasable vs erasable
Speed of access
Access time
Memory cycle time
Transfer rate
Capacity
Word size
Number of words
Method of access
Random
Sequential
Direct
Associative
Organization
Location of bits in random access memory
Registers
Cache
Virtual memory
Transfer unit
Word
Block
Page

All of these characteristics of the design of a memory system will influence its performance. The location of the memory in relationship to the processor and the closer it is to the processor the faster it will be. Another key characteristic is the unit of transfer that is dependent on the size of the bus from the memory unit with increased size of the transfer unit increasing performance. Increased number of accessible units will often decrease the performance of memory, and this is remedied by increasing the word size. The first three physical characteristics have more influence on performance than do the last two. The RAID standard has been adopted by industry, and it provides a mechanism for substituting large capacity disks with multiple smaller-capacity disk drives and causes the data to be distributed in such a way as to allow the simultaneous access to data from the multiple drives. Adoption of the RAID standard provides disk access performance enhancements and provides a mechanism for allowing incremental increases in overall storage capacity of the disk system. The design of the memory system is critical to the performance of a computer system.

E. Enhancement Using Parallel Processors

Computer professionals have strived to achieve the ultimate of computer design by interconnecting many existing simple computers together. This is the basis for a computer using multiprocessors. Multiprocessors are computers that have two or more processors that have a common access to memory. Customers desire scalable multiprocessors that allow them to order the current required number of processors and then add additional processors as needed. It is also desirable to have the performance increase proportionately to the number of the processors in the system, and to have the computer system continue to function properly when one or more of the processors fail. The typical range of processors in a multiprocessor computer is between 2 and 256.

The design of the multiprocessor computer is a key factor in the performance of the resulting computer system and the following are some of the key issues that must be addressed:

• The mechanism for sharing memory between processors must be defined
• The coordination of processors must be determined
• The number of processors must be established
• The interconnection between the processors must be defined

The most common mechanism for sharing memory is through a single address space. This provides each processor with the opportunity to access any memory location through their instruction set. Shared memory multi-

processor may be symmetric multiprocessor (SMP) processors, in which the access to a word of memory takes the same time no matter which processor accesses it or which word is accessed, and the nonuniform memory access multiprocessors, which does not have the equal access characteristic. The alternative to the shared memory multiprocessor is one whose processors have private memory, and coordination is then achieved through the passing of messages. A key element for memory is coherency, which requires data validity among the different processors. Multiple processors are interconnected either by a bus or network.

The multiprocessor area of computer architecture is highly active. It is difficult to maintain awareness of current examples but the World Wide Web is an excellent source of information with links to several types of multiprocessor examples at http://www.mkp.com/books_catalog/cod2/ch9links.htm.

VIII. CONCLUSIONS

Computer architecture has been and will continue to be one of the most researched areas of computing. It is about the structure and function of computers. Although there is explosive change in computing, many of the basic fundamentals remain the same. It is the current state of technology along with the cost/performance requirements of the customers that changes how the principles are applied. The need of high performance computers has never been greater than it is now, but the difficulty of designing the ideal system has never been more difficult because of the rapid increase in speed of processors, memory, and interconnection of components. The difficulty is in designing a balanced system according to Amdahl's Law.

Professionals from every area related to computing should understand architecture principles and appropriate aspects of hardware and software. It is important for those professionals to understand the interrelationship between the assembly language, organization of a computer, and the design of a computer system. It is important to be able to sort out those concepts that are the basis for modern computers and to be able to understand the relationship between hardware and software.

SEE ALSO THE FOLLOWING ARTICLES

COMPILERS • COMPUTER NETWORKS • MICROCOMPUTER BUSES AND LINKS • OPERATING SYSTEMS • PROCESS CONTROL SYSTEMS • SOFTWARE ENGINEERING • SOFTWARE MAINTENANCE AND EVOLUTION

BIBLIOGRAPHY

Adams, D. R., and Wagner, G. E. (1986). "Computer Information Systems: An Introduction," South-Western Publishing, Cincinnati, OH.

Bell, C., and Newell, A. (1978). "Computer Structures: Readings and Examples," McGraw-Hill, New York.

Bitter, G. G. (1984). "Computers in Today's World," Wiley, New York.

Chen, P., Lee, E., Gibson, G., Katz, R., and Patterson, D. (1994, June). "RAID: High Performance, Reliable Secondary Storage," ACM Computing Surveys.

Campbell-Kelly, M., and Aspray, W. (1996). "Computer: A History of the Information Machine," Basic Books, New York.

Digital Equipment Corporation. (1983). "PDP-11 Processor Handbook," DEC, Maynard, MA.

Digital Equipment Corporation. (1983). "VAX Hardware Handbook," DEC, Maynard, MA.

Gimarc, C., and Milutinovic, V. (1987). "A Survey of RISC Processors and Computers of the Mid-1980s," IEEE Computer, Long Beach, CA.

Hennessy, J. L., and Patterson, D. A. (19**XX**). "Computer Architecture: A Quantitative Approach," 2nd ed., Morgan Kaufmann, San Francisco.

Kidder, T. (1981). "Soul of a New Machine," Little, Brown, New York.

Patterson, D., and Hennessy, J. (1998), "Computer Organization and Design—The Hardware/Software Interface," 2nd ed., Morgan Kaufmann, San Francisco.

Stallings, W. (1999). "Computer Organization and Architecture," 5th ed., Macmillan, New York.

Stone, H. (1993). "High-Performance Computer Architecture," Addison-Wesley, Reading, MA.

Tanenbaum, A. S. (1995). "Distributed Operating Systems," Prentice-Hall, Englewood Cliffs, NJ.

Tanenbaum, A. S. (1999). "Structured Computer Organization," 4th ed., Prentice-Hall, Englewood Cliffs, NJ.

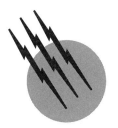

Computer-Based Proofs of Mathematical Theorems

C. W. H. Lam
Concordia University

GLOSSARY

Axiom A statement accepted as true without proof.
Backtrack search A method of organizing a search for solutions by a systematic extension of partial solutions.
Computer-based proof A proof with a heavy computer component; one which is impossible to do by hand.
Computer programming A process of creating a sequence of instructions to be used by a computer.
Enumerative proof A proof method by exhibiting and analyzing all possible cases.
Monte Carlo method A method of estimation by performing random choices.
Optimizing a computer program A process of fine-tuning a computer program so that it runs faster.
Predicate A statement whose truth value depends of the values of its arguments.
Proof A demonstration of the truth of a statement.
Proposition A statement which is either true or false, but not both.
Search tree A pictorial representation of the partial solutions encountered in a backtrack search.

EVER SINCE the arrival of computers, mathematicians have used them as a computational aid. Initially, they were used to perform tedious and repetitious calculations. The tremendous speed and accuracy of computers enable mathematicians to perform lengthy calculations without fear of making careless mistakes. Their main application, however, has been to obtain insight into various mathematical problems, which then led to conventional proofs independent of the computer. Recently, there was a departure from this traditional approach. By exploiting the speed of a computer, several famous and long-standing problems were settled by lengthy enumerative proofs, a technique

Encyclopedia of Physical Science and Technology, Third Edition, Volume 3
Copyright © 2002 by Academic Press. All rights of reproduction in any form reserved.

only suitable for computers. Two notable examples are the four-color theorem and the nonexistence of a finite projective plane of order 10. Both proofs required thousands of hours of computing and gave birth to the term "a computer-based proof." The organization of such a proof requires careful estimation of the necessary computing time, meticulous optimization of the computer program, and prudent control of all possible computer errors. In spite of the difficulty in checking these proofs, mathematicians are starting to accept their validity. As computers are getting faster, many other famous open problems will be solved by this new approach.

I. MATHEMATICAL THEORIES

What is mathematics? It is not possible to answer this question precisely in this short article, but a generally acceptable definition is that *mathematics is a study of quantities and relations using symbols and numbers.* The starting point is often a few undefined objects, such as a *set* and its *elements.* A mathematical theory is then built by assuming some axioms which are statements accepted as true. From these basic components, further properties can then be derived.

For example, the study of geometry can start from the undefined objects called *points.* A *line* is then defined as a set of points. An axiom may state: "Two distinct lines contain at most one common point." This is one of the axioms in Euclidean geometry, where it is possible to have parallel lines. The complete set of axioms of Euclidean geometry was classified by the great German mathematician David Hilbert in 1902. Using these axioms, further results can be derived. Here is an example:

Two triangles are congruent if the three sides of one are equal to the three sides of the other.

To show that these derived results follow logically from the axioms, a system of well-defined principles of mathematical reasoning is used.

A. Mathematical Statements

The ability to demonstrate the truth of a statement is central to any mathematical theory. This technique of reasoning is formalized in the study of *propositional calculus.*

A *proposition* is a statement which is either true or false, but not both. For example, the following are propositions:

- $1 + 1 = 2$.
- 4 is a prime number.

The first proposition is true, and the second one is false.

The statement "$x = 3$" is not a proposition because its truth value depends on the value of x. It is a *predicate,* and its truth value depends of the value of the *variable* or *argument* x. The expression $P(x)$ is often used to denote a predicate P with an argument x. A predicate may have more than one argument, for example, $x + y = 1$ has two arguments. To convert a predicate to a proposition, values have to be assigned to all the arguments. The process of associating a value to a variable is called a *binding.*

Another method of converting a predicate to a proposition is by the quantification of the variables. There are two common quantifiers: *universal* and *existential.* For example, the statement

$$\text{For all } x, x < x + 1$$

uses the universal quantifier "for all" to provide bindings for the variable x in the predicate $x < x + 1$. If the predicate is true for every possible value of x, then the proposition is true. The symbol $\forall$ is used to denote the phrase "for all." Thus, the above proposition can also be written as

$$\forall x, x < x + 1.$$

The set of possible values for x has to come from a certain *universal* set. The universal set in the above example may be the set of all integers, or it may be the set of all reals. Sometimes, the actual universal set can be deduced from context and, consequently, not stated explicitly in the proposition. A careful mathematician may include all the details, such as

$$\forall \text{ real } x > 1, x > \sqrt{x}.$$

The choice is often not a matter of sloppiness, but a conscious decision depending on whether all the exact details will obscure the main thrust of the result.

An existential quantifier asserts that a predicate $P(x)$ is true for one or more x in the universal set. It is written as

$$\text{For some } x, P(x),$$

or, using the symbol $\exists$, as

$$\exists x, P(x).$$

This assertion is false only if for all x, $P(x)$ is false. In other words,

$$\neg[\exists x, P(x)] \equiv [\forall x, \neg P(x)].$$

B. Enumerative Proof

A proof of a mathematical result is a demonstration of its truth. To qualify as a proof, the demonstration must be absolutely correct and there can be no uncertainty nor ambiguity in the proof.

Many interesting mathematical results involve quantifiers such as ∃ or ∀. There are many techniques to prove quantified statements, one of which is the enumerative proof. In this method, the validity of $\forall x, P(x)$ is established by investigating $P(x)$ for every value of x one after another. The proposition $\forall x, P(x)$ is only true if $P(x)$ has been verified to be true for all x. However, if $P(x)$ is false for one of the values, then it is not necessary to consider the remaining values of x, because $\forall x, P(x)$ is already false. Similarly, the validity of $\exists x, P(x)$ can be found by evaluating $P(x)$ for every value of x. The process of evaluation can stop once a value of x is found for which $P(x)$ is true, because $\exists x, P(x)$ is true irrespective of the remaining values of x. However, to establish that $\exists x, P(x)$ is false, it has to be shown that $P(x)$ is false for all values of x.

To ensure a proof that is finite in length, enumerative proofs are used where only a finite number of values of x have to be considered. It may be a case where x can take on only a finite number of values, or a situation where an infinite number of values can be handled by another proof technique, leaving only a finite number of values for which $P(x)$ are unknown.

An enumerative proof is a seldom-used method because it tends to be too long and tedious for a human being. A proof involving a hundred different cases is probably the limit of the capacity of the human mind. Yet, it is precisely in this area that a computer can help, where it can evaluate millions or even billions of cases with ease. Its use can open up new frontiers in mathematics.

C. Kinds of Mathematical Results

Mathematicians love to call their results *theorems*. Along the way, they also prove lemmas, deduce corollaries, and propose conjectures. What are these different classifications? The following paragraphs will give a brief answer to this question.

Mathematical results are classified as lemmas, theorems, and corollaries, dependent on their importance. The most important results are called theorems. A *lemma* is an auxiliary result which is useful in proving a theorem. A *corollary* is a subsidiary result that can be derived from a theorem. These classifications are quite loose, and the actual choice of terminology is based on the subjective evaluation of the discoverer of the results. There are cases where a lemma or a corollary have, over a period of time, attained an importance surpassing the main theorem.

A *conjecture*, on the other hand, is merely an educated guess. It is a statement which may or may not be true. Usually, the proposer of the conjecture suspects that it is highly probable to be true but cannot prove it. Many famous mathematical problems are stated as conjectures. If a proof is later found, then it becomes a theorem.

D. What Makes an Interesting Theorem

Of course, a theorem must be interesting to be worth proving. However, whether something is interesting is a subjective judgement. The following list contains some of the properties of an interesting result:

- Useful
- Important
- Elegant
- Provides insight

While the usefulness of some theorems is immediately obvious, the appreciation of others may come only years later. The importance of a theorem is often measured by the number of mathematicians who know about it or who have tried proving it. An interesting theorem should also give insights about a problem and point to new research directions. A theorem is often appreciated as if it is a piece of art. Descriptions such as "It is a beautiful theorem" or "It is an elegant proof" are often used to characterize an interesting mathematical result.

II. COMPUTER PROGRAMMING

Figure 1 shows a highly simplified view of a computer. There are two major components: the central processing unit (CPU) and memory. Data are stored in the memory. The CPU can perform operations which changes the data. A program is a sequence of instructions which tell the CPU what operations to perform and it is also stored in the computer memory.

Computer programming is the process of creating the sequence of instructions to be used by the computer. Most programming today is done in high-level languages such as Fortran, Pascal, or C. Such languages are designed for the ease of creating and understanding a complex program by human beings. A program written in one of these languages has to be translated by a compiler before it can be used by the computer.

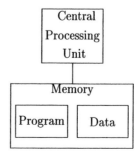

FIGURE 1 A simplified view of a computer.

Due to the speed with which a computer can perform operations it is programmed to do and its capacity in dealing with huge amounts of information, it is well suited to do lengthy or repetitive tasks that human beings are not good at doing.

III. COMPUTER AS AN AID TO MATHEMATICAL RESEARCH

Ever since the arrival of computers, mathematicians have used them as a computational aid. Initially, computers were used to perform tedious and repetitive calculations. The tremendous speed and accuracy of computers enable mathematicians to perform lengthy calculations without fear of making careless mistakes. For example, π can be computed to several billion digits of accuracy, and prime numbers of ever-increasing size are being found continually by computers.

Computer can also help mathematicians manipulate complicated formulae and equations. MACSYMA is the first such symbol manipulation program developed for mathematicians. A typical symbol manipulation program can perform integration and differentiation, manipulate matrices, factorize polynomials, and solve systems of algebraic equations. Many programs have additional features such as two- and three-dimensional plotting or even the generation of high-level language output.

Even though these programs have been used to prove a number of results directly, their main application has been to obtain insight into the behavior of various mathematical objects, which then has led to conventional proofs. Mathematicians still prefer computer-free proofs, if possible. A proof that involves using the computer is difficult to check and so its correctness cannot be determined absolutely. The term "a computer-based proof" is used to refer to a proof where part of it involves extensive computation, for which an equivalent human argument may take millions of years to make.

IV. EXAMPLES OF COMPUTER-BASED PROOFS

In the last 20 years, two of the best known mathematical problems were solved by lengthy calculations: the four-color conjecture and the existence question of a finite projective plane of order 10.

A. Four-Color Conjecture

The *four-color conjecture* says that four colors are sufficient to color any map drawn in the plane or on a sphere

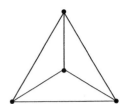

FIGURE 2 The complete graph K_4.

so that no two regions with a common boundary line are colored with the same color.

Francis Guthrie was given credit as the originator of this problem while coloring a map of England. His brother communicated the conjecture to DeMorgan in October 1852. Attempts to prove the four-color conjecture led to the development of a major branch of mathematics: graph theory.

A graph $G(V, E)$ is a structure which consists of a set of vertices $V = \{v_1, v_2, \ldots\}$ and a set of edges $E = \{e_1, e_2, \ldots\}$; each edge e is incident on two distinct vertices u and v. For example, Fig. 2 is a graph with four vertices and six edges. The technical name for this graph is the *complete graph* K_4. It is *complete* because there is an edge incident on every possible pair of vertices. Figure 3 is another graph with six vertices and nine edges. It is the *complete bipartite* graph $K_{3,3}$. In a *bipartite* graph, the set of vertices is divided into two classes, and the only edges are those that connect a vertex from one class to one of the other class. The graph $K_{3,3}$ is complete because it contains all the possible nine edges of the bipartite graph. A graph is said to be *planar* if it can be drawn on a plane in such a way that no edges cross one another, except, of course, at common vertices. The graph K_4 in Fig. 2 is drawn with no crossing edges and it is obviously planar. The graph $K_{3,3}$ can be shown to be not planar, no matter how one tries to draw it.

One can imagine that a planar graph is an abstract representation of a map. The vertices represent the capitals of the countries in the map. Two vertices are joined by an edge if the two countries have a common boundary. Instead of coloring regions of a map, colors can be assigned to the vertices of a graph representing the map. A graph is *k-colorable* if there is a way to assign colors to the graph such that no edge is incident on two vertices with the same

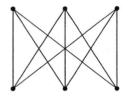

FIGURE 3 The complete bipartite graph $K_{3,3}$.

color. So, the four-color conjecture can also be stated as every planar graph is 4-colorable.

The graph K_4 is not 3-colorable. This can be proved by contradiction. Suppose it is 3-colorable. Since there are four vertices, two of the vertices must have the same color. Since the graph is complete, there is an edge connecting these two vertices of the same color, which is a contradiction.

Kempe in 1879 published an incorrect proof of the four-color conjecture. Heawood in 1890 pointed out Kempe's error, but demonstrated that Kempe's method did prove that every planar graph is 5-colorable. Since then, many famous mathematicians have worked on this problem, leading to many significant theoretical advances. In particular, it was shown that the validity of the four-color conjecture depended only on a finite number of graphs. These results laid the foundation for a computer-based proof by Appel and Haken in 1976. Their proof depended on a computer analysis of 1936 graphs which took 1200 h of computer time and involved about 10^{10} separate operations. Finally, the four-color conjecture became the four-color theorem.

B. Projective Plane of Order 10

The question of the possible existence of a projective plane of order 10 was also settled recently by using a computer. A *finite projective plane of order n*, with $n > 0$, is a collection of $n^2 + n + 1$ lines and $n^2 + n + 1$ points such that

1. Every line contains $n + 1$ points.
2. Every point is on $n + 1$ lines.
3. Any two distinct lines intersect at exactly one point.
4. Any two distinct points lie on exactly one line.

The smallest example of a finite projective plane is one of order 1, which is a triangle. The smallest nontrivial example is one of order 2, as shown in Fig. 4. There are seven points labeled from 1 to 7. There are also seven lines labeled $L1$ to $L7$. Six of them are straight lines, but $L6$ is represented by the circle through points 2, 6, and 7.

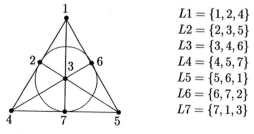

$L1 = \{1, 2, 4\}$
$L2 = \{2, 3, 5\}$
$L3 = \{3, 4, 6\}$
$L4 = \{4, 5, 7\}$
$L5 = \{5, 6, 1\}$
$L6 = \{6, 7, 2\}$
$L7 = \{7, 1, 3\}$

FIGURE 4 The finite projective plane of order 2.

	1	2	3	4	5	6	7
L1	1	1	0	1	0	0	0
L2	0	1	1	0	1	0	0
L3	0	0	1	1	0	1	0
L4	0	0	0	1	1	0	1
L5	1	0	0	0	1	1	0
L6	0	1	0	0	0	1	1
L7	1	0	1	0	0	0	1

FIGURE 5 An incidence matrix for the plane of order 2.

The earliest reference to a finite projective plane was in an 1856 book by von Staudt. In 1904, Veblen used the projective plane of order 2 as an exotic example of a finite object satisfying all the Hilbert's axioms for geometry. He also proved that this plane of order 2 cannot be drawn using only straight lines. In a series of papers from 1904 to 1907, Veblen, Bussey, and Wedderburn established the existence of most of the planes of small orders. Two of the smallest orders missing are 6 and 10. In 1949, the celebrated Bruck-Ryser theorem gave an ingenious theoretical proof of the nonexistence of the plane of order 6. The nonexistence of the plane of order 10 was established in 1988 by a computer-based proof.

In the computer, lines and points are represented by their incidence relationship. The *incidence matrix* $A = [a_{ij}]$ of a projective plane of order n is an $n^2 + n + 1$ by $n^2 + n + 1$ matrix where the columns represent the points and the rows represent the lines. The entry a_{ij} is 1 if point j is on line i; otherwise, it is 0. For example, Fig. 5 gives the incidence matrix for the projective plane of order 2. In terms of an incidence matrix, the properties of being a projective plane are translated into

1. A has constant row sum $n + 1$.
2. A has constant column sum $n + 1$.
3. The inner product of any two distinct rows of A is 1.
4. The inner product of any two distinct columns of A is 1.

These conditions can be encapsuled in the following matrix equation:

$$AA^T = nI + J,$$

where A^T denotes the transpose of the matrix A, I denotes the identity matrix, and J denotes the matrix of all 1's.

As a result of intensive investigation by a number of mathematicians, it was shown that the existence question of the projective plane of order 10 can be broken into four starting cases. Each case gives rise to a number of geometric configurations, each corresponding to a partially completed incidence matrix. Starting from these partial matrices, a computer program tried to complete them to a full plane. After about 2000 computing hours on a CRAY-1A supercomputer in addition to several years of computing on a number of VAX-11 and micro-VAX computers, it was shown that none of the matrices could be completed, which implied the nonexistence of the projective plane of order 10. About 1012 different subcases were investigated.

V. PROOF BY EXHAUSTIVE COMPUTER ENUMERATION

The proofs of both the four-color theorem and the nonexistence of a projective plane order 10 share one common feature: they are both enumerative proofs. This approach to a proof is often avoided by humans, because it is tedious and error prone. Yet, it is tailor-made for a computer.

A. Methodology

Exhaustive computer enumeration is often done by a programming technique called *backtrack search*. A version of the search problem can be defined as follows:

Search problem:

Given a collection of sets of candidates $C_1, C_2, C_3, \ldots, C_m$ and a boolean *compatibility* predicate $P(x, y)$ defined for all $x \in C_i$ and $y \in C_j$, find an *m-tuple* $(x_1, \ldots, x_m)$ with $x_i \in C_i$ such that $P(x_i, x_j)$ is true for all $i \neq j$.

A *m*-tuple satisfying the above condition is called a *solution*.

For example, if we take $m = n^2 + n + 1$ and let C_i be the set of all candidates for row i of the incidence matrix of a projective plane, then $P(x, y)$ can be defined as

$$P(x, y) = \begin{cases} \text{true} & \text{if } \langle x, y \rangle = 1 \\ \text{false} & \text{otherwise,} \end{cases}$$

where $\langle x, y \rangle$ denotes the inner product of rows x and y. A solution is then a complete incidence matrix.

In a backtrack approach, we generate *k-tuples* with $k \leq m$. A *k-tuples* $(x_1, \ldots, x_k)$ is a *partial solution* at level k if $P(x_i, x_j)$ is true for all $i \neq j \leq k$. The basic idea of the backtrack approach is to *extend* a partial solution at level k to one at level $k + 1$, and if this extension is impossible, then to go back to the partial solution at level $k - 1$ and attempt to generate a different partial solution at level k.

For example, consider the search for a projective plane of order 1. In terms of the incidence matrix, the problem is to find a 3×3 (0,1)-matrix A satisfying the matrix equation

$$AA^T = I + J.$$

Suppose the matrix A is generated row by row. Since each row of A must have two 1's, the candidate sets C_i are all

equal to $\{[110], [101], [011]\}$. A partial solution at level 1 can be formed by choosing any of these three rows as x_1. If x_1 is chosen to be $[110]$, then there are only two choices for x_2 which would satisfy the predicate $P(x_1, x_2)$, namely, $[101]$ and $[011]$. Each of these choices for x_2 has a unique extension to a solution.

A nice way to organize the information inherent in the partial solutions is the backtrack *search tree*. Here, the empty partial solution ( ) is taken as the root, and the partial solution $(x_1, \ldots, x_k)$ is represented as the child of the partial solution $(x_1, \ldots, x_{k-1})$. Following the computer science terminology, we call a partial solution a *node*. It is also customary to label a node $(x_1, \ldots, x_k)$ by only x_k because this is the choice made at this level. The full partial solution can be read off the tree by following the branch from the root to the node in question. Figure 6 shows the search tree for the projective plane of order 1. The possible candidates are labeled as r_1, r_2, and r_3. The right-most branch, for example, represents choosing r_3 or $[011]$ as the first row, r_2 as the second row, and r_1 as the third row.

It is often true that the computing cost of processing a node is independent of its level k. Under this assumption, the total computing cost of a search is equal to the number of nodes in the search tree times the cost of processing a node. Hence, the number of nodes in the search tree is an important parameter in a search. This number can be obtained by counting the nodes in every level, with α_i defined as the node count at level i of the tree. For example, in the search tree for the projective plane of order 1 shown in Fig. 6, $\alpha_1 = 3$, $\alpha_2 = 6$, and $\alpha_3 = 6$.

B. Estimation

It is very difficult to predict *a priori* the running time of a backtrack program. Sometimes, one program runs to completion in less than 1 sec, while other programs seem to take forever. A minor change in the strategy used in the

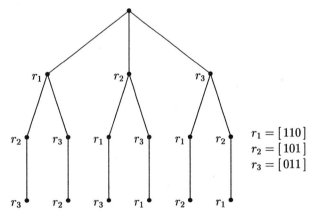

FIGURE 6 Search tree for the projective plane of order 1.

backtrack routine may change the total running time by several orders of magnitude. Some "minor improvements" may speed up the program by a factor of 100, while some other "major improvements" actually slow down the program. It is useful to have a simple and reliable method to estimate the running time of such a program. It can be used to

1. Decide whether the problem can be solved with the available resources.
2. Compare the various methods of "improvement."
3. Design an optimized program to reduce the running time.

A good approximation to the running time of a backtrack program is to multiply the number of nodes in the search tree by a constant representing the time required to process each node. Hence, a good estimate of the running time can be obtained by finding a good estimate to the number of nodes in the search tree. The size of the tree can be approximated by summing up the estimated α_i's, the node counts at each level.

An interesting solution to the estimation problem is a *Monte Carlo* approach developed by Knuth. The idea is to run a number of experiments, with each experiment consisting of performing the backtrack search with a randomly chosen candidate at each level of the search. Suppose we have a partial solution $(x_1, \ldots, x_k)$ for $0 \le k < n$, where n is the depth of the search tree. We let

$$C'_{k+1} = \{x_{k+1} \in C_{k+1} | P(x_1, \ldots, x_{k+1}) \text{ is true}\}$$

be the set of acceptable candidates which extend $(x_1, \ldots, x_k)$. We choose x_{k+1} at random from C'_{k+1} such that each of the $|C'_{k+1}|$ possibilities are equally likely to be chosen. We let $d_k = |C'_k|$ be the number of elements in C'_k for $k = 1, \ldots, n$. Then the node count at level i, α_i, can be estimated by

$$\alpha_i \approx d_1 \ldots d_i.$$

Now, the total estimated size of the search tree is

$$\sum_{k=1}^{n} d_1 \ldots d_k.$$

The cost of processing a node can be estimated by running a few test cases, counting the nodes, and dividing the running time by the number of nodes.

These estimated values of the node counts can best be presented by plotting the logarithm of α_i, or equivalently the number of digits in α_i, as a function of i. A typical profile is shown in Fig. 7. The value of i for which log α_i is maximum is called the *bulge* of the search tree. A backtrack search spends most of its time processing nodes near the bulge.

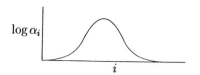

FIGURE 7 Typical shape of a search tree.

C. Optimization

A computer-based proof using backtracking may take a large amount of computing time. Optimization is a process of fine-tuning the computer program so that it runs faster. The optimization methods are divided into two broad classes:

1. Those whose aim is to reduce the size of the search tree
2. Those whose aim is to reduce the cost of the search tree by processing its nodes more efficiently

As a general rule, methods that reduce the size of the search tree, a process also called *pruning* the search tree, can potentially reduce the search by many orders of magnitude, whereas improvements obtained by trying to process nodes more efficiently are often limited to 1 or 2 orders of magnitude. Thus, given a choice, one should first try to reduce the size of the search tree.

There are many methods to prune the search tree. One possibility is to use a more effective compatibility predicate, while preserving the set of solutions. In a search tree, there are many branches which do not contain any solution. If these branches can be identified early, then they can be eliminated, hence reducing the size of the tree.

Another method to prune the search tree is by *symmetry pruning*. Technically, a symmetry is a property preserving operation. For example, two columns of a matrix A can be interchanged without affecting the product AA^T. Consider again the search tree for a projective plane of order 1. The interchange of columns 1 and 2 will induce a relabeling of r_2 as r_3 and vice versa. Thus, after considering the partial solution $x_1 = [101]$, there is no need to consider the remaining partial solution $x_1 = [011]$, because its behavior will be a duplicate of the earlier case. In combinatorial problems, the size of symmetries tends to be large and symmetry pruning can be very effective.

Methods to reduce the cost of processing a node of the search tree are often adaptions of well-known methods in computer programming. For example, one can use a better algorithm such as replacing a linear search by a binary search. We can also replace the innermost loop by an assembly language subroutine, or a faster computer can be used.

One common optimization technique is to move invariant operations from the inside of a loop to the outside. This idea can be applied to a backtrack search in the following manner. We try to do as little as possible for nodes near the bulge, at the expense of more processing away from the bulge. For example, suppose we have a tree of depth 3 and that $\alpha_1 = 1$, $\alpha_2 = 1000$, and $\alpha_3 = 1$. If the time required to process each node is 1 sec, then the processing time for the search tree is 1002 sec. Suppose we can reduce the cost of processing the nodes at level 2 by a factor of 10 at the expense of increasing the processing cost of nodes at levels 1 an 3 by a factor of 100. Then, the total time is reduced to 300 sec.

D. Practical Aspects

There are many considerations that go into developing a computer program which runs for months and years. Interruptions, ranging from power failures to hardware maintenance, are to be expected. A program should not have to restart from the very beginning for every interruption; otherwise, it may never finish. Fortunately, an enumerative computer proof can easily be divided into independent runs. If there is an interruption, just look up the last completed run and restart from that point. Thus, the disruptive effect of an untimely interrupt is now limited to the time wasted in the incomplete run. Typically, the problem is divided into hundreds or even millions of independent runs in order to minimize the time wasted by interruptions.

Another advantage of dividing a problem into many independent runs is that several computers can be used to run the program simultaneously. If a problem takes 100 years to run on one computer, then by running on 100 computers simultaneously the problem can be finished in 1 year.

E. Correctness Considerations

An often-asked question is, "How can one check a computer-based proof?" After all, a proof has to be absolutely correct. The computer program itself is part of the proof, and checking a computer program is no different from checking a traditional mathematical proof. Computer programs tend to be complicated, especially the ones that are highly optimized, but their complexity is comparable to some of the more difficult traditional proofs.

The actual execution of the program is also part of the proof, and the checking of this part is difficult or impossible. Even if the computer program is correct, there is still a very small chance that a computer makes an error in executing the program. In the search for a projective plane of order 10, this error probability is estimated to be 1 in 100,000.

So, it is impossible to have an absolute error-free, computer-based proof! In this sense, a computer-based proof is an experimental result. As scientists in other disciplines have long discovered, the remedy is an independent verification. In a sense, the verification completes the proof.

VI. RECENT DEVELOPMENT: RSA FACTORING CHALLENGE

Recently, there has been a lot of interest in factorizing big numbers. It all started in 1977 when Rivest, Shamir, and Adleman proposed a public-key cryptosystem based on the difficulty of factorizing large numbers. Their method is now known as the RSA scheme. In order to encourage research and to gauge the strength of the RSA scheme, RSA Data Security, Inc. in 1991 started the RSA Factoring Challenge. It consists of a list of large composite numbers. A cash prize is given to the first person to factorize a number in the list. These challenge numbers are identified by the number of decimal digits contained in the numbers. In February 1999, the 140-digit RSA-140 was factorized. In August 1999, RSA-155 was factorized. The best known factorization method divides the task into two parts: a sieving part to discover relations and a matrix reduction part to discover dependencies. The sieving part has many similarities with an enumerative proof. One has to try many possibilities, and the trials can be divided into many independent runs. In fact, for the factorization of RSA-155, the sieving part took about 8000 MIPS years and was accomplished by using 292 individual computers located at 11 different sites in 3 continents. The resulting matrix had 6,699,191 rows and 6,711,336 columns. It took 224 CPU hours and 3.2 Gbytes of central memory on a Cray C916 to solve. Fortunately, there is never any question about the correctness of the final answer, because one can easily verify the result by multiplying the factors together.

VII. FUTURE DIRECTIONS

When the four-color conjecture was first settled by a computer, there was some hesitation in mathematics circles to accept it as a proof. There is first the question of how to check the result. There is also the aesthetic aspect: "Is a computer-based proof elegant?" The result itself is definitely interesting. The computer-based proof of the nonexistence of a projective plane of order 10 again demonstrated the importance of this approach. A lengthy enumerative proof is a remarkable departure from traditional thinking in terms of simple and elegant proofs. As computers are getting faster, many other famous open problems may be solved by this approach. Mathematicians are slowly coming to accept a computer-based proof as another proof technique. Before long, it will be treated as just another tool in a mathematician's tool box.

SEE ALSO THE FOLLOWING ARTICLES

COMPUTER ARCHITECTURE • DISCRETE MATHEMATICS AND COMBINATORICS • GRAPH THEORY • MATHEMATICAL LOGIC • PROBABILITY

BIBLIOGRAPHY

Cipra, B. A. (1989). "Do mathematicians still do math?" Res. News, Sci. **244,** 769–770.

Kreher, D. L., and Stinson, D. R. (1999). "Combinatorial Algorithms, Generation, Enumeration, and Search," CRC Press, Boca Raton, FL.

Lam, C. W. H. (1991). "The search for a finite projective plane of order 10," *Am. Math. Mon.* **98,** 305–318.

Lam, C. W. H. (1989). "How reliable is a computer-based proof?" *Math. Intelligencer* **12,** 8–12.

Odlyzko, A. M. (1985). Applications of Symbolic Algebra to Mathematics, In "Applications of Computer Algebra" (R. Pavelle, ed.), pp. 95–111, Kluwer-Nijhoff, Boston.

Saaty, T. L., and Kainen, P. C. (1977). "The Four-Color Problem," McGraw-Hill, New York.

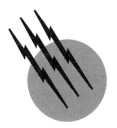

Computer-Generated Proofs
of Mathematical Theorems

David M. Bressoud
Macalester College

GLOSSARY

Algorithm A recipe or set of instructions that, when followed precisely, will produce a desired result.

Binomial coefficient The binomial coefficient $\binom{n}{k}$ is the coefficient of x^k in the expansion of $(1+x)^n$. It counts the number of ways of choosing k objects from a set of n objects.

Computer algebra system A computer package that enables the computer to do symbolic computations such as algebraic simplification, formal differentiation, and indefinite integration.

Diophantine equation An equation in several variables for which only integer solutions are accepted.

Hypergeometric series A finite or infinite summation, $1 + a_1 + \cdots + a_k + \cdots$, in which the first term is 1 and the ratio of successive summands, a_{k+1}/a_k, is a quotient of polynomials in k.

Hypergeometric term A function of, say, k, that is the kth summand in a hypergeometric series.

Proof certificate A piece of information about a mathematical statement that makes it possible to prove the statement easily and quickly.

Proper hypergeometric term A function of two variables such as n and k that is of the following form: a polynomial in n and k times $x^k y^n$ for some fixed x times a product of quotients of factorials of the form $(an + bk + c)!$ where a, b, and c are fixed integers.

Rising factorial A finite product of numbers in an arithmetic sequence with difference 1. It is written as $(a)_n = a(a+1)(a+2)\cdots(a+n-1)$.

IN A COMPUTER-BASED PROOF, the computer is used as a tool to help guess what is happening, to check cases, to do the laborious computations that arise. The person who is creating the proof is still doing most of the work. In contrast, a computer-generated proof is totally automated. A person enters a carefully worded mathematical statement for which the truth is in doubt, hits the RETURN key, and within a reasonable amount of time the computer responds either that the statement is true or that it is false. A step beyond this is to have the computer do its own searching for reasonable statements that it can test.

Such fully automated algorithms for determining the truth or falsehood of a mathematical statement do exist.

With Doron Zeilberger's program EKHAD, one can enter the statement believed or suspected to be correct. If it is true, the computer will not only tell you so, it is capable of writing the paper ready for submission to a research journal. Even the search for likely theorems has been automated. A good deal of human input is needed to set parameters within which one is likely to find interesting results, but computer searches for mathematical theorems are now a reality.

The possible theorems to which this algorithm can be applied are strictly circumscribed, so narrowly defined that there is still a legitimate question about whether this constitutes true computer-generated proof or is merely a powerful mathematical tool. What is not in question is that such algorithms are changing the kinds of problems that mathematicians need to think about.

I. THE IDEAL VERSUS REALITY

A. What Cannot Be Done

Mathematics is frequently viewed as a formal language with clearly established underlying assumptions or axioms and unambiguous rules for determining the truth of every statement couched in this language. In the early decades of the twentieth century, works such as Russell and Whitehead's *Principia Mathematica* attempted to describe all mathematics in terms of the formal language of logic. Part of the reason for this undertaking was the hope that it would lead to an algorithmic procedure for determining the truth of each mathematical statement. As the twentieth century progressed, this hope receded and finally vanished. In 1931, Kurt Gödel proved that no axiomatic system comparable to that of Russell and Whitehead could be used to determine the truth or falsehood of every mathematical statement. Every consistent system of axioms is necessarily incomplete.

One broad class of theorems deals with the existence of solutions of a particular form. Given the mathematical problem, the theorem either exhibits a solution of the desired type or states that no such solution exists. In 1900, as the tenth of his set of twenty-three problems, David Hilbert challenged the mathematical community: "Given a Diophantine equation with any number of unknown quantities and with rational integral numerical coefficients: To devise a process according to which it can be determined by a finite number of operations whether the equation is solvable in rational integers." A well-known example of such a Diophantine equation is the Pythagorean equation, $x^2 + y^2 = z^2$, with the restriction that we only accept integer solutions such as $x = 3$, $y = 4$, and $z = 5$. Another problem of this type is Fermat's Last Theorem. This theorem asserts that no such positive integer solutions exist for

the equation $x^n + y^n = z^n$ when n is an integer greater than or equal to 3. We know that the last assertion is correct, thanks to Andrew Wiles.

For a Diophantine equation, if a solution exists then it can be found in finite (though potentially very long) time just by trying all possible combinations of integers, but if no solution exists then we cannot discover this fact just by trying possibilities. A proof that there is no solution is usually very hard. In 1970, Yuri Matijasevič proved that Hilbert's algorithm could not exist. It is impossible to construct an algorithm that, for every Diophantine equation, is able to determine whether it does or does not have a solution.

There have been other negative results. Let E be an expression that involves the rational numbers, π, $\ln 2$, the variable x, the functions sine, exponential, and absolute value, and the operations of addition, multiplication, and composition. Does there exist a value of x where this expression is zero? As an example, is there a real x for which

$$e^x - \sin(\pi \ln 2) = 0?$$

For this particular expression the answer is "yes" because $\sin(\pi \ln 2) > 0$, but in 1968, Daniel Richardson proved that it is impossible to construct an algorithm that would determine in finite time whether or not, for every such E, there exists a solution to the equality $E = 0$.

B. What Can Be Done

In general, the problem of determining whether or not a solution of a particular form exists is extremely difficult and cannot be automated. However, there are cases where it can be done. There is a simple algorithm that can be applied to each quadratic equation to determine whether or not it has real solutions, and if it does, to find them.

That $x^2 - 4312x + 315 = 0$ has real solutions may not have been explicitly observed before now, but it hardly qualifies as a theorem. The theorem is the statement of the quadratic formula that sits behind our algorithm. The conclusion for this particular equation is simply an application of that theorem, a calculation whose relevance is based on the theory.

But as the theory advances and the algorithms become more complex, the line between a calculation and a theorem becomes less clear. The Risch algorithm is used by computer algebra systems to find indefinite integrals in Liouvillian extensions of difference fields. It can answer whether or not an indefinite integral can be written in a suitably defined closed form. If such a closed form exists, the algorithm will find it. Most people would still classify a specific application of this algorithm as a calculation, but it is no longer always so clear-cut. Even definite integral evaluations can be worthy of being called theorems.

Freeman Dyson conjectured the following integral evaluation for positive integer z in 1962:

$$(2\pi)^{-n} \int_0^{2\pi} \cdots \int_0^{2\pi} \prod_{1 \le j < k \le n} \left| e^{i\theta_j} - e^{i\theta_k} \right|^{2z} d\theta_1 \cdots d\theta_n$$
$$= \frac{(nz)!}{(z!)^n}.$$

Four proofs have since been published. Dyson's conjecture cannot be proven by the Risch or any other general integral evaluation algorithm because the dimension of the space over which the integral is taken is a variable, but its proof is now close to the boundary of what can be totally automated.

Most of this article will focus on the WZ method developed by Wilf and Zeilberger in the early 1990s. Given a suitable hypergeometric series, the WZ method will determine whether or not it has a closed form. If it does, the algorithm will find it. It can even be used to find new hypergeometric series that can be expressed in closed form. Again, the important mathematics is the theory that is used to create and justify the algorithm, but specific applications now look very much like theorems. One example of a result that can be proved by the WZ method is the following identity, discovered and proved by J. C. Adams in the nineteenth century. Let $P_n(x)$ be the Legendre polynomial defined by

$$P_n(x) := \frac{1}{2^n} \sum_{k=0}^n \binom{n}{k}^2 (x-1)^k (x+1)^{n-k},$$

and let $A_k = \binom{2k}{k}$, then

$$\int_{-1}^1 P_m(x) P_n(x) P_{m+n-2k}(x)\, dx$$
$$= \frac{1}{(m+n+1/2-k)} \cdot \frac{A_k A_{m-k} A_{n-k}}{A_{m+n-k}}. \quad (1)$$

Note that the term-by-term integration is not difficult for a computer algebra system. What distinguishes this particular identity is that the number of terms in each summation is left as a variable.

Given a hypergeometric series, the WZ method can be used to find the closed expression that it equals, provided such an expression exists. We take as an example

$$f(n) = \sum_{0 \le k \le n/3} 2^k \frac{n}{n-k} \binom{n-k}{2k}.$$

The algorithm produces a recursion satisfied by $f(n)$:

$$f(n+3) - 2f(n+2) + f(n+1) - 2f(n) = 0.$$

This is a particularly nice example because the coefficients are constants and standard techniques can be applied to discover that

$$\sum_{0 \le k \le n/3} 2^k \frac{n}{n-k} \binom{n-k}{2k}$$
$$= 2^{n-1} + \frac{1}{2}(i^n + (-i)^n), \qquad n \ge 2.$$

In general, the coefficients in the recursion will be polynomials in n. In 1991 Marko Petkovšek created an algorithm that will find a closed form solution for such a recursion, or prove that no such formula exists. The combination of the WZ method with Petkovšek's algorithm gives an automated proof that a particular type of solution cannot exist, or else it finds such a solution. As an example, there is a computer-generated proof of the fact that

$$\sum_{k=0}^n \binom{n}{k}^2 \binom{n+k}{k}^2$$

cannot be written as a linear combination of hypergeometric terms in n.

The WZ method combined with Petkovšek's algorithm is producing fully automated proofs of results that, until recently, have required considerable human ingenuity. Significantly, it replies not just with a statement that a particular identity is true, but also with a proof certificate, a critical insight that enables anyone with pencil and paper and a little time to verify that this identity is correct. At the very least, these algorithms have moved the line of demarcation between what constitutes a proof and what is only a computation.

II. HYPERGEOMETRIC SERIES IDENTITIES

A. What Is a Hypergeometric Series?

A series, $1 + a_1 + a_2 + a_3 + \cdots$, is called hypergeometric if the ratio of consecutive terms, a_{n+1}/a_n, is a rational function of n, say $a_{n+1}/a_n = P(n)/Q(n)$, where P and Q are polynomials. Most of the commonly encountered power series are hypergeometric or can be expressed in terms of hypergeometric series (see Fig. 1). A hypergeometric term is a function of n that is a summand of a hypergeometric series indexed by n. In particular, a hypergeometric term is of the form

$$a_k = \prod_{n=0}^{k-1} \frac{a_{n+1}}{a_n} = \prod_{n=0}^{k-1} \frac{P(n)}{Q(n)},$$

for some pair of polynomials P and Q.

If we factor P and Q,

$$P(n) = c_1(n+\alpha_1)(n+\alpha_2) \cdots (n+\alpha_m),$$

$$Q(n) = c_2(n+\beta_1)(n+\beta_2) \cdots (n+\beta_{n+1}),$$

Exponential function:

$$e^x = 1 + \sum_{n=1}^{\infty} \frac{x^n}{n!}, \quad \frac{a_{n+1}}{a_n} = \frac{x}{n+1},$$

Sine function:

$$\sin x = x \left(1 + \sum_{n=1}^{\infty} \frac{(-1)^n x^{2n}}{(2n+1)!}\right), \quad \frac{a_{n+1}}{a_n} = \frac{-x^2}{4(n+1)(n+3/2)},$$

Bessel function of the first kind:

$$J_k(x) = \frac{x^k}{\Gamma(k+1)} \left(1 + \sum_{n=1}^{\infty} \frac{(-1)^n x^{2n}}{4^n \, n! \, (k+1)_n}\right), \quad \frac{a_{n+1}}{a_n} = \frac{-x^2}{4(n+1)(n+k+1)},$$

Error function:

$$\text{erf}(x) = \frac{2x}{\sqrt{\pi}} \left(1 + \sum_{n=1}^{\infty} \frac{(-1)^n x^{2n}}{(2n+1) \, n!}\right), \quad \frac{a_{n+1}}{a_n} = -x^2 \frac{(2n+1) \, n}{(2n+3)(n+1)}.$$

FIGURE 1 Examples of common functions expressed in terms of hypergeometric series.

then the hypergeometric term can be written as

$$a_k = c \frac{(\alpha_1)_k (\alpha_2)_k \cdots (\alpha_m)_k}{(\beta_1)_k (\beta_2)_k \cdots (\beta_{n+1})_k},$$

where $c = c_1/c_2$ and $(\alpha)_k$ is the rising factorial:

$$(\alpha)_k = \alpha(\alpha+1)(\alpha+2)\cdots(\alpha+k-1).$$

B. The Chu–Vandermonde Identity

A large part of the impetus behind the development of the WZ method and the reason why it has become such an influential tool is that there is a rich and ever-expanding store of useful identities for hypergeometric series. These recur throughout mathematics, playing important roles in the solutions of both theoretical and applied problems.

The binomial theorem was the first and is the most fundamental of these identities. It is the foundation upon which all others are proved. Mathematicians have been building upon the binomial theorem for many years. In 1303, Chu Shih-Chieh wrote *Precious Mirror of the Four Elements (Ssu Yü Chien)*, in which he may have been the first person to state the fundamental result:

$$\sum_{i=0}^{\infty} \binom{a}{i} \binom{b}{k-i} = \binom{a+b}{k}. \tag{2}$$

In Chu's identity, a, b, and k are positive integers. Note that all summands will be zero once i is greater than a or k. Equation (2) is easily derived from the binomial theorem by comparing the coefficients of x^k in

$$(1+x)^a (1+x)^b = \sum_{i=0}^{a} \binom{a}{i} x^i \sum_{j=0}^{b} \binom{b}{j} x^j,$$

and

$$(1+x)^{a+b} = \sum_{k=0}^{a+b} \binom{a+b}{k} x^k.$$

Equation (2) was rediscovered by Alexandre Vandermonde in 1772 and is today known as the *Chu–Vandermonde Identity*.

The ratio of successive terms in the summation is

$$\binom{a}{n+1} \binom{b}{k-n-1} \bigg/ \binom{a}{n} \binom{b}{k-n}$$

$$= \frac{(n-a)(n-k)}{(n+1)(n+b-k+1)}.$$

If we divide both sides of Eq. (2) by the first summand, $\binom{b}{k}$, it can be rewritten in terms of rising factorials as

$$1 + \sum_{n=1}^{\infty} \frac{(-a)_n (-k)_n}{n!(b-k+1)_n} = \frac{(a+b)! \, (b-k)!}{(a+b-k)! \, b!}. \tag{3}$$

In 1797, Johann Friedrich Pfaff showed that, subject only to convergence conditions, Eq. (3) holds for complex values, in which case it can be expressed as

$$1 + \sum_{n=1}^{\infty} \frac{(\alpha)_n (\beta)_n}{n!(\gamma)_n} = \frac{\Gamma(\gamma - \alpha - \beta)\Gamma(\gamma)}{\Gamma(\gamma - \alpha)\Gamma(\gamma - \beta)}. \tag{4}$$

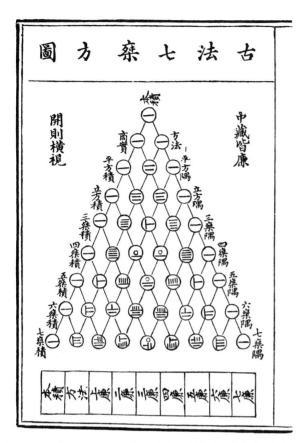

FIGURE 2 The representation of "Pascal's" triangle in Chu's *Precious Mirror of the Four Elements* of 1303. (Reprinted with the permission of Cambridge University Press.)

Pfaff's student, Carl Friedrich Gauss, used hypergeometric series in his astronomical work and advanced their study. Among his contributions, he found sharp criteria for whether or not a hypergeometric series converges. Throughout the nineteenth and twentieth century, a great number of identities for hypergeometric series were discovered, many of which were collected in the Bateman Manuscript Project published as *Higher Transcendental Functions* in 1953–1955.

C. Standardized Notation

Most hypergeometric series can be written as sums of rational products of binomial coefficients, but this representation is problematic because it is not unique. As an example,

$$\sum_{n=0}^{m} 2^{m-k-2n} \binom{m}{n} \binom{m-n}{n+k} = \binom{2m}{m+k}$$

appears to be different from the Chu-Vandermonde identity [Eq. (2)]. But if we look at the ratio of consecutive summands, it is

$$\frac{1}{4} \frac{(m-2n-k-1)(m-2n-k)}{(n+1)(n+k+1)}$$
$$= \frac{(n+(k+1-m)/2)(n+(k-m)/2)}{(n+1)(n+k+1)}.$$

This is simply the Chu-Vandermonde identity with $\alpha = (k+1-m)/2$, $\beta = (k-m)/2$, and $\gamma = k+1$.

There is clearly an advantage to using the rising factorial notation, in which case we write

$$_m F_n \binom{\alpha_1, \ldots, \alpha_m}{\beta_1, \ldots, \beta_n}; x) := 1 + \sum_{k=1}^{\infty} \frac{(\alpha_1)_k \cdots (\alpha_m)_k}{k!(\beta_1)_k \cdots (\beta_n)_k} x^k.$$

Even with this standardized notation, there are equivalent identities that look different because there are nontrivial transformation formulas for hypergeometric series. As an example, provided the series in question converge, we have that

$$_2 F_1 \binom{a, b}{2a}; x) = \left(1 - \frac{x}{2}\right)^{-b}$$

$$\times {}_2 F_1 \binom{b/2, (b+1)/2}{a + 1/2}; \left(\frac{x}{2-x}\right)^2).$$

This is why, even if all identities for hypergeometric series were already known, it would not be enough to have a list of them against which one could compare the candidate in question. Just establishing the equivalence of two identities can be a very difficult task. This makes the WZ method all the more remarkable because it is independent of the form in which the identity is given and can even be used to verify (or disprove) a conjectured transformation formula.

III. THE WZ METHOD

A. Sister Celine's Technique

The WZ method for finding and proving identities for hypergeometric series builds on a succession of developments that began with the Ph.D. thesis of Sister Mary Celine Fasenmyer at the University of Michigan in 1945. We consider a sum of the form

$$f(n) = \sum_k F(n, k),$$

where $F(n, k)$ is a proper hypergeometric term. This means that it is a polynomial in n and k times $x^k y^n$, for fixed x and y, times a product of quotients of factorials of the form $(an+bk+c)!$, where a and b, and c are fixed integers. As an example,

$$\sum_{0 \le k \le n/3} 2^k \frac{n}{n-k} \binom{n-k}{2k} = \sum_{0 \le k \le n/3} n \cdot 2^k \cdot \frac{(n-k-1)!}{(2k)! \, (n-3k)!}.$$

is such a series.

Every such sum of proper hypergeometric terms will satisfy a finite recursion of the form

$$\sum_{j=0}^{J} a_j(n) f(n+j) = 0.$$

Sister Celine showed how to reduce the problem of finding these coefficients to one of solving a system of linear equations. It was Doron Zeilberger who realized that this gives us an algorithm for proving hypergeometric series identities because we need only verify that each side satisfies the same recursion and the same initial conditions. The problem with using Sister Celine's approach is that her particular algorithm for finding the coefficients is slow. Later developments would speed it up considerably, though in the process would lose the easy generalization of Sister Celine's technique to summations over several indices.

B. Gosper's algorithm

In 1977 and 1979, R. W. Gosper, Jr., took a different approach and became one of the first people to use computers to discover and check identities for hypergeometric series. Given a proper hypergeometric term $F(n, k)$, Gosper showed how to automate a search for a proper hypergeometric term $G(n, k)$ with the property that

$$G(n, k+1) - G(n, k) = F(n, k).$$

If such a G could be found, then

$$f(n) = \sum_{k=0}^{n} (G(n, k+1) - G(n, k))$$
$$= G(n, n+1) - G(n, 0).$$

An example of the application of this algorithm is the computer-generated proof of an identity discovered and first proved by J. S. Lomont and John Brillhart: Let $1 \le m \le n$, where $n \ge 2$ and $1 \le s \le \min(m, n-1)$, then

$$\sum_{j=0}^{s} \left[(-1)^j (m+n-2j) \binom{m}{j} \binom{m-j}{m-s} \binom{n}{j} \binom{n-j}{n-s} \right.$$
$$\left. \times \binom{m+n}{j} \binom{m+n-s-j-1}{s-j} \middle/ \binom{s}{j}^2 \right] = 0.$$

Given this conjecture, the program EKHAD replies with the proof certificate:

$$-s, \; j * (m+n-s-j)/(m+n-2*j).$$

This means that if $F(n, j)$ is the summand in the conjectured identity, then

$$-sF(n, j) = G(n, j+1) - G(n, j),$$

where $G(n, j) = j(m+n-s-j)F(n, j)/(m+n-2j)$. The sum over j of $G(n, j+1) - G(n, j)$ telescopes, and therefore the original summation equals $[G(n, s+1) - G(n, 0)]/(-s) = 0$.

Gosper's algorithm is a fertile approach that is often applicable, but it is limited by the fact that such a G does not always exist.

C. Wilf and Zeilberger

Major progress was made by Doron Zeilberger who, starting in 1982, began to combine the ideas of Sister Celine and William Gosper. In the early 1990s, Herbert Wilf joined Zeilberger in extending and refining these methods into a fully automated proof machine that is now known as the WZ method. If $F(n, k)$ is a proper hypergeometric term, then there always is a proper hypergeometric term $G(n, k)$ such that $G(n, k+1) - G(n, k)$ is equal to a linear combination of $\{F(n+j, k) \mid 0 \le j \le J\}$ for some explicitly computable J,

$$\sum_{j=0}^{J} a_j(n) F(n+j, k) = G(n, k+1) - G(n, k), \quad (5)$$

where the $a_j(n)$ are polynomials in n. If $f(n) = \sum_{k=0}^{K} F(n, k)$, then we can sum both sides of Eq. (5) over $0 \le k \le K$. The right side telescopes, and we are left with the recursive formula

$$\sum_{j=0}^{J} a_j(n) f(n+j) = G(n, K+1) - G(n, 0).$$

Gosper's technique—which is very fast—can be used to find the function G. The coefficients $a_j(n)$ are then found by solving a system of linear equations.

Gosper's algorithm is the special case of the WZ method in which $J = 0$. The other case of particular interest is when $J = 1$ and $a_1 = -a_0 = 1$. Consider the conjectured identity:

$$\sum_{k} \binom{n}{2k} \binom{2k+1}{k} \frac{n+2}{2k+1} 2^{n-2k-1} = \binom{2n+1}{n}. \quad (6)$$

If we divide each side by $\binom{2n+1}{n}$, this can be rewritten as

$$f(n) = \sum_{k} \binom{n}{2k} \binom{2k+1}{k} \frac{n+2}{2j+1} 2^{n-2k-1} \middle/$$
$$\binom{2n+1}{n} = 1.$$

If this is true, then $f(n)$ satisfies the recursion $f(n + 1) - f(n) = 0$. Let $F(n, k)$ be the summand,

$$F(n, k) = \binom{n}{2k}\binom{2k + 1}{k}\frac{n + 2}{2j + 1}2^{n-2k-1}\bigg/\binom{2n + 1}{n}.$$

If we could find a proper hypergeometric term $G(n, k)$ for which

$$F(n + 1, k) - F(n, k) = G(n, k + 1) - G(n, k), \quad (7)$$

then it would follow that $f(n + 1) - f(n) = 0$, and so $f(n)$ would be constant. It would be enough to check that $f(0) = 1$.

In fact, such a G does exist. The WZ method finds it. The proof certificate is the rational function,

$$\frac{G(n, k)}{F(n, k)} = \frac{4k(k + 1)}{(2n + 3)(2k - n - 1)}.$$

To check Eq. (6), we only need to verify that F and G, which we now know, do indeed satisfy Eq. (7).

In general, the WZ method returns either the ratio $G(n, k)/F(n, k)$ [if the recursion is of the form given in Eq. (7)], or it returns the actual recursive formula satisfied by $f(n)$. The only drawback to the WZ method is that the number of terms in the recursive formula may be too large for practical use.

D. Petkovšek and Others

In his Ph.D. thesis of 1991, Marko Petkovšek showed how to find a closed form solution—or to show that such a solution does not exist—for any recursive formula of the form

$$\sum_{j=0}^{J} a_j(n)f(n + j) = g(n),$$

in which $g(n)$ and the $a_j(n)$ are polynomials in n. By closed form, we mean a linear combination of a fixed number of hypergeometric terms.

Combined with the WZ method, Petkovšek's algorithm implies that in theory if not in practice, given any summation of proper hypergeometric terms, there is a completely automated computer procedure that will either find a closed form for the summation or prove that no such closed from exists. A full account of the WZ method and Petkovšek's algorithm is given in the book $A = B$ by Petkovšek, Wilf, and Zeilberger.

Others have worked on implementing and extending the ideas of the WZ method. One of the centers for this work has been a group headed by Peter Paule at the University of Linz in Austria. Ira Gessel has been at the forefront of those who have used this algorithm to implement computer searches that both discovered and proved a large number of new identities for hypergeometric series.

IV. EXTENSIONS, FUTURE WORK, AND CONCLUSIONS

A. Extensions and Future Work

All of the techniques described in this article have been extended to q-hypergeometric series such as

$$1 + \sum_{k=0}^{n} q^{k^2}\frac{(1 - q^n)(1 - q^{n-1})\cdots(1 - q^{n-k+1})}{(1 - q)(1 - q^2)\cdots(1 - q^k)},$$

in which the ratio of consecutive summands is a rational function of q^k.

Many general determinant evaluations can be reduced to problems that can be solved using the WZ method. Work is progressing on fully automating proofs of such results.

There are other theorems that appear to be amenable to an automated computer attack. These include results on real closed fields using techniques of George Collins and geometrical theorems proved using algebraic techniques such as Gröbner bases.

B. Conclusions

The net effect of the algorithms that prove identities for hypergeometric series is that a piece of mathematics that once could only be done by those with cleverness and insight has been turned into a purely mechanical calculation. Rather than limiting the scope of mathematics, the WZ method has widened it. Problems that had once been intractable are now within reach. The situation is different in degree but not in kind from the invention of calculus. This was the discovery of mechanical procedures that enabled scientists to shift their attention away from laborious and ingenious techniques for finding areas and tangent lines and to begin addressing the really interesting questions.

Perhaps this will always be the fate of computer-generated proofs. That one class of problems has been moved into the category of those that can be solved by computers means that we are freed to direct our attention to those questions that are most important.

Web Sites for the WZ Method and Related Algorithms

Home page for the book $A = B$:
    http://www.cis.upenn.edu/~wilf/AeqB.html
Wilf and Zeilberger's programs:
    http://www.cis.upenn.edu/~wilf/progs.html
Programs of the RISC group at the University of Linz:
    http://www.risc.uni-linz.ac.at/research/combinat/risc/

SEE ALSO THE FOLLOWING ARTICLES

COMPUTER ALGORITHMS • COMPUTER ARCHITECTURE • DISCRETE MATHEMATICS AND COMBINATORICS • GRAPH THEORY • MATHEMATICAL LOGIC • PROBABILITY

BIBLIOGRAPHY

Collins, G. E. (1975). *In* "Automata Theory and Formal Languages," Lecture Notes in Computer Science, (H. Brakhage, ed.) Vol. 33, pp. 134–183, Springer, Berlin.

Hilbert, D. (1902). "Mathematical Problems," *Bulletin of the American Mathematical Society* **8**, pp. 1–34. (Translation of the original German article.)

Nemes, I., Petkovšek, M., Wilf, H., and Zeilberger, D. (1997). "How to do MONTHLY problems with your computer," *American Mathematical Monthly* **104**, 505–519.

Petkovšek, M., Wilf, H., and Zeilberger, D. (1996). "*A = B*," A K Peters, Wellesley, MA.

Wilf, H., and Zeilberger, D. (1992). "An algorithmic proof theory for hypergeometric (ordinary and "*q*") multisum/integral identities," *Inventiones Mathematicae* **108**, 575–633.

Computer Networks

Jordanka Ivanova
Michael Jurczyk
University of Missouri-Columbia

GLOSSARY

Computer network System consisting of an interconnected collection of computers that are able to exchange information.

Network protocol Encapsulated module of services with defined interfaces, providing communication functions to other protocols or applications.

Network architecture Structured collection of protocols and layers with specification of the services provided in each layer.

Quality of Service (QoS) Description of the quality of services provided by a computer network. QoS is characterized by parameters such as bandwidth, delay, and jitter.

Network security service Service that enforces the security policy to provide a certain level of network security.

World Wide Web (WWW) Large-scale, on-line repository of information in the form of linked documents, web pages, and web sites.

Internet The global collection of networks (internet) that uses TCP/IP protocols.

I. DEFINITIONS

For the past 10 years, computer networks have become a major part of modern communications, data processing, business, culture, and everyday life. The properties that make them so popular are access to remote data, resource sharing, and human communication. Network users can access and use remote data and machine resources as if they were local. Computer resources such as printers and database servers can be shared by many remote users, making the computing process more efficient. Resource sharing and remote data access through networking are also used to increase the reliability, accessibility, and uptime of many applications. For example, if a bank has a database server and a backup server, in case of failure of the main server the backup server can then provide service to the users, preventing loss of data accessibility.

Apart from sharing data and resources, computer networks offer a human communication medium. This is very important for large companies, which have offices around the country, or international organizations, which have offices all over the world. The efficient and convenient means of communication like e-mail, video conferencing, and the Internet as a global information environment are used to link people together. The Internet as a worldwide network brings new business opportunities, where products and services find their way to the clients much faster and with less effort. E-commerce is a growing and very successful business.

Networks are as important for personal use as for corporate use. Information access is one of the major motivations. Users can access information from millions of servers available through the Internet containing articles, publications, on-line books, manuals, magazines (e-zines), and newspapers. Video and audio, virtual reality, entertainment, and network games are other popular applications used over the Internet. On-line shopping is not just popular, but it is becoming a major competitor for conventional shopping. Personal communications is another important feature of computer networks. E-mail, on-line pagers, and Internet telephony are all network based. Newsgroups and chat rooms are interactive ways, in real time, to communicate with other people. A new approach in the educational system is distance learning. Based on on-line courses and programs, it offers course material combined with multimedia to the students. This distance learning is especially helpful for people holding full-time jobs or people with disabilities.

Computer network is a system consisting of an interconnected collection of computers that are able to exchange information. Computers connected in a network are usually called nodes. *Hosts* are nodes used by users and are usually running applications. Apart from hosts, network nodes such as routers, bridges, or gateways that execute network support functions (e.g., messages routing) might also be present. Physical links that interconnect nodes can be copper wire, fiber optics, microwave, radio, infrared, or satellite connections. There are two connectivity technologies independent of the physical link type: multiple-access and point-to-point links. A *multiple-access link* allows more than two nodes that share that link to be attached to it. In contrast, a *point-to-point link* connects two nodes only. A network that

uses multi-access links and allows messages sent by a node to be received by all other nodes is called a *broadcast network*. Broadcasting is the operation of sending a message/packet to all destinations. A *point-to-point network* or *switched network* consists of many point-to-point links in an organized way forwarding data from one link to another. Usually, smaller networks tend to be broadcast networks, while larger ones are point-to-point networks.

Networks can be circuit switched or packet switched. *Circuit-switched networks* establish a connection from the sender to the receiver first. After the connection is up, the sender will send data in a large chunk over the connection. After the sending process is finished, the connection is torn down. In *packet-switched networks*, a message is split into packets. Each packet consists of a body (the actual data) and a header that includes information such as the destination address. A connection is not established, but each packet is routed through the network individually. Packet-switched networks can provide connection-oriented or connectionless services. *Connection-oriented services* establish a connection from the sender to the receiver by specifying the exact route for all the packets through the network. All packets use this route and will be received by the receiver in order. In contrast, *connectionless services* do not establish a fixed route for the packets. Packets belonging to the same connection might travel over different routes through the network and might be received out of order.

Size is an important characteristic of networks. A key factor is not the number of nodes a network connects, but the time needed for the data to propagate from one end of the network to the other. When the network is restricted in size, the upper bound of transmission time is known. In larger networks, this time is generally unknown.

Smallest networks covering usually a single room are termed *system area networks* (SAN). They are used to interconnect PCs or workstation clusters forming server systems and to connect data vaults to the system. Smaller networks, typically spanning less than 1 km, are called *local area networks* (LAN). LANs are generally used to connect nodes within a few buildings and are privately owned. They are used to connect workstations for resource sharing (printer, database server) and for information exchange among each other and often employ multi-access links to connect all of the computers. LANs can run at speeds from 10 megabits per second (Mbps) up to few hundreds Mbps and have low delay and usually low bit error rates. The next in size are *metropolitan area networks* (MAN), which usually extend up to tens of kilometers and serve nearby corporate offices or interconnect a city. The last type, *wide area networks* (WAN), has no size limit and can be worldwide. WAN spans a large geographical area, often a country or continent. Because of the great number of hosts and the distance among them, WANs are in general not broadcast-type networks, but they can provide broadcast operations. WANs use the point-to-point principle and employ a set of nodes dedicated to internetwork communication. These nodes are usually called *routers*, and the set of routers is called *subnet*. When a host wants to communicate with another host, it sends the message, often divided into several packets, to the closest router. That router stores the packets and forwards them to a router that is closer to the destination host. This repeats until the packets reach a router that is connected to the destination node. This router will then deliver the packets to the destination or to the LAN the destination is connected to.

Another common classification of networks is based on the topology of how the nodes and the links are interconnected. The topology of the network is represented by a graph in which nodes are vertices and links are edges. The network graph can be symmetric or irregular. Symmetric ones are used mainly in LANs, such as the star topology, in which all the nodes are connected to one center node, the bus topology, in which all nodes are attached to a single cable/bus, the ring network, in which nodes are connected to form a ring, or the fully connected topology, in which every node is connected to every other node. WANs are likely to have irregular topology, since they have too many links to enforce a symmetric topology.

II. NETWORK PROTOCOLS

A. Overview

One of the main objectives when designing a computer network is to keep it as simple as possible. To achieve this, different abstractions for services and objects can be used. The idea of encapsulating components of the system into an abstraction and providing interfaces that can be used by other system components hides the complexity. In networks, this strategy leads to the concept of *layering* of the network software. The general rule is to start from the underlying hardware and build a sequence of software layers on top, each one providing a higher level of service. Layering provides two features: less complexity and modularity. Instead of constructing a network with one solid mass of services, it is much easier to separate them into manageable components. Modularity gives the assurance that as long as the interface and the functions are consistent and compatible with the original specification, the layer can be modified at any time and new functions can be added. Each layer provides certain services to the layer above, implemented in terms of services offered by the

layer below it. A layer on one machine communicates with exactly the same layer on another machine. Corresponding layers are usually called *peers*. Peers, together with rules and conventions that they have for communication, are called *protocol*. Each protocol defines the interface to its peers and the service interface to the other layers. Peer interface describes the specifications and format of the messages exchanged between the peers. Service interface specifies the functions provided by the layer. The set of layers and protocols is called *network architecture*. In fact, only the hardware level peers communicate with each other directly, all other layers pass messages to the next lower layer. If two applications on different computers want to communicate, the initiator sends a message containing data and some special formatting information usually called *header*. The next lower layer receives the message, adds its own header, and passes it to next lower level and so on until the message is transmitted over the link (see Fig. 1). On the receiver machine, this message will progress up the different layers until it reaches the application involved in this communication. Each layer will strip the corresponding layer header from the message, will apply its services, and will pass the message to its next higher layer. Viewing the group of protocols in a machine or a network as a stack can be helpful for visualization, but the protocols need not be in a linear configuration. More than one protocol can exist on the same level, providing similar yet different services. The *protocol graph* is used to describe the protocols and their dependencies in a network system. A chain of protocols relying on each other's services used by a specific application or a system is called *protocol stack* and is a single path in the protocol graph. The use of the term protocol is somehow ambiguous, referring sometimes to the abstract interface and sometimes to the actual implementation of the interface. To distinguish this, we will refer to the latter as *protocol specification*.

Network architectures and protocols are described in reference models. In the following subsections, a few very important reference models will be discussed: the OSI reference model, the TCP/IP reference model, and the B-ISDN ATM reference model.

B. ISO OSI Reference Model

One of the first standards for computer communications was proposed and developed by the *International Standards Organization* (ISO) in the early 1980s. This network architecture model, the *open systems interconnection* (OSI) reference model shown in Fig. 1, describes a network through seven layers. On any of these layers, one or more protocols can implement the functions specified for the layer. Some protocol specifications based on this

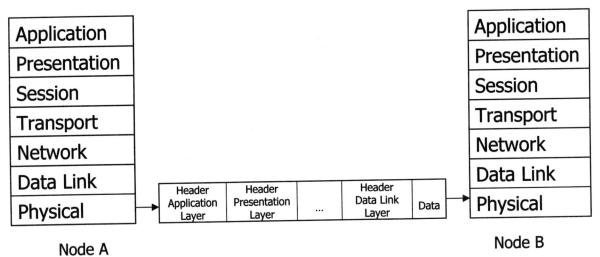

FIGURE 1 An OSI model and example of data transmission from node A to node B.

model, the "*X dot*" series X.25, X.400, etc., were specified by the *International Telecommunications Union* (ITU). Even though most of today's protocols do not follow this reference model, it captures very well the concepts of service, protocol, and interface. OSI design is general and quite complex, but it is able to present the functionality of a network in a way it can be used as a guide for designing new networks.

There are seven layers in the OSI model, starting with the *physical layer* handling the raw data transmission over a physical medium. The most common transmission media are twisted pair (copper wires), coaxial cable, and fiber optics. The *data link layer*, usually implemented in the network adaptors, is above the physical layer and is concerned with the organization of data into *frames* and the reliable transportation of these frames over a direct link. The specific problems of multi-access links such as channel allocation and collision detection are handled by the data link sub-layer called *medium access control* (MAC). Reliable frame delivery, frame ordering, and frame retransmission are provided in the layer by *sliding window protocols*. This is a set of protocols for full-duplex data frame transmission, in which the sender and the receiver both keep windows of frame acknowledgements and send frames only if a certain number of already sent frames were acknowledged by the receiver. The data link layer also includes some error detection and correction functions such as parity bit code and cyclic redundancy code (CRC).

The next higher layer is called *network layer*, and it addresses the problem of finding a route through the network from the source to the destination. This layer also addresses the problems of interconnecting different networks. Protocols in this layer are used to glue together heterogeneous networks into one scalable internetwork or

internet. Mind the difference between Internet and internet, where the former is the global, widely used internetwork and the latter is the general term for a logical network consisting of a collection of physical networks. The *transport layer* takes care of the efficient and reliable delivery of data from the source to the destination node. Protocols in this layer are sometimes called end-to-end protocols. Modified versions of the sliding window protocols ensure reliable delivery, synchronization, and flow control. The fifth layer, the *session layer*, is the one that manages traffic direction and synchronization. The last two layers, the presentation and application layers, are mainly application-oriented service layers. The *presentation layer* is responsible for data representation and data coding, while the *application layer* offers a variety of services for particular applications, such as e-mail or file transfer. Security services such as encryption/decryption, authentication, and integrity are usually implemented in the last two layers.

C. TCP/IP Reference Model

While the OSI model was carefully designed, standardized, and then implemented, the *Transmission Control Protocol* (TCP)/*Internet Protocol* (IP) architecture was implemented first in the early 1980's and modeled afterwards. Well accepted in the university circles first, and later in the industry, TCP/IP, also called the Internet architecture, is nowadays one of the major architectures. However, the TCP/IP reference model is not general and consistent, but rather implementation bound, and it does not have a clear concept of services and protocols. The IP protocol serves as a joining point for many different networks, providing them with a method for communication.

Layers   Protocols

Application	FTP	DNS	TELNET	SMTP	HTTP
Transport	TCP			UDP	
Internet	IP				
Host-to-Network	Ethernet		SONET	ATM	

FIGURE 2 Protocols and layers in the TCP/IP model. Applications can either use TCP or UDP, depending on their implementation.

TCP/IP consists of four layers, as shown in Fig. 2, which can be matched to some of the OSI model layers. The *host-to-network layer* is similar to the physical and data link layers. The *internet layer* with the IP protocol corresponds to the network layer. The main concept for TCP/IP, implemented on this layer, is a connectionless packet-switched network.

Above the internet layer is the *transport layer* with two main protocols, TCP and *User Datagram Protocol* (UDP). TCP is a reliable and connection-oriented, end-to-end protocol. TCP has error detection/correction, data retransmission, and flow control mechanisms and takes care of out-of-order messages. UDP in contrast is an unreliable, connectionless protocol providing support for audio and video applications. The last layer, the *application layer*, hosts a variety of high-level protocols such as File Transfer Protocol (FTP) or Simple Mail Transfer Protocol (SMTP).

D. B-ISDN ATM Reference Model

The Asynchronous Transfer Mode (ATM) technology is the core of the Broadband Integrated Services network (B-ISDN). ATM is a connection-oriented and packet-switched technology, which emerged in the 1980s. Data in ATM is transmitted in small packets, called *cells*, which have a fixed size of 48 bytes payload and 5 bytes header. To set up a connection, a path through the network, a *virtual circuit*, is established first. This connection setup process is called *signaling*. After network resources are reserved and allocated to that virtual circuit, the source can start sending data cells. All cells transported over a specific virtual circuit take the same path/route through the network so that cells belonging to a specific connection are always received in order. Several virtual circuits can be combined into a single *virtual path*. A virtual path identifier and a virtual circuit identifier are present in each cell header to map a cell to a specific route through the network. If congestion occurs inside the network, cells

are temporarily buffered in the ATM switches. If a switch buffer overflows, cells are discarded inside the switch and are lost. Cell switching is one of the biggest advantages of ATM. During the signaling process, not only appropriate routes are found but also resources are allocated in the switches. In ATM, it is therefore possible to make quality of services (QoS) reservations for a specific virtual circuit. QoS parameters, such as sustained cell rate, peak cell rate, maximum burst tolerance, cell delay, and cell delay jitter, characterize the services provided.

ATM link speed of 155 Mbps to a few gigabits per second and the capability to guarantee QoS are very important for real-time applications like audio, video, and multimedia. ATM networks are increasingly used as backbones for WANs. One example is the very high speed backbone network (vBNS), which is currently used as one of the backbones for the next-generation Internet (Internet-2). vBNS runs TCP/IP over ATM. ATM can also be used to implement high-speed LANs.

As depicted in Fig. 3, the B-ISDN ATM reference model has different layers and a different structure as compared to the OSI model. While the OSI model is two dimensional, the ATM model is three dimensional. ATM's physical layer corresponds to both the physical and data link layer of the OSI model. The physical layer deals with the physical transmission of the bit stream and therefore depends on the physical medium used. Copper cable or fiber optics can be used for ATM. Above the physical layer is the ATM layer that deals with flow control, virtual circuit management, and cell header generation. The ATM layer is functionally equivalent to the OSI data link and network layers. On top of the ATM layer sits the ATM adaptation layer (AAL) that supports the different ATM services. AAL lays somewhere between the transport and session layers in the OSI model and provides assembly and reassembly of packets that are larger than a cell. Four different services are currently defined for ATM, resulting in four different AAL classes. AAL1 supports circuit

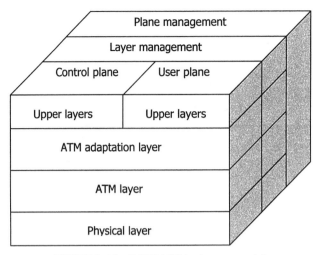

FIGURE 3 The B-ISDN ATM reference model.

emulation over ATM networks and is well suited for transporting constant bit rate (CBR) traffic. AAL2 is designed to support variable bit rate (VBR) traffic, where a timing relationship between the source and the destination is required. AAL3/4 is used to transfer data that is sensitive to loss but not to delay. Finally, AAL5 supports data traffic with no real-time constraints and is currently the most-used adaptation layer in industry.

Above the AAL layer are upper layers that are divided into a user and a control plane. The user plane is concerned with flow control, data transport, and error correction, while the control plane deals with connection management such as call admission and QoS. The layer and plane management functions in the third dimension are responsible for interlayer coordination and resource management.

To achieve high-speed data rates of up to a few gigagits per second, ATM networks use fiber optics. The physical layer standard most often used with ATM is the *synchronous optical network* (SONET), which was released in the late 1980s. SONET provides a common signaling standard with respect to wavelength, timing, framing, etc. It also offers a way to multiplex multiple digital channels together and provides support for network operation, administration, and maintenance.

III. NETWORK CLASSES

A. System Area Network (SAN)

SANs are used to interconnect PC clusters or workstation clusters forming server systems and to connect data vaults and other I/O subsystems to the overall system. SANs are also used to connect the individual processor nodes

in a parallel computer (these networks are also termed *interconnection networks*). It is very common to connect SANs to LANs and WANs, and they are usually at the leading edge in terms of performance.

B. Local Area Network (LAN)

LANs have become the most popular form of computer networks; they are inexpensive and widely available. The primary topologies used in LANs are star, ring, and bus.

One bus-based standard is *Ethernet*, where computers share a single transmission medium. Ethernet segments are limited to 500 m. The speed of Ethernet starts from 10 Mbps and goes up to 1 Gbps, in the latest version called Gigabit Ethernet. If a node has something to send over the bus, it listens to the bus first. If there is no data currently on the bus, the node starts sending its data. Because there is no coordination among the nodes, two or more nodes might start sending data at the same time, resulting in a collision. To deal with collisions, *carrier sense multiple access with collision detection* (CSMA/CD) is used in Ethernet as the MAC protocol. For collision detection, every node monitors the bus while sending data. If the data on the bus is different to the data a node is currently sending, a collision is detected. After detecting a collision, the nodes involved stop sending, wait a random time of up to d seconds, and then try again. In case of a new collision, the parameter d is doubled, and the nodes try sending again. This mechanism is called *binary exponential backoff*.

An example for a ring network is the IBM *Token Ring*. In this network, computers are connected in a loop and have a medium access mechanism called *token passing*. Token is a special message that circulates in the network. Whenever a node receives the token, that node has the right to transmit data. After sending the data, it will send the token to the next neighbor node to allow that node to send data. This scheme guarantees freedom of collision and fair access to the network for all nodes. The token ring protocol can also be implemented in a bus network, resulting in a *token bus* network. In this case, a virtual ring structure is established in which each node has a fixed address and a node sends the token to the node with the next higher address.

Recently, ATM found its way into LANs. An ATM switch can be used to centrally connect up to 32 computers or other LANs, providing bit rates starting at 155 Mbps.

LANs provide a good combination of capacity, price, and speed, but they have a distance limitation of 500 m. This limitation is primarily due to the properties of the copper wire used. To extend the span of a LAN, devices such as repeaters and bridges can be used. These devices are introduced in Section IV.

C. Metropolitan Area Network (MAN)

MANs employ technologies similar to the one used in LANs. A standard called *distributed queue dual bus* (DQDB) has been defined for MANs that does not have the LAN's cable length limitations and performance problems when connecting thousands of nodes. DQDB uses two parallel, unidirectional optical buses that connect all nodes. To transmit data, a node has to know whether the destination node is to the left or to the right of it. If it is to the left, it uses bus A, while if it is to the right, it uses bus B. At the head of each bus, a steady stream of empty data frames is produced. These frames in conjunction with counters in each node are used for fair bus arbitration. DQDB networks span a distance of up to 160 km at a data rate of 45 Mbps.

D. Wide Area Network (WAN)

WAN technologies are able to provide good performance for large-size networks accommodating a huge number of nodes/connections. Usually, WANs are packet switched and use point-to-point links that interconnect the routers/switches and connect the WAN to MANs and LANs. A hierarchical addressing scheme is used that makes routing decisions easier. In most cases, *next-hop forwarding* is implemented in which routers decide to which router a packet is to be forwarded. The information about destinations and next-hop relations is kept in a routing table within each router. There are two types of routing mechanisms: static and dynamic routing. Static routing protocols use network topology information only to decide on a route, resulting in simple, low-overhead protocols. However, static protocols do not adapt to changing traffic conditions and link failures in the network. Dy-

namic routing protocols, on the other hand, take the current traffic conditions into account as well and are able to deal with network/link failures by routing traffic around the failed links. Most WANs use dynamic routing protocols for increased flexibility and efficiency. As mentioned before, ATM is increasingly used to implement WANs.

IV. NETWORK COMPONENTS

Components most often used in networks are repeaters, switches, routers, bridges, and gateways. The main distinction among these components is the layer it is operating on and if they change data that they relay. Networks using copper wires have a distance limitation. This limitation is primarily due to the properties of the copper wire used. A signal on a copper wire becomes weaker and weaker as it travels along the wire, which limits the distance it can travel. This is a fundamental problem with the technologies used in LANs. A common technology to overcome this signal loss is the use of repeaters. A *repeater* is an electronic device operating at the physical layer that takes the signal received on one side of the cable and transmits it amplified on the other side. To limit the propagation delay in Ethernet, a maximum of four repeaters can be used, which extends the Ethernet distance to up to 2500 m. Another device used to extend LANs is a *bridge*. A bridge operates on the data link layer and connects two LAN segments like a repeater does but it helps to isolate interference and other problems as it forwards only correct frames and ignores the corrupted ones. Bridges do not change frame information. In addition, most bridges perform frame filtering to transmit only frames that have their destination located in the LAN on the other side of the bridge. Adaptive bridges learn node locations automatically by observing source and destination addresses in the frame headers. Bridges can be used to extend LANs between two buildings or even over longer distances. Furthermore, satellite connections or fiber links can be used between two filtering bridges with buffering capabilities to further increase the LAN span.

Routers work on the network level, selecting the most convenient route based on traffic load, link speed, cost, and link failures. They route packets between potentially different networks. *Switches* also operate on the network level, transferring packets from one link to another, implementing packet-switched networks. Sometimes a device might combine router, switch, and bridge functions. A *gateway* is a device that acts as a translator between two systems that do not use the same communication protocols and/or architectures.

V. WIRELESS NETWORKS

Wireless networks have penetrated LAN and WAN systems. *Wireless LANs* provide low mobility and high-speed data transmission within a confined region such as a building/office. A base station connected to the wireline network is placed in a central spot in a building, while mobile users use a wireless transmitter/receiver with their computer/laptop (e.g., a PCMCI card) to communicate with the base station. The range of a wireless connection is around 100 m, depending on the base station location and building/office characteristics such as building material and floorplan. Recent technological developments have led to several *wireless WAN* systems aimed at personal communications services (PCS). Although the major application is voice transmission, current-generation wireless WAN systems are also capable of data communications in the range of 100 k bits per second (bps), and future-generation wireless systems are expected to provide a T3 data rate of 45 Mbps, which will enable multimedia communications including data, text, audio, image, video, animation, graphics, etc.

In general, wireless networks can be divided into two classes: (1) cellular networks and (2) ad-hoc networks. In *cellular networks*, a geographic area is divided into overlapping cells (as in cellular phone networks). A fixed base station (BS) is situated in each cell, and mobile hosts (MHs) (e.g., users with a laptop) roam these cells. Each base station is connected to a wireline network and is responsible for broadcasting data received by the wireline network to the mobile hosts currently present in the cell and for receiving data from the individual mobile hosts and forwarding the data over the wireline network to their destinations. Mobile hosts cannot directly exchange data among each other but must use the base station. Thus, to send data from MH1 to MH2, MH1 transmits the data to its corresponding base station BS1. Assuming that MH1 and MH2 are currently in different cells, BS1 will forward the data to base station BS2 of the cell MH2 resides in using the wireline network. BS2 then broadcasts the data to MH2.

In *ad-hoc networks*, there is no cellular structure and there are no fixed base stations. Instead, a mobile host communicates directly with another mobile host that is a single radio hop away. A mobile host can also function as an intermediate node relaying data between MHs that are more than a single hop apart. Ad-hoc mobile networks allow spontaneous LANs to be created anywhere and anytime, supporting nomadic collaborative computing. A route in an ad-hoc network comprises the source, the destination, and intermediate nodes. Movements by any of these nodes may affect the route. If any of the nodes involved in a specific route moves out of the range of its neighbor nodes, the route becomes invalid and the communication terminates. MHs that are able to communicate with each other (using intermediate nodes) form a group. On the one hand, independent groups that are currently not connected among each other might become connected through MH movements forming larger groups. On the other hand, a group may also break up into smaller, independent groups, depending on the node movements. The unpredictability of node movements and group memberships call for sophisticated and efficient routing protocols for ad-hoc wireless networks.

Both cellular and ad-hoc networks rely on high-speed wireless data transmission. The main problems in providing high-speed wireless connections are (1) the significantly higher bit error rate as compared to wireline networks, (2) frequency reuse in cellular wireless networks, and (3) handoff from one base station to the next in cellular networks. Most network protocols such as TCP or ATM were designed under the assumption of low bit error rates. For example, if a packet is lost under TCP, network congestion rather than poor channel conditions are assumed, and the packet will be retransmitted. If TCP would be used over a wireless channel, degraded channels would lead to a massive retransmission, resulting in very poor network performance. In cellular networks, a cell overlaps with its neighboring cell to guarantee complete network coverage over an area. Thus, if a certain frequency is used in a cell for a connection, this frequency cannot be used in any neighbor cell to avoid data collisions at the cell boundaries. This frequency restriction limits the number of different frequencies that can be used in a single cell, which limits the number of concurrent users in a cell. If a mobile host moves in a cellular network and crosses cell boundaries by leaving cell A and entering cell B, a handoff of that communication from the base station of cell A to the base station of cell B has to be performed. Thus, the connection has to be rerouted in the wireline network connecting those base stations. This rerouting has to be done in a way that the connection experiences minimized packet loss and minimized packet delay jitter during the handoff. In addition, the probability of connection termination during handoff (e.g., due to insufficient available bandwidth within the wireline network or within the new cell) should be minimal.

VI. INTERNET AND APPLICATIONS

A. Fundamentals of the Internet

The Internet was first introduced in the late 1960s. At that time, the Advanced Research Projects Agency (ARPA, later DARPA) launched an experimental network called

ARPANET. ARPANET was dedicated for university research teams working on DoD projects, allowing them to exchange data and results. This network was the first one to use store-and-forward packet switching mechanisms, which are still used in today's Internet. Starting with four nodes at 1969, ARPANET was spanning the United States, connecting more than 50 nodes just a few years later. The invention of TCP/IP (1974) and its integration into Berkeley UNIX helped ARPANET to grow to hundreds of hosts in 1983. The fact that ARPANET was not open to all academic centers led the National Science Foundation (NSF) to design a high-speed successor to the ARPANET, which was called NSFNET. In the mid 1980s, NSFNET and ARPANET were interconnected and many regional networks joined them. The name Internet was then associated with this collection of networks. Growing exponentially since 1990, the Internet now connects millions of hosts and tens of millions of users, and these numbers double every year. For a host to be "on-line" the only requirements are to run TCP/IP and to have a valid IP address. The IP address can either be permanently assigned to a node or a temporary address can be used (many users/machines are on-line only for a limited time while connected to their Internet provider so that a permanent address is not necessary). The topology of the Internet is unsymmetric, unstructured, and constantly changing. The Internet architecture is based on the TCP/IP reference model, as depicted in Fig. 4. Applications can bypass individual layers and can even access the network directly.

B. Tools and Applications

1. Domain Name System (DNS)

In the Internet, each host is assigned a unique binary IP address. In the header of each IP packet there are fields for the IP addresses of the source and the destination nodes that are used to find a route through the network from the source to the destination. For scalability reasons, the address space is divided into domains and sub-domains,

resulting in an address with domain numbers separated by periods, such as 128.206.21.57. Each field in the address specifies a domain/sub-domain the node is in. This address format simplifies the routing algorithm running on each IP router. However, these addresses are usually hard to remember by users. Therefore, each node is also assigned an alphanumeric name, such as mackerel.cecs.missouri.edu that can be easier interpreted by a user. The alphanumeric names are usually used by applications to address nodes. Thus, an address translation is needed to translate the alphanumeric name into its corresponding IP address. The *Domain Name System* (DNS) is implementing this translation.

DNS is a hierarchical, domain-based naming scheme and a distributed database system that implements this naming scheme. For example, consider the node name *mackerel.cecs.missouri.edu*. The left-most segment represents the computer name (*mackerel*), the next two segments represent sub-domains: Computer Engineering and Computer Science Department (*cecs*) at the University of Missouri (*missouri*), while the right-most segment specifies the top-level domain (in our case *edu*, an educational institution). In general, the top-level domain includes names for each country (e.g., *us* for U.S.A. or *de* for Germany) or generic names such as *edu* for educational institutions, *com* for commercial, *gov* for the U.S. federal government, *int* for certain international organizations, *mil* for the U.S. armed forces, *org* for non-profit organizations, and *net* for network providers. To obtain a domain name, an organization or a user needs to register with the Internet authority. Once a domain name is given, the organization can subdivide the domain and create its own hierarchy. There are no standards for this interdomain structure, and the names of the computers belonging to this organization may not follow the same pattern. This gives the organizations freedom to change, coordinate, or create names in their own domain. The entire DNS system operates as a large distributed database. A root server occupies the top level of the naming hierarchy. The root server does not know all possible domain names, but it knows how to contact other servers to resolve an address. If a node needs an address translation, it sends a DNS request message to a local name server (the IP address of this server has to be known in advance by the node). If the server is able to resolve the name, it sends the translation back to the requesting node. Otherwise, this server contacts another name server higher up in the hierarchy to resolve the name.

2. E-mail

One of the first and very popular Internet applications was electronic mail or e-mail. It was implemented for the

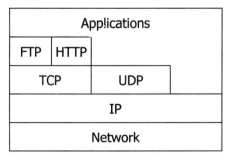

FIGURE 4 Internet architecture.

ARPANET to provide the users with the ability to send each other messages over the network. This application has three components: message format standard, transfer protocol, and user interface (mail reader). The standard for message format, RFC 822, divides an e-mail message into a header and a body part. While the header contains simple information such as the names and e-mail addresses of the sender and the recipient, time of sending, and message subject, the body of the message can be more complicated. Initially, RFC 882 defined the body to contain ASCII text only. In 1993, the *Multi Purpose Internet Mail Extensions* (MIME) was introduced, which made it possible to also transmit binary files such as executables, images, audio, or video files. In MIME, the binary data is first encoded into plain ASCII text and then sent over the network. The protocol for transferring e-mail messages over the Internet is called *Simple Mail Transfer Protocol* (SMTP). SMTP requires a mailer daemon to run on the hosts. To send a message, the mailer daemon on the sending machine connects to the destination machine's daemon using SMTP over TCP and transmits the message, as well as receives messages if there are any. Received messages are then moved to the user's inbox. SMTP as the name suggests is a simple, ASCII-based protocol that uses commands of the type HELLO, MAIL, etc. This only works if both the sender and the receiver nodes are directly connected to the Internet. In the case that a node has no direct access to the Internet, it can still send and receive e-mail by using an e-mail server that is connected to the network. E-mail is received by the server and stored in a remote mailbox on the server. The destination node then connects to the server and fetches the e-mail from that server. Two simple protocols can be used for e-mail fetching: POP3 and IMAP. The *Post Office Protocol* (POP3) allows a user to connect to the mail server to download received messages. Once a message is downloaded, it is deleted from the server. The *Interactive Mail Access Protocol* (IMAP) transforms the e-mail server into an e-mail repository. A user can access the server and can read the messages without downloading them to the local machine. This protocol allows users to access their e-mail from different machines.

The last component necessary for e-mail, the mail reader or client program, organizes e-mail messages into folders and does the MIME encoding/decoding. Some popular mail readers are Qualcomm's Eudora, Netscape Browser, Microsoft Outlook, etc.

3. File Transfer Protocol (FTP) and Telnet

File Transfer Protocol is one of the oldest Internet applications and is still widely used. First implemented for the ARPANET, the FTP was designed to transfer/copy files from one host to the other over the Internet. FTP first establishes a TCP connection between the two hosts and requests an authorization from the user. After the user supplies a valid user name and password, an FTP control session starts. Simple commands such as append, open, send, and rename are used to transfer files. The user can either supply these commands directly or a file transfer application program can be used. Data transfer can be done in two modes, binary and ASCII mode, allowing the transfer of text files as well as binary files such as executables, images, or video/audio files. One or more FTP connections can exist concurrently between any pair of computers.

The telnet application is used to log-on to a remote computer. It enables a user to access and work on a remote node as if the user were directly connected to that node.

4. World Wide Web (WWW)

Applications like e-mail, FTP, and Telnet are popular and brought many people to use the Internet, but the real Internet boom started with the introduction of the World Wide Web (WWW). It started in 1989 with a proposal from the physicist Tim-Berners Lee from the European Center for Nuclear Research, CERN. CERN had several large international teams and many research projects with complex experiments. Since team members were distributed across several countries, it was very difficult for them to constantly exchange reports, drawings, plans, and other documents. The Web was intended to satisfy their needs by creating a web of linked documents easy to view and modify. Soon after 1993, when the first graphical interface, Mosaic, was released, it became very popular—so popular that it brought $1.5 billion in stock for its successor Netscape. A year later, MIT and CERN created a WWW Consortium, an organization to further develop the Web, standardizing protocols and providing control over the Web. The main reason for the Web's popularity is the easy-to-use graphical interface and the enormous amount of information available to anyone. Not only is information available, but many companies offer services over the Internet such as on-line shopping, banking, education, and entertainment. The new possibilities for business are a very important factor; it is estimated that by 2003 the business-to-business commerce over the Internet will reach $1.3 trillion. The Web is a set of servers, clients, and linked documents, and it uses the *Hypertext Transfer Protocol* (HTTP). Linked documents can contain text, image, data, audio, video, and, of course, links to other documents or parts of documents. A single document is usually called a page, and a set of linked pages is called a site. Web servers have a process listening on TCP port 80 for clients. Clients (Web browsers) request linked documents from a server and view them on the client's computer. The most common operations are GET and POST, to fetch a Web site and to append a web page, respectively. HTTP, like SMTP, is a text-oriented protocol with human understandable commands. An HTTP message has a header and a body. The message body is the data requested by the client or is empty in request messages. A message header contains information similar to the one in e-mail headers, such as the web page address. Instead of using the IP address of the web page, a *uniform resource locator* (URL) is used. The URL is a compact representation of the location and access method for a resource such as a web page. The URL provides a pointer to any object (files, file directories, documents, images, video clips, etc.) that is accessible on any machine connected to the Internet. Each URL consists of two parts: access method and location. For example, in the URL *http://www.cecs.missouri.edu, http* indicates that the HTTP protocol is to be used to access the web page located on the machine *www.cecs.missouri.edu.* Other access methods are ftp, mailto, news, telnet, etc.

The first version of HTTP1.0 establishes one connection per client and allows only one request per connection. This means that if a client wants to retrieve more than one web page it needs to set up multiple connections to the server. This is improved in the latest HTTP version, which allows persistent connections. Caching web pages is done either on the client side or on proxy servers to reduce Internet traffic and web page access time. HTTP supports caching by providing an expiration field in the web page header. A client can use the cached copy of the page until the date specified in this field expires, in which case the client has to reacquire the page from the same node.

5. Multimedia Applications

The success of the World Wide Web and the HTTP protocol led to the introduction of many new applications, such as multimedia. Multimedia applications combine text, image, video, and audio. While the transmission of text and images over the Internet is non-critical, video and audio transmissions are more problematic because of the relatively high transmission bandwidth requirements, especially for video.

Two general types of audio/video applications are streaming and conferencing. Streaming applications typically have servers delivering audio and/or video streams to the client. Examples are video broadcast, radio broadcast, and video on demand. It is common for these applications to broadcast the streams so that multiple clients are able to receive them. The second type, conferencing applications (e.g., video conferencing), is more interactive. Although different in nature, these applications need high bandwidth, low delay, and low delay *jitter* (variance of the delay).

Another important characteristic of multimedia applications is the need for multicast. During a video broadcast, a video stream has to be delivered to multiple destinations. The same is true for video conferencing applications. During a multicast, a sender could send the multicast stream multiple times to each individual destination. This, however, would result in a high traffic load in the network. Thus, the network itself should be able to distribute/copy packets of a single video stream. In IP networks such as the Internet, *IP Multicast* was therefore introduced and relies on special multicast routers. A multicast router is responsible for copying incoming packets belonging to a multicast stream to specific router outputs to generate the multicast distribution tree. An announce/listen mechanism was adopted in which multicast senders periodically send session announce messages. Nodes that want to join a certain multicast group will listen to these messages and will send a message announcing their join request. After joining the session, each node periodically sends announce messages. The multicast routers will listen to the announce messages and will generate the multicast tree according to the current session membership. If a node leaves the session or a temporary network failure occurs, the nearest router implementing the session will stop receiving the node's announce messages and will stop sending any multicast packets to that node. Thus, changes in session membership are handled locally in the network, resulting in scalability of the IP multicast.

Another important mechanism for multimedia communication is the handling of temporary network congestion during a multimedia transmission. The *real-time transport protocol* (RTP) in conjunction with the *real-time control protocol* (RTCP) is most often used today to deliver streaming video over the Internet that can be used to react on network congestion. RTP is not a protocol layer, but rather a tree-based hierarchy of interdependent protocols supporting different video and audio coding schemes. Among the services provided by RTP are delivery monitoring, time stamping and sequence numbering of packets for packet loss detection and time reconstruction and synchronization of multimedia data streams, payload and coding scheme identification. RTP/RTCP runs over UDP/IP and uses IP multicast. RTCP is responsible for generating periodic sender and receiver reports that can be used by the multimedia application to detect temporary network congestion. A source could, for example, adjust the video coder rate to the current network state (e.g., decrease the rate during network congestion), resulting in a varying quality of the received video that depends on the current network conditions. The *real-time streaming protocol* (RTSP), a client-server multimedia presentation protocol used with RTP/RTCP, provides a "VCR-style" remote control functionality to audio/video streams. RTSP provides functions such as pause, fast forward, rewind, absolute positioning in a stream, etc.

VII. SECURITY

Computer networks have a very important feature— sharing. Sharing resources and data as well as extensive access to the Internet imply the need of some security mechanisms. While very intuitive, computer security is a quite ambiguous term. It is well known that a system (computer network) is as secure as its weakest element. However, defining security in general is quite hard. The intrinsic problem is that in order to define what a secure system is, the system's assets have to be evaluated. Since a computer network is a dynamic structure, the security should be a dynamic rather than a static process. The description of the levels and types of security services a network needs is called *security policy*. Some basic services to be provided in a network are authentication, authorization, integrity, and confidentiality. In the real world, we have established mechanisms to implement and enforce similar services as well. For example, passport and/or signature are enough to authenticate a person. In computer networks, analogous mechanisms are deployed. Usually, user name and password are sufficient to verify and accept a valid user of the network, relying on the authentication protocol. Preventing security holes and attacks is necessary, but sometimes detection and recovery from a system failure is equally or more important.

A. Security Services

Security services are intended to protect a system from security attacks, to prevent attacks, or both by utilizing different security mechanisms. *User authentication* is the process of verifying the identity of a user. In the case of a user-to-user communication, both users have to be checked. Traditionally, in the client–server domain, the authentication is focused on the client side, since the system should be protected from users and not vice versa. However, for some applications such as e-commerce, server authentication is equally important to ensure that it is the correct server a customer is communicating with. *Data authentication* describes the verification of a particular data or message origin.

Authorization refers to the restriction of access to data and/or nodes. A user can be accepted into the network/node through authentication, but he/she might not have access to all of the files. Restriction lists or access lists and membership control are generally provided by the operating system. Another important service is *integrity*.

It protects transmitted data from changes, duplication, or destruction. Modifications due to an error or intruder can usually be detected and fixed by the network protocol. If the data is sensitive, then integrity is combined with confidentiality service. *Confidentiality* is a service that protects all user data transmitted over a network. Even if data is intercepted by a third party, that third party will be unable to read the data. The *non-repudiation* service prevents a sender or receiver from denying a transmitted message (e.g., for on-line purchase proof).

Security attack is defined as any action, intended or not, that compromises the security of the information and/or system. Attacks can generally be passive or active. Passive attacks can be the copying of information or a traffic analysis. Active attacks involve some modification of the original data or fabrication of new data, such as replay or interruption of data. Security mechanisms are designed to prevent, protect, and recover from security attacks. Since no technique is able to provide full protection, the designers and/or system administrators of a network are responsible for choosing and implementing different security mechanisms.

B. Security Building Blocks

To incorporate and enforce the security policy, appropriate mechanisms are needed. Some very fundamental mechanisms are encryption/decryption, security management tools, firewalls, and detection and recovery tools. The choice of a specific mechanism depends on the level of security needed.

1. Cryptography

Cryptography is one of the oldest and most powerful security-providing technologies. The word cryptography, science of information security, comes from the Greek word *kryptos*, meaning hidden. It is mainly associated with *encryption*, the process of scrambling data with a secret parameter called an encryption key into *ciphertext*. The opposite process, *decryption*, takes the ciphertext and converts it back to plain text with the help of a decryption key. The mechanism of breaking ciphers without having the decryption key is called *cryptanalysis*. This is usually done by an *intruder* who has access to the data transmission channel.

There are few different types of encryption methods: substitution, transposition, and one-time pad cipher. One of the oldest ciphers is *substitution cipher*, known as Caesar cipher, which substitutes every symbol with another from its group. For example A with D, B with G, and so on. This is very easy to analyze and break with common letter statistics. *Transposition cipher* preserves the symbol meanings but reorders them in a new way depending on the key word. *One-time pad* is a theoretically unbreakable cipher. It combines (e.g., using *exclusive or*) two strings, the plaintext and a secret key. The result is a cipher, which holds no information about the plaintext, and every plaintext is an equally probable candidate for a specific ciphertext. One general rule is that the more secure the method is, the more redundancy information is required. However, the more data that is available for a cryptanalyst, the higher the chances are to break the code. Modern cryptography used in computer systems applies the basic techniques described here with a tendency toward very complex algorithms and short secret keys. There are two main classes of encryption algorithms: symmetric and asymmetric.

Symmetric algorithms use one key to encrypt and decrypt the data. Therefore, the key should be distributed between the communicating parties in a safe way. An example is the *data encryption standard* (DES) algorithm. This cipher developed by IBM and standardized by the National Security Agency (NSA) was initially used with 56-bit keys. Unfortunately, this key size results in around 7.2×10^{16} different key combinations only, and it is possible to break a DES cipher using exhaustive search within a few hours using powerful PCs. To increase the security level of DES, a 128-bit key is used now. Using this key size, it requires around 100 million years to break a DES cipher with exhaustive search.

The problem of symmetric algorithms is the secure distribution of the secret keys among nodes. This distribution problem is solved by *asymmetric algorithms*. The idea is to use two different keys, one encryption (public) key, which is available to everyone, and one decryption (private) key, which is known only by the user owning the key. This scheme is also called *public key cryptography*. One efficient method developed in 1978 by a group at MIT is the *RSA algorithm* named after the authors Rivest, Shamir, and Adleman. The RSA algorithm uses the product of very large prime numbers to generate the public and private keys. In order to decrypt a cipher, an intruder knows the public key but has to calculate the private key by factorizing the public key. However, factorization of very large numbers is not trivial (for example, for a 200-digit public key, a billion years of computational time is required to calculate the private key). This makes the RSA algorithm very secure.

2. Authentication Protocols and Digital Signatures

Authentication protocols can be based on shared secret key, public key, key distribution center, or the *Kerberos protocol*. The protocol based on *shared secret key* requires users A and B to share a secret key in order to use the protocol. The protocol consists of five message exchanges. A first sends a communication initiation message to B. B does not know whether this message is from A or from an intruder, so B sends a very large random number to A. To prove its identity, A then will encrypt the number with the shared secret key and return it to B. B will decrypt the message to obtain the original number back. Because only A and B know the secret key, B now knows that the message is coming from A. Next, A sends a challenge (large random number) to B, B encrypts the number with the secret key and sends it back to A, and A decrypts the number to find out whether the message actually came from B. After this, the real communication can start. A problem with the secret key authentication is the secure distribution of the secret key.

The *public key authentication protocol* uses two keys per node, a public key for encryption and a private key for decryption. Everybody has access to the public key of a node, while the private key is secret. During authentication, random numbers are generated and exchanged, similar to the shared secret key protocol. The only difference is that the public key of the receiving node is used by the sending node to encrypt the random number, while the secret key of the receiving node is used to decrypt the received number. A disadvantage of this protocol is the non-trivial distribution of the public keys.

Another authentication method uses trusted *key distribution centers* (KDC). Each user has only one key that is shared with the distribution center. Whenever A wants to communicate with B, it generates a session key, encrypts the key with its own secret key, and sends it to the distribution center. The center knows A's secret key and is able to decrypt the session key. It then encrypts the session key with B's secret key and sends it to B, which is able to decrypt the session key again. The session key is then used for secure communication between A and B. To avoid replay attacks where intruders copy messages and resend them at a later time, time stamps or unique numbers are included in the messages to detect the message resending.

The *Kerberos authentication protocol* consists of a set of two additional servers, the *authentication server* (AS) and the *ticket-granting server* (TGS). The AS is similar to a key distribution center in that it shares a secret key with each user. To start a communication between users A and B, A contacts the AS. The server will send back a session key KS and a TGS ticket to A, both encrypted with A's secret key. The TGS ticket can later be used as proof that the sender is really A whenever A requests another ticket from the TGS server. A sends the request to communicate with B to the TGS. This request contains KS and the TGS ticket that A received from the AS. This ticket was encrypted by the AS using a secret TGS key. By decrypting the ticket, the TGS is therefore able to validate that it is communicating with user A. The TGS then creates a session key KAB for the A/B communication and sends two versions back to A: one version encrypted with KS and one version encrypted with B's secret key. A decrypts the first version by using its session key to obtain KAB. It then sends the second version to B, which is also able to decrypt KAB. Now A and B both have the same secret session key KAB to start a secure communication.

Authentication protocols authenticate users only. In many applications, such as financial transactions or e-commerce, messages themselves have to be authenticated as well. Digital signatures were therefore introduced. *Digital signatures* are used to verify the identity of the sender, to protect against repudiation of the message by the sender later on, and to detect if a receiver has concocted a message himself. In general, any public key authentication algorithm can be used to produce digital signatures. For example, if user A wants to send a message with a signature to user B, A first generates the signature (by applying a cryptographic function such as a hash function to the message) and encrypts it with his private key and then with B's public key as shown in Fig. 5. After receiving the message and the signature, user B will decrypt the signature first with his private key and then with A's public key. After decrypting the actual message, B can generate a message signature and can compare it with the decrypted signature to verify the message and its sender. The *Digital Signature Standard* (DSS) introduced in 1991 uses a similar mechanism.

3. Firewalls

A *firewall* is a device barrier between a secure intranet and the outside world (e.g., the Internet). Firewalls are typically implemented as "screening routers." A *screening router* is a router or gateway that examines the incoming/outgoing traffic and selectively routes packets based on various criteria. A firewall may range from impermeable (allowing little or no traffic in or out) to porous (allowing most or all traffic in or out). For example, a typical screening router may block inbound traffic traveling on any TCP/IP port except port 80 (generally used for WWW services). An "absolute" protection against flooding the intranet of packets from an external host, called denial-of-service attack, can be achieved by forbidding any inbound traffic but allowing outbound traffic. Given that some other security measures exist in the network and that users will want to use the network to share and exchange data, this measure is extreme in most cases. Some firewalls have the ability to provide dynamic port selection. This is helpful in cases when the ports are specified during the transmission. A second common type of firewall is a proxy-based firewall. *Proxy*, in general, is a

process/computer situated between a client and a server. When a client sends a request to a server, or vice versa, the request is actually sent to the proxy, and the proxy passes the request to the server. These proxies can help to enforce security policies by examining the packets and dynamically deciding which packets to forward and which packets to drop. Proxies have to understand the application layer protocol used (e.g., HTTP, Telnet, FTP). Firewalls add a layer of protection to the whole network connected to it. Depending upon the setup of the firewall, it can be used for prevention (stop the traffic in particular direction and port), protection, and recovery. Protection can be implemented by analyzing the traffic streams. If the traffic behavior deviates from the normal/anticipated behavior, the traffic source can be localized, and access for this host can be prevented. Recording of traffic information and data transmitted is possible at a firewall and can be later used for recovery. Although firewalls are useful for protection from external users, they cannot protect the system from malicious internal users. Another vulnerability of firewalls is the use of mobile codes or tunneling techniques. While the inspected packets look innocent to the firewall, they might carry programs threatening the security of the system. Thus, firewalls should always be used together with security support services available inside the intranet.

VIII. QUALITY OF SERVICE IN COMPUTER NETWORKS

Many applications such as multimedia need some network guarantees to deliver a certain level of quality independent of the current or future network state. For example, to deliver a high-quality video, the network should provide a video connection with a guaranteed bandwidth of, e.g., 4 Mbps. Thus, a network should provide a certain level of *quality of service* (QoS). QoS is characterized by QoS parameters such as bandwidth, delay, and delay jitter. The user/application negotiates a QoS contract with the network by specifying QoS parameters that the network will guarantee during the lifetime of a connection.

In ATM networks, QoS guarantees are explicitly included through resource reservation, while IP networks are best-effort networks without any QoS guarantees. To enable QoS in IP networks such as the Internet, QoS mechanisms were introduced in IP. The two most important ones are *Integrated Services* (IntServ) and *Differentiated Services* (DiffServ). IntServ provides two different classes of services over the Internet: (1) the guaranteed service and (2) the controlled-load service. The *guaranteed service* class guarantees bandwidth and delay requirements for a connection. This service can be used for real-time audio/video connections with hard delay requirements. The

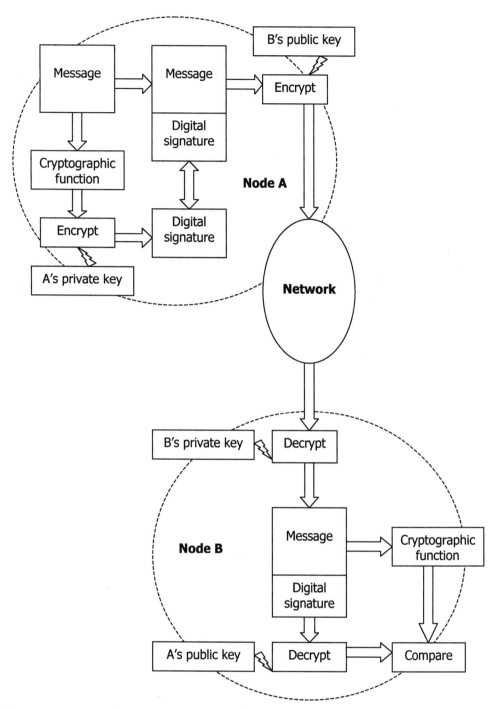

FIGURE 5 Example of public key encryption and digital signature. A sends B an encrypted message signed with A's digital signature. B decrypts the message and the signature and checks if the signature corresponds to the right message and right person.

controlled-load service class does not guarantee anything but tries to minimize packet loss and delay of a connection. This service can be used for adaptive audio/video applications.

IntServ relies on resource reservation and uses the *resource reservation protocol* (RSVP). RSVP is a signaling protocol that supports IP Multicast and permits receiver-driven resource reservation of an IP packet flow without establishing an explicit connection. To accomplish this, special RSVP routers are needed. The main problem with RSVP is that its applicability and scalability over large networks such as the Internet are limited. Because RSVP implements per-flow resource reservation, it requires very large state tables to be maintained in the core network routers, which limits its scalability.

The current approach for supporting IP QoS is *differentiated services*. Instead of reserving resources for each individual flow, DiffServ classifies these flows at the network edge and applies a per-class service in the network. Each service class is associated with a per-hop behavior (PHB) in the network. Networks supporting DiffServ therefore need two kinds of routers: boundary routers at the edge of the network that classify and shape/police incoming traffic and interior routers that apply PHBs to the different classes. Currently, three classes are supported: (1) best effort (regular IP traffic), (2) *expedited forwarding* (EF), and (3) *assured forwarding* (AF). The EF service supports low loss, low delay, and low delay jitter connections with a guaranteed peak bandwidth. EF emulates a point-to-point virtual leased line. The AF service defines four relative classes of service with each service supporting three levels of packet drop precedence. If a router encounters congestion, packets with higher drop precedence will be dropped ahead of those with a lower precedence. No specific bandwidth or delay constraints are defined for the different AF classes. EF and AF services are implemented through internal router queue management and scheduling such as random early detection and weighted fair queuing.

IX. TRENDS

The backbones of today's networks are telephone lines, coaxial cables, and optical fibers. This is going to change with the growing number of cell phones (predicted to outnumber PCs by 2005) that will result in more people being connected to the Internet via wireless than wired connections by 2008. The cell phone manufacturers are adopting new protocols for wireless communications. The current lack of interoperability (European cell phones do not work in the United States and vice versa) and the low bandwidth of current wireless connections are going to be changed by the new cell phone standard 3G (third generation). By 2004, 3G promises to bring interoperability, increase the

bandwidth, and drop the average cost of making a cell phone call. The bandwidth predictions are for 144-Kbps access for automobile passengers and for up to 2-Mbps connectivity for stationary users for as low as 2 cents per minute. This will allow home users to receive high-quality movies and business travelers on the road to receive their multimedia e-mail. The future ubiquitous Internet (Ubi-Net) may evolve into a network of networks, all running IP over a combination or wired and wireless networks, and might be a reality by as early as 2008.

SEE ALSO THE FOLLOWING ARTICLES

ARTIFICIAL NEURAL NETWORKS • COMPUTER ARCHITECTURE • DATABASES • SOFTWARE ENGINEERING • TELECOMMUNICATION SWITCHING AND ROUTING • WIRELESS COMMUNICATION • WWW (WORLD-WIDE WEB)

BIBLIOGRAPHY

Comer, D. (1999). "Computer Networks and Internets," 2nd ed., Prentice-Hall PTR, Upper Saddle River, NJ.

Foster, I., and Kesselman, C., eds. (1999). "The Grid: Blueprint for a New Computing Infrastructure," Morgan Kauffman, San Francisco, CA.

Händel, R., Huber, M., and Schröder, S. (1995). "ATM Networks: Concepts, Protocols, Applications," Addison-Wesley, Berkeley, CA.

Jurczyk, M., Siegel, H. J., and Stunkel, C. (1999). Interconnection Networks for Parallel Computers. In "Wiley Encyclopedia of Electrical and Electronics Engineering" (J. G. Webster, ed.), Vol. 10, pp. 555–564, Wiley, New York.

Kuo, F., Effelsberg, W., and Garcia-Luna-Aceves, J., eds. (1998). "Multimedia Communications," Prentice-Hall PTR, Upper Saddle River, NJ.

Lewis, T. (1999). "UbiNet: The ubiquitous Internet will be wireless," *IEEE Computer* **3210,** 127–128.

Peterson, L., and Davie, B. (2000). "Computer Networks: A System Approach," 2nd ed., Morgan Kauffman, San Francisco, CA.

Prycker, M. (1995). "Asynchronous Transfer Mode, Solution for Broadband ISDN," 3rd ed., Prentice-Hall PTR, Upper Saddle River, NJ.

Stallings, W. (1999). "Cryptography and Network Security: Principles and Practice," 2nd ed., Prentice-Hall PTR, Upper Saddle River, NJ.

Tanenbaum, A. (1996). "Computer Networks," 3rd ed., Prentice-Hall PTR, Upper Saddle River, NJ.

Toh, C.-K. (1997). "Wireless ATM and Ad-hoc Networks," Kluwer Academic, Boston, MA.

Computer Viruses

Ernst L. Leiss
University of Houston

GLOSSARY

Data integrity Measure of the ability of a (computer) system to prevent unwanted (unauthorized) changes or destruction of data and software.

Data security Measure of the ability of a (computer) system to prevent unwanted (unauthorized) access to data and software.

Logical bomb Code embedded in software whose execution will cause undesired, possibly damaging, actions.

Subversion Any action that results in the circumvention of violation of security principles.

Worm Self-contained program that is usually not permanently stored as a file and has the capacity of self-replication and of causing damage to data and software.

Virus Logical bomb with the ability of self-replication. It usually is a permanent part of an existing, permanently stored file and has the capability of causing damage to data and software.

I. BACKGROUND AND MOTIVATION

In the years before 1988, a number of incidents suggested the potential for major problems related to the organized and widespread subversion of (networked) computer sys-

tems, accompanied by the possibility of massive destruction of data and software. While until then these concerns were considered rather remote, the Internet attack of 1988 shattered this complacency. In the intervening decade, computer viruses have attained significant visibility in the computer-literate population, rivalling the notoriety of Y2K-related problems but with substantially greater staying power.

The reason for the attention attracted by these intruders lies in their potential for destruction of data and software. With the exception of some highly secured systems related to defense and national security, virtually all larger computer systems are connected via computer networks, commonly referred to as the Internet. Personal computers, if they are not permanently linked into these networks, have at least the capability of linking up to them intermittently through a variety of Internet service providers. Networks are systems that allow the transmission of digitally encoded information (data, software, messages, as well as still images, video, and audio) at relatively high speeds and in relatively convenient ways from one system to another. Subverting the function of a network may therefore result in the subversion of the computers linked by it. Consequently, a scenario is very plausible in which a program may be transmitted that is capable of destroying large amounts of data in all the computers in a given network.

Encyclopedia of Physical Science and Technology, Third Edition, Volume 3

The case that such a scenario is plausible has been made for many years, starting with F. Cohen's demonstration of a computer virus in 1983.

In 1988, such a scenario was played out for the first time on a worldwide scale. Since then, numerous incidents have reinforced the public's sense of vulnerability to attacks by insidious code fragments on software and data stored in all kinds of computers. While earlier virus attacks spread via diskettes and later via electronic bulletin boards (in ways that required some user participation through loading infected programs), in recent years, the World Wide Web and more sophisticated e-mail systems have provided transmission channels that facilitated the worldwide spread of the attackers at a unprecedented speed. Moreover, infection which earlier required some explicit action by the victim has become much more stealthy, with the advent of viruses that become activated through the opening (or even previewing) of an apparently innocent attachment to an e-mail document.

The destruction of data and software has obvious economic implications. The resulting malfunctioning of computer systems may also affect safety-critical systems, such as air-traffic control systems or control systems for hydro-electric dams or nuclear power plants. Futhermore, the potential for disruption can be damaging: a bomb threat can conceivably be more paralyzing that the explosion of a small bomb itself. Protection against such treats may be either impossible or unacceptable to usesrs in the necessarily resulting reduction of functionality and ease of use of computer systems. It must be borne in mind that by necessity, the notion of user friendliness of a computer system of communications network is antithetical to the notions of data security and data integrity.

II. VIRUSES, WORMS, AND SO FORTH

From a technical point of view, the most alarming aspect of the attackers under discussion in this article is that they are self-replicating. In other words, the piece of software that performs the subversion has the ability of making copies of itself and transmitting those copies to other programs in the computer or to other computers in the network. Obviously, each of these copies can now wreak havoc where it is and replicate itself as well! Thus, it may be sufficient to set one such program loose in one computer in order to affect all computers in a given network. Since more and more computers, including personal computers, are interconnected, the threat of subversion can assume literally global dimensions. Let us look at this in greater detail. First, we define a few important terms.

A logical bomb is a piece of code, usually embedded in other software, that is only activated (executed) if a certain condition is met. It does not have the capability of self-replication. Activation of the logical bomb may abort a program run or erase data or program files. If the condition for execution is not satisfied at all times, it may be regarded as a logical time bomb. Logical bombs that are activated in every invocation are usually not as harmful as time bombs since their actions can be observed in every execution of the affected software. A typical time bomb is one where a disgruntled employee inserts into complex software that is frequently used (a compiler or a payroll system, for example) code that will abort the execution of the software, for instance, after a certain date, naturally chosen to fall after the date of the employee's resignation or dismissal.

While some programming errors may appear to be time bombs (the infamous Y2k problem certainly being the best known and most costly of these), virtually all intentional logical bombs are malicious.

A computer virus is a logical bomb that is able to self-replicate, to subvert a computer system in some way, and to transmit copies of itself to other hardware and software systems. Each of these copies in turn may self-replicate and affect yet other systems. A computer virus usually attaches itself to an existing program and thereby is permanently stored.

A worm is very similar to a computer virus in that it is self-replicating and subverts a system; however, it usually is a self-contained program that enters a system via regular communication channels in a network and then generates its own commands. Therefore, it is frequently not permanently stored as a file but rather exists only in the main memory of the computer. Note that a logical bomb resident in a piece of software that is explicitly copied to another system may appear as a virus to the users, even though it is not.

Each of the three types of subversion mechanisms, bombs, viruses, and worms, can, but need not, cause damage. Instances are known in which bombs and viruses merely printed out some brief message on the screen and then erased themselves, without destroying data or causing other disruptions. These can be considered as relatively harmless pranks. However, it must be clearly understood that these subversion mechanisms, especially the self-replicating ones, most definitely have enormous potential for damage. This may be due to deliberate and explicit erasure of data and software, or it may be due to far less obvious secondary effects. To give one example, consider a worm that arrives at some system via electronic mail, thereby activating a process that handles the receiving of mail. Typically, this process has a high priority; that is, if there are any other processes executing, they will be suspended until the mail handler is finished. Thus, if the system receives many mail messages, a user may get the impression that the system is greatly slowed down. If these mail messages are all copies of the same worm, it is clear that the system can easily be saturated and thereby damage can be done, even though no data or programs are erased.

This is what happened in the historic case study cited above. On November 2, 1988, when a worm invaded over 6000 computers linked together by a major U.S. network that was the precursor to the present-day Internet, including Arpanet, Milnet, and NSFnet. Affected were computers running the operating system Berkeley Unix 4.3. The worm took advantage of two different flaws, namely, a debugging device in the mail handler (that most centers left in place even though it was not required any longer after successful installation of the mail handler) and a similar problem in a communications program. The worm exploited these flaws by causing the mail handler to circumvent the usual access controls in a fairly sophisticated way; it also searched users' files for lists of trusted users (who have higher levels of authority) and used them to infiltrate other programs. The worm's means of transmission between computer was the network. Because infiltrated sites could be reinfiltrated arbitrarily often, systems (especially those that were favorite recipients of mail) became saturated and stopped performing useful work. This was how users discovered the infiltration, and this was also the primary damage that the worm caused. (While it did not erase or modify any data, it certainly was capable of doing this had it been so designed.) The secondary damage was caused by the efforts to remove the worm. Because of the large number of sites affected, this cost was estimated to have amounted to many years of work, even though it was relatively easy to eliminate the worm by rebooting each system because the worm was never permanently stored.

One reason this worm made great waves was that it caused the first major infiltration of mainframe computers. Prior to this incident, various computer viruses (causing various degrees of damage) had been reported, but only for personal computers. Personal computers are typically less sophisticated and originally had been designed for personal use only, not for networking; for these reasons they had been considered more susceptible to attacks from viruses. Thus, threats to mainframes from viruses were thought to be far less likely than threats to personal computers. The November 2, 1988, incident destroyed this myth in less than half a day, the time it took to shut down Internet and many computer systems on it.

Since then, a wide variety of attackers have appeared on the scene, substantially aided by the explosive growth of the World Wide Web. Not surprisingly, given the dominance that Microsoft's operating systems have in the market, most recent viruses exist within the context of that company's operating systems. Many of these viruses use the increasingly common use of attachments to be transmitted surreptitiously; in this case, opening an attachment may be all that is required to get infected. In fact, users may not even be aware that an attachment was opened, because it occurred automatically (to support more sophisticated mail functions, such as previewing or mail sorting according to some user-specified criterion). Frequently, the resulting subversion of the mail system facilitates further distribution of the virus, using mailing lists maintained by the system.

III. PREVENTION AND DETECTION

There are two different approaches to defending against viruses and worms. One is aimed at the prevention or detection of the transmission of the infiltrator; the other tries to prevent or detect damage that the infiltrator may cause by erasing or modifying files. The notable 1988 worm incident illustrates, however, that even an infiltrator that does not alter any files can be very disruptive.

Several defense mechanisms have been identified in the literature; below some of the more easily implementable defenses are listed. One should, however, keep in mind that most of them will be implemented by more software, which in turn could be subject to infiltration. First, however, it is necessary to state two fundamental principles:

1. No infection without execution.
2. Detection is undecidable.

The first principle refers to the fact that infection cannot occur unless some type of (infected) software is executed. In other words, merely looking at an infected program will not transmit a virus. Thus, simple-minded e-mail programs that handle only flat ASCII files are safe since no execution takes place. As we pointed out earlier, the execution involved can be virtually hidden from the user (e.g., in the case of previewing attachments), but in every case, the user either enabled the execution or could explicitly turn it off. The second principle has major implications. Essentially, it states that it is probably impossible to design a technique that would examine an arbitrary program and determine whether it contains a virus. This immediately raises the question of how virus detections software functions. Let us make a small excursion first. We claim that any half-way effective virus must have the ability of determining whether a program has already been infected by it. If it did not have this capability,it would reinfect an already infected program. However, since a virus is a code fragment of a certain length, inserting that same code fragment over and over into the same program would result in a program that keeps on growing until it eventually exceeds any available storage capacity, resulting in immediate detection of the virus. Returning now to our question of how virus detection software works, we can say that it does exactly the same that each virus does. This ofcourse implies trivially that that test can be carried out only if the virus is known. In other words, virus detection software will never be able to find any virus; it will only be able to detect viruses that were known to the authors of the detection software at the time it was written. The upshot is that old virus detection software is virtually worthless since it will not be able to detect any viruses that appeared since the software was written. Consequently, it is crucial to update one's virus detection software frequently and consistently.

While many virus detection programs will attempt to remove a virus once it is detected, removal is significantly trickier and can result in the corruption of programs. Since viruses and worms typically have all the access privileges that the user has, but no more, it is possible to set the permissions for all files so that writing is not permitted, even for the owner of the files. In this way, the virus will not be able to write the files, something that would be required to insert the virus. It is true that the virus could subvert the software that controls the setting of protections, but to date (February 2000), no virus has ever achieved this. (Whenever a user legitimately wants to write a file, the user would have to change the protection first, then write the file, and then change the protection back.) The primary advantage of this method is that it is quite simple and very effective. Its primary disadvantage is that users might find it inconvenient. Other, more complicated approaches include the following:

1. Requirement of separate and explicit approval (from the user) for certain operations: this assumes as interactive environment and is probably far too combersome for most practical use. Technically, this can be implemented either as software that requires the user to enter an approval code or as hardware addendum to the disk drive that prevents any unapproved writes to that disk. Note, however, that a good deal of software, for example, compilers, legitimately write to disk, even though what is written may be already infiltrated.
2. Comparison with protected copy: another way of preventing unauthorized writes is to have a protected copy (encrypted or on a write-once disk, see method 6) of a program and to compare that copy with the conventionally stored program about to be executed.
3. Control key: a control key can be computed for each file. This may be a type of check sum, the length of the file, or some other function of the file. Again, it is important that this control key be stored incorruptible.

4. Time stamping: many operating systems store a date of last modification for each file. If a user separately and incorruptibly stores the date of last modification for important files, discrepancies can indicate possible infiltrations.
5. Encryption: files are stored in encrypted format (that is, as cipher text). Before usage, a file must be decrypted. Any insertion of unencrypted code (as it would be done by a virus trying to infiltrate a program) will give garbage when the resulting file is decrypted.
6. Write-once disks: certain codes (immutable codes, balanced codes) can be used to prevent (physically) the change of any information stored on write once (or laser) disks.

All methods attempt to prevent unauthorized changes in programs or data. Methods 2 (protected copy), 3 (control key), and 6 (write-once disk) work primarily if no changes at all are permitted. Methods 4 (time stamping) and 5 (encryption) work if changes are to be possible.

None of these methods guarantees that attempted infiltrations will be foiled. However, these methods may make it very difficult for a virus to defeat the security defenses of a computer system. Note, however, that all are aimed at permanently stored files. Thus, a worm as in the November 2, 1988, incident may not be affected at all by any of them. As indicated, this worm took advantage of certain flaws in the mail handler and a communications program.

IV. CONCLUSION

Viruses and worms are a threat to data integrity and have the potential for endangering data security; the danger of these attackers is magnified substantially by their ability of self-replication. While the general mechanisms of viruses have not changed significantly over the past decade, the useage patterns of computers have, providing seemingly new ways of attacking data and software. Certain defense mechanisms are available; however, they do not guarantee that all attacks will be repulsed. In fact, technical means in general are insufficient to deal with the threats arising from viruses and worms as they use commonly accepted and convenient means of communication to infiltrate systems.

Using reasonable precautions such as restricting access; preventing users from running unfamiliar, possibly infiltrated software; centralizing software development and maintenance; acquiring software only from reputable vendors; avoiding opening attachments (at least from unknown senders); using virus detection software in a systematic and automated way (e.g., every log-on triggers a virus scan); and most importantly, preparing, internally disseminating, and strictly adhering to a carefully thought-out disaster recovery plan (which must function even if the usual computer networks are not operational!), it is likely that major damage can be minimized.

SEE ALSO THE FOLLOWING ARTICLE

COMPUTER NETWORKS

BIBLIOGRAPHY

Anti-Virus Emergency Response Team, hosted by Network Associates (http://www.avertlabs.com).

Bontchev, V. (199). "Future Trends in Virus Writing," Virus Test Center, University of Hamburg, Germany (http://www.virusbtn.com/other papers/Trends).

Computer Emergency Response Team (CERT). Registered Service mark of Carnegie-Mellon University, Pittsburgh (http://www.cert.org).

Department of Energy Computer Incident Advisory Capability (http://www.ciac.org).

European Institute for Computer Anti-Virus Research (http://www.eicar.com).

Kephart, J. O., Sorkin, G. B., Chess, D. M., and White, S. R. (199). "Fighting Computer Viruses," Sci. Am., New York (http://www.sciam.com/ 1197issue/1197kephart.html).

Leiss, E. L. (1990). "Software Under Siege: Viruses and Worms," Elsevier, Oxford, UK.

Polk, W. T., and Bassham, L. E. (1994). "A Guide to the Selection of Anti-Virus Tools and Techniques," Natl. Inst. Standards and Technology Computer Security Divison (http://csrc.ncsl.nist.gov/nistpubs/select).

Concrete, Reinforced

R. Park
University of Canterbury

I. Introduction
II. Cement and Concrete
III. Development of Reinforced Concrete
IV. Design Approaches
V. Modern Reinforced Concrete Theory
VI. Prestressed Concrete
VII. Ferrocement and Fiber-Reinforced Concrete
VIII. General Applications of Structural Concrete

GLOSSARY

Aggregrate Inert granular material such as sand, gravel, or crushed stone that is mixed with Portland cement and water to form concrete.

Beam Horizontal structural member subjected primarily to loads producing flexure.

Cast-in-place concrete Concrete cast in its final position in the structure.

Column Vertical structural member subjected primarily to compressive loads.

Composite concrete members Concrete members of precast and cast-in-place concrete elements constructed in separate placements but so interconnected that all the elements respond to forces as a unit.

Deformed reinforcement Steel bars with surface deformations used for reinforcement in concrete.

Modulus of elasticity Ratio of normal stress to corresponding strain for tensile or compressive stresses below the limit of proportionality of the material.

Precast concrete Concrete cast in other than its final position in the structure.

Prestressed concrete Concrete in which tensioned steel elements are used to introduce internal stresses of such magnitude and distribution that the stresses resulting from external loading are counteracted to a desired degree.

Reinforced concrete Concrete containing steel reinforcement and designed and detailed so that the two materials act together in resisting forces.

Service loads Loads comprising the dead weight of the structure, live load representing external working loads, and other loads, such as wind, earthquake, and earth pressure, expected during normal working conditions.

Stirrup Reinforcement located perpendicular to and around the longitudinal reinforcement in beams used to resist shear, torsion, and buckling of the longitudinal bars and to confine the compressed concrete; similar to hoop and tie.

Strength design Members of structures are designed taking inelastic strains into account to reach ultimate (maximum) strength when an ultimate load equal to the sum of each service load multiplied by its respective load factor is applied to the structure.

Encyclopedia of Physical Science and Technology, Third Edition, Volume 3

Structure Assemblage of members that provide resistance to forces acting on a building or bridge.

Ultimate strength design Similar to Strength design.

Working stress design Members of structures are designed using elastic theory so that at the service loads the stresses in the reinforcing steel and the concrete do not exceed specified allowable stresses, which are fixed proportions of the ultimate or yield strength of the materials.

CONCRETE is a mixture of Portland cement or other hydraulic cement, fine aggregate, coarse aggregate, and water, with or without admixtures. It is a material that is relatively strong in compression but weak in tension. The successful use of concrete in structures has come about from the addition of steel reinforcement to provide tensile resistance. Reinforced concrete is designed on the assumption that the concrete and steel reinforcement act together in resisting forces. Reinforced concrete is now one of the most common materials from which structures are built, being widely used for the construction of buildings, bridges, and many other types of structures. Reinforced concrete structures may be of cast-in-place construction or formed from precast concrete components generally acting compositely with cast-in-place concrete. Advances in the analysis, design, and construction of reinforced concrete structures have been based on analytical and experimental studies and on practical experience.

I. INTRODUCTION

Concrete is relatively strong in compression but weak in tension. The compressive strength of concrete in structures is typically 3000–9000 psi (20–60 MPa), and the tensile strength is normally about 10–15% of the compressive strength. The successful use of concrete in structures has come about from the addition of metal reinforcement to provide tensile resistance. When properly reinforced, concrete is transformed from a relatively brittle material into a composite material with a reasonable and calculable strength in compression, tension, flexure, and shear. The use of reinforced concrete in structures has not come about as a result of a single vital discovery, but rather its use has increased gradually through the years as knowledge and experience have accumulated. The first use of iron and steel as reinforcement in concrete was made by practical engineers and was not based on the considerations of theorists. The success of many of these early reinforced concrete structures was based on sound judgment. However, there were many failures as well, due to the lack of an analytical approach and to errors in detailing reinforce-

ment. Confidence in and development of reinforced concrete design and construction through the years have been achieved by combining the ideas of practitioners with the growing knowledge of the behavior of reinforced concrete obtained from analysis and experiment. Concrete is now one of the most common materials from which structures are built.

II. CEMENT AND CONCRETE

The development of concrete as a construction material has required the ready availability of a reliable cement. Cementing materials of various types were used by the ancient Egyptians, Greeks, and Romans. After a decline of knowledge in the Middle Ages, John Smeaton provided the first significant advance when he was commissioned in England in 1756 to rebuild the Eddystone Lighthouse off the Cornish Coast. For the construction of the lighthouse he used a pozzolana mixed with lime as a mortar to bind the stones together, having determined that the capacity of certain limes to harden under water was dependent on their clay contents. The first artificial cement was produced by the English mason and building contractor Joseph Aspin, who took out a patent in 1824 for cement made by burning a mixture of clay and limestone together. Aspin referred to this product as Portland cement because of its resemblance to Portland stone. The scientific basis for the production of modern Portland cement can be credited to Isaac Charles Johnston, a compatriot of Aspin, who in 1844 in England patented the process of continuing the burning process until sintering and, by experiment, obtained the best mixture of clay and chalk. Modern Portland cement is generally made from aluminum and silica as found in clay or shale and from a calcareous material such as limestone or chalk. The process of manufacture consists of grinding the raw materials, mixing them in certain proportions, burning in a large rotary kiln to a temperature approaching 1400°C until the material sinters and fuses into clinker, and grinding the clinkers when cool into a fine powder with some gypsum added to give Portland cement. In addition to ordinary Portland cement, a number of special-purpose cements are now made, for example, rapid-hardening Portland cement, high-alumina cement, and sulfate-resisting cement.

Concrete is a mixture of cement, fine aggregate, coarse aggregate, and water, with or without admixtures. The aggregate is inert and normally occupies a large proportion of the volume of the concrete. In concrete mix design, the distribution of the aggregate particle sizes and the relative proportion of cement, aggregate, and water are determined. These variables are normally expressed as the aggregate grading, aggregate/cement ratio, and water/cement ratio. In concrete mix design, values for these variables are sought, bearing in mind that the mix should be as economical as possible, that the fresh concrete should be sufficiently workable to be properly placed, and that the hardened concrete must be durable and most of it must attain a specified compressive strength. A mean strength greater than the specified strength must be the aim of the mix design because of the inevitable variations in the actual strengths achieved, and this requires the application of statistical and probability theory.

The most important variables affecting the strength of concrete at a given age are the water/cement ratio and the degree of compaction. When concrete is fully compacted, its compressive strength is inversely proportional to the water/cement ratio, as was discovered by Duff Abrams in the United States in 1919. Abrams's finding was a special case of a more general law formulated by Feret in France in 1896 that expressed the concrete strength in terms of the cement/total voids ratio. The presence of air voids in concrete causes a reduction in the compressive strength, and hence adequate workability for proper compaction during placing is very important. The use of concrete in large-scale works was made possible by the development of power-driven mixers and was aided by the discovery by Freyssinet in France in 1917 of the value of mechanical high-frequency vibration for the compaction of concrete.

III. DEVELOPMENT OF REINFORCED CONCRETE

The principle of combining concrete and iron as a composite structural material received considerable attention in the latter half of the nineteenth century. In 1848 Lambot in France built a concrete rowing boat that was reinforced by a rectangular mesh of iron rods. In 1849 the Paris gardener Joseph Monier made for the first time concrete tubs with a reinforcement of wire mesh. The first patent for the use of reinforced concrete as a composite structural material was taken out in England in 1854 by Wilkinson. The patent was for embedding in floors or beams of concrete (either arched or flat) a network of flat iron bars or secondhand wire rope raised over the supports and sagging near the bottom of the beam or slab at midspan. Little was done in England to follow Wilkinson's lead. In France in 1852, Francois Coignet had made a building by encasing an iron skeleton framework in concrete. In 1855 Coignet took out an English patent that included the statement that floors could be made "by burying beams, iron planks or a square mesh of rods in concrete on falsework." Coignet foresaw most uses of reinforced concrete, but he produced no theory on which to base design. In the United States in 1878 a patent was granted to Thaddeus Hyatt for construction in reinforced concrete. Hyatt recognized

the importance of the near equality of the coefficients of thermal expansion of concrete and iron and advocated the use of deformed bars for reinforcement. However, it was mainly through the expansion of Monier's activity that the engineering world at large became acquainted with reinforced concrete. Monier's numerous patents taken out from 1867 onward involved containers, floors, beams, pipes, bridges, and other elements. It is notable that Monier and the other early pioneers did not use any theoretical calculations to determine the dimensions of the concrete or reinforcement.

The first theoretical analysis of the behavior of reinforced concrete beams was published in 1886 in Germany by Koenen, who assumed that the steel alone took the tensile stresses, the neutral axis was at the middepth of the section, and the concrete above the neutral axis had a straight-line distribution of compressive stress. It was P. Neumann who first recognized the importance of the relationship between the moduli of elasticity of the steel and concrete on the position of the neutral axis. However, it was Edmund Coignet (son of Francois Coignet) and N. de Tedesco who in 1894 in France first presented the elastic theory for flexure in a form similar to that used today. Their theory assumed that plane sections before bending remained plane after bending, the material obeyed Hooke's law, the ratio of the moduli of elasticity of steel and concrete was a constant, and all tension in the concrete was ignored. There were many rival theories in the 1890s because it was known that concrete did not strictly obey Hooke's law, and not all theorists were convinced that it was reasonable to neglect the tensile resistance of concrete (Fig. 1).

Reinforced concrete construction developed particularly rapidly in France. In 1895 Considère had begun an extensive experimental study of reinforced concrete beams and columns, which led to a number of important concepts—for example, the use of spiral reinforcement in columns as a means of reducing the cross section of heavily loaded columns and the anchorage of the ends of main bars in beams by forming a hook with a diameter five times that of the bar. However, the greatest influence on reinforced concrete construction was that of Francois Hennebique (1843–1921), whose genius resulted in many innovations. For example, Hennebique is credited with having introduced the T beam, in which the cross-sectional shape makes the best use of the materials making up the composite section (Fig. 2a). Hennebique also combined columns, beams, and floors into completely monolithic structures. The Hennebique system was introduced to Britain in 1897. Figure 2b shows a typical continuous beam from his patent of that year. Hennebique's work marked the beginning of a new era of reinforced concrete construction.

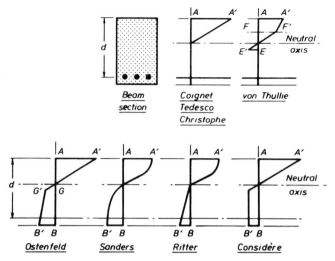

FIGURE 1 Some rival theories around 1900 of the distribution of longitudinal stress down the depth of a reinforced concrete section (*AA'*, maximum compressive stress in concrete; *BB'*, maximum tensile stress in concrete).

Other significant early contributions made by Europeans were those of Freyssinet in France, Emperger in Austria, and E. Mörsch in Germany. Mörsch's contributions included a classical textbook on reinforced concrete construction. The pioneer in the application of reinforced concrete in the United States was Ransome, who erected his first fully concrete-framed building in 1903.

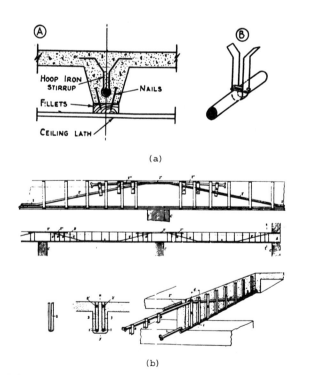

FIGURE 2 (a) Hennebique's T Beam of 1892 and details of the stirrup and cotters. (b) Hennebique's main British patent of 1897.

Early structural forms in reinforced concrete commonly followed traditional concepts. For example, some early floors were clearly direct imitations of floors made entirely of wood or of wood supported on iron or steel beams. However, new structural forms that owed their origin to the particular properties of reinforced concrete were developed. An example is the flat slab floor (a floor without beams), which was mainly an American innovation, being patented in the United States in 1902 by O. W. Norcross. The use of the flat slab floor for heavy loading was made possible by the addition of the drop panel (a slab thickening around the column) by A. R. Lord and the enlarged mushroom head to the column by C. A. P. Turner, both of the United States. It is of interest that there was no acceptable form of analysis available when the first flat slab floor was built by Turner in 1906 in Minneapolis. The building was erected at Turner's risk and load-tested before the owner would accept it. Other applications for which concrete proved suitable were frames, walls, and shell roofs for buildings, tanks, bridges, and other structures.

There has been a steady development in the theory of reinforced concrete since the start of the present century. The properties of the materials are now much better understood and have been greatly improved. The stresses in and the modes of failure of the common range of reinforced concrete structural elements and assemblages have received considerable investigation, both theoretically and experimentally. As a result calculations can now be made with increasing confidence, both in the service load range using elastic theory and at the ultimate load of the structure using ultimate strength theory.

The theory of ultimate flexural strength considers the behavior of a member when the concrete has reached its compressive (crushing) strength and the reinforcing steel has generally reached its yield strength. Charles Whitney in the United States in 1942 suggested replacing the actual curved shape of the distribution of compressive stress in the concrete by an equivalent rectangle, resulting in a convenient analytical simplification when determining the ultimate flexural strength. Flexural strength calculations can be carried out with accuracy. Unfortunately, the theories developed for the ultimate strength in shear and torsion are still significantly dependent on empirical data, as are aspects of bond, anchorage, and cracking behavior.

Reinforced concrete structures have, if well constructed, shown very good durability and fire resistance through the years. The demonstrated durability is due mainly to the high alkalinity of Portland cement concrete, which inhibits the corrosion of embedded steel reinforcement, provided that a reasonable thickness of well-compacted concrete cover is present over the steel and that the crack widths are limited. A reasonable thickness of concrete cover over reinforcement also leads to very good fire resistance, since the thermal conductivity of concrete is relatively poor. Fire ratings have commonly been stated in terms of the concrete cover thickness necessary to keep the steel beneath a critical temperature for the prescribed duration of a fire.

IV. DESIGN APPROACHES

A. Working Stress Design

Several of the early studies of the flexural behavior of reinforced concrete members can be likened to ultimate strength theories, because they assume nonlinear distributions of concrete stress down the depth of the section (see, e.g., Fig. 1). However, when reinforced concrete theory was first developed, the elastic (straight-line) theory of Coignet and Tedesco became generally accepted, mainly because elastic theory was the conventional method of design for other materials and also because the straight-line distribution of stress led to mathematical simplification. In addition, tests had shown that the use of elastic theory with carefully chosen values for the working stresses led to a structure displaying satisfactory behavior at the service loads and having an adequate margin of safety against collapse. Thus elastic theory has formed the basis of reinforced concrete design for many years.

In the "working stress" design approach the stresses in the structure at the service loads as calculated by elastic theory are limited to allowable stresses, which are specified proportions of the concrete compressive strength and the steel yield strength. Typically the allowable stresses are taken to be about one-half of the material strengths. For example, in the building code of the American Con-

crete Institute (ACI) the allowable stresses for reinforced concrete beams designed by this method are $0.45 f_c'$ for concrete in compression and 0.4 to $0.5 f_y$ for steel reinforcement in tension, where f_c' is the specified compressive strength of concrete and f_y the specified yield strength of the reinforcement.

B. Ultimate Strength Design

Recently there has been renewed interest in ultimate strength theory as a basis of design. After more than half a century of practical experience and laboratory tests, the knowledge of the actual behavior of structural concrete has vastly increased and the deficiencies of the working stress (elastic theory) design method have become increasingly apparent. Design based on ultimate strength first became accepted as an alternative to working stress design in the building codes for reinforced concrete of the ACI in 1956 and of Britain in 1957. In "ultimate strength" design (or "load factor" design) the sections of the members are designed taking the inelastic (plastic) strains into account so as to have sufficient dependable ultimate strength to resist the design ultimate actions arising from the service loads factored so as to give an adequate margin of safety against collapse.

The design requirement can be written

$$U \leqq \phi S_u \tag{1}$$

where U is the design ultimate action at the section, ϕ the strength reduction factor, and S_u the nominal ultimate strength of the section. These terms are described more fully below.

In the building code of the ACI, the design ultimate actions are found from the service loads using load factors that are intended to ensure adequate safety. The design ultimate actions are found from the worst of the following load cases (the service loads are dead load D, live load L, and wind load W):

$$U = 1.4D + 1.7L \tag{2}$$

$$U = 0.75(1.4D + 1.7L + 1.7W) \tag{3}$$

where L may have its full value or zero, and

$$U = 0.9D + 1.3W \tag{4}$$

If design is to include earthquake loading E, the above load combinations apply, except that $1.1E$ is substituted for W. The nominal ultimate strength of a section of a member, S_u, is obtained from theory predicting failure behavior of the section and on given member dimensions and specified material strengths. The dependable ultimate strength is given by ϕS_u, where ϕ is the strength reduction factor to allow for approximations in calculations and variations in the material strengths, workmanship, and dimensions.

Values for ϕ recommended by the ACI code depend on the importance of the variable quantities. For the design of beams for flexure, $\phi = 0.9$. For the design of columns for flexure, if spirally reinforced, $\phi = 0.75$; otherwise, $\phi = 0.7$; except that for columns with axial compression approaching zero, $\phi \to 0.9$. For the design of members for shear and torsion, $\phi = 0.85$.

C. Trend toward Ultimate Strength Design

There has been a gradual trend toward the use of ultimate strength design for reinforced concrete since the 1950s. The issues behind this trend are numerous.

A main issue has concerned the modular ratio (ratio of the modulus of elasticity of the steel to that of the concrete) used in working stress design. The stress–strain behavior of concrete is time dependent, and the creep strains of concrete that occur during sustained stress may be several times the initial elastic strain. Therefore, any one value chosen for the modular ratio can be only a crude approximation. Creep strains can cause a substantial redistribution of stress in reinforced concrete sections, and this means that the stresses that actually exist at the service loads often bear little relation to the design stresses calculated by elastic theory for a particular value of the modular ratio. For example, the compression reinforcement in columns may reach the yield strength during the sustained application of service loads, although this occurrence is not evident from an elastic theory analysis using a normally recommended value for the modular ratio. Ultimate strength design does not require a value for the modular ratio.

Another issue concerns the economy of materials. Ultimate strength design utilizes reserves of strength resulting from a more efficient distribution of stresses allowed by plastic strains in the concrete and reinforcing steel, and at times it indicates the working stress method to be very conservative. For example, the steel reinforcement in a structural concrete beam usually reaches the yield strength at the ultimate load, but elastic theory may indicate a lower steel stress when the concrete is crushed. Also, ultimate strength design makes more efficient use of high-strength reinforcement, and smaller beam depths can be used without compression reinforcement.

A third issue centers around the proper assessment of risk against collapse. Elastic theory cannot reliably predict the ultimate strength of the reinforced concrete members because plastic strains are not taken into account. For structures designed by the working stress method, therefore, the exact load factor (ultimate load/service load) is unknown and varies from structure to structure. Also, ultimate strength design allows a rational selection of load factors, with values depending on the particular load combinations.

Finally, ultimate strength design is attractive because it also allows the designer to assess the ductility of the structure in the postelastic range. The available ductility is an important consideration, affecting the possible redistribution of bending moments in the design for gravity, earthquake, or blast loading.

The building code of the ACI permits the use of either working stress or ultimate strength design. However, the ACI code emphasizes the ultimate strength design approach. The working stress method is referred to as the alternative design method and has been relegated to an appendix of the code. In this alternative design method elastic theory is used to design beams for flexure, but factored-down ultimate strength equations are used to design members for all other actions. It is evident that this alternative method has been retained only in an attempt to keep what has been in the past the conventional design approach. Future ACI codes may omit this alternative procedure completely. Also of interest is a change in the terminology in ACI codes. The word *ultimate* rarely appears. For example, "ultimate strength" is written as "strength."

It is also worth noting that ultimate strength theory has been used for proportioning sections in the Russia and some other European countries for many years.

D. Limit State Design and Strength and Serviceability Design

It has been recognized internationally that the design approach for reinforced concrete ideally should combine the best features of ultimate strength and working stress design. This approach is desirable because, if the members are proportioned by ultimate strength requirements alone, there is a danger that, although the load factor is adequate, the cracking and the deflections at the service loads may be excessive.

In 1964 the European Concrete Committee produced its recommendations for an international code of practice for reinforced concrete. This document introduced the concept of limit state design, proposing that a structure be designed with reference to several limit states. The strength characteristics of the materials and the load factors were based on probability theory. The most important limit states were strength at ultimate load, deflections at service load, and crack widths at service load. This approach is being adopted by many codes.

The strength design procedure emphasized by the building code of the ACI could more properly be referred to as strength and serviceability design, since although the members are proportioned by strength requirements it is also required that the service load performance of the structure be shown to be satisfactory. Deflections at service load are controlled by using adequately stiff members or by calculating the deflections by elastic theory and ensuring that limiting values are not exceeded. Crack widths at service load are controlled by using well-distributed arrangements of reinforcing steel. Thus the ACI strength and serviceability design approach is similar in principle to the European limit state design approach.

V. MODERN REINFORCED CONCRETE THEORY

Some examples of reinforced concrete theory are reviewed in this section. It is not intended that the review be comprehensive, but it will indicate some procedures for analysis and design.

A. Strength of Sections with Flexure and Axial Load

Figure 3a shows an eccentrically loaded, symmetrically reinforced concrete column section when the flexural strength is reached under the action of the ultimate load P_u and ultimate moment $M_u = P_u e$. The area of tension reinforcement is A_s and that of compression reinforcement is A'_s, and $A_s = A'_s$. The neutral axis depth is c. It is assumed that plane sections before bending remain plane after bending, and the tensile strength of the concrete is ignored. Values for the strain ε_c at the extreme compression fiber of the concrete at the flexural strength (maximum moment) have been found by experimental and analytical studies. The strain ε_c is commonly taken to be 0.0035

in European practice and 0.003 in North American and Australasian practice. The difference between these two ε_c values is of no great significance since most sections maintain nearly maximum moment capacity over curvatures corresponding to those strains. The actual curved shape of the concrete compressive stress distribution is commonly replaced by an equivalent rectangle, as originally suggested by Charles Whitney. Shown in Fig. 3a is the North American and Australasian equivalent rectangular stress distribution, which has a mean stress of $0.85 f'_c$, where f'_c is the concrete compressive cylinder strength, and a depth a, where the ratio a/c depends on the value of f'_c. Typically $a/c = \beta_1 = 0.85$ for $f'_c \leq 30$ MPa (4350 psi) and is smaller for higher values of f'_c. These parameters were found so that the actual and equivalent distributions of concrete compressive stress have the same area and position of centroid. European practice also proposes an equivalent compressive stress distribution, generally either parabolic–rectangular or rectangular in shape.

The force and moment equilibrium equations that give the ultimate load and ultimate moment in terms of the internal stresses are

$$P_u = 0.85 f'_c ab + f'_s A'_s - f_s A_s \tag{5}$$

$$
\begin{aligned}
M_u = P_u e &= 0.85 f'_c ab(0.5h - 0.5a) \\
&\quad + f'_s A'_s(0.5h - d') + f_s A_s(0.5h - d')
\end{aligned}
\tag{6}
$$

where the strains in the compression and tension reinforcement are obtained from the strain diagram of Fig. 3a as

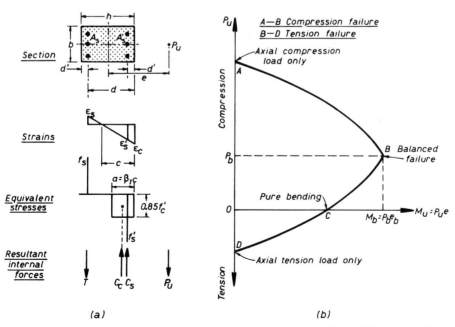

(a)   (b)

FIGURE 3 (a) Reinforced concrete short column at ultimate eccentrically applied load; (b) interaction diagram showing combinations of load and moment that cause failure.

$$\varepsilon'_s = \varepsilon_c(a - d'\beta_1)/a \tag{7}$$

$$\varepsilon_s = \varepsilon_c(d\beta_1 - a)/a \tag{8}$$

and the stresses of f'_s and f_s in the compression and tension reinforcement are obtained from those strains and the stress–strain relationships for the steel.

For a given ultimate load P_u, the depth of the equivalent rectangular stress distribution a can be found from Eqs. (5), (7), and (8) and the stress–strain relations for the steel, and then the ultimate moment M_u corresponding to that load can be found from Eq. (6). Alternatively, the P_u and M_u values corresponding to a range of a values can be found from Eqs. (5) to (8) and the stress–strain relationships for the steel, resulting in the interaction diagram shown in Fig. 3b. Any combination of load and moment that gives a point on the interaction diagram will cause failure of the column; any combination of load and moment giving a point within the area of the interaction diagram can be supported without failure.

If the ultimate load P_u is smaller than a particular load P_b (see Fig. 3b), a tension failure occurs in which the tension reinforcement yields ($f_s = f_y =$ steel yield strength), followed eventually by crushing of the concrete. If $P_u > P_b$, a compression failure occurs in which the tension steel does not yield ($f_s < f_y$). This behavior is a consequence of the increase in neutral axis depth and hence the reduction in tension steel strain, with increase in compressive load. The ductility displayed by a column at the ultimate load can be considerable during a tension failure but negligible during a compression failure. The ductility during a compression failure can be improved by the presence of closely spaced transverse reinforcement in the form of hoops or spirals, which improve the compression strain capacity of the concrete.

When the column is loaded concentrically in compression, the axial load strength (point A in Fig. 3b) is the sum of the compressive strengths of the concrete and steel. The axial tensile strength as a tie member (point D in Fig. 3b) is simply the yield strength of the steel multiplied by the total area of the longitudinal reinforcement.

B. Strength of Sections with Flexure

If the member acts as a beam, the ultimate moment M_u is found from Eqs. (5) to (8) with $P_u = 0$ substituted (point C in Fig. 3b). This gives, for a beam without compression reinforcement ($A'_s = 0$) and for the case of a tension failure ($f_s = f_y$), an ultimate moment of

$$M_u = 0.85 f'_c ab \left(d - 0.59 \frac{A_s f_y}{f'_c b} \right) \tag{9}$$

When compression reinforcement is not present in a beam, a ductile failure can be achieved by limiting the quantity

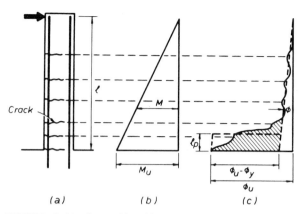

FIGURE 4 Member with ultimate moment and curvature reached. (a) Member; (b) bending moment diagram; (c) curvature diagram (full lines, actual; dashed lines, idealized; shaded area, plastic rotation).

of tension reinforcement present to ensure that it reaches the yield strength before the concrete crushes.

C. Rotations and Deflections of Members Due to Flexure

Figure 4 shows a reinforced concrete member that, under transverse load, has reached the ultimate moment at the critical section. The region of inelastic curvature is spread over a length of member, this region being at least that over which the bending moment exceeds the moment when the tension steel first yields in the section. Local peaks of curvature occur at the cracks, because between the cracks the flexural rigidity of the member is increased as a result of the tension carried by the uncracked concrete. The actual distribution of curvature can be idealized as having elastic and inelastic regions (Fig. 4c).

In the elastic range the effective (average) flexural rigidity of a cracked reinforced concrete member is partway between the cracked transformed section and the gross (uncracked) section value. The smaller the first cracking moment is compared with the maximum moment in the member, the more closely the effective flexural rigidity of the sections approaches the cracked transformed section value. That is, the tension stiffening caused by the uncracked concrete between cracks becomes less important at high moments. For example, if the moment at first cracking is one-half of the maximum moment in a typical member at a particular load, the assumption of all concrete cracked will lead to less than 15% error in the deflection calculated in the elastic range.

The shaded area in Fig. 4c gives the inelastic rotation that can occur at a "plastic hinge" in the vicinity of the critical section. The shaded area can be replaced by an equivalent area of height $\phi_u - \phi_y$ and length l_p, where ϕ_y is the curvature when the tension reinforcement first yields, ϕ_u the curvature when the compressed concrete reaches its maximum strain capacity, and l_p the equivalent length of the plastic hinge. When diagonal tension (inclined) cracks are present in a plastic hinge region, the inelastic curvature is spread farther along the member than the bending moment diagram implies. Also, any deterioration of bond between concrete and reinforcement occurring in a plastic hinge region tends to spread the yielding. Therefore, values for l_p have been proposed on the basis of experimental evidence. The value for l_p is at least 0.4 of the member depth and generally greater. The available plastic hinge rotation is given by the shaded area of Fig. 4c and is

$$\theta_p = (\phi_u - \phi_y)l_p \qquad (10)$$

Deflections can be calculated from the curvatures.

D. Strength of Members with Shear

The combination of flexure and significant shear force in reinforced concrete members causes inclined cracks, referred to as diagonal tension cracks. Figure 5a shows part of a simply supported beam without shear reinforcement to one side of a diagonal tension crack. The shear force V is resisted by a shear force V_{cz} carried across the compression zone, a dowel force V_d transmitted across the crack by the flexural reinforcement, and the vertical component V_{av} of the force due to the shear stress transmitted across the

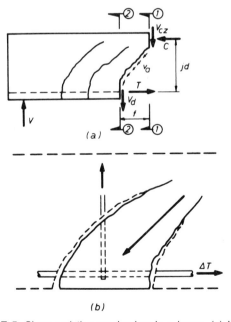

FIGURE 5 Shear-resisting mechanism in a beam. (a) Internal forces resisting shear without web reinforcement present; (b) concrete cantilever acting as strut when web reinforcement is present.

inclined crack by means of interlocking of the aggregate particles. Thus

$$V = V_{cz} + V_d + V_{av} \qquad (11)$$

If the resultant aggregate interlock force is assumed to pass through the intersection of the forces C and V_{cz} and if the contribution of V_d is neglected as being unreliable because of possible longitudinal splitting of concrete along the bars, then $M = T j d$, which means that the tensile force T at section 2 is due to the moment M at section 1. This shift in magnitude of the tensile force causes some spread of flexural yielding along the member and influences the plastic hinge length, as discussed previously. It must also be taken into account when the curtailment of longitudinal bars is being considered along spans. For example, a bar should be extended by at least a beam effective depth beyond the point at which the bending moment diagram indicates that the remaining bars can carry the moment.

In reinforced concrete beams of normal proportions without shear reinforcement, shear failure can occur by breakdown of the concrete cantilevers between the diagonal tension cracks due to bond forces arising from beam action. Although extensive analytical and experimental work has improved the understanding of the behavior of reinforced concrete members subjected to shear, particularly in recent years, no well-established theoretical equations for the ultimate shear strength have emerged. Therefore, relatively simple semiepirical design equations for the shear strength have been adopted by most codes. The shear force carried by the resisting mechanisms of Fig. 5a at shear failure can be written $v_c b_w d$, where b_w is the width of the beam web, d is the effective depth of the beam, and the nominal shear stress v_c is a function of the major variables affecting the shear resistance expressed in Eq. (11). These major variables are the tensile strength of the concrete, which is a function of the square root of the concrete compressive strength; the crack control, as measured by the ratio of longitudinal tension steel $A_s/b_w d$; and the shear span/depth ratio, as measured by M/Vd. A commonly assumed simplified value of v_c used in North America and Australasia is $2\sqrt{f_c'}$ psi ($0.17\sqrt{f_c'}$ MPa), where f_c' is the concrete compressive cylinder strength in pounds per square inch (or megapascals). Axial compression (as in columns) increases the shear carried by the concrete because of better crack control and the larger neutral axis depth.

Shear reinforcement, commonly in the form of steel stirrups, can carry shear force by truss action (Fig. 5b). In the truss analogy, first postulated by E. Mörsch in 1908, the web of the truss is considered to consist of stirrups in tension and concrete struts running parallel to the diagonal tension cracks. The top and bottom chords of the truss are formed by the compressed concrete due to flexure and the longitudinal tension reinforcement. The number of stirrups perpendicular to the axis of the member crossing an assumed 45° crack is approximately d/s, where s is the stirrup spacing. Therefore, the shear force carried by perpendicular stirrups is approximately $f_y A_v d/s$, where f_y is the stirrup steel yield strength and A_v the area of each stirrup.

The ultimate shear strength of a member with perpendicular stirrups is found by summing the contribution from the concrete mechanisms and the shear reinforcement as

$$V_u = v_c b_w d + f_y A_v d/s \qquad (12)$$

E. Bond and Anchorage

The attainment of satisfactory bond between concrete and steel reinforcement is essential for the satisfactory performance of reinforced concrete. Local bond stresses are developed by the change in the force in reinforcing bars along their length. It has been observed, however, that, provided sufficient anchorage length is available at the ends of bars, failure originating from local bond stress does not occur. The anchorage requirement is that a bar must extend beyond any section by sufficient distance to transmit the force at that section to the concrete by bond. Existing code requirements for anchorage are entirely empirical. A straight anchorage length of 40 bar diameters, commonly used for many years for plain, round mild steel bars, has been much reduced by the use of bars with a deformed surface, which improves the bond resistance. For example, according to the building code of the ACI, the basic development length for a deformed bar of 1-in. (25.4-mm) diameter in concrete of compressive strength $f'_c = 3000$ psi (20.7 MPa) is 23 bar diameters if the steel yield strength f_y is 40,000 psi (275 MPa) or 34 bar diameters if $f_y = 60,000$ psi (414 MPa).

A bond failure of deformed bars generally occurs by concrete splitting due to the wedging action of the deformations, and hence the required anchorage length is a function mainly of the bar force at yield, the tensile strength of the concrete, the position of the bar in the member (top-cast bars perform adversely when compared with bottom-cast bars because of the effect of water gain and sedimentation under the top bars), lateral spacing between bars, concrete cover thickness, and the quantity of confining transverse steel present. Hooks formed at the end of bars can significantly reduce the required anchorage length.

F. Cracking at Service Load

The occurrence of cracks in reinforced concrete is inevitable because of the low tensile strength of concrete. With high service load steel stresses, particularly as a result of the use of high-strength reinforcement, significant cracking may occur at the service load. The cracking of a reinforced concrete structure at the service load should not be such as to spoil the appearance of the structure or lead to corrosion of the reinforcement. For example, the distribution of reinforcement permitted by the building code of the ACI is aimed at ensuring that the surface crack widths at service load will not exceed 0.016 in. (0.41 mm) in the case of members with interior exposure and 0.013 in. (0.33 mm) in the case of members with exterior exposure.

A number of semitheoretical and empirical approaches have been developed for the determination of the most probable maximum surface width of flexural cracks. Simplistically, the crack width can be postulated to be the elongation of the reinforcement between two adjacent cracks. Statistical studies of test results have shown crack widths to be primarily a function of the stress in the steel, the thickness of concrete cover, the distribution of the steel reinforcement, and the relative distances from the neutral axis of the reinforcing steel and the extreme concrete tension fiber. Crack widths are also influenced by shrinkage of concrete and other time-dependent effects and by repeated loading. Crack width measurements are inherently subject to large scatter, even in careful laboratory work. Thus great accuracy cannot be expected from existing equations for crack widths.

The best crack control at the service load is obtained when reinforcing bars are well distributed in the zone of concrete tension. Thus a number of small-diameter bars are preferable to the use of a few larger-diameter bars. It should also be emphasized that concrete with a low water/cement ratio and a reasonable thickness of well-compacted good-quality concrete are also essential for durable concrete structures.

G. Determination of Design Actions at Ultimate Load using Linear Elastic Structural Analysis

The main approach used in design to determine the bending moments and forces in a reinforced concrete structure at the ultimate load is linear elastic structural analysis. The advantages of using this approach are that the moments and forces can be found using relatively simple and well-established structural theory, that the stresses in the concrete and reinforcement at the service loads are kept as low as possible, thus minimizing the width of cracks and deflections, and that only a small amount of moment redistribution is required at the ultimate load. The last aspect means that the plastic rotations required at the critical sections will be small. Some redistribution of moments will almost always be necessary because, unless the structural analysis is based on the final complex distribution of flexural stiffnesses after cracking, redistribution of moments will have to occur before all the critical sections can attain their ultimate flexural strength.

Most codes also allow the moments and forces derived from linear elastic structural analysis to be modified for a moderate amount of moment redistribution, the amount permitted being limited to ensure that the crack widths at the service loads are not excessive and that the critical sections are adequately ductile to achieve the required redistribution at the ultimate load. For example, the negative moments at the supports of continuous reinforced concrete flexural members may be increased or decreased by up to 20%, according to the building code of the ACI, the exact amount depending on the section ductility, provided that the positive moments within the spans are adjusted to maintain static equilibrium between internal forces and external loads.

H. Limit Design of Frames

The limit design approach for reinforced concrete frames, analogous to the "plastic theory" design of structural steel frames, allows any distribution of bending moments at the ultimate load to be used, provided that the plastic rotation capacity is sufficient to permit the assumed distribution of moments to be developed, and hence that the collapse mechanism can form at the ultimate load, and that the cracking and deflections at the service load are not excessive.

The major difficulty with this approach lies in the calculation of the required plastic rotations. This calculation is necessary because reinforced concrete sections, unlike structural steel, may in some cases fail in a brittle manner at a lower ultimate load before the collapse mechanism develops. A method that appears to have the most promise for reinforced concrete is that of A. L. L. Baker. To calculate the plastic rotations actually required at the plastic hinges to develop a particular limit moment diagram. Baker considers the structure to be in the state immediately before ultimate load, where sufficient plastic hinges have formed to make it statically determinate and the last plastic hinge or hinges are about to form and convert the structure to a collapse mechanism. In this state the structure has known ultimate moments at the plastic hinges and at the plastic hinge or hinges about to form, and the members are in the elastic range between the hinges. The rotations required of the plastic hinges can be calculated by the flexibility coefficient method. For a structure that is n times statically indeterminate with known ultimate moments $M_1, M_2, M_3, \ldots, M_n$ at the plastic hinges, the plastic rotation θ_i at hinge i is

$$-\theta_i = \delta_{i0} + \sum M_k \delta_{ik} \qquad (13)$$

where δ_{i0} is rotation at hinge i due to the external ultimate loads acting with zero hinge moments, M_k the ultimate moment at hinge k, and δ_{ik} the rotation at hinge i due to $M_k = 1$ acting alone at hinge k. Thus the plastic rotations required for a selected limit moment diagram can be calculated and checked against the permissible rotations computed from Eq. (10) to ensure that full moment redistribution can take place. Baker's method is unwieldy in all but the simple cases, however, and although it has been recommended by some design codes, it can hardly be regarded as a design office technique in its present form.

I. Limit Design of Slabs

Reinforced concrete slabs are usually lightly reinforced because the span/depth ratios are normally governed by requirements of adequate stiffness to avoid excessive deflections. Hence the sections generally have sufficient plastic rotation capacity to allow full moment distribution to take place at ultimate load. It is therefore commonly assumed that limit design can be applied to reinforced concrete slabs without the need to check whether sufficient plastic rotation capacity is available. Lower- or upper-bound approaches can be used to obtain the design ultimate moments, assuming that a flexural failure occurs. The shear strength is checked separately.

The lower-bound method of reinforced concrete slab design involves finding a distribution of moments that satisfies the equilibrium equations and the boundary conditions when the ultimate load is applied and providing sufficient reinforcement to carry these moments without the ultimate moment being exceeded anywhere in the slab. This procedure can be fulfilled by using in design the distribution of moments given by thin-plate elastic theory. Alternatively, a commonly used lower-bound approach is Hillerborg's strip method. The simple strip method obtains the distribution of moments and shears by replacing the slab by two systems of strips, normally running in two directions at right angles, which share the external loading. The strips are not considered to carry any load by torsion, and the design bending moments are found by simple statics. A. Hillerborg has also introduced the advanced strip method, which features rectangular corner-supported elements, as well as triangular and rectangular edge-supported elements, for use in the design of beamless slabs.

The upper-bound method of reinforced concrete slab design is known as yield line theory and is due mainly to K. W. Johansen. Collapse mechanisms for slabs are known as yield line patterns and are composed of segments of slabs separated by lines of plastic hinging. The ultimate load of the slab is calculated from the postulated yield line pattern, and the position of the yield lines that gives the smallest ultimate load is sought. The correct yield line patterns for all common cases are now well known. In design the procedure is to provide adequate reinforcement to ensure that the required ultimate load can be reached.

J. Complete Behavior of Structural Systems

The full analysis to determine the complete behavior of reinforced concrete structures at all levels of loading, where flexural deformations predominate, can be undertaken with the aid of a computer with large storage. For example, a given frame or slab system can be divided into small elements of the thickness of the member or slab, each with its known moment–curvature relation, including the effect of tension stiffening between the cracks. The distribution of moments throughout the structure as the external load is increased incrementally from zero can be determined using the stiffness method. At each level of external load an iterative procedure is used to obtain the distribution of moments that is compatible with the boundary conditions, the equilibrium equations, and the stiffnesses of the elements as given by the nonlinear moment–curvature relations. Eventually, with increasing load, the elements at the critical sections commence to reach their ultimate flexural strength, and plastic hinges spread throughout the structure. The ultimate load is reached either when a collapse mechanism forms or when a brittle failure occurs at a critical section due to lack of plastic rotation capacity. A computer program capable of this analysis procedure could be used in design only on a trial-and-error basis, but it would be a powerful tool for evaluating or checking structural performance over the full range of loading, including behavior at the service and ultimate loads. Geometric changes in the structure during loading could also be included in such a program to allow for the effect of deflections on internal actions and to indicate any instability effects. The full analytical approach is feasible now and some special-purpose computer programs are available, but are not commonly used.

VI. PRESTRESSED CONCRETE

A. Introduction

Prestressed concrete is defined as concrete that has been prestressed so that the induced internal actions counteract the external loading to a desired degree. Prestressed concrete members can be placed in one of two categories: pretensioned or posttensioned. Pretensioned prestressed concrete members are produced by stretching steel tendons between external buttresses before the concrete is placed. When the concrete has reached the required strength, the tendons are released from the buttresses and the prestressing force is transferred to the concrete of the member by the bond between the tendons and the concrete. In the case of posttensioned prestressed concrete members, steel tendons are stretched in ducts in the concrete, after the concrete has hardened and gained sufficient strength, by jacking against the concrete member itself, and then the tendons are anchored at the ends of the member.

The general principles of prestressing concrete were known for many years before it became a practical method of construction. Early attempts to prestress concrete were unsuccessful, mainly because high-strength steel was not used for the tendons and the low initial steel stress was lost with time as a result of the shrinkage and creep of concrete. The general concepts of prestressed concrete were first formulated in the 1880s by Doehring of Germany and Jackson of the United States. The modern development of prestressed concrete is credited to Freyssinet of France, who clearly showed the effect of concrete creep and shrinkage and in approximately 1928 demonstrated that by the use of high-tensile steel the losses in prestress could be kept down to reasonable proportions. The practical development of pretensioning was due mainly to Hoyer of Germany. The wide application of prestressed concrete became possible when reliable and economical methods of tensioning and end anchorage were developed. In 1939 Freyssinet developed double-acting jacks and conical end anchorages. In 1940 Magnel of Belgium developed metal wedges.

B. Basic Behavior

The usual concept of prestressed concrete is that of a material weak in tension that is compressed by prestress to avoid the occurrence of tension due to external loading. If the tensile strength of the concrete is not exceeded, the stresses can be determined by standard elastic theory for uncracked sections. Figure 6a shows a prestressed concrete beam with a symmetrical cross section and with a prestressing tendon placed in a curved profile with variable eccentricity e with respect to the centroid of the section of the member. The extreme fiber concrete stresses at section 1 due to prestress alone are

$$f = \frac{F}{A} \pm \frac{Fe}{Z} \tag{14}$$

where F is the prestressing force, A is the section area of the member, and Z is the section modulus of the member. When external loading is applied, the extreme fiber concrete stresses due to the external moment M at section 1 are $\pm M/Z$, and these stresses are added to those due to prestress. Figure 6c shows the distribution of concrete stresses at section 1 due to prestress alone and due

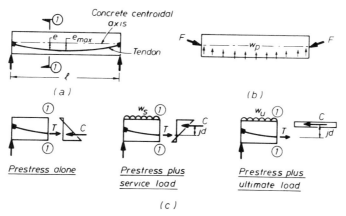

FIGURE 6 Behavior of a prestressed concrete member. (a) Beam and tendon; (b) forces acting on concrete due to parabolic tendon alone; (c) internal couple at a section when external moment at that section increases from zero to the ultimate moment.

to prestress plus service load (before cracking). In design a small tensile stress in the concrete, insufficient to cause cracking, is often allowed.

Another concept of prestressed concrete is that of a combination of steel and concrete in which the tensile force in the prestressing steel and the resultant compressive force in the concrete form an internal couple that resists the bending moment due to the external loading. Figure 6c illustrates the concept. The compressive force in the concrete C acts at the centroid of the compressive stress distribution. The tensile force in the tendon is T. For equilibrium, C must equal T at all stages. When there is no external loading on the member, C and T have the same line of action; that is, the internal couple is zero. When external loading is applied, the centroid of the compressive stress distribution shifts up the section because of the change in concrete compressive stress distribution, and the internal couple is equal to Cjd or Tjd, which exactly equals the bending moment acting at the section due to the external loading. As the external load varies within the service load range, T remains practically constant in a prestressed concrete member because a high initial steel stress is used and the increase in steel stress due to bending strains in the concrete represents only a small percentage of change in steel force. Hence as the service load varies, C and T may be considered to remain constant, but jd varies. By comparison, in a reinforced concrete member, as the external load varies within the service load range C and T vary but jd remains constant. If the external load exceeds the service load, the concrete will crack when it reaches its tensile strength. Cracking will result in an increase in the tensile stress in the steel and in the compressive stress in the concrete as the stresses readjust to carry the external moment. With further loading the concrete and steel enter the inelastic range and eventually the ultimate moment of the section is reached (Fig. 6c).

Reinforced concrete cannot effectively use very high tensile steel because the high steel stresses associated with the efficient use of such steel mean that large tensile strains must be induced in the steel, and this would cause the surrounding concrete to crack seriously at service load. In prestressed concrete this problem is overcome because the steel is prestretched. Also, high-strength concrete can be used to advantage. Hence prestressed concrete makes effective use of high-strength materials.

Prestressed concrete can also be regarded as an attempt to balance the external loading by the actions induced by the prestressing tendon. Figure 6a shows a simply supported beam prestressed by a tendon with force F and a parabolic profile. The tendon applies an upward uniform load w_p per unit length to the concrete (Fig. 6b), since it is forced to remain in the shape of a parabola, and from simple statics

$$w_p = 8Fe_{max}/l^2 \qquad (15)$$

where e_{max} is the eccentricity of the tendon at midspan. Thus external loading on a member can be balanced by the upward loading from the tendon. If the external load is exactly balanced by the tendon, the beam will not be deflected and the concrete will be subjected to uniform stress F/A, since the moment due to external loading will be exactly balanced by the moment due to the prestress along the beam. Similarly, the net shear force will be zero. In practice it is not generally possible or desirable to balance the total service loads fully, mainly because such a design would be uneconomical, requiring a large amount of prestress, and because the live load fluctuates between zero and its full value during service conditions. However, by suitable design of the prestressing force, it is possible to offset a good deal of deflection due to external load. This allows the use of more slender spans in prestressed concrete than is possible with reinforced concrete. The load balance approach was introduced by T. Y. Lin in 1963. It is a useful concept for design, especially in the case of posttensioned flat slab floor systems.

C. Statically Indeterminate Structures

When statically indeterminate structures are posttensioned during construction, the members deform elastically and secondary prestressing moments may arise as a result of the restraints offered by the support conditions. The secondary prestressing moments so induced are often not secondary in magnitude and must be added to the primary prestressing moments due to tendon eccentricity to find the resulting (total) prestressing moments. Concordant tendon profiles, for which the secondary prestressing moments are zero, can be found for continuous beams but are often uneconomical in design since full use cannot be made of the available section eccentricity. However, efficient nonconcordant (general) tendon profiles can readily be incorporated in design. In the case of a general tendon profile the resulting prestressing moment can be calculated directly, using a method of structural analysis, from the equivalent loads acting on the structure caused by the prestressing tendons. The equivalent loads are due to end anchorages and to the angular changes of the tendons along the members. Much of the pioneering theoretical work on the analysis of statically indeterminate structures was that of Y. Guyon, whose volumes still form a major reference source on prestressed concrete.

D. Design Approaches for Prestressed Concrete

Prestressed concrete structures are generally designed with reference to the limit states of allowable stresses and deflection at service loads using elastic theory, and adequate ultimate load using ultimate strength theory. The allowable concrete tensile stress at service load may be zero or a small value that will not lead to cracking, according to the concept of prestressing advocated by Freyssinet.

An alternative design approach that has been advocated in more recent years permits tensile stress to develop in the concrete, which may lead to cracking when the full service live load is applied. The acceptance of cracked prestressed concrete sections at service loads leads to the concept of partially prestressed concrete, as advocated by Abeles. The use of a lesser amount of prestressing steel in a partially prestressed concrete structure generally means that some nonprestressed steel reinforcement must also be present to help control the cracking at service load and to provide the additional flexural strength necessary to achieve an adequate ultimate load.

VII. FERROCEMENT AND FIBER-REINFORCED CONCRETE

Providing tensile resistance in concrete has conventionally involved embedded steel reinforcing bars or prestressing tendons. Nevertheless, significant progress has been made in obtaining a composite material formed from cement mortar or concrete acting with layers of small-diameter wire mesh or with short, randomly oriented fibers.

Ferrocement is the name given to the composite material formed by plastering several layers of steel wire mesh with cement mortar. The earliest recorded use of ferrocement was Lambot's concrete rowing boat, which was built in France in 1848. For many years ferrocement continued to be used mainly for boats. The most important developments in ferrocement were due to P. L. Nervi in Italy, who constructed several vessels in the early postwar years and was the first to make significant use of ferrocement in buildings. In recent years ferrocement has been used widely for boats and architectural elements of buildings. The close spacing of small-diameter wires results in very well controlled cracking of concrete, and only a small thickness of mortar cover is necessary for protection against corrosion. Also, the impact resistance is high. Complicated shapes can be readily constructed because the mortar is plastered directly onto the wire mesh, and formwork is unnecessary.

The idea of endowing concrete with a greater tensile strength and ductility by the use of fibers was proposed at the turn of the century. However, the practical use of short, randomly oriented fibers in concrete is relatively new. Much of the pioneering work on steel fiber reinforcement was conducted by J. P. Romualdi and G. B. Batson in the United States in the late 1950s and early 1960s. Ordinary glass fibers are unsuitable, since glass is attacked by the highly alkaline environment of concrete, but alkaline-resistant glass fibers have now been developed. The failure mechanism of fiber-reinforced concrete involves mainly fiber pull-out, and it is possible to improve the bond strength by deforming the fibers in various ways. The maximum size and quantity of aggregate in the concrete must be limited, because the fibers can only reinforce the matrix. The presence of fibers reduces the workability of the mix, and hence a maximum of $\sim$3% by volume of fibers can be added when mixing. The inclusion of fibers improves the toughness of concrete, and the load-carrying capacity can be sustained at relatively large tensile and compressive strains, well beyond peak load conditions. Fiber-reinforced concrete has possible applications when the toughness and energy-absorbing characteristics are important, such as for pavements and linings. Glass-fiber reinforced concrete has also been used as spray-on cladding for structures.

VIII. GENERAL APPLICATIONS OF STRUCTURAL CONCRETE

A. Buildings

Concrete floors are generally either of all-cast-in-place construction or formed from precast members, generally acting compositely with cast-in-place concrete.

Three basic types of all-cast-in-place reinforced concrete floors are shown in Fig. 7. Beams may or may not be present. The flat plate (Fig. 7a) with uniform thickness is the simplest form of beamless slab construction. The flat slab (Fig. 7b) has capitals at the tops of the columns and/or drop panels, which are thickened areas of slab surrounding each column, in order to provide the additional slab stiffness and shear strength necessary for larger spans and heavier live loads. The two-way slab (Fig. 7c) is supported on beams in two directions on the column lines. The beams can be replaced by walls. If the span between the column lines is large, secondary beams may also be present between the column lines in order to reduce the spans of the slabs. If the span of the slab in one direction is more than twice that in the other direction, the greater part of the slab loading will be carried in the direction of the short span, and the slab can be regarded as a one-way slab. One-way slab action will also occur if the beams or walls are present only in one direction.

Waffle slabs, containing recesses formed in the lower surface, can be used to reduce the dead load. Waffle slabs can also be used as a means of increasing the effective depth of a slab without an accompanying increase in dead load. Hollow clay tiles or other fillers can be employed in place of recesses as a further variation of a waffle slab.

The use of prestressing in flat slab floors has become common in many parts of the world, the posttensioned tendons allowing excellent crack and deflection control and longer spans.

Floor slabs may be cast in place on formwork in their final position. Some flat plates are built using lift–slab construction, in which all the slabs are cast in a stack at ground level and then lifted into their final positions after being cured.

The choice of slab type depends on the magnitude of the live load, the spans, the cost of materials and labor during construction, and the local preference. Cast-in-place two-way slab floors are more labor intensive during construction but require less reinforcing steel because of the large effective depths of the beams compared with beamless floors with the same spans between columns and live loads.

Precast concrete one-way floors consisting of double-T, hollow core plank, or other sections spanning beams or walls can be used as an alternative to an all-cast-in-place floor. Precast floors are generally used with a cast-in-place concrete topping slab. The precast units are either of reinforced or of pretensioned prestressed concrete. Precast floors are widely used in many parts of the world.

Concrete floors also connect parts of the vertical structural system to form a structure capable of transferring both gravity and horizontal loads to the foundations. The vertical structural system commonly comprises either continuous frames consisting of columns and beams rigidly connected together, or structural walls, which support the floors. Buildings can include various combinations of these vertical structural systems. Examples of these basic

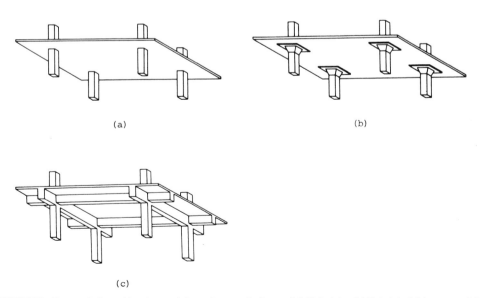

(a)

(b)

(c)

FIGURE 7 Types of all-cast-in-place reinforced concrete floors. (a) Flat plate; (b) flat slab; (c) two-way slab.

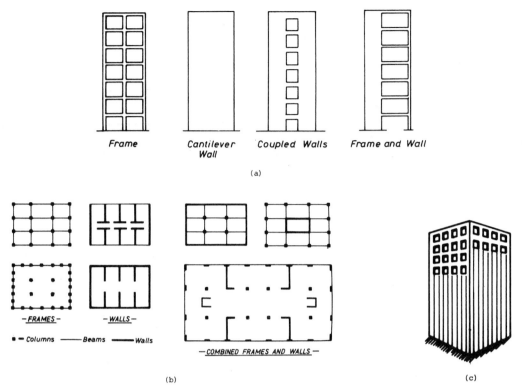

Frame   Cantilever   Coupled Walls   Frame and Wall
          Wall

(a)

■ = Columns ——— Beams ——— Walls

— FRAMES —   — WALLS —

— COMBINED FRAMES AND WALLS —

(b)   (c)

FIGURE 8 Basic structural concrete systems for resisting gravity and horizontal loads. (a) Frames and walls; (b) some building plans; (c) framed tube.

structural systems and combinations of them are shown in Fig. 8a,b.

A problem with relying entirely on frame action for strength and stiffness is that interstory horizontal deflections due to horizontal wind or earthquake loading may be significant. Codes generally require calculations to check that the horizontal drift is within tolerable limits, which should ensure that the discomfort caused to occupants and the damage caused to nonstructural components are not excessive. Structural walls, because of their greater stiffness, can be effectively combined with frames to help control behavior under horizontal loads. Indeed, several well-placed structural walls in a building can resist a large portion of the horizontal load and limit horizontal drift, leaving the more flexible frames to carry the majority of the gravity loading. Thus, combined systems of walls and frames can often be used to advantage.

Some buildings have frames with closely spaced columns around the perimeter, which simulate a perforated rectangular hollow tube, sometimes referred to as a framed tube (Fig. 8c). Cantilever walls and coupled walls can also make up tubes. Tubes are an efficient means of resisting lateral loading since part of the tube wall at right angles to the direction of lateral loading can act as the "flange" to the tube wall in the direction of the loading.

An effective structural system for tall buildings is a tube within a tube, consisting of an inner tube formed by the walls enclosing the service core and an outer tube formed by frames with closely spaced columns around the perimeter of the building.

Design for gravity and wind loading is essentially a matter of ensuring that the building structure can withstand the greatest likely load applications without failure. However, it would be uneconomical to design a structure to withstand the greatest likely horizontal earthquake loading without damage. Dynamic analyses of building structures responding elastically to ground shaking recorded during severe earthquakes have shown that the theoretical response inertia loads are generally significantly greater than the static design horizontal loads recommended by codes. Hence structures designed for the horizontal earthquake loads normally recommended by codes can survive very severe earthquakes only if they have sufficient ductility to absorb and dissipate seismic energy by inelastic deformations. Hence buildings constructed in earthquake zones require special attention in design to ensure that in the event of a major earthquake they can deform in the inelastic range in a ductile manner. In practice, this means designing ductile plastic hinge regions in members and ensuring that failure does not occur by shear, bond, or column failure. In frames a strong column–weak beam concept is used, and potential plastic hinge regions are made ductile by ensuring that the tension reinforcement content is not too high, that adequate compression steel is present, and that sufficient transverse reinforcement is available to prevent shear failure and compression bar buckling and to confine the concrete.

Conventionally, most concrete building structures have been of cast-in-place monolithic construction. However, the cost of skilled labor to build formwork and the need for speed of construction have made the use of precast concrete components, prefabricated off site, an attractive alternative to cast-in-place construction. The precast concrete components have followed the traditional patterns of beams, slabs, and columns. More recently they have also taken the form of precast bearing walls or panels and precast slab units. Care is needed in incorporating precast units into the design to ensure that the connections between them have both adequate strength and ductility. The progressive collapse of the Ronan Point building in England in 1968 after a domestic gas explosion was due to poor connections between precast concrete walls and floors and a poor design concept.

The first tall concrete building to be built with a reinforced concrete frame was the 16-story Ingalls Building, which was constructed in 1903 in Cincinnati, Ohio. The world's tallest concrete building has been Water Tower Place, built in 1975 in Chicago. This building is 859 ft (262 m), 76 stories tall and has high-strength 9000-psi (62-MPa) concrete at its base. The tower of the building consists of structural walls and closely spaced perimeter columns forming a tubular structure. Lightweight concrete has been used to reduce the dead weight of some tall concrete buildings. Figure 9 shows the 715-ft (218-m), 60-story-tall One Shell Plaza building built in Houston, Texas, in 1969 using high-strength 6500-psi (45-MPa) lightweight concrete. The CN Tower, in Toronto, completed in 1975, is the world's tallest free-standing tower (Fig. 10). The CN Tower serves for communications and observations and has a concrete structure 1464 ft (446 m) high topped by a steel structure to give a total height of 1815 ft (553 m).

As well as floors and vertical structural systems, concrete has had significant application for shell roof construction. The first concrete shell roof was built in Jena, Germany, in 1924. Since then, many developments in form have occurred and spans increased. For example, the roof of the building of the Centre Nationale des Industries et Technologies in Paris, completed in 1958, consists of three corrugated fan-shaped half-arched shells of reinforced concrete. This roof structure is supported at three points that form in plan an equilateral triangle, the side length of which is 675.8 ft (206 m). The Kings County

FIGURE 9 One Shell Plaza, Houston, Texas.

Stadium, situated in Seattle, Washington, was a dome of 661-ft (201.5-m) diameter consisting of a thin concrete shell stiffened by concrete ribs.

B. Bridges

Concrete bridges are generally either arch or beam in structural type. The beam is by far the most common structural element used in bridges. An arch carries the load of the bridge, mainly by axial compressive thrust in the member developed by the resistance against spread of the arch at the abutments. Some flexure and shear must also be resisted by the arch since the line of thrust for the loading generally does not follow the centroidal axis of the arch exactly. Beams are used in many structural configurations in bridges and carry load by flexure and shear. The beam can have a shaped cross section, such as an I or T section or a hollow box, and the depth can be varied along the length of the span to allow efficient use of materials. In multispan bridges either a series of simply supported beams can be used or the beam can be continuous. Cable-stayed bridges, in which cables attached to the beam transfer loading to towers at the bridge piers, are also used for long spans.

The first large-scale use of concrete, made with Portland cement, in bridges was probably the long series of concrete arches forming the Grand Maitre Aqueduct constructed in

FIGURE 10 CN Tower, Toronto, Canada.

of Switzerland. Significant use of reinforced concrete in bridge construction followed. The next era in bridge design and construction occurred after the Second World War when Eugene Freyssinet, with considerable courage and ingenuity, introduced the use of prestressed concrete. The first of Freyssinet's prestressed concrete bridges was completed in 1949 across the River Marne at Esbly, France, and was an arch of 243-ft (74-m) span. The ability to balance external loads with prestressing forces has given designers considerable scope for innovation in form and has led to slender, graceful structures. The St. James Park Footbridge in England built in 1958 (Fig. 11) is an excellent example of the elegance made possible by prestressed concrete. The 70-ft (21-m) main span of this bridge has a depth of only 14 in. (0.36 m) at midspan.

Bridge designs in many countries have led to the development of standard precast, prestressed concrete beams that are normally designed to act compositely with a cast-in-place reinforced concrete deck slab. For bridges with short spans the precast sections are normally voided rectangular, channel, or T shape. For bridges with medium spans of up to ~130 ft (~40 m), the precast sections are commonly of I shape. For long spans, segmental prestressed concrete hollow box girders are generally used. Typically in this form of construction the piers are cast in place. Then construction of the spans proceeds in each direction from the pier by the balanced cantileyer method, which involves placing precast or cast-in-place segments, each posttensioned to the previously completed construction. Alternatively, the span-by-span method can be used where the placing of the segments commences at one end of the structure and proceeds continuously to the other end and the segments are supported during construction by movable falsework. A box-girder bridge during segmental construction is shown in Fig. 12. Cable-stayed prestressed concrete box girders are also being used competitively for long spans. A novel form of bridge suitable for light loads and long spans is the stressed-ribbon bridge, consisting of

France around 1870. The first concrete bridge in Britain was an arch of 75-ft (23-m) span built at Earls Court, London, in 1867 by John Fowler. An early British reinforced concrete beam bridge is the Homersfield Bridge of 50-ft (15-m) span built in 1870 in Suffolk.

However, it was the French engineer Hennebique who had the greatest influence on early reinforced concrete bridge construction. Examples of Hennebique's work are his bridge built at Chatellerault in France in 1899, which is three arch spans of 131, 164, and 131 ft (40, 50, and 40 m), each with rise/span ratios of ~0.1, and his Risorgimento Bridge built over the Tiber in Rome in 1911, which is a cellular arch of 328-ft (100-m) span with a rise/span ratio of 0.1. Although many of these early arch bridges are outstanding structures, the use of reinforced concrete in other bridge types was not particularly notable until the Swiss engineer Maillart (1872–1940) designed and built some outstanding bridge structures in the mountainous terrain

FIGURE 11 St. James Park Footbridge, London, England.

FIGURE 12 Knight Street Bridge, Vancouver–Richmond, Canada, during balanced cantilever construction.

a concrete deck supported directly on steel cables. Fatigue has not proved to be a problem in concrete bridges.

The world's longest span in concrete has been the Brotonne Bridge over the Seine in France, which is a cable-stayed structure with a prestressed concrete deck and a main span of 1050 ft (320 m). The longest span concrete arch bridge is the Gladesville Bridge in Australia, completed in 1964, which has a span of 1000 ft (305 m) and a rise of 134 ft (40.8 m). The longest main span for a cantilever bridge is 790 ft (241 m) and occurs in the Koror–Babelthuap Bridge in the Palau Islands in the Trust Territory of the Pacific Islands, which was completed in 1977. The longest main span of a stressed-ribbon bridge is 710 ft (216 m) and is the Holderbank–Wildegg Bridge in Switzerland built in 1964. It is technically feasible for these spans to be increased. For example, T. Y. Lin of the United States has proposed a scheme to build a bridge across the Bering Strait from Alaska to Siberia involving 250 precast prestressed cable-stayed spans, each with a length of 1200 ft (366 m).

C. Other Structures

Concrete has had significant use in such marine structures as docks, locks, barrages, barges, and towers. Perhaps the most interesting application in recent years has been the large concrete offshore towers that are in use in the North Sea for the drilling, production, and storage of gas or oil. These concrete towers are built in sheltered construction sites by land and then towed floating into position and sunk into place, leaving a working platform above water. The method of construction minimizes the amount of construction required in the open sea. These offshore towers are known as gravity structures because they are not pinned to the sea bottom by piles but simply sit on the sea bed. They have sufficient size and weight to give stability

against overturning. The technology for concrete offshore structures has developed rapidly. The first structure of this type was the Ekofisk oil storage tank, with a capacity of 1 million barrels, which was installed in 230 ft (70 m) of water in the North Sea in 1973 (Fig. 13). In the period from 1973 to 1978, 16 offshore concrete towers were installed in the North Sea, in water depths ranging from 50 to 500 ft (15 to 152 m) with base diameters varying from 160 to 460 ft (50 to 140 m). These structures are indeed large in all respects. The Doris Ninian Tower required $\sim$185,000 yd^3 ($\sim$142,000 m^3) of concrete. Increasing use has also been made of reinforced and prestressed concrete for floating vessels. An example is a floating storage facility for liquid petroleum gas (LPG), 528 ft (161 m) long by 138 ft (42 m) wide, which was constructed in Tacoma, Washington, and towed almost 10,000 miles (16,000 km) to its working site in the Java Sea. On the whole, well-designed and -constructed concrete has proved to be a durable material in fixed and floating offshore structures.

Dams are impressive because of their size. The world's highest concrete dam is the Grande Dixence gravity dam of Switzerland, which was completed in 1962 and has a height of 935 ft (285 m). The Hoover Dam on the Colorado River was completed in 1936 and, with its height of 726 ft (221 m), was for 22 years the world's highest dam. The Hoover Dam is of the arch–gravity type, its structural action relying partly on arching between the valley walls and partly on self weight, and was a generation ahead of other large concrete dams.

Concrete is also widely used in airport and road pavements, foundation structures, cooling towers, tanks, tunnels, and other structures.

FIGURE 13 Efofisk oil storage tank under tow to the Ekofish oil field in the North Sea.

SEE ALSO THE FOLLOWING ARTICLES

BONDING AND STRUCTURE IN SOLIDS ● FRACTURE AND FATIGUE ● MASONRY ● MECHANICS OF STRUCTURES

BIBLIOGRAPHY

Abeles, P. W. (1967). *J. Am. Concr. Inst.* **64** (10).

American Concrete Institute (1999). "ACI 318-99 and 318R-99 Building Code Requirements for Structural Concrete and Commentary," CRC Press, Boca Raton, FL.

Carpinteri, A. (ed.) (1999). "Minimum Reinforcement in Concrete Members," Elsevier, Amsterdam.

Elfgren, L. (ed.) (1989). "Fracture Mechanics of Concrete Structures," Chapman and Hall, London.

Hamilton, S. B. (1956). "A Note on the History of Reinforced Concrete in Buildings," Natl. Build. Stud., Spec. Rep. No. 24, Dep. Sci. Ind. Res., H.M. Stationery Office, London.

Hanna, A. S. (1998). "Concrete Formwork Systems," Vol. 2, Civil and Environmental Engineering, Marcel Dekker, New York.

Hillerborg, A. (1975). "Strip Method of Design" (transl. from Swedish), Cement and Concrete Association, London.

Hopkins, H. J. (1970). "A Span of Bridges," David & Charles, Newton Abbot, England.

Institution of Civil Engineers Research Committee (1962). "Ultimate Load Design of Concrete Structures," *Proc. Inst. Civ. Eng.*, Vol. 21. ICE, London.

Johansen, K. W. (1962). "Yield Line Theory" (transl. from Danish), Cement and Concrete Association, London.

Lin, T. Y. (1963). "Design of Prestressed Concrete Structures," Wiley, New York.

Mehta, P. (1996). "Concrete: Microstructure, Properties, and Materials," 2nd ed., Primis Custom Publishing.

Neville, A. M. (1975). "Properties of Concrete," Pitman, London.

Oehlers, D. J., and Brodford, M. A. (1995). "Composite Steel and Concrete Structural Members," Elsevier, Amsterdam.

Park, R., and Gamble, W. L. (1980). "Reinforced Concrete Slabs," Wiley, New York.

Park, R., and Paulay, T. (1975). "Reinforced Concrete Structures," Wiley, New York.

Radomski, W. (2000). "Bridge Rehabilitation," World Scientific, Singapore.

Tanabe, T. (1999). "Comporative Performances of Seismic Design Codes for Concrete Structures," 2 vols., Elsevier, Amsterdam.

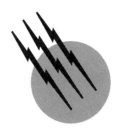

Constitutive Models for Engineering Materials

Kaspar J. Willam

University of Colorado at Boulder

GLOSSARY

Elastic damage mechanics Degradation of elastic stiffness properties due to progressive damage.

Elasticity Linear and nonlinear stress–strain relations preserving reversibility in the small (hypoelasticity) and in the large (hyperelasticity).

Failure analysis Loss of stability, loss of uniqueness, and loss of ellipticity (localization) at the constitutive level.

Plasticity Yield constraint of stress (strain, energy) trajectory leading to irreversible deformations due to plastic flow.

Triaxial constitutive models Algebraic, differential, and integral equations relating three-dimensional stress and strain tensors.

CONSTITUTIVE MODELS describe the response behavior of natural and manufactured materials under different mechanical and environmental conditions. The focus of this overview is the mechanical performance of engineering materials expressed in terms of stress, strain, and internal state variables which describe the effect of the previous load history on the current properties. Thereby, the domain of traditional constitutive models encompasses the continuum concepts of elasticity, plasticity, viscosity, and their extension to include thermal and other environmental effects through continuum thermodynamics.

The constitutive model introduces or describes the physical properties of a given material. It connects the kinematic with the kinetic descriptions of motion, thus closing the formulation of the initial boundary value problem. In this context it is important to keep in mind the needs of modern computational analysis techniques, such as the finite element method, which demand triaxial constitutive models for realistic model-based simulations. In fact, it is the range of the underlying material models which delimits the predictive value of large-scale simulations nowadays involving thousands of degrees of freedom. In short, our survey will concentrate on continuum-based material formulations currently used in engineering practice, in research and education, as well as in commercial finite element software packages for stress and deformation analyses. Primary applications cover life-cycle performance assessment of civil, mechanical, and aeronautical structures ranging from dams, bridges, and containment vessels to air- and spacecraft components as well as automotive crash simulations and micro-electro-mechanical systems.

I. INTRODUCTORY REMARKS

Traditionally, "material science" studies the behavior of materials at different scales in order to observe and quantify the chemo-physical processes at the underlying micromechanical, molecular, and atomistic levels. Multiscale material modeling upscales these processes onto the macroscopic level. The first step is to decompose the entire range of scales into several subranges. From the viewpoint of characterizing and designing engineering materials, we distinguish among the four scales illustrated in Fig. 1:

- *Meter level*: Practical problems in civil, mechanical, and aerospace structures such as the analysis and design of dams and containment vessels are solved at this level.
- *Millimeter level*: Most material properties are obtained from laboratory specimens at this level. In our terminology, this constitutes the macroscale—a level at which engineering materials may be treated as homogeneous continua after homogenizing the effect of the microstructural constituents into so-called *effective* properties.
- *Micrometer level*: Microstructural features such as microdefects, the grain size of polycrystals, and hydra-

tion products in cement-based materials are observed at this scale. In current terminology this constitutes the mesoscale—a level at which materials may be treated as heterogeneous composites (e.g., metal matrix and concrete composites) in which particle inclusions are bonded to the matrix by cohesive/frictional interface layers.

- *Nanometer level*: Molecular and atomistic processes take place at this level, which includes the molecular chaining of polymers and the behavior of single crystals. Many diffusion mechanisms for moisture and aggressive chemicals are considered to be active at this level, which includes, for example, the transport process of ions and chemical compounds. It should be noted that cause–effect relations in many cases reach beyond Newtonian mechanics, especially when subatomic processes are considered at the level of quantum mechanics.

Historically, the current thinking of materials dates back to the "corpuscular" natural philosophy advanced by René Descartes (1596–1650), who postulated that the properties of matter emerge from multilevel microstructures comprised of molecules and voids. As early as 1722, René de Réaumur developed a fairly realistic picture of the anatomy and multilevel morphology of quench-hardened steel in his treatise on iron at different length scales. On a separate line of materials engineering, Joseph Aspdin patented in 1824 hydraulic Portland cement for manufacturing mortar and concrete. In the historical context, the great challenge is to bridge the gap between the atomistic thinking of the classical Greek school of natural philosophers around Democritus (460–370 BC), and the continuum world of differential calculus by Isaac Newton (1642–1727) and Gottfried Wilhelm Leibniz (1646–1716), which provides the framework of modern continuum mechanics founded by Augustin Cauchy (1789–1857).

Traditionally, engineering materials are considered to be macroscopically homogeneous (and often isotropic). While in most applications this approach may be adequate, progressive degradation processes can only be explained properly by considering microstructural features of the material. This requires characterization of each constituent and the interface bond conditions, in addition to the morphology of the specific meso- and microstructures, respectively. These types of studies are still very demanding in terms of manpower and computing power, in spite of the rapid development of computer simulations to investigate degradation processes at the micro-, meso-, and macrolevels in three-dimensional space and time.

In the long history of alchemy an ancient dream is to create precious materials by synthesizing and transmuting low-cost constituents through innovative chemo-physical processes. Translated into the world of high-tech materials

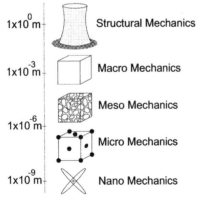

FIGURE 1 Multiscale material mechanics.

of today, this endeavor is at the core of "materials engineering," where optimization and material systems engineering meet well-defined performance objectives. Consequently, the design of new high-performance materials requires thorough understanding of the properties of each constituent and their interaction in chemo-physical processes. Thus, the role of constitutive relations is to quantify the performance of the resulting material compounds under mechanical and environmental load histories. For engineering purposes the length scale under consideration is the macroscale of traditional material testing laboratories, which provide the test data to calibrate the material parameters of the constitutive relation according to specifications of the American Society for Testing and Materials (ASTM) or other professional engineering societies.

In what follows, we will review established constitutive models for engineering materials, which may be broadly classified into three groups depending on their mathematical constructs:

1. Algebraic constitutive relations, such as linear and nonlinear elastic models
2. Differential constitutive relations, such as plastic and viscoplastic models
3. Integral constitutive relations, such as hereditary viscoelastic models

It is understood that the constitutive models are linear for performance analyses under service load conditions and highly nonlinear for limit state studies, when material failure is considered in the form of discontinuous fracture processes. For the sake of simplicity, this review will be restricted to small strains and local constitutive models in order to confine our attention to the interrelationship of stresses and strains, which are symmetric second-order tensors.

The review is organized as follows: Section II reviews primarily isotropic elastic models and examines three families of nonlinear elastic constitutive formulations using representation theorems of scalar and tensor functions. Section III summarizes the flow theory of elastoplasticity with a dicussion of volumetric–deviatoric coupling in the case of one-, two-, and three-invariant formulations. Section IV examines material failure at the constitutive level and introduces localization analysis with an application for loading in simple shear. Section V briefly reviews recent developments in continuum damage mechanics. Section VI concludes the state-of-the art report with seminal remarks on current research activities. Appendix 1 summarizes a few results of tensor algebra, while Appendix 2 contains some remarks on elastoplastic failure analysis subject to the kinematic constraint of plane strain.

II. ELASTIC MODELS

Linear elasticity is the main staple of material models in solids and structures. The statement "ut tensio sic vis," attributed to Robert Hooke (1635–1703), characterizes the behavior of a linear spring in which the deformations increase proportionally with the applied forces according to the anagram "ceiiinossstuv." The original format of Hooke's law included the geometric properties of the wire test specimens; therefore, the spring constant did exhibit a pronounced size effect. The definition of the modulus of elasticity E, where

$$\sigma = E\epsilon \qquad (1)$$

is attributed to Thomas Young (1773–1829). He expressed the proportional material behavior through the notion of a normalized force density and a normalized deformation measure, though the original formulation also did not entirely eliminate the size effect.

The tensorial character of stress was established by Cauchy, who defined the triaxial state of stress by three traction vectors using the celebrated tetraeder argument of equilibrium. The state of stress is described in terms of Cartesian coordinates by the second-order tensor:

$$\sigma(x, t) = \begin{bmatrix} \sigma_{11} & \sigma_{12} & \sigma_{13} \\ \sigma_{21} & \sigma_{22} & \sigma_{23} \\ \sigma_{31} & \sigma_{32} & \sigma_{33} \end{bmatrix} \qquad (2)$$

The conjugate state of strain is a second-order tensor with Cartesian coordinates,

$$\epsilon(x, t) = \begin{bmatrix} \epsilon_{11} & \epsilon_{12} & \epsilon_{13} \\ \epsilon_{21} & \epsilon_{22} & \epsilon_{23} \\ \epsilon_{31} & \epsilon_{32} & \epsilon_{33} \end{bmatrix} \qquad (3)$$

which is normally expressed in terms of the symmetric part of the displacement gradient, if we restrict our attention to infinitesimal deformations. In the case of nonpolar media we may confine our attention to stress measures, which are symmetric according to the axiom of L. Boltzmann,

$$\sigma = \sigma^t \quad \text{or} \quad \sigma_{ij} = \sigma_{ji} \qquad (4)$$

and the conjugate strain measures,

$$\epsilon = \epsilon^t \quad \text{or} \quad \epsilon_{ij} = \epsilon_{ji} \qquad (5)$$

where $i = 1, 2, 3$ and $j = 1, 2, 3$. As a result, the eigenvalues are real-valued and constitute the set of principal stresses and strains with zero shear components in the principal eigendirections of the second-order tensor. In contrast, nonsymmetric stress and strain measures may exhibit complex conjugate principal values and maximuum normal stress and strain components in directions with nonzero shear components characteristic for micropolar Cosserat continua.

Restricting this exposition to symmetric stress and strain tensors, they may be cast into vector form using the Voigt notation of crystal physics:

$$\boldsymbol{\sigma}(\boldsymbol{x}, t) = [\sigma_{11} \quad \sigma_{22} \quad \sigma_{33} \quad \tau_{12} \quad \tau_{23} \quad \tau_{31}]^t \quad (6)$$

and

$$\boldsymbol{\epsilon}(\boldsymbol{x}, t) = [\epsilon_{11} \quad \epsilon_{22} \quad \epsilon_{33} \quad \gamma_{12} \quad \gamma_{23} \quad \gamma_{31}]^t \quad (7)$$

where $\tau_{ij} = \sigma_{ij}, \gamma_{ij} = 2\epsilon_{ij}, \forall i \neq j$. The vector form of stress and strain will allow us to formulate material models in matrix notation used predominantly in engineering, (some of the properties of second-order tensors and basic tensor operations are expanded in Appendix 1).

A. Linear Elastic Material Behavior

Generalization of the scalar format of Hooke's law is based on the notion that the triaxial state of stress is proportional to the triaxial state of strain through the linear transformation,

$$\boldsymbol{\sigma} = \mathcal{E} : \boldsymbol{\epsilon} \quad \text{or} \quad \sigma_{ij} = \mathcal{E}_{ijkl}\epsilon_{kl} \quad (8)$$

Considering the symmetry of the stress and strain, the elasticity tensor involves in general 36 elastic moduli which may be further reduced to 21 elastic constants if we invoke major symmetry of the elasticity tensor; that is,

$$\mathcal{E} = \mathcal{E}^t \quad \text{or} \quad \mathcal{E}_{ijkl} = \mathcal{E}_{klij}$$

$$\text{with} \quad \mathcal{E}_{ijkl} = \mathcal{E}_{ijlk} \quad \text{and} \quad \mathcal{E}_{ijkl} = \mathcal{E}_{jikl} \quad (9)$$

The task of identifying 21 elastic moduli is simplified if we consider specific classes of symmetry, thus orthotropic elasticity involves nine, and transversely anisotropic elasticity five elastic moduli.

1. Isotropic Linear Elasticity

In the case of isotropy, the fourth-order elasticity tensor has the most general representation,

$$\mathcal{E} = a_o \mathbf{1} \otimes \mathbf{1} + a_1 \mathbf{1} \bar{\otimes} \mathbf{1} + a_2 \mathbf{1} \underline{\otimes} \mathbf{1}$$

$$\text{or} \quad \mathcal{E}_{ijkl} = a_o \delta_{ij}\delta_{kl} + a_1 \delta_{ik}\delta_{jl} + a_2 \delta_{il}\delta_{jk} \quad (10)$$

where $\mathbf{1} = [\delta_{ij}]$ stands for the second-order unit tensor. The three-parameter expression may be recast in terms of symmetric and skew symmetric fourth-order tensor components as:

$$\mathcal{E} = a_o \mathbf{1} \otimes \mathbf{1} + b_1 \mathcal{I} + b_2 \mathcal{I}^{skew} \quad (11)$$

where the symmetric fourth-order unit tensor reads:

$$\mathcal{I} = \frac{1}{2}[\mathbf{1} \bar{\otimes} \mathbf{1} + \mathbf{1} \underline{\otimes} \mathbf{1}] \quad \text{or} \quad \mathcal{I}_{ijkl} = \frac{1}{2}[\delta_{ik}\delta_{jl} + \delta_{il}\delta_{jk}] \quad (12)$$

and the skewed symmetric one,

$$\mathcal{I}^{skew} = \frac{1}{2}[\mathbf{1} \bar{\otimes} \mathbf{1} - \mathbf{1} \underline{\otimes} \mathbf{1}] \quad \text{or} \quad \mathcal{I}^{skew}_{ijkl} = \frac{1}{2}[\delta_{ik}\delta_{jl} - \delta_{il}\delta_{jk}] \quad (13)$$

Because of the symmetry of stress and strain the skewed symmetric contribution is inactive, $b_2 = 0$, and the isotropic linear elasticity of the material behavior is fully described by two independent elastic constants. In short, the fourth-order material stiffness tensor reduces to

$$\mathcal{E} = \Lambda \mathbf{1} \otimes \mathbf{1} + 2G\mathcal{I}$$

$$\text{or} \quad \mathcal{E}_{ijkl} = \Lambda \delta_{ij}\delta_{kl} + G[\delta_{ik}\delta_{jl} + \delta_{il}\delta_{jk}] \quad (14)$$

where the two elastic constants Λ and G are named after Gabriel Lamé (1795–1870).

$$\Lambda = \frac{E\nu}{[1 + \nu][1 - 2\nu]} \quad (15)$$

denotes the cross modulus, and

$$G = \frac{E}{2[1 + \nu]} \quad (16)$$

designates the shear modulus which have a one-to-one relationship with the modulus of elasticity and Poisson's ratio, E, ν. In the absence of initial stresses and initial strains due to environmental effects, the linear elastic relation reduces to

$$\boldsymbol{\sigma} = \Lambda[tr\boldsymbol{\epsilon}]\mathbf{1} + 2G\boldsymbol{\epsilon} \quad \text{or} \quad \sigma_{ij} = \Lambda\epsilon_{kk}\delta_{ij} + 2G\epsilon_{ij} \quad (17)$$

Here, the trace operation is the sum of the diagonal entries of the second-order tensor corresponding to double contraction with the identity tensor $tr\boldsymbol{\epsilon} = \epsilon_{kk} = \mathbf{1} : \boldsymbol{\epsilon}$.

2. Matrix Form of Elastic Stiffness ($\boldsymbol{\sigma} = E\boldsymbol{\epsilon}$)

Isotropic linear elastic behavior may be cast in matrix format using the Voigt notation of symmetric stress and strain tensors and the engineering definition of shear strain, $\gamma_{ij} = 2\epsilon_{ij}$. The elastic stiffness matrix may be written for isotropic behavior as:

$$E = \begin{bmatrix} \Lambda + 2G & \Lambda & \Lambda & & & \\ \Lambda & \Lambda + 2G & \Lambda & & 0 & \\ \Lambda & \Lambda & \Lambda + 2G & & & \\ \hline & & & G & & \\ & 0 & & & G & \\ & & & & & G \end{bmatrix} \quad (18)$$

3. Matrix Form of Elastic Compliance ($\epsilon = C\sigma$)

In the isotropic case, the normal stress σ_{11} gives rise to three normal strain contributions, the direct strain $\epsilon_{11} = \frac{1}{E}\sigma_{11}$ and the normal strains $\epsilon_{22} = -\frac{\nu}{E}\sigma_{11}$ and $\epsilon_{33} = \frac{\nu}{E}\sigma_{11}$ because of the cross effect attributed to Siméon Denis Poisson (1781–1840). Using the principle of superposition, the additional strain contributions due to σ_{22} and σ_{33} enter the compliance relation for isotropic elasticity in matrix format:

$$\begin{bmatrix} \epsilon_{11} \\ \epsilon_{22} \\ \epsilon_{33} \end{bmatrix} = \frac{1}{E} \begin{bmatrix} 1 & -\nu & -\nu \\ -\nu & 1 & -\nu \\ -\nu & -\nu & 1 \end{bmatrix} \begin{bmatrix} \sigma_{11} \\ \sigma_{22} \\ \sigma_{33} \end{bmatrix} \quad (19)$$

In the isotropic case, the shear response is entirely decoupled from the direct response of the normal components. Thus, the compliance matrix expands into the partitioned form:

$$C = \frac{1}{E} \left[\begin{array}{ccc|ccc} 1 & -\nu & -\nu & & & \\ -\nu & 1 & -\nu & & 0 & \\ -\nu & -\nu & 1 & & & \\ \hline & & & 2[1+\nu] & & \\ & 0 & & & 2[1+\nu] & \\ & & & & & 2[1+\nu] \end{array} \right] \quad (20)$$

where isotropy entirely decouples the shear response from the normal stress–strain response. This cross effect of Poisson is illustrated in Fig. 2, which shows the interaction of lateral and axial deformations under axial compression. It is intriguing that in his original work a value of $\nu = 0.25$ was proposed by Poisson based on molecular considerations. The elastic compliance relation reads in direct and indicial notations:

$$\mathcal{C} = -\frac{\nu}{E}\mathbf{1} \otimes \mathbf{1} + \frac{1}{2G}\mathcal{I}$$

$$\text{or} \quad C_{ijkl} = -\frac{\nu}{E}\delta_{ij}\delta_{kl} + \frac{1+\nu}{2E}[\delta_{ik}\delta_{jl} + \delta_{il}\delta_{jk}] \quad (21)$$

4. Canonical Format of Isotropic Elasticity

Decomposing the stress and strain tensors into spherical and deviatoric components,

$$s = \sigma - \sigma_{vol}\mathbf{1} \quad \text{where} \quad \sigma_{vol} = \frac{1}{3}[tr\sigma] \quad (22)$$

$$e = \epsilon - \epsilon_{vol}\mathbf{1} \quad \text{where} \quad \epsilon_{vol} = \frac{1}{3}[tr\epsilon] \quad (23)$$

leads to the stress deviator

$$s(x, t) = \left[\begin{array}{ccc} \dfrac{2\sigma_{11} - \sigma_{22} - \sigma_{33}}{3} & \sigma_{12} & \sigma_{13} \\ \sigma_{21} & \dfrac{2\sigma_{22} - \sigma_{33} - \sigma_{11}}{3} & \sigma_{23} \\ \sigma_{31} & \sigma_{32} & \dfrac{2\sigma_{33} - \sigma_{11} - \sigma_{22}}{3} \end{array} \right] \quad (24)$$

and the strain deviator

$$e(x, t) = \left[\begin{array}{ccc} \dfrac{2\epsilon_{11} - \epsilon_{22} - \epsilon_{33}}{3} & \epsilon_{12} & \epsilon_{13} \\ \epsilon_{21} & \dfrac{2\epsilon_{22} - \epsilon_{33} - \epsilon_{11}}{3} & \epsilon_{23} \\ \epsilon_{31} & \epsilon_{32} & \dfrac{2\epsilon_{33} - \epsilon_{11} - \epsilon_{22}}{3} \end{array} \right] \quad (25)$$

which have the property $trs = 0$ and $tre = 0$. The decomposition decouples the volumetric from the distortional response, because of the underlying orthogonality of the spherical and deviatoric partitions, $s : [\sigma_{vol}\mathbf{1}] = 0$ and $e : [\epsilon_{vol}\mathbf{1}] = 0$. The decoupled response reduces the elasticity tensor to the scalar form,

$$\sigma_{vol} = 3K\epsilon_{vol} \quad \text{and} \quad s = 2Ge \quad (26)$$

in which the bulk and the shear moduli,

$$K = \frac{E}{3[1-2\nu]} = \Lambda + \frac{2}{3}G$$

$$\text{and} \quad G = \frac{E}{2[1+\nu]} = \frac{3}{2}[K - \Lambda] \quad (27)$$

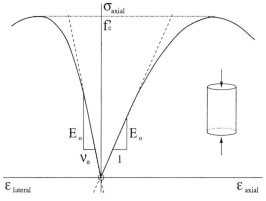

FIGURE 2 Poisson effect in axial compression.

define the volumetric and the deviatoric material stiffness. Consequently, the internal strain-energy density expands into the canonical form of two decoupled contributions,

$$2W = \sigma : \epsilon = [\sigma_{vol}\mathbf{1}] : [\epsilon_{vol}\mathbf{1}] + s : e = 9K\epsilon_{vol}^2 + 2Ge : e \tag{28}$$

such that the positive strain energy argument delimits the range of possible values of Poisson's ratio to $-1 \le \nu \le 0.5$.

5. Isotropic Elasticity under Initial Volumetric Strain

In the case of isotropic material behavior, with no directional properties, the size of a representative volume element may change due to thermal effects or shrinkage and swelling, but it will not distort. Consequently, the expansion is purely volumetric (i.e., identical in all directions). Using direct and indicial notation, the additive decomposition of strain into elastic and initial volumetric components, $\epsilon = \epsilon_e + \epsilon_o$, leads to the following extension of the elastic compliance relation:

$$\epsilon = -\frac{\nu}{E}[tr\sigma]\mathbf{1} + \frac{1}{2G}\sigma + \epsilon_o\mathbf{1}$$

$$\text{or} \quad \epsilon_{ij} = -\frac{\nu}{E}\sigma_{kk}\delta_{ij} + \frac{1}{2G}\sigma_{ij} + \epsilon_o\delta_{ij} \tag{29}$$

where $\epsilon_o = \epsilon_o\mathbf{1}$ denotes the initial volumetric strain (e.g., due to thermal expansion). The inverse relation reads:

$$\sigma = \Lambda[tr\epsilon]\mathbf{1} + 2G\epsilon - 3\epsilon_o K\mathbf{1}$$

$$\text{or} \quad \sigma_{ij} = \Lambda\epsilon_{kk}\delta_{ij} + 2G\epsilon_{ij} - 3\epsilon_o K\delta_{ij} \tag{30}$$

Rewriting this equation in matrix notation, we have:

$$\begin{bmatrix} \sigma_{11} \\ \sigma_{22} \\ \sigma_{33} \end{bmatrix} = \begin{bmatrix} K+\frac{4}{3}G & K-\frac{2}{3}G & K-\frac{2}{3}G \\ K-\frac{2}{3}G & K+\frac{4}{3}G & K-\frac{2}{3}G \\ K-\frac{2}{3}G & K-\frac{2}{3}G & K+\frac{4}{3}G \end{bmatrix} \begin{bmatrix} \epsilon_{11} \\ \epsilon_{22} \\ \epsilon_{33} \end{bmatrix}$$

$$-3K\epsilon_o \begin{bmatrix} 1 \\ 1 \\ 1 \end{bmatrix} \tag{31}$$

Considering the special case of plane stress, $\sigma_{33}=0$, the stress–strain relations reduce in the presence of initial volumetric strains to:

$$\begin{bmatrix} \sigma_{11} \\ \sigma_{22} \end{bmatrix} = \frac{E}{1-\nu^2}\begin{bmatrix} 1 & \nu \\ \nu & 1 \end{bmatrix}\begin{bmatrix} \epsilon_{11} \\ \epsilon_{22} \end{bmatrix} - \frac{E}{1-\nu}\epsilon_o\begin{bmatrix} 1 \\ 1 \end{bmatrix} \tag{32}$$

where the shear components are not affected by the temperature change in the case of isotropy.

6. Free Thermal Expansion

Under stress-free conditions, the thermal expansion $\epsilon_o = \alpha[T-T_o]\mathbf{1}$ leads to $\epsilon = \epsilon_o$; that is,

$$\epsilon_{11} = \alpha[T - T_o] \tag{33}$$

$$\epsilon_{22} = \alpha[T - T_o] \tag{34}$$

$$\epsilon_{33} = \alpha[T - T_o] \tag{35}$$

Thus, the change of temperature results in free thermal expansion, while the mechanical stress remains zero under zero confinement, $\sigma = \mathcal{E} : \epsilon_e = \mathbf{0}$.

7. Thermal Stress under Full Confinement

In contrast to the unconfined situation above, the thermal expansion is equal and opposite to the elastic strain $\epsilon_e = -\epsilon_o$ under full confinement, when $\epsilon = 0$. In the case of plane stress, the temperature change $\Delta T = T - T_o$ leads to the thermal stresses:

$$\sigma_{11} = -\frac{E}{1-\nu}\alpha[T - T_o] \tag{36}$$

$$\sigma_{22} = -\frac{E}{1-\nu}\alpha[T - T_o] \tag{37}$$

B. Nonlinear Elasticity

In linear elasticity, the resulting constitutive relations do not depend on the point of departure, except for the symmetry argument, which is derived most naturally from the energy argument. In contrast, nonlinear elastic material models strongly depend on the setting of the constitutive formulation. Considering for a moment the uniaxial stress–strain relation shown in Fig. 3, it appears that the nonlinear material law may be formulated in terms of any one of the following three possibilities:

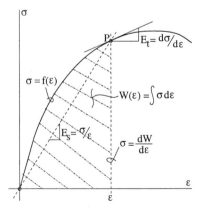

FIGURE 3 Conceptual material models of nonlinear elasticity.

1. *Algebraic format* ($\sigma = f(\epsilon)$): The algebraic format extends Hooke's law $\sigma = \mathcal{E} : \epsilon$ into the nonlinear regime. The triaxial generalization of the scalar-valued algebraic function leads in the simplest case to the concept of a secant stiffness, $\sigma = E_s : \epsilon$. In the triaxial situation, this leads to the question of how to represent a symmetric tensor-valued function of the symmetric strain tensor, $\sigma = f(\epsilon)$. This type of constitutive model is limited to rate- and history-independent material behavior, such as linear and nonlinear elasticity, which do not exhibit hysteretic effects.

2. *Integral format* ($\sigma = \frac{\partial W}{\partial \epsilon}$): The integral format is the repository of functional representations which include linear and nonlinear viscoelasticity in which creep and relaxation are considered in the form of fading memory effects, $\sigma(t) = \mathcal{F}_{t_o \leq \tau \leq t} \frac{\partial W(\epsilon(\tau), q)}{\partial \epsilon}$. As long as the rate sensitivity does not include any history effects, $q = 0$, the functional representation does not exhibit hysteretic effects. In this case and in the absence of rate dependence, we recover the format of instantaneous hyperelasticity, in which the stress is derived from the gradient of an elastic potential, $\sigma = \frac{\partial W}{\partial \epsilon}$.

3. *Differential format* ($d\sigma = E_t : d\epsilon$): In its elementary format the differential form reduces to the tangential stress–strain relationship. In a broader sense it provides also a repository for internal variables, which memorize inelastic changes of the material properties in plasticity and viscoplasticity. In this case, hysteretic effects are included to account for material dissipation and energy release in the case of fracture. Restricting our attention to nonlinear hypolasticity, the differential format leads to the question of how to represent the stress rate in terms of a tensor-valued function of two symmetric tensors, $\dot{\sigma} = g(\sigma, \dot{\epsilon})$.

In what follows we will see that the three versions of nonlinear elasticity lead to constitutive formulations which exhibit fundamental differences when we consider triaxial conditions.

1. Algebraic Format

The so-called "total" stress–strain relation expresses stress in terms of a nonlinear function of strain. As an alternative, the secant relation, $\sigma = \mathcal{E}_s : \epsilon$ provides a pseudolinear format, which simply redefines the nonlinear function $f(\epsilon)$ in terms of the nonlinear secant modulus $\mathcal{E}_s$.

The triaxial setting, however, leads to two algebraic formats of nonlinear elasticity which differ in a fundamental manner:

a. Cauchy elasticity ($\sigma = f(\epsilon)$). In this case, the triaxial state of stress is a nonlinear tensor function of the strain tensor—in indicial notation, $\sigma_{ij} = f_{ij}(\epsilon_{kl})$. Using the representation theorems of second-order symmetric tensor functions the possibilities are restricted to a small set of possible choices when isotropy is invoked. In this case, the most general format of Cauchy elasticity may have one of the two representations,

$$\sigma = \Phi_1 \mathbf{1} + \Phi_2 \epsilon + \Phi_3 \epsilon^2 \ \text{ or } \ \sigma = \Psi_1 \mathbf{1} + \Psi_2 \epsilon + \Psi_3 \epsilon^{-1} \tag{38}$$

because of the Cayleigh–Hamilton theorem. Thus, the three response functions Φ_i and Ψ_i are scalar functions of the three invariants of either stress or strain. It is important to keep in mind that Cauchy elasticity is based on a second-order symmetric tensor function of a second-order symmetric tensor.

b. Secant or pseudoelasticity ($\sigma = \mathcal{E}_s : \epsilon$). In the pseudoelastic format of elasticity, the nonlinearity is incorporated into the fourth-order secant stiffness tensor. This format has been used early on in different engineering disciplines to develop nonlinear extensions of simple classes of linear elasticity. The so-called variable stiffness models retain the format of linear elasticity and simply replace the two elastic constants of linear elasticity by nonlinear functions. We will see later on that the nonlinear K–G models are theoretically sound if the nonlinear response decouples the volumetric from the deviatoric response; that is, $K = K(\epsilon_{vol})$ and $G = G(tre^2)$ where $tre^2 = e : e$.

It is not very surprising that elastic damage models did start from this secant format of nonlinear elasticity using arguments of effective stiffness properties, which are in some sense equivalent to distributed micromechanical defects. In fact, the original proposal of scalar damage was nothing but a nonlinear pseudoelastic model in which the secant stiffness is in matrix notation:

$$\sigma = E_s \epsilon \quad \text{where} \quad E_s = [1 - d]E_o \quad \text{and} \quad d = 1 - \frac{E_s}{E_o} \tag{39}$$

In its basic format, the secant matrix of elastic damage retains the structure of the initial elastic stiffness except for the factor $[1 - d]$:

$$E_s = \frac{[1 - d]E_o}{[1 + v_o][1 - 2v_o]}$$

$$\times \begin{bmatrix} 1 - v_o & v_o & v_o & & & \\ v_o & 1 - v_o & v_o & & \mathbf{0} & \\ v_o & v_o & 1 - v_o & & & \\ \hline & & & \frac{1 - 2v_o}{2} & & \\ & \mathbf{0} & & & \frac{1 - 2v_o}{2} & \\ & & & & & \frac{1 - 2v_o}{2} \end{bmatrix}$$

$$\tag{40}$$

which measures the remaining integrity of the material when the damage variable increases from zero to one, $0 \le d \to 1$. Restricting damage to the format of isotropic linear elasticity, it is natural to decompose degradation into volumetric and deviatoric damage components such that:

$$K_s = [1 - d_{vol}]K_o \quad \text{and} \quad G_s = [1 - d_{dev}]G_o \quad (41)$$

The corresponding secant relation of elastic damage reads in matrix notation

$$E_s = \begin{bmatrix} K_s + \frac{4}{3}G_s & K_s - \frac{2}{3}G_s & K_s - \frac{2}{3}G_s & & \\ K_s - \frac{2}{3}G_s & K_s + \frac{4}{3}G_s & K_s - \frac{2}{3}G_s & & 0 \\ K_s - \frac{2}{3}G_s & K_s - \frac{2}{3}G_s & K_s + \frac{4}{3}G_s & & \\ \hline & & & G_s & \\ & 0 & & & G_s & \\ & & & & & G_s \end{bmatrix}$$

$$(42)$$

From this expression we observe that the secant format of isotropic elastic damage is very simple because of the underlying decoupling of volumetric from deviatoric degradation. As long as we are only interested in a given state of damage, this isotropic pseudoelastic formulation suffices to describe the response behavior using effective material properties based on the principle of stress or strain equivalence. However, the constitutive formulation becomes far more intricate when a thermodynamically consistent damage formulation is needed for progressive damage simulations. In this case, loading and unloading have to be defined for the general case of triaxial conditions, and evolution laws are required to describe the two independent processes of volumetric and deviatoric damage which will be described in Section 5 in more detail.

2. Integral Format

The integral description of elasticity starts from the postulate of a strain energy function from which the stress is derived by differentiation. This hyperelastic format of elasticity dates back to the original work of George Green (1793–1841).

a. Green elasticity ($\sigma = \frac{\partial W}{\partial \epsilon}$). Given a strain energy density potential, $W = W(\epsilon)$, the hyperelastic stress–strain relation is simply the gradient of that potential function with respect to strain,

$$\sigma = \frac{\partial W}{\partial \epsilon} \quad \text{or} \quad \sigma_{ij} = \frac{\partial W}{\partial \epsilon_{ij}} \quad (43)$$

The compliance relation derives from the dual complementary strain energy potential. The important aspect of this integral formulation is the path-independence of the line integral which defines the internal strain energy,

$$W(\epsilon) = \int_\epsilon \sigma : d\epsilon = \int_\epsilon \frac{\partial W}{\partial \epsilon} : d\epsilon = \int_\epsilon dW \quad (44)$$

in terms of the total differential dW. The traditional notions of elasticity, such as reversibility and lack of energy dissipation under closed cycles of strain, are a direct consequence of path independence; that is,

$$W(\epsilon) = \oint_\epsilon dW = 0 \quad (45)$$

In other terms, the hyperelastic material description is nondissipative and preserves energy under arbitrary strain histories.

The corresponding hyperelastic stiffness tensor is a measure of the curvature of the strain energy function involving the second derivatives of $W = W(\epsilon)$,

$$\dot{\sigma} = \mathcal{E}_t : \dot{\epsilon} \quad \text{where} \quad \mathcal{E}_t = \frac{\partial^2 W}{\partial \epsilon \otimes \partial \epsilon} \quad (46)$$

Consequently, the elasticity tensor is symmetric, $\mathcal{E}_t = \mathcal{E}_t^t$, if $W(\epsilon)$ is sufficiently smooth. This reduces the 36 elastic constants to 21 in the general case of anisotropic elasticity, and to two in case of isotropy. The positive definiteness of the hyperelastic tangent operator, $\det \mathcal{E}_t > 0$, is directly connected to the convexity of the strain energy functional and the uniqueness argument of boundary value problems in elasticity.

b. Isotropic hyperelastic models. In the case of isotropy, frame objectivity requires that the strain energy density function must be independent from the coordinate system of an observer. Thus, the potential function must be a function of strain invariants and combinations thereof. In the past, two different families of nonlinear isotropic hyperelastic material models have been proposed; one leads to the format of nonlinear K–G models, while the other results in a pseudo-orthotropic format based on principal coordinates of stress and strain.

1. *Hyperelastic K–G models*: Starting from the invariant representation of the strain energy density function, $W(\epsilon) = W(I_1, I_2, I_3)$, the stress–strain relation follows from the chain rule of differentiation as:

$$\sigma = \frac{\partial W}{\partial \epsilon} = \frac{\partial W}{\partial I_1}\frac{\partial I_1}{\partial \epsilon} + \frac{\partial W}{\partial I_2}\frac{\partial I_2}{\partial \epsilon} + \frac{\partial W}{\partial I_3}\frac{\partial I_3}{\partial \epsilon} \quad (47)$$

Using the moment invariants for the sake of convenience, where $I_i = \frac{1}{i}tr\epsilon^i$, the derivatives simplify to $\frac{\partial I_i}{\partial \epsilon} = \epsilon^{i-1}$. Consequently, the nonlinear isotropic stress–strain relation results in the general hyperelastic format for nonlinear isotropic behavior,

$$\sigma = W_1 \mathbf{1} + W_2 \epsilon + W_3 \epsilon^2 \quad (48)$$

The three scalar functions $W_i = \frac{\partial W}{\partial I_i} = W_i(I_j)$ describe the degree of nonlinearity in terms of three irreducible invariants, while the tensorial properties are characterized by the three irreducible basis tensors $\mathbf{1}$, ϵ, ϵ^2. Note the interrelationship between the scalar functions,

$$\frac{\partial W_i}{\partial I_j} = \frac{\partial W_j}{\partial I_i} \quad \text{since} \quad \frac{\partial^2 W}{\partial I_i \partial I_j} = \frac{\partial^2 W}{\partial I_j \partial I_i} \quad (49)$$

when they are derived from a sufficiently differentiable scalar potential, $W = W(I_1, I_2, I_3)$. In contrast, the scalar functions Φ_i and Ψ_i of the pseudoelastic formulation do not satisfy this form of reciprocity if the underlying integrability conditions are not enforced. On the issue of volumetric–deviatoric coupling we observe that the trace operation leads to:

$$tr\sigma = 3W_1 + W_2 tr\epsilon + W_3 tr\epsilon^2 \quad (50)$$

Considering a simple shear deformation, with $\epsilon_{12} = 0.5\gamma$, W_3 and thus the dependence of $W = W(I_1, I_2, I_3)$ on the third invariant are responsible for volumetric-deviatoric interaction. On another note, the quadratic expansion of the strain energy density function leads exactly to the two Lamé constants of linear isotropic elasticity since $W_3 = 0$. Thus, it was Green hyperelasticity which firmly established the bimodular theory of isotropic elasticity in contrast to the unimodular theory which was widely accepted by the French school of engineering scientists in the last century. In fact, this bimodular format also holds for nonlinear hyperelasticity if the strain energy density function can be decomposed into two independent functions, one defining the volumetric and the other the deviatoric behavior:

$$W(\epsilon) = W_{vol}(tr\epsilon) + W_{dev}(tr\epsilon^2) \quad (51)$$

This infers, however, that the influence of the third invariant I_3 remains negligible, since it is this term which is responsible for coupling the volumetric and deviatoric response behavior. The decomposition of the strain energy function leads to the appealing concept of nonlinear $K–G$ models because of their inherent simplicity which retains the two-modular form of linear elasticity. Figures 4 and 5 illustrate the secant stiffness relations, which may be expressed best in the form of the octahedral components of stress σ_o, τ_o and strain ϵ_o, γ_o:

$$\begin{bmatrix} \sigma_o \\ \tau_o \end{bmatrix} = \begin{bmatrix} 3K_s & 0 \\ 0 & 2G_s \end{bmatrix} \begin{bmatrix} \epsilon_o \\ \gamma_o \end{bmatrix} \quad (52)$$

where $K_s = K(tr\epsilon)$ and $G_s = G(tr\epsilon^2)$. The so-called $K–G$ models shown in Figs. 4 and 5 play a prominent role for modeling nonlinear material behavior. Combining the volumetric and deviatoric scalar relations leads to the secant stiffness relation:

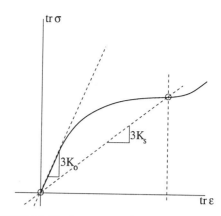

FIGURE 4 Nonlinear volumetric response behavior.

$$\sigma = \mathcal{E}_s : \epsilon \quad \text{where} \quad \mathcal{E}_s = \left[K_s - \frac{2}{3}G_s\right]\mathbf{1} \otimes \mathbf{1} + 2G_s\mathcal{I} \quad (53)$$

where the elastic material constants have been simply replaced by their secant values. The corresponding tangential stiffness relation, however, introduces an additional fourth-order tensor term because of the chain rule of differentiation when taking derivatives of $G_s = G_s(\sqrt{e : e})$,

$$\dot{\sigma} = \mathcal{E}_t : \dot{\epsilon} \quad \text{where} \quad \mathcal{E}_t = \left[K_t - \frac{2}{3}G_s\right]\mathbf{1} \otimes \mathbf{1} + 2G_s\mathcal{I} + \frac{4[G_t - G_s]}{tr e^2}e \otimes \mathbf{e} \quad (54)$$

The dyadic tensor product $e \otimes e$ originates from the nonlinearity and reflects the state of the deviator strain on the incremental material law. This term is the source of strain-induced anisotropy, which characterizes the tangential stiffness relations of all elastoplastic and elastic damage models.

2. Hyperelastic model in principal coordinates: In this case, the strain energy function is expressed in terms of the principal strain values,

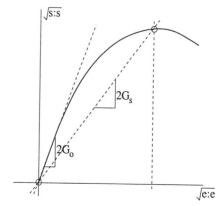

FIGURE 5 Nonlinear deviatoric response behavior.

$$W = W(\epsilon_1, \epsilon_2, \epsilon_3) = W(\epsilon_2, \epsilon_3, \epsilon_1) = W(\epsilon_3, \epsilon_1, \epsilon_2)$$
$$(55)$$

whereby the underlying isotropy infers cyclic permutation of indices. Because of the underlying coaxiality of the principal axes of stress and strain, the stress–strain relation involves only their principal values; that is,

$$\sigma_i = \frac{\partial W(\epsilon_1, \epsilon_2, \epsilon_3)}{\partial \epsilon_i} \qquad (56)$$

The tangential stress–strain relation leads to the following matrix representation of the principal coordinates:

$$\begin{bmatrix} \dot{\sigma}_1 \\ \dot{\sigma}_2 \\ \dot{\sigma}_3 \end{bmatrix} = \begin{bmatrix} \frac{\partial^2 W}{\partial \epsilon_1 \partial \epsilon_1} & \frac{\partial^2 W}{\partial \epsilon_1 \partial \epsilon_2} & \frac{\partial^2 W}{\partial \epsilon_1 \partial \epsilon_3} \\ \frac{\partial^2 W}{\partial \epsilon_2 \partial \epsilon_1} & \frac{\partial^2 W}{\partial \epsilon_2 \partial \epsilon_2} & \frac{\partial^2 W}{\partial \epsilon_2 \partial \epsilon_3} \\ \frac{\partial^2 W}{\partial \epsilon_3 \partial \epsilon_1} & \frac{\partial^2 W}{\partial \epsilon_3 \partial \epsilon_2} & \frac{\partial^2 W}{\partial \epsilon_3 \partial \epsilon_3} \end{bmatrix} \begin{bmatrix} \dot{\epsilon}_1 \\ \dot{\epsilon}_2 \\ \dot{\epsilon}_3 \end{bmatrix} \quad (57)$$

This illustrates the symmetry of the tangential stiffness properties if the strain energy function is sufficiently smooth. Moreover, the tangential stiffness properties are positive definite if the strain energy function remains convex. Although the tangential relation appears to be anisotropic in principal coordinates, the nonlinear stress–strain rate relation maintains coaxiality between $\sigma - \epsilon$ as long as the tangential shear terms satisfy the condition (Ogden, 1984):

$$\begin{bmatrix} \dot{\tau}_{12} \\ \dot{\tau}_{23} \\ \dot{\tau}_{31} \end{bmatrix} = \begin{bmatrix} \frac{\sigma_2 - \sigma_1}{\epsilon_2 - \epsilon_1} & 0 & 0 \\ 0 & \frac{\sigma_3 - \sigma_2}{\epsilon_3 - \epsilon_2} & 0 \\ 0 & 0 & \frac{\sigma_1 - \sigma_3}{\epsilon_1 - \epsilon_3} \end{bmatrix} \begin{bmatrix} \dot{\gamma}_{12} \\ \dot{\gamma}_{23} \\ \dot{\gamma}_{31} \end{bmatrix} \quad (58)$$

In the past, the principal coordinate representation of nonlinear stress–strain relations has been popularized under the name of "orthotropic" material models (see Chen and Han, 1988), though strictly speaking the constitutive format is isotropic in which the strain-induced orthotropy appears because of the different levels of nonlinearity in the principal coordinates.

3. Differential Format

The differential description of elasticity starts from representation theorems of nonlinear tensor functions which lead in the simplest case of grade-one materials to an incrementally linear format of tangential stiffness. The hypoelastic terminology is attributed to Clifford Truesdell (1919–2000), although the linear format of incremental elasticity is normally referred to as the tangential stiffness model.

a. Truesdell elasticity ($\dot{\sigma} = g(\sigma, \dot{\epsilon})$). In the differential format of elasticity, the stress rate is expanded into

a symmetric tensor function of two second-order tensors. In the case of a stress-based formulation, the two independent arguments of the tensor function are the stress and the rate of strain tensors. Invoking the argument of material objectivity and frame indifference, the irreducible set of base tensors encompasses the following terms:

$$\mathbf{1}, \sigma, \sigma^2, \dot{\epsilon}, \dot{\epsilon}^2, (\sigma \cdot \dot{\epsilon} + \dot{\epsilon} \cdot \sigma), (\sigma \cdot \dot{\epsilon}^2 + \dot{\epsilon}^2 \cdot \sigma),$$
$$(\sigma^2 \cdot \dot{\epsilon} + \dot{\epsilon} \cdot \sigma^2), (\sigma^2 \cdot \dot{\epsilon}^2 + \dot{\epsilon}^2 \cdot \sigma^2) \qquad (59)$$

Conseqently, an isotropic tensor function of two symmetric tensors involves in the most general case nine response functions, $\phi_1, \ldots, \phi_9$, which depend in turn on the six moment invariants of the stress and strain rate tensors below,

$$\bar{I}_1^\sigma = tr\sigma, \qquad \bar{I}_2^\sigma = tr\sigma^2, \qquad \bar{I}_3^\sigma = tr\sigma^3;$$
$$\bar{I}_1^{\dot{\epsilon}} = tr\dot{\epsilon}, \qquad \bar{I}_2^{\dot{\epsilon}} = tr\dot{\epsilon}^2, \qquad \bar{I}_3^{\dot{\epsilon}} = tr\dot{\epsilon}^3 \qquad (60)$$

as well as on the four joint invariants:

$$\bar{J}_1 = tr(\sigma \cdot \dot{\epsilon}), \qquad \bar{J}_2 = tr(\sigma^2 \cdot \dot{\epsilon}),$$
$$\bar{J}_3 = tr(\sigma \cdot \dot{\epsilon}^2), \qquad \bar{J}_4 = tr(\sigma^2 \cdot \dot{\epsilon}^2) \qquad (61)$$

The general format results in the general stress-strain rate relation,

$$\dot{\sigma} = \phi_1 \mathbf{1} + \phi_2 \sigma + \phi_3 \sigma^2 + \phi_4 \dot{\epsilon} + \phi_5 \dot{\epsilon}^2$$
$$+ \phi_6 (\sigma \cdot \dot{\epsilon} + \dot{\epsilon} \cdot \sigma) + \phi_7 (\sigma \cdot \dot{\epsilon}^2 + \dot{\epsilon}^2 \cdot \sigma)$$
$$+ \phi_8 (\sigma^2 \cdot \dot{\epsilon} + \dot{\epsilon} \cdot \sigma^2) + \phi_9 (\sigma^2 \cdot \dot{\epsilon}^2 + \dot{\epsilon}^2 \cdot \sigma^2) \quad (62)$$

For rate independence, the expansion must be homogeneous of order one, thus the rate terms of the tensor function must be restricted to the first order. In other terms, the hypoelastic material law is rate independent if and only if:

$$g(\sigma, \alpha \dot{\epsilon}) = \alpha g(\sigma, \dot{\epsilon}) \qquad (63)$$

Among the numerous possibilities, two classes of hypoelastic constitutive models may be distinguished:

1. *Incrementally linear hypoelastic models*: The linear restriction of the hypoelastic stress–strain relations leads to the classical tangential stiffness format:

$$\dot{\sigma} = \mathcal{E}_t : \dot{\epsilon} \quad \text{where} \quad \mathcal{E}_t = \mathcal{E}(\sigma) \qquad (64)$$

This stress-based format is reversible in the small, but not in the large. In other words, the classical hypoelastic formulation leads to path-dependence:

$$\sigma = \int_\epsilon \mathcal{E}_t(\sigma) : \frac{d\epsilon}{dt} dt \qquad (65)$$

This infers energy dissipation and irreversible behavior for arbitrary load histories as opposed to hyperelasticity.

In fact, the hyperelastic property of path independence is recovered only if appropriate integrability conditions are satisfied which assure that the stress is the gradient of a single potential function (i.e., $\sigma = \frac{\partial W}{\partial \epsilon}$). The most general format of the hypoelastic tangent operator involves 12 hypoelastic response functions which depend in general on all ten stress invariants, $C_i = C_i(I_j, J_k)$. The tensorial structure involves 12 fourth-order tensor products between the second-order unit tensor and stress tensors up to the fourth power:

$$\mathcal{E}_t =$$
$$
\begin{bmatrix}
C_1 \mathbf{1} \otimes \mathbf{1} & + C_2 \sigma \otimes \mathbf{1} & + C_3 \sigma^2 \otimes \mathbf{1} \\
+ C_4 \mathbf{1} \otimes \sigma & + C_5 \sigma \otimes \sigma & + C_6 \sigma^2 \otimes \sigma \\
+ C_7 \mathbf{1} \otimes \sigma^2 & + C_8 \sigma \otimes \sigma^2 & + C_9 \sigma^2 \otimes \sigma^2 \\
+ C_{10}[\mathbf{1} \bar{\otimes} \mathbf{1} + \mathbf{1} \underline{\otimes} \mathbf{1}] & + C_{11}[\sigma \bar{\otimes} \mathbf{1} + \mathbf{1} \underline{\otimes} \sigma] & + C_{12}[\sigma^2 \bar{\otimes} \mathbf{1} + \mathbf{1} \underline{\otimes} \sigma^2]
\end{bmatrix}
$$
$$(66)$$

Consequently, the tangential stiffness operator of the nonlinear K–G model in Section II.B.2 forms a very limited subclass as it activates only three out of the 12 terms in the general framework of incrementally linear hypoelasticity. Under the name of variable moduli models, a good number of simplified hypoelastic material models have been proposed and are still used in structural and geotechnical engineering.

2. Incrementally nonlinear hypoelastic models: Another rate-independent restriction leads to a class of incrementally nonlinear models which have been proposed under the name of hypoplastic models. Because of the incremental nonlinearity they are capable of distinguishing between different loading and unloading stiffness properties in analogy to the endochronic time model introduced by Valanis (1975). In the absence of a loading function, it is understood that the irreversible contribution leads to continuous energy dissipation under repeated load cycles in contrast to unload–reload cycles in elastoplasticity.

III. ELASTOPLASTIC MODELS

There are several points of departure to develop a consistent theory for elastoplastic constitutive models. Traditionally, we start from the underlying rheological model shown in Fig. 6 which lends itself to develop the elastoplastic stress-strain relations for uniaxial conditions. The basic concepts may be generalized using the internal variable theory of Coleman and Gurtin (1967) based on continuum thermodynamics which provides an elegant repository for history-dependent inelasticity. In the sequel, we will confine our discussion to rate-independent material behavior. In this case, the starting point of the internal

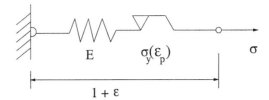

FIGURE 6 Elastoplastic serial model of spring and friction element.

variable models is the assumption that the stress tensor may be represented as a function of the current strain tensor and a finite number of parameters, $q_1, q_2, \ldots, q_n$:

$$\sigma = f(\epsilon, q_1, q_2, \ldots, q_n) \qquad (67)$$

These internal variables $q_1, q_2, \ldots, q_n$ represent the material state that depends on the process history. Their current values is defined in terms of evolution equations which form in general a system of ordinary first-order differential equations:

$$\dot{q}_i(t) = g_i(\sigma, \epsilon, q_1(t), q_2(t), \ldots, q_n(t))$$
$$\text{where} \quad i = 1, 2, \ldots, n \quad (68)$$

The evolution equations specify the temporal change of the internal variables depending on their current value and input histories either in the form of stress or strain control. For strain control, the general integral of the functional

$$q_i(t) = \mathcal{F}_{t_o \leq \tau \leq t}(\epsilon(\tau), q_1(t_o), q_2(t_o), \ldots, q_n(t_o)) \quad (69)$$

needs to be evaluated numerically. The internal variables represent the memory properties of the inelastic material behavior. Their physical meaning is to represent the dynamical process which take place because of inelasticity at the material microstructure and to make visible their macroscopic effects. In plasticity, they constitute the plastic strain tensor and the reduced variables for isotropic and kinematic hardening. In elastic damage mechanics they form the damage tensor and the scalar damage variables for volumetric and deviatoric elastic stiffness degradation. Thus, the evolution equations are additional constitutive equations which augment the stress–strain relation.

In the sequel we start from the uniaxial relations of the the rheological model and extend them to triaxial conditions using an engineering approach. For illustration we consider the one-invariant von Mises plasticity model and its extension to the two-invariant Drucker and Prager (1952) model as well as the three-invariant extension along the line of the five parameter model by Willam and Warnke (1975). We will conclude this section with the example problem of simple shear which will illustrate the dramatic effect of volumetric–deviatoric interaction on the overall response behavior.

A. Elastoplastic Rheological Model

The serial arrangement of an elastic spring and a friction element dates back to Reuss and the additive decomposition of strain into elastic and plastic components. The combination of an elastic spring and the friction element leads to two formats of elastoplasticity, one is the "deformation theory" of Hencky, and the other is the "rate theory" of Prandtl–Reuss. The deformation theory is essentially a secant-type formulation of plasticity along the nonlinear K–G model which is augmented by a plastic load-elastic unload condition. This leads to discontinuities in the transition region from elasticity to plasticity under repeated loading and unloading cycles. Thus, we will concentrate on the flow theory of plasticity, although the continuity leads to overly stiff elastic response predictions when plastic loading takes place to the side of a smooth yield surface without corners.

In small deformation problems, the flow theory is based on the additive decomposition of the total strain rate into elastic and plastic components,

$$\dot{\epsilon} = \dot{\epsilon}_e + \dot{\epsilon}_p \tag{70}$$

The serial arrangement of an elastic spring and a plastic friction element infers stress equivalence in the two elements. In other terms, the state of stress in the elastic spring is limited by the slip conditions of the friction element. The behavior in the elastic response regime $\sigma \leq \sigma_y$ is described by Hooke's law:

$$\dot{\epsilon}_e = \frac{\dot{\sigma}}{E} \tag{71}$$

while the plastic response is active when the stress reaches the yield condition, such that under persistent plastic loading when $\sigma = \sigma_y$,

$$\dot{\epsilon}_p = \frac{\dot{\sigma}}{E_p} \tag{72}$$

Consequently,

$$\dot{\epsilon} = \frac{\dot{\sigma}}{E} + \frac{\dot{\sigma}}{E_p} = \frac{\dot{\sigma}}{E_{ep}} \tag{73}$$

with the elastoplastic tangent stiffness,

$$E_{ep} = \frac{EE_p}{E + E_p} \tag{74}$$

Figure 7 illustrates the elastoplastic response for a load–unload–reload input history. Note that the tangent modulus $E_t = E_{ep}$ may range from positive to negative values

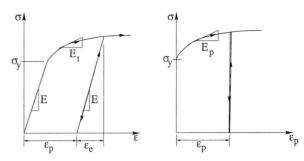

FIGURE 7 Load–unload–reload response of elastoplastic hardening solid.

according to:

$$E_{ep} \begin{cases} >0, & \text{hardening} \\ =0, & \text{perfectly plastic} \\ <0, & \text{softening} \end{cases}$$

with the understanding that a critical value for softening is reached when $E_p^{crit} = -E$.

We also note the indeterminancy of the plastic strain rate for perfectly plastic behavior, when the plastic modulus is zero, $\dot{\epsilon}_p = \frac{0}{0}$. This indicates that strain rather than stress control must be used to maintain uniqueness for elastic–perfectly plastic conditions as well as for strain softening:

$$\dot{\epsilon}_p = \frac{\dot{\sigma}}{E_p} = \frac{E}{E + E_p} \dot{\epsilon} \tag{75}$$

This assures that the plastic strain rate increases with increasing strain rate and vice versa. Moreover, the intuitive loading condition must be replaced by a more precise yield condition of the form of the scalar-valued yield function of stress:

$$F(\sigma) = |\sigma| - \sigma_y = 0 \tag{76}$$

which acts as a threshold condition when the stress demand $|\sigma|$ reaches the yield capacity of the elastoplastic material. We note that plastic loading requires that (1) the stress path has to reach the yield strength, which is here assumed to be the same in uniaxial tension and compression; and (2) the rate of strain must be such that under persistent plastic action $\epsilon \dot{\epsilon} > 0$ under strain control.

B. General Format of the Flow Theory of Elastoplasticity

Starting from the small strain setting, where the strain tensor is the symmetric part of the displacement gradient, $\epsilon = \frac{1}{2}[\nabla u + \nabla^t u]$, the serial arrangement of the rheological model in Fig. 6 decomposes the strain into the sum of elastic and plastic parts:

$$\epsilon = \epsilon_e + \epsilon_p \tag{77}$$

1. Elastic Behavior

Extending the hyperelastic concept of a strain energy potential, we start this time from the free energy function Ψ according to Helmholtz, in which the plastic strains and the internal variable κ account for the hardening behavior:

$$\Psi = \Psi(\epsilon, \epsilon_p, \kappa) \tag{78}$$

With the classical arguments of Clausius–Duhem, the inequality of the second law of thermodynamics yields the stress tensor σ as a thermodynamically conjugate variable to the elastic strains,

$$\sigma = \frac{\partial \Psi}{\partial \epsilon_e} \quad \text{and} \quad \dot{\sigma} = \mathcal{E} : [\dot{\epsilon} - \dot{\epsilon}_p] \tag{79}$$

Thereby the relation between the stress rates and the elastic strain rates is defined in terms of the fourth-order elasticity tensor $\mathcal{E}$ discussed earlier on in Section II.

2. Plastic Yield Condition

The yield function

$$F(\sigma, \kappa) = f(\sigma) - r_y(\kappa) \le 0 \quad \text{with} \quad n = \frac{\partial F}{\partial \sigma} \tag{80}$$

delimits the elastic domain with the normal n with respect to the tangent plane at the yield surface. The geometric interpretation of the yield function in the form of a yield surface helps to visualize the elastic region which may expand or shrink according to the underlying hardening/softening model. Here, the yield resistance r_y is conjugate to the internal variable for isotropic hardenening κ:

$$r_y = \frac{\partial \Psi}{\partial \kappa} \quad \text{and} \quad \dot{r}_y = H_p \dot{\kappa}, \tag{81}$$

Thus, the relation between the rate of the yield resistance and the hardening variable defines the hardening modulus H_p.

3. Plastic Flow Rule

The evolution of plastic strains is governed by the plastic flow rule,

$$\dot{\epsilon}_p = \dot{\lambda} m \quad \text{with} \quad m = \frac{\partial Q}{\partial \sigma} \tag{82}$$

whereby m denotes the normal to the plastic potential Q which differs from the yield function F in the nonassociated case. In the above equation, $\dot{\lambda}$ denotes the plastic multiplier, and $Q = Q(\sigma)$ is the plastic flow potential. We speak of associated flow when $m \parallel n$ which infers normality of plastic flow in the direction of the gradient of the yield function. In the case of nonassociated flow, $Q \ne F$, one speaks of loss of normality.

4. Plastic Consistency Condition

For plastic behavior not only must the yield condition hold ($F = 0$), but the plastic consistency condition must also be satisfied under persistent plastic action. The so-called Prager-consistency condition $\dot{F} = 0$ enforces the stress path to remain on the yield surface (at least in the differential sense).

The set of constitutive equations is completed by the Kuhn–Tucker condition for plastic loading:

$$F \le 0 \quad \dot{\lambda} \ge 0 \quad F\dot{\lambda} = 0 \tag{83}$$

Together with the plastic consistency condition,

$$\dot{F} = 0. \tag{84}$$

the plastic multiplier results in:

$$\dot{\lambda} = \frac{1}{h_p} n : \mathcal{E} : \dot{\epsilon} \quad \text{with} \quad h_p = H_p + n : \mathcal{E} : m \tag{85}$$

5. Elastoplastic Stiffness Relation

Substituting the plastic multiplier into the elastoplastic stress–strain relation yields:

$$\dot{\sigma} = \mathcal{E} : [\dot{\epsilon} - \dot{\lambda} m] = \mathcal{E} : \left[\dot{\epsilon} - m \frac{n : \mathcal{E} : \dot{\epsilon}}{H_p + n : \mathcal{E} : m} \right] \tag{86}$$

Rearranging results in the elastoplastic tangent operator $\mathcal{E}_{ep}$ which relates the stress and strain rates:

$$\dot{\sigma} = \mathcal{E}_{ep} : \dot{\epsilon} \tag{87}$$

in the form of a rank-one update of the elastic material operator,

$$\mathcal{E}_{ep} = \mathcal{E} - \frac{1}{h_p} \mathcal{E} : m \otimes n : \mathcal{E} = \mathcal{E} - \frac{1}{h_p} \bar{m} \otimes \bar{n} \tag{88}$$

where $\bar{m} = \mathcal{E} : m$ and $\bar{n} = n : \mathcal{E}$. The update notation emphasizes that the elastic reference tensor is being reduced by the plastic rank-one modification $\bar{m} \otimes \bar{n}$. Note that the critical softening modulus is reached when $h_p^{crit} = 0$ or, in other words, when $H_p^{crit} = -n : \mathcal{E} : m$. We also note the loss of symmetry (i.e., $\mathcal{E}_{ep} \ne \mathcal{E}_{ep}^t$) for nonassociated flow when $n \ne m$.

C. Special case of J_2-Plasticity

In J_2-plasticity we combine the von Mises condition of plastic yielding with the plastic flow rule and the additive decomposition of Prandtl–Reuss. In view of the deviatoric setting of plasticity we may restrict the development of J_2-plasticity to a deviatoric stress–strain relation. For the sake of simplicity we describe the resistance of the material by the yield strength in uniaxial tension, $r_y = \sigma_y$, and we assume elastic perfectly plastic behavior to eliminate the need for internal variables to describe hardening/softening.

1. *J_2-yield function*:

$$F(s) = \frac{1}{2}s : s - \frac{1}{3}\sigma_y^2 = 0 \qquad (89)$$

2. *Associated plastic flow rule*:

$$\dot{\epsilon}_p = \dot{\lambda}s \quad \text{where} \quad m = \frac{\partial F}{\partial s} = s \qquad (90)$$

3. *Plastic consistency condition*:

$$\dot{F} = \frac{\partial F}{\partial s} : \dot{s} = s : \dot{s} = 0 \qquad (91)$$

The deviatoric stress–strain rates are related by:

$$\dot{s} = 2G[\dot{e} - \dot{e}_p] = 2G[\dot{e} - \dot{\lambda}s] \qquad (92)$$

Substituting into the consistency condition,

$$\dot{F} = 2Gs : [\dot{e} - \dot{\lambda}s] = 0 \qquad (93)$$

leads to the plastic multiplier:

$$\dot{\lambda} = \frac{s : \dot{e}}{s : s} \qquad (94)$$

4. *Deviatoric stress–strain relation*: The deviatoric stress–strain rate expression may be written as:

$$\dot{s} = 2G\left[\dot{e} - \frac{s : \dot{e}}{s : s}s\right] = 2G\left[\mathcal{I} - \frac{s \otimes s}{s : s}\right] : \dot{e} \qquad (95)$$

or in short form as:

$$\dot{s} = \boldsymbol{\mathcal{G}}_{ep} : \dot{e} \quad \text{with} \quad \boldsymbol{\mathcal{G}}_{ep} = 2G\left[\mathcal{I} - \frac{s \otimes s}{s : s}\right] \qquad (96)$$

5. *Tangent stiffness operator of J_2-elastoplasticity*: Including the elastic volumetric response, $[tr\dot{\sigma}] = 3K[tr\dot{\epsilon}]$, the elastoplastic stress–strain relationship leads to:

$$\dot{\sigma} = \frac{1}{3}[tr\dot{\sigma}]\mathbf{1} + \dot{s} = K[tr\dot{\epsilon}]\mathbf{1} + \boldsymbol{\mathcal{G}}_{ep} : \dot{e} \qquad (97)$$

Considering

$$\dot{\sigma} = K[tr\dot{\epsilon}]\mathbf{1} + \boldsymbol{\mathcal{G}}_{ep} : \left[\dot{\epsilon} - \frac{1}{3}[tr\dot{\epsilon}]\mathbf{1}\right] \qquad (98)$$

the combined elastoplastic tangent relationship results in:

$$\dot{\sigma} = \boldsymbol{\mathcal{E}}_{ep} : \dot{\epsilon} \quad \text{with} \quad \boldsymbol{\mathcal{E}}_{ep} = \Lambda\mathbf{1} \otimes \mathbf{1} + 2G\left[\mathcal{I} - \frac{s \otimes s}{s : s}\right] \qquad (99)$$

which has the same constitutive structure as the nonlinear K–G model in Eq. (54).

D. Example Problem of Simple Shear

In frictional materials the so-called Reynolds effect of shear dilatancy is one of the central mechanisms responsible for the interaction of the volumetric and deviatoric components. In the case of plastic loading, the elastoplastic tangential stiffness relationship may be written as

$$\boldsymbol{\mathcal{E}}_{ep} = \boldsymbol{\mathcal{E}} - \frac{1}{h_p}\boldsymbol{\mathcal{R}}_p \quad \text{where} \quad \boldsymbol{\mathcal{R}}_p = \boldsymbol{\mathcal{E}} : m \otimes n : \boldsymbol{\mathcal{E}} \quad (100)$$

Here, the plastic dyadic tensor product $\boldsymbol{\mathcal{R}}_p$ is the sole repository of stress- or strain-induced anisotropy and for coupling between the direct and shear behavior in the case of isotropic elasticity.

For definiteness let us examine the relation between shear strains and normal stresses under simple shear, a strain-controlled test illustrated in Fig. 8. The objective is to determine the source of coupling of the direct stresses and shear strains in the case of elastoplastic formulations of increasing complexity.

Figure 9 illustrates the fundamentally different response predictions of the pressure-independent von Mises and a pressure-sensitive and dilatant two-invariant plasticity formulation in terms of $I_1 = [tr\,\sigma]$ and $J_2 = \frac{1}{2}[tr\,s^2]$ as proposed by Drucker and Prager (1952). A parabolic generalization of the original two-parameter model may be developed in terms of the friction angle α_F and the cohesive resistance $r_y = \tau_y$ to define the shear strength in the form:

$$F(I_1, J_2) = J_2 + \alpha_F I_1 - \tau_y^2 = 0 \qquad (101)$$

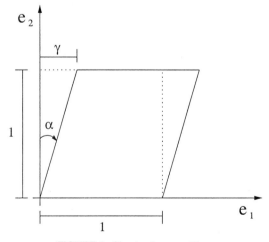

FIGURE 8 Simple shear problem.

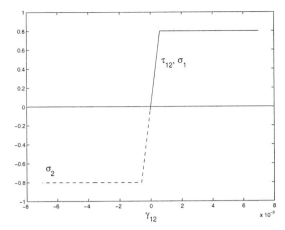
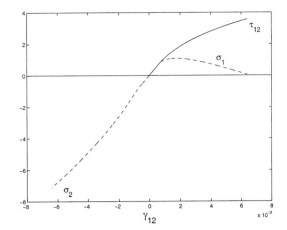

FIGURE 9 Simple shear response behaviour of one-invariant von Mises model (left) and two-invarient parabolic Drucker–Prager model (right).

The two parameters may be identified from uniaxial tension and compression test data f_t' and f_c' as follows:

$$\alpha_F = \frac{f_c' - f_t'}{3} \quad \text{and} \quad \tau_y^2 = \frac{f_c' f_t'}{3} \qquad (102)$$

The von Mises yield condition in shear, $\sqrt{J_2} = \tau_y = 0.8$ *ksi*, reproduces the elastic–perfectly plastic input, whereby the shear stress–strain relation coincides with the principal stress response. Note, strain control coincides with the stress-controlled situation of pure shear, $\tau_{12} = \sigma_1 = -\sigma_2 = \tau_y$, if there is no coupling between the volumetric and deviatoric response behavior. In contrast, the simple shear test exhibits apparent hardening, when an associated flow rule is used in conjunction with the parabolic Drucker–Prager yield condition. Calibration of the two-parameter representation with the uniaxial tensile and compressive strength values $f_c' = 4$ *ksi* and $f_t' = 0.6$ *ksi* leads to the same incipient shear strength as the von Mises model. However, in contrast to the pressure-independent plasticity model, the parabolic Drucker–Prager model exhibits very different behavior in the plastic regime because of the apparent hardening under persistent plastic flow in spite of the underlying assumption of perfectly plastic behavior. We observe a large increase of shear stress, which is accompanied by an equivalent increase of the minor principal stress $-\sigma_2 \gg \tau_y$ at the cost of reducing the major principal tensile stress from tension into compression.

In view of these fundamental differences of pressure-sensitive vs. pressure-insensitive plasticity models, we further investigate the source of these discrepancies. To this end we systematically examine the influence of the three invariants in the plastic yield condition on the simple shear response and the shear dilatancy in particular.

1. One-Invariant Plasticity Formulation

To start with, we consider the effect of the second invariant in the yield condition, $F = F(\rho)$, where the deviator $\rho = \sqrt{s : s}$ defines the radial distance of the yield condition from the hydrostatic axis of the cylindrical Haigh–Westergaard coordinates ξ, ρ, θ of stress. In this case, the expression for the gradient of the yield function defines the normal as:

$$\boldsymbol{n} = \frac{\partial F}{\partial \boldsymbol{\sigma}} = \frac{\partial F}{\partial \rho} \frac{\partial \rho}{\partial \boldsymbol{\sigma}} = \frac{F_\rho}{\rho} \boldsymbol{s} \quad \text{where} \quad \frac{\partial \rho}{\partial \boldsymbol{\sigma}} = \frac{1}{\rho} \boldsymbol{s} \quad (103)$$

The partial derivative, $F_\rho = \partial F / \partial \rho$, is a positive scalar which determines the magnitude of the normal shown in Fig. 10.

The plastic flow direction is defined by the gradient of the plastic potential. In the case of pressure-insensitive plastic flow

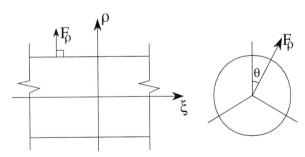

FIGURE 10 Gradient of generic single-invariant yield surface.

$$\boldsymbol{m} = \frac{\partial Q}{\partial \boldsymbol{\sigma}} = \frac{\partial Q}{\partial \rho}\frac{\partial \rho}{\partial \boldsymbol{\sigma}} = \frac{Q_\rho}{\rho}\boldsymbol{s} \qquad (104)$$

The expressions for $\boldsymbol{n}$ and $\boldsymbol{m}$ simplify considerably for loading in simple shear when $\dot{\gamma}_{12} > 0$ and when all other strain components remain zero, $\epsilon_{11} = \epsilon_{22} = \epsilon_{33} = \gamma_{23} = \gamma_{13} = 0$. At incipient yielding the elastic shear stress is the only deviatoric component which is nonzero, $s_{12} = \tau_{12} = G\gamma_{12}$. With $\rho = \sqrt{\boldsymbol{s}:\boldsymbol{s}} = \sqrt{2}\tau_{12}$, the gradients of the yield function and plastic potential $\boldsymbol{n}$ and $\boldsymbol{m}$ in vector notation reduce to:

$$\boldsymbol{n} = \begin{bmatrix} 0 & 0 & 0 & \frac{1}{\sqrt{2}}F_\rho & 0 & 0 \end{bmatrix}^t$$

$$\text{and} \quad \boldsymbol{m} = \begin{bmatrix} 0 & 0 & 0 & \frac{1}{\sqrt{2}}Q_\rho & 0 & 0 \end{bmatrix}^t \qquad (105)$$

For plastic loading in simple shear the elastoplastic tangent matrix has the simple format:

$$\begin{Bmatrix} \dot{\sigma}_{11} \\ \dot{\sigma}_{22} \\ \dot{\sigma}_{33} \\ \cdots \\ \dot{\tau}_{12} \\ \dot{\tau}_{23} \\ \dot{\tau}_{13} \end{Bmatrix} =$$

$$\begin{bmatrix} K+\frac{4}{3}G & K-\frac{2}{3}G & K-\frac{2}{3}G & \vdots & 0 & 0 & 0 \\ K-\frac{2}{3}G & K+\frac{4}{3}G & K-\frac{2}{3}G & \vdots & 0 & 0 & 0 \\ K-\frac{2}{3}G & K-\frac{2}{3}G & K+\frac{4}{3}G & \vdots & 0 & 0 & 0 \\ \cdots & \cdots & \cdots & \cdots & \cdots & \cdots & \cdots \\ 0 & 0 & 0 & \vdots & G\left[1-\frac{1}{2h_p}GF_\rho Q_\rho\right] & 0 & 0 \\ 0 & 0 & 0 & \vdots & 0 & G & 0 \\ 0 & 0 & 0 & \vdots & 0 & 0 & G \end{bmatrix} \begin{Bmatrix} 0 \\ 0 \\ 0 \\ \cdots \\ \dot{\gamma}_{12} \\ 0 \\ 0 \end{Bmatrix}$$

$$(106)$$

For von Mises plasticity the dyadic tensor product adds only a single term to E_{44}^{ep} on the principal diagonal. Note that the off-diagonal partitions of the elastoplastic tangent operator are zero, similar to isotropic elasticity exhibiting no coupling between the normal stresses and shear strains and vice versa.

For completeness let us examine under what conditions the elastoplastic tangent operator $\boldsymbol{E}_{ep}$ becomes singular. To this end, we partition the elastoplastic tangent matrix into the four components and use the Schur theorem of determinants which states:

$$\det \boldsymbol{E}_{ep} = \det \boldsymbol{E}_{ss}\det\left(\boldsymbol{E}_{nn} - \boldsymbol{E}_{ns}\boldsymbol{E}_{ss}^{-1}\boldsymbol{E}_{sn}\right) \qquad (107)$$

A necessary and sufficient condition for $\det \boldsymbol{E}_{ep} = 0$ is reached when either $\det \boldsymbol{E}_{ss} = 0$ or when the determinant of the so-called Schur complement is zero, $\det\left(\boldsymbol{E}_{nn} - \boldsymbol{E}_{ns}\boldsymbol{E}_{ss}^{-1}\boldsymbol{E}_{sn}\right) = 0$. In the case of no coupling,

$\boldsymbol{E}_{ns} = \boldsymbol{E}_{sn} = \boldsymbol{0}$, and with the positive partition of the elastic stiffness, $\det \boldsymbol{E}_{nn} > 0$, the only possibility that the elastoplastic tangent operator turns singular arises when $\det \boldsymbol{E}_{ss} = 0$. This condition is true if and only if the diagonal term vanishes (i.e., $E_{44}^{ep} = 0$). Substituting the normals $\boldsymbol{n}$ and $\boldsymbol{m}$ in Eq. (105) into the denominator $h_p = H_p + \boldsymbol{n}:\boldsymbol{E}:\boldsymbol{m} = H_p + \frac{1}{2}GF_\rho Q_\rho$, the plastic degradation of the elastic stiffness results in:

$$E_{44}^{ep} = G\left[1 - \frac{1}{1+\frac{2H_p}{GF_\rho Q_\rho}}\right] \qquad (108)$$

As the shear modulus is strictly positive, $G > 0$, and as $F_\rho Q_\rho > 0$ in the case of the von Mises yield cylinder, the diagonal shear stiffness goes to zero, $E_{44}^{ep} \to 0$, only if the hardening parameter goes to zero, $H_p \to 0$. Consequently,

$$\det \boldsymbol{E}_{ep} = 0 \quad \text{iff} \quad H_p = 0 \qquad (109)$$

2. Two-Invariant Plasticity Formulation

In the second stage, we examine the dilatational stress relation of a two-invariant Drucker–Prager yield condition, which includes both hydrostatic and deviatoric plastic effects as shown in Fig. 11.

The normal gradient of the two-invariant formulation reads in the case of two-invariant plasticity:

$$\boldsymbol{n} = \frac{\partial F}{\partial \boldsymbol{\sigma}} = \frac{\partial F}{\partial \xi}\frac{\partial \xi}{\partial \boldsymbol{\sigma}} + \frac{\partial F}{\partial \rho}\frac{\partial \rho}{\partial \boldsymbol{\sigma}} = \frac{F_\xi}{\sqrt{3}}\boldsymbol{1} + \frac{F_\rho}{\rho}\boldsymbol{s} \qquad (110)$$

where $\xi = \frac{1}{\sqrt{3}}\boldsymbol{\sigma}:\boldsymbol{1}$ denotes the dependence on the first stress invariant $I_1 = [tr\,\boldsymbol{\sigma}]$. The scalar derivative is in this case $F_\xi = \partial F/\partial \xi$, while the tensor derivatives are $\partial \xi/\partial \boldsymbol{\sigma} = \frac{1}{\sqrt{3}}\boldsymbol{1}$.

In analogy to the normal $\boldsymbol{n}$, the gradient of the nonassociated plastic potential is expressed as:

$$\boldsymbol{m} = \frac{\partial Q}{\partial \boldsymbol{\sigma}} = \frac{\partial Q}{\partial \xi}\frac{\partial \xi}{\partial \boldsymbol{\sigma}} + \frac{\partial Q}{\partial \rho}\frac{\partial \rho}{\partial \boldsymbol{\sigma}} = \frac{Q_\xi}{\sqrt{3}}\boldsymbol{1} + \frac{Q_\rho}{\rho}\boldsymbol{s} \qquad (111)$$

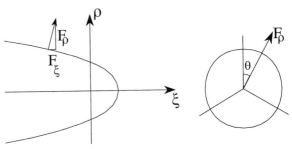

FIGURE 11 Gradient of generic two-invariant yield surface.

In vector form, the two gradients read for simple shear:

$$n = \left[\frac{F_\xi}{\sqrt{3}}, \frac{F_\xi}{\sqrt{3}}, \frac{F_\xi}{\sqrt{3}}, \frac{1}{\sqrt{2}}F_\rho, 0, 0\right]^t$$

$$\text{and} \quad m = \left[\frac{Q_\xi}{\sqrt{3}}, \frac{Q_\xi}{\sqrt{3}}, \frac{Q_\xi}{\sqrt{3}}, \frac{1}{\sqrt{2}}Q_\rho, 0, 0\right]^t \quad (112)$$

The dyadic product $\mathcal{R}_p = \mathcal{E} : m \otimes n : \mathcal{E}$ of the plastic tangent stiffness leads to loss of symmetry if $F_\rho = Q_\rho$ and/or $F_\xi \neq Q_\xi$. The factor 2 in the normal-shear partition is due to the use of the technical shear strain definition in matrix notation.

$$E_{44}^{ep} = G\left[1 - \frac{1}{1 + \frac{2H_p + 6KF_\xi Q_\xi}{GF_\rho Q_\rho}}\right] \quad (115)$$

The elastic bulk and shear moduli (K, G) and the derivatives F_ξ, F_ρ and Q_ξ, Q_ρ are all positive in the conical region of the yield surface and the plastic potential. Consequently, the diagonal term will remain positive, $E_{44}^{ep} > 0$, and the simple shear response exhibits "apparent" hardening in the case of perfectly plastic behavior when $H_p = 0$. The tangent operator becomes singular only when the plastic modulus reaches the limiting softening value:

$$R_p = \begin{bmatrix} 3K^2 F_\xi Q_\xi & 3K^2 F_\xi Q_\xi & 3K^2 F_\xi Q_\xi & \vdots & \sqrt{\frac{3}{2}}KGF_\rho Q_\xi & 0 & 0 \\ 3K^2 F_\xi Q_\xi & 3K^2 F_\xi Q_\xi & 3K^2 F_\xi Q_\xi & \vdots & \sqrt{\frac{3}{2}}KGF_\rho Q_\xi & 0 & 0 \\ 3K^2 F_\xi Q_\xi & 3K^2 F_\xi Q_\xi & 3K^2 F_\xi Q_\xi & \vdots & \sqrt{\frac{3}{2}}KGF_\rho Q_\xi & 0 & 0 \\ \cdots & \cdots & \cdots & \cdots & \cdots & \cdots & \cdots \\ \sqrt{\frac{3}{8}}KGF_\xi Q_\rho & \sqrt{\frac{3}{8}}KGF_\xi Q_\rho & \sqrt{\frac{3}{8}}KGF_\xi Q_\rho & \vdots & \frac{1}{2}G^2 F_\rho Q_\rho & 0 & 0 \\ 0 & 0 & 0 & \vdots & 0 & 0 & 0 \\ 0 & 0 & 0 & \vdots & 0 & 0 & 0 \end{bmatrix} \quad (113)$$

The hardening parameter in the denominator is comprised of three terms:

$$h_p = H_p + \frac{1}{2}GF_\rho Q_\rho + 3KF_\xi Q_\xi \quad (114)$$

whereby the third contribution introduces an additional term, which is positive as long as $F_\xi Q_\xi > 0$. As expected, now there is coupling between the normal stresses and the shear strain. This shows that plastic loading in the two-invariant formulation results in normal stresses, which are negative as long as $F_\rho Q_\xi > 0$ and vice versa, $F_\xi Q_\rho > 0$. In other terms, pressure-sensitive plastic loading in simple shear induces compromise confinement in the case of Einematic constraints. Thus, the volumetric–deviatoric interaction depends to a large extent on the volumetric component of the plastic flow rule, which is positive in the conical region of the plastic potential and which diminishes to zero (i.e., $Q_\xi \to 0$) when the plastic potential approaches the von Mises condition.

The singularity of the elastoplastic tangent operator E_{ep} is again determined in terms of singular partitions using the Schur theorem of determinants. The diagonal format of the shear partition suggests examination of the diagonal term:

$$H_p^{limit} = -3KF_\xi Q_\xi \quad (116)$$

In summary, softening is needed to compensate for the apparent hardening effect of the volumetric dependence of the yield function and the plastic potential. Therefore, the singularity depends critically on the volumetric components of the yield condition and the plastic potential (i.e., $H_p \to 0$ as $F_\xi Q_\xi \to 0$).

3. Three-Invariant Plasticity Formulation

In the third and final study we generalize the plasticity formulation and include the effect of the third invariant in the yield function and the plastic potential. The effect of the third invariant leads to the triple-symmetric shape of the yield function in the deviatoric plane as illustrated in Fig. 12.

In this case, the gradient operators are becoming considerably more involved in view of the definition of the third invariant, which is here expressed in terms of the angle of similarity:

$$\theta = \frac{1}{3}\cos^{-1}\left[\frac{3\sqrt{3}J_3}{2J_2^{1.5}}\right] \quad (117)$$

where $J_3 = \det s$.

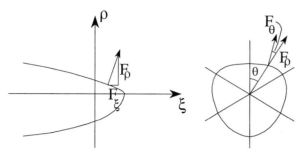

FIGURE 12 Gradients of generic three-invariant yield surface.

The gradients n, m are comprised of three contributions; that is,

$$n = \frac{\partial F}{\partial \sigma} = \frac{\partial F}{\partial \xi}\frac{\partial \xi}{\partial \sigma} + \frac{\partial F}{\partial \rho}\frac{\partial \rho}{\partial \sigma} + \frac{\partial F}{\partial \theta}\frac{\partial \theta}{\partial \sigma}$$

$$= \frac{F_\xi}{\sqrt{3}}\mathbf{1} + \frac{F_\rho}{\rho}s + F_\theta\frac{\partial \theta}{\partial \sigma} \qquad (118)$$

and

$$m = \frac{\partial Q}{\partial \sigma} = \frac{\partial Q}{\partial \xi}\frac{\partial \xi}{\partial \sigma} + \frac{\partial Q}{\partial \rho}\frac{\partial \rho}{\partial \sigma} + \frac{\partial Q}{\partial \theta}\frac{\partial \theta}{\partial \sigma}$$

$$= \frac{Q_\xi}{\sqrt{3}}\mathbf{1} + \frac{Q_\rho}{\rho}s + Q_\theta\frac{\partial \theta}{\partial \sigma} \qquad (119)$$

The additional scalar derivatives include $F_\theta = \frac{\partial F}{\partial \theta}$ and $Q_\theta = \frac{\partial Q}{\partial \theta}$, while the tensor derivative is complicated:

$$\frac{\partial \theta}{\partial \sigma} = \frac{\sqrt{3}}{2\sqrt{1-\cos^2 3\theta}}\left[\frac{1}{J_2^{1.5}}\frac{\partial J_3}{\partial \sigma} - \frac{3J_3}{2J_2^{2.5}}\frac{\partial J_2}{\partial \sigma}\right]$$

with $\dfrac{\partial J_3}{\partial \sigma} = s \cdot s - \dfrac{2}{3}J_2\mathbf{1}$ and $\dfrac{\partial J_2}{\partial \sigma} = s$ (120)

In vector form, the gradients for simple shear appear as:

$$n = \left[\frac{F_\xi}{\sqrt{3}} + \frac{F_\theta}{2\sqrt{3}\tau_{12}}, \frac{F_\xi}{\sqrt{3}} + \frac{F_\theta}{2\sqrt{3}\tau_{12}}, \frac{F_\xi}{\sqrt{3}}\right.$$

$$\left. - \frac{F_\theta}{\sqrt{3}\tau_{12}}, \frac{1}{\sqrt{2}}F_\rho, 0, 0\right]^t \qquad (121)$$

and

$$m = \left[\frac{Q_\xi}{\sqrt{3}} + \frac{Q_\theta}{2\sqrt{3}\tau_{12}}, \frac{Q_\xi}{\sqrt{3}} + \frac{Q_\theta}{2\sqrt{3}\tau_{12}}, \frac{Q_\xi}{\sqrt{3}}\right.$$

$$\left. - \frac{Q_\theta}{\sqrt{3}\tau_{12}}, \frac{1}{\sqrt{2}}Q_\rho, 0, 0\right]^t \qquad (122)$$

In short, the plastic contributions of the third invariant enter the dyadic product $\mathcal{R}_p = \mathcal{E}:m\otimes n:\mathcal{E}$, affecting the coupling partitions $\mathcal{R}_{ns}^p$ and $\mathcal{R}_{sn}^p$ as well as the direct partition $\mathcal{R}_{nn}^p$. To understand their effect, we study the individual terms in the different 3×3 partitions below:

$$\mathbf{R}_{nn}^p = \begin{bmatrix} R_{11}^p & R_{12}^p & R_{13}^p \\ R_{21}^p & R_{22}^p & R_{23}^p \\ R_{31}^p & R_{32}^p & R_{33}^p \end{bmatrix} \qquad (123)$$

where

$$R_{11}^p = R_{22}^p = R_{12}^p = R_{21}^p = 3K^2 F_\xi Q_\xi$$

$$+ \frac{G^2 F_\theta Q_\theta}{3\tau_{12}^2} + \frac{KG}{\tau_{12}}(F_\theta Q_\xi + F_\xi Q_\theta) \quad (124)$$

$$R_{33}^p = 3K^2 F_\xi Q_\xi + \frac{4G^2 F_\theta Q_\theta}{3\tau_{12}^2}$$

$$- \frac{2KG}{\tau_{12}}(F_\theta Q_\xi + F_\xi Q_\theta) \qquad (125)$$

$$R_{13}^p = R_{23}^p = 3K^2 F_\xi Q_\xi - \frac{2G^2 F_\theta Q_\theta}{3\tau_{12}^2}$$

$$+ \frac{KG}{\tau_{12}}(F_\xi Q_\theta - 2F_\theta Q_\xi) \qquad (126)$$

$$R_{31}^p = R_{32}^p = 3K^2 F_\xi Q_\xi - \frac{2G^2 F_\theta Q_\theta}{3\tau_{12}^2}$$

$$+ \frac{KG}{\tau_{12}}(F_\theta Q_\xi - 2F_\xi Q_\theta) \qquad (127)$$

The individual terms indicate that the dependence on the third invariant, $F_\theta Q_\theta$, introduces different terms in the in-plane and out-of-plane stiffness contributions as opposed to the dependence on the first invariant, $F_\xi Q_\xi$. In contrast, the shear partition maintains the same format as the two-invariant model,

$$\mathbf{R}_{ss}^p = \begin{bmatrix} \frac{1}{2}G^2 F_\rho Q_\rho & 0 & 0 \\ 0 & 0 & 0 \\ 0 & 0 & 0 \end{bmatrix} \qquad (128)$$

The coupling partitions exhibit a distinct difference between in-plane and out-of-plane stiffness contributions in simple shear:

$$\mathbf{R}_{ns}^p = \begin{bmatrix} \sqrt{\frac{3}{2}}KGF_\rho Q_\xi + \frac{G^2}{\sqrt{6}}\frac{F_\rho Q_\theta}{\tau_{12}}, 0, 0 \\ \sqrt{\frac{3}{2}}KGF_\rho Q_\xi + \frac{G^2}{\sqrt{6}}\frac{F_\rho Q_\theta}{\tau_{12}}, 0, 0 \\ \sqrt{\frac{3}{2}}KGF_\rho Q_\xi - \frac{2G^2}{\sqrt{6}}\frac{F_\rho Q_\theta}{\tau_{12}}, 0, 0 \end{bmatrix} \quad (129)$$

and

$$\boldsymbol{R}_{sn}^{p} = \frac{1}{2} \begin{bmatrix} \sqrt{\frac{3}{2}} KGF_{\xi}Q_{\rho} + \frac{G^2}{\sqrt{6}} \frac{Q_{\rho}F_{\theta}}{\tau_{12}}, 0, 0 \\ \sqrt{\frac{3}{2}} KGF_{\xi}Q_{\rho} + \frac{G^2}{\sqrt{6}} \frac{Q_{\rho}F_{\theta}}{\tau_{12}}, 0, 0 \\ \sqrt{\frac{3}{2}} KGF_{\xi}Q_{\rho} - \frac{2G}{\sqrt{6}} \frac{Q_{\rho}F_{\theta}}{\tau_{12}}, 0, 0 \end{bmatrix}^{t} \quad (130)$$

We note that the entries of the dyadic product are no longer the same as in the two-invariant formulation. In fact, the negative sign in the third entry introduces additional structure and distinguishes in-plane from out-of-plane action, which reduces significantly the out-of-plane confining stress σ_{33} under simple shear. We also observe the loss of symmetry in all partitions except for the shear partition, when the plastic dilatancy differs from the frictional resistance of the yield condition, $Q_{\xi} \neq F_{\xi}$. It might surprise the reader that the shear partition $\boldsymbol{R}_{ss}^{p}$ remains unchanged except for the denominator h_p which is augmented by an apparent hardening term from the contribution of the third invariant; that is,

$$h_p = H_p + \frac{1}{2} GF_{\rho}Q_{\rho} + 3KF_{\xi}Q_{\xi} + \frac{3}{2} K \frac{F_{\theta}Q_{\theta}}{\tau_{12}^2} \quad (131)$$

Analysis of the singularity in simple shear leads to the question of when the diagonal term vanishes (i.e., $E_{44}^{ep} \to 0$). The answer follows the previous argument of the two invariant formulation. In fact, $\det \boldsymbol{E}_{ep} = 0$ when the plastic softening modulus reaches the limiting value:

$$H_p^{limit} = -3K \left[F_{\xi}Q_{\xi} + \frac{1}{2} \frac{F_{\theta}Q_{\theta}}{\tau_{12}^2} \right] \quad (132)$$

This indicates that dependence on the third invariant further stabilizes the elastoplastic tangent operator beyond the limit point condition of the two-invariant format.

E. Concrete Plasticity Model

To illustrate the observations above we resort to numerical simulation of the simple shear test with the help of the three-invariant plasticity model by Kang and Willam (1999). In this case, the curvilinear loading surface, $F(\xi, \rho, \theta, q_h, q_s) = 0$, is C^1-continuous except at the vertex in equitriaxial tension (see Fig. 13a. The triaxial concrete model exhibits pressure sensitivity of the deviatoric strength as a function of the third stress-invariant, inelastic dilatancy during shearing and brittle–ductile transition from fragile behavior in tension to ductile behavior with increasing confinement.

The plastic loading function is comprised of three components: the triaxial failure envelope and hardening or softening contributions, which are mobilized alternatively in the pre- and post-peak regions:

$$F(\xi, \rho, \theta, q_h, q_s) = F(\xi, \rho, \theta)_{\text{fail}} + F(\xi, \rho, \theta, k(q_h))_{\text{hardg}}$$
$$+ F(\xi, \rho, \theta, c(q_s))_{\text{softg}} \quad (133)$$

The failure envelope,

$$F(\xi, \rho, \theta)_{\text{fail}} = \frac{\rho r(\theta, e)}{f_c'} - \frac{\rho_1}{f_c'} \left[\frac{\xi - \xi_o}{\xi_1 - \xi_o} \right]^{\alpha} = 0 \quad (134)$$

fixes the triaxial strength in stress space in terms of a curvilinear triple-symmetric cone. Figure 13a depicts the failure envelope in terms of the meridians in triaxial compression and extension, and Fig. 13b shows the failure envelope in terms of the deviatoric traces at different levels of hydostatic stress.

The simple shear response is shown in Fig. 14 for the three-invariant concrete model. In comparison to the two-invariant formulation shown in Fig. 9, the shear strength

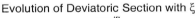

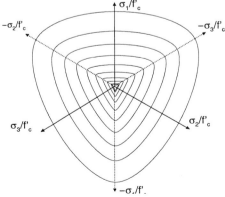

FIGURE 13 (a) Tension and compression meridians, (b) deviatoric contours of triaxial concrete envelope. [From Kang, H., and Willam, K. (1199). *ASCE J. Eng. Mech.* **125**, 941–950. With permition.]

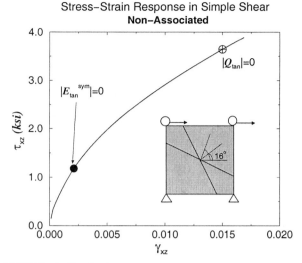

FIGURE 14 Simple shear response of three-invariant concrete model. [From Kang, H., and Willam, K. (1999). *ASCE J. Eng. Mech.* **125,** 941–950. With permission.]

and shear ductility are reduced, because of the diminished out-of-plane confinement of the three-invariant model.

Note that the nonassociated plasticity model exhibits loss of stability ($\det \mathcal{E}_{ep}^{sym} = \det \mathcal{E}_{tan}^{sym} = 0$) and discontinuous bifurcation in the form of localized failure ($\det \boldsymbol{Q}_{ep} = \det \boldsymbol{Q}_{tan} = 0$) in the apparent hardening regime before a limit point condition is reached. For an understanding of the different failure diagnostics, the reader is referred to Section IV and to Appendix 2 for additional comments on elastoplastic failure in plane strain.

IV. ANALYSIS OF MATERIAL FAILURE

In this section, we examine different criteria for initiating failure at the constitutive level of materials. The traditional approach in strength of materials is to probe demand ver-

sus resistance with the aid of a limit state condition. This leads to the geometrical visualization of the triaxial state of stress and its proximity to the envelope condition according to Otto Mohr (1835–1918).

Note that the geometrical interpretation of the triaxial strength hypothesis $F = F(\boldsymbol{\sigma}) = 0$ leads to a critical combination of normal and shear tractions σ, σ_T on a critical failure plane $\pm\theta$ which determines the mode of failure. Thus, the Mohr envelope criterion tells us not only when failure takes place, but also what kind of failure mode develops. The main drawback is that it does not depend on the intermediate principal stress since the major principal circle is governed by the maximum and minimum values of stress, $R_{max} = \frac{1}{2}|\sigma_1 - \sigma_3|$. Representative strength criteria for cohesive-frictional materials include the maximum shear stress conditions of Mohr–Coulomb, Tresca, and Leon, as well as the maximum normal stress condition of Rankine for tensile cracking. Alternatively to strength of materials, failure has also been described by deformation hypotheses $F(\epsilon) = 0$ such as the Saint Venant criterion of maximum normal strain for tensile cracking. Moreover, energy criteria $F(\boldsymbol{\sigma} : \boldsymbol{\epsilon}) = 0$ have been proposed to describe material failure such as the Beltrami condition of maximum strain energy and the Huber criterion of maximum distortional energy. In fact, fracture criteria of a critical stress intensity factor or an equivalent strain energy release rate may also be included in this list of failure criteria, $F(K_I, K_{II}, K_{III}) = 0$ and $F(G_f) = 0$, though discrete fracture mechanics normally starts from the existence of a crack or a notched defect and defines crack propagation in terms of fracture concepts. In this sense, failure criteria differ from fracture criteria in a fundamental fashion, the former indicate initiation while the latter monitor propagation.

In the original envelope concept of Mohr (Fig. 15), the actual mechanism behind material failure was left open in terms of its kinematic or static manifestation. There

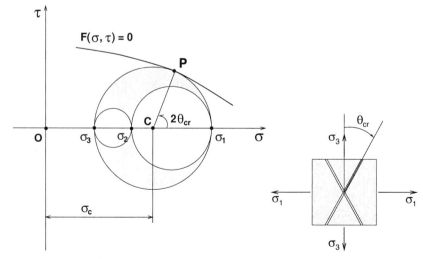

FIGURE 15 Mohr concept of universal failure envelope.

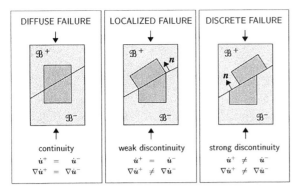

FIGURE 16 Kinematic deterioration of the continuum into a discontinuum.

was the conceptual distinction between tensile cracking and separation of material interfaces in mode I, and frictional slip among adjacent material interfaces in mode II. However, nothing was said about what happens after failure initiates. In fact, failure initiation was assumed to be critical and trigger collapse of the entire structure.

In view of recent progress in localization analysis it is important to keep in mind that failure is a process of events which start small at the material level and lead to progressive deterioration of the continuum into a discontinuum. Figure 16 illustrates how his process leads to a progression of kinematic deterioration. It initiates in the form of diffuse failure and leads through localized, weakly discontinuous failure to discrete or strongly discontinuous failure. The following discussion will be restricted to the first two stages of failure, whereby diffuse failure is characterized by loss of material stability and loss of uniqueness, while localized failure results in the formation of weak discontinuities, which signal the onset of kinematic degradation of the continuum into a discontinuum synonymous with loss of ellipticity.

A. Loss of Stability/Material Instability

The loss of material stability was identified early on by Drucker and Hill with the loss of positive internal work. In fact, the exclusion functional of positive second-order work density,

$$d^2W = \frac{1}{2}\dot{\sigma} : \dot{\epsilon} = \frac{1}{2}\dot{\epsilon} : \mathcal{E}_t : \dot{\epsilon} > 0, \qquad \forall \dot{\epsilon} \neq 0 \quad (135)$$

is widely accepted as a sufficient condition for material stability.

In the case of nonassociated elastoplasticity, where $\mathcal{E}_{ep} \neq \mathcal{E}_{ep}^t$, this criterion leads to:

$$d^2W = \frac{1}{2}\dot{\epsilon} : \mathcal{E}_{ep} : \dot{\epsilon} = \frac{1}{4}\dot{\epsilon} : [\mathcal{E}_{ep} + \mathcal{E}_{ep}^t] : \dot{\epsilon}$$

$$= \frac{1}{2}\dot{\epsilon} : \mathcal{E} : \dot{\epsilon} - \frac{1}{4h_p}\dot{\epsilon} : [\bar{m} \otimes \bar{n} + \bar{n} \otimes \bar{m}] : \dot{\epsilon} \quad (136)$$

In this context, it is important to recall that the energy functional extracts only the symmetric part of the tangent operator which leads to the following observation based on the Bromwich eigenvalue bounds of nonsymmetric matrices:

$$\lambda_{min}\left(\mathcal{E}_{ep}^{sym}\right) \leq \Re(\lambda_{min}(\mathcal{E}_{ep})) \cdots \leq \lambda_{max}\left(\mathcal{E}_{ep}^{sym}\right) \quad (137)$$

In other terms, the instability argument coincides with the loss of positive definiteness of the symmetric material operator,

$$\det \mathcal{E}_{ep}^{sym} \overset{!}{=} 0 \rightarrow \lambda_{min}\left(\mathcal{E}_{ep}^{sym}\right) = 0 \quad (138)$$

This argument leads to the critical value of plastic hardening for stability (see Maier and Hueckl, 1979):

$$H_p^{stabil} = \frac{1}{2}[\sqrt{(\boldsymbol{n} : \mathcal{E}_o : \boldsymbol{n})(\boldsymbol{m} : \mathcal{E}_o : \boldsymbol{m})} - \boldsymbol{n} : \mathcal{E}_o : \boldsymbol{m}] \quad (139)$$

Therefore, the sufficient condition for stability may be turned around into an upper-bound argument of loss of stability in terms of the maximum plastic modulus $H_p^{stabil} \geq H_p^{limit}$.

B. Diffuse Failure/Loss of Uniqueness

According to Fig. 16, diffuse failure maintains continuity in the rate of displacement and the displacement gradient fields. It corresponds to a stationary stress state which defines a limit point on the response path of the material:

$$\dot{\sigma} = \boldsymbol{0} \quad (140)$$

For incrementally linear materials with $\dot{\sigma} = \mathcal{E}_t : \dot{\epsilon}$, this infers that:

$$\mathcal{E}_t : \dot{\epsilon} = \boldsymbol{0} \quad (141)$$

Thus, the condition for the loss of uniqueness is equivalent to a singular behavior of the tangent operator $\mathcal{E}_t$,

$$\det \mathcal{E}_t \overset{!}{=} 0 \rightarrow \lambda_{min}(\mathcal{E}_t) = 0 \quad (142)$$

Recall that the elastoplastic tangent operator involves a rank-one update of the elastic material operator,

$$\mathcal{E}_{ep} = \mathcal{E} - \frac{1}{h_p}\bar{m} \otimes \bar{n} \quad (143)$$

whereby the update tensors were defined as follows:

$$\bar{m} = \mathcal{E} : m \quad (144a)$$

$$\bar{n} = \boldsymbol{n} : \mathcal{E} \quad (144b)$$

Preconditioning of the elastoplastic tangent operator with the inverse elasticity tensor, the following relation appears.

$$\mathcal{E}^{-1} : \mathcal{E}^{ep} = \mathcal{I} - \mathcal{E}^{-1} : \frac{\bar{m} \otimes \bar{n}}{h_p} \qquad (145)$$

Due to the rank-one update structure of the elastoplastic tangent operator, the critical eigenvalue λ_{min} of the generalized eigenvalue problem may be evaluated in closed form. This motivates the introduction of the scalar-valued measure of material integrity $d_{\mathcal{E}}$,

$$\lambda_{min}(\mathcal{E}^{-1} : \mathcal{E}^{ep}) = 1 - d_{\mathcal{E}} \quad \text{with} \quad d_{\mathcal{E}} := \frac{n : \mathcal{E} : m}{H_p + n : \mathcal{E} : m}. \qquad (146)$$

From the above definition, a necessary condition for loss of uniqueness leads to the critical hardening modulus $H_p^{limit} = 0$ associated with full loss of integrity:

$$1 - d_{\mathcal{E}} \overset{!}{=} 0 \qquad (147)$$

The criterion presented above can only be understood as a necessary condition for the loss of uniqueness. For nonsymmetric elastoplastic tangent operators, which arise in nonassociated plasticity formulations, loss of stability in the form of $\det \mathcal{E}_{ep}^{sym} \overset{!}{=} 0$ provides a lower bound condition according to the Bromwich bounds. Consequently, loss of stability which is synonymous with a singularity of the symmetric tangent operator takes place before the limit point condition in Eq. (142) is reached.

C. Localized Failure/Loss of Ellipticity

The localization condition is based on the early works of Hadamard and Hill. In contrast to the diffuse failure mode described in the previous section, localized failure of weak discontinuities is characterized through a discontinuity in the rate of the displacement gradient, while the field of the displacement rate itself is still continuous. According to the Maxwell compatibility condition, the jump in the rate of the displacement gradient may be expressed in terms of a scalar-valued jump amplitude α, the unit jump vector M, and the unit normal vector to the discontinuity surface N shown in Fig. 16:

$$[\![\nabla \dot{u}]\!] = \alpha M \otimes N \rightarrow [\![\dot{\varepsilon}]\!] = \alpha [M \otimes N]^{sym} \qquad (148)$$

Equilibrium along the discontinuity surface requires that the traction vectors are equal and opposite on both sides of the discontinuity:

$$[\![\dot{t}]\!] = \dot{t}^+ - \dot{t}^- = 0 \qquad (149)$$

According to Cauchy's theorem, we have

$$[\![\dot{t}]\!] = N \cdot [\![\dot{\sigma}]\!] = N \cdot [\![\mathcal{E}_t : \dot{\varepsilon}]\!] = 0 \qquad (150)$$

With the assumption of a linear comparison solid, $[\![\mathcal{E}_t]\!] = \mathcal{E}_t^+ - \mathcal{E}_t^- = 0$, the localization condition may be expressed in the form of an eigenvalue problem:

$$\alpha \, Q_t \cdot M = 0 \quad \text{with} \quad Q_t = N \cdot \mathcal{E}_t \cdot N \qquad (151)$$

whereby Q_t denotes the tangential localization tensor. The necessary condition for the onset of localization is thus characterized through the singularity of the localization tensor which happens to coincide with the acoustic tensor in wave propagation:

$$\det Q_t \overset{!}{=} 0 \rightarrow \lambda_{min}(Q_t) = 0 \qquad (152)$$

The elastoplastic localization tensor may be expressed as a rank-one update of the elastic acoustic tensor,

$$Q_{ep} = Q - \frac{1}{h_p} e_m \otimes e_n \qquad (153)$$

whereby the following abbreviations have been introduced for the update vectors e_m and e_n:

$$e_m = N \cdot \mathcal{E} : m \qquad (154a)$$

$$e_n = n : \mathcal{E} \cdot N \qquad (154b)$$

Instead of probing the lowest eigenvalue of the localization tensor, Eq. (152), we resort to the generalized eigenvalue problem $\det[Q^{-1} \cdot Q^{ep}] \overset{!}{=} 0$ in order to detect whether the localization tensor turns singular (see Ottosen and Runesson, 1991). Preconditioning of the elastoplastic acoustic tensor with the inverse of the elastic acoustic tensor leads to:

$$Q^{-1} \cdot Q_{ep} = 1 - Q^{-1} \cdot \frac{e_m \otimes e_n}{h_p} \qquad (155)$$

The closed form solution for the lowest eigenvalue λ_{min} introduces a scalar-valued measure of integrity with respect to localization in the form of:

$$\lambda_{min}(Q^{-1} \cdot Q_{ep}) = 1 - d_Q \quad \text{with} \quad d_Q = \frac{e_n \cdot Q^{-1} \cdot e_m}{H_p + n : \mathcal{E} : m} \qquad (156)$$

It defines a necessary condition for localization as well as the critical hardening modulus H_p^{loc} indicating loss of ellipticity, when

$$1 - d_Q \overset{!}{=} 0 \rightarrow H_p^{loc} = e_n \cdot Q^{-1} \cdot e_m - n : \mathcal{E} : m \qquad (157)$$

For nonsymmetric elastoplastic tangent operators in nonassociated plasticity, loss of strong ellipticity in terms of a vanishing determinant of the symmetric part of the tangent acoustic operator, $\det Q_{ep}^{sym} \leq \det Q_{ep} \overset{!}{=} 0$, provides a lower bound condition according to the Bromwich theorem. Consequently, loss of strong ellipticity, which is synonymous with a singularity of the symmetric localization operator, may take place before the limit point condition

Eq. (142) is reached. But, most importantly, the critical hardening modulus for localization may still be positive (compare Rudnicki and Rice, 1975) before a limit point is reached because of the symmetric features of the second-order localization tensor when compared to the fourth-order material tensor.

D. Geometric Localization Condition

The analytical localization condition, Eq. (152), may be illustrated geometrically in the form of an envelope condition in analogy to the contact condition of Mohr. In the case of the two-invariant plasticity model, the localization condition plots as an ellipse in the $\sigma - \sigma_T$ Mohr coordinates. The localization envelope was developed by Benallal and Comi (1996) for elastoplastic material models.

In the following, the geometric localization format will be established reproducing some results from Kuhl *et al.* (2000) to illustrate the model problem of simple shear. To this end we briefly summarize the basic ideas of the geometric localization analysis for the nonassociated parabolic Drucker–Prager model introduced in the previous section. Eq. (157) indicates the onset of localization when

$$H_p^{loc} + \boldsymbol{n} : \mathcal{E} : \boldsymbol{m} = \boldsymbol{e}_n \cdot \boldsymbol{Q}^{-1} \cdot \boldsymbol{e}_m \qquad (158)$$

Herein, $\mathcal{E}$ denotes the fourth-order elasticity tensor, while $\boldsymbol{Q}^{-1}$ is the inverse of the elastic acoustic tensor which can be determined analytically:

$$\mathcal{E} = \Lambda \boldsymbol{1} \otimes \boldsymbol{1} + 2G\mathcal{I} \qquad (159a)$$

$$\boldsymbol{Q}^{-1} = \frac{1}{G}\boldsymbol{1} - \frac{1}{2G[1-\nu]}\boldsymbol{N} \otimes \boldsymbol{N} \qquad (159b)$$

In the following, the localization condition, Eq. (158), is recast in the Mohr coordinates σ and σ_T:

$$\sigma := \boldsymbol{N} \cdot \boldsymbol{\sigma} \cdot \boldsymbol{N} \qquad (160a)$$

$$\sigma_T^2 := [\boldsymbol{\sigma} \cdot \boldsymbol{N}] \cdot [\boldsymbol{\sigma} \cdot \boldsymbol{N}] - \sigma^2 \qquad (160b)$$

Combination of these terms defines the localization ellipse in the Mohr coordinates,

$$\frac{[\sigma - \sigma_o]^2}{A^2} + \frac{\sigma_T^2}{B^2} = 1 \qquad (161)$$

where σ_o locates the center, while A and B determine the half axes of the ellipse in the normal and tangential stress components, respectively. In the case of the parabolic Drucker–Prager two-invariant model, the normal to the yield surface $\boldsymbol{n}$ and the normal to the plastic potential $\boldsymbol{m}$ take the form:

$$\boldsymbol{n} = \frac{\partial F}{\partial \boldsymbol{\sigma}} = \boldsymbol{s} + \alpha_F \boldsymbol{1} \qquad (162a)$$

$$\boldsymbol{m} = \frac{\partial Q}{\partial \boldsymbol{\sigma}} = \boldsymbol{s} + \alpha_Q \boldsymbol{1} \qquad (162b)$$

The abbreviations introduced in Eq. (154) result in the following expressions:

$$\boldsymbol{e}_n = 2G\boldsymbol{s} \cdot \boldsymbol{N} + 3K\alpha_F \boldsymbol{N} \qquad (163a)$$

$$\boldsymbol{e}_m = 2G\boldsymbol{s} \cdot \boldsymbol{N} + 3K\alpha_Q \boldsymbol{N} \qquad (163b)$$

The localization ellipse is defined in terms of the parameters,

$$\sigma_o = \frac{1}{3}I_1 - \frac{1+\nu}{2[1-2\nu]}[\alpha_F + \alpha_Q]$$

$$A^2 = 2\frac{1-\nu}{1-2\nu}B^2 \qquad (165)$$

$$B^2 = \frac{1}{4G}H_p + J_2 + \frac{[1+\nu]^2}{8[1-2\nu][1-\nu]}[\alpha_F + \alpha_Q]^2$$
$$+ \frac{1+\nu}{1-\nu}\alpha_F\alpha_Q$$

where α_F, α_Q denote the friction and dilatancy parameters of the parabolic Drucker–Prager model.

The geometrical properties of the localization ellipse are illustrated in Fig. 17. Note that the center and the shape of the ellipse are not influenced by the plastic hardening modulus H_p. The hardening modulus only influences the size of the ellipse. Figure 17 also depicts the major Mohr circle of stress,

$$[\sigma - \sigma_c]^2 + \sigma_T^2 = R^2 \qquad (166)$$

which characterizes the actual stress state. Herein, σ_c and R denote the center and the radius of the circle, respectively, in terms of the principal stresses σ_1 and σ_3:

$$\sigma_c = \frac{\sigma_1 + \sigma_3}{2} \qquad (167a)$$

$$R = \frac{\sigma_1 - \sigma_3}{2} \qquad (167b)$$

The tangency condition results in a quadratic equation, which defines the analytical solutions for critical failure

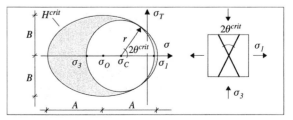

FIGURE 17 Mohr representation of localization ellipse and major stress circle.

angle θ^{crit} and the critical hardening modulus H_p^{loc}. For the nonassociated Drucker–Prager plasticity model, the critical failure angle may be expressed as follows:

$$\tan^2 \theta^{crit}$$
$$= \frac{R - [[1 - 2\nu][\sigma_c - I_1/3] + [1 + \nu][\alpha_F + \alpha_Q]/2]}{R + [[1 - 2\nu][\sigma_c - I_1/3] + [1 + \nu][\alpha_F + \alpha_Q]/2]}$$
(168)

The critical failure angle is strongly influenced by the friction coefficient α_F and the dilatancy parameter α_Q. Furthermore, the failure angle is also influenced by Poisson's ratio. The analytical solution for the critical localization modulus takes the following form:

$$H_p^{loc} = 4G\left[[1 - 2\nu]\left[\sigma_c - \frac{1}{3}I_1 + \frac{1 + \nu}{2[1 - 2\nu]}[\alpha_F + \alpha_Q]\right]^2 \right.$$
$$+ R^2 - J_2 - \frac{[1 + \nu]^2}{8[1 - 2\nu][1 - \nu]}[\alpha_F + \alpha_Q]^2$$
$$\left. - \frac{[1 + \nu]}{[1 - \nu]}\alpha_F \alpha_Q\right]$$
(169)

which may be positive only in the case of nonassociative flow when $\alpha_F \neq \alpha_Q$.

E. Model Problem of Simple Shear

Under strain control the direct shear experiment exhibits different levels of dilatancy depending on the dyadic product in the plastic portion of the elastoplastic operator assuming that the elastic behavior is isotropic. The confinement of full strain control introduces failure modes that vary from a ductile shear failure mode under high confinement to brittle failure under zero confinement. An intriguing consequence of the Reynolds effect is the apparent hardening of the perfectly plastic material model which arises because of the constrained dilatancy.

In the numerical simulation of simple shear (see Fig. 18) the in-plane shear strain γ_{12} increases monotonically. The critical directions of localized failure are studied by means of the nonassociated parabolic Drucker–Prager plasticity

TABLE I Critical Failure Angle of Parabolic Drucker–Prager Model, Simple Shear Problem

$f_c':f_t'$	$\nu = 0.000$ (°)	$\nu = 0.125$ (°)	$\nu = 0.250$ (°)	$\nu = 0.375$ (°)	$\nu = 0.499$ (°)
1:1	45.00	45.00	45.00	45.00	45.00
3:1	35.26	33.99	32.69	31.36	30.01
5:1	29.45	27.24	24.90	22.38	19.64
8:1	22.20	18.26	13.37	5.39	0.00
12:1	11.78	0.00	0.00	0.00	0.00

formulation when $\alpha_Q = 0$, which corresponds to incompressible plastic flow. Therefore, the influence of lateral confinement is increased gradually by increasing the contrast ratio of compressive to tensile strength, $f_c':f_t'$. This ratio directly affects the value of the friction coefficient α_F, whereas the parameter α_Q is kept zero according to a plastic potential of the von Mises type.

Table I summarizes the critical failure angle θ^{crit} for different Poisson ratios and strength ratios $f_c':f_t'$. The critical failure angles range from $0°$ to $45°$, thus including failure modes which range from pure mode I decohesive failure to mode II slip failure. As expected, the critical failure angle decreases with an increase of lateral confinement which is caused by increasing the contrast of the compressive to tensile strength. Moreover, the influence of lateral confinement increases for larger values of Poisson's ratio. Figures 19 to 21 show the result of the corresponding geometric localization analysis for a Poisson's ratio of $\nu = 0.2$. The first figure is for a strength ratio of $f_c':f_t' = 1:1$. This choice corresponds to a vanishing friction coefficient, $\alpha_F = 0.0$, representing the classical yield function of the von Mises type. For this analysis, two critical directions are found under $\theta^{crit} = 45°$ and $\theta^{crit} = 135°$ indicating mode II shear failure. This mode is typically

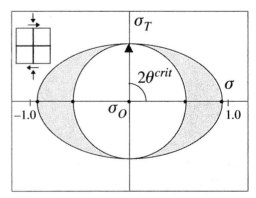

FIGURE 19 Geometric localization analysis, $f_c':f_t' = 1:1$. [From Kuhl, E. *et al.* (2000). *Int. J. Solids Structures* **37**, 48–50. With permission.]

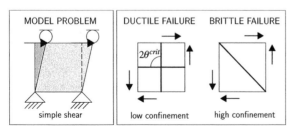

FIGURE 18 Model problem, simple shear.

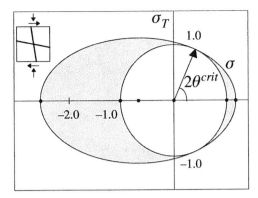

FIGURE 20 Geometric localization analysis, $f_c':f_t' = 3{:}1$. [From Kuhl, E. *et al.* (2000). *Int. J. Solids Structures* **37**, 48–50. With permission.]

observed in pressure-insensitive metals exhibiting Lüders bands.

Figure 20 depicts the result of localization analysis for the strength ratio $f_c':f_t' = 3{:}1$ representative for cast iron. The corresponding friction coefficient of $\alpha_F = 0.667$ induces relatively low confinement. Again, two critical directions are found, which have rotated slightly towards the direction of maximum principal strain at 45°. The corresponding critical failure angles of $\theta = 33.211°$ and $\theta = 146.79°$ indicate mixed shear-compression failure.

Finally, the compressive strength is assumed to be 12 times the tensile strength, $f_c':f_t' = 12{:}1$, which introduces high confinement. The friction coefficient takes a very large value of $\alpha_F = 3.667$, which is representative of cementitious materials such as concrete. For this class of materials with a very large contrast between compressive and tensile strength values, a brittle mode of failure is observed. Figure 21 illustrates mode I decohesive failure for which the critical direction corresponds to the direction of maximum principal stress, where $\theta^{crit} = 0°$.

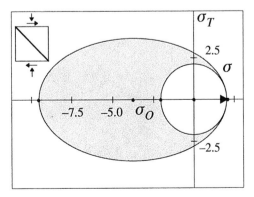

FIGURE 21 Geometric localization analysis, $f_c':f_t' = 12{:}1$. [From Kuhl, E. *et al.* (2000). *Int. J. Solids Structures* **37**, 48–50. With permission.]

V. ELASTIC DAMAGE MODELS

Elastic degradation formulations have received increasing attention since their inception by Lazar Markovich Kachanov (1914–1993). Nowadays, damage models are used regularly to describe the reduction of the elastic stiffness properties in quasibrittle materials, such as concrete, rocks, ceramics, etc. (see Lemaitre, 1992). For improved realism, additional features have been incorporated beyond the basic format of a scalar damage factor in order to account for different stiffness properties in tension and compression, to capture stiffness recovery due to microcrack closure, and to model the interaction of damage and plasticity processes.

As indicated in Section II.B.1, the basic secant formulation encompasses the simplest format of isotropic damage, which has emerged in the literature. Most scalar damage models start from the basic notion of the effective area concept by Kachanov,

$$A_{eff} = A - A_d = [1 - d]A, \quad \text{where} \quad d = \frac{A_d}{A}$$
$$\text{and} \quad 0 \le d \le 1 \tag{170}$$

according to which the internal stresses are transferred by the intact material skeleton. In other words, distributed microdefects and stress concentrators reduce the effective load-bearing area as compared to the nominal area. This argument leads directly to the effective stress concept through equilibrium considerations, whereby:

$$\sigma A = \sigma_{eff} A_{eff} \quad \text{leads to} \quad \sigma = [1 - d]\sigma_{eff} \tag{171}$$

The strain equivalence concept $\epsilon_{eff} = \epsilon$ of a parallel system of damaging elements leads to the classical scalar format of elastic damage,

$$\sigma_{eff} = E\epsilon_{eff} \quad \text{such that} \quad \sigma = [1 - d]E\epsilon \tag{172}$$

The $[1 - d]$ factor reduces the elastic stiffness properties when $d = 0 \to 1$. In summary, the elementary damage model leads to the secant stress–strain relationship,

$$\sigma = E_s\epsilon, \quad \text{where} \quad E_s = [1 - d]E$$
$$\text{and} \quad d = 1 - \frac{E_s}{E} \tag{173}$$

For a more differentiated isotropic approach, the degradation distinguishes between volumetric and deviatoric damage. In the case of anisotropic degradation, a number of representations have been proposed, starting from vector to second-, fourth-, and eighth-order tensor formulations. Thus, the most popular concept is the second-order damage tensor $[d_{ij}] = \boldsymbol{d}$ and its equivalent integrity tensor,

$$\boldsymbol{\phi} = \mathbf{1} - \mathbf{d} \quad \text{or} \quad \phi_{ij} = \delta_{ij} - d_{ij} \tag{174}$$

Whereas the secant formulation of elastic stiffness degradation is well established for isotropic as well as anisotropic damage, the constitutive formulation of the damage process requires a loading function, a damage rule, and some hardening/softening laws that describe the degradation process in terms of a reduced number of internal variables. In the spirit of the internal variable theory, they need to be defined in the space of conjugate thermodynamic forces, which turns out to be the strain energy density function when a scalar damage variable is used.

A. Basic Format of Elastic Scalar Damage

In the traditional $[1 - d]$ model of scalar damage, the isotropic stiffness and compliance tensors are replaced by their secant values,

$$\mathcal{E}_s = [1 - d]\mathcal{E}_o \quad \text{and} \quad \mathcal{C}_s = \frac{1}{1 - d}\mathcal{C}_o \quad (175)$$

Differentiation of the secant relations leads to:

$$\dot{\mathcal{E}}_s = -\dot{d}\mathcal{E}_o \quad \text{and} \quad \dot{\mathcal{C}}_s = \frac{\dot{d}}{1 - d}\mathcal{C}_s \quad (176)$$

The compliance term suggests changing variables and using the logarithmic scalar damage variable,

$$\dot{\ell} = \frac{\dot{d}}{1 - d}, \quad \text{where} \quad \ell = \int \dot{\ell} = ln\left(\frac{1}{1 - d}\right)$$
$$\text{and} \quad d = 1 - e^{-\ell} \quad (177)$$

This permits us to rewrite the secant tensors and the derivative of the compliance in the form of:

$$\mathcal{E}_s = e^{-\ell}\mathcal{E}_o \quad \text{and} \quad \mathcal{C}_s = e^{\ell}\mathcal{C}_o \quad \text{such that} \quad \dot{\mathcal{C}}_s = \dot{\ell}\mathcal{C}_s \quad (178)$$

In analogy to plastic associativity we need to examine the thermodynamic force conjugate to the damage variable in the dissipation inequality. The recoverable part of the elastic energy density is a function of the current secant stiffness or compliance properties; that is,

$$2W = \epsilon : \mathcal{E}_s : \epsilon = \sigma : \mathcal{C}_s : \sigma \quad (179)$$

Differentiation leads to the following energy exchange:

$$\dot{W} = \sigma : \dot{\epsilon} - \frac{1}{2}\sigma : \dot{\mathcal{C}}_s : \sigma \quad (180)$$

where $\sigma : \dot{\epsilon}$ is the external energy supply and $\dot{W}$ the increase of elastic strain energy density. Thus, the difference defines the dissipation rate, which must remain positive according to the second thermodynamic law:

$$\dot{\mathcal{D}} = \sigma : \dot{\epsilon} - \dot{W} = \frac{1}{2}\sigma : \dot{\mathcal{C}}_s : \sigma \geq 0 \quad (181)$$

The thermodynamic force $\mathcal{Y}$, which is taken to be conjugate to the basic change of the secant compliance, is

$$\dot{\mathcal{D}} = -\mathcal{Y} :: \dot{\mathcal{C}}_s \quad \text{where} \quad -\mathcal{Y} = \frac{1}{2}\sigma \otimes \sigma \quad (182)$$

In the scalar format of logarithmic damage, the thermodynamic conjugate force reduces to the elastic strain energy density function W, as:

$$\dot{\mathcal{D}} = -\mathcal{Y} :: \mathcal{C}_s\dot{\ell} = \frac{1}{2}\sigma : \mathcal{C}_s : \sigma\dot{\ell} = W\dot{\ell} \quad (183)$$

For associativity, the damage function needs to be expressed in terms of the conjugate thermodynamic force, which delimits the undamaged response regime of the strain energy resistance function r_d in analogy to the yield condition of plasticity,

$$F_d = F(W, \ell) = W - r_d(\ell) = 0 \quad (184)$$

In other terms, the resistance function $r_d = W$ is the internal strain energy of a specific test environment following the classical argument of Beltrami, or its deviatoric component in the case of the distortional energy criterion of Huber.

The derivation follows the elastoplastic format developed by Carol *et al.* (2001) for elastic dagradation. Starting from the decomposition of strain rate into an elastic stress producing and a degrading strain component, $\dot{\epsilon} = \dot{\epsilon}_e + \dot{\epsilon}_d$, illustrated in Fig. 22, the current secant stiffness is being used to differentiate between the elastic stress-producing deformations and the inelastic damage stress. This results in zero permanent deformations under load–unload cycles, though dissipation will take place according to the degree of nonlinearity during loading as opposed to linear unloading.

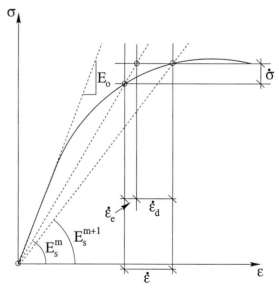

FIGURE 22 Response behavior of elastic degradation.

The set of rate equations which describes progressive degradation follows the individual steps of the flow theory of plasticity:

1. *Stress–strain rate relation*, $\sigma = \mathcal{E}_s : \epsilon$ and $\dot{\sigma} = \mathcal{E}_s : [\dot{\epsilon} - \dot{\epsilon}_d]$, where the rate of damage strain defines the deviation from linearity with regard to the linear secant stiffness as indicated in Fig. 22.

2. *Damage rule*, $\dot{\epsilon}_d = \dot{\lambda} m_d$, where $m_d = \mathcal{C}_s : \sigma = \epsilon$ defines the directional properties of elastic degradation and where the damage multiplier $\dot{\lambda}$ defines the magnitude of the inelastic damage strain.

3. *Damage threshold condition*, $F_d = f(\mathcal{Y}) - r_d(\lambda) = 0$, where $f(\mathcal{Y})$ denotes the demand in terms of a scalar-valued representation of the conjugate force in the dissipation inequality, and where r_d defines the threshold resistance function.

4. *Consistency condition for persistent elastic damage*, $\dot{F}_d = n_d \dot{\sigma} - H_d \dot{\lambda} = 0$, where $n_d = \frac{\partial F_d}{\partial \sigma}$ is the gradient of the damage function and where $H_d = -\frac{\partial F_d}{\partial \lambda}$ denotes the hardening/softening modulus of the resistance function.

5. *Damage multiplier*, $\dot{\lambda} = \frac{1}{h_d} n : \mathcal{E}_s : \dot{\epsilon}$, where the strain-driven format of damage introduces the denominator $h_d = H_d + n_d : \mathcal{E}_s : m_d$ which is subject to the same constraint, $h_d > 0$, as the analogous denominator term in elastoplasticity. This leads to the critical softening modulus for elastic damage $H_d^{crit} = -n_d : \mathcal{E}_s : m_d$.

6. *Tangential scalar damage properties*, $\mathcal{E}_{ed} = \mathcal{E}_s - \frac{1}{h_d} \mathcal{E}_s : m_d \otimes n_d : \mathcal{E}_s$, where the main difference from elastoplasticity is the secant reference stiffness. The constitutive format has the same structure as elastoplasticity except that $\mathcal{E}_s \neq const.$

In the isotropic case of the logarithmic scalar damage model, $\dot{\lambda} = \dot{\ell}$ and both the damage rule as well as the gradient of the damage function are associative as long as the damage function is expressed in terms of the conjugate force,

$$F_d = W - r_d(\ell) = 0 \quad \text{with} \quad n_d = \frac{\partial F_d}{\partial \sigma} = \mathcal{C}_s : \sigma = \epsilon \tag{185}$$

The hardening/softening modulus H_d at constant stress is defined as:

$$H_d = -\frac{\partial F_d}{\partial \lambda} = \frac{\partial r_d}{\partial \ell} \tag{186}$$

From these terms, the tangent stiffness of elastic degradation may be assembled in the standard manner which yields:

$$\mathcal{E}_{ed} = e^{-\ell} \mathcal{E}_s - \frac{1}{h_d} \sigma \otimes \sigma \tag{187}$$

where the denominator $h_d = \frac{\partial r_d}{\partial \ell} + \sigma : \epsilon$ must remain strictly positive under strain control. The symmetric format of the tangent stiffness operator is a consequence of the associative damage function, in contrast to damage functions such as the strain-based damage model of Mazars and his coworkers.

B. Simple Shear Response of Logarithmic Scalar Damage Model

In what follows we compare the results of the scalar-valued damage formulation with the corresponding elastoplastic results in Section III.D in order to assess their volumetric–deviatoric interaction. Analogous to plasticity, the elastic damage matrix may be partitioned into four submatrices which correspond to the normal and shear components of stress and strain. In the case of isotropic elasticity, there is no coupling between elastic normal and shear stresses and normal and shear strains in the initial stage before damage takes place.

Strain control of the simple shear test results in $\tau_{12} = G_0 \gamma_{12}$ at the onset of damage. Substituting into the elastic damage operator leads to the rate equations:

$$\begin{Bmatrix} \dot{\sigma}_{11} \\ \dot{\sigma}_{22} \\ \dot{\sigma}_{33} \\ \cdots \\ \dot{\tau}_{12} \\ \dot{\tau}_{23} \\ \dot{\tau}_{13} \end{Bmatrix} = e^{-\ell} \begin{bmatrix} K_o + \frac{4}{3}G_o & K_o - \frac{2}{3}G_o & K_o - \frac{2}{3}G_o & \vdots & 0 & 0 & 0 \\ K_o - \frac{2}{3}G_o & K_o + \frac{4}{3}G_o & K_o - \frac{2}{3}G_o & \vdots & 0 & 0 & 0 \\ K_o - \frac{2}{3}G_o & K_o - \frac{2}{3}G_o & K_o + \frac{4}{3}G_o & \vdots & 0 & 0 & 0 \\ \cdots & \cdots & \cdots & \cdots & \cdots & \cdots & \cdots \\ 0 & 0 & 0 & \vdots & G_o - \frac{e^\ell}{h_d}\tau_{12}^2 & 0 & 0 \\ 0 & 0 & 0 & \vdots & 0 & G_o & 0 \\ 0 & 0 & 0 & \vdots & 0 & 0 & G_o \end{bmatrix} \begin{Bmatrix} 0 \\ 0 \\ 0 \\ \cdots \\ \dot{\gamma}_{12} \\ 0 \\ 0 \end{Bmatrix} \tag{188}$$

The coupling partitions are zero, showing no interaction between the normal stresses and the shear strains, when the material experiences damage in analogy to the relation in Eq. (106) for J_2-plasticity.

1. Singularity of Tangent Damage Operator

Possible singularities of $\boldsymbol{E}_{ed}$ may be studied again with the Schur theorem of zero subdeterminant. We note that the normal partition of the stiffness matrix $\boldsymbol{E}_{nn}^{ed}$ simply reflects the degradation of the intact elastic partition $\boldsymbol{E}_{nn}^0$ by the integrity factor $e^{-\ell} = 1 - d$ as $\ell \to \infty$. The second possibility, which could cause the tangent operator to become singular, is the shear partition, $\boldsymbol{E}_{ss}^{ed}$. The two diagonal terms E_{55}^{ed} and E_{66}^{ed} are affected by degradation of the elastic properties; they will induce a singularity only when $e^{-\ell} \to 0$. The diagonal term E_{44}^{ed}, however, can diminish to zero before the material is fully damaged, since:

$$E_{44}^{ed} = e^{-\ell} G_o - \frac{1}{h_d} \tau_{12}^2 = 0 \qquad (189)$$

when

$$h_d = \frac{\tau_{12}^2}{e^{-\ell} G_o} = \frac{\tau_{12}^2}{G_s} = \tau_{12}\gamma_{12} \qquad (190)$$

The damage denominator $h_d = H_d + \boldsymbol{\sigma} : \boldsymbol{\epsilon}$ infers that the hardening modulus must be zero; that is,

$$H_d = \frac{\partial r}{\partial \ell} = 0 \quad \text{when} \quad h_d = \tau_{12}\gamma_{12} \qquad (191)$$

for any value of damage, ℓ. Thus, for perfectly damaging behavior, when $H_d = 0$ we have $E_{44}^{ed} = 0$, and consequently the tangent operator turns singular, $\det \boldsymbol{E}_{ed} = 0$.

This behavior is very much analogous to pressure-insensitive von Mises plasticity and holds independently of the load condition, when elastic damage is restricted to the deviatoric response behavior.

C. Damage Models under Simple Shear

To illustrate the performance of damage models let us consider again the strain-controlled load case of simple shear. To start with, we examine the relationship between scalar damage mechanics and the earlier development of one-invariant von Mises plasticity. Comparing the governing tangent operators in Eqs. (88) and (188) we recognize the same structure except for the integrity factor $e^{-\ell}$ in the case of scalar damage. Using the equivalent energy density value to calibrate the damage threshold $r_d = \frac{1}{2G}\tau_y^2$, Fig. 23a shows that the response behavior is identical to that of elastoplasticity shown for monotonic loading in Fig. 9. The basic difference between damage and plasticity shows up when we consider cyclic loading which takes place along the degraded secant stiffness, and which does not distinguish between tension and compression because of the underlying Beltrami condition of a maximum strain energy threshold. There is no confinement effect and the shear strength in this strain-controlled problem coincides with the tensile strength. Since the simple shear response may be viewed as an equi-biaxial tension-compression test with no direct-shear coupling, all damage stresses degrade exponentially according to the logarithmic damage evolution, in tension as well as in compression and shear.

In contrast, we consider the recent anisotropic damage formulation of Carol *et al.* (2001) to examine the influence of anisotropy on the volumetric–deviatoric interaction.

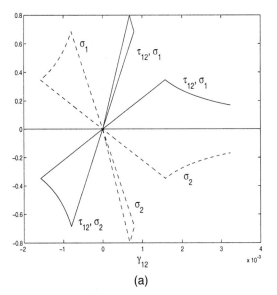

(a)

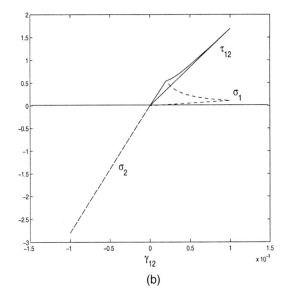

(b)

FIGURE 23 Simple shear response: (a) scalar damage model, (b) anisotropic damage model. [From Carol, I. *et al.* (2001). *Int. J. Solids Structures* **38**, 491–546. With permission.]

In this case, a second-order integrity tensor $\phi = 1 - d$ is adopted in extension of the pseudologarithmic scalar format above. Resorting to the so-called energy equivalence of Cordebois and Sidoroff, where neither effective stress nor effective strain coincides with its nominal counterpart, we develop a fourth-order damage effect tensor,

$$\sigma_{eff} = \alpha : \sigma \quad \text{and} \quad \epsilon = \alpha^t : \epsilon_{eff} \tag{192}$$

Assuming linear elastic behavior at the level of effective stress and effective strain, the resulting secant compliance retains symmetry:

$$\mathcal{C}_s = \alpha^t : \mathcal{C} : \alpha \tag{193}$$

In the principal axes of damage, the damage effect tensor reduces to diagonal form:

$$\alpha = \begin{bmatrix} \phi_1 & 0 & 0 & & & \\ 0 & \phi_2 & 0 & & 0 & \\ 0 & 0 & \phi_3 & & & \\ \hline & & & \sqrt{\phi_1 \phi_2} & & \\ & 0 & & & \sqrt{\phi_2 \phi_3} & \\ & & & & & \sqrt{\phi_3 \phi_1} \end{bmatrix} \tag{194}$$

Consequently, the corresponding secant compliance of anisotropic damage has the format

$$\mathcal{C}_s = \frac{1}{E} \begin{bmatrix} \phi_1^2 & -\nu\phi_1\phi_2 & -\nu\phi_1\phi_3 & & & \\ -\nu\phi_2\phi_1 & \phi_2^2 & -\nu\phi_2\phi_3 & & 0 & \\ -\nu\phi_3\phi_1 & -\nu\phi_3\phi_2 & \phi_3^2 & & & \\ \hline & & & 2[1+\nu]\phi_1\phi_2 & & \\ & 0 & & & 2[1+\nu]\phi_2\phi_3 & \\ & & & & & 2[1+\nu]\phi_3\phi_1 \end{bmatrix} \tag{195}$$

where the secant stiffness involves five independent material parameters as opposed to nine in orthotropic elasticity. For damage evolution we need to define a damage criterion. For associated damage the demand function $F_d = f(\mathcal{Y}) - r_d(\ell) = 0$ is based on the conjugate force which, in this case, is the second-order energy tensor $\mathcal{Y} = \sigma \cdot \epsilon$. In this case, we resort to the Rankine criterion of maximum principal energy where the scalar damage function is $F_d = \mathcal{Y}_{max} - r_d(\ell) = 0$. Adopting a no-damage switch in compression, the simple shear response results in the response behavior shown in Fig. 23b. In comparison to the scalar damage format in Fig. 23a, the simple shear response of the anisotropic damage model in Fig. 23b shows some very interesting features: (1) after reaching the damage threshold at $\tau_{12} = 0.8 \, ksi$, the principal tensile stress diminishes again exponentially; and (2) the shear stress and the compressive principal stress increase without bounds because of volumetric–deviatoric interaction which introduces increasing confinement. Moreover, there

is no damage in compression, and unloading occurs at the secant stiffness in tension, leaving no permanent deformation, while the compressive stiffness remains intact.

In contrast to the scalar damage model, we notice that the anisotropic damage model does exhibit volumetric–deviatoric coupling. Thus, the shear stress increases far beyond the initial shear strength threshold $\tau_y = 0.8 \, ksi$ in the presence of out-of-plane confinement. At the same time, the nominal tensile stress is subject to softening due to progressive tensile damage according to the logarithmic degradation law.

In summary, isotropic scalar damage models do not exhibit shear dilatancy, while anisotropic models do reproduce the Reynolds dilatancy effect of frictional materials similarly to pressure-sensitive plasticity.

VI. CONCLUSIONS

This state-of-the-art report has focused on major developments in rate-independent elasticity and inelasticity. Many topics have not been covered for the sake of brevity—most notably, nonlinear hardening and softening formats in plasticity and damage (Lubliner, 1990); the general form of the internal variable description (Halphen and Nguyen, 1975); the microstructural features thereof (Kröner, 1963); the size and gradient effects (Bažant and Planas, 1997); and last but not least the computational aspects of inelastic analysis.

The exposition presented an apercu of linear and nonlinear elastic constitutive models as well as plastic and elastic damage models. Specifically, we have examined the interaction of normal and shear components which is directly related to the population in the constitutive gradients n, m responsible for stress- and/or strain-induced anisotropy in the tangent operator.

ACKNOWLEDGMENTS

The author wishes to acknowledge the help of Dr. Eric Hansen and former coworkers Dr. Ellen Kuhl and Dr. Hong Kang for their assistance and for preparing some of the figures. This report was partially sponsored by the National Science Foundation grant CMS 9622940 to the University of Colorado, Boulder. Opinions expressed in this paper are those of the writer and do not reflect those of the sponsor.

APPENDIX 1: SOME RESULTS OF MATRIX AND TENSOR ANALYSIS

The mathematical tools behind material models are housed in linear algebra and tensor analysis, in particular. For this reason, we revisit a few well-established relationships which may provide additional insight beyond the mere mechanics of elementary tensor manipulation.

Stress and Strain Tensors

Stress and strain tensors are a set of objects $\sigma, \epsilon \in \Re^3$ which may be represented as (3×3) matrices whose coefficients are the coordinates of a set of orthonormal base vectors.

1. Scalar (inner) product of $\sigma, \epsilon \in \Re^3$ (double contraction):
This product operation generates a scalar value:

$$w = \sigma : \epsilon = \sigma_{ij}\epsilon_{ij} \qquad (196)$$

The direction between two tensors is

$$\cos\theta = \frac{\sigma : \epsilon}{\sigma \epsilon} \qquad (197)$$

where $0 \leq \theta \leq \pi/2$, and $\sigma = (\sigma : \sigma)^{\frac{1}{2}}$ and $\epsilon = (\epsilon : \epsilon)^{\frac{1}{2}}$ are the Euclidean lengths of the two vectors. *Note*: The internal strain energy is directly related to the scalar product of stress and strain.

2. Outer product of $\sigma, \epsilon \in \Re^3$ (single contraction): This product operation generates a second order tensor:

$$w = \sigma \cdot \epsilon = \sigma_{ij}\epsilon_{jk} \qquad (198)$$

Note: The thermodynamic force in the form of a tensor product of stress and strain is conjugate to the second order damage tensor.

3. Dyadic or tensor product of $\sigma, \epsilon \in \Re^3$ (no contraction): This product operation generates a fourth-order rank-one tensor:

$$W = \sigma \otimes \epsilon = \sigma_{ij}\epsilon_{kl} \qquad (199)$$

Note: The tensorial order of the dyadic product is the sum of the tensorial order of each factor; e.g., the dyadic product of two second-order tensors generates a fourth-order tensor, etc.

4. Inverse of rank-one modification of unit tensor I:

$$B = I + a \otimes b \quad \text{then} \quad B^{-1} = I - \frac{1}{1+a \cdot b}a \otimes b \qquad (200)$$

as long as $a \cdot b \neq -1$.

5. Sherman–Morrison formula of inversion:

$$B = A + X \otimes Y \qquad (201)$$

then

$$B^{-1} = A^{-1} - \frac{1}{1 + YA^{-1}X}A^{-1}X \otimes YA^{-1} \qquad (202)$$

6. Outer products do not commute:

$$A \cdot B \neq B \cdot A; \quad \text{however,} \quad B \cdot A = [A^t \cdot B^t]^t \qquad (203)$$

This also holds for tensor products.

Determinant

The determinant of a second-order tensor is a single number which summarizes the tensorial property in the form of a multilinear functional.

1. Determinant of tensor products:

$$\det(A \cdot B) = \det A \det B \qquad (204)$$

2. The Schur theorem states that the determinant of partitions in tensor A may be written in terms of the determinant product of partitions:

$$\begin{aligned} \det A &= \det A_{22} \det A_{11}^{compl} \\ &= \det A_{22} \det \left(A_{11} - A_{12}A_{22}^{-1}A_{21}\right) \end{aligned} \qquad (205)$$

Note: The Schur complement couples the partitions and leads to the bound $\det A \leq \det A_{22} \det A_{11}$ as long as $x \cdot A_{22} \cdot x > 0$ is positive definite.

Eigenvalues and Eigenvectors

There exists a nonzero vector x such that the linear transformation $\sigma \cdot x$ is a multiple of x:

$$\sigma \cdot x = \lambda x \qquad (206)$$

Note: The eigenvector x_i spans the triad of principal directions, and the eigenvalues λ_i define the three principal values of stress.

1. Characteristic polynomial: The eigenvalue problem is equivalent to stating that:

$$(\sigma - \lambda I)x = 0 \qquad (207)$$

For a nontrivial solution $x \neq 0$, then $(\sigma - \lambda_i I)$ must be singular. Consequently,

$$\det(\sigma - \lambda I) = 0 \qquad (208)$$

generates the characteristic polynomial

$$p(\lambda) = \det(\sigma - \lambda I) \qquad (209)$$

the roots of which are the eigenvalues $\lambda(\sigma)$. According to the fundamental theorem of algebra, a polynomial of degree 3 has exactly 3 roots, thus each matrix $\sigma \in \Re^3$ has 3 eigenvalues. *Note*: All three eigenvalues are real as long as $\sigma = \sigma^t$ is symmetric, which is the case for nonpolar materials because of conjugate shear stresses $\sigma_{ij} = \sigma_{ji}$.

2. Cayley–Hamilton theorem: This theorem states that every square matrix satisfies its own characteristic equation. In other words, the scalar polynomial $p(\lambda) = \det(\lambda I - \sigma)$ also holds for the stress polynomial $p(\sigma)$. One important application of the Cayley–Hamilton theorem is to express powers of the stress tensor σ^k as a linear combination of the irreducible bases I, σ, σ^2 for $k \geq 2$.

3. Spectral properties of rank-one update of unit tensor: Spectral analysis of a square matrix generated by a rank-one update of the unit tensor of second-order

$$B = I + a \otimes b \qquad (210)$$

reduces to the eigenvalue analysis,

$$[(b \cdot a - \lambda I)\alpha]a = 0 \qquad (211)$$

The eigenvalues and eigenvectors of $B = I + a \otimes b$ are related to the eigenpairs of the update matrix; that is,

$$\lambda(B) = 1 + \lambda \quad \text{and} \quad x(B) = x \qquad (212)$$

In the case of a single rank-one update of the unit matrix, we find $\lambda_1(B) = 1 + \lambda$ and $\lambda_k(B) = 1 \forall k = 2, 3, \ldots, n$ with the determinant $\det(B) = \det(I + a \otimes b) = 1 + a \cdot b$ and $\lambda = a \cdot b$.

APPENDIX 2: PLANE STRAIN CONSTRAINT OF ELASTOPLASTIC LIMIT POINT

On a final note, we observe that the kinematic constraint of zero out-of-plane deformations delays and may suppress altogether the formation of a limit point associated with peak strength in pressure-sensitive plasticity. We recall the shear response does not reach a limit point under plane strain. The reason for this puzzling observation is the kinematic constraint that restricts the formation of a limit point when:

$$\dot{\sigma} = \mathcal{E} : [\dot{\epsilon} - \dot{\epsilon}_p] = 0 \quad \text{where} \quad \dot{\epsilon}_p = \dot{\lambda} m \qquad (213)$$

In the plastic flow rule, $\dot{\lambda}$ denotes the plastic multiplier, which is strictly positive under sustained plastic flow, and $m = \frac{\partial Q}{\partial \sigma}$ is the direction of plastic flow in terms of the gradient of the plastic potential. Consequently, the only possibility to form a limit point at peak occurs if the strain rate equals the plastic strain rate, $\dot{\epsilon} = \dot{\epsilon}_p$, and the elastic strain rate remains zero, $\dot{\epsilon}_e = 0$. The plane strain constraint infers that the out-of-plane strain rate must vanish; that is,

$$\dot{\epsilon}_{33} = \dot{\lambda} m_{33} = \dot{\lambda} \frac{\partial Q}{\partial \sigma_{33}} = 0 \qquad (214)$$

In the three-invariant plasticity formulation above, this constraint implies that:

$$m_{33} = \frac{1}{\sqrt{3}} \left(Q_\xi - \frac{Q_\theta}{\tau_{12}} \right) = 0 \quad \text{or} \quad Q_\xi = \frac{Q_\theta}{\tau_{12}} \qquad (215)$$

In short, pressure-sensitive plastic flow, $Q_\xi \neq 0$, does not permit formation of a limit point as shown in Figs. 9 and 14. And, even if this would be the case, the shear stress would never reach the peak value because loss of stability and localization take place much earlier in the ascending hardening regime.

SEE ALSO THE FOLLOWING ARTICLES

BONDING AND STRUCTURE IN SOLIDS • ELASTICITY • FRACTURE AND FATIGUE • MATERIALS CHEMISTRY • PLASTICITY

BIBLIOGRAPHY

Bažant, Z. P., and Planas, J. (1997). "Fracture and Size Effect in Concrete and Other Quasibrittle Materials," CRC Press, Boca Raton, FL.

Benallal, A., and Comi, C. (1996). "Localization analysis via a geometrical method," *Int. J. Solids Structures* **33**, 99–119.

Carol, I., Rizzi, E., and Willam, K. (2001). "On the formulation of anisotropic elastic degradation. Part I: Theory based on a pseudo-logarithmic damage tensor rate; Part II: Generalized pseudo-Rankine model for tensile damage," *Int. J. Solids Structures* **38**, 491–546.

Chen, W. F., and Han, D. J. (1988). "Plasticity for Structural Engineers," Springer-Verlag, New York.

Coleman, B. D., and Gurtin, M. E. (1967). "Thermodynamics with internal state variables," *J. Chem. Phys.* **47**, 597–613.

Drucker, D. C., and Prager, W. (1952). "Soil mechanics and plastic analysis of limit design," *Q. Appl. Math.* **10**, 157–175.

Halphen, B., and Nguyen, Q. S. (1975). "Sur les matériaux standards généralisés," *J. de Mécanique* **14**(1), 39–63.

Kang, H., and Willam, K. (1999). "Localization characteristics of a triaxial concrete model," *ASCE J. Eng. Mech.* **125**, 941–950.

Kröner, E. (1963). "Dislocation: a new concept in the continuum thery of platicity," *J. Math. Phys.* **42**, 27–37.

Kuhl, E., Ramm, E., and Willam, K. (2000). "Failure analysis of elastoplastic materials at different levels of observation," *Int. J. Solids Structures* **37**, 48–50; 7259–7280.

Lemaitre, J. (1992). "A Course on Damage Mechanics," Springer-Verlag, Berlin.

Lubliner, J. (1990). "Plasticity Theory," Macmillan, New York.

Maier, G., and Hueckel, T. (1979). "Nonassociated and coupled flow rules of elastoplasticity for rock-like materials," *Int. J. Rock Mech. Min. Sci.* **16**, 77–92.

Ogden, R. W. (1984). "Non-Linear Elastic Deformations," Ellis Horwood, Chichester, U.K.

Ottosen, N. S., and Runesson, K. (1991). "Properties of discontinuous bifurcation solutions in elasto-plasticity," *Int. J. Solids Structures*, **27**, pp. 401–421.

Rudnicki, J. W., and Rice, J. R. (1975). "Conditions for the localization of deformation in pressure-sensitive dilatant materials," *J. Mech. Phys. Solids*, **23**, 371–394.

Valanis, K. C. (1975). "On the foundations of the endochronic theory of viscoplasticity," *Arch. Mech. Stos.* **27**, 857–868.

Willam, K. J., and Warnke, E. P. (1975). "Constitutive model for the triaxial behaviour of concrete" *Proc. Int. Assoc. Bridge Structural Engineers*, Report 19, Section III, Zürich, Switzerland, pp. 1–30.

Continental Crust

Walter D. Mooney
U.S. Geological Survey

I. Methods of Studying the Crust
II. Deep Structure of the Continental Crust
III. Global Crustal Thickness
IV. Continental Fault Zones
V. Fluids in the Crust
VI. Composition of Continental Crust
VII. Origin of Continental Crust

GLOSSARY

Geologic province Large region of the crust characterized by similar geologic history and development, e.g., orogens, platforms, continental arcs, rifts, and extended crust.

Lithosphere The lithosphere is the crust and subcrustal uppermost mantle (continental lithosphere may be up to 150 km thick and oceanic lithosphere can reach a value of 80–100 km).

Moho Seismic boundary between the crust and uppermost mantle; the lower crust typically has a compressional wave velocity of 6.8–7.4 km/sec and the uppermost mantle a velocity greater than 7.6 km/sec, with an average value of 8.07 km/sec.

Plate tectonics A theory of the global scale dynamics involving the relative movements of many rigids lithospheric plates. Along the edges of these lithospheric plates are zones of seismic and tectonic activity.

Pn Compressional wave velocity of the uppermost mantle, i.e., directly below the Moho. Pn velocities show a range of values from 7.6 to 8.8 km/s with an average of 8.07 km/s.

Seismic refraction profiles Seismic data recorded with widely spaced (100–5000 m) geophones and long offsets (100–300 km) between sources and receivers. The data provide excellent constraints on seismic velocity within the lithosphere.

Seimic reflection profiles Seismic data recorded with closey spaced (5–100 m) geophones and sources that yield a high-resolution image of the crust, but generally do not constrain seismic velocities in the middle or lower crust.

THE CONTINENTAL CRUST, the exterior layer of the earth, has been thoroughly studied by earth scientists due to its accessibility. Indeed, since the crust was extracted from the mantle, much of what we infer about the evolution of the entire planet is based on our knowledge of the continental crust. It is also of great practical importance as our primary source of mineral resource wealth, all arable land, and fresh water.

The crust is defined as the layer of the earth above a very distinct seismic boundary that lies at a depth of about 12 km beneath the oceans and from 20 to 80 km beneath

Encyclopedia of Physical Science and Technology, Third Edition, Volume 3

TABLE I Physical Characteristics of the Continental Crust[a]

Area	2.10×10^8 km^2
Fraction of earth's area	0.412
Mean thickness	41 km
Range of thickness	10–75 km
Average thickness at sea level	28 km
Volume	8.61×10^9 km^3
Mean density	2.83 g/cm^3
Mass	2.44×10^{25} g
Percent of earth's mass	0.40%
Percent of earth's mass, excluding core	0.62%
Percent of area above mean sea level	71.3%
Mean elevation	125 m
Mean elevation lying above 200 m isobath	690 m
Mean compressional wave seismic velocity (V_p)	6.45 km/sec
Mean shear wave seismic velocity (V_s)	3.65 km/sec
Mean poisson's ratio (σ)	0.27

[a] Modified from McLennan, S. M., and Taylor, S. R. (1999). *In* "Encyclopedia of Geochemistry" (C. P. Marshall and R. W. Fairbridge, eds.), pp. 145–150, Kluwer Academic, New York.

continents. This boundary is called the Mohorovicic seismic discontinuity, or "Moho" for short, in honor of the man who discovered it in 1909. The Moho is defined as the boundary (or transition zone) where the compressional wave seismic velocity, as measured by seismic refraction data, increases to greater than or equal to 7.6 km/s. Although the seismic definition of the crust is a limited one, in fact the change in seismic velocity at the Moho is so pronounced that it must also correspond to a fundamental geologic change, specifically from low-density, silica-rich rocks to higher-density olivine-rich rocks. The Moho has been found wherever seismic measurements have been made, and thus can be considered as a universal boundary, much like the boundary that separated the earth's silicate mantle from its iron core. The mean thickness of the continental crust is 41 km (Table I). Not all of the continental crust is above sea level. It extends offshore on the continental shelf up to the slope break, where the bathymetry abruptly increases and oceanic crust forms the basement.

The crust does not "float" in the underlying mantle. The crust and the uppermost mantle together comprise the lithosphere, which is defined as the cold and strong, outmost portion of the earth that deforms in an essentially elastic manner. The lithosphere is also sometimes defined as the portion of the earth through which heat from the interior is transported solely by conduction. This latter definition implies a lithospheric thickness beneath continents of about 50–350 km depending on the age and recent tectonic history of the lithosphere. Thus, it is essential to consider the evolution of the entire lithosphere when discussing the evolution of the continental crust.

The continental crust is in general significantly older than the oceanic crust, a fact that is well understood in the context of plate tectonics. The oldest oceanic crust is only about 200 million years old because the oceanic crust is continuously recycled back into the mantle as an essential part of the plate tectonic process. In contrast, more than 60% of the continental crust is 0.6–3.0 billion (600–3000 million) years old, with a small amount of continental crust being 3.0–4.0 billion years old. The oldest portions of the continental crust are the stable cratons, which are further divided into regions with exposed crystalline rocks (shields) and regions covered with younger sediments (platforms; Fig. 1).

Our initial ideas regarding the continental crust came from geologic mapping. The origins of modern geologic thinking can be traced to James Hutton, who published his thesis "Theory of the Earth with Proof and Illustrations" in 1795. A key concept that was pioneered by Hutton is the cyclical view of the earth, with active mountain building balanced by erosion and destruction of the crust. In the past 200 years the entire earth has been mapped in considerable detail, and the distribution of rock types and tectonic settings is well established.

Geophysical studies of the earth as a whole can be traced to at least 1600. In this year William Gilbert, the physician to Queen Elizabeth, published *De Magnete*, a landmark treatise on magnetism that concluded that the earth itself was a large magnet. Geophysical studies of the crust are more than 100 years old. The first seismographs were developed in Italy (F. Ceccgi, 1875), Japan (by J. Ewing of England), and Germany (Potsdam Observatory, 1889). The basis for the modern electromagnetic seismograph was developed on 1910 in Russia by B. B. Galitzen, the father of seismometry. By 1940, H. Jeffreys and K. E. Bullen published a complete model for the earth's interior, one of the great scientific achievements of the twentieth century. The second half of the twentieth century saw extensive geophysical exploration of the continents, both for basic research and for mineral and hydrocarbon exploration. Geophysical data, especially regarding seismicity, marine magnetic anomalies, and paleomagnetism, also provided the critical evidence needed to establish the principles of plate tectonics.

In comparison with relatively older fields of geology and geophysics, geochemical studies of the earth are a twentieth-century phenomenon. Geochemistry was provided a firm basis in 1908 when F. W. Clarke of the U.S. Geological Survey published the first edition of his landmark book, *The Data of Geochemistry*. V. I. Vernadsky of the former USSR was a key figure in the application of geochemistry to geologic problems, and was the first to recognize the importance of living organisms in geologic and geochemical processes. In the first half of the 1900s,

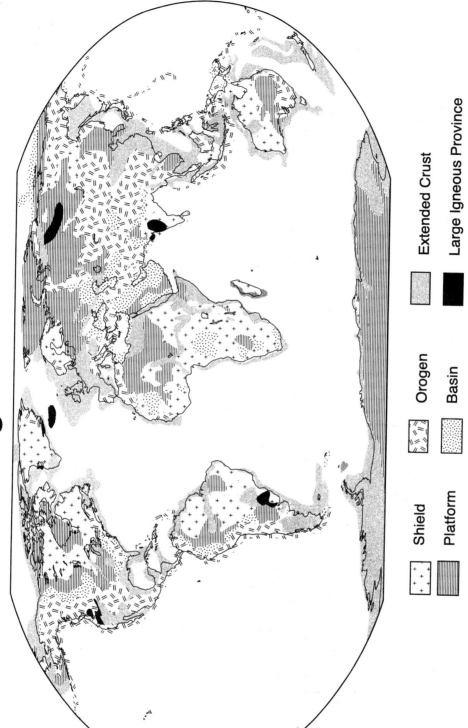

Geologic Province

FIGURE 1 Continental geologic provinces: shield (exposed stable crust, also called cratonic crust); platform (cratonic crust with flat-lying sedimentary cover); orogen (area of recent or ancient mountain building; basin (sedimentary accumulations of variable thickness); extended crust (region subjected to horizontal extension, accompanied by thinning of the crust); large igneous province (region covered by extensive igneous extrusions usually 1–5 km in thickness).

Shield
Platform
Orogen
Basin
Extended Crust
Large Igneous Province

V. M. Goldschmidt (Germany and Norway) established the basic principle of geochemistry that ionic radius and charge permits isomorphous substitution of elements within crystals. In recent decades a large database has been developed regarding the distribution of chemical elements in the continental crust. This in turn provides important constraints on the composition and evolution of the crust.

I. METHODS OF STUDYING THE CRUST

A. Seismic Studies

1. Deep Seismic Refraction and Reflection Data

Seismic techniques provide the highest-resolution measurements of the structure of the crust, and have been conducted on a worldwide basis (Fig. 2). Seismic studies of the deep continental crust that utilize man-made sources are classified into two categories, seismic refraction and reflection data, depending on their field acquisition parameters. Refraction data provide reliable information regard-

ing the gross layering of seismic velocities within the crust, and are very effective in mapping crustal thickness (i.e., the depth to Moho). In contrast, seismic reflection data provide an image of the crust at a finer scale (50 m in the vertical and horizontal dimensions). As a result, reflection and refraction data have complementary strengths: reflection data provide a structural image of the crust, whereas refraction data provide an estimate of the seismic velocity distribution in the crust.

The seismic velocities measured with refraction data are primarily determined by five factors: mineralogical composition, confining pressure, temperature, anisotropy, and pore fluid pressure. Thus, in order to draw inferences about the mineralogical composition of the deep crust we must first estimate the contribution of the other four properties. Confining pressure can be calculated from depth of burial, and temperature can be estimated from the surface heat flow. The roles of seismic anisotropy and pore fluid pressure are more difficult to estimate. The inherent uncertainty of this procedure can be reduced when both compressional (V_p) and shear wave (V_s) velocities are

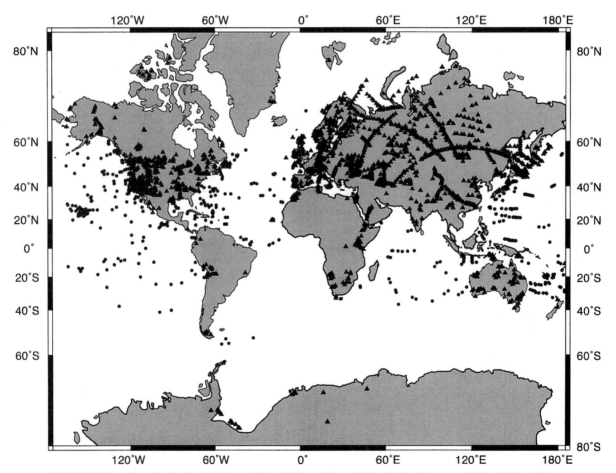

FIGURE 2 Location of seismic refraction profiles within continents (triangles) and oceans (circles). [From Mooney, W. D., Laske, G., and Masters, T. G. (1998). *J. Geophys. Res.* **103,** 727–747.]

measured. The relation between V_p and V_s is expressed by Poisson's ration, which varies from 0.23 to 0.32 for most minerals but quartz has a value of only 0.08. Thus, the measurement of Poisson's ratio offers the means of distinguishing between felsic (quartz-rich) and mafic (quartz-poor) rocks.

2. Seismic Surface Waves and Teleseismic Data

An earthquake near the earth's surface will generate seismic surface waves of the Rayleigh and Love types. The amplitude of motion for these waves decreases exponentially with depth in the earth (hence the term surface waves), with longer-wavelength waves being sensitive to the seismic velocity structure at greater depth than shorter-wavelength waves. This wavelength dependence of seismic velocity gives rise to the dispersion of surface waves and makes them a valuable tool for studying the vertical velocity structure of the crust. The analysis of seismic surface waves for crustal structure has been applied on a global basis and is particularly effective in determining the shear-wave structure of the crust. Thus, seismic surface waves provide complementary information to seismic refraction data that normally provide information on the compressional-wave structure of the crust.

Local and distant (teleseismic) earthquake data can be used to determine crustal structure. One method, seismic tomography, uses earthquake arrivals at a network of seismic stations to determine crustal structure by examining the arrival times of many criss-crossing paths between the earthquakes and seismometers. This technique, which is similar to medical tomography, provides a three-dimensional picture of the crust. A second technique uses teleseismic data and searches for seismic arrivals that are produced by the phases conversions at seismic boundaries. This technique has the advantage of needing only a single high-quality seismic station to produce reliable results, and thus has become an increasingly valuable source of information in the past 20 years.

B. Nonseismic Methods

1. Gravity, Magnetic, Electrical, and Paleomagnetic Methods

All rock types have a variety of distinct, albeit nonunique, physical properties that includes density, magnetic susceptibility, and conductivity. Geophysical surveying techniques to measure these properties are highly developed, and it is now possible to make detailed maps of lateral changes in rock density, magnetic properties, or conductivity. Advanced digital processing of such data can also be used to reliably infer these rock properties at depth, al-though the resolution of such inferences decreases steadily with depth. Such studies are widely used and have the capability to distinguish between competing geologic models of the structure of the crust.

Rocks commonly retain a magnetism that originates from the time of their formation. Measurement of this paleomagnetic direction can be used to determine the latitude at which the rock was created. In many cases this latitude will differ from the rock's present latitude, and the difference can be ascribed to the distance the rock has moved, due to continental drift, since its formation. Such measurements provided the first quantitative estimates of relative plate motions, and today provide a detailed picture of the coalescing and dispersion of continents.

2. Geologic Mapping, Petrologic and Geochemical Studies, and Deep Drilling

The continental crust can only be understood from the surface down—that is, all models for the deep structure and composition must rigorously satisfy the constraint that it is consistent with exposed sections of crust. One such constraint is that the upper crust has a dominantly silica-rich ("felsic") composition, whereas exposures of lower crustal sections reveal mafic rocks that have been metamorphosed to high metamorphic grade and have a low content of water. Additional information regarding the deep crust comes from samples of rocks that are carried to the surface by volcanic rocks. These so-called xenoliths (Greek for "foreign rock") provide valuable petrologic and geochemical information on rocks at depth within the crust. Direct sampling deep within the upper crust has also been achieved. Deep scientific drilling has reached depths of about 10–12 km in southern Germany and the Kola Peninsula of northwestern Russia. Samples taken from these drill holes, as well as measurements made within the drill holes, provide invaluable information on the *in situ* properties of the deep crust.

3. Heat Flow

Radioactive decay of isotopes of Uranuim (U), Thorium (Th), and Potasium (K) generates a considerable amount of heat within the earth, particularly the crust and mantle. These three elements are strongly concentrated in the continental crust because the crust has been extracted from the mantle by a melt that is rich in these elements. Silicic rocks, such as granites, are significantly more radiogenic than mafic rocks, such as basalt. Thus, the analysis of heat flow data can provide significant constraints on crustal composition. For example, a 40-km-thick crust that is purely granitic in composition would contain abundant U, Th, and K throughout the 40 km thick crust. It would therefore generate an excessive geothermal heat flow in comparison with what is typically measured. Likewise, a purely mafic crust would have too little U, Th, and K, and generate too little heat flow. Heat flow data indicate that a more suitable crustal composition consists of a granitic upper crust that grades into a mafic lower crust.

Heat flow data also provide the most reliable information regarding temperatures within the crust. Current estimates are that the temperature at the base of the crust (40 km) ranges from about 400 to 900°C. Crustal temperatures are significant because they effect the strength (rheology) of the rocks within the crust.

4. Laboratory Measurements of Rock Properties

Field measurements of rock properties would be of limited use without complementary laboratory measurements to calibrate them and to permit inferences of deep crustal composition. Extensive databases are available for all rock properties, including measurements at elevated pressures and temperatures. These include measurements of density, seismic velocity, conductivity, magnetic susceptibility, and porosity.

These measurements document the important effects of pressure and temperature on seismic velocity. As confining pressure increases to about 200 Mpa, microcracks close, causing P- and S-velocities to increase rapidly (about 0.5–1.0 km/sec/100 Mpa); at higher pressures, the increase is slight (about 0.02–0.06 km/sec/100 Mpa for most rock types). As temperature increases, both compressional and shear-wave velocities decrease, with typical coefficients of about 2.0–6.0×10^{-4} km/sec/°C, depending on rock type. Since pressure and temperature both increase with depth in the earth, their effects on velocity compete. For a typical continental geotherm of about 15°C/km, a crust of uniform lithology will have a constant seismic velocity at middle- and lower-crustal depths, while in high heat-flow provinces (25–35°C/km) the temperature effect dominates, causing negative velocity gradients.

II. DEEP STRUCTURE OF THE CONTINENTAL CRUST

A. Seismic Velocities

Seismic refraction data have provided information on the seismic velocity structure of the crust on a global basis (Fig. 2). The crystalline crust that underlies sedimentary deposits generally has a compressional wave velocity between 5.7 and 7.4 km/sec. In many cases it is possible to subdivide the seismic structure of the crust into three distinct layers (Fig. 3), although it should be emphasized

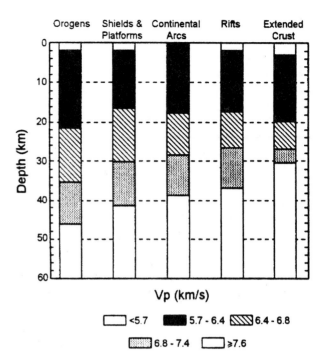

FIGURE 3 Average crustal structure for five geologic provinces, from thickest (orogens) to thinnest (extended crust). The crystalline crust (i.e., below sedimentary deposits) is subdivided into three layers that correspond to granitic/gneissic rocks (upper crust, 5.7–6.4 km/sec), dioritic/amphibolitic rocks (middle crust, 6.4–6.8 km/sec), and gabbroic/granulitic rocks (6.8–7.4 km/sec). These crustal columns are only averages. Each of these five geologic provinces shows considerable diversity. [From Christensen, N. I., and Mooney, W. D. (1995). *J. Geophys. Res.* **100,** 9761–9788.]

that such a subdivision is not intended to imply simply layering within the crust. Exposed crustal sections show that in fact the crust is complexly layered at all scales.

It has long been recognized that there is an important correlation between crustal structure and geologic province. We have therefore divided them into five primary geologic provinces: orogens, shields and platforms, continental arcs, rifts, and extended (stretched) crust. Large igneous provinces are regions where voluminous volcanic outpourings cover basement rocks, which can consist of any of the five primary geologic provinces. Shields and platforms occupy by far the largest area of continental crust (Fig. 1). Orogens include the young, active mountain belts of the Alps, Andes, and Tibet, and ancient orogens such as the Urals, Appalachians, and the Tien Shan, China. Continental arcs include the trans-Mexican volcanic belt, Cascades of North America, and active volcanic belts of the western Pacific. Extended crust includes such regions as the Basin and Range of the western United States and much of western Europe. Rifts include East Africa, Lake Baikal, and the Rio Grande Rift. We estimate the following proportions of continental crust by area: 69% shield and platform; 15% old and young orogens; 9% extended crust; 6% magmatic arc; 1% rift.

Shields and platforms have an average crustal thickness of 41.5 km, a value that that is close to the worldwide average crustal thickness (39.17 km). Extended crust, as the name implies, has been thinned by extension, and has an average thickness of 30.5 km. Orogens show a wide range of crustal thicknesses, ranging from about 30 to 72 km. Rifts, both active and inactive, also show a broad range, from 18 to 46 km. It should be emphasized that some of these ranges are obtained within a single tectonic province. For example, the crustal thickness within the Alps of southern Europe varies from about 35 to as much as 60 km, and the Kenya rift shows crustal variations along strike of the rift that amounts to 20–36 km.

The thickest crust occurs beneath mountain belts (orogens) and the thinnest crust beneath highly extended crust (Fig. 3). The upper crust is commonly 15–20 km thick and has seismic velocities in the range 5.7–6.4 km/sec. The middle crust is commonly 8–15 km thick and has velocities in the range 6.4–6.8 km/sec. The lower crust is highly variable in thickness (0–12 km) and has seismic velocities in the range 6.8–7.4 km/sec. The range of continental crustal thickness measured globally (Fig. 4A) is 16–72 km, with an average of 39.7 km. Crust that is thinner than about 25 km, or thicker than 60 km, is very rare.

Average crustal velocity (Fig. 4B) is an important property because it is a weighted average of the bulk crustal composition. The average seismic velocity of the crust ranges from 6.0 km/sec (corresponding to a felsic bulk composition) to about 6.9 km/sec (dominantly mafic composition), with a global average of 6.45 km/sec (corresponding to a bulk composition that is intermediate between felsic and mafic, and equivalent to a diorite).

The top of the uppermost mantle is defined as that depth where the seismic velocity exceeds 7.6 km/sec. The seismic velocity of the uppermost mantle is frequently referred to as the *Pn* velocity, for "normal P (compressional) wave." *Pn* velocities show a range of values from 7.6 to 8.8 km/sec, the average is 8.07 km/sec with a standard deviation of 0.21 km/sec (Christensen and Mooney, 1995).

B. Crustal Reflectivity

Seismic reflection data (Fig. 5) provide the highest resolution information on the *in situ* structure of the lower continental crust and Moho. The seismic properties that are most readily obtained from reflection data are reflectivity patterns, and these correlate with distinct geologic settings. When reflectivity patterns are interpreted with complementary seismic velocity and nonseismic crustal parameters, inferences regarding the composition and evolution of the lower crust can be made.

Various hypotheses have been advanced for the origin of lower crustal reflectivity. The origin of this reflectivity must, in fact, be multigenetic, a point that is unambiguously demonstrated by the correlation of reflection data with boreholes and by tracing reflection horizons to outcrop. Causes of crustal reflectivity that have been demonstrated by such direct evidence include igneous intrusions (sills), fine-scaled metamorphic layering, shear (mylonite) zones, and lithologic layering.

Recent reflection data in Precambrian shields reveal strong crustal reflectivity, including subhorizontal laminae in the lower crust. Many of the Precambrian reflectivity patterns appear to preserve structures that date from ancient collisional events, which implies that this reflectivity is not due to post-orogenic ductile shear, igneous intrusions, or fluids. Rather, reflectivity is due to primary lithologic and metamorphic layering, and Precambrian shear zones that were formed during ancient compressional orogenies.

In extended crust, the origin of crustal reflectivity is probably depth dependent, with ductile shear-enhancing reflectivity that has its primary origin in lithologic and metamorphic layering in the lower crust, and igneous layering within the 3–5 km thick Moho transition zone. Layered zones of high pore pressure (up to several kilometers thick and due to metamorphic dewatering of the lower crust) may sometimes be present at the top of the lower crust, as indicated by low seismic refraction velocities and high electrical conductivity.

The observation of widespread crustal reflectivity leads to the suggestion that a common process acts to promote its existence. It is likely that regional and global plate stresses act to enhance reflectivity by inducing lower crustal ductile flow that produces subhorizontal lamination. This ordering process would enhance reflectivity that has its primary cause in igneous intrusions, or compositional, metamorphic, or mineralogic layering. This ordering process requires elevated temperatures to promote ductile flow, and the dipping reflection patterns in Precambrian terranes appear to be primary patterns associated with Proterozoic microplate collisions, rather than secondary reflectivity associated with ductile flow. Thus, the Precambrian crust appears to possess considerable long-term thermal and tectonic stability.

III. GLOBAL CRUSTAL THICKNESS

A contour map of global crustal thickness (Fig. 6) shows the bimodal division of the earth's crustal thickness. Ocean basins have 6–7 km thick crust (not including 4–5 km of water) and continents have an average thickness of 39.7 km. The crust is typically 30 km thick at the ocean-continent margin and gradually increases toward the

(A)

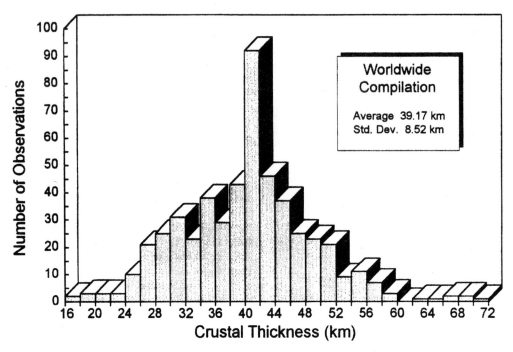

(B)

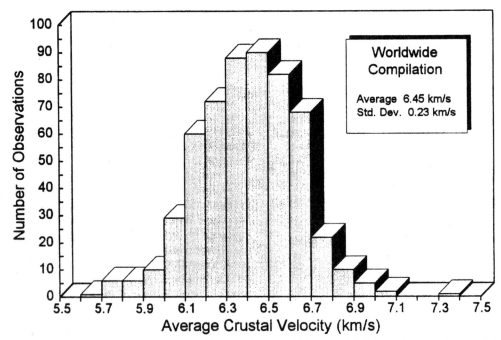

FIGURE 4 (A) Histogram of continental crustal thickness. Average thickness is 39.17 km, with a standard deviation of 8.52 km. Weighted for land area, the average thickness is 41 km (cf. Table I). Ninety percent of continental crust has a thickness between 24 and 54 km. (B) Histrogram of average crustal compressional-wave velocity, which is a proxy for bulk crustal composition and density. The average is 6.45 km/sec, corresponding to a bulk crustal composition equivalent to a diorite, and a density of 2.84 g/cc. [From Christensen, N. I., and Mooney, W. D. (1995). *J. Geophys. Res.* **100,** 9761–9788.]

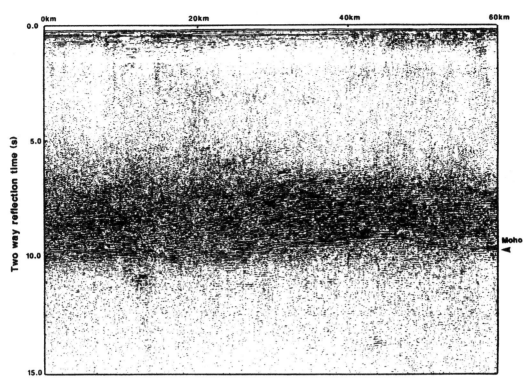

FIGURE 5 Seismic reflection profile recorded in shallow water offshore of Britain showing a highly reflective lower crust and relatively transparent upper crust and mantle. [From Warner, M. (1990). *Tectonophys.* **173**, 163–174.]

continental interior to 40–45 km. Crust that is thicker than 50 km is restricted to a few regions, including the Tibetan Plateau of western China, the Andes of western South America, and the Precambrian shield of southern Finland. The contour map (Fig. 6) only shows large-scale crustal features, thus some regions of locally thick crust are not visible. The crust does not show a pattern of increased thickness with age, as would be the case if it were repeatedly subjected to igneous intrusions from the underlying mantle. For example, the crust of western Australia is older (3000 Ma.) than central Australia (less than 2000 Ma.), yet the crust is at least 10 km thinner in western Australia. Indeed, the crust with a thickness in excess of 50 km is almost always a young, active mountain belt. These regions have high topography and are subject to rapid erosion. Continental extension and rifting is also an important geologic process and results in thinned crust. Examples include western Europe and the western U.S.A.

IV. CONTINENTAL FAULT ZONES

Continental fault zones are important because they preserve the history of crustal deformation (such as thrusting of one rock unit over another during a compressional

event), and because destructive earthquakes occur on continental faults, such as the San Andreas and North Anatolian faults. The deformation of rocks due to faulting causes changes in lithology, pore pressure, and seismic velocity, all of which are evident from geophysical measurements. The deformation of rocks by faulting ranges from intragrain microcracking to severe alteration. Saturated microcracked and mildly fractured rocks do not exhibit a significant reduction in velocity, but from borehole measurements, densely fractured rocks do show significantly reduced velocities, the amount of reduction generally proportional to the fracture density. Highly fractured rock and thick fault gouge along the creeping portion of the San Andreas fault are evidenced by a pronounced seismic low-velocity zone (LVZ), which is either very thin or absent along locked portions of the fault. Thus there is a correlation between fault slip behavior and seismic velocity structure within the fault zone; high pore pressure within the pronounced LVZ may be conductive to fault creep.

The internal properties of fault zones provide critical information on the manner in which deformation takes place within continental interiors, and is an important constraint on the physical basis for slip behavior of faults. Furthermore, fault zones are likely regions for future successful

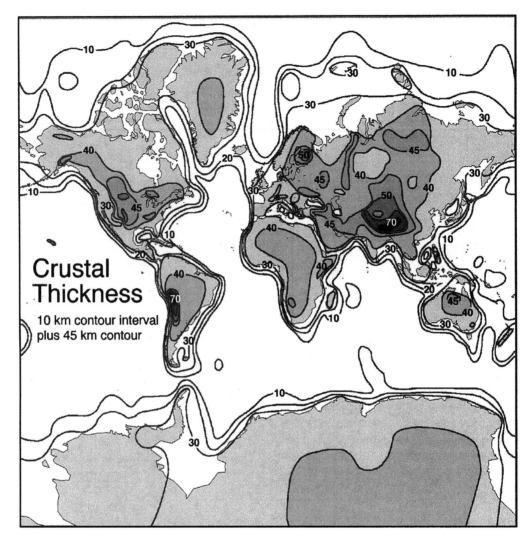

FIGURE 6 Mercator projection of contour map of crustal thickness based on seismic refraction data shown in Fig. 2. The average crustal thickness of oceanic crust is 6–7 km (excluding the 4–5 km of water) and the average thickness of continental crust is about 40 km. [From Mooney, W. D., Laske, G., and Masters, T. G. (1998). *J. Geophys. Res.* **103**, 727–747.]

earthquake-prediction measurements, since they may potentially exhibit premonitory phenomena.

V. FLUIDS IN THE CRUST

The presence of fluids in the crust is of great importance to many basic geologic processes. Fluids can lower the melting temperature of rocks by several hundred degrees, thereby giving rise to magmatism. Indeed, the clear layering of lower-velocity felsic rocks above higher-velocity mafic rocks (Fig. 3) may be indicative of magmatic differentiation of the crust at some point in its history. Fluids also reduce the total pore pressure in rocks, and therefore will enhance faulting. The primary evidence for fluids in the deep continental crust comes from interpretations of electromagnetic data, which in many places (particularly in Phanerozoic crust) require a zone of high conductance in the lower crust. Saline fluids in rocks with a porosity of a few percent would account for the high conductivity. The effect that such porosity would have on seismic velocities depends on the pore pressure of the fluids: fluids at low pore pressure will have a negligible effect on seismic velocity, while pore pressures near lithostatic pressure will significantly reduce seismic velocity. For the low-porosity metamorphic and igneous rocks expected in the lower crust, porosity of several percent would have to be held open by high pore pressure, a condition that is unlikely to exist for long time periods. It is therefore generally considered that the middle and lower crust is dry, that is, has a very low percent concentration of fluids.

VI. COMPOSITION OF CONTINENTAL CRUST

An estimate of bulk crustal composition must satisfy at least four demanding constraints: (1) provide a match to the average seismic velocity structure of the crust; (2) agree with exhumed deep crustal sections; (3) agree with the composition of deep crustal xenoliths; and (4) provide the appropriate amount of radiogenic heat production to match the average surface heat flow of 57 mW/m^2 (as discussed in the section entitled Heat Flow). The bulk composition of the continental crust has been established by these methods to be intermediate in silica content (i.e., ~60% SiO$_2$ between basalt and rhyolite). The most common extrusive rock with an intermediate composition is an andesite, named after the Andean volcanism of South America; the equivalent intrusive rock is a diorite.

The mineralogical composition of the upper crust is given in Table II. This table was determined from numerous surface samples. Felsic minerals predominate, notably quartz, plagioclase, and orthoclase. The bulk chemical composition of the crust can be summarized with a dozen oxides (Table III). We note for comparison that the weight percent of SiO$_2$ in the mantle is 45% (vs 61.7% in the crust) and the weight percent of MgO in the mantle is 38% (vs 3% in the crust). Thus the mantle is significantly more mafic in composition compared with the crust. The mantle also has a density (3.3 g/cc at a depth of 50 km) that is significantly greater than the average crustal density (2.84 g/cc). Table IV lists the weight percent for the

TABLE III Bulk Crustal Chemical Composition[a]

Oxide	Weight%
SiO$_2$	61.7
TiO$_2$	0.9
Al$_2$O$_3$	14.7
Fe$_2$O$_3$	1.9
FeO	5.1
MgO	3.1
CaO	5.7
MnO	0.1
Na$_2$O	3.6
K$_2$O	2.1
P$_2$O$_5$	0.2
H$_2$O	0.8

[a] From Christensen, N. I., and Mooney, W. D. (1995). *J. Geophys. Res.* **100,** 9761–9788.

nine most abundant elements that occur in the continental crust.

A hypothetical cross section through the continental crust (Fig. 7) summarizes many of the previously discussed conclusions regarding the structure and composition of the crust. Whereas the crust is grossly stratified compositionally into a felsic upper crust, intermediate middle crust, and mafic lower crust, Fig. 7 emphasizes the complex interfingering of all rock types throughout the crust. The abrupt change in crustal thickness on the left-hand side of the figure depicts a strike-slip fault that separates 50-km-thick crust from 30-km-thick crust. Such a boundary is found between the central Europe platform and shield (Poland and Ukraine) and the extended crust of western Europe (Germany and Austria).

TABLE II Mineralogical Composition (%) of the Upper Continental Crust[a]

	Weight %	
	Upper crust	Exposed crust
Quartz	23.2	20.3
Plagioclase	39.9	34.9
Glass	0.0	12.5
Orthoclase	12.9	11.3
Biotite	8.7	7.6
Muscovite	5.0	4.4
Chlorite	2.2	1.9
Amphibole	2.1	1.8
Pyroxene	1.4	1.2
Olivene	0.2	0.2
Oxides	1.6	1.4
Other	3.0	2.6
Total	100.2	100.1

[a] From Nesbitt, H. W., and Young, G. M. (1984), *Geochim. Cosmochim. Acta* **48,** 1523–1534.

TABLE IV Nine Most Abundant Elements and Their Average Concentration in Continental Crust[a]

Element	Weight %
O	45.5
Si	26.8
Al	8.4
Fe	7.1
Ca	5.3
Mg	3.2
Na	2.3
K	0.9
Ti	0.5

[a] Modified from McLennan, S. M., and Taylor, S. R. (1999). *In* "Encyclopedia of Geochemistry" (C. P. Marshall and R. W. Fairbridge, eds.), pp. 145–150, Kluwer Academic, New York.)

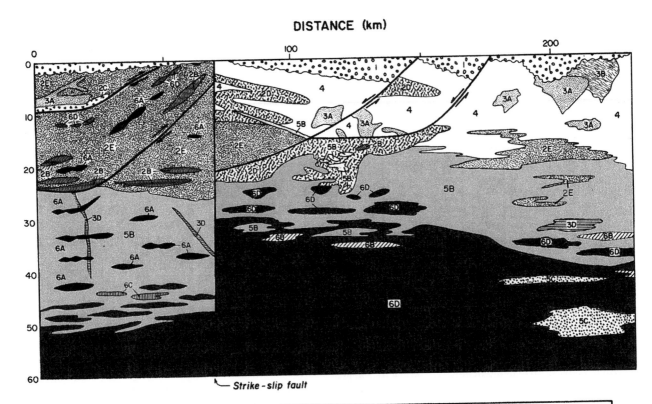

DISTANCE (km)

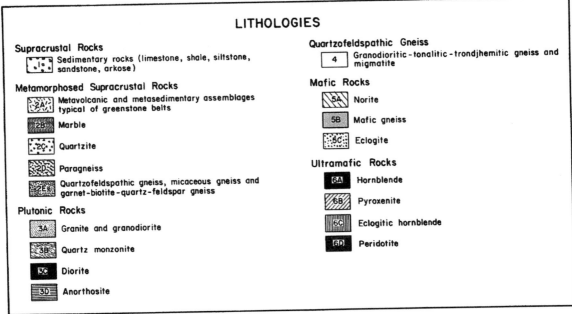

LITHOLOGIES

Supracrustal Rocks

Sedimentary rocks (limestone, shale, siltstone, sandstone, arkose)

Metamorphosed Supracrustal Rocks

2A — Metavolcanic and metasedimentary assemblages typical of greenstone belts

2B — Marble

2C — Quartzite

2D — Paragneiss

2E — Quartzofeldspathic gneiss, micaceous gneiss and garnet-biotite-quartz-feldspar gneiss

Plutonic Rocks

3A — Granite and granodiorite

3B — Quartz monzonite

3C — Diorite

3D — Anorthosite

Quartzofeldspathic Gneiss

4 — Granodioritic-tonalitic-trondjhemitic gneiss and migmatite

Mafic Rocks

5A — Norite

5B — Mafic gneiss

5C — Eclogite

Ultramafic Rocks

6A — Hornblende

6B — Pyroxenite

6C — Eclogitic hornblende

6D — Peridotite

FIGURE 7 Cross section of the continental crust and uppermost mantle to a depth of 60 km. Lithologies are indicated in key. In this hypothetical section, the crust decreases from a thickness of 50 km on the left to 30 km to the right of the strike-slip fault. This section illustrates the complexity of the crust, with numerous upper-crustal, low-angle faults and isolated sills of mafic and ultramafic rocks in the lower crust. [From Fountain, D. M., and Christensen, N. I. (1989). "Geological Society of America Memoir 172," pp. 711–742, Boulder, CO.]

VII. ORIGIN OF CONTINENTAL CRUST

The problem of the origin of the continental crust can be resolved into two fundamental questions: (1) the location and mechanisms of initial mantle extraction of the primitive crust and (2) the processes by which this primitive crust is converted into the stable, cratonic continental crust. There is general agreement that the continental crust that formed in the past 2.0 Ga was created by plate tectonic processes. Some investigators extend the time that this process was active to 3.0–4.0 Ga, but this is a matter of considerable debate.

We can confidently state that for the past 2.0 Ga continental crust has had its primary origin at convergent plate boundaries where subduction of oceanic lithosphere gives rise to hydration and melting of the mantle above the subducting plate. The result is the formation of island arcs for ocean–ocean convergence, and a continental magmatic arc for ocean–continent convergence. Examples of island arcs are the Aleutians of southern Alaska and Japan; continental magmatic arcs include the volcanoes of Mexico, the western U.S.A. (Cascades), and western South America. The basic process of crustal growth consists of melts rising from the mantle to form new continental material. However, the process is not simple, since both island arc and continental arc magmatism (collectively referred to here as "arcs") produces a melt that is more basaltic than the average bulk composition of continental crust. Thus, new crust created by the arcs must undergo further evolution to provide typical continental crust. For example, if the dense root of the arc is detached and sinks back into the mantle (a process referred to as delamination), the upper two-thirds of the arc will have a composition close to typical continental crust. However, it is usually difficult to document crustal delamination using geophysical data, and the importance of this process is therefore subject to debate.

Arcs are not the only sites of extraction of voluminous melt from the mantle. Oceanic plateaus are large volcanic features that are found in all ocean basins. They are believed to be the result of largely submarine volcanism above a hot mantle upwelling, known as a mantle plume. Examples include the Kerguelen Plateau of the South Indian Ocean and the submerged Ontong-Java Plateau of the Southwest Pacific. The larger oceanic plateaus are too thick and buoyant to subduct, and thus accrete to the continental margin at an ocean–continent converent margin. The basaltic crust that comprises western Oregon and Washington State is believed to be an example of an accreted oceanic plateau. However, like arcs, plateaus have a bulk basaltic composition, and thus they also must undergo further evolution in composition to form typical continental crust. One such process is remelting and igneous differentiation, with silicic melts rising in the crust and denser, more mafic residuum sinking to the lowermost crust.

As mentioned above, it is not agreed how crustal formation has evolved over the past 4.0 Ga. However, crust older than about 2.5 Ga (Archean) is compositionally distinct from younger continental crust. Archean igneous rocks are dominantly bimodal, consisting of mafic thoeleiite plus dacitic, and are depleted in heavy rare earth elements, in contrast to the dominantly unimodal, roughly andesitic calc-alkaline magmatism that forms younger crust. These compositional differences may be due to different mechanisms of crustal extraction from the mantle or to different mechanisms of differentiation and alteration of newly formed continental crust. For example, we may speculate that two processes were more important in crustal formation during the Archean: (1) higher mantle temperatures during the Archean may have given rise to abundant mantle–plume magmatism (analogous to present-day volcanism in Hawaii and Iceland), and/or (2) high mantle temperatures resulted in the melting of the basaltic layer in subducted oceanic crust, producing the observed bimodal magmatism. In contrast, crust younger than 2.0 Ga is largely the product of arc volcanism. The question of the evolution of continental crust over the past 4.0 Ga remains an open question, and will be a focus for research in years to come.

SEE ALSO THE FOLLOWING ARTICLES

EARTHQUAKE MECHANISMS AND PLATE TECTONICS • EARTHQUAKE PREDICTION • EARTH SCIENCES, HISTORY OF • EARTH'S CORE • ENVIRONMENTAL GEOCHEMISTRY • GLOBAL SEISMIC HAZARDS • HEAT FLOW • MINERALOGY AND INSTRUMENTATION • OCEANIC CRUST • PLATE TECTONICS • STRESS IN THE EARTH'S LITHOSPHERE

BIBLIOGRAPHY

Christensen, N. I., and Mooney, W. D. (1995). "Seismic velocity structure and composition of the continental crust: A global view." *J. Geophys. Res.* **100,** 9761–9788.

Fountain, D. M., and Christensen, N. I. (1989). Geological Society of America Memoir 172, pp. 711–742, Boulder, CO.

McLennan, S. M., and Taylor, S. R. (1999). *In* "Encyclopedia of Geochemistry," (C. P. Marshall and R. W. Fairbridge, eds.) pp. 145–150, Kluwer Academic, New York.

Mooney, W. D., Laske, G., and Masters, T. G. (1998). "CRUST 5.1: A global crustal model at 5° × 5°. *J. Geophys. Res.* **103,** 727–747.

Nesbitt, H. W., and Young, G. M. (1984). "Prediction of some weathering trends of plutonic and volcanic rocks based on thermodynamic and kinetic considerations," *Geochim. Cosmochim. Acta* **48,** 1523–1534.

Warner, M. (1990). *Tectonophys.* **173,** 163–174.

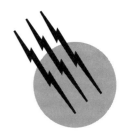

Controls, Adaptive Systems

I.D. Landau

Centre National de la Recherche Scientifique and Laboratorie d'Aromatique de Grenoble

GLOSSARY

Adaptation mechanism Component of an adaptive system that drives the adjustable parameters of the system.

Adaptive predictor Adjustable predictor of the output or the state of the plant whose parameters are driven by a parameter adaptation algorithm.

Direct adaptive control The parameters of the controller are directly adapted by the adaptation mechanism that processes the plant-model error.

Index of performance Measure of the performance of a system.

Indirect adaptive control Adaptation of the parameters of the controllers is done in two steps: (1) estimation of the process parameters and (2) computation of the controller parameters based on the current process estimated parameters.

Model reference adaptive control Direct adaptive control technique that uses a filter called *reference model* for specifying the desired control performance.

On-line parameter identification Recursive parameter estimation techniques based on the use of an adjustable predictor and a parameter adaptation algorithm that processes sequentially the input-output data as they are available.

Parameter adaptation algorithm Recursive algorithm for updating the adjustable parameters that is used in the adaptation mechanism.

Plant-model error, prediction error, adaptation error Measures of performance error used directly or indirectly in parameter adaptation algorithms.

Self-tuning control Indirect adaptive control technique that uses an on-line estimation of the process parameters for the computation of the controller parameters.

ADAPTIVE CONTROL is a set of techniques for automatic adjustment in real time of controllers in order to achieve or maintain a desired level of performance of a control system when the process parameters are unknown or change with time. These techniques have been

Encyclopedia of Physical Science and Technology, Third Edition, Volume 3

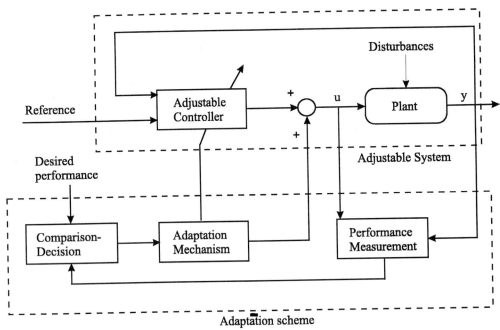

FIGURE 1 Basic configuration of an adaptive control system.

developed in order to design high-performance control systems when *a priori* knowledge of current values of the process parameters is not available. These techniques are also used for automatic tuning of control systems. Research in the area started at the end of the 1950s. After many years of research and experimentation and the advent of the microprocessors, the application of these techniques is expanding. Dedicated software and adaptive regulators are available.

I. ADAPTIVE CONTROL SYSTEMS: BASIC PRINCIPLES

The unknown and unmeasurable variations of process parameters (or insufficient knowledge of them) degrade the performance of control systems. These variations (or errors) are termed *parameter disturbances*. The disturbances acting on a control system can be classified as (1) disturbances acting on the controlled variables and (2) parameter disturbances acting on the performance of the control system.

Feedback is used in the conventional control systems basically to eliminate the effect of disturbances on the controlled variables and to bring them back to their desired values according to a certain index of performance (IP).

A similar conceptual approach can be considered for the problem of achieving and maintaining the desired performance of a control system in the presence of parameter disturbances. We must define an IP for the performance of a control system (e.g., the damping factor), then we mea-

sure this IP. The measured IP is compared with the desired IP, and the difference is fed into an adaptation mechanism. The output of the adaptation mechanism acts on the parameters of the controller and/or on the control signal in order to modify the system performance accordingly. A basic configuration of an adaptive control system is given in Fig. 1.

An adaptive control system can be interpreted as a feedback system where the controller variable is the IP. As shown in Fig. 1, the adaptive control loop is an additional feedback loop to be used when it is necessary to monitor the performance of the basic feedback control system.

II. BASIC ADAPTIVE CONTROL TECHNIQUES

A. Open-Loop Adaptive Control (Gain Scheduling)

In addition to closed-loop adaptive control systems, simple open-loop systems have also been developed. An example is the gain scheduling technique. A block diagram of such a system is shown in Fig. 2.

This technique assumes the existence of a rigid relationship between some measurable variables of the environment and the parameters of the process model. It is possible using this relationship to reduce (or eliminate) the effect of parameter variations by changing accordingly the parameters of the controllers. The modification of system performance by means of gain scheduling is not

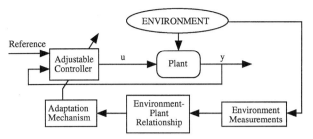

FIGURE 2 Open-loop adaptive control system (gain scheduling).

monitored. Therefore, this system can fail if for some reason the rigid relationship between the environment measurement and process parameters changes. We shall consider next the two most commonly used closed-loop adaptive control techniques: direct adaptive control (DAC) and indirect adaptive control (IAC).

B. Direct Adaptive Control (Model Reference Adaptive Control)

Figure 3 illustrates the basic philosophy of designing a linear controller (in a deterministic environment). The design assumes a knowledge of the plant dynamic model and of the desired performances. In most cases the desired performances of the feedback control system can be specified in terms of the characteristics of a dynamic system that is a "realization" of the desired behavior of the closed-loop system. For example, a tracking objective can be specified in terms of the desired input-output behavior by a given transfer function. A regulation objective can be specified in terms of the evolution of the output starting from an initial disturbed value by specifying the desired pole location of the closed loop (i.e., by a given transfer function). The controller is designed such that for a given plant model the closed-loop control system has the characteristics of the desired dynamic system.

Since the desired performance corresponds in fact to the evolution of the desired dynamic system, which is prespecified, the design problem can be equivalently reformulated, as in Fig. 4. The reference model in Fig. 4 is a realization of the system with the desired performance.

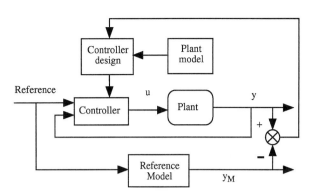

FIGURE 4 Design of a linear controller using a reference model.

The controller is now designed such that (1) the error between the output of the plant and the output of the reference model is identically zero for identical conditions, and (2) an initial error will vanish with a certain dynamics.

When the plant parameters are unknown or change with time, in order to achieve the desired performances an adaptive control approach has to be considered for the linear control schemes given in Figs. 3 and 4. Figure 5 shows the adaptive control scheme known as model reference adaptive control (MRAC), which is an extension of the linear control scheme presented in Fig. 4. This scheme is based on the observation that the difference between the output of the plant and the output of the reference model (subsequently called *plant-model error*) is a measure of the difference between the real and the desired performance. This information is used through the adaptation mechanism (which also receives other information) to adjust the parameters of the controller automatically in order to force asymptotically the plant-model error to zero.

C. Indirect Adaptive Control (Self Tuning Control)

Figure 6 shows an indirect adaptive control scheme known as self-tuning control (STC), which is an extension of the

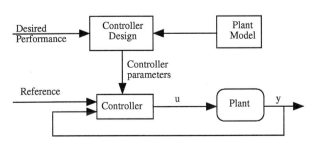

FIGURE 3 Design of a linear controller (deterministic environment).

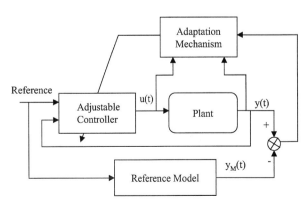

FIGURE 5 Model reference adaptive control system (direct adaptive control).

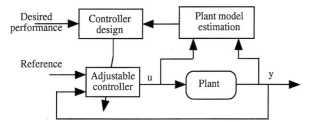

FIGURE 6 Indirect adaptive control (self-tuning control).

linear control scheme presented in Fig. 3. The basic idea is that a suitable controller can be designed if a model of the plant is estimated on-line based on input and output available data. To understand how this adaptive control schemes operates, we must first consider in more detail the on-line estimation of a plant model.

The basic scheme for the operation of all types of on-line parameter estimation algorithms is shown in Fig. 7. The main idea in on-line parameter estimation is to build up an adjustable predictor for the output of the plant. The error between the plant output and the predicted output (prediction error) is used by a recursive estimation (adaptation) algorithm to adjust the parameters of the adjustable predictor. The objective is to force the prediction error asymptotically to zero (in a deterministic environment). This type of scheme is in fact an adaptive predictor, which enables one to obtain asymptotically an estimated model giving a correct input–output description of the plant for a given sequence of inputs.

Recursive identification schemes can also be interpreted in terms of a model reference adaptive system (MRAS). The plant to be identified can be interpreted as the reference model. The parameters of the adjustable system (the adjustable predictor) are driven by the adaptation mechanism in order to bring asymptotically to zero the error between the output of the plant and the output of the adjustable model.

Introducing the block diagram for the parameter estimation algorithms given in Fig. 7 into the scheme of Fig. 6, one obtains the general configuration of the indi-

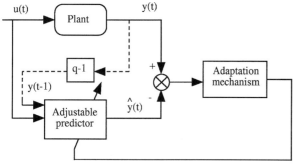

FIGURE 7 Basic scheme for on-line parameter estimation.

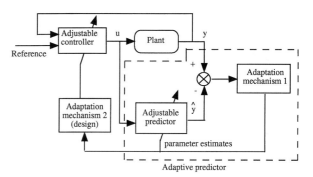

FIGURE 8 Indirect adaptive control (detailed scheme).

rect adaptive control scheme given in Fig. 8. Using this basic scheme, one can obtain a large variety of adaptive control algorithms by combining various recursive parameter estimation algorithms (adaptation algorithm) with various strategies for designing the controller based on a knowledge of the parameters of the plant model (adaptation mechanism). However, the behavior of these schemes should be analyzed carefully. Unfortunately, analytical results describing the behavior of such adaptive control schemes are available only for a limited choice of parameters estimation algorithms and control strategies.

One of the most popular IAC is the adaptive pole placement scheme.

D. Direct and Indirect Adaptive Control

Comparing the MRAC shown in Fig. 5 with the IAC shown in Fig. 8 one observes an important difference. In the scheme of Fig. 5, the parameters of the controller are directly adapted (estimated) by the adaptation mechanism. In the scheme of Fig. 8, adaptation mechanism 1 adapts the parameters of an adjustable predictor, and these parameters are then used to compute the controller parameters. The MRAC shown in Fig. 5 belongs to the class of direct adaptive control schemes, while the scheme shown in Fig. 8 belongs to the class of indirect adaptive control schemes.

In many cases, however, by an appropriate parametrization of the adjustable predictor (reparametrization of the predictor model in terms of controller parameters) the parameter adaptation (estimation) algorithm in Fig. 8 will directly estimate the parameters of the controller, yielding a direct adaptive control scheme. In such cases, the adaptive mechanism 2 in Fig. 8 disappears and the connection with explicit MRAC is obvious since in both schemes one directly adjusts the parameters of the controller.

A number of well-known indirect adaptive control schemes (minimum variance self-tuning control, etc.) are in fact DAC since in these schemes one directly estimates the parameters of the controller.

Direct adaptive control is a very appealing approach because the implementation is simpler and it avoids the numerical problems associated in many cases with the computing controller parameters from plant parameters estimates. Unfortunately, its application is limited to plant models that have stable zeros.

E. Basic Design of Adaptive Controllers

The concepts and configurations in a deterministic environment, as discussed above, can be extended in a straightforward manner to operations in a stochastic environment. When designing a control system in a stochastic environment, one has to consider, in addition to the plant model, the model for the stochastic disturbance. When the disturbance is modeled as an autoregressive moving average (ARMA) process, no matter what kind of linear controller will be used, the output of the plant will also be an ARMA process (in regulation). Therefore, the control performance can be specified in terms of an ARMA model that has the desired statistical properties (mean value, variance). The controller will be designed such that the output behaves as the desired ARMA model. Note that an ARMA process is obtained by sending a Gaussian white-noise sequence through a pole-zeros filter, the properties of the resulting ARMA process being defined entirely by the filter transfer function. Therefore, a stochastic reference model can be defined. However, in a stochastic environment one cannot specify exactly the desired future values of the plant output. The best one can achieve is a design in which the error between the desired predicted output and the plant output is an innovation sequence (or a white-noise sequence). For this reason, only a prediction reference model can be built.

When the parameters are known, the design is such that the error between the plant output and the desired predicted output is an innovation sequence (or a white-noise sequence). This sequence can be used when necessary to excite the stochastic part of the prediction reference model. When the parameters of the plant model and disturbance model are unknown or vary, the error between the output of the plant and the predicted output given by the reference model is no longer an innovation sequence. Then, as indicated in Figs. 5 and 7, an adaptation mechanism that uses this error will drive the parameters of the controller such that the plant model prediction error is forced to become an innovation sequence asymptotically.

In Figs. 1, 5, and 8 one observes that, despite the fact that the plant can be assumed to be linear but with unknown parameters, the adaptive control scheme is in fact nonlinear. The parameters of the controller depend on the plant variables through the adaptation mechanism.

Since adaptive control systems are nonlinear, global stability in the deterministic environment and global convergence in the stochastic environment are basic properties that such systems should have. In both deterministic and stochastic environments it is possible to cast the analysis (and design) of DAC into a stability problem for a system disturbed from its equilibrium. In the deterministic case, the equilibrium corresponds to a null plant-model error; in the stochastic case, the equilibrium corresponds to the situation when the plant-model error is an innovation sequence.

The steps in the design of adaptive control systems are summarized as follows:

1. Design of the linear controller assuming that the parameters of the plant model and disturbance model are known.
2. Definition of the structure of the adaptive control loop.
3. Design of the parameter adaptation algorithms.

III. DESIGN OF DIRECT ADAPTIVE CONTROLLERS

A. Known-Parameters Case

Consider a continuous-time plant characterized by the transfer function:

$$H(s) = Ge^{-s\tau}/(1 + sT) \qquad (1)$$

which one would like to control through a computer. The pulsed transfer function using a zero-order hold corresponding to $H(s)$ is given by:

$$H(z^{-1}) = B(z^{-1})/A(z^{-1}) = \frac{b_1 z^{-1} + b_2 z^{-2}}{1 + a_1 z^{-1}} \qquad (2)$$

where:

$$a_1 = -e^{-\Delta/T} \qquad (3)$$

$$b_1 = G\left(1 - e^{-(\tau-\Delta)/T}\right) \qquad (4)$$

$$b_2 = Ge^{-\Delta/T}(e^{\tau/T} - 1) \qquad (5)$$

Here, Δ represents the sampling period, τ is the pure time delay (in this example, $\tau < \Delta$), and z is a complex variable.

The discretized plant model characterized by the discrete transfer function $H(z^{-1})$ can be represented in the time domain by the recursive equation:

$$y(t + 1) = -a_1 y(t) + b_1 u(t) + b_2 u(t - 1) \qquad (6)$$

where $u(t)$ and $y(t)$ are the input and output of the plant, respectively, at the normalized sampled time t (real time/sampling period).

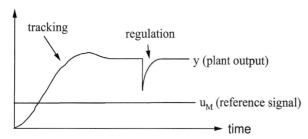

FIGURE 9 Tracking and regulation objectives.

Using the delay operator q^{-1}:

$$q^{-1}y(t+1) = y(t) \qquad (7)$$

we can rewrite Eq. (6) as:

$$\left(1 + a_1 q^{-1}\right)y(t+1) = b_1 u(t) + b_2 u(t-1) \qquad (8)$$

The tracking and regulation objectives are illustrated in Fig. 9 and can be specified as follows.

Regulation [null reference signal, $u_M(t) \equiv 0$]: Compute the control $u(t)$ such that an initial disturbance is eliminated with a dynamics defined by:

$$P(q^{-1})y(t+1) = \left(1 + p_1 q^{-1}\right)y(t+1)$$
$$= 0, \ \forall t > 0 \qquad (9)$$

Taking, for example, $p_1 = -0.5$, one has from Eq. (9) $y(t+1) = 0.5y(t)$ and $y(t) = (0.5)^t y(0)$, respectively. One observes that p_1 defines the speed of disturbance rejection.

Tracking: Compute the control $u(t)$ such that the process output satisfies a relation of the form:

$$A_m(q^{-1})y(t+1) = B_m(q^{-1})u_M(t) \qquad (10)$$

where $A_m(q^{-1})$ is an asymptotically stable polynomial (all roots are inside the unit circle) and $u_M(t)$ is the sequence of reference signals. The transfer function:

$$z^{-1}B_m(z^{-1})/A_m(z^{-1}) \qquad (11)$$

(obtained when replacing the delay operator q^{-1} by the complex variable z^{-1}) defines the reference model. By effectively defining a reference model that generates the reference trajectory $y_M(t)$ as:

$$A_m(q^{-1})y_M(t+1) = B_m(q^{-1})u_M(t+1) \qquad (12)$$

the control objectives defined above can be summarized as follows. Find $u(t)$ such that:

$$P(q^{-1})[y(t+1) - y_M(t+1)] = 0, \ \forall t > 0 \qquad (13)$$

To compute $u(t)$ we introduce $u(t)$ in Eq. (13). From Eq. (7), by adding to both sides $p_1 y(t)$, we obtain:

$$P(q^{-1})y(t+1) = \left(1 + p_1 q^{-1}\right)y(t+1) = (p_1 - a_1)y(t)$$
$$+ b_1 u(t) + b_2 u(t-1) \qquad (14)$$

Introducing Eq. (14) in Eq. (13) and using the notation:

$$r_0 = p_1 - a_1 \qquad (15)$$

we obtain the expression:

$$u(t) = \frac{1}{b_1}\left[P(q^{-1})y_M(t+1) - r_0 y(t) - b_2 u(t-1)\right] \qquad (16)$$

The block diagram for the resulting scheme is shown in Fig. 10, where:

$$B(q^{-1}) = b_1 + b_2 q^{-1} + \cdots + b_{n_B} q^{-n_B} \qquad (17)$$
$$= q^{-1} B^*(q^{-1}) \qquad (18)$$
$$A(q^{-1}) = 1 + a_1 q^{-1} + \cdots + a_{n_A} q^{-n_A} \qquad (19)$$
$$= 1 + q^{-1} A^*(q^{-1}) \qquad (20)$$
$$R(q^{-1}) = r_0 + r_1 q^{-1} + \cdots + r_{n_r-1} q^{-n_r+1};$$
$$r_i = p_{i+1} - a_{i+1} \qquad (21)$$
$$P(q^{-1}) = 1 + p_1 q^{-1} + \cdots + p_{n_p} q^{-n_p} \qquad (22)$$

In Fig. 10, the variable:

$$\varepsilon^0(t) = P(q^{-1})[y(t) - y_M(t)] \qquad (23)$$

is identically null if the controller is correctly tuned. In other words, this variable is a measure of the difference between the desired and real performance and will be used later for the automatic adaptation of the controller parameters.

Eq. (16), defining the controller, can be rewritten:

$$P(q^{-1})y_M(t+1) = \theta_c^T \phi(t) \qquad (24)$$

where θ_c is the vector of controller parameters defined by:

$$\theta_c^T = [r_0, b_1, b_2] \qquad (25)$$

and $\phi(t)$ is the measurement vector defined by:

$$\phi(t)^T = [y(t), u(t), u(t-1)] \qquad (26)$$

The structure of Eq. (24) remains the same for the general case corresponding to a plant model of any order and with

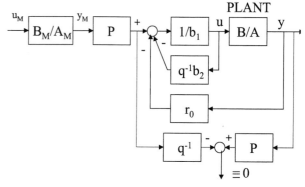

FIGURE 10 Tracking and regulation with independent objectives (known plant parameters).

pure time delay $> \Delta$ (only the dimension of θ_c and $\phi(t)$ will change).

Note that the same controller is obtained if one considers as the control objective the following. Find $u(t)$ such that the quadratic criterion:

$$
\begin{aligned}
J(t + 1) &= \varepsilon^0(t + 1)^2 \\
&= P(q^{-1})[y(t + 1) - y_M(t + 1)]^2 \quad (27)
\end{aligned}
$$

is minimized; that is, find $u(t)$ such that $[\partial J(t + 1)] / \partial u(t) = 0$.

The transfer function from the reference signal to the output of the plant is given by:

$$
H(z^{-1}) = \frac{z^{-1} B_m(z^{-1})}{A_m(z^{-1})} \quad (28)
$$

The use of this control strategy requires the assumption that $B(z^{-1})$ has all its roots inside the unit circle—i.e., the zeros of the plant model discrete transfer function are stable (the controller cancels the plant model zeros). In this example, the asymptotic stability of $B(z^{-1})$ requires $|b_2| < |b_1|$ which implies that $\tau < 0.5\Delta$.

B. Direct Adaptive Control

In the case of unknown plant parameters one would like to minimize at each step the criterion of Eq. (27) in order to achieve asymptotically:

$$
\lim_{k \to x} \varepsilon^0(t + 1) = \lim_{t \to x} P(q^{-1})[y(t + 1) - y_M(t + 1)] = 0 \quad (29)
$$

The structure of the controller will remain the same, as in the case of known parameters, but the constant parameters will be replaced by adjustable parameters,

$$
\begin{aligned}
u(t) = \frac{1}{\hat{b}_1(t)} [P(q^{-1})y_M(t + 1) \\
- \hat{r}_0(t)y(t) - \hat{b}_2(t)u(t - 1)] \quad (30)
\end{aligned}
$$

where $\hat{r}_0(t)$, $\hat{b}_1(t)$ and $\hat{b}_2(t)$ are the estimates at time t of the parameters of the controller. Defining the adjustable parameter vector as:

$$
\hat{\theta}(t)^T = [\hat{r}_0(t), \hat{b}_1(t), \hat{b}_2(t)] \quad (31)
$$

we can rewrite the equation of the adjustable controller as:

$$
P(q^{-1})y_M(t + 1) = \hat{\theta}^T(t)\phi(t) \quad (32)
$$

where $\phi(t)$ is given by Eq. (26).

The next step is the design of the parameter adaptation algorithm (PAA) in a recursive form, in order to implement it in real time. The structure of such a recursive PAA is

$$
\begin{aligned}
\hat{\theta}(t + 1) &= f[\hat{\theta}(t), \phi(t), \varepsilon^0(t + 1)] \\
&= \hat{\theta}(t) + f'[\hat{\theta}(t), \phi(t), \varepsilon^0(t + 1)] \quad (33)
\end{aligned}
$$

In other words, the estimated values of the parameters at time $t + 1$ depend on the estimated values of the parameters at time t, on the measurements at time t, and on the performance error $\varepsilon^0(t + 1)$, which itself depends on $\hat{\theta}(t)$. The objective is to find a PAA of the form of Eq. (33), which minimizes at each step the criterion of Eq. (34) [obtained from Eq. (27) when using Eq. (32)]:

$$
\begin{aligned}
J(t + 1) &= \varepsilon^0(t + 1)^2 \\
&= P(q^{-1})[y(t + 1) - \hat{\theta}(t)^T \phi(t)]^2 \quad (34)
\end{aligned}
$$

In this case, the structure of the control law is fixed and the minimization is done with respect to $\hat{\theta}(t)$. Using the recursive gradient optimization procedure, one finds:

$$
\begin{aligned}
\frac{1}{2} \frac{\partial J(t + 1)}{\partial \hat{\theta}(t)} &= \frac{\partial \varepsilon(t + 1)}{\partial \hat{\theta}(t)} \varepsilon^0(t + 1) \\
&= -\phi(t)\varepsilon^0(t + 1) \quad (35)
\end{aligned}
$$

and the corresponding PAA will have the form:

$$
\begin{aligned}
\theta(t + 1) &= \theta(t) + F\phi(t)\varepsilon^0(t + 1) \\
F &= \alpha I, \ \alpha > 0 \quad (36)
\end{aligned}
$$

where I is the unit diagonal matrix. The matrix F is called the *adaptation gain*. The geometric interpretation of algorithm (36) is given in Fig. 11 (note that $F = \alpha I$ is sometimes replaced by a positive definite matrix $F > 0$).

To ensure the convergence of the adaptation, α in Eq. (36) should be very small. A detailed analysis of the stability of PAA indicates that, in order to ensure the convergence of the algorithm for any positive definite F, the correction term should be normalized by dividing by $1 + \phi(t)^T F\phi(t)$, which gives:

$$
\hat{\theta}(t + 1) = \hat{\theta}(t) + \frac{F\phi(t)\varepsilon^0(t + 1)}{1 + \phi(t)^T F\phi(t)} \quad (37)
$$

An indirect adaptive control approach can be also used for this case to design an adaptive control system. However, by an appropriate reparametrization of the plant model one can directly estimate the controller parameters, yielding a direct adaptive control scheme.

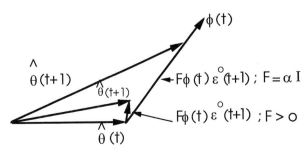

FIGURE 11 Geometric interpretation of parameter adaptation algorithms.

IV. DESIGN OF INDIRECT ADAPTIVE CONTROLLERS

A. Known Parameters Case

We will now consider the pole-placement control strategy, which can be applied for the control of plants characterized by discrete transfer functions with unstable zeros.

The plant model is given by:

$$H(z^{-1}) = \frac{z^{-d}B(z^{-1})}{A(z^{-1})} \qquad (38)$$

where the polynomials $A(z^{-1})$ and $B(z^{-1})$ have been defined in Eqs. (17) and (19) and d is the pure time delay expressed as a number of sampling periods. One assumes that $A(z^{-1})$ and $B(z^{-1})$ do not have commun factors.

The desired closed loop poles are defined by the polynomial:

$$P(z^{-1}) = 1 + p_1 z^{-1} + \cdots + p_{n_p} z^{-n_p} \qquad (39)$$

The controller is given by:

$$u(t) = -\frac{R(q^{-1})}{S(q^{-1})}y(t) + \frac{T(q^{-1})}{S(q^{-1})}y_M(t) \qquad (40)$$

where $y_M(t)$ is the output of the reference model defined in Eq. (12).

The controller polynomials $R(q^{-1})$, $S(q^{-1})$ are obtained as solution of the Bezout equation:

$$P(q^{-1}) = A(q^{-1})S(q^{-1}) + z^{-d}B(q^{-1})R(q^{-1}) \qquad (41)$$

and the polynomial $T(q^{-1})$ is given by:

$$T(q^{-1}) = \frac{P(q^{-1})}{B^*(1)} \qquad (42)$$

The transfer function from the reference signal to the output is given by:

$$H_{pp}(z^{-1}) = \frac{z^{-d}B_M(z^{-1})B(z^{-1})}{A_M(z^{-1})B^*(1)} \qquad (43)$$

(i.e., the plant zeros are not canceled).

Consider the same examples as for the direct adaptive control given by Eq. (2) but with $\Delta > \tau > 0.5\Delta$. In this case, direct adaptive control cannot be applied since $B(z^{-1})$ is unstable. The pole placement is the convenient control strategy. In this case:

$$A(q^{-1}) = 1 + a_1 q^{-1}; \quad B(q^{-1}) = b_1 q^{-1} + b_2 q^{-2} \qquad (44)$$

The desired closed loop poles are given by:

$$P(q^{-1}) = 1 + p_1 q^{-1} + p_2 q^{-2} \qquad (45)$$

Selecting $R(q^{-1})$ and $S(q^{-1})$ as:

$$R(q^{-1}) = r_0; \qquad S(q^{-1}) = 1 + s_1 q^{-1} \qquad (46)$$

Eq. (41) becomes:

$$(1 + a_1 q^{-1})(1 + s_1 q^{-1}) + (b_1 q^{-1} + b_2 q^{-2})r_0$$
$$= 1 + p_1 q^{-1} + p_2 q^{-2} \qquad (47)$$

which is equivalent to a set of linear equations (each one for a power of q) written in a matrix form as:

$$\begin{bmatrix} 1 & 0 & 0 \\ a_1 & 1 & b_1 \\ 0 & a_1 & b_2 \end{bmatrix} \begin{bmatrix} 1 \\ s_1 \\ r_0 \end{bmatrix} = \begin{bmatrix} 1 \\ p_1 \\ p_2 \end{bmatrix} \qquad (48)$$

and which can be rewritten as:

$$M(\theta) \cdot \theta_c = p \qquad (49)$$

where $M(\theta)$ is a matrix depending on the plant parameter vector θ_p defined as:

$$\theta_p^T = [a_1, b_1, b_2] \qquad (50)$$

θ_c is the vector of the controller parameter:

$$\theta_c^T = [1, s_1, r_0] \qquad (51)$$

and

$$p^T = [1, p_1, p_2] \qquad (52)$$

contains the coefficients of $P(q^{-1})$.

The vector of controller parameters will be given by:

$$\theta_c = M(\theta)^{-1} p \qquad (53)$$

In the adaptive case, the plant parameters are unknown.

B. Indirect Adaptive Control

The adaptation of the parameters of the controller is done in two steps: (1) process parameter estimation and (2) computation of the controller parameters based on the process-estimated parameters.

1. Estimation of the Plant Parameters

The plant is characterized by:

$$y(t + 1) = -a_1 y(t) + b_1 u(t) + b_2 u(t - 1) \qquad (54)$$

where the parameters a_1, b_1, and b_2 are now unknown. An adjustable predictor of the process output is naturally obtained from Eq. (54) by replacement of the true parameters with their estimates at time t. One obtains:

$$\hat{y}^0(t + 1) = -\hat{a}_1(t)y(t) + \hat{b}_1(t)u(t) + \hat{b}_2(t)u(t)$$
$$= \hat{\theta}_p^T(t)\phi(t) \qquad (55)$$

where

$$\hat{\theta}_p^T(t) = [\hat{a}_1(t), \hat{b}_1(t), \hat{b}_2(t)] \qquad (56)$$

and

$$\phi^T(t) = [-y(t), u(t), u(t - 1)] \qquad (57)$$

The term $y^0(t+1)$ is the *a priori* output of the adjustable predictor. The *a priori* prediction error is given by:

$$\varepsilon^0(t+1) = y(t+1) - \hat{y}^0(t+1)$$
$$= y(t+1) - \hat{\theta}_p^T(t)\phi(t) \qquad (58)$$

One must now find a PAA of the form of Eq. (33) for $\hat{\theta}_p(t)$ that will minimize a quadratic criterion in terms of the prediction error:

$$J(t+1) = \varepsilon^0(t+1)^2 \qquad (59)$$

Applying the gradient procedure as above, one gets, after normalization:

$$\hat{\theta}_p(t+1) = \hat{\theta}_p(t) + \frac{F\phi(t)\varepsilon^0(t+1)}{1 + \phi(t)^T F\phi(t)} \qquad (60)$$

2. Computation of the Controller Parameters

Taking in account Eqs. (48) and (53) which give the controller parameters in the case of known parameters, one must, in the adaptive case, compute at each instant t a new value of the controller parameters based on the current estimation of the plant-model parameters. This leads to:

$$\hat{\theta}_c(t) = M(\theta_p(t))^{-1} p \qquad (61)$$

where

$$\hat{\theta}_c(t) = [1, \hat{s}_1(t), \hat{r}_0(t)] \qquad (62)$$

and $M(\theta_p(t))$ corresponds to $M(\theta_p)$ in Eq. (53) where the constant parameters are replaced by their estimates at time t.

The polynomial $T(q^{-1})$ in Eq. (42) will be given, at each instant, by:

$$T(t, q^{-1}) = \frac{1 + p_1 q^{-1} + p_2 q^{-2}}{\hat{b}_1(t) + \hat{b}_2(t)} \qquad (63)$$

V. GENERAL FORM OF PARAMETER ADAPTATION ALGORITHMS

A general form for the PAA is the following:

$$\hat{\theta}(t+1) = \hat{\theta}(t) + F(t)\phi(t)\nu(t+1) \qquad (64)$$

$$F(t+1)^{-1} = \lambda_1(t)F(t)^{-1} + \lambda_2(t)\phi(t)\phi(t)^T$$

$$0 < \lambda_1(t) \leq 1; \quad 0 \leq \lambda_2(t) < 2; \quad F(0) > 0 \quad (65)$$

$$\nu(t+1) = \frac{\nu^0(t+1)}{1 + \phi(t)^T F(t)\phi(t)} \qquad (66)$$

In Eq. (64), the term $\nu(t+1)$ is the *a posteriori* adaptation error, which can be computed using Eq. (66) (dividing the *a priori* adaptation algorithm error by the normalization term). The adaptation error sometimes corresponds to the prediction error or to the performance error (as in the examples discussed previously) and sometimes corre-

sponds to filtered values of these quantities. The *a priori* adaptation error depends on $\hat{\theta}(i)$ up to $i = t$.

The PAA includes in the general case a time-varying adaptation gain matrix, the inverse of which is updated recursively using Eq. (65). However, using the matrix inversion lemma, a recursive formula for F itself is obtained:

$$F(t+1) = \frac{1}{\lambda_1(t)} \cdot \left[F(t) - \frac{F(t)\phi(t)\phi(t)^T F(t)}{\frac{\lambda_1(t)}{\lambda_2(t)} + \phi(t)^T F(t)\phi(t)} \right] \qquad (67)$$

The terms $\lambda_1(t)$ and $\lambda_2(t)$ have opposite effects on the variation of the adaptation gain; $\lambda_1(t) < 1$ tends to increase the adaptation gain [$\lambda_1(t)$ is also called the *forgetting factor*]; and $\lambda_2(t) > 0$ tends to decrease the adaptation gain [the bound $\lambda_2(t) < 2$ results from a stability analysis]. For $\lambda_1(t) \equiv 1$, $\lambda_2(t) \equiv 0$, one obtains the gradient-type algorithm (constant adaptation gain). For $\lambda_1(t) \equiv 1$, $\lambda_2(t) \equiv 1$, one obtains the decreasing gain algorithm (least square type). Other laws of variations can be obtained by taking advantage of the variations allowed for $\lambda_1(t)$ and $\lambda_2(t)$ (variable forgetting factor, constant trace, etc.).

VI. APPLICATIONS

The typical tasks achieved by adaptive control systems are as follows:

1. Automatic tuning of controllers (which reduces the tuning time and improves the performance of the control systems).
2. Building up a gain schedule for various operation points.
3. Maintaining the system performance when the process parameters vary.

Adaptive control systems make the following possible:

1. The use of controllers that are more complex than PID (proportional + integral + derivative actions) and offer better performance.
2. The capacity to monitor the process parameters and to predict and detect failure.
3. The design of new technological processes that include from the beginning adaptive control loops.

A list of basic references for the field of adaptive control is provided below.

SEE ALSO THE FOLLOWING ARTICLES

CONTROLS, BILINEAR SYSTEMS • CONTROLS, LARGE-SCALE SYSTEMS • CONTROL SYSTEMS, IDENTIFICATION

● CYBERNETICS AND SECOND ORDER CYBERNETICS
● HYBRID SYSTEMS CONTROL ● SELF-ORGANIZING SYS-
TEMS ● SYSTEM THEORY

BIBLIOGRAPHY

Aström, K. J. (1995). "Adaptive Control," 2nd ed., Addison-Wesley, Reading, MA.

Aström, K. J., and Wittenmark, B. (1984). "Computer Controlled Systems: Theory and Design," Prentice Hall, Englewood Cliffs, NJ.

Goodwin, G., and Sin, K. S. (1984). "Adaptive Filtering Prediction and Control," Prentice Hall, Englewood Cliffs, NJ.

Ioannou, P., and Sun, J. (1996). "Robust Adaptive Control: A Unified Approach," Prentice Hall, New York.

Landau, I. D. (1979). "Adaptive Control: The Model Reference Approach," Dekker, New York.

Landau, I. D. (1990). "System Identification and Control Design," Prentice Hall, New York.

Landau, I. D., Lozano, R., and M'Saad, M. (1997). "Adaptive Control," Springer Verlag, London.

Controls, Bilinear Systems

Ronald R. Mohler
Oregon State University

GLOSSARY

Admissible control Class of functions that is acceptable to the analyst or designer. Here, assumed to be at least piecewise continuous with time and bounded in magnitude.

Bang–bang control Piecewise constant policy that takes on only two values (maximum and minimum allowed).

Controllable system System that can be controlled from any given initial state to any given final state, in finite time with an admissible control.

Lvapunov function Positive definite function V whose negative time derivative $-\dot{V}$ is also positive definite.

Singular control Control function that causes a zero switching function for some finite time duration.

Switching function Function that determines the zeros of a bang–bang policy.

A BILINEAR SYSTEM (BLS) is a nonlinear mathematical model utilizing differential equations to represent the control of a dynamic process. Bilinear systems are characterized by differential equations that are linear in the state of the system and linear in the control, but not jointly linear in both.

I. BILINEAR SYSTEMS ANALYSIS

A. Introduction

Mathematical models of dynamic processes and their control are of increasing significance in high technology. While most of these processes are nonlinear in nature, perturbation theory and linear-system mathematics generally have been the foundation of their analysis and design. The nonlinear class called *bilinear systems* is both prevalent in nature and useful for system design and control. Bilinear system dynamics and control have such diverse applications as immunology, national defense, nuclear power generation, solar energy, automobile dynamics, and socioeconomics. Bilinear models also represent approximations to more highly nonlinear processes, and bilinear control can offer significantly improved performance.

B. Definitions

Linear systems are characterized by linear differential equations, that is, ordinary differential equations that are linear in the dependent variables, linear in their derivatives with respect to the independent variable (time), and linear

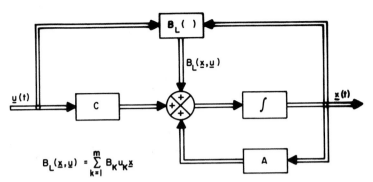

FIGURE 1 Bilinear state diagram.

in the input function or control. With respect to a vector or state formulation, the system is given by:

$$dx(t)/dt = Ax(t) + Cu(t) \qquad (1)$$

where $\mathbf{x}(t_0) = \mathbf{x}_0$, where t is time and x_0 is the initial state; $\mathbf{x}(t)$ is an n-dimensional state vector of elements $x_1(t), \ldots, x_n(t)$; $\mathbf{u}(t)$, is an m-dimensional control vector of elements $u_1(t), \ldots, u_m(t)$; and A, C are matrices of appropriate dimensions whose elements are assumed to be constant here. Such systems of equations have been thoroughly analyzed and applied to system design and control. This has been the case more for mathematical convenience than for accuracy and performance.

Attempting to keep the process amenable to some of the mathematical tools, yet providing an adaptive feature or variable structure leads to a BLS state formulation of the form:

$$\frac{d\mathbf{x}(t)}{dt} = A\mathbf{x}(t) + \sum_{k=1}^{m} B_k u_k(t)\mathbf{x}(t) + C\mathbf{u}(t) \qquad (2)$$

where $\mathbf{x}(t_0) = \mathbf{x}_0$ and B_k are constant matrices of appropriate dimension; $A, C, \mathbf{x}(t), \mathbf{u}(t)$ are as defined above for Eq. (1). The BLS state diagram is given in Fig. 1. Note that the multiplicative term in the feedback provides the adaptive feature of BLSs that makes them so prevalent in nature and so useful for control system design.

For cases with the system dynamics still more highly nonlinear, BLS analysis represents another order of approximation beyond that of linear systems analysis. Perturbation analyses, using the Taylor series, may be used to derive the BLS model by keeping the first-order products of control and state perturbations that are dropped in the linear system approximation.

For the nonlinear system $d\mathbf{x}/dt = \mathbf{f}(\mathbf{x}, u)$, with single input $u(t)$ and state vector $\mathbf{x}(t)$, the Taylor series with respect to variation $\delta\mathbf{x}(t)$ from a reference $\mathbf{x}^*$ (assumed constant here, with $\dot{\mathbf{x}}^* = 0$) becomes:

$$\delta\dot{\mathbf{x}} = A\delta\mathbf{x} + \mathbf{c}\delta u + B\delta u\delta\mathbf{x} + O^2(\delta u, \delta\mathbf{x})$$

where $O^2(\delta u, \delta\mathbf{x})$ refers to second- and higher-order terms, $\delta\mathbf{x} = \mathbf{x} - \mathbf{x}^*$, $\delta u = u - u^*$, $A = [\partial\mathbf{f}/\partial\mathbf{x}]_*$, $\mathbf{c} = \partial\mathbf{f}/\partial u|_*$, and $B = [\partial^2\mathbf{f}/\partial\mathbf{x}\partial\mathbf{u}]_*$ (* designates values taken at the reference $\mathbf{x}^*, u^*$).

Obviously, linear systems are a special class of BLS, but otherwise the simplest example is given by:

$$dx/dt = ux \qquad (3)$$

where x is a scalar function of time and u a scalar control function. In demography, the control u might be birth rate minus death rate plus migration rate into a region of population x.

II. PHYSICAL APPLICATIONS

A. Engineering

1. Automobiles and Aircraft

The frictional force between an automobile brake shoe and drum (Fig. 2) is nearly proportional to the product of the orthogonal force u_1 between the surfaces and their relative velocity. Though actually involving Coulomb friction and velocity terms, the frictional force generated by the mechanical brake is commonly approximated by:

$$f_b = c_b u_1 \dot{x}$$

Then, by a summation of engine force u_2 with inertial, braking, and other frictional forces, it is seen from

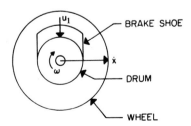

FIGURE 2 Bilinear mechanical brake.

Newton's second law that the state of the vehicle is given by:

$$dx/dt = A\mathbf{x} + u_1 B\mathbf{x} + \mathbf{c}u_2 \qquad (4)$$

where $\mathbf{x}$ is composed of x, position, and $\dot{x}$, velocity:

$$A = \begin{bmatrix} 0 & 1 \\ 0 & -kc_f/m \end{bmatrix}; \qquad B = \begin{bmatrix} 0 & 0 \\ 0 & -kc_b/m \end{bmatrix};$$

$$c = \begin{bmatrix} 0 \\ 1/m \end{bmatrix}$$

Here, k is a proportionality constant, c_f a vehicle frictional constant, c_b a brake constant, and m vehicle mass. As in Eq. (3), function arguments hereafter are dropped for convenience; for example, $x = x(t)$. It is interesting that the brake pedal introduces parametric control and the accelerator introduces additive control. A similar phenomenon is realized in the landing mode of a high-performance aircraft. During this phase of flight, the aircraft drag coefficient is controlled by means of relatively large wing "flaps." Indeed, one can feel the change in damping during this phase. While the parametric control allows improved state controllability, the additive controls in the form of thrust reversers and the frictional brakes are necessary to bring the aircraft to equilibrium at the state-space origin. Such structural changes frequently allow systems to reach otherwise unattainable states, while the additive control is usually required for local controllability to equilibrium, as further explained by early results for BLS controllability. The landing aircraft, of course, is more highly nonlinear due to attitude and mach changes. However, the BLS approximation is much closer to reality than a linear perturbation model.

2. Nuclear Fission

Similar to biological-population and cell fission models, the nuclear fission BLS model evolves according to:

$$\frac{dn}{dt} = \frac{u(1-\beta)-1}{l} n + \sum_{i-1}^{6} \lambda_i c_i$$

$$\frac{dc_i}{dt} = \frac{u\beta_i}{l} n - \lambda_i c_i, \qquad i = 1, \dots, 6 \qquad (5)$$

where u is neutron multiplication, n is neutron population, c_i ($i = 1, \dots, 6$ for ^{235}U) are precursor populations (or delayed neutrons). l is average neutron generation time, and λ_i and β_i are decay constants and proportionality constants associated with the ith group such that:

$$\sum_{i=1}^{6} \beta_i = \beta$$

In heat-exchange dynamics, BLSs arise if it is assumed that conduction is manipulated or that the convective coefficient is changed such as by altering coolant mass flow rate. For example, consider an energy balance on a perfectly insulated cylinder (such as a reactor core element) of average power generation q [proportional to n from Eq. (5)], temperature T_c, and mass heat capacity c_1 with negligible axial conduction. Coolant flows through the cylinder and has average temperature T_g, mass heat capacity c_2, specific heat c_p, and mass flow rate w. First, assume that the weighted average coolant temperature is given by:

$$T_g = (T_i + \theta T_o)/(1 + \theta) \qquad (6)$$

where θ is a constant and T_i and T_o are inlet and outlet temperatures, respectively. Then, the energy balance yields:

$$d\mathbf{x}/dt = Bw\mathbf{x} + Cu \qquad (7)$$

where $\mathbf{x}' = [T_c, T_g]$ (i.e., $\mathbf{x}$ is composed of T_c and T_g, with the prime designating row vector or matrix transpose):

$$B = \begin{bmatrix} -c_h/c_1 & C_h/c_1 \\ c_h/c_2 & (\alpha + c_h/c_2) \end{bmatrix}, \qquad C = \begin{bmatrix} c_1^{-1} & 0 \\ 0 & \alpha T_i \end{bmatrix} \qquad (8)$$

Here, c_h is a convective heat-transfer parameter (nearly constant and design dependent), and $\alpha = c_p(1 + \theta)/\theta c_2$. Although numerous assumptions are made, more general finite-difference models result in lumped BLSs, and spacially distributed derivations yield bilinear partial differential equations.

3. Heating Systems

Heating, ventilating, and air-conditioning systems, including solar panels, storage tanks, and heat pumps, can be similarly modeled by BLSs. Such analyses result in significant energy savings over the traditional linear optimal control approach.

Bilinear system models arise naturally for heat-transfer processes when heat-transfer coefficients or coolant mass flow rates are considered as control variables. Of course, nonlinear valve characteristics and nonlinear heat-transfer characteristics such as radiation would introduce more highly nonlinear terms that might be approximated in turn by a BLS or coupled BLS. The latter may approximate polynomial nonlinearities as well as others. Linear models, on the other hand, are not very accurate for any appreciable variations in temperature.

A two-tank, solar-assisted, heat-pump system that should be efficient for winter heating in northern latitudes provides a good illustrative example. By an energy balance, the heat-rate input Q_c to the system from the collector is described approximately by:

$$Q_c = h_c A'_c (T_p - T_g)$$
$$= F_c A_c [S\alpha - h_L(T_p - T_a)] - C_{pm}\dot{T}_p \qquad (9)$$

where the dot signifies differentiation with respect to time, Q_c is useful heat delivered by the collector, F_c a heat removal factor, A_c effective collector area. S incident solar radiation (kJ/hr × m² received on a tilted collector), α plate absorptance, h_L collector loss coefficient to the environment, T_p average absorber plate temperature, T_a ambient temperature, C_{pm} effective mass heat capacity associated with the plate, h_c transfer coefficient to the low-temperature (LT) coolant, A'_c effective transfer area to the LT coolant, and T_g an appropriately weighted average of T_i, collector inlet temperature, and outlet T_o given by Eq. (6). For a preliminary description, it might be assumed that $T_o = T_c$, LT coolant temperature.

Usually, C_{pm} is relatively small and neglected. Depending on the design; however, H_L and h_c may be functions of u_3, coolant mass flow rate through the collector. If heat losses in the pipe between the collector inlet and the LT storage tank are neglected, it can be assumed that $T_i = T_o = T_g$, which simplifies the model. Otherwise, the middle equality in Eq. (8) must be set equal to the increase in enthalpy of the circulating coolant with mass heat capacity negligible. Still, the model, with the heat-transfer coefficient as a control or with the coefficient a linear function of u_3 (a control), is a BLS.

The collected solar energy can be stored as sensible heat in liquid or solid or as latent heat in selected salts. The latter stores heat in the form of crystallization but has not been as thoroughly developed as the former. For sensible heat storage Q_s,

$$Q_s = C_{ms}\Delta T \qquad (10)$$

where C_{ms} is the storage mass heat capacity and ΔT its temperature range. Again, heat (Q_L) may be lost to the environment mainly by conduction from the storage tank, with

$$Q_L = a_L(T_L - T_a) \qquad (11)$$

where a_L is a heat-transfer constant (actually, conductance times area, nearly constant). Collector efficiency is improved by means of low and high-temperature (HT) storage tanks since the LT tank keeps the collector temperature lower. Also, the coefficient of performance of the heat pump is boosted since the input temperature is higher than ambient. The LT tank energy balance yields:

$$C_{mL}\dot{T}_L = Q_c - Q_L - Q_{hp} \qquad (12)$$

where Q_{hp} is the rate at which heat is taken out of the LT tank by the heat pump and C_{mL} is the LT-stored coolant mass heat capacity. It can be shown that:

$$Q_{hp} = u_2[COP(T_h, T_L) - 1] \qquad (13)$$

where u_2 is the heat-pump electrical input; T_h is the HT storage temperature; and $COP(T_h, T_L)$, the coefficient of performance, is approximated by a linear function of T_h and T_L.

The heat delivered to the building, of enclosure temperature T_b, is given approximately by:

$$Q_b = c_b u_1(T_h - T_b) = C_{mb}\dot{T}_b + Q_1 \qquad (14)$$

where c_b is the appropriate specific heat, C_{mb} is enclosure mass heat capacity, and u_1 is the building pump mass flow rate (out of the HT tank). The heat load on the building is

$$Q_1 = a_1(T_b - T_a) \qquad (15)$$

where a_1 is approximately a heat-transfer constant. Apparently, the required auxiliary heat input rate to the building is

$$Q_{aux} = Q_b - Q_1 \geq 0 \qquad (16)$$

Here, $Q_{aux} = 0$ if the building load can be provided from storage.

Typical control variables can be taken from pump mass-flow rates u_1, u_3 and heat-pump electrical input u_2. State variables can be selected as the most important temperatures (e.g., building and storage temperatures). Again, a complicated BLS model arises.

For northern latitudes, the heat-pump and two-tank solar systems seem to be an especially effective combination, since the solar system is most efficient during electrical peak hours, and the heat pump would be most efficient during off-peak hours. Also, the two-tank system is more efficient than the one-tank system.

4. Electrical and Neural Networks

It is readily seen that a field-controlled d.c. motor can be modeled as a BLS that is similar to parametrically controlled circuits. Also, a BLS model has been derived for a frequency-controlled induction motor, as well as thin-film transistors, field-effect tetrodes, and field-effect transistors (JFET and MOSFET).

Switched-impedance electrical networks are quite common in applications and include a conventional ignition sparking system for internal-combustion engines for which the distributor switch (control) either opens (infinite resistance) or closes (small resistance) in parallel with a capacitor and battery. Analogously, the automobile's mechanical suspension system can provide smoother motion by controlled damping. Flexible a.c. transmission systems (FACTS) use thyristor-controlled series capacitors or other manipulated-impedance devices (parametric control) to transmit significantly increased electrical power over existing lines. Such designs can result in significant economic and ecologic benefits. Similarly, thin-film transistors, field-effect tetrodes, and field-effect transistors (JFET and MOSFET), when used for network impedance control, result in BLS.

In neural networks, ion currents (primarily sodium and potassium) are commonly modeled as electrical currents with controlled impedances. As in many applications, however, the closing and opening of different ion channels (changing impedances) may depend on the network state and result in a more highly nonlinear closed-loop system. Artificial neural networks (ANN) may be synthesized in this manner and result in combinations of BLS coupled together by the traditional squashing functions. In some cases, generation of the neural action potential (spike) may be modeled by a bang–bang controlled BLS (i.e., piecewise time-invariant and linear). Similarly, ANN of many BLS elements may use heaviside (a specific squashing) coupling functions to approximate nonlinear dynamics in a simple manner. It is interesting that the neural spike seems to be generated analogous to that of the electrical spark in the internal-combustion engine.

B. Biological Control

1. Cell Division

Similar to nuclear fission and population equations, cellular division and differentiation can be derived as cascades of BLSs. Certain genetic information is stored in the cell to control the development of various chains developing into organs and processes. Even the membrane permeability of the cell wall is parametrically controlled by enzyme concentration to allow nutrients to feed the cell or, in the case of a cancer cell, to allow fluid to leak in and eventually burst it. The latter action takes place in the form of enzyme–protein reactions in the complement part of the human immune system. This enzyme–protein cascade of nine stages appears mathematically as a bilinear amplifier: increasing enzyme (complement) concentration until it "drills" the cell (lysis). Antibody is necessary to trigger the complement process. This mechanism is portrayed in Fig. 3.

2. Immune System

The overall BLS synthesis of the major portion of the immune process is shown in Fig. 4. The process includes the cooperation of antibody (Ab), cytotoxic T cells (T_c), macrophages (M_ϕ), and complement (C_3, C_8) to destroy the alien matter [antigen (Ag) for bacteria, virus, tumor, or

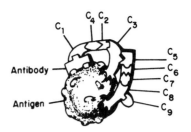

FIGURE 3 Complement cascade with tumor lysis.

chemical contaminant]. The humoral process includes the generation of different classes of B cells leading to plasma cells, which secrete antibody molecules of different structures. The cell-mediated immune system, like the humoral process, has stem cells (white blood cells) as its source and leads to the generation of macrophages and various classes of T cells. T helper cells add positive reactivity to "turn on" the humoral process, while T suppressor cells "turn off" the same process according to the concentration and structure of the alien. The collaboration and stimulation, represented by the nonlinear nondynamic block in Fig. 4, are not well defined. Indeed, most of immunological experimentation is focused in this direction. It is hoped that system theory will be helpful.

Immunological experimentation demonstrates the multiplicative immune modes for bacteria destruction in serum. Plotting the survival percentage (on a log scale) versus time for uncoated, antibody-coated, and antibody-plus-complement-coated bacteria results in nearly straight lines of decreasing descent. Consequently, a simple first-order homogeneous BLS, with the control multiplying all three concentrations, is suggested. Actually, the humoral immune system is modeled quite accurately by a BLS with state variables, including concentrations of immunocompetent cells, memory cells, plasma cells, immune

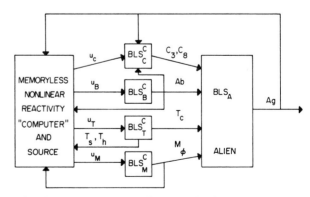

FIGURE 4 Immune system bilinear synthesis, where u refers to control reactivity, subscripts C to complement, B to B cells, T to T cells, M to macrophages, h to helpers, s to suppressors; BLS^C refers to coupled bilinear systems; C_3, C_8 refer to complement stages; and Ab and Ag refer to antibody and antigen, respectively.

complexes, and antigen (alien). The immunocompetent cells are sensitized lymphocyte cells with particular surface receptors for antigen according to chemical affinity. They may differentiate into plasma cells or into memory cells. The latter, which may further divide, enter the pool of immunocompetent cells; plasma cells are nonreproducing offspring of stimulated immunocompetent cells. Immune complexes individually include bound antibody and antigen. Antigen triggers the response mechanism. Additive controls include antigen inoculation and the bone marrow source of stem cells. Reactivity (e.g., by stimulation or differentiation) coefficients comprise multiplicative controls.

The additive control from bone marrow is independent of the multiplicative control variables and is significant in immunotherapy, since certain cancers (e.g., leukemia) have been treated by implantation of healthy bone marrow cells into the patient. The model can be used to analyze the effectiveness of such treatment by consideration of deterioration of the infected process or organs as an added component to the model. Though this source of stem cells is distributed according to affinity k (usually assumed to be Poisson or Gaussian), an average k seems representative in most practical cases.

The other additive control, inoculation of antigen, is independent of the other control variables and has significance in disease prevention (or, more correctly, disease control) by vaccination as well as in simulation of experiments whereby certain animals are inoculated with antigens of particular characteristics.

A similar T-cell model and its cooperation with macrophages and B cells has been suggested. Other BLS models of various complexities (including antibody switchover, distributed affinity, organ distribution, and infection) have been studied.

3. Metabolism and Hormone Levels

Biochemical processes concerned with metabolism take place within the cell itself. In this case, the law of mass action yields the following bilinear approximation:

$$dx_1/dt = a_1 x_2 + bu x_1$$
$$dx_2/dt = a_2 x_1 + bu x_2 \tag{17}$$

where a_1 and a_2 are reaction rate constants, u is enzyme concentration, x_1 is substrate concentration, and x_2 is the concentration of some complex that decomposes into an end product and enzyme.

Various control schemes have been hypothesized to describe the manner by which the bilinear control (enzyme concentration) is varied to control the reaction. It is known that the amount of enzymes is determined according to DNA information. Again, any closed-loop control system

with $u(x)$ would be nonlinear, but this does not preclude the utility of bilinear concepts in analyzing the process and the quality of control.

A similar bilinear model can be utilized to describe the regulation of circulating thyroxin in the human body. For this process, the law of mass action yields a bilinear system similar to Eq. (17) with x_1 being free thyroxin concentration, x_2 protein-bound thyroxin concentration, and u free protein concentration.

4. Respiratory System

So many physiological systems involve biochemical reactions of some sort that it is not unreasonable to expect bilinear processes to appear frequently in the study of physiology. The regulation of carbon dioxide (CO_2) in the respiratory system is a good example of a slightly more complex process. The bilinear model, as derived by Grodins of the University of Southern California, presupposes that the respiratory system consists of a lumped tissue reservoir from CO_2 and a ventilated lung reservoir which are connected by two unidirectional lines to represent arterial and venous circulatory systems. It is further assumed that the arterial blood and venous blood are in chemical equilibrium with the air in the lungs and the tissue CO_2 concentration, respectively. It is also assumed that tissue and blood have the same CO_2 absorption characteristic. Then, the CO_2 continuity equation states that the rate of change of reservoir CO_2 is equal to the difference between the rate of CO_2 inflow and outflow. In this manner, fractional CO_2 concentrations in the lungs, x_1, and in the tissue, x_2, are described by:

$$d\mathbf{x}/dt = u_1 B_1 \mathbf{x} + u_2 B_2 \mathbf{x} + C\mathbf{u} + \mathbf{d} \tag{18}$$

where:

$$\mathbf{x} = \begin{bmatrix} x_1 \\ x_2 \end{bmatrix}; \qquad \mathbf{u} = \begin{bmatrix} u_1 \\ u_2 \end{bmatrix}; \qquad \mathbf{d} = \begin{bmatrix} 0 \\ m_r/v_1 \end{bmatrix}$$

$$B_1 = \begin{bmatrix} -1/v_1 & 0 \\ 0 & 0 \end{bmatrix}; \qquad B_2 = \begin{bmatrix} -p_b \alpha/v_1 & 1/v_1 \\ p_b \alpha/v_2 & -1/v_2 \end{bmatrix}$$

$$C = \begin{bmatrix} f/v_1 & -\beta/v_1 \\ 0 & \beta/v_2 \end{bmatrix}$$

In terms of physical quantities, u_1 is lung volumetric ventilation rate; u_2 is volumetric cardiac output; v_1 and v_2 are lung tissue volumes, respectively; m_r is the volumetric metabolic rate of CO_2 production; P_b is barometric pressure, α and β are the slope and intercept of the linear absorption graph, respectively (arterial CO_2 concentration versus x_1); and f is the fractional CO_2 concentration in inspired air. Though metabolic rate m_r causes an additive defined term (not a control normally), a simple change of variables can be made to put Eq. (18) in the usual bilinear form, since m_r can be assumed constant for various periods of time.

For this system, the bilinear control mode arises from the dependence CO_2 flow rate on concentration; this is true despite the fact that the blood and air can be pumped at controllable rates.

In this process, the control variables, ventilation rate, and cardiac output are regulated in response to changes in metabolic rate such as take place during exercise. As always, the closed-loop control system with $u(x)$ is no longer bilinear.

5. Other Body Regulators

It is interesting that such a multiplicative nonlinearity also appears in the Grodins model of the cardiac regulator and in the description of the generator potential in a photoreceptor. The bilinear hydraulic cardiovascular model involves the systemic–arterial, systemic–venous, pulmonary–arterial, and pulmonary–venous circuits with circuit pressures as state variables, circuit susceptances as bilinear controls, and ventrical outputs as additive controls.

The following formulation shows that the functioning of the human thermoregulation system is roughly a bilinear process. For this purpose, the thermal plant of the body is represented by a two-lump approximation that consists of a core body of average temperature x_1, which is covered by a skin of average temperature x_2. A simple heat balance with the system in the so-called comfort mode of operation again yields Eq. (18), with:

$$B_1 = \begin{bmatrix} -1/c_{m1} & 1/c_{m1} \\ 1/c_{m2} & -1/c_{m1} \end{bmatrix}; \qquad B_2 = \begin{bmatrix} 0 & 0 \\ 0 & -1/c_{m2} \end{bmatrix}$$

$$C = \begin{bmatrix} 0 & 0 \\ 0 & x_\alpha/c_{m2} \end{bmatrix}; \qquad \mathbf{d} = \begin{bmatrix} q/c_{m1} \\ 0 \end{bmatrix}$$

In terms of physical quantities, u_1 is the thermal conductance of the core; u_2 is the thermal conductance of the skin; c_{m1} and c_{m2} are thermal capacitances of the core and wall, respectively; x_α is ambient temperature; and q (causing an independent additive control in this case) is heat-generated by metabolic rate, shivering, exercise, and so on. The conductances are varied to control the system by such mechanisms as vasomotor control of circulation and the formation of "goose bumps." Slightly more complicated models include sweating and heat-generating muscles, which are combined into a third lump.

Even the dynamic behavior of the temperature feedback receptors located in the skin exhibits a variable structure that suggests a bilinear model. Of course, the closed-loop

control system, in this case via the hypothalamus, is again nonbilinear.

It is apparent that variable-structure processes are quite common in physiology. For example, certain parameters may deteriorate due to disease, but the body seems to adapt by creating new conditions that may be optimum in some new sense. Also, muscle control with controlled use of an unstable mode suggests a bilinear model.

Typical of population growth discussed above, cells seem to increase in number according to a controlled divergent mode that is so common to bilinear systems. This behavior is exhibited by body growth itself, which seems to be roughly exponential with a controlled time constant that eventually yields the appropriate near-equilibrium size. A better understanding of the mathematical behavior of such processes may very well lead to a better understanding of body diseases eventually; such models may be used in diagnosis and even in drug prescriptions.

Many compartmental BLSs appear in biology as a result of process distribution to various body organs or suborgans. Since it is usually impossible to measure significant states directly, tracer experiments are frequently required with minimal-state realizations determined by compartmental accessibility (insertion and observation).

C. Other Fields

In economics, BLSs arise in numerous ways, such as parametric control via capital investment and tax rates. Numerous bilinear models have been studied in ecology and agriculture. These include grassland models in which additive control includes sunlight, temperature, and moisture, with parametric control being modeling harvest rates and grazing. Other models include the generation of beef with feed-rate control and pest-management control by means of pesticides, pheromones, and predators.

Deterministic and stochastic bilinear population models have been developed, the simplest example being given by Eq. (3). In addition to human demography, these models apply, for example, to general biological species, biological cells, chemical molecules, atomic particles, and manufacturing and distribution of products. In such processes, there may be birth rates, death rates, migration rates between compartments, and sometimes differentiation rates that are proportional to the population themselves. Control is introduced by the manipulation of appropriate coefficients such as the birth-rate coefficient, cellular or neutron multiplication, and migration-attractiveness multipliers or membrane permeability.

Normally, human population is not considered to be determined by a control process, but attempts are being made to control the population of some countries by influencing the population coefficient u. Though the problem is somewhat academic at this time, an optimal control process $u(t)$ might be computed to control the population in such a manner that the national economy can be optimized in some sense. Once the optimal policy is known, however, the government in question would still have to solve the extremely difficult problem of implementation, which might involve massive educational efforts, mass distribution of pills, and other techniques. Of course, humans must also compete with microorganisms for survival, and our control of such alien populations is implemented by altering their population coefficient, again a bilinear control function.

Other BLS models that come to mind include electromagnetic separation of isotopes, sunspot activity, and earthquakes. Similar to the immune defense model above, BLSs can be used to model national defense, hierarchical communication, and command and control, such as generalizations of Lanchester models for combating forces.

1. Structure

Bilinear systems have an apparent adaptive structure that allows them to exhibit better performance and to model nonlinear physical phenomena more accurately than linear systems. It is interesting that, if they are defined, the product and quotient of two scalar homogeneous BLS outputs result in a BLS. Of course, certain generalizations follow. Using the fact that, if $\mathbf{x}$ satisfies a linear equation, then $\mathbf{x}'\mathbf{x}$ satisfies one also, it is readily shown that if $\mathbf{x}$ satisfies a homogeneous BLS [Eq. (2) with zero C matrix], then $\mathbf{y}' = [x_1^2, x_1, x_2, x_1, x_3, \ldots, x_n^2]$ satisfies another homogeneous BLS with the same control with the new matrices derived from A and B_k ($k = 1, \ldots, m$). Similarly, any homogeneous BLS with output multilinear in the state can be similarly realized by a BLS with a linear-in-state output. It is readily seen that this result generalizes even further, since systems that involve transcendental functions can be represented by multilinear systems of increased order.

The phase-lock loop in communications is a good example in which the $\sin \theta$ and $\cos \theta$ as states lead to a BLS. Indeed, this generalization of BLSs makes them and some of their associated methodologies amenable to a larger class of nonlinear systems that also includes certain rigid-body dynamics.

Within each BLS itself there evolves a certain hierarchical structure that is associated with a "canonical" decomposition such as is convenient for its Volterra representation (assuming its existence, of course). In this manner, the BLS of Eq. (2), with output $\mathbf{y} = D\mathbf{x}$, can be generated by:

$$\mathbf{y} = D\sum_{i=1}^{x}\mathbf{x}_i \approx D\sum_{i=1}^{N}\mathbf{x}_i \tag{19}$$

where:

$$\dot{\mathbf{x}}_1 = A\mathbf{x}_1 + Cu$$

$$\dot{\mathbf{x}}_2 = A\mathbf{x}_2 + \sum_{k=1}^{m} B_k u_k \mathbf{x}_1 + C\mathbf{u}$$

$$\vdots \tag{20}$$

$$\dot{\mathbf{x}}_l = A\mathbf{x}_l + \sum_{k=1}^{m} B_k u_k \mathbf{x}_{l-1} + C\mathbf{u}$$

$$\vdots$$

with $\mathbf{x}(0) = 0$ for convenience. It is readily seen that x_i, $i = 1, 2, \ldots$ correspond to the terms in the Volterra series, and the corresponding kernels are generated as a "nesting" of linear-system impulse responses according to Eq. (20). In this manner, the Volterra series of the homogeneous time-invariant BLS with the output $\mathbf{y} = D\mathbf{x}$ is given by:

$$\mathbf{y} = De^{At}\mathbf{x}_0 + \int_0^t De^{A(t-\tau)}C\mathbf{u}(\tau)\,d\tau$$

$$+ \sum_{i=1}^{l}\sum_{1=k_1,\ldots,k_i}^{m}\cdots\sum \int_0^t \int_0^{\tau_1}\cdots\int_0^{\tau_{i-1}}$$

$$\times De^{A(t-\tau_1)}B_{k_1}\cdots B_{k_i}e^{A\tau_i}\mathbf{x}_0 \prod_{l=1}^{i} u_{k_l}(\tau_l)\,d\tau_l$$

$$+ \sum_{i=1}^{x}\sum_{1=k_1,\ldots,k_{i-1}}^{m}\cdots\sum \int_0^t \int_0^{\tau_1}\cdots\int_0^{\tau_1} De^{A(t-\tau_1)}$$

$$\times B_{k_1}\cdots B_{k_t}e^{A(\tau_1-\tau_{l-1})}\mathbf{c}_{k_i-1}\prod_{l=1}^{i-l} u_{k_l}(\tau_l)\,d\tau_l \tag{21}$$

where $C = [\mathbf{c}_1\cdots\mathbf{c}_m]$, $\mathbf{x}(0) = \mathbf{x}_0$, $\tau_i \geq \tau_{i-1} \geq 0$, $i = 1, 2, \ldots$, and $k_i = 1, 2, \ldots, m$. The state transition matrix e^{At} is the solution to the matrix equation $\dot{X} = AX$, with $X(t_0) = I$, the identity matrix. The Volterra kernel synthesis (nth kernel) involves linear system operation on a product of controls and previous term outputs [from the $(n-1)th$ kernel]. For strongly bilinear systems, this takes the form of an infinite sequence of "nesting" and multiplication operations starting with the first kernel or linear-system impulse response. The Volterra series for BLSs is sometimes approximated by a finite number of terms [Eq. (20)] and, in the case of so-called weakly bilinear systems, can be exactly represented by a finite number or decomposed into a finite number of linear systems with outputs multiplied together to form inputs to successive linear systems according to Eq. (20). Consequently, a physical system approximated (or given) by a finite hierarchical structure associated with Eq. (20) is conveniently analyzed by linear-system theory.

III. CONTROL

It is obvious that the variable-structure feature of BLSs is a main reason for their utility in modeling so many natural processes. As a result of its bilinear control mode these systems are adaptive, and generally they are more controllable than linear systems. While stability is a keynote of classical controller design, it is the controlled use of instability that allows BLSs to attain a large set of states. This is made more apparent in the following section.

A. Controllability

Roughly put, the measure of a control system's capacity to attain certain states with an admissible control in finite time is called *controllability*, and it is a very basic property of dynamic systems in general.

For physiological processes, controllability or the lack of controllability might be quite useful in the diagnosis of disease and other malfunctions. Several rather trivial conclusions can be made without a detailed analysis. For example, the state-space set of attainability for the thermal regulator can deteriorate with circulatory ailments. If the system state were disturbed into a region from which the comfort-mode model could not control the process to the desired terminal set, a new control mode such as sweating might be employed. By means of controllability, the effectiveness of enzymes on biochemical reactions is quite apparent, a fact that is now being exploited by the detergent industry. Also, the respiratory chemostat model shows the significance of cardiac output by consideration of controllability. Even the effectiveness of the treatment of cancer might be analyzed by a study of controllability of cell division and, as shown above, is amenable to bilinear modeling.

While the bilinear control mode is useful for varying the dynamic structure of the process, it is the additive control terms that effect control of equilibrium conditions. Consequently, it is reasonable to expect that a better mathematical synthesis of the automobile might include an additive braking force (torque reversal) and a bilinear engine force with, perhaps, system damping increasing with auto velocity. While controllability enhancement may not be sufficient reason to warrant the implementation of such mechanisms for conventional automobiles, it is interesting to note the use of jet engine thrust reversers, which admit an additive braking force to aircraft. Again, the conventional braking flaps cause a change in the aerodynamic drag coefficient for bilinear control of the aircraft. Also, aircraft with variable wing geometry can be analyzed as a bilinear control system.

For certain pursuit systems, such as high-performance aircraft or antiballistic missile (ABM), it may be necessary

to attain an extremely large region in state space. While it is physically impossible to reach all states in finite time, the idea is useful from a mathematical standpoint. If any finite state can be reached from any other finite state in finite time with an admissible control, here the system is said to be completely controllable. Certainly, then, complete controllability would be the strongest type of controllability, and appropriate necessary and sufficient conditions for complete controllability could be useful in control system design or even more general systems analysis.

In this regard, it has been shown that the bilinear system of Eq. (2) with $\mathbf{u} \in U$ is completely controllable if the following are true (here U is a closed, bounded set): (1) Control values $\mathbf{u}^+$ and $\mathbf{u}^-$ exist such that the real parts of the eigenvalues of the system matrix are positive and negative, respectively, and equilibrium states $\mathbf{x}^e(\mathbf{u}^+)$, $\mathbf{x}^e(\mathbf{u}^-)$ are contained in a connected component of the equilibrium set. (2) For each $\mathbf{x}$ in the equilibrium set with an equilibrium control $\mathbf{u}^e(\mathbf{x}) \in U$, such that $\mathbf{f}[\mathbf{x}, \mathbf{u}^e(x)] = \mathbf{0}$, there exists an m-dimensional $\mathbf{v}$ such that the n vectors $D_\mathbf{v}, ED\mathbf{v}, E^2 D\mathbf{v}, \ldots, E^{n-1} D\mathbf{v}$ are linearly independent, where U is the compact control set:

$$E = A + \sum_{k=l}^{m} u_k e(\mathbf{u}) B_k$$

and

$$D = (B_1 \mathbf{x} B_2 \mathbf{x} \cdots B_m \mathbf{x}) + C$$

Though these conditions appear somewhat formidable to apply, they frequently simplify. For example, Eq. (2) is satisfied if all the eigenvalues of the system matrix:

$$A + \sum_{k=1}^{m} u_k B_k$$

can be shifted across the imaginary axis of the complex plane without passing through zero as $\mathbf{u}$ ranges continuously over a subset of U. Also, for canonical phase-variable systems ($x_1 = x$, $x_2 = \dot{x}$, $\ldots$, $x_n = x^{n-1}$), condition (2) is always satisfied if C is a nonzero matrix.

That BLSs with magnitude-constrained control can be completely controllable may in itself be a useful property for synthesis and analysis. Linear systems seldom possess this property. For example, consider the autonomous linear system $\dot{x} = A\mathbf{x}$, where real parts of the eigenvalue of A have negative real parts. Since this linear system is stable, there exists a positive-definite quadratic form (a Lyapunov function) $V(\mathbf{x})$, with $\dot{V}(x)$ a negative-definite quadratic form. Then, the surface $V(\mathbf{x}) = \text{const}$ are hyperspheriods that enclose the origin in state space.

Now consider the same system with an added bounded linear control term in Eq. (1). For this linear system, the following derivation shows that a Lyapunov hyperspheroid $V(\mathbf{x}) = M$ exists such that $\dot{\mathbf{x}}$ points into the interior of the volume that $V(\mathbf{x}) = M$ bounds, for every bounded control $\mathbf{u}$:

$$\dot{\mathbf{x}} \cdot \frac{\partial V}{\partial \mathbf{x}} = A\mathbf{x} \cdot \frac{\partial V}{\partial \mathbf{x}} + C\mathbf{u} \cdot \frac{\partial V}{\partial \mathbf{x}} = \dot{V} + C\mathbf{u} \cdot \frac{\partial V}{\partial \mathbf{x}}$$

Since the last term is linear in $\mathbf{x}$, and since $V(\mathbf{x})$ is a negative-definite quadratic form, this sum is negative for every bounded $\mathbf{u}$ with $\mathbf{x}$ (and therefore M) sufficiently large. Consequently, the linear system [Eq. (1)] with $\mathbf{u}$ bounded is not completely controllable if the eigenvalues of A have negative real parts. For example, the linear system:

$$\ddot{x} + a\dot{x} + x = u$$

with $a > 0$ and u bounded is not completely controllable. However, by adding a bilinear control so that

$$\ddot{x} + (a + u)\dot{x} + a = u$$

with $a > 0$, the system can be made completely controllable on the $(x, \dot{x})$ state space if $a < 1$ by the previous result. For example, $a = 0.5$ and $u = \pm 0.6$ results in the superposition of a stable focus (complex eigenvalues with negative real parts) and an unstable focus (complex eigenvalues with positive real parts).

For many systems, complete controllability with respect to some isolated region is all that is required. The neutron kinetics, with only positive states admitted and with control constraints, is a good example. Frequently, the neutron kinetics [Eq. (5)] is approximated by:

$$\begin{aligned} \frac{dn}{dt} &= \frac{u - \beta}{l} n + \lambda c \\ \frac{dc}{dt} &= \frac{\beta}{l} n - \lambda c \end{aligned} \qquad (22)$$

Here, one average precursor group approximates the six groups in Eq. (5), and the control becomes reactivity (excess multiplication $< \beta$).

It is readily seen from Eq. (22) that the bilinear neutron kinetics is locally controllable about any equilibrium point x_0 except the origin if $(\beta - 1)/l \neq \lambda$. Of course, this inequality is always valid in practice since $\beta \ll 1$ and $\lambda > 0$.

For $u = 0$, there is an eigenvector:

$$\mathbf{x}^0 = \begin{bmatrix} 1 \\ (1 + \lambda)/\lambda l \end{bmatrix}$$

in the first quadrant of the state plane. However, $\lambda l \ll 1$, so that

$$\mathbf{x}^0 \approx \begin{bmatrix} 1 \\ 1/\lambda l \end{bmatrix}$$

Then, as shown in Fig. 5, the slope of $\mathbf{x}^0$ is greater than the slope of $\mathbf{b}$ (which is β/λ) since $\beta \ll l$. Hence, with the con-

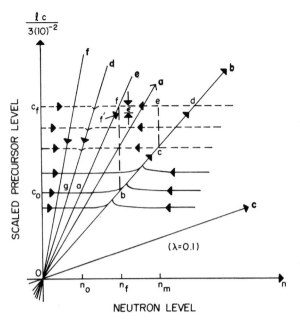

FIGURE 5 Time-optimal neutronic state trajectories, where e is the equilibrium line ($c = 0$, $n = 0$ with $u = 0$); eigenvectors $\mathbf{a}$, $\mathbf{b}$, $\mathbf{d}$, $\mathbf{c}$, $\mathbf{f}$ for $u = 0.25\beta$, $\pm 0.5\beta$, and 0.9β, respectively.

trol $u = 0$ (minimum admissible value) and $u = \alpha$ (maximum admissible value), it is readily seen from Eq. (22) that any state between the equilibrium line $\mathbf{e}$ in Fig. 5 and the eigenvector $\mathbf{x}^0$ can be reached from the equilibrium line in finite time. (Note, however, that it requires infinite time to reach the eigenvector.) Similarly, it can be shown that there is an eigenvector $\mathbf{x}^+$ in the first quadrant with slope less than that of the equilibrium line such that all states between the equilibrium and the $\mathbf{x}^+$ eigenvector are attainable from the equilibrium line. Then, any state between the eigenvectors $\mathbf{x}^0$ and $\mathbf{x}^+$ can be reached from any other state in this region since the equilibrium line can be reached from any state between x^0 and $\mathbf{x}^+$ [from Eq. (22) and Fig. 5] and since the system is locally controllable about the equilibrium line (except the origin). Hence, the region between $\mathbf{x}^+$ and $\mathbf{x}^0$ in Fig. 5 is controllable. Typical parameter values are used for Fig. 5, with $\beta = 0.0065$, $\lambda = 0.1$ sec^{-1}, and $l = 10^{-4}$ sec.

In practice, however, the control is rate constrained as well as magnitude constrained, but the equilibrium line can still be reached in finite time from any finite state between $\mathbf{x}^+$ and $\mathbf{x}^0$ (with the origin a point exception).

B. Design

If the cost function, as well as the plant, is linear in control, it is seen from Pontryagin's maximum principle that "maximum allowable effort" controls (e.g., bang–bang control pieced together in time with state-constraint and singular trajectories) are optimal for a broad class of BLS optimal control problems. Optimal switching times for such trajectories are computed by means of a gradient algorithm, which is taken with respect to a switching-time vector and has been termed the *switching-time variation method*. This method has been applied to time-optimal and fuel-optimal control of vehicle position and speed and to nuclear reactor kinetics as well as other processes.

Application of the maximum principle without state constraint shows immediately that the form of optimal control of Eq. (2) with $u \in U$ is a bang–bang process. If U is defined by $|u_k| \le 1$ and $s_k(t)$ is the switching function, the control has the form:

$$u_k = \operatorname{sgn} s_k(t), \qquad k = 1, \ldots, m \qquad (23)$$

where $\operatorname{sgn} s_k = s_k/|s_k|$ for $s_k \ne 0$. If $s_k(t) = 0$ for some finite time, the controls are said to be singular, and these must be considered.

The minimum-time reactivity to control the neutron process from some initial equilibrium state to some terminal state is computed in this section. The process is given by Eq. (22) or by:

$$d\mathbf{x}/dt = A(u)\mathbf{x} = \mathbf{f}(\mathbf{x}, u) \qquad (24)$$

where $A(u)$ is defined by:

$$A = \begin{bmatrix} u - \beta/l & \lambda \\ \beta/l & -\lambda \end{bmatrix}$$

$x_1 = n$, $x_2 = c$.

The first problem is to find the admissible control $|u(t)| \le \alpha$ from the class of piecewise continuous functions that drives the system from an initial equilibrium (n_0, c_0) to a terminal equilibrium (n_f, c_f) in minimal time.

Following the maximum principle, the optimal control must maximize the inner product:

$$H(n, c, \mathbf{p}, u) = \langle f(\mathbf{x}, u), \mathbf{p} \rangle$$

$$= (u/l)np_1$$

$$+ (p_2 - p_1)[(\beta/l)n - \lambda c] \qquad (25)$$

where $\mathbf{p}(t)$ is a continuous nonzero n vector with:

$$d\mathbf{p}/dt = -A'(u)\mathbf{p} \qquad (26)$$

$\mathbf{p}(t_0) = \mathbf{p}_0$. Consequently, since $n(t)$ is positive, the optimal control maximizes Eq. (25) and has the form:

$$u = \alpha \operatorname{sgn} p_1(t) \qquad (27)$$

Substitution of $u = \pm \alpha$ into Eq. (22) with $(\beta/l)n_0 = \lambda c_0$ for initial equilibrium yields:

$$n = \frac{[\rho_2 \mp (\alpha/l)]n_0 \exp[\rho_1(t - t_0)]}{\rho_2 - \rho_1}$$

$$\frac{-[\rho_1 \pm (\alpha/l)]n_0 \exp[\rho_2(t - t_0)]}{\rho_2 - \rho_1} \qquad (28)$$

and

$$c = \frac{\rho_2 c_0 \exp[\rho_1(t - t_0)] - \rho_1 c_0 \exp[\rho_2(t - t_0)]}{\rho_2 - \rho_1} \qquad (29)$$

where the eigenvalues ρ_1, ρ_2 are given by the following, with $u = \pm \alpha$,

$$\rho_1 \cdot \rho_2 = \frac{(u - \beta - \lambda l) \pm [(u - \beta - \lambda l)^2 + 4u\lambda l]^{1,2}}{2l}$$

and corresponding eigenvectors have directions θ_1, θ_2 such that

$$\tan \theta_i = [l\rho_i - (u - \beta)]/\lambda l, \qquad i = 1, 2$$

Since the two eigenvalues are real and of opposite sign for $u = \alpha$ and are negative and real for $u = -\alpha$, the bang–bang-controlled neutron kinetics are characterized in the state plane by a saddle-point portrait superimposed on a stable-node portrait with a singular point at the origin. It can be seen by elimination of time from Eqs. (28) and (29) that the state trajectories are located by:

$$\left[\frac{(\rho_2 + \lambda)c - (\beta/l)n}{(\rho_2 + \lambda)c_0 - (\beta/l)n_0} \right]^{-\rho_2}$$

$$= \left[\frac{(\rho_1 + \lambda)c - (\beta/l)n}{(\rho_1 + \lambda)c_0 - (\beta/l)n_0} \right]^{-\rho_1} \qquad (30)$$

where n_0 and c_0 locate any given state. Certain such combinations of time-optimal trajectories are given in Fig. 5. State behavior of the adjoint system of Eq. (26), with $u = 0.9\beta$ for positive p_1 and $u = -0.9\beta$ for negative p_1, is described in the right half of the state plane by a saddle-point portrait, while that in the left half-plane is typical of an unstable-node portrait.

Since it is required that $\mathbf{p}(t)$ be continuous, it is seen directly that $p_1(t)$, the switching function in Eq. (26), cannot have more than one zero in finite time. Since the constant-u trajectories are unique solutions of the differential equations of the system, Fig. 5 indicates that there is only one possible trajectory that joins any reachable state from some initial state with a maximum of one switching; therefore, such a trajectory must be optimal (e.g., trajectory a–d–f has $u = \alpha$ from a to d and $u = -\alpha$ from d to f). Also, the principle of optimality (any portion of an optimal trajectory is optimal) shows that if a desired terminal state can be reached from a given initial state with no switchings and $|u| = \alpha$, then the joining trajectory is optimal.

An example of an optimal startup trajectory linking an initial steady state (n_0, c_0) with a desired terminal steady state (n_f, c_f) is represented by trajectory a–d–f in Fig. 5. Initially, the system is at equilibrium with zero reactivity. Then reactivity is made equal to the maximum positive constant α while $p_1(t)$ is positive. At point d, $p_1(t)$ goes through zero and becomes negative as reactivity is switched to the minimum constraint. When the desired state point f is reached, reactivity is returned to zero to maintain quiescence.

The switching point for the time-optimal startup process is conveniently determined from Eq. (30) by the neutron level n (or the neutron precursor level c) at point d. In general, Eq. (30) can be used to generate a switching line for a time-optimal startup to a desired terminal state (with n_0 replaced by n_f and c_0 by c_f). Below this line, $u = \alpha$; and above the switching line, $u = -\alpha$. If n and c are known, an on-line computation can be made (e.g., digitally or by means of an analog function generator and a comparator) to determine the switching point.

Since the trajectory is asymptotic to the eigenvector b in Fig. 5, for any significant difference $n_f - n_0$, the switching point is nearly independent of the initial state. In many physical systems, it might be desirable to switch to a simple control law that is a continuous function of the system state within some small predetermined region of the desired terminal state. The size and shape of this region could be determined by the accuracy of the equations in defining the physical process, the accuracy of the measuring devices, and the capability of the closed-loop control. Hence, the exact switching point might not be critical.

Though the state control has utility (e.g., during calibration of instrumentation when time is available), it is frequently desirable to traverse from some initial equilibrium state to a terminal equilibrium neutron level in minimal time without the terminal precursor level necessarily being in equilibrium. This problem can be specified by merely defining the terminal set to be the vertical line $n = n_f$ in the n versus c state plane shown in Fig. 5. Here, the projection of the costate vector on state space must be perpendicular to this terminal line at the terminal time t_f.

With $p_2(t_f) = 0$ and time reversed, it is seen that $p_1(t)$ can have no zeros for $t < t_f$. Similar plots, with the eigenvectors in the same quadrants, could be obtained for other values of α, hence the time-optimal control with a reactivity constraint for $t_0 < t < t_f$. That this process requires no switching on $(t_{0,f})$ can also be shown by the solution to the adjoint system with time reversed, which is similar to the neutronics solution.

At the terminal time, the neutron precursor level is not yet at equilibrium. Therefore, reactivity must be different from zero to maintain constant neutron level $n(t) = n_f$ for $t > t_f$. From Eq. (22), it is seen that

$$u = l\dot{c}/n_f = \beta - \lambda l c(t)/n_f$$

is the required terminal reactivity. From Eqs. (28) and (29) it is readily shown that this required terminal control does belong to the admissible set with $|u(t)| \leq \alpha$.

If the switching function of the control equation vanishes for any finite time, the control is said to be singu-

lar. For Eq. (27) a singular control exists on some time interval if $p_1(t)$ is zero on this time interval. It is seen from Eq. (26), however, that this would require $p_2(t) = 0$. Hence, $\mathbf{p}(t) = \mathbf{0}$ on this finite interval, which would contradict the maximum principle since $\mathbf{p}(t)$ is nonzero. Therefore, the problem does not have a singular solution. Along with the uniqueness of the bang–bang-controlled state trajectories, it is obvious that the extremal neutronic policy derived here is indeed the unique time-optimal control.

If six precursor groups are considered, the state is described by Eq. (5) and $u(t) = 1$ can be assumed in the precursor equation. The time-optimal control is still obtained from Eq. (27), but in this case the switching function $p_1(t)$ satisfies the adjoint equation of (5). Again, the problem of most physical interest is to control the fission process from an initial point to a terminal hyperplane, $n = n_f$, in a seven-dimensional state space. The transversality condition then requires that:

$$p_i(t_f) = 0 \quad \text{for} \quad i = 1$$

From this condition, it is readily seen from the homogeneity of the costate equations that $p_1(t)$ cannot have a zero in finite time. Consequently, $u(t)$ does not switch on (t_0, t_f).

Again, the terminal control for $t > t_f$ must compensate for the decaying precursors. In this case, the terminal control is

$$u = \frac{l}{n_f} \sum_{i-1}^{6} \frac{dc_i}{dt}$$

It can be seen from the state equations that this terminal control is admissible, just as for the one-precursor case.

It can be shown that a small error in measurement to estimate the required terminal control results in a divergence of $n(t)$ away from n_f. A multiple bang–bang or so-called dither or sliding control that holds neutron level essentially constant can be synthesized. The idea is to allow $n(t)$ to vary within certain small prescribed limits about n_f and to drive the control at maximum effort about the theoretical in order to correct $n(t)$ when its deviation exceeds these limits.

The key to the synthesis of the time-optimal neutron level control appears to be the ability to maintain neutron density essentially constant while the precursor level is not near steady state.

Although the bang–bang neutronic control is synthesized very simply, the required terminal control is slightly more complicated. In many cases, a conventional type of closed-loop control is satisfactory and may even approach the performance of the optimal process. Even for these cases, however, the optimization analysis provides a yardstick of performance. One convenient technique of approximating the optimal process is to synthesize a continuous feedback control system that is fast acting but limited in control variations. Initially, the system is at equilibrium. Then maximum allowable reactivity is added until the neutron level is within some region of the desired value. At that time, a proportional plus integral type of feedback controller replaces the constant control process. The controller was introduced at that time to limit the magnitude of the controller integral signal. (Log-power control or a separate integral controller limit would allow the loop to be closed for the entire startup.) With high gain, the reactivity stays at its value until the power level is close to the desired state, at which time the proportional plus integral control takes effect.

For many power reactors there is another complication due to the accumulation of poison buildups, which may offer severe constraints to reactor restarts after shutdown. Two alternatives are to let the poison concentration decay sufficiently or to provide sufficient reactivity to override the poison. To minimize losses due to excessive shutdown periods, optimal control can play a significant role.

Similar to other regenerative processes, poison concentration can be described by a bilinear model. In this case,

$$dx_1/dt = \gamma_1 u - \lambda_1 x_1 \tag{31}$$

and

$$dx_2/dt = \gamma_2 u + \lambda_1 x_1 - \lambda_2 x_2 - \sigma_2 u x_2$$

where $0 \le u(t) \le u_b$, $x_1(t) = [^{135}\text{I}]/\Sigma_f$, and $x_2(t) = [^{135}\text{Xe}]/\Sigma_f$; u is the thermal neutron flux, [I] the iodine concentration, [Xe] the xenon concentration, Σ_f the macroscopic fission cross section, γ_1 the fractional yield of ^{135}I equal to 0.061, γ_2 the fractional yield of ^{134}Xe equal to 0.002, λ_1 the ^{135}I decay constant equal to 2.9×10^{-5} sec^{-1}, λ_2 the decay constant of ^{135}Xe equal to 2.1×10^{-5} sec^{-5}, and σ_2 the microscopic capture cross section of ^{135}Xe for thermal neutrons equal to 2.7×10^{-18} cm^2. The dynamics of this process are so slow that neutron flux rather than reactivity can be used as a control variable.

A sensible optimal control problem here is to change states in minimum time from $x(t_0) = x_0$ to $x(t_f)$ that belongs to some terminal set Γ_f so that xenon concentration is constrained by $x_2 \le x_{2m}$. The equilibrium set for equilibrium flux values on the xenon–iodine state plane is shown in Fig. 6. Here, $x_{2m} = 5 \times 10^{16}$ cm^{-2}, and the terminal set Γ_f is the zero-flux shutdown line, which does not violate this constraint but just reaches it.

From the maximum principle, it is obvious that, while the state is not on the constraint boundary, the extremal control is bang–bang, with

$$u = \begin{cases} 0 & \text{for } S < 0 \\ u_b & \text{for } S > 0 \end{cases} \tag{32}$$

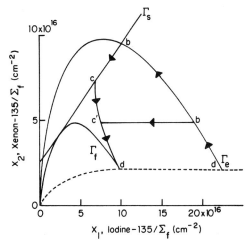

FIGURE 6 Reactor shutdown trajectories in the iodine–xenon state plane.

where the switching function is

$$S = -\gamma_1 p_1 + p_2(\sigma_2 x_2 - \gamma_2)$$

and $\mathbf{p}(t)$ is described by the usual adjoint equations of (31).

From the state equation, the target set Γ_t can be described by:

$$x_2 = g(x_1) = \left[\frac{x_{2m}}{(\alpha x_{2m})^\alpha} + \left(\frac{\alpha x_{2m}}{1 - \alpha} \right)^{1-\alpha} \right] x_1^\alpha - \frac{x_1}{1 - \alpha}$$

where $\alpha = \lambda_2/\lambda_1$. Then the orthogonal transversality condition requires that $(p_1 + p_2) dg/dx_1 = 0$, where:

$$\frac{dg}{dx_1} = \alpha \left[\frac{\alpha x_{2m}}{(\alpha x_{2m})^\alpha} + \left(\frac{\alpha x_{2m}}{1 - \alpha} \right)^{1-\alpha} \right] x_1^{\alpha-1} - \frac{1}{1 - \alpha}$$

For this problem, it is possible for the switching function $S(t)$ to vanish on some finite time interval. In such a case, the control is singular and must only force the state to satisfy $S(t) = 0$ and $dS/dt = 0$. From these conditions, a singular set Γ_s can be computed so that, whenever trajectories intersect such a set, the singular solution must be considered. Hence, Γ_s is defined by:

$$x_2 = x_1 + \frac{\gamma_1 + \gamma_2}{\sigma_2} - \frac{\gamma_2 \gamma_2}{\sigma_2 \lambda_1}$$
$$= x_1 + 2.28 \times 10^{16} \text{ cm}^{-2} \tag{33}$$

This singular set is graphed in Fig. 6. It is shown that if the [Xe] constraint were relaxed, the optimal trajectory a–b–c–d would involve a singular control on the b–c arc, where point a corresponds to equilibrium with $u = 10^{14}$ neutrons/(cm^2 sec). The optimal trajectory is computed from the equations above and involves $u = 0$ on

a–b, $u = u_b = 10^{14}$ neutrons/(cm^2 sec) on c–d, and, for the singular portion,

$$u = \frac{x_1(2\lambda_1 - \lambda_2) - \lambda_2 h}{\sigma_2 x_1 + 2\gamma_1 - (\lambda_2 \gamma_2 / \gamma_1)}$$

on b–c, where from Eq. (33) $h = 2.28 \times 10^6$ cm^{-2}. [It can be shown readily that $u(t)$ on b–c is less than u_b.]

With the state constraint on x_2, however, the flux to maintain $x_2(t) = x_{2m}$ is computed from Eq. (31). Thus,

$$u = (\gamma_2 x_{2m} - \lambda_1 x_1)/(\gamma_2 - \sigma_2 x_{2m}) \qquad (34)$$

which also is less than u_b.

From these necessary conditions, it is not too difficult to show that singular control is not needed with such a low x_{2m}, and the trajectory a–b'–c'–d shown in Fig. 6 is optimal with the following policy: $u = 0$ on a–b', $u = u_m$ on c'–d, and $u(t)$ is computed by Eq. (34) on b'–c'. At point d, u should return to zero to maintain the desired terminal set. The time consumed in traversing the trajectory a–b'–c'–d is 865 min. With such a slow response, the neutron response time for reactivity changes is certainly negligible.

As another example of the maximum principle, consider the following second-order regulator with classical position feedback and controlled rate feedback:

$$\begin{aligned} \dot{x}_1 &= x_2 \\ \dot{x}_2 &= -x_1 - \alpha(1 + u)x_2 \end{aligned} \qquad (35)$$

where x_1 is a position error and $|u| \leq 1$. The optimal control is to transfer an initial state to a small circle of radius ρ about the origin in such a manner as to minimize:

$$J(x) = \int_0^{t_f} \left[x_1^2(t) + x_2^2(t) \right] dt \qquad (36)$$

where t_f is the terminal time.

The value of $u(t)$ along an extremal, as deduced from the maximum principle, is

$$u(t) = -\text{sgn}[p_2(t)x_2(t)] \qquad (37)$$

where $\mathbf{p}(t)$ satisfies the adjoint equation to Eq. (35) with $\mathbf{p}(t_f) = c\mathbf{x}(t_f)$, $c = $ const. Now the switching curves can be computed by solving Eqs. (35) and (37) and the adjoint equations backward in time from the terminal circle. These switching lines approach eigenvectors with switchings used to bypass slow response modes.

In the limit, the target is just the origin, but this point cannot be reached in finite time. However, the switching curves obtained by the contraction process approach a well-defined limit, becoming just the radial line with slope $-\lambda_1$. The plane is divided into sectors R^+ and R^-, with $u(x) = +1$ on R^+ and $u(x) = -1$ on R^-.

The result can be given a very simple interpretation. Without the bilinear mode of control (i.e., for the equa-

tion $\ddot{x} + 2\alpha\dot{x} + x = 0$), all trajectories would approach the origin asymptotic to the direction of the "slow" eigenvector $\mathbf{w}^2$. This would result in relatively poor performance. Moreover, as α becomes large the performance deteriorates. since $\lambda_2 \to 0$. With a bilinear mode of control, however, the direction of $\mathbf{w}^2$ lies entirely in the sectors R^-, and the effect of the "slow mode" is nullified. All extremals approach the origin along the "fast" eigenvectors $\mathbf{w}^1$. As the damping α becomes large, the eigenvalue λ_1, becomes large and the performance improves monotonically. It can be shown that the limiting synthesis is not quite optimal.

It is of interest to compare the performance of the second-order BLS with that of a linear system when the input is other than a step. Hence, a sinusoidal input with variable frequency is utilized for the bilinear system:

$$\ddot{x} + \alpha[1 + u(x - x_d, \dot{x})]\dot{x} + x = x_d \qquad (38)$$

where x is the system response for a demanded input x_d. The simple switching law from Eq. (37) is incorporated for $u(x - x_d, \dot{x})$. This law was obtained as the limit of syntheses for step function inputs and was shown to be suboptimal even for this class of inputs. Therefore, no claim is made regarding optimality for the results that are presented.

The control law, however, has two pleasant properties. In the first place, it is very easy to synthesize in practice. Also, the nonlinearity is strictly independent of radius in the phase plane, which means that solutions of the resulting system have the important scalar multiplication property. That is, if the forced system of Eq. (38) has the response $x(t)$ for an input $x_d(t)$, it has response $kx(t)$ for the $kx_d(t)$. Observe that because the switching lines are radial,

$$u(x - x_d, \dot{x}) = u(kx - kx_d, k\dot{x}),$$

and therefore

$$k\ddot{x} + \alpha[1 + u(kx - kx_d, k\dot{x})]k\dot{x} + kx = kx_d$$

is satisfied.

The solid curves of Fig. 7 represent normalized mean-squared tracking error as a function of frequency for the BLS with $\alpha = 1, 2$, and 4. Because of the second property discussed previously, these are universal curves that do not depend on the amplitude of $x_d(t)$. The dashed curves represent mean-square error for the linear system obtained by setting $u = 1$ in Eq. (38), with $\alpha = 1, 2$, and 4. The performance of the BLS is clearly superior, and the trend for increasing α is reversed from that of the linear system.

It can be shown that under certain broad assumptions, BLSs may be stabilized by appropriate quadratic feedback. Of even more significance, however, is the ability to improve the state-space region of stabilization and controllability by means of bilinear control in Eq. (2). Still,

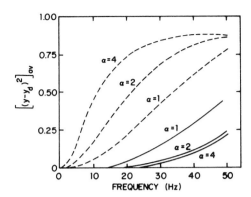

FIGURE 7 Mean-square error comparison of linear and bilinear servosystems for sinusoidal inputs.

the success of traditional linear control, the last term in Eq. (2), for more local control must not be overlooked.

IV. CONCLUSIONS AND PROJECTIONS

Bilinear systems with their adaptive structure are prevalent in nature and are useful for engineering design and control. They offer higher performance and controllability than linear systems in many cases. Bilinear system models for nuclear reactors, heat transfer, and immunology have been discussed in some detail.

The capacity of the immune system to control a replicating alien (e.g., virus) is a consequence of its controlled use of instability associated with cascades of BLSs. One example is the humoral immune process leading to the generation of antibodies which is turned on by T helper cells and turned off by T suppressor cells. Similarly, complement is synthesized as a series of bilinear amplifiers of enzyme concentration; nuclear fission in a reactor behaves in an analogous manner.

In their quest for high performance, humans can learn a lesson from nature. While engineering systems are traditionally designed from linear theory in a conservative fashion, breakthroughs might be achieved by the methodologies presented here. The advent of the microprocessor makes such designs even more feasible and almost dictates a new approach to control-system design.

As in bilinear control, numerical algorithms can be generated that are adaptive in step size and/or predictor slope. In signal processing, bilinear filters already have been shown to offer higher performance than classical linear filters or even traditional nonlinear filters in most cases that have been investigated. However, this is still a relatively new area of study. With regard to computers, bilinear automata and biochips may be on the horizon. To envision the information available via bionetworks, consider the humoral immune system with some 10^{12} lymphocytes

and 10^{20} antibody molecules of 10^7 different chemical affinities in the unstimulated human body.

Most recent research indicates that BLSs can play a significant role in signal processing as well as adaptive control. For example, nongaussian signals arise as a result of nonlinear media distortions or from naturally nonlinear processes, and bilinear models may arise directly from the physical phenomena or as valid approximations. In general, windowing can be conceptualized as a bilinear "control" term which multiplies a function of time which is used in an estimation procedure. If this is done in the frequency domain, the result is equivalent to convolution in the time domain. Even simple autoconvolution of signals generally enhances signal coherence while suppressing noise distortions.

Nonlinear adaptive control by means of BLS with appropriate nonlinear feedback can be useful for a variety of applications. Unfortunately, the problems are much tougher than for linear adaptive control. It should be noted, however, that even gain-manipulated linear adaptive control amounts to a bilinear control process, with the controller gains becoming the control variables multiplying state feedback.

Addressing the nonlinear adaptive control problem by means of BLS feedback-control theory is being researched for modern automobile comfort, safety, and roadability by means of controlled suspension. Similar methods are being applied to high-performance, high-angle-of-attack aircraft control and to large-scale electric power networks such as a flexible a.c. transmission system (FACTS). Self-tuning and model reference schemes have been derived which utilize bilinear autoregressive moving average (BARMA) models.

For the control of cancerous tumors by immunotherapy, experimentation has confirmed the significance of interleukin-2 (IL-2). This molecular substance is naturally secreted by T helper cells (T_h) and stimulates the generation of cytotoxic T cells (T_c) which can effectively destroy tumor cells. In the immune system, their concentration represents a bilinear (parametric) control of the T_c-generation rate, which in turn becomes a parametric control of the tumor death rate. IL-2 can be synthesized artificially to stimulate T_c generation, or genetically engineered T_h (or their appropriate precursors) could produce more IL-2. Computer simulations and experiments suggest great potential for such treatment.

Artificial neural networks (ANN) are a promising area of research for more efficient computing, communication, and control. BLS research may provide a more solid scientific foundation for its success, as a consequence of the logical coupled BLS synthesis of ANN and biological neurons. For certain BLS, such as those represented by a finite Volterra series and for certain rigid-body dynamics, Lie algebra presents a most convenient analytical tool utilizing combinations of linear system solutions.

SEE ALSO THE FOLLOWING ARTICLES

ARTIFICIAL NEURAL NETWORKS • CONTROLS, ADAPTIVE SYSTEMS • CONTROLS, LARGE-SCALE SYSTEMS • CYBERNETICS AND SECOND ORDER CYBERNETICS • HYBRID SYSTEMS CONTROL • SELF-ORGANIZING SYSTEMS • SIGNAL PROCESSING

BIBLIOGRAPHY

Asachenkov, A., Marchuk, G., Mohler, R., and Zuev, S. (1994). "Disease Dynamics," Birkhauser, Boston and Basel.

Baciotti, A. (1990). Constant feedback stabilizability of bilinear systems. *In* "Realization and Modeling in System Theory" (M., Kaashoek, J., Van Schippen, and A., Ran, eds.), pp. 357–367, Birkhauser, Basel.

Celikovsky, S. (1993). "On the stabilization of homogeneous bilinear systems." *Syst. Control Lett.* **21,** 503–510.

Khapalov, A., and Mohler, R. (1996). "Reachable sets and controllability of bilinear time-invariant systems: A qualitative approach." *IEEE Trans. Auto. Control* **AC41,** 1342–1347.

Khapalov, A., and Mohler, R. (1997). "Asymptotic stabilization of the bilinear time-invariant system via piecewise constant feedback." *Syst. Control Lett.* **33,** 47–54.

Mohler, R. (1991). "Nonlinear Systems": Vol. 1, "Dynamics and Control"; Vol. 2, "Applications to Bilinear Control," Prentice Hall, Englewood Cliffs, NJ.

Mohler, R., and Khapalov, A. (2000). "Bilinear-system control and application to flexible a.c. transmission." *J. Optimization Theory Applic.* **106**(3).

Sachkov, Y. L. (1997). "On positive orthant controllability of bilinear systems in small codimensions." *SIAM J. Control Opt.* **35,** 29–35.

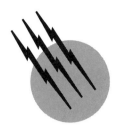

Controls, Large-Scale Systems

M. Jamshidi
NASA ACE Center, The University of New Mexico

GLOSSARY

Aggregation Mathematical process of contracting or reducing the total number of a system's states.

Boundary layer Initial transient history of a fast variable lost due to strong coupling.

Centrality Notion pertaining to the case in which all the components of a system are close to one another.

Coordination Process of putting together subsystem solutions, checking for overall system feasibility, and redirecting the subsystems to another round of solutions or stop; this coordination process is executed by the "coordinator."

Coupling Nature of interconnection between two or more subsystems; it can be either "weak," meaning that the system is decomposed into two or more disjoint subsystems, or "strong," meaning that the decomposition produces a reduction in the model as well.

Decentralized Notion pertaining to a system in which centrality does not hold.

Descriptive (descriptor) variable System variable used to describe a system fully (e.g., force on a spring).

Fixed modes Modes (eigenvalues) of systems that are not influenced by decentralized (local) feedback control.

Hierarchical Term describing a multilevel structure in which each member of a level "coordinates" all the members of the level immediately below it, while it is being coordinated along with others at the same level from above; also called "multilevel" or "hierarchical."

Interaction Influence of all the remaining subsystems on a given subsystem.

Perturbation Mathematical process in which a system's variables are designated "slow" or "fast" in time-scale variations.

A SYSTEM is sometimes considered to be large scale if it can be partitioned or decoupled into a number of subsystems, that is, small-scale systems. Another viewpoint is that a system is termed large scale if its dimensions are so great that conventional techniques of modeling, analysis, control, design, optimization, estimation, and computation fail to give reasonable solutions with reasonable efforts. A third definition is based on the notion of *centrality*. Until the advent of large-scale systems, almost all control systems analysis and design procedures were limited to having system components and information flow from one point to another, localized or centralized in one geographical location or center, such as a laboratory. Thus, another definition of a large-scale system is a system in which the concept of centrality fails. This can be due to a lack of either centralized computing capability or a centralized information structure. Large-scale systems appear in such diversified fields as sociology, business, management, economics, the environment, energy, data networks, computer networks, power systems, flexible space structures, Internet-based systems, transportation, aerospace, and navigational systems.

I. HISTORICAL BACKGROUND

Since the 1950s, system theory has evolved from a semi-heuristic discipline directed toward the design and analysis of electronic and/or aerospace systems consisting of a handful of components into a very sophisticated theory capable of treating complex and large systems with myriad applications. The theory must deal not only with electronic and aerospace systems, the complexities of which have increased by several orders of magnitude, but also with a vast number of real-life systems in society, the economy, industry, and government.

Initially, system engineers attempted to cope with this increasing system complexity through the development of sophisticated numerical techniques in order to apply classical systems theory to large systems. This approach, however, soon reached a point of diminishing returns, and it became apparent that new theoretical techniques would be necessary to handle large and complex systems. Although many such techniques are still being developed and require a great deal of fine-tuning, it is generally accepted that a key to the successful treatment of a large-scale system is to exploit fully its structural interconnection. This exploitation traditionally takes place in two ways: through the full use of, say, sparse matrix techniques or through the "decomposition" of a larger system into a finite number of smaller systems.

II. MODELING AND MODEL REDUCTION

The first step in any scientific or technological study of a system is to design a mathematical model of the real problem. In any modeling tasks, two often conflicting factors prevail: simplicity and accuracy. On the one hand, if a system model is oversimplified, presumably for the sake of computational effectiveness, incorrect conclusions may be drawn from it in representing an actual real system. On the other hand, a highly detailed model can lead to unnecessary complications, and, even if a feasible solution is attainable, the amount of detail generated may be

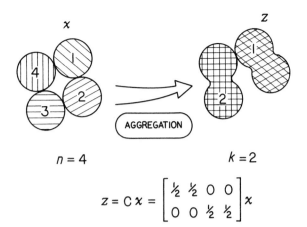

FIGURE 1 Pictorial representation of the aggregation process.

so vast that further investigations on the system behavior are impossible and the practical value of the model becomes questionable. Clearly, a mechanism by which a compromise can be made between a complex, more accurate model and a simple, less accurate model is needed. Creating such a mechanism is not a simple undertaking.

In the area of large-scale systems there have been three general classes of modeling techniques: *aggregation, perturbation,* and *descriptive variable* schemes. An aggregate model of a system is described by a "coarser" set of variables. The underlying reason for aggregating a system model is to retain the key qualitative properties of the system, such as stability, which can be viewed as a natural process through the second method of Lyapunov. In other words, the stability of a system described by several state variables is fully represented by a single variable— the Lyapunov function. Figure 1 presents a representation of the aggregation process. The system on the left is described by four variables (circles), and the system on the right represents an aggregated model in which two variables now describe the system. Variable 1, called z1, is an average of the first two variables of the full model, while the second aggregated variable, $z2$, is an average of the third and fourth variables.

Another approach for large-scale system modeling is perturbation, which is based on ignoring certain interactions of the dynamic or structural nature of a system. Here again, however, the key system properties must not be sacrificed for the sake of reduced computations. Although both perturbation and aggregation schemes tend to reduce the computations needed and perhaps provide a simplification of structure, there has been no firm evidence that they are the most desirable for large-scale systems.

A new type of large-scale system modeling is the *descriptive variable* scheme. Here, the fundamental principle is that the accuracy of a given large-scale system model is most likely preserved if the system is represented by the actual physical or economical variables that describe the operation of the system—hence, the name "descriptive variable."

This section is devoted to an examination of aggregation and perturbation, methods viewed as modeling alternatives for large-scale systems.

A. Aggregation

Aggregation has long been a technique for analyzing static economic models. The modern treatment of aggregation is based on the formulations of Malinvaud, which are shown in Fig. 2. In this diagram, $\mathcal{X}$, $\mathcal{Y}$, $\mathcal{Z}$, and $\mathcal{V}$ are topological (or vector) spaces, and f represents a linear continuous map between the exogenous variable $x \,\varepsilon\, \mathcal{X}$ and endogenous variable $y \,\varepsilon\, \mathcal{Y}$. The aggregation procedures $h\colon \mathcal{X} \to \mathcal{Z}$ and $g\colon \mathcal{Y} \to \mathcal{V}$ lead to aggregated variables $z \in \mathcal{Z}$ and $v \,\varepsilon\, \mathcal{V}$. The map $k\colon Z \to \mathcal{V}$ represents a simplified or an aggregated model. The aggregation is said to be "perfect" when k is chosen such that the relation:

$$gf(x) = kh(x) \qquad (1)$$

holds for all $x \in \mathcal{X}$. The notion of perfect aggregation is an idealization at best, and in practice it is approximated through two alternative procedures, according to econometricians: (1) to impose some restrictions on f, g, and h while leaving $\mathcal{X}$ unrestricted, and (2) to require Eq. (1) to hold on some subset of $\mathcal{X}$.

1. Balanced Aggregation

One of the main shortcomings of model reduction methods is the lack of a strong numerical tool to go with the well-developed theory. For example, the minimal realization theory of Kalman offers a clear understanding of the internal structure of linear systems. The associated discussions on controllability, observability, and minimal realization often illustrate the points, but numerical algorithms are adequate only for low-order textbook examples. Furthermore, there has been little connection made between minimal realization, controllability and observability, and model reduction. In this section, we propose the *principle component analysis* of statistics along with some algorithms for the computation of "singular value

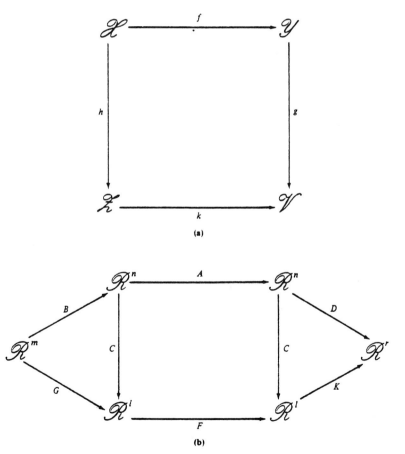

FIGURE 2 Pictorial representation of aggregation: (a) static system; (b) dynamic linear system.

decomposition" of matrices to develop a model reduction scheme that makes the most controllable and observable modes of the system transparent. Under a certain matrix transformation, the system is said to be "balanced," and the most controllable and observable modes would become prime candidates for reduced-order model states.

Consider an asymptotically stable, controllable, and observable linear time invariant system (A, B, C) defined by:

$$\dot{x}(t) = Ax(t) + Bu(t) \tag{2a}$$

$$y(t) = Cx(t) \tag{2b}$$

where x, u, y are state, input, and output vectors, and A, B, and C are $n \times n$ system, $n \times m$ input, and $r \times n$ output matrices, respectively. The balanced matrix method is based on the simultaneous diagonalization of the positive definite controllability and observability Gramians of Eq. (2), which satisfy the following Lyapunov-type equations:

$$G_c A^T + A G_c + B B^T = 0 \tag{3}$$

$$G_0 A + A^T G_0 + C^T C = 0 \tag{4}$$

The balanced approach to model reduction is essentially the computation of a similarity transformation matrix S such that both G_c and G_0 become equal and diagonal, that is, balanced. This transformation matrix is given by:

$$S = VDP\Sigma^{-1/2} \tag{5}$$

where orthogonal matrices V and P satisfy the following symmetric eigenvalue/eigenvector problems:

$$V^T G_c V = D^2 \tag{6}$$

and

$$P^T \left[(VD)^T G_0 (VD) \right] P = \Sigma^2 \tag{7}$$

$$\Sigma = S^T G_0 S = S^{-1} G_c (S^{-1})^T$$

$$= diag(\sigma_1, \sigma_2, \ldots, \sigma_n) \tag{8}$$

Here, D is a diagonal matrix like Σ. The diagonal elements of Σ have the property that $\sigma_1, \geq \sigma_2, \ldots, \sigma_n > 0$ and are called *second-order modes* of the system. Using the transformation $\hat{x} = S^{-1}x$, one obtains the following full-order equivalent system:

$$\dot{\hat{x}} = \hat{A}\hat{x} + \hat{B}u = \left[\begin{array}{c|c} F & \hat{A}_{12} \\ \hline \hat{A}_{21} & \hat{A}_{22} \end{array} \right] \hat{x} + \left[\begin{array}{c} G \\ \hline \hat{B}_2 \end{array} \right] u \tag{9}$$

$$y = \hat{C}\hat{x} = [H \mid \hat{C}_2]\hat{x} \tag{10}$$

where:

$$\hat{A} = S^{-1}AS, \quad \hat{B} = S^{-1}B, \quad \text{and} \quad \hat{C} = CS. \tag{11}$$

Now, if $\sigma_r \gg \sigma_r + 1$ for a given r, an internally dominant, reduced-order model of order r can be obtained from Eqs. (9) and (10) by:

$$\dot{z} = F_z = Gu \tag{12}$$

$$y = Hz \tag{13}$$

where (F, G, K) matrices represent the desired reduced-order model.

Although this partitioning of second-order models leading to a reduced and residual model are somewhat arbitrary, grouping the most controllable and observable modes together does provide a reasonable criterion for model reduction.

B. Perturbation

The basic concept of perturbation methods is the approximation of a system's structure by neglecting certain interactions within the model that lead to lower order. From a large-scale system modeling viewpoint, perturbation methods can be considered to be approximate aggregation techniques.

Two basic classes of perturbation are applicable to large-scale system modeling: "weakly coupled" models and "strongly coupled" models. This classification is not universally accepted, but a great number of authors have adapted it; others refer to these classes as nonsingular (regular) and singular perturbations.

1. Weakly Coupled Models

In many industrial control systems, certain dynamic interactions are neglected to reduce the computational burden for system analysis, design, or both. This is practiced in chemical process control and space guidance, for example, where different subsystems are designed for flow, pressure, and temperature control in an otherwise coupled process or for each axis of a three-axis altitude-control system. The computational advantages of neglecting weakly coupled subsystems, however, are offset by a loss of overall system performance. In this section, weakly coupled models for large-scale linear systems are introduced.

Consider the following linear large-scale system split into two subsystems:

$$\begin{bmatrix} \dot{x}_1 \\ \dot{x}_2 \end{bmatrix} = \begin{bmatrix} A_1 & \varepsilon A_{12} \\ \varepsilon A_{21} & A_2 \end{bmatrix} \begin{bmatrix} x_1 \\ x_2 \end{bmatrix} + \begin{bmatrix} B_1 & \varepsilon B_{12} \\ \varepsilon B_{21} & B_2 \end{bmatrix} \begin{bmatrix} u_1 \\ u_2 \end{bmatrix} \tag{14}$$

It is clear that when $\varepsilon = 0$, this system decouples into two subsystems:

$$\dot{\hat{x}}_2 = A_1\hat{x}_1 + B_1\hat{u}_1$$
$$\dot{\hat{x}}_2 = A_2\hat{x}_2 + B_2\hat{u}_2 \tag{15}$$

which correspond to two approximate aggregated models, one for each subsystem. In this way, the computation

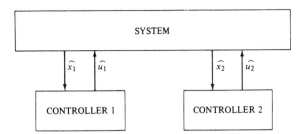

FIGURE 3 Decentralized control structure for two weakly coupled subsystems.

associated with simulation and design will be reduced drastically, especially for large-system order n and k greater than two subsystems. In view of the decentralized structure of large-scale systems (Section IV), these two subsystems can be designed separately in a decentralized fashion, as shown in Fig. 3.

Research on weakly coupled systems has followed two main lines. The first is to set $\varepsilon = 0$ in Eq. (2) and try to find a quantitative measure of the resulting approximation when in fact $\varepsilon \neq 0$ under actual conditions. Such measures usually correspond to a loss of performance for a linear optimal control problem. Our focus here is not on the loss of optimality due to decomposition; rather, our object is to introduce conditions under which a system can be considered weakly coupled.

The second line of research is to exploit such a system in an algorithmic fashion in order to find an approximate optimal feedback gain through a MacLaurin series expansion of the accompanying Riccati matrix in the coupling parameter ε. It has been shown that retaining k terms of the Riccati matrix expansion would give an approximation of order $2k$ to the optimal cost.

III. STRONGLY COUPLED MODELS

Strongly coupled models are those with variables of highly distinct speeds. Such models are based on the concept of *singular perturbation*, which differs from regular perturbation (weakly coupled systems) in that perturbation is to the left of the system's state equation, that is, a small parameter multiplying the time derivative of the state vector. In practice, many systems, most of them large in dimension, possess fast-changing variables displaying a singularly perturbed characteristic. For example, in a power system, the frequency and voltage transients vary from a few seconds in generator regulators, shaft-stored energy, and speed governor motion to several minutes in prime mover motion, stored thermal energy, and load voltage regulators.

Similar time-scale properties prevail in many other practical systems and processes, such as industrial control

systems (e.g., cold-rolling mills), biochemical processes, aircraft and rocket systems, and chemical diffusion reactions. In fact, some of the "order reduction" techniques that were discussed can be explained in terms of singular perturbation.

Consider a singularly perturbed system described by:

$$\dot{x}(t) = A_1 x(t) + A_{12} z(t) + B_1 u(t) \tag{16a}$$
$$x(t_0) = x_0$$

$$\varepsilon \dot{z}(t) = A_{21} x(t) + A_2 z(t) + B_2 u(t) \tag{16b}$$
$$z(t_0) = z_0$$

If A_2 is nonsingular, as $\varepsilon \to 0$, Eq. (16) becomes:

$$\dot{x}(t) = \left(A_1 - A_{12}^{-1} A_{21}\right)\hat{x} + \left(B_1 - A_{12}^{-1} A_2\right)\hat{u} \tag{17}$$

$$\hat{z}(t) = -A_2^{-1}\hat{x} - A_2^{-1} B_2 \hat{u} \tag{18}$$

Equation (17) is an approximate aggregated model in which the n eigenvalues of the original system are, in effect, approximated by the l eigenvalues of the $(A_1 - A_{12} A^{-1} A_{21})$ matrix in Eq. (17). This observation follows the same line of argument in discussions of conditions for weakly coupled systems. A very important phenomenon associated with a singularly perturbed system is the existence of so-called *boundary layers*. In going from Eq. (16) to Eq. (17), the initial condition of $z(t)$ is lost and the values of $\hat{z}(t_0)$ and $z(t)_0 = z_0$ are in general different; the difference is termed a *left-side boundary layer*, which corresponds to the fast transients of Eq. (16). Figure 4 shows the boundary layer phenomenon for the fast state $z(t)$.

IV. HIERARCHICAL CONTROL

A large-scale system, as briefly discussed in Section I, can be described as a complex system composed of a number of constituents or smaller subsystems serving particular functions and shared resources and governed by interrelated goals and constraints. Although interaction

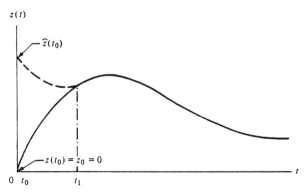

FIGURE 4 Boundary layer correction for fast-state $z(t)$. $-\cdot-$, $\hat{z}(t)$; $——$, $\hat{z}(t) + \eta(t)$.

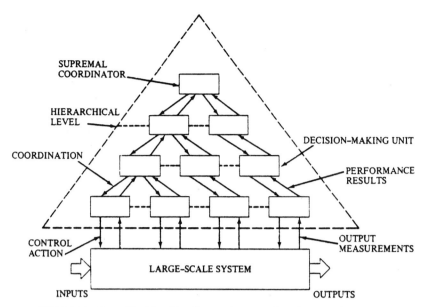

FIGURE 5 Hierarchical (multilevel) control strategy for a large-scale system.

among subsystems can take many forms, one of the most common is hierarchical, which appears to be somewhat natural in economic management and in organizational and complex industrial systems such as those of steel, oil, and paper. Within the hierarchical structure, the subsystems are positioned on levels with different degrees of hierarchy. A subsystem at a given level controls or coordinates the units on the level below it and is, in turn, controlled or coordinated by the unit on the level immediately above it. Figure 5 shows a typical hierarchical (multilevel) system. The highest level coordinator, sometimes called the *supremal coordinator,* can be thought of as the board of directors of a corporation, while another level's coordinators may be the president, vice-president, and directors. The lower levels are occupied by the plant manager, shop managers, and so on, while the large-scale system is the corporation itself. In spite of this seemingly natural representation of a hierarchical structure, its exact behavior is not well understood, mainly because relatively little quantitative work has been done on these large-scale systems.

There is no unique or universally accepted set of properties associated with hierarchical systems; however, the following are some of the key properties:

1. A hierarchical system consists of decision-making components structured in a pyramidal shape (Fig. 5).
2. The system has an overall goal that may or may not be in harmony with all of its individual components.
3. The various levels of hierarchy in the system exchange information (usually vertically) among themselves iteratively.

4. As the level of hierarchy increases, the time horizon increases; that is, the lower level components are faster than the higher level ones.

On the basis of the discussion so far, one can tentatively conclude that the successful operation of a hierarchical system is best described by two processes: *decomposition* and *coordination* (Fig. 6). In summary, the basic aims of hierarchical control are (1) to decompose a given large-scale system into a number of small-scale subsystems, and (2) through the coordination of these subsystems' solutions to achieve feasibility and optimality of the overall solution by means of a multilevel iterative algorithm.

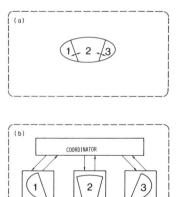

FIGURE 6 Pictorial representation of system decomposition and coordination: (a) interconnected system; (b) hierarchically structured system.

A. Goal Coordination: Interaction Balance

Consider a linear large-scale system described by a set of differential equations, commonly known as the state equation:

$$\dot{x} = Ax + Bu, \qquad x(t_0) = x_0 \qquad (19)$$

with its cost function to be minimized:

$$J = \frac{1}{2}x^T(t_f)F_x(t_f) + \frac{1}{2}\int_{t_0}^{t_f}(x^T Q x + u^T R u)\,dt \quad (20)$$

where F, $Q \geq 0$; $R > 0$; t_0 and t_f are initial and final values of time, respectively; and x_0 is the initial state. All the remaining terms correspond to the usual linear systems theory. Assume that the order n of system (19) is too large (say, $n > 200$) for this problem to be solved as is. It is assumed that Eq. (19) can be decomposed into N subsystems described by:

$$\dot{x}_i = A_i x_i + B_i u_i + z_i, \qquad x_i(t_0) = x_{i0}$$
$$i = 1, \ldots, N \quad (21)$$

where z_i is defined by:

$$z_i = \sum_{j=1, j \neq i}^{N} G_{ij} x_j \qquad (22)$$

which describes the ith subsystem with the remaining $N-1$ subsystems. The original system's optimal control problem is reduced to the optimization of N subsystems that collectively satisfy Eqs. (21) and (22) while minimizing:

$$J = \sum_{i=1}^{N} \left\{ \frac{1}{2}x_i^T(t_f)F_i x_i(t_f) + \frac{1}{2}\int_{t_0}^{t_f} \left[x_i^T(t)Q_i x_i(t) \right. \right.$$
$$\left. \left. + u_i^T(t)R_i u_i(t) + z_i^T(t)S_i z_i(t) \right] dt \right\} \qquad (23)$$

where F_i and Q_i are $n_i \times n_i$ positive semidefinite matrices, and R_i and S_i are $m_i \times m_i$ and $n_i \times n_i$ positive definite matrices with:

$$n = \sum_{i=1}^{N} n_i; \qquad m = \sum_{i=1}^{N} m_i \qquad (24)$$

The physical interpretation of the last term in the integrand of Eq. (23) is difficult at this point. In fact, the purpose of introducing this term is to avoid singular controls.

In this decomposition of a large interconnected linear system, the common coupling factors among its N subsystems are the "interaction" variables $z_i(t)$, which, along with Eqs. (21) and (22), constitute the "coupling" constraints. This formulation is called *global* and is denoted by S_G. The following assumption is considered to hold. The global problem is replaced by a family of N subproblems coupled together through a parameter vector $\alpha = (\alpha_1, \ldots, \alpha_N)^T$ and denoted by $s_i(\alpha)$, $i = 1, 2, \ldots, N$. In other words, the global system problem is "imbedded" into a family of subsystem problems through an imbedding parameter α in such a way that for a particular value of α^* the subsystems $s_i(\alpha^*)$, $i = 1, 2, \ldots, N$ yield the desired solution to S_G. In terms of hierarchical control notation, this imbedding concept is nothing but the notion of coordination, but in mathematical programming problem terminology it is known as the *master problem*. As will be seen later, α represents a vector of Lagrange multipliers in an optimization problem setting. Figure 7 shows a two-level control structure of a large-scale system. Under this strategy, each local controller i receives α_i^l from the coordinator (second-level hierarchy), solves $s_i(\alpha_i^l)$, and transmits (reports) some function y_i^l of its solution to the coordinator. The coordinator, in turn, evaluates the next updated value of α; that is,

$$\alpha^{l+1} = \alpha^l + \varepsilon^l d^l \qquad (25)$$

where ε^l is the lth-iteration step size, and the update term d is defined from the following conjugate gradient interaction:

$$d^{l+1} = e^{l+1}(t) + \gamma^{l+1} d^l(t) \qquad 0 \leq t \leq t_f \qquad (26)$$

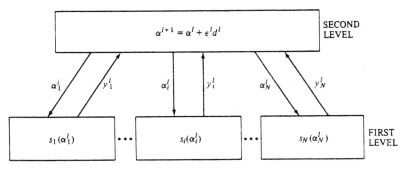

FIGURE 7 Two-level goal-coordination structure for dynamic systems.

where:

$$\gamma^{l+1} = \frac{\int_0^{l_f} [e^{l+1}(t)]^T e^{l+1}(t)\, dt}{\int_0^{l_f} (e^l)^T e^l\, dt} \tag{27}$$

and $d^0 = e^0$. The vector $d(t)$ is commonly known as the *interaction error*, defined by:

$$e_i[\alpha(t), t] = z_i[\alpha(t), t] - \sum_{\substack{j=1 \\ j \neq 1}}^{N} G_{ij} x_j[\alpha(t), t] \tag{28}$$

The relations expressed in Eqs. (25) to (27) constitute the so-called second-level, or coordinator, problem. The first-level, or subsystem, problem is represented by Eqs. (21) to (23), which constitute a linear regulator problem and can be handled by a Riccati formulation. By virtue of Fig. 7, each time the coordinator passes down a new (updated) coordination vector $\alpha^* = (\alpha_N^{*T} \cdots \alpha_N^{*T})$, the N subsystems need to solve N near-optimum control problems defined by Eqs. (21) to (23) until the *interaction balance* is achieved, that is, the normalized interaction error.

$$\text{Error} = \left[\sum_{i=1}^{N} \int_0^{l_f} \left(z_i - \sum_{j=1}^{N} G_{ij} x_j \right)^T \right.$$
$$\left. \times \left(z_i - \sum_{j=1}^{N} G_{ij} x_j \right) dt \right] \Big/ \Delta t \tag{29}$$

is sufficiently small. Here, Δt is the step size of integration.

B. Interaction Prediction

An alternative approach in optimal control of hierarchical systems that has both open- and closed-loop forms is the interaction prediction method, which avoids second-level, gradient-type iterations. Consider a large-scale, linear, interconnected system that is decomposed into N subsystems, each of which is described by:

$$\dot{x}_i(t) = A_i x_i(t) + B_i u_i(t)$$
$$x_i(0) = x_{io} \qquad i = 1, 2, \ldots, N \tag{30}$$

where the interaction vector z_i is

$$z_i(t) = \sum_{j=1}^{N} G_{ij} x_j(t) \tag{31}$$

The optimal control problem at the first level is to find a control $u_i(t)$ that satisfies Eqs. (30) and (31) while minimizing a quadratic cost function:

$$J_i = \frac{1}{2} x_i^T(t_f) Q_i x_i(t_f)$$
$$+ \frac{1}{2} \int_0^{l_f} \left[x_i^T(t) Q_i x_i(t) + u_i^T(t) R_i(t) u_i(t) \right] dt \tag{32}$$

This problem can be solved by first introducing a set of Lagrange multipliers $\alpha_i(t)$ and costate vectors $p_i(t)$ to augment the "interaction" equality constraint [Eq. (31)] and subsystem dynamic constraint [Eq. (18)] to the integrand of the cost function, that is, the ith subsystem Hamiltonian, defined by:

$$H_i = \frac{1}{2} x_i^T(t) Q_i x_i(t) + \frac{1}{2} u_i^T(t) R_i u_i(t) + \alpha_i^T z_i$$
$$- \sum_{j=1}^{N} \alpha_j^T G_{ji} x_i + p_i^T (A_i x_i + B_i u_i + C_i z_i) \tag{33}$$

Then, utilizing the necessary conditions of optimality, using a Riccati formulation,

$$p_i(t) = K_i(t) x_i(t) + g_i(t) \tag{34}$$

and simplifying the resulting two-point boundary value (TPBV) problem, one obtains

$$\dot{K}_i(t) = -K_i(t) A_i = A_i^T J i(t) + K_i(t) S_i K_i(t) - Q_i \tag{35}$$

$$\dot{g}_i(t) = -[A_i - S_i K_i(t)]^T g_i(t) - K_i(t) z_i(t)$$
$$+ \sum_{j=1, j \neq 1}^{N} G_{ji}^T \alpha_j^T(t) \tag{36}$$

the final conditions of which, $K_i(t_f)$ and $g_i(t_f)$, follow from:

$$p_i(t_f) = \partial \left[\tfrac{1}{2} x_i^T(t_f) Q_i x_i(t_f) \right] \Big/ \partial x_i(t_f) = Q_i x_i(t_f) \tag{37}$$

and Eq. (34); that is,

$$K_i(t_f) = Q_i; \qquad g_i(t_f) = 0 \tag{38}$$

Following this formulation, the first-level optimal control becomes:

$$u_i(t) = -R_i^{-1} B_i^T K_i(t) x_i(t) - R_i^{-1} B_i^T g_i(t) \tag{39}$$

which has a partial feedback (closed-loop) term and a feedforward (open-loop) term. Two points are made here. First, the solution of the differential matrix Riccati equation, which involves $(n_i + l) n_i / 2$ nonlinear scalar equations is independent of the initial state $x_i(0)$. Second, unlike $K_i(t)$, $g_i(t)$ in Eq. (36), by virtue of $z_i(t)$, is dependent on $x_i(0)$.

The second-level problem is essentially updating the new coordination vector. For this purpose, using partial derivatives, the additively separable Lagrangian would result in the following second-level coordination procedure at the lth iteration:

$$\begin{bmatrix} \alpha_i(t) \\ z_i(t) \end{bmatrix}^{l+1} = \begin{bmatrix} -p_i(t) \\ \sum_{j=1, j\neq 1}^{N} G_{ji}x_j(t) \end{bmatrix}^l \quad (40)$$

The interaction prediction method is formulated by the following algorithm.

Step 1. Solve N independent differential matrix Riccati equations [Eq. (35)] with the final condition (38) and store $k_i(t), i = 1, 2, \ldots, N$.

Step 2. For initial α_i^l, z_i^l, solve the "adjoint" equation [Eq. (36)] with the final condition (38). Evaluate and store $g_i(t), i = 1, 2, \ldots, N$.

Step 3. Solve the state equation:

$$\dot{x}_i(t) = [A_i - S_i K_i(t)]x_i(t) - S_i g_i(t) + z_i(t)$$
$$x_i(0) = x_{i0} \quad (41)$$

and store $x_i(t), i = 1, 2, \ldots, N$.

Step 4. At the second level, use the results of steps 2 and 3 and Eq. (40) to update the coordination vector.

Step 5. Check for convergence at the second level by evaluating the overall interaction error:

$$e(t) = \sum_{j=1}^{N} \int_0^{t_f} \left[z_i(t) - \sum_{j=1}^{N} G_{ij}x_j(t) \right]^T$$
$$\times \left[z_i(t) - \sum_{j=1}^{N} G_{ij}x_j(t) \right] dt \Big/ \Delta t \quad (42)$$

It must be noted that, depending on the type of digital computer and its operating systems, subsystem calculations can be done in parallel and the N matrix Riccati equations at step 1 are independent of $x_i(0)$; hence, they must be computed once regardless of the number of second-level interactions in the interaction prediction algorithm [Eq. (40)]. It must be further noted that, unlike the goal coordination methods, no $z_i(t)$ term is needed in the cost function, which is intended to avoid singularities.

V. DECENTRALIZED CONTROL

As mentioned in Section I, in many large-scale systems the notion of centrality does not hold. Under such conditions, the system remains structurally intact while its output information is shared among N controllers, referred to as *decentralized,* which collectively control the system. Therefore, the basic difference between decentralized and hierarchical control is the following. In hierarchical control, the system is *decomposed* into subsystems, the solutions of which are *coordinated* by a higher level controller

(e.g., a supervisory type of control). In decentralized control, on the other hand, the system's output measurement is shared among a finite number of controllers that collectively help to control the system.

The main motivation behind decentralized control is the failure of conventional methods of centralized control theory. Some fundamental techniques of the latter theory, such as pole placement, state feedback, optimal control, and state estimation, require complete information from all system sensors for the sake of feedback control. This scheme is clearly inadequate for feedback control of large-scale systems. Due to the physical configuration and often high dimensionality of such systems, a centralized control is neither economically feasible nor even necessary. Therefore, in many applications of feedback control theory to linear large-scale systems, some degree of restriction is assumed to prevail on the transfer of information. In some cases, a total decentralization is assumed; that is, every local control u_i is obtained from the local output y_i and possible external input v_i. In others, an intermediate restriction on the information is possible.

In this section, one major problem related to the decentralized structure of a large-scale control system is addressed. This problem has to do with finding a state or an output feedback gain whereby the closed-loop system has all its poles in specified locations. This problem is commonly known as *feedback stabilization.* Alternatively, the closed-loop poles of a controllable system can be preassigned through the state or output feedback.

A. Stabilization Problem

Consider a large-scale linear time-invariant (TIV) system with N local control stations (channels):

$$\dot{x}(t) = Ax(t) + \sum_{i=1}^{N} B_i u_i(t) \quad (43)$$

$$y_i(t) = C_i x, \quad i = 1, 2, \ldots, N \quad (44)$$

where x *is* an $n \times 1$ state vector, and u_i and y_i are $m_i \times 1$ and $r_i \times 1$ control and output vectors associated with the ith control station, respectively. The original system control and output orders m and r are given by:

$$m = \sum_{i=1}^{N} m_i; \quad r = \sum_{i=1}^{N} r_i \quad (45)$$

The decentralized stabilization problem is defined as follows. Obtain N local output control laws, each with its independent dynamic compensator,

$$u_i(t) = H_i z_i(t) + K_i y_i(t) + L_i v_i(t) \quad (46)$$

$$\dot{z}_i(t) = F_i z_i(t) + S_i y_i(t) + G_i v_i(t) \quad (47)$$

so that the system in Eqs. (43) and (44) in its closed-loop form is stabilized. In Eqs. (46) and (47) $\dot{z}_i(t)$ is the $n_i \times 1$ output vector of the ith compensator, $v_i(t)$ is the $m_i \times 1$ external input vector for the ith controller, and matrices H_i, K_i, L_i, F_i, S_i, and B_i are $m_i \times n_i$, $m_i \times r_i$, $m_i \times m_i$, $n_i \times n_i$, $n_i \times r_i$, and $n_i \times m_i$, respectively. Alternatively, the problem can be restated as follows. Find matrices H_i, K_i, L_i, F_i, S_i, and B_i, $i = 1, 2, \ldots, N$, so that the resulting closed-loop system described by Eqs. (43) and (44) has its poles in a set $\mathcal{L}$, where $\mathcal{L}$ is a nonempty, symmetric, open subset of the complex s plane. It is clear that the membership of a closed-loop system pole $\lambda \in \mathcal{L}$ implies its complex conjugate $\lambda^* \in \mathcal{L}$ in a prescribed manner.

B. Fixed Modes and Polynomials

The notions of fixed polynomials and fixed modes are generalizations of the "centralized" system pole-placement problem, in which any uncontrollable and unobservable mode of the system must be stable to the decentralized case. The idea of fixed modes for decentralized control has been used extensively with regard to the general servomechanism problem.

Consider the decentralization stabilization problem described by Eqs. (43) to (47). The dynamic compensator [Eqs. (46) and (47)] can be rewritten in compact form:

$$u(t) = Hz(t) + Ky(t) + Lv(t)$$
$$\dot{z}(t) = F_z(t) + S_y(t) + G_v(t) \tag{48}$$

where:

$$H \stackrel{\Delta}{=} \text{block diag}\{H_1, H_2, \ldots, H_N\}$$
$$K \stackrel{\Delta}{=} \text{block diag}\{K_1, K_2, \ldots, K_N\}$$
$$L \stackrel{\Delta}{=} \text{block diag}\{L_1, L_2, \ldots, L_N\}$$
$$F \stackrel{\Delta}{=} \text{block diag}\{F_1, F_2, \ldots, F_N\} \tag{49}$$
$$S \stackrel{\Delta}{=} \text{block diag}\{S_1, S_2, \ldots, S_N\}$$
$$G \stackrel{\Delta}{=} \text{block diag}\{G_1, G_2, \ldots, G_N\}$$

and

$$u^T(t) \stackrel{\Delta}{=} \{u_1^T(t) \vdots \cdots s \cdots \vdots u_N^T(t)\}$$
$$z^T(t) \stackrel{\Delta}{=} \{z_1^T(t) \vdots \cdots \vdots z_N^T(t)\}$$
$$y^T(t) \stackrel{\Delta}{=} \{y_1^T(t) \vdots \cdots \vdots y_N^T(t)\} \tag{50}$$
$$v^T(t) \stackrel{\Delta}{=} \{v_1^T(t) \vdots \cdots \vdots v_N^T(t)\}$$

If the control [Eqs. (48) and (49)] is applied to Eqs. (43) and (44), the following augmented system results:

$$\begin{bmatrix} \dot{x}(t) \\ \hdashline \dot{z}(t) \end{bmatrix} = \begin{bmatrix} A + BKC & BH \\ \hdashline SC & F \end{bmatrix} \begin{bmatrix} x(t) \\ \hdashline z(t) \end{bmatrix} + \begin{bmatrix} BL \\ \hdashline G \end{bmatrix} v(t) \tag{51}$$

where

$$B = \underbrace{\left[B_1 \vdots \cdots \vdots B_1 \right]}_{m_1 \qquad m_N} \quad \text{and} \quad C = \begin{bmatrix} C_1 \\ \cdots \\ \vdots \\ \cdots \\ C_N \end{bmatrix} \begin{matrix} \}r_1 \\ \\ \\ \}r_N \end{matrix} \tag{52}$$

As mentioned earlier, the problem is to find the control laws [Eqs. (48) and (49)] so that the overall augmented system [Eq. (51)] is asymptotically stable. In other words, by way of local output feedback, the closed-loop poles of the decentralized system are required to lie on the left half of the complex s plane. The following definitions and theorem provide the ground rules for this problem.

Definition 1. Consider the system (C, A, B) describing Eqs. (43) and (44) and integers $m_i, r_i, i = 1, 2, \ldots, N$ in Eq. (45). Let the $m \times r$ gain matrix K be represented as a member of the following set of block-diagonal matrices:

$$\mathbf{K} = \left(K \mid K = \begin{matrix} m_1\{ \end{matrix} \begin{bmatrix} \overbrace{K_1}^{r_1} & & & \\ & K_2 & & \\ & & \ddots & \\ & & & K_N \\ & & & \underbrace{}_{r_N} \end{bmatrix} \begin{matrix} \\ \\ \\ \}m_N \end{matrix} \right) \tag{53a}$$

where $\dim(K_i) = m_i \times r_i$, $i = 1, 2, \ldots, N$. Then the "fixed polynomial" of (C, A, B) with respect to $\mathbf{K}$ is the greatest common divisor (gcd) of the set of polynomial $|\lambda I - A - BKC|$ for all $K \in \mathbf{K}$ and is denoted by:

$$\phi(\lambda; C, A, B, K) = \text{gcd}\{|\lambda I - A - BKC|\} \tag{53b}$$

Definition 2. For the system (C, A, B) and the set of output feedback gains K given by Eq. (53), the set of "fixed modes" of (C, A, B) with respect to K is defined as the intersection of all possible sets of the eigenvalues of matrix $(A + BKC)$; that is,

$$\Lambda(C, A, B, K) = \bigcap_{K \in \mathbf{K}} \lambda(A + BKC) \tag{54}$$

where $\lambda(\cdot)$ denotes the set of eigenvalues of $(A + BKC)$. Note also that K can take on the null matrix; hence, the set of fix modes $\Lambda(\cdot)$ is contained in $\lambda(A)$. In view of definition 1, the members of $\Lambda(\cdot)$ (that is, the "fixed modes") are the roots of the "fixed polynomials" $\phi(\cdot\,;\cdot)$ in Eq. (53b):

$$\Lambda(C, A, B, K) = \{\lambda \mid \lambda \in s \text{ and } \phi(\lambda, C, A, B, K) = 0\} \tag{55}$$

where s denotes a set of values on the entire complex s plane.

The fixed modes of centralized system (C, A, B, K), where K is $m \times r$, correspond to the uncontrollable and unobservable modes of the system. The following theorem provides the necessary and sufficient conditions for the stabilizability of a decentralized closed-loop system.

Theorem 1. For the system (C, A, B) in Eqs. (43) and (44) and the class of block-diagonal matrices **K** in Eq. (53a), the local feedback laws [Eqs. (46) and (47)] will asymptotically stabilize the system if and only if the set of fixed modes of (C, A, B, K) is contained in the open, left-half s plane; that is, $\Lambda(C, A, B, K) \in s^-$, where s^- is the open, left-half s plane.

C. Stabilization via Dynamic Compensation

One of the earliest efforts in dynamically compensating centralized systems made use of output feedback; the problem can be briefly stated as follows: Consider a linear TIV system:

$$\dot{x}(t) = Ax(t) + Bu(t) \tag{56}$$

$$y(t) = Cx(t) \tag{57}$$

We wish to find a dynamic compensator:

$$\dot{z}(t) = Fz(t) + Sy(t) \tag{58}$$

$$u(t) = Hz(t) + Ky(t) \tag{59}$$

so that the closed-loop system,

$$\dot{x}(t) = (A + BKC)x(t) + BHz(t) \tag{60}$$

has a prescribed set of poles.

For the case of finding the dynamic compensator, let n_o and n_c be the smallest integers such that:

$$\text{rank } [B, AB, \ldots, A^n cB] = n$$

$$\text{rank } [C^T, A^T C^T, \ldots, A^{Tn} oC^T] = n \tag{61}$$

Now, for convenience purposes, let $\eta = \min(n_c, n_o)$ and $\Lambda_\eta = (\lambda_1, \ldots, \lambda_{n+\eta})$ be a set of arbitrary complex numbers with the only restriction being that, for each λ_i with $\text{Im}(\Lambda_i) \neq 0$, a complex conjugate pair $\Lambda_i = \text{Re}(\lambda_i) \pm j\,\text{Im}(\lambda_i)$ is contained in Λ_η. Let us define the following augmented triplet (C_η, A_η, B_η):

$$A_\eta = \left[\begin{array}{c|c} A & 0 \\ \hline 0 & 0 \end{array}\right] \begin{array}{l} \}n \\ \}\eta \end{array} \qquad B_\eta = \left[\begin{array}{c|c} B & 0 \\ \hline 0 & I \end{array}\right] \begin{array}{l} \}n \\ \}\eta \end{array}$$

$$C_\eta = \left[\begin{array}{c|c} C & 0 \\ \hline 0 & I \end{array}\right] \begin{array}{l} \}r \\ \}\eta \end{array} \tag{62}$$

The following theorem determines the existence of an output feedback gain for proper pole placement.

Theorem 2. Let (C, A, B) be a controllable and observable system, and let the triple (C_η, A_η, B_η) be defined by Eq. (62) with $\eta = \min(n_c, n_o)$ and a set of prescribed poles $\Lambda_\eta = (\lambda_1, \ldots, \lambda_{n+\eta})$. Then, there exists a gain matrix K such that the eigenvalues $(A_\eta, +B_\eta K C_\eta)$ are exact elements of A_η.

This canonical structure theorem and the decentralized stabilization problem 1 of Section IV.A have been used to find a dynamic decentralized control. Consider that the set of N dynamic compensators [Eqs. (46) and (47)] and the triplet in Eq. (62) define a real constant $(m + \eta) \times (r + \eta)K_\eta$ matrix:

$$K_\eta = \begin{bmatrix} K_1 & & 0 & H_1 & & 0 \\ & K_2 & & & H_2 & \\ & & \ddots & & & \\ 0 & & K_N & 0 & & H_N \\ S_1 & & & F_1 & & \\ & S_2 & 0 & & F_2 & \\ & 0 & & & & \ddots \\ & & S_N & & & F_N \end{bmatrix} \begin{array}{l} \}m_1 \\ \}m_2 \\ \vdots \\ \}m_N \\ \}\eta_1 \\ \}\eta_2 \\ \vdots \\ \}\eta_N \end{array} \tag{63}$$

where K_i, H_i, S_i, and F_i are $m_i \times r_i$, $m_i \times \eta_i$, $\eta_i \times r$, and $\eta_i \times \eta_i$ submatrices, respectively, defined in Eqs. (46) and (47); m and r are defined in Eq. (45); and $\eta = \sum_{i=1}^N \eta_i$. The following proposition summarizes the decentralized control pole-placement problem.

Proposition 1. Considering the triplets (C, A, B) and (C_η, A_η, B_η) defined earlier and the set of block-diagonal **K** in Eq. (53a), for any set of integers $\eta_1, \ldots, \eta_N$ with $\eta_1 \geq 0$, the following two "fixed polynomials" are identical:

$$\phi(\lambda; C, A, B, K) = \phi(\lambda; C_\eta A_\eta B_\eta K_\eta) \tag{64}$$

where C_η, A_η, B_η, and K_η have already been defined. In other words, the greatest common divisor of $\det(\lambda I - A - BKC)$ and $\det(\lambda I - A - B_\eta K_\eta C_\eta)$ are the same. The result of this proposition and a matrix identity can be used to place poles in decentralized output feedback controllers through dynamic compensation.

SEE ALSO THE FOLLOWING ARTICLES

CONTROLS, ADAPTIVE SYSTEMS • CONTROLS, BILINEAR SYSTEMS • MATHEMATICAL MODELING • SYSTEM THEORY

BIBLIOGRAPHY

Davison, E. J. (1976). *In* "Direction in Large-Scale Systems" (Y. C. Ho and S. K. Mitter, eds.), pp. 303–323. Plenum, New York.

Ho, Y. C., and Mitter, S. K., eds. (1976). "Directions in Large Scale System," pp. v–x. Plenum, New York.

Institute of Electrical and Electronics Engineers (1982). "International Large-Scale Systems Symposium," IEEE, Virginia Beach, VA.

International Federation of Automatic Control (1976). "Theory and Applications of Large-Scale Systems," IFAC, Udine, Italy.

International Federation of Automatic Control (1979). "Theory and Applications of Large-Scale Systems," IFAC, Toulouse, France.

International Federation of Automatic Control (1982). "Theory and Applications of Large-Scale Systems," IFAC, Warsaw, Poland.

International Federation of Automatic Control (1985). "Theory and Applications of Large-Scale Systems," IFAC, Zurich, Switzerland.

Jamshidi, M. (1983). "Large-Scale Systems: Modeling and Control," Elsevier/North-Holland, New York.

Jamshidi, M. (1996). "Large-Scale Systems: Modeling, Control and Fuzzy Logic," Prentice Hall, Englewood Cliffs, NJ.

Luenberger, D. G. (1978). *IFAC J. Autom.* **14,** 473–480.

Mahmoud, M. S., and Singh, M. G. (1981). "Large Scale Systems Modeling," Pergamon, Oxford.

Mesarovic, M. D., Macko, D., and Takahara, Y. (1970). "Theory of Hierarchical Multilevel Systems," Academic Press, New York.

Saeks, R., and DeCarlo, R. A. (1981). "Interconnected Dynamical Systems," Dekker, New York.

Sage, A. P. (1977). "Methodologies for Large-Scale Systems," McGraw-Hill, New York.

Siljak, D. D. (1978). "Large-Scale Dynamic Systems," Elsevier/North-Holland, New York.

Singh, M. G. (1980). "Dynamical Hierarchical Control," rev. ed. North-Holland, Amsterdam.

Singh, M. G., and Titli, A. (1978). "Systems," Pergamon, Oxford.

Stengel, D. N., Luenberger, D. G., Larson, R. E., and Cline, T. S. (1979). "A Descriptor Variable Approach to Modeling and Optimization of Large-Scale Systems," Rep. No. CONS-2858-T1. U.S. Dept. of Energy, Oak Ridge, TN.

Wang, S. H., and Davison, E. J. (1973). *IEEE Trans. Autom. Control* **AC-18,** 473–478.

Control Systems, Identification

R. K. PEARSON
ETH Zürich

I. Building Dynamic Models
II. Linear Model Identification
III. Nonlinear Model Identification
IV. Practical Issues
V. Keeping Up with Recent Progress

GLOSSARY

ARMA models or *autoregressive, moving-average models* are linear, discrete-time dynamic models in which the output at each time instant depends on past outputs (defining the autoregressive terms) and past inputs (defining the moving average terms).

Bias refers to *systematic errors* in parameter estimation, often caused by violation of working assumptions (e.g., the presence of *colored noise* in algorithms based on the assumption that model errors are *white*).

Colored noise refers to an error sequence $\{e_k\}$ in which errors e_k and e_j are statistically correlated for $k \neq j$; contrast with *white noise*.

FIR models or *finite impulse response models* are generally linear dynamic models characterized by finite-order moving average representations, implying that their responses to impulse inputs go to zero after a finite number of time steps, equal to the model's memory length.

Identification or *system identification* refers to the fitting of mathematical models to observed input/output data.

Linear models satisfy the *principle of superposition:* The response of the model to a linear combination of inputs is equal to the corresponding linear combination of the responses to each individual input.

Nonlinear models are any models that *do not satisfy* the *principle of superposition* discussed in connection with *linear models*; an important practical observation is that the class of nonlinear models is infinitely larger and more varied than the class of linear models.

Outliers are *anomalous data points* that are inconsistent with the bulk of the available data; unless they are handled appropriately, outliers can cause significant *biases* in estimated model parameters.

Persistently exciting input sequences are those for which the estimated autocorrelation matrix is nonsingular; persistence of excitation is a sufficient condition for the existence of solutions for linear least squares model identification procedures.

Prediction error methods are generally system identification procedures for linear models, based on minimizing some measure of the model prediction errors; minimizing the sum of squared prediction errors leads to a *least squares* procedure.

State-space models are systems of coupled differential equations (in the continuous-time case) or systems of coupled difference equations (in the discrete-time case) that completely describe the evolution of a system, whether linear or nonlinear.

Subspace methods of system identification construct state-space models directly from observed input/output data.

White noise refers to an error sequence $\{e_k\}$ in which e_k and e_j are statistically independent and therefore uncorrelated if $k \neq j$; contrast with *colored noise*.

I. BUILDING DYNAMIC MODELS

The term *system identification* refers to the development of *dynamic models* relating one or more *input variables* to one or more *output variables*. In this context, a dynamic model is a mathematical model that relates the time evolution of the input variables to that of the output variables, typically consisting of differential equations or difference equations. Usually, system identification refers to an *empirical procedure* in which observed input/output datasets are fit to a particular model structure, but variations on this theme are possible and some of them are discussed briefly in this article. Most commonly, system identification is concerned with *linear dynamic models*, which satisfy the *principle of superposition*: The response $\mathcal{L}[au + bv]$ to a linear combination $au + bv$ of inputs is simply the corresponding linear combination $a\mathcal{L}[u] + b\mathcal{L}[v]$ of the individual responses to these inputs. Conversely, the problem of *nonlinear model identification* is one of growing practical importance, so it is also discussed here.

Ogunnaike and Ray (1994, p. 645) distinguish the following three important problems involving a set $\mathcal{U}$ of inputs, a set $\mathcal{Y}$ of outputs, and a dynamic model $\mathcal{M}$:

1. *Simulation:* Given $\mathcal{U}$ and $\mathcal{M}$, determine $\mathcal{Y}$.
2. *Control:* Given $\mathcal{Y}$ and $\mathcal{M}$, determine $\mathcal{U}$.
3. *Identification:* given $\mathcal{U}$ and $\mathcal{Y}$, determine $\mathcal{M}$.

In *simulation problems*, we are interested in the responses of a known model $\mathcal{M}$ to a specified set of inputs or stimuli $\mathcal{U}$, for a variety of reasons including improved understanding of an existing physical system (e.g., improved understanding of the operation of an existing manufacturing process) or the design of a new system (e.g., performance prediction of a proposed automotive engine design). In *control problems*, we are interested in determining one or more inputs $\mathcal{U}$ that cause a known system model $\mathcal{M}$ to achieve a desired response, specified by the output set $\mathcal{Y}$. Many different solutions to the control problem have been proposed, and a discussion of these solutions is well beyond the scope of this article, but an essential point is that many control strategies require a process model $\mathcal{M}$ that is of a particular structure. As a specific example, *linear*

model predictive control is an approach that has enjoyed fairly great success in the process industries (particularly the petrochemical industry) (Allgöwer and Zheng, 1998; Ogunnaike and Ray, 1994), and it is typically based on linear FIR models, a class defined and discussed in Section I.B.2 of this article. The *system identification problem* is the task of constructing a model $\mathcal{M}$ that is compatible with a given set $\mathcal{U}$ of inputs and a given set $\mathcal{Y}$ of outputs. Often, the underlying motivation for system identification is the solution of some control problem, and this point is important since it strongly influences some of the solutions that have been proposed for the system identification problem and it can introduce some important complications (e.g., the problem of closed-loop identification discussed briefly in Section IV.D).

A. Modeling Approaches

The task of building dynamic models arises both in simulation and control applications, and it may be approached from many different perspectives. It is useful to divide these approaches into the following four basic categories, discussed by Sjöberg and Astrom (1995, p. 1691):

1. White-box, glass-box or fundamental models
2. Gray-box models obtained via physical modeling
3. Gray-box models obtained via semiphysical modeling
4. Black-box or purely empirical models

The terms *white-box model* and *glass-box model* refer to mathematical models whose structure has a direct physical interpretation. These models are typically developed by writing balance equations for mass, momentum, energy, or any other species (e.g., electrical charge) the conservation of which represents a fundamental physical requirement. In addition, these models typically include various constitutive relations relating some of the physical variables that enter the model, and these equations are often algebraic, so the resulting model often takes the form of a set of coupled differential algebraic equations (DAEs). Further, these models are frequently quite complex, involving dozens, hundreds, or even thousands of individual component equations. As a specific example, Chucholowoski *et al.* (1999) describe the development of simulation models for automotive vehicle dynamics, noting that typical models for this purpose are highly structured index 3 DAEs, which are difficult to handle numerically; as a simpler alternative, they describe a model involving only ordinary differential equations (ODEs), but the scale of this model is representative: 24 ODEs are required for the vehicle body and axle dynamics, 8 ODEs for the dynamics of the tires, 19 ODEs for the power train dynamics, and 5 ODEs for the dynamics of the steering mechanism, for a total of 56 coupled ODEs. A number of chemical process examples are discussed by Pearson (1999b, Sec. 1.1), each involving DAE systems with several hundred equations. Similarly, in his opening plenary talk at the 1997 IFAC Symposium on System Identification, Akaike notes that 276 differential equations are required to simulate the planar motion of an athlete's long-jump takeoff (Sawaragi and Sagara, 1997, p. 1).

At the opposite extreme, *black-box models* are those for which no direct physical interpretation of the variables is possible. Although this lack is widely recognized as a disadvantage, the principal advantage of black-box models is that their structure can be chosen to facilitate either subsequent controller design or parameter estimation, or both. An extremely important example, both in theory and in practice, is the class of *finite-dimensional, linear, time-invariant* models, which are frequently used to describe the dynamic behavior of physical systems exhibiting small amplitude variations around some specified steady-state operating point. A detailed treatment of this problem is given in Section II, together with a number of references to much more comprehensive treatments. Unfortunately, it has been observed by various authors that pure black-box modeling is an ill-posed problem, in the sense that the solution is not generally unique and often depends discontinuously on the input–output data from which it is computed. This observation is particularly pertinent for nonlinear dynamic models, but it also holds in the case of linear models, an important point discussed further in Section IV.A.

Because both white-box and black-box models exhibit important practical limitations, there is increasing interest in *gray-box models*, which attempt to combine fundamental knowledge about the physical system to be modeled with empirical input–output data. This idea may be implemented in various ways, and Sjöberg and Astrom (1995, p. 1691) draw a useful distinction between *physical gray-box models* in which the *model structure* is chosen on physical grounds and *model parameters* are estimated from observed data, and *semiphysical gray-box models* in which physical insights are used to suggest certain nonlinear combinations of variables to be incorporated in an empirical model. As a simple example of physical gray-box modeling, Nowak and Deuflhard (see Deuflhard and Hairer, 1983) describe the problem of estimating kinetic parameters in large systems of chemical reactions, outlining a procedure that exploits the considerable chemical structure of this problem (in particular, both a positivity constraint on the parameters and a modification of the usual least squares fitting criterion). Other examples, many similar in spirit, are discussed in other papers presented at the same conference; also, Chucholowoski *et al.* (1999) note that automobile dynamics are a strong function of

tire forces, and their fundamental model incorporates as a component a tire model involving about 80 parameters per tire, all of which must be measured or estimated experimentally. Empirical submodel components also enter the fundamental models discussed by Pearson (1999b, Sec. 1.1).

A variation of the semiphysical gray-box modeling strategy that illustrates the basic notion is described by Pearson and Pottmann (2000): Motivated by the observation that steady-state behavior of complex systems is often better understood than dynamic behavior (e.g., from steady-state design models), model structures are chosen that can exactly match the known steady-state behavior of the physical system (these structures are discussed in Sections I.B.3 and III.A), reducing the nonlinear model identification problem to a constrained parameter estimation problem for a *linear* dynamic model that is incorporated into the overall nonlinear model as a subsystem. Finally, note that both physical and semiphysical gray-box modeling approaches are necessarily somewhat application-specific because it is difficult to incorporate fundamental knowledge into a general problem framework. Conversely, a few semiphysical gray-box problems may be formulated in fairly general terms, such as matching steady-state behavior, imposing stability constraints, or requiring that monotone inputs yield monotone responses.

The rest of this article presents a collection of somewhat more detailed discussions of a variety of specific approaches to black-box and semiphysical gray-box modeling problems where the techniques of system identification are most often employed. Because these techniques depend strongly on the model structure considered—specifically, continuous time vs. discrete time, state-space vs. input/output vs. other structures (e.g., block-oriented structures), and linear vs. nonlinear—the rest of this section is devoted to a brief introduction to these different model structures. Before proceeding to these discussions, however, one final point is worth noting: Many of the most popular approaches to system identification are those that ultimately reduce the initial problem to one of regression analysis, to which standard techniques may be applied (Draper and Smith, 1998). In its simplest formulation, this problem is of the form:

$$\mathbf{y} = \mathbf{X}\theta + \mathbf{e} \qquad (1)$$

where $\mathbf{y}$ is a vector of N observations of some *dependent variable*, $\mathbf{X}$ is an $N \times p$ matrix of observations of one or more *independent variables*, θ is a p-vector of unknown model parameters (or related quantities) to be determined, and $\mathbf{e}$ is a vector of N model prediction errors. Standard approaches seek estimates $\hat{\theta}$ of the unknown parameters that minimize some measure of $\mathbf{e}$. A particularly popular choice is the *method of least squares*, in which $\hat{\theta}$ is chosen to minimize:

$$J = \sum_{k=1}^{N} e_k^2 \tag{2}$$

This method dates back to Gauss and Legendre around the end of the eighteenth century and part of its current popularity lies in the fact that a simple explicit solution exists, together with many useful associated results (Draper and Smith, 1998):

$$\hat{\theta} = (\mathbf{X}^T\mathbf{X})^{-1}\mathbf{X}^T\mathbf{y}. \tag{3}$$

Conversely, despite its continued popularity, this approach is not without its limitations, a topic discussed further in Section IV.C.

B. Model Structures

1. Continuous vs. Discrete Time

Physical systems typically evolve more or less smoothly in continuous time, and fundamental models that attempt mechanistic descriptions of these systems generally consist of collections of ordinary differential equations, partial differential equations, integral equations, and so forth. Conversely, computer control is generally implemented at discrete time instants, based on measurements made at discrete time instants. Consequently, computer-based control is commonly based on discrete-time dynamic models, possibly obtained as approximate discretizations of continuous-time models. Also, a point that is particularly relevent to this discussion is that discrete-time models are generally much easier to identify from input–output data than continuous-time models. In the case of simple linear models, an exact correspondence between continuous-time and discrete-time models is possible, but this correspondence does not extend to the nonlinear case.

As an important example, consider the first-order linear dynamic model in continuous-time, described by the following ordinary differential equation:

$$\frac{dy(t)}{dt} = -\frac{1}{\tau}y(t) + \frac{G}{\tau}u(t) \tag{4}$$

Here, $u(t)$ represents an input variable, $y(t)$ is the corresponding output variable, τ is the *time constant* for the system, and G is the *steady-state gain* of this system. If the variables $u(t)$ and $y(t)$ are sampled uniformly in time and we write $u_k = u(t_k)$ and $y_k = y(t_k)$ where $t_k = t_0 + kT$, it is possible to obtain the following *difference equation* relating u_k and y_k, also known as a *discrete-time dynamic model*:

$$y_k = \alpha y_{k-1} + \beta u_{k-1} \tag{5}$$

The basis on which Eq. (5) is derived from Eq. (4) is the assumption that $u(t)$ is piecewise constant and does not change between sampling instants, a reasonable assumption in cases where the sequence $\{u_k\}$ is computer generated and a reasonable approximation in many other cases. Under this assumption, the coefficients of the continuous-time and discrete-time models are related by:

$$\alpha = e^{-T/\tau}, \quad \beta = G(1 - e^{-T/\tau}) \tag{6}$$

Similar equivalences exist between higher order linear ordinary differential equations and linear difference equations (Pearson, 1999b, p. 12), but it is also important to note that the transformation just described from the continuous-time model, Eq. (4), to the discrete-time model, Eq. (5), imposes significant restrictions on the parameter α. In particular, for a *stable* first-order linear model, the response time τ is necessarily positive; one consequence is that responses to step changes in $u(t)$ from u_0 for $t \leq 0$ to u_+ for $t > 0$ result in responses that settle out essentially to their asymptotic value Gu_+ in ∼3 to ∼5 time constants τ. In contrast, if $\tau < 0$ in Eq. (4), the step response is unstable, growing exponentially without bound as $t \to \infty$. This stability range $(0 < \tau < \infty)$ corresponds to the range $0 < \alpha < 1$ for the parameter α in the discrete-time model, whereas unstable models with $\tau < 0$ correspond to $\alpha > 1$ in the discrete-time model. It is important to note that the discrete-time model also exhibits a stable response for $-1 < \alpha < 0$, but this response is oscillatory, corresponding to behavior that is only possible for linear continuous-time models of order 2 or higher. This result illustrates that continuous-time and discrete-time models can behave somewhat differently even for simple linear models where these differences are, relatively speaking, quite small.

For nonlinear models, the differences between continuous-time and discrete-time behavior are much more pronounced. This point may be seen by considering the following simple nonlinear reactor model, discussed in more detail by Pearson (1999b, Sec. 8.5):

$$\frac{dy(t)}{dt} = -hy^2(t) - \frac{y(t)u(t)}{V} + \frac{\delta u(t)}{V} \tag{7}$$

where h, V, and δ are constants. Assuming as in the previous example that $u(t)$ is piecewise constant and defining $\mu_k = u(t_k)/2hV$ ultimately lead to the following exact discretization:

$$y_k = \frac{[1 - \tau_{k-1}\mu_{k-1}]y_{k-1} + 2\delta\tau_{k-1}\mu_{k-1}}{1 + \tau_{k-1}[y_{k-1} + \mu_{k-1}]}$$

$$\tau_{k-1} = \frac{\tanh\left[hT\sqrt{\mu_{k-1}^2 + 2\delta\mu_{k-1}}\right]}{\sqrt{\mu_{k-1}^2 + 2\delta\mu_{k-1}}}$$

The key point here is that, although the order remains the same on discretization (i.e., a first-order nonlinear differential equation results in a first-order nonlinear difference equation), the *forms* of these two equations are nothing alike; the simple polynomial appearing on the right-hand side of the differential equation is mapped into a messy combination of hyperbolic functions, square roots, and fractions.

It is, of course, possible to *approximate* a set of nonlinear differential equations by a structurally similar set of nonlinear difference equations. As a specific example, the *Euler approximation* simply replaces derivatives with finite differences; $dy(t)/dt \simeq [y_k - y_{k-1}]/T$, but this approximation is not without its difficulties. As a specific example, consider the Euler discretization of the second-order nonlinear differential equation:

$$\frac{dx(t)}{dt} = -\alpha x^2(t) \rightarrow x_k = x_{k-1} - \alpha T x_{k-1}^2. \quad (8)$$

Comparing the responses of these equations to the initial condition $x(t) = x_0$ at $t = 0$ illustrates an important difference: the exact solution of the continuous-time equation is $x(t) = x_0/(1 + \alpha x_0 t)$, which is stable for all $x_0 > 0$ and all $\alpha > 0$, decaying asymptotically to zero like $1/\alpha t$ as $t \rightarrow \infty$. In contrst, the discretized model can be shown to be unstable if $\alpha x_0 T > 1$. A comparison of the continuous-time solution and the approximate discrete-time solution is shown in Fig. 1 for $\alpha x_0 T = 1.026$.

Because they are necessary for computer-based control applications and because the range of techniques available for discrete-time model identification appears to be wider than that for continuous-time model identification, discrete-time system identification methods appear to be more popular in practice and are discussed more extensively in the literature. Consequently, the remainder of this article restricts consideration to discrete-time model identification. For a general overview and some useful references on the continuous-time case, consult the survey paper by Unbehauen and Rao in Sawaragi and Sagara (1997, pp. 973–999); this survey deals with both linear and nonlinear continuous-time dynamic models, and other papers in the same conference proceedings provide useful illustrations of these and related ideas, especially those presented in the session on continuous-time identification (Sawaragi and Sagara, 1997, pp. 1281–1310). In addition, the discussion given here is also restricted to *time-invariant* or *shift-invariant* linear models, again because these models are most widely used in practice. The topic of *time-varying* linear model identification is also treated in a dedicated session at the SYSID'97 meeting, beginning with a useful overview by Forssell and Ljung (Sawaragi and Sagara, 1997, pp. 1655–1678).

2. State-Space vs. Input–Output Models

The underlying notion of a state-space model is the following: If $\mathbf{x}(t_0)$ represents the *state* of the system at time

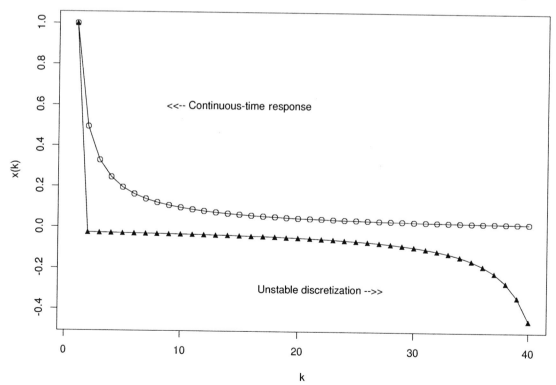

FIGURE 1 Continuous- vs. discrete-time model responses.

t_0, this vector contains all information necessary to completely predict the future evolution of the system for all $t > t_0$, and the basis for these predictions is the *state-space model*. In continuous-time, this model can often be expressed in the general form:

$$\dot{\mathbf{x}}(t) = \mathbf{f}(\mathbf{x}(t), \mathbf{u}(t)), \quad \mathbf{y}(t) = \mathbf{g}(\mathbf{x}(t), \mathbf{u}(t)) \quad (9)$$

Here, $\mathbf{u}(t)$ represents a vector of m real-valued input signals $\{u_i(t)\}$ and $\mathbf{y}(t)$ is a vector of r real-valued output signals $\{y_i(t)\}$. The n-dimensional vector $\mathbf{x}(t)$ is the *state vector* for the system, and the functions $\mathbf{f}: R^n \times R^m \rightarrow R^n$ and $\mathbf{g}: R^n \times R^m \rightarrow R^r$ define how the state vector and the output vector, respectively, evolve in time. If these functions are *analytic* (implying the existence of derivatives of all orders with respect to all arguments), then it is possible to expand the right-hand side of Eq. (9) in a Taylor series about some steady-state solution. Sufficiently small deviations from this steady-state solution then evolve approximately according to the *linearized model*, which may be written in matrix notation as:

$$\dot{\mathbf{x}}(t) = \mathbf{F}\mathbf{x}(t) + \mathbf{G}\mathbf{u}(t), \quad \mathbf{y}(t) = \mathbf{H}\mathbf{x}(t) + \mathbf{M}\mathbf{u}(t) \quad (10)$$

As in the simple first-order linear example discussed in Sec. I.B.1, this continuous-time model may be discretized under the assumption that the inputs are piecewise constant to obtain a discrete-time state-space model of the same dimension n:

$$\mathbf{x}_{k+1} = \mathbf{A}\mathbf{x}_k + \mathbf{B}\mathbf{u}_k, \quad \mathbf{y}_k = \mathbf{C}\mathbf{x}_k + \mathbf{D}\mathbf{u}_k \quad (11)$$

The subspace-based methods for linear system identification discussed in Sec. II.C apply to linear discrete-time state-space models of this form, estimating the unknown matrices $\mathbf{A}$, $\mathbf{B}$, $\mathbf{C}$, and $\mathbf{D}$ from observed input–output data.

Alternatively, linear dynamic models may also be represented in other ways, of which the following three are particularly important. First, for a linear state-space model defined by Eq. (11) with $m = 1$ and $r = 1$ (i.e., a single-input, single-output, or SISO, model), one equivalent representation is the *autoregressive moving average* model, often designated *ARMA(p, q)*. This model structure is an *input–output representation* defined by the following equation:

$$y(k) = \sum_{i=1}^{p} a_i y(k-i) + \sum_{i=0}^{q} b_i u(k-i) \quad (12)$$

and the linear state-space model, Eq. (11), may be represented as an *ARMA(p, q)* model with $0 \leq q \leq p = n$. One advantage of this representation is that it lends itself nicely to regression formulations, a point discussed further in Section II.A. Another equivalent representation for this model is obtained by applying the z-transform to Eq. (12) (Ljung, 1999; Pearson, 1999b) to obtain the *transfer function*:

$$H(z) = \frac{\sum_{i=0}^{q} b_i z^{-i}}{1 - \sum_{i=1}^{p} a_i z^{-i}} \quad (13)$$

Evaluating this function at $z = \exp j\omega T$ leads to an expression for the complex angular *frequency response* $H(\omega)$, where T is the uniform sampling rate for the sequences $\{u_k\}$ and $\{y_k\}$, and $\omega = 2\pi f$ varies between 0 and the Nyquist limit $\omega = \pi$; this limit follows from the fact that $f = 1/2T$ represents the highest frequency component that can be present in a continuous-time signal $u(t)$ if it is to be sampled uniformly every T time units without aliasing (Ljung, 1999, p. 445). This characterization also lends itself to the development of linear system identification algorithms, and the ideas behind these methods are discussed further in Section II.B. Finally, a third general representation for linear discrete-time dynamic models is the *impulse response* or *convolution* model:

$$y_k = \sum_{i=0}^{\infty} h_i u_{k-i} \quad (14)$$

Direct estimation of the impulse response coefficients $\{h_k\}$ is possible, but this approach is less popular than others, partly because an infinite number of impulse response coefficients is required for a complete representation and partly because the estimation of these parameters is more difficult than it sounds (Soderstrom and Astrom, 1995, p. 1836). Alternatively, for any *stable* linear dynamic model, the impulse response decays asymptotically to zero, providing one motivation for considering *finite impulse response* (FIR) models, obtained by truncating the sum in Eq. (14) to a finite number of terms. In fact, FIR models are extremely popular in process control applications, providing an important practical basis for linear model predictive control implementations (Ogunnaike and Ray, 1994).

These representations are almost equivalent in the sense that interconversion between them is generally possible, but the system identification algorithms required to obtain model parameters in the different representations are *not* generally equivalent, as these algorithms normally make considerable use of the representation for which they were developed. More specifically, any time-invariant, linear discrete-time dynamic model may be represented as a convolution model and this representation is unique. Similarly, *most* linear models encountered in practice may be represented by the ARMA structure, often with conveniently small values for the *order parameters* p and q, although exceptions do arise, as in the case of distributed parameter models that exhibit extremely slow (nonexponential) decays; for a discussion of this topic, refer to Pearson (1999b, ch. 2). The ARMA model structure and the transfer function are completely equivalent, being related via the z-transform, and any strictly proper rational transfer function $H(z)$ (i.e., any transfer function of the form of Eq. (13) with finite $q < p$) may be represented as a state-space model of dimension p.

3. Linear vs. Nonlinear Models

Strictly speaking, the term "nonlinear" is usually interpreted to mean "not necessarily linear," implying only that the principle of superposition noted at the beginning of this article cannot be assumed to hold. As a consequence, the term "nonlinear" really tells us nothing about how the system *does* behave, in marked contrast to the term "linear," which tells us enough to construct explicit representations (specifically, every linear, time-invariant, discrete-time dynamic model may be completely characterized in terms of its impulse response sequence $\{h_i\}$ (Pearson, 1999b, p. 64). Consequently, most useful discussions of nonlinear dynamic models begin with some *structural* description of a particular nonlinear model class. One important example is the very broad class of NARMAX models, introduced by Billings and Voon (1986) and defined by the following nonlinear generalization of the linear *ARMA(p, q)* model discussed in Section I.B.2:

$$y_k = \Phi(y_{k-1}, \ldots, y_{k-p}, u_k, \ldots, u_{k-q},$$
$$e_{k-1}, \ldots, e_{k-r}) + e_k \qquad (15)$$

where $\{e_k\}$ represents a sequence of model prediction errors. Although parameter estimation algorithms exist for various classes of functions $\Phi(\cdot)$ in this general model structure, many applications benefit from the use of restricted special cases of this structure, for a number of reasons discussed further in Section III of this article. One practically important special case of this general model structure is the *Hammerstein model*, obtained by taking $\Phi(\cdot)$ to be of the following form:

$$\Phi(y_{k-1}, \ldots, y_{k-p}, u_k, \ldots, u_{k-q})$$
$$= \sum_{i=1}^{p} a_i y_{k-i} + \sum_{i=0}^{q} b_i g(u_{k-i}), \quad r = 0 \quad (16)$$

where $g(\cdot)$ is any real-valued function. This model also exhibits the block diagram representation shown in Fig. 2, and this representation is one of the contributing factors to the popularity of the Hammerstein model since it combines the popular linear dynamic model with a simple memoryless nonlinearity.

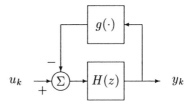

FIGURE 3 The Lur'e model structure.

It is important to emphasize that the way in which these components are interconnected has a profound influence on the qualitative behavior of the resulting nonlinear model, on its possible representations, and on the algorithms used in estimating model parameters from input–output data. For example, the *Lur'e model* combines the same two elements as the Hammerstein model, but in the feedback structure shown in Fig. 3. This nonlinear model structure has the following input–output representation:

$$y_k = \sum_{i=1}^{p} a_i y_{k-i} + \sum_{i=0}^{q} b_i [u_{k-i} - g(y_{k-i})] \qquad (17)$$

which may be derived directly from the block diagram. As illustrated in Section III.B, the qualitative behavior of the Hammerstein and Lur'e models is radically different, illustrating the general difference in behavior between models with nonlinear moving average terms such as $g(u_{k-i})$ appearing in the Hammerstein model and models with nonlinear autoregressive terms such as $g(y_{k-i})$ appearing in the Lur'e model.

Still a third way of combining these two components is to form the *Wiener model* shown in Fig. 4, consisiting of a similar cascade connection to that defining the Hammerstein model, but connected in the opposite order. In the case of linear dynamic models, the order of interconnection makes no difference, but with nonlinear models, this order does make a difference, as may be seen by considering input–output representations for the Wiener model. If the function $g(\cdot)$ is invertible, it is possible to derive the following input–output representation:

$$y_k = g\left(\sum_{i=1}^{p} a_i g^{-1}(y_{k-i}) + \sum_{i=1}^{p} b_i u_{k-i}\right) \qquad (18)$$

but for the case where $g(\cdot)$ is not invertible, no input–output representation exists, in general (Pearson, 1999b, p. 173). In addition, the invertibility of the function $g(\cdot)$ plays an important role in the parameter estimation

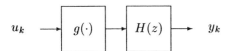

FIGURE 2 The Hammerstein model structure.

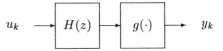

FIGURE 4 The Wiener model structure.

problem for Wiener models, in marked contrast to the case of Hammerstein models; this point is discussed further in Section III.B. Conversely, one other case where an input–output representation is always possible is the case of a Wiener model based on a FIR linear subsystem. In particular, the input–output representation is then simply:

$$y_k = g\left(\sum_{i=0}^{q} h_i u_{k-i}\right) \qquad (19)$$

Finally, it is instructive to conclude this discussion of nonlinear model structures with one more example, that of *bilinear models* (Pearson, 1999b, ch. 3). In the single-input, single-output case, these models may be defined in two different, nonequivalent ways. The first is the state-space representation,

$$\mathbf{x}_k = \mathbf{A}\mathbf{x}_k + \mathbf{b}u_k + u_k\mathbf{N}\mathbf{x}_k, \quad y_k = \mathbf{c}^T\mathbf{x}_k + du_k \quad (20)$$

where **b** and **c** are n-vectors, d is a constant, and **N** is an $n \times n$ matrix. The second way of defining bilinear models is via the input–output representation:

$$y_k = \sum_{i=1}^{p} a_i y_{k-i} + \sum_{i=1}^{q} b_i u_{k-i} + \sum_{i=1}^{P}\sum_{j=1}^{Q} c_{ij} y_{k-i} u_{k-j} \qquad (21)$$

In some cases, it is possible to convert bilinear models between these representations, but not in the general case (Pearson, 1999b, p. 94). This observation is important because, as noted earlier, algorithms for estimating the model parameters in either of these representations necessarily exploit this representation.

II. LINEAR MODEL IDENTIFICATION

Without question, the system identification methods that are most widely used in practice are those for linear models. Further, because this field has been an active area of research for some time, many different methods exist, and these methods may be classified in many different ways. The following sections present a brief overview of some of these approaches; because of both space limitations and the breadth of the field, this overview cannot be complete, but the following discussions attempt to give a representative flavor of some of the most important ideas in linear system identification. Three broad classes of methods are described in Sections II.A, B, and C below, organized by representations introduced in Section I.B.2: time-domain methods based on the ARMA model representation (Section II.A), frequency-domain methods based on the transfer function representation (Section II.B), and subspace methods based on the state-space representation (Section II.C). For more complete discussions of these topics,

refer to the book of Ljung (1999), the SYSID'97 conference proceedings (Sawaragi and Sagara, 1997) the *Automatica* special issue on system identification (Soderstrorn and Astrom, 1995), or the related references cited by any of these authors.

A. Time-Domain Methods

Time-domain methods typically employ the ARMA model representation discussed in Section I.B.2, constructing a *prediction model* of the general form:

$$\hat{y}_k = \sum_{i=1}^{p} \alpha_i y_{k-i} + \sum_{i=0}^{q} \beta_i u_{k-i} \qquad (22)$$

Prediction error methods of linear system identification then determine the parameter vector,

$$\theta = [\alpha_1, \ldots, \alpha_p, \beta_0, \ldots, \beta_q]^T \qquad (23)$$

to minimize some measure of the prediction errors:

$$e_k = y_k - \hat{y}_k = y_k - \sum_{i=1}^{p} \alpha_i y_{k-i}$$

$$- \sum_{i=0}^{q} \beta_i u_{k-i} \equiv y_k - \phi_k^T \theta \qquad (24)$$

where $\phi_k^T = [y_{k-1}, \ldots, y_{k-p}, u_k, \ldots, u_{k-q}]$. Generally, prediction error methods choose θ to minimize a criterion of the form:

$$J(\theta) = \frac{1}{N} \sum_{k=1}^{N} \rho[e_k] \qquad (25)$$

where $\rho[x]$ is some nonnegative function that measures the "size" of the errors e_k. A particularly common choice is $\rho[x] = x^2$, leading to a *least squares* identification procedure. Specifically, we then have

$$J(\theta) = \frac{1}{N} \sum_{k=1}^{N} \left[y_k - \phi_k^T \theta\right]^2 \qquad (26)$$

Because of its simple form, this optimization problem may be solved explicitly for the least squares solution:

$$\hat{\theta} = \left[\frac{1}{N} \sum_{k=1}^{N} \phi_k \phi_k^T\right]^{-1} \left[\frac{1}{N} \sum_{k=1}^{N} y_k \phi_k\right] \qquad (27)$$

Note that this equation involves a matrix (the first term in brackets) of dimension $(p+q+1) \times (p+q+1)$ and a vector of dimension $p+q+1$. For the solution $\hat{\theta}$ to exist, it is necessary that this matrix inverse exist, and a sufficient condition for the invertability of this matrix is that the input sequence be *persistently exciting* (Ljung, 1999, Sec. 13.2), a notion discussed further in Section IV.B. If, in addition, the components of the vector ϕ_k are uncorrelated with the errors e_k, it follows that the parameter estimate $\hat{\theta}$ will be *unbiased*, converging to the unknown true parameter vector θ as $N \to \infty$. Conversely, if the components of ϕ_k are correlated with these errors, the estimated parameter vector $\hat{\theta}$ will be *biased*. As Ljung (1999, p. 205) notes, ϕ_k and e_k will be uncorrelated, yielding unbiased estimates, under either of two conditions:

1. The sequence $\{e_k\}$ is white (more properly, *independent, identically distributed*, or i.i.d., meaning that e_k and e_j are statistically independent if $k \neq j$), or
2. The vector ϕ_k involves only inputs u_{k-j}, meaning that $p = 0$ in Eq. (22).

It is interesting to note that this latter case corresponds to the use of FIR prediction models, and it was noted previously that these models are quite popular in industrial practice as a basis for model predictive control applications.

Often, the error sequence $\{e_k\}$ is *colored* (i.e., e_k and e_j are correlated for $k \neq j$), causing the least squares procedure just described to yield biased parameter estimates. In such cases, the following observation is sometimes useful. Suppose the input sequence $\{u_k\}$ and the output sequence $\{y_k\}$ are both passed through a linear *prefilter* $\mathcal{L}$ with transfer function $L(z)$. To see the effects of this prefilter, consider the frequency-domain representation of the prediction model defined in Eq. (22), obtained by taking the z-transform of this equation:

$$Y(z) = G(z, \theta)U(z) + E(z) \qquad (28)$$

where $Y(z)$, $U(z)$, and $E(z)$ represent the z-transforms of the data sequences $\{y_k\}$, $\{u_k\}$, and $\{e_k\}$, respectively. It follows that the effect of the prefiltering operation is to modify Eq. (28) to

$$L(z)Y(z) = G(z, \theta)L(z)U(z) + E(z)$$

$$\Rightarrow Y(z) = G(z, \theta)U(z) + L^{-1}(z)E(z) \qquad (29)$$

where $L^{-1}(z)$ denotes the inverse of the filter $\mathcal{L}$. Hence, the effect of the prefiltering is only to modify the character of the noise model. In particular, if $L(z)$ can be chosen to match the power spectrum of the sequence $\{e_k\}$, the effect will be to replace this original error sequence with an *uncorrelated* sequence $\{w_k\}$. If the errors are Gaussian (a common assumption in practice, although refer to Section IV.C for further discussion of this point), it follows that the resulting sequence $\{w_k\}$ is white. As a practical matter, the use of prefilters can introduce added complications (this point is discussed by Ljung, 1999, p. 207), but if the error

sequence can be modeled adequately as an *autoregressive process*:

$$e_k = \sum_{i=1}^{r} \gamma_i e_{k-i} + w_k \Rightarrow \mathcal{L}x_k = x_k - \sum_{i=1}^{r} \gamma_i x_{k-i} \qquad (30)$$

prefiltering may be built into the least squares prediction error method by simply augmenting the order parameters p and q appearing in the prediction model to $p + r$ and $q + r$.

Another alternative in cases where the simple least squares procedures yields biased estimates is the *instrumental variables method*, based on the following idea. Suppose e_k is correlated with the components of ϕ_k; the instrumental variables method attempts to find another vector sequence ξ_k that is of the same dimension as ϕ_k and correlated with this vector, but is uncorrelated with the prediction errors e_k. This objective is accomplished by requiring that

$$\frac{1}{N} \sum_{k=1}^{N} \xi_k [y_k - \phi_k^T \theta] = 0 \qquad (31)$$

If we can find such a sequence, this equation may be solved for the *instrumental variables estimate* $\tilde{\theta}$:

$$\tilde{\theta} = \left[\frac{1}{N} \sum_{k=1}^{N} \xi_k \phi_k^T \right]^{-1} \left[\frac{1}{N} \sum_{k=1}^{N} \xi_k y_k \right] \qquad (32)$$

provided that ξ_k is also chosen such that the required matrix inverse exists. For a further discussion of this idea, including choices of the *instruments* ξ_k and some extensions of the basic approach, refer to Ljung (1999, Sec. 7.6).

B. Frequency-Domain Methods

The linear model identification approaches just described exhibit two fundamental characteristics. First, they are *time-domain* approaches, taking uniformly sampled input and output sequences $\{u_k\}$ and $\{y_k\}$ and using them to construct a difference equation that describes the evolution of the system in time. Second, these approaches are *parametric*, meaning that the computational objective is to estimate an unknown parameter vector θ that has a finite number of components. *Frequency-domain methods* generally also take the time-sampled sequences $\{u_k\}$ and $\{y_k\}$ as a starting point, but the objective is to obtain an estimate of a frequency-domain characterization, either the frequency response $H(\omega)$ as a function of the continuous frequency variable $\omega = 2\pi f$ or the transfer function $H(z)$ as a ratio of two polynomials, as defined in Eq. (13). Because it depends on the continuous variable ω and is not described by a finite number of parameters, the frequency response represents a *nonparametric* characterization of the linear system; conversely, since it is described by a finite set of poles (the roots of the denominator polynomial) and zeros (the roots of the numerator polynomial), the transfer function defined in Eq. (13) is a parametric description, exactly equivalent to the time-domain descriptions considered in the previous section.

The *discrete Fourier transform* (DFT) of a sequence $\{x_k\}$ is defined as:

$$X(\omega) = \sum_{k=-\infty}^{\infty} x_k e^{-i2\pi f kT} \qquad (33)$$

and this transform provides the basis for the following frequency-domain characterization of a linear system. Define $U(\omega)$ as the DFT of the input sequence $\{u_k\}$, $Y(\omega)$ as the DFT of the output sequence $\{y_k\}$, and $H(\omega)$ as the DFT of the impulse response sequence $\{h_k\}$ defining the linear system. In general, note that all of these quantities are complex valued and may therefore be characterized by a magnitude and a phase. It is a standard result that the following equation holds for linear models:

$$Y(\omega) = H(\omega)U(\omega) \Rightarrow H(\omega) = \frac{Y(\omega)}{U(\omega)} \qquad (34)$$

If the input sequence consists of a single sinusoid at frequency f_0, it follows that the response of the linear system will also be a single sinusoid of frequency f_0 with amplitude $|H(\omega_0)|$ and phase shift $\angle H(\omega_0)$, where $|H|$ denotes the magnitude of the complex quantity H, $\angle H$ represents its phase, and $\omega_0 = 2\pi f_0$. Consequently, one possible approach to linear system identification would be to determine the magnitude and phase of the system's response to a sequence of sinusoids of known amplitude and varying frequency. In fact, this is roughly how spectrum analyzers used in electronic system characterization work. Conversely, if $\{u_k\}$ is a Gaussian white-noise sequence, the response of the linear system will be a Gaussian colored-noise sequence and the difference between the character of the input and the output may again be used to estimate the frequency response $H(\omega)$. More generally, if $\{u_k\}$ is a sequence of Gaussian random variables, it is completely characterized by its autospectrum $S_{uu}(\omega)$, and the response $\{y_k\}$ of any linear system is another Gaussian sequence that is completely characterized by its autospectrum $S_{yy}(f)$. The relationship between these signals is characterized by the cross-spectrum $S_{uy}(\omega)$, which is related to the frequency response of the system by:

$$S_{uy}(\omega) = H(\omega)S_{uu}(\omega) \Rightarrow H(\omega) = \frac{S_{uy}(\omega)}{S_{uu}(\omega)} \qquad (35)$$

This observation provides the basis for nonparametric frequency response estimation procedures: The autospectrum $S_{uu}(\omega)$ and the cross-spectrum $S_{uy}(\omega)$ are each estimated from the input and output data sequences, and these estimates are substituted into Eq. (35) to obtain an estimate

of $H(\omega)$. It is important to emphasize that there is considerable practical art associated with obtaining reliable auto- and cross-spectral estimates, as the next example illustrates; for this reason, those interested in pursuing this approach should consult references such as Chave *et al.* (1987) for a detailed discussion of the practical considerations involved.

Another useful observation is that the autospectrum $S_{yy}(\omega)$ of the output sequence is related to that of the input sequence by:

$$S_{yy}(\omega) = |H(\omega)|^2 S_{uu}(\omega) \qquad (36)$$

Combining this observation with Eq. (35) yields the result that the *magnitude-squared coherence*

$$\gamma_{uy}^2(\omega) = \frac{|S_{uy}(\omega)|^2}{S_{uu}(\omega)S_{yy}(\omega)} \qquad (37)$$

satisfies $\gamma_{uy}^2(\omega) = 1$ for all ω if $\{u_k\}$ and $\{y_k\}$ are the input and output, respectively, of *any* linear system. Carefully computed, this quantity can provide a useful indication of the presence of noise or nonlinearity perturbing this linear relationship, resulting in *coherence suppression*: $\gamma_{uy}^2(\omega) < 1$ over the frequency range where this perturbation is influential. Conversely, if sufficient care is *not* exercised in these computations, the results can be completely meaningless. As a specific example, the simplest possible spectral estimator is the periodogram (Ljung, 1999, p. 30):

$$P_{xx}(\omega) = \frac{1}{N} \left| \sum_{k=1}^{N} x_k e^{-ik\omega T} \right|^2$$

$$= \frac{1}{N} \sum_{k=1}^{N} \sum_{\ell=1}^{N} x_k x_\ell e^{-i(k-\ell)\omega T} \qquad (38)$$

which may be extended to a *cross-periodogram* between two sequences $\{x_k\}$ and $\{y_k\}$:

$$P_{xy}(\omega) = \frac{1}{N} \sum_{k=1}^{N} \sum_{\ell=1}^{N} x_k y_\ell e^{-i(k-\ell)\omega T} \qquad (39)$$

An apparently reasonable estimator of $\gamma_{xy}^2(\omega)$ is that obtained by replacing the auto- and cross-spectra in Eq. (37) with their periodogram estimates P_{uu}, P_{yy}, and P_{uy}. This procedure, however, leads to the estimate $\hat{\gamma}_{xy}^2(\omega) = 1$ for all ω, for any two sequences $\{x_k\}$ and $\{y_k\}$, regardless of how or whether they are related. It is possible to overcome this difficulty by replacing the raw periodogram $P_{xy}(\omega)$ with a better spectral estimator, but the point of this example is to illustrate that care is required in forming nonparametric estimates of quantities such as $H(\omega)$ or $\gamma_{uy}^2(\omega)$.

Given an estimate of the frequency response $\hat{H}(\omega)$, it is possible to obtain parametric transfer function estimates by choosing the parameters in the transfer function model (i.e., the parameter vector θ defined in Eq. (23), defining the poles and zeros of the transfer function) so that the frequency response $\tilde{H}(\omega)$ computed from this transfer function represents a reasonable approximation of $\hat{H}(\omega)$. This approximation usually involves a frequency weighting so that the resulting transfer function captures the most important features of the frequency response (e.g., behavior near the crossover frequency for control system design applications). In addition, this approximation approach usually results in a nonlinear least squares problem, although variants that lead to linear regression problems do exist. For further discussion of these ideas, refer to Ljung (1999, Sec. 7.7).

C. Subspace Methods

Subspace methods for linear system identification are based on the state-space representation defined by Eq. (11), and the objective of these methods is to estimate the system matrices $\mathbf{A}$, $\mathbf{B}$, $\mathbf{C}$, and $\mathbf{D}$ from observed input–output data. A particular advantage of these methods is that, because they directly employ the state-space representation for the system, they extend easily to multiple-input, multiple-output (MIMO) systems. Conversely, a significant issue that must be addressed in all such methods is that state-space realizations are not unique. That is, if $(\mathbf{A}, \mathbf{B}, \mathbf{C}, \mathbf{D})$ defines an n-dimensional state-space model and $\mathbf{T}$ is any nonsingular $n \times n$ matrix, the original state-space model is equivalent to the transformed model $(\mathbf{TAT}^{-1}, \mathbf{TB}, \mathbf{CT}^{-1}, \mathbf{D})$ with respect to input–output behavior; in particular, these models are related by a change of basis for the state-space defined by $\mathbf{z}_k = \mathbf{Tx}_k$. Further, if a particular linear system can be realized by a state-space model of dimension n, it can also be realized by state-space models of all higher dimensions $r > n$. These higher dimensional models necessarily exhibit either uncontrollable or unobservable modes, however, meaning that either the controllability Grammian Ω_n or the observability Grammian Γ_n is singular, where:

$$\Omega_n = [\mathbf{B}, \mathbf{AB}, \ldots, \mathbf{A}^{n-1}\mathbf{B}], \quad \text{and} \quad \Gamma_n = \begin{bmatrix} \mathbf{C} \\ \mathbf{CA} \\ \vdots \\ \mathbf{CA}^{n-1} \end{bmatrix} \quad (40)$$

This singularity has both theoretical and practical consequences that are undesirable, but it may be avoided by identifying a *minimal realization*, defined as a state-space model of the minimum dimension n that is capable of matching the observed input–output behavior.

Viberg (Soderstrom and Astrom, 1995, pp. 1835–1851) presents a good overview of the important and currently very popular class of *subspace-based state-space system identification* algorithms, commonly known by the acronym 4SID (pronounced "forsid"). Most of these algorithms are closely related to the following basic idea. First, define $\mathbf{h}_i$ as the $r \times m$ matrix of impulse response coefficients at time step i and suppose that we are able to observe or estimate a finite collection of these coefficients, $\{\mathbf{h}_i\}$ for $i = 0, 1, 2, \ldots, 2n + 1$. It follows directly from the state-space representation that these coefficients are related to the matrices $\mathbf{A}$, $\mathbf{B}$, $\mathbf{C}$, and $\mathbf{D}$ via

$$\mathbf{h}_i = \begin{cases} 0 & i < 0 \\ \mathbf{D} & i = 0 \\ \mathbf{CA}^{i-1}\mathbf{B} & i > 0 \end{cases} \quad (41)$$

Hence, the matrix $\mathbf{D}$ may be obtained directly from $\mathbf{h}_0$ and the later impulse response coefficients contain information about the other three matrices defining the state-space representation of the system. To obtain these matrices, first form the $(n + 1)rm \times (n + 1)rm$ *Hankel matrix*:

$$\mathbf{H} = \begin{bmatrix} \mathbf{h}_1 & \mathbf{h}_2 & \mathbf{h}_3 & \cdots & \mathbf{h}_{n+1} \\ \mathbf{h}_2 & \mathbf{h}_3 & \mathbf{h}_4 & \cdots & \mathbf{h}_{n+2} \\ \mathbf{h}_3 & \mathbf{h}_4 & \mathbf{h}_5 & \cdots & \mathbf{h}_{n+3} \\ \vdots & \vdots & \vdots & \vdots & \vdots \\ \mathbf{h}_{n+1} & \mathbf{h}_{n+2} & \mathbf{h}_{n+3} & \cdots & \mathbf{h}_{2n+1} \end{bmatrix} \quad (42)$$

This matrix can be factored into the product of the extended controllability and observability Grammians, each of order $n + 1$:

$$\mathbf{H} = \Gamma_{n+1}\Omega_{n+1} \quad (43)$$

Various full-rank factorization methods exist (e.g., the singular value decomposition, or SVD), which may be used to determine Γ_{n+1} and Ω_{n+1}, given $\mathbf{H}$. For a system with m inputs and r outputs, the first m rows of Γ_{n+1} and the first r columns of Ω_{n+1} are simply equal to the unknown matrices $\mathbf{C}$ and $\mathbf{B}$, respectively, as may be seen in Eq. (40). Further, it is possible to determine the $n \times n$ system matrix $\mathbf{A}$, as follows. Define $\Gamma_{1:n}$ as the matrix obtained from Γ_{n+1} by deleting the bottom r rows (corresponding to the matrix $\mathbf{CA}^n$) and $\Gamma_{2:(n+1)}$ as the matrix obtained by deleting the top r rows of Γ_{n+1} (corresponding to the matrix $\mathbf{C}$). It then follows that $\Gamma_{1:n}$ is the nonsingular controllability Grammian (that is, the Grammian Γ_n defined in Eq. (40)), and these two new matrices are related via:

$$\Gamma_{2:(n+1)} = \Gamma_{1:n}\mathbf{A} \Rightarrow \mathbf{A} = \Gamma_{1:n}^{-1}\Gamma_{2:(n+1)} \quad (44)$$

As noted previously, the state-space representation is not unique, a consequence of the freedom we have in choosing a basis for the state-space; in procedures of the type just outlined, this basis is determined by the choice of the full-rank factorization procedure used to compute Γ_{n+1} and Ω_{n+1} from $\mathbf{H}$. This observation illustrates one of the inherent disadvantages of the 4SID methods: Many characteristics of the methods and the models identified by these methods are not easily determined, arising as in this case through specific algorithmic choices made when the methods are applied (see, for example, the discussion of 4SID methods by Ljung and McKelvey in the SYSID'97 conference proceedings (Sawaragi and Sagara, 1997, pp. 1075–1079).

Viberg notes that an important practical difficulty with the approach just outlined is the difficulty of obtaining accurate estimates for the impulse response coefficients $\mathbf{h}_i$, and he divides 4SID methods into two types: *realization-based methods* that address this issue and *direct methods* that avoid forming the Hankel matrix $\mathbf{H}$. These latter methods appear to be more popular and Ljung (1999, pp. 208–211, 340–351) describes several of them in some detail. Essentially, the basic idea is to first observe that *if the state vector* $\mathbf{x}_k$ *were known*, we could apply standard linear regression procedures to the following equation:

$$\mathbf{Y}_k = \Theta \Phi_k + \mathbf{E}_k \qquad (45)$$

where the matrices and vectors appearing in this equation are defined as:

$$\mathbf{Y}_k = \begin{bmatrix} \mathbf{x}_{k+1} \\ \mathbf{y}_k \end{bmatrix} \qquad \Theta = \begin{bmatrix} \mathbf{A} & \mathbf{B} \\ \mathbf{C} & \mathbf{D} \end{bmatrix}$$

$$\Phi_k = \begin{bmatrix} \mathbf{x}_k \\ \mathbf{u}_k \end{bmatrix} \qquad \mathbf{E}_k = \begin{bmatrix} \mathbf{w}_k \\ \mathbf{v}_k \end{bmatrix}$$

The vectors $\mathbf{v}_k$ and $\mathbf{w}_k$ appearing in the matrix $\mathbf{E}_k$ correspond to noise terms that enter the state-space model; specifically, this model is assumed to be of the form:

$$\mathbf{x}_{k+1} = \mathbf{A}\mathbf{x}_k + \mathbf{B}\mathbf{u}_k + \mathbf{w}_k, \qquad \mathbf{y}_k = \mathbf{C}\mathbf{x}_k + \mathbf{D}\mathbf{u}_k + \mathbf{v}_k$$

In practice, we do not know the state vector $\mathbf{x}_k$, but it may be estimated from prediction models of the general form:

$$\hat{y}(k + j - 1|k - 1) = \sum_{i=1}^{s_1} \alpha_i y_{k-i} + \sum_{i=1}^{s_2} \beta_i u_{k-i} \quad (46)$$

for $j = 1, 2, \ldots, \ell$ for some ℓ that is at least as large as the system order n. Specifically, for $p \geq n$, there exists a matrix $\mathbf{L}$ of dimension $n \times rp$ such that $\mathbf{x}_k = \mathbf{L}\hat{\mathbf{Y}}_k$ where $\hat{\mathbf{Y}}_k$ is the matrix of predictions:

$$\hat{\mathbf{Y}}_k = \begin{bmatrix} \hat{y}(1|0) & \cdots & \hat{y}(N|N-1) \\ \vdots & \vdots & \vdots \\ \hat{y}(p|0) & \cdots & \hat{y}(N+p-1|N) \end{bmatrix} \quad (47)$$

Note that the coefficients α_i and β_i appearing in Eq. (46) may be obtained using standard regression procedures,

choosing them to minimize some measure of the prediction errors $\hat{y}(k + j - 1|k - 1) - y_{k+j-1}$. Also, note that the general 4SID approach outlined here involves a number of parameters (e.g., j, s_1, and s_2 appearing in Eq. (46) and p appearing in Eq. (47)), with different choices essentially corresponding to different subspace-based algorithms. For a much more detailed treatment of these algorithms, refer to the discussion by Ljung (1999, pp. 340–351) and the original references cited either there or in the review of Viberg (Soderstrom and Astrom, 1995, pp. 1835–1851).

III. NONLINEAR MODEL IDENTIFICATION

As noted previously, the term "system identification" most often refers to the identification of linear dynamic models, but fundamental (i.e., white-box) models are almost always nonlinear, and there is a growing need for approximate models that capture more of the dynamic behavior of these models than their linearizations can. For example, at the 1998 International Symposium on Nonlinear Model Predictive Control (NMPC) in Ascona, Switzerland, two authors emphasized that the lack of suitable nonlinear dynamic models was one of the key obstacles to more extensive practical adoption of the NMPC methodology. Specifically, Qin and Badgewell (Allgöwer and Zheng, 1998, pp. 128–145) present a survey of NMPC applications in industry, concluding that:

There is no systematic approach for building nonlinear dynamic models for NMPC.

Similarly, Lee (Allgöwer and Zheng, pp. 91–107) begins his survey of modeling and identification requirements for NMPC by noting the marked contrast between the widespread industrial application of *linear* MPC and the extremely limited industrial application of *nonlinear* MPC, and cites the following reason for this disparity:

Asked to choose the most important, however, most of us will, without hesitation, point the arrow to one common factor: *the inability to construct a nonlinear model on a reliable and consistent basis.*

One of the basic reasons for this state of affairs is that the class of nonlinear dynamic models is extremely large and heterogeneous compared with its linear subset, containing many distinct families that are as different from each other in behavior as they are from linear models. The following discussion briefly examines this important point; for a more detailed discussion, refer to the book by Pearson (1999b).

TABLE I Nine Forms of Nonlinear Qualitative Behavior

i	Symbol	Qualitative behavior class $\mathcal{B}_i$
1	HARM	Harmonic generation from sinusoidal inputs
2	SUB	Subharmonic generation from sinusoidal inputs
3	ASYM	Asymmetric responses to symmetric inputs
4	IDS	Input-dependent stability
5	SSM	Steady-state multiplicity (input or output)
6	CHAOS	Chaotic responses to simple inputs
7	HOM	Nonlinear homogeneous behavior
8	PHOM	Nonlinear positive-homogeneous behavior
9	SL	Nonlinear static-linear behavior

A. Which Kind of Nonlinear Model Do We Need?

Despite their many advantages, linear models are incapable of exhibiting certain types of qualitative behavior. Hence, if we wish to model systems in which this behavior is important, a nonlinear dynamic model is required. Table I lists nine forms of qualitative behavior, the first six of which are inherently nonlinear, and the last three of which represent relaxations of linearity (Pearson, 1999b). The key point here is that, if we want our dynamic model to describe any of these observable forms of input–output behavior, a nonlinear model is necessary. In addition, observation of specific forms of nonlinear behavior can also provide useful insight into which type of nonlinear model structure is necessary to describe this behavior (e.g., Hammerstein, Wiener, Lur'e, or something else?).

This point is illustrated in Table II, which summarizes the ability of three model classes—Hammerstein mod-

TABLE II Qualitative Behavior for Three Model Classes

Class Index i	Behavior Class $\mathcal{B}_i$	Hammerstein Models	Bilinear Models	Lur'e Models	Comments
1	HARM	Yes	Yes	Yes	Typical nonlinear behavior
2	SUB	No	No[a]	Yes	
3	ASYM	Yes	Yes	Yes	Requires asymmetric nonlinearity
4	IDS	No	Yes[b]	Yes	
5	SSM	Input	No	Output	
6	CHAOS	No	No	Yes	
7	HOM	No	No	No	
8	PHOM	YES[c]	No	Yes[c]	
9	SL	No	Yes	No	

[a] Possible for transient responses only.
[b] Characteristic behavior.
[c] Piecewise-linear models.

els, bilinear models (specifically, the input–output bilinear models defined by Eq. (21)), and Lur'e models—to exhibit each of these forms of behavior. In some cases (e.g., harmonic generation and asymmetric responses to symmetric inputs), all three of these model classes are capable of exhibiting the indicated behavior, whereas in other cases (e.g., nonlinear homogeneous responses), none of these model classes is. Most cases are intermediate, with some models capable of the specified behavior and other model classes incapable of it. Such observations are useful in model structure selection since if a particular form of qualitative behavior is important in the application under consideration, it is desirable to at least select a model class that is capable of this behavior and possibly to constrain the model parameters in a way that guarantees any identified model exhibits this behavior. The following three examples illustrate this idea.

1. Hammerstein vs. Wiener Models

Some of the subtle differences between the structurally similar Hammerstein and Wiener models were noted in Section I.B.3. The behavioral consequences of these differences are illustrated in Fig. 5, which compares the responses of these two models, built from the same two subsystems. Specifically, the curve marked with the open circles in this figure shows the response of the Hammerstein model, whereas the curve marked with the solid triangles shows the response of the Wiener model. In both cases, the model components are

$$y_k = 0.8y_{k-1} + 0.2u_{k-1}, \quad g(x) = x^{1/3}$$

The input generating both of these responses consists of two steps, from $u_k = 0$ to $u_k = 1$ at $k = 50$ and from $u_k = 1$ to $u_k = 0$ at $k = 130$. For the upward step at $k = 50$, the Wiener model response is slightly faster than that of the Hammerstein model, whereas for the downward step at $k = 130$, the Wiener model response is *much slower*. This directional dependence of the dominant response time is a clear indication of the nonlinearity of the Wiener model, but the key point here is that this behavior is not possible for the Hammerstein model; there, the dominant time constant is determined completely by the linear model, independent of the static nonlinearity $g(\cdot)$. In fact, in this particular case, the cube-root nonlinearity has no effect at all on the pair of unit input steps since $g(0) = 0$ and $g(1) = 1$; hence, the Hammerstein model responses are precisely the same as those of the linear model on which it is based. For a different choice of input amplitudes, the nonlinearity of the Hammerstein model would be apparent from the amplitude dependence of the steady-state responses, but the primary point of this example is to illustrate that the general type of qualitative behavior exhibited by these

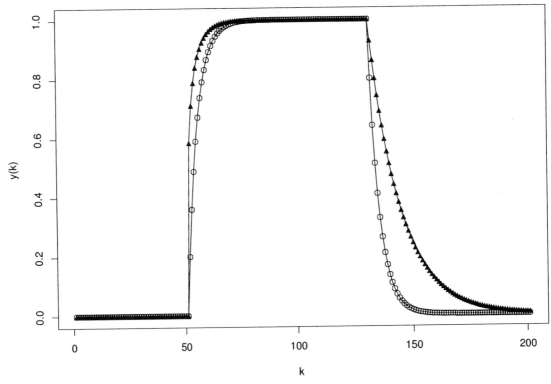

FIGURE 5 Step responses, Hammerstein vs. Weiner models.

two nonlinear models can be quite different, despite their strong structural relationship.

2. Bilinear Models

Figure 6 shows the response of the following bilinear model to a sequence of nine successive step inputs, each increasing in value by 0.6:

$$y_k = 0.8y_{k-1} + 0.2u_{k-1} - 0.4y_{k-1}u_{k-1} \qquad (48)$$

(Because this model is a first-order difference equation, it belongs to both the bilinear state-space model class and the bilinear input–output model class.) Note the pronounced amplitude dependence of the dynamic character of these step responses: At low amplitudes, these responses are relatively slow and monotonic but, as the input amplitude increases, the response becomes faster, then exhibits an overshoot, then sustained oscillations, and ultimately oscillatory instability. In fact, this behavior is characteristic of bilinear models. To avoid these instabilities, it is necessary to either limit the range of the inputs considered (a practical option in some circumstances) or avoid the bilinear model class altogether if this behavior is unacceptable.

3. Lur'e Models and Recurrent Artificial Neural Networks

It was noted in Section I.B.3 that the inclusion of nonlinear autoregressive terms in a model often leads to rather exotic behavior. This point is illustrated in Fig. 7, which shows four different model responses; the top two plots illustrate the responses of the following Lur'e model to a sinusoidal input:

$$y_k = 0.8y_{k-1} + 0.2[u_{k-1} - f(y_{k-1})], \qquad f(x) = -g\sin x$$

$$(49)$$

and the bottom two plots illustrate the responses of an unrealistically simple three-layer artificial neural network (ANN) structure (Ljung, 1999; Pearson, 1999b). Specifically, this network consists of a two-node hidden layer, having the general structure $u_k \rightarrow h_1(k), h_2(k) \rightarrow y_k$ and defined by the equations:

$$h_1(k) = \tanh[-0.5y_{k-1} + u_{k-2}]$$

$$h_2(k) = \tanh[1.3y_{k-2} + u_{k-2}]$$

$$y_k = \tanh[3h_1(k) - h_2(k)]$$

The key point is that the inclusion of nonlinear autoregressive terms (also called *recurrent* or *recursive* terms) often leads to exotic behavior like that shown here.

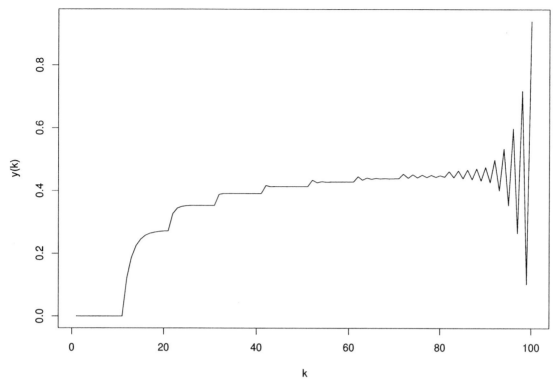

FIGURE 6 Input-dependent dynamics of the bilinear model.

B. Parameter Estimation: Hammerstein vs. Wiener Models

The discussion in Section III.A.1 illustrated some of the behavioral differences between Hammerstein and Wiener models; the following discussion illustrates some of the important differences that arise in attempting to identify these two closely related model structures from input–output data. In some cases, the steady-state behavior of the system of interest is known, either from the steady-state solution of a detailed fundamental model or on the basis of detailed empirical observations. In such cases, the dynamic behavior is often less well known and it may then be reasonable to identify an empirical model subject to the constraint that this model exactly match the known steady-state behavior (Pearson and Pottmann, 2000). In the case of both Hammerstein and Wiener models, if the steady-state gain of the linear subsystem is constrained to be 1, the steady-state response map for the overall nonlinear model is simply given by the nonlinear function: $y_s = g(u_s)$. Hence, this nonlinear function may be chosen to match the steady-state behavior, reducing the nonlinear model identification problem to that of empirically determining the constants a_i and b_i in the $ARMA(p, q)$ representation of the linear subsystem. An important consequence of this approach, however, is that these linear model parameters must be estimated subject to the constraint that the steady-state gain of the linear subsystem is 1, implying,

$$\sum_{i=1}^{p} a_i + \sum_{i=0}^{q} b_i = 1$$

This constraint leads to a *restricted least squares problem*, for which the solution is known (Draper and Smith, 1998, p. 122).

For the Hammerstein case, the resulting identification problem is relatively straightforward: Given $g(\cdot)$, we form the sequence of transformed inputs $v_k = g(u_k)$ and apply the restricted least squares approach to the data pairs (v_k, y_k) to obtain the model parameters a_i and b_i. Conversely, the Wiener solution is somewhat more involved because of the way in which the nonlinear function $g(\cdot)$ enters the problem. In particular, even when the function $g(\cdot)$ is known, the estimation problem becomes much more complicated when this function is not invertible, in contrast to the Hammerstein problem just described, where invertibility plays no role. In the case where $g^{-1}(\cdot)$ exists, we may proceed analogously to the Hammerstein case, first forming the transformed *output* sequence $z_k = g^{-1}(y_k)$ and then estimating the model parameters a_i and b_i from the data pairs (u_k, z_k) via the restricted least squares procedure. Unfortunately, even here there is a complication, as it is possible to obtain poor fits between the

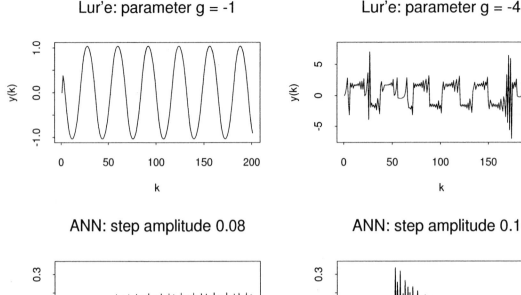

Lur'e: parameter g = -1   Lur'e: parameter g = -4

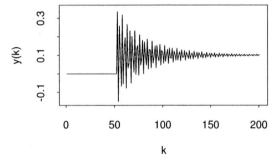

ANN: step amplitude 0.08   ANN: step amplitude 0.16

FIGURE 7 Nonlinear autoregressive model responses.

observed outputs y_k and their Wiener model predictions. In particular, the least squares fit criterion penalizes prediction errors for the transformed variable z_k and not the system output. These differences can be quite significant, motivating the use of *weighted least squares*, where the weights are based on the known function $g(\cdot)$ (Pearson and Pottmann, 2000).

In cases where the nonlinear function $g(\cdot)$ is unknown, a popular algorithm for Hammerstein model identification is that due to Narendra and Gallman (Pearson, 1999b). This algorithm starts with an expansion of the unknown nonlinearity in terms of *known* basis functions $\psi_j(x)$, most commonly polynomials:

$$g(x) = \sum_{j=0}^{m} c_j \psi_j(x) \qquad (50)$$

reducing the determination of $g(\cdot)$ to a problem of estimating the constants $\{c_j\}$. Combining this expansion with the input–output representation for the Hammerstein model then leads to an expression of the form:

$$y_k = \sum_{i=1}^{p} a_i y_{k-i} + \sum_{i=0}^{q} \sum_{j=0}^{m} b_i c_j \psi_j(u_{k-i}) \qquad (51)$$

The Narendra–Gallman algorithm proceeds by first fixing the coefficients $c_j = c_j^0$, thus fixing the static nonlinearity and reducing the problem to an unconstrained linear model identification problem. This problem may be solved by standard methods (in particular, least squares) for estimates a_i^1 and b_i^1 of the linear model parameters. Next, these parameters are fixed at the values a_i^1 and b_i^1, again reducing the model (Eq. (51)) to one that is linear in the unknown parameters c_j, which are then estimated by standard methods. Iterating this procedure normally yields reasonable estimates of the model parameters a_i, b_i, and c_j, although it is known that the algorithm can fail to converge if the initial parameter estimates c_j^0 are too far from the true values. Finally, note that the nice structure of the Narendra–Gallman algorithm for Hammerstein models (i.e., it may be treated as iterated pairs of linear least squares problems) does not extend to the Wiener case. For example, consider the case of a Wiener model based on a linear FIR subsystem; combining the input–output representation, Eq. (19), with the basis expansion, Eq. (50), for the nonlinear function then yields the model:

$$y_k = \sum_{j=0}^{m} c_j \psi_j \left(\sum_{i=0}^{q} h_i u_{k-i} \right)$$

Although this model is linear in the parameters c_j, the parameters h_i now enter nonlinearly, forcing us to alternate between linear and nonlinear least squares problems at each stage of the iteration procedure. Consequently, the advantages of this approach over simply solving the nonlinear least squares problem for the parameters c_j and h_i simultaneously are not clear, in marked contrast to the Hammerstein case where the Narendra–Gallman algorithm represents a way of avoiding the complexities of explicit nonlinear least squares problems.

Finally, it is worth noting that various other algorithms for estimating both Hammerstein and Wiener model parameters have been proposed. Space limitations do not permit detailed discussions of these algorithms, but two fairly recent results illustrate how developments in linear system identification are being extended to various nonlinear model classes. Specifically, the subspace methods discussed in Section II.C for linear system identification have been extended to both Hammerstein and Wiener model identification (Verhaegen and Westwick, 1996; Westwick and Verhaegen, 1996). As in the case of linear model identifications, one of the advantages of these algorithms is that they permit the identification of multiple-input, multiple-output models.

C. Some General Observations

It is useful to conclude this discussion of nonlinear dynamic model identification with a few general observations. First, it is important to emphasize that nonlinear model identification is generally more difficult than linear model identification, and this factor provides a partial explanation for the difficulties encountered in developing models for NMPC noted at the beginning of this section. As a simple but specific example, binary input sequences can be extremely effective in linear system identification, but they are generally ineffective in the identification of Hammerstein models (Pearson, 1999b, p. 408). Another extremely important practical point is that the term "nonlinear" can mean an uncountably infinite number of different things. For example, the differences in qualitative behavior between Hammerstein and Lur'e models are much greater than the differences between linear models and Hammerstein models, even if the Hammerstein and Lur'e models are constrained to exhibit the same steady-state behavior and are constructed from the same linear subsystems. As a consequence, nonlinear model identification algorithms tend to be highly structure-specific, as the Hammerstein/Wiener comparison presented in Section III.B illustrates. Finally, note that the area of nonlinear model identification is a rapidly evolving one, so it is worth checking the recent literature for useful results that may be relevent to any particular model identification effort; see Section V for a further discussion of the identification literature.

IV. PRACTICAL ISSUES

Much of the system identification literature focuses on the problem of *parameter estimation*, based on the assumptions that a reasonable model structure has been chosen and that representative input–output datasets are available. In practice, empirical model identification generally involves several iterations through the following sequence of steps:

1. Specify a model structure $\mathcal{S}$.
2. Based on this choice, design a suitable input sequence $\{u_k\}$.
3. Apply this sequence to the physical system and observe the response sequence $\{y_k\}$.
4. Determine the model or models from structure class $\mathcal{S}$ that best fit the input–output data.
5. Assess the reasonableness of these results, both with respect to goodness-of-fit and other criteria.

The following sections briefly discuss some of the important practical aspects of each of these steps.

A. Structure Specification and Qualitative Behavior

The first and last steps in the iterative procedure just described are more closely related than might be apparent. In particular, model structure choices typically dictate much about the possible range of model behavior. The traditional measure of model adequacy is goodness-of-fit, but in practice reasonable goodness-of-fit is only a *necessary condition* for a good model, *not* a sufficient condition. That is, different models exhibiting comparable goodness-of-fit may differ enormously in terms of important qualitative behavior such as stability, monotonicity of step responses, or the tendency to exhibit chaotic responses to simple inputs. A specific and illuminating example is that of Tulleken (1993), who developed a large set of linear models for an industrial distillation process. Paying careful attention to issues such as input sequence specification and using standard model identification procedures, Tulleken obtained a large collection of models which he subsequently examined in terms of important qualitative behavior; specific criteria included stability, minimum- or nonminimum-phase behavior, monotonicity or nonmonotonicity of step responses, signs of the steady-state gains, and general agreement with dominant settling times. Ultimately, he rejected almost all of the identified models as unsuitable because they failed to match the behavior of the physical system with respect to one or more of these criteria. To overcome these problems, Tulleken developed model parameter constraints to guarantee the stability of the identified model; models identified subject to these constraints were found to exhibit much more reasonable behavior with respect to the other criteria considered. The key point of this example is that standard model identification procedures maximize goodness-of-fit but they *cannot* guarantee "reasonable qualitative behavior" without explicit constraints to force such agreement. The following two examples illustrate ways in which such constraints may be developed and imposed.

One of Tulleken's criteria was monotonicity of step responses, and it was noted in Section I.B.1 that this behavior is characteristic of first-order linear discrete-time models, *provided the coefficient α is positive*. In fact, a similar result holds for the entire class of linear systems. It can be shown (Pearson, 1999, p. 341) that a linear model exhibits monotonically increasing responses to all monotonically increasing inputs if and only if the impulse response coefficients $\{h_i\}$ are nonnegative. Hence, if we are identifying linear FIR models and want them to exhibit such preservation of monotonicity, it is best to explicitly constrain the model identification procedure to guarantee $h_i \geq 0$ for all i. One of Tulleken's other qualitative behavior criteria was rough agreement with known steady-state behavior (i.e., he rejected models for which the signs of certain steady-state gains were incorrect). For both linear models and some classes of nonlinear models, it is possible to impose constraints that guarantee an exact match to known steady-state behavior. This point was discussed in Section III.B, and a much more detailed discussion appears in the paper by Pearson and Pottmann (2000). More generally, Tulleken's experience and these two examples emphasize the point made by Lindskog and Ljung (Sawaragi and Sagara, 1997, p. 693):

Don't estimate what you already know!

B. Input Sequence Considerations

An extremely important issue in practical system identification is the choice of input sequence $\{u_k\}$ to use in generating input–output data. One way of seeing this importance is to note that constant input sequences yield constant responses that contain no information about the system dynamics to be identified. In the case of linear system identification, it was noted earlier that the inputs should be *persistently exciting* (Ljung, 1999, p. 412), a condition that may be satisfied using many different classes of input sequences, including random or pseudorandom binary sequences, white-noise sequences of arbitrary distribution, multisines, chirps, random step sequences, and many others. More generally, we have the following four "design

variables" at our disposal in developing effective identification inputs (Pearson, 1999b, Sec. 8.4.2):

1. Sequence length N
2. Range of variation $[a, b]$
3. Distribution of $\{u_k\}$ over the range $[a, b]$
4. *Shape* or *frequency content* of the sequence

All of these design variables are important, but practical considerations generally impose different constraints on them. For example, at fixed sampling time T, the length N of a sequence used for model identification determines the time required to obtain it. Generally, the best results are obtained with input sequences designed to elicit information from the process, rather than from historical operating data, although preliminary examination of historical operating data can provide useful insights into the general qualitative behavior of the system (e.g., it may be possible to at least roughly infer many of the qualitative criteria considered by Tulleken from such data). Because identification sequences often do perturb the physical system significantly, identification experiments tend to be somewhat invasive and their total duration is often constrained by the patience of those operating the system, limiting the length of the available sequences; this point is emphasized by Tulleken (1993). Similarly, the range of input variation is also typically limited, as excessive excursions may result in significant degradation in process performance (e.g., the manufacture of unsalable product) or result in unsafe operation. A general rule of thumb is to obtain as much data as possible, maximizing sequence length N and covering as wide an operating range as is reasonable.

The other two design variables—the sequence distribution and frequency content—are generally more at the discretion of the model developer. The frequency content is particularly important, both for linear and nonlinear model identification, as it essentially determines how well the input sequence satisfies persistence of excitation conditions. More detailed discussions of the options available for input sequence design and how to select among these options are available in the Bibliography. For linear model identification, Ljung (1999, Sec. 13.3) gives a particularly good treatment, covering the popular binary sequences, multisines, and swept sinusoids and discussing the advantages and disadvantages of each. For a discussion of some of the options available for nonlinear model identification and a comparison of the results obtained with different sequences for a simple example, refer to Pearson (1999b, ch. 8).

C. Noise vs. Outliers

Different possible noise assumptions were discussed briefly in Section II in connection with the various linear model identification approaches. Usually, the term "noise" refers to highly irregular, short-term fluctuations that are sufficiently *well-behaved* to be reasonably approximated by a standard random variable model, most often assumed to be Gaussian. Because least squares procedures correspond to maximum likelihood estimators for the case of additive, independent Gaussian random errors (Ljung, 1999, p. 217), the Gaussian assumption has some important theoretical advantages, but this assumption is not always reasonable in practice. In particular, *outliers* may be defined informally as "data points that are inconsistent with our expectations, based on the behavior of the bulk of the data." This notion is illustrated in Fig. 8, which shows four real-world datasets, each containing visually obvious outliers.

The consequences of outliers in both linear and nonlinear system identification can be severe. As a specific example, the influence of five outliers in a data sequence of length 100 for first-order linear model identification can cause errors of ~40% to ~100% in the estimated model parameters (Pearson, 1999a). Further, outliers also tend to seriously degrade the results of model order selection or structure determination procedures such as cross-validation (Pearson, 1999b, Sec. 8.2). Space limitations do not permit a detailed treatment of this topic here, but more

extensive discussions may be found in Pearson (1999a,b) and the further references cited there. The primary point here is that outliers do occur frequently in practice and they cannot be ignored; at very least, glaring outliers should be replaced temporarily with even crude estimates of more reasonable values and the model identification process should be performed both with and without this replacement. Comparing the results obtained by these two procedures can then provide some idea of the influence of these anomalous points; if large differences are observed, other qualitative criteria such as those discussed in Section IV.A should be invoked to determine which result is more reasonable.

D. Open-Loop vs. Closed-Loop Identification

Because system identification is often used in the development of models for closed-loop control, it is important to note that the presence of existing controllers can complicate the system identification problem significantly. Further, this problem arises frequently in practice, for a variety of reasons, including the necessity of operating open-loop unstable processes with a stabilizing controller, the frequent requirement to perform identification experiments without interfering excessively with

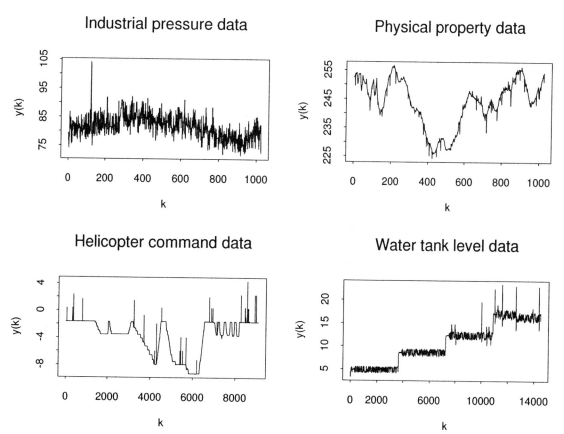

FIGURE 8 Four datasets containing outliers.

normal process operation, or the possibility that unsuspected feedback mechanisms exist in an incompletely understood complex physical, chemical, or biological system.

Ljung (1999, Sec. 13.4) gives a fairly detailed discussion of the additional complications that arise in doing closed-loop model identification, including the following example. Suppose the first-order linear model discussed in Section I.B.1 is to be identified from input–output data, but under the proportional feedback law $u_k = -fy_k$ for some *known* feedback gain f. This assumption leads to the model identification equations:

$$y_k = \alpha y_{k-1} + \beta u_{k-1} + e_k,$$

$$u_k = -fy_k, \Rightarrow y_k = (\alpha + \beta f)y_{k-1} + e_k$$

It follows that the estimates $\hat{\alpha} = \alpha + \gamma f$ and $\hat{\beta} = \beta - \gamma$ of the model parameters α and β fit the data equally well for *any* value of the free parameter γ. As a consequence, it is not possible to identify the model parameters α and β in this situation, even if the feedback law f is known and even if the input $\{u_k\}$ satisfies persistence of excitation conditions that are sufficient to guarantee identifiability in the open-loop case. Conversely, this problem may be overcome by suitably restricting the model structure to reduce the number of parameters that must be estimated; a particularly interesting observation in light of the discussion presented in Section IV.A is that one such suitable restriction would be to fix the steady-state gain $G = \beta/(1 - \alpha)$ of the open-loop system.

Ljung (1999, Sec. 13.5; Sawaragi and Sagara, 1997, pp. 155–160) also describes three general approaches to closed-loop linear model identification:

1. *Direct Methods*, based on input–output data from the system of interest, effectively ignoring the presence of the feedback controller
2. *Indirect methods*, based on input–output data from the *closed-loop system* (i.e., the controller reference input r_k is used instead of the system input u_k)
3. *Joint input–output methods*, which consider both the system output y_k and the system input u_k as responses of a system driven by the reference input r_k

Of these three methods, Ljung argues strongly in favor of the first, noting that it is applicable even in cases where the controller is extremely complicated (as is often the case in industrial settings) or not completely known (e.g., if *assumed* controller tunings are incorrect). Further, Ljung argues that prediction error methods (including the least squares procedure described in Section II.A) appear to give the best results in closed-loop identification, provided an accurate noise model is used. Conversely, other meth-

ods which work well for open-loop identification may perform poorly in closed-loop, including both instrumental variable methods and subspace methods.

E. Software

Few areas evolve more rapidly than software, so it is probably not reasonable to devote too much space to this topic, but because software underlies essentially all practical applications of system identification, it is important to say *something* about it here. The following examples are not meant to be exhaustive, but they are representative of what is currently available in specialized identification packages. Also, these examples are *not* intended as a product recommendation, despite the fact that they are all based on the MATLAB software package (a product of Mathworks, Inc; Natik, MA), which is quite widely used in the system identification community. It is worth reiterating, however, that many problems can be cast as linear regression problems of the form discussed at the end of Section I.A, and many other standard software packages have efficient built-in solution procedures for these problems.

A reasonably detailed discussion of Mathworks' System Identification Toolbox is given by its primary developer in his system identification book (Ljung, 1999, ch. 17). This software package provides reasonably comprehensive support for linear system identification, including procedures to support data handling (e.g., plotting, segmentation, trend removal, etc.), frequency-domain identification, time-domain identification, and a variety of characterizations of the resulting linear model (e.g., simulation of model responses, determination of poles and zeros, generation of frequency responses, etc.). Descriptions of other MATLAB-based system identification software packages are available in the SYSID'97 conference proceedings (Sawaragi and Sagar, 1997) including one for the development of nonlinear global models based on operating regime decompositions (pp. 943–946), one for the identification of nonlinear models based on neural networks (pp. 931–936), one for closed-loop linear model identification (pp. 937–942), and one for frequency-domain identification of linear models (pp. 943–946). Finally, a list of five other system identification software packages is given in Ljung (1999, p. 522), together with further references for more detailed descriptions.

V. KEEPING UP WITH RECENT PROGRESS

In general, it is fair to say that the subject of system identification is vast and growing rapidly. Some very useful, broad surveys may be found in the *Automatica* special issue on system identification (Soderstrom and Astrom, 1995), which contains the following 10 papers on various different aspects of system identification:

1. Sjöberg, J. *et al.*, "Nonlinear black-box modeling in system identification: a unified overview," pp. 1691–1724.

2. Juditsky, A. *et al.*, "Nonlinear black-box models in system identification: mathematical foundations," pp. 1725–1750.

3. van den Hof, P. M. J., and Scharma, R. J. P., "Identification and control—closed-loop issues," pp. 1751–1770.

4. Ninnes, B., and Goodwin, G. C., "Estimation of model quality," pp. 1771–1797.

5. Mäkilä, P. M., Partington, J. R., and Gustafsson, T. K., "Worst-case control-relevant identification," pp. 1799–1819.

6. van den Hof, P. M. J., Heuberger, P. S. C., and Bokor, J., "System Identification with generalized orthonormal basis functions," pp. 1821–1834.

7. Viberg, M., "Subspace-based methods for the identification of linear time-invariant systems," pp. 1835–1851.

8. van Overschee, P., and de Moor, B., "A unifying theorem for three subspace system identification algorithms," pp. 1853–1864.

9. Deistler, M., Peternell, K., and Scherrer, W., "Consistency and relative efficiency of subspace methods," pp. 1865–1875.

10. van Overschee, P., and de Moor, B., "Choice of state-space basis in combined deterministic-stochastic subspace identification," pp. 1877–1883.

It is clear from this list of titles that many of the topics addressed in this article are discussed in more detail in the papers listed here. More generally, journals such as *Automatica*, *IEEE Transactions on Automatic Control*, *IEEE Transactions on Signal Processing*, *IEEE Transactions on Circuits and Systems*, and *International Journal of Control* regularly publish current research results in the area of system identification. Finally, it is also useful to note the existence of the IFAC (International Federation of Automatic Control) symposium series on system identification, which are international conferences held every three years. The IFAC'97 symposium proceedings consists of three volumes, numbering 1724 pages and including 278 papers.

SEE ALSO THE FOLLOWING ARTICLES

CONTROLS, ADAPTIVE SYSTEMS • CONTROLS, BILINEAR SYSTEMS • CONTROLS, LARGE-SCALE SYSTEMS • CYBERNETICS AND SECOND ORDER CYBERNETICS • HYBRID SYSTEMS CONTROL • SELF-ORGANIZING SYSTEMS • SIGNAL PROCESSING, GENERAL • SYSTEM THEORY

BIBLIOGRAPHY

Allgöwer, F., and Zheng, A., eds. (1998). "Preprints, International Symposium on Nonlinear Model Predictive Control: Assessment and Future Directions," Ascona, Switzerland, June 3–5.

Billings, S. A., and Voon, W. S. F. (1986). "A prediction-error and stepwise-regression estimation algorithm for non-linear systems," *Int. J. Control* **44**, 803–822.

Chave, A. D., Thomson, D. J., and Ander, M. E. (1987). "On the robust estimation of power spectra, coherences, and transfer functions," *J. Geophys. Res.* **92**(B1), 633–648.

Chucholowoski, C., Vögel, M., Von stryk, O., and Wolter, T. M. (1999). "Real time simulation and online control for virtual test drives of cars." *In* "High Performance Scientific and Engineering Computing" (H.-J. Bungartz, F. Durst, and C. Zenger, eds.), Springer-Verlag, Berlin.

Deuflhard, P., and Hairer, E., eds. (1983). "Numerical Treatment of Inverse Problems in Differential and Integral Equations," Birkhäuser, Basel.

Draper, N. R., and Smith, H. (1998). "Applied Regression Analysis," 3rd ed., John Wiley & Sons, New York.

Ljung, L. (1999). "System Identification: Theory for the User," 2nd ed., Prentice Hall, New York.

Ogunnaike, B. A., and Ray, W. H. (1994). "Process Modeling, Dynamics, and Control," Oxford University Press, New York.

Pearson, R. K. (1999a). "Data cleaning for dynamic modeling and control," paper BP-3, No. 6, In *Proc. Eur. Control Conf.*, Karlsruhe, Germany, September (CD-ROM).

Pearson, R. K. (1999b). "Discrete-Time Dynamic Models," Oxford University Press, New York.

Pearson, R. K., and Pottmann, M. (2000). "Gray-box identification of block-oriented nonlinear models," *J. Process Control* **10**, 301–315.

Sawaragi, Y., and Sagara, S., eds. (1997). "Proc. IFAC Symposium System Identification (SYSID'97) (3 volumes)," Kitakyusha, Fukuoka, Japan, July 8–11 .

Soderstrom, T., and Astrom, K. J., eds. (1995). "Special issue on trends in system identification," *Automatica* **31**(12).

Tulleken, H. J. A. F. (1993). "Grey-box modelling and identification using physical knowledge and Bayesian techniques," *Automatica* **29**(2), 285–308.

Verhaegen, M., and Westwick, D. (1996). "Identifying MIMO Hammerstein systems in the context of subspace model identification methods," *Int. J. Control* **63**, 331–349.

Westwick, D., and Verhaegen, M. (1996). "Identifying MIMO Wiener systems using subpace model identification methods," *Signal Processing* **52**, 235–258.

Conversion of Step Response to Frequency Response

Kamran Forouhar

Boeing Corporation, Inc.

GLOSSARY

Closed-loop system System in which the output affects the signal implemented on the plant.

Frequency response Steady-state response of a system to a sinusoidal input.

Open-loop system System in which the output does not affect the signal implemented on the plant (dynamic of the system).

Steady state Behavior of the response of a system when time approaches infinity.

Step response Response of a system to a step input when all the initial conditions are zero.

THIS ARTICLE PRESENTS a technique for converting the step response of a linear system to its frequency response, that is, gain and phase versus frequency or gain versus phase, and so on. This technique can be programmed such that the step response sequence of a linear system can be used as the input data. The computer program will then generate the corresponding frequency response plots. It is also possible to obtain the open-loop frequency response from the transient response of a closed-loop feedback control system.

To obtain the closed-loop frequency response with this technique there is no need for any information about the dynamic or structure of the system. Instead, a finite sequence of numbers, which is usually the first few seconds of the system step response, is the only required information about the system. This sequence of numbers should approach the steady state with a constant slope. Open-loop frequency response can also be obtained if we assume unity feedback control systems or canonical feedback control systems with known error functions.

The programs based on this technique have a variety of applications—for example, signal processing, aerospace engineering, control and communication systems, and other fields of engineering. The use of the programs based on this technique will reduce the time and expense of costly experimental testing programs.

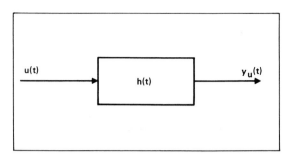

FIGURE 1 Linear system.

I. MATHEMATICAL PRELIMINARIES

Consider a single-input, single-output continuous linear system with the impulse response function $h(t)$ in Fig. 1, where $u(t)$ is the step input and $y_u(t)$ the step response of the system. It is known that the impulse response function is the time derivative of the step response:

$$h(t) = dy_u(t)/dt \qquad (1)$$

Hypothetical diagrams of the step response and its resulting impulse response are given in Figs. 2 and 3, respectively. According to Eq. (1) and the assumption that the step response approaches the steady state with approximately zero slope at some time t_1 shown in Fig. 2, the impulse response will be approximately equal to zero after time t_1, as shown in Fig. 3; that is,

$$h(t) \simeq 0 \quad \text{for} \quad t > t_1 \qquad (2)$$

The transfer function of a linear system is related to its impulse response function by:

$$H(s) = \int_0^\infty h(t)e^{-st}\, dt \qquad (3)$$

Now, let

$$s = J\omega \qquad (4)$$

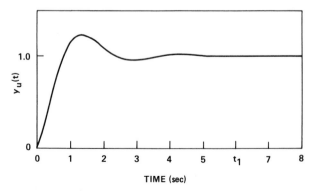

FIGURE 2 Hypothetical step response of a linear system.

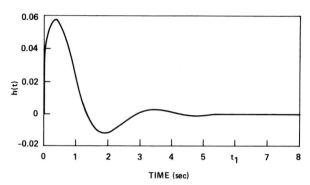

FIGURE 3 Impulse response.

Then,

$$H(J\omega) = \int_0^\infty h(t)e^{-J\omega t}\, dt \qquad (5)$$

Because

$$e^{-J\omega t} = \cos \omega t - J \sin \omega t \qquad (6)$$

we have

$$H(J\omega) = \int_0^\infty h(t)\cos \omega t\, dt - J \int_0^\infty h(t) \sin \omega t\, dt \qquad (7)$$

From Eq. (2), Eq. (7) can be written:

$$H(J\omega) \simeq \int_0^{t_1} h(t)\cos \omega t\, dt - J \int_0^{t_1} h(t) \sin \omega t\, dt \qquad (8)$$

The gain and phase of the frequency transfer function $H(J\omega)$ are obtained from Eq. (8) and are given by the following relationships:

$$\text{Gain} = |H(J\omega)| = \{A^2(\omega) + B^2(\omega)\}^{1/2} \qquad (9)$$

$$\text{Phase} = \angle H(J\omega) = \tan^{-1}[-B(\omega)/A(\omega)] \qquad (10)$$

where:

$$A(\omega) = \int_0^{t_1} h(t)\cos \omega t\, dt \qquad (11)$$

$$B(\omega) = \int_0^{t_1} h(t)\sin \omega t\, dt \qquad (12)$$

A similar approach can be used for discrete linear systems. Let us assume that $y(kT)$, $k = 0, 1, 2, \ldots$ is the step response sequence and $h(kt)$, $k = 0, 1, 2, \ldots$ is the impulse response sequence of a discrete linear system where T is the sampling interval. Then, $h(kT)$ is obtained by the relationship:

$$h(kT) = y_u(kT) - y_u(kT - T)^* \qquad (13)$$

(The proof of this equation is given in Section IV.)

If $H(z)$ is the pulse transfer function of the discrete system, it is related to the impulse response function by the z-transform relationship:

$$H(z) = \sum_{k=0}^{\infty} h(kT)z^{-k} \qquad (14)$$

In the steady state, $h(kT)$ is approximately zero, and this occurs at some finite time nT; that is,

$$h(kT) \simeq 0 \quad \text{for} \quad k > n \qquad (15)$$

Now, let us substitute $z = e^{J\omega T}$ in Eq. (14):

$$H(e^{J\omega T}) = \sum_{k=0}^{\infty} h(kT)e^{-kJ\omega T} \qquad (16)$$

If we substitute the identity (6) into Eq. (16) and take Eq. (15) into consideration, $H(e^{J\omega T})$ becomes:

$$H(e^{J\omega T}) \simeq \sum_{k=0}^{n} h(kT)\cos k\omega T - J \sum_{k=0}^{n} h(kT)\sin k\omega T \qquad (17)$$

The gain and phase of the frequency transfer function are calculated as:

$$\text{Gain} = |H(e^{J\omega T})| = \{C^2(\omega T) + D^2(\omega T)\}^{1/2} \qquad (18)$$

$$\text{Phase} = \angle H(e^{J\omega T}) = \tan^{-1}[-D(\omega T)/C(\omega)] \qquad (19)$$

where:

$$C(\omega T) = \sum_{k=0}^{n} h(kT)\cos k\omega T \qquad (20)$$

and

$$D(\omega T) = \sum_{k=0}^{n} h(kT)\sin k\omega T \qquad (21)$$

The values of t_1 and n can be prespecified as part of the input data if the step response is already known, or it can be found by a criterion coded in the computer program. The criterion is given as follows. Let us define α as:

$$\alpha = |y_u(t) - y_u(t - \Delta t)| \qquad (22)$$

in which the value of α is approximately zero when the step response reaches the steady state with zero slope. In most problems of interest the step response comes very close to steady state after a few seconds. Then, the time t_1 is found when α at time t_1 is less than a prespecified small value. For example,

$$|y_u(t_1) - y_u(t_1 - \Delta t)| = \alpha < 10^{-6} \qquad (23)$$

Note that in a damped sinusoidal response that has not yet reached the steady state, the inequality of Eq. (23) may be satisfied at local minima or local maxima of the step response. To solve this problem, additional statements in the computer program can be added to Eq. (23):

$$\beta = |y_u(t_1) - y_u(t_1 - 2\Delta t)| \qquad (24)$$

and

$$\max(\alpha, \beta) < 10^{-6} \qquad (25)$$

Similarly, for discrete systems let

$$\gamma = |y_u(kT) - y_u(kT - T)| \qquad (26)$$

and

$$\delta = |y_u(kT) - y_u(kT - 2T)| \qquad (27)$$

Then, the stopping criterion is

$$\max(\gamma, \delta) < 10^{-6} \qquad (28)$$

After k points when the inequality of Eq. (28) is satisfied, n is equal to k.

A computer program was written based on this technique and was used to obtain the results for the following example:

EXAMPLE 1. The step response of a discrete linear system is known and is illustrated in Fig. 4. As shown in the figure, after ~ 12 sec, the step response with the sampling interval $T = 0.002$ sec reaches the steady state in which $n = 6000$ points. By means of the program, the frequency response of the system was obtained and is shown in Fig. 5 as solid lines.

The diagram in Fig. 4 is the output due to a unit step response of a fourth-order discrete linear system modeled by two cascaded second-order subsystems. These subsystems are represented by the difference equations (29) and (30) with the sampling interval $T = 0.002$ sec:

$$y(k) = 1.996405y(k - 1) - 0.998002y(k - 2)$$

$$+ 0.000399w(k) + 0.000798w(k - 1)$$

$$+ 0.000399w(k - 2) \qquad (29)$$

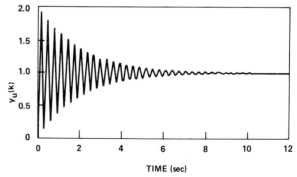

FIGURE 4 Step response of the discrete linear system of Example 1.

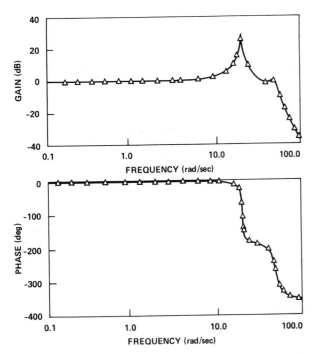

FIGURE 5 Frequency response of the discrete linear system of Example 1.——, Frequency response obtained from the conversion technique; △, frequency response obtained directly from the transfer function.

$$w(k) = 1.970370w(k-1) - 0.980247w(k-2)$$
$$+ 0.002469x(k) + 0.004938x(k-1)$$
$$+ 0.002469x(k-2) \qquad (30)$$

The frequency response diagram was also plotted directly from the pulse transfer function of the same system, which can be represented by:

$$H(z) = \frac{0.000399(z^2 + 2.0z + 1.0)}{z^2 - 1.996405z + 0.998002}$$
$$\frac{0.002469(z^2 + 2.0z + 1.0)}{z^2 - 1.970370z + 0.980247} \qquad (31)$$

These curves have been overlaid on the curves in Fig. 5 and are represented by the △ symbols.

Notice that the proposed technique with 12 sec or 6000 points of the step response generates a very accurate result for this example.

II. DERIVATION OF THE OPEN-LOOP FREQUENCY RESPONSE

A. Unity Feedback Systems

In this case, the structure of the system is assumed to be that of a unity feedback (Fig. 6), but the structure or transfer function of the open-loop system G does not have to be known. The closed-loop transfer function of this system is

$$Y_u/U = G/(1 + G) \qquad (32)$$

where U and Y_u are the Laplace or z transform of step input and step response, respectively. For convenience, the arguments z or s are dropped.

Let

$$M = Y_u/U \qquad (33)$$

Then,

$$G = M/(1 - M) \qquad (34)$$

From Section I, the frequency transfer function $M(J\omega)$ for continuous systems or $M(e^{J\omega T})$ for discrete systems can be found, and in turn the gain and phase of the open-loop systems can be obtained as follows:

$$\text{Gain} = |G| = |M/(1 - M)| = |M|/|1 - 1M| \qquad (35)$$

and

$$\text{Phase} = \angle G = \angle M - \angle 1 - M \qquad (36)$$

Now, assume that the gain and phase of the closed-loop systems are defined as:

$$r = |M| \qquad (37)$$

and

$$\phi = \angle M \qquad (38)$$

Hence,

$$M = re^{J\phi} = r\cos\phi + Jr\sin\phi \qquad (39)$$

By substituting Eq. (39) into Eqs. (35) and (36), the gain and phase of the open-loop systems are derived in terms of closed-loop gain and phase as follows:

$$\text{Gain} = |G| = \frac{r}{[(1 - r\cos\phi)^2 + r^2\sin^2\phi]^{1/2}} \qquad (40)$$

$$\text{Phase} = \phi - \tan^{-1}\left(\frac{r\sin\phi}{1 - r\cos\phi}\right) \qquad (41)$$

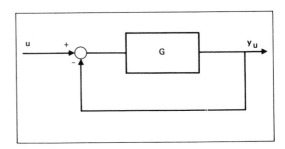

FIGURE 6 Unity feedback system.

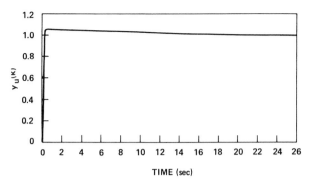

FIGURE 7 Step response of a unity feedback system of Example 2.

Note that the instability of the open-loop systems does not create any difficulty in applying the above technique.

EXAMPLE 2. The step response of a discrete unity feedback system is known and is plotted with a sampling interval $T = 0.01$ sec in Fig. 7. As one observes, the closed-loop step response has approached the steady state after almost 22 sec in which $n = 2200$ points.

By extension of the program to unity feedback systems (Section II.A), the open-loop frequency response of this system was obtained and is shown in Fig. 8. The output of this system in the time domain was computed by using the

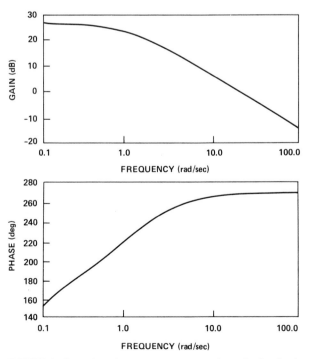

FIGURE 8 Open-loop frequency response of a unity feedback system of Example 2.

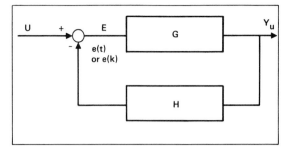

FIGURE 9 Canonical feedback system.

second-order difference equation (42). A unit step input was applied, and the sampling interval was 0.010 sec.

$$y(k) = 1.826397y(k-1) - 0.826488y(k-2)$$
$$+ 0.0913449x(k) + 0.0000456x(k-1)$$
$$- 0.0912933x(k-2) \quad (42)$$

Note that the open-loop transfer function of this system can be derived from the closed-loop representation of this system [Eq. (42)] as:

$$G(z) = \frac{0.100527z^2 + 0.000051z - 0.100477}{z^2 - 2.010050z + 1.010050} \quad (43)$$

This transfer function represents an unstable open-loop system, but since the feedback loop stabilizes the closed-loop system, the technique can be applied. This stabilization approach can be used to extend this technique to a special group of unstable systems.

B. Canonical Feedback Systems

In this case, it is assumed that the general structure of the system can be represented by a block diagram (Fig. 9), and the error function $e(t)$ or $e(kT)$ due to a step input is known or can be measured.

The open-loop transfer function of the system is GH. This system can be restructured according to the block diagram in Fig. 10 such that the open-loop transfer function is

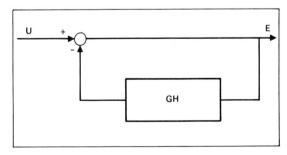

FIGURE 10 Restructured version of the canonical feedback system of Fig. 9.

still GH but the output is the error function of the canonical system shown in Fig. 9. The closed-loop transfer function of the system in Fig. 10 is

$$E/U = 1/(1 + GH) \tag{44}$$

Now, similar to unity feedback systems (Section II.A) let

$$N = E/U \tag{45}$$

If the error function of the system in Fig. 9 is known, the frequency transfer function $N(J\omega)$ or $N(e^{J\omega T})$ can be found. Then the open-loop frequency transfer function GH can be derived as a function of N:

$$GH = (1 - N)/N \tag{46}$$

Hence,

$$\text{Gain} = |GH| = \left| \frac{1 - N}{N} \right| = \frac{|1 - N|}{|N|} \tag{47}$$

and

$$\text{Phase} = \angle GH = \angle 1 - N - \angle N \tag{48}$$

Similar to Section II.A, let

$$N = r_1 e^{J\phi_1} \tag{49}$$

where:

$$r_1 = |N| \tag{50}$$

and

$$\phi_1 = \angle N \tag{51}$$

By substituting Eq. (49) into Eqs. (46) and (47), the open-loop gain and phase are found.

EXAMPLE 3. The error sequence $e(kT)$, $k = 0, 1, 2, \ldots$ of a discrete canonical feedback linear system to a unit step sequence with sampling interval $T = 0.04$ is known and is shown in Fig. 11. After almost 10 sec or $n = 250$ points, the error reaches the steady state. By ex-

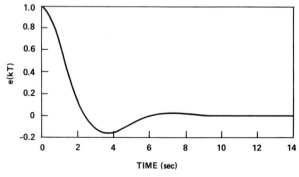

FIGURE 11 Error response sequence of Example 3.

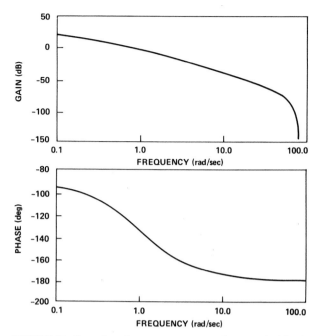

FIGURE 12 Open-loop frequency response of a canonical feedback system of Example 3.

tending the program to the canonical feedback systems technique (Section II.B), the Bode plot in Fig. 12 was obtained.

The error function of this system in response to a unit step sequence was generated by the canonical system in which

$$G(z) = \frac{0.0196078(z + 1)}{z - 0.960784} \tag{52}$$

$$H(z) = \frac{0.02(z + 1)}{z - 1} \tag{53}$$

III. SYSTEMS WITH UNBOUNDED STEP RESPONSE

For a special case in which the step response approaches a ramp function, it can be decomposed to a transient and a ramp sequence. For example, if $y_u(kT)$, $k = 0, 1, 2, \ldots$, is the step response sequence of a system, we have

$$y_u(kT) = y_{u1}(kT) + y_{u2}(kT), \quad k = 0, 1, 2, \ldots \tag{54}$$

where $y_{u1}(kT)$ is the transient part and $y_{u2}(kT)$ the ramp sequence. If $H(e^{j\omega T})$ is the frequency transfer function of the linear systems, we have

$$H(e^{J\omega T}) = H_1(e^{J\omega T}) + H_2(e^{J\omega T}) \tag{55}$$

where $H_1(e^{J\omega T})$ corresponds to $y_{u1}(kT)$, and $H_2(e^{J\omega T})$ corresponds to $y_{u2}(kT)$. The technique presented in Section I is used to find $H_1(e^{J\omega T})$. Also, $H_2(e^{J\omega T})$ is found as:

$$H_2(e^{J\omega T}) = KZ\left[\frac{1}{s^2}\right]\Bigg|_{Z=e^{J\omega T}} \qquad (56)$$

where $Z[1/s^2]$ is the Z-transform of the unit ramp function and K is the slope of the unit ramp function and K is the slope of the ramp. As one alternative, Tustin approximation can be used for the ramp in the discrete domain as:

$$H_2(e^{J\omega T}) \approx K\left[\frac{T}{2}\frac{z+1}{z-1}\right]^2\Bigg|_{Z=e^{J\omega T}} \qquad (57)$$

The value of sampling interval T is an effective factor for the accuracy of gain and phase.

Note that the same approach is used for continuous systems. $H_2(J\omega)$, which is the frequency response of a ramp function, is obtained as:

$$H_2(J\omega) = \frac{K}{s^2}\Bigg|_{s=J\omega} \qquad (58)$$

EXAMPLE 4. The step response of a continuous linear system approaches a ramp function with the slope equal to 1.0 (Fig. 13). By extending the program to the technique described in Section III, this response was converted to the frequency response shown in Fig. 14. The step response was generated through a system represented by the differential equation:

$$\frac{d^2y(t)}{dt^2} + \frac{dy(t)}{dt} = x(t) \qquad (59)$$

to a unit step input.

In signal processing and digital filtering, we may be interested in group delay instead of phase. In this case, we can obtain the group delay by differentiating the phase with respect to ω; that is,

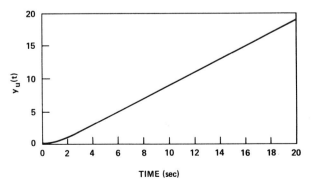

FIGURE 13 Step response of a continuous linear system of Example 4.

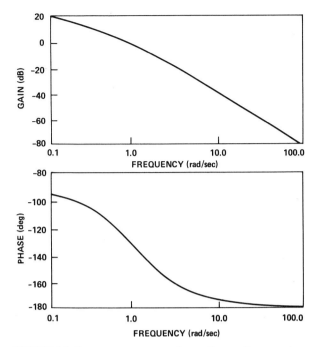

FIGURE 14 Frequency response of a continuous linear system of Example 4.

$$\tau_g(e^{J\omega T}) = -\frac{d\phi}{d\omega} = -\frac{d}{d\omega}\left(\tan^{-1}\frac{-C(\omega T)}{D(\omega T)}\right) \qquad (60)$$

where $C(\omega T)$ and $D(\omega T)$ are given by Eqs. (20) and (21).

There are cases when the step response does not reach the steady state with constant slope in a sufficiently small amount of time. In these cases, the frequency response for higher frequencies can be obtained in a relatively short period in the time domain, but for low frequencies it will be distorted by using the same technique. The results obtained in these cases will be useful if we are interested in higher frequencies such as in the study of stability of flexible vehicles.

IV. APPENDIX

The proof of Eq. (13) is as follows. In the z domain, the input–output relation of a discrete linear system is

$$Y(z) = H(z)X(z) \qquad (A-1)$$

The step response of the system $y_u(z)$ is obtained by substituting Eq. (A-2) in Eq. (A-1):

$$X(z) = \frac{1}{1-z^{-1}} \qquad (A-2)$$

Hence,

$$Y_u(z) = H(z)\frac{1}{1 - z^{-1}} \qquad \text{(A-3)}$$

and

$$H(z) = Y_u(z) - z^{-1}Y_u(z) \qquad \text{(A-4)}$$

Since $H(z)$ is the z transform of $h(kT)$, by taking the inverse z transform of Eq. (A-4) we get:

$$h(kT) = y_u(kT) - y_u(kT - T) \qquad \text{(A-5)}$$

SEE ALSO THE FOLLOWING ARTICLES

CONTROL SYSTEMS, IDENTIFICATION • SIGNAL PROCESSING

BIBLIOGRAPHY

Ackenhusen, J. G. (2000). "Real Time Signal Processing Design and Implementation of Signal Processing Systems," Advanced Information Systems Group, ERIM International, Prentice Hall, New York.

Deller, J. R., Jr., Hansen, J. H. L., and Proakis, J. G. (2000). "Discrete-Time Processing of Speech Signals," Signal Processing Society, Institute of Electrical and Electronics Engineers, Virginia Beach, VA.

Haykin, S., and Kosko, B. (2001). "Intelligent Signal Processing," Institute of Electrical and Electronics Engineers, Virginia Beach, VA.

Mecklenbräuker, W., and Hlawatsch, F. (1997). "The Wigner Distribution: Theory and Applications in Signal Processing," Elsevier, Amsterdam.

Porat, B. (1996). "A Course in Digital Signal Processing," John Wiley & Sons, New York.

Poularikas, A., ed. (1999). "The Handbook of Tables and Formulas for Signal Processing," CRC Press, Boca Raton, FL.

Convex Sets

A. C. Thompson
Dalhousie University

GLOSSARY

Affine map The composition of a linear map and a translation; i.e., if T is linear, then $A(x) := T(x + x_0) = T(x) + T(x_0)$ is an affine map.

Compact set A closed and bounded set in $\mathbb{R}^n$; in a metric space, a set C such that if (x_k) is a sequence of elements of C there is a subsequence that converges to a point of C.

Dual space The collection of all linear functions from a vector space to the set of real numbers. These functions can be added and multiplied by real numbers in a point-by-point fashion which makes this collection into another vector space.

Hyperplane A *level set* of a function in the dual space; i.e., if f is a linear function from a vector space X to $\mathbb{R}$ and if α is a number, then $H_f^\alpha := \{x \in X : f(x) = \alpha\}$ is a typical hyperplane.

Line segment The set of points (vectors) $\{x : x = a + \lambda(b - a),\ 0 \le \lambda \le 1\} = \{x : x = (1 - \lambda)a + \lambda b,\ 0 \le \lambda \le 1\}$ is called the *line segment* joining a and b and is denoted by $[a, b]$.

Linear map (transformation) A function T between vector spaces that respects the vector operations of addition and multiplication by numbers; i.e., $T(\alpha x + \beta y) = \alpha T(x) + \beta T(y)$.

$\mathbb{R}^n$ The most usual vector spaces consisting of n-tuples of real numbers that are added and multiplied by numbers in a coordinate-by-coordinate fashion.

Vector space A collection of things called *vectors* or *points* that can be added (via a parallelogram law) and multiplied by numbers (also called *scalars*). Here the numbers will be real but in other contexts complex or other number systems are possible.

Encyclopedia of Physical Science and Technology, Third Edition, Volume 3

I. INTRODUCTION

Relatively few shapes in the natural world are convex. When they do occur— for example, soap bubbles, drops of dew, smoothly worn stones on the beach, single crystals of amethyst and salt— we find them to be esthetically pleasing. Among manufactured objects, rectangles, circles, hexagons, cubes, cylinders, and cones are quite ubiquitous. We first encounter them as children and enjoy the shapes of wooden building blocks and colored tiles.

Convexity is the study of these shapes. Two-dimensional convex shapes (circles, ellipses, triangles, polygons) and the regular Platonic solids have been objects of mathematical study for a very long time. The study of convexity as a specific mathematical topic dates back only to the end of the 19th century. The primary influence was the pioneering work of Minkowski for which one should consult his collected works.[3] For a good historical summary, see the article by Peter Gruber in the *Handbook of Convex Geometry*.[2] This chapter covers only some of the topics that fall under the heading of convexity. Other aspects can be found in articles in Reference 2. There is not space to mention the many interactions of convexity with other branches of mathematics. The book by Roger Webster[5] is a readable, elementary introduction to the subject and has some interesting applications.

To define convex sets precisely we need an ambient space in which they may exist. This space requires one type of mathematical structure and usually comes equipped with another.

The structure that is required is that of a *vector* or *linear space*; the most familiar vector spaces are those that we denote by $\mathbb{R}^2$, $\mathbb{R}^3$, and, in general, $\mathbb{R}^n$. This space consists of n-tuples of numbers $(\xi_1, \xi_2, \ldots, \xi_n)$ whose entries are called the *coordinates* of the vector. It is customary to write single vectors as *rows*, but matrices (and other functions that operate on vectors) are usually written to the left of the vector. This means that the vectors should 'really' be viewed as *columns*.

If a vector y is expressed in the form:

$$y = \alpha_1 x_1 + \alpha_2 x_2 + \cdots + \alpha_k x_k$$

then it is said to be a *linear combination* of the vectors $\{x_i : i = 1 \ldots, k\}$. The usual basis for $\mathbb{R}^n$ consists of the vectors:

$$e_1 = (1, 0, 0, \ldots, 0), \quad e_2 = (0, 1, 0, \ldots, 0), \quad \ldots,$$

$$e_n = (0, 0, 0, \ldots, 1)$$

where the ith vector has a 1 as the ith coordinate and a 0 elsewhere. Any vector $x = (\xi_1, \xi_2, \ldots, \xi_n)$ can be expressed uniquely as:

$$x = \xi_1 e_1 + \xi_2 e_2 + \cdots + \xi_n e_n.$$

A set of vectors $\mathcal{B}$ with the property that every vector has a unique representation as a finite linear combination of the vectors in $\mathcal{B}$ is called a *basis* for X. Each vector space has a basis, and all bases for the same space have the same number of elements. That number is called the *dimension* of the space. The dimension of $\mathbb{R}^n$ is n. The space is said to be *finite dimensional* if it has a basis with a finite number of elements and is *infinite dimensional* otherwise. We shall be concerned almost entirely with finite dimensional spaces.

A linear map is called an *isomorphism* if it is one to one and onto. If X is finite dimensional, then, corresponding to a basis $(x_1, x_2, \ldots, x_n)$, there is an isomorphism, T, of X onto $\mathbb{R}^n$ defined as follows: each $x \in X$ has a unique representation as $x = \sum \alpha_i x_i$; set $T(x) := (\alpha_1, \alpha_2, \ldots, \alpha_n)$. In this sense, there is no real loss of generality if, when dealing with finite dimensional spaces, we restrict attention to $\mathbb{R}^n$.

The linear mappings from X to $\mathbb{R}$ are given the special name of *linear functionals*. The set of all of them is called the *dual space* of X and denoted by X^*. If X is finite dimensional and if it is given a basis, then the linear functionals can be represented by $1 \times n$ matrices (i.e., row vectors of the same size as the column vectors from X). Thus, in the finite dimensional case, the dimension of X^* is the same as that of X.

In $\mathbb{R}^n$, the length of a vector $x = (\xi_1, \xi_2, \ldots, \xi_n)$ is denoted by $\|x\|$ and is defined to be the number:

$$\|x\| := \left(\sum_{i=1}^{n} |\xi_i|^2 \right)^{1/2}.$$

The structure that is usually present in discussions of convex sets in addition to the vector space structure just outlined is that of a *topological space*. Most frequently the topology derives from a metric or distance. This is the case in $\mathbb{R}^n$ where the distance between two vectors x and y, $d(x, y)$, is defined to be the length of $x - y$:

$$d(x, y) := \|x - y\|.$$

The topological structure allows one to talk about convergence, continuity, and such concepts as open sets, closed sets, connected sets, and compact sets. The Heine-Borel Theorem asserts that C is a compact subset of $\mathbb{R}^n$ if and only if it is both closed and bounded. This is no longer true in infinite dimensional spaces.

II. DEFINITIONS

Definition. A set K in a vector space X is said to be *convex* if whenever $x, y \in K$ then the line segment $[x, y]$ is contained in K. Thus, in Fig. 1, the set (a) is convex while (b) is not.

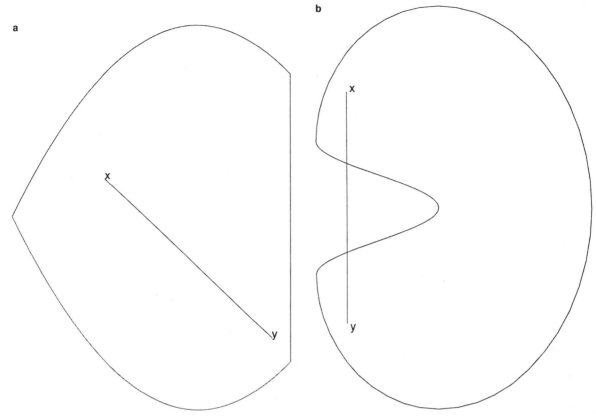

FIGURE 1

A point of the line segment $[a, b]$— a point of the form:

$$a + \lambda(b - a) = (1 - \lambda)a + \lambda b, \ 0 \le \lambda \le 1,$$

is said to be a convex combination of a and b.

A related notion is that of being *star shaped*. A set S is star shaped about a point x_0 if for all x in S the line segment $[x, x_0]$ is contained in S. In Fig. 2, the set (a) is star shaped about x but not about y, and (b) is not star shaped about any point. A set is convex if and only if it is star shaped about every point. Each convex set (and each star-shaped set) is connected.

In the one-dimensional space $\mathbb{R}$ the collections of convex sets, star-shaped sets, and connected sets coincide. Each of these is the class of intervals (closed, open, half-open, bounded, and unbounded). Therefore, in order to have an interesting theory of convexity, the dimension of the underlying space should be at least 2.

There is also a definition of convexity as an adjective that applies to functions rather than sets:

Definition. A real valued function f defined on a convex set K is said to be *convex* if, for all x and y in K,

$$f(\lambda x + (1 - \lambda)y) \le \lambda f(x) + (1 - \lambda)f(y).$$

Note that the convexity of K is needed to ensure that $\lambda x + (1 - \lambda)y$ is in the domain of f when x and y are.

The relation between the definitions of convexity for a function and for a set is twofold. The *graph* of f is a subset of $K \times \mathbb{R}$ and is defined as:

$$\text{graph}(f) := \{(x, \eta) : f(x) = \eta\}.$$

Extending this idea, the *epigraph* of f is the set that lies above the graph of f:

$$\text{epigraph}(f) := \{(x, \eta) : f(x) \le \eta\}.$$

Then, f is convex (as a function) if and only if epigraph(f) is convex as a set. Secondly, if f is convex then, for each (extended) real number α, the sets $\{x : f(x) \le \alpha\}$ and $\{x : f(x) < \alpha\}$ are convex. This illustrates the connection between convexity and certain types of inequality.

A function is sublinear if it is both subadditive—$f(x + y) \le f(x) + f(y)$—and *non-negatively homogeneous*—$f(\alpha x) = \alpha f(x)$ for all $\alpha \ge 0$. From now on, "homogeneity" will mean "non-negative homogeneity." All linear functionals are sublinear and all sublinear functions are convex. Hence, if f is sublinear then sets of the form $\{x : f(x) \le \alpha\}$ are convex.

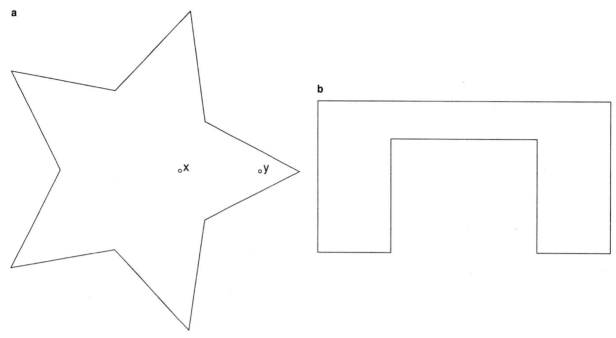

FIGURE 2

III. EXAMPLES

If K is a convex set and if A is an affine mapping, then $A(K) := \{Ax : x \in K\}$ is also convex. Affine mappings include translations, rotations, dilations, and reflections. Therefore, it is often convenient to think of examples as being located at some particular point in space (centered at the origin, for example) or with some particular orientation or with some particular scaling. These are not relevant to the property of being convex and may not be very relevant to other properties of the convex sets.

    1. A single point is a convex set.

    2. Lines, line segments (with or without the end points), and rays (sets of the form $\{x : x = a + \lambda b, \text{ with } \lambda \geq 0\}$) are convex sets. As indicated in Section II, in $\mathbb{R}$ the only convex sets are intervals.

    3. The ball of radius 1 centered at the origin (briefly, the unit ball),

$$B := B[0, 1] := \{x : \|x\| \leq 1\}$$

is convex. While this is geometrically clear in two and three dimensions, the proof (the same in all dimensions) is not so immediate. We need to show that the norm is a sublinear functional. That it is homogeneous is easy; the fact that it is subadditive,

$$\|x + y\| \leq \|x\| + \|y\|,$$

is usually referred to as the *triangle inequality* and is a consequence of the Cauchy-Schwarz inequality.

Any closed ball $B[x_0, r] := \{x : \|x - x_0\| \leq r\}$ of center x_0 and radius r is convex. An open ball, $B(x_0, r) := \{x : \|x - x_0\| < r\}$ is also convex (in general, the interior of a convex set is also convex). Note that we may consider, for example, a two-dimensional ball (a disc) as a subset of $\mathbb{R}^3$ or a higher dimensional space. It is still convex, but whereas as a subset of $\mathbb{R}^2$ it has interior points, in $\mathbb{R}^3$ it does not. We shall say more about the idea of relative interior in Section IV. Closed and bounded convex sets are called *convex bodies*. Some authors use this term to also imply that the set has a non-empty interior.

    4. The image of the unit ball under any invertible affine map is convex. This gives the important class of convex bodies known as *ellipsoids*.

    5. The *unit cube* is defined to be the set $\{x = (\xi_1, \xi_2, \ldots, \xi_n) : 0 \leq \xi_i \leq 1\}$. If a cube centered at the origin is required, we often consider one that is dilated by a factor of 2 and call it the *standard cube*, C_n:

$$C_n := \{x = (\xi_1, \xi_2, \ldots, \xi_n) : -1 \leq \xi_i \leq 1\}.$$

This set is a convex body in $\mathbb{R}^n$. The image of a standard cube under an invertible affine map is called a *parallelotope*.

    6. The *standard simplex*, S_n, is defined by the following equation:

$$S_n := \left\{x = (\xi_1, \xi_2, \ldots, \xi_n) : \xi_i \geq 0 \text{ and } \sum \xi_i \leq 1\right\}.$$

A general n-simplex is the image of S_n under an invertible affine map.

7. The *standard cross-polytope* in $\mathbb{R}^n$, $\mathcal{O}_n$, is defined by:

$$\mathcal{O}_n := \left\{ x = (\xi_1, \xi_2, \ldots, \xi_n) : \sum |\xi_i| \leq 1 \right\}.$$

The letter $\mathcal{O}$ is used because in $\mathbb{R}^3$ this set is a regular octahedron. Note that $\mathcal{O}_2$ and C_2 are both squares but they are oriented differently.

8. A *hyperplane* $H_f^\alpha := \{x : f(x) = \alpha\}$ (where f is a linear functional) is convex. A (closed) *half-space* is a set of the form $H_f^{\alpha-} := \{x : f(x) \leq \alpha\}$ and is described as one side of the hyperplane H_f^α. This set is convex as is its interior (the open half-space for whose definition $\leq$ is replaced by $<$). One also has the half-space $H_f^{\alpha+} := \{x : f(x) \geq \alpha\} = H_{-f}^{-\alpha-}$.

9. In addition to the standard Euclidean ball B in $\mathbb{R}^n$ we may consider the ℓ_p ball $B(p)$ which is defined by:

$$B(p) := \left\{ x : \sum |\xi_i|^p \leq 1 \right\}.$$

If $1 \leq p$, then this set is convex.

There will be more examples of convex sets in Sections IV and V, where we discuss general methods of constructing them and of getting new sets from old ones.

IV. DESCRIPTIONS OF CONVEX SETS

How should a convex set be specified? How does one decide whether a given point is inside a particular convex set or not? There are more sophisticated computational versions of these questions: What are the most efficient algorithms for describing a convex set or for deciding whether a point is inside or not? These are difficult questions which will not be tackled here. Instead, we give a variety of answers to the more general questions, any one of which may be the best for a particular situation.

The intersection of an arbitrary collection of convex sets is convex. Since any set is contained in at least one convex set (the whole vector space in which it sits), it follows that any set, A, is contained in a smallest convex set, namely the intersection of *all* the convex sets that contain A. It is called the *convex hull* of A and is written co$A$. Thus,

$$\text{co}A := \bigcap K$$

where the intersection is taken over all convex sets K with $A \subseteq K$. One visualizes the convex hull as the set obtained by stretching a rubber sheet around the set A or, in two dimensions, an elastic band (see Fig. 3).

Therefore, one may describe a convex set either as the intersection of certain simpler convex sets or as the convex hull of some simpler set.

The convex hull of a finite set is called a *convex polytope*. Sometimes more general types of polytopes may be

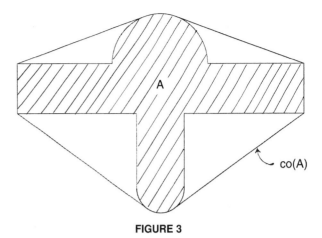

FIGURE 3

considered but here the word will always denote a closed, convex set and extra adjectives will be dropped.

Alternatively, a polytope may be described as the intersection of finitely many closed half-spaces. There is a difference between the two notions. The convex hull of finitely many points is always bounded; the intersection of half-spaces may not be. A bounded polytope that has an interior may be described either by the points of which it is the convex hull or by the bounding hyperplanes. This is the first example of the duality relationship discussed in Section V.

Examples. The standard simplex is the convex hull of the finite set $\{0, e_1, e_2, \ldots, e_n\}$. The standard octahedron is the convex hull of the finite set $\{\pm e_1, \pm e_2, \ldots, \pm e_n\}$. The standard cube is the intersection of the following half-spaces: $\{x : f_i(x) \leq 1\}$ and $\{x : -f_i(x) \leq 1\}$, where $f_i(x) = \xi_i$ and $i = 1, 2, \ldots, n$.

Here we digress to discuss the dimension of a convex set. An affine combination of vectors $\{x_1, x_2, \ldots, x_k\}$ is a linear combination $\alpha_1 x_1 + \alpha_2 x_2 + \cdots + \alpha_k x_k$ in which $\sum \alpha_i = 1$. If K is a convex set, then the set of all affine linear combinations of elements of K is called the *affine hull* of K. If $0 \in K$, then the affine hull of K is the same as the set of all linear combinations (because one can add a suitable multiple of 0 to make the coefficients sum to 1). Hence, in this case, the affine hull of K is a subspace (containing K). In the general case, the affine hull of K is a flat (the older term is *affine variety*)—i.e., the translate of a subspace. Subspaces and hence flats have a well-defined dimension. The dimension of a convex set is the dimension of its affine hull.

Related to this notion are the following concepts. First, a finite set of points with n elements is said to be in general position if its affine hull (or, equivalently, its convex hull) has dimension $(n - 1)$ which is the maximum possible. Two distinct points are always in general position, three points if they are not collinear, four points if they are not coplanar, and so on. The convex hull of $(n + 1)$ points in general position (necessarily in $\mathbb{R}^m$ with $m \geq n$) is an n-simplex. To see that there is an affine map of this set onto S_n observe that there is a translation that takes one point to 0 and then a linear map that takes the remaining n to the usual basis vectors. Second, the relative interior of K is the interior when K is regarded as a subset of its affine hull. The relative interior of a non-empty convex set is always non-empty.

Definition. A *face* F of a convex set K is a convex subset of K with the property that if y is in F and if y can be represented in the form $y = \alpha x_1 + (1 - \alpha)x_2$ with x_1 and x_2 in K, then, in fact, x_1 and x_2 are in F. A more geometrical description is that if an open line segment in K contains points of F then the whole line segment lies in F. If P is an n-dimensional polytope in $\mathbb{R}^n$, then its 0-dimensional faces are called *vertices*, its one-dimensional faces are called *edges*, and the $(n-1)$ dimensional faces will be called *facets* (this latter term is not universally used).

Examples. The faces of the standard cube C_3 in $\mathbb{R}^3$ are the cube itself, the six facets of the cube (the sets where one fixed coordinate has a prescribed value from $\{-1, 1\}$), the 12 edges of the cube (the sets where two fixed coordinates have prescribed values from $\{-1, 1\}$), and the eight vertices (the sets where all three coordinates have prescribed values from $\{-1, 1\}$).

The faces of the standard simplex S_3 in $\mathbb{R}^3$ are S_3; the four facets of the simplex (the intersections of S_3 with the planes $\{x : \xi_i = 0\}$ and $\{x : \sum \xi_i = 1\}$); the six edges (the line segments $[0, e_i]$, $[e_1, e_2]$, $[e_2, e_3]$, $[e_3, e_1]$); and the four vertices $\{0\}$, $\{e_i\}$.

Faces of a convex set K that are single points $\{z\}$ are called *extreme points* of K.

Theorem (Minkowski). A convex body in $\mathbb{R}^n$ is the convex hull of its extreme points.

There is an infinite dimensional extension of this theorem due to Krein and Milman which states that each compact convex set in a normed linear space is the closed convex hull of its extreme points.

A face of a convex set that is a ray is called an *extreme ray*. Another generalization of Minkowski's theorem is that any closed convex set that contains no lines is the convex hull of its extreme points and extreme rays.

The set of extreme points of a compact set need not be closed. This is shown by the following example.

Example. In $\mathbb{R}^3$ let K be the convex hull of the circle $\{x : \xi_1^2 + \xi_2^2 = 1\}$ and the points $(1, 0, 1)$ and $(1, 0, -1)$. K looks like a double slanted cone. The extreme points are

$(1, 0, 1)$, $(1, 0, -1)$, and all the points of the circle except $(1, 0, 0)$.

The next way to describe a convex set is to first suppose that the origin is an interior point of the set. This effectively means that the set has interior points because one can always either translate the set or choose the origin appropriately. Then one describes the set by saying how far in any direction the boundary is from the origin.

Definition. If K is a convex set with 0 as an interior point, then the *radial function* of K, $r_K(x)$, is defined by:

$$r_K(x) := \sup\{\lambda : \lambda x \in K\}.$$

A slight variant of this is to say how much K has to be dilated to contain a given vector.

Definition. If K is a convex set with 0 as an interior point, then the *gauge function* (or Minkowski functional) of K, $g_K(x)$, is defined by:

$$g_K(x) := \inf\{\lambda \geq 0 : x \in \lambda K\}.$$

It is evident that $1/r_K(x) = g_K(x)$ (with $1/\infty = 0$ if K is unbounded). The function g_K has the advantage of being both homogeneous and subadditive (i.e., sublinear). If K is bounded, then the radial function is finite for all $x \neq 0$ and the gauge function is non-zero for $x \neq 0$. If, in addition, K is symmetric about the origin so that $g_K(-x) = g_K(x)$, then the gauge function has the properties of a norm. This leads to the use of convexity in the study of normed spaces.

If K is closed, it is described as $K = \{x : g_K(x) \leq 1\}$. The boundary of K (denoted by ∂K) is the set of points for which $g_K(x) = r_K(x) = 1$.

Instead of thinking of the boundary of K as a set of points, we may also regard it as an envelope of half-spaces (in the same way that a polytope was described as a finite intersection of half-spaces).

Definition. A hyperplane H_f^α is said to be a *support hyperplane* for a convex body K if (1) $K \cap H_f^\alpha \neq \emptyset$, and (2) K is contained in one of the half-spaces $H_f^{\alpha+}$, $H_f^{\alpha-}$.

If K is a convex body, if f is a linear functional, and if $\alpha := \max\{f(x) : x \in K\}$, then H_f^α is a support hyperplane of K. This leads to the idea of the support function of K.

Definition. If K is a closed convex set then the *support function* of K, $h_K(f)$, is defined by:

$$h_K(f) := \sup\{f(x) : x \in K\}.$$

The support function is a sublinear function. If K is bounded, then h_K is finite for all x and we may replace sup by max. If $0 \in K$, then h_K is non-negative, and if 0 is an interior point, then h_K is strictly positive. If K is symmetric, then h_K is a norm on the dual space. There are generalizations of this idea to infinite dimensional spaces, but there

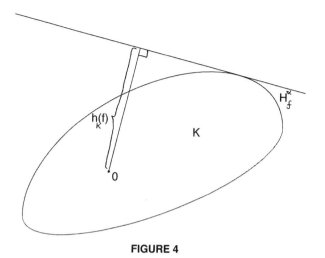

FIGURE 4

the dual space has a slightly different connotation—one must restrict attention to continuous linear functionals.

Most books on convexity that restrict attention to finite dimensional spaces identify the dual space $(\mathbb{R}^n)^*$ with $\mathbb{R}^n$ by means of the inner product. By that we mean that for each linear functional f there is a vector y_f such that $f(x) = \langle y_f, x \rangle$ where the symbol $\langle ., . \rangle$ denotes the inner product. In this case, the definition of support function is given in terms of the inner product.

To interpret the support function geometrically, we restrict our attention to linear functionals f with length 1. In this special case, the α that appears in the equation of the hyperplane:

$$f(x) = \alpha$$

represents the perpendicular distance from the origin to the hyperplane. Therefore, the support function (restricted to linear functions of length 1) represents the perpendicular distance from the origin to the supporting hyperplane of K that is in the direction f (see Fig. 4).

The following theorems are important.

Theorem. If f is a sublinear functional on $\mathbb{R}^n$, then there is a convex body K such that $f = h_K$.

Theorem. If K is a non-empty convex body, then $K = \{x : f(x) \le h_K(f) \text{ for all linear functionals } f\}$.

The proof of the second theorem requires the separation theorem from Section VII. There are a variety of proofs of the first theorem (see Schneider[4]).

V. NEW CONVEX SETS FROM OLD

A. The Convex Hull

We begin this section with a second look at the operation of the convex hull, the intersection of all convex sets con-

taining A. Rather than cut down to the convex hull from outside the set, we may also build up the convex hull from inside.

If $x_1, x_2, \ldots, x_k$ is a finite set of vectors, then y is said to be a *convex combination* of these vectors if:

$$y = \lambda_1 x_1 + \lambda_2 x_2 + \cdots + \lambda_k x_k \text{ with } \lambda_i \ge 1 \text{ and } \sum \lambda_i = 1.$$

The definition of convexity uses only two points but, by induction, one shows that if K is convex then every convex combination of points in K is also in K. If A is not convex, every convex combination of points from A is in co$A$. Finally, one shows that the set of all convex combinations of points from A is a convex set. Therefore, co$A$ is precisely the set of all convex combinations of points from A.

Note that k, the length of the convex combination, is arbitrary but Carathéodory's Theorem in Section VII shows that in finite dimensional spaces this is really not so.

Theorem. The convex hull of a compact set is compact and of an open set is open.

Theorem. If A is a non-empty bounded set in $\mathbb{R}^n$, then the diameter of co$A$ is the same as the diameter of A. (Here, "diameter" means the supremum of the distances between points of A.)

A nice application of the ideas here is the Gauss-Lucas Theorem. A short and elegant proof can be found in Webster.[5]

Theorem. If $p(z)$ is a non-constant polynomial, then the roots of $p'(z)$ (the derivative of p) are contained in the convex hull of the roots of $p(z)$.

B. The Polar of a Convex Set

By definition, the convex hull operator assigns to each set a convex set. The next very important operation does the same thing.

Definition. If A is a non-empty subset of X, then the *polar* of A, denoted by A° (or A^*), is defined by:

$$A^\circ := \{f \in X^* : f(x) \le 1 \text{ for all } x \in A\}.$$

If $A \subseteq X$, then $A^\circ \subseteq X^*$. Often, the distinction between X and X^* is obscured and the inner product is used in the definition of A°. Similarly, for a non-empty set B in X^* we have:

$$B^\circ := \{x \in X : f(x) \le 1 \text{ for all } f \in B\}.$$

Repeated applications of this operation are indicated without parentheses, thus $A^{\circ\circ}$, $B^{\circ\circ}$, and so on. For all sets A (in either X or X^*) we have $A \subseteq A^{\circ\circ}$. The operation reverses inclusions: If $A_1 \subseteq A_2$, then $A_2^\circ \subseteq A_1^\circ$. It follows that $A^\circ = A^{\circ\circ\circ}$ always.

The definition of B° reveals it to be an intersection of closed half-spaces so it is always a closed, convex set in X that contains 0. The same holds for A° in X^*. If $A = \{0\}$, then $A^\circ = X^*$ and $X^\circ = \{0\}$. If A is a single point other than 0, then A° is a half-space. Thus, the duality between points and half-spaces (encountered in Section IV) is implemented by this operation. However, if A is a half-space, then A° is not a singleton but a line segment joining the expected point to 0. Because A° is always a closed, convex set containing the origin, in order to get an exact duality we must restrict attention to this class of sets.

Theorem. If K is a closed, convex set with $0 \in K$, then $K = K^{\circ\circ}$.

This is a most important theorem whose proof relies on the separation theorem in Section VII. For an arbitrary set A, $A^{\circ\circ}$ is the closed convex hull of A and 0.

On the collection of closed convex sets that contain 0, the polar mapping is one to one and maps this collection onto the corresponding collection in X^*. If 0 is an interior point of K, then K° is compact. If K is compact, then K° has 0 as an interior point. Thus, the polar operation also maps the class of compact convex sets with 0 as an interior point onto the same class in X^*. If K is also symmetric about 0, then so is K°. In this last case, K plays the role of the unit ball in a normed space and K° is the dual ball in the dual space X^*.

Examples. If B is the unit ball in X, then B° is the unit ball in X^*.

If C_n is the standard cube in $\mathbb{R}^n$, then C_n° is the standard cross-polytope and conversely.

If S_n is the standard simplex, then S_n° is the following unbounded set. If $f = (\phi_1, \phi_2, \ldots, \phi_n)$, then $f \in S_n^\circ$ if and only if $\phi_i \leq 1$ for all i.

Finally, this duality also connects the gauge (and radial function) of K with the support function of K°.

Theorem. If K is a convex body with 0 as an interior point, then the gauge function of K° is the support function of K, and the gauge function of K is the support function of K°.

Since the radial function is the reciprocal of the gauge function, this is readily rewritten using the radial functions.

C. The Collection of Convex Sets as a Lattice

The intersection of two convex sets is again a (possibly empty) convex set. It is the largest convex set contained in both of them. The union of two convex sets need not be convex, but the convex hull of the union is the smallest convex set that contains both of them. Therefore, we define the following binary operations:

$$K_1 \wedge K_2 := K_1 \cap K_2$$

$$K_1 \vee K_2 := \mathrm{co}(K_1 \cup K_2).$$

With these operations, the collection of convex sets is a *lattice*. The underlying order relation is that of inclusion.

There are two important sublattices: the collection of closed convex sets that contain 0 and the collection of compact convex sets with 0 as an interior point. On each of these sublattices, the polar map is a bijection that reverses inclusion and hence reverses the lattice operations:

$$(K_1 \wedge K_2)^\circ = (K_1 \cap K_2)^\circ = K_1^\circ \vee K_2^\circ = \mathrm{co}\big(K_1^\circ \cup K_2^\circ\big)$$

$$(K_1 \vee K_2)^\circ = (\mathrm{co}(K_1 \cup K_2))^\circ = K_1^\circ \wedge K_2^\circ = K_1^\circ \cap K_2^\circ.$$

D. Algebraic Operations on Convex Sets

In addition to the operations just discussed, vector operations can be performed on the collection of convex sets. These are done "elementwise."

Definition. If K_1 and K_2 are convex sets in a vector space X and λ is a non-negative scalar, then:

$$K_1 + K_2 := \{x : x = x_1 + x_2 \text{ with } x_i \in K_i\}$$

and

$$\lambda K_1 := \{\lambda x : x \in K_1\}.$$

The reason for the restriction to $\lambda \geq 0$ is twofold. First, although we may define $(-1)K$, it may be (if K is symmetric) that $(-1)K = K$. Furthermore, the distributive law:

$$(\lambda + \mu)K = \lambda K + \mu K$$

only holds generally if $\lambda, \mu \geq 0$. Hence, there is no such thing as $K + (-1)K = \{0\}$. With the restriction to non-negative scalars, all the usual algebraic identities are valid. Despite this, we shall occasionally have need to talk about $-K := (-1)K$ below. Any non-negative linear combination of convex sets is again convex. This is the basis of what is called the *Brunn–Minkowski Theory* of convex sets (see Schneider[4]).

Examples. The sum of a convex set K with a single point $\{x_0\}$ is the same as the translation of K by x_0. In general, $K + L$ is the union of translates of K by points in L (or vice versa). Hence, one way to think of this operation is to visualize one of the convex sets K with its boundary ∂K and the other one L with the origin as a reference point somewhere in L. First translate L so that the reference point lies on ∂K and then slide L round K (just

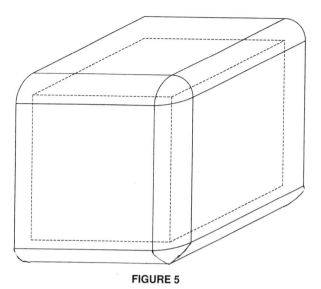

FIGURE 5

by translation) so that the reference point stays on ∂K and eventually has traversed all of it. The convex body swept out in this process is $K + L$.

In $\mathbb{R}^2$, the sum of the line segments $[-e_1, e_1]$ and $[-e_2, e_2]$ is the square C_2. If, in $\mathbb{R}^3$, we now add the line segment $[-e_3, e_3]$, we get the standard cube in $\mathbb{R}^3$. In general, the cube C_n is the sum of the line segments $[-e_i, e_i]$ for $i = 1, 2, \ldots, n$.

In $\mathbb{R}^2$, if we add the line segment $[(-1, -1), (1, 1)]$ to the square C_2 we get a hexagon with vertices at $(2, 0)$, $(2, 2)$, $(0, 2)$ and their negatives.

The sum of a finite number of line segments is a special type of polytope called a *zonotope*. These have a number of significant properties (see Section X). The sum of the cube C_3 and a multiple λB of the unit ball is shown in Fig. 5.

One of the reasons that the support function is so important is that it behaves well for these operations:

$$h_{K_1+K_2} = h_{K_1} + h_{K_2} \text{ and } h_{\lambda K} = \lambda h_K.$$

For the gauge and radial functions, we have:

$$g_{\lambda K} = \lambda^{-1} g_K \text{ and } r_{\lambda K} = \lambda r_K.$$

For the polar operation, likewise:

$$(\lambda K)^\circ = \lambda^{-1}(K^\circ).$$

Finally, these operations can be defined for arbitrary sets, and then the convex hull operation is also well behaved:

$$\mathrm{co}(A + B) = \mathrm{co}(A) + \mathrm{co}(B) \text{ and } \mathrm{co}\lambda A = \lambda \mathrm{co}(A).$$

E. Operations that Increase Dimension

So far we have been concerned with operations that take place within one fixed vector space X or from that space to its dual X^*. We now consider two vector spaces X and Y and their Cartesian product $X \times Y$, which has dimension equal to the sum of the dimensions of X and Y (assuming all are finite dimensional). There is a natural embedding of

X into $X \times Y$ that sends a vector x to $(x, 0)$ and similarly for Y. Then $X \times Y$ is the direct sum of these embedded spaces.

If K is a convex set in X and L is one in Y, then $K \times L$ (the Cartesian product) is a convex set in $X \times Y$. However, if we apply the embedding maps to K and L so that both can be thought of as lying in $X \times Y$, then $K \times L = K + L$.

In this sense, the cube C_n can be thought of either as a sum of line segments (as above) or as the Cartesian product of the interval $[-1, 1]$ with itself n times.

Examples. The product of the unit ball in $\mathbb{R}^2$ and the interval $[-1, 1]$ in $\mathbb{R}$ is the standard circular cylinder in $\mathbb{R}^3$. Likewise, the product of any convex set K with a line segment is a cylinder. In the case when K is a polytope, the word "prism" is often used instead of cylinder.

Another operation that can be performed on K and L as above is the suspension operation. As just indicated, we may think of both K and L as embedded in $X \times Y$ and then $K * L$, the suspension of K and L, is defined to be $\mathrm{co}(K \cup L) = K \vee L$.

Example. The standard cross-polytope is built up from the interval $[-1, 1]$ by repeated suspension operations.

If we are careful to interpret the polar operations in the appropriate spaces, we then get:

$$(K + L)^\circ = K^\circ * L^\circ \text{ and } (K * L)^\circ = K^\circ + L^\circ.$$

F. Symmetrizing Operations

The question of symmetry is an important one. A convex set is said to be *symmetric* about the origin if $K = -K$. It is often useful to operate on a convex set in such a way as to make it more symmetrical. For example, to prove certain inequalities where some quantity is maximized by a ball, one may be able to show that the quantity increases under a symmetrizing operation.

The first of these is quite simple. We define the *difference set* of K to be the convex set $D(K) := K - K := K + (-1)K$. The set $D(K)$ is always symmetric about 0. One readily checks that the support function of $-K$ is given by $h_{-K}(f) = h_K(-f)$, hence $h_{D(K)}(f) = h_K(f) + h_K(-f)$. For linear functionals f of norm 1, this last quantity is defined to be the *width* of K in the direction f and is denoted by $w_K(f)$. It represents the distance between parallel supporting hyperplanes with normal direction f (see Fig. 6).

A set is said to be of constant width if $w_K(f)$ is constant for all f with $\|f\| = 1$; that is, if $h_{D(K)}$ (restricted to those f with $\|f\| = 1$) is constant, which is so if and only if $D(K) = B$. We shall say more about these sets in Section X. For now, note that the only convex set of constant width that is symmetric about 0 is the ball.

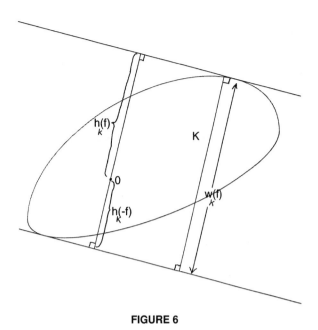

FIGURE 6

Examples. From some points of view, the most asymmetric convex set is the standard simplex. The difference body is rather regular. The difference body of the standard two-simplex S_2 is an affine regular hexagon and that of S_3 is an affine regular cuboctahedron (see Fig. 7).

The most well-known and simplest example of a non-symmetric set of constant width is the Reuleaux triangle (see Section X for a definition of this set; see Fig. 8 for a representation of it.)

The second symmetrizing operation is more complicated and requires an inner product and hence a notion of orthogonality. It is due to Jakob Steiner, is called *Steiner symmetrization*, and deals with symmetry with respect to a hyperplane.

Let K be a compact convex set in X. We wish to symmetrize K with respect to a hyperplane H in X. For each point $x \in H$ let $\ell(x)$ denote the line through x orthogonal to H. The intersection $\ell(x) \cap K$ is a line segment, a point, or is empty. If it is a line segment, let $S(x)$ be the translation of $\ell(x) \cap K$ whose midpoint is at x; if the intersection is a point, then $S(x) = x$, otherwise, $S(x)$ is empty. The Steiner symmetral of K with respect to H, $S_H(K)$, is defined by the equation $S_H(K) := \cup S(x)$.

It turns out that $S_H(K)$ is convex. If K has interior, then so does $S_H(K)$. If K and L are compact convex sets, then $S_H(K) + S_H(L) \subseteq S_H(K + L)$. The most important property of this symmetrization is that, given any convex body K, one may, by symmetrizing successively with a suitable sequence of hyperplanes, obtain a convex body that approximates a ball to within any degree of accuracy. Although we have not defined these measures, we also point out here that Steiner symmetrization leaves volume unchanged and does not increase either the surface area or the diameter of the set.

The third such operation is similar but reverses the roles of line and hyperplane. If ℓ is a line, for each point $x \in \ell$ let $H(x)$ be the hyperplane through x orthogonal to ℓ. If $H(x) \cap K \neq \emptyset$, let $B(x)$ be the $(n-1)$-dimensional ball centered at x and lying in $H(x)$ whose $(n-1)$-dimensional volume (see Section VIII) is the same as that of $H(x) \cap K$. Let $S_\ell(K) := \cup B(x)$. This object is also convex. It is called the *Schwarz symmetral* of K.

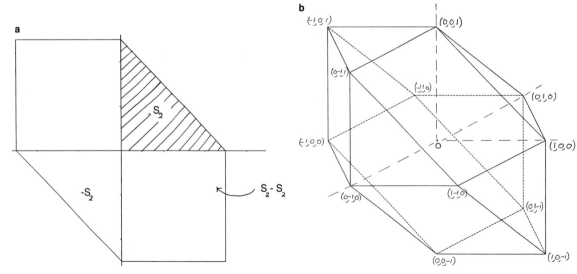

FIGURE 7

G. Other Operations

Two more ways to get new convex sets from old ones ought to be mentioned. The first is the projection body operator Π. There are several ways to define this operation; the simplest way is to use orthogonal projections. If K is a convex body in X, for each unit vector u in X consider the projection of K onto the hyperplane H_u orthogonal to u. This projection is the set $p_u(K)$ consisting of those points $x \in H_u$ such that the line $x + \lambda u$ has a non-empty intersection with K. Now let $h(u)$ be the $(n-1)$-dimensional volume (see Section VIII) of $p_u(K)$. It turns out that this function is subadditive. If it is extended to all of X by (positive) homogeneity, then it is sublinear and hence is the support function of a convex body in X^* called the *projection body* of K, $\Pi(K)$.

This is a most interesting object. It is always symmetric with respect to the origin. If K is a polytope, then $\Pi(K)$ is not just a polytope but is a zonotope (a sum of line segments) which has all of its faces centrally symmetric.

Examples. The construction (explained in Section X) of projection bodies of polytopes is relatively easy. The projection body of C_3 is (up to a scaling) C_3 itself. The projection body of $\mathcal{O}_3$ is a rhombic dodecahedron. This object has eight vertices that coincide with those of C_3 and six that are at $\pm 2e_1, \pm 2e_2, \pm 2e_3$. It has 12 facets that are all alike and are rhombi, hence the name.

The second operation is (in some not entirely clear way) a dual construction to Π. We begin with a convex body K such that $0 \in K$. Let f be a norm 1 linear functional in X^*. Let H_f be the hyperplane that is the kernel of f. Let $i_f(K) := K \cap H_f$ and let $r(f)$ be the $(n-1)$-dimensional volume (see Section VIII) of $i_f(K)$. Now let $I(K)$ be the set in X^* consisting of all the line segments $[0, r(f)f]$. In other words, by again extending r, we make it the radial function of $I(K)$. It is an important theorem (whose proof is difficult) of Busemann that if K is symmetric about 0 then $I(K)$ is convex. It is called the *intersection body* of K. For more general convex sets, $I(K)$ is star shaped. Lutwak has shown that the "proper" setting for this operation is the class of star-shaped sets.

VI. SPACES OF CONVEX SETS

In the last section, various operations on the collection of all convex sets were considered. With the algebraic operations of addition and multiplication by (non-negative) scalars, the collection of convex sets has many of the attributes of a vector space. In this section, we show that the collection of compact convex sets is also a metric space. In fact, there are two metrics that can be considered. One is defined on all compact convex sets in a given vector space

regardless of their dimension. The second distinguishes between points, line-segments, two-dimensional convex sets, and so on.

A. The Hausdorff Metric

If K and L are two compact convex sets, then they are bounded. Hence, there exist non-negative scalars λ, μ such that:

$$K \subseteq L + \lambda B \text{ and } L \subseteq K + \mu B.$$

Now let λ_0 be the infimum (in fact, the minimum) of all such λ's and similarly for μ_0. The numbers λ_0 and μ_0 may also be defined directly in terms of the norm (or distance) on X by:

$$\lambda_0 = \min_{x \in L} \max_{y \in K} \|x - y\|$$

and similarly for μ_0.

Definition. The *Hausdorff metric* δ on the set of compact convex sets is defined by the equation:

$$\delta(K, L) := \max\{\lambda_0, \mu_0\}.$$

Theorem. The function δ is a metric on the set of all compact convex sets in X.

If the support functions of convex sets are restricted to the unit ball in X^*, then the support function of a multiple λB of the unit ball is just the constant λ; hence, $K \subseteq (L + \lambda B)$ if and only if $h_K \leq h_L + \lambda$, which implies that the Hausdorff metric between the sets is the same as the uniform metric between the support functions. All the operations discussed in the previous section are continuous with respect to this metric as are several important functions that are defined in later sections.

The most important fact about this metric is a compactness result, known as the *Blaschke selection theorem* (the name refers to the selection of a convergent subsequence from a given bounded sequence).

Theorem (Blaschke). The set of compact convex sets contained in some ball $B[x, r]$ and equipped with the Hausdorff metric is compact.

Various collections of convex sets, for example the collection of all ellipsoids, form closed subsets in this metric. Therefore, a bounded sequence of ellipsoids has a subsequence that converges to another ellipsoid (provided we allow degenerate cases such as points and line segments).

Other important classes are dense in this metric. For example, the collection of all polytopes is dense. Therefore, any convex body can be approximated as closely as we please (with respect to the Hausdorff metric) by a polytope. This, together with the continuity of various functions, means that the proof of a theorem can often be accomplished by first proving the result for polytopes and then extending it to all convex bodies "by continuity."

A convex set is said to be strictly convex if its boundary does not contain any line segment (of positive length). A convex set is said to be smooth if there is a unique supporting hyperplane at each point of its boundary. The sets of smooth, of strictly convex, and of both smooth and strictly convex bodies are all dense in the set of all convex bodies.

B. The Banach-Mazur Metric

It is sometimes appropriate to consider that convex sets of different dimension are infinitely far apart. This section is concerned with metrics of this sort. We limit our attention to convex bodies with 0 as an interior point.

If K and L are two convex bodies with 0 in their interiors, then there are scalars λ and μ such that:

$$K \subseteq \lambda L \text{ and } L \subseteq \mu K.$$

As before, we can now take λ_0 and μ_0 to be the minimal such λ and μ. Then set $\Phi(K, L) := \lambda_0 \mu_0$. The fundamental result is now John's Theorem.

Theorem (John). If K is a centrally symmetric convex body in an n-dimensional space X and if 0 is an interior point of K, then there is an ellipsoid E such that:

$$E \subseteq K \subseteq \sqrt{n}E$$

The standard cube C_n and cross polytope $\mathcal{O}_n$ show that $\sqrt{n}$ cannot be improved. If we remove the condition that K be centrally symmetric, then we must replace $\sqrt{n}$ by n. This bound is attained by the simplex S_n.

The ellipsoid E that appears in this result is of considerable interest. It is the ellipsoid of maximal volume contained in K and is called the *Löwner-John ellipsoid*. It occurs in linear and nonlinear programming in Khachiyan's polynomial time algorithm and in Schor's algorithm.

The functional Φ is not a metric for two reasons. First, the construction is multiplicative rather than additive; e.g., $\Phi(K, K) = 1$ rather than 0. If we want a genuine metric we must take the logarithm of Φ. Since not all authors do so, one should be careful when reading the literature to see the precise definition.

Second, $\log(\Phi(K, \alpha K)) = 0$; Therefore, one should consider equivalence classes of multiples of K. However, it is more appropriate to enlarge the equivalence classes and say that the sets K_1 and K_2 are equivalent if there is an invertible linear map T such that $T(K_1) = K_2$. Finally, the definition of the Banach-Mazur metric is $\Delta(K, L) := \inf\{\log[\Phi(K, T(L))] : T \text{ is an invertible linear map}\}$.

An infimum is used here because it is also a useful definition in the case of infinite dimensional spaces. Restricted to the finite dimensional case, the infimum is attained. The functional Δ is a metric on the equivalence classes of convex sets under the above equivalence relation. If we consider the norms generated by K and L, then the equivalence relation is one of isometry between normed spaces and Δ measures the distance between equivalence classes of normed spaces.

John's Theorem now says that the distance between the equivalence class containing K and the set of ellipsoids (which is the equivalence class containing B) is no more than $(\log n)/2$. Hence, the distance between any two equivalence classes is no more than $\log n$. If we allow non-symmetric sets, then these numbers are doubled. It is surprising that the exact diameter of these metric spaces is only known in the case of two dimensions.

VII. BASIC THEOREMS

A. Separation and Support Theorems

The notion of separation involves placing a hyperplane between two convex sets. There are varying degrees of separation that can be considered.

Definition. A hyperplane $H = H_f^\alpha$ is said to *separate* the convex sets K and L if K lies in one of the closed half-spaces determined by H, and L lies in the other. The separation is *proper* if it is not the case that both sets lie in H. The separation is *strict* if one can replace "closed" by "open" in the first sentence. Finally, the separation is *strong* if there exist α and β with $\alpha < \beta$ and $K \subseteq H_f^{\alpha-}$ and $L \subseteq H_f^{\beta+}$ (or vice versa).

Separation Theorem. Let K be a convex set in $\mathbb{R}^n$ and suppose x is not in K, then K and x can be separated. If K is closed, then K and x can be strongly separated.

It follows from the second statement that every closed convex set can be represented as an intersection of closed half-spaces. It follows from the first statement that if K has a non-empty interior then at each point x of the boundary of K there is a supporting hyperplane obtained by separating x from the interior of K.

There is a converse to this theorem. If A is a closed set with a non-empty interior and if, at each boundary point, there is a supporting hyperplane, then A is convex.

The separation theorem can be generalized to the following: If K and L are two convex sets whose relative interiors are disjoint, then K and L can be separated. If one set is closed and the other is compact, then the separation is strong.

The proof of the seemingly more general statement follows from the earlier one because K and L can be (strongly) separated if and only if $K - L$ and $\{0\}$ can be (strongly) separated. Also, if one set is compact and the other closed, then $K - L$ is closed. This is *not* true in general for two closed sets.

Example. In $\mathbb{R}^2$ the sets $K := \{(x, y) : y = 0\}$ and $L := \{(x, y) : x \geq 0 \text{ and } y \geq 1/x\}$ are both closed and convex. They cannot be strongly (or even strictly) separated, and the set $K - L$ is open.

These theorems have very important analogs in infinite dimensional spaces. In that setting, they are all consequences of the Hahn-Banach Theorem, which is one of the most important theorems of functional analysis.

B. Carathéodory's Theorem and Its Relatives

The following theorems are all closely related, but the Carathéodory result appears the most fundamental.

Theorem (Carathéodory). If A is a subset of an n-dimensional space and if $x \in \mathrm{co}A$, then x can be expressed as a convex combination of $(n + 1)$ or fewer points.

Other ways of phrasing the conclusion is to say that x is a convex combination of a set of points in general position. Another is to say that x lies in a simplex whose vertices are in A. Thus, when constructing the convex hull, the length of the convex combinations needed is bounded (in finite dimensional spaces).

Theorem (Radon). If a finite set F of points in an n-dimensional space is *not* in general position, then it may be decomposed into two disjoint subsets F_1 and F_2 such that $\mathrm{co}(F) \cap \mathrm{co}(G) \neq \emptyset$.

In particular, this is true of any set of at least $(n + 2)$ points.

Theorem (Helly). Let $K_1, K_2, \ldots, K_m$ be a finite family of convex sets in an n-dimensional space ($m \geq n + 1$). If every subfamily with exactly $(n + 1)$ members has a non-empty intersection, then the whole family has a non-empty intersection.

Eggleston[1] shows how Helly's Theorem can be derived from Carathéodory's and conversely.

Since any family of compact sets has a non-empty intersection if every finite subfamily does, there is an easy extension to infinite families of compact convex sets. If an arbitrary family of compact convex sets in an n-dimensional space is such that every subfamily with $(n + 1)$ members has a non-empty intersection, then so does the whole family. A transversal for a family of sets is a line that meets every member of the family.

Corollary. If $\mathcal{F}$ is a finite family of parallel line segments in $\mathbb{R}^2$ such that every three of them has a transversal, then the whole family has a transversal.

Corollary. Let $\mathcal{F}$ be a finite family of convex sets in $\mathbb{R}^n$ and let K be a convex set. If for every finite subfamily with $(n + 1)$ elements there is a translate of K that intersects each member of the subfamily, then there is a single translate of K which intersects every member of the whole family.

Theorem (Kirchberger). Let F_1 and F_2 be finite sets in an n-dimensional space such that $F_1 \cup F_2$ has at least $(n + 2)$ elements. Suppose that for every subset F of $F_1 \cup F_2$ with exactly $(n + 2)$ points the sets $F \cap F_1$ and $F \cap F_2$ can be strictly separated, then F_1 and F_2 can be strictly separated.

Webster[5] gives an elegant proof of Jung's theorem also based on Helly's theorem.

Theorem (Jung). Every set A in $\mathbb{R}^n$ with diameter 1 is contained in a closed ball of radius no more than $\sqrt{n/(2n + 2)}$.

Finally, we give Krasnosel'skii's Theorem (sometimes called the "Art Gallery" Theorem).

Theorem (Krasnosel'skii). Let A be a compact subset of $\mathbb{R}^n$. If, for every $(n + 1)$ points, $a_1, a_2, \ldots, a_{n+1}$ of A there is a point x of A such that the line segments $[x, a_i]$ all lie in A, then A is star shaped.

In other words, in an art gallery, if for every finite set of $(n + 1)$ pictures there is a point in the gallery from which one can see all $(n + 1)$, then there is a point from which one can see the whole art gallery.

All of these theorems have been much generalized. For one collection of such results, see the article by J. Eckhoff in the *Handbook of Convex Geometry*.[2]

VIII. VOLUMES AND MIXED VOLUMES

For many of the more interesting properties of convex sets it is necessary to measure them in some way. Such measurement requires more advanced ideas than were presented in Section I. The basic concept is that of volume in an n-dimensional space. There are several approaches which coincide for compact convex sets but may not for more general types of sets.

The most straightforward approach is Eggleston's,[1] which says that the volume of a convex set in $\mathbb{R}^n$ is its n-dimensional Lebesgue measure. The volume of a set in $\mathbb{R}^n$ will be denoted by $V_n(K)$. One-dimensional volume is usually called *length*, and V_2 is usually called *area*. In

n-dimensional space, V_{n-1} is also frequently referred to as area. No confusion should arise.

The volume functional takes values in the extended real numbers and has a number of very important properties:

1. It is non-negative—$V_n(K) \geq 0$ for all convex sets K in $\mathbb{R}^n$—and is strictly positive if K has a non-empty interior (has dimension n).
2. It is countably additive, in the sense that, if $\{K_m : m = 1, 2, \ldots, \}$ is a sequence of disjoint convex sets, then $V_n(\cup_m K_m) = \sum_m V_n(K_m)$.
3. It is finite for compact sets; $V_n(K) < \infty$ if $K \subseteq \mathbb{R}^n$ is compact.
4. It is continuous with respect to the Hausdorff metric.
5. It is monotonic; if $K \subseteq L$, then $V_n(K) \leq V_n(L)$ (this follows from (property 2)).
6. It is translation invariant; $V_n(K + x) = V_n(K)$.
7. For linear transformations T, it has the property that $V_n(T(K)) = \det T \, V_n(K)$.
   In particular:
8. $V_n(\lambda K) = \lambda^n V_n(K)$ and also V_n is invariant under rigid motions.

The important fact from the theory of Haar measure is that, up to a scalar factor, there is only one functional that has properties 1, 2, 3, and 6.

However, if we relax property 2 a little, then a number of important functions are relevant to the theory of convex sets.

A real-valued function v defined on a collection of sets S is said to be a valuation if:

$$v(K \cup L) + v(K \cap L) = v(K) + v(L)$$

whenever K, L, $K \cup L$, and $K \cap L$ are all in S.

It follows that for finitely many disjoint sets, a valuation is additive, hence, if its values are non-negative, it is monotonic. Volume is a valuation.

The n-dimensional volume of a convex set may be calculated by integrating the $(n-1)$-dimensional volumes of its cross sections in some particular direction. Thus, if f is a linear functional and if K is a convex set and if we let $K_\alpha := K \cap H_f^\alpha$, then:

$$V_n(K) = \int_{-\infty}^{\infty} V_{n-1}(K_\alpha) d\alpha.$$

Examples. The above formula can be used inductively to prove that:

$$V_n(B) = \frac{\pi^{n/2}}{\Gamma(1 + n/2)}.$$

This number is abbreviated ϵ_n (many authors use κ_n).

There is a more general formula for the n-dimensional p-ball $B(p)$:

$$V_n(B(p)) = \frac{2^n \Gamma(1 + 1/p)^n}{\Gamma(1 + n/p)}.$$

The volume of the standard cube C_n is $V_n(C_n) = 2^n$.

The volume of the standard simplex S_n is $V_n(S_n) = 1/n!$.

The volume of the standard cross-polytope $\mathcal{O}_n$ is $V_n(\mathcal{O}_n) = 2^n/n!$.

If K is a convex set such that $K \subseteq H_f^\alpha$ with $\alpha \neq 0$ and $\|f\| = 1$, then the pyramid with vertex at 0 and base K is the convex hull of K and 0. The volume of this pyramid is $\alpha V_{n-1}(K)/n$.

The previous formula can be generalized. If P is a polytope with 0 as an interior point and if F_i, $i = 1, 2, \ldots, m$, denote the facets of P and if α_i is the perpendicular distance from 0 to the hyperplane containing F_i, then $V_n(P) = (1/n) \sum_1^m \alpha_i V_{n-1}(F_i)$.

With a notion of surface area, this can be further generalized to arbitrary convex sets:

$$V_n(K) = \frac{1}{n} \int_{S^{n-1}} h_K(f) d\sigma_K(f)$$

where σ_K is the "surface area measure" induced on the surface of the dual unit ball S^{n-1} by K (see Schneider[4] for details).

Properties (7) and (8) show how V_n behaves for scalar multiples and under linear maps. The basic question, first considered by Brunn and Minkowski toward the end of the 19th century, is how does V_n behave with respect to the addition of sets; i.e., how is property 6 extended from singletons to general convex sets?

To see how this might work, consider the special case of $K + \lambda K$; then:

$$V_n(K + \lambda K) = V_n((1 + \lambda)K) = (1 + \lambda)^n V_n(K)$$

$$= \sum \binom{n}{i} \lambda^i V_n(K);$$

that is, we get a polynomial in λ whose coefficients are multiples of $V_n(K)$. In fact, we always get a polynomial.

Theorem. If K and L are convex bodies, then $V_n(K + \lambda L)$ is a polynomial in λ of degree n whose coefficients are written in the following way:

$$V_n(K + \lambda L) = \sum_0^n \binom{n}{i} V(K, n-i; L, i) \lambda^i.$$

The numbers $V(K, n-i; L, i)$ are called the *mixed volumes* of K and L. The numbers $n-i$ and i are inserted in this notation because the mixed volumes are functions of n variables and here K and L occur $n-i$ times and i times, respectively. Since this is true for a summand with two terms, it can be extended inductively to sums with an arbitrary number of terms. The essential feature is that the volume of a linear combination $(\sum \lambda_i K_i)$ is a homogeneous polynomial in the λ_i's of degree n whose coefficients are functions of precisely n of the K_i's.

Example. Referring again to Figure 5, one sees that

$$V_3(C_3 + \lambda B) = V_3(C_3) + A(C_3)\lambda$$
$$+ (3\pi)L(C_3)\lambda^2 + V_3(B)\lambda^3$$

where $A(C_3)$ is the sum of the areas of the facets of the cube and $L(C_3)$ is the length of an edge of the cube and measures the width of the cube. Thus, in this instance, we have $V(C_3, C_3, C_3) = V_3(C_3)$, $V(C_3, C_3, B) = A(C_3)/3$, $V(C_3, B, B) = \pi L(C_3)$, and $V(B, B, B) = V_3(B)$.

Mixed volumes have the following properties:

1. They are non-negative.
2. They are monotonic in each variable.
3. They are homogeneous (for non-negative scalars) in each variable. Recall that with the notation $V(K, n - i; L, i)$ the variable K occurs $n - i$ times.
4. They are additive in each variable.
5. They are translation invariant (in each variable).
6. They are continuous with respect to the Hausdorff metric.
7. $V(K, K, \ldots, K) = V_n(K)$.

The example shows that mixed volumes of the form $V(K, n - i; B, i)$ are closely related to the geometry of K. For this, the notation is modified.

Definition. The *quermassintegrals* or *cross-sectional measures* or *Minkowski functionals* of K are written $W_i(K)$ and are defined by $W_i(K) := V(K, i; B, n - i)$ (note the change of place of i).

Then, $W_0(K) = V_n(K)$, $W_1(K) = A(K)/n$, and $W_{n-1}(K) = \epsilon_n b(K)/2$, where $A(K)$ is the surface area of K and $b(K)$ is the mean width of K. These last two equations may either be taken as the definition of these quantities or, if they are defined in other ways, then as theorems. Perhaps surprisingly, even $W_n(K)$ has relevance to the set K: $W_n(K) = \epsilon_n \chi(K)$, where $\chi(K)$ is the Euler characteristic of K which (for convex sets) is always 1.

Since W_1 is the coefficient of λ in a certain polynomial, it (and hence $A(K)$) can be obtained via differentiation in the usual way.

Theorem. The surface area $A(K)$ of K is obtained by the formula:

$$A(K) = \lim_{\lambda \to 0} \frac{V_n(K + \lambda B) - V_n(K)}{\lambda}.$$

There are also integral formulas for both $A(K)$ and $b(K)$:

Theorem (Cauchy). If K is a convex body in $\mathbb{R}^n$, then:

$$A(K) = \frac{1}{n\epsilon_n} \int_{S^{n-1}} V_{n-1}(p_u(K)) d\omega(u)$$

where S^{n-1} is the surface of the unit ball, $d\omega$ is surface area measure (Lebesgue measure) on S^{n-1}, u is a unit vector, and $p_u(K)$ is the projection of K onto the subspace orthogonal to u.

Theorem. If K is a convex body in $\mathbb{R}^n$, then:

$$b(K) = \frac{2}{n\epsilon_n} \int_{S^{n-1}} h_K(u) d\omega(u).$$

(This is often taken as the definition of $b(K)$).

There are various relationships between the functionals W_i expressed as integral formulas.

All the functionals W_i are valuations and, in a certain precise sense, these are *all* the valuations.

Theorem (Hadwiger). If v is a valuation on the collection of convex bodies that is invariant under rigid motions and is either continuous or monotonic, then:

$$v(K) = \sum_0^n \alpha_i W_i(K)$$

where the coefficients α_i are real if v is continuous and non-negative if v is monotonic. The first straightforward proof of this theorem was given in 1995 by Dan Klain.

IX. INEQUALITIES

There is a huge variety of inequalities that relate to convex sets in one way or another. We present a brief sample. The reader is referred to the work of Erwin Lutwak and, in particular, his article in the *Handbook of Convex Geometry*.[2] We do not always give the most general result (which may need more notation or definitions); many extensions can be found in Schneider.[4] The word "homothetic" often occurs in the conditions for equality. The sets A and B are homothetic if $A = \lambda B + a$ ($\lambda > 0$); i.e., A is the image of B under a translation and a (positive) dilation.

We begin with the fact that the nth root of the volume is a concave function.

Theorem (Brunn-Minkowski). If K and L are convex bodies with interior in $\mathbb{R}^n$ and if $0 \leq \alpha \leq 1$, then:

$$V_n^{1/n}(\alpha K + (1 - \alpha)L) \geq \alpha V_n^{1/n}(K) + (1 - \alpha)V_n^{1/n}(L)$$

with equality if and only if K and L are homothetic.

Theorem (Minkowski inequality for mixed volumes). If K and L are convex bodies in $\mathbb{R}^n$, then:

$$V(K, n-1; L, 1)^n \geq V_n(K)^{n-1} V_n(L)$$

with equality if and only if K and L are homothetic.

The power of this inequality is shown by the fact that, substituting the unit ball for L, one immediately gets the isoperimetric theorem for convex sets.

Theorem (Isoperimetric). If K is a convex body in $\mathbb{R}^n$ with prescribed volume v, then $A(K) \geq A(B_v)$, where B_v is the dilation of B with volume v. Moreover, equality holds if and only if K is a translate of B_v.

Corollary (Isoperimetric inequality). If K is a convex body in $\mathbb{R}^n$, then:

$$\frac{A(K)^n}{V_n(K)^{n-1}} \geq \frac{A(B)^n}{V_n(B)^{n-1}} = n^n \epsilon_n.$$

Theorem (Urysohn). If K is a convex body in $\mathbb{R}^n$, then:

$$\frac{b(K)^n}{V_n(K)} \geq \frac{b(B)^n}{V_n(B)} = 2^n/\epsilon_n.$$

Since $b(K) \leq D(K)$ (the diameter of K), we get the following corollary.

Corollary (Isodiametric inequality). If K is a convex body in $\mathbb{R}^n$, then:

$$\frac{D(K)^n}{V_n(K)} \geq \frac{D(B)^n}{V_n(B)} = 2^n/\epsilon_n.$$

In the last three inequalities, equality holds if and only if K is a ball.

The next set of inequalities relate the volume functional to the operations given in Section V. However, they are also affine inequalities because the quantities involved are unchanged by affine transformations.

Theorem (Blaschke-Santaló). If K is a convex body in $\mathbb{R}^n$, then:

$$V_n(K)V_n(K^\circ) \leq V_n(B)V_n(B^\circ) = \epsilon_n^2$$

with equality if and only if K is an ellipsoid.

The quantity $V_n(K)V_n(K^\circ)$ is called the *volume product* of K and is an affine invariant. There is a conjecture of Mahler on the lower bound for the volume product. It has been proved for zonoids (the closure, in the Hausdorff metric, of the set of zonotopes).

Theorem (Reisner). If K is a zonoid in $\mathbb{R}^n$, then:

$$V_n(K)V_n(K^\circ) \geq V_n(C_n)V_n(C_n^\circ) = V_n(\mathcal{O}_n)V_n(\mathcal{O}_n^\circ) = 4^n/n!$$

with equality if and only if $K = C_n$ ($\mathcal{O}_n$ is not a zonoid).

Theorem (Rogers-Shephard). If K is a convex body in $\mathbb{R}^n$ then

$$2^n V_n(K) \leq V_n(D(K)) \leq \binom{2n}{n} V_n(K)$$

with equality on the left if and only if K is symmetric and on the right if and only if K is a simplex. (The left-hand inequality is trivial; it is the other one that is due to Rogers and Shephard.)

Theorem (Busemann intersection inequality). If K is a convex body in $\mathbb{R}^n$ with 0 as an interior point, then:

$$\frac{V_n(I(K))}{V_n(K)^{n-1}} \leq \frac{V_n(I(B))}{V_n(B)^{n-1}} = \frac{\epsilon_{n-1}^n}{\epsilon_n^{n-2}}$$

with equality if and only if K is an ellipsoid.

Theorem (Petty projection inequality). If K is a convex body in $\mathbb{R}^n$, then:

$$V_n(K)^{n-1} V_n([\Pi(K)^\circ]) \leq V_n(B)^{n-1} V_n([\Pi(B)^\circ])$$
$$= (\epsilon_n/\epsilon_{n-1})^n$$

with equality if and only if K is an ellipsoid.

Here, we insert results dealing with two problems that generated a great deal of research in the latter part of the 20th century: the Busemann-Petty problem and its "dual," the Shephard problem. Both deal with convex bodies symmetric about 0 because otherwise it is easy to give a negative answer to both questions.

Question (Busemann-Petty). If K and L are two centrally symmetric convex bodies in $\mathbb{R}^n$ and if, for every linear functional f, we have:

$$V_{n-1}(K \cap H_f^0) \leq V_{n-1}(L \cap H_f^0)$$

does it follow that $V_n(K) \leq V_n(L)$?

The breakthrough on this problem came when Lutwak showed that the answer is "yes" if the body K is an intersection body (in a broader sense than our definition), otherwise, there will be a convex body L yielding a counter example. The work of many authors and finally of Zhang, Gardner, and Koldobsky showed that the answer to the problem is "no" in all dimensions ≥ 5 but that for $n = 2, 3, 4$ the answer is "yes."

Question (Shephard). If K and L are two centrally symmetric convex bodies in $\mathbb{R}^n$ and if, for every direction u, we have:

$$V_{n-1}(p_u(K)) \leq V_{n-1}(p_u(L))$$

does it follow that $V_n(K) \leq V_n(L)$?

Petty and Schneider showed that the answer is "no" in general but "yes" if the body L is a projection body. Since all symmetric convex bodies in $\mathbb{R}^2$ are projection bodies, the answer is "yes" in $\mathbb{R}^2$ but "no" in all higher dimensions.

Finally we mention two well-known inequalities that relate to the finite (and infinite) dimensional ℓ_p-spaces.

Theorem (Hölder's inequality). If the numbers $p \geq 1$ and q are related by the equation $p^{-1} + q^{-1} = 1$ and if $x = (\xi_i)$ and $y = (\eta_i)$ are two vectors in $\mathbb{R}^n$, then:

$$\left| \sum \xi_i \eta_i \right| \leq \left(\sum |\xi_i|^p \right)^{1/p} \left(\sum |\eta_i|^q \right)^{1/q}.$$

A consequence of this inequality is that of Minkowski.

Theorem (Minkowski's inequality). If x and y are vectors in $\mathbb{R}^n$ and if $\|x\|_p := (\sum |\xi_i|^p)^{1/p}$, then

$$\|x + y\|_p \leq \|x\|_p + \|y\|_p.$$

Therefore, the functional $\|x\|_p$ is a norm and the set $B(p)$ defined in Example 9 in Section III is convex.

X. SPECIAL CLASSES OF CONVEX SETS

A. Polytopes

In this section we deal with bounded polytopes. Recall that such a polytope is a convex body that may be regarded either as the convex hull of finitely many points or, dually, as the intersection of finitely many half-spaces. Polygons and polyhedra have been studied since the beginnings of mathematics. The existence theorem of Minkowski is very important.

Theorem (Minkowski). If $\{u_1, u_2, \ldots u_k\}$ is a set of dual unit vectors which do not all lie in a hyperplane and if $\{\alpha_1, \alpha_2, \ldots, \alpha_k\}$ are positive real numbers such that:

$$\sum_i^k \alpha_i u_i = 0$$

then there is a polytope (unique up to translation) that has k facets whose areas are given by the α_i's and whose "normals" are given by the u_i's.

There is a generalization of this theorem to general convex bodies determined by general "surface area measures" (see Schneider[4]).

By the facial structure of a polytope, we mean the collection of all of its faces classified by their dimension. If P is an n-dimensional polytope, then there is precisely one n-dimensional face, P itself. The empty set is usually included as the unique face of dimension -1. The 0-dimensional faces are the vertices, the one-dimensional faces are the edges, and the $(n-1)$-dimensional faces are the facets. The f-vector of P is the vector $(f_0, f_1, f_2, \ldots, f_{n-1})$ where f_i is the number of faces of dimension i. The faces of P form a lattice under the inclusion relation (this is why P and $\emptyset$ are included as faces). The lattice of faces of P° forms the dual lattice.

A key combinatorial problem is to characterize those vectors that are f-vectors of some polytope. Only in three dimensions has this problem been solved. The primary necessary condition for a vector to be an f-vector is that it satisfies the so-called Euler relation.

Theorem. If f_i denotes the number of faces of a polytope of dimension i, then

$$\sum_0^n (-1)^i f_i = 1.$$

The number 1 is the Euler characteristic of P.

Corollary. In three-dimensional space, the number of vertices f_0 of edges f_1 and of facets f_2 satisfy the equation $f_2 - f_1 + f_0 = 2$.

Theorem (Steinitz). In three-dimensional space, (f_0, f_1, f_2) is the f-vector of a polyhedron if and only if it satisfies, in addition to the Euler relation, the following inequalities:

1. $4 \leq f_0 \leq 2f_2 - 4$
2. $4 \leq f_2 \leq 2f_0 - 4$

Such conditions are not completely known in higher dimensions but there are many partial results, especially for $n = 4$ (see the survey article by Bayer and Lee[2]). However, it is known that the Euler relation is the only affine relation satisfied by all f vectors.

Of particular appeal are the regular figures. If $n = 2$, then there is an infinite family of regular polygons. Up to rigid motions and dilations, there is precisely one for each number k of vertices (edges). If $n = 3$, then there are precisely five regular polyhedra (the Platonic solids). There are many proofs that there can be no more. We outline one. Suppose that the facets are p-gons and that q meet at each vertex. Then the Euler relation implies that $p^{-1} + q^{-1} > 1/2$. Since $p, q \geq 3$, only the values $(p, q) = (3, 3)$, $(3, 4)$, $(4, 3)$, $(3, 5)$ and $(5, 3)$ are possible.

In all dimensions it is possible to construct a regular cube C_n, a regular cross-polytope $\mathcal{O}_n$, and a regular simplex (a linear image of S_n). For $n \geq 5$, these are the only regular polytopes. When $n = 4$, there are three more with, respectively, 24 octahedral facets, 120 dodecahedral facets, and 600 tetrahedral facets.

A great variety of polytopes have some degree of regularity; for example, the cuboctahedron has six square facets and eight that are equilateral triangles and whose vertices are all alike. Its dual is the rhombic dodecahedron, which has all its facets alike (they are all rhombi).

B. Zonotopes

This special class of polytopes was first brought to attention by Federov in connection with crystallography. They are centrally symmetric but, more than that, they are characterized by having all their two-dimensional faces centrally symmetric. A theorem of Alexandrov implies that *all* of the faces are centrally symmetric. Thus, when $n = 3$, the zonohedra are those polyhedra with centrally symmetric facets. For $n > 3$, there are polytopes whose facets are centrally symmetric but are not zonotopes.

From the point of view of the Brunn-Minkowski theory based on sums and scalar multiples of convex sets, zonotopes are the simplest objects because they are the ones that can be expressed as sums of line segments. To simplify matters, the line segments can be centered at 0 so that the zonotope is also symmetric about 0. The support function of a line segment $[-x, x]$ is $h_{[-x,x]}(f) = |f(x)|$ and, since support functions respect addition, if $Z := \sum_i [-x_i, x_i]$, then $h_Z(f) = \sum_i |f(x_i)|$.

In addition to crystallography, these objects are important in several areas of mathematics. First, if we restrict the domain of the projection body operator Π to the polytopes, its range is the set of zonotopes. Moreover, if the domain is restricted to the centrally symmetric polytopes, then Π is one–one.

If P is a polytope, then the action of Π on P is described in the following way. Let the facets of P be denoted by $F_1, F_2, \ldots, F_m$. Each F_i is contained in a hyperplane $H_{f_i}^{\alpha_i}$ with $\|f_i\| = 1$. The functional f_i is called the unit *normal* to the facet F_i; the sign is usually chosen so that it is directed outwards but that is immaterial here. The f_i's are in the directions of the vertices of P°. If μ_i denotes the area of the facet F_i, then:

$$\Pi(P) = 1/2 \sum_i^m \mu_i/2[-f_i, f_i].$$

One factor of $1/2$ is to make the line segments be of length μ_i, the other is because the projection of P in some direction is covered twice by the projections of the facets; if P is symmetric, the sum can be taken over half the facets, one from each pair. This construction reveals that Π maps objects in X to objects in the dual space X^*.

Examples. For the standard cube C_3, the normals are the usual basis vectors (and their negatives). The areas are all 4, hence, $\Pi(C_3) = \sum 2[-e_i, e_i] = 2C_3$.

For the standard octahedron $\mathcal{O}_3$, the normals are

$$3^{-1/2}(1, 1, 1), \ 3^{-1/2}(1, -1, 1) \ 3^{-1/2}(-1, -1, 1)$$
$$3^{-1/2}(-1, 1, 1)$$

and the areas are all $\sqrt{3}/2$. Another way to find the sum of the corresponding line segments is as the convex hull of the vectors:

$$1/4(\pm(1, 1, 1) \pm (1, -1, 1) \pm (-1, -1, 1) \pm (-1, 1, 1))$$

where all (16) possible choices of signs are taken. Of these, 14 are vertices and the other two are interior points (they are 0). These vertices are those of the rhombic dodecahedron, as stated in Section V (with an extra factor of $1/2$).

For the standard simplex, $\Pi(S_3)$ is also a sum of four line segments and is combinatorially the same as $\Pi(\mathcal{O}_3)$ but it has different proportions. However, the two are affinely equivalent showing that Π is not one to one on the set of all polytopes.

There are two other useful characterizations of zonotopes. If P is a centrally symmetric polytope with k pairs of opposite vertices, then P is a projection of $\mathcal{O}_k$. Similarly, if a zonotope is a sum of k line segments, then it is a projection of C_k. The dual characterization is that the dual of a zonotope is a central cross section of $\mathcal{O}_k$ for some suitable k.

A problem of great antiquity is that of characterizing and classifying the tilings of space. A more tractable problem is to ask what objects can be used as a single tile that will tile space by translation. In $\mathbb{R}^3$ these are known. There are five such objects and all are zonohedra. In higher dimensions, it is known (Venkov) that such objects are centrally symmetric polytopes with centrally symmetric facets. However, for $n > 3$ it was noted above that these need not be zonotopes. For results on tilings, see the article of that title by Schulte.[2]

C. Zonoids

Whereas the polytopes form a dense set among the convex bodies in $\mathbb{R}^n$, this is far from true of the zonotopes. The class of zonoids is defined as the closure of the set of zonotopes in the Hausdorff metric (in some fixed vector space). Thus, every zonoid is the limit of a sequence of zonotopes. Since the map from convex body to support function is continuous, it follows that the support function of a zonoid is the limit of a sequence of functions of the form $\sum_i |f(x_i)|$. Such limits can be written as an integral. Thus, a convex body Z is a zonoid if and only if its support function has the form:

$$h_Z(f) = \int |f(x)| d\rho(x)$$

where the integral is over the surface of the unit ball and ρ is an even measure on that surface.

The set of zonoids is also the range of the projection body operator Π when its domain is the set of all convex bodies. This operator is one to one on the set of centrally symmetric bodies. An important context in which zonoids arise is that of vector measures. The range of any non-atomic vector measure is a zonoid.

In finite-dimensional normed spaces, there are several ways in which area can be defined. For that due to Holmes and Thompson the solution to the isoperimetric problem is always a zonoid. This fact makes that definition especially suitable for integral geometry. An interesting open question is whether there are non-Euclidean normed spaces for which the solution to the isoperimetric problem is the unit ball. For further results about zonoids, consult the article by Goodey and Weil.[2]

D. Ellipsoids

An ellipsoid in $\mathbb{R}^n$ was defined in Section III to be an affine image of the unit ball. Here, we restrict attention to ellipsoids with the origin as center and as an interior point; that is, the image of the unit ball under an invertible linear map T. Alternatively, if S is a positive definite symmetric matrix, the ellipsoid E_S is defined (using the inner product) by:

$$E_S := \{x : \langle Sx, x \rangle \leq 1\}.$$

Thus, the unit ball B is E_I where I is the identity matrix. The two approaches are connected by the fact that if T is an invertible matrix then $S := T^{-1*}T^{-1}$ is positive definite symmetric and $T(B) = T(E_I) = E_S$. Conversely, a positive definite matrix S has a positive definite square root $S^{1/2}$ and

$$E_S = S^{-1/2}(E_I) = S^{-1/2}(B).$$

Many properties of ellipsoids may be regarded as facts about positive definite symmetric matrices and conversely. If an ellipsoid E is given in the form $E = T(B)$, then $V_n(E) = \det T \, \epsilon_n$. Similarly, $V_n(E_S) = \epsilon_n / \sqrt{\det S}$. Useful here is the fact that the determinant of a matrix is the product of its eigenvalues.

The gauge function of a centrally symmetric convex body with 0 as an interior point is a norm. Ellipsoids are precisely those bodies whose corresponding norms come from an inner product—a Euclidean norm and (in infinite dimensions) a Hilbert space norm. Therefore, theorems that characterize ellipsoids among convex bodies also characterize inner product spaces among normed spaces (and conversely). Theorems of this type are extremely numerous in the literature. Here we give a small sample.

The most well-known characterization of inner product spaces is via the parallelogram law and is due to Jordan and von Neumann: a norm $\| . \|$ comes from an inner product if and only if:

$$\|x + y\|^2 + \|x - y\|^2 = \|x\|^2 + \|y\|^2$$

for all vectors x and y. Thus, a convex body is an ellipsoid if and only if its gauge function satisfies the above equation. Another characterization of inner product spaces is that of Kakutani and is closely related to the Blaschke result below.

Theorem (Kakutani). If $(X, \| . \|)$ is a normed space of dimension $n \geq 3$ and if for every two-dimensional subspace H of X there is a projection of X onto H which does not increase the norm of any vector, then the norm comes from an inner product.

Ellipsoids are characterized by the equality case in a number of the inequalities in Section IX. A few more geometric results are the following.

Theorem (Bertrand, Brunn). Let K be a convex body. For every vector x, consider the chords of K that are parallel to x. The midpoints of these chords all lie on a hyperplane if and only if K is an ellipsoid with center on that hyperplane.

Definition. If K is a convex body, if $x \neq 0$ is a vector, and if L_x is the line spanned by x, then the shadow boundary of K in the direction x is the set $S_x := \partial(K + L_x) \cap \partial K$.

The image here is of a beam of light shining on K in the direction x. Part of the body is illuminated and part is in shadow. The set S_x is the "edge" of the shadow. If K has some flat parts to its boundary, the edge may be broad.

Theorem (Blaschke). If $n \geq 3$ and if K is a convex body, then the shadow boundary S_x lies in a hyperplane for all x if and only if K is an ellipsoid.

Theorem (Shaĭdenko, Goodey). If $\lambda \geq 3$ and if K and L are two convex bodies such that for all translates $L + x$ of L (except if $L + x = K$) the set $\partial K \cap \partial(L + x)$ is contained in a hyperplane, then K and L are homothetic ellipsoids.

Lastly, we recall the Löwner-John ellipsoids (see Section VI). For every convex body K, there is a *unique* ellipsoid of maximal volume inscribed to K and a *unique* ellipsoid of minimal volume circumscribed to K.

E. Simplices

In many respects, among convex bodies simplices stand at the opposite extreme from ellipsoids. Whereas ellipsoids are among the most symmetric of bodies, simplices are among the most asymmetric. In some of the inequalities in Section IX, ellipsoids are precisely at one extreme and simplices at the other.

The place where many students and users of mathematics meet the word "simplex" is in the terms *simplex method* or *simplex algorithm*; see Webster.[5]

Choquet observed that if K is a convex body such that its intersection with any homothetic image is again a homothetic image then K is a simplex. He used this idea to define a simplex in infinite-dimensional spaces. Moreover, if K is of dimension $(n-1)$ and is contained in a hyperplane H_f^1 in $\mathbb{R}^n$ (*not* a hyperplane through 0), then one can construct a cone $C_K := \{\lambda x : x \in K, \ \lambda \geq 0\}$ which induces an order relation on $\mathbb{R}^n$ by:

$$x \leq y \text{ if and only if } y - x \in C_K.$$

This order relation makes $\mathbb{R}^n$ into a lattice if and only if K is a simplex. This is the starting point of a far-reaching theory of infinite-dimensional spaces initiated by Choquet.

Finally, we give a characterization due to Martini that nicely connects three of the operations of Section V.

Theorem (Martini). A polytope is a simplex if and only if $D(K) = \lambda \Pi(K)^\circ$ for some $\lambda > 0$.

Other characterizations of (and information about) simplices can be found in the article by Heil and Martini in the *Handbook of Convex Geometry*.[2]

F. Sets of Constant Width

The width of a convex set in a direction f, $w_K(f)$, was defined in Section V and sets of constant width were also introduced there. In Section V, it was shown that K is of constant width if and only if $D(K)$ is a ball. If K_1 and K_2 are each of constant width, then so is $K_1 + K_2$.

For each point x in ∂K, if we define the diameter of K at x to be $d_K(x) := \sup\{\|x - y\| : y \in K\}$, then K is of constant width if and only if $d_K(x)$ is constant.

If the convex body K has diameter d, we may define the spherical hull of K, $\mathrm{sh}(K)$, to be the intersection of all balls of radius d with center in K:

$$\mathrm{sh}(K) := \bigcap\{B[x, d] : x \in K\}.$$

Then $K \subseteq \mathrm{sh}(K)$ and equality occurs if and only if K is of constant width. Sets of constant width are precisely those sets of diameter d such that if $y \notin K$ then $\mathrm{diam}(K \cup \{y\}) > d$. Moreover, every set K of diameter d is contained in a set of constant width d.

The inradius of a convex body K is the maximal radius of a ball contained in K, and the circumradius is the min-

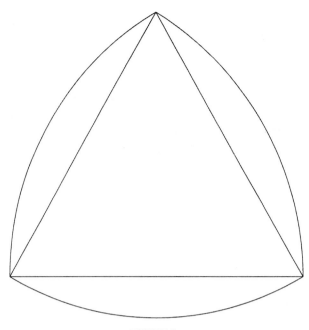

FIGURE 8

imal radius of a ball that contains K. If K has inradius r and circumradius R and is of constant width w, then there is a unique inscribed ball of radius r and a unique circumscribed ball of radius R, these balls are concentric, and $r + R = w$.

There is an inequality relating the volume and surface area of bodies of constant width w in R^n:

$$V_n(K)/V_{n-1}(\partial K) \leq w/2n.$$

Two-dimensional sets of constant width have been much studied. A Reuleaux triangle of width w is the intersection of three discs of radius w centered at the vertices of an equilateral triangle (see Fig. 8). Reuleaux polygons with any odd number of sides may be constructed similarly. Discs $B[x, w/2]$ are the sets of constant width w whose area is maximal.

Theorem (Blaschke, Lebesgue). If K is a two-dimensional set of constant width w, then:

$$V_2(K) \geq \left(\pi - \sqrt{3}\right)w^2/2$$

with equality if and only if K is a Reuleaux triangle.

The corresponding problem in higher dimensions is unsolved. Constructions analogous to the Reuleaux polygons can be made in higher dimensions but not so easily.

Cauchy's formula for surface area (Section IX) yields an elegant result for bodies of constant width in $\mathbb{R}^2$. The length of the perimeter of such a body is πw where w is the width.

SEE ALSO THE FOLLOWING ARTICLES

FUNCTIONAL ANALYSIS • LINEAR OPTIMIZATION • SET THEORY • TOPOLOGY, GENERAL

BIBLIOGRAPHY

Eggleston, H. G. (1958). "Cambridge Tracts in Mathematics and Physics," Vol. 47 "Convexity," Cambridge University Press, Cambridge, U.K.

Gruber, P. M., and Wills, J. M. (eds.) (1993). "Handbook of Convex Geometry" (two vols.), North-Holland, Amsterdam.

Minkowski, H. (1911). "*Gesammelte Abhandlungen*," Vol. II, Teubner, Leipzig.

Schneider, R. (1993). "Encyclopedia of Math. and Its Applications," Vol. 44, "Convex bodies: the Brunn-Minkowski Theory," Cambridge University Press, Cambridge, U.K.

Webster, R. J. (1994). "Convexity," Oxford University Press, London.

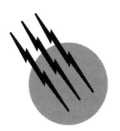

Coordination Compounds

R. D. Gillard

University of Wales, Cardiff, Wales

GLOSSARY

Brönsted base Proton acceptor.

Chelation Binding of a single ligand to a single metal ion through two (or more) Brönsted donor atoms, making ring structures.

Chrysotherapy Treatment of disease with gold compounds.

Complex Species formed by combination between a metal ion and a ligand.

Coordination Formation of a link between a metal ion and a ligand by the donation of electrons from the ligand to the metal ion.

Coordination number Number of ligand atoms directly attached to a central metal ion in a complex. These directly attached ligands make up the first coordination sphere.

Enantiomers Isomers that are related as an object and its mirror image.

Isoelectric Pertaining to species (whether molecules, ions, or radicals) that have the same total number of electrons occupying the same orbitals around the same number of nuclei. For example, AlF_6^{3-}, SiF_6^{2-}, PF_6^-, and SF$_6$ are isoelectronic with each other, as are $[AuCl_2]^-$, $HgCl_2$, and $[TlCl_2]^+$.

Isomerism Existence of more than one distinct substance with the same composition, for example, ammonium cyanate, NH_4NCO, and urea, $(H_2N)_2CO$, both of which are CH_4N_2O in stoichiometry, or potassium fulminate, $K(CNO)$, and potassium cyanate, $K(NCO)$. The distinct substances are called isomers.

Labile Attaining equilibrium rapidly.

Ligand Any moiety bonded to a metal ion in a complex; often denoted by L in formulas and equations.

Resolution Separation of enantiomers from one another.

Stability constant Thermodynamic equilibrium constant measuring the ease of formation of a complex ion from its constituents [the metal ion and the ligand(s)].

CHEMISTRY DEALS WITH changes (reactions) of substances such as the conversion of organic matter (e.g., sugar) with dioxygen to carbon dioxide and water:

$$C_6H_{12}O_6 + 6O_2 \rightarrow 6CO_2 + 6H_2O$$

Metal complexes are formed from Brönsted bases (substances with many electrons) and Lewis acids (usually metal cations, which are positively charged and therefore interact favorably with electron donors). A typical metal complex compound is $[Cu(NH_3)_4](SO_4) \cdot H_2O$, which is made by allowing to crystallize the dark blue solution obtained from adding an excess of strong ammonia to aqueous copper sulfate.

The interaction of ligand with metal ion is often called *coordination*, and the solid compounds that contain metal ions complexed by ligands are called coordination compounds. Typical examples are $K_2[PtCl_4]$, $K_2[PtCl_6]$, $[Pt(NH_3)_4]Cl_2$, and $[Pt(NH_3)_6]Cl_4$.

I. INTRODUCTION

Whenever any metal salt and any Brönsted base (an anion or other molecule with electronegative atoms, such as nitrogen or oxygen, that has the capacity to donate electrons) come into contact, coordination is likely to occur to give a complex compound. For example, when solid nickel chloride (yellow) reacts with a stream of ammonia gas, it is converted to purple hexamminenickel(II) chloride:

$$NiCl_2 \text{ (solid)} + 6NH_3 \text{ (g)} \rightarrow [Ni(NH_3)_6]Cl_2 \text{ (solid)}$$

Similarly, on the addition of strong ammonia, a base (in water, $NH_3 + OH_2 \rightleftharpoons NH_4^+ + OH^-$; all four species are present), little by little to a blue aqueous solution of copper(II) sulfate, the final soluble species is the complex tetraammine–copper(II) ion; its stability is clear from the fact that it is formed by the dissolution [Eq. (2)] of the intermediate solid [Eq. (1)]. This solid is basic copper(II) sulfate, known also as several minerals (brocchantite, langite, wroewulfite) in oxidized sulfide ore zones. The word *basic* in the name simply reflects the presence of the hydroxide ion. OH^-, the basic constituent of water.

$$4Cu^{2+} \text{ (aq)} + SO_4^{2-} + 6OH^- \rightarrow \{Cu_4(SO_4)(OH)_6\} \tag{1}$$

$$\{Cu_4(SO_4)(OH_6)\} + 16NH_3 \rightarrow 4[Cu(NH_3)_4]^{2+}$$
$$+ SO_4^{2-} + 6OH^- \tag{2}$$

The dark blue solution of the tetraammine–copper(II) species in strong aqueous ammonia (Schweitzer's solution) will dissolve cellulose. On acidification, the ammonia is neutralized (protonated forming ammonium ion, NH_4^-) and the cellulose is reformed (at one time, through spinnerets, to make a commercial fiber).

When silver halide emulsions are used as photographic films, development of the parts exposed to light to give black silver is followed by "fixing," which is simply metal complexing. The unchanged (nonimaged) silver halide (usually chloride and bromide) must be removed before

the negative can be handled in daylight. The ligand used is thiosulfate in the form of an aqueous (and therefore ionized) solution of its sodium salt $Na_2S_2O_3 \cdot 5H_2O$, so-called photographer's hypo. The silver halide dissolves:

$$AgX + 2(S_2O_3)^{2-} \rightarrow [Ag(S_2O_3)_2]^{3-} + X^- \tag{3}$$

The sodium salt of the anion, bisthiosulfatoargentate(I), can be crystallized and was said by its discover. Herschel, to have a sweet taste. Oddly, the corresponding compound of the heaviest of the "coinage metals" (copper, silver, gold). $Na_3[Au(S_2O_3)_2] \cdot 2H_2O$ (**1**), also has biological properties, being used under the name Sanochrysine in treating rheumatoid arthritis by chrysotherapy.

1

The gold ion is linearly bonded by two sulfur atoms.

Leather is a complicated material but contains much protein; it is always a good ligand, because it contains peptide linkages, from which either oxygen or nitrogen can donate electrons to a metal ion, forming a coordinate bond, and in which the many functional groups of the side chains of the amino acids may also interact with metal ions (as shown by the atoms underlined in structure **2**).

2

This extra element of cross-linking of the polymer (polypeptide) chains by metal ions—in practice, chiefly chromium(III) giving "chrome-leather"—underlies the utility of leather tanning. It is the formatin of coordinated chromium(III) that gives the tan.

Metal ions coordinated by ligands are common, giving rise to effects of striking beauty and great importance. The changes in the color of blood on oxygenation arise because of a change in one ligand on the iron ion in the coordination compound hemoglobin. Many examples of chemotherapy depend on the formation of coordinated metal ions, as in the removal of the excess of copper from patients with Wilson's disease (hepatolenticular degeneration) using penicillamine (**3**) or triethylenetetramine (**4**). The latter is systematically named 1,4,7,11-tetraazaundecane.

3

4

II. FORMATION

A. Equilibria between Cations and Donors in Water

When a donor molecule (a Brönsted base, i.e., a proton acceptor) forms a bond by donating electron density to a positive center (a Lewis acid, i.e., an electron acceptor), coordination is said to occur, as shown in Eq. (4).

$$AlCl_3 + Cl^- \rightarrow [AlCl_4]^- \tag{4}$$

More commonly, coordination occurs in solution (often aqueous), to give overall processes of metal complexing such as

$$Ni^{2+}\,(aq) + 6NH_3 \rightarrow Ni(NH_3)_6^{2+}\,(aq) \tag{5}$$

In fact, the six water ligands (W) are being replaced, one by one, stepwise, by six ammonia ligands:

100 %

5

The replacement of one water in **5** by ammonia (A) gives the monosubstituted complex **6**:

100 %

6

The position marked with an asterisk across from A in **6** (*trans* to A) is unique. The replacement of a second water in **6** by ammonia gives the disubstituted complex; there are in **6** four positions next (*cis*) to A, which gives **7**, and only one across from A (*trans*), which gives **8**:

(cis), 80 %   (trans), 20 %
7   **8**

If a third W is replaced in **7** or **8** by A, the *trans* (**8**, 20% of the whole) can give only **9**, but the 80% that was *cis* (**7**), gives 40% *mer* (**9**) and 40% *fac* (**10**) for a final result of 60% *mer*-**9** and 40% *fac*-**10**.

(mer), 60 %   (fac), 40 %
9   **10**

A fourth replacement gives, from the 40% *fac* (**10**), only *cis*-A_4W_2 (since all three Ws are equivalent), but from *mer*, the position marked with an asterisk is unique, giving *trans*-W_2A_4 for one-third of the replacements into mer (A_3W_3) (i.e., 20% of the total, since *mer* was 60% of the whole). Thus, we have 20% *trans*-$[MA_4W_2]$ and 80% *cis*-$[MA_4W_2]$. The fifth substitution, of course, gives MA_5W.

There is an obvious symmetry to this series of replacements: We may start from MW_6 and replace W by A, or vice versa [Eq. (6)]:

$$MA_6 + W \rightleftharpoons MA_5W_1 \rightleftharpoons MA_4W_2 \rightleftharpoons MA_3W_3$$
$$\rightleftharpoons MA_2W_4 \rightleftharpoons MA_1W_5 \rightleftharpoons MW_6 + A \tag{6}$$

Notice that six-coordinate complex ions $[MA_nB_{6-n}]$, where $2 \le n \le 4$, may exist in two forms. In the example above, A is NH_3 and B is H_2O, but the occurrence of two forms (**7** as against **8**, or **9** as against **10**) differing in the arrangement of ligands A and B about the center is quite general.

B. Stability Constants and Relationships among Them

The general form (**7**) of the overall stability constant for the coordination

$$pM^{n+} + qL + rH^+ \rightleftharpoons [M_pL_qH_r] \qquad (7)$$

is written

$$\beta_{pqr} = [M_pL_qH_r]/[M]^p[L]^q[H^+]^r \qquad (8)$$

For example,

$$Ag^{2+}(aq) + 4C_5H_5N(aq) \rightleftharpoons [Ag(C_5H_5N)_4]^{2+}(aq) \quad (9)$$

$$\beta_{140} = [Ag(C_5H_5N)_4]/[Ag^{2+}][C_5H_5N]^4 \qquad (10)$$

$$\log \beta_{140} = 25 \qquad (11)$$

These *overall* stability constants describe only the equilibrium of a given stochiometric *composition* of formation of a complex and do not, for example, distinguish between the forms of [MW$_4$A$_2$] above [the *trans* (**7**) and the *cis* (**8**)]. These are macroconstants and are a weighted sum of the individual microconstants. For example, the macroconstant β_{120} for Eq. (12) is made up of the two microconstants $\beta_{120(cis)}$ [Eq. (13)] and $\beta_{120(trans)}$ [Eq. (14)]:

$$Ni^{2+}(aq) + 2NH_3 \rightleftharpoons [Ni(NH_3)_2(H_2O)_4]^{2+} \qquad (12)$$

$$Ni^{2+}(aq) + 2NH_3 \rightleftharpoons cis\text{-}[Ni(NH_3)_2(H_2O)_4]^{2+} \qquad (13)$$

$$Ni^{2+}(aq) + 2NH_3 \rightleftharpoons trans\text{-}[Ni(NH_3)_2(H_2O)_4]^{2+} \quad (14)$$

The constant describing the equilibrium [Eq. (15)] between the forms with the same stoichiometry is simply Eq. (16).

$$cis\text{-}[M(NH_3)_2(H_O)_4]^{2+} \rightleftharpoons trans\text{-}[M(NH_3)_2(H_2O)_4]^{2-}$$
$$\qquad (15)$$

$$K_{isom} = [trans]/[cis] = \beta_{120(trans)}\beta_{120(cis)} \qquad (16)$$

Reverting to the ammine system above, if everything else were unchanged (if A and W had an equal chance of being attached to M), then the *statistical* proportions shown would apply. Furthermore, if the probability were the same of attaching an ammonia to the metal and removing it from the metal, then we would expect the probability of complexing to be 36 times greater for the first substitution of water by ammonia [Eq. (17)]:

$$MW_6 + NH_3 \rightleftharpoons M(NH_3)W_5 + W \qquad (17)$$

(six equivalent ways forward, only one back) than for replacing the final water ligand by a sixth ammonia [Eq. (18)] (only one possible place where the water can be replaced by ammonia and six ways back where any of the six ammonia ligands can be replaced by water). On these grounds, the *stepwise* equilibrium constants for the

formation of M(NH$_3$)$_6$ decrease (electrical charge on ions is omitted):

$$MW(NH_3)_5 + NH_3 \rightleftharpoons M(NH_3)_6 + W \qquad (18)$$

$$M + L \rightleftharpoons ML \qquad (19)$$

$$K_{110} = [ML]/[M][L] \qquad (20)$$

$$ML + L \rightleftharpoons ML_2 \qquad (21)$$

$$K_{120} = [ML_2]/[ML][L] \qquad (22)$$

$$ML_2 + L \rightleftharpoons ML_3 \qquad (23)$$

$$K_{130} = [ML_3]/[ML_2][L] \qquad (24)$$

So $K_{110} > K_{120} > K_{130}$ on these statistical grounds.

Furthermore, if we consider a similar series [Eqs. (25)–(28)] in which the ligand is an anion, carrying a negative charge, we can see a similar trend due to the effect of charge neutralization (here the ionic charges *are* printed):

$$M^{n+} + L^- \rightleftharpoons (ML)^{(n-1)+} \qquad (25)$$

$$K_{110} = \{[ML]^{(n-1)+}\}/\{[M]^{n+}\}[L^-] \qquad (26)$$

$$(ML)^{(n-1)+} + L^- \rightleftharpoons (ML_2)^{(n-2)+} \qquad (27)$$

$$K_{120} = \{[ML_2]^{(n-2)+}\}/\{[ML]^{(n-1)+}\}[L^-]$$
$$\qquad (28)$$

The first stepwise complexation, in Eq. (25), involves the neutralization, by mutual Coulombic attraction, of the full $n+$ charge of the metal ion by the single negative charge of L$^-$. However, for the second step, in Eq. (27), the Coulombic interaction is smaller, now being between the same single negative charge on L$^-$ and the less intensely charged (ML) species, with only $(n-1)$ charge. Typically, if M^{n-} were Al^{3-}, and L$^-$ were F$^-$, the attraction in Eq. (25) would be between a 3+ and a 1− pair, but that in Eq. (27) would be between the 2− of [AlF]$^{2-}$ and the 1− of the fluoride ligand.

In the final stages of stepwise coordination, the negatively charged anionic ligand is being added to an already negatively charged metal species. The Pauling electroneutrality principle reminds us that ligands will add readily only until the affinity for electrons of the central metal ion is essentially satisfied.

Thus, on the grounds both of statistics and of charge interaction, the order $K_{11} > K_{12} > K_{13} > \cdots K_{1n}$ is expected (and almost always observed). Table I lists some examples.

If there is a departure from this order or a sharp, distinct irregularity, something in the system has changed, often the coordination number of the metal ion. For example, with Cl$^-$ + Fe^{3+}, K_{11}, K_{12}, and K_{13} decrease steadily but K_{14} shows a sudden drop. This suggests that the octahedral

TABLE I Aqueous Stability Constants (log β_{1n0})[a]

M^{n+}	Ligand	β_{110}	β_{120}	β_{130}	β_{140}
Cu^{2+}	NH_3	4.2	7.7	10.6	12.7
Ni^{2+}	NH_3	2.8	5.04	6.77	7.97
Al^{3+}	F^-	6.1	11.1	14.9	17.7

[a] Note that the overall constants β_{1n0} are the product of stepwise constants $K_{110} \times K_{120} \times \cdots K_{1n0}$, e.g., $\beta_{140} = K_{110} \times K_{120} \times K_{130} \times K_{140}$.

six-coordinated $Fe(OH_2)_6^{3+}$ combines (stepwise) with one, then a second, then a third Cl^- without change in coordination number but that, at the next addition of chloride, a major change occurs, probably

$$[Fe(OH_2)_3(Cl)_3]^0 + Cl^- \rightleftharpoons [FeCl_4]^-$$

Note that no such discontinuity occurs with fluoride forming complexes with iron(III).

The coordination number of a given metal ion, such as Co^{2+} or Al^{3+} (which indicates the number of ligands attached to it), often varies from ligand to ligand. For example, whereas for aluminum, aqueous chloride forms $[AlCl_4]^-$ and the addition of another chloride ligand (a fifth and a sixth) is not easy, up to six fluoride ligands coordinate rather readily, forming AlF_6^{3-}.

C. Irving–Williams Series

The past discussion has dealt with variation of stability for the coordination compounds of a given metal ion. What about variation for a particular ligand with a range of metal ions? Again, the simple ideas of Coulombic interaction take us a long way. The greater the charge and the smaller, the size of a cation (Lewis acid), the better it will interact with an electron donor (Brönsted base). Thus, for a given ligand, say glycinate (gly-O),

$$K_{110} \text{ for } Na^+ < K_{110} \text{ for } Mg^{2+} < K_{110} \text{ for } Al^{3+}$$

Similarly, for the change in cationic size with constant charge, as in the lanthanides (where the "lanthanide contraction" means that the elements with higher atomic numbers give the smallest 3+ ions) where *only* the metal ion changes, we generally find the stability of a series of coordination compounds to *increase* from lanthanum (atomic number 57) to lutetium (atomic number 71).

The similar decrease in size through the transition series (where the ionic radii of the doubly charged ions M^{2+} decrease from V^{2+} to Zn^{2+}) leads to an increase in stability for coordinated compounds (Fig. 1). The order shown for β_{110} for the second half of the first transition series (Mn < Fe < Co < Ni ≤ Cu > Zn) is the Irving–Williams series.

The "double-humped" plot is famous; it summarizes many properties of analogous sets of coordination compounds. Notice that it is the number of d electrons that dictates this shape. The upper dashed plot is for the triply charged metal ions. Naturally, the interaction between M^{3+} and the electrons of a given ligand is better than that of M^{2+}, so the dashed curve of Fig. 1, representing stability constants for triply charged metal ions, like Fe^{3+}, lies above the full curve for doubly charged ions like Fe^{2+}.

D. Inverse Complexing

In the crystal lattice of many solid binary compounds of the type AgI, the metal ion (here Ag^+) will have several anions (here four I^-) surrounding it. This is akin to coordination of metal ion by iodide, giving the complex anions AgI_2^-, AgI_3^{2-}, AgI_4^{3-}, and so on. Conversely, each iodide in solid silver iodide will be surrounded by silver ions. In some cases, chiefly for large central *anions*, complex cations like IAg_2^+, IAg_3^{2+}, and $TeAg_6^{4+}$ exist in aqueous solutions. The formation of a cationic complex by the addition of several positive ions to a central negative ion is called inverse complexing (or sometimes metallocomplexing). Inverse complexes are a chief factor in the dissolution of silver halides in excess of soluble silver salts such as nitrate [Eq. (29)].

$$AgI + AgNO_3 \rightarrow [IAg_2]^+ + NO_3^- \qquad (29)$$

and central inverse complexed units such as $(OM_3)^{y+}$, $(NM_3)^{x+}$ are fairly common. These include $[OIr_3]^{8+}$, in Lecoq de Boisbaudran's compound, $Ir_3O(SO_4)_4$, named for the discoverer of the element gallium; $(OM_3)^{7+}$ in the basic chromium(III), iron(III), and ruthenium(III) acetates $[M_3O(CH_3COO)_6]^+$; and $(Hg_2N)^+$ in Millon's base.

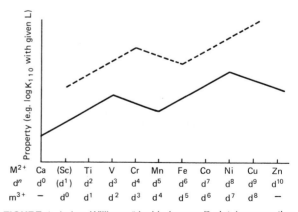

FIGURE 1 Irving–Williams "double-humped" plot (representing the change in properties of the doubly charged ions M^{2+}) for thermodynamic properties of ions of the first transition series. The dashed plot show the related changes for the triply charged M^{3+}.

The phenomena are closely akin to the remarkable dissolution by aqueous solutions of beryllium and zinc salts of their own water, insoluble oxides [Eq. (30)]:

$$2H_2O + MO + 2M^{2+} \rightleftharpoons M_3(OH)_3^{3+} + H^+$$

$$(M = Be, Zn) \quad (30)$$

Indeed, whenever the coordination number of a simple anion in a complex is greater than 1, there is an element of "inverse complexing," and interesting properties result. The following are examples:

1. The "oxo-bridged" species, where either O^{2-} or OH^- links two metal ions, as in $[Cl_5RuORuCl_5]^{4-}$; Durrant's anion $[(O_4C_2)_2Co(OH, OH)Co(C_2O_4)_2]^{4-}$; or the famous Werner's hexol $[Co\{(OH)_2Co(NH_3)_4\}_3]^{6+}$. Such bridging ligands carry the symbol μ.

2. Halide bridges. These are common, so that there are many species with two-coordinated halide ligands, as in Pt(II)–Cl–Pt(II), often in the same complex species as one-coordinated "terminal halides." The properties of the terminal and bridging halide ions are quite different.

E. Chelation

Both in nature and in synthetic chemistry, there are many cases of polydonor molecules, each containing several atoms that can coordinate to a metal ion. Often, these several atoms may be so distributed in space that they coordinate to the same metal ion. This is shown in Eq. (31):

11

$$(31)$$

The underlined nitrogen and oxygen atoms of the glycinate ion bind to the same metal ion (here Ni^{2+}), giving a new cyclic structure (**11**) with the five-membered "chelate" ring (NiNCCO). Such di- or polydonor ligands are called chelating. Cyclic molecules (carbocyclic, aliphatic as in cyclohexane; aromatic as in benzene; heterocyclic, aromatic as in pyridine; or alicyclic as in piperidine) are most common when five- or six-membered. The same is true for chelate cycles in coordination compounds. These are formed most readily when five- or six-membered (**12–14**; M indicates a metal ion).

12

13

14

The presence of chelate rings stabilizes a molecule, for example, in Eq. (32), where en = 1,2-diaminoethane:

$$[(H_2O)_4Ni(NH_3)_2]^{2+} + en \rightarrow [(H_2O)_4Ni(en)]^{2+} + 2NH_3$$

$$(32)$$

Omitting ionic charges and the four water molecules that remain attached to the nickel ion throughout, we obtain

$$K = [Ni(en)][NH_3]^2/[Ni(NH_3)_2][en] = 10^3 \quad (33)$$

Note that this is a macroconstant. Clearly, the relevant equilibrium constant is actually for joining *cis*-nitrogen ligands in pairs [Eq. (34); W indicates a water ligand].

$$(34)$$

This stability of chelated coordination is very important in chemical analysis. Two examples of chelated coordination are as follows:

1. The Tchugaev reaction [Eq. (35)] of 2,3-butanedionedioxime (dimethylglyoxime, abbreviated here as H₂DMG) with nickel ions, where the chelated product is a bright red precipitate.

Ni(HDMG)₂

The remarkable stability of this red solid stems from the chelation and the further factors of the short, strong intramolecular hydrogen bonds (O–H---O) and an out-of-plane intermolecular interaction.

2. Complexing, as in Eq. (36), of metal ions by 1,2-diaminoethane-N,N,N',N'-tetraacetate (**15**) from so-called ethylenediaminetetraacetic acid, H_4(EDTA), giving a highly chelated product (**16**).

15 **16**

$$\tag{36}$$

When the divalent ions, here M^{2+}, are of calcium or magnesium, as in "hard" water, they are sufficiently complexed ("sequestered") by the $EDTA^{4-}$ ligands to render the water "softer."

F. Oxidation States and Their Stability

Coordinating a metal ion (say, Fe^{2+}) to a ligand alters its stability. Coordinating the same ligand to a differently charged ion (e.g., Fe^{3+}) of the same metal (i.e., in a different oxidation state) alters the stability of that ion as well, but usually to a different degree. Such a situation can be analyzed by means of the Nernst equation [Eq. (37)],

$$E = E^0 - 2.303 \frac{RT}{nF} \log \frac{[M^{n+}]}{[M^{(n+1)+}]} \tag{37}$$

or at equilibrium,

$$E^0 = 2.303 \frac{RT}{nF} \log \frac{[Fe^{2+}]}{[Fe^{3+}]} \tag{38}$$

The concentrations $[Fe^{2+}]$ and $[Fe^{3+}]$ here are those of the free, uncomplexed metal ions (i.e., those in a solvent environment). Now, what happens if a coordinating ligand X is added? Both M^{n+} and $M^{(n+1)+}$ (e.g., Fe^{2+} and Fe^{3+}) are bound, but to differing degrees. This depresses the concentration of the two free (aquated) ions differentially. Such binding by the ligand is described by stability constants for the two oidation states, β_{110}^{II} and β_{110}^{III}, so that

$$E = E^0 - 2.303 \frac{RT}{nF} \log \frac{[Fe(II)X]}{\beta_{110}^{II}} \frac{\beta_{110}^{III}}{[Fe(III)X]} \tag{39}$$

If we add a ligand that will bind more tightly with the oxidized ion, $M^{(n+1)+}$ than with M^{n+}, then the potential E will shift strongly.

An extreme example is that, on the addition of cyanide salts (e.g., KCN) to a solution of a cobalt(II) salt in water, cobalt(II) becomes so greatly destabilized relative to its oxidized ion, cobalt(III) (whose cyano complexes are remarkably favored), that the potential for $Co^{3+} + e^- \rightarrow Co^{2+}$ (where $E^0 = +1.84$ V in water in the absence of coordinating agents) becomes -0.82 V. That is, the cobalt couple (extremely strongly oxidizing in water) in now so reducing that it will drive electrons onto protons in the water to give ($H^+ + e^- \rightarrow \frac{1}{2} H_2$) dihydrogen gas.

In a similar way, though silver ion in commonly a good oxidant (i.e., the half-cell $Ag^+ + e^- \rightarrow Ag$ is favored relative to $H^+ + e^- \rightarrow H_2$), in the presence of iodide ions—which diminish the value of $[Ag^+]$ by virtue of the gross insolubility of silver iodide, AgI—silver metal dissolves in hydriodic acid, HI, to give dihydrogen gas.

G. Stability Constants and pH

Most ligands are bases (having lone pairs of electrons), and many examples of coordination may be viewed as competition [e.g., Eq. (40)] for these lone pairs between solvated protons (acid–base equilibria) and other solvated cations:

17 **18**

$$\tag{40}$$

Clearly, the addition of metal ion (in the form of its salts) to an aqueous solution of the ligand, here

TABLE II Equilibrium Constants K for Coordination of Hydroxide[a]

Metal	Ligand	n	x	$\log K$
Pt	2,2-Bipyridyl	2	2	4.3
Pt	5,5'-DMB[b]	2	2	4.8
Pd	5,5'-DMB[b]	2	2	5.5
Pd	HDMG[c]	2	0	5.5

[a] Equivalent to $r = -1$ in β_{pqr}:

$$(ML_n)^{x+} + OH^- \rightarrow [(ML_n)OH]^{(x-1)+}$$

[b] DMB, Dimethyl-2,2'-bipyridyl.
[c] HDMG, "Dimethylglyoxime," 2,3-butanedionedioxime. Equation (35) shows the structure of the analogous $[Ni(DMG)_2]$.

8-hydroxyquinoline (**17**) or *H*-oxinate, displaces some protons, causing a fall in pH. Conversely, the addition of protonic acids to metal complexes such as **18** will reverse this formation and cause the dissociation of the coordinate bond. One common way of measuring stability constants is to set up such competitive equilibria as shown in Eq. (41):

$$2HL + M^{2+} \rightleftharpoons ML_2 + 2H^+ \tag{41}$$

For example,

$$2H_3NCH_2COO + Cu^{2+} \rightleftharpoons M(H_2NCH_2COO)_2 + 2H^+ \tag{42}$$

This is done in the presence of concentrations of metal ion from zero to levels comparable with those of ligands.

For aqueous equilibria, where the species M^{n-}, L^-, and OH^- are present, the utility of K_{pqr} is clear. Coordination compounds may contain protons (particularly on polydentate ligands). Values of r (0 in examples so far) represent the involvement of protons in complex formation. Of course, when r is negative, this may arise from loss of a proton from somewhere in the coordination species or, because $[H^+][OH^-] = K_w$, that is, $[OH^-] = K_w[H^+]^{-1}$, it may signify the gain of a hydroxide, as in the examples in Table II.

III. ELECTRONIC CONFIGURATIONS

There are some 70 metallic elements. All metal ions (Lewis acids) form coordination compounds. At present, the coordination compounds of the 27 transition metals are the most widely studied and applied, and this section refers to them. In the periodic table, at the onset of each of the transition series, the energies of the n s, n p, and $(n-1)$ d [or $(n-2)$ f if appropriate] orbitals are so close that they are made to interchange fairly readily. For example, the ground state of the barium atom (atomic

number 56, at the start of the third transition series) is $6s^2$, and the atom before it, cesium, has the $6s^1$ ground state. However, for cesium, the next electronic state lies not far from the ground state, and indeed on compression, the conductance of cesium changes sharply, as the d orbital is squeezed below the s orbital. Barium shows a similar transition with pressure, corresponding to a change in configuration from $6s^2$ to $6s^1 5d^1$. In view of the angled hybrids (s $\mp$ d) given by this configuration, the angular (bent) structures as monomers in the vapor phase of MX_2 for M = Ca, X = F; M = Sr, X = Cl, F; M = Ba, X = F, Cl, Br, I are examples of transition metal chemistry. In the same way, although the ground states of the *atoms* of vanadium and nickel are $4s^2 3d^3$ and $4s^2 3d^8$, the shrinkage on ionization (which can be regarded as equivalent to the effect of huge pressure) squeezes the 3d orbitals to lower energies than the 4s, so that the ground states of the ions are $3d^3$ and $3d^8$, respectively.

The concept that isoelectronic d^n configurations have related properties is most valuable. In particular, such properties as color, magnetism, and rates of chemical reaction, which depend rather directly on numbers of d-electrons, can be rationalized and predicted. For example, the metal ions whose chiral coordination ions have enough kinetic inertness to be separated (resolved) into long-lasting enantiomers most commonly have six d electrons (n d^6 configurations).

A. Splitting Diagrams for Octahedral Coordination

When six ligands surround a metal ion to give octahedral coordination, the situation for two of the five d orbitals of the metal is shown in Fig. 2. These two orbitals are representative of others like themselves.

1. The d_{yz} orbital is typical of those that "point between" the axes (defined by the six ligands): These three

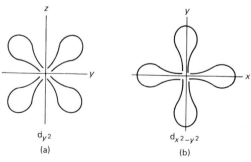

FIGURE 2 Differing spatial distribution of the orbitals (a) d_{yz} (between axes) and (b) $d_{x^2-y^2}$ (along axis).

[d_{yz} (Fig. 2a, **19**), d_{xz} (**20**), and d_{xy} (**21**)] have electron density as far from

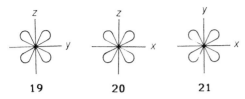

19	**20**	**21**

the ligand electron density as possible, and hence are nonbonding.

2. The $d_{x^2-y^2}$ orbital (Fig. 2b) typifies those that point directly at the ligands: These two ($d_{x^2-y^2}$, d_{z^2}) can overlap with ligand orbitals and form good bonds along the axes. From these two metal-centered d orbitals and the composite ligand–donor orbitals are formed two bonding and two antibonding combinations, as shown in Fig. 3, representing the overlap of the s, p (three of these), and d (five) orbitals of the central metal ion with the six equivalent donor orbitals (lone pairs) of the six equivalent ligands of the octahedral complex. Now there are six donor orbitals (two electrons each) on the six ligands, so those 12 electrons will be accommodated in the levels lying up to and including $d_{(\sigma)}$. That diagram is then appropriate to *any* octahedral system, whether or not central atom d electrons are involved. For example, the isoelectronic series AlF_6^{3-}, SiF_6^{2-}, PF_6^-, SF_6 (where the central atom makes bonds using the orbitals 3s, 3p, and 3d) have the electronic structure shown, and in the same way $InCl_6^{3-}$, $SnCl_6^{2-}$, and $SbCl_6^-$ (using the 5s, 5p, and 5d atomic orbitals of the metals) have the same diagram filled, up to and including $d_{(\sigma)}$.

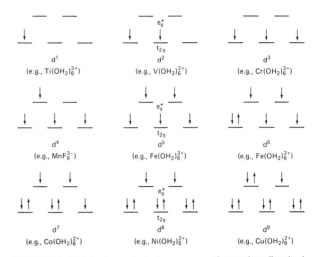

FIGURE 4 Distributions of electrons among the nonbonding (t_{2g}) set and the antibonding (e_g^*) set for ions with various d^n configurations, assuming small values for $\delta E (=e_g^* - t_{2g})$.

B. Spin States

For the transition elements, the metal ion has n d electrons, so that those over and above the bonding framework of the 12 originating from the ligands require that the t_{2g} and e_g sets be occupied. This is shown for the d^1–d^9 configurations in Fig. 4 (note that only the nonbonding t_{2g} orbitals, originally d_{xy}, d_{yz}, and d_{zx}, and the antibonding e_g pair are shown, that is, the five molecular orbitals in Fig. 3 enclosed in a solid-line box). This, set of distributions (high spin) assumes Hund's rule. What if Hund's rule is not obeyed? Then, instead of the maximum spin multiplicity (greatest number of unpaired, i.e., parallel, spins, "spin free") as drawn in Fig. 4, the fact that the energy of the e_g^* set is higher than that of t_{2g} will lead to full population of the lower-energy t_{2g} set, with consequent spin pairing ("spin-paired," "low spin"). This will happen where the energy separation ($e_g^* - t_{2g}$) is greatest, which is where the bonds (L → M *along* axes x, y, and z) are strongest, since $e_g^* - e_g$ will then be greatest (the nonbonding t_{2g} serves as an unmoving marker between the bonding and antibonding e_g orbitals).

Thus, for d^4–d^7, there are the possible "low-spin" electronic configurations of Fig. 5, quite different from those in Fig. 4, the "high-spin" ones. Table III summarizes the different numbers of unpaired electrons for high- and low-spin complex compounds.

C. Simple Magnetic Properties

One view of the interaction of matter with magnetic fields stems from the electrons acting as revolving charges, setting up a magnetic dipole. Electrons paired in orbitals cancel one another (the small residual effects are lumped

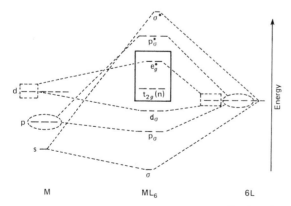

FIGURE 3 Bonding in an octahedral complex, ML_6^{n+}, viewed as a combination (overlap in space) of the s, the p (x, y and z), and the d (xy, yz, zx and $x^2 - y^2$, z^2) orbitals of metal (M) with one, three, and two, respectively, of the six donor orbitals on the six ligands (L). Shown here are the combinations of only those d orbitals that point at ligand lone pairs, that is, the $d_{x^2-y^2}$ and d_{z^2} with selected ligand orbitals. The remaining three of the five d orbitals of the metal (d_{xy}, d_{yz}, d_{zx}) are shown as being unaffected in energy when the complex forms, that is, nonbonding (t_{2g}).

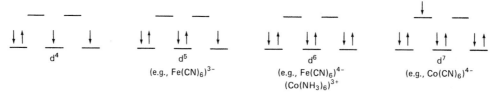

FIGURE 5 Distributions of electrons for ions with d^4–d^7 configurations, where $\Delta E\,(e_g^* - t_{2g})$ is large.

together under the name *diamagnetism*). Lone (unpaired) electrons give large magnetic moments $\mu > 0$ (paramagnetism) interacting strongly with an applied magnetic field,

$$\mu = [n(n + 2)]^{1/2} \tag{43}$$

where n is the number of unpaired electrons: μ is in units of Bohr magnetons (BM). Taking iron(III) as an example, the high-spin (weak-field; see Fig. 4) configuration $(t_{2g})^3(e_g)^2$ with five unpaired electrons as in ferric alum, where the coordination sphere of the iron is $[Fe(OH_2)_6]^{3+}$, gives $\mu = (5 \times 7)^{1/2} = 35^{1/2} = 5.9$ BM per iron. For the low-spin (strong-field; see Fig. 5) ferricyanide ion, where the iron(III) has the configuration $(t_{2g})^5(e_g)^0$, with only one unpaired electron, $\mu = 3^{1/2} = 1.73$ BM. A more systematic name for this spin-paired ion is hexacyanoferrate(III).

D. Colors

As with any other class of matter, if an energy gap in a complex compound matches the energy of an incident photon, absorption occurs. In many coordination compounds, the quantum required for excitation from the ground state is of visible light, so that these compounds are of many colors, often of great beauty. Two chief selection rules decide which of all the possible transitions occur:

TABLE III Occupancy of Orbitals in High- and Low-Spin Octahedral Coordination Compounds[a]

	d^1	d^2	d^3	d^4	d^5	d^6	d^7	d^8	d^9
High spin									
e_g^*	—	—	—	—	—	1	2	3	4
t_{2g}	1	2	3	4	5	5	5	5	5
Unpaired electrons (HS)	1	2	3	4	5	4	3	2	1
S	2	3	4	5	6	5	4	3	2
Low spin									
e_g	—	—	—	—	—	—	1	2	3
t_{2g}	1	2	3	4	5	6	6	6	6
Unpaired electrons (LS)	1	2	3	2	1	0	1	2	1
S	2	3	4	3	2	1	2	3	2
Difference (HS – LS)	0	0	0	2	4	4	2	0	0

[a] HS, High spin; LS, low spin, S denotes spin multiplicity.

1. s $\leftrightarrow$ p, p $\leftrightarrow$ d (etc.) atomic transitions are allowed, but p $\leftrightarrow$ p, d $\leftrightarrow$ d are not; in more general terms, $g \leftrightarrow u$ is allowed, but $g \leftrightarrow g$, $u \leftrightarrow u$ are forbidden.

2. The spin multiplicity does not change in an allowed transition. It is for this reason that all high-spin d^5 systems are at best weakly colored: There is only the one possible spin-parallel arrangement of five electrons in five d orbitals, so *any* transition must be to a state with different spin multiplicity. Coordination compounds of Mn^{2+} ($3d^5$) are usually in consequence very pale. Notice that this doubly forbidden character is true for any octahedral six-coodinated *high-spin* ion with five d-electrons. Examples include $[Fe(OH_2)_6]^{3+}$ in alums, which is a weak field environment, but not, of course, $[Fe(CN)_6]^{3-}$, which has its five d-electrons paired.

From the first rule (sometimes called the Laporte rule), any d–d transition in a centrosymmetric molecule is weak. Compared with dyestuffs (often with molar extinction coefficient $\varepsilon \sim 50,000$), complex compounds are poor absorbers of photons.

So far, only the electronic configurations for regular octahedral coordination have been given, and color has been described in terms of excitation of an electron by a quantum of visible light from the ground sate to an upper state. The size of the gap E in energy separating the ground and upper states controls the color ($E = h\nu$). The larger the gap, the greater the energy of the photon required to promote an electron across it. The gap depends on the strength of the M $\leftarrow$ L bonds along the axes (see Fig. 3).

In general, descending a triad of transition elements, like cobalt, rhodium, iridium [where the M^{3+} ions have the configurations $(3d)^6$, $(4d)^6$, $(5d)^6$, respectively], the energy gap ΔE in analogous species, say $M(NH_3)_6^{3+}$, increases, as in Table IV. This means, of course, that the compounds become less obviously colored; even the lowest-energy spin-allowed d–d transitions ($5d \rightarrow 5d$) for molecules containing third-row elements go into the ultraviolet.

Taking one metal ion, variation of color arises from the same cause, the variation of the energy gap. For the electronic configuration d^3 in the coordination complexes of chromium(III), for the lowest-energy transition arising

TABLE IV Energy Gaps in Analogous Species Containing Cobalt, Rhodium, and Iridium

Compound	Configuration	Transition	Energy (kK)[a]	Color
$Co(NH_3)_6^{3+}$	$(3d)^6$	$(t_{2g})^6 \rightarrow (t_{2g})^5(e_g)^1$	21.3	Orange-yellow
$Rh(NH_3)_6^{3+}$	$(4d)^6$	$(t_{2g})^6 \rightarrow (t_{2g})^5(e_g)^1$	33.3	White
$Ir(NH_3)_6^{3+}$	$(5d)^6$	$(t_{2g})^6 \rightarrow (t_{2g})^5(e_g)^1$	40.0	White
$trans$-$[Co(py)_4Cl_2]^{+}$ [b]	$(3d)^6$	$(t_{2g})^6 \rightarrow (t_{2g})^5(e_g)^1$	16.0	Green
$trans$-$[Rh(py)_4Cl_2]^{+}$ [b]	$(4d)^6$	$(t_{2g})^6 \rightarrow (t_{2g})^5(e_g)^1$	24.5	Yellow
$trans$-$[Ir(py)_4Cl_2]^{+}$ [b]	$(5d)^6$	—	—	White

[a] 1 kK (kilokayser) = 1000 cm^{-1}.

[b] All are the *trans*-dichloro isomers; py = pyridine, C_5H_5N.

from $(t_{2g})^3 \rightarrow (t_{2g})^2(e_g)^1$, the result of increasing bond strength (i.e., stability) is to increase the ligand field splitting ΔE (Table V).

The order of increasing splitting of the d levels by the ligands is the spectrochemical series: $F^- < OH_2 < NH_3 < NO_2^- < CN^-$. The last entry in Table V shows an example of the baricenter (center of gravity) rule. If the frequency of a particular transition in an environment (MA_6) is ν_1 and that for MB_6 is ν_2, then for $[MA_nB_{6-n}]$ it is the weighted average [Eq. (44)]. Such mixed coordination

$$(n\nu_2 + (6-n)\nu_2)/6 \qquad (44)$$

spheres, where there is more than one type of molecule acting as ligand, are very common. The effects of substituting ammonia by chloride for cobalt(III) $[3d^6]$ and chromium(III) $[(3d)^3]$ are shown in Table VI.

E. Photochemistry

The photochemistry (chemical reactions of molecules in excited electronic states, made by irradiating with light of the appropriate energy to promote them from their ground states) of complex compounds is not as useful as might be expected. In general, the most effective reactions are

brought about not by d–d excitation but by other light absorptions, such as charge transfer. For example, on the irradiation of trisoxalatoferrate(III) ion with UV light, the process shown in Eq. (45) takes place through the photochemical excitation shown in Eq. (46).

$$[Fe(C_2O_4)_3]^{3-} \rightarrow [Fe(C_2O_4)_2]^{2-} + CO + CO_2^- \quad (45)$$

$$\left[Fe(III) \leftarrow (C_2O_4)^{2-}\right] + h\nu \rightarrow \left[Fe(II) - (C_2O_4)^-\right]^+$$
$$(46)$$

F. Solvent Effects

Nearly all the basic notions of the chemistry of complex ions are derived from aqueous systems, and striking departures often occur when other solvents are used. For example, the coordination of iron(II) by diimine ligands changes its complexion as follows. In water $\beta_{110} > \beta_{120} \ll \beta_{130}$. The tris species, for example, $[Fe(phen)_3]^{2+}$, in water or isolated from water as salts with chloride and so on is *diamagnetic* (d^6 spin-paired). In *water*, where anions such as chloride are well solvated, paramagnetic solids such as $[Fe(phen)_2Cl_2]$ dismute, as in Eq. (47).

$$6H_2O + 3[Fe(phen)_2Cl_2] \rightarrow 2[Fe(phen)_3]^{2+}$$
$$+ Fe(OH_2)_6^{2+} + 6Cl^- \quad (47)$$

In acetone and similar nonprotic solvents, the bis species is perfectly stable, and it is now the tris species that is unstable [i.e., $\beta_{120} \gg \beta_{130}$, the opposite of water; see Eq. (48)]. To speak of the spin pairing of the tris species as ligand field stabilization is incorrect, since that stabilization is effective only in water. The remarkable, and often quoted, reversal of the stability constants for adding the second and third chelating ligands to iron(II) ions must actually stem rather from changes in solvation energies of the anions present.

$$[Fe(phen)_3]^{2+} + 2Cl^- \rightarrow [Fe(phen)_2Cl_2] + phen \quad (48)$$

TABLE V Colors of d^3 Coordination Complexes

Complex	λ^a	$n (=E/h)^b$	Color
$[CrF_6]^{3-}$	671	14.9	Green
$[Cr(OH_2)_6]^{3+}$	575	17.4	Violet
$[Cr(NH_3)_6]^{3+}$	464	21.55	Orange-yellow
$[Cr(CN)_6]^{3-}$	376	26.6	Pale yellow
$[Cr(NC_5H_5)_3Cl]_3^c$	629	15.9	Green
$[Cr(OH_2)_4Cl_2]^{+d}$	635	15.75	Green

[a] Wavelength of absorption maximum in nanometers.

[b] Frequency of absorption maximum in kilokaysers (1 kK = 1000 cm^{-1}).

[c] *Mer* isomer; NC_5H_5 is pyridine.

[d] *Trans* isomer.

TABLE VI   Mixed Coordination Spheres on Cobalt(III) and Chromium(III)

Compound	ν^a log ε				Color
	Cobalt		**Chromium**		
$[M(NH_3)_6]^{3+}$	20.7	1.8	21.0	1.6	Orange
$[M(NH_3)_5Cl]^{2+}$	19.4	1.75	19.5	1.6	Pink
trans-$[M(NH_3)_4Cl_2]^+$	16.1	1.4	16.5	1.4	Green
cis-$[M(NH_3)_4Cl_2]^+$	19.1	1.9	19.2	1.9	Green

^{a} In kilokaysers (1 kK = 1000 cm^{-1}).

G. Kinetic Properties of Coordination Compounds

Reactions of coordination compounds can be divided into several classes, depending on whether the oxidation state of any atom changes during the transformation of starting materials (factors) into products. No change gives a reaction such as Eq. (49).

$$[Co(NH_3)_5Cl]^{2+} + OH_2 \rightarrow [Co(NH_3)_5(OH_2)]^{3+} + Cl^- \quad (49)$$

Many formations and decompositions or other equilibrations of coordination compounds are extremely rapid. The half-life of a reaction such as the replacement ("substitution") by ammonia of water coordinated to nickel(II) ions is typically microseconds to milliseconds, and there is indeed a convenient distinction (due to Taube) for reactions in solution between kinetically labile and kinetically inert systems. On mixing 0.1 M aqueous solutions of the reagents, labile equilibria are fully established within 1 min, whereas inert systems take longer. Many of the ions of the heavier (second and third row) transition elements in several oxidation states (e.g., both Pt^{2+} and Pt^{4+}) are inert, as are many spin-paired d^6 ions (Fe^{2+}, Co^{3+}, Ni^{4+}) and chromium(III) in the first row. Kinetic lability in solution is the rule for coordination compounds containing main group metals. Reactions of solid coordination compounds (like most other solid-state changes) are usually slow. It is this kinetic inertness that has led to the isolation of so many metastable coordination compounds.

The decomposition (via substitution) of hexaamminecobalt(III) salts in acidic water [Eq. (50)] is thermodynamically very favorable; that is, K in Eq. (51) is very large, but the rate is extremely small. Solutions of such hexaamminecobalt(III) salts in dilute acid are unchanging for weeks.

$$[Co(NH_3)_6]^{3+} + 6H_3O^+ \rightleftharpoons [Co(OH_2)_6]^{3+} + 6NH_4^+ \quad (50)$$

$$K = [Co(OH_2)_6](NH_4^+)^6 / [Co(NH_3)_6](H_3O^+)^6 \quad (51)$$

Isolating the less thermodynamically stable of two interconvertible forms of the same composition (whether it be coordination compounds, allotropes of elements, or any other chemical composition) can be done only if the *rate* of reaching equilibrium is so slow as to render the conversion of the *metastable* isomer to the stable one very protracted. This is a form of Ostwald's law of metastable intermediates. Such rates are slow (minutes $< t_{1/2} <$ years) for the equilibrations of coordinated cobalt(III), chromium(III), a few (spin-paired) d^6 ferrous compounds, such as salts of the ferroin **22a**, tris-1,10-phenanthrolineiron(II) cation, and a few octahedral nickel(II) species with strong field ligands (3d^8, e.g., the tris-1,10-phenanthrolinenickel(II) ion **22b**).

a: M = Fe^{2+} and b: M = Ni^{2+}

Each N⌒N represents the ligand:

22

Oxidations and reductions are common and important, chiefly because coordination compounds of the transition metals may readily pass (often rapidly) from one oxidation state to another and because one-electron changes are common (whereas elsewhere in the periodic table, this would involve free-radical formation). For example, several named organic reagents utilize half-cells based on coordination compounds.

Typically, Decker's reagent is alkaline ferricyanide [hexacyanoferrate(III)], which may be used to oxidize the pseudobase **23** to the pyridone **24**, as in Eq. (52).

            23   **24**

Many other such reactions occur with changes by a combination of oxidation–reduction and substitution. When Tollens's reagent, $[Ag(NH_3)_2]^+$, oxidizes aldehydes, the product is silver metal. In Fehling's solution (which he originally called Barreswils's solution) the oxidant (for reducing sugars) is a complex compound of copper(II) with tartrate ions, and in Benedict's solution it is a complex with citrate ions. Sarett's reagent for oxidizing alcohols to aldehydes is a coordination compound of chromium(VI), $Cr(C_5H_5N)_2O_3$, where the chromium(VI) becomes reduced. There are many minor variations of this reaction.

Often, the compound to be oxidized is made a ligand, and the oxidation can then be intramolecular, as in the Dow phenol process [Eq. (53)], where the benzoate of basic copper(II) benzoate is oxidized to salicylate by hydroxide and then carbon dioxide is eliminated to yield phenol.

$$C_6H_5COOCuOH \rightarrow C_6H_5OH + CO_2 + Cu^0 \quad (53)$$

A degradation of coordinated salicylate [25, Eq. (54)] is also typical. This chelating contraction [25 (six-membered ring) → 26 (five-membered ring)] occurs readily in acid

$$(54)$$

permanganate. Here is another side to the coin: The metal ion, in this case cobalt(III), somehow prevents the oxalate ion bound to it in 26 from undergoing its normal ("high school") oxidation by permanganate. Such "shielding" effects (loss of normal reactions) on coordination are well known.

Such modifications of the chemistry of ligands attached to metal ions are becoming increasingly important. Not all are oxidation or reduction.

The high-field, spin-paired nitrosopentacyanoferrate ion $[Fe(CN)_5(NO)]^{2-}$, often used as an aqueous solution of its salt—so-called sodium nitroprusside, $Na_2[Fe(CN)_5NO] \cdot 2H_2O$—undergoes many reactions of analytical importance, chiefly as qualitative tests, in which the nitroso ligand becomes modified, usually without detaching from the iron. Examples are the Gmelin test for sulfur in organic matter (Lassaigne sodium fusion to give sulfide, which gives a strong purple color with the FeNO unit) and the Bodlander reaction with sulfite to give a bright red color.

Evens so simple a reaction as acid–base equilibrium in the ligands is strongly modified by metal ions. In Eq. (55),

$$[L_5M(OH_2)]^{n+} \rightleftharpoons [L_5M(OH)]^{(n-1)+} + H^+ \quad (55)$$

the acid dissociation to form proton and the conjugate base may be strong (e.g., for $L = H_2O$, $M = Fe$, $n = 3$,

$pK_a = 2.7$, comparable to that of monochloracetic acid!). Even the acid dissociation of ammonia to give its conjugate base is said to become perceptible when it is attached to a highly charged metal ion, as in Eq. (56).

$$[Pt(NH_3)_6]^{4+} (aq) \rightleftharpoons [Pt(NH_3)_5(NH_2)]^{3+} + H^+ (aq)$$
$$(56)$$

Certainly such proton transfers have extremely large rate constants.

In water, nearly all substitutions into a coordination sphere (i.e., of one ligand for another) have a very simple rate equation:

$$Rate = k[complex] \quad (57)$$

That is true whether the rate constant k is large (labile) or small (inert), but tells us nothing about the detailed mechanism.

There are a few situations in which the rate equation is a little more interesting than Eq. (57); for example:

1. Substitution at platinum(II) $[(5d)^8]$; four-coordinated square species. Here, often,

$$Rate = k[complex][incoming\ nucleophile] \quad (58)$$

2. "Base hydrolysis" (substitution by hydroxide ion) of coordination compounds of cobalt(III) with ligands containing an N–H group. The unusual rate equation is

$$Rate = k[complex][OH^-] \quad (59)$$

This is commonly thought to imply the presence of a reactive conjugate base, typically via Eqs. (60)–(63):

$$[Co(NH_3)_5Cl]^{2+} + OH^-$$
$$\rightarrow [Co(NH_3)_4(NH_2)Cl]^+ + H_2O \quad (60)$$
$$[Co(NH_3)_4(NH_2)Cl]^+ \rightarrow [Co(NH_3)_4(NH_2)]^{2+} + Cl^-$$
$$(61)$$
$$[Co(NH_3)_4(NH_2)]^{2+} \rightarrow [Co(NH_3)_4(NH_2)(OH_2)]^{2+}$$
$$(62)$$
$$[Co(NH_3)_4(NH_2)(OH_2)]^{2+} + H_2O$$
$$\rightarrow [Co(NH_3)_5(OH_2)]^{3+} + OH^- \quad (63)$$

This gives Eq. (64) [with the loss of chloride from the conjugate base, Eq. (61), as the slow-rate-determining step]:

$$Rate = k[Co(NH_3)_4(NH_2)Cl]^{2+} \quad (64)$$
$$= kK[Co(NH_3)_5Cl][OH^-] \quad (65)$$

The base hydrolyses of the much studied cobalt(III) compounds are dominated by this type of equation, whereas the coordination compounds of the equally polarizing ions chromium(III) and rhodium(III) with ligands capable of forming conjugate bases do not seem to exhibit this behavior.

3. Base hydrolysis (and in a very few cases attack by cyanide or other nucleophiles) on a number of coordination compounds with the ligand C=N. Often, the ligand is part of an aromatic ring. For instance, for all the base hydrolyses such as Eq. (66) (substitution of an N-heterocyclic ligand LL, usually 2,2'-bipyridyl or 1,10-phenanthroline, by hydroxide), although there can obviously be no conjugate base formed by protonic dissociation (there are *no* acidic protons) the rate equations are nevertheless as in Eq. (67):

$$M(LL)_3^{n+} + OH^- \rightarrow [M(LL)_2(OH)(OH_2)]^{(n-1)+} + LL \tag{66}$$

$$\text{Rate} = [M(LL)_3]\left(k_0 + k_1[OH^-] + k_2[OH^-]^2\right) \tag{67}$$

In a similar way, the rate equations for substitution by cyanide ion are as in Eq. (68):

$$\text{Rate} = [M(LL)_3]\left(k_0 + k_1[CN^-] + k_2[CN^-]^2\right) \tag{68}$$

In the latter case, a typical example is the ready reaction of ferroin with cyanide in water to give the Schilt–Barbieri compound:

$$[Fe(phen)_3]^{2+} + 2CN^- \rightarrow [Fe(phen)_2(CN)_2] + phen \tag{69}$$

The most reasonable interpretation (there have been many) is to consider the hydroxide or cyanide as forming first an sp^3-hybridized carbon atom (a pseudobase or Reissert-type adduct, respectively) and then being transmitted from carbon to metal ion. In other words, the change in reactivity of an N-heterocycle on coordination to a metal ion is akin to that of the same N-heterocycle on classical quaternization by an organic agent such as methyl iodide. The unusual rate equation [Eq. (67) or (68)] involving the nucleophile's concentration in first- and second-order terms arises because the rates of these reactions (apparently hydrolysis or substitution by cyanide at the metal ion) are actually controlled by rates of reaction at the ligand ($27 \rightarrow 28$).

27   28

products

Certainly, the presence of an imine-like carbon atom adjacent to the coordinated nitrogen seems necessary for values of k_1 in rate equations (67) or (68) to be large.

4. The unusually rapid coordination to initially inert aquo metal ions of a few oxo anions (XO_{ny}^{y-}), for example, $X = N$, $n = 2$, $y = 1$ for ONO^-, may involve a rate-controlling reaction at the oxygen of water (path **29**) rather than the metal (path **30**).

29   30

H. Catalysis

If one defines a catalyst as "a species whose activity appears to a higher power in the rate equation than in the stoichiometric equation," many kinds of transformation of coordination compounds may be subject to catalysis. For example, the replacement of fluoride in Eq. (70) has the stoichiometric and rate equations given in Eqs. (71) and (72), respectively. The proton is said to be a catalyst, probably through the intermediate compound (**31**) with hydrogen fluoride as ligands:

$$[Co(NH_3)_5F]^{2+} + H_2O \rightarrow [Co(NH_3)_5(OH_2)]^{3+} + F^- \tag{70}$$

$$K = \frac{[Co(NH_3)_5(OH_2)][F]}{[Co(NH_3)_5F]} \tag{71}$$

$$\text{Rate} = k[Co(NH_3)_5F][H^+] \tag{72}$$

Other ligands that are the conjugate bases of weak Brönsted acids (e.g., NO_2^-) show similar catalysis by a proton. In the case of coordinated nitrite, the proposed protonated intermediates (**32**) may actually be isolated in solid salts [here the nitrate $[Co(NH_3)_5(HONO)](NO_3)_3$]. In both **31** and **32**. A represents the ligand NH_3.

31   32

Genesis of the famous Zeise's salt. $K[Pt(C_2H_4)Cl_3]$, made by the slow reaction of potassium tetrachloroplatinate(II) with ethene in water (for ~ 5 days) is dramatically

$$trans\text{-}[Cr(OH_2)_4Cl_2]^+ + (EDTA)^{4-} \longrightarrow [Cr(EDTA)]^- + 2Cl^- \tag{73}$$

Zn | Green

Purple

$[Cr(OH_2)_4Cl_2]^+$

$$[Cr^{II}(OH_2)_4Cl_2]^0 + (EDTA)^{4-} \longrightarrow [Cr(EDTA)]^{2-} \tag{74}$$

catalyzed by a small amount of tin(II) chloride; in its presence, the Zeise's salt forms quickly.

By far the most important catalytic reactions of coordination compounds are those based on oxidation and reduction. Typically, a kinetically labile oxidation state, usually for the first transition series, $3d^{n+1}$, is added to or generated (in situ) from an inert one, $3d^n$. An example is shown in Figs. (73) and (74).

Reaction (73) has a very large value of $K = [Cr(EDTA)^-]/[[Cr(OH_2)_4Cl_2]^+][(EDTA)^{4-}]$, so the formation of the purple complex chelated product should be strongly favored, but the half-life is in fact several hours at 15°C. Zinc metal (merely a reducing agent, a source of electrons) rapidly reduces chromium(III) to chromium(II), which is (with its $(3d)^4$ electronic configuration) kinetically labile, and so comes to equilibrium with the $(EDTA)^{4-}$ ligand rapidly ($t_{1/2} = 10^{-7}$ s), forming a quite stable chromium(II)–EDTA compound [Eq. (74); $\beta_{110} \approx 10^{14}$]. The product is now oxidized rapidly [Eq. (75)] by electron transfer:

$$[Cr(EDTA)]^{2-} + trans\text{-}[Cr(OH_2)_4Cl_2]^+$$
$$\rightarrow [Cr(EDTA)]^- + [Cr(OH_2)_4Cl_2]^0 \tag{75}$$

The same result can be achieved by adding a salt of chromium(II) rather than by forming it in situ. Table VII gives some similar redox-catalyzed substitutions. Note that, as in Table VII, whereas the first-row metal ions

(here, chromium and cobalt) often have stable oxidation states separated by a single electron [cf. Fe(III) and Fe(II); Cu(II) and Cu(I)], their heavier congeners (here, rhodium and platinum) often have states differing by two charges [cf. Pd(IV) and Pd(II); Au(III) and Au(I)]. The principle of complementarity indicates that, for these heavier metals, such two-electron reductants as primary or secondary alcohols will readily give the catalytic-reduced states.

Many of the most important catalytic activities of coordination compounds and metal ions (particularly iron and copper) are in the electron transport chains of cellular metabolism, where they act as catalysts for the oxidation of organic intermediates. Several other transition metal ions (including vanadium and molybdenum) have important metabolic roles in a variety of organisms. Indeed, recent discoveries suggest that even such metals as chromium and nickel have biological functions.

IV. SHAPE

A. Coordination Number

The formula of a complex compound is established once we know the oxidation state of the metal in the ion and the (coordination) number of ligands attached to it. What can we say about the latter quantity?

The coordination number (i.e., the number of ligand atoms in direct contact with the metal atom) of metal ions

TABLE VII Syntheses via Catalysis

Metal	Inert oxidation state	Electronic configuration	Catalyst	Labile oxidation state	Electronic configuration	Example
Chromium	III	$3d^3$	Zinc	II	$3d^4$	$[Cr(EDTA)]^-$, $[Cr(en)_3]^{3+}$
Cobalt	III	$3d^6$	Charcoal[a]	II	$3d^7$	$Co(NH_3)_6^{3+}$
Rhodium	III	$4d^6$	$R'_3CCHROH$[b]	I	$4d^8$	$[Rh(NH_3)_5Cl]Cl_2$
Platinum	IV	$5d^6$	—	II	$5d^8$	Halo substitution, $[Pt(S_5)_3]^{2-}$

[a] Acts, rather like a graphite electrode, as a source of electrons, i.e., a reducing agent.

[b] The hydrogen underlined here is thought to act as the two-electron source (i.e., $H^- \rightarrow H^+ + 2e^-$).

TABLE VIII Coordination Numbers and Associated Shapes

Coordination number	Shape	Point group	Examples
2[a]	Linear	$C_{\infty h}$	$[CuCl_2]^-$
			$[Ag(NH_3)_2]^+$
4	Tetrahedral	T_d	$[BeF_4]^{2-}$
			$[CoCl_4]^{2-}$
			$[Zn(NH_3)_4]^{2+}$
	Plannar	D_{4h}	$PtCl_4^{2-\,b}$
			$AuCl_4^{-\,b}$
			ICl_4^-
5	Pyramidal	C_{4v}	$[OV(OH_2)_4]^{2+}$
	Bipyramidal	D_{3h}	$[InCl_5]^{2-}$
6	Octahedral	O_h	$[AlF_6]^{3-}$
			$[RhCl_6]^{3-}$

[a] On the whole, two coordination is found for the lower oxidation state (I) of the coinage metals (copper, gold, silver) and among their isoelectronic neighbors [e.g., $Hg(OH_2)_2)^{2+}$, $Tl(OH_2)_2^{3+}$].

[b] Note that these are isoelectronic (so the same structure is expected).

in simple coordination compounds varies (as it does in binary and ternary crystalline structures) from 2 up to ~ 12. The most common (all others up to 12 are known) are 2, 4, 5, 6, and (for some of the larger ions—barium and radium; thorium; zirconium, hafnium, and some lanthanides and actinides) 8. Table VIII gives examples of coordination numbers and the associated shapes.

There are many rules of thumb for rationalizing *changes* in coordination number for a particular metal and among metals in general. For a given metal ion with a particular ligand in a particular solvent—usually water—such changes are manifested by sudden discontinuities in properties. With Hg^{2+} in water, the successive stepwise stability constants with chloride are $K_{110} > K_{120} \ggg K_{130} \cdots$. Whereas the first two chloride ligands attach to mercury very well, giving successively $(HgCl)^+$ and $(HgCl_2)^0$, the third one has little affinity. Presumably, the stable linear two-coordinated structure is being altered to a three-coordinated $HgCl_3^-$ (triangular) or four-coordinated $Hg(OH_2)Cl_3^-$ (tetrahedral).

In general, if a particular metal ion in a particular oxidation state manifests, under different circumstances (i.e., with a variety of ligands), more than one coordination number, the changes (based on the Pauling electroneutrality principle) are as follows:

1. Anions have lower coordination numbers than cations:

$$[CoCl_4]^{2-} \quad \text{versus} \quad [Co(OH_2)_6]^{2+}$$

$$[MnCl_4]^{2-} \quad \text{versus} \quad [Mn(OH_2)_6]^{2+}$$

2. The more polarizable the ligand, the lower is the coordination number: $[AlCl_4]^-$, $[AlF_6]^{3-}$.

For any one metal, the ions in increasingly positive oxidation states become smaller [i.e., rFe^{3+} (spin free) $(3d^5) < rFe^{2+}$ (spin free) (d^6); $rCu^{2+}(d^9) < rCu^+(d^{10})$]. The coordination number with a given ligand tends to increase with this shrinkage, perhaps because the more highly charged cation has much increased electron attachment enthaply, requiring more of the same ligands to become electroneutral. Examples (which abound) are $TlCl_6^{3-}$, $TlCl_4^-$; ICl_4^-, ICl_2^-; CuF_6^{3-}, CuF_4^{2-}; $PtCl_6^{2-}$, $PtCl_4^{2-}$; $SnCl_6^{2-}$, $SnCl_3^-$; $Ag(C_5H_5N)_4^{2+}$, $Ag(C_5H_5N)_2^+$; $(AuCl_4)^-$, $(AuCl_2)^-$. In the cases of the electronic configurations d^8 and d^6, for the same metal the interconversion of one state to the other is often called oxidative addition (or reductive elimination in the opposite direction):

$$MCl_4^{2-} + Cl_2 \rightarrow MCl_6^{2-} \qquad (\text{e.g., M = Pd, Pt}) \quad (76)$$

For the same ligand with varying metal ions there is (in crystalline binary compounds) a general tendency for the coordination numbers of the cations to increase on going down a eutropic family (as in SiO_2, GeO_2, SnO_2, PbO_2). This is not as true of isolated complexes in coordination compounds. The sizes of corresponding ions do increase down the three transition series, but this increase is often swamped by the sharing of the effect among several ligands. For example, the bond lengths M $\leftarrow$ N for $Co(NH_3)_6^{3-}$, $Rh(NH_3)_6^{3+}$ and $Ir(NH_3)_6^{3+}$ are sufficiently alike that many triads of their analogous salts, for example, $[M(NH_3)_6](NO_3)_3 \cdot HONO_2$ or $[M(NH_3)_5](OH_2)](NO_3)_3 \cdot HONO_2$, are isostructural for M = Co, Rh, Ir.

The growth in size of the s-block ions is well known, as in $rMg^{2+} < rCa^{2+}$ (1.06 Å) $< rSr^{2+}$ (1.33 Å) $< rBa^{2+} < rRa^{2+}$, and the coordination numbers with *like* ligands do tend to increase down these series. However, the metal ions [other than the very small ones of the higher oxidation states, Mn(VII) and the like] of atomic number greater than 19 are all large enough to accommodate the higher coordination numbers (8). For the electropositive elements at the start of the transition series, the lanthanide contraction ensures that $(5d)^n$ ions are about the same size as their $4d^n$ congeners but both are larger than $3d^n$ (e.g., $Ti^{4+} < Zr^{4+} \simeq Hf^{4+}$; $V^{3+} < Nb^{3+} \simeq Ta^{3+}$; $Cr^{3+} < Mo^{3+} \simeq W^{3+}$). Typical complex compounds contain TiF_6^{2-}, ZrF_8^{4-}, and HfF_8^{4-}.

B. Structures and Their Symmetries

The coordination number 4 is fairly common, and there are two limiting shapes, as listed in Table VIII: the planar and the tetrahedral. For main group metals, tetrahedral ions such as $[BeF_4]^{2-}$, $[AlCl_4]^-$, and $[Zn(NH_3)_4]^{2+}$ are common. For the first transition series, they are quite common among anions and with very polarizable ligands, as in the blue ion $[CoBr_4]^{2-}$ and in the blue β form (so-called for historical reasons: names used to distinguish polymorphic varieties are unsystematic) of $[Co(C_5H_5N)_2Cl_2]$. There is another form of this dimorphic compound, the pink α, which is polymeric in the crystal (through chlorine bridging two cobalt ions) and contains octahedral cobalt(II) ions. The very highly polarizing ions of the highest oxidation states of the central transition elements often have essentially tetrahedral shapes. Some examples are VO_4^{3-}, CrO_4^{2-}, MnO_4^- these being an isoelectronic set; MnO_4^{2-}; FeO_4^{2-} (so-called ferrate); $[CrO_3Cl]^-$, $[TcO_4]^-$, $[ReO_4]^-$, $[OsO_4N]^-$ (osmiamate); and many others.

The planar four-coordinated molecules are of great rarity in main group chemistry (ICl_4^-, XeF_4) but dominate the simple coordination chemistry of spin-paired d^8 species. These crop up among the second and third row platinum and coinage metals, and there are a few examples for $(3d)^8$, chiefly for nickel(II) with strong-field (polarizable) ligands: $[Ni(CN)_4]^{2-}$, $[Pd(NH_3)_4]^{2+}$, $[Pt(NH_3)_2Cl_2]$, $[AuCl_4]^-$.

C. Isomerism

Kinetically labile copper(II) ions, in water, combine with an excess of L-alaninate ligands (LAla-O)$^-$ to give a solution (at pH 8.3) that contains only the uncharged species $[Cu(LAla-O)_2]$. This species, planar about the copper ion, may exist as *cis* (**33**) or *trans* (**34**) isomers. The equilibrium

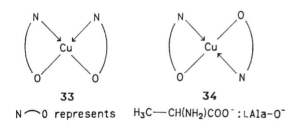

33   **34**

N⌒O represents H_3C—$CH(NH_2)COO^-$: LAla-O$^-$

ratio in solutions has been said to be

$$K = [trans]/[cis] = 1.5$$

Trans and *cis* isomers must be in labile equilibrium, the copper(II) ion being so rapid in its reactions. Nevertheless, both *trans* and *cis* forms can be obtained as crystals from the solution. A concentrated solution initially deposits Cambridge blue crystals of the *trans* isomer **34**, but if left for a few days these are gradually replaced by the more stable Oxford blue crystals of the *cis* isomer **33**.

TABLE IX   Properties of *Cis* and *Trans* Isomers of Square $[PtA_2B_2]$

Geometry				
A	B	Isomer	μ^a	Other properties
NH_3	Cl	*Cis*	—	$H_2SO_4 \rightarrow$ adduct Drug
NH_3	Cl	*Trans*	—	$H_2SO_4 \rightarrow$ no adduct Not drug
$P(C_2H_5)_3$	Cl	*Cis*	10.6	White
$P(C_2H_5)_3$	Cl	*Trans*	0	Yellow

^{a} Dipole moment (in debye units).

The subtlety of metal complex chemistry is manifest in the apparently similar bisglycinatocopper(II) system, the behavior of which is actually opposite to that of alaninate: The initial aqueous solution gives first the metastable crystals of the *cis* isomer, which is slowly replaced by the *trans*.

For the kinetically inert spin-paired d^8 species, such as palladium(II) and platinum(II), this *cis–trans* (geometric) isomerism occurs frequently. Table IX shows some examples. For such compositions as those in Table IX, there is often another kind of isomerism (ionization isomerism) in which the same stoichiometry Pt–$2NH_3$–$2Cl$ is given by such salts as $[Pt(NH_3)_4][PtCl_4]$ and $[Pt(NH_3)_3Cl][Pt(NH_3)Cl_3]$. The former, Magnus's green salt, has columns or chains of platinum ions running through the structure (**35**, A = NH_3).

35

Related structures may give rise to desirable electronic properties, such as one-dimensional conductance (along the Pt–Pt axis).

Far and away the most common coordination number is 6, with the ligands arranged at the corners of an octahedron. For metals from atomic number 11 (sodium) upward, more complex ions and compounds contain six-coordinate metal ions than not. Examples of ions are $[Cr(\underline{N}CS)_6]^{3-}$, where the underlining of the N indicates that it is this constituent of the ligand, rather than the sulfur atom, that is directly attached to the metal, $[SnCl_6]^{2-}$, $[Ni(H_2NCH_2CH_2NH_2)_3]^{2+}$, $[Ru(bipy)_3]^{2+}$, and $[Pt(S_5)_3]^{2-}$.

Isomers are particularly common among six-coordinated compounds, although there are many cases for other coordination numbers. The extraordinarily selective synthesis form the dianion of the natural amino acid L-cysteine as a *bi*dentate ligand at cobalt(III), to give dark green tris-L-cysteinato-(*N,S*)-cobaltate(III), illustrates the chief causes of isomerism among coordination compounds.

L-Cysteinate (the doubly charged anion) is represented by structure **36**.

36

The three atoms underlined are all possible donor sites, but since only two are used to form each of the three chelate rings, this ligand could be attached through (1) nitrogen and oxygen (chelated α-amino acidate, with a pendant unattached thiolate, $-CH_2S^-$), (2) sulfur and oxygen (a chelated β-thiocarboxylate, with unattached NH_2), or (3) sulfur and nitrogen (chelated 1-amino-2-thiolate, with pendant carboxylate). This variability of mode of attachment of a ligand is called linkage isomerism. Other ligands with such multiplicity of possible binding sites may be unidentate and include (among several) nitrite attached via nitrogen, for example, [Co(NH$_3$)$_5$(NO$_2$)], so-called nitropentamminecobalt(III) ion, a yellow ion; nitrite attached via oxygen, [Co(NH$_3$)$_5$(ONO)]$^{2+}$, so-called nitritopentamminecobalt(III) ion, a red ion; thiocyanate attached via sulfur, for example, [Rh(NH$_3$)$_5$(SCN)]$^{2+}$; or "isothiocyanate," nitrogen-bonded [Rh(NH$_3$)$_5$(NCS)]$^{2+}$.

Representing the bidentate L-cysteinate as N–S, three of these may be coordinated as in **37** or **38**.

(fac)   (mer)
37   **38**

In the discussion of stepwise stability constants [see Eq. (5)], the more symmetric facial and less symmetric meridional isomers of [MA$_3$B$_3$] were distinguished from their probabilities of formation (*mer*-**9**, 60%; *fac*-**10**, 40%). In the present case, only the *fac* isomer is formed. Such geometric isomers have the same composition but different molecular symmetries.

D. Optical Activity

Finally, there are for the *fac* isomer **37** (point group C_3) (and indeed the *mer* isomer **38**, C_1, or for any other geometry that has only axial symmetry, D_3, C_2, and so on) two forms of the molecule that differ, as do right- and left-handed propellers (**39** and **40**). Such an object (molecule) cannot be superimposed on its image in a mirror. The labels attached (by arbitrary convention) to propeller molecules, such as

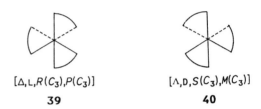

$[\Delta, L, R(C_3), P(C_3)]$ $[\Lambda, D, S(C_3), M(C_3)]$
39   **40**

are shown [from minus (M), plus (P), sinistral (S), rectal (R); Λ refers to left-helicity about the threefold axis, Δ to right; D and L are for dextro (right) and levo (left)].

E. Stereoselectivity

To distinguish right from left requires an all-one-handed agent (either all right or all left). For an all-right-distinguishing agent, the differing interaction combinations are right–right and right–left. This is the same difference as between shaking hands and holding hands. Isomers formed from differing combinations of right- and left-handed (chiral) parts of molecules are known as diastereoisomers. Natural (optically active) L-cysteine selects the $S(C_3)$ propeller of cobalt(III). Such selection among the possible diastereoisomers is known as stereoselectivity.

The sole product, from among all the possible isomers, is $S(C_3)$-*fac*-[Co(L-cysteinate-N,S)$_3$]$^{3-}$, which can be readily crystallized as its potassium salt. Such selective syntheses (where one among several possible reactions occurs preferentially rather than at random) are now more readily achieved than hitherto.

Tartaric acid (or its conjugate bases) can select between the right and left hands (**41** and **42**) of DL-[Co(en)$_2$(CO$_3$)]$^+$, giving the very stable diastereoisomer **43**. This effectively takes all the L-cobalt centers out of commission, so that on adding symmertric reagents (e.g., NO$_2^-$), the only place for reactions is the D-cobalt of **44**, giving the D-[Co(en)$_2$(NO$_2$)$_2$]$^+$ ion **45a**. The overall ionic charges on the complex ions **41**–**46** are omitted, but all the complexed cobalt centers are cobalt(III).

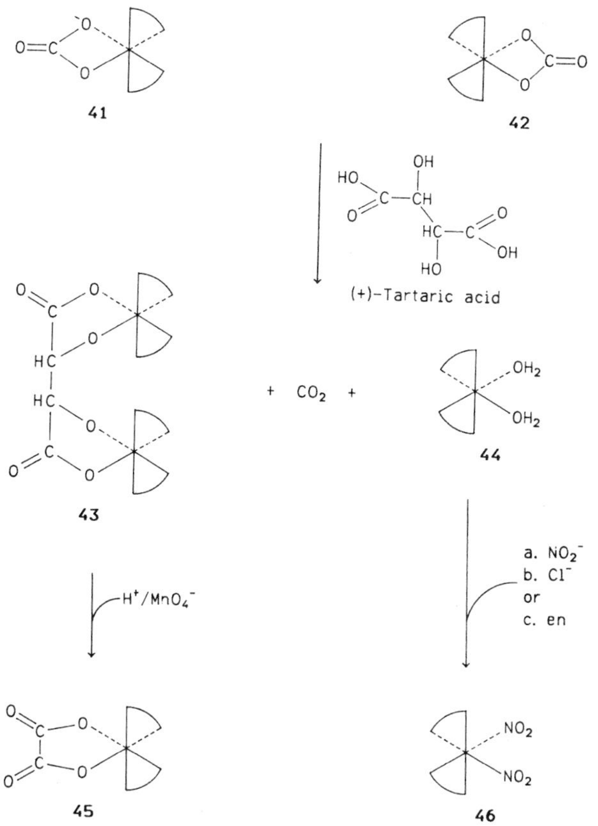

41

42

(+)-Tartaric acid

+ CO_2 +

44

43

a. NO_2^-
b. Cl^-
or
c. en

$-H^+/MnO_4^-$

45

46

This is a neat dissymmetric synthesis of the product, *cis*-dinitrobis(1,2-diaminoethane)cobalt(III) (**46**), which, in principle, can be done in one reaction vessel.

A similar convenient "one-pot" synthesis of an optically active coordination compound from "off-the-shelf" reagents involves reducing aqueous hexachlororuthenate(IV) ion with dissymmetric (+)-tartaric acid (a natural product) and presumably forming preferentially one of the several possible diastereoisomeric (+)-tartratoruthenium species. On the addition of excess of 2,2'-bipyridyl, it substitutes stereoselectively, giving an excess of (+)-[Rubipy$_3$]$^{2+}$ over its (−) enantiomer. This product, when racemic, has been used as a photocatalyst in many attempts to catalyze reaction (77), the photolysis of water (as a solar energy utilizer),

$$2H_2O + h\nu \rightarrow 2H_2 + O_2 \qquad (77)$$

A great deal of effort is being made to modify the ligands in this *N*-heterocyclic complex, to improve or alter solubility and oxidation-reduction properties. The final oxidative degradation by acid permanganate of the tartrate ligand in the stable diastereoisomer **43** (this bridging two cobalt ions) gives breakage of the carbon–carbon bond but no rupture of any bond to chiral cobalt(III). Clearly, the handedness of the propeller about cobalt in the product

(**45**) is the same as that in the reactant (**43**). This chemical transformation proved that the two chiralities (handedness of propeller) are related and is called a chemical correlation of configuration. Other examples include Eqs. (78)–(80):

$$(+)\text{-}cis\text{-}[Co(en)_2(\underline{NCS})_2]^+$$
$$\rightarrow (+)\text{-}cis\text{-}[Co(en)_2(\underline{NH_3})_2]^{3+} \qquad (78)$$

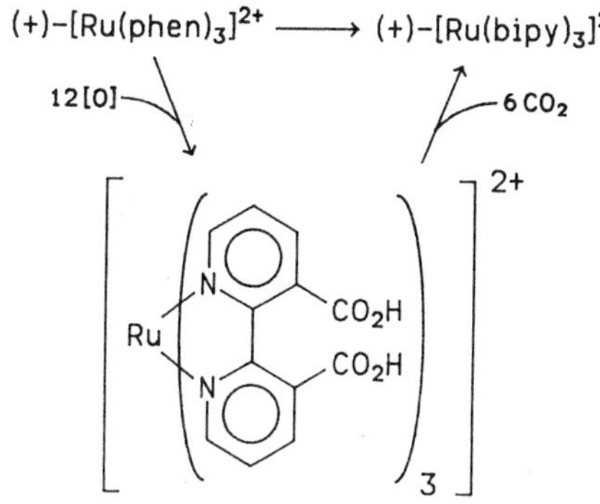

$$(+)\text{-}[Ru(phen)_3]^{2+} \longrightarrow (+)\text{-}[Ru(bipy)_3]^{2+}$$

12[O]   −6 CO_2

$$(79)$$

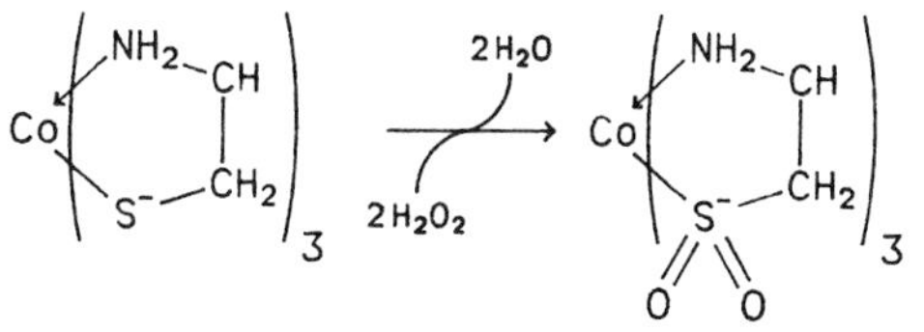

$$(80)$$

Equation (80) represents the transformation of coordinated thiolate in the dark green triscysteinate complex of cobalt(III) by oxidation using hydrogen peroxide to the yellow sulfinate. In full, it can be written

$$fac\text{-}[Co(LCys)_3]^{3-} + 6H_2O_2$$
$$\rightarrow \text{L-}fac\text{-}[Co(LCysu\text{-}N,S)_3]^{3-} + 6H_2O \quad (81)$$

The cysteine sulfinate is abbreviated Cysu, and this dissymmetric yellow anion is a splendid resolving agent for triply charged racemic cations such as [M(en)$_3$]$^{3+}$ (M=Cr, Co, Rh) or [M(bipy)$_3$]$^{3+}$ (M=Co, Cr). All such chemical correlations of configuration are reactions of coordinated ligands.

F. Resolutions

Selection among diastereoisomers is not, of course, restricted to such intramolecular cases as the tris-(*N*,*S*)-cysteinatocobaltate(III), where all the chiral centers (at the metal ion and in the organic molecules) are in the same molecule. Wherever a single-handed molecule or chiral influence (say, right) interacts with a racemic (50 : 50 right–left) mixture, there are the diastereoisomeric pair of unequal possibilities right–right for the right-handed half of

TABLE X Diastereoisomer Salt Formation

Racemate (X)	Resolving agent (A)	Solvent	LSD^a
$(NH_4)_2[Pt(S_5)_3]$	$(-)$-$[Ru(bipy)_3](ClO_4)_2$	Acetone	$(-)A(-)X$
$[Co(EDTA)]^-$	L-Histidinium	Water–ethanol	$(+)A(+)X$
$Na[Rh(\overset{\overset{H}{N}}{\underset{\underset{H}{N}}{}}SO_2)_2(H_2O)_2]^-$	$(+)$-$H_3CCH(C_6H_5)NH_2$	Water	$(+)A(+)X$
$[Ru(bipy)_3]^{3+}$	$(+)$-$[Co(LCysu)_3]^{3-}$	Water	$(+)_{350}X(+)A$
$[Co(en)_2(NO_2)_2]^+$	$(+)$-$[Co(EDTA)]^-$	Water	$(-)X(+)A$

^{a} This column describes the less soluble diastereoisomeric (LSD) slat formed from the resolving agent (A) and one hand of the racemate. It gives (sign of rotation of the cation) (sign of rotation of the anion). This measurement was done at the yellow lines of sodium, except for the $[Rh(biby)_3]^{3+}$ ion, where the rotation is at 350 nm, as indicated by the subscript $(+)_{350}$.

the racemic mixture and right–left for the left-handed half. Table X shows some examples of resolutions.

Pasteur found for organic racemates involving asymmetric tetrahedral carbon atoms that microorganisms (themselves made up of chiral molecules all of one hand; i.e., bacteria are very stereospecific reagents) metabolize one hand much more rapidly than the other. In much the same way, microorganisms show at least stereoselectivity and occasionally, apparently, stereospecificity (100% selection) for those few octahedral chelated compounds that have been studied. For example, racemic *mer*-trisglycinatocobalt(III) serves as a nitrogen source for *Pseudomonas stutzeri*, but only the D hand is consumed; the L is unaffected. This is the best way to obtain this uncharged complex compound in optically active form.

G. Jahn–Teller Effect

For the common four- and six-coordinated shapes, lowering of symmetry by di- or trisubstitution of one ligand by another leads to geometric isomerism (where the shape of the field around the metal ion differs between the two isomers, e.g., fac and mer). Similarly, for the six-coordinated octahedron (O_h), lowering of symmetry by removing reflection elements (center, plane, or improper axis), as in **47**, gives the

a

```
    a          a
a.  |  .c   c.  |  .a
 \  | /      \  | /
  \ |/        \ |/
b――――c      c――――b
  / |          / |
 b             b
 |             |
 D             L
      47
```

possibility of enantiomerism. This is usually done by chelating. Examples of racemates D-**47** + L-**47** are known.

Of the several claimed separations of D-**47** from L-**47** (optical resolution) at least that of $[Co(NH_3)_2(NO_2)_2(CN)_2]$ seems authentic.

Quite apart from such modification of essentially symmetric shapes, there is another general effect, causing distortion of symmetric shapes, that is particularly common among coordination compounds of some transition metals (but also known for other species, such as the first electronic excited state of benzene). This is the Jahn–Teller effect. For nonlinear assemblies of nuclei and electrons, unequal occupancy of degenerate orbitals is unstable, and there will be a lower energy state of different geometry in which the nuclear framework has been distorted. Taking the electronic configurations of spin-free d^4 [e.g., Cr(II), Mn(III)] and any d^9 [e.g., Cu(II), Ag(II)] as examples, in octahedral ligand fields the electron occupancy is shown in **48** and **49**,

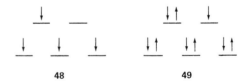

respectively. The Jahn–Teller theorem says that the nuclei will move to lift the degeneracy (i.e., $x = y \neq z$), as in Fig. 6. There are many cases of such distortion. Indeed, either (or both) the ground state or the first excited state of all configurations d^n ($1 \leq n \leq 9$) is subject to Jahn–Teller distortion. Particularly noticeable are departures from regular shape in which the unequally occupied degenerate orbitals are antibonding; some examples are given in Table XI. The energy level diagram for T_d for a particular d^n is the inverse of that for O_h, for the same d^n shown in Figs. 4 and 5.

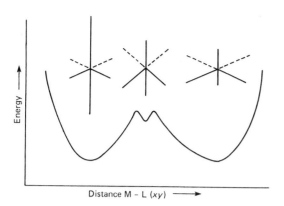

FIGURE 6 Two possible distortions (along normal coordinates) of an octahedral compound MX$_6$ with all six MX distances equal (center point) to "four short, two long" MX$_4^s$X$_2^1$ or "two short, four long" MX$_s^2$X$_4^1$.

V. BIOLOGICAL METAL COMPLEXING

The number of metals known to be essential to a range of living species has increased markedly during the twentieth century, so that the "biological periodic table" is now a large fraction of the periodic table itself. Most of the metal requirements are as trace elements, such as vanadium in mushrooms. The vanadium (in the famous *Amanita muscaria*, the red-capped mushroom with white spots) occurs in the same coordination compound throughout the mushroom. This compound, amavanadin, originally isolated from a Black Forest species, is said to have structure **50**.

50

Many natural coordination compounds exist—for example, vitamin B$_{12}$, in which the central metal is

cobalt(III), with six ligands, five of which are imine-like or heterocyclic nitrogen, and the sixth an alkyl group (or in the well-known artifact, now called cyanocobalamin, formed during the original isolations, cyanide).

The striking fact about biological metal complex chemistry is its novelty. When first isolated, metal-containing molecules and metal-in-volving systems from living cells almost always turn out to have features that have no real counterpart in known synthetic (*in vitro*) coordination compounds. Much effort has been expended on developing "model systems" to reproduce these natural biochemical ingenuities. For example, the stable carbon–cobalt(III) bond in vitamin B$_{12}$ has been mimicked in the bis-1,2-dionedioximatocobalt(III) moiety (**51**) (the so-called cobaloximes), in which the nitrogen donors of the oxime ligands (whose imine function is very similar to that in N-heterocycles) are held planar by strong intramolecular hydrogen bonds.

51

One reminder of the power of natural syntheses is the range of iron-binding molecules that have been found in bacteria. Several functional groups have emerged by human design as useful ligands to form stable complex compounds in aqueous media specifically with iron(II) or iron(III). These are shown in Table XII. How wonderful that the same groups are used by all living things. For example, the diimine molecules **52**, caerulomycin A, and **53** (ferropyrimine) of Table VII are very closely akin to the 2,2'-bipyridyl compounds so common in analytical chemistry.

VI. THERAPEUTIC METAL COMPLEXES

Whether a solution of a coordination compound is ingested or injected, it enters an aqueous medium and, depending

TABLE XI Jahn–Teller Distorted Structures

General shape	Configuration	Ion	Example
O_h, antibonding ($d_{x^2-y^2}$, d_{z^2})	$(3d)^4$	Mn(III)	MnF$_3$
	$(3d)^4$	Cr(II)	KCrF$_3$
	$(3d)^9$	Cr(II)	K$_2$CuF$_4$
T_d, antibonding (d_{xy}, d_{yz}, d_{zx})	$(3d)^8$	—	NiCr$_2$O$_4$
	$(3d)^8$	—	Cs$_2$[CuCl$_4$]

TABLE XII Some "Specific" Complexing Agents for Aqueous Iron and Natural Analogs

Compound (oxidation state)	Ligand	Natural example
$Fe(OH_2)_6^{3+}$ (III)	Salicylate (2-hydroxybenzoate)	—
		—
	Hydroxamic acids	Ferrioxamine
$Fe(OH_2)_6^{2+}$ (II)	2,2′-Bipyridyl	

[a] Structure **52** and derivatives known as caerulomycins B, C, and D (all based on the 2,2′-bipyridyl-6-aldoxime moiety) are isolated from *Streptomyces caeruleus*.

[b] Ferropyrimine (**53**) is isolated from the organism *Erwinia rhapontici*, the cause of crown rot in rhubarb, which is also a rich source of the chelating agent $(C_2O_4)^{2-}$, oxalate, or ethane 1,2-dioate.

on concentration, temperature, ionic strength, pH, and so forth, will come to equilibrium with the environment. Establishing the actual nature (*speciation*) of the metal-containing ions and molecules (*species*) that are present in such a given environment is obviously important. It is best done, at present, by a combination of potentiometry [described in Section II.G, Eqs. (41) and (42)] and spectroscopic analysis. The potentiometry defines ranges of possible speciation, and spectroscopy decides among them. Characteristic therapeutic effects [other than those of the separated constituent parts, metal ion and ligand(s) added separately] are therefore to be expected for kinetically inert compounds. Synergic effects are, of course, conceivable for kinetically labile metal ions with particular ligands.

The two best-known therapies involving kinetically inert coordination compounds are chrysotherapy (treatment of rheumatoid arthritis with gold compounds) and the relatively new treatment of certain cancers with platinum compounds.

In chrysotherapy, following earlier uses of "potable gold" and of "colloidal gold," the intact complex compounds (with trade names) used in the form of aqueous solutions are $Na_3[Au(S_2O_3)_2]$ (sanochrysine), sodium gold thiomalate (myocrisin), and a phosphine complex (Solganol). These may have grave side effects but are commonly effective in restoring expression to locked joints. The detailed mechanism of action is unknown, but an intriguing fact is that when two enantiomeric joints are examined, one affected, the other not (such as two elbows), gold is found only in the rheumatic or arthritic joint.

Platinum therapy, discovered by Rosenberg, used *cis*-dichlorodiammineplatinum(II) (see Table IX). This is sold under the names Platinol, *cis*-platin, or Neo-platin. It is thought to act by interfering with nucleic acid replication. Platinosis (a sensitivity, revealed as an allergy, to certain complex compounds of platinum) has become a notifiable industrial disease in France. It has been known since the early years of this century among workers in factories producing platinum chemicals. The novel utility of this simplest of coordination compounds, *cis*-$[Pt(NH_3)_2Cl_2]$, is certainly a major breakthrough in cancer chemotherapy.

Ligands by themselves are often effective drugs or detoxificants. For example, D-penicil-lamine (**3**, a substituted cysteine) is used to mobilize copper deposited in reducing tissues in patients with Wilson's disease (hepatolenticular degeneration), a hereditary defect in copper metabolism. The copper transport protein (ceruloplasmin) of blood plasma is faulty and bonds copper ions less effectively than it should. The enantiomeric L-penicillamine is ineffective as a treatment. If (as may happen) D-penicillamine is either inactive or gives rise to intense nausea, triethylenetetraamine (trien; **4**) is often used.

SEE ALSO THE FOLLOWING ARTICLES

CATALYSIS, HOMOGENEOUS • CATALYSIS, INDUSTRIAL • ELECTRON TRANSFER REACTIONS • KINETICS (CHEMISTRY) • LIGAND FIELD CONCEPT • NOBLE METALS • RARE EARTH ELEMENTS AND MATERIALS

BIBLIOGRAPHY

Kauffman, G. B. (ed.) (1994). "Coordination Chemistry, A Century of Progress," ACS Symposium Series 565; American Chemical Society, Washington, D.C.

Wilkinson, G., Gillard, R. D., and McCleverty, J. A. (1987). "Comprehensive Coordination Chemistry," Vols. 1–7, Pergamon Press, Oxford.

Corrosion

Samuel A. Bradford
University of Alberta

GLOSSARY

Anodic protection Technique for reducing corrosion by polarizing a metal into its passive region.

Cathodic protection Technique for corrosion reduction by making a metal the cathode of an electrochemical cell and supplying it with electrons.

Concentration cell Electrolytic cell deriving its electromotive force from a difference in concentration of some electrolyte component.

Crevice corrosion Corrosion occurring in regions not fully exposed to the environment, such as in screw threads, under absorbent gaskets, or under rust.

Erosion corrosion Chemical–mechanical corrosion caused by the rapid movement of a corrosive environment over a metal surface.

Galvanic corrosion Corrosion of a metal when electrically connected to a less reactive metal or nonmetal conductor.

Hot corrosion Metal oxidation accelerated by a surface layer of fused salt.

Hydrogen embrittlement Reduction of ductility of metals by atomic hydrogen absorbed in the metal crystals. The hydrogen may cause blistering, internal cracking, formation of metal hydrides, or simply loss of ductility.

Inhibitor Chemical added to the environment to reduce corrosion without significant reaction with the environment.

Intergranular corrosion Corrosion concentrating at metal grain boundaries where precipitates or impurity atoms have accumulated.

Passivation Reaction of a metal with an oxidizing environment to produce an extremely thin and protective film of corrosion products on the metal surface.

Pitting Localized corrosion due to incomplete passivation of a metal. A small area remains anodic and thereby continues to corrode in the form of cavities.

Polarization Electrode potential shift toward the cathodic direction when a metal is made an anode (or toward the anodic when the metal is made a cathode).

Rust Corrosion product of iron or steel consisting of hydrated iron oxides, primarily hydrated ferric oxide.

Stress-corrosion cracking Cracking of a metal under the combined action of sustained tensile stress and a corrosive environment.

I. CORROSION PRINCIPLES

A. Electrochemical Corrosion Cells

Corrosion is primarily an electrochemical process. Michael Faraday established this principle in the early nineteenth century, and it is still fundamental to an understanding of the problem and to corrosion prevention.

All electrochemical corrosion cells must have four components: (1) an anode (the corroding metal), (2) a cathode (metal, graphite, or semiconducting electron conductor), (3) an electrolyte containing a reducible species, and (4) an electron-conducting connection between the electrodes. If any component is missing in the cell, then electrochemical corrosion will not occur.

Corrosion reactions can be split into anode and cathode half-cell reactions for better understanding of the process. The anode reaction is simple: The anode metal corrodes and goes into solution as positive metal ions:

$$M \rightarrow M^{n+} + ne^- \tag{1}$$

The electrons produced remain on the corroding metal where they would quickly halt corrosion if no cathode reaction were available to remove them.

In electrochemical corrosion, the electrons produced at the anode are consumed by reaction of the electrolyte on the cathode surface. Since a wide variety of corrosives can attack metals, several cathode reactions can exist. The most common is the one occurring in nature and in neutral or basic solutions, the reduction of dissolved oxygen:

$$O_2 + 2H_2O + 4e^- \rightarrow 4OH^- \tag{2}$$

In acids, the cathode reaction is the reduction of hydrogen ions:

$$2H^+ + 2e^- \rightarrow H_2\uparrow \tag{3}$$

Not so common, but very corrosive, is the cathode process for oxidizing acids or aerated acids:

$$O_2 + 4H^+ + 4e^- \rightarrow 2H_2O \tag{4}$$

In some chemical processes, the solutions contain oxidizing agents that can be reduced in valence; for example, ferric ions (Fe^{3+}) can be reduced to ferrous ions (Fe^{2+}), or cupric ions can be reduced to metallic copper:

$$Cu^{2+} + 2e^- \rightarrow Cu\downarrow \tag{5}$$

B. Polarization

If a metal is made an anode in a corrosion cell, then it will polarize so that its electrode potential shifts in the noble (cathodic) direction, making it a less efficient anode as more current flows from the metal surface into the electrolyte. Similarly, if the metal is made a cathode by impressing current onto it, its potential drops, making it a less efficient cathode.

In corrosion reactions, polarization has two principal causes. Activation polarization results from a sluggish step in the electrode reaction that cannot respond instantly to the supply or demand for electrical current. Tafel has shown that activation polarization obeys the equation:

$$\eta = a \pm b \log i \tag{6}$$

where η is the polarization; i is the current density (A/m^2), either anodic (+) or cathodic (−); and a and b are constants. If the electrode potential is within about ± 50 mV of equilibrium, the Tafel equation is not obeyed, however. Within this range, the electrode has both anodic and cathodic reactions occurring on it, with the net current being the difference between current to the electrode and current from the electrode.

The second cause of polarization is concentration polarization, which is commonly only a problem for the cathode. The cathode reaction is continually consuming dissolved oxygen, hydrogen ions, or other reducible species from the corrosive solution. Large current demands on the cathode will deplete the reactants in solution near the cathode. The cathode reaction then proceeds only as fast as reactants diffuse through the depleted layer of solution to reach the cathode surface. In time, the entire corrosion process may become limited by this diffusion until an increase in cathodic polarization cannot accelerate the diffusion of the reacting species any further.

C. Passivity

Some metals, notably the stainless steels, titanium, aluminum, and chromium, corrode to form a thin film of corrosion product on their surfaces that greatly protects them from further attack. Carbon steels form a passive film that is only partially protective and is easily damaged.

To passivate, a metal must first corrode at an extremely high rate, with the anodic current density reaching a critical value, until the passive film forms over the entire surface. When the metal has passivated, the Tafel equation no longer holds; the corrosion current drops sharply to a low value as potential is increased. If the potential of a passivated metal is decreased, or the solution becomes less oxidizing, then the metal may perform well for a while but may suddenly begin corroding at a disastrous rate if the passive film becomes damaged.

D. Mixed Potential Theory

The cathode and anode currents must be equal in a corrosion cell. The cathode reaction can proceed only when it receives electrons from the anode. If the anode tries to supply electrons faster than the cathode can consume them, then the excess electrons choke off the anode reaction.

In a corrosion cell, the cathode and anode potentials also must be equal because the two electrodes are directly connected. Initially, the cathode reaction potential is much greater than the anode potential, but as corrosion begins and current starts flowing, polarization of both the anode and cathode brings them to the corrosion potential where the corrosion current flows from anode to cathode through the electrolyte.

II. ENVIRONMENTS

A. Aqueous Solutions

1. Effect of Concentration

A nonpassivating metal corrodes more rapidly as the corrosive concentration increases, as a general rule. The rule only applies, however, if sufficient water remains in the solution to ionize the corrosive and to solvate the metal ions as they leave the anode. Steel, for example, corrodes severely in sulfuric acid concentrations <60% but is the common material of construction for storage tanks and piping that hold concentrated 95% sulfuric acid.

Figure 1 shows a pipe corroded much worse at the waterline, where the maximum amount of corrosive, in this case oxygen, can provide the most efficient cathode reaction [Eq. (2)].

A passivating metal corrodes more rapidly as the corrosive concentration increases, up to a point. When the corrosion rate increases enough, the metal spontaneously passivates so that the corrosion rate drops to a very low value, perhaps several orders of magnitude. This low corrosion persists even when the corrosive concentration is increased further, although for most passivated metals the

FIGURE 1 Wastewater pipe normally half full. Steel has severely corroded at the waterline where oxygen is plentiful.

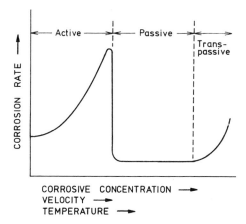

FIGURE 2 Effect of solution variables on corrosion rate (idealized). A particular metal in a given solution may follow only part of the curve, that is, show only active corrosion, an active-to-passive transition, and so forth.

concentration can eventually be increased enough that the passive film itself is attacked. At extremely high concentrations, then, a transpassive region exists where corrosion rates again become severe. Figure 2 illustrates the possible effects that concentration may have.

2. Effect of Solution Velocity

For an active (nonpassivating) metal, increasing velocity will help overcome concentration polarization, increasing the cathode reaction rate that is limiting the process and thus increasing corrosion. At a very high solution velocity, concentration polarization may be entirely overcome so that further increase in velocity has no effect on the corrosion rate. Corrosion systems tend to be diffusion-controlled at low velocities but reaction-controlled at high velocities.

An active–passive metal may be active at low velocities but passive at high velocities where enough oxidizer reaches the metal surface to passivate it. In that case, the corrosion rate increases greatly as velocity increases, up to a velocity sufficient for passivation, whereupon the corrosion decreases to a very low rate and is not further affected by velocity. For such a metal the corrosion rate may be acceptable if the velocity is maintained above a critical velocity, but the corrosion could be severe in crevices and under gaskets where velocity is very low (see Fig. 2).

3. Effect of Temperature

Both chemical reaction rates and diffusion rates increase with increasing temperature. Consequently, polarization decreases.

For activation-controlled corrosion, polarization is less at higher temperatures because the electrode reactions can respond quickly to an extra electron demand or supply. Both anode and cathode reactions are speeded so that the corrosion current, or corrosion rate, greatly increases.

For concentration-controlled corrosion (that is, where concentration polarization is rate limiting), increasing the temperature increases diffusion so that the effect is like increasing solution velocity (see Fig. 2).

4. Differential Temperature Cells

Especially important with heat exchangers, boilers, immersion heaters, and nuclear reactor cooling systems, differential temperature cells occur wherever metal surfaces cannot remain at a uniform temperature. Because hot metal is more reactive than cool metal, a hot spot that is anodic to a large cool area will corrode very seriously. In some cases, if the temperature difference is not great ($<75°C$), then the potential difference between anode and cathode will be small and no serious corrosion cell would exist but for the fact that the hotter metal passivates. In this case, severe corrosion then occurs on the cooler, active metal while the hot, passive metal serves as the cathode (see Fig. 2).

Differential temperature cells persist for the life of the plant. They may be unimportant, lying dormant until the environment changes in some way, such as an increase in conductivity or aggressive ion concentration. When activated, the differential temperature cells can increase corrosion tremendously.

B. Natural Environments

Atmospheric corrosion occurs only when water is present as a surface film on the metal. The relative humidity and temperature thus are critical variables. As they recur in a somewhat cyclic manner, it is commonly found that atmospheric corrosion can be expressed as $C = At^n$, where t is the exposure time and A and n are constants that depend on the metal and environmental conditions. With this equation, steel exposure tests of 3–4 years can be reasonably extrapolated to 20–30 years. Although an increase in temperature will increase reaction and diffusion kinetics, it will decrease relative humidity and reduce oxygen solubility in the water film. Rainfall, although wetting the metal surface, provides benefit by leaching corrosives (such as sulfate and chloride ions) out of the deposited corrosion products. Atmospheric environments are commonly classified as rural, urban, industrial, marine, and so forth, in an imprecise attempt to indicate the degree of contamination with SO_2, $NaCl$, and the like.

Corrosion of metals buried in soils depends in a very complex manner on moisture content, dissolved ions, acidity, and the presence of corrosive bacteria. The first two of these variables can be evaluated surprisingly well by simply measuring the soil resistivity. A resistivity of $<1000 \, \Omega$ cm indicates highly corrosive soil; $10,000 \, \Omega$ cm, slightly corrosive soil; and $>100,000 \, \Omega$ cm, noncorrosive soil. Soil acidity between pH 4 and 8.5 has little effect on corrosion but more acidic or more alkaline soils are highly corrosive.

C. Microbial Corrosion

Microbial colonies (e.g., fungi, bacteria) may grow to form a barrier to oxygen, which causes the metal under the colony to corrode while the surrounding metal acts as a cathode. In addition, microbes may increase the reaction rate of either anode or cathode, reduce surface-film resistance, or produce a corrosive metabolic product.

Anaerobic bacteria grow in the absence of oxygen—for example, in swamps, pipelines, and cooling systems. The most corrosive bacteria in soils are the sulfate-reducing types that feed on organic matter, reducing polarization, forming cathodic sulfide, and producing a highly corrosive metabolic product. Under colonies of other bacteria and within their slimy, protective biofilms, these bacteria generate their own anaerobic conditions even in aerated soils.

Among aerobic bacteria, the iron bacteria cause the most serious problem. They feed on ferrous ions and exude ferric ions that form a thin ferric oxide shell (tubercule) over the colony, setting up a barrier to oxygen. Sulfate-reducing bacteria can then also grow within the tubercule.

Sulfur-oxidizing bacteria are another aerobic type present in soils around sulfur storage, oil fields, and sewage disposal facilities. They convert elemental sulfur to sulfuric acid, reducing the pH to <1 within the colony. The strong acid rapidly deteriorates metals and concrete. Microbial corrosion is not stopped by most corrosion inhibitors, but once the type of microbe is identified a suitable biocide can be completely effective.

III. TYPES OF CORROSION

A. Uniform Attack

On a tonnage basis, more metal corrodes by uniform attack than by any other mechanism. A common example is the rusting of an unpainted shovel. All the surface is rusting so anodes must be everywhere, but at the same time cathode reactions must be occurring at some locations that continuously alternate with the anode reactions. Thus, while oxidation and reduction processes are obviously occurring, the anode and cathode of a corrosion cell cannot be distinguished.

Although certainly a common process, this uniform attack poses no unexpected problem for the engineer. The engineer can estimate the corrosion rate and combat it—for example, by choosing suitable materials and coatings or by using inhibitors or cathodic protection. Situations where unexpected corrosion cells appear create much more serious corrosion problems.

B. Galvanic Corrosion

Two different metals electrically connected in a corrosive solution set up an electrochemical cell with the more reactive metal as the anode.

A list of metals in a galvanic series is commonly used in corrosion work to determine which metal is more reactive in a particular environment (e.g., seawater, soil). Metals that can be both active and passive are listed both ways. Potentials are usually not listed because corrosion does not have any standard conditions; that is, potentials change with solution concentration, minor variations in alloy composition, stirring rate, aeration, and so forth. A galvanic series in seawater is given in Table I.

The more reactive metal will be the anode when coupled to a less reactive (more noble) metal. The two metals usually share the corrosion task: One provides surface for

TABLE I Galvanic Series in Seawater

(Cathodic or Noble End)
Graphite
Platinum
Nickel alloys B, C, 825
Titanium
Nickel-copper alloys
Austenitic stainless steels (passive) (e.g., "20," 304)
Silver
Nickel and nickel-chromium alloy 600
Lead
Cupronickels
Ferritic, martensitic stainless steels (passive) (e.g., 430, 410)
Bronzes
Copper
Brasses
Stainless steels (active)
Carbon steels, cast iron
Cadmium
Aluminum and aluminum alloys
Zinc
Magnesium and magnesium alloys
(Anodic or Active End)

the cathode reaction and the other corrodes much more rapidly than if it were alone. If a galvanic cell cannot be avoided, the choice of two metals close together in the galvanic series minimizes the corrosion.

Cathode reactions are usually slow and require a large cathode surface for reaction because they involve adsorption of the reactants onto the metal surface. Consequently, a large cathode continues to consume all the electrons released by the anode, but a small cathode limits the galvanic corrosion. As a general rule, if galvanic corrosion cannot be avoided, then the cathode should be as small as possible.

Galvanic corrosion is worst where the two metals make contact, because the electrolyte path for ions is shortest there and cell resistance the least. Therefore, one way of mitigating galvanic corrosion is to separate the two metals as far as is practical. If copper and steel water pipes are separated by an insulating gasket inside the pipes, even though they are electrically connected externally, the high resistance of the water often reduces galvanic corrosion to a minor problem.

High-purity metals corrode much less than commercially pure metals. Likewise, one-phase alloys with all elements in solid solution corrode less than two-phase alloys. Thus, even microscopic galvanic corrosion cells greatly speed up the attack.

Galvanic cell potentials $\geq 100\,\text{mV}$ will develop between stressed (anodic) and unstressed metal. Cold-worked metal at a bend will therefore corrode in preference to the neighboring unworked metal. Even the crystal boundaries within metals, their grain boundaries, have a higher energy than the bulk of the grains and tend to corrode preferentially.

C. Intergranular Corrosion

Grain boundaries, besides being high-stress areas, are sites for accumulation of precipitates and impurity atoms, setting up microscopic galvanic cells. An important example occurs in the welding of some stainless steels containing carbon. Chromium carbide can precipitate in grain boudaries adjacent to the weld, where temperatures have been in the 500–800°C range. A galvanic cell with a large cathode to anode area ratio is set up between the chromium-depleted regions immediately next to the boundaries and the high-chromium stainless steel in the bulk of the grains. Severe corrosion is concentrated along the boundaries so that a trench corrodes out in the heat-affected zone on both sides of the weld.

An example of intergranular corrosion in the heat-affected zone of a welded aluminum pipe is shown in Fig. 3. Note that many grains have become completely separated from the parent metal.

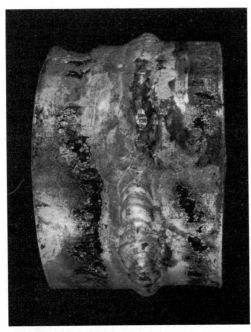

FIGURE 3 Intergranular corrosion in the heat-affected zone of a weld on an aluminum-alloy pipe buried in soil.

D. Selective Leaching of Alloys

In certain environments, an alloy may corrode by dissolving away one component and leaving another untouched. One example is dezincification of high-zinc brasses in which both the zinc and copper corrode, but copper immediately plates back on the metal because it is much more noble than the zinc. The result is a very porous, weak, pink copper area in the yellow brass.

Selective leaching also occurs with gray cast-iron pipe in soils. The iron corrodes, leaving the cathodic graphite flakes in place. The graphite with the compact rust still has the shape and appearance of a pipe and if left in well-packed soil will continue to carry water. But the graphitized cast-iron is weak and will fracture under any sudden pressure change or stress.

E. Crevice Corrosion

Crevice corrosion occurs where stagnant solution is in contact with the metal in crevices. such as in threaded connections, beneath absorbent gaskets, and under deposits of dirt and rust. Stainless steels in chloride solutions are especially susceptible because their passive film becomes damaged in the crevice, thus setting up a cell between the active crevice and the passive stainless steel outside the crevice.

When the metal first contacts the solution, uniform corrosion occurs all over the metal surface, with the cathode reaction usually involving the reduction of dissolved oxy-

gen molecules. The oxygen quickly becomes depleted in the crevice but is not replaced if the crevice is tight enough to limit diffusion. The cathode reaction then must stop in the crevice while the anode reaction continues, but outside the crevice the cathode reaction continues because O_2 can easily diffuse to the metal. In this way the O_2 concentration cell restricts the corrosion to only the crevice, with the cathode reaction outside the crevice serving to support the corrosion.

Chloride ions intensify corrosion by diffusing into the crevice, attracted by the metal ions produced there. The resulting metal chloride hydrolyzes to form insoluble hydroxide or oxide while the chloride ions catalyze the process:

$$MCl + H_2O \rightarrow MOH + H^+ + Cl^- \qquad (7)$$

F. Pitting

Pitting causes more *unexpected* losses than any other type of corrosion. It occurs when a passivated metal is almost, but not entirely, resistant to attack. Attack begins at a weak spot in the passive film but the pit fails to repassivate. The solution inside the pit becomes acidic, usually with a high Cl^- ion concentration, just as in crevice corrosion.

Metals susceptible to pitting are also susceptible to crevice corrosion, although the converse is not necessarily true. Both types of corrosion concentrate the attack at one small location; therefore, weight-loss measurements do not indicate the extent of the damage.

High velocity of solution is likely to wash the concentrated, corrosive solution out of the pit, since a pit is much more open and exposed than a tight crevice is. Even in fairly stagnant solutions, many pits develop, grow briefly, and then repassivate because convection currents have flushed out the acid solution in the pits.

Figure 4 shows severe pitting of steel by water. The flat-bottomed pits are indicative of attack by carbon dioxide dissolved in the water, forming carbonic acid.

IV. CHEMICAL–MECHANICAL CORROSION

A. Erosion Corrosion

Erosion corrosion is caused by rapid movement of a corrosive environment over the metal surface. If attack is localized, especially in the early stages, then it appears as elongated, comet-shaped pits pointing downstream. If attack is more general, then a pattern of grooves or waves develops, leaving a smooth, wavy surface (see Fig. 5). The welds, which are harder and slightly noble with respect to the pipe, are unattacked. The pipe shown in the figure was

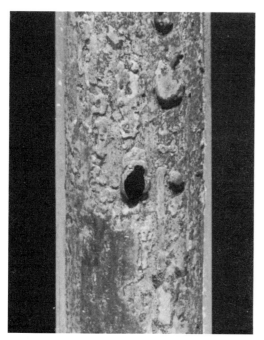

FIGURE 4 Pitting of a steel water pipe by water containing carbon dioxide.

evidently run only partly full; the directional pits at the top indicate the earlier stage of erosion corrosion.

Most metals are susceptible, but especially the passivated metals that may have their passive film damaged.

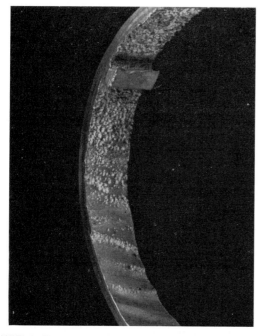

FIGURE 5 Erosion corrosion of a chromium–nickel stainless steel. The weld is more resistant than the pipe. Directional pitting at the top indicates the first stage of erosion corrosion.

Hard, strong metals, although resistant to wear, are not resistant to erosion corrosion if their corrosion resistance is low or if their passive film is weak.

Attack is particularly severe in high-velocity slurries because of the abrasive action of the suspended solids. Turbulence of the solution is a common cause of erosion corrosion; laminar flow seldom causes corrosion. Impingement, where the liquid flow strikes the metal surface and is forced to change direction, is extremely damaging.

B. Cavitation

With high-velocity, turbulent liquids (as found in pumps and turbines), cavitation can be a great problem, sometimes occurring in combination with erosion corrosion. Cavitation is caused by momentary vaporization of liquid at a low-pressure spot in the turbulent liquid, instantly followed by implosion of the vapor bubble. Shock waves from the implosions exceed the metal's yield strength and "hammer" the metal, which can often be heard as a popping or crackling sound. Cavitation damage in its early stage appears as a patch of closely spaced pits on the metal surface. In time, the pits change to a roughened spot and then to a deep gouged-out area.

C. Hydrogen Damage

The first steps in the hydrogen cathode reaction are the adsorption of the hydrogen ion onto the metal surface and the reception of an electron from the metal to form a hydrogen atom. Instead of two hydrogen atoms combining to molecular H_2 and leaving the surface, the small hydrogen atoms may diffuse into the metal, particularly if the surface is poisoned by sulfide. If the atoms diffuse to internal voids or interfaces between phases, they can form H_2 gas there, creating internal pressure and even blistering the metal. In addition, atomic hydrogen, collected at internal surfaces such as grain boundaries, may be responsible for stress-corrosion cracking by lowering the internal surface energy. This decreases the additional fracture energy required to create the high-energy crack surfaces. Internal atomic hydrogen can also form brittle hydrides within some metals, notably titanium and zirconium, making crack propagation easy.

D. Stress-Corrosion Cracking

Cracking of a metal under the combined action of sustained tensile stress and a corrosive environment is termed *stress-corrosion cracking* (SCC). The cracks usually have a fine, branched appearance and may follow grain boundaries (e.g., steel in caustic) or may be transgranular (austenitic stainless steel in chlorides). The corrosive

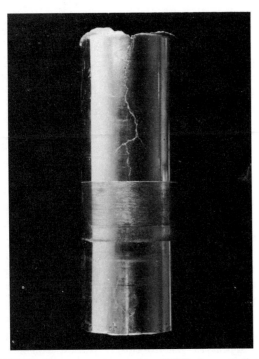

FIGURE 6 Stress-corrosion cracking (SCC) of a vertical pump shaft sleeve in a salt slurry. Branched cracking of the stainless steel is typical.

environment and the stress must be present simultaneously for synergistic action.

Figure 6 shows an example of SCC of an austenitic, stainless steel vertical pump shaft sleeve in a chloride environment.

Some metals appear to have a threshold stress—a lower limit of stress necessary for cracks to propagate—but in many metals the cracks can extend under extremely low stress or stress intensity. Stress-corrosion cracking is sometimes considered a brittle failure, but the cracks propagate more slowly than brittle cracks and the alloys that crack may be inherently ductile.

It is now apparent that SCC is a general term for several different environment-stress interactions. The most important mechanisms are

1. *Hydrogen embrittlement:* Acidic conditions generated within the crack produce atomic hydrogen that diffuses ahead of the crack, embrittling the metal. This is the case for high-strength steels.
2. *Film rupture:* A passive film or tarnish cracks under the applied stress, setting up an active/passive cell that accelerates corrosion and repassivates the metal. The corrosion creates a notch that acts to increase stress repeatedly to rupture the film. This is probably the SCC mechanism for metals such as brass and aluminum.

3. *Preexisting crack paths:* Weak or highly corrodible paths already exist within some metals. For example, age-hardened aluminum alloys have a weak, precipitate-free zone adjacent to grain boundaries; austenitic stainless steels normally crack transgranularly in chlorides but, if sensitized, they crack along the corrodible grain boundaries.
4. *Adsorption:* Certain specific species adsorb at the crack tip, lowering the surface energy of new surfaces and consequently lowering the fracture stress. This is considered to be the cracking mechanism in liquid metal environments and possibly in organic liquids.

Stress-corrosion cracking can be prevented by selecting materials with low residual stresses and microstructures not susceptible to intergranular attack, or by using protective coatings in stressed areas. Operating temperatures should be kept as low as possible. Cathodic protection can stop SCC, but overprotection can charge the metal with hydrogen. The environment can be modified by inhibitor additions or by removal of the species responsible for SCC, such as Cl^- for stainless steels and ammonia for brass.

E. Corrosion Fatigue

Fatigue, the cracking of a metal under cyclic or alternating stress, is greatly accelerated by corrosion. With corrosion present, the fatigue has no endurance limit, that is, a limiting stress below which the crack will not extend. In corrosion fatigue, the crack surface usually shows some corrosion product deposits, unlike the clean, smooth surface of a fatigue crack. Almost all metals are susceptible to both fatigue and corrosion fatigue, with the cracks being transgranular and usually unbranched.

Corrosion fatigue can be prevented by retarding the corrosion with inhibitors or cathodic protection, or by reducing the chances of fatigue: selecting ductile metals with low internal stresses, minimizing cyclic stresses and vibration, eliminating factors that increase stress, or applying surface compressive stresses by methods such as shot peening or sandblasting.

F. Fretting and Corrosive Wear

Fretting, also called *friction oxidation*, occurs when highly loaded metal surfaces vibrate or slip slightly against one another. Unlike wear, fretting reaches a maximum at $\sim$7.5 μm amplitude, and seldom is a problem above $\sim$25 μm. Bolted, riveted, and keyed joints, splines, couplings, and bearings are commonly attacked.

Fretting destroys close tolerances, ruins bearing surfaces, and may initiate fatigue cracks. It may wear away protective coatings to initiate galvanic or crevice corrosion.

The mechanism of fretting is a combination of oxidation and wear in the sequence: (1) metal surfaces oxidize, (2) vibration between rubbing surfaces breaks oxide at contact points, (3) bare metal rubbing against bare metal causes friction welding at the contact points, (4) continued vibration breaks these microscopic welds, and (5) the process repeats. Any metal fragments from the fracture will oxidize and act as an abrasive between the loaded surfaces.

Fretting corrosion, with a mechanism combining wet corrosion and wear, is a problem in nuclear reactors at contact points between the fuel element and pressure tube. Vibration comes from turbulent flow of the coolant or pulsing of the coolant from the pump.

Corrosive wear continually removes oxide or other protective corrosion products from metal surfaces, allowing the corrosion to continue unhindered while the broken particles of product often act as an abrasive. Corrosion usually produces ionic corrosion products that are harder than the underlying metal. If the product layer is very hard and thin, it may actually protect the metal from wear, but if it is thick it usually is brittle and breaks off. A few corrosion products, particularly the salts of fatty acids, are soft and ductile, acting as lubricants and reducing both corrosion and wear.

V. CORROSION MONITORING

A. Corrosion Rate Expressions

Corrosion rates are described either by weight loss or by depth of penetration. Weight loss has been most commonly given as mdd (milligrams of metal lost per square decimeter), but many other metric, English, and bastard units have been used. The preferred metric (SI) unit is $g \cdot m^{-2} \cdot d^{-1}$.

Expressing corrosion as weight loss is only valid for uniform attack, since for localized attack (such as pitting) the weight loss is meaningless.

Either uniform or localized attack can be expressed as depth of penetration. The favorite penetration unit has been mpy (mils per year, a mil being 0.001 inch), whereas the metric unit is mm/year. If pitting is present, then the penetration in mm/year usually refers to the deepest pit observed.

B. Coupon Tests

Where visual inspection is difficult in plant equipment and pipelines, corrosion is commonly monitored with metal coupons inserted into the stream. Coupons are also used in laboratory and field testing to evaluate different materials, protective coatings, and inhibitors. The coupon test is a long-time averaging technique, applicable in any environment, and is the universally accepted standard method. But it requires strict attention to sample preparation and cleaning.

The composition, heat treatment, and surface condition of the coupons should be identical to those of the actual equipment monitored. Operating stresses, velocity effects, and heat transfer conditions are difficult to duplicate and are often ignored, although they may be extremely important in some situations.

Each specimen is marked for identification and its position in the holder is recorded. Stamped identification numbers may corrode severely or cause cracking. If uniform stress conditions are desired, then the specimens can be marked with drilled holes, edge notches, or the like. A minimum of three replicate specimens is recommended to detect errors and establish the variation to be expected. Duplicate tests usually give approximately ±10% reproducibility.

Monitoring of plant equipment is an ongoing process with coupons removed at regular intervals. Tests of new materials or coatings in plant tests are continued as long as possible, with the hours of testing ≥50 ÷ corrosion rate in mm/year. The test duration is reported along with weight loss or depth of penetration after the samples have been cleaned of all corrosion products. Microscopic and visual examination assess pitting and cracking. The coupons may also be subjected to mechanical tests where strength is sensitive to corrosion.

C. Linear Polarization Probes

Instantaneous corrosion rates of equipment can be measured with electronic probes inserted into the corrosive solution. The signals from the probes can be used for troubleshooting or continuous equipment protection. In some cases, the probes have proved to be so accurate and so fast in response that they can be used for process control.

Linear polarization probes are based on the principle that for polarization within ~10 mV of the corrosion potential the applied current is a *linear* function of the polarization. The probe consists of two electrodes made of the same metal as the equipment concerned. (Some probes have a reference electrode also.) Typically, a 10-mV potential difference is applied between the two working electrodes, and the current flowing between them is read on an ammeter connected between the electrodes. The current is directly proportional to the corrosion rate.

The method has proved to be accurate in many industrial situations and is superior to coupon tests if corrosion rates are very low. But, as with any test method, it can be misleading in some situations. It does not accurately detect pitting and cannot be used if the corrosive medium is not an electrolyte.

D. Electrical Resistance Probes

A probe that can be used in any environment, whether or not an electrolyte, is the electrical resistance probe. It is simply a hairpin wire of the metal of interest, electrically insulated from surrounding metal and inserted in the corrosive medium. As corrosion proceeds, the diameter of the wire decreases and the electrical resistance of the wire increases. This method averages corrosion over long periods of time and does not detect short-term corrosion excursions. Because of the small surface area of the wire, pitting may be missed, but if the wire does pit its electrical resistance greatly increases. Scaling or fouling can short-circuit the wire and give erroneous readings. Electrical resistance is very temperature dependent, so correction must always be made for temperature fluctuations.

E. Hydrogen Probes

Hydrogen probes are designed to detect hydrogen charged into the metal by corrosion, thus anticipating the dangers of hydrogen embrittlement and hydrogen blistering of steel.

The hydrogen patch probe monitors the amount of hydrogen charged into the actual equipment. A small electrochemical cell is clamped to the outside wall of the vessel or pipe, with a palladium foil anode in contact with the steel wall. Atomic hydrogen in the steel is oxidized to H^+ by a potential applied to the auxiliary cell. The H^+ ions diffuse through the electrolyte in the probe to a cathode, where they are reduced to H_2 gas. The cell current is monitored as an indirect measure of the corrosion inside the equipment.

VI. CORROSION CONTROL

When faced with a corrosion problem, the engineer can solve it by any of five ways:

1. The material or coating may be changed to a more suitable one.
2. The environment chemistry may be changed (e.g., pH, inhibitors).
3. The operating conditions may be changed (e.g., temperature, velocity).
4. The design of the apparatus may be changed.
5. The electrical potential of the metal may be changed (i.e., cathodic or anodic protection).

A. Materials Selection

The lists of commercially available alloys, plastics, ceramics, composite materials, glass, concrete, and wood are almost endless. The engineer chooses materials based primarily on the various mechanical, electrical, and chemical properties required, as well as availability and cost. However, three groups of metals are particularly important in corrosion: carbon steels, stainless steels, and nickel alloys.

Steels are not corrosion resistant, but they are used extensively because of their excellent mechanical properties and relatively low cost. If the environment is highly corrosive, then steels are protected by coatings or cathodic protection.

Stainless steels are ferrous alloys containing >10% chromium. They passivate well in oxidizing environments that would destroy ordinary carbon steels, but chlorides can pit and crack the stainless alloys.

Nickel alloys, commonly containing copper or molybdenum or chromium and iron, are much more resistant to SCC than are high-strength steels or stainless steels. But the nickel alloys are also much more expensive. Most of the nickel alloys are formulated to be highly resistant to a specific corrodant such as HCl or HF.

Certain metals are so well suited for certain common environments that they are the first materials considered for those applications. Table II lists some of these compatible combinations. Every metal also has its weaknesses to corrosion that are sometimes overlooked. Especially incompatible combinations are also shown in the table.

TABLE II Compatible and Incompatible Combinations

Metal group	Compatible	Incompatible
Carbon steels	Concentrated sulfuric acid	Dilute sulfuric acid
Stainless steels	Nitric acid	Hydrochloric acid
Aluminum	The atmosphere	Alkalis
Nickel and nickel alloys	Caustic	Hot, sulfur-containing gases
Copper and copper alloys	Water	Ammonia
Titanium	Strongly oxidizing solutions	Molten caustic
Lead	Dilute sulfuric acid	Concentrated sulfuric acid
Tin	Food	Strong acids and bases

B. Protective Coatings

1. Metallic Coatings

Corrosion resistance can be imparted to a metal by coating it with another metal. Metallic coatings are either sacrificial or noble with respect to their substrate. Sacrificial coatings, such as zinc or cadmium on steel, will corrode in preference to any underlying metal exposed at pores, scratches, or cut edges. The thicker the coating, the longer this small-scale cathodic protection continues. Noble coatings, such as nickel or chromium on steel, depend on the high corrosion resistance of the coating for protection, but if the coating is damaged, then the corrosion of the substrate is accelerated by galvanic action.

Protective metallic coatings can be applied in a wide variety of ways. The most common method is electroplating on a base metal, which is most often low- or medium-strength steel. Hot dipping in a molten bath of metal is generally the least expensive method for applying a relatively thick coating, but the method is limited to the low-melting metals: zinc (galvanizing), aluminum, lead, and tin. Diffusion coatings are produced to alloy the metal surface, usually for oxidation resistance. Examples of diffusion coatings are chromizing by heating chromium powder rolled onto steel and aluminizing in aluminum vapor. Aluminum coatings are often produced by flame-spraying molten aluminum droplets on the base metal surface. Chemical reduction is used to produce electroless nickel coatings in which the part to be coated is immersed in a nickel salt solution that decomposes to deposit nickel on the metal surface.

2. Conversion Coatings

A metal can be protected by converting its surface, either chemically or electrochemically, to a corrosion-resistant compound. Steel is often given a thin phosphate coating before painting to improve paint adherence and provide some corrosion protection. Treatment in the phosphate bath for alonger time produces a thick coating that effectively absorbs oil, thereby improving lubrication and corrosion resistance.

Chromate coatings are often produced on aluminum, magnesium, and zinc. Such coatings provide a good base for paints and serve as inhibitors by gradually releasing soluble chromate during exposure. Blueing of steel, by treatment with hot alkali and oxidizer, produces an attractive layer of Fe_3O_4 that is not very protective but does effectively absorb oil and wax.

Anodizing, usually on aluminum or magnesium, involves making the metal an anode in an electrolytic cell. The resulting oxide layer readily absorbs dyes and can be sealed in boiling water or steam to produce a colorful, highly corrosion-resistant surface.

3. Ceramic Coatings

Vitreous enamels are applied on the metal surface as a powdered glass frit (commonly alumino-borosilicate), which is then fired in a furnace to soften the glass and bond it to the metal. The resulting layer is virtually impenetrable to water and oxygen, and it provides a smooth, attractive surface that is easily cleaned. Vitreous enamel coatings are very resistant to atmospheric exposure, can be formulated to resist strong acids or mild alkalies, and are often used for protection from high-temperature gases. These enamel coatings are sensitive to mechanical and thermal shock.

Portland cement coatings have the advantages of low cost, a thermal expansion similar to steel, and easy application and repair. They can be applied by troweling, spraying, or centrifugal casting for pipe linings. Cement linings are commonly used in acid tanks, waste treatment tanks, and process vessels. The linings are susceptible to mechanical and thermal shock.

4. Organic Coatings

Coatings are arbitrarily defined as <20 mils thick (0.5 mm) to differentiate them from linings. A typical paint coat is 2–4 mils (50–100 μm). For corrosion protection it must shield the metal from the environment, but it is never a perfect barrier. Oxygen and water can always permeate the multitude of pores in the film; however, as long as the pores are fine enough and long enough to restrict ionic permeation, giving a coating resistance $\geq 10^8$ Ω/cm^2, corrosion will be prevented. In addition, some coatings reduce the potential of the corrosion cell by inhibition if they contain anodic inhibitors, such as red lead or zinc chromate, or a more reactive metal, such as zinc dust.

All paints consist of vehicle (liquid) and pigment (solid). The vehicle contains the binder that will transform to a continuous solid film by polymerization or by evaporation of solvent. The pigment imparts color, reinforces the film, protects the binder from sunlight, and may aid corrosion-resistance by inhibitive or galvanic action. Other additives are usually present as stabilizers, driers, plasticizers, wetting agents, and flow-control agents.

Poor surface preparation causes ~75% of premature failures of paint coatings. Even the best coatings do not perform well on a poor surface, but a poor coating on a good surface can give good service. Moisture, dirt, oil, grease, mill scale, rust, chemicals, and old coatings on the metal surface can cause poor adhesion of the prime coat. Furthermore, the more corrosionresistant coatings (e.g., vinyls, chlorinated rubbers, and epoxies) have poor adherence. Poor application of the coating is another cause of many failures. Common application problems include dry spray (paint particles that never flow together), usually caused by holding the spray gun too far from the work, and application under bad weather conditions: high humidity, high or low temperature, or high wind.

C. Inhibitors

Inhibitors are chemical substances added in small concentrations to a corrosive environment to decrease the corrosion rate. They are usually proprietary mixtures of substances formulated for specific environments, along with surfactants and additives to increase inhibitor solubility. Inhibitors can be classified by their inhibiting mechanism as oxidizers, filming inhibitors, or a variety of other processes.

Passivators, or oxidizers, cause the anode in the corrosion cell to become passive. Direct passivators, which include chromates, nitrates, and molybdates, primarily treat cooling waters but must always be present in excess of the critical concentration necessary for passivation or they will severely increase corrosion. The critical concentration depends on chloride ion concentration and water temperature, but it is usually approximately 10^{-4} to 10^{-3} M. Indirect passivators are alkaline compounds, such as caustic, phosphates, and silicates, that improve the adsorption of dissolved oxygen on the metal, thus relying on the oxygen in the environment to passivate the corrosion.

Filming inhibitors form a barrier between the metal surface and the environment. Organic adsorption inhibitors are long-chain molecules with a polar end that adsorbs on the metal surface. Examples are amines, thioureas, and aldehydes that fasten to the surface and present their organic end to the aqueous environment to repel it. Inorganic adsorption inhibitors precipitate an insoluble layer on the surface. Examples are bicarbonates that precipitate a carbonate layer and the widely used phosphates that precipitate ferrous and ferric phosphates on steel.

Other types of inhibitors include the poisons, which interfere with the cathode H^+ or O_2 reduction. Examples are arsenic and antimony compounds that prevent adsorbed hydrogen atoms from leaving the metal surface as molecular H_2 bubbles. Scavengers inhibit corrosion by removing dissolved oxygen from solution. For example, Na_2SO_3 reacts with O_2 to form a stable sulfate. Neutralizing inhibitors react with acids to remove the hydrogen ions that could serve as cathode reactants. For example, ammonia is added to boiler water to raise its pH. Scale inhibitors are not corrosion inhibitors in a strict sense, but they do prevent crevice corrosion under boiler scale by forming soft, nonadherent precipitates instead of $CaCO_3$ scale.

D. Cathodic Protection

Cathodic protection supplies electrons to the metal to be protected (usually steel), thus stifling the anodic areas. The electron source can be either a more reactive metal that corrodes preferentially or an external electrical power supply. Cathodic protection can stop any type of corrosion, uniform or localized, but cannot prevent mechanical damage; for example, corrosion fatigue can be stopped but not the fatigue. SCC can be eliminated, but overprotection can charge the metal with hydrogen and cause failure by hydrogen embrittlement. Cathodic protection requires an electrolyte (aqueous solution or soil) through which current can flow to the protected metal surface; thus, attack by gas or vapor cannot be stopped.

The sacrificial anode method connects the corrodible metal to an even more reactive anode such as magnesium, magnesium alloy, or zinc. In sea water, aluminum does not passivate and therefore is most commonly used as an anode. The method is inexpensive, needs no external power, and distributes a fairly uniform protection current. Its current output is low, however, which limits the amount of protection it can provide, but it is also unlikely to cause stray currents that damage other metal nearby. The anodes gradually corrode away and must be replaced after several years.

The impressed current method requires an external dc power source connected between the corrodible metal (cathode) and a chemically inert anode. The method is extremely versatile in providing whatever protection is needed if the resistivity of the electrolyte varies, but overprotection can disbond coatings or charge hydrogen into the metal being protected. The large currents often required for cathodic protection may also damage nearby steel that is not connected to the protection system.

Permanent anodes most commonly used for impressed currents are graphite, highsilicon cast iron, magnetite, platinum-clad titanium, and mixed-metal oxides sintered on titanium. Rectifiers usually supply the dc-protection current necessary but, in remote areas where power lines are impractical, solar cells and batteries, engine generators, thermoelectric generators, and vapor turbogenerators have been successfully used.

For impressed-current cathodic protection of buried pipelines, the anodes are commonly placed in remote ground beds located at 100 ft to several miles from the pipeline. Deep-well ground beds, usually >300 ft deep, are often used to protect well casings and can also protect pipelines and the bottoms of aboveground storage tanks. Distributed anodes buried in a row parallel to the pipeline and usually at least 12 ft from it, produce a minimum of interference with neighboring systems in cities. The distributed anodes, all connected to each other and to the rectifier, can be fine-tuned to provide precisely the amount of current needed along the line.

Combining cathodic protection with a coating on the metal is very effective and requires protection current only at flaws or damaged spots in the coating. However, cathodic protection is ineffective in areas shielded so that the protection current does not reach them. One example is the failure of an external anode to throw current down the inside of a pipe; all protection current short-circuits to the pipe wall within a length of approximately four times the diameter. Another example is the failure to protect cathodically a pipe buried in cinders. The cinders, electrically conductive, pick up all the protection current and short-circuit it to the pipe, rather than allowing the current to flow all over the pipe surface to protect it.

Stray current hazards, previously mentioned, require carefully designed and controlled cathodic protection systems. An anode ground bed protecting a pipeline tends to throw current through the soil in all directions. If some of the current flows to a foreign metal structure (e.g., a buried tank as shown in Fig. 7), then it must leave the tank somewhere and flow on to the pipeline. The discharge areas on the tank will greatly corrode and may require that the tank have its own cathodic protection or be tied into the pipeline system.

E. Anodic Protection

Anodic protection is a technique to reduce corrosion on a metal by polarizing it into its passive region and maintaining it there. It is used primarily on carbon steels, stainless steels, nickel alloys, and titanium in environments such as sulfuric and phosphoric acids, sulfite digesters, and aqueous ammonia. Anodic protection cannot be used if the passive film is too readily damaged or is not very protective,

as may occur in chloride solutions or at high temperatures. The large currents required temporarily to passivate and repassivate the metal correspond to high corrosion rates, but the initially high corrosion is followed by low corrosion in the passive region.

The electronic control necessary to passivate and hold the metal at the required potential is the potentiostat, which maintains a constant potential with respect to a reference electrode. Any electronic conductor (e.g., steel, graphite) can serve as the cathode, while the metal to be protected is the anode.

The applied current is proportional to the corrosion rate of the protected system, making anodic protection less expensive than cathodic protection for extremely corrosive environments. Other advantages over cathodic protection include the operating conditions that can be set precisely by laboratory experiments and the current distribution, or "throwing power," that is extremely uniform. Anodic protection can also be achieved by attaching galvanic cathodes of platinum or titanium, without the need of a potentiostat, in solutions that readily passivate steel or stainless steel.

VII. DESIGNING FOR CORROSION

A. Allowance for Corrosion

After the proper material has been selected for the corrosive conditions, the common method of combating uniform corrosion is to make sufficient allowance for corrosion in the design so that the metal thickness lasts for the required life of the equipment. No modern engineering project is designed to last forever; a typical chemical plant may be designed to last 6–8 years. If it lasts longer than that, it is probably obsolete. However, allowing the metal to corrode is sometimes not a suitable solution, especially if (1) close tolerances must be kept, (2) corrosion is localized by unexpected electrochemical cells, or (3) the corrosion products could create problems such as contaminating the product, exerting stresses with their large volumes, and clogging filters.

B. Avoidance of Corrosion Cells

A simple design principle, often overlooked, is simply to keep corrodents away from the metal, or at least to limit the contact time as much as possible. For example, tank bottoms should be sloped to drain completely, while corrugated decks and channels should be equipped with drain holes if they cannot be sloped. Pipe insulation that may absorb moisture from high-humidity atmospheres or from drips or spray should be protected with an impervious

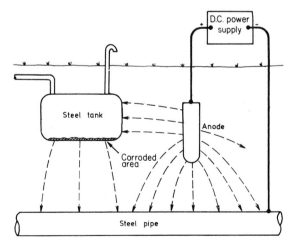

FIGURE 7 Stray-current corrosion of a tank by an impressed-current cathodic protection system.

waterproof coating. Penetration of imperfect plastic wrappings in such cases is the most common cause of SCC in chemical plants.

Most design engineers, aware of galvanic corrosion, know that they should avoid combining dissimilar metals in a corrosive environment, but they sometimes underestimate the corrodibility while trying to find metals with the best wear resistance, conductivity, formability, or any other properties needed. Too often the engineers rely completely on gaskets and seals to keep out the corrosive; any misassembly or failure of a seal completely destroys the careful design. Where galvanic cells are unavoidable, the three rules to follow are (1) metals should be chosen with a small potential difference, (2) the cathode metal should be small, and (3) the two metals should be placed as far apart as possible.

Careful design can eliminate most crevice corrosion by designing out the crevice (for example, by welding instead of riveting) or the crevice can be protected galvanically, such as by using zinc-coated bolts. A common crevice problem is the use of absorbent gaskets. Metals in a corrosive environment should not contact open sponge rubber, fiberglass blankets, plastic foams, cardboard, or wood.

Erosion corrosion is usually combated by proper design. Pipe diameters and bend radii can be increased and inlets and outlets streamlined to reduce flow rates. Pipe elbows are often made thicker than the pipes to allow for erosion corrosion, or replaceable impingement plates may be inserted in the elbow to deflect the liquid stream. Tube-type heat exchangers, which often suffer erosion corrosion in the first few inches of the turbulent, inlet end of the tubes, may have their lives doubled by extending the tubes ~4 in. beyond the tube sheet, thus moving the erosion corrosion out to a position where leaks are less serious.

High stresses must be avoided in the design of equipment. These even include stresses from machining that can create a corrosion cell with unstressed metal as the cathode. Assembly stresses are a common corrosion problem, particularly stresses created by welding. If an assembly must operate at high temperatures, then the materials should be matched for thermal expansion; otherwise, large stress cells may be created. Ideally, the design should keep the total stresses, both operating and residual, below the stress intensity that initiates SCC.

"Hot spots" are especially serious. The local heating produces stresses along with high local corrosion rates. Hot spots may evaporate the corrodent and concentrate it. A differential temperature cell may develop an active/passive situation and the mechanical strength of the metal may decrease. Altogether, a hot spot is an ideal situation for either severe localized attack or SCC.

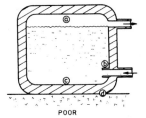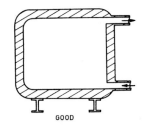

FIGURE 8 Design errors in chemical equipment: (a) vapor space permits splashing; (b) pipe extension leaves areas difficult to clean; (c) tank cannot be drained completely; and (d) crevice under tank traps spilled liquid.

C. Inspection and Maintenance

Proper design should always provide ready accessibility to critical areas so that complete cleaning and inspection can be performed. The design must also consider easy replacement of parts. This is often overlooked even in as carefully designed a machine as an automobile. Where visual inspection is not practical, as in a buried pipeline, provision should be made to position monitoring probes or to insert and remove coupons.

Figure 8 shows a design error that prevents thorough cleaning by making some areas of the equipment inaccessible to brushes. Also shown in the figure are features that leave corrosive in contact with the metal longer than necessary. The tank can neither be completely filled nor completely drained. Too often the design engineer assumes that corrosion considerations end with the selection of materials.

VIII. REACTION OF METALS WITH GASES

A. Oxidation Principles

The oxidation of metals in high-temperature, gaseous environments is an oxidation-reduction reaction, with electron exchange between metal and gas, much like electrochemical corrosion processes in liquids. All metals, except gold, are thermodynamically unstable in air at room temperature; they are driven to react with air by the Gibbs energy of formation of their oxides. Nevertheless, metals can be used in gases, even at high temperatures if their reaction rates are sufficiently slow. Diffusion of the gas to the metal or the metal to the gas usually controls oxidation kinetics.

One of the first attempts to apply a mechanism theory to oxidation of metals was the proposal of the Pilling–Bedworth ratio. Pilling and Bedworth observed that most metals quickly react and form a protective scale of oxide that serves as a barrier layer between metal and gas, greatly slowing the reaction as the scale grows thicker. However, some metals, particularly the alkali and alkaline earth metals, do not decrease oxidation rate with time, even though they form very dense oxides. The oxides are so dense, in fact, that their volumes are smaller than the volume of metal reacted. The Pilling–Bedworth ratio was thus defined as the volume of oxide divided by the volume of metal reacted. If the ratio is <1, then the oxide presumably cannot cover all the metal surface and thus leaves bare metal exposed to continue reaction directly with the gas (assuming no lateral diffusion).

If the ratio is ≥1, then the oxide completely covers the metal surface and allows the reaction to continue only after solid-state diffusion through the oxide, a very slow process. As the scale grows, it further separates metal from gas, slowing the diffusion even more. However, if the Pilling–Bedworth ratio is high (>2), then internal stresses in the oxide build up until the oxide may eventually spall off and expose bare metal once again for rapid oxidation. Although the ratio is by no means precise, it does indicate the mechanisms controlling the oxidation.

Once a barrier layer of oxide protects the metal surface, oxygen molecules from the gas adsorb on the oxide, dissociate to oxygen atoms, and pick up electrons from the oxide to become O^{2-} ions. Metal oxide can grow only if (1) electrons can pass through its structure and O^{2-} ions can diffuse inward through the oxide to the metal surface, (2) metal ions diffuse outward to the adsorbed O^{2-} ions, or (3) both ions diffuse toward each other.

Since oxides allow the passage of electrons through their structure, most can be classed as semiconductors. Semiconductors have either positive (p-type) charge carriers or negative (n-type) charge carriers, or both. Wustite ($Fe_{<1}O$) is a good example of a p-type oxide. As the formula indicates, some Fe^{2+} cations are missing from the structure. Electrical charge neutrality is maintained in the oxide by the presence of some Fe^{3+} cations; that is, some iron ions contribute an additional low-energy electron to satisfy the demands of the oxygen in the structure. These empty low-energy electron positions, termed electron "holes," allow the movement of electrical charge as an electron from one Fe^{2+} ion moves into the hole of an Fe^{3+} ion. The hole, considered to have a positive charge, moves to the position vacated by the moving electron. Wustite thus allows current to pass by hole conduction and allows ionic diffusion of Fe^{2+} through its cation vacancies.

The n-type semiconductors may be either an ion deficient or cation excessive. ZrO_2 is an example of the anion-deficient type. A small fraction of the cations are Zr^{2+} instead of Zr^{4+}, allowing a few O^{2-} anions to be absent from the oxide structure while still maintaining electrical neutrality. The anion vacancies allow O^{2-} ions to diffuse through the structure, while the two extra electrons on the

Zr^{2+} ions are easily detached to move about freely in the structure to carry most of the current.

ZnO is an example of a cation-excessive, n-type semiconducting oxide. The Zn^{2+} ion is small enough to fit interstitially between the ions in the ZnO crystal structure. As the metal oxidizes, the Zn surface releases Zn^{2+} ions and free electrons into the ZnO layer. The free electrons are the negative charge carriers and the Zn^{2+} interstitial ions diffuse from one interstitial position to another toward the O^{2-} ions at the outer oxide surface.

B. Kinetics of Oxide Growth

For nonprotective oxides such as those with Pilling–Bedworth ratios <1, those that are cracked or porous, or those that melt, the oxidation rate remains constant with time. This leads to a linear rate law:

$$w = k_L t \qquad (8)$$

where w is the weight of oxide formed, k_L is the rate constant, and t is the oxidizing time (see Fig. 9a).

Most commonly observed is the diffusion controlled situation in which diffusion of ions through the oxide

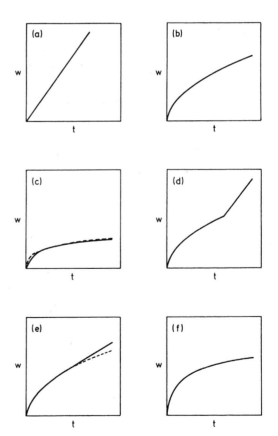

FIGURE 9 Observed variations in oxidation with time at constant temperature: (a) linear; (b) parabolic; (c) thin films, log (——), universe log (– – –); (d) breakaway; (e) paralinear; (f) logarithmic.

barrier slows the oxidation as the scale thickens, because the diffusion distances increase. The rate is inversely proportional to the diffusion distance, leading to the parabolic equation:

$$w^2 = k_p t \qquad (9)$$

The oxide layer protects the metal so that the oxidation rate quickly decreases. If the parabolic rate constant k_p is small, then the metal may perform quite satisfactorily in the environment even though it oxidizes (see Fig. 9b).

Parabolic oxidation is not observed in the first stages of oxidation, however. In this thin-film range, an electric field is set up between the adsorbed O^{2-} ions on the outer surface of the oxide and the metal ions on the metal surface. At film thicknesses <10 nm, the field is so strong that ions are pulled through the film much more rapidly than by normal diffusion. The rate-controlling step in the oxidation then becomes the electron transfer across the film, which appears to follow either a logarithmic or inverse logarithmic rate law:

$$w = k_e \log(At + 1) \quad \text{or}$$
$$1/w = \beta - k_i \log t \qquad (10)$$

Depending on how the constants A and B are chosen, and the rate constants k_e and k_i, experimental measurements fit the two equations about equally well. A variety of thin-film theories have been proposed that predict one equation or the other. Some theories predict a logarithmic equation in the earliest stages or at low temperatures which then changes to an inverse logarithmic equation later or at higher temperatures (see Fig. 9c).

Breakaway oxidation can occur where compressive stresses exceed the adhesive forces of the scale. The protective scale, growing parabolically, cracks or spalls off and exposes bare metal that oxidizes linearly. In gases where oxidation is very rapid, the exposed metal may again be protected by an oxide layer until the scale breaks again somewhere on the metal, initiating a series of parabolic steps. Breakaway oxidation is common if the metal is alternately heated and cooled; this creates high stresses in the scale because thermal expansion of the scale is much less than that of the metal (see Fig. 9d).

If a metal forms two oxides, the inner one protective but gradually changing to an outer unprotective layer, then paralinear oxidation occurs. The outer oxide may be porous or may sublime, while the inner layer forms a protective layer according to the parabolic equation. As the growth of the inner oxide gradually slows, it eventually equals the rate of transformation to the outer oxide, and the thickness of the inner oxide then remains constant.

The original parabolic growth curve therefore changes to a tangential linear region as the oxidation rate becomes constant (see Fig. 9c).

Some metals are observed to oxidize according to a logarithmic equation even though the scales are far too thick for any thin-film mechanism to apply. This occurs in protective scales that decrease their diffusion area as oxidation proceeds.

Three situations are known to reduce the diffusion area with time:

1. Cation vacancies in the scale may collect near the metal surface and form voids, which then restrict further cation diffusion.
2. O^{2-} ions diffusing inward may create stresses that cause the scale to crack parallel to the metal surface or may detach the scale from the metal in a few places.
3. Alloys may have more than one metal oxidizing, with particles of one oxide forming within the other oxide layer. If diffusion is very slow through the oxide particles, they may reduce diffusion almost as effectively as voids. The result is that oxidation is slowed more rapidly than normal parabolic behavior so that a logarithmic equation closely approximates the scale growth (see Fig. 9f).

Oxidation may be a catastrophe in cases where it is linear but more especially for metal powder or foil that heats up from the exothermic oxidation process itself. In those cases, the metal literally burns. However, the term *catastrophic oxidation* is reserved for situations in which a liquid oxide forms. A few oxides are notorious for melting at low temperatures, particularly V_2O_5, MoO_3, and PbO. If one of these oxides forms from impurities either in the metal or the gas, the liquid will penetrate the protective scale along pores, cracks, or grain boundaries until it gets between the metal and the scale, detaches the scale, and fluxes it off. Thus, a stainless steel exhaust stack, protected by its Cr_2O_3 layer, will be destroyed if fuel oil containing vanadium is burned. The V_2O_5 in the fly ash, melting at 675°C, will cause catastrophic oxidation.

C. Alloy Oxidation

When an alloy is oxidized, what oxide or oxides form and where they form depend on several factors: the thermodynamic stability of the oxides, diffusion rates in the metal and in the oxides, relative volumes of the various oxides, solubilities of the oxides with each other, and possibility of forming ternary oxides.

The simplest situation can illustrate the various possibilities. Metal A, which would oxidize to AO, is alloyed with metal B, which forms BO (BO being more stable thermodynamically than AO). On oxidation, the following products have been observed. An internal oxide of BO may form if oxygen diffuses rapidly into the metal. BO appears as a layer of fine precipitates just below the alloy surface and grows inward parabolically. It can even form if the partial pressure of oxygen in the gas is less than the dissociation pressure of BO.

A single external scale of BO may form if AO is unstable in the environment or if the concentration of the A is too low in the alloy. Below a critical concentration of A, the reaction $AO + B \rightarrow BO + A$ is favored, so any AO that might momentarily form is reduced by the high concentration of B available. Similarly, only an AO scale may form if the concentration of B in the alloy is less than its critical value.

Most commonly observed, however, are separate layers of AO and BO. The faster growing oxide (e.g., AO) first forms a layer, leaving the metal surface depleted in A. Then a layer of BO forms.

Occasionally precipitates of one oxide may be seen in the scale of the other. The slower growing oxide is stable enough to form while being surrounded by the faster growing one.

Ternary oxides may form in some A–B–O systems. If AO and BO are at least partially miscible, an oxide of variable composition, $(A, B)O$, can form. Ternary compounds, too, may form, the so-called spinels being particularly important. Spinels have both divalent and trivalent cations with the formula $AO \cdot B_2O_3$ and the same crystal structure as the mineral spinel ($MgO \cdot Al_2O_3$). Many spinels have few ion vacancies and so grow quite slowly, providing a very protective oxide scale on the metal.

D. Hot Corrosion

Hot corrosion is an accelerated oxidation caused by a gas, usually combustion products, forming a layer of molten salt, usually a sulfate, on the metal or oxide surface. The problem has been studied most extensively in jet engines of aircraft flying over seawater but has also appeared in gas turbines in ships, fast breeder reactors, and municipal waste incinerators.

During an initiation period, which may take thousands of hours, the scale-forming element becomes depleted in the alloy, oxides partially dissolve in the salt, and growth stresses increase in the scale. Once the salt penetrates the oxide scale, hot corrosion becomes extremely rapid and often catastrophic. The scale may change to a porous, honeycomb structure or may be undermined and completely removed in some places. The oxide layer ceases to provide an effective barrier to further oxidation by the gas.

E. Oxidation Prevention

None of the refractory (high-melting) pure metals can be used unprotected in air at high temperatures. Molybdenum and tungsten form molten oxides, the platinum metals all form gaseous oxides, and niobium and tantalum oxides have extremely high Pilling–Bedworth ratios (>2.5). Alloying these elements, although it can improve oxidation resistance, always worsens their mechanical properties so that the only way found to use them in high-temperature reactive gases is to coat them.

A coating must oxidize slowly, if at all, and must adhere well. Related to adherence is a wide range of properties that the coating should possess: resistance to impact, erosion, and abrasion; ability to withstand creep and plastic deformation; and the lack of formation of compounds that would worsen mechanical properties. In addition, it is desirable for the coating to be self-healing and easy to apply.

Coatings can be inert, such as ceramics and noble metals, or reactive with the environment, such as reactive metals and compounds (e.g., silicides, carbides, borides). Reactive metals have been very successful as protective coatings on the refractory metals. For example, zinc coatings on niobium diffuse into the substrate to form a series of niobium–zinc intermetallic compound layers. These compounds oxidize to ZnO, which severely limits Zn^{2+} diffusion even at high temperatures. Aluminide coatings, applied to niobium and tantalum by hot-dipping, oxidize to α-Al_2O_3, which forms an excellent diffusion barrier. Molybdenum and tungsten are often protected by silicate coatings that slowly oxidize to SiO_2.

Ceramic coatings are used extensively in gas turbines. In addition to being unreactive in the environment, they also act as a thermal barrier, thus allowing higher gas inlet temperatures and increasing engine efficiency. A plasma-sprayed zirconia–yttria layer 0.1–0.5 mm thick typically is applied over a 0.1-mm bond layer. Thermal cycling is accomodated by high thermal expansion of the stabilized zirconia.

SEE ALSO THE FOLLOWING ARTICLES

ELECTROCHEMISTRY • ELECTROLYTE SOLUTIONS, THERMODYNAMICS • ELECTROLYTE SOLUTIONS, TRANSPORT PROPERTIES • EMBRITTLEMENT, ENGINEERING ALLOYS • FRACTURE AND FATIGUE

BIBLIOGRAPHY

Atkinson, J. T. N., and Van Droffelaar. (1995). "Corrosion and Its Control: An Introduction to the Subject," 2nd ed., NACE, Houston, TX.

Bradford, S. A. (1993). "Corrosion Control," Van Nostrand-Reinhold, New York.

Bradford, S. A. (1998). "Practical Self-Study Guide to Corrosion Control," CASTI Publishing, Edmonton, Canada.

Bradford, S. A. (2000). "CASTI Handbook of Corrosion in Soils," CASTI Publishing, Edmonton, Canada.

Craig, B., and Anderson, D., eds. (1995). "Handbook of Corrosion Data," 2nd ed., ASM International, Materials Park, OH.

Davis, J. R., ed. (2000). "Corrosion: Understanding the Basics," ASM International, Materials Park, OH.

During, E. D. (1997). "Corrosion Atlas," 3rd ed., Elsevier Science, New York.

Jones, D. A. (1996). "Principles and Prevention of Corrosion," Prentice Hall International, Englewood Cliffs, NJ.

Mattsson, E. (1996). "Basic Corrosion Technology for Scientists and Engineers," The Institute of Materials, London.

Schweitzer, P. A. (1996). "Corrosion Engineering Handbook," Dekker, New York.

Talbot, D., and Talbot, J. (1997). "Corrosion Science and Technology," CRC Press, Boca Raton, FL.

Timmins, P. F. (1996). "Predictive Corrosion and Failure Control in Process Operations," ASM International, Materials Park, OH.

von Baeckmann, W., Schwenk, W., and Prinz, W. (1997). "Handbook of Cathodic Protection: Theory and Practice of Electrochemical Protection Processes," 3rd ed., Gulf Pub., Houston, TX.

Cosmic Inflation

Edward P. Tryon

Hunter College and Graduate Center of The City University of New York

GLOSSARY

Comoving Moving in unison with the overall expansion of the universe.

Cosmic microwave background (CMB) Emitted when the universe was ~300,000 years old, this electromagnetic radiation fills all of space and contains detailed information about the universe at that early time.

Cosmic scale factor The distance between any two comoving points is a constant fraction of the cosmic scale factor $a(t)$, which grows in unison with the cosmic expansion. The finite volume of a closed universe is $V_u(t) = 2\pi^2 a^3(t)$.

Cosmological constant Physical constant Λ in the field equations of general relativity, corresponding to an effective force that increases in proportion to distance and is repulsive for positive Λ. Its effects are equivalent to those of constant vacuum energy and pressure.

Critical density Curvature of space is positive, zero, or negative according to whether the average density of mass-energy is greater than, equal to, or less than a critical density determined by the Hubble parameter and cosmological constant.

Curvature constant A constant k that is $+1$, 0, or -1 for positive, zero, or negative spatial curvature, respectively.

Doppler effect Change in observed frequency of any type of wave due to motion of the source relative to the observer: motion of the source away from an observer decreases the observed frequency by a factor determined by the relative velocity.

False vacuum Metastable or unstable state predicted by grand unified theories, in which space has pressure P and mass-energy density ρ related by $P = -c^2\rho$; the gravitational effects mimic a positive cosmological constant.

Field Type of variable representing quantities that may, in principle, be measured (or theoretically analyzed) at any point of space–time; hence, fields possess values (perhaps zero in some regions) at every point of space–time.

Higgs fields In unified theories, field quantities that, when all are zero, have positive energy density and negative pressure giving rise to false vacuum. At least one Higgs field must be nonzero in the state of minimum energy (true vacuum), which breaks an underlying symmetry of the theory.

Horizon distance Continually increasing distance beyond which neither light nor any causal influence could yet have traveled since the universe began.

Hubble parameter Ratio H of the recessional velocity of a comoving object to its distance from a comoving observer; by Hubble's law, this ratio has the same value for all such objects and observers. Its present value is called *Hubble's constant.*

Inflaton fields In theories of elementary particles, the generic term for fields with a false vacuum state that could give rise to inflation; e.g., Higgs fields.

Riemannian geometry The geometry of any curved space or space-time that is approximately flat over short distances.

Robertson–Walker (RW) metric: In homogeneous, isotropic, cosmological models, describes space–time and its evolution in terms of a curvature constant and (evolving) cosmic scale factor.

Second-rank tensor A 4×4 array (i.e., matrix) of quantities whose values depend, in a characteristic way, on the reference frame in which they are evaluated.

Stress-energy tensor The source of gravity in general relativity, a second-rank tensor whose components include the energy and momentum densities, pressure, and other stresses.

Work-energy theorem Any change in energy of a system equals the work done on it; as a corollary, the energy of an isolated system is constant.

COSMIC INFLATION refers to a conjectured, early period of exponential growth, during which the universe is believed to have increased in linear dimensions by a factor exceeding $\sim 10^{30}$. The inflationary period ended before the universe was $\sim 10^{-30}$ seconds old, after which cosmic evolution proceeded in accord with the standard big bang model. Inflation is based on the general theory of relativity together with grand unified (and several other) theories of the elementary particles. The latter predict a peculiar state called the *false vacuum*, which rendered gravitation repulsive for a very early and brief time, thereby causing inflation. Several long-standing cosmic puzzles are resolved by this conjecture.

Standard theories of inflation predict that our universe is spatially flat. This prediction is strikingly supported by recent studies of the cosmic microwave background (CMB). Inflation also makes quantitative predictions about tiny deviations from perfect uniformity at very early times, the slight concentrations of matter that later grew into galaxies and larger structures. These predictions also appear to be confirmed by recent, detailed studies of the CMB.

I. EINSTEIN'S THEORY OF GRAVITATION

In Newton's theory of gravitation, mass is the source of gravity and the resulting force is universally attractive. The empirical successes of Newton's theory are numerous and familiar. All observations of gravitational phenomena prior to this century were consistent with Newtonian theory, with one exception: a minor feature of Mercury's orbit. Precise observations of the orbit dating from 1765 revealed an advance of the perihelion amounting to 43 arc sec per century in excess of that which could be explained by the perturbing effects of other planets. Efforts were made to explain this discrepancy by postulating the existence of small, perturbing masses in close orbit around the sun, and this type of explanation retained some plausibility into the early part of the twentieth century. Increasingly refined observations, however, have ruled out such a possibility: there is not sufficient mass in close orbit around the sun to explain the discrepancy.

In 1907, Einstein published the first in a series of papers in which he sought to develop a relativistic theory of gravitation. From 1913 onward, he identified the gravitational field with the metric tensor of Riemannian (i.e., curved) space–time. These efforts culminated in his general theory of relativity (GTR), which was summarized and published in its essentially final form in 1916. The scope and power of the theory made possible the construction of detailed models for the entire universe, and Einstein published the first such model in 1917. This early model was subsequently retracted, but the science of mathematical cosmology had arrived.

The GTR reproduces the successful predictions of Newtonian theory and furthermore explains the perihelion advance of Mercury's orbit. Measurement of the bending of starlight in close passage by the sun during the solar eclipse of 1919 resulted in further, dramatic confirmation of GTR, which has since been accepted as the standard theory of gravitation. Additional successes of GTR include correct predictions of gravitational time dilation and, beginning in 1967, the time required for radar echoes to return to Earth from Mercury and artificial satellites.

There are no macroscopic phenomena for which the predictions of GTR are known (or suspected) to fail. Efforts to quantize GTR (i.e., to wed GTR with the principles of quantum theory) have met with continuing frustration, primarily because the customary perturbative method for constructing solutions to quantum equations yields divergent integrals whose infinities cannot be canceled against one another (the quantized version of GTR is not renormalizable). Hence there is widespread belief that microscopic gravitational phenomena involve one or more features differing from those of GTR in a quantized form. In the *macroscopic* realm, however, any acceptable theory of gravitation will have to reproduce the successful predictions of GTR. It is reasonable (though not strictly necessary) to suppose that all the macroscopic predictions of GTR, including those that remain untested, will be reproduced by any theory that may eventually supersede it.

For gravitational phenomena, the obvious borderline between macroscopic and microscopic domains is provided by the *Planck length* (L_P):

$$L_P \equiv (Gh/c^3)^{1/2} = 4.0 \times 10^{-33} \text{ cm} \qquad (1)$$

where G is Newton's constant of gravitation, h is Planck's constant, and c is the speed of light. Quantum effects of gravity are expected to be significant over distances comparable to or less than L_P. According to the wave–particle duality of quantum theory, such short distances are probed only by particles whose momenta exceed Planck's constant divided by L_P. Such particles are highly relativistic, so their momenta equal their kinetic energies divided by c. Setting their energies equal to a thermal value of order kT (where k is Boltzmann's constant and T the absolute temperature), the *Planck temperature* (T_P) is obtained:

$$T_P \equiv ch/kL_P = 3.6 \times 10^{32} \text{ K} \qquad (2)$$

where "K" denotes kelvins (degrees on the absolute scale). For temperatures comparable to or greater than T_P, effects of quantum gravity may be important.

The fundamental constants of quantum gravity can also be combined to form the *Planck mass* (M_P):

$$M_P \equiv (hc/G)^{1/2} = 5.5 \times 10^{-5} \text{ g} \qquad (3)$$

The Planck mass and length give rise to the *Planck density* ρ_P:

$$\rho_P \equiv M_P / L_P^3 = 8.2 \times 10^{92} \text{ g/cm}^3 \qquad (4)$$

For mass densities comparable to or greater than ρ_P, the effects of quantum gravity are expected to be significant. (Definitions of L_P, T_P, M_P, and ρ_P sometimes differ from those given here by factors of order unity, but only the orders of magnitude are important in most applications.)

According to the standard big bang model, the very early universe was filled with an extremely hot, dense gas of elementary particles and radiation in (or near) thermal equilibrium. As the universe expanded, the temperature and density both decreased. Calculations indicate that the temperature fell below T_P when the uni-

verse was $\sim 10^{-45}$ sec. old, and the density fell below ρ_P at $\sim 10^{-44}$ sec. Hence currently unknown effects of quantum gravity may have been important during the first $\sim 10^{-44}$ sec or so of cosmic evolution. For all later times, however, the temperature and density were low enough that GTR should provide an adequate description of gravitation, and our theories of elementary particles together with thermodynamics and other established branches of physics should be applicable in a straightforward way. We shall assume this to be the case.

As remarked earlier, GTR identifies the gravitational field with a second-rank tensor, the metric tensor of curved space–time. The source of gravitation in GTR is therefore not simply mass (nor mass-energy), but another second-rank tensor: the stress-energy tensor. Hence stresses contribute to the source of gravitation in GTR, the simplest example of stress being pressure.

It is commonly said that gravitation is a universally attractive force, but such a sweeping statement is not supported by the field equations of GTR. It is true that the kinds of objects known (or believed) to constitute the present universe attract each other gravitationally, but GTR contains another possibility: granted an extraordinary medium wherein pressure is negative and sufficiently large, the resulting gravitational field would be repulsive. This feature of GTR lies at the heart of cosmic inflation: it is conjectured that during a very early and brief stage of cosmic evolution, space was characterized by a state called a false vacuum with a large, negative pressure. A false vacuum would mimic the effects of a positive cosmological constant in the field equations, as will be explained in some detail in Section IV.G. Cosmic dimensions would have increased at an exponential rate during this brief period, which is called the *inflationary era*.

The concept of negative pressure warrants a brief explanation. A medium characterized by positive pressure, such as a confined gas, has an innate tendency to expand. In contrast, a medium characterized by negative pressure has an innate tendency to contract. For example, consider a solid rubber ball whose surface has been pulled outward in all directions, stretching the rubber within. More generally, for a medium characterized by pressure P, the internal energy U contained within a variable volume V is governed by $\Delta U = -P\Delta V$ for adiabatic expansion or contraction (a result of the work-energy theorem, where ΔU and ΔV denote small changes in U and V, respectively). If U increases when V increases, then P is negative.

The possibility of an early false vacuum with resulting negative pressure and cosmic inflation was discovered in 1979 by Alan H. Guth, while he was studying grand unified theories (GUTs) of elementary particles and their interactions. (The relevant features of GUTs will be described in Section VII.C.) The resulting inflation is believed to have increased cosmic linear dimensions by a factor of $\sim 10^{30}$ in the first $\sim 10^{-30}$ sec, and probably by very much more.

Over the two decades since Guth's initial discovery, it has been noted that inflation is possible in many different versions of GUTs. Furthermore, superstring theory and several other types of particle theory contain features similar to those of GUTs, and give rise to comparable inflation. Inflation therefore encompasses a *class* of theories, including at least fifty distinct versions that have been proposed since Guth's initial model.

All the theories of elementary particles that predict inflation remain speculative, as does inflation itself. We shall see, however, that inflation would explain several observed features of the universe that had previously defied explanation. Inflation also makes a quantitative prediction about deviations from perfect uniformity in the early universe, deviations that later gave rise to the clumping of matter into galaxies and clusters of galaxies. This prediction has recently been strikingly confirmed by detailed studies of the cosmic background radiation. Empirical evidence for an early period of cosmic inflation is therefore quite substantial, whatever the details of the underlying particle theory might prove to be.

In common with the initial model of Guth, the great majority of inflationary theories predict that the average curvature of space is (virtually) zero, corresponding to *flat* universes. Such theories are called "standard inflation." A few theories predict nonzero, negative (*hyperbolic*) spatial curvature, which, if small, cannot be ruled out by present observations. The latter theories are sometimes called "open inflation" (a potentially misleading name, because flat and hyperbolic universes are both infinite, and traditionally have both been called "open"). Most predictions of standard and open inflation are virtually the same, however, so we shall only distinguish between them where they differ.

We shall next enumerate and briefly describe some heretofore unexplained features of the universe that may be understood in terms of cosmic inflation, and then we will proceed to a detailed presentation of the theory.

II. COSMOLOGICAL PUZZLES

A. The Horizon Problem

Both the special and general theories of relativity preclude the transmission of any causal agent at a speed faster than light. Accepting the evidence that our universe has a finite age, each observer is surrounded by a *particle horizon*: no comoving particle beyond this horizon could have yet made its presence known to us in any way, for there has not yet been sufficient time for light (or anything else) to have traveled the intervening distance. Regions of the universe outside each others' horizons are causally disconnected, having had no opportunity to influence each other. The distance to the horizon naturally increases with the passage of time. In the standard big bang model, an order-of-magnitude estimate for the horizon distance is simply the speed of light multiplied by the time that has elapsed since the cosmos began. (Modifications to this simple estimate arise because the universe is expanding and because space–time is curved; see Section VI.A.)

Astronomical observations entail the reception and interpretation of electromagnetic radiation, that is, visible light and, in recent decades, radiation from nonvisible parts of the spectrum such as radio waves, X-rays, and γ-rays. We shall refer to that portion of the cosmos from which we are now receiving electromagnetic radiation as the *observable universe*; it is that fraction of the whole which may be studied by contemporary astronomers.

The universe was virtually opaque to electromagnetic radiation for times earlier than $\sim 300,000$ yr (as will be explained). Therefore, the present radius of the observable universe is less than the present horizon distance— $\sim 300,000$ ly less in the standard big bang model, or very much less if cosmic inflation occurred at the very early time characteristic of inflationary models. The cosmos is currently ~ 14 billion years old, and the present radius of the observable universe is ~ 50 billion light-years. (Sources of light have been moving away from us while the light traveled toward us.) Approximately 10^{11} galaxies lie within the observable universe. About 10% of these galaxies are found in groups called *clusters*, some of which form larger groups called *superclusters* (i.e., clusters of clusters).

When averaged over suitably large regions, the universe appears to be quite similar in all directions at all distances, i.e., isotropic and homogeneous. The distribution of matter is considerably less homogeneous than was supposed only twenty years ago, however. Over the last two decades, continually improving technology has made possible far more detailed surveys of the sky, and surprising features have been discovered.

Over distances of roughly 100 million light-years, the concentration of matter into galaxies, clusters, and superclusters is roughly a fractal pattern. No fractal pattern holds over larger distances, but new and larger structures have become apparent. Clusters and superclusters are arrayed in filaments, sheets, and walls, with nearly empty voids between them. Diameters of the voids typically range from 600 to 900 million light-years. One sheet of galaxies, the "Great Wall," is about 750 million light-years long, 250 million light-years wide, and 20 million light-years thick. The *largest* voids are nearly 3 *billion* light-years across. In very rough terms, the distribution of matter resembles an irregular foam of soap bubbles, except that many of the voids are interconnected as in a sponge. No pattern is evident in the locations or sizes of the walls and voids—their distribution appears to be random. In a *statistical* sense, the universe still appears to be isotropic and homogeneous.

More dramatic and precise evidence for isotropy is provided by the cosmic background radiation, whose existence and properties were anticipated in work on the early universe by George Gamow and collaborators in the late 1940s. These investigators assumed that the early universe was very hot, in which case a calculable fraction of the primordial protons and neutrons would have fused together to form several light species of nuclei [^{2}H (*deuterium*), ^{3}He, ^{4}He, and ^{7}Li] during the first few minutes after the big bang. (Heavier nuclei are believed to have formed later, in the cores of stars and during stellar explosions.) The abundances predicted for these light nuclei agreed with observations, which confirmed the theory of their formation.

After this early period of nucleosynthesis, the universe was filled with a hot plasma (ionized gas) of light nuclei and electrons, together with a thermal distribution of photons. When the universe had expanded and cooled for $\sim$300,000 yr, the temperature dropped below 3000 K, which enabled the nuclei and electrons to combine and form electrically neutral atoms that were essentially transparent to radiation. (At earlier times the plasma was virtually opaque to electromagnetic radiation, because the photons could travel only short distances before being scattered by charged particles: the observable universe extends to the distance where we now see it as it was at the age of $\sim$300,000 years. This is called the *time of last scattering*.) The thermal photons that existed when the atoms formed traveled freely through the resulting gas, and later through the vast reaches of empty space between the stars and galaxies that eventually formed out of that gas. These photons have survived to the present, and they bathe every observer who looks toward the heavens. (If a television set is tuned to a channel with no local station and the brightness control is turned to its lowest setting, $\sim$1% of the flecks of "snow" on the screen are a result of this cosmic background radiation.)

The photons now reaching earth have been traveling for $\sim$14 billion yr and hence were emitted by distant parts of the universe receding from us with speeds near that of light. The Doppler effect has redshifted these photons, but preserved the blackbody distribution; hence the radiation remains characterized by a temperature.

Extending earlier work with Gamow, R. A. Alpher and R. C. Herman described the properties of this cosmic background radiation in 1948, estimating its present temperature to be $\sim$5 K (with its peak in the *microwave* portion of the spectrum). No technology existed, however,

for observing such radiation at that time. In 1964, Arno A. Penzias and Robert W. Wilson accidently discovered this radiation (at a single wavelength) with a large horn antenna built to observe the Echo telecommunications satellite. Subsequent observations by numerous workers established that the spectrum is indeed blackbody, with a temperature of 2.73 K. The present consensus in favor of a hot big bang model resulted from this discovery of the relic radiation.

The cosmic background radiation is isotropic within $\sim$one part in 10^5 (after the effect of earth's velocity with respect to distant sources is factored out). This high degree of isotropy becomes all the more striking when one realizes that the radiation now reaching earth from opposite directions in the sky was emitted from regions of the cosmos that were over 100 times farther apart than the horizon distance at the time of emission, assuming the horizon distance was that implied by the standard big bang model. How could different regions of the cosmos that had never been in causal contact have been so precisely similar?

Of course the theoretical description of any dynamical system requires a stipulation of initial conditions; one might simply postulate that the universe came into being with perfect isotropy as an initial condition. An initial condition so special and precise begs for an explanation, however; this is called the *horizon problem*.

B. The Flatness Problem

In a sweeping extrapolation beyond the range of observations, it is often assumed that the entire cosmos is homogeneous and isotropic. This assumption is sometimes called the *Cosmological Principle*, a name that is perhaps misleading in suggesting greater sanctity for the assumption than is warranted either by evidence or by logic. (Indeed, current scenarios for cosmic inflation imply that the universe is distinctly different at some great distance beyond the horizon, as will be explained in Section VIII.B.) The existence of a horizon implies, however, that the observable universe has evolved in a manner independent of very distant regions. Hence the evolution of the observable universe should be explicable in terms of a model that assumes the universe to be the same everywhere. In light of this fact and the relative simplicity of models satisfying the Cosmological Principle, such models clearly warrant study.

In 1922, Alexandre Friedmann discovered three classes of solutions to the field equations of GTR that fulfill the Cosmological Principle. These three classes exhaust the possibilities (within the stated assumptions), as was shown in the 1930s by H. P. Robertson and A. G. Walker. The three classes of Friedmann models (also called

Friedmann–Robertson–Walker or Friedmann–Lemaître models) are distinguished from one another by the curvature of space, which may be positive, zero, or negative. The corresponding spatial geometries are *spherical*, *flat*, and *hyperbolic*, respectively. Einstein's field equations correlate the spatial geometry with the average density ρ of mass-energy in the universe. There is a critical density ρ_c such that spherical, flat, and hyperbolic geometries correspond to $\rho > \rho_c$, $\rho = \rho_c$, and $\rho < \rho_c$, respectively. A spherical universe has finite volume (space curves back on itself), and is said to be "closed"; flat and hyperbolic universes have infinite volumes and are said to be "open."

Astronomical observations have revealed no departures from Euclidean geometry on a cosmic scale. Furthermore, recent measurements of the cosmic microwave background reveal space to be so nearly flat that $\rho = (1.0 \pm 0.1)\rho_c$. This is a striking feature of our universe, given that the ratio ρ/ρ_c need only lie in the range from zero to infinity. The ratio can be arbitrarily large in closed universes, which furthermore can be arbitrarily small and short-lived. The ratio can be arbitrarily close to zero in hyperbolic universes, with arbitrarily little matter. Flat universes are quite special, having the highest possible value for ρ/ρ_c of any infinite space. Our universe is remarkably close to flat, and the question of why is called the *flatness problem* (first pointed out, in detail, by Robert H. Dicke and P. James E. Peebles in 1979).

C. The Smoothness Problem

Although the observable universe is homogeneous on a large scale, it is definitely lumpy on smaller scales: the matter is concentrated into stars, galaxies, and clusters of galaxies. The horizon problem concerns the establishment of large-scale homogeneity, whereas the smoothness problem concerns the formation of galaxies and the chain of events that resulted in the observed sizes of galaxies. That a problem exists may be seen in the following way.

A homogeneous distribution of matter is rendered unstable to clumping by gravitational attraction. Any chance concentration of matter will attract surrounding matter to it in a process that feeds on itself. This phenomenon was first analyzed by Sir James Jeans, and is called the *Jeans instability*. If the universe had evolved in accord with the standard big bang model since the earliest moment that we presume to understand (i.e., since the universe was 10^{-44} sec old), then the inevitable thermal fluctuations in density at that early time would have resulted in clumps of matter *very* much larger than the galaxies, clusters, and superclusters that we observe. This was first noted by P. J. E. Peebles in 1968, and the discrepancy is called the *smoothness problem*.

D. Other Problems

In addition to the questions or problems thus far described, there are others of a more speculative nature that may be resolved by cosmic inflation. We shall enumerate and briefly explain them here. The first is a problem that arises in the context of GUTs.

Grand unified theories predict that a large number of magnetic monopoles (isolated north or south magnetic poles) would have been produced at a very early time (when the universe was $\sim 10^{-35}$ sec old). If the universe had expanded in the simple way envisioned by the standard big bang model, then the present abundance of monopoles would be vastly greater than is consistent with observations. (Despite extensive searches, we have no persuasive evidence that *any* monopoles exist.) This is called the *monopole problem* (see Sections VII.C and VIII.A).

The remaining questions and problems that we shall mention concern the earliest moment(s) of the cosmos. In the standard big bang model, one assumes that the universe began in an extremely dense, hot state that was expanding rapidly. In the absence of quantum effects, the density, temperature, and pressure would have been infinite at time zero, a seemingly incomprehensible situation sometimes referred to as the *singularity problem*. Currently unknown effects of quantum gravity might somehow have rendered finite these otherwise infinite quantities at time zero. It would still be natural to wonder, however, why the universe came into being in a state of rapid expansion, and why it was intensely hot.

The early expansion was not a result of the high temperature and resulting pressure, for it can be shown that a universe born stationary but with arbitrarily high (positive) pressure would have begun immediately to contract under the force of gravity (see Section IV.D). We know that any initial temperature would decrease as the universe expands, but the equations of GTR are consistent with an expanding universe having no local random motion: the temperature could have been absolute zero for all time. Why was the early universe hot, and why was it expanding?

A final question lies at the very borderline of science, but has recently become a subject of scientific speculation and even detailed model-building: How and why did the big bang occur? Is it possible to understand, in scientific terms, the creation of a universe *ex nihilo* (from nothing)?

The theory of cosmic inflation plays a significant role in contemporary discussions of all the aforementioned questions and/or problems, appearing to solve some while holding promise for resolving the others. Some readers may wish to understand the underlying description of cosmic geometry and evolution provided by GTR. Sections

III and IV develop the relevant features of GTR in considerable detail, using algebra and, occasionally, some basic elements of calculus. Section V describes the content, age, and future of the universe. Section VI explains the horizon and flatness problems in detail. Most of the equations are explained in prose in the accompanying text, so that a reader with a limited background in mathematics should nevertheless be able to grasp the main ideas. Alternatively, one might choose to skip these sections (especially in a first reading) and proceed directly to Section VII on unified theories and Higgs fields, without loss of continuity or broad understanding of cosmic inflation (the glossary may be consulted for the few terms introduced in the omitted sections).

III. GEOMETRY OF THE COSMOS

A. Metric Tensors

General relativity presumes that special relativity is valid to arbitrarily great precision over sufficiently small regions of space–time. GTR differs from special relativity, however, in allowing for space–time to be *curved* over extended regions. (Curved space-time is called *Riemannian*, in honor of the mathematician G. F. B. Riemann who, in 1854, made a seminal contribution to the theory of curved spaces.) Another difference is that while gravitation had been regarded as merely one of several forces in the context of special relativity, it is conceptually not a force at all in GTR. Instead, all effects of gravitation are imbedded in, and expressed by, the curvature of space–time.

In special relativity, there is an *invariant interval* s_{12} separating events at t_1, $\mathbf{r}_1$ and t_2, $\mathbf{r}_2$ given by

$$(s_{12})^2 = -c^2(t_1 - t_2)^2 + (x_1 - x_2)^2 + (y_1 - y_2)^2$$
$$+ (z_1 - z_2)^2 \qquad (5)$$

where x, y, and z denote Cartesian components of the position vector $\mathbf{r}$. If $t_1 = t_2$, $|s_{12}|$ is the distance between $\vec{r}_1$ and $\vec{r}_2$. If $\vec{r}_1 = \vec{r}_2$, then $|s_{12}|/c$ is the time interval between t_1 and t_2. [Some scientists define the overall sign of $(s_{12})^2$ to be opposite from that chosen here, but the physics is not affected so long as consistency is maintained.]

The remarkable relationship between space and time in special relativity is expressed by the fact that the algebraic form of s_{12} does not change under *Lorentz transformations*, which express the space and time coordinates in a moving frame as linear *combinations* of space and time coordinates in the original frame. This mixing of space with time implies a dependence of space and time on the reference frame of the observer, which underlies the counter-intuitive predictions of special relativity. The *flatness* of the space-time of special relativity is implied by the existence of coordinates t and $\vec{r}$ in terms of which s_{12} has the algebraic form of Eq. (5) *throughout* the space-time (called *Minkowski space–time*, after Hermann Minkowski).

The most familiar example of a curved space is the surface of a sphere. It is well-known that the surface of a sphere cannot be covered with a coordinate grid wherein all intersections between grid lines occur at right angles (the lines of longitude on earth are instructive). Similarly, a curved space–time of arbitrary (nonzero) curvature cannot be spanned by any rectangular set of coordinates such as those appearing in Eq. (5). One can, however, cover any *small* portion of a spherical surface with *approximately* rectangular coordinates, and the rectangular approximation can be made arbitrarily precise by choosing the region covered to be sufficiently small. In a like manner, any small region of a curved space–time can be spanned by rectangular coordinates, with the origin placed at the center of the region in question. The use of such coordinates (called *locally flat* or *locally inertial*) enables one to evaluate any infinitesimal interval ds between nearby space–time points by application of Eq. (5).

To span an extended region of a curved space–time with a single set of coordinates, it is necessary that the coordinate grid be curved. Since one has no *a priori* knowledge of how space–time is curved, the equations of GTR are typically expressed in terms of arbitrary coordinates x^μ ($\mu = 0, 1, 2,$ and 3). The central object of study becomes the *metric tensor* $g_{\mu\nu}$, defined implicitly by

$$(ds)^2 = g_{\mu\nu}\, dx^\mu\, dx^\nu \qquad (6)$$

together with the assumption that space–time is *locally* Minkowskian (and subject to an axiomatic proviso that $g_{\mu\nu}$ is symmetric, i.e., $g_{\mu\nu} = g_{\nu\mu}$).

In principle, one can determine $g_{\mu\nu}$ at any given point by using Eq. (5) with locally flat coordinates to determine infinitesimal intervals about that point, and then requiring the components of $g_{\mu\nu}$ to be such that Eq. (6) yields the same results in terms of infinitesimal changes dx^μ in the arbitrary coordinates x^μ. The values so determined for the components of $g_{\mu\nu}$ will clearly depend on the choice of coordinates x^μ, but it can be shown that the variation of $g_{\mu\nu}$ over an *extended* region also determines (or is determined by) the curvature of space–time. It is customary to express models for the geometry of the cosmos as models for the metric tensor, and we shall do so here.

B. The Robertson–Walker Metric

Assuming the universe to be homogeneous and isotropic, coordinates can be chosen such that

$$(ds)^2 = -c^2(dt)^2 + a^2(t)\left\{\frac{(dr)^2}{1 - kr^2}\right.$$

$$\left. + r^2[(d\theta)^2 + \sin^2\theta(d\varphi)^2]\right\} \qquad (7)$$

as was shown in 1935 by H. P. Robertson and, independently, by A. G. Walker. The ds of Eq. (7) is called the *Robertson–Walker* (RW) *line element*, and the corresponding $g_{\mu\nu}$ (with coordinates x^μ identified as ct, r, θ, and φ) is called the *Robertson–Walker metric*. The cosmological solutions to GTR discovered by Friedmann in 1922 are homogeneous and isotropic, and may therefore be described in terms of the RW metric.

The spatial coordinates appearing in the RW line element are *comoving*; that is, r, θ, and φ have constant values for any galaxy or other object moving in unison with the overall cosmic expansion (or contraction). Thus t corresponds to proper (ordinary clock) time for comoving objects. The radial coordinate r and the constant k are customarily chosen to be dimensionless, in which case the *cosmic scale factor* $a(t)$ has the dimension of length. The coordinates θ and φ are the usual polar and azimuthal angles of spherical coordinates. Cosmic evolution in time is governed by the behavior of $a(t)$. [Some authors denote the cosmic scale factor by $R(t)$, but we shall reserve the symbol R for the curvature scalar appearing in Section IV.A.]

The constant k may (in principle) be any real number. If $k = 0$, then three-dimensional space for any single value of t is Euclidean, in which case neither $a(t)$ nor r is uniquely defined: their product, however, equals the familiar radial coordinate of flat-space spherical coordinates. If $k \neq 0$, we may define

$$\bar{r} \equiv |k|^{1/2}r \quad \text{and} \quad \bar{a}(t) \equiv a(t)/|k|$$

in terms of which the RW line element becomes

$$(ds)^2 = -c^2(dt)^2 + \bar{a}^2(t)\left\{\frac{(d\bar{r})^2}{1 \mp \bar{r}^2}\right.$$

$$\left. + \bar{r}^2[(d\theta)^2 + \sin^2\theta(d\varphi)^2]\right\} \qquad (8)$$

where the denominator with ambiguous sign corresponds to $1 - (k/|k|)\bar{r}^2$. Note that Eq. (8) is precisely Eq. (7) with $a = \bar{a}$, $r = \bar{r}$, $k = \pm 1$. Hence there is no loss of generality in restricting k to the three values 1, 0, and -1 in Eq. (7), and we shall henceforth do so (as is customary). These values for k correspond to positive, zero, and negative spatial curvature, respectively.

C. The Curvature of Space

If $k = \pm 1$, then the spatial part of the RW line element differs appreciably from the Euclidean case for objects with r near unity, so that the kr^2 term cannot be neglected. The proper (comoving meter-stick) distance to objects with r near unity is of order $a(t)$. Hence $a(t)$ is roughly the distance over which the curvature of space becomes important. This point may be illustrated with the following examples.

The proper distance $r_P(t)$ to a comoving object with radial coordinate r is

$$r_p(t) = a(t)\int_0^r \frac{dr'}{(1 - kr'^2)^{1/2}} \qquad (9)$$

The integral appearing in Eq. (9) is elementary, being $\sin^{-1}r$ for $k = +1$, r for $k = 0$, and $\sinh^{-1}r$ for $k = -1$. Thus the radial coordinate is related to proper distance and the scale factor by $r = \sin(r_p/a)$ for $k = +1$, r_p/a for $k = 0$, and $\sinh(r_p/a)$ for $k = -1$.

Two objects with the same radial coordinate but separated from each other by a small angle $\Delta\theta$ are separated by a proper distance given by Eq. (7) as

$$\Delta s = ar\Delta\theta \qquad (10)$$

For $k = 0$, $r = r_p/a$ and we obtain the Euclidean result $\Delta s = r_p\Delta\theta$. For $k = +1$ (or -1), Δs differs from the Euclidean result by the extent to which $\sin(r_p/a)$ [or $\sinh(r_p/a)$] differs from r_p/a. Series expansions yield

$$\Delta s = r_p\Delta\theta\left\{1 - \frac{k}{6}(r_p/a)^2 + \mathbb{O}\big[(r_p/a)^4\big]\right\} \qquad (11)$$

where "$\mathbb{O}[x^4]$" denotes additional terms containing 4th and higher powers of "x" that are negligible when x is small compared to unity. Equation (11) displays the leading-order departure from Euclidean geometry for $k = \pm 1$, and it also confirms that this departure becomes important only when r_p is an appreciable fraction of the cosmic scale factor. Standard inflation predicts the scale factor to be enormously greater than the radius of the observable universe, in which case no curvature of space would be detectable even if $k \neq 0$. Open inflation could produce a wide range of values for the scale factor, with no upper limit to the range.

Equation (7) implies that the proper area A of a spherical surface with radial coordinate r is

$$A = 4\pi a^2 r^2 \qquad (12)$$

and that the proper volume V contained within such a sphere is

$$V = 4\pi a^3 \int_0^r dr' \frac{r'^2}{(1 - kr'^2)^{1/2}} \qquad (13)$$

The integral appearing in Eq. (13) is elementary for all three vlaues of k, but will not be recorded here.

When $k = 0$, Eqs. (9), (12), and (13) reproduce the familiar Euclidean relations between A, V, and the proper radius r_p. For $k = \pm 1$, series expansions of A and V in terms of r_p yield

$$A = 4\pi r_p^2 \left\{ 1 - \frac{k}{3}(r_p/a)^2 + \mathcal{O}\left[(r_p/a)^4\right] \right\} \quad (14)$$

$$V = \frac{4\pi}{3} r_p^3 \left\{ 1 - \frac{k}{5}(r_p/a)^2 + \mathcal{O}\left[(r_p/a)^4\right] \right\} \quad (15)$$

For any given r_p, the surface area and volume of a sphere are smaller than their Euclidean values when $k = +1$, and larger than Euclidean when $k = -1$.

The number of galaxies within any large volume of space is a measure of its proper volume (provided the distribution of galaxies is truly uniform). By counting galaxies within spheres of differing proper radii (which must also be measured), it is possible in principle to determine whether proper volume increases less rapidly than r_p^3, at the same rate, or more rapidly; hence whether $k = +1, 0$, or -1. If $k \neq 0$, detailed comparison of observations with Eq. (15) would enable one to determine the present value of the cosmic scale factor.

Other schemes also exist for comparing observations with theory to determine the geometry and, if $k \neq 0$, the scale factor of the cosmos. Thus far, however, observations have revealed no deviation from flatness. Furthermore, recent precise measurements of the cosmic microwave background indicate that space is flat or, if curved, has a scale factor at least as large as the radius of the observable universe ($\sim$50 billion light-years, explained in Section VI.A).

For $k = 0$ and $k = -1$, the coordinate r spans the entire space. This cannot be true for $k = +1$, however, because $(1 - kr^2)$ is negative for $r > 1$. To understand this situation, a familiar analog in Euclidean space may be helpful. Consider the surface of a sphere of radius R, e.g., earth's surface. Let r, θ, and φ be standard spherical coordinates, with $r = 0$ at Earth's center and $\theta = 0$ at the north pole. Longitudinal angle then corresponds to φ. Line elements on the surface can be described by $R\,d\theta$ and $R\sin\theta\,d\varphi$, or by $Rd\rho/\sqrt{1 - \rho^2}$ and $R\rho\,d\varphi$, where $\rho = \sin\theta$. In the latter case, the radial line element is singular at $\rho = 1$, which corresponds to the equator at $\theta = \pi/2$. This singularity reflects the fact that all lines of constant longitude are parallel at the equator. It also limits the range of ρ to $0 \le \rho \le \pi/2$, the northern hemisphere.

The Robertson-Walker metric for $k = 1$ is singular at $r = 1$ because radial lines of constant θ and φ become parallel at $r = 1$. In fact this three-dimensional space is precisely the 3-D "surface" of a hypersphere of radius $a(t)$

in 4-D Euclidean space. With $r = 0$ at the "north pole," the "equator" lies at $r = 1$, so the coordinate r only describes the "northern hemisphere." A simple coordinate for describing the *entire* 3-D space is $u \equiv \sin^{-1} r$, which equals $\pi/2$ at the "equator" and π at the "south pole." The analogs of Eqs. (7)–(15) may readily be found by substituting $\sin u$ for r, wherever r appears. The entire space is then spanned by a finite range of u, namely $0 \le u \le \pi$.

Such a universe is said to be *spherical* and *closed*, and $a(t)$ is called the *radius of the universe*. Though lacking any boundary, a spherical universe is finite. The proper circumference is $2\pi a$, and the proper volume is $V(t) = 2\pi^2 a^3(t)$.

In a curved space, lines as straight as possible are called *geodesics*. Such lines are straight with respect to *locally* flat regions of space surrounding each segment, but curved over longer distances. In spaces described by a Robertson-Walker metric, lines with constant θ and φ are geodesics (though by no means the only ones). On earth's surface, the geodesics are great circles, e.g., lines of constant longitude. Note that adjacent lines of longitude are parallel at the equator, intersect at both poles, and are parallel again at the equator on the opposite side. A spherical universe is very similar. Geodesics (e.g., rays of light) that are parallel in some small region will intersect at a distance of one-fourth the circumference, in either direction, and become parallel again at a distance of half the circumference. If one follows a geodesic for a distance *equal* to the circumference, one returns to the starting point, from the opposite direction.

For $k = 0$ and $k = -1$, space is infinite in linear extent and volume; such universes are said to be *open*. As remarked earlier, a universe with $k = 0$ has zero spatial curvature for any single value of t; such a universe is said to be *Euclidean* or *flat*. The RW metric with $k = 0$ does not, however, correspond to the flat space–time of Minkowski. The RW coordinates are comoving with a spherically symmetric expansion (or contraction), so that the RW time variable is measured by clocks that are moving relative to one another. Special relativity indicates that such clocks measure time differently from the standard clocks of Minkowski space–time. In particular, a set of comoving clocks does not run synchronously with any set of clocks at rest relative to one another. Furthermore, the measure of distance given by the RW metric is that of comoving meter sticks, and it is known from special relativity that moving meter sticks do not give the same results for distance as stationary ones.

Since both time and distance are measured differently in comoving coordinates, the question arises whether the differences compensate for each other to reproduce Minkowski space–time. For $k = 0$, the answer is negative: such a space–time can be shown to be curved, provided the scale factor is changing in time so that comoving instruments actually are moving.

If $k = -1$, space has negative curvature: geodesics (e.g., light rays) that are parallel nearby diverge from each other as one follows them away. An imperfect analogy is the surface of a saddle; two "straight" lines that are locally parallel in the seat of the saddle will diverge from each other as one follows them toward the front or back of the saddle. (The analogy is imperfect because the curvature of space is presumed to be the same everywhere, unlike the curvature of a saddle, which varies from one region to another.) A universe with $k = -1$ is called *hyperbolic*.

D. Hubble's Law

It is evident from Eq. (9) that the proper distance r_p to a comoving object depends on time only through a multiplicative factor of $a(t)$. It follows that

$$\dot{r}_p / r_p = \dot{a}/a \equiv H(t) \tag{16}$$

$$\dot{r}_p = Hr_p \tag{17}$$

where dots over the symbols denote differentiation with respect to time, for example, $\dot{f} \equiv df/dt$. Thus $\dot{r}_p$ equals the velocity of a comoving object relative to the earth (assuming that earth itself is a comoving object), being positive for recession or negative for approach. An important proviso, however, arises from the fact that the concept of velocity relative to earth becomes problematic if the object is so distant that curvature of the intervening space–time is significant. Also, r_p is (hypothetically) defined in terms of comoving meter sticks laid end to end. All of these comoving meter sticks, except the closest one, are moving relative to earth, and are therefore perceived by us to be shortened because of the familiar length contraction of special relativity. Thus r_p deviates from distance in the familiar sense for objects receding with velocities near that of light. For these reasons, $\dot{r}_p$ only corresponds to velocity in a familiar sense if r_p is less than the distance over which space–time curves appreciably and $\dot{r}_p$ is appreciably less than c.

A noteworthy feature of Eq. (17) is that $\dot{r}_p$ exceeds c for objects with $r_p > c/H$. There is no violation of the basic principle that c is a limiting velocity, however. As just noted, $\dot{r}_p$ only corresponds to velocity when small compared to c. Furthermore, a correct statement of the principle fo limiting velocity is that no object may *pass by* another object with a speed greater than c. In Minkowski space–time the principle may be extended to the relative velocity between distant objects, but no such extension to distant objects is possible in a curved space–time.

Equations (16) and (17) state that if $\dot{a} \neq 0$, then all comoving objects are receding from (or approaching) earth with velocities that are proportional to their distances from us, subject to the provisos stated above. One of the most direct points of contact between theoretical formalism and observation resides in the fact that, aside from local random motions of limited magnitude, the galaxies constituting the visible matter of our universe are receding from us with velocities proportional to their distances. This fact was first reported by Edwin Hubble in 1929, based on determinations of distance (later revised) made with the 100-in. telescope that was completed in 1918 on Mt. Wilson in southwest California. Velocities of the galaxies in Hubble's sample had been determined by V. M. Slipher, utilizing the redshift of spectral lines caused by the Doppler effect.

The function $H(t)$ defined by Eq. (16) is called the Hubble parameter. Its present value, which we shall denote by H_0, is traditionally called *Hubble's constant* (a misnomer, since the "present" value changes with time, albeit very slowly by human standards). Equation (17), governing recessional velocities, is known as *Hubble's law*.

The value of H_0 is central to a quantitative understanding of our universe. Assuming that redshifts of spectral lines result entirely from the Doppler effect, it is straightforward to determine the recessional velocities of distant galaxies. Determinations of distance, however, are much more challenging.

Most methods for measuring distances beyond the Milky Way are based on the appearance from earth of objects whose intrinsic properties are (believed to be) known. Intrinsic luminosity is the simplest property used in this way. Apparent brightness decreases at increasing distance, so the distance to any (visible) object of known luminosity is readily determined. Objects of known intrinsic luminosity are called "standard candles," a term that is sometimes applied more generally to *any* kind of distance indicator.

Given a set of identical objects, the relevant properties of (at least) *one* of them must be measured in order to use them as standard candles. This can be a challenge. For example, the brightest galaxies in rich clusters are probably similar, representing an upper limit to galactic size. Since they are luminous enough to be visible at great distances (out to $\sim 10^{10}$ ly), they would be powerful standard candles. Determining their intrinsic luminosity (i.e., *calibrating* them) was difficult, however, because the nearest rich cluster is (now known to be) $\sim 5 \times 10^7$ light-years away (the *Virgo cluster*, containing ~ 2500 galaxies). This distance is much too great for measurement by trigonometric parallax, which is ineffective beyond a few thousand light-years. Also note that even when calibrated, the brightest galaxies are useless for measuring distances less that $\sim 5 \times 10^7$ light-years, the distance to the nearest rich cluster.

In order to calibrate sets of standard candles spanning a wide range of distances, a *cosmic distance ladder* has been constructed. Different sets of standard candles form successive rungs of this ladder, with intrinsic luminosity increasing at each step upward. The bottom rung is based on parallax and other types of stellar motion. The resulting distances have been used to determine the intrinsic luminosities of main-sequence stars of various spectral types, which form the second rung. Every rung above the first is calibrated by using standard candles of the rung immediately below it, in a bootstrap fashion. Since any errors in calibration are compounded as one moves up the ladder, it is clearly desirable to minimize the number of rungs required to reach the top.

A key role has been played by *variable stars* (the third rung), among which *Cepheids* are the brightest and most important. Cepheids are supergiant, pulsating stars, whose size and brightness oscillate with periods ranging from (roughly) 1 to 100 days, depending on the star. In 1912, a discovery of historic importance was published by Henrietta S. Leavitt, who had been observing 25 Cepheids in the Small Magellenic Cloud (a small galaxy quite near the Milky Way).

Leavitt reported that the periods and apparent brightnesses of these stars were strongly correlated, being roughly proportional. Knowing that all of these stars were about the same distance from earth, she concluded that the *intrinsic* luminosities of Cepheids are roughly proportional to their periods. She was unable to determine the constant of proportionality, however, because no Cepheids were close enough for the distance to be measured by parallax.

Over the next decade, indirect and laborious methods were used by others to estimate the distances to several of the nearest Cepheids. Combined with the work of Leavitt, a *period-luminosity relation* was thereby established, enabling one to infer the distance to any Cepheid from its period and apparent brightness. Only a small fraction of stars are Cepheids, but every galaxy contains a great many of them. When Cepheids were observed (by Hubble) in Andromeda in 1923, the period-luminosity relation revealed that Andromeda lies far outside the Milky Way. This was the first actual proof that other galaxies exist.

Until recently, telescopes were only powerful enough to observe Cepheids out to ~8 million light-years. At this distance (and considerably beyond), random motions of galaxies obscure the systematic expansion described by Hubble's law. Two more rungs were required in the distance ladder in order to reach distances where motion is governed by the Hubble flow. (This partly explains why Hubble's original value for H_0 was too large by ~700%.) Different observers favored different choices for the higher rungs, and made different corrections for

a host of complicating factors. During the two decades prior to launching of the Hubble Space Telescope (HST) in 1990, values for H_0 ranging from 50 to 100 km/sec/Mpc were published (where Mpc denotes *megaparsec*, with 1 Mpc $= 3.26 \times 10^6$ ly $= 3.09 \times 10^{19}$ km). The reported uncertainties were typically near ± 10 km/sec/Mpc, much too small to explain the wide range of values reported. Unknown systematic errors were clearly present.

The initially flawed mirror of the HST was corrected in 1993, and the reliability of distance indicators has improved substantially since then. Our knowledge of Cepheids, already good, is now even better. Far more important, the HST has identified Cepheids in galaxies out to ~80 million light-years, nearly 10 times farther than had previously been achieved. This tenfold increase in the range of Cepheids has placed them next to the highest rung of the distance ladder. Every standard candle of greater luminosity and range has now been calibrated with Cepheids, substantially improving their reliabilities.

Ground-based astronomy has also become much more powerful over the last decade. Several new and very large telescopes have been built, with more sensitive light detectors (charge-coupled devices), and with computer software that minimizes atmospheric distortions of the images. Large areas of the sky are surveyed and, when objects of special interest are identified, the HST is also trained on them. Collaboration of this kind has made possible the calibration of a new set of standard candles, supernovae of type Ia (SNe Ia, or SN Ia when singular).

Type SNe Ia occur in binary star systems containing a white dwarf and a bloated, nearby giant. The dwarf gradually accretes matter from its neighbor until a new burst of nuclear fusion occurs, with explosive force. The peak luminosity equals that of an *entire galaxy*, and is nearly the same for all SNe Ia. Furthermore, the rate at which the luminosity decays is correlated with its value in a known way, permitting a measurement of distance within $\pm 6\%$ for *each* SN Ia. This precision exceeds that of any other standard candle.

Because of their enormous luminosities, SNe Ia can be studied out to very great distances (at least 12 billion light-years with present technology). In 1997, observations of SNe Ia strongly suggested that our universe is expanding at an *increasing* rate, hence that $\Lambda > 0$ (where Λ denotes the cosmological constant). Subsequent studies have bolstered this conclusion. An accelerating expansion is the most dramatic discovery in observational cosmology since the microwave background (1964), and perhaps since Hubble's discovery of the cosmic expansion (1929).

Many calculations in cosmology depend on the value of H_0, which still contains a range of uncertainty (primarily due to systematic errors, implied by the fact that different standard candles yield somewhat different results). It is therefore useful (and customary) to characterize H_0 by a dimensionless number h_0 (or simply h), of order unity, defined by

$$H_0 = h_0 \times 100 \, \text{km/sec/Mpc} \tag{18}$$

An HST Key Project team led by Wendy L. Freedman, Robert C. Kennicutt, Jr., and Jeremy R. Mould has used nine different secondary indicators whose results, when combined, yield $h_0 = 0.70 \pm 0.07$. Type SNe Ia were one of their nine choices and, by themselves, led to $h_0 = 0.68 \pm 0.05$. The preceding values for h_0 would be reduced by ∼0.05 if a standard (but debatable) adjustment were made to correct for differences in metallicity of the Cepheids used as primary indicators.

As previously suggested, there are many devils in the details of distance measurements, and highly respected observers have dealt with them in different ways. For several decades Allan Sandage, with Gustav A. Tammann and other collaborators, has consistently obtained results for h_0 between 0.5 and 0.6. Using SNe Ia as secondary indicators, their most recent result is $h_0 = 0.585 \pm 0.063$. Several other teams have also obtained h_0 from SNe Ia, with results spanning the range from 0.58 to 0.68.

In light of the preceding results and other studies far too numerous to mention, it is widely believed that h_0 lies in the range

$$h_0 = 0.65 \pm 0.10 \tag{19}$$

Before leaving this section on cosmic geometry, we note that a space–time may contain one or more finite regions that are internally homogeneous and isotropic while differing from the exterior. Within such a homogeneous region, space–time can be described by an RW metric (which, however, tells one nothing about the exterior region). This type of space–time is in fact predicted by theories of cosmic inflation; the homogeneity and isotropy of the observable universe are not believed characteristic of the whole.

IV. DYNAMICS OF THE COSMOS

A. Einstein's Field Equations

A theory of cosmic evolution requires dynamical equations, taken here to be the field equations of GTR:

$$R_{\mu\nu} - \tfrac{1}{2} g_{\mu\nu} R + \Lambda g_{\mu\nu} = 8\pi G c^{-4} T_{\mu\nu} \tag{20}$$

where $R_{\mu\nu}$ denotes the Ricci tensor (contracted Riemann curvature tensor), R denotes the curvature scalar (contracted Ricci tensor), Λ denotes the cosmological constant, and $T_{\mu\nu}$ denotes the stress-energy tensor of all forms of matter and energy excluding gravity. The Ricci tensor and curvature scalar are determined by $g_{\mu\nu}$ together with its first and second partial derivatives, so that Eq. (20) is

a set of second-order (nonlinear) partial differential equations for $g_{\mu\nu}$. As with all differential equations, initial and/or boundary conditions must be assumed in order to specify a unique solution. (A definition of the Riemann curvature tensor lies beyond the scope of this article. We remark, however, that some scientists define it with an opposite sign convention, in which case $R_{\mu\nu}$ and R change signs and appear in the field equations with opposite signs from those above.)

All tensors appearing in the field equations are symmetric, so there are 10 independent equations in the absence of simplifying constraints. We are assuming space to be homogeneous and isotropic, however, which assures the validity of the RW line element. The RW metric contains only one unknown function, namely the cosmic scale factor $a(t)$, so we expect the number of independent field equations to be considerably reduced for this case.

Since the field equations link the geometry of space–time with the stress-energy tensor, consistency requires that we approximate $T_{\mu\nu}$ by a homogeneous, isotropic form. Thus we imagine all matter and energy to be smoothed out into a homogeneous, isotropic fluid (or gas) of dust and radiation that moves in unison with the cosmic expansion. Denoting the proper mass-energy density by ρ (in units of mass/volume) and the pressure by P, the resulting stress-energy tensor has the form characteristic of a "perfect fluid":

$$T_{\mu\nu} = (\rho + P/c^2) U_\mu U_\nu + P g_{\mu\nu} \tag{21}$$

where U_μ denotes the covariant four-velocity of a comoving point: $U_\mu = g_{\mu\nu} \, dx^\nu / d\tau$, with τ denoting proper time ($d\tau = |ds|/c$).

Under the simplifying assumptions stated above, only two of the 10 field equations are independent. They are differential equations for the cosmic scale factor and may be stated as

$$\left(\frac{\dot{a}}{a}\right)^2 = \frac{8\pi G}{3}\rho + \frac{c^2 \Lambda}{3} - k\left(\frac{c}{a}\right)^2 \tag{22}$$

$$\frac{\ddot{a}}{a} = -\frac{4\pi G}{3}\left(\rho + \frac{3P}{c^2}\right) + \frac{c^2 \Lambda}{3} \tag{23}$$

where $\dot{a}$ and $\ddot{a}$ denote first and second derivatives, respectively, of $a(t)$ with respect to time. Note that the left side of Eq. (22) is precisely H^2, the square of the Hubble parameter.

B. Local Energy Conservation

A remarkable feature of the field equations is that the *covariant divergence* (the curved–space-time generalization of Minkowski four-divergence) of the left side is identically zero (*Bianchi identities*). Hence the field equations imply that over a region of space–time small enough to be flat, the Minkowski four-divergence of the stress-energy tensor is zero. A standard mathematical argument then leads to the conclusion that energy and momentum are conserved (locally, which encompasses the domain of experiments known to manifest such conservation). Hence within the context of GTR, energy and momentum conservation are not separate principles to be adduced, but rather are consequences of the field equations to the full extent that conservation is indicated by experiment or theory.

If a physical system is spatially localized in a space–time that is flat (i.e., Minkowskian) at spatial infinity, then the total energy and momentum for the system can be defined and their constancy in time proven. The condition of localization in asymptotically flat space–time is not met, however, by any homogeneous universe, for such a universe lacks a boundary and cannot be regarded as a localized system. The net energy and momentum of an unbounded universe defy meaningful definition, which precludes any statement that they are conserved. (From another point of view, energy is defined as the ability to do work, but an unbounded universe lacks any external system upon which the amount of work might be defined.)

In the present context, local momentum conservation is assured by the presumed homogeneity and isotropy of the cosmic expansion. Local energy conservation, however, implies a relation between ρ and P, which may be stated as

$$\frac{d}{dt}(\rho a^3) = -\frac{P}{c^2}\frac{d}{dt}(a^3) \qquad (24)$$

[That Eq. (24) follows from the field equations may be seen by solving Eq. (22) for ρ and differentiating with respect to time. The result involves $\ddot{a}$, which may be eliminated by using Eq. (23) to obtain a relation equivalent to Eq. (24).]

To see the connection between Eq. (24) and energy conservation, consider a changing spherical volume V bounded by a comoving surface. Equation (13) indicates that V is the product of a^3 and a coefficient that is independent of time. Multiplying both sides of Eq. (24) by that coefficient and c^2 yields

$$\frac{d}{dt}(\rho c^2 V) = -P\frac{dV}{dt} \qquad (25)$$

The Einstein mass-energy relation implies that ρc^2 equals the energy per unit volume, so $\rho c^2 V$ equals the energy U internal to the volume V. Hence Eq. (25) is simply the work-energy theorem as applied to the adiabatic expansion (or contraction) of a gas or fluid: $\Delta U = -P\Delta V$. Since energy changes only to the extent that work is done, (local) energy conservation is implicit in this result.

By inverting the reasoning that led to Eq. (24), it can be shown that the second of the field equations follows from the first and from energy conservation; that is, Eq. (23)

follows from Eqs. (22) and (24). [To reason from Eqs. (23) and (24) to (22) is also possible; k appears as a constant of integration.] Hence evolution of the cosmic scale factor is governed by Eqs. (22) and (24), together with an equation of state that specifies the relation between ρ and P.

C. Equation for the Hubble Parameter

The present expansion of the universe will continue for however long H remains positive. Since $H \equiv \dot{a}/a$, Eq. (22) may be written as

$$H^2 = \frac{8\pi G}{3}\rho + \frac{c^2\Lambda}{3} - k\left(\frac{c}{a}\right)^2 \qquad (26)$$

The density ρ is inherently positive. If $\Lambda \geq 0$ (as seems certain), then $H > 0$ for all time if $k \leq 0$. Open universes, flat or hyperbolic, therefore expand forever. A closed, finite universe may (or may not) expand forever, depending on the relation between its total mass M_u and Λ (as will be discussed).

If $P = 0$, Eq. (24) implies that ρ varies inversely with the cube of $a(t)$, which simply corresponds to a given amount of matter becoming diluted as the universe expands:

$$P = 0: \qquad \rho = \rho_0\left(\frac{a_0}{a}\right)^3 \qquad (27)$$

where ρ_0 and a_0 denote present values.

The density ρ has been dominated by matter under negligible pressure (i.e., $P \ll \rho c^2$) since the universe was $\sim 10^6$ years old, so Eq. (27) has been a good approximation since that early time. Recalling again that $H \equiv \dot{a}/a$, Eqs. (26) and (27) imply

$$\dot{a}^2 = \frac{8\pi G\rho_0 a_0^3}{3a} + \frac{c^2\Lambda a^2}{3} - kc^2 \qquad (28)$$

The scale factor $a(t)$ increases as the universe expands. Open universes expand forever, and their scale factors increase without limit. If $\Lambda = 0$, Eq. (28) implies that $\dot{a}^2 \to -kc^2$ as $t \to \infty$. From Eqs. (9) and (16), we see that every comoving object has $\dot{r}_p \to 0$ in a flat ($k = 0$) universe, while $\dot{r}_p \to c \times \sinh^{-1} r$ in a hyperbolic ($k = -1$) universe.

If $\Lambda > 0$ in an open universe, then $\dot{a}/a \to c\sqrt{\Lambda/3}$ as $a \to \infty$. In this limit,

$$\dot{a} = \Gamma a \qquad (\text{with } \Gamma \equiv c\sqrt{\Lambda/3}) \qquad (29)$$

The solution is

$$a(t) = a(t_{in})\exp[\Gamma(t - t_{in})] \qquad (30)$$

where t_{in} denotes any initial time that may be convenient. Equation (29) will be satisfied to good approximation after some minimum time $t_{\min}$ has passed, so $a(t)$ will be given for all $t \geq t_{\min}$ by Eq. (30), with $t_{in} = t_{\min}$.

Equation (30) implies an exponential rate of growth for r_p in the final stage of expansion, with a doubling time t_d given by

$$t_d = \frac{\ln 2}{c} \sqrt{\frac{3}{\Lambda}} \qquad (31)$$

Equation (29) then implies exponential growth for $\dot{r}_p$ as well, with the same doubling time.

Next we consider the closed, $k = +1$ case in some detail, where $a(t)$ is called the *radius* of the universe. The total mass of such a universe is finite and given by $M_u = \rho V_u = \rho(2\pi^2 a^3)$. Assuming zero pressure for simplicity, M_u is constant, and Eq. (26) implies

$$H^2 = \frac{4GM_u}{3\pi a^3} + \frac{c^2 \Lambda}{3} - \frac{c^2}{a^2} \qquad (32)$$

If $\Lambda = 0$, H will shrink to zero when the radius $a(t)$ reaches a maximum value

$$a_{\max} = \frac{4GM_u}{3\pi c^2} \qquad (33)$$

The time required to reach $a_{\max}$ can be calculated, and is given by

$$t(a_{\max}) = \frac{2GM_u}{3c^3} \qquad (34)$$

It is remarkable that both $a_{\max}$ and the time required to reach it are determined by M_u. The dynamics and geometry are so interwoven that the rate of expansion does not enter as an independent parameter.

After reaching $a_{\max}$, the universe will contract ($H < 0$) at the same rate as it had expanded. Equation (23) ensures that the scale factor will not remain at $a_{\max}$, and Eq. (28) implies that the magnitude of H is determined by the value of $a(t)$.

If the universe is closed and $\Lambda > 0$, a more careful analysis is required. A detailed study of Eq. (32) reveals that the present expansion will continue forever if

$$\Lambda \geq \left(\frac{\pi c^2}{2GM_u} \right)^2 \qquad (35)$$

If the *in*equality holds, the size of the universe will increase without limit. As for the open case, $\dot{a}/a \to c\sqrt{\Lambda/3}$ as $a \to \infty$. After some time t_0 has passed, Eq. (30) will describe $a(t)$, and the final stage of expansion will be the same as for the open case. Both r_p and $\dot{r}_p$ will grow at exponential rates, with a doubling time given by Eq. (31).

If the *equality* in Eq. (35) holds, then the universe will expand forever but approach a maximum, limiting size $a_{\max}$ as $t \to \infty$, with $a_{\max} = 2GM_u/\pi c^2$ (exactly 3/2 as large as for $\Lambda = 0$.) If the inequality (35) is *violated* (small Λ), then the universe will reach a maximum size in a finite time, after which it will contract (as for the special case where $\Lambda = 0$).

We note in passing that despite the name *hyperbolic*, a completely empty universe with $\Lambda = 0$ and $k = -1$ has precisely the flat space-time of Minkowski. Empty Minkowski space-time is homogeneous and isotropic, and therefore must correspond to an RW metric. With $\rho = \Lambda = 0$, Eq. (28) implies that $k = -1$ [and that $\dot{a} = c$, in which case Eq. (23) implies that $P = 0$, as does Eq. (24).] The difference in appearance between the Minkowski metric and the equivalent RW metric results from a fact noted earlier, namely that RW coordinates are comoving. The clocks and meter sticks corresponding to the RW metric are moving with respect to each other, so they yield results for time and distance that differ from those ordinarily used in a description of Minkowski space-time.

To make explicit the equivalence between empty, hyperbolic space-time and Minkowski space-time, we recall that Eq. (28) implies $\dot{a} = c$. Since $a = 0$ at $t = 0$, it follows that $a(t) = ct$ for the hyperbolic case in question. It may readily be verified that the Minkowski metric transforms into the corresponding RW metric with $r_M = rct$, $t_M = t(r^2 + 1)^{1/2}$, where r_M and t_M denote the usual coordinates for Minkowski space-time.

D. Cosmic Deceleration or Acceleration

We saw in Section IV.B that Eq. (23) contains no information beyond that implied by Eq. (22) and local energy conservation. Equation (23) makes a direct and transparent statement about cosmic evolution, however, and therefore warrants study. Recalling again that the proper distance $r_p(t)$ to a comoving object depends on time only through a factor of $a(t)$, we see that Eq. (23) is equivalent to a statement about the acceleration $\ddot{r}_p$ of comoving objects (where again the dots refer to time derivatives):

$$\ddot{r}_p = \left\{ -\frac{4\pi G}{3} \left(\rho + \frac{3P}{c^2} \right) + \frac{c^2 \Lambda}{3} \right\} r_p \qquad (36)$$

Note that $\ddot{r}_p$ is proportional to r_p, which is essential to the preservation of Hubble's law over time. A positive $\ddot{r}_p$ corresponds to acceleration *away* from us, whereas a negative $\ddot{r}_p$ corresponds to acceleration *toward* us (e.g., deceleration of the present expansion). The same provisions that conditioned our interpretation of $\dot{r}_p$ as recessional velocity apply here; $\ddot{r}_p$ equals acceleration in the familiar sense only for objects whose distance is appreciably less than the scale factor (for $k \neq 0$) and whose $\dot{r}_p$ is a small fraction of c.

As a first step in understanding the dynamics contained in Eq. (34), let us consider what Newton's laws would predict for a homogeneous, isotropic universe filling an infinite space described by Euclidean geometry. In particular, let us calculate the acceleration $\ddot{r}_p$ (relative to a comoving observer) of a point mass at a distance r_p away from the observer. (The distinction of "proper" distance is superfluous in Newtonian physics, but the symbol r_p is being used to maintain consistency of notation.)

In Newton's theory of gravitation, the force F between two point masses m and M varies inversely as the square of the distance d between them: $F = GmM/d^2$. If more than two point masses are present, a principle of superposition states that the net force on any one mass is the vector sum of the individual forces resulting from the other masses.

While it is not obvious (indeed, Newton invented calculus to prove it), the inverse–square dependence on distance together with superposition implies that the force exerted by a uniform sphere of matter on any exterior object is the same as if the sphere's mass were concentrated at its center. It may also be shown that if a spherically symmetric mass distribution has a hollow core, then the mass exterior to the core exerts no net force on any object inside the core. (These theorems may be familiar from electrostatics, where they also hold because Coulomb's law has the same mathematical form as Newton's law of gravitation.)

The above features of Newtonian gravitation may be exploited in this study of cosmology by noting that in the observer's frame of reference, the net force on an object of mass m (e.g., a galaxy) at a distance r_p from the observer results entirely from the net mass M contained within the (hypothetical) sphere of radius r_p centered on the observer:

$$M = \left(\tfrac{4}{3}\pi r_p^3\right)\rho \tag{37}$$

The force resulting from M is the same as if it were concentrated at the sphere's center, and there is no net force from the rest of the universe exterior to the sphere. Combining Newton's law of gravitation with his second law of motion (that is, acceleration equals F/m) yields

$$\ddot{r}_p = -GM/r_p^2 = -\frac{4\pi G}{3}\rho r_p \tag{38}$$

where Eq. (35) has been used to eliminate M, and the minus sign results from the attractive nature of the force.

Note that Eq. (36) agrees precisely with the contribution of ρ to $\ddot{r}_p$ in the Einstein field equation (36); this part of relativistic cosmology has a familiar, Newtonian analog. In GTR, however, stresses also contribute to gravitation, as exemplified by the pressure term in Eq. (36). If $\Lambda \neq 0$, then a cosmological term contributes as well.

In ordinary circumstances, pressure is a negligible source of gravity. To see why, we need only compare the gravitational effect of mass with the effect of pressure filling the same volume. For example, lead has a density $\rho = 11.3 \times 10^3 \, \text{kg/m}^3$. In order for pressure filling an equal volume to have the same effect, the pressure would need to be $P = c^2\rho/3 = 3.39 \times 10^{20} \, \text{N/m}^2$ (where $1 \, \text{N} = 0.225 \, \text{lb}$). This is $\sim 3 \times 10^{15}$ times as great as atmospheric pressure, and $\sim 10^9$ times as great as pressure at the center of the earth.

The density ρ appearing in the field equations is positive, so its contribution to $\ddot{r}_p$ in Eq. (36) is negative: the gravitation of mass-energy is attractive. Positive pressure also makes a negative contribution to $\ddot{r}_p$, but pressure may be either positive or negative. Equation (36) implies that negative pressure causes gravitational repulsion.

If $\Lambda \neq 0$, its contribution to $\ddot{r}_p$ has the same sign as Λ. If $\Lambda > 0$ and a time arrives when

$$\Lambda > \frac{4\pi G}{c^2}\left(\rho + \frac{3P}{c^2}\right) \tag{39}$$

then $\ddot{r}_p > 0$, which corresponds to an *accelerating* expansion. The density and pressure both decrease as the universe expands, whereas Λ is constant. After the first moment when the inequality (39) is satisfied, the universe will expand forever, at an accelerating rate. A time will therefore come when Eq. (29) is satisfied, after which r_p and $\dot{r}_p$ will both grow at exponential rates, in accordance with Eqs. (29)–(31). There is recent evidence that our universe has in fact entered such a stage, as will be discussed in Section V.E.

The *name* of the big bang model is somewhat misleading, for it suggests that the expansion of the cosmos is a result of the early high temperature and pressure, rather like the explosion of a gigantic firecracker. As just explained, however, positive pressure leads not to acceleration "outward" of the cosmos, but rather to deceleration of any expansion that may be occurring. Had the universe been born stationary, positive pressure would have contributed to immediate collapse. Since early high pressure did not contribute to the expansion, early high temperature could not have done so either. (It is perhaps no coincidence that the phrase *big bang* was initially coined as a term of derision, by Fred Hoyle—an early opponent of the model. The model has proven successful, however, and the graphic name remains with us.)

E. The Growth of Space

It is intuitively useful to speak of ρ, P, and Λ as leading to gravitational attraction or repulsion, but the geometrical role of gravitation in GTR should not be forgotten. The geometrical significance of gravitation is most apparent for a closed universe, whose total proper volume is finite and equal to $2\pi^2 a^3(t)$. The field equations describe the change in time of $a(t)$, hence of the size of the universe. From a naive point of view, Hubble's law describes cosmic expansion in terms of recessional velocities of galaxies; but more fundamentally, galaxies are moving apart because the space between them is expanding. How else an increasing volume for the universe?

The field equation for $\ddot{a}(t)$ implies an equation for $\ddot{r}_p$, which has a naive interpretation in terms of acceleration and causative force. More fundamentally, however, $\ddot{a}$ and

$\ddot{r}_p$ represent the derivative of the rate at which space is expanding. Our intuitions are so attuned to Newton's laws of motion and causative forces that we shall often speak of velocities and accelerations and of gravitational attraction and repulsion, but the geometrical nature of GTR should never be far from mind.

The interpretation of gravity in terms of a growth or shrinkage of space resolves a paradox noted above, namely, that positive pressure tends to make the universe collapse. Cosmic pressure is the same everywhere, and uniform pressure pushes equally on all sides of an object leading to no net force or acceleration. Gravity, however, affects the rate at which space is growing or shrinking, and thereby results in relative accelerations of distant objects. One has no *a priori* intuition as to whether positive pressure should lead to a growth or shrinkage of space; GTR provides the answer described above.

F. The Cosmological Constant

For historical and philosophical reasons beyond the scope of this article, Einstein believed during the second decade of this century that the universe was static. He realized, however, that a static universe was dynamically impossible unless there were some repulsive force to balance the familiar attractive force of gravitation between the galaxies. The mathematical framework of GTR remains consistent when the cosmological term $\Lambda g_{\mu\nu}$ is included in the field equations (20), and Einstein added such a term to stabilize the universe in his initial paper on cosmology in 1917.

Although the existence of other galaxies outside the Milky Way was actually not established until 1923, Einstein (in his typically prescient manner) had incorporated the assumptions of homogeneity and isotropy in his cosmological model of 1917. With these assumptions, the field equations reduce to (22) and (23). The universe is static if and only if $\dot{a}(t) = 0$, which requires (over any extended period of time) that $\ddot{a}(t) = 0$ as well. Applying these conditions to Eqs. (22) and (23), it is readily seen that the universe is static if and only if

$$\Lambda = \frac{4\pi G}{c^2}\left(\rho + \frac{3P}{c^2}\right) \tag{40}$$

$$k\left(\frac{c}{a}\right)^2 = 4\pi G\left(\rho + \frac{P}{c^2}\right) \tag{41}$$

Since our universe is characterized by positive ρ and positive (albeit negligible) P, the model predicts positive values for Λ and k. Hence the universe is closed and finite, with the scale factor (radius of the universe) given by Eq. (37) with $k = 1$. For example, if $\rho \sim 10^{-30}$ g/cm^3 (roughly the density of our universe) and $P = 0$, then Eq. (41) yields a cosmic radius of $\sim 10^{10}$ ly, $\sim 1/5$ the size of our observable universe.

The static model of Einstein has of course been abandoned, for two reasons. In 1929, Edwin Hubble reported that distant galaxies are in fact receding in a systematic way, with velocities that are proportional to their distance from us. On the theoretical side, it was remarked by A. S. Eddington in 1930 that Einstein's static model was intrinsically unstable against fluctuations in the cosmic scale factor: if $a(t)$ were momentarily to decrease, ρ would increase and the universe would collapse at an accelerating rate under the newly dominating, attractive force of ordinary gravity. Conversely, if $a(t)$ momentarily increased, ρ would decrease and the cosmological term would become dominant, thereby causing the universe to expand at an accelerating rate. For this combination of reasons, Einstein recommended in 1931 that the cosmological term be abandoned (i.e., set $\Lambda = 0$); he later referred to the cosmological term as "the biggest blunder of my life." From a contemporary perspective, however, only the static model was ill-conceived. The cosmological term remains a viable possibility, and two recent observations indicate that $\Lambda > 0$ (as will be discussed in Section V.E).

We have seen that Einstein was motivated by a desire to describe the cosmos when he added the $\Lambda g_{\mu\nu}$ term to the field equations, but there is another, more fundamental reason why this addendum is called the cosmological term. Let us recall the structure of Newtonian physics, whose many successful predictions are reproduced by GTR. Newton's theory of gravity predicts that the gravitational force on any object is proportional to its mass, while his second law of motion predicts that the resulting acceleration is inversely proportional to the object's mass. It follows that an object's mass cancels out of the acceleration caused by gravity. Hence the earth (or any other source of gravity) may be regarded as generating an *acceleration field* in the surrounding space. This field causes all objects to accelerate at the same rate, and may be regarded as characteristic of the region of space.

The GTR regards gravitational accelerations as more fundamental than gravitational forces, and focuses attention on the acceleration field in the form of $g_{\mu\nu}$ as the fundamental object of study. The acceleration field caused by a mass M is determined by the field equations (20), wherein M contributes (in a way depending on its position and velocity) to the stress-energy tensor $T_{\mu\nu}$ appearing on the right side. The solution for $g_{\mu\nu}$ then determines the resulting acceleration of any other mass that might be present, mimicking the effect of Newtonian gravitational force. The term $\Lambda g_{\mu\nu}$ in Eq. (20), however, does not depend on the position or even the existence of any matter. Hence the influence of $\Lambda g_{\mu\nu}$ on the solution for $g_{\mu\nu}$ cannot be interpreted in terms of forces between objects.

To understand the physical significance of the cosmological term, let us consider a special case of the field equations, namely their application to the case at hand:

a homogeneous, isotropic universe. As we have seen, the field equations are then equivalent to Eqs. (26) and (34). From the appearance of Λ on the right sides, it is clear that Λ contributes to the recessional velocities of distant galaxies (through H), and to the acceleration of that recession. This part of the acceleration cannot be interpreted in terms of forces acting among the galaxies for, as just remarked, the cosmological term does not presume the existence of galaxies.

The proper interpretation of the cosmological term resides in another remark made earlier: that the recession of distant galaxies should be understood as resulting from growth of the space between them. We conclude that the cosmological term describes a growth of space that is somehow intrinsic to the nature of space and its evolution in time. Such a term clearly has no analog in the physics predating GTR.

G. Equivalence of Λ to Vacuum Energy and Pressure

An inspection of Eqs. (22) and (23) reveals that the contributions of Λ are precisely the same as if all of space were characterized by constant density and pressure given by

$$\rho_V = \frac{c^2\Lambda}{8\pi G} \qquad (42)$$

$$P_V = -\frac{c^4\Lambda}{8\pi G} = -c^2\rho_V \qquad (43)$$

From a more basic point of view, Eqs. (42), (43), and (21) yield

$$T_{\mu\nu} = P_V g_{\mu\nu} = -c^2\rho_V g_{\mu\nu} \qquad (44)$$

Comparing Eq. (44) with the field equations in their most fundamental form, i.e., Eq. (20), the equivalence between Λ and the above ρ_V and P_V is obvious. We also note that local energy conservation is satisfied by a constant ρ_V with $P_V = -c^2\rho_V$, as may be seen from Eq. (24). In fact all known principles permit an interpretation of Λ in terms of the constant ρ_V and P_V given by Eqs. (42) and (43).

Is there any reason to suppose that $T_{\mu\nu}$ might contain a term such as that described above, equivalent to a cosmolgical constant? Indeed there is. In relativistic quantum field theory (QFT), which describes the creation, annihilation, and interaction of elementary particles, the vacuum state is not perfectly empty and void of activity, because such a perfectly ordered state would be incompatible with the uncertainty principles of quantum theory. The vacuum is defined as the state of minimum energy, but it is characterized by spontaneous creation and annihilation of particle–antiparticle pairs that sometimes interact before disappearing.

For most purposes, only changes in energy are physically significant, in which case energy is only defined within an arbitrary additive constant. The energy of the vacuum is therefore often regarded as zero, the simplest convention. In the context of general relativity, however, a major question arises. Does the vacuum of QFT really contribute nothing to the $T_{\mu\nu}$ appearing in Einstein's field equations, or do quantum fluctuations contribute and thereby affect the metric $g_{\mu\nu}$ of spacetime?

Let us consider the possibility that quantum fluctuations act as a source of gravity in the field equations of GTR. These fluctuations occur throughout space, in a way governed by the local physics of special relativity. The energy density of the vacuum should therefore not change as the universe expands. Denoting the vacuum density by ρ_V [with $\rho_V = (\text{energy density})/c^2$], the corresponding pressure P_V can be deduced from Eq. (24). Assuming that ρ_V is constant, Eq. (24) implies $P_V = -c^2\rho_V$, in perfect accordance with Eq. (43).

If $\Lambda \geq 0$ (as seems certain on empirical grounds), then the corresponding vacuum has $\rho_V \geq 0$ and $P_V \leq 0$. Negative pressure has a simple interpretation in this context. Consider a (hypothetical) spherical surface expanding in unison with the universe, so that all parts of the surface are at rest with respect to comoving coordinates. Assuming the vacuum energy density to be constant, the energy inside the sphere will increase, because of the increasing volume. Local energy conservation implies that positive work is being done to provide the increasing energy, hence that an outward force is being exerted on the surface.

No such force would be required unless the region inside the sphere had an innate tendency to contract, which is precisely what is meant by negative pressure. The required *outward* force on the surface is provided by negative pressure in the *surrounding* region, which exactly balances the internal pressure (as required in order for every part of the surface to remain at rest with respect to comoving coordinates). A constant ρ_V and the P_V of Eq. (43) describe this situation perfectly. It is not yet known how to calculate ρ_V, so the value of ρ_V and the corresponding Λ remain empirical questions.

The theory of cosmic inflation rests on the assumption that during a very early period of cosmic history, space was characterized by an unstable state called a "false vacuum." Over the brief period of its existence, this false vacuum had an enormous density $\rho_{FV} \sim 10^{80}$ g/cm^3, with $P_{FV} \cong -c^2\rho_{FV}$. This caused the cosmic scale factor to grow at an exponential rate [as in Eqs. (29)–(31)], with a doubling time $t_d \sim 10^{-37}$ sec.

Even if the false vacuum lasted for only 10^{-35} sec, about 100 doublings would have occurred, causing the universe to expand by a factor of $\sim 2^{100} \approx 10^{30}$. Recall that in curved spaces, departures from flatness at a distance r_p are of order $(r_p/a)^2$ (as explained in Section III.C). Standard inflation predicts that even if space *is* curved, the scale factor is now so large that the *observable* universe is virtually flat. [Hyperbolic ("open") inflation is assumed to have occurred in a space with negative curvature. The inflation, while substantial, may not have lasted long enough to render the observable universe ∼ flat.]

V. CONTENTS, AGE, AND FUTURE OF THE COSMOS

A. The Critical Density

The field equation (26) implies a relation between H, ρ, Λ, and the geometry of space, since the latter is governed by k. For present purposes, it is convenient (and customary) to express Λ in terms of the equivalent vacuum density ρ_V, given by Eq. (42). The sum of all *other* contributions will be denoted by ρ_M. ("M" is short for matter *and radiation*. The latter is negligible at present, but dominated ρ_M at very early times.) We next consider Eq. (26) with $\Lambda = 0$ and $\rho = \rho_M + \rho_V$.

Setting $k = 0$ in Eq. (26) and solving for ρ, we obtain ρ for the intermediate case of a *flat* universe. Its value is called the *critical density*, and is presently given by

$$\rho_c = \frac{3H_0^2}{8\pi G} = (1.88 \times 10^{-29} \text{g/cm}^3)h_0^2 \qquad (45)$$

where h_0 is defined by Eq. (18). Inspection of Eq. (26) reveals that k is positive, zero, or negative if ρ is greater than, equal to, or less than ρ_c, respectively. We note in passing that ρ_c is extremely small. For $h_0 \cong 0.65$, ρ_c is comparable to a density of five hydrogen atoms per cubic meter of space. This is far closer to a perfect vacuum than could be achieved in a laboratory in the foreseeable future.

To facilitate a comparison of individual densities with the critical density, it is customary to define a set of numbers Ω_i by

$$\rho_i = \Omega_i \rho_c \qquad (46)$$

where the index i can refer to any of the particular kinds of density. For example, $\rho_M = \Omega_M \rho_c$, so $\Omega_M = \rho_M/\rho_c$ is the fraction of ρ_c residing in matter (and radiation). The *total* density is described at present by

$$\Omega_0 \equiv \Omega_M + \Omega_V \qquad (47)$$

Equation (26) now implies that $k > 0$, $k = 0$, or $k < 0$ if $\Omega_0 > 1$, $\Omega_0 = 1$, or $\Omega_0 < 1$, respectively. If $k \neq 0$, the present value of the scale factor is readily seen to be

$$k = \pm 1: \qquad a_0 = \frac{c}{H_0}\sqrt{\frac{k}{\Omega_0 - 1}} \qquad (48)$$

(Recall from Section III.B that if $k = 0$, then a_0 has no *physical* significance, and can be chosen arbitrarily.)

Standard inflation does not predict the geometry of the *entire* universe, only that the *observable* universe is *very close* to flat. As was shown in Section III.C, this would be true for either type of curved space if a_0 were very much larger than the radius of the observable universe. Equation (48) reveals that even if $k = \pm 1$, the observable universe can be arbitrarily close to flat if Ω_0 is sufficiently close to 1.

Note that inflation does not predict whether the universe is *finite* ($k = +1$) or *infinite* ($k \leq 0$), for reasons stated above and also because the observable universe is not regarded as representative of the whole. In the latter case, the RW metric (upon which our discussion is based) would not be applicable to the entire universe.

B. Kinds of Matter

The nature and distribution of matter are complex subjects, only partially understood. Broadly speaking, the principal sources of ρ_M are ordinary matter and other (*exotic*) matter. Electrons and atomic nuclei (made of protons and neutrons) are regarded as ordinary, whether they are separate or bound together in atoms. The universe has zero net charge, so electrons and protons are virtually identical in their abundance. (Charged particles are easily detected, and all other charged particles that we know of are short-lived.)

Protons and neutrons have nearly equal masses, and are ∼1800 times as massive as electrons. Virtually all the mass of ordinary matter therefore resides in nuclei. Protons and neutrons are members of a larger group of particles called baryons, and cosmologists often refer to ordinary matter as *baryonic matter* (despite the fact that electrons are not baryons). All other baryons are very short-lived and rare, so any significant amount of other matter is *nonbaryonic*. We shall denote the densities of baryonic and nonbaryonic matter by ρ_B and ρ_{NB}, respectively. The total, present density of matter (and radiation) may now be expressed as $\rho_M \cong \rho_B + \rho_{NB}$, where the approximation is valid within a tiny fraction of 1%.

We have no evidence that stars and planets are constructed of anything but ordinary matter. It follows that any other matter whose density is a significant fraction of ρ_c must be electrically neutral, and also immune to the strong nuclear force (the *strong interaction*): its presence would otherwise be conspicuous in stars and planets. Such matter can interact only weakly with ordinary matter, through gravity and perhaps through the weak nuclear force (the *weak interaction*), and conceivably through some other weak force that remains to be identified.

Neutrinos are particles with the above properties. They are also abundant, having been copiously produced during the early period of nucleosynthesis studied by Gamow and coworkers (*big bang nucleosynthesis*, or BBN). Their rest masses were long thought to be zero (a convoluted piece of jargon meaning that, like photons, they always travel at the speed of light). The expansion and cooling of the universe would have reduced the energy of such relic neutrinos to a negligible amount, as has happened with the relic photons that now make up the cosmic microwave background. Recent experiments have revealed, however, that neutrinos have small but nonzero rest masses. Even with very small masses, neutrinos are so abundant that their density ρ_ν could be a significant fraction of ρ_c. [The Greek letter ν (nu) is a standard symbol for neutrinos.]

There is gravitational evidence (to be described) for the existence of electrically neutral, weakly interacting particles *other* than neutrinos. No such particles have yet been observed in laboratories, so their identity and detailed properties remain unknown. Such matter is therefore called *exotic*, and its density will be denoted by ρ_X (however many varieties there may be). We then have $\rho_{NB} = \rho_\nu + \rho_X$. (Neutrinos are sometimes regarded as exotic matter, but the present distinction is useful.)

C. Amounts of Matter

Stars are made of baryonic matter, as are interstellar and intergalactic clouds of gas and dust. Bright stars (brighter than red dwarfs, the dimmest of main sequence stars) are easily seen and studied, and are found to contain only $\sim 0.5\%$ of ρ_c. Clouds of gas and dust contain several times as much, but the amount has been difficult to measure. In fact our best estimate for the total amount of baryonic matter is based on the relative abundances of nuclei formed during the early period of BBN.

The fraction of primordial baryonic matter converted into ^{2}H (deuterium) during BBN depended strongly on the density of baryons at that early time, in a (presumably) known way. Intergalactic clouds of gas have never been affected by nucleosynthesis in stars, so they should be pure samples of the mixture produced by BBN. By studying light from distant quasars that has passed through intergalactic clouds, the relative abundance of deuterium was measured in 1996 by David R. Tytler, Scott Burles, and coworkers. On the basis of that and subsequent measurements, it is now believed that

$$\Omega_B = (0.019 \pm 0.001)/h_0^2 = 0.045 \pm 0.009 \qquad (49)$$

The amount of baryonic matter is therefore ~ 9 times greater than appears in bright stars.

It has become apparent in recent decades that the universe contains a great deal of *dark* (nonluminous) *matter*. As indicated above, only $\sim 1/9$ of ordinary matter is in

bright stars, and most of the remainder is cold (or dilute) enough to be dark. There is also *gravitational* evidence for ~ 7 times as much dark matter as there is baryonic matter. The evidence resides primarily in the dynamics of galaxies and galactic clusters, which are held together by gravitational forces. The motions of stars in galaxies, and of galaxies in clusters, reveal the strength of gravity within them. The total amount of mass can then be inferred.

We first consider orbital motion around some large, central mass, such as the motion of planets around our sun. Under the simplifying assumption of circular orbits, Newton's laws of gravitation and motion (the latter being $\vec{F} = m\vec{a}$, with centripetal acceleration given by $a = v^2/r$) imply that

$$v = \sqrt{GM_\odot/r} \qquad (50)$$

where v and r denote the speed around, and radius of, the orbit, and $M_\odot$ denotes the mass of the sun.

We can test this model for the solar system by seeing whether orbital speeds and radii of the nine planets are related by the proportionality $v \propto 1/\sqrt{r}$, as predicted by Eq. (50). The orbits pass this test (making allowances for the fact that orbits are actually ellipses). Larger orbits are indeed traversed at smaller speeds, in the way predicted by Newton's laws. Since Eq. (50) implies that $M_\odot = v^2 r/G$, the sun's mass can be determined from the speed and radius of a *single* orbit, with confidence that all orbits yield the same answer (because $v \propto 1/\sqrt{r}$). This is how we know the value of $M_\odot$.

The Milky Way has a small nucleus that is very densely populated with stars (and may contain a black hole of $\sim 10^6 M_\odot$). Centered about this nucleus is a "nuclear bulge" that is also densely populated, with a radius of $\sim 10^4$ ly. The visible disk has a radius of $\sim 5 \times 10^4$ ly, and it becomes thinner and more sparsely populated as one approaches the edge. For the preceding reasons, it was long supposed that most of the galaxy's mass is less than 2×10^4 ly from the center. For stars in circular orbits outside this central mass, orbital speeds should then be a decreasing function of radius, roughly satisfying $v \propto 1/\sqrt{r}$ (as for planets in the solar system).

Beginning in the 1970s, Vera Rubin and others began systematic studies of orbital speeds, out to the edge of our galaxy and beyond (where a relatively small number of stars are found). To everyone's surprise, the speeds did *not* decrease with increasing distance. In fact the speeds at 5×10^4 ly (the visible edge) are $\sim 10\%$ *greater* than at 2×10^4 ly. Furthermore, the speeds are even larger (by a small amount) at 6×10^4 ly, *beyond* the visible edge. Studies of this kind indicate that beyond $\sim 2 \times 10^4$ ly, the amount of mass within a distance r of the center is roughly proportional to r, out to 10^5 ly or so (twice the radius of the visible disk).

Studies of other spiral galaxies have revealed similar properties. The Milky Way and other spirals are now believed to have roughly spherical distributions of dark matter, containing most of the mass and extending far beyond the disk, with densities falling off (roughly) like $1/r^2$ beyond the nuclear bulge. Most of this dark matter is probably nonbaryonic (as will soon be explained).

The largest clusters contain thousands of galaxies, so it is plausible (but not certain) that little dark matter extends beyond their visible boundaries. There are three distinct ways to estimate the total masses of large clusters. The original method (first used by Fritz Zwicky in 1935) is based on the motions of galaxies within a cluster, from which the strength of gravity and total mass can be inferred. The second method exploits the presence of hot, X-ray emitting, intracluster gas. The high temperature presumably results from motion caused by gravity, so the total mass can be inferred from the spectrum of the X-rays. The third method is based on gravitational lensing. The light from more distant galaxies bends as it passes by (or through) a cluster, and detailed observations of this effect enable one to estimate the total mass. All three methods yield similar results.

The total luminosities of nearby clusters are easily determined, and the mass-to-luminosity ratios are about the same for all large clusters that have been studied. If one assumes this ratio to be representative of *all* galaxies (only $\sim 10\%$ of which reside in clusters), then the average density of matter is readily obtained from the total luminosity of all galaxies within any large, representative region of space. This line of reasoning has been applied to the relevant data, yielding $\Omega_M = 0.19 \pm 0.06$. This value for Ω_M would be an underestimate, however, if the matter inside large clusters were more luminous than is typical of matter. This would be true if an appreciable amount of dark matter extends beyond the visible boundaries of large clusters, or if an atypically large fraction of baryonic matter inside of large clusters has formed bright stars. Both possibilities are plausible, and at least one appears to be the case.

In addition to emitting X-rays, the hot intracluster gas slightly alters (by scattering) the cosmic microwave radiation passing through it (the Sunyaev-Zel'dovich effect). Since only charged matter generates or affects radiation, the X-rays and slightly diminished CMB reaching earth both contain information about the amount of baryonic matter. The most *precise* inference from such studies is the *ratio* of baryonic to nonbaryonic mass in large clusters; namely $\rho_B/\rho_M = (0.075 \pm 0.002)/h_0^{3/2}$. Assuming this ratio to be typical of matter everywhere, it can be combined with Eq. (49) to obtain $\Omega_M = (0.253 \pm 0.02)/\sqrt{h_0} = 0.31 \pm 0.03$. This is regarded as the most reliable value for Ω_M. There is a long

history of underestimating uncertainties in this field, however, so it seems appreciably safer to say that

$$\Omega_M = 0.31 \pm 0.06 = (6.9 \pm 1.9)\Omega_B \qquad (51)$$

In terms of mass, only $\sim 1/7$ of matter is baryonic, and only $\sim 1/60$ is in bright stars. Luminous matter is quite special, even less revealing than the tips of icebergs (which display $\sim 1/40$ of the whole, and reveal the composition of the remainder).

D. Nonbaryonic Matter

By mass, $\sim 6/7$ of matter is nonbaryonic. Neutrinos have been studied for decades, and are part of the "standard model" of elementary particles. As mentioned previously, there is recent evidence that neutrinos have nonzero rest masses (the Super-Kamiokande detection of neutrino oscillations, beyond the scope of this article). The present data place a lower limit on Ω_ν, namely $\Omega_\nu \geq 0.003$. Laboratory experiments have not ruled out the possibility that neutrinos account for *all* the nonbaryonic matter, but there are reasons to believe otherwise.

Indirect evidence for the nature of the nonbaryonic matter arises in computer models for the progressive concentration of matter into galaxies, clusters of galaxies, superclusters, sheets (or *walls*) of galaxies, and voids. All of these structures evolved as gravity progressively amplified small deviations from perfect uniformity, beginning at a very early time. (The original inhomogeneities are believed to have been microscopic quantum fluctuations that were magnified by inflation.)

When the temperature dropped below $\sim 3,000$ K, the electrons and nuclei were able to remain together as electrically neutral atoms, forming a transparent gas. This happened $\sim 300,000$ yr after the big bang, the *time of last scattering*. The photons then present have been traveling freely ever since, and now form the cosmic microwave background. Studies of the CMB reveal that deviations from perfect uniformity at the time of last scattering were on the order of $\delta\rho_M/\rho_M \sim 10^{-5}$. Those slight inhomogeneities have since been amplified by gravity into all the structures that we see today.

Computer models for the growth of structure require assumptions about the exotic dark matter. This matter is usually assumed to consist of stable elementary particles that are relics from a very early time. Because they are weakly interacting, only gravity should have affected them appreciably (gravity is the only weak force acting over large distances).

Since space was virtually transparent to exotic particles, their speeds would have played a key role in structure formation. High speeds would have inhibited their being captured by growing concentrations of matter, unless those concentrations were unusually extended and massive. Computer simulations indicate that fast weakly-interacting particles would only have formed very large concentrations, of cluster and even supercluster size, whereas slow particles would first have formed smaller concentrations of galactic size. The scenario wherein concentrations of cluster size preceded the formation of galaxies is called "top down" growth of structure. If galaxies formed first and later clumped into clusters, the growth is called "bottom up."

Since light travels at a finite speed, astronomers see things as they were in the increasingly distant past as they look out to increasingly greater distances. The evidence is very clear that galaxies formed first, with only $\sim 10\%$ of them clumping later into clusters. Structure formation occurred from the bottom up. It follows that nonbaryonic particles were predominately slow. Of course "fast" and "slow" are relative terms, and the average speeds of *all* types of massive particles decreased as the universe expanded. These simple labels are useful in a brief discussion, however, and they have fairly definite meanings among workers in the field. In further jargon, fast nonradiating (uncharged) particles are referred to as "hot dark matter," and slow ones as "cold dark matter." Bottom-up structure formation implies that dark matter is predominately cold.

There are three species of neutrino (electron, muon, and tau-neutrinos). All three species should have reached thermal equilibrium with other matter and photons before the universe was one second old. (The early, high density of matter more than compensated for the weakness of neutrino interactions.) For this and other reasons, the present abundance of relic neutrinos is believed to be known, and Ω_ν can be expressed as

$$\Omega_\nu = \sum_i \frac{m_i c^2}{93.5 h_0^2 \, \mathrm{eV}} = \sum_i \frac{(0.025 \pm 0.005) m_i c^2}{1 \, \mathrm{eV}} \quad (52)$$

where the index i refers to the three species (1 eV = 1 electronvolt = 1.60×10^{-19} joule).

If only *one* species had a nonzero mass, with $m_i c^2 \cong 39$ eV, the result would be $\Omega_\nu \cong 1$. (For comparison, the lightest particle of ordinary matter is the electron, with $m_e c^2 = 5.11 \times 10^5$ eV.) Since $\Omega_{NB} \cong 0.26$, the upper limit on neutrino mass-energies is on the order of 10 eV. It is the combination of weak interactions with very small masses that has frustrated efforts to determine the mass of even one species of neutrino. The Super-Kamiokande experiment is only sensitive to *differences* in mass between different species. The observed mass difference implies $\Omega_\nu \geq 0.003$, where the equality would hold if only one species had nonzero mass.

Because neutrinos are so light and were in thermal equilibrium at an early, hot time, they are the paradigm of "hot dark matter." They slowed as the universe expanded, but at any *given* temperature, the lightest particles were the fastest. Computer simulations rule out neutrinos as the predominate form of nonbaryonic matter. Recalling the lower limit discussed previously, we conclude that

$$0.003 \leq \Omega_\nu < \tfrac{1}{2}\Omega_{NB} \quad (53)$$

The two most favored candidates for exotic matter are *axions* and *neutralinos*. These arise in speculative, but plausible, theories that go beyond the standard model of elementary particles. (Neutralinos occur in *supersymmetric* theories, which receive tentative support from a very recent experiment on the precession of muons.) For quite disparate reasons, both qualify as cold dark matter.

Axions are predicted to be much lighter than neutrinos. Unlike neutrinos, however, they would never have been in thermal equilibrium, and would *always* have moved slowly (for abstruse reasons beyond the scope of this article). In sharp contrast, neutralinos would have been in thermal equilibrium when the universe was very young and hot. Neutralinos (if they exist) are much more massive than protons and neutrons, however, and their *large masses* would have rendered them (relatively) slow-moving at any given temperature. Other candidates for exotic "cold dark matter" have been proposed, but will not be mentioned here.

E. Vacuum Energy

In 1998, two teams of observers, one led by Saul Perlmutter and the other by Brian P. Schmidt, announced that the cosmic expansion appears to be *accelerating*. This revolutionary claim was based on observations of type SNe Ia over a wide range of distances, extending out to $\sim 6 \times 10^9$ ly. Distances approximately twice as great have since been explored using SNe Ia as standard candles.

Redshifts are usually described in terms of a red-shift parameter $z = \lambda_0/\lambda_e - 1$, where λ_0 denotes the observed wavelength and λ_e the emitted wavelength. With spacetime described by a RW metric, it is readily shown that $z = a(t_0)/a(t_e) - 1$, where t_0 and t_e denote the (present) time of observation and the earlier time of emission, respectively. Supernovae of type Ia have been observed with redshifts as large as $z \approx 1$, corresponding to $a(t_0) \approx 2a(t_e)$. Such light was emitted when the universe was only half its present size, and roughly half its present age.

Since the intrinsic luminosity of SNe Ia is known, the apparent brightness reveals the distance of the source, from which the time of emission can be inferred (subject to possible corrections for spatial curvature). The evolution over time of $a(t)$ has been deduced in this way, and indicates an accelerating expansion. Since ρ_M has decreased over time while ρ_V remained constant, they have affected $a(t)$ in different ways. Thus Ω_M and Ω_V can both be deduced from detailed knowledge of $a(t)$. Present SNe Ia data indicate that $\Omega_M \approx 0.3$ and $\Omega_V \approx 0.7$, with uncertainties in both cases on the order of ± 0.1.

The CMB radiation provides a snapshot of the universe at the time of last scattering, $\sim 300{,}000$ yr after the big bang. The CMB reaching us now has come from the outermost limits of the observable universe, and has a redshift $z \sim 1100$. It has an extremely uniform temperature of $T = (2.728 \pm 0.002)$K in all directions, except for tiny deviations on the order of $\delta T/T \sim 10^{-5}$ over small regions of the sky. These deviations are manifestations of slight inhomogeneities in density of order $\delta \rho_M / \rho_M \sim 10^{-5}$ at the time of last scattering. Gravity produced these slight concentrations of matter from much smaller ones at earlier times, in ways that are believed to be well understood. In particular, the *spectrum* (probability distribution) of different *sizes* of inhomogeneity is calculable in terms of much earlier conditions.

The angle subtended by an object of known size at a known distance depends on the curvature of space, as discussed in Section III.C. (For example, the circumference of a circle differs from $2\pi r$ in a curved space.) In the present context, the "objects of known size" were the *largest* concentrations of matter, whose diameters were governed by the distance that density waves could have traveled since the big bang (the *sound horizon* at the time of last scattering).

The speed of density waves is a well-understood function of the medium through which they travel, and the sound horizon at last scattering has been calculated with (presumably) good accuracy. The angular diameters of the largest concentrations of matter, manifested by $\delta T/T$, therefore reveal the curvature of space. Within observational uncertainties, the data confirm the foremost prediction of standard inflation, namely that *space is very close to flat*. There is no evidence of curvature. The CMB data indicate that space is flat or, if curved, that $a_0 \geq 3c/H_0 \sim 45$ billion light-years. Equation (48) then implies that

$$\Omega_0 = 1.0 \pm 0.1 \qquad (54)$$

From Eq. (51) and the SNe Ia result cited above, it follows that

$$\Omega_V = 0.69 \pm 0.12 \qquad (55)$$

Spatial flatness and vacuum energy are the most important and dramatic discoveries in observational cosmology since the CMB (1964).

There has been some speculation that the mass-energy density filling all of space might not be *strictly* constant, but change slowly over time. (Such an energy field is sometimes called "quintessence.") Denoting the density of such a hypothetical field by ρ_Q, observations would imply a large, negative pressure, $P_Q < -0.6c^2\rho_Q$. Present data are wholly consistent with the simplest possibility, however, the one described here in detail: a constant ρ_V with $P_V = -c^2\rho_V$, perfectly mimicking a cosmological constant.

F. Age and Future of the Cosmos

We now consider the evolution over time of a flat universe, under the simplifying assumption that $P_M = 0$ (an excellent approximation since the first 10 million years or so). Equations (22), (27), (42), and (43) then imply

$$t = \frac{2}{3H_0} \int_0^{x(t)} \frac{dx'}{\sqrt{\Omega_V x'^2 + \Omega_M}} \qquad (56)$$

where t denotes the time since the big bang, $x(t) \equiv [a(t)/a_0]^{3/2}$, denotes the present value of the cosmic scale factor $a(t)$, and Ω_M and Ω_V are *present* values, with $\Omega_M + \Omega_V = 1$.

The scale of time is set by H_0, where $t_H \equiv 1/H_0$ is called the *Hubble time*. To understand its significance, suppose that every distant galaxy had always been moving away from us at the same speed it has now. For motion at constant speed v, the time required to reach a distance d is $t = d/v$. According to Hubble's law, $v = H_0 d$ [Eq. (17)], so the time required for *every* galaxy to have reached its present distance from us would be $d/v = 1/H_0 = t_H$. Thus t_H would be the age of the universe if galactic speeds had never been affected by gravity. (It is sometimes called the "Hubble age.") With H_0 given by Eqs. (18) and (19),

$$t_H \equiv \frac{1}{H_0} = (15.1 \pm 2.4) \times 10^9 \text{ yr} \qquad (57)$$

Denoting the *actual* age of the universe by t_0 (the *present* time), we note that $x(t_0) = 1$. For the simple case where $\Omega_M = 1$ (and $\Omega_V = 0$), Eq. (56) yields $t_0 = \frac{2}{3}t_H \sim 10 \times 10^9$ yr (less than the age of the oldest stars). For $\Omega_V > 0$, Eq. (56) implies

$$\frac{t}{t_H} = \frac{2}{3\sqrt{\Omega_V}} \ln\left(\frac{\sqrt{\Omega_V(a/a_0)^3} + \sqrt{\Omega_V(a/a_0)^3 + \Omega_M}}{\sqrt{\Omega_M}} \right) \qquad (58)$$

expressing t/t_H as a function of a/a_0. Using the observed value $\Omega_M = 0.31 \pm 0.06$ (with $\Omega_V = 1 - \Omega_M$) and setting $a = a_0^4$, we obtain the present age

$$t_0 = (0.955 \pm 0.053)t_H = (14.4 \pm 2.4) \times 10^9 \text{ yr} \qquad (59)$$

The oldest stars are believed to be $(12 \pm 2) \times 10^9$ years old, and they are expected to have formed $\sim 2 \times 10^9$ yr after the big bang. The discovery of substantial vacuum energy (or something very similar) has increased the theoretical age of the universe comfortably beyond that of the oldest stars, thereby resolving a long-standing puzzle.

The time when the cosmic expansion began accelerating is readily determined. Replacing Λ in Eq. (23) with the ρ_V and P_V of Eqs. (42) and (43), we obtain

$$\frac{\ddot{a}}{a} = -\frac{4\pi G}{3}\left(\rho_M - 2\rho_V + \frac{3P_M}{c^2}\right) \qquad (60)$$

With $P_M \cong 0$, the acceleration ($\ddot{a} > 0$) began when ρ_M fell below ρ_V. Using Eq. (27) for ρ_M, this happened when $(a/a_0)^3 = \Omega_M/2\Omega_V \simeq 0.22$. Denoting the corresponding time by t_a, Eq. (58) yields $t_a \simeq 0.52 t_H \simeq 0.55 t_0$.

When a/a_0 reaches 2, ρ_M will have shrunk (by dilution) to ~6% of ρ_V. This will happen at $t \simeq 1.7 t_H \simeq 1.8 t_0$. From roughly that time onward, the universe will grow at an exponential rate, as described by Eqs. (29)–(31) and (42). The doubling time will be

$$t_d = \frac{\ln 2}{\sqrt{\Omega_V}} t_H \approx 13 \times 10^9 \text{ yr} \qquad (61)$$

The ratio $a(t)/a_0$ is displayed in Fig. 1, which should be reliable for $t/t_0 \geq 10^{-3}$.

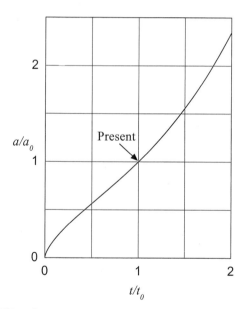

FIGURE 1 Cosmic expansion: the distance between any pair of comoving objects is proportional to the cosmic scale factor $a(t)$. Its present value is denoted by a_0, and the present time by t_0. For example, two objects were half as far apart as now when $a/a_0 = 1/2$, which occurred at $t \approx 0.4 t_0$. They will be twice as far apart as now when $a/a_0 = 2$, which happen at $t \approx 1.8 t_0$. The graph is only valid for $t/t_0 \geq 10^{-3}$ (after inflation).

VI. HORIZON AND FLATNESS PROBLEMS

A. The Horizon Problem

The greatest distance that light could (in principle) have traveled since the universe began is called the *horizon distance*. Since no causal influence travels faster than light, any two regions separated by more than the horizon distance could not have influenced each other in any way. Observations reveal the universe to be flat (or nearly so) out to the source of the CMB now reaching us. We shall assume for simplicity that space is precisely flat ($k = 0$, $\Omega = 1$).

Every light path is a "null geodesic," with $(ds)^2 = 0$ in Eq. (7). With $k = 0$, $c\,dt = \pm a(t)\,dr$. Let us place ourselves at $r = 0$, and determine the proper distance $r_p = a_0 r$ to an object whose light is now reaching us with redshift $z = (a_0/a) - 1$. Since $dt = da/\dot{a}$ and $\dot{a} = aH$,

$$r_p = a_0 c \int_a^{a_0} \frac{da'}{a'^2 H(a')} \qquad (62)$$

where $H(a)$ is given by Eq. (26) in terms of $\rho(a)$. We shall initially assume the pressure of matter and radiation to be negligible, in which case $\rho \propto 1/a^3$. Equation (26) then implies

$$H^2 = H_0^2\left[\Omega_M\left(\frac{a_0}{a}\right)^3 + \Omega_V\right] \qquad (63)$$

where Λ is being represented by ρ_V as in Eq. (42) (and, as before, Ω_M and Ω_V denote *present* values). Equations (62) and (63) imply

$$r_p(z) = \frac{c}{H_0} f(z),$$

$$f(z) \equiv 2 \int_{1/\sqrt{z+1}}^{1} \frac{dy}{\sqrt{\Omega_M + \Omega_V y^6}} \qquad (64)$$

where $y \equiv \sqrt{a/a_0}$.

The integral defining $f(z)$ cannot be evaluated in terms of elementary functions. For small z, however, one can define $x \equiv 1 - y$, expand the integrand in powers of x, and thereby obtain

$$f(z) = z\left[1 - \tfrac{3}{4}\Omega_M z + \mathbb{O}(z^2)\right] \qquad (65)$$

where $\mathbb{O}(z^2)$ denotes corrections of magnitude ~z^2. The two terms displayed in Eq. (65) are accurate within 1% for $z \leq 1$ and $\Omega_M \sim 0.3$. (Despite appearances, Eq. (65) involves Ω_V, through our assumption that $\Omega_V = 1 - \Omega_M$.)

Next we note that $f(\infty) - f(z)$ is twice the integral over y from 0 to $1/\sqrt{z+1}$. For large z, y remains small, and the integrand can be expanded as a power series in y. We thereby obtain

$$f(z) = f(\infty) - \frac{2}{\sqrt{\Omega_M(z+1)}}\left\{1 - O\left[\frac{\Omega_V}{14\Omega_M(z+1)^3}\right]\right\} \quad (66)$$

where the correction shown is *exact* below order $1/(z+1)^6$. If one keeps this correction term, Eq. (66) is valid within 0.2% for $z \geq 1$ and $\Omega_M \sim 0.3$. The value of $f(\infty)$ is readily determined by numerical integration of Eq. (64).

For times earlier that $t_{EQ} \sim 3 \times 10^5$ yr (the time of *equality* between low and high-pressure ρ_M), radiation and highly relativistic particles contribute more to ρ_M than does matter with negligible pressure. (It appears to be coincidental that this is close to the time of last scattering, when the universe became transparent and the CMB set forth on its endless journey.) The transition is, of course, a smooth one. For simplicity, however, we assume that $P_M = 0$ for $t > t_{EQ}$, in which case Eq. (64) would be valid. For $t < t_{EQ}$, we assume that $P_M = c^2\rho_M/3$, which is the pressure exerted by radiation and highly relativistic particles.

We also impose continuity on a and $\dot{a}$ at t_{EQ} or, equivalently, on a and H. Finally, the CMB was emitted at $t \sim t_{EQ}$, and has $z_{CMB} \sim 1100$. The value of z_{EQ} can only be estimated, but it seems certain that $z_{EQ} \geq 100$. It follows that $\rho_V \leq 10^{-6}\rho_M$ at t_{EQ}, and we shall neglect it for $t \leq t_{EQ}$.

For $P_M = c^2\rho_M/3$, Eq. (24) implies that $\rho_M \propto 1/a^4$. It then follows from Eq. (26) that $2a\dot{a} = da^2/dt$ is constant for $t \leq t_{EQ}$. We are presently concerned with the standard big bang (without inflation), wherein $\rho \propto 1/a^4$ all the way back to $t = 0$ (when $a = 0$). We therefore have

$$t \leq t_{EQ}: \quad \left(\frac{a}{a_{EQ}}\right)^2 = \frac{t}{t_{EQ}}, \quad H \equiv \frac{\dot{a}}{a} = \frac{1}{2t} \quad (67)$$

where $a_{EQ} \equiv a(t_{EQ})$.

Equation (67) holds at early times *regardless* of what the present universe is like, an extraordinary fact of great simplicity and usefulness.

Equation (67) implies that a^2H is constant for $t \leq t_{EQ}$, so the integral in Eq. (60) is trivial for $t \leq t_{EQ}$. Imposing continuity on H, Eq. (63) can be used to express H_{EQ} in terms of H_0, Ω_M, and a_0 (neglecting Ω_V at t_{EQ}). We thereby obtain

$$z \geq z_{EQ}:$$

$$r_p(z) = \frac{c}{H_0}\left[f(\infty) - \frac{1}{\sqrt{\Omega_M(z_{EQ}+1)}}\frac{(z+z_{EQ}+2)}{z+1}\right] \quad (68)$$

For $\Omega_M = 0.31$ (with $\Omega_V = 0.69$), a numerical integration yields $f(\infty) = 3.26$. As z_{EQ} ranges from 100 to the (un-

realistic) value of ∞, $r_p(\infty)$, varies by less than 6%, and $r_p(z_{EQ})$ differs from $r_p(\infty)$ by less than 6%. Note that $r_p(\infty)$ is the present *standard horizon* distance $d_{SH}(t_0)$, i.e., the horizon distance if there had been no inflation. For $z_{EQ} \sim 1000$, we obtain

$$d_{SH}(t_0) = r_p(\infty) \approx 3.2\frac{c}{H_0} \approx 48 \times 10^9 \text{ ly} \quad (69)$$

A 20% uncertainty in Ω_M would only render the factor of 3.2 uncertain by $\sim$10%. It may seem puzzling that $d_{SH}(t_0) \sim 3 \times$ (age of universe), but the cosmic expansion has been moving the source away from us while the light traveled toward us. This effect has been amplified by acceleration of the expansion.

The distance to the source of the CMB is of particular interest, for reasons that will become apparent. Since $z_{CMB} \approx 1100$, the percentage difference between $r_p(z_{CMB})$ and $d_{SH}(t_0)$ is small, and we have seen that it varies little with any choice of $z_{EQ} \geq 100$. For $z_{EQ} = z_{CMB} = 1100$, Eq. (68) yields $r_p(z_{CMB}) \approx 3.1(c/H_0) \approx 47 \times 10^9$ ly. The CMB sources in opposite directions from us are therefore separated by $\sim$90 $\times 10^9$ ly. Proper distances between comoving points scale like $a/a_0 = z + 1$, so sources in opposite directions were $\sim$8 $\times 10^7$ ly apart when they emitted the CMB we now observe. This is a fact of extreme interest, as we shall see.

The CMB reaching us now began its journey at $t_{LS} \approx 300,000$ yr, where t_{LS} denotes the time of last scattering. We wish to determine $d_{SH}(t_{LS})$, the (standard) horizon distance at that time. We consider light emitted from $r = 0$ at $t = 0$, and consider the proper distance $d_H(t) = a(t)r(t)$ it has traveled by time t:

$$d_H(t) = ca(t)\int_0^t \frac{dt'}{a(t')} \quad (70)$$

Using Eq. (67), we obtain

$$t \leq t_{EQ}: \quad d_{SH}(t) = 2ct, \quad (71)$$

a wonderfully simple result. If $t_{LS} \leq t_{EQ}$, Eq. (71) would imply that $d_{SH}(t_{LS}) \approx 600,000$ ly.

For a considerable time later than t_{EQ}, a/a_0 remained small enough that $\rho_V \ll \rho_M$. (For example, with $\Omega_V/\Omega_M \approx 7/3$, $\rho_V \leq 10^{-2}\rho_M$ for $a/a_0 \leq \frac{1}{6}$. This is satisfied for $t \leq 0.7t_0 \approx 10^{10}$ year, as may be seen from Fig. 1.) With $\rho \propto 1/a^3$, Eq. (63) implies that $da^{3/2}/dt$ is constant, hence that $a^{3/2} = \alpha t + \beta$ for constant α and β. Continuity of a and H at t_{EQ} determines these constants, with the results

$$t \geq t_{EQ}, \rho_V = 0: \quad a(t) = a_{EQ}\left(\frac{3t + t_{EQ}}{4t_{EQ}}\right)^{2/3}, \quad (72)$$

$$d_{SH}(t) = c(3t + t_{EQ})\left\{1 - \left[\frac{t_{EQ}}{2(3t + t_{EQ})}\right]^{1/3}\right\} \quad (73)$$

Now recall that if $t_{LS} \leq t_{EQ}$, Eq. (71) yields $d_{SH}(t_{LS}) \approx$ 600,000 ly. If $t_{LS} \geq t_{EQ}$, Eq. (73) yields $d_{SH} \approx 630,000$ ly (within 5%), for any $t_{EQ} \geq 50,000$ yr (a very conservative lower bound). In either case, $d_{SH}(t_{LS})$ was less than 1% of 80 million ly, the distance between sources of the CMB in opposite directions from us when the CMB was omitted. This is the *horizon problem* described in Section II.A.

Next consider a simple inflationary model. Let t_b and t_e denote the times when inflation begins and ends, respectively, with $t_e \ll 1$ sec. For $t \leq t_b$, $(a/a_b)^2 = t/t_b$, with $H = 1/2t$ as in Eq. (67). For $t_b \leq t \leq t_e$, $a = a_b \exp\Gamma(t - t_b)$, with $H = \Gamma$, a constant [recall Eqs. (29) and (30)]. Continuity of H implies $\Gamma = 1/2t_b$. The doubling time t_d is related to Γ by $\Gamma = \ln 2/t_d$, so $t_d = (2 \ln 2)t_b$.

Inflation ends at t_e with $a_e = a_b \exp\Gamma\Delta t$ and $H = \Gamma$, where $\Delta t \equiv t_e - t_b$. The density then reverts to its earlier behavior $\rho \propto 1/a^4$, and Eq. (26) again implies that da^2/dt is constant. The general solution is $a^2 = \xi t + \eta$, where ξ and η are fixed by continuity of a and H at t_e. One readily finds that $a^2 = a_e^2[2\Gamma(t - t_e) + 1]$, but a more elegant solution is at hand.

From the end of inflation onward, the universe evolves precisely like a model without inflation, but which had the same conditions at t_e. Denoting time in the latter model by t', $H = 1/2t'$, and $H = \Gamma$ at $t'_e = 1/2\Gamma$. The scale factor was zero at $t' = 0$ in this model, so $(a/a_e)^2 = t'/t'_e$. Since $t'_e = 1/2\Gamma$ when $t = t_e$, the relation is $t' = (t - t_e) + t_d/\ln 4$.

Using $dr = c\, dt/a(t)$, it is straightforward to calculate $r(t_b)$, $r(t_e) - r(t_b)$, and $r(t') - r(t'_e)$. We thereby obtain

$$\frac{d_H(t')}{2ct'} = (e^{\Gamma\Delta t} - 1)\sqrt{\frac{2t_d}{t' \ln 2} + 1} \qquad (74)$$

(presuming the *actual* horizon distance d_H is given by the inflationary model). The horizon problem is solved if $d_H(t_{LS}) \gg r_p(z_{CMB})/z_{CMB}$. This inequality needs to be a strong one, in order to explain how the CMB arriving from opposite directions can be so *precisely* similar. Combined with Eq. (74), this requires inflation by a factor

$$e^{\Gamma\Delta t} \gg \frac{r_p(z_{CMB})}{z_{CMB}c\sqrt{t_d t_{LS}}} \qquad (75)$$

where factors of order unity have been omitted. For $t_d \sim 10^{-37}$ sec,

$$e^{\Gamma\Delta t} \gg 10^{28} \qquad (76)$$

is required. Resolution of the horizon problem is displayed in Figure 2.

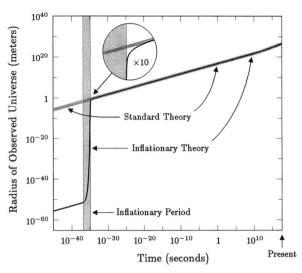

FIGURE 2 Solution to the horizon problem: the size of the observed universe in the standard and inflationary theories. The vertical axis shows the radius of the region that evolves to become the presently observed universe, and the horizontal axis shows the time. In the inflationary theory, the universe is filled with a false vacuum during the inflationary period, marked on the graph as a vertical gray band. The gray line describing the standard hot big bang theory is drawn slightly thicker than the black line for the inflationary theory, so that one can see clearly that the two lines coincide once inflation has ended. Before inflation, however, the size of the universe in the inflationary theory is far smaller than in the standard theory, allowing the observed universe to come to a uniform temperature in the time available. The inset, magnified by a factor of 10, shows that the rapid expansion of inflation slows quickly but smoothly once the false vacuum has decayed. (The numerical values shown for the inflationary theory are not reliable, but are intended only to illustrate how inflation can work. The precise numbers are highly uncertain, since they depend on the unknown details of grand unified theories.) [From "The Inflationary Universe" by Alan Guth. Copyright © 1997 by Alan Guth. Reprinted by permission of Perseus Books Publishers, a member of Perseus Books, L.L.C.]

B. The Flatness Problem

Let us now define

$$\Omega(t) \equiv \rho(t)/\rho_c(t) \qquad (77)$$

where $\rho_c(t)$ is the critical density given by Eq. (45) with H_0 replaced by $H(t) = \dot{a}/a$. As emphasized by Robert Dicke and P. James E. Peebles in 1979, any early deviation of Ω from unity would have grown very rapidly with the passage of time, as Figure 3 displays. Replacing Λ by the equivalent ρ_V, Eq. (26) implies

$$\frac{\Omega - 1}{\Omega} = \frac{3kc^2}{8\pi G\rho a^2} \leq \frac{3kc^2}{8\pi G\rho_M a^2} \qquad (78)$$

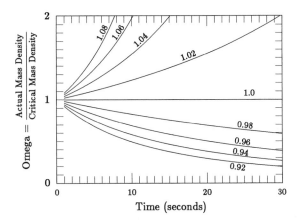

FIGURE 3 The evolution of omega for the first 30 sec. The curves start at one second after the big bang, and each curve represents a different starting value of omega, as indicated by the numbers shown on the graph. [From "The Inflationary Universe" by Alan Guth. Copyright © 1997 by Alan Guth. Reprinted by permission of Perseus Books Publishers, a member of Perseus Books, L.L.C.]

where the inequality holds for $\rho_V \geq 0$. Since $\rho = (\Omega/\Omega_M)\rho_M$,

$$\frac{\Omega_0 - 1}{\Omega_0} = \left(\frac{\Omega_{M0}}{\Omega_0}\right)\frac{3kc^2}{8\pi G\rho_{M0}a_0^2} \quad (79)$$

where the subscript "0" denotes present values.

The pressure P_M of matter (and radiation) became negligible when the universe was $\sim 10^6$ years old. Since that early time, $\rho_M \simeq \rho_{M0}(a_0/a)^3$ [as in Eq. (27)]. Equations (78) and (79) therefore imply

$$\frac{\Omega - 1}{\Omega} \leq \frac{\Omega_0 - 1}{\Omega_{M0}}\left(\frac{a}{a_0}\right) \quad (80)$$

which actually holds for *all* times since inflation, because the left side decreases even more rapidly than the right side as one goes back into the period when $P_M > 0$. As explained in Section V, it appears that $\Omega_{M0} \geq 0.2$ and $|\Omega_0 - 1| \leq 0.2$, in which case

$$\frac{|\Omega - 1|}{\Omega} \leq \frac{a}{a_0} \quad (81)$$

for all times since inflation. The CMB began its journey at $t_{LS} \approx 300,000$ yr, and now has $z_{CMB} \approx 1100$. It follows that $|\Omega_{LS} - 1| < 10^{-3}$.

We do not receive any light emitted before t_{LS}. To determine Ω at earlier times, we must study the time-dependence of the scale parameter a. We begin by noting that Eq. (26) (with $\Lambda = 0$) can be written as $H^2 = \Omega H^2 - k(c/a)^2$, which implies that $k(c/a)^2 = [(\Omega - 1)/\Omega] \times \Omega H^2$. The k-term in Eq. (26) is therefore negligible when $|\Omega - 1| \ll 1$, and we shall drop it for $t \leq t_{LS}$.

For all times earlier than t_{LS}, ρ_V was completely negligible. It is also known that for times earlier than

$t_{ER} \sim 40,000$ yr (the *end* of the *radiation* era), radiation and/or relativistic particles dominated ρ_M, exerting a pressure $P_M \simeq \rho_M c^2/3$. Thus $\rho \simeq \rho_{ER}(a_{ER}/a)^4$, and Eq. (78) implies

$$t \leq t_{ER}: \quad \frac{\Omega - 1}{\Omega} = \frac{\Omega_{ER} - 1}{\Omega_{ER}}\left(\frac{a}{a_{ER}}\right)^2 \quad (82)$$

We are here considering the standard model (without inflation), so $a^2 \propto t$.

Neither matter nor radiation dominated for times between t_{LS} and t_{ER}. Since $t_{ER} < t_{LS}$, however, $|\Omega_{ER} - 1| < |\Omega_{LS} - 1| < 10^{-3}$. Equation (82) then implies

$$t \leq t_{ER}: \quad |\Omega - 1| < 10^{-3}\frac{t}{t_{ER}} \quad (83)$$

Since $t_{ER} = 1.3 \times 10^{12}$ sec, it follows that $|\Omega - 1| < 10^{-15}$ at $t = 1$ sec.

It is striking that Ω was so close to unity at 1 sec, but there is no physical significance to the time of 1 sec. Had there been no inflation, the inequality (83) should have been valid all the way back to the Planck time $t_P \equiv L_P/c \sim 10^{-43}$ sec. Quantum gravity would have been significant at earlier times, and is only partially understood, but Eq. (83) implies

$$t = t_{Pl}: \quad |\Omega - 1| < 10^{-58} \quad (84)$$

We have no reason to suppose that quantum gravity would lead to a value for Ω near unity, let alone so *extremely* close. The present age of the universe is $\sim 10^{60} \times t_{Pl}$, and it is a profound mystery why Ω now differs from unity (if at all) by less than 20%. This is the "flatness problem."

The Hubble parameter $H \equiv \dot{a}/a$ remains constant during any period of exponential growth [recall Eqs. (29) and (30)]. If $a(t)$ increased by a factor of 10^N during a period of exponential growth, a generalization of Eq. (48) to arbitrary times would imply that $|\Omega - 1|$ *decreased* by a factor of 10^{2N}. To remove the necessity for Ω to have been close to unity at the Planck time, inflation by a factor $\geq 10^{30}$ is required. It seems unlikely that such growth would have been the *bare minimum* required to explain why Ω is near unity after $10^{60} \times t_{Pl}$, so standard inflation predicts growth by *more* than 30 powers of 10 (already required to solve the horizon problem), probably *many* more. It would follow that Ω_0 is extremely close to unity, which standard inflation predicts.

VII. UNIFIED THEORIES AND HIGGS FIELDS

A. Quantum Theory of Forces

In quantum field theory (QFT), forces between particles result from the exchange of *virtual quanta*, i.e., transient, particlelike manifestations of some field that gives rise to the force. Electromagnetic forces arise from the exchange of virtual photons, where *photon* is the name given to a quantum or energy packet of the electromagnetic field. The weak nuclear force results from the exchange of *weak vector bosons*, of which there are three kinds called W^+, W^-, and Z. The strong nuclear force results, at the most fundamental level, from the exchange of eight *colored gluons* (a whimsical, potentially misleading name in that such particles are without visible qualities; *color* refers here to a technical property). From the viewpoint of QFT, gravitation may be regarded as resulting from the exchange of virtual *gravitons*. The four types of force just enumerated encompass all of the fundamental interactions currently known between elementary particles.

It is well-known that a major goal of Einstein's later work was to construct a unified field theory in which two of the existing categories of force would be understood as different aspects of a single, more fundamental force. Einstein failed in this attempt; in retrospect, the experimental knowledge of elementary particles and their interactions was too fragmentary at the time of his death (1955) for him to have had any real chance of success. The intellectual heirs of Einstein have kept his goal of unification high on their agendas, however. With the benefit of continual, sometimes dramatic experimental advances, and mathematical insights from a substantial number of theorists, considerable progress in unification has now been achieved. A wholly unexpected by-product of contemporary unified theories is the possibility of cosmic inflation. To understand how this came about, let us briefly examine what is meant by a unification of forces.

Each of the four categories of force enumerated above has several characteristics that, taken together, serve to distinguish it from the other types of force. One obvious characteristic is the way in which the force between two particles depends on the distance between them. For example, electrostatic and (Newtonian) gravitational forces reach out to arbitrarily large distances, while decreasing in strength like the inverse square of the distance. Forces of this nature (there seem to be only the two just named) are said to be *long range*. The weak force, in contrast, is effectively zero beyond a very short distance ($\sim 2.5 \times 10^{-16}$ cm). The strong force between protons and neutrons is effectively zero beyond a short range of $\sim 1.4 \times 10^{-13}$ cm, but is known to be a residual effect of the more fundamental force between the quarks (elementary, subnuclear particles) of which they are made. It is the force between quarks that is mediated by colored gluons (in a complicated way with the remarkable property that, under special conditions, the force between quarks is virtually independent of the distance between them).

The range of a force is intimately related to the mass of the virtual particle whose exchange gives rise to the force. Let us consider the interaction of particles A and B through exchange of virtual particle P. For the sake of definiteness, suppose that particle P is emitted by A and later absorbed by B, with momentum transmitted from A to B in the process.

The emission of particle P appears to violate energy conservation, for P did not exist until it was emitted. This process is permitted, however, by the time-energy uncertainty principle of quantum theory: energy conservation may be temporarily violated by an amount ΔE for a duration Δt, provided that $\Delta E \times \Delta t \approx \hbar$ (where $\hbar$ denotes Planck's constant divided by 2π). If particle P has a rest-mass m_p, then $\Delta E \geq m_p c^2$, so the duration is limited to $\Delta t \leq \hbar / m_p c^2$. The farthest that particle P can travel during this time is $c\Delta t \leq \hbar / m_p c$. This limits the range of the resulting force to a distance of roughly $\hbar / m_p c$ (called the *Compton wavelength* of particle P).

The preceding discussion is somewhat heuristic, but is essentially correct. If a particle being exchanged has $m > 0$, then the resulting force drops rapidly to zero for distances exceeding its Compton wavelength; conversely, long-range forces (i.e., electromagnetism and gravitation) can only result from exchange of "massless" particles. (Photons and gravitons are said to be massless, meaning zero rest-mass. Calling them "massless" is a linguistic convention, because such particles always travel at the speed of light and hence defy any measurement of mass when at rest. They may have arbitrarily little energy, however, and they behave as theory would lead one to expect of a particle with zero rest-mass.)

There are other distinguishing features of the four types of force, such as the characteristic strengths with which they act, the types of particles on which they act, and the force direction. We shall not pursue these details further here, but simply remark that two (or more) types of force would be regarded as unified if a single equation (or set of equations) were found, based on some unifying principle, that described all the forces in question. Electricity and magnetism were once regarded as separate types of force, but were united into a single theory by James Clerk Maxwell in 1864 through his equations which show their interdependence. From the viewpoint of QFT, electromagnetism is a single theory because both electrical and magnetic forces result from the exchange of photons; forces are currently identified in terms of the virtual quanta or particles that give rise to them.

B. Electroweak Unification via the Higgs Mechanism

When seeking to unify forces that were previously regarded as being of different types, it is natural to exploit any features that the forces may have in common. Einstein hoped to unify electromagnetism and gravitation, in considerable measure because both forces shared the striking feature of being long-range. As we have noted, Einstein's efforts were not successful. In the 1960s, however, substantial thought was given to the possibility of unifying the electromagnetic and *weak* interactions.

The possibility of unifying the weak and electromagnetic interactions was suggested by the fact that both interactions affected all electrically charged particles (albeit in distinctly different ways), and by the suspicion that weak interactions might result from exchange of a hypothetical set of particles called "weak vector bosons." (*Vector bosons* are particles with intrinsic angular momentum, or "spin," equal to $\hbar$, like the photon.)

The electromagnetic and weak forces would be regarded as unified if an intimate relationship could be found between the photon and weak vector bosons, and between the strengths of the respective interactions. (More specifically, in regrettably technical terms, it was hoped that the photon and an uncharged weak vector boson might transform into one another under some symmetry group of the fundamental interactions, and that coupling strengths might be governed by eigenvalues of the group generators. The symmetry group would then constitute a unifying principle.)

The effort to unify these interactions faced at least two obstacles, however. The most obvious was that photons are massless, whereas the weak interaction was known to be of very short range. Thus weak vector bosons were necessarily very massive, which seemed to preclude the intimate type of relationship with photons necessary for a unified theory. A second potential difficulty was that while experiments had suggested (without proving) the existence of electrically charged weak bosons, now known as the W^+ and W^-, unification required a third, electrically neutral weak boson as a candidate for (hypothetical) transformations with the neutral photon. There were, however, no experimental indications that such a neutral weak boson (the Z) existed.

Building on the work of several previous investigators, Steven Weinberg (in 1967) and Abdus Salam (independently in 1968) proposed an ingenious unification of the electromagnetic and weak interactions. The feature of principle interest to us here was the remarkable method for unifying the massless photon with massive weak bosons. The obstacle of different masses was overcome by positing that, at the most fundamental level, there was no mass difference. Like the photon, the weak vector bosons were assumed to have no *intrinsic* mass. How, then, could the theory account for the short range of the weak force and corresponding large effective masses of the weak bosons? The theory did so by utilizing an ingenious possibility discovered by Peter Higgs in 1964, called the *Higgs mechanism* for (effective) mass generation.

A QFT is defined by the choice of fields appearing in it and by the *Lagrangian* chosen for it. The Lagrangian is a function of the fields (and their derivatives) that specifies the kinetic and potential energies of the fields, and of the particles associated with them (the *field quanta*). In more general terms, the Lagrangian specifies all intrinsic masses, and all interactions among the fields and particles of the theory. The Langrangian contains two kinds of physical constants that are basic parameters of the theory: the masses of those fields that represent massive particles, and *coupling constants* that determine the strengths of interactions.

A theory wherein the Higgs mechanism is operative has a scalar field variable $\Phi(\mathbf{r}, t)$ (a *Higgs field*). This Φ appears in the Lagrangian in places where a factor of mass *would* appear for certain particles, if those particles *had* intrinsic mass. The potential energy of the Higgs field is then defined in such a way that its state of minimum energy has a nonzero, constant value Φ_0. This state of lowest energy corresponds to the *vacuum*. Particles with no *intrinsic* mass thereby acquire *effective* masses proportional to Φ_0. These effective masses also contain numerical factors that depend on the type of particle, giving rise to the differences in mass that are observed.

The Higgs mechanism seems like a very artificial way to explain so basic a property as mass. When first proposed, it was regarded by many as a mathematical curiosity with no physical significance. The Higgs mechanism plays a central role in the Weinberg–Salam electroweak theory, however, for the weak bosons are assumed to acquire their large effective masses in precisely this way. Furthermore, the Weinberg–Salam theory has been experimentally confirmed in striking detail (including the existence of the neutral weak Z boson, with just the mass and other properties predicted by the theory). Verification of the theory led to Weinberg and Salam sharing the 1979 Nobel Prize in Physics with Sheldon Lee Glashow, who had laid some of the ground-work for the theory.

Only a single Higgs field has been mentioned thus far, but the Higgs mechanism actually requires at least two: one that takes on a nonzero value in the vacuum, plus one for each vector boson that acquires an effective mass through the Higgs mechanism. [There are four real (or two complex) Higgs fields in the Weinberg–Salam theory, since the W^+, W^-, and Z weak bosons all acquire effective masses in this way.] Furthermore, there are many different states that share the same minimum energy, and hence many possible vacuum states for the theory.

As a simple example, consider a theory with two Higgs fields Φ_1 and Φ_2 and a potential energy–density function V given by

$$V(\Phi_1, \Phi_2) = b\left(\Phi_1^2 + \Phi_2^2 - m^2\right)^2 \qquad (85)$$

where b and m are positive constants. (We shall see that m corresponds to the Φ_0 mentioned previously.) The minimum value of V is zero, which occurs when $\Phi_1^2 + \Phi_2^2 = m^2$. This is the equation for a circle of radius m, in a plane where the two axes measure Φ_1 and Φ_2, respectively. Every point on this circle corresponds to zero energy, and could serve as the vacuum.

A trigonometric identity states that $\cos^2 \psi + \sin^2 \psi \equiv 1$, for arbitrary angle ψ. The circle of zero energy in the Φ_1, Φ_2, plane is therefore described by

$$\begin{aligned} \Phi_1 &= m \cos \psi \\ \Phi_2 &= m \sin \psi \end{aligned} \tag{86}$$

where ψ ranges from 0 to 360°. Any *particular* angle ψ_0 corresponds to a particular choice of vacuum state. It is useful to consider a pair of Higgs fields Φ_1' and Φ_2' defined by

$$\begin{aligned} \Phi_1' &\equiv \Phi_1 \sin \psi_0 - \Phi_2 \cos \psi_0 \\ \Phi_2' &\equiv \Phi_1 \cos \psi_0 + \Phi_2 \sin \psi_0 \end{aligned} \tag{87}$$

Comparing Eq. (86) for $\psi = \psi_0$ with Eq. (87), we see that $\Phi_1' = 0$ and $\Phi_2' = m$. Hence one may say that "the vacuum" is characterized by $\Phi_1' = 0$ and $\Phi_2' = m$, so long as one bears in mind that some *particular* vacuum corresponding to a *particular* value of ψ has been selected by nature, from a continuous set of possible vacuums. Some intrinsically massless particles acquire *effective* masses from the fact that $\Phi_2' = m$ under ordinary conditions, as discussed previously.

Theories containing Higgs fields are always constructed so that the Higgs fields transform among one another under the action of a symmetry group, i.e., a group of transformations that leaves the Langrangian unchanged. [For the simple example just described, with V given by Eq. (85), the symmetry group is that of rotations in a plane with perpendicular axes corresponding to Φ_1 and Φ_2: the group is called $U(1)$.]

Higgs fields are always accompanied in a theory by (fundamentally) massless vector bosons that transform among one another under the same symmetry group. The process whereby nature chooses some particular vacuum (corresponding in our simple example to a particular value for ψ) is essentially random, and is called *spontaneous symmetry breaking*. Once the choice has been made, the original symmetry among all the Higgs fields is broken by the contingent fact that the actual vacuum realized in nature has a nonzero value for some particular Higgs field but not for the others. At the same time, the symmetry among the massless vector bosons is broken by the generation of effective mass for one (or more) of them, which in turn shortens the range of the force resulting from exchange of one (or more) of the vector bosons.

Theories involving the Higgs mechanism are said to possess a *hidden symmetry*. Such theories contain a symmetry that is not apparent in laboratories, where nature has *already* selected some *particular* vacuum, thereby concealing the existence of other possible vacuums. Furthermore, every possible vacuum endows a subset of intrinsically massless particles with effective masses, thereby obscuring their fundamental similarity to other particles that remain massless. Exchanges of the effectively massive particles result in forces of limited range ($\sim \hbar / mc$), whereas any particles that remain massless generate long-range forces (electromagnetism being the only one, except for gravity). Extraordinary imagination was required to realize that electromagnetism and the weak interaction might be different aspects of a single underlying theory. Actual construction of the resulting *electroweak theory* may reasonably be regarded as the work of geniuses.

It seems possible, even likely, that in other regions of space–time far removed from our own, the choice of vacuum has been made differently. In particular, there is no reason for the selection to have been made in the same way in regions of space–time that were outside each others' horizons at the time when selection was made. The fundamental laws of physics are believed to be the same everywhere, but the spontaneous breaking of symmetry by random selections of the vacuum state should have occurred differently in different "domains" of the universe.

We have remarked that the state of minimum energy has a nonzero value for one (or more) of the Higgs fields. If we adopt the convention that this state (the vacuum) has zero energy, it follows that any region of space where all Higgs fields vanish must have a mass-energy density $\rho_H > 0$.

Our example with the Higgs potential described by Eq. (85) would have

$$\rho_H = bm^4/c^2 \tag{88}$$

in any region where $\Phi_1 = \Phi_2 = 0$. Work would be required to expand any such region, for doing so would increase the net energy. Such a region is therefore characterized by negative pressure P_H that, by the work-energy theorem, is related to ρ_H by $P_H = -c^2 \rho_H$ [as may be seen from Eq. (25) for constant ρ and P].

Of all the fields known or contemplated by particle theorists, only Higgs fields have positive energy when the fields themselves vanish. This feature of Higgs fields runs strongly against intuition, but the empirical successes of the Weinberg-Salam theory provide strong evidence for Higgs fields. If all Higgs fields were zero throughout the cosmos, the resulting ρ_H and negative P_H would perfectly mimic a positive cosmological constant, causing an exponential rate of growth for the universe. The original theory of cosmic inflation is based on the premise that all Higgs fields were zero at a very early time. The rate of growth would have been enormous, increasing the size of the universe by many powers of ten during the first second of cosmic history. This extraordinary period of *cosmic inflation* ended when the Higgs fields assumed their present values, corresponding to our vacuum.

The Higgs fields relevant to inflation arise in *grand unified theories* (GUTs) of elementary particles and their interactions. The possibility of inflation was first discovered while analyzing a GUT, and many of the subsequent models for inflation are based on GUTs. Inflation also occurs in *supersymmetric* and *superstring* theories of elementary particles, where Higgs fields have analogs called *inflaton fields* (inflation-causing fields). The Higgs fields of GUTs have most of the essential features, however, and we shall discuss them first.

C. Grand Unified Theories (GUTs)

As evidence was mounting for the Weinberg–Salam theory in the early 1970s, a consensus was also emerging that the strong force between quarks is a result of the exchange of eight colored gluons, as described by a theory called *quantum chromodynamics*. Like the photon and the weak bosons, gluons are vector bosons, which makes feasible a unification of all three forces. [Gravitons are not vector bosons (they have a spin of twice $\hbar$), so a unification including gravity would require a different unifying principle.]

A striking feature of the standard model is that the strengths of all three forces are predicted to grow closer as the energy increases, becoming *equal* at a collision energy of $\sim 2 \times 10^{16}$ GeV (somewhat higher than reported earlier, because of recent data). This merging of all three strengths at a single energy suggested very strongly that the forces are unified at higher energies, by some symmetry that is spontaneously broken at $\sim 2 \times 10^{16}$ GeV. A symmetry of this kind would also explain why electrons and protons have precisely equal electrical charges (in magnitude). Experiments have established that any difference is less than one part in 10^{21}, a fact that has no explanation unless electrons are related by some symmetry to the quarks of which protons are made.

Guided by these considerations and by insights gained from the earlier electroweak unification, Howard M. Georgi and Sheldon Glashow proposed the first GUT of all three forces in 1974 [a group called SU(5) was assumed to describe the underlying symmetry]. The Higgs mechanism was adduced to generate effective masses for all particles except photons in the theory (gravitons were not included). In the symmetric phase (all Higgs fields zero), electrons, neutrinos, and quarks would be indistinguish-

able. The theory also predicted an (effectively) supermassive X boson, whose existence was required to achieve the unification.

Following the lead of Georgi and Glashow, numerous other investigators have proposed competing GUTs, differing in the underlying symmetry group and/or other details. There are now many such theories that reproduce the successful predictions of electroweak theory and quantum chromodynamics at the energies currently accessible in laboratories, while making different predictions for very high energies where we have little or no data of any kind. Thus we do not know which (or indeed whether any) of these GUTs is correct. They typically share three features of significance for cosmology, however, which we shall enumerate and regard as general features of GUTs.

Grand unified theories involve the Higgs mechanism (with at least 24 Higgs fields), and contain numerous free parameters. The energy where symmetry is broken may be set at the desired value ($\sim 2 \times 10^{16}$ GeV) by assigning appropriate values to parameters of the theory. This energy determines the gross features of the Higgs potential, and also the approximate value of ρ_H (the mass-energy density when all Higgs fields are zero). The precise value of ρ_H is somewhat different for different GUTs, and also depends on some parameters whose values can only be estimated. Numerical predictions for ρ_H and the resulting inflation therefore contain substantial uncertainties. We shall see, however, that the observable predictions of inflation are scarcely affected by these uncertainties.

The density ρ_H is given roughly (within a few powers of ten) by

$$\rho_H \sim (10^{16} \text{ GeV})^4/(\hbar^3 c^5) \sim 10^{80} \text{ g/cm}^3 \qquad (89)$$

This density is truly enormous, $\sim 10^{47}$ times as great as if our entire sun were compressed into a cubic centimeter. In fact ρ_H is comparable to the density that would arise if all the matter in the observable universe were compressed into a volume the size of a hydrogen atom. As indicated by Eq. (89), this enormous value for ρ_H results from the very high energy at which symmetry appears to be broken.

If all Higgs fields were zero throughout the cosmos, then Eqs. (42) and (89) would imply an enormous, effective cosmological constant

$$\Lambda_H \sim 10^{57} \text{ m}^{-2} \qquad (90)$$

Any universe whose expansion is governed by a positive Λ will grow at an exponential rate. From Eq. (31), we see that the doubling time corresponding to Λ_H is

$$t_d \sim 10^{-37} \text{ sec} \qquad (91)$$

At this rate of growth, only 10^{-35} sec would be required for 100 doublings of the cosmic scale factor (which governs distances between comoving points, i.e., points moving in unison with the overall expansion). Note that $2^{100} \sim 10^{30}$. If inflation actually occurred, it is quite plausible that our universe grew by a factor much larger than 10^{30}, during a very brief time.

We have remarked that the value of ρ_H is uncertain by a few powers of ten. If ρ_H were only 10^{-10} times as large as indicated by Eq. (89), then t_d would be 10^5 times larger than given by Eq. (91). Even with $t_d \sim 10^{-32}$ sec, however, 100 doublings would occur in 10^{-30} sec. Whether 10^{-30} sec were required, or only 10^{-35} sec, makes no observable difference.

A second important feature of GUTs concerns *baryon number*. Baryon number is a concept used to distinguish certain types of particles from their antiparticles. Protons and neutrons are assigned a baryon number of $+1$, in contrast with -1 for antiprotons and antineutrons. Grand unified theories do not conserve baryon number, which means that matter and antimatter need not be created (or destroyed) in equal amounts.

Nonconservation of baryon number is a high-energy process in GUTs, because it results from emission or absorption of X, bosons, whose mc^2 energy is $\sim 10^{16}$ GeV. This is far beyond the range of laboratories, so it cannot be studied directly. When the universe was very young, however, it was hot enough for X bosons to have been abundant. A universe created with equal amounts of matter and antimatter could have evolved at a very early time into one dominated by matter, such as our own. Such a process is called *baryogenesis*.

In 1964, Steven Weinberg presciently remarked that there was no apparent reason for baryon number to be exactly conserved. He had recently shown that the QFT of any long-range force is inconsistent, unless the force is generated by a strictly conserved quantity. Electromagnetism is just such a force, which "explains" why electric charge is precisely conserved. Gravitation is another (the only other) such force, which "explains" why energy is conserved (more precisely, why $\partial^\mu T_{\mu\nu} = 0$). There is no long-range force generated by a baryon number, however, so it would be puzzling if it were exactly conserved. Weinberg discussed the possibility of baryogenesis in the early universe in general terms, but was limited by the absence of any detailed theory for such processes. Andrei Sakharov published a crude model for baryogenesis in 1967, but no truly adequate theory was available before GUTs.

In 1978, Motohiko Yoshimura published the first detailed model for baryogenesis based on a GUT, and many others soon followed. It was typically assumed that the universe contained equal amounts of matter and antimatter prior to the spontaneous breaking of symmetry, and that baryogenesis occurred as thermal energies were dropping below the critical value of $\sim 10^{16}$ GeV. Such calculations seem capable of explaining how matter became dominant over antimatter, and also why there are $\sim 10^9$ times as many photons in the cosmic microwave background as there are baryons in the universe.

The apparent success (within substantial uncertainties) of such calculations may be regarded as indirect evidence for GUTs (or theories with similar features, such as supersymmetric and superstring theories). More direct and potentially attainable evidence would consist of observing proton decay, which is predicted by various GUTs to occur with a half-life on the order of 10^{29}–10^{35} yr (depending on the particular GUT and on the values of parameters that can only be estimated). No proton decay has yet been observed with certainty, however, and careful experiments have shown the proton half-life to exceed 10^{32} yr (which rules out some GUTs that have been considered).

A third feature of GUTs results from the fact that there is no single state of lowest energy but rather a *multiplicity* of possible vacuum states. The choice of vacuum was made by nature as the universe cooled below $\sim 10^{29}$ K, when thermal energies dropped below $\sim 10^{16}$ GeV at a cosmic age of $\sim 10^{-39}$ sec. Since different parts of the universe were outside each others' horizons and had not yet had time to interact in any way, the choice of vacuum should have occurred differently in domains that were not yet in causal contact. These mismatches of the vacuum in different domains would correspond to surfacelike defects called *domain walls* and surprisingly, as was shown in 1974 by G. t' Hooft and (independently) by A. M. Polyakov, also to pointlike defects that are magnetic monopoles.

If the universe has evolved in accord with the standard big bang model (i.e., without inflation), then such domain walls and magnetic monopoles should be sufficiently common to be dramatically apparent. None have yet been observed with certainty, however, which may either be regarded as evidence against GUTs or in favor of cosmic inflation: inflation would have spread them so far apart as to render them quite rare today.

D. Phase Transitions

The notion of phase is most familiar in its application to the three states of ordinary matter: solid, liquid, and gas. As heat is added to a solid, the temperature gradually rises to the melting point. Addition of further energy equal to the latent heat of fusion results in a phase transition to the liquid state, still at the melting point. The liquid may then be heated to the boiling point, where addition of the latent heat of vaporization results in a transition to the gaseous phase. The potential energy resulting from intermolecular forces is a minimum in the solid state, distinctly greater in the liquid state, and greater still for a gas. The latent heats of fusion and vaporization correspond to the changes in potential energy that occur as the substance undergoes transition from one phase to another.

In QFT, the state of minimum energy is the vacuum, with no particles present. If the theory contains Higgs fields, then one of them is nonzero in the vacuum. The vacuum (if perfect) has a temperature of absolute zero, for any greater temperature would entail the presence of a thermal (blackbody) distribution of photons. A temperature of absolute zero is consistent, of course, with the presence of any number of massive particles.

If heat is added to some region of space it will result in thermal motion for any existing particles, and also in a thermal distribution of photons. If the temperature is raised to a sufficiently high level, then particle–antiparticle pairs will be created as a result of thermal collisions: the mc^2 energies of the new particles will be drawn from the kinetic energies of preexisting particles. The Higgs field that was a nonzero constant in the vacuum would retain that value over a wide range of temperatures, however, and this condition may be regarded as defining a phase: it implies that certain vector bosons have effective masses, so that the underlying symmetry of the theory is broken in a particular way.

If heat continues to be added, then the temperature and thermal energy will eventually rise to a point where the Higgs fields no longer remain in (or near) their state of lowest energy: they will undergo random thermal fluctuations about a central value of zero. When this happens, the underlying symmetry of the theory is restored: those vector bosons that previously had effective masses will behave like the massless particles they fundamentally are. This clearly constitutes a different phase of the theory, and we conclude that nature undergoes a phase transition as the temperature is raised above a critical level sufficient to render all Higgs fields zero.

Roughly speaking, this symmetry-restoring phase transition occurs at the temperature where the mass-energy per unit volume of blackbody radiation equals the ρ_H corresponding to the vanishing of all Higgs fields; an equipartition of energy effects the phase transition at this temperature when latent heat equal to $\rho_H c^2$ is added. For the ρ_H given by Eq. (89), the phase transition occurs at a temperature near 10^{29} K. Like all phase transitions, it can proceed in either direction, depending on whether the temperature is rising or falling as it passes through the critical value.

The relation between phase transitions and symmetry breaking may be elucidated by noting that ordinary solids break a symmetry of nature that is restored when the solid melts. The laws of physics have rotational symmetry, as do

liquids: there is no preferred direction in the arrangement of molecules in a liquid. When a liquid is frozen, however, the rotational symmetry is broken because solids are crystals. The axes of a crystal may be aligned in any direction; but once a crystal has actually formed, its particular axes single out particular directions in space.

It is also worth noting that rapid freezing of a liquid typically results in a moderately disordered situation where different regions have their crystal axes aligned in different directions: domains are formed wherein the rotational symmetry is broken in different ways. We may similarly expect that domains of broken-symmetry phase were formed as the universe cooled below 10^{29} K, with the symmetry broken in different ways in neighboring domains. In this case the horizon distance would be expected to govern the domain size.

There is one further aspect of phase transitions that is of central importance for cosmic inflation, namely, the phenomenon of *supercooling*. For a familiar example, consider water, which normally freezes at 0°C. If a sample of very pure (distilled) water is cooled in an environment free of vibrations, however, it is possible to cool the sample more than 20°C below the normal freezing point while still in the liquid phase. Liquid water below 0°C is obviously in an unstable state: any vibration is likely to trigger the onset of crystal formation (i.e., freezing), with consequent release of the latent heat of fusion from whatever fraction of the sample has frozen.

When supercooling of water is finally terminated by the formation of ice crystals, *reheating* occurs: release of the crystals' heat of fusion warms the partially solid, partially liquid sample (to 0°C, at which temperature the remaining liquid can be frozen by removal of its heat of fusion). Such reheating of a sample by released heat of fusion is to be expected whenever supercooling is ended by a phase transition (though not all systems will have their temperature rise as high as the normal freezing point).

Some degree of supercooling should be possible in any system that undergoes phase transitions, including a GUT with Higgs fields. A conjecture that supercooling (and reheating) occurred with respect to a GUT phase transition in the early universe plays a crucial role in theories of cosmic inflation, which we are now prepared to describe in detail.

VIII. SCENARIOS FOR COSMIC INFLATION

A. The Original Model

We now assume that some GUTs with the features described earlier (in Section VII.C) provide a correct description of particle physics, and consider the potential consequences for the evolution of the cosmos. According to the standard big bang model, the temperature exceeded 10^{29} K for times earlier than 10^{-39} sec. Hence the operative GUT would initially have been in its symmetric phase wherein all Higgs fields were undergoing thermal fluctuations about a common value of zero.

As the temperature dropped below the critical value of $\sim 10^{29}$ K, there are two distinct possibilities (with gradations between). One possibility is that the GUT phase transition occurred promptly, in which case the expansion, cooling, and decrease in mass-energy density of the universe would have progressed in virtually the same way as in the standard model. The GUT symmetry would have been broken in different ways in domains outside each others' horizons, and one can estimate how many domain walls and magnetic monopoles would have resulted. A standard solution to the field equations (Section VI.A) tells one how much the universe has expanded since then, so it is a straightforward matter to estimate their present abundances.

No domain walls have yet been observed, nor magnetic monopoles with any confidence (there have been a few candidates, but too few for other explanations to be ruled out). The abundances predicted by the preceding line of reasoning vastly exceed the levels consistent with observations. For example, work by John Preskill and others indicated in 1979 that monopoles should be $\sim 1\%$ as common as protons. Their magnetic properties would make them easy to detect, but no search for them has succeeded. Furthermore, monopoles are predicted to be $\sim 10^{16}$ times as massive as protons. If they were $\sim 1\%$ as common as protons, their average density in the universe would be $\sim 10^{12}$ times the critical density. In reality, the observed density of monopoles is vastly less than predicted by GUTs with a prompt phase transition. This was called the "monopole problem."

The monopole problem was discovered by elementary particle theorists soon after the first GUT was introduced, and appeared for several years to constitute evidence against GUTs. In 1980, however, Alan H. Guth proposed a remarkable scenario based on GUTs that held promise for resolving the horizon, flatness, monopole, and smoothness problems, and perhaps others as well. Guth conjectured that as the universe expanded and cooled below the GUT critical temperature, supercooling occurred with respect to the GUT phase transition. All Higgs fields retained values near zero for a period of time after it became thermodynamically favorable for one (or a linear combination) of them to assume a nonzero value in the symmetry-breaking phase.

Supercooling would have commenced when the universe was $\sim 10^{-39}$ sec old. The mass-energy density of all forms of matter and radiation other than Higgs fields would have continued to decrease as the universe expanded, while that of the Higgs fields retained the constant value ρ_H corresponding to zero values for all Higgs fields. Within $\sim 10^{-38}$ sec, ρ_H and the corresponding negative pressure $P_H = -\rho_H c^2$ would have dominated the stress-energy tensor, which would then have mimicked a large, positive cosmological constant. The cosmic scale factor would have experienced a period of exponential growth, with a doubling time of $\sim 10^{-37}$ sec. Guth proposed that all this had happened, and called the period of exponential growth the *inflationary era*. (The numbers estimated by Guth were somewhat different in 1980. Current values are used here, with no change in the observable results.)

The Higgs fields were in a metastable or unstable state during the inflationary era, so the era was destined to end. Knowledge of its duration is obviously important: How many doublings of the scale factor occurred during this anomalous period of cosmic growth? The answer to this question depends on the precise form of the Higgs potential, which varies from one GUT to another and unfortunately depends, in any GUT, on the values of several parameters that can only be estimated.

Guth originally assumed that the Higgs potential had the form suggested by Fig. 4a. For such a potential, a zero Higgs field corresponds to a local minimum of the energy density, and the corresponding state is called a *false vacuum*. (The *true vacuum* is of course the state wherein the Higgs potential is an absolute minimum; that is the broken-symmetry phase where one of the Higgs fields has a nonzero constant value.)

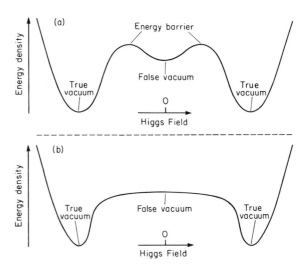

FIGURE 4 Higgs potential energy density for (a) original inflationary model, and (b) new inflationary model. For a theory with two Higgs fields, imagine each curve to be rotated about its vertical axis of symmetry, forming a two-dimensional surface. A true vacuum would then correspond to any point on a circle in the horizontal plane.

In the classical (i.e., nonquantum) form of the theory, a false vacuum would persist forever. The physics is analogous to that of a ball resting in the crater of a volcano. It would be energetically favorable for the ball to be at a lower elevation, hence outside the volcano at its base. In classical physics there is no way for the ball to rise spontaneously over the rim and reach the base, so the ball would remain in the crater for all time.

In quantum physics, the story changes: a ball in the crater is represented by a wave packet, which inevitably develops a small extension or "tail" outside the crater. This feature of quantum mechanics implies a finite probability per unit time that the ball will "tunnel through" the potential energy barrier posed by the crater's rim and materialize outside at a lower elevation, with kinetic energy equal to the decrease in potential energy. Such quantum-mechanical tunneling through energy barriers explains the decay of certain species of heavy nuclei via the spontaneous emission of alpha particles, and is a well-established phenomenon.

If there exists a region of lower potential energy, for either alpha particles or the value of a Higgs field, then "decay" of the initial state via tunneling to that region is bound to occur sooner or later. The timing of individual decays is governed by rules of probability, in accordance with the statistical nature of quantum-mechanical predictions.

The decay of the false vacuum was first studied and described by Sidney R. Coleman. Initially all Higgs fields are zero, but one or another of them tunnels through the energy barrier in a small region of space and acquires the nonzero value corresponding to the true vacuum, thereby creating a "bubble" of true vacuum in the broken symmetry phase.

Once a bubble of true vacuum has formed, it grows at a speed that rapidly approaches the speed of light, converting the surrounding region of false vacuum into true vacuum as the bubble expands. As the phase transition proceeds, energy conservation implies that the mass-energy density ρ_H of the preexisting false vacuum is converted into an equal density of other forms of mass-energy, which Guth hoped would become hot matter and radiation that evolved into our present universe.

A low rate of bubble formation corresponds to a high probability that any particular region of false vacuum will undergo many doublings in size before decay of the false vacuum ends inflation for that region. Guth noted that if the rate of bubble formation were low enough for inflation to have occurred by a factor of $\sim 10^{30}$ or more, then the observable portion would have emerged from small enough a region for all parts to have been in causal contact before inflation. Such causal contact would have resulted in a smoothing of any earlier irregularities via processes leading toward thermal equilibrium, thereby explaining (one hopes) the homogeneity and isotropy now observed: the horizon problem would be solved (and perhaps the smoothness problem as well).

It was furthermore noted by Guth that comparable growth of the cosmic scale factor could solve the flatness problem. Recall the definition of Ω in Section VI.B: $\Omega \equiv \rho/\rho_c$, where ρ and ρ_c denote the actual and critical densities of mass-energy, respectively. If the scale factor had grown by a factor of 10^{30} or more during the period of inflation, any difference between Ω and unity *prior* to inflation would have been reduced by a factor of 10^{60} or more by its *end*. [Equation (48) holds at all times]. R. Dicke and P. J. E. Peebles have shown that Ω could not have differed from unity by more than one part in 10^{15} when the universe was one second old, if the present density of matter lies between 20 and 200% of the critical density (as it does, with $\rho_M \approx 0.3\rho_c$). Perhaps of even greater interest, they showed that if Ω had differed from unity by more that one part in 10^{14} at one second, then *stars would never have formed*. (For $\Omega > 1$, the universe would have collapsed too soon. For $\Omega < 1$, matter would have been diluted too soon by the expansion.) This need for extraordinary "fine-tuning" at early times is the "flatness problem." Inflation by a factor $\geq 10^{30}$ could explain why Ω was so near unity at 1 sec.

If such inflation has occurred, the cosmic scale factor a_0 would be 10^{30} (or more) times greater than in the standard model. Even if space *is* curved, the curvature only becomes significant over distances comparable to the value of a_0 (as explained in Section III.C). Inflation could render a_0 so much larger than the radius of the observable universe that no curvature would be apparent.

The rate at which bubbles form depends on the particular GUT and, within any GUT, on parameters including some whose values can only be estimated. Guth assumed that inflation *is* the reason why our universe is nearly flat, hence that bubble formation was slow enough for ~ 100 or more doublings to occur ($2^{100} \sim 10^{30}$). With a doubling time $t_d \sim 10^{-37}$ sec, this would have required only $\sim 10^{-35}$ sec.

It would seem coincidental if 100 doublings had occurred but not, for example, 105 or more, which would have reduced $|\Omega - 1|$ by another factor of $2^{10} \approx 1000$ (or more). Standard inflation predicts that considerably more doublings have occurred than are required to explain the obvious features of our universe (e.g., the presence of stars and us). Standard inflation therefore predicts enough doublings to render the observed universe indistinguishable from flat. Even 10^7 doublings would require only $\sim 10^{-30}$ sec, and this would be inflation by a factor of $\sim 10^{3,000,000}$.

Guth sought to understand whether a region now the size of the observable universe was more likely to have evolved from few or many bubbles, and he encountered a major shortcoming of his model. The role played by bubbles in the phase transition led to a bizarre distribution of mass-energy, quite unlike what astronomers observe. The problem may be described as follows.

We have remarked that the ρ_H of the false vacuum is converted into an equal amount of mass-energy in other forms as the phase transition proceeds. It can be shown, however, that initially the mass-energy is concentrated in the expanding walls of the bubbles of true vacuum, in the form of moving "kinks" in the values of the Higgs fields. Conversion of this energy into a homogeneous gas of particles and radiation would require a large number of collisions between bubbles of comparable size. Bubbles are forming at a constant rate, however, and grow at nearly the speed of light. If enough time has passed for there to be many bubbles, there will inevitably be a wide range in the ages and therefore sizes of the bubbles: collisions among them will convert only a small fraction of the energy in their walls into particles and radiation. In fact it can be shown that the bubbles would form finite clusters dominated by a single largest bubble, whose energy would remain concentrated in its walls.

One could explain the homogeneity of the observable universe if it were contained within a single large bubble. There remains, however, the problem that the mass-energy would be concentrated in the bubble's walls: the interior of a single large bubble would be too empty (and too cold at early times) to resemble our universe. In the best spirit of science, Guth acknowledged these problems in his initial publication, and closed it with an invitation to readers to find a plausible modification of his inflationary model, one that preserved its strengths while overcoming its initial failure to produce a homogeneous, hot gas of particles and radiation that might have evolved into our present universe.

We note in passing that Andre D. Linde, working at the Lebedev Physical Institute in Moscow, had discovered the principal ingredients of inflation prior to Guth. Furthermore, Linde and Gennady Chibisov realized in the late 1970s that supercooling might occur at the critical temperature, resulting in exponential growth. To their misfortune, however, they were not aware of the horizon and flatness problems of standard cosmology. When they realized that bubble collisions in their model would lead to gross inhomogeneities in the universe, they saw no reason to publicize their work.

Alexei Starobinsky, of the Landau Institute for Theoretical Physics in Moscow, proposed a form of inflation in a talk at Cambridge University in 1979. His focus was on avoidance of an initial singularity at time zero, however, and no mention was made of the horizon, flatness, or monopole problems. Although well received, his talk did not generate enough excitement for word of it to reach the United States. By the time it was published in 1980, Guth had already given many lectures emphasizing how inflation might solve problems of standard cosmology. It was this emphasis that made the potential significance of Starobinsky's and Linde's work apparent to a wide audience.

B. The New Inflationary Model

A major improvement over Guth's original model was developed in 1981 by A. D. Linde and, independently, by Andreas Albrecht with Paul J. Steinhardt. Success hinges on obtaining a "graceful exit" from the inflationary era, i.e., an end to inflation that preserves homogeneity and isotropy and furthermore results in the efficient conversion of ρ_H into the mass-energy of hot particles and radiation.

The new inflationary model achieves a graceful exit by postulating a modified form for the potential energy function of Higgs fields: the form depicted in Fig. 4b, which was first studied by Sidney Coleman in collaboration with Erick J. Weinberg. A Coleman–Weinberg potential has no energy barrier separating the false vacuum from true vacuum; instead of resembling the crater of a volcano, the false vacuum corresponds to the center of a rather flat plateau. (Though not a local minimum of energy, the plateau's center is nevertheless called a false vacuum for historical reasons.)

Supercooling of a Coleman–Weinberg false vacuum results in the formation of contiguous, causally connected domains throughout each of which the Higgs fields gradually evolve in a simultaneous, uniform way toward a single phase of true vacuum. The initial sizes of these domains are comparable to the horizon distance at the onset of supercooling, $\sim 10^{-28}$ cm. The mass-energy density of the false vacuum was that of null Higgs fields, $\rho_H \sim 10^{80}$ g/cm^3. Despite this enormous density, the initial mass of the domain in which our entire universe lies was $\sim 10^{-4}$ g.

Within each domain the evolution of a Higgs field away from its initial value of zero is similar to the horizontal motion of a ball that, given a slight nudge, rolls down the initially flat but gradually increasing slope of such a plateau. The equations describing evolution of the Higgs fields have a term corresponding to a frictional drag force for a rolling ball; in addition to the initial flatness of the plateau, this "drag term" serves to retard the evolution of a Higgs field away from zero toward its true-vacuum, broken-symmetry value. The result is called a "slow-rollover" transition to the true vacuum.

As always, there are different combinations of values of Higgs fields that have zero energy and qualify as true vacua; different domains undergo transitions to different true vacua, corresponding to the possibility of a ball rolling off a plateau in any one of many different directions.

As a Higgs field (or linear combination thereof) "rolls along" the top of the potential energy plateau, the mass-energy density remains almost constant despite any change in the volume of space. The corresponding pressure is large and negative, so inflation occurs. The doubling time for the cosmic scale factor is again expected to be $\sim 10^{-37}$ sec. The period of time required for a Higgs field to reach the edge of the plateau (thereby terminating inflation) can only be estimated, but again inflation by a factor of 10^{30}, or by very much more, is quite plausible. As in the "old inflationary model" (Guth's original scenario), new "standard" inflation assumed that the actual inflation greatly exceeded the amount required to resolve the horizon and flatness problems.

When a Higgs field "rolls off the edge" of the energy plateau, it enters a brief period of rapid oscillations about its true-vacuum value. In accordance with the field-particle duality of QFT, these rapid oscillations of the field correspond to a dense distribution of Higgs particles. The Higgs particles are intrinsically unstable, and their decay results in a rapid conversion of their mass-energy into a wide spectrum of less massive particles, antiparticles, and radiation. The energy released in this process results in a high temperature that is less than the GUT critical temperature of $\sim 10^{29}$ K only by a modest factor ranging between 2 and 10, depending on details of the model. This is called the "reheating" of the universe.

The reheating temperature is high enough for GUT baryon-number–violating processes to be common, which leads to matter becoming dominant over antimatter. Reheating occurs when the universe is very young, probably less than 10^{-30} seconds old, depending on precisely how long the inflationary era lasts. From this stage onward, the observable portion evolves in the same way as in the standard big bang model. (A slow-rollover transition is slow relative to the rate of cooling at that early time, so that supercooling occurs, and slow relative to the doubling time of $\sim 10^{-37}$ sec; but it is extremely rapid by any macroscopic standard.)

The building blocks of the new inflationary model have now been displayed. As with the old inflationary model, there is no need to assume that any portion of the universe was initially homogeneous and isotropic. It is sufficient to assume that at least one small region was initially expanding, with a temperature above the GUT critical temperature. The following chain of events then unfolded.

As long as the temperature exceeded the critical temperature of $\sim 10^{29}$ K, all Higgs fields would have undergone thermal fluctuations about a common value of zero, and the full symmetry of the GUT would have been manifest. The assumed expansion caused cooling, however, and the temperature fell below 10^{29} K when the universe was $\sim 10^{-39}$ sec old. Supercooling began, and domains formed with diameters comparable to the horizon distance at that time, $\sim 10^{-28}$ cm, with initial masses of $\sim 10^{-4}$ g.

As supercooling proceeded, the continuing expansion caused dilution and further cooling of any preexisting particles and radiation, and the stress-energy tensor became dominated at a time of roughly 10^{-37} sec by the virtually constant contribution of Higgs fields in the false vacuum state. With the dominance of false-vacuum energy and negative pressure came inflation: the expansion entered a period of exponential growth, with a doubling time of $\sim 10^{-37}$ sec. The onset of inflation accelerated the rate of cooling; the temperature quickly dropped to $\sim 10^{22}$ K (the *Hawking temperature*), where it remained throughout the latter part of the inflationary era because of quantum effects that arise in the context of GTR.

For the sake of definiteness, assume that the domain in which we reside inflated by a factor of 10^{50}, which would have occurred in $\sim 2 \times 10^{-35}$ sec. With an initial diameter of $\sim 10^{-28}$ cm and mass of $\sim 10^{-4}$ g, such a domain would have spanned $\sim 10^{22}$ cm when inflation ended (as in Fig. 2), with a total mass of $\sim 10^{146}$ g. *All but one part in $\sim 10^{150}$ of this final mass was produced during inflation, as the mass-energy of the false vacuum grew in proportion to the exponentially increasing volume.* (Guth has called this "the ultimate free lunch.") The radius of our presently observable universe would have been ~ 60 cm when inflation ended (based on the methods of Section VI.A, assuming that $t_{EQ} \geq t_{LS}$).

Inflation ended with a rapid and chaotic conversion of ρ_H into a hot ($T \geq 10^{26}$ K) gas of particles, antiparticles, and radiation. Note that if baryon number were strictly conserved, inflation would be followed for all time by equal amounts of matter and antimatter (which would have annihilated each other almost completely, with conversion of their mass-energy into radiation). The observable universe, however, is strongly dominated by matter. Hence inflation requires a GUT (or a theory sharing key features with GUTs) not only to explain an accelerating expansion, but also to explain how the symmetry between matter and antimatter was broken: through baryon-number–violating processes at the high temperature following reheating.

C. New, Improved Inflation

In the standard big bang model, thermal fluctuations in density at the Planck time would gradually have been amplified, by the concentrating effect of gravity, into present clumps of matter much larger than anything we see. This is the *smoothness problem*, first noted by P. J. E. Peebles in 1968. When inflation was proposed in 1980, Guth and others hoped that it would solve this problem. The end of inflation creates the initial conditions for the ensuing big bang, so the density perturbations at the end of inflation were the basic issue. During the Nuffield Workshop on the Very Early Universe in the summer of 1982, however, it became clear that the simplest GUT, the SU(5) model of Georgi and Glashow, was unacceptable: it gave rise to values for $\delta\rho/\rho$ that were too large by a factor of $\sim 10^6$.

As other GUTs were analyzed with similar results, a consensus gradually emerged that *no* symmetry-breaking Higgs field in a GUT could give rise to acceptable inflation. The size of $\delta\rho/\rho$ at the end of inflation depends critically on details of the Higgs potential. The potential of a GUT Higgs field, like that in Figure 4b, would have to be wider and flatter by factors $\sim 10^{13}$ in order for the resulting $\delta\rho/\rho$ to have given rise to the stars and galaxies that we see.

Another problem requiring a much wider and flatter potential became apparent. As the early universe was cooling toward the GUT critical temperature, all Higgs fields would indeed have undergone thermal fluctuations about *average* values of zero, because of cancellations between positive and negative values. For Higgs potentials that meet the requirements of GUTs, however, the average *magnitudes* would have been comparable to those in a *true* vacuum. The creation of a false vacuum by cooling to the critical temperature would have been extremely unlikely. "Unlikely" is not "impossible," but plausibility of the theory would be undermined by the need for a special initial condition.

In contrast, if the potential were 10^{13} times wider and flatter, cooling below the critical temperature should have resulted in a false vacuum over *some* tiny region of space, which might then have inflated into a universe like that in which we find ourselves. We are therefore led to consider theories of this kind.

The term *inflaton* (*in*-fluh-*tonn*) refers to any hypothetical field that could cause inflation. We are interested, of course, in inflatons that might have produced *our* universe. A scalar field (having no spatial direction) ϕ seems best suited (perhaps two of them). The underlying particle theory must convert false-vacuum energy into a hot soup of appropriate particles at the end of inflation, in a graceful exit producing density perturbations of the required kind. Finally, the particle theory must contain a mechanism for baryogenesis, making matter dominant over antimatter at an early time.

The potential $V(\phi)$ could have an extremely broad, almost flat region surrounding $\phi = 0$, with $V(0) > 0$. An attractive alternative was proposed by Andrei Linde, however, in 1983. The inflaton potential of Linde's *chaotic inflation* has the shape of a very wide, upright bowl, with the true vacuum at its center ($\phi = 0$). Space-time could have been in a highly disordered, chaotic state at the Planck time. The theory only requires that a single, infinitesimal region had a value for ϕ far enough up the wall to have produced the desired inflation while "rolling" slowly toward the center. Like chaotic inflation, most inflaton models assume that inflation began very near the Planck time, in part to explain why the initial microcosm did not collapse before inflation began.

A broad range of particle theories can be constructed with suitable inflaton fields, including supersymmetric theories, supergravity theories, and superstring theory. Further discussion would take us beyond the range of this article. Improved models employing a wide range of inflatons share all the successes of the "new inflationary model" of 1982, however, and those predicting a flat universe will be included when we speak of "standard inflation."

D. Successes of Inflation

Our universe lies inside a domain with a uniform vacuum state, corresponding to a particular way in which any (now) hidden symmetry of the underlying particle theory was broken. Standard inflation predicts that this domain inflated by a factor much greater than 10^{30}, in which case our universe (almost certainly) lies *deeply* inside. Neither domain walls nor magnetic monopoles would then be found within the observable universe (except for a few that might have resulted from extreme thermal fluctuations after inflation ended). Standard inflation therefore solves the monopole problem. The horizon problem is solved as well, as illustrated in Fig. 2.

The flatness problem has an interesting history. When inflation was proposed in 1980, the matter yet found by astronomers corresponded to a density of matter ρ_M that was $\sim 10\%$ of the critical density ρ_c. Astronomers had by no means completed their search for matter, however. When inflation was proposed, it seemed possible that future discoveries would bring ρ_M up to ρ_c, as required by standard inflation's prediction of flatness. With the passage of time, the presence of additional (dark) matter was indicated by observations, but there has never been evidence that ρ_M is even half of ρ_c. The currently observed value is $\rho_M = (0.31 \pm 0.06)\rho_c$, and it seems virtually certain that $\rho_M \leq \rho_c/2$ (Section V).

By the early 1990s, the apparent discrepancy between ρ_M and ρ_c led to a serious consideration of "open" inflationary models, wherein $\rho_M < \rho_c$ and space has negative curvature. Such models are ingenious and technically viable, but require special assumptions that seem rather artificial.

As evidence mounted that $\rho_M < \rho_c$, however, standard inflation acquired a bizarre feature of its own. Its prediction of flatness implied the existence of *positive vacuum energy*, in order to bring the total density up to ρ_c. This would be equivalent to a positive cosmological constant (Section IV.F and G), a concept that Einstein abandoned long ago and few had been tempted to resuscitate. Such vacuum energy would have caused an *acceleration* of the cosmic expansion, beginning when the universe was roughly half its present age (Section V.F).

In 1998 two teams of astronomers, one led by Saul Perlmutter and the other by Brian P. Schmidt, reported an astonishing discovery. They had observed supernovae of type Ia (SNe Ia) over a wide range of distances out to ~12 billion light-years (redshift parameter $z \approx 1$, see Section VI.A). Their data strongly indicated that the cosmic expansion is, indeed, accelerating. This conclusion has been confirmed by subsequent studies of SNe Ia. Furthermore, the *amount* of vacuum energy required to explain the acceleration appears to be just the amount required ($\pm 10\%$) to make the universe flat: $\rho_V \approx 0.7\rho_c$, hence $(\rho_M + \rho_V) \approx \rho_c$ (details in Section V.E).

The vacuum energy implied by supernovae data has been strikingly confirmed by studies of the CMB, which has been traveling toward us since the time of last scattering: $t_{LS} \approx 300,000$ yr. The CMB is extremely uniform, but contains small regions with deviations from the average temperature on the order of $\delta T/T \sim 10^{-5}$. These provide a snapshot of temperature perturbations at t_{LS}, which were strongly correlated with density perturbations.

Prior to t_{LS}, baryonic matter was strongly ionized. The free electrons and nuclei were frequently scattered by abundant photons, which pressurized the medium. Density waves (often called *acoustic waves*) will be excited by perturbations in any pressurized medium, and the speed of such waves is determined (in a known way) by the medium's properties. At any given time before t_{LS}, there was a maximum distance that such waves could yet have traveled, called the *acoustic horizon distance*.

As the acoustic horizon expanded with the passage of time, it encompassed a growing number of density perturbations, which excited standing waves in the medium. The fundamental mode acquired the greatest amplitude, spanning a distance comparable to the acoustic horizon distance. This is called the *first acoustic peak*. Perturbations in the CMB should contain this peak, with an angular diameter reflecting the acoustic horizon distance at t_{LS}.

The angular diameters of perturbed regions in the CMB depend not only on the linear dimensions of their sources, but also on the curvature (if any) of space. Sources of the CMB are presently a distance $r_p(z_{CMB}) \approx 47 \times 10^9$ ly from us (Section VI.A). If space is curved, angular diameters of perturbations in the CMB would be noticeably af-

fected unless the cosmic scale factor a_0 satisfied (roughly) $a_0 \geq r_p(z_{CMB})$ (Section III.B). From Eqs. (48) and (69), we see that this would require $|\Omega_0 - 1| \leq 0.11$.

BOOMERanG (Balloon Observations of Millimetric Extragalactic Radiation and Geomagnetics) results published in 2000 are portrayed in Fig. 5. The index ℓ refers to a multipole expansion (spherical harmonics, averaged over m for each ℓ). If space were flat, the angular diameter of the first acoustic peak would be $\approx 0.9°$, corresponding to $\ell_{peak} \approx 180°/0.9° \approx 200$.

With $\rho_M \approx 0.3\rho_c$ but no vacuum energy, space would be negatively curved, giving rise to $\ell_{peak} \approx 500$. This is clearly ruled out by the data. For Ω_0 near unity, the prediction can be expressed as $\ell_{peak} \approx 200/\sqrt{\Omega_0}$, and the data in Fig. 5 indicate $\ell_{peak} = (197 \pm 6)$. At the 95% confidence level, the authors concluded that $0.88 \leq \Omega_0 \leq 1.12$. The best-fitting Friedmann solution for spacetime (Section IV.A) has $\rho_M = 0.31\rho_c$ and $\rho_V = 0.75$, reproducing the SNe Ia result for vacuum energy within the margin of error. A more resounding success for standard inflation is not easily imagined.

A sweeping, semiquantitative prediction about density perturbations is made by inflation. In the absence of quantum fluctuations, inflation would rapidly drive the false vacuum to an extremely uniform state. Quantum fluctuations are ordinarily microscopic in size and very short-lived, but rapid inflation would negate both of these usual properties.

Inflation would multiply the diameters of quantum fluctuations, and furthermore, do this so rapidly that opposite sides would lose causal contact with each other before the

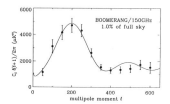

FIGURE 5 Evidence for flatness and vacuum energy (cosmological constant). Angular distribution of $(\delta T/T)^2$ of the cosmic microwave background, measured by instruments in balloons launched from McMurdo Station, Antarctica, by the BOOMERanG Project. Curve indicates best-fitting Friedmann solution for spacetime, with cosmological parameters $\Omega_M = 0.31$, $\Omega_V = 0.75$, and Hubble parameter $h_0 = 0.70$. At the 95% confidence level, the study concluded that $0.88 \leq \Omega_0 \leq 1.12$, where $\Omega_0 \equiv \Omega_M + \Omega_V = 1$ for flat space. The positive Ω_V means positive vacuum energy, corresponding to a cosmological constant that would accelerate the cosmic expansion. The index ℓ refers to a multipole expansion in terms of spherical harmonics $Y_{\ell m}(\theta, \varphi)$, where m has been summed over for each ℓ. [Adapted from P. de Bernardis, *et al.* Reprinted by permission from *Nature* **404**, 955 (2000) MacMillan Magazines Ltd.]

fluctuation vanished. The resulting density perturbations would be "frozen" into the expanding medium, thereby achieving permanency. By inflation's end, the sizes of perturbations would span an *extremely* broad range. The earliest fluctuations would have been stretched enormously, the latest very little, and all those from intermediate times by intermediate amounts. The magnitudes of the resulting $\delta\rho/\rho$ are very sensitive to details of the inflaton potential, but the mechanism described above results in a distribution of sizes that depends rather little on details of the potential.

The kind of probability distribution described above should be apparent in the CMB, since density perturbations correspond closely to perturbations in the radiation's temperature. The Cosmic Background Explorer satellite, better known as COBE (*co*-bee), was launched in 1989 to study the CMB with unprecedented precision. Its mission included measuring deviations from uniformity in the CMB's temperature. Over two years were required to gather and analyze the data displayed in Fig. 6.

These data were first presented at a cosmology conference in Irvine, CA, in March of 1992. The long awaited results struck many attendees as a historic confirmation of inflation, and none have since had reason to feel otherwise. Many subsequent studies have confirmed and refined the data in Fig. 6, but we present the earliest results because of their historical significance.

The density perturbations predicted by inflation are similar to those required to explain the evolution of structure formation in our universe. Indeed, one may reasonably hope that some *particular* inflaton potential will be wholly successful in describing further details of the CMB soon to be measured, and also in explaining details of the structures we observe.

Inflation provides the only known explanations for several puzzling, quite special features of our universe. Standard inflation has also made two quantitive predictions, both of which have been strikingly confirmed (at least within present uncertainties). Only time can tell, but the prognosis for inflation seems excellent, in some form broadly similar to that described here.

E. Eternal Inflation, Endless Creation

Throughout inflation, the inflation field ϕ is on a downhill slope of the potential $V(\phi)$. The false-vacuum density ρ_{FV} is therefore not strictly constant, but gradually decreases as ϕ approaches the end of inflation. Since $t_d \propto 1/\sqrt{\rho_{FV}}$ [Eqs. (31) and (42)], the rate of exponential growth also tends to decrease, all else being equal. In our discussion of density perturbations, however, we noted that quantum fluctuations in ϕ result in perturbations $\delta\rho$ that are frozen

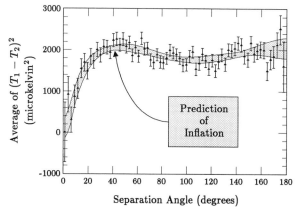

FIGURE 6 COBE nonuniformity data. The data from the COBE satellite gives the temperature T of the background radiation for any direction in the sky. The experimental uncertainties for any one direction are large, but statistically meaningful quantities can be obtained by averaging. The COBE team considered two directions separated by some specific angle, say 15°, and computed the square of the temperature difference, $(T_1 - T_2)^2$, measured in microkelvins (10^{-6} K). By averaging this quantity over all pairs of directions separated by 15°, they obtained a statistically reliable number for that angle. The process was repeated for angles between 0 and 180°. The computed points are shown as small triangles, with the estimated uncertainty shown as a vertical line extending above and below the point. The gray band shows the theoretically predicted range of values corresponding to the scale-invariant spectrum of density perturbations arising from inflation. The effect of the earth's motion through the background radiation has been subtracted from both the data and the prediction, as has a specific angular pattern (called a quadrupole) which is believed to be contaminated by interference from our own galaxy. Since inflation determines the shape of the spectrum but not the magnitude of density perturbations, the magnitude of the predicted gray band was adjusted to fit the data. [From "The Inflationary Universe" by Alan Guth. Copyright © 1997 by Alan Guth. Reprinted by permission of Perseus Books Publishers, a member of Perseus Books, L.L.C.]

into permanency by the rapid expansion. A fascinating question therefore arises.

For simplicity, suppose that ϕ is precisely uniform throughout space at the beginning of inflation. Quantum fluctuations will immediately result in perturbations that become frozen into space. In regions where $\delta\rho > 0$, the expansion rate will be slightly greater than average, and such regions will fill an increasing fraction of space. Within these growing regions, later fluctuations will produce *new* regions where $\delta\rho > 0$, compounding the original deviation from average density and rate of expansion. One expects negative $\delta\rho$ as often as positive, but a string of positives causes that region of space to grow more rapidly, increasing the volume of space where a randomly positive $\delta\rho$ would lengthen the string. This can continue indefinitely.

Of course the *average* ρ is decreasing while the region described above continues rising above the average, so the outcome is unclear. Will inflation come to an end in such a region, sometime later than average, or will there be regions where ρ_{FV} never shrinks to zero and inflation never ends?

The answer clearly depends on the size, frequency, and magnitude of quantum fluctuations, and also on details of the inflaton potential. Alexander Vilenkin answered this question for new inflationary models (Section VIII.B) in 1973. His extraordinary conclusion was that once "new inflation" has begun, there will *always* be regions where it continues. Andrei Linde established this result for a wide class of "improved" inflationary models, including his own "chaotic inflation," in 1986. This phenomenon is called *eternal inflation*, and appears to be characteristic of virtually all inflationary models that might be relevant to our universe.

Eternal inflation is accompanied by endless formulation of "pockets" of true vacuum. Each is surrounded by eternally inflating false vacuum, containing within it all the other pockets yet created. A detailed analysis reveals that the *number* of pockets of true vacuum *grows at an exponential rate*, with a doubling time comparable to that of the surrounding false vacuum (typically *much* less than 10^{-30} sec).

If standard inflation is correct, *our observable universe lies deep within a single such pocket*. The number of such pockets would double a minimum of 10^{30} times each second. This corresponds to the number of pockets (and universes contained therein) increasing by a *minimum* factor of $(10^{1,000,000,000})^3$, every second, for eternity. The possibility of truth being stranger than fiction has risen to a new level.

Given that standard inflation implies endless creation, human curiosity naturally wonders whether such inflation had any *beginning*. A definitive answer remains elusive. Under certain very technical but plausible assumptions, however, Arvind Borde and Alexander Vilenkin proved in 1994 that inflation cannot extend into the infinite *past*. A beginning is required, which renews an ancient question.

F. Creation *ex nihilo*

We have seen that the observable universe may have emerged from an extremely tiny region that experienced inflation and then populated the resulting cosmos with particles and radiation created from the mass-energy of the false vacuum. An ancient question arises in a new context: How did that tiny region come into being, from which the observable universe emerged? Is it possible to understand the creation of a universe *ex nihilo* (from nothing)?

Scientific speculation about the ultimate origin of the universe appears to have begun in 1973, with a proposal by Edward P. Tryon that the universe arose as a *vacuum fluctuation*, i.e., a spontaneous quantum transition from an empty state to a state containing tangible matter and energy. The discovery of this theoretical possibility shares two features with Guth's discovery of inflation. The first is that neither investigator was considering the question to which an answer was found. Guth was a particle theorist, seeking to determine whether GUTs failed the test of reality by predicting too many monopoles. He had not set out to study cosmology, let alone to revolutionize it.

In Tryon's case, he had never imagined it *possible* to explain how and why the big bang occurred. Universes with no beginning had been quite popular, in large measure because they nullified (by fiat) the issue of creation. When the steady-state model was undermined by discovery of the CMB in 1965, the eternally oscillating model replaced it as the favored candidate, despite the mystery of how a universe could have bounced back from each collapse and furthermore retained its uniformity through infinitely many bounces. The only alternative was a universe of finite age, which seemed incomprehensible from a scientific point of view.

Interests in cosmology, quantum theory, and particle physics arose in Tryon's undergraduate years. Graduate studies commenced in 1962 at the University of California, Berkeley, where his courses included quantum field theory and general relativity. Classical and quantum theories of gravitation were the subjects of his dissertation.

It had long been a mystery why mass plays two quite different roles. It describes resistance to acceleration, or *inertia*, in Newton's laws of motion. In Newton's law of gravitation, $F = GmM/r^2$, mass plays the role of gravitational "charge" (analogous to electric charge in Coulomb's law). This puzzle had led Ernst Mach in the 1880s to conjecture that inertia might somehow be a result of interaction with the rest of the universe (*Mach's Principle*). Tryon found this possibility tantalizing, and took a break from his thesis project one afternoon to see if he could invent a quantitative theory.

Denoting inertial mass by m_I and gravitational mass (or "charge") by m_G, Newton's gravity becomes $F = m_G M_G / r^2$ in suitably chosen units. The two types of mass are presumably related by

$$m_I = f(\vec{r}, t) m_G, \qquad (92)$$

where f is a scalar field expressing the inertial effect of all matter in the universe. An obvious constraint on any such theory is the empirical fact that $f \simeq 1/\sqrt{G}$ near earth.

Guided by dimensional analysis and simplicity, Tryon guessed that f might be the magnitude of a relativistic analog to gravitational potential, divided by c^2:

$$f = |V_G|/c^2, \qquad (93)$$

where the gravitational potential energy (GPE) of a mass m would be

$$\text{GPE} = -m_G|V_G|, \qquad (94)$$

since $V_G < 0$. In the nonrelativistic (Newtonian) limit, one would then have $\nabla^2 f = -4\pi\rho_G/c^2$, where ρ_G denotes the density of gravitational mass or "charge," and ∇^2 is the Laplacian differential operator. The desired analog is

$$f\square^2 f = \frac{4\pi}{c^2}T^\mu_\mu, \qquad (95)$$

where $\square^2$ is the invariant d'Alembertian of general relativity, $T^\mu_\mu \equiv g^{\mu\nu}T_{\mu\nu}$, and $T_{\mu\nu}$ is the *inertial* stress-energy tensor (so that ρ_G appears in $T_{\mu\nu}/f$).

For a simple test of the theory, Tryon used the $T_{\mu\nu}$ of Eq. (21) and the critical density ρ_c given by Eq. (45), assuming zero pressure. (The actual density was known to be comparable to ρ_c, perhaps even equal, and calculations were simpler in a flat space). In a rough estimate for f, the universe was treated as static. Since no causal influence could have reached us from beyond the horizon, the volume integral over ρ_c was cut off at the horizon distance $d_{SH} = 2c/H_0$ [Eqs. (62) and (64)]. The result was a cut-off version of Newton's gravity:

$$f^2 \approx \left(M_I/c^2\right) \times \langle 1/r \rangle, \qquad (96)$$

where M_I denotes the total (inertial) mass within the horizon, and $\langle 1/r \rangle = 3/2d_{SH}$ is the average value of $1/r$.

With M_I given by ρ_c, Eq. (96) yields $f \approx \sqrt{3/G}$, which is the result desired within a factor of $\sqrt{3}$. It would be difficult to exaggerate the adrenaline that was produced by this discovery. Given the crudity of the calculation, it seemed likely that a more careful analysis would yield the *exact* result desired for *some* density near ρ_c, perhaps the density of our own universe.

When Tryon met with his mentor the next day, he was disappointed to learn that this approximate relation between ρ_c and G was already known. In fact Carl Brans and Robert Dicke had published a similar (but far more sophisticated) theory a few years earlier (1961). One difference was that unlike Tryon's crude theory, theirs contained an adjustable parameter ω, and became identical with GTR in the limit as $\omega \to \infty$.

Active interest in the Brans-Dicke theory was soon undermined by observations, all of which were consistent with standard GTR. The Brans-Dicke theory could not be ruled out by such data, since ω could always be chosen large enough to obtain agreement. In the absence of any discrepancy between observations and GTR, however, GTR remained the favored candidate. Tryon followed these developments closely, and gradually abandoned the Machian interpretation of the relation ρ_c between and G. He retained a deep conviction, however, that the nearly

critical density of our universe was no coincidence. He believed it to be a cryptic manifestation of some truth of fundamental importance, and he resolved to revisit the issue from time to time, hoping eventually to decode the message.

Tryon's research centered on the strong interactions for several years, but he occasionally spent a few hours wrestling with the significance of a density near ρ_c. One afternoon in 1973, an epiphany occurred. To his utter surprise and astonishment, his "mind's eye" saw a brilliant flash of light, and he realized instantly that he was witnessing the birth of the universe. In that same instant, he realized how and why it happened: the universe is a vacuum fluctuation, with its cosmic size and duration explained by the relation between ρ_c and G that he had struggled for so long to understand.

For pedagogical purposes, textbooks on quantum field theory discuss a few simple models which, for a variety of reasons, fail to describe physical reality. One such model contains a field ψ with a potential $V(\psi) = \lambda\psi^3$, where λ is a constant. This model is *immediately* rejected, because it has no state of minimum energy. (For $\lambda = \pm|\lambda|$, $V(\psi) \to -\infty$ as $\psi \to \mp\infty$). The existence of a state of minimum energy, called the *vacuum*, is traditionally regarded as essential to any realistic quantum field theory of elementary particles.

Consider now a distribution of particles with total mass M. Adding Einstein's Mc^2 to Newtonian approximations for kinetic and potential energies, one obtains a total energy

$$E = M\left(c^2 + \tfrac{1}{2}\langle v^2 \rangle\right) - \tfrac{1}{2}GM^2 \times \langle 1/r \rangle, \qquad (97)$$

where angled brackets denote average values (of velocity-squared and inverse separation, respectively). Note that E has no minimum value: $E \to -\infty$ as $M \to \infty$. *Gravitational potential energy is precisely the kind that has traditionally been rejected in quantum field theory*, because such a theory has no state of minimum energy. An initially empty state with $E = M = 0$ would spontaneously undergo quantum transitions to states with $E = 0$ but *arbitrary* $M > 0$, subject only to the constraint that

$$\langle 1/r \rangle = \frac{2\left(c^2 + \tfrac{1}{2}\langle v^2 \rangle\right)}{GM} \qquad (98)$$

Quantum field theory has no stable vacuum when gravity is included.

A vacuum fluctuation from $M = 0$ to a state with $M > 0$ would require a form of quantum tunneling, since $E > 0$ for values of M between 0 and the final value. Such tunneling is a completely standard feature of quantum theory, however. We have already considered one example, the inevitable tunneling of a Higgs field from the false vacuum of Fig. 4a to the true vacuum.

In quantum field theory, pairs of particles (e.g. electrons and positrons) are continuously appearing in the most empty space possible (Section IV.G). Energy conservation is violated by an amount ΔE for a duration Δt. This is permitted (actually *implied*) by quantum mechanical uncertainties, however, subject to the constraint that $\Delta E \Delta t \approx \hbar$. Eventually, by chance, enough particles would spontaneously appear close enough together for them to have zero net energy, because of cancellation between positive mass-energy and negative gravitational energy. Such a state would have $\Delta E = 0$, and could last forever.

Now recall Tryon's guess that the f in Eq. (92) might be the magnitude of the cosmic gravitational potential. This would require that $f \simeq 1/\sqrt{G}$ but, in a crude calculation assuming the density to be ρ_c, he had obtained $f \simeq \sqrt{3/G}$. Equations (92)–(94) then imply that $m_I c^2 + \text{GPE} \approx 0$ for every piece of matter, i.e., *a universe with density near ρ_c could have zero energy*. Such a universe could have originated as a spontaneous vacuum fluctuation. Conversely, such an origin would explain *why* our universe has a density near ρ_c.

More generally, creation *ex nihilo* (from nothing) would imply that our universe has zero net values for all conserved quantities ("the quantum numbers of the vacuum"). The only apparent challenge to this prediction arises from the fact that our universe is strongly dominated by matter (i.e., antimatter is quite rare). Laboratory experiments have revealed that the symmetry between matter and antimatter is not exact, however, and most theories of elementary particles going beyond the standard model contain mechanisms for matter to have become dominant over antimatter at a very early time (*baryogenesis*, see Section VII.C).

The preceding discussion has used Newtonian approximations for energy, and these are only suggestive. We note, however, that Newtonian theory resembles GTR far more than one might guess. For a matter-dominated universe, Newton's and Einstein's physics predict the same rate of slowing for the expansion, as was shown in Section IV.D. As another example, the escape velocity from the surface of a planet of mass M and radius R is given by Newtonian mechanics as $v_{esc} = \sqrt{2GM/R}$. For $v_{esc} = c$, the relation between R and M is precisely that of GTR describing a black hole (where R is called the *Schwarzschild radius*).

The field equation (22) is of particular interest. With $a(t)$ denoting the cosmic scale factor and $\dot{a} \equiv da/dt$, Eq. (22) implies

$$(kc^2 + \dot{a}^2) - \frac{8\pi G(\rho a^3)}{3a} = 0 \qquad (99)$$

Now consider a closed, finite universe ($k = +1$). The proper volume of such a universe (Section III.C) is

$V(t) = 2\pi^2 a^3(t)$, and the total mass-energy is $M(t) = \rho(t)V(t)$. [M is constant for zero pressure, but otherwise not; see Eq. (24).] With $r_p(t)$ denoting proper distance (Section III.C) and $\dot{r}_p \equiv dr_p/dt$ corresponding to velocity, Eq. (97) multiplied by M yields

$$M\left(c^2 + 0.36\langle \dot{r}_p^2 \rangle\right) - 0.55 GM^2 \times \langle 1/r_p \rangle = 0, \qquad (100)$$

where $\dot{a}^2$ and $1/a$ have been expressed in terms of

$$\langle \dot{r}_p^2 \rangle = \frac{1}{V}\int dV\left(\dot{r}_p^2\right) = \frac{2\dot{a}^2}{\pi}\int_0^\pi du(u \sin u)^2, \quad (101)$$

$$\langle 1/r_p \rangle = \frac{1}{V}\int dV \frac{1}{r_p} = \frac{2}{\pi a}\int_0^\pi du \frac{\sin^2 u}{u} \qquad (102)$$

(see Section III.C). Note that the density, pressure, and total mass M are arbitrary in the above analysis. [In particular, Eqs. (99)–(102) are valid in both standard and inflationary models.]

Apart from slight differences in numerical coefficients, the left side of Eq. (100) is identical in form to the right side of Eq. (97). This suggests that, in some sense, all closed universes have zero energy. This interpretation is buttressed by the fact that such universes are finite and self-contained, so that no gravitational flux lines could extend outward from them into any surrounding space in which they might be embedded. The energy of a universe has no rigorous definition within the context of GTR (Section IV.B), but the preceding observations are nevertheless quite suggestive.

We also note that if dominated by matter, a closed universe expands to a maximum size in a finite time and then collapses back to zero size (Section IV.C). This is the typical behavior of a vacuum fluctuation. For the above reasons, and also because physical processes are expected to give rise to finite systems, Tryon's conjecture of 1973 was accompanied by a prediction that our universe is closed and finite.

It is a remarkable fact that creation *ex nihilo* could have been proposed at any time since the early 1930s. All the theoretical ingredients had been in place. There had been a long period when the magnitude and grandeur of the cosmos had befogged the normally clear vision of physicists, however. A few examples will illustrate this point.

Before Hubble's discovery of the cosmic expansion in 1929, Einstein had believed the universe to be infinitely old and static. This view was incompatible with the physics of even his time, however, in at least three respects.

1. In 1823, Heinrich Olbers had pointed out that the entire night sky should be as bright as the surface of a typical star in such a model. (*Olbers' paradox*: His reasoning was that every line of sight would end on a star, a valid argument. In more modern terms, the

entire universe would have reached a state of thermal equilibrium, in which case all of space would be filled by light as intense as that emitted from the surfaces of stars.)

2. Conservation of energy would be grossly violated by stars that shone forever.

3. Over an infinite span of time, gravitational instabilities would have caused all stars to merge into enormous, dense concentrations of matter, as was evident from work published in 1902 by Sir James Jeans (the *Jeans instability*).

Even Einstein, with few equals in the history of science, felt these considerations to be outweighed by the philosophical appeal of an infinitely old and static universe.

Hubble's discovery of the cosmic expansion resulted in a period of intense speculation about the nature of the universe. In the 1930s, several inventors of cosmological models based their conjectures on a so-called "cosmological principle," namely that the universe must (or *should*) be homogeneous and isotropic. This was extended in the late 1940s to the "perfect cosmological principle," meaning that the universe must (or *should*) also be eternal and unchanging. The steady-state model of cosmology was based on this "principle," and was widely favored in professional circles until empirical evidence undermined it in the 1960s. We remark here (as occasional skeptics did at the time) that both of the aforementioned "principles" were of a highly dubious nature, because (1) nothing supported them except observations which they pretended to explain, and (2) they had no *generality*, referring only to a single system (the entire universe).

The preceding examples are not meant as criticisms of the brilliant scientists who held such views, but rather as illustrations of how the mystique of the cosmos exempted it from the objective standards routinely applied to lesser systems.

Early in his graduate studies (about 1963), Tryon attended a colloquium given by a 30-year-old professor then unknown to him. In straightforward and compelling terms, this professor elucidated several ways in which neutrinos might play critical roles in astrophysics. The lecture inspired Tryon, and transformed his scientific worldview. For the first time, he realized that enormous structures, perhaps even the entire universe, could (and *should*) be analyzed in terms of the same principles that govern phenomena on earth, *including the principles of microscopic physics*. The professor's name was Steven Weinberg. It was Tryon's great good fortune that Weinberg later accepted him as a thesis student. In retrospect, Tryon has long felt deeply indebted to Weinberg for opening his mind to the possibility that the entire universe might have arisen as a spontaneous vacuum fluctuation.

We come now to the second striking feature shared by the psychological and theoretical breakthroughs of Guth and Tryon. In his highly engaging history and survey of inflation (see bibliography), Guth describes how he became involved in the research that culminated in his discovery.

In early 1979, Guth was a young research physicist at Cornell University, preoccupied with issues unrelated to his later work. He has written that in April of 1979,

"my attitude was changed by a visit to Cornell by Steven Weinberg.... Steve gave two lectures on how grand unified theories can perhaps explain why the universe contains an excess of matter over antimatter.... For me, Weinberg's visit had a tremendous impact. I was shown that a respectable (and even a highly respected) scientist could think about things as crazy as grand unified theories and the universe at 10^{-39} sec. And I was shown that really exciting questions could be asked."

The day after Weinberg's visit ended, Guth immersed himself in a study of GUTs in the early universe. By year's end, he had discovered the possibility of cosmic inflation, and he knew immediately that he should take the possibility seriously.

In truth, Weinberg influenced and inspired a generation of young physicists, the first generation entirely liberated from the disorienting mystique of the cosmos. Through widespread lectures, writing, and the extraordinary power and clarity of his vision, Weinberg emboldened a generation to apply the principles of microscopic physics to the early universe.

We have now reached the extraordinary position where the following are regarded as plausible, even likely in the minds of many: (1) the universe originated as a spontaneous quantum fluctuation, (2) quantum fluctuations in the false vacuum lead to eternal inflation and endless creation, and (3) quantum fluctuations in the false vacuum were the primordial deviations from homogeneity that evolved into galaxies and the largest structures in the universe. These and related ideas are sometimes referred to as "quantum cosmology," which is surely among the most striking paradigm shifts in the history of science. Steven Weinberg may fairly be regarded as the spiritual father of quantum cosmology.

One can only speculate about the stage upon which our universe was born. If it was infinite in extent (in space and/or "time"), however, vacuum fluctuations giving rise to universes like ours would be inevitable if the probability were different from zero (no matter how small) on a finite stage. Indeed, such creations would be more numerous than any finite number—such would be the nature of an infinite stage. It is nevertheless true that this quantum theory of creation became much more widely accepted after Guth's proposal of inflation, because inflation explains how even a microscopic fluctuation could balloon into a cosmos.

A mathematically detailed scenario for creation *ex nihilo* was published in 1978 by R. Brout, F. Englert, and E. Gunzig. This model assumed a large, negative pressure for the primordial state of matter, giving rise to exponential growth that converted an initial microsciopic quantum fluctuation into an open universe. This model was a significant precursor to inflationary scenarios based on GUTs, as well as being a model for the origin of the universe.

During the years since Guth's proposal of inflation, numerous creation scenarios have been published, differing primarily in their details. An especially interesting and simple one is that of Alexander Vilenkin (1982). Vilenkin sought to eliminate the need for *any* kind of stage upon which creation occurred, i.e., he sought a more literal creation from *nothing*.

Vilenkin noted that a closed universe has finite volume: $V(t) = 2\pi^2 a^3(t)$. For $a = 0$, the volume is zero and, if such a universe is completely empty, it is then a candidate for true "nothingness"—a mere abstraction. He proposed that such an abstraction could experience quantum tunneling to a microscopic volume filled with false vacuum, which would then inflate to cosmic proportions. We do not (and may never) know whether our universe was born within some larger physical reality or emerged from a pure abstraction, but both scenarios seem plausible.

Creation *ex nihilo* suggests (without implying) a radical idealism, in the sense of ideals being abstract principles. To illustrate the point, it may be helpful to consider the alternative (and usual) view.

Well before a child acquires language and the ability to think abstractly, it has become aware of tangible matter in its surroundings. As the child develops, it furthermore discerns patterns of regularity—splinters hurt, chocolate tastes good, and day alternates with night. Given that matter exists, it must behave in *some* way, and one learns from experience and perhaps the sciences how very regular that behavior is. The unarticulated hierarchy in one's mind is

(1) matter exists, (2) it must do *something*, and (3) it seems to follow some set of rules. Matter is the *fundamental* reality, without which it would make no sense for there to be principles governing it.

The maximal version of creation *ex nihilo* inverts this hierarchy: abstract principles (*ideals*) are the fundamental reality. Tangible matter and energy are *manifestations of a deeper reality*, the principles that gave rise to their existence. This remains, of course, an open question, but it is one of the many issues that quantum cosmology has given us to contemplate.

SEE ALSO THE FOLLOWING ARTICLES

CELESTIAL MECHANICS • COSMIC RADIATION • COSMOLOGY • DARK MATTER IN THE UNIVERSE • GALACTIC STRUCTURE AND EVOLUTION • HEAT TRANSFER • QUANTUM THEORY • RELATIVITY, GENERAL • RELATIVITY, SPECIAL • STELLAR STRUCTURE AND EVOLUTION • UNIFIED FIELD THEORIES

BIBLIOGRAPHY

Guth, A. (1997). "The Inflationary Universe," Perseus Books, New York.

Krauss, L. M. (2000). "Quintessence: The Mystery of the Missing Mass in the Universe," Basic Books, New York.

Liddle, A. R. (1999). "An Introduction to Modern Cosmology," Wiley, New York.

Liddle, A. R., and Lyth, D. H. (2000). "Cosmological Inflation and Large-Scale Structure," Cambridge Univ. Press, Cambridge, U.K.

Linde, A. D. (1990). "Inflation and Quantum Cosmology," Academic Press, San Diego.

Peacock, J. A. (1999). "Cosmological Physics," Cambridge Univ. Press, Cambridge, U.K.

Peebles, P. J. E. (1993). "Principles of Physical Cosmology," Princeton Univ. Press, Princeton, New Jersey.

Weinberg, S. (1972). "Gravitation and Cosmology: Principles and Applications of the General Theory of Relativity," Wiley, New York.

Weinberg, S. (1977). "The First Three Minutes," Basic Books, New York.

Weinberg, S. (1992). "Dreams of a Final Theory," Pantheon Books, New York.

Cosmic Radiation

Peter L. Biermann

*Max-Planck Institute for Radioastronomy
and University of Bonn*

Eun-Suk Seo

University of Maryland

GLOSSARY

Active galactic nuclei When massive black holes accrete, their immediate environment, usually thought to consist of an accretion disk and a relativistic jet, emits a luminosity often far in excess of the emission of all stars in the host galaxy put together; this phenomenon is called an active galactic nucleus.

Antimatter All particles known to us have antiparticles, with opposite properties in all measures, such as charge.

Big Bang Our universe is continuously expanding, and its earliest stage reachable by our current physical understanding is referred to as the Big Bang.

Black holes Compressing a star to a miniscule size, in the case of our sun to a radius of 3×10^5 cm, makes it impossible for any radiation to come out; all particles and radiation hitting such an object disappear from this world. This is called a black hole.

Chemical elements In atoms the number of protons Z in the nucleus, equal to the number of electrons in the surrounding shell, determines the chemical element.

Cosmic ray airshower When a primary particle at high energy, either a photon or a nucleus, comes in to the upper atmosphere, the sequence of interactions and cascades forms an air shower.

Cosmic ray ankle At an energy of 3×10^{18} eV, or 3 EeV, there is an upturn in the spectrum, to an approximate spectral index of 2.7 again.

Cosmic ray GZK cutoff The interaction with the cosmic microwave background is predicted to produce a strong cutoff in the observed spectrum at 5×10^{19} eV called the GZK cutoff. This cutoff is not seen.

Cosmic ray knee At about 5×10^{15} eV, or 5 PeV, there is a small bend downward in the cosmic ray spectrum by about 0.4 in spectral index, from 2.7 to 3.1.

Cosmic ray spectrum The number of particles at a certain energy E within a certain small energy interval dE is called the spectrum. Flux is usually expresssed as the number of particles coming in per area, per second,

Elementary particles The natural constituents of normal matter are the proton, neutron, and electron.

Gamma ray bursts Bursts of gamma ray emission coming from the far reaches of the universe, and almost certainly the result of the creation of a stellar mass black hole.

Interstellar matter The medium between the stars in our Galaxy, which is composed of very hot gas (order 4×10^6 K), various stages of cooler gas, down to about 20 K, dust, cosmic rays, and magnetic fields.

Magnetic monopoles The physics of electric and magnetic fields contains electric charges but no magnetic charges. In the context of particle physics it is likely that monopoles, basic magnetically charged particles, also exist.

Microwave background The very high temperature of the Big Bang is still visible in the microwave background, a universal radiation field of 2.73 K temperature.

Our Galaxy Our Galaxy is a flat, circular, disklike distribution of stars and gas enmeshed with interstellar dust and embedded in a spheroidal distribution of old stars.

Nuclear collisions When elementary particles collide with each other or with a photon, other particles can be created that are often unstable.

Radio galaxies In the case that an active galactic nucleus produces a very powerful and visible jet, often with lobes and hot spots, most readily observable at radio wavelengths, it is called a radio galaxy.

Spallation The destruction of atomic nuclei in a collision with another energetic particle such as another nucleus or, commonly, a proton.

Supernovae All stars above an original mass of more than 8 solar masses are expected to explode at the end of their lifetime after they have exhausted nuclear burning; the observable effect of such an explosion is called a supernova.

Topological defects It is generally believed that in the very early times of the universe, when the typical energies were far in excess of what we can produce in accelerators, there was a phase when the typical energies of what might be called particles was around 10^{24} eV, usually referred to as topological defects, or relics.

Units: Energy Electron volts (eV); 1 eV is of the order of what is found in chemical reactions; $1 \text{ eV} = 1.6 \times 10^{-12}$ erg.

Units: length Centimeters; the radius of the earth is 6.4×10^8 cm and the radius of the Sun is 7×10^{10} cm.

Units: mass Grams; the sun has a mass of 2×10^{33} g; the earth has a mass of 6×10^{27} g.

Units: time Seconds; the travel time of light from the sun to the earth is about 8 min $= 480$ sec; the number of seconds in a year is 3.15×10^7.

ENERGETIC PARTICLES, traditionally called *cosmic rays*, were discovered nearly a 100 years ago and their origin is still uncertain. Their main constituents are the normal nuclei as in the standard cosmic abundances of matter, with some enhancements for the heavier elements; there are also electrons. Information on isotopic abundances shows some anomalies as compared with the interstellar medium. There is also antimatter, such as positrons and antiprotons. The known spectrum extends over energies from a few hundred MeV to 300 EeV ($= 3 \times 10^{20}$ eV), and shows a few clear spectral signatures: There is a small spectral break near 5 PeV ($= 5 \times 10^{15}$ eV), commonly referred to as the *knee*, where the spectrum turns down; and there is another spectral break near 3 EeV ($= 3 \times 10^{18}$ eV), usually called the *ankle*, where the spectrum turns up again. Due to interaction with the microwave background arising from the Big Bang, there is a strong cutoff expected near 50 EeV ($= 5 \times 10^{19}$ eV), which is, however, not seen; this expected cutoff is called the GZK cutoff after its discoverers, Greisen, Zatsepin, and Kuzmin. The spectral index α is near 2.7 below the knee, near 3.1 above the knee, and again near 2.7 above the ankle, where this refers to a differential spectrum of the form $E^{-\alpha}$. We will describe the various approaches to understanding the origin and physics of cosmic rays.

I. INTRODUCTION AND HISTORY

Cosmic rays were discovered by Hess and Kohlhörster in the beginning of the 20th century through their ionizing effect on airtight vessels of glass enclosing two electrodes with a high voltage between them. This ionizing effect increased with altitude during balloon flights, and therefore the effect must come from outside the earth, so the term *cosmic rays* was coined. The earth's magnetic field acts on energetic particles according to their charge, and hence they are differently affected coming from east and west, and so their charge was detected, proving that they are charged particles; at high energies near 10^{18} eV or 1 EeV, there is observational evidence that a small fraction of the particles are neutral and in fact neutrons. From around 1960 onward there has been evidence of particles at or above 10^{20} eV, with today about two dozen such events known. After the cosmic microwave background was discovered in the early 1960s, it was noted only a little later by Greisen, Zatsepin, and Kuzmin that near and above an energy of 5×10^{19} eV (called the GZK cutoff) the interaction with the microwave background would lead to strong losses if these particles were protons, as is now believed on the basis of detailed air shower data. In such an interaction, protons see the photon as having an energy of above the pion mass, and so pions can be produced in the reference frame of the collision, leading to about a 20% energy loss of the proton in the observer frame. Therefore for an assumed cosmologically homogeneous distribution of sources for protons at extreme energies, a spectrum at earth is predicted which shows a strong cutoff at 5×10^{19} eV, the GZK cutoff. This cutoff is not seen, leading to many speculations as to the nature of these particles and their origin.

Cosmic rays are measured with instruments on balloon flights, satellites, the Space Shuttle, the International Space Station, and with ground arrays. The instrument chosen depends strongly on what is being looked for and the energy of the primary particle. One of the most successful campaigns has been with balloon flights in Antarctica, where a balloon can float at about 40 km altitude and circumnavigate the South Pole once and possibly even several times during one Antarctic summer. For very high precision measurements very large instruments on the Space Shuttle or the International Space Station are used, such as for the search for antimatter.

Critical measurements are the exact spectra of the most common elements, hydrogen and helium, the fraction of antiparticles (antiprotons and positrons), isotopic ratios of elements such as neon and iron, the ratio of spallation products such as boron to primary nuclei such as carbon as a function of energy, the chemical composition near the knee, at about 5×10^{15} eV, and beyond, and the spectrum and nature of the particles beyond the ankle, at 3×10^{18} eV, with special emphasis on the particles beyond the GZK cutoff, at 5×10^{19} eV.

II. PHYSICAL CONCEPTS

Here we expand upon the terms explained briefly in the Glossary.

- *Big Bang*. Our universe is continuously expanding, and its earliest stage reachable by our current physical understanding is referred to as the Big Bang, when energy densities were extremely high. Within the first 3 min the chemical elements such as hydrogen and helium were produced, and minute amounts of deuterium (an isotope of hydrogen), ^{3}He, an isotope of helium, and ^{7}Li, an isotope of the third element, lithium.
- *Microwave background*. The very high temperature of the Big Bang is still visible in the microwave background, a universal radiation field of 2.73 K. There is a corresponding cosmic bath of low-energy neutrinos. Both

photons and neutrinos have a density of a few hundred per cubic centimeter.
- *Units*: *length* Centimeters; the radius of the earth is 6.4×10^8 cm and the radius of the sun is 7×10^{10} cm. The distance from earth to sun is 1.5×10^{13} cm. One pc = parsec = 3.086×10^{18} cm, 1 kpc = 10^3 pc, 1 Mpc = 10^6 pc; 1 pc is about 3 light-years, the distance traveled by light in 3 years; the speed of light c is 3×10^{10} cm/sec, and is the same in any inertial reference frame. The basic length scale of the universe is about 4000 Mpc.
- *Units*: *mass*. Grams; the sun has a mass of 2×10^{33} g. The earth has a mass of 6×10^{27} g. A typical galaxy like our own has a mass of order 10^{11} solar masses. A proton has a mass of 1.67×10^{27} g.
- *Units*: *time* Seconds; the travel time of light from the sun to the earth is about 8 min = 480 sec; the number of seconds in a year is 3.15×10^7. The age of the solar system is about 4.5×10^9 years, and the age of our Galaxy and also of our universe is about 1.5×10^{10} years; our galaxy is younger than the universe, but we do not know the two ages well enough to determine the difference with any reliability.
- *Units*: *energy*. Electron volts (eV); 1 eV is of the order of what is found in chemical reactions, 1 eV = 1.6×10^{-12} erg; 1 MeV = 10^6 eV, 1 GeV = 10^9 eV, 1 TeV = 10^{12} eV, 1 PeV = 10^{15} eV, 1 EeV = 10^{15} eV.
- *Elementary particles*. The natural constituents of matter are the proton, neutron, and electron. Protons have a mass of about 938 MeV, neutrons of about 940 MeV, and electrons of about 0.511 MeV. This is in energy units using Einstein's equivalence $E = mc^2$, where E is the energy, m the rest mass, and c the speed of light. The proton is positively charged, the electron negatively charged, with the same charge as the proton, and the neutron is neutral. All atomic nuclei are built from protons and neutrons, where the number of protons determines the chemical element, and the number of neutrons determines the various isotopes of each chemical element. The surrounding shell of electrons has for the neutral atom exactly the same number of electrons as the nucleus has protons. Photons are another primary stable constituent, have no rest mass, no charge, and always travel at the speed of light, in any frame of reference. Neutrinos come in three varieties, and appear to continuously change among themselves; they have a very low mass.
- *Antimatter*. All known particles have antiparticles, with opposite properties in all measures, such as charge. The collision of a particle and its antiparticle always leads to a burst of radiation, when both particles are annihilated. In cosmic rays we observe antiprotons, and positrons, the antiparticles to electrons. The search for antinuclei has not been successful; any detection of even a single antinucleus, such as antihelium, would provide extremely strong constraints on the physics of matter in the universe.

The instrument AMS, first on the Space Shuttle, and then the International Space Station, will search for antimatter particles.

- *Nuclear collisions.* When elementary particles collide with each other or with a photon, other particles can be created that are often unstable, i.e., they decay into other particles and continue to do so until they reach a state where only stable particles result. Such a process is called a cascade. Common intermediate and final particles are the pion, the muon, the photon, and the neutrino. The pion comes in various forms, charged and uncharged, the muon is always charged, and the neutrino is always neutral. The neutrino is characterized by a very small interaction cross section with matter. The pions have a mass, again in energy equivalents, of about 140 MeV, the muons of about 106 MeV, and the neutrinos of about 0.03 eV, a still rather uncertain number.

- *Chemical elements.* In atoms the number of protons Z in the nucleus, equal to the number of electrons in the surrounding shell, determines the chemical element. The number of neutrons in the nucleus is approximately equal to the number of protons. Chemical elements are now known to beyond Z of 110. The first and most common elements are hydrogen ($Z = 1$), helium (2), carbon (6), oxygen (8), neon (10), magnesium (12), silicon (14), sulfur (16), calcium (20), and iron (26). The intermediate elements lithium ($Z = 3$), beryllium (4), and boron (5) are very rare. The odd-Z elements are commonly rarer than the even-Z elements. The overall abundance by mass is about 73% for hydrogen, 25% for helium, and the rest all other elements combined, with the most abundant among these being carbon and oxygen.

- *Cosmic ray airshower.* When a primary particle at high energy, either a photon or a nucleus, comes into the upper atmosphere, the sequence of interactions and cascades forms an airshower. These airshowers are dominated by electrons and photons generated in electromagnetic subshowers. Cerenkov light, a bluish light, is emitted in a narrow cone around the shower direction, produced when particles travel at a speed (always less than or equal to c; particles have exactly the speed of light at zero rest mass, such as photons) higher than the speed of light c divided by the local index of refraction (which is 4/3 in water, for instance, and about 1.0003 in air). Observing this bluish light allows observations of high-GeV-to-TeV photon sources in the sky because Cerenkov light allows good pointing. For particles such as protons or atomic nuclei at high energy, air fluorescence can be used as a means of observation, advantageous because it is omnidirectional in its emission and gives enough light at high energy: Such fluorescence occurs when normal emission lines of air molecules are excited. The airshower includes a pancake of secondary electrons and positrons as well as muons. The ratio between Cerenkov light and fluorescence light is almost constant and independent of particle energy. Most modern observations of very high energy cosmic rays are done either by observing the air fluorescence (arrays such as Fly's Eye, HIRES, or AUGER) or by observing the secondary electrons and positrons (in arrays such as Haverah Park, AGASA, Yakutsk, or also AUGER). In the future such observations may be possible from space by observing the air fluorescence or the reflected Cerenkov light from either the International Space Station or from dedicated satellites. Fly's Eye was and HIRES is in Utah, AUGER is in Argentina, AGASA is in Japan, Yakutsk is in Russia, and Haverah Park was in the United Kingdom.

- *Spallation.* Spallation is the destruction of atomic nuclei in a collision with another energetic particle, such as another nucleus, or commonly a proton. In this destruction many pieces of debris can be formed, with one common result being the stripping of just one proton or neutron, and another common result a distribution of smaller unit nuclei. Since the proton number determines the chemical element, such debris is usually other nuclei, such as boron.

- *Cosmic ray spectrum.* The number of particles at a certain energy E within a certain small energy interval dE is called the spectrum. As a function of energy E this is usually described by power laws, such as $E^{-2.7} dE$; the exponent is called the spectral index, here 2.7. Heat radiation from a normal object has a very curved spectrum in photons. Cosmic rays usually have a power-law spectrum, which is called a nonthermal behavior. Flux is usually expressed as the number of particles coming in per area, per second, per solid angle in steradians (all-sky is 4π), and per energy interval.

- *Cosmic ray knee.* At about 5×10^{15} eV, or 5 PeV, there is a small bend downward in the cosmic ray spectrum, by about 0.4 in spectral index, from 2.7 to 3.1. This feature is called the *knee*. There is some evidence that it occurs at a constant energy-to-charge ratio for different nuclei. There is also considerable evidence that toward and at the knee the chemical composition slowly increases in favor of heavier nuclei such as iron. A similar somewhat weaker feature is suggested by new AGASA and HIRES data near 3×10^{17} eV, where again the spectrum turns down a bit more.

- *Cosmic ray ankle.* At an energy of 3×10^{18} eV, or 3 EeV, there is an upturn in the spectrum, to an approximate spectral index of 2.7 again. At the same energy there is evidence that the chemical composition changes from moderately heavy to light, i.e., back to mostly hydrogen and helium.

- *Cosmic ray GZK cutoff.* The interaction with the cosmic microwave background should produce a strong cutoff in the observed spectrum at 5×10^{19} eV called the GZK cutoff; this is provided that (a) these particles are protons (or neutrons) and (b) the source distribution is homogeneous in the universe. This cutoff is not seen; in fact, no cutoff is seen at any energy, up to the limit of data, at 3×10^{20} eV, or 300 EeV. This is one of the most serious problems facing cosmic ray physics today.

• *Black holes.* Compressing a star to a miniscule size, in the case of our sun, to a radius of 3×10^5 cm, makes it impossible for any radiation to come out; all particles and radiation hitting such an object disappear from this world. This is called a black hole. It is now believed that almost all galaxies have a massive black hole at their center, with masses sometimes ranging up 10^{10} solar masses, but usually much less. There are also stellar mass black holes, but their number is not well known, probably many thousands in each galaxy.

• *Our Galaxy.* Our Galaxy is a flat, circular, disklike distribution of stars and gas enmeshed with interstellar dust and embedded in a spheroidal distribution of old stars. The age of this system is about 15 billion (=15×10^9) years; its size is about 30 kpc across, and its inner region is about 6 kpc across. At its very center there is a black hole with 2.6×10^6 solar masses. The gravitational field is dominated in the outer parts of the Galaxy by an unknown component, called dark matter, which we deduce only through its gravitational force. In the innermost part of the Galaxy normal matter dominates. The mass ratio of dark matter to stars to interstellar matter in our Galaxy is about 100:10:1. Averaged over the nearby universe these ratios are shifted in favor of gas, with gas dominating over stars probably, but with dark matter still dominating over stars and gas by a large factor.

• *Interstellar matter.* The medium between the stars in our Galaxy is composed of very hot gas (order 4×10^6 K), various stages of cooler gas, down to about 20 K, dust, cosmic rays, and magnetic fields. All three components, gas, cosmic rays, and magnetic fields, have approximately the same energy density, which happens to be also close to the energy density of the microwave background, about 1 eV/cm^3. The average density of the neutral hydrogen gas, of temperature a few 10^{10} K to a few 10^3 K, is about 1 partcile/cm^3, but highly clumped, in a disk of thickness about 100 pc (=3×10^{20} cm). The very hot gas extends much farther from the symmetry plane, about 2 kpc on either side.

• *Supernovae.* All stars above an original mass of more than 8 solar masses are expected to explode at the end of their lifetime after they have exhausted nuclear burning; the observable effect of such an explosion is called a supernova. When they explode, they emit about 3×10^{53} ergs in neutrinos and also about 10^{51} erg in visible energy, such as in shock waves in ordinary matter, the former stellar

envelope, and interstellar gas. These neutrinos have an energy in the range of a few MeV to about 20 MeV. When stars are in stellar binary systems, they can also explode at low mass, but this process is believed to give only 10% or less of all stellar explosions. The connection to gamma ray bursts is not clear. It is noteworthy that above an original stellar mass of about 15 solar masses, stars also have a strong stellar wind, which for original masses above 25 solar masses becomes so strong that it can blow out a large fraction of the original stellar mass even before the star explodes as a supernova. The energy in this wind, integrated over the lifetime of the star, can attain the visible energy of the subsequent supernova, as seen in the shock wave of the explosion.

• *Gamma ray bursts.* Bursts of gamma ray emission come from the far reaches of the universe, almost certainly the result of the creation of a stellar-mass black hole. The duration of these bursts ranges from a fraction of a second to usually a few seconds, and sometimes hundreds of seconds. Some gamma ray bursts have afterglows in other wavelengths like radio, optical, and X rays, with an optical brightness which very rarely comes close to being detectable with standard binoculars. The emission peaks near 100 keV in observable photon energy, and appears to have an underlying power-law character, suggesting nonthermal emission processes.

• *Active galactic nuclei.* When massive black holes accrete, their immediate environment, usually thought to consist of an accretion disk and a relativistic jet (i.e., where the material flies with a speed very close to the speed of light), emits a luminosity often far in excess of the emission of all stars in the host galaxy put together. There are such black holes of a mass near 10^8 solar masses, with a size of order the diameter of the earth orbit around the sun and a total emission of 1000 times that of all stars in their host galaxy.

• *Radio galaxies.* In the case that an active galactic nucleus produces a very powerful and visible jet, often together with lobes and hot spots, most readily observable at radio wavelengths, it is called a radio galaxy. The radio image of such a galaxy can extend to 300 kpc or more, dissipating the jet in radio hot spots embedded in giant radio lobes. The frequency of such radio galaxies with powerful jets, hot spots, and lobes is rare, less than 1/1000 of all galaxies, but in the radio sky they dominate due to their extreme emission.

• *Topological defects.* It is generally believed that in the very early times of the universe, when the typical energies were far in excess of what we can produce in accelerators, there was a phase when the typical energies of what might be called particles was around 10^{24} eV, usually referred to as topological defects, or relics. It is conceivable that such particles have survived to today, and some of them decay, emitting a copious number of neutrinos, photons, and also protons. These particles then themselves would have near 10^{24} eV initially, but interact strongly with the cosmic radiation field.

• *Magnetic monopoles.* The physics of electric and magnetic fields contains electric charges but no magnetic charges. In the context of particle physics it is likely that monopoles, basic magnetically charged particles, also exist. Such monopoles are a special kind of topological defect. The basic property of magnetic monopoles can be described as follows: (a) Just as electrically charged particles short-circuit electric fields, monopoles short-circuit magnetic fields. The observation of very large scale and permeating magnetic fields in the cosmos shows that the universal flux of monopoles is very low. (b) Monopoles are accelerated in magnetic fields, just as electrically charged particles are accelerated in electric fields. In cosmic magnetic fields, the energies which can be attained are of order 10^{21} eV or more. Any relation to the observed high-energy cosmic rays is uncertain at present.

III. ENERGIES, SPECTRA, AND COMPOSITION

The solar wind prevents low-energy charged articles from entering the inner solar system due to interaction with the magnetic field in the solar wind, a steady stream of gas going out from the sun into all directions, originally discovered in 1950 from the effect on cometary tails: they all point outward, at all latitudes of the sun, and independent of whether the comet actually comes into the inner solar system or goes outward, in which case the tail actually precedes the head of the comet. This prevents us from knowing anything about interstellar energetic particles with energies lower than about 300 MeV. Above about 10 GeV per charge unit Z of the particle, the effect of the solar wind becomes negligible. Since cosmic ray particles are mostly fully ionized nuclei (with the exception of electrons and positrons), this is a strong effect.

Our Galaxy has a magnetic field of about 6×10^{-6} G in the solar neighborhood; the energy density of such a field corresponds approximately to 1 eV/cm^3, just like the other components of the interstellar medium. In such a magnetic field charged energetic particles gyrate with a radius of gyration called the Larmor radius, which is proportional to the momentum of the particle perpendicular to the magnetic field direction. For highly relativistic particles this entails that around 3×10^{18}-eV protons, or other nuclei of the same energy-to-charge ratio, no longer gyrate in the disk of the Galaxy, i.e., their radius of gyration is larger than the thickness of the disk. Thus, they cannot possibly originate in the Galaxy, and must come from out-

side; indeed, at that energy there is evidence for a change both in chemical composition and in the slope of the spectrum.

The energies of these cosmic ray particles that we observe range from a few hundred MeV to 300 EeV. The integral flux ranges from about 10^{-5} per cm^2, per sec, per sterad at 1 TeV per nucleus for hydrogen or protons, to 1 particle per sterad per km^2 and per century around 10^{20} eV, a decrease by a factor of 3×10^{15} in integral flux, and a corresponding decrease by a factor of 3×10^{23} in the spectrum, i.e., per energy interval, which means in differential flux. Electrons have only been measured to a few TeV.

The total particle spectrum spectrum is about $E^{-2.7}$ below the knee and about $E^{-3.1}$ above the knee, at 5 PeV, and flattens again to about $E^{-2.7}$ beyond the ankle, at about 3 EeV. Electrons have a spectrum which is similar to that of protons below about 10 GeV, and steeper near $E^{-3.3}$ above this energy. The lower energy spectrum of electrons is inferred from radio emission, while the steeper spectrum at the higher energies is measured directly (Fig. 1).

The chemical composition is rather close to that of the interstellar medium, with a few strong peculiarities relative to that of the interstellar medium: (a) hydrogen and helium are less common relative to silicon. Also, the ratio of hydrogen to helium is smaller. (b) lithium, beryllium, and boron, the odd-Z elements, as well as the sub-iron elements (i.e., those with Z somewhat less than iron) are all enhanced relative to the interstellar medium (Fig. 3). (c) Many isotope ratios are quite different, while some are identical. (d) Among the cosmic ray particles there are radioactive isotopes, which give an age of the particles since acceleration and injection of about 3×10^7 years. (e) Toward the knee and beyond the fraction of heavy elements appears to continuously increase, with moderately heavy to heavy elements almost certainly dominating beyond the knee, all the way to the ankle, where the composition becomes light again. This means that at that energy we observe a transition to what appears to be mostly hydrogen and helium nuclei. At much higher energies we can only show consistency with a continuation of these properties; we cannot prove unambiguously what the nature of these particles is.

The fraction of antiparticles, ie., positrons and antiprotons, is a few percent for the positron fraction and a few 10^4 for antiprotons. No other antinuclei have been found (Figs. 3–5).

There is no significant anisotropy of cosmic rays at any energy, not even at the highest energies, beyond the GZK cutoff. Only at those highest energies is there a persistent hint that events at quite different energies occasionally cluster into pairs and triplets in the sky in the arrival direction, and this is hard to understand in almost any model.

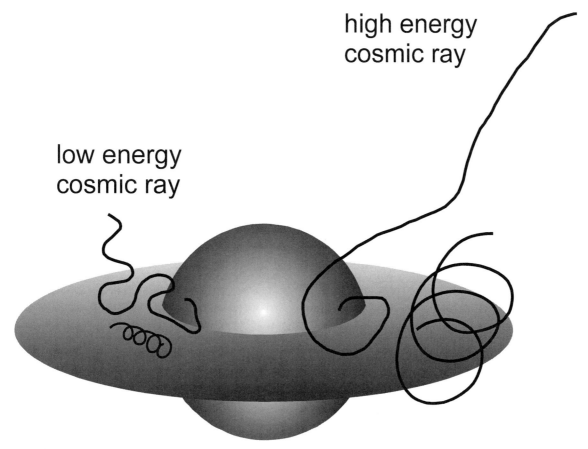

high energy
cosmic ray

low energy
cosmic ray

FIGURE 1 The Galaxy with some sample orbits of low energy and high energy cosmic ray particles. The Galaxy is shown with the disk and the spherodial component of older stars.

IV. ORIGIN OF GALACTIC COSMIC RAYS

It has been long surmised that supernova explosions provide the bulk of the acceleration of cosmic rays in the Galaxy. The acceleration is thought to be a kind of ping-pong effect between the two sides of the strong shock wave sent out by the explosion of a star. This ping-pong effect is a repeated reflection via the magnetic resonant interaction between the gyromotion of the energetic charged particles and waves of the same wavelength as the Larmor motion in the magnetic thermal gas. Since the reflection is usually thought to be a gradual diffusion in direction, the process is called diffusive shock acceleration, or, after its discoverer, Fermi acceleration.

For a shock wave sent out directly into the interstellar gas this kind of acceleration easily provides particle energies up to about 100 TeV. While the detailed injection mechanism is not quite clear, the very fact that we observe the emission of particles at these energies in X rays provides a good case and a rather direct argument for highly energetic electrons. Even though protons are by a factor of about 100 more abundant than electrons at energies near

1 GeV, we cannot yet prove directly that supernova shocks provide the acceleration; only the analogy with electrons can be demonstrated.

However, we observe what ought to be galactic cosmic rays up to energies near the knee, and beyond to the ankle, i.e., 3 EeV.

The energies, especially for particles beyond 100 TeV, can be provided by several possibilities, with the only theory worked out to a quantitative level suggesting that those particles also get accelerated in supernova shock waves, in those which run through the powerful stellar wind of the predecessor star. Then it can be shown that energies up to 3 EeV per particle are possible (mostly iron). An alternate possibility is that a ping-pong effect between various supernova shock waves occurs, but in this case seen from outside. In either (or any other) such theory it is a problem that we observe a knee, i.e., a downward bend of the spectrum at an energy-to-charge ratio which appears to be fairly sharply defined. The concept that stellar explosions are at the origin entails that all such stars are closely similar in their properties, including their magnetic field, at the time of explosion; this implies a specific

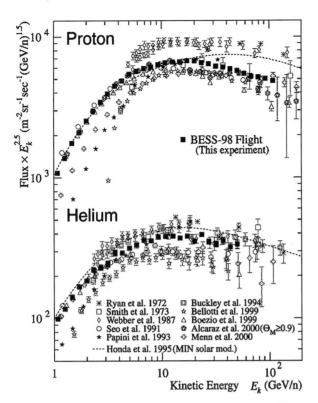

FIGURE 2 The proton and helium spectra at low energy. [From Suzuki, T., *et al.* (2000). "Precise measurements of cosmic-ray proton and helium spectra with BESS spectrometer," *Astrophys. J.* **545**, 1135–1142; The AMS Collaboration. (2000). "Protons in near earth orbit," *Phys. Lett.* B **472**, 215–226; The AMS Collaboration. (2000). "Helium in near earth orbit," *Phys. Lett.* B **494**, 193–202.]

mic rays at least in the GeV to many GeV energy range is always the same in various locations in a galaxy and also in different galaxies. During this travel inside a galaxy the cosmic rays interact with the interstellar gas, and in this interaction produce gamma ray emission from pion decay, positrons, and also neutrons, antiprotons, and neutrinos. The future gamma ray emission observations of the galactic disk will certainly provide very strong constraints on this aspect of cosmic rays. In these interactions secondary nuclei and isotopes are also produced by spallation. This spallation also gives secondary isotopes such as ^{10}Be, which is radioactive and decays with a half-life of 1.6×10^6 years, allowing models of cosmic ray propagation to be tested.

One kind of evidence about where cosmic rays come from and what kind of stars and stellar explosions really dominate among their sources is the isotopic ratios of various isotopes of neon, iron, and other heavy elements; these isotope ratios suggest that at least one population is indeed the very massive stars with strong stellar winds; however, whether these stars provide most of the heavier elements, as one theory proposes, is still quite an open question.

Antimatter as observed today can all be produced in normal cosmic ray interactions. However, even the detection of a single antinucleus of an element such as helium would constitute proof that the universe contains antimatter regions and would radically change our perception of the matter–antimatter symmetries in our world.

There is some evidence that just near EeV energies there is one component of galactic cosmic rays which is spatially associated in arrival direction with the two regions of highest activity in our Galaxy, at least as seen from earth: the Galactic Center region as well as the Cygnus region show some weak enhancement. Such a directional association is only possible for neutral particles, and since neutrons at that energy can just about travel from those regions to here before they decay (only free neutrons decay, neutrons bound into a nucleus do not decay), a production of neutrons in cosmic ray interactions is conceivable as one explanation of these data.

length scale in the explosion, connected to the thickness of the matter of the wind snowplowed together by the supernova shock wave. While this is certainly possible, we have too little information on the magnetic field of pre-supernova stars to verify or falsify this. In the case of the other concept it means that the transport through the interstellar gas has change in properties also at a fairly sharply defined energy-to-charge ratio, indicating a special scale in the interstellar gas, for which there is no other evidence.

Galactic cosmic rays get injected from their sources with a certain spectrum. While they travel through the Galaxy from the site of injection to escape or to the observer, they have a certain chance to leak out from the hot magnetic disk of several kpc full-width thickness of the Galaxy; occasionally this thick disk is referred to as the halo. The escape of cosmic rays becomes easier with higher energy. As a consequence their spectrum steepens, as shown by comparing source and observed spectrum. The radio observations of other galaxies show consistency with the understanding that the average spectrum of cos-

V. THE COSMIC RAYS BETWEEN 3 AND 50 EeV, THE EXPECTED GZK CUTOFF

The cosmic rays between the ankle and the expected GZK cutoff are readily explained by many possible sources, almost all outside our Galaxy.

Some, but not all of these proposals can also explain particles beyond the GZK cutoff, discussed in a separate section.

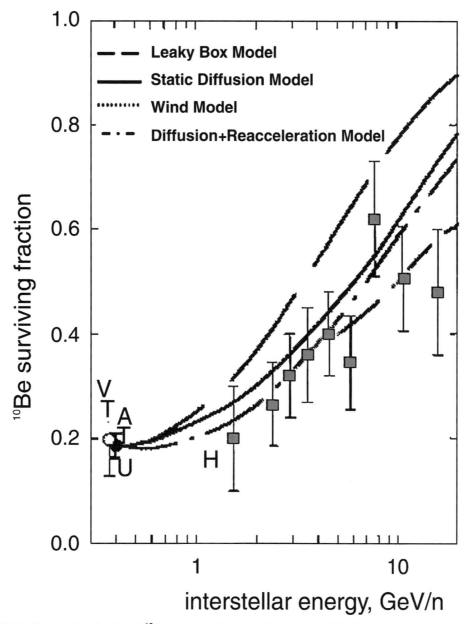

FIGURE 3 The surviving fraction of ^{10}Be in data and a comparison with models of cosmic ray propagation. [From Ptuskin, V. S. (2000). "Cosmic ray transport in the galaxy," In "ACE-2000 Symposium" (R. A. Mewaldt *et al.*, eds.), pp. 390–395, AIP, Melville, NY.]

Pulsars, especially those with very high magnetic fields, called magnetars, can almost certainly accelerate charged particles to energies of 10^{21} eV. There are several problems with such a notion, one being the adiabatic losses on the way from close to the pulsar out to the interstellar gas, and another one the sky distribution, which should be very anisotropic given the distribution and strength of galactic magnetic fields. On the other hand, if this concept could be proven, it would certainly provide a very easy expla-

nation for why there are particles beyond the GZK cutoff: for galactic particles the interaction with the microwave background is totally irrelevant, and so no GZK cutoff is expected.

Another proposal is gamma ray bursts. However, since we know that the bulk of gamma ray bursts arise from stellar explosions at cosmological distances this notion is hard to maintain. The frequency of nearby gamma ray bursts is too small to provide any appreciable flux. However, since

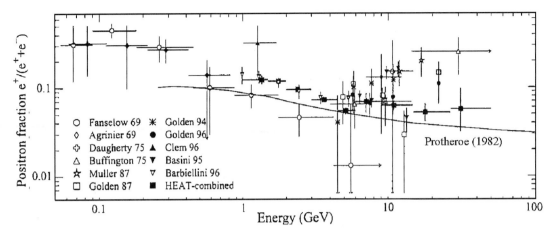

FIGURE 4 The positron fraction compared with various models of cosmic ray interaction. [From Coutu, St., *et al.* (1999). "Cosmic-ray positrons: Are there primary sources?" *Astrophys. J.* **11**, 429–435; The AMS Collaboration. (2000). "Leptons in near earth orbit," *Phys. Lett. B* **484**, 1022; Wiebel-Sooth, B., and Biermann, P. L. (1999). "Cosmic rays," In "Landolt-Börnstein," Vol. VI/3c, pp. 37–90, Springer-Verlag, Berlin.]

ultimately we do not yet know what constitutes a gamma ray burst, their contribution cannot be settled with full certainty.

Shock waves running through a magnetic and ionized gas accelerate charged particles, as we know from *in situ* observations in the solar wind, and this forms the basis of almost all theories to account for galactic cosmic rays. The largest shock waves in the universe have scales of many tens of Mpc, and have shock velocities of around 1000 km/sec. These shock waves arise in cosmological large-scale structure formation, seen as a soap-bubble-like distribution of galaxies in the universe. The accretion flow to enhance the matter density in the resulting sheets and filaments is continuing, and causes shock waves to exist all around us. In such shock waves, which also have been shown to form around growing clusters of galaxies, particles can be accelerated and can attain fairly high energies. However, the maximum energies can barely reach the energy of the GZK cutoff.

The most conventional and easiest explanation is radio galaxies, of which provide the hot spots an obvious acceleration site: These hot spots are giant shock waves, often of a size exceeding that of our entire Galaxy. The shock speeds may approach several percent, maybe even several tens of percent, of the speed of light. Therefore, in this interpretation, these radio galaxy hot spots provide a very straightforward acceleration site. Integrating over all known radio galaxies readily explains the flux and spectrum as well as chemical composition of the cosmic rays in this energy range, and quite possibly also beyond.

VI. PARTICLES BEYOND THE GZK CUTOFF

Since we do not observe the expected GZK cutoff, we need to look for particles which defy the interaction with the microwave background or for a source distribution which reduces the time for interaction with the microwave background substantially.

Here we emphasize those concepts which can explain the events beyond the GZK cutoff; these proposals do not necessarily also explain those particles below the expected cutoff.

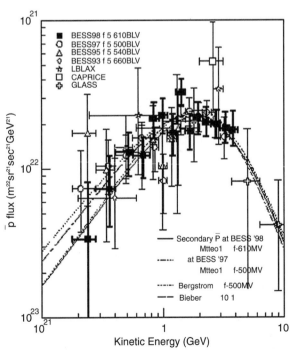

FIGURE 5 The antiproton spectrum. [From Yoshimura, K., *et al.* (2000). "Cosmic-ray antiproton and antinuclei," *Advan. Space Res.*, in press.]

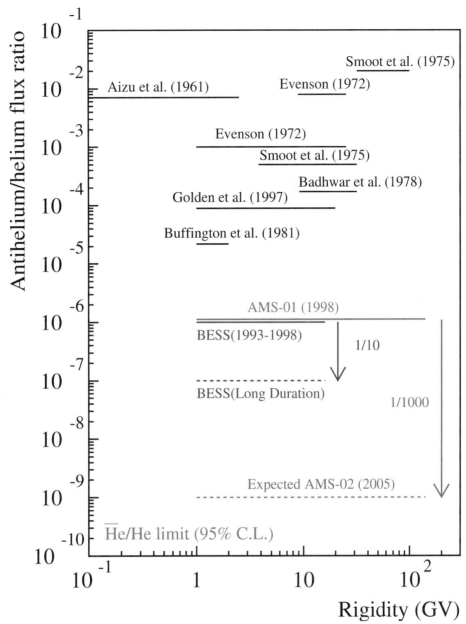

FIGURE 6 The antihelium limits at present. [From Yoshimura, K., *et al.* (2000). "Cosmic-ray antiproton and antinuclei," *Advan. Space Res.*, in press.]

A source distribution which is closely patterned after the actual galaxy distribution in the nearby universe greatly enhances the expected flux of events near the GZK cutoff and may allow an interpretation of the observed events given a properly biased version of the galaxy distribution and a source population like a special class of galaxies.

However, the data are also compatible with a spectrum which suggests a rise beyond the expected GZK cutoff, and this strongly hints at a totally different and new event class. This supports an interpretation as the result of the decay of primordial relics, with an energy of 10^{24} eV. Such a decay produces a large number of highly energetic neutrinos, photons, and only 3% of ultimately protons. Since these protons start with a much flatter spectrum than normal cosmic rays, and also have such extreme initial energies, the lack of a GZK cuotff can be understood. There is, however, a strong prediction: The large number of photons produced yields a cascade that gives rise to a strong contribution to the gamma ray background. This gamma ray background adds to all the gamma rays produced by active galactic nuclei. It is not finally resolved whether observations exclude this option because of an excessive gamma ray background or whether the actual spectrum of the gamma ray background in fact supports it.

There is an interesting variant of this idea, and that is to consider neutrinos which come from large distances in the universe and interact with the local relic neutrinos (the population of Big Bang neutrinos predicted by standard Big Bang theory, akin to the microwave background). In such a theory, the low density of relic neutrinos may be a serious problem.

Obviously, if we could understand the sky distribution of events at these extreme energies, then the option of using pulsars would become very attractive. For particles with a single charge as protons, the anisotropies in arrival directions would be severe, but for iron nuclei this constraint would be much alleviated. Therefore, the option of considering pulsars entails an interpretation as iron particles, inconsistent at the transition from galactic to extragalactic cosmic rays at about 3 EeV, but conceivable at a possible transition from one source population to another at the expected GZK cutoff.

On the other hand, these particles could be something very different, as occasionally speculated, particles suggested by some versions of an extension of particle physics that may not interact with the microwave background and yet interact in the atmosphere very similarly to protons. In this case, we may ask what sources produce so extreme particles. The class of compact radio galaxies has a powerful jet, and hot spots that live currently inside a very large amount of local interstellar gas, and so provide both accelerator and beam dump, i.e., make a particle physics experiment in the sky. However, whether anything detectable comes down to us is open to question.

Finally, it has been shown that radio galaxy hot spots can in fact accelerate protons to the required energies. Here the difficulty is to identify the single most important source for the events at the highest energies. One idea, originally suggested around 1960, has been to consider the nearby radio galaxy M87. It has several advantages, and also disadvantages: First, the main problem with this notion is that the events do not show any clustering in direction with the direction to M87, which is in the nearby Virgo cluster, and so magnetic scattering or bending is required. On the other hand, M87 clearly can accelerate protons to the required energies, and in fact one needs such protons to explain its nonthermal spectrum. Intergalactic magnetic fields and also the fields in our galactic halo can bend and maybe even scatter the orbits of very energetic charged particles. These intergalactic magnetic fields clearly play the key role here, and have not been fully understood.

In conclusion, the origin of the events beyond the expected GZK cutoff remains a unsolved problem in modern high-energy physics.

VII. OUTLOOK

The origin of cosmic rays with observed particle energies to 300 EeV remains an unsolved problem.

A number of efforts related to cosmic rays will surely help our understanding:

- The determination of gamma ray spectra and maps of the Galaxy at high energy.
- More information on stellar evolution including rotation and magnetic fields. Are massive stars all converging to a small set of final states, which includes the magnetic field?
- The search to determine the strength and structure of cosmic magnetic fields, both in the halo of galaxies and in the web of the cosmological galaxy distribution. Clearly this is the key to understand the propagation of high-energy cosmic rays.
- The search for possible correlations in arrival directions of high-energy cosmic rays with astronomical sources. If there were such a subset of events, it would provide very strong constraints on the nature of the particles, as already is the case with some events near EeV energies. At higher energies it might provide new constraints on models for the fundamental properties of particles.

Progress in our understanding of cosmic rays will be mainly determined by much better data:

- Very accurate spectra, at low energies, such as now available from the first AMS data.
- Very accurate data on antimatter, positrons, antiprotons, and antinuclei, also from AMS and extended campaigns of balloon flights.
- Very accurate data on isotopic ratios, from a new generation of balloon flights.
- Accurate chemical composition near the knee and beyond.
- Accurate spectra, sky distribution, and information on the nature of particles in the three energy ranges from the knee to the ankle, through the energy range from 3 EeV to 50 EeV, the expected GZK cutoff, and beyond.

The search for the origin of cosmic rays promises to remain at the focus of research in physics in the 21st century.

ACKNOWLEDGMENTS

The authors would like to thank Ray Protheroe and Todor Stanev for very helpful comments. We thank Sera Markoff and Ramin Sina for help with the graphs.

SEE ALSO THE FOLLOWING ARTICLES

COSMIC INFLATION • COSMOLOGY • GAMMA-RAY AS-TRONOMY • INFRARED ASTRONOMY • NEUTRINO AS-TRONOMY • PARTICLE PHYSICS, ELEMENTARY • RADI-ATION PHYSICS • SOLAR PHYSICS • SUPERNOVAE

BIBLIOGRAPHY

Bhattacharjee, P., and Sigl, G. (2000). "Origin and propagation of extremely high-energy cosmic rays," *Phys. Rep.* **327,** 109–247.

Biermann, P. L. (1997a). "The origin of the highest energy cosmic rays," *J. Phys. G* **23,** 1–27.

Biermann, P. L. (1997b). "Supernova blast waves and pre-supernova winds: Their cosmic ray contribution," *In* "Cosmic Winds and the Heliosphere," (J. R. Jokipii *et al.*, eds.), pp. 887–957, University of Arizona Press, Tucson, AZ.

Clay, R., and Dawson, B. (1998). "Cosmic Bullets," Addison-Wesley, Reading, MA.

Diehl, R., Kallenbach, R., Parizot, E., and von Steiger, R. (eds.) (2001). "The Astrophysics of Galactic Cosmic Rays," Kluwer, Dordrecht.

Kronberg, P. P. (2000). "Magnetic fields in the extragalactic universe, and scenarios since recombination for their origin—A review," *In* "Proceedings the Origins of Galactic Magnetic Fields," Manchester, England.

Longair, M. S. (1992/1994). "High Energy Astrophysics," Vols. 1 and 2; 2nd ed., Cambridge Univesity Press, Cambridge.

Nagano, M., and Watson, A. A. (2000). "Observations and implications of the ultra-high energy cosmic rays," *Rev. Mod. Phys.* **72,** 689–732.

Piran, T. (1999). "Gamma-ray bursts and the fireball model," *Phys. Rep.* **314,** 575.

Wiebel-Sooth, B., and Biermann, P. L. (1999). "Cosmic rays," *In* "Landolt-Börnstein," Vol. VI/3c, pp. 37–90, Springer-Verlag, Berlin.

Cosmology

John J. Dykla

Loyola University Chicago and Theoretical Astrophysics Group, Fermilab

GLOSSARY

Background radiation Field quanta, either photons or neutrinos, presumably created in an early high-temperature phase of the universe. The photons last interacted strongly with matter before the temperature fell below 3000 K and electrons combined with nuclei to form neutral, transparent hydrogen and helium atoms. Although shifted to the radio spectrum by the expansion of the universe, this "light" has been moving independently since this decoupling (in the standard model, ~700,000 years after the "big bang"), providing direct evidence of how different the universe was at this early time from its appearance now. The measured homogeneity and isotropy of the photon background are very difficult to account for in models, such as the "steady-state" theories, which avoid a "big bang."

Black hole Region of space-time with so large a curvature (gravitational field) that, classically, it prevents any matter or radiation from leaving. The "one-way" surface, which allows passage into the black hole but not out, is called an *event horizon*. Black holes are completely characterized by their mass, charge, and angular momentum.

Cosmological principle The assumption that at each moment of proper time an observer at any place in the universe would find the large-scale structure of surrounding matter and space the same as any other observer (homogeneity) and appearing the same in all directions (isotropy). Combined with the general theory of relativity, this forms the basis of the standard model in cosmology. Observational evidence for the expansion of the universe, combined with the principle of conservation of mass–energy, indicates that the large-scale average density, though everywhere the same at a given time, is decreasing as the universe evolves. The perfect cosmological principle seeks to avoid a "big bang" by asserting that the homogeneous and isotropic density of matter at all points in space is also unchanging in time. Such an assumption could be reconciled with an expanding universe only by postulating the continuous creation of matter and appears untenable in light of the microwave background radiation.

Dark matter problem The discrepancy by approximately an order of magnitude between the ordinary matter that is directly visible through astronomical observations and the mass inferred from dynamic theory applied to the observed motions of galaxies and clusters of galaxies. The calculated average density is roughly the critical value for a flat universe, which is also strongly favored by "inflationary" particle physics scenarios. Thus, various exotic forms of astronomical objects and elementary particles have been proposed to account for the dark matter assumed present.

Doppler shift Change in frequency (or, inversely because of invariant speed of propagation, wavelength) of radiation due to the relative motion of source and observer. For motion purely along the line of sight, recession implies increased wavelength (redshift) and approach implies decreased wavelength (blueshift), complicated by nonlinear relativistic effects when the relative velocity of source and observer is an appreciable fraction of the velocity of light. Purely transverse motion is associated with a relativistic and intrinsically nonlinear redshift.

Event Point in space at a moment of time. The generalization, in the four-dimensional space–time continuum of relativity theory, of the concept of a point in a space as a geometric object of zero "size."

General theory of relativity Understanding of gravitation invented by Albert Einstein as the curvature of space-time, described mathematically as a four-dimensional geometric manifold. His field equations postulate the way in which the distribution of mass–energy tells space–time how to curve and imply the equations by which curved space–time tells mass–energy how to move. With the cosmological principle, it forms the basis of the standard model in cosmology.

Nebula Latin for "cloud," this old term in observational astronomy refers to a confusing variety of extended objects easily distinguished from the pointlike stars. Some are now known to be dust or gas clouds associated with the births or deaths of stars in the Milky Way, but others are galaxies far outside the Milky Way.

Perfect fluid Idealization of matter as continuous and completely characterized by three functions of space–time: mass–energy density, pressure, and temperature. At each event, these functions are related through a thermodynamic equation of state.

Proper time Measure of time by a standard clock in the reference frame of a hypothetical observer. In describing periods in the evolution of the entire universe, the reference frame is chosen as one that at each event is comoving with the local expansion. The proper time since the "big bang" singularity is called the *age of the universe*.

Singularity Event or set of events at which some physical quantities that are generally measurable, such as density and curvature (gravitational field), are calculated to have infinite values. It represents a limit to the validity of a field theory.

COSMOLOGY as a scientific endeavor is the attempt to construct a comprehensive model of the principal features of material composition, geometric structure, and temporal evolution of the entire physically observable universe. The primary tools of the modern cosmologist are those of observational astronomy and theoretical physics. Signals are gathered over a very broad range of wavelengths of the electromagnetic spectrum, from radio waves collected by the 300-m-diameter dish at Arecibo, Puerto Rico, through visible quanta recorded by charge-coupled devices attached to the twin 10-m Keck telescopes atop Mauna Kea, Hawaii, to γ rays detected by instruments orbiting above the earth's atmosphere. In recent years, experiments in particle physics have become increasingly relevant to cosmological questions. Examples include measurements of the solar neutrino flux which challenge our understanding of energy generation in the stars and attempts to detect the superpartners, axions, strings, and other exotic objects predicted by various models in high-energy physics and possible solutions to the missing-mass problem.

The standard theoretical model in cosmology for the past several decades has been based on Einstein's general theory of relativity, supplemented by assumptions about the homogeneity and isotropy of space–time, and data from spectroscopy, consistent with understanding gained from nuclear physics, about the distribution of ordinary matter among various species. Recently, progress in understanding the early universe has been made by fusing this theory with the standard $SU(3)_C \times SU(2)_L \times U(1)_Y$ model in particle physics, which seems to describe very accurately the interactions of quarks and leptons at energies up to $\sim 2 \times 10^3$ GeV, the current limit of terrestrial accelerators. Traditional analytical techniques of the theorist are now often supplemented by the raw-number-crunching power of increasingly rapid and sophisticated computers, especially in applications to otherwise intractable calculations involving real or speculative space–times or quantum fields. Further synthesis is clearly dependent on continued advances in observational astronomy, high-energy experimentation, and theoretical physics.

Research in cosmology is subject to one obvious but fundamental limitation that no other field of scientific inquiry shares. By definition, there is only one entire physical reality that we can observe and attempt to understand. Since we cannot acquire data from outside all this, it is impossible in principle to perform a controlled experiment to test a truly comprehensive theory of the cosmos. The best we can hope for are ever more refined models, constrained by data from increasingly diverse observational sources, which become more broadly successful as we continue to measure and to think.

I. ISSUES FROM PRESCIENTIFIC INQUIRIES

A. Philosophical Speculations and Myths

Cosmological thought recorded in diverse cultures during past millennia displays a continued fascination with a relatively small number of fundamental questions about the physical universe and our place within it. Many attempts to provide satisfying answers have been made, It is not appropriate to present here a comparative survey of the myriad world views that have been proposed, but there is value in understanding those issues raised by early thought that are still at the core of modern scientific cosmology.

Logically, the first issue is that of order versus chaos. Indeed, the word *cosmology* implies an affirmative answer to the question of whether the totality of physical experience can be comprehended in a meaningful pattern. We should be aware, however, that adherents of the view that nature is fundamentally inscrutable or capricious have presented their arguments in many cultures throughout recorded history.

Is matter everywhere in the universe of the same type as that which is familiar to us on earth, or are there some distinctively "celestial" substances? Speculations favoring one view or the other have been stated with assurance as long as humans have engaged in intellectual debate.

Is the space of the universe finite or infinite, bounded or unbounded? To this, also, pure thought unfettered by empirical evidence has, in many places and times, offered dogmatic pronouncements on either side. The logical independence between the issues of finiteness and boundedness was not fully clarified until the invention of non-Euclidean geometrics in the nineteenth century. Thus, we now see that the opening question of this paragraph actually comprises two distinct but related questions. Of the issues discussed in this section, this is perhaps the only one in which progress in pure thought has raised questions not addressed by ancient philosophers and myth makers.

Is the universe immutable in its total structure, or does it evolve with the passage of cosmic time? If the latter, did it have a creation in time, or has it always existed? Similarly, will a changing cosmos continue forever, or will there be an end to time that we can experience? Such questions have sometimes been distinguished from the structural questions of cosmology and said to be of a different order called *cosmogony*. We shall see that in scientific cosmology such a separation is artificial and unproductive. Indeed, all the questions posed in this section are, in all current scientific models, inextricably interrelated.

B. Value of Some Early Questions

All modern science is predicated on the philosophical assumption that its subject is comprehensible. Although even Albert Einstein remarked that this is the most incomprehensible thing about the universe, it is difficult to see how this can be a question for scientific debate. The presumption of discoverable regularities, not meaningless chaos, is a necessary underpinning of any scientific endeavor. Although twentieth-century quantum mechanics has compelled reappraisal of earlier determinism, the work of physicists is founded upon belief in objective laws correlating observations of natural phenomena and confirmed by successful predictions.

Philosophical speculations in earlier ages about the substance of the cosmos often assumed an insurmountable qualitative distinction between the "stuff" of terrestrial experience and the matter of the celestial spheres. The beginnings of modern science included an explicit rejection of this view in favor of one in which all the matter of the universe is of fundamentally the same type, subject to laws that hold in all places and at all times. It is interesting that some recently proposed solutions to the dark matter problem in astrophysics suggest that the vast majority of the matter of the universe may be in one or more exotic forms that are predicted by elementary particle theory but for which we do not yet have experimental evidence. This is not a return to prescientific notions. It is still assumed that, no matter what the substance of the universe, all physical reality is subject to one set of laws. Most important, it is assumed that such laws can be discovered with the aid of data obtained from terrestrial experiments as well as astronomical observations.

Although most early cosmologists imagined a finite and bounded universe, some thinkers (e.g., the Greek philosopher Democritus in the fifth century B.C.) argued eloquently for an infinite and unbounded expanse of space. Until the latter part of the nineteenth century, however, a finite space was presumed synonymous with a bounded one. The non-Euclidean geometries of Riemann, Gauss, and Lobachevsky first presented the separation of the questions of extension and boundedness in the form of logically possible examples amenable to mathematical study. Originally, generalizations of the theorem of Pythagoras to curved spaces assumed that the square of the distance between two distinct points is greater than zero, described by a positive definite metric tensor. In relativity theory, time is treated as an additional "geometric" dimension, distinguished by the fact that the square of the interval between two events may be a quantity of either sign (depending on whether there is a frame of reference in which they are at the same place at different times or a frame of reference in which they are at different places at the same time) or zero (if they can be related only by the passage of a signal at the speed of light). Gravitation is understood to be the relation between the curvature of space–time and the mass–energy contained and moving within it. Modern research on the asymptotic structure of space–times is based on the analysis of four-dimensional geometries of indefinite metric within the framework of relativistic gravitational theory. All viable models consistent with minimal assumptions about our lack of privileged position describe spaces that are without a boundary surface at each moment of time, although some are finite in volume.

Myths of creation, whether based on purely philosophical speculations or carrying the authority or religious conviction, are commonplace in almost all cultures of which we have any knowledge. Some suppose the birth of a static universe in a past instant, but most describe a process of formation or evolution of the cosmos into its present appearance. In many eras, some individuals or groups have championed an eternal universe. Most of these models of the universe are static or cyclic, but occasionally they represent nature with infinite change, never repeating but with no beginning. In earlier times, a beginning in time was almost always taken to imply finiteness in a bounded space. The new geometries mentioned earlier have made possible serious consideration of a universe that began at a definite moment in the past, is not bounded in space, and may be finite or infinite in extent. The question of whether there will be an end to time or whether there will be eternal evolution is intimately related to the question of finite or infinite spatial extent in these models. At present, the empirical evidence about whether time will end is inconclusive. The "inflationary scenarios" that are currently in vogue favor a universe remarkably close to the marginally infinite and thus eternally unwinding "asymptotically flat" model, but allow the possibility of a miniscule difference from flatness of either sign, resulting in ultimate closure or eternal expansion. Although the observed density of visible matter appears much too small to halt the present expansion, there are numerous hypothetical candidates for the dark matter.

II. ORIGINS OF MODERN COSMOLOGY IN SCIENCE

A. Copernican Solar System

Centuries before the Christian era, several Greek philosophers had proposed heliocentric models to coordinate the observed motions of the sun, moon, and planets in relation to the "fixed" stars. However, the computational method of Alexandrian astronomer Claudius Ptolemy which he developed during the second century A.D. using a structure of equants and epicycles based on the earth as the only immovable point also made successful predictions of apparent motions. Although additional epicycles had to be added from time to time to accommodate the increasingly precise observations of later generations, this system became dominant in the Western world. Its assumption of a dichotomy between terrestrial experience and the laws of the "celestial spheres" stifled fundamental progress until the middle of the sixteenth century. In 1543, the Polish cleric Mikolaj Kopernik, better known by the Latin form of his name, Nicholas Copernicus, cleared the way for modern astronomy with the publication of *De Revolutionibus Orbium Coelestium* (*Concerning the Revolutions of the Celestial Worlds*). Appearing in his last year, this work summarized the observations of a lifetime and presented logical arguments for the simplicity to be gained by an analysis based on the sun as a fixed point.

More important, this successful return to the heliocentric view encouraged the attitude that a universal set of laws governs the earth and the sky. In the generations following Copernicus, the formulation of physical laws and their empirical testing were undertaken in a manner unprecedented in history: Modern science had begun. The Dane Tycho Brahe observed planetary orbits to an angular precision of 1 arc min. The German Johannes Kepler used Brahe's voluminous and accurate data to formulate three simple but fundamental laws of planetary motion. The Italian Galileo Galilei performed pioneering experiments in mechanics and introduced the use of optical telescopes in astronomy. From these beginnings, a scientific approach to the cosmological questions that humanity had been asking throughout its history developed.

B. Newtonian Dynamics and Gravitation

The first great conceptual synthesis in modern science was the creation of a system of mechanics and a law of gravitation by the English physicist Isaac Newton, published in his *Principia* (*Mathematical Principles of Natural Philosophy*) of 1687. His "system of the world" was based on a universal attraction between any two point objects described by a force on each, along the line joining them, directly proportional to the product of their masses and inversely proportional to the square of the distance between them. He also presented the differential and integral calculus, mathematical tools that became indispensable to theoretical science.

On this foundation, Newton derived and generalized the laws of Kepler, showing that orbits could be conic sections other than ellipses. Nonperiodic comets are well-known examples of objects with parabolic or hyperbolic orbits. The constant in Kepler's third law, relating the squares of orbital periods to the cubes of semimajor axes, was found to depend on the sum of the masses of the bodies attracting one another. This was basic to determining the masses of numerous stars, in units of the mass of the sun, through observations of binaries. As Cavendish balance experiments provided an independent numerical value for the constant in the law of gravitation, stellar masses in kilograms could be calculated. Later, analysis of observed perturbations in planetary motions led to the prediction of previously unseen planets in our solar system. Studies of the stability of three bodies moving under their mutual gravitational influence led to the discovery of two clusters of asteroids sharing the orbit of Jupiter around the sun. These successes fostered confidence in the view of one boundless Euclidean space, the preferred inertial frame of reference, as the best arena for the description of all physical activity. Lacking direct evidence to the contrary, theoretical cosmologists at first assumed that the space of the universe was filled, on the largest scales, with matter distributed uniformly and of unchanging density. Analysis soon disclosed that Newtonian mechanics implied instability to gravitational collapse into clumps for an initially homogeneous and static universe. This stimulated the observational quest for knowledge of the present structure of the cosmos outside the solar system.

C. Technology of Observational Astronomy

Galileo's first telescopes had revealed many previously unseen stars, particularly in the Milky Way, where they could resolve numerous individual images. Based on elementary observations and careful reasoning, the ancient concept of a sphere of fixed stars centered on the earth had been supplanted by a potentially infinite universe with the earth at a position of no particular distinction. Further progress in understanding the distribution of matter in space was dependent on actual measurements of the distances to the stars. The concept of stellar parallax, variation in the apparent angular positions of nearby stars in relation to more distant stars as seen from earth at various points in its orbit around the sun, became useful when the observational precision of stellar location surpassed the level of 1 arc sec. An important contribution was the development of the micrometer, first used by William Gascoigne, later independently by Christiaan Huygens, and systematically appearing in the telescope sights of Jean Picard and others after the latter part of the seventeenth century. The first parallax measurements were reported by Bessel in Germany, Henderson in South Africa, and Struve in Russia, in rapid succession in 1838 and 1839. By 1890, distances were known for nearly 100 stars in the immediate neighborhood of the sun, and the limits of the trigonometric parallax approach (~100 parsec, based on limits to the

useful size of an optical telescope looking through the earth's atmosphere) were being approached.

As long as sufficient light can be gathered from a star to be dispersed for the study of its spectrum, much can be learned about it even if one is not certain how far away it is. Spectroscopic studies of stars other than the sun were first undertaken by William Huggins and Angelo Secchi in the early 1860s. In 1868, Huggins announced the measurement of a Doppler redshift in the lines of Sirius indicating a recession velocity of 47 km/sec, and Secchi published a catalog of 4000 stars divided into four classes according to the appearance of their spectra. The introduction of the objective prism by E. C. Pickering in 1885 tremendously increased the rate at which spectra could be obtained. The publication of the Draper catalogs in the 1890s, based on the spectral classification developed by Pickering and Antonia Maury, provided the data that led to the rise of stellar astrophysics in the twentieth century.

D. Viewing the Island Universe

Some hazy patches among the fixed stars, such as the Andromeda nebula or the Magellanic Clouds, are clearly visible to the unaided human eye. When Galileo observed the band called the Milky Way through his telescope and was surprised to find it resolved into a huge number of faint stars, he concluded that most nebulosities were composed of stars. Early in the eighteenth century, the Swedish philosopher Emanuel Swedenborg described the Milky Way as a rotating spherical assembly of stars and suggested that the universe was filled with such spheres. The English mathematician and instrument maker Thomas Wright also considered the Milky Way to be one among many but supposed its shape to be a vast disk containing concentric rings of stars. By 1855, Immanuel Kant had further developed the disk model of the Milky Way by applying Newtonian mechanics, explaining its shape through rotation. He assumed that the nebulae are similar "island universes."

Early in the nineteenth century, however, William Herschel discovered the planetary nebulae, stars in association with true nebulosities. This reinforced a possible interpretation of nebulae as planetary systems in the making, in agreement with the theory of the origin of the solar system developed by Pierre Simon de Laplace. John Herschel's observations of 2500 additional nebulae, published in 1847, emphasized that they were mostly distributed away from the galactic plane. The "zone of avoidance" was used by some to argue for the physical association of the nebulae with the Milky Way. (It is now known that dust and gas in the plane of the galaxy diminish the intensity of light passing through it, whether from sources inside it or beyond.) In 1864, Huggins studied the Orion nebula and found it to display a bright line spectrum similar to that of a hot gas. Also, photographs of Orion and the Crab nebula did not show resolution into individual stars. As the twentieth century began, the question of the nebulae was intricately linked with that of the structure of the Milky Way. The size of nebulae was still quite uncertain in the absence of any reliable distance indicators, but most astronomers believed the evidence favored considering the nebulae part of the Milky Way, thus reducing the universe to this one island of stars.

III. REVOLUTIONS OF THE TWENTIETH CENTURY

A. New Theoretical Models

The two great conceptual advances of theoretical physics in the twentieth century, relativity and quantum mechanics, have had profound implications for cosmology. The general theory of relativity provides the basis for the evolving space–time of the now standard model, and fundamental particle identities and interactions are keys to understanding the composition of matter in the universe.

In 1905, Albert Einstein established the foundations of the special theory of relativity, which connects measurements of space and time for observers in all possible inertial frames of reference. He devoted the next decade to developing a natural way of including observers in accelerated frames of reference, using as a fundamental principle the fact that all test masses undergo the same acceleration in a given gravitational field. The result was Einstein's theory of space–time and gravitation, the general theory of relativity (GTR), completed in 1916. It was immediately apparent that the GTR would have a substantial impact on cosmological questions. Although many model space–times have been studied in the GTR, in 1922 Alexander Friedmann showed that the assumptions of spatial homogeneity and isotropy can be embodied in only three. They are presented here in the modern notation of the Robertson–Walker metric with the choice of units commonly used in theoretical research. The intimate relationship of space and time in relativistic theories is recognized through units of measure in which the speed of light is numerically 1, eliminating the letter c representing its value in units such as those of SI. We may think of the units as geometrized (e.g., time measured in meters) or chronometrized (e.g., length measured in seconds), where $1 \sec = 299, 792, 458$ m (exactly, by definition).

The invariant element of separation ds between two neighboring events in a homogeneous and isotropic universe can be expressed by:

$$ds^2 = -dt^2 + R^2[dr^2/(1-kr^2) + r^2(d\theta^2 + \sin^2\theta \, d\phi^2)]$$

$$(1)$$

where $R(t)$ is the scale factor that evolves with cosmic time and k is the curvature constant, which may be positive, negative, or zero. The invariant separation depends on k only through the pure number kr^2, where r is a comoving space coordinate, for which matter is locally at rest. Without loss in generality, we may choose k (and hence also r) dimensionless and set the scale of r so that k is $+1$, -1, or 0. Although R carries the units of spatial length, it is not possible to interpret it as a "radius of the universe" if k is not positive, since the volume of a model with $k = 0$ or $k = -1$ is infinite.

To ensure that the physics is described in a manner independent of arbitrary choices, it is useful to introduce the scale change rate, also known as the Hubble parameter, defined as $H = (dR/dt)/R$. Since H has dimensions of velocity divided by length, $1/H$ is a characteristic time for evolution of the model. For an expanding empty universe, $1/H$ would be the time since the beginning of the expansion. Assuming that the distribution of matter on a large scale can be described in terms of a perfect fluid of total density ρ and pressure p, the coupled evolution of space-time and matter in the GTR are determined by the Friedmann equations,

$$H^2 = 8\pi G\rho/3 - k/R^2 \qquad \text{and} \qquad d(\rho R^3) = -p \, d(R^3)$$

$$(2)$$

where G is the Newtonian gravitational constant. (Note that the term *density* can be taken as either mass per volume or energy per volume in units where $c = 1$.)

Although the GTR implies that a homogeneous and isotropic space generally expands or contracts, at the time the theory was formulated the common presumption held that the real universe was static. This led Einstein to modify his original simple theory by introducing a "cosmological term" into the field equations. The "cosmological constant" in this term was chosen to ensure a stable static solution. When the evidence for an expanding universe became apparent in the next decade, Einstein dropped consideration of this additional term, calling its introduction the biggest blunder of his career. We shall leave it out of the presentation in this section but later remark on recent motivations for a possible reinstatement of the cosmological constant with a value other than that which makes the universe static. Thus, there are only three possible homogeneous and isotropic universes in the GTR, the Friedmann models. In view of Eq. (2), which determines the evolution of the Hubble parameter, the curvature constant can be positive only if the density exceeds the value $\rho_{\text{crit}} = 3H^2/8\pi G$. Thus, an expanding universe in which $k = +1$ must halt its growth and begin to collapse before its mean density drops below this critical value. Such a space-time is finite in volume at each moment but has no boundary. While this model is expanding, the Robertson–Walker coordinate time since the expansion began is less than $2/3H$, depending on how much the density exceeds the critical value. The classical GTR implies that a finite proper time passes between its beginning in a singularity of zero volume and its return to such a singularity. (However, Charles Misner has argued that a logarithmic time scale based on the volume of the universe is more appropriate as a measure of possible change that may occur, giving even this closed cosmology a potentially "infinite" future and past.) If the density is less than the critical value, space–time has negative curvature and will have a positive velocity of expansion into the infinite future of proper time, when H approaches zero. The time since expansion began is greater than $2/3H$, depending on how much the density is less than the critical value, but never exceeds $1/H$. Although such an open universe still had a beginning in proper time at a singularity, its volume is always infinite and it contains an infinite mass–energy. If the density precisely equals the critical value, then the curvature vanishes. Such a flat universe has been expanding for a time $2/3H$, it has infinite mass–energy in an infinite volume described by Euclidean geometry at each moment, and its expansion velocity as well as Hubble parameter will approach zero asymptotically as proper time in comoving coordinates continues toward the infinite future. Which of these three models best describes the universe we inhabit is an empirical as well as theoretical question that remains central to current research in cosmology.

B. New Observational Evidence

Beginning in the late 1920s, Vesto Slipher, Edwin Hubble, and Milton Humason used the Hooker telescope at Mt. Wilson, California, then the largest optical instrument in the world, to measure Doppler shifts in the spectra of nebulae that they realized were outside the Milky Way. Convincing reports of a relationship between recession velocities v deduced from Doppler redshifts and estimates of the distances d to these galaxies were published in 1929 and 1931. Hubble pointed out that the velocities, in kilometers per second, were directly proportional to the distances, in megaparsecs (1 Mpc $= 3.08 \times 10^{19}$ km). The observational relation is "smooth" only when the distances considered are at least of the order of a few megaparsecs, allowing the averaging of the distribution of galaxies to produce an approximately homogeneous and isotropic density. Since the ratio v/d is clearly the current value of the scale change rate H in a Friedmann universe, this was the first observational evidence in support of GTR cosmology.

The parameter H was at first called the Hubble constant but is now recognized as a misnomer, because the early data did not extend far enough into space to correspond to looking back over a sustantial fraction of the time since the expansion began. Hubble's initial value of H was so large that the time scale for expansion derived from it was substantially less than geological estimates of the age of the earth, $\sim 4.7 \times 10^9$ years. Subsequently, the numerical value of the Hubble parameter has undergone several revisions due to reevaluations of the cosmic distance scale, yielding a universe older than originally supposed and thus a much larger volume from which signals at the speed of light can be observed at our position. Allowing for present uncertainties, H is now believed to be between 50 and 80 (km/sec)/Mpc. Thus, the time scale of the cosmic expansion, $1/H$, is between approximately 1.2 and 2×10^{10} years.

When early large values of the Hubble parameter were in conflict with geological and astrophysical estimates of the age of the solar system, it was suggested that the simple GTR cosmologies might not be viable. One solution was to reinstate the cosmological constant, not to halt the universal expansion but merely to slow it down in a "coasting" phase. Recently, some observations of supernovae suggest an extragalactic distance scale with H greater than 70 (km/sec)/Mpc. If we accept this and also believe that stellar evolutionary theory is sufficiently established to yield precise ages of globular cluster stars in excess of 1.3×10^{10} years, then invoking a nonzero cosmological constant would be a way of avoiding the unacceptable conclusion that the universe contains stars older than its expansion time. Models studied by George Lemaitre and others after 1927 often had unusual features, such as a closed space of positive curvature that could continue expanding for an infinite proper time. A more radical idea, which had other motivations as well, was to abandon conservation of energy in favor of a "steady-state" cosmology based on the continuous creation of matter throughout an infinite past. In such a theory, introduced by Hermann Bondi, Thomas Gold, and Fred Hoyle in 1948, there is no initial singularity, or "big bang," to mark the beginning of the expansion. In 1956, George Gamow predicted that a residual electromagnetic radiation at a temperature of only a few kelvins should fill "empty" space as a relic of the high temperatures at which primordial nucleosynthesis occurred soon after the initial singularity in a GTR cosmology. In 1965, engineers Arno Penzias and Robert Wilson detected a microwave background coming from all directions in space into a communications antenna they had designed and built for Bell Laboratories in Holmdel, New Jersey. Cosmologists Robert Dicke and P. J. E. Peebles at nearby Princeton quickly explained the significance of this 2.7 K blackbody spectrum to them and thus dealt a severe blow to the viability of any cosmological model that avoids a "hot big bang."

A direct determination of whether the universe is spatially closed or open would necessitate precisely measuring density over volumes of many cubic megaparsecs, detecting the sign of departures from Euclidean geometry in an accurate galactic census over distances exceeding several tens of megaparsecs, or finding the change in the Hubble parameter associated with looking back toward the beginning over distances exceeding several hundred megaparsecs. Unfortunately, these conceptually simple observations appear to be somewhat beyond the scope of our present technology. However, there are various indirect ways to estimate, or at least place bounds on, the present mean density of the cosmos. To eliminate the influence of uncertainties in the Hubble parameter on the precise value of the critical density, it is now common to describe the resulting estimates or bounds in terms of the dimensionless quantity $\Omega = \rho/\rho_{crit}$. Clearly, $\Omega > 1$ corresponds to a closed, finite universe and $\Omega \leq 1$ corresponds to an infinite universe. On the basis of the amount of ordinary visible matter observed as stars and clouds in galaxies, we conclude that $\Omega > 0.01$. Requiring the present density to be low enough so that the age of the universe (which depends inversely on H and, through a monotonically decreasing function, on Ω) is at least as great as that of the oldest stars observed, estimated to be 10^{10} years, yields an upper bound $\Omega < 3.2$. Of the infinite range of conceivable values, it is quite remarkable that the universe can so easily be shown to be nearly "flat." Further evidence and theoretical insight suggest that near coincidence (within one or two powers of 10) of the density and the critical density is not an accident.

The most severe constraints on the contribution of ordinary matter to Ω presently come from demanding that primordial synthesis of nuclei during the "hot big bang" of a standard Friedmann model produces abundance ratios in agreement with those deduced from observation. In nuclear astrophysics, it is customary to specify temperatures in energy units which corresponds to setting the Boltzmann constant equal to 1. Thus, the relation between the megaelectronvolt of energy and the kelvin of absolute temperature is

$$1 \text{ MeV} = 1.1605 \times 10^{10} \text{ K}$$

Primordial nucleosynthesis models are based on the assumption that the temperature of the universe was once higher than 10 MeV, so that complex nuclei initially could not exist as stable structures but were formed in, and survived from, a brief interval as the universe expanded and cooled to temperatures of less than 0.1 MeV, below which nucleosynthesis does not occur. Computing all relevant nuclear reactions throughout this temperature range to determine the final products is a formidable undertaking. Of the various programs written since Gamow suggested the idea of primordial nucleosynthesis in 1946, the one published by Robert Wagoner in 1973 has become the accepted standard. With updates of reaction rates by several groups since then, the numerical accuracy of the predicted abundances is now believed to be $\sim$1%. Since the weakly bound deuteron is difficult to produce in stars and easily destroyed there, its abundance, 1×10^{-5} relative to protons as determined in solar system studies and from ultraviolet absorption measurements in the local interstellar medium, is generally accepted as providing a lower bound to its primordial abundance and hence an upper bound of 0.19 to the contribution of baryons to Ω. Analogous arguments may be applied to establishing a concordance between predicted and observed abundances of ^{3}He, ^{4}He, and ^{7}Li, the only other isotopes calculated to be produced in significant amounts during this primordial epoch. The resulting constraint on the contribution of baryons to the critical density, $0.010 \leq \Omega_B \leq 0.080$, shows clearly that baryons alone cannot close the universe. Of course, this does not eliminate the possibility that the universe may have positive curvature due to the presence of less conventional forms of as yet unseen matter.

IV. UNSOLVED CLASSICAL PROBLEMS

A. Finite or Infinite Space

Although concordance between the standard model and observed nuclear abundances limits baryon density to well below the critical value needed for closure of a Friedmann universe, the question of finite or infinite space remains observationally undecided owing to other complications. In principle, it should be possible to determine the curvature constant k by direct measurement of the deviation from Hubble's simple linear relationship between velocity and distance. Unfortunately, substantial deviation is not expected, until sources at distances of the order of $1/H$ are studied, and galaxies are too faint to have their spectra measured adequately at such distances by present technology. Furthermore, observing galaxies at such distances implies seeing them at earlier times, and estimates of their distance could be subject to systematic errors due to unknown evolution of galactic luminosity.

Since the discovery of quasars by Maarten Schmidt at the Palomar Observatory in 1963, cosmologists have hoped that these most distant observed objects could be used to extend the Hubble relationship to the nonlinear regime and decide the sign of the curvature. Quasars are now known with spectral features up to 6.0 times their terrestrial wavelengths, corresponding to Doppler recession velocities up to 96% of the speed of light, placing them at substantial fractions of the distance from us to the horizon of the observable universe. However, uncertainties in estimating very large distances in the universe due to insufficient understanding of the evolution of galaxies, not to mention the structure of quasars, have prevented unambiguous determination of the sign of the curvature. In fact, Hubble's law is still used to estimate distances to the quasars, rather than they being used to determine both distance and redshift and thus test their relationship. The question of whether space is finite or infinite remains unresolved by observation at this time.

B. Eternal Expansion or an End to Time

In the Friedmann cosmologies of the GTR, a finite space implies an end to proper time in the future, but this is not required in some nonstandard models. Assuming the Robertson–Walker form of the metric for a homogeneous and isotropic space–time, it is convenient to discuss the future evolution of any such expanding universe in terms of a dimensionless deceleration parameter, defined as:

$$q = -R(d^2R/dt^2)/(dR/dt)^2$$

Throughout most of its history, the dynamics of the universe have been dominated by matter in which the average energy density is very much greater than the pressure. Neglecting the pressure of nongravitational fields, space will reverse its expansion and collapse in finite proper time if and only if $q > \frac{1}{2}$. From the Friedmann equation for the Hubble parameter, it is easy to show that $q = \Omega/2$ in a space–time described by the GTR with zero cosmological constant. In nonstandard models, the deceleration parameter depends on the density, the cosmological constant, and the Hubble parameter in more complicated ways. For some choices of cosmological constant, it is even possible to have an accelerating universe ($q < 0$) with a positive density. However, a cosmological constant whose magnitude substantially exceeded the critical energy density would produce detectable local effects that are not observed. Thus, we can conclude that a sufficiently large density must imply an end to time.

Applications of the virial theorem of Newtonian mechanics to galactic rotation and the dynamics of clusters indicate that masses often exceed those inferred from visible light by about an order of magnitude. Such results push at the upper bound of the baryon density inferred from nucleosynthesis but are still far short of supporting $q > \frac{1}{2}$. However, if only one-tenth of the ordinary matter in galaxies and clusters may be visible to us, is it not possible that there exists mass–energy, in as yet undetected forms or places, of sufficient quantity to produce deceleration exceeding the critical value? Primordial black holes,

formed before the temperature of the universe had dropped to 10 MeV, would not interfere with nucleosynthesis in the standard model but could substantially increase the value of Ω. (Notice that black holes due to the collapse of stars are made of matter that contributed to Ω_B during the time of primordial nucleosynthesis and thus are limited by the nuclear abundance data.) Calculations of the primordial black hole mass spectrum, such as those by Bernard Carr, have demonstrated that present observational data are insufficient to decide whether the contribution of black holes to the density will reverse expansion. More conventional astronomical candidates for the dark matter including brown dwarfs (examples have been detected in the halo of the Milky Way through gravitational lensing) and dead (radio-quiet) pulsars (almost certain to exist but not as yet detected) are unfortunately constrained by the nucleosynthesis limits on baryon density. Some speculative ideas in high-energy particle physics present other exotic candidates for dark matter that may dominate the gravitational dynamics of the universe as a whole. These include massive neutrinos (although observations of electron neutrinos from SN1987A [Supernova Shelton] constrain the electron neutrino mass to probably less than 10 eV, tau and mu neutrinos could be heavier), the axions required to banish divergences in grand unified theories (GUTs) (as yet undetected), and the supersymmetric partners of all known fermions and bosons (none yet found experimentally). Empirically, whether there will be an end to time remains an unresolved question.

C. Observations and Significance of Large-Scale Structure

The measured homogeneity and isotropy of the cosmic microwave background radiation temperature ($\Delta T/T \sim 10^{-5}$) is strong evidence that the observable universe is rather precisely homogeneous and isotropic on the largest scale ($1/H$ is ~3000 Mpc). However, it is well known that on only slightly smaller scales, up to 120 Mpc, the universe today is very inhomogeneous, consisting of stars, galaxies, and clusters of galaxies. For example, the variation in density divided by the average density of the universe, $\delta\rho/\rho$, is of the order of 10^5 for galaxies. The density of visible matter in large "voids" recently discovered is typically less than average by about one order of magnitude. Since gravitational instability tends to enhance any inhomogeneity as time goes on, the difficulty is not in creating inhomogeneity but rather in deviating from perfect homogeneity at early times in just the way that can account for the structural length scales, mass spectra, and inferred presence of dark matter that are so obvious in the universe today. Opinions about the structure of the universe at early times have run the gamut from Misner's chaotic "mixmaster" to Peebles's quite precisely homogeneous and isotropic space-times. The issue of the origin of structure in the universe on the largest scales remains unresolved, because conflicting scenarios that adequately account for galaxies and clusters can involve so many tunable parameters that it is difficult to distinguish among competing models observationally.

The three-dimensional map of more than 1.1×10^4 galaxies within a sphere of diameter 400 Mpc centered on the Milky Way, created by Huchra and Geller on the basis of more than 5 years' data gathered with an earth-bound telescope comparable in aperture to the Hubble Space Telescope (HST returned its first images May 1990), shows voids and filamentary structures on scales up to about 120 Mpc. They strain "top down" scenarios of the evolution of cosmic structure, since it is difficult to have gravitationally bound structures so large which "later" fragmented to form quasars when the universe was less than 7% of its present age. Advocates of such pictures have questioned the statistical significance of a sample only on the order of 10^{-7} of the galaxies within our horizon. The HST, despite initial difficulties in forming optimally sharp images (corrected in 1993), has gathered spectra of a sample from roughly an order of magnitude deeper into space (implying roughly 10^3 times as many galaxies). Combined with evidence of galaxy formation at early times visible in the Hubble deep fields, this enlarged sample supports a two-component model of structure formation. Hot dark matter provides a gravitational field which fragments on the large scale of galaxy clusters, while cold dark matter pockets act as seeds for the concurrent formation of subunits of galaxies. HST will spend part of its remaining life in orbit further improving statistics to refine the interpretation of an enlarged sample. It might finally gather data that might unravel the contributions of galactic evolution from that of the cosmological deceleration to the redshift-distance relation of the most distant active galaxies and quasars.

D. Viability of Nonstandard Models

The assumption that the universe is homogeneous and isotropic on a sufficiently large scale has been called the cosmological principle. This principle, applied to the Riemannian geometry of space-time using the methods of group theory, leads to the Robertson–Walker form of the invariant separation between events. This mathematically elegant foundation for theoretical cosmology is independent of a particular choice of gravitational field equations. However, the evolution of the scale factor and the relation of the curvature constant to the matter distribution are, of course, intimately related to the structure of the gravitational theory that is assumed.

Previous mention was made of the logical possibility of complicating the equations of the GTR by introducing a cosmological constant. Though hints of a possible conflict between ages of galactic halo stars and the Hubble time might be resolved by a nonzero cosmological constant, there is abundant evidence that it cannot be significantly larger than the critical density. Attempts to derive a value from quantum theories of fundamental interactions yield estimates too large by many orders of magnitude. This embarrassing failure leaves advocates of a cosmological constant only the unpleasant option of arbitrarily adjusting it to a suitable small value. Critics then ask why it should not be chosen to vanish exactly.

Numerous alternatives to the GTR consistent with the special theory of relativity have been proposed. Some may be eliminated from further consideration by noncosmological tests of gravitational theory. For example, theories based on a space–time metric that is conformally flat, such as that published by Gunnar Nordstrom in 1912, are untenable because they fail to predict the deflection of light rays in the gravitational field of the sun, first measured by Dyson, Eddington, and Davidson during a solar eclipse in 1919. Others, such as the scalar–tensor theory published by Carl Brans and Robert Dicke in 1961, may be made consistent with current observational data by suitable adjustment of a parameter. Their cosmological consequences present distinct challenges to the standard model for some times during the evolution of the universe. However, sufficiently close to singularities, the predictions of most such theories become indistinguishable from those of the GTR. For example, in 1971 Dykla, Thorne, and Hawking showed that gravitational collapse in the Brans–Dicke theory inevitably leads to the "black hole" solutions of the GTR. Hence, cosmological models based on these theories make predictions very much like those of the GTR for the strong fields near the "big bang" and the nearly flat space-time of today. If they are very different in some intermediate regime, perhaps one important to the formation of structures such as galaxies, there currently appear to be no crucial tests that could cleanly decide in their favor. Thus, by an application of Occam's razor, most contemporary models are constructed assuming the validity of the GTR.

An influential exception to the dominance of GTR models was the "steady-state" cosmology of Bondi, Gold, and Hoyle. The philosophical foundation of this work was the extension of the cosmological principle to the "perfect cosmological principle," which asserted that there should be uniformity in time as well as spatial homogeneity and isotropy. As remarked earlier, observations of the microwave background presently render this model untenable. It is also unable to account for the abundances of various light nuclei synthesized when the universe was much hotter and denser than it is now. The apparent overthrow of the perfect cosmological principle encouraged questioning of the assumption of spatial homogeneity and isotropy. Since the empirical evidence in favor of spatial uniformity on very large scales is quite strong, most cosmologists today would like to deduce the cosmological principle rather than assume it. That is, we would like to demonstrate that, starting with arbitrary initial conditions, inhomogeneities and anisotropies are smoothed out by physical processes in a small time compared with the age of the universe. A serious difficulty in attempts to derive the cosmological principle was first emphasized by Misner in 1969. Relativistic space–times of finite age have particle horizons, so that at any moment signals can reach a given point only from limited regions, and parts of the universe beyond a certain distance from one another have not yet had any possibility of communicating. The correlation in conditions at distant regions that is asserted by the cosmological principle can be derived only if chaotic initial conditions smooth themselves out through an infinite number of processes, such as the expansions and contractions along different axes in "mixmaster" universes. In the standard model, the cosmological principle is regarded as an unexplained initial condition.

V. INTERACTION OF QUANTUM PHYSICS AND COSMOLOGY

A. Answers from Particle Physics

The interaction between elementary particle physics and cosmology has increased greatly since 1980, to the benefit of both disciplines. Several initial conditions of classical cosmology are given a tentative explanation in "inflationary universe" scenarios, proposed by Alan Guth in 1981 and modified by Linde, Albrecht, and Steinhardt in 1982. These models assume a time when the energy density of a "false vacuum" in a grand unified theory (GUT) dominated the dynamics of the universe. Since the density was essentially constant throughout this period, the Robertson–Walker scale factor grew exponentially in time, allowing an initially tiny causally connected region (even smaller than the small value of $1/H$ at the start of inflation) to grow until it included all of the space that was to become the currently observable universe. The original version (1981) assumed that this occurred while the universe remained trapped in the false vacuum.

Unfortunately, such a universe that inflated sufficiently never made a smooth transition to a radiation-dominated, early Friedmann cosmology. In the "new inflationary" models (1982), the vacuum energy density dominates while the relevant region of the universe inflates and evolves toward the true vacuum through the spontaneous breaking of the GUT symmetry by nonzero vacuum expectation values of the Higgs scalar. The true vacuum is reached in a rapid and chaotic "phase transition" when the universe is of the order of 10^{-35} sec old, resulting in the production of a large number and variety of particles (and antiparticles) at a temperature of the order of 10^{14} GeV. It is supposed that the universe evolves according to the standard model after this early time.

Since the entire observable universe evolves from a single causally connected region of the quantum vacuum, inflationary models obviously avoid horizon problems. The homogeneity and isotropy of the present observable universe are a consequence of the dynamic equilibrium in the tiny region. Since the density term in the Friedmann equation for the evolution of the Hubble parameter remains essentially constant throughout the inflationary era, while the curvature term is exponentially suppressed, these scenarios also offer a natural explanation for the present approximate flatness of the universe. In fact, plausible suppression of the curvature greatly exceeds that required by any astrophysical observations. Only by artificially contrived choices of parameters in an inflationary universe could we avoid the conclusion that any difference between the current value of Ω and unity is many orders of magnitude less than 1. If this result and the bounds on Ω_B from primordial nucleosynthesis are both true, we must conclude that dark matter of as yet undetermined form dominates the dynamics of the universe. Since GUTs predict the nonconservation of B (baryon number), C (charge conjugation), and CP (product of charge conjugation and parity), the decay of very heavy bosons far from thermodynamic equilibrium offers a way of dynamically generating the predominance of matter over antimatter rather than merely asserting it as an initial condition. In the absence of observation of the decay of the proton or accelerators capable of attaining energies at which GUTs predict convergence of coupling "constants," the apparent baryon asymmetry of the universe is perhaps the best empirical support for some sort of unified theory of quarks and leptons.

B. Constraints from Cosmology

Even as elementary particle theory solves some problems of cosmology, it is subject to limitations derived from cosmological data involving energies far beyond the 2×10^3 GeV limit of existing terrestrial accelerators. An important example involves the production of magnetic monopoles in the early universe. In 1931, P. A. M. Dirac showed that assuming the existence of magnetic monopoles led to a derivation of the quantization of magnetic and electric charge and a relation between them implying that magnetic charges would have to be very large. However, other properties of the hypothetical monopoles, such as mass and spin, were undetermined in his theory. In 1974, Gerhard t'Hooft and Alexander Polyakov showed that monopoles must be produced in gauge theory as topological defects whenever a semisimple group breaks down to a product that contains a $U(1)$ factor, for example,

$$SU(5) \to SU(3) \times SU(2) \times U(1)$$

All proposed GUTs, which attempt to unify the strong and the electroweak interactions, are examples of gauge theories in which monopoles are required and have masses of the order of the vacuum expectation value of the Higgs field responsible for the spontaneous symmetry breaking. The present experimental lower limit, of the order of 10^{33} years, for the mean life of the proton implies a lower limit for this mass of the order of 10^{18} GeV. Only within a time of at most 10^{-35} sec after the "big bang" was any place in the universe hot enough to produce particles of so great a mass, either as topological "knots" or as pairs of monopoles and antimonopoles formed in the energetic collisions of ordinary particles.

In addition to their enormous masses and relatively large magnetic charges, monopoles are predicted by GUTs to serve as effective catalysts for nucleon decay. Thus, if present in any appreciable abundance in the universe today, monopoles should make their presence obvious by doing some conspicuously interesting things. If monopoles were made in about the same abundance as baryons, their density alone would exceed the critical value for a closed universe by a factor of $\sim 10^{11}$. Monopoles would use up the potential energy of stationary magnetic fields, such as that of the Milky Way, by converting it to increases in their own kinetic energy. Collecting in a star throughout its history, they would render its collapsed "final state" short-lived by catalyzing nucleon decay. The observational upper bound on Ω, lower limits on the galactic magnetic field, and the life spans of neutron stars place severe constraints on the flux of monopoles at present and hence on their rate of production during the spontaneous symmetry-breaking era. In fact, unacceptably large magnetic monopole production in the simplest GUTs was one of the primary motivations for the development of the new inflationary universe models, which solve the problem through an exponential dilution of monopole density that leaves very roughly one monopole in the entire observable universe at present. While such a scenario appears to make the experimental search for monopoles essentially futile, it is important not to ascribe too much quantitative significance to this result, because the predicted number is exponentially sensitive to the ratio of the monopole mass to the highest temperature reached in the phase transi-

tion at the end of spontaneous symmetry breaking. Thus, an uncertainty of a mere factor of 10 (theoretical uncertainties are at least this large) in this ratio changes the predicted number of monopoles by a factor of the order of 10^8.

On the experimental front, Blas Cabrera claimed the detection of a magnetic monopole on February 14, 1982, after 150 days of searching with a superconducting quantum interferometer device (SQUID). If this observation were correct and even approximately corresponded to the typical distribution of monopoles in space, then neither the excessive production of a naive GUT model nor the extreme scarcity of a new inflationary model could be credible. Confirmation of this monopole detection would leave current theory totally at a loss to explain the monopole abundance, but neither Cabrera nor other observers have yet claimed another detection. After more than 3000 days had passed, most workers were of the opinion that the single "event" was due to something less exotic than a magnetic monopole.

The apparent smallness of the cosmological constant is a fact that has not yet been explained in any viable theory of particle physics or gravitation. Below some critical temperature in electroweak theory or GUTs, the effective potential function of the Higgs fields behaves like a cosmological constant in contributing a term equal to this potential function times the space–time metric to the stress–energy–momentum tensor of the universe. Empirical bounds on the vacuum energy density today imply that this potential at the spontaneous symmetry-breaking minimum was already less than 10^{-102} times the effective potential of the false vacuum. There is no derivation of this extremely small dimensionless number within the framework of GUTs. In fact, the assumption that the cosmological constant is negligible today is an unexplained empirical constraint on the otherwise undetermined scale of the effective potential in a gauge theory.

Another possible success of inflationary models is the natural development of nearly scale-independent density inhomogeneities from the quantum fluctuations in the Higgs field of GUTs during inflation. Inhomogeneities should later evolve by gravitational clumping into galaxies and clusters of galaxies. This opens the possibility of calculation from first principles of the spectrum of later structural hierarchies. Comparison of the results of such calculations with the observed large-scale structure of the universe may provide the most stringent constraints on new inflationary models.

C. Singularities and Quantum Gravity

This survey has looked back nearly 10^{18} sec from the present to the "big bang" with which the standard cosmological model claims the universe began. The attempt to understand its evolution reveals a number of significant eras. Let us review them from the present to the initial singularity in reverse chronological order, which is generally the order of decreasing direct experimental evidence and thus increasing tentativeness of conclusions. At a time of the order of 10^{13} sec after the universe began, when the temperature was ~ 0.3 eV, the photons that now compose the cosmic microwave radiation background last appreciably interacted with matter, which then "recombined" into transparent neutral atoms of hydrogen, helium, and lithium and began to form the large-scale structures familiar to us: stars, galaxies, and clusters of galaxies. At a time of the order of 10^{-2} sec after the beginning, when the temperature was ~ 10 MeV, the free neutrons and some of the free protons underwent the primordial synthesis that formed the nuclei of these atoms, and the cosmic background neutrinos ceased having significant interactions with matter. At a time of the order of 10^{-5} sec after the beginning and a temperature ~ 300 MeV, quarks became "confined" to form the hadrons as we now know them. At a time of the order of 10^{-12} sec after the beginning, the temperature was $\sim 10^3$ GeV, the present limit of terrestrial accelerators. The distinction between electromagnetic and weak interactions was not significant before then. The reconstruction of earlier history is of necessity much more tentative. The spontaneous symmetry breaking of the grand unification of electroweak and strong interactions is thought to have occurred at a time of roughly 10^{-34} sec and a temperature of the order of 10^{14} GeV. During the "inflation" preceding this epoch the baryon asymmetry of the universe may have been generated by fluctuations from thermal equilibrium in GUTs, and magnetic monopoles may have been produced by symmetry breaking. Any attempt to analyze events at substantially earlier times must address the unfinished program of constructing a quantum theory that unifies gravitation with the strong and electroweak interactions.

In the absence of complete understanding, the time and temperature scales of quantum gravitational effects can be estimated by dimensional analysis applied to the fundamental constants that must appear in any such theory. These are the quantum of action, the Newtonian gravitational constant, the speed of light, and the Boltzmann constant. The results, a time of the order of 10^{-44} sec and a temperature of the order of 10^{19} GeV, delineate conditions so near the classically predicted singularity of infinite density and curvature at the "big bang" that the very concept of a deterministic geometry of space–time breaks down. Violent fluctuations of space–time should generate particles in a manner analogous to that which occurs in the vicinity of a collapsing black hole, as first studied by Stephen Hawking in 1974. Such processes could conceivably be

the source of all existing matter, and the possible removal of an infinite-density singularity at time zero would make it scientifically meaningful to ask what the universe was doing before the "big bang."

Not only the existence and structure of matter but even the topology and dimensionality of space–time become properties to be derived rather than postulated in the quantum gravity era. In 1957, John Wheeler suggested that space–time need not necessarily be simply connected at the Planck scale (of the order of 10^{-35} m) but could have a violently fluctuating topology. If so, its description in terms of a smooth continuum would not be appropriate and would have to be replaced by some other mathematical model. The first viable unification of electromagnetism and gravitation, proposed by Theodor Kaluza in 1921 and independently by Oskar Klein in 1926, used a five-dimensional space–time with an additional "compact" spatial dimension subject to constraints that reduced their model to a sterile fusion of Maxwell's and Einstein's field equations. Removing these constraints allows Kaluza–Klein spaces to be used for alternate formulations of gauge field theories. Models with a total of 11 dimensions are currently being actively explored in relation to supersymmetry theories, which seek to provide a unification of bosons and fermions and all interactions among them. Another approach, string field theory, involves replacing the pointlike particles of conventional quantum field theory with fundamental objects with extent in one spatial dimension. It has been demonstrated that topology-changing processes can be explicitly realized in a Kaluza–Klein superstring theory, encouraging the hope that this could be the long-sought basis for a theory of everything. The full implications of such studies for particle physics and for cosmology are not yet clear.

VI. THOUGHTS ON SOURCES OF FUTURE PROGRESS

A. Emerging Observational Technologies

Since the pioneering research of Galileo, advances in telescope capabilities have been the source of more and richer data to constrain cosmological speculation. The current generation of astronomers has seen photographic techniques increasingly augmented by electronic image intensifiers. Improvements in photon detectors, such as charge-coupled devices, are beginning to approach their limits, so that achieving substantial gains will involve increasing the aperture in the next generation of optical and infrared instruments. The twin Keck telescopes, which can be used as an optical interferometer of 85-m baseline, are the largest general-purpose telescopes on earth. A consortium of nations have built Gemini, a matched pair of 8-m telescopes, one in Hawaii and one in Chile. Arizona astronomers have built the Large Binocular Telescope, which will carry two 8.4-m mirrors in a single mounting. The European Southern Observatory has completed the first of four 8.2-m telescopes that will eventually observe as a single Very Large Telescope from the Andes in Chile. Even larger telescopes are planned, and the use of adaptive optics will endow modern telescopes with "seeing" much better than that possible in the past.

As more powerful telescopes looking farther into outer space observe signals from earlier in the history of the universe, higher energy accelerators probing farther into inner space measure particle behavior under conditions simulating earlier times during the "big bang." The installation of superconducting magnets in the tevatron at the Fermi National Accelerator Laboratory at Batavia, Illinois, enabled it to produce protons with an energy of 10^3 GeV in 1984. In 1986, it made available a total energy of 2×10^3 GeV by accommodating countercirculating beams of protons and antiprotons that are made to collide. Since the pioneering effort toward the detection of gravitational waves by Joseph Weber in the late 1960s, astrophysicists have eagerly anticipated the maturing of this technology to open a new window on the universe. Long experience with the Hulse–Taylor binary pulsar is strong evidence that gravitational waves do indeed exist and have the properties predicted by Einstein's general theory of relativity. The development of large dedicated facilities such as the laser interferometer gravitational-wave observatory (LIGO) at last promises to soon move gravitational wave astronomy from a curiosity with isolated applications to a tool for the exploration of any cosmic environment involving strong gravitational fields. As usual in the opening of a new area of science, the unexpected discoveries will surely be the most exciting. Continued experiments at energies of several thousand GeV will lead to important new insights into the structure of matter on a scale of less than 10^{-19} m.

B. Concepts and Mathematical Tools

The search for a viable extension of GUTs to a theory of everything (TOE), which would include quantum gravity as well as the strong and electroweak interactions, is an active area of particle theory research. The energies at which such unification was achieved in nature are presumably even higher than those for GUTs. Thus, no data from accelerators, even in the most optimistic projections of foreseeable future technology, can serve to constrain speculation as well as does information from cosmology. The currently fashionable attempts to derive a TOE are based on "supersymmetry," which is the idea that the fundamental Lagrangian contains equal numbers of Bose and Fermi fields and that they can be transformed into each other by a supersymmetry. This immediately doubles the particle spectrum, associating with each particle thus far observed (or predicted by GUTs) a "superpartner" of opposite quantum statistics. There is at present no experimental evidence for the existence of any of these superpartners, inviting doubt as to the necessity of the supersymmetry assumption. However, supersymmetric theories have the potential to address one otherwise unanswered issue of cosmology: Why is the cosmological constant so small, perhaps precisely zero? Supersymmetric theories are the only known quantum field theories that are sensitive to the vacuum energy level. This appears to imply that a derivation of the cosmological constant from first principles should be possible within a supersymmetric theory, but the problem remains unsolved.

As the number of degrees of freedom being considered in field theories increases, increased computing speed and power become more important on working out the consequences of various proposed models. Some new hardware architectures such as concurrent processing appear to be a means of achieving performance beyond the limits of any existing machines but require the further development of software exploiting their distinctive features to achieve their full potential. Progress in discrete mathematics is of benefit to both computer science and pure mathematics. In the past, fundamental insights have often been derived from mathematical analysis without the benefit of "number crunching," and there is no reason to expect that this process has come to an end. It is, of course, impossible to predict what new closed-form solution of a recalcitrant problem may be discovered tomorrow, or what impact such a discovery may have.

In the last analysis, any attempt to predict the direction of progress in cosmology further than the very near future seems futile. By the nature of the questions that cosmology seeks to answer, the scope of potentially relevant concepts and information is limitless. It is entirely possible that within a decade carefully reasoned thought, outrageous unexpected data, or some combination of the two may overthrow some of today's cherished "knowledge." Aware of the questions still unanswered and of the possibility that some of the right questions have not yet been asked, we can only hope that future discoveries, anticipated or unforeseen, will result in ever greater insights into the structure, history, and destiny of the universe.

SEE ALSO THE FOLLOWING ARTICLES

CELESTIAL MECHANICS ● CHAOS ● COSMIC INFLATION ●
DARK MATTER IN THE UNIVERSE ● GALACTIC STRUCTURE

AND EVOLUTION • PARTICLE PHYSICS, ELEMENTARY •
QUASARS • RELATIVITY, GENERAL • RELATIVITY, SPE-
CIAL • STELLAR STRUCTURE AND EVOLUTION • UNIFIED
FIELD THEORIES

BIBLIOGRAPHY

Auborg, E., Montmerle, T., Paul, J., and Paul, P., eds. (2000). *Texas
    Symp. on Relativistic Astrophys. and Cosmology, Nuclear Physics B
    (Proc. Suppl.)* 80.
Bothun, G. (1998). "Modern Cosmological Observations and Problems,"
    Taylor & Francis, London.
Gribbon, J., and Rees, M. (1989). "Cosmic Coincidences," Bantam
    Books, New York.
Kolb, E., Turner, M., Lindley, D., Olive, K., and Seckel, D., eds. (1986).
    "Inner Space/Outer Space," Univ. of Chicago Press, Chicago.
Misner, C., Thorne, K., and Wheeler, J. (1973). "Gravitation," Freeman,
    New York.
Peebles, P. J. E. (1980). "The Large-Scale Structure of the Universe,"
    Princeton Univ. Press, Princeton, NJ.
Isham, C., Penrose, R., and Sciama, D., eds. (1981). "Quantum Gravity
    II," Oxford Univ. Press, London.
Rowan-Robinson, M. (1985). "The Cosmological Distance Ladder,"
    Freeman, New York.

CP (Charge Conjugation Parity) Violation

John F. Donoghue
Barry R. Holstein
University of Massachusetts

I. CP Symmetry
II. Observation of CP Violation
III. Mechanisms of CP Violation
IV. Recent Progress
V. Future Areas of Research

GLOSSARY

B meson Elementary particle containing one heavy quark called a b quark as well as a lighter antiquark, or a $\bar{b}$ antiquark, and a light quark. The B mesons are about five times more massive than a proton.

Charge conjugation invariance Invariance of physical laws under the process of interchanging particle and antiparticle.

Decay Elementary particles can transform into other combinations of particles as long as energy, momentum, charge, etc., are conserved. The process of an isolated particle transforming into several lighter particles is called *decay*.

Electric dipole moment Classically, an electric dipole moment is a separation of charges so that, although the whole system is electrically neutral, the distribution of charge has a region of positive charge and a region of negative charge separated along some axis.

Kaon Elementary particle with a mass of about one-half of that of the proton. The kaon is the lightest of the particles, with a quantum number called "strangeness" and containing a "strange" quark or antiquark.

Parity invariance Invariance of physical laws under the process of reversing spatial coordinates. If accompanied by a 180° rotation, this is equivalent to a reflection in a mirror.

CHARGE CONJUGATION PARITY (CP) violation is said to occur when two processes, which differ by the combined action of charge conjugation and parity reversal, do not occur at the same rate. This phenomenon is rare, but it has been observed in the decay of the neutral K meson system. The origin of this slightly broken symmetry is not currently understood, and it may tell us more about the structure of the fundamental interactions.

I. CP SYMMETRY

From ancient times, the concept of symmetry has commanded a powerful influence upon our view of the universe. However, many symmetries are only approximate, and the way in which they are broken can reveal much about the underlying dynamics of physical law. Perhaps the earliest example of a broken symmetry was the required modification of the presumed perfect circular orbits of the outer planets by epicycles—circles on circles—in order to explain the observation that occasionally the trajectories of theses planets through the sky double back on themselves. This breaking of perfect symmetry, although small, forced scientists to search more deeply into the basic forces responsible for celestial orbits, leading ultimately to Newton's law of universal gravitation.

More recently, in the mid-1950s, the concept of parity invariance—left-right symmetry—was found to be violated by the weak interaction, that is, the force responsible for such processes as nuclear beta decay. The concept of right or left in such a process is realized by particles whose direction of spin is respectively parallel or antiparallel to the particle momentum. Wrapping one's hand around the momentum vector with fingers pointing in the direction of rotation, as in Fig. 1, the particle is said to be right-left-handed if the right/left thumb points in the direction of the momentum vector. Parity invariance would require the absence of handedness, that is, the emission of equal numbers of both right- and left-handed particles in the de-

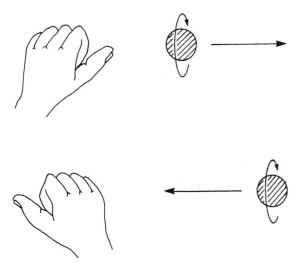

FIGURE 1 A spinning particle is described as either left-handed or right-handed, depending on which hand, when wrapped around the direction of motion with fingers pointing in the direction of the spin, has the thumb pointing in the direction of travel. Thus, the top figure indicates left-handed motion and the bottom figure shows right-handed motion.

cay process. Beta decay processes occur with both electron (e^-) and positron (e^+) emission:

$$A \rightarrow B + e^- + \bar{v}_e$$
$$B' \rightarrow A' + e^+ + \bar{v}_e$$

where v_e ($\bar{v}_e$) is the accompanying neutrino (antineutrino). In such decays, the neutrinos or antineutrinos are found to be completely left- or right-handed, indicating a maximal violation of parity. We now understand this as being due to the left-handed character of the particles that mediate the weak interactions, the W and Z bosons.

Even after the overthrow of parity in 1957, it was believed that a modified remnant of the symmetry remained, that of CP. Here P designates the parity operation, while C signifies charge conjugation, which interchanges particle and antiparticle. Thus, CP invariance requires equality of the rates for:

$$A \rightarrow B + e^- + \bar{v}_e (right)$$

and

$$\bar{A} \rightarrow \bar{B} + e^+ + \bar{v}_e (left)$$

Here $\bar{A}$, $\bar{B}$ are the antiparticles of A, B. Such CP invariance occurs naturally in the theories that have been developed to explain beta decay. It would then also be expected to extend to other weak interaction processes, such as the decays of other elementary particles.

There exists also a related symmetry called *time reversal*, T. In this case, the symmetry corresponds to the replacement of the time t by $-t$ in all physical laws (plus the technical addition of using the complex conjugate of the transition amplitude). Pictorially, this consists of taking a film, say, of a scattering amplitude $A + B \rightarrow C + D$ and then running the film backwards to obtain $C + D \rightarrow A + B$ (see Fig. 2). Time reversal invariance requires that the two processes occur with equal probability. In addition, there is a very powerful and fundamental theorem, called the *CPT theorem*, that asserts that in all of the currently known class of field theories the combined action of CP and T transformations must be a symmetry. Of course, this CPT invariance is also being subjected to experimental scrutiny and may in fact be violated in a new class of theories called *string theory* in which the fundamental units are not particles but elementary strings. In most reactions, both CP and T appear to be true symmetries and it is only in exotic reactions that any violation is possibly manifest.

It should be noted that there is an extremely important way in which the world is not CP invariant. This concerns the observed contents of the universe. When we look throughout the visible world, we see mostly electrons but very few positrons and mostly protons but very

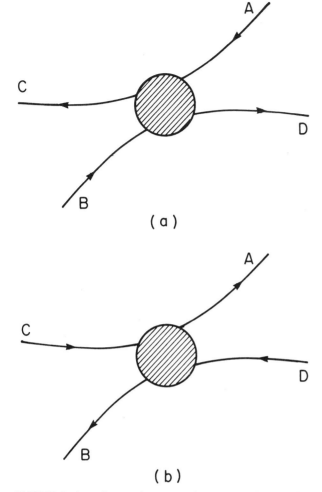

FIGURE 2 According to time-reversal symmetry, the reaction $A + B \rightarrow C + D$ indicated in (a) must be identical to the reaction $C + D \rightarrow A + B$ shown in (b).

few antiprotons. Thus, the matter that exists around us is not CP invariant! This may very well be an indication of CP violation in the early history of the universe. It is natural to assume that at the start of the "big bang" equal numbers of particles and antiparticles were produced. In this case, one requires a mechanism whereby the interactions of nature created a slight preference of particles over antiparticles, such that an excess of particles can remain at the present time. It can be shown that this scenario requires CP-violating interactions. Thus, it can be said that our existence is very likely due to the phenomenon of CP violation.

II. OBSERVATION OF CP VIOLATION

In 1964, a small breaking of CP symmetry was found in a particular weak interaction. In order to understand

how this phenomenon was observed, we need to know that under a left–right transformation a particle can have an intrinsic eigenvalue, either $+1$ or -1, since under two successive transformations one must return to the original state. Pi mesons, for example, are spinless particles with a mass of around 140 MeV/c^2. Via study of pion reactions, they are found to transform with a negative sign under parity operations. Pions exist in three different charge states—positive, negative, and neutral—and under charge conjugation, the three pions transform into one another:

$$C|\pi^+\rangle = |\pi^-\rangle$$
$$C|\pi^0\rangle = |\pi^0\rangle$$
$$C|\pi^-\rangle = |\pi^+\rangle$$

(Note that the π^0 is its own antiparticle.) Under CP, then, a neutral pion is negative:

$$CP|\pi^0\rangle = -|\pi^0\rangle$$

so that a state consisting of two or three neutral pions is, respectively, even or odd under CP. The argument is somewhat more subtle for charged pions, but it is found that spinless states $|\pi^+\pi^-\rangle$ and a symmetric combination of $|\pi^+\pi^-\pi^0\rangle$ are also even and odd, respectively, under the CP operation.

Before 1964, it was believed that the world was CP invariant. This had interesting implications for the system of K mesons (spinless particles with a mass of about 498 MeV/c^2) which decay into 2π and 3π by means of the weak interaction. There are two neutral K species, K^0 and $\bar{K}^0$, particle and antiparticle, with identical masses and opposite strangeness quantum numbers. We now understand these particles in the quark model, where K^0 is a bound state of a down quark and a strange antiquark, while $\bar{K}^0$ contains a down antiquark and a strange quark. Under CP, we have:

$$CP|K^0\rangle = |\bar{K}^0\rangle$$
$$CP|\bar{K}^0\rangle = |K^0\rangle$$

Although neither K^0 nor $\bar{K}^0$ is a CP eigenstate, one can form linear combinations:

$$|K_1\rangle = \frac{1}{\sqrt{2}}(|K^0\rangle + |\bar{K}^0\rangle) \qquad CP = +1$$

$$|K_2\rangle = \frac{1}{\sqrt{2}}(|K^0\rangle - |\bar{K}^0\rangle) \qquad CP = -1$$

If the world were CP invariant, then the particle that decays into a two-pion final state must itself be an eigenstate of CP with CP $= +1$, while that which decays into a three-pion final state must have CP $= -1$. Therefore, we expect that the decay scheme is

$$K_1 \rightarrow \pi^+\pi^-, \pi^0\pi^0$$
$$K_2 \rightarrow \pi^+\pi^-\pi^0, \pi^0\pi^0\pi^0$$

In fact, precisely this phenomenon is observed. The neutral kaons that decay weakly into these pionic channels are different particles (K_L, K_S) with different lifetimes, labeled L, S for long and short:

$$\tau_S \approx 10^{-10} \text{ sec}$$
$$\tau_L \approx 10^{-8} \text{ sec}$$

In the limit of CP conservation, $K_S = K_1$ and $K_L = K_2$. One can even observe strangeness oscillations in the time development of the neutral kaon system, which is a fascinating verification of the quantum mechanical superposition principle at work, but it is not our purpose to study this phenomenon here.

Rather, we return to 1964 when an experiment at Brookhaven National Laboratory by Christenson, Cronin, Fitch, and Turlay observed that the same particle—the longer-lived kaon—could decay into both 2π and 3π channels. The effect was not large—for every 300 or so $K_L \rightarrow 3\pi$ decays, a single $K_L \rightarrow 2\pi$ was detected—but it was definitely present. Since these channels possess opposite CP eigenstates, it was clear that a violation of CP symmetry had been observed.

III. MECHANISMS OF CP VIOLATION

The phenomenon of CP violation is of interest because we do not yet understand its origin. It is possible, but not yet proven, that it could be a manifestation of the Standard Model, which is the current theory of the fundamental interactions that appears to describe most of what we see in particle physics. However, because the breaking of CP symmetry is so small, it is also possible that it is a manifestation of some new type of interaction that is not part of our current Standard Model. In this case, the phenomenon is our initial indication of a deeper theory that will tell us yet more about the workings of nature. The goal of present and future research in this field is to identify the origin of the CP-violating interaction.

The mechanism that allows CP violation within the Standard Model was first articulated by Kobayashi and Maskawa (KM). It makes use of the fact that the interaction of the quarks with the charged bosons that mediate the weak force $W^\pm$ have different strengths. Generalizing from work by Cabibbo in the 1960s, the strength of the couplings can be described by angles, and therefore obey a "unitarity" constraint which is a generalization of the relation $\cos^2\theta + \sin^2\theta = 1$. Kobayashi and Maskawa noted that this can be generalized to *complex* angles (i.e.,

including complex phases) as long as a unitarity condition is satisfied. This has the form:

$$\sum_j V_{ij}^* V_{ji} = 1$$

where V_{ij} are the elements of a 3×3 matrix (the KM matrix); i, j refer to the different types of quarks, and $*$ denotes complex conjugation. It is the addition of complex phases to these couplings that allows for the existence of CP violation. Normally the phase of an amplitude $A = |A|e^{i\phi}$ is not observable, since the decay probability is given by the square of the absolute value of the amplitude $|A|^2$. However, relative phases can sometimes be observed, and kaon mixing can involve the two different amplitudes, $K^0 \rightarrow \pi\pi$ and the mixing-induced amplitude $K^0 \rightarrow \bar{K}^0 \rightarrow \pi\pi$, which can have different relative phases. For example, if the mass eigenstates were K_1 and K_2 described previously, then these two relative phases would be observable:

$$A(K^0 \rightarrow \pi\pi) = |A|e^{i\phi}$$
$$A(\bar{K}^0 \rightarrow \pi\pi) = |A|e^{-i\phi}$$
$$A(K_1 \rightarrow \pi\pi) = \sqrt{2}|A|\cos\phi$$
$$A(K_2 \rightarrow \pi\pi) = \sqrt{2}|A|\sin\varphi$$

In the KM scheme, this phase resides in the coupling of light quarks (u, d, and s quarks) to heavy quarks (c, b, and t quarks) and this makes the effect naturally small. The origin of this phase in the heavy quark couplings is not well understood and its magnitude is not predicted, but its existence is compatible with the theory.

There are other theories that have been proposed to explain the phenomenon of CP violation. Indeed, one of these, the superweak model of Wolfenstein, predates the Standard Model. This theory proposes a new, very weak force which can mix K_1 and K_2. In the modern framework of gauge theory, this would involve the exchange of a very heavy particle. Because it is so weak, however, it is very unlikely to be seen in any other effect besides the mixing of the neutral kaons. There are also other mechanisms. For example, in the theory of supersymmetry, which postulates a symmetry between fermions and bosons, there are many complex phases in addition to the one of the KM model, and these can lead to a rich variety of CP-violating processes. Likewise, if there exist extra Higgs bosons, which are spinless particles often postulated in new theories, their couplings will almost always involve CP-violating phases.

IV. RECENT PROGRESS

The most important recent progress involves the observation of an effect that is clearly *not* simply the mixing of

K_1 and K_2. This is then a new effect, often referred to as *direct* CP violation. This emerges from the study of the two different charged states that can emerge from kaon decay. Conventionally, we describe the ratio of two rates by two parameters, ε and ε', defined:

$$\frac{A(K_L \rightarrow \pi^+\pi^-)}{A(K_S \rightarrow \pi^+\pi^-)} = \varepsilon + \varepsilon'$$

$$\frac{A(K_L \rightarrow \pi^0\pi^0)}{A(K_S \rightarrow \pi^0\pi^0)} = \varepsilon - 2\varepsilon'$$

If the mixing of the neutral kaons were the only phenomenon contributing to this process, then both these ratios would be identical and $\varepsilon' = 0$. The superweak model leads to this prediction. On the other hand, the KM theory has a mechanism such that the two decays can differ by a small amount. The prediction of this difference is very difficult because of the need to calculate decay amplitudes within the theory of the strong interactions. In fact, the range of predictions in the literature encompasses an order of magnitude $\varepsilon'/\varepsilon = 0.0003 \rightarrow 0.003$.

Very beautiful and precise experiments have been carried out over the last decade at CERN (the European Laboratory for Particle Physics) in Geneva and Fermilab (near Chicago) which now agree on a value $\varepsilon'/\varepsilon = 0.0022 \pm 0.0003$. This result is a major advance for the field. It offers convincing proof that direct CP violation exists. Since not all effects are in the mixing mechanism, it rules out the superweak theory. The result also appears compatible with the Standard Model within the present range of theoretical uncertainty. However, further theoretical work is required in order to refine the prediction if this is to become a firm test.

V. FUTURE AREAS OF RESEARCH

Despite a long history of investigation, CP violation has only been detected in the neutral kaon system. The recent observations of ε' have been extremely important but have not decisively identified the mechanism responsible for this phenomenon. Clearly, in order to understand the origin of CP non-conservation, additional experimental observations are required. This is recognized as an important problem in the field and work is underway around the world that may help to clarify this situation.

The most focused effort at present is in the area of B meson decay. B mesons are particles that carry a heavy (b) quark, as well as a lighter (u, d, or s) antiquark. These heavy particles are produced only in high-energy reactions and decay with a lifetime of about 10^{-12} seconds. The neutral particles in this family—called B_d^0 and B_s^0, where the subscript labels the antiquark flavor—undergo mixing with their corresponding antiparticles in a fash-

ion similar to the neutral kaon. This mixing, together with possible direct CP violation in the decay amplitude, leads to a possibility of CP non-conservation in the decay of B mesons. However, because they are heavier there exist far more channels open for B meson decay than are possible for kaons, so the experimental exploration is both richer and more difficult.

There are a few decay channels for which the Standard Model yields precise predictions. The most accessible of these is the reaction, $B_d^0 \rightarrow \Psi K_S^0$, where the symbol Ψ denotes the bound state of a charmed quark and a charmed antiquark. The signal being looked for is the difference between the decay to this state, as a function of time, of B_d^0 and its antiparticle $\bar{B}_d^0$. In the ratio of these decay rates, the magnitude of the decay amplitude cancels out, leaving only a well-defined combination of the KM angles, as described above. In addition, this decay mode is experimentally accessible. Both the Ψ and the K_S^0 are readily identified by the particle detectors, and indeed the decay rate relevant for this process has already been measured. Experimentally, the most stringent requirement is the observation of the time dependence of the decay, for which the asymmetric B factories are needed (see below). At the time of this writing, there exist preliminary indications for a CP-violating asymmetry, although the present precision is not sufficient to know if it agrees with the Standard Model prediction. This asymmetry is a valuable test of the Standard Model mechanism and by itself could signal the need for new nonstandard interactions. Moreover, there are many other decay modes that may exhibit CP violation. The overall pattern of such decays will allow a thorough study of the mechanism of CP violation.

The experiments on these heavy particle decays are being carried out at all of the present high-energy accelerator facilities, but most especially at dedicated B factories. These are specialized accelerators that are designed to provide the maximum number of B mesons in an environment that gives experimenters the clearest access to the relevant decay channels. There are three B factories operating in the world as of this writing: at Cornell University, Stanford Linear Accelerator Center, and the KEK laboratory in Japan. The latter two are "asymmetric" machines, where the energies of the two colliding beams are not equal. This design requirement was specifically chosen in order to facilitate the observation of the time dependence of the decay asymmetries. It is expected that these B factories will soon provide preliminary results on CP asymmetries for some of the more accessible modes, to be followed up by a multiyear precision exploration of all facets of heavy quark physics.

A second important area of current and future research is that of measurement of electric dipole moments of

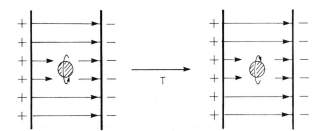

FIGURE 3 Under time reversal, a spinning particle placed between capacitor plates, as shown, will reverse its direction of spin, but the direction of the electric field stays the same. Thus, an interaction of the form $\vec{S} \cdot \vec{E}$ violates time-reversal invariance.

particles. This refers to the interaction between a particle and an applied electric field of the form:

$$H \propto \vec{S} \cdot \vec{E}$$

where $\vec{S}$ is the spin of the particle and $\vec{E}$ is the electric field. Imagine a spinning particle placed in an electric field set up by two oppositely charged capacitor plates (Fig. 3). It is easy to see that if time is reversed, the spin reverses but not the electric field, so that an interaction of this form violates time reversal and, hence, by the CPT theorem also violates CP. There is a long history of experiments that have looked for the possible electric dipole moment of the neutron. The use of a neutral particle is a necessity since a charged particle would accelerate out of the experimental region under the influence of the electric field. At the present time, the experimental upper limit of a possible neutron electric dipole moment is at the level of several times 10^{-26} e-cm. This is an incredible sensitivity. If one imagines a neutron expanded to the size of the earth, the above limit corresponds to a charge separation of only one micron! Similar searches for a nonzero electric dipole moment are being performed with atoms. While no elec-

tric dipole moment has yet been found, the interpretation of any such result in terms of a limit is made uncertain by the shielding of the nucleus from the full effect of the electric field because of the shifting of the electron cloud. The Standard Model mechanism for CP violation predicts a dipole moment which is many orders of magnitude too small to be seen by present experiments. However, many other models predict electric dipole moments in the range under investigation, and this may prove to be a powerful indication of new physics.

SEE ALSO THE FOLLOWING ARTICLES

ACCELERATOR PHYSICS AND ENGINEERING • PARTICLE PHYSICS, ELEMENTARY • RADIATION PHYSICS • RADIOACTIVITY

BIBLIOGRAPHY

Bigi, I., and Sanda, A. (2000). "CP Violation," Cambridge University Press, Cambridge, U.K.

Donoghue, J. F., Golowich, E., and Holstein, B. R. (1994). "Dynamics of the Standard Model," Cambridge University Press, Cambridge, U.K.

Georgi, H. (1984). "Weak Interactions and Modern Particle Theory," Benjamin-Cummings, Redwood City, CA.

Gottfried, K., and Weisskopf, V. F. (1984). "Concepts of Particle Physics," Oxford University Press, London.

Halzen, F., and Martin, A. D. (1984). "Quarks and Leptons: An Introductory Course in Modern Particle Physics," Wiley, New York.

Perkins, D. H. (1982). "Introduction to High Energy Physics," 2nd ed., Addison–Wesley, Reading, MA.

Quinn, H., and Witherall, M. (1998). "The asymmetry between matter and antimatter," *Sci. Am.* **Oct.**, 76.

Winstein, B., and Wolfenstein, L. (1993). "The search for direct CP violation," *Rev. Mod. Phys.* **65**, 1113.